Advances in Plant Breeding Strategies

Volume 11

Series Editors

Jameel M. Al-Khayri, Agricultural Biotechnology, King Faisal University, Al-Hassa, Saudi Arabia

Shri Mohan Jain, Department Agricultural Sciences, University of Helsinki, Helsinki, Finland

This book series is intended to provide a platform for plant scientists to present the latest advances in both innovative and traditional methodologies applied in plant breeding for the development of new crop varieties under different environmental conditions leading to sustainable food production, food and nutrition security, food supply chain, germplasm conservation and utilization. This series is an excellent reference source for plant breeders, geneticists, biotechnologists, and molecular biologists using molecular tools, bioinformatics, artificial intelligence, and other emerging technologies. The subject matter of all volumes will relate to postgraduate students specializing in agriculture, biotechnology, and molecular breeding as well as policymakers, seed companies, researchers, and professionals interested in plant genetic improvement.

Khaled F. M. Salem • Jameel M. Al-Khayri
Shri Mohan Jain
Editors

Breeding and Biotechnology of Leaf, Fruit, and Seed Fiber Crops

Springer

Editors
Khaled F. M. Salem
Department of Biological Sciences
Shaqra University
Ad-Dawadimi, Saudi Arabia

Department of Plant Biotechnology,
Genetic Engineering and Biotechnology
Research Institute (GEBRI)
University of Sadat City
Sadat City, Egypt

Jameel M. Al-Khayri
Department of Agricultural Biotechnology
King Faisal University
Al-Ahsa, Saudi Arabia

Shri Mohan Jain
Department Agricultural Sciences
University of Helsinki
Helsinki, Finland

ISSN 3004-8737 ISSN 3004-8745 (electronic)
Advances in Plant Breeding Strategies
ISBN 978-3-032-00406-2 ISBN 978-3-032-00407-9 (eBook)
https://doi.org/10.1007/978-3-032-00407-9

This Springer imprint is published by the registered company Springer Nature Switzerland AG
The registered company address is: Gewerbestrasse 11, 6330 Cham, Switzerland

Preface of the Series

Contemporary plant breeders rely on conventional and emerging methodologies to provide sustainable food production worldwide. However, the human population is growing at an alarming rate, especially in developing countries, and food availability to feed new mouths could gradually become a serious problem. Agriculture production is negatively impacted by environmental pollution, rapid industrialization, water scarcity, soil erosion of fertile topsoil, limited areas of potentially arable land, lack of improvement of local plant types, erosion of genetic diversity, and dependence on just a few major crop species for the world's food supplies. Breeding programs of these crops have been very much dependent on the ready availability of genetic variation, whether spontaneous or induced. Plant breeders and geneticists are under constant pressure to sustain food production by using innovative breeding strategies and enhancing minor crops that are well adapted to the marginal land and source of nutrition as well as abiotic and biotic stresses. An inherent limitation of traditional breeding is that the introduction of one or few genes into a cultivar requires several generations of backcrossing, making it a time-consuming process.

With the development of new molecular tools, molecular marker-assisted breeding helped rapid introgression of transgene and reduction of linkage drag. The continued development and adaptation of plant biotechnology approaches including genetic engineering, genome editing, molecular markers, genomics, proteomics, bioinformatics, artificial intelligence, and nanotechnology has established new tools for the creation, analysis, and manipulation of genetic variation for the development of improved cultivars to achieve agricultural sustainability and food security. A multi-disciplinary approach combining traditional plant breeding, mutation breeding, plant biotechnology, and molecular biology would be strategically ideal for developing improved crop varieties to ensure global food security, provide natural pharmaceutical sources, and supply fiber and ornamental crops for industry, esthetic and environmental purposes. However, breeders continue to face challenged such as environmental stress, the limitation of arable land expansion, and erosion of genetic resources particularly under the impact of global climate change.

This book series is intended to provide a platform for plant scientists to present the latest advances in both innovative and traditional methodologies applied in plant breeding for the development of new crop varieties under different environmental conditions leading to sustainable food production, food and nutrition security, food supply chain, germplasm conservation, and plant sustainable utilization. This series is an excellent reference source for plant breeders, geneticists, biotechnologists, and molecular biologists using molecular tools, bioinformatics, artificial intelligence, and other emerging technologies. The subject matter relates to postgraduate students specializing in agriculture, biotechnology, and molecular breeding as well as policymakers, seed companies, researchers, and professionals interested in plant genetic improvement.

Al-Ahsa, Saudi Arabia
Helsinki, Finland

Jameel M. Al-Khayri
Shri Mohan Jain

Preface

Fiber crops serve various purposes in agriculture, industry, medical supplies, and the clothing and textile industries. Additionally, certain fiber crops' plant parts are essential for producing oil for human and animal consumption. In order to meet consumer demands for new, desired natural plant traits such as morphological traits, physiological traits, maturity traits, yield traits, and fiber quality traits like color, and improvements are necessary. It is projected that from 2025 to 2034, the global market for plant natural fibers will expand at a compound annual growth rate (CAGR) of over 7.2%, from an estimated USD 5.5 billion in 2024. One of the agricultural sectors experiencing rapid growth is the apparel, textile, and plant waste industries that are derived from them. Due to innovation driving this business, there is a growing need for plant breeders in both the public and private sectors to meet customer demands. Breeding mutations and creating novel fiber-specific features that benefit the apparel and textile industries are ideal tasks for biotechnological techniques such as genetic transformation, genomics, nanotechnology, gene editing, and induced mutagenesis. In addition, digital farming, artificial intelligence, and vertical farming are becoming increasingly important in this field.

The main idea of this book is to explore how traditional and novel plant breeding techniques can be applied to create new fiber crop varieties that are suitable for different environmental conditions in a changing global climate, in order to achieve sustainable production for use in the apparel and textile industries. In addition to traditional breeding methods, this book focuses on the latest advancements in molecular tools and biotechnology for a variety of leaf, fruit, and seed fiber crops, and their application in plant breeding.

The book contains 21 chapters supported by 182 color figures and 70 tables. The chapters are grouped into three parts. Part I, titled Leaf Fiber Crops, covers the genera *Musa*, *Agave*, *Trifolium*, *Phoenix*, *Hyphaene*, *Ananas*, *Metroxylon*, *Arenga*, and *Yucca*. Part II, titled Fruit Fiber Crops, covers the genera *Citrus*, *Cocos*, and *Elaeis*. Part III, titled Seed Fiber Crops, covers the genera *Gossypium* and *Ceiba*. Each chapter discusses breeding techniques for specific species or a group of related species. Topics covered include current cultivation methods and challenges, biodiversity, germplasm conservation, breeding methods including conventional, molecular, tissue culture, genetic engineering, gene editing, hybridization, and mutation breeding. Each chapter concludes with future research directions and a comprehensive list of references for further reading. Sixty-nine globally recognized scientists from 12 different nations contributed to the chapters based on their expertise in the subject. The chapters underwent a thorough review process to ensure scientific accuracy and presentation quality.

The book is a valuable resource for professional plant breeders, geneticists, researchers, entrepreneurs, and advanced undergraduate and graduate students interested in biotechnology, molecular breeding, and agriculture.

The editors would like to thank the staff of Springer for the opportunity to publish this book and the contributing authors for their unwavering dedication and hard work.

Ad-Dawadimi, Saudi Arabia — Khaled F. M. Salem
Al-Ahsa, Saudi Arabia — Jameel M. Al-Khayri
Helsinki, Finland — Shri Mohan Jain

Acknowledgment

The editors extend their appreciation to the support of the Deanship of Scientific Research, Vice Presidency for Graduate Studies and Scientific Research, King Faisal University, Saudi Arabia (Project No.KFU253808).

The original version of the frontmatter has been revised. An Acknowledgement has been added to the book's front matter.

Contents

Part III Seed Fiber Crops

Editors and Contributors

About the Editors

Khaled F. M. Salem is affiliated with Biological Sciences Department, Faculty of Science and Humanities, Shaqra University, Ad-Dawadimi, Saudi Arabia, and Plant Biotechnology Department, Genetic Engineering and Biotechnology Research Institute, University of Sadat City, Egypt. He earned his BS in Crop Science with excellent honor degree in 1993 and MS with focus on rice breeding from Menoufiya University, Egypt, in 1998. He conducted his MS at Rice Research and Training Center, Field Crops Research Institute, Agriculture Research Center, Giza, Ministry of Agriculture, Egypt. In 2004, he completed his PhD in Plant Breeding and Genetics, Martin Luther University Halle-Wittenberg, Germany. He conducted his PhD project at IPK, Gatersleben, Germany. From 2001 to 2004 he worked as visiting scientist at Gene Bank Department and Gene and Genome Mapping Group, IPK, Germany. Additionally, he attended several training courses as visiting scientist in China, Italy, Spain, Turkey, and CIMMYT. He serves as reviewer for 50 journals and editor for several journals. He has published 78 articles, 30 book chapters, and 25 conference proceedings. In addition, he has edited plant gene bank reference book. His research has focused on cereal, fiber, and vegetable crops biotechnology for the past three decades. He has also mentored and supervised master's and PhD students.

Jameel M. Al-Khayri is affiliated with the Department of Agricultural Biotechnology, King Faisal University, Saudi Arabia. He received BS in Biology in 1984 from the University of Toledo, MS in Agronomy in 1988, and PhD in Plant Science in 1991 from the University of Arkansas. He is the National Correspondent of the International Association of Plant Tissue Culture and Biotechnology. He has authored 150 research articles and 86 chapters and has edited several journal special issues. In addition, he has edited 42 reference books on plant biotechnology, genetic resources, breeding, genomics, and nanotechnology. In addition to teaching, advising, and research, he held administrative responsibilities as the Assistant Director of Date Palm Research Center, Head of Department of Plant Biotechnology, and Vice Dean for Development and Quality Assurance. He served as a Member of Majlis Ash-Shura (Saudi Legislative Council) for the 2009–2012 term. He is interested in agricultural innovations and sustainability.

Shri Mohan Jain (Adjunct) is currently associated with the Department of Agricultural Sciences, University of Helsinki, Helsinki, Finland. He received MPhil in 1973 and PhD in 1978 from Jawaharlal Nehru University, New Delhi, India. He served as a Technical Officer, Plant Breeding, and Genetics, International Atomic Energy Agency (IAEA), Vienna, Austria, from 1999 to 2005 when the agency was awarded Nobel Peace Prize, 2005. He is on the Editorial Board of *Euphytica*, *In Vitro*, and *Propagation of Ornamental Plants* and a reviewer in Plant Cell Tissue and Organ Culture, *Plant Cell Reports*, and *Plants*. He has published 225 peer-reviewed journals articles, book chapters, and conference proceedings. Additionally, he has edited 82 books; has been an invited speaker; and served as Chairperson in several international conferences worldwide. He is a former consultant to the IAEA, the European Union, the Government of Grenada, the Iranian Private Company, and the Egyptian government.

Contributors

Amal M. E. Abdel-Hamid Department of Biological and Geological Sciences, Faculty of Education, Ain Shams University, Cairo, Egypt

Ahmed M. Abdelmoghny Cotton Research Institute (CRI), Agricultural Research Center (ARC), Giza, Egypt

Eman H. Afifi Fruit Breeding, Ornamental Plants and Woody Tress Department, Horticulture Research Institute (HRI), Agriculture Research Centre (ARC), Giza, Egypt

Norma Almaraz-Abarca Centro Interdisciplinario de Investigación para el Desarrollo Integral Regional unidad Durango (CIIDIR Durango), Instituto Politécnico Nacional, Durango, Mexico

Abeer S. Arafa Cotton Research Institute (CRI), Agricultural Research Center (ARC), Giza, Egypt

Dwi Asmono Department of Research and Development, PT Sampoerna Agro Tbk, Sampoerna Strategic Square, North Tower, Jakarta, Indonesia

José Antonio Ávila-Reyes Centro Interdisciplinario de Investigación para el Desarrollo Integral Regional unidad Durango (CIIDIR Durango), Instituto Politécnico Nacional, Durango, Mexico

Asmah Awal Faculty of Plantation and Agrotechnology, UiTM Cawangan Melaka, Merlimau, Melaka, Malaysia

Sandeep Bagri ICAR-Central Institute for Cotton Research, Regional Station, Sirsa, Haryana, India

Mochamad Hasjim Bintoro Department of Agronomy and Horticulture, Faculty of Agriculture, IPB University, IPB Dramaga Campus, Bogor, West Java Province, Indonesia

Imam Civi Cartealy Research Center for Computing, Research Organization for Electronics and Informatics, National Research and Innovation Agency (BRIN), Science and Technology Park of Soekarno, Cibinong, West Java Province, Indonesia

Subhash Chand Division of Crop Improvement, ICAR-Indian Grassland and Fodder Research Institute, Jhansi, Uttar Pradesh, India

Jingjing Chen Key Laboratory of Tropical Fruit Biology, Ministry of Agriculture and Rural Affairs, South Subtropical Crops Research Institute, Chinese Academy of Tropical Agricultural Sciences, Zhanjiang, China

Xiaohong Dai Key Laboratory of Tropical Fruit Biology, Ministry of Agriculture and Rural Affairs, South Subtropical Crops Research Institute, Chinese Academy of Tropical Agricultural Sciences, Zhanjiang, China

Hargopal Dutta ICAR-Indian Agricultural Research Institute, New Delhi, India

Suman Dutta Ramakrishna Mission Vivekananda Educational & Research Institute, Howrah, West Bengal, India

Luis E. Eguiarte Laboratory of Molecular and Experimental Evolution, Department of Evolutionary Ecology, Institute of Ecology, University National Autonomous of Mexico, Mexico City, Mexico

Karima Mohamed El-Absy Biology Department, University College of Tayma, University of Tabuk, Tabuk, Saudi Arabia
Eco-physiology Unit, Plant Ecology and Ranges Department, Desert Research Center, Cairo, Egypt

Essam Fathy El-Hashash Department of Agronomy, Faculty of Agriculture, Al-Azhar University, Cairo, Egypt

Yusuf Sigit Ahmad Fauzan Research Center for Applied Botany, Research Organization for Life Sciences and Environment, National Research and Innovation Agency (BRIN), Science and Technology Park of B.J. Habibie, South Tangerang City, Banten Province, Indonesia

Abdul Ghaffar Department of Agronomy, MNS-University of Agriculture, Multan, Pakistan

Reham H. A. O. Gibely Cotton Research Institute (CRI), Agricultural Research Center (ARC), Giza, Egypt

Federico A. Gutiérrez-Miceli Plant Tissue Culture Laboratory, National Technological Institute of Mexico/IT of Tuxtla Gutierrez, Graduate Studies and Research Division, Tuxtla Gutiérrez, Chiapas, Mexico

Venus E. Hasan Department of Biotechnology, General Commission for Scientific Agricultural Research, Latakia, Syria

Gulfam Hassan Institute of Agricultural Extension, Education and Rural Development, Faculty of Social Sciences, University of Agriculture Faisalabad, Faisalabad, Pakistan

H. Mary Henrietta Department of Mathematics, Saveetha Engineering College (Autonomous), Chennai, Tamil Nadu, India

Muhammad Iftikhar Institute of Agricultural Extension, Education and Rural Development, Faculty of Social Sciences, University of Agriculture Faisalabad, Faisalabad, Pakistan

Albertus Fajar Irawan Faculty of Animal and Agricultural Science, Diponegoro University, Jl Prof. Jacub Rais. Diponegoro University Campus, Tembalang, Semarang, Central Java Province, Indonesia

José G. Joya-Dávila Institute of Agricultural Sciences, Autonomous University of Baja California (ICA-UABC), Nuevo León, Baja California, Mexico

Sohail Kamaran Department of Plant Breeding and Genetics, Faculty of Agriculture, University of Agriculture, Faisalabad, Faisalabad, Pakistan

Karyanti Research Center for Estate Crops, Research Organization for Agriculture and Food, National Research and Innovation Agency (BRIN), Science and Technology Park of Soekarno, Cibinong, West Java Province, Indonesia

Asif Ali Khan Institute of Horticultural Sciences, University of Agriculture, Faisalabad, Pakistan

Iqrar Ahmad Khan Institute of Plant Breeding and Biotechnology, MNS-University of Agriculture, Multan, Pakistan

Zulqurnain Khan Institute of Plant Breeding and Biotechnology, MNS-University of Agriculture, Multan, Pakistan

Plosha Khanum Institute of Plant Breeding and Biotechnology, MNS-University of Agriculture, Multan, Pakistan

Neeraj Kharor Forage Section, Department of Genetics and Plant Breeding, Chaudhary Charan Singh Haryana Agricultural University, Hisar, Haryana, India

Pummy Kumari Forage Section, Department of Genetics and Plant Breeding, Chaudhary Charan Singh Haryana Agricultural University, Hisar, Haryana, India

Antonio G. Lalusin Institute of Crop Science, College of Agriculture and Food Science, University of the Philippines, Los Banos, Laguna, Philippines

Carlos A. Lecona-Guzmán Plant Tissue Culture Laboratory, National Technological Institute of Mexico/IT of Tuxtla Gutierrez, Graduate Studies and Research Division, Tuxtla Gutiérrez, Chiapas, Mexico

Dongliang Li Key Laboratory of Tropical Fruit Biology, Ministry of Agriculture and Rural Affairs, South Subtropical Crops Research Institute, Chinese Academy of Tropical Agricultural Sciences, Zhanjiang, China

Meghdad Kamali Moghaddam Department of Textile Engineering, University of Bonab, Bonab, Iran

Nazatul Asikin Muda Faculty of Plantation and Agrotechnology, UiTM Cawangan Melaka, Merlimau, Melaka, Malaysia

K. U. K. Nampoothiri ICAR-Central Plantation Crops Research Institute, Kasargod, Kerala, India

Nidhi ICAR-Central Institute for Cotton Research, Regional Station, Sirsa, Haryana, India

Debashis Paul ICAR-Central Institute for Cotton Research, Regional Station, Sirsa, Haryana, India

Devit Purwoko Research Center for Applied Botany, Research Organization for Life Sciences and Environment, National Research and Innovation Agency (BRIN), Science and Technology Park of B.J. Habibie, South Tangerang City, Banten Province, Indonesia

Muhammad Ahsan Qureshi Institute of Horticultural Sciences, University of Agriculture, Faisalabad, Faisalabad, Pakistan

S. N. Rahana ICAR-Indian Institute of Oil Palm Research, Regional Station, Palode, Thiruvananthapuram, Kerala, India

M. K. Rajesh ICAR-Regional Station, ICAR-Central Plantation Crops Research Institute, Vittal, Karnataka, India

Illy Kamaaina Ramle Faculty of Plantation and Agrotechnology, UiTM Cawangan Melaka, Merlimau, Melaka, Malaysia

Hafiz Amjad Ali Rana Department of Outreach and Continuing Education, Faculty of Social Sciences and Humanities, Muhammad Nawaz Shareef University of Agriculture, Multan, Pakistan

Sheila J. Reyes-Zambrano Plant Tissue Culture Laboratory, National Technological Institute of Mexico/IT of Tuxtla Gutierrez, Graduate Studies and Research Division, Tuxtla Gutiérrez, Chiapas, Mexico

Imron Riyadi Indonesian Oil Palm Research Institute, Bogor Unit, Bogor, West Java Province, Indonesia

Karen Y. Ruiz-Mondragón Laboratory of Molecular and Experimental Evolution, Department of Evolutionary Ecology, Institute of Ecology, University National Autonomous of Mexico, Mexico City, Mexico

Khaled F. M. Salem Department of Biological Sciences, Shaqra University, Ad-Dawadimi, Saudi Arabia

Department of Plant Biotechnology, Genetic Engineering and Biotechnology Research Institute (GEBRI), University of Sadat City, Sadat City, Egypt

Samir Samah Biogenetix-Lab S. de R.L. de C.V., Ixtapaluca, State of Mexico, Mexico

Satpal Forage Section, Department of Genetics and Plant Breeding, Chaudhary Charan Singh Haryana Agricultural University, Hisar, Haryana, India

Dibyendu Seth Department of Genetics and Plant Breeding, Chaudhary Charan Singh Haryana Agricultural University, Hisar, Haryana, India

Muhammad Shahzad Department of Agriculture Extension Education, Faculty of Agriculture & Environment, The Islamia University of Bahawalpur, Bahawalpur, Punjab, Pakistan

B. L. Sharma Forage Section, Department of Genetics and Plant Breeding, Chaudhary Charan Singh Haryana Agricultural University, Hisar, Haryana, India

K. Sunilkumar ICAR-Indian Institute of Oil Palm Research, Regional Station, Palode, Thiruvananthapuram, Kerala, India

Teuku Tajuddin Research Center for Applied Botany, Research Organization for Life Sciences and Environment, National Research and Innovation Agency (BRIN), Science and Technology Park of B.J. Habibie, South Tangerangm City, Banten Province, Indonesia

Eman Tawfik Botany and Microbiology Department, Faculty of Science, Helwan University, Helwan, Egypt

Ernestina Valadez-Moctezuma Universidad Autónoma Chapingo, Departamento de Fitotecnia, Chapingo, State of Mexico, Mexico

Sugey Vásquez-Hernández Plant Tissue Culture Laboratory, National Technological Institute of Mexico/IT of Tuxtla Gutierrez, Graduate Studies and Research Division, Tuxtla Gutiérrez, Chiapas, Mexico

Surender Kumar Verma ICAR-Indian Institute of Wheat and Barley Research, Seed Research Farm, Hisar, Haryana, India
ICAR-Central Institute for Cotton Research, Regional Station, Sirsa, Haryana, India

Luli Wang Key Laboratory of Tropical Fruit Biology, Ministry of Agriculture and Rural Affairs, South Subtropical Crops Research Institute, Chinese Academy of Tropical Agricultural Sciences, Zhanjiang, China

Waleed M. B. Yehia Cotton Research Institute, Agriculture Research Center (ARC), Giza, Egypt

Aysen Yumurtaci Marmara University, Institute for Graduate Studies in Pure and Applied Sciences, Biology Department, Goztepe Campus, Istanbul, Turkey

Part I
Leaf Fiber Crops

Chapter 1
Advancements in Abaca (*Musa textilis* Nee.) Breeding and Biotechnology for Ensuring the Sustainability of the Philippine Abaca Industry

Antonio G. Lalusin

Abstract Abaca (*Musa textilis* Nee.) is an endemic crop to the Philippines and thrives very well in various regions across the country. It is a member of the *Musaceae* family and ranks second in importance among leaf fiber types. Abaca fiber, unlike many others, is obtained from the petioles or pseudostems of the plant's leaves. The abaca industry is a significant source of dollar revenue and serves as a vital export crop for the nation. However, the local abaca sector's future faces substantial challenges due to viral diseases-primarily bunchy top, mosaic, and bract mosaic, with bunchy top being the most severe. Introducing high-yield, virus-resistant abaca hybrids as an alternative to traditional varieties holds considerable promise for addressing this critical issue within the Philippine abaca industry. This chapter also explores market dynamics, industry challenges, distribution, cultivation, genetic resources, and various breeding methodologies for abaca. Advances in tissue culture, mutation breeding, genetic engineering, transcriptomics, and genomics have been utilized to demonstrate the successful transfer of resistance genes from donor species to traditional abaca varieties. Noteworthy achievements in abaca breeding include the development of Dioscoro 1 formerly named Backcross Abaca with Native and Desirable Accessions to Lift-up the Abaca industry (BANDALA) variety at the University of the Philippines Los Baños (UPLB). Dioscoro 1 or BANDALA is the result of wide hybridization between wild *Musa balbisiana* (Pacol) and abaca. Through a series of backcrossing and marker-assisted selection, Pacol, serving as the donor parent for abaca bunchy top virus disease resistance, was incorporated into traditional abaca varieties. BANDALA is intended for use in textiles, handicrafts, pulp, and paper production. This chapter outlines the progress in abaca breeding in the Philippines and the specific strategies used to generate various types of abaca varieties.

A. G. Lalusin (✉)
Institute of Crop Science, College of Agriculture and Food Science, University of the Philippines, Los Banos, Laguna, Philippines
e-mail: aglalusin@up.edu.ph

K. F. M. Salem et al. (eds.), *Breeding and Biotechnology of Leaf, Fruit, and Seed Fiber Crops*, Advances in Plant Breeding Strategies 11,
https://doi.org/10.1007/978-3-032-00407-9_1

Keywords Abaca (*Musa textilis* Nee.) · Abaca bunchy top virus · BANDALA · Breeding · Resistance

1.1 Introduction

Abaca (*Musa textilis* Nee.), a plant native to the Philippines, is well known for its strong fiber, commonly referred to as Manila hemp. It belongs to the Musaceae family and has a chromosome count of 2n = 20, in contrast to the banana's 2n = 22. Abaca fiber is highly versatile and is widely used in making ropes, textiles, pulp, specialty papers, and handicrafts. These specialty papers include currency notes, filter papers, stencil papers, and tea bags (Lalusin 2010). As a significant contributor to the Philippine economy, the abaca industry is experiencing growth, driven by the increasing global demand for biodegradable materials and sustainable forestry practices. The continuous discovery of new applications for abaca enhances its industrial relevance, thanks to its exceptional natural properties. Philippine abaca plays a crucial role in advancing environmental sustainability and forest conservation on a global scale.

With the rising preference for eco-friendly and organic products, manufacturers, particularly in the US and Europe, are turning to natural, biodegradable materials like abaca. Known for its strength and durability, abaca outperforms synthetic alternatives in various industries. It is particularly favored in naval and commercial shipping, as well as oil exploration, due to its non-slip qualities and minimal impact on marine ecosystems. Similarly, the pulp and paper industries value abaca pulp for its resilience and recyclability. Paper made from abaca fibers can be recycled far more frequently than those derived from wood or other natural sources, making it an ideal substitute for coniferous pulp in paper manufacturing. Given that the global pulp and paper industry consumes approximately 200 million metric tons of wood pulp annually, integrating abaca as a raw material could significantly help reduce deforestation. With its superior characteristics and environmentally friendly nature, abaca stands out as a sustainable alternative for industries worldwide.

Abaca plants can be harvested for fiber within 2 years of planting, allowing for continuous production cycles that contribute to global tree conservation. The crop is well-suited for reforestation farming systems, enhancing biodiversity when intercropped with coconut palms and other tree species in former monoculture plantations and rainforest restoration projects. Additionally, abaca cultivation helps mitigate soil erosion and sedimentation in coastal areas, benefiting marine ecosystems. Its ability to improve soil water retention plays a vital role in preventing floods and landslides, while its minimal soil depletion and low land requirements make it an environmentally sustainable crop. Abaca waste materials also serve as organic fertilizer, further enhancing soil health and sustainability. However, the Philippine abaca industry faces challenges due to the crop's susceptibility to diseases, leading to genetic erosion. To address this, large-scale rehabilitation efforts are underway in key abaca-growing regions to ensure long-term industry sustainability. Breeding programs are being revitalized with more adaptive strategies to meet evolving demands for abaca fiber (Lalusin and Villavicencio 2015). Advanced plant breeding techniques are being employed to develop improved varieties with higher fiber yield, superior quality, and increased resistance to major diseases affecting the crop.

These efforts integrate conventional breeding methods with cutting-edge advances in molecular biology and biotechnology to create targeted solutions for disease management. Biotechnological approaches enable the isolation of resistance genes from abaca varieties or their wild relatives. With advanced insights into plant-pathogen interaction mechanisms, the development of abaca varieties with durable resistance is expected to progress rapidly. These improved varieties could be directly used for commercial cultivation or serve as vital germplasm for creating other high-yielding, disease-resistant strains. This chapter discusses the advances in abaca breeding in the Philippines and the breeding techniques employed in the development of these abaca varieties.

1.1.1 Current Cultivation and Production of Abaca

The Philippine abaca sector maintains a robust presence in both global and local markets, yielding an annual revenue of US$100M (PSA 2024). Dominating the worldwide abaca trade, the Philippines supplies approximately 86.1 percent of global demand for abaca fiber, while Ecuador and Costa Rica contribute 13.7 percent and 0.2 percent, respectively. As of 2022, abaca cultivation covered 155,455 hectares, yet only 86,854 hectares were effectively utilized, resulting in a fiber output of 17,640 metric tons (MT) (PSA 2024). The structure of abaca farms primarily comprises smallholdings owned and operated by individual farmers, averaging 0.69 hectares per farmer. Cooperative-managed farms, spanning 10–100 hectares, are scarce in comparison.

The Philippines maintains its position as the world's foremost abaca producer, capturing an 85% share of global imports. Concurrently, domestic demand is rising at a rate of 5.7%. The vital role of abaca is emphasized by its significant influence on both the economy and society, notably seen in the prevalence of smallholder farmers, who typically possess about 2 hectares of land. Cultivation-wise, abaca dominates in the Eastern Visayas and Bicol regions, accounting for 36% and 30% of the total planted land area, respectively.

Over the last decade, abaca fiber production in the Philippines has maintained a relatively stable average of 65,701 metric tons (MT) annually (Fig. 1.1). Focused on three main regions, the Bicol Region, Eastern Visayas, and Davao Region, these areas collectively contribute 79% to the nation's total abaca output over the period. Notably, Catanduanes remains the leading abaca producer among provinces, producing 18,971 MT in 2021, representing 33% of the region's total production. Leyte closely follows with 12%, while Northern Samar holds the third position with 7%. However, abaca production in the Bicol region has faced fluctuations due to recurring typhoons, notably impacting Catanduanes, the primary production hub.

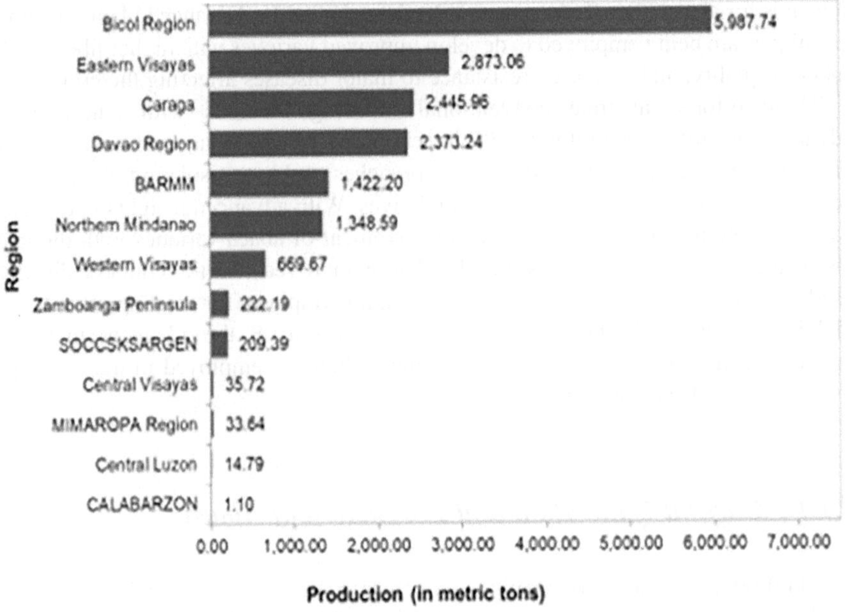

Fig. 1.1 Distribution of abaca production by region in the Philippines from April to June 2023. (Source Philippine Statistics Authority (PSA))

Abaca has been cultivated in the Philippines for centuries and was widely known to Filipinos long before Spanish colonization. The crop was grown, processed, and traded, with its products even used as a form of tax payment. When the Spaniards arrived on Cebu Island in 1521, they observed locals wearing garments made from abaca fibers, indicating that weaving was already a well-established practice. However, the commercial and export potential of abaca was not immediately recognized.

Historical records indicate that in 1820, a U.S. Navy lieutenant introduced abaca fiber to the United States, leading to its first export in 1825. Over time, abaca became renowned for its exceptional tensile strength and durability in saltwater, making it one of the strongest materials for marine cordage. Efforts to cultivate abaca in other regions, including India (1822), Borneo, German East Africa, the West Indies, and Florida, proved either unsuccessful or commercially unviable (Copeland 1911).

By the twentieth century, abaca had become one of the Philippines' top export commodities. To secure a stable supply of marine cordage for the U.S. Navy, the U.S. government promoted abaca cultivation in regions with climates similar to that of the Philippines. In 1923, abaca was introduced to several Latin American countries with tropical, humid, and warm climates, including Panama, Ecuador, Costa Rica, Guatemala, Honduras, the Dominican Republic, Brazil, British and French Guiana, Cuba, Jamaica, Puerto Rico, Martinique, Guadeloupe, Trinidad, Mexico, St. Vincent, Bolivia, Peru, Nicaragua, and El Salvador (Dempsey 1963). Additionally, in 1925, abaca seed pieces from the Philippines were used to establish plantations in Sumatra, as well as in British Borneo and Malaya.

Abaca was also introduced to New Caledonia and Queensland (Torres and Garrido 1939), and an abaca plant was reportedly discovered growing in a garden in New Zealand. In 1958, the crop was introduced to Vietnam using seed pieces sourced from Costa Rica. After World War II, a Japanese plantation owner in Davao, Philippines, conducted formal field tests and successfully cultivated abaca in Ecuador. Today, Ecuador is a key producer, supplying approximately 15% of the global abaca market.

In the Philippines, abaca is cultivated in nearly all regions except Ilocos, Cagayan Valley, parts of Region 3, and the provinces of Cavite and Batangas in Region 4. Small-scale experimental plantings have also been conducted in Ilocos Norte, Rizal, Sultan Kudarat, and Tawi-Tawi. Abaca thrives in areas with abundant rainfall, high humidity (78–88%), and average temperatures below 27 °C, with no prolonged dry seasons.

Commercially known as Manila hemp, abaca has expanded beyond its traditional use in rope and paper production to applications in the fashion and automotive industries. Philippine designers craft embroidered barong shirts from abaca sinamay, while Japanese martial artists use makiwara wrapped with abaca rope for training. Its exceptional tensile strength, folding durability, and high porosity make it ideal for a wide range of products, including currency paper (Philippine peso bills contain 20% abaca, while Japanese yen bills contain 70%), furniture, home décor, textiles, cosmetics, cigarette papers, surgical masks, tea bags, and coffee filters.

Globally, the Philippines and Ecuador dominate abaca production, yielding approximately 60,000 MT and 10,000 MT of fiber annually, respectively. The Philippines contributes about 85% of the world's supply, focusing primarily on exporting pulp over raw fiber. Annual exports—including raw fiber, pulp, cordage, and fiber crafts generate approximately USD 76.8 million for the country. The Philippines cultivates abaca mainly through smallholder farms, while Ecuador relies on large estates. Indonesia, a minor producer, is gradually expanding its abaca cultivation.

Ranked ninth among the Philippines' top agricultural exports, abaca trails products such as coconut oil, bananas, pineapples, and tuna. Key export markets include the United Kingdom, Japan, and the United States for abaca pulp, with Germany serving as the largest buyer, followed by Japan and the UK. The USA, Singapore, and Canada are primary buyers of abaca cordage. However, foreign exchange earnings from abaca exports have declined by 2.8% annually, except for abaca pulp, which has seen a 2.6% annual growth. Meanwhile, earnings from raw fiber, cordage, yarns, fabrics, and fiber crafts have experienced downward trends (Fig. 1.2).

1.1.2 Importance of Abaca in the Philippine Economy

Abaca is a vital fiber crop in the Philippine economy, supported by a complex supply chain that connects local farmers to international markets. The process begins with

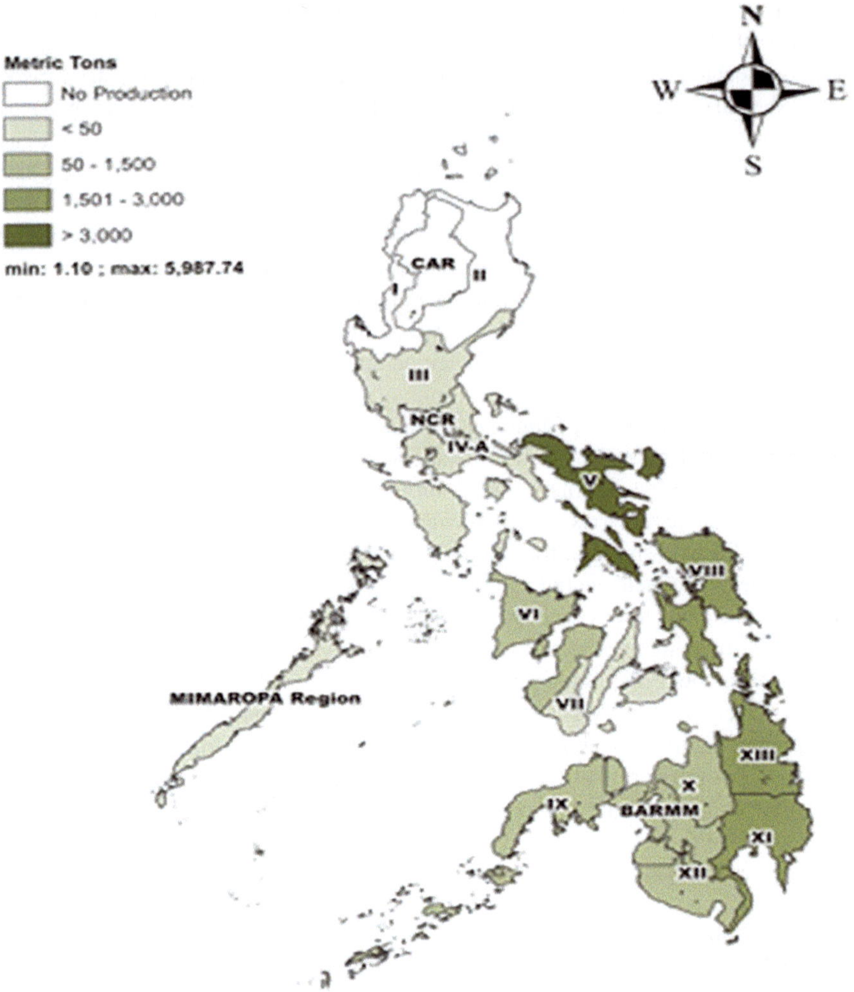

Fig. 1.2 Map of the Philippines showing the provinces where abaca is being produced from April to June 2023. (Source: Philippine Statistics Authority (PSA))

producers, primarily farmers and fiber strippers, who sell their abaca fibers in an "all-in" condition meaning the fibers are ungraded directly to barangay (village) dealers. At this local level, the fibers are aggregated for further processing. From there, the fibers are sold to town or city dealers, who typically carry out basic sorting and grading. In some instances, farmers bypass these intermediaries and sell directly to exporters or grading and baling establishments (GBEs). GBEs are pivotal in the supply chain, responsible for grading and compressing the fibers into bales suitable for both domestic and international markets. To enhance efficiency and reduce reliance on intermediaries, some farmers' cooperatives and associations have established direct partnerships with domestic processors. These cooperatives help farmers secure better prices and more stable markets by streamlining the supply chain (PhilFIDA 2020).

The graded and baled abaca fibers eventually reach manufacturers, who transform them into a wide variety of products such as ropes, textiles, and specialty paper. While this multi-tiered trading system facilitates the movement of fibers across the market, it also introduces challenges for farmers, including issues related to pricing, market access, and maintaining quality standards.

1.1.2.1 Local Consumption

Domestic processors utilize approximately 66% of the Philippines' total abaca fiber production, playing a crucial role in the country's abaca industry. The steady increase in local demand, averaging 3.2% annually, has contributed significantly to economic growth by generating higher income, creating employment opportunities, and fostering the production of high-value goods.

Among the various sectors utilizing abaca fiber, the pulp industry has emerged as the fastest-growing segment, driven by the increasing global demand for specialty paper products. These include meat casings, tea bags, cigarette papers, and high-quality filtration materials. Over the past decade, the pulp sector has accounted for approximately 57% of total domestic consumption, experiencing an annual growth rate of 6.9%. The strong performance of this sector is attributed to the superior qualities of abaca pulp, such as high tensile strength, durability, and recyclability, making it an essential raw material in the paper manufacturing industry.

The cordage sector, historically a major consumer of abaca fiber, has faced challenges in recent years. While it has contributed around 31% of total domestic fiber usage, its demand has declined by an average of 1.4% per year over the last decade. This decline is primarily due to increasing competition from synthetic cordage materials, which offer lower production costs and greater uniformity. Despite this, abaca cordage remains preferred in specific applications, such as marine and naval industries, where its superior durability, non-slipping properties, and minimal environmental impact provide a distinct advantage over synthetic alternatives.

As global markets continue shifting toward sustainable and biodegradable materials, the abaca industry, particularly the pulp sector, is well-positioned for further expansion. Strategic investments in processing technology, research on improving fiber quality, and efforts to promote abaca-based products internationally can further strengthen the industry's long-term growth and competitiveness.

1.1.2.2 Export

The Philippines' abaca industry plays a vital role in the country's economy, generating an average of US$79 million annually from exports of both processed abaca products and raw fiber. The majority of these earnings—approximately 82% (US$65 million)—come from processed products such as pulp, cordage, and fiber crafts, while the remaining 18% (US$14 million) is derived from raw fiber exports. Among processed abaca products, abaca pulp is the largest contributor to export revenues, accounting for 39% of total earnings. Fiber crafts, which include handmade paper products, woven goods, and decorative items, follow at 22%, showcasing the industry's growing diversification. Meanwhile, cordage and allied products contribute 14%, while yarns and fabrics make up 0.7% of total export value.

The United States remains the primary market for abaca cordage, largely due to its applications in the maritime, shipping, and oil exploration industries. Japan and Germany are the leading importers of abaca pulp, primarily for their specialty paper industries. Over the past decade, demand for abaca pulp has expanded to France, Taiwan, Korea, and China, reflecting the increasing global preference for sustainable and biodegradable materials. Similarly, the market for fiber crafts is strongest in the United States, Germany, Japan, and Australia, where consumers favor eco-friendly, handmade products. In terms of raw fiber exports, the Philippines ships an average of 12,887 metric tons (MT) per year, with the United States, the United Kingdom, and Japan accounting for 91% of total shipments. However, raw fiber exports have been gradually declining at a rate of 2.3% per year, as the industry shifts its focus toward exporting higher-value processed products. This shift has helped sustain overall growth, ensuring a positive long-term trajectory for the Philippine abaca industry.

1.1.2.3 Sectors of the Abaca Industry

The abaca industry is made up of five major sectors: farming, fiber stripping, trading, fiber exporting and processing.

Farming Sector

The abaca industry plays a vital role in the Philippines' agricultural sector, sustaining the livelihoods of more than 77,500 farmers who cultivate around 127,258 hectares of land. Most abaca farms are small-scale, with an average size of two hectares per farmer.

However, productivity remains a challenge, as the current national average fiber yield is only 850 kg per hectare, significantly below the potential yield of 2000 kg per hectare. This disparity highlights the need for better cultivation methods, disease-resistant abaca varieties, and stronger support systems to enhance production and maximize the industry's growth.

Several challenges hinder productivity. Farmers often lack access to high-quality planting materials, modern farming technologies, and sustainable practices. Diseases like abaca bunchy top virus (ABTV) and mosaic diseases significantly reduce yields, while inadequate infrastructure, such as poor post-harvest facilities and limited transportation, further restrict profitability. Additionally, climate change and extreme weather, including typhoons, pose significant risks to abaca farming.

To address these issues and boost productivity, interventions are needed. Introducing improved abaca varieties resistant to pests and diseases, training farmers in modern techniques, and strengthening government and private sector support can enhance yields and economic returns. Investments in post-harvest infrastructure and the development of climate-resilient farming systems are also critical. By addressing these challenges, the Philippines can maximize the potential of its abaca industry and improve the livelihoods of its farmers, ensuring long-term sustainability and growth.

Fiber Stripping Sector

Abaca fiber stripping in the Philippines is carried out manually or using mechanical methods, with about 80% of fibers hand-stripped and the remaining 20% produced through spindle stripping machines. Hand-stripping, a traditional and labor-intensive technique, is primarily practiced in the Bicol region and certain parts of Leyte and Samar provinces, whereas spindle stripping is used in Mindanao and some areas of Leyte.

The process of fiber stripping involves three main steps: cutting down or tumbling of the mature abaca stalks, followed by tuxying or removing the stalks' outer layers to expose fiber bundles and then air-drying the stripped fibers to meet quality standards (Fig. 1.3). Fiber strippers are compensated through two systems: cash payment or a sharing arrangement. In the sharing system, strippers receive 50%, 60%, or 70% of the harvest's value, based on locally agreed practices. This approach ensures fair compensation for workers while accounting for the labor-intensive nature of stripping. Hand-stripping is widely practiced due to its ability to produce high-quality fibers, though mechanical methods are gaining ground in some regions for efficiency (Fig. 1.4). The combination of traditional and modern practices

Fig. 1.3 Manual stripping of abaca fibers (**a**), the stripped abaca fibers were collected (**b**). (Source: AG Lalusin, ICropS, CAFS, the University of the Philippines Los Baños (UPLB))

Fig. 1.4 Harvested abaca fibers were being dried under the sun near the river (**a**) and along the street (**b**) in Catanduanes Province, Philippines. Mr. Vicente M. Borabo, an expert in abaca fiber, inspects the quality of the fibers. (Source: AG Lalusin, ICropS, CAFS, the University of the Philippines Los Baños (UPLB))

supports the abaca industry's sustainability while providing employment and income for thousands of workers.

Trading Sector

The Philippine abaca trading system is a multi-tiered structure that facilitates the movement of abaca fibers from small-scale farmers to larger markets. Trading occurs at different levels, determined by the farmers' locations and fiber accumulation points. These levels include barrio, town, provincial, city, and regional traders, with a total of 617 traders operating nationwide.

At the barrio level, traders work closest to farmers, purchasing abaca directly from producers. These fibers are then sold to towns and provincial traders, who consolidate larger volumes from multiple sources (Fig. 1.5). Finally, city and regional traders accumulate fibers from various provinces for bulk sales to domestic and international markets. Each trading level applies a markup to the price of abaca to account for services like transportation, storage, and logistics. While this system ensures efficient distribution, it also means that prices increase with each level, potentially limiting the income of farmers at the start of the chain. To improve profitability for farmers, interventions such as streamlining the supply chain or creating direct market access opportunities could reduce intermediaries and

Fig. 1.5 Transport of extracted abaca fibers from the farmer's field to the buyer in Catanduanes Province, Philippines. (Source: AG Lalusin, ICropS, CAFS, the University of the Philippines Los Baños (UPLB))

associated costs. Strengthening cooperatives or farmer groups may also enhance their bargaining power and ability to retain a greater share of the final market price.

Fiber Exporting Sector

Fiber exporters, along with grading and baling establishments (GBEs), are essential in the abaca supply chain, ensuring fibers are processed and prepared for both domestic and international markets. These establishments are primarily located in major abaca-producing regions but maintain liaison offices in Metro Manila to facilitate business operations and communication.

GBEs employ classifiers, whose responsibility is to ensure that the abaca fibers meet the government-mandated standards for quality. This step is vital for maintaining the reputation of Philippine abaca in global markets and ensuring consistency in fiber quality. A key function of GBEs is the operation of pressing machines used for bailing fibers. This process compacts the fibers into standardized bales, facilitating storage, transportation, and trade. The standard bale weighs 125 kilograms and measures 100 cm × 55 cm × 60 cm, providing a uniform unit for trading purposes.

Through these activities, GBEs serve as the final processing point in the supply chain before fibers reach buyers, ensuring that abaca products meet both local and international standards. Their role is essential for maintaining quality control, improving marketability, and sustaining the competitiveness of the Philippine abaca industry in the global market.

Processing Sector

Pulp Mills

In the Philippines, there are six abaca pulp companies that play a significant role in the global market. Together, these companies have a collective rated capacity of 16,180 metric tons (MT) per year for producing abaca pulp, which is primarily used in the paper and specialty products industries.

These companies have developed strong market networks, focusing largely on international demand. The global demand for abaca pulp, known for its strength and durability, is primarily driven by industries such as high-quality paper production, textiles, and other specialty products. The Philippines, being the world's largest producer of abaca, holds a competitive advantage in supplying these markets. The ability of these companies to maintain high production capacity and access global markets helps sustain the Philippine abaca industry, providing significant economic benefits, especially to those involved in the processing and export sectors. This also enhances the country's position as a key player in the global pulp market, ensuring that the demand for abaca pulp continues to be met with high-quality, sustainably sourced materials.

Cordage Manufacturers

Six cordage firms operate across key regions in the Philippines, including Metro Manila, Cebu, and Davao. These companies primarily use abaca as raw material to manufacture various products, such as ropes, cordage, and twine (Fig. 1.6). To meet specific buyer requirements, the companies sometimes blend abaca with other natural fibers, such as maguey, enhancing the properties of the final products.

Fig. 1.6 Abaca cordage, crafted from abaca fibers, is a durable and robust type of rope known for its exceptional strength and resistance to saltwater. Its reliability makes it a preferred choice for marine and fishing uses, as well as industrial applications like rigging and towing. (Source: AGLalusin, ICropS, CAFS, the University of the Philippines Los Baños (UPLB))

These cordage companies have a combined rated capacity of approximately 21,350 metric tons (MT) per year, highlighting the industry's significant scale. Abaca's strength and durability make it an ideal material for manufacturing robust and high-quality cordage products, which are in demand for various applications, including maritime, agriculture, and construction industries.

The production of cordage products not only supports local economies but also positions the Philippines as a key global supplier of natural fiber-based ropes and twine. By utilizing abaca and other natural fibers, these companies contribute to the sustainable use of the country's agricultural resources while tapping into both domestic and international markets.

Fiber Craft Manufacturers

The fiber craft sector in the Philippines, encompassing handmade papermaking and carpet manufacturing, is predominantly cottage-based and operates mainly in rural areas, particularly in the central regions of the country (Fig. 1.7). This sector serves as an important source of livelihood, employing many, particularly women and out-of-school youth. The work in this sector involves traditional, artisanal techniques that produce distinctive, high-quality products. Handmade paper and fiber-based carpets are often crafted with creativity, showcasing the unique designs and craftsmanship of local artisans. These products are valued for their functional qualities as well as their artistic appeal, making them attractive to both local and international markets. Over time, many of these manufacturers have successfully penetrated international markets, gaining recognition for their eco-friendly and sustainable

Fig. 1.7 (**a**, **b**) Abaca handicrafts (hats, rugs, baskets and flower vases) are artisanal products made from the fibers of the abaca plant. Known for its strength, durability, and flexibility, abaca fiber is one of the most sustainable materials for crafting. Abaca handicrafts are highly valued for their eco-friendliness and natural aesthetic, often incorporating traditional weaving techniques and cultural artistry. (Source: AGLalusin, ICropS, CAFS, the University of the Philippines Los Baños (UPLB))

products. This has allowed the sector to expand its reach, boosting the livelihoods of rural communities and creating economic opportunities for artisans.

In addition to offering economic benefits, the fiber craft sector helps preserve traditional skills and promotes the use of natural, renewable fibers, contributing to sustainable production practices in the Philippines. The sector's success highlights the potential of combining artisanal expertise with global market demand for unique and environmentally conscious products.

Textiles/Fabrics

The textile and fabric sector in the Philippines specializes in producing handwoven abaca fabrics, known for their versatility and wide range of applications. These fabrics are used as raw materials for creating novelty and household items, décor, wrapping materials, and fashion wear and accessories. To cater to diverse markets, the sector incorporates various design elements, such as metallic threads or polyester blends in some weaves, while others feature striped and ethnic patterns to suit different consumer preferences.

The textile/fabric industry is primarily concentrated in regions like Western Visayas, the Bicol Region, and Southern Mindanao, where indigenous communities, particularly in T'nalak weaving, are integral to the production of these abaca fabrics (Fig. 1.8). Tinalak, a traditional handwoven fabric, holds cultural significance and is woven by local artisans, often using intricate patterns passed down through generations.

In contrast, the development of new product lines, especially in fashion wear, accessories, and specialty or novelty items, is mainly concentrated in Metro Manila. This urban hub serves as the focal point for design innovation and market expansion, linking traditional weaving techniques with contemporary fashion trends to cater to both local and international markets. This integration of craftsmanship with modern designs has helped the sector thrive and diversify its product offerings.

1.2 Challenges of the Abaca Industry

Despite its historical significance, the Philippine abaca industry continues to struggle with persistent challenges, particularly in farm productivity and fiber quality. Since the 1980s, abaca production has significantly declined, largely due to the absence of improved varieties and the widespread impact of three major viral diseases: abaca bunchy top (ABT), abaca mosaic (AM), and abaca bract mosaic (BM). Although resistant lines have been identified in the abaca germplasm collection, they often yield fibers of lower quality. Despite ongoing disease management efforts, controlling these viruses remains a major challenge. Other diseases further compound the challenges to abaca production. Developing superior hybrids with resistance to viral and pathogenic diseases through conventional breeding and the use of biotechnology

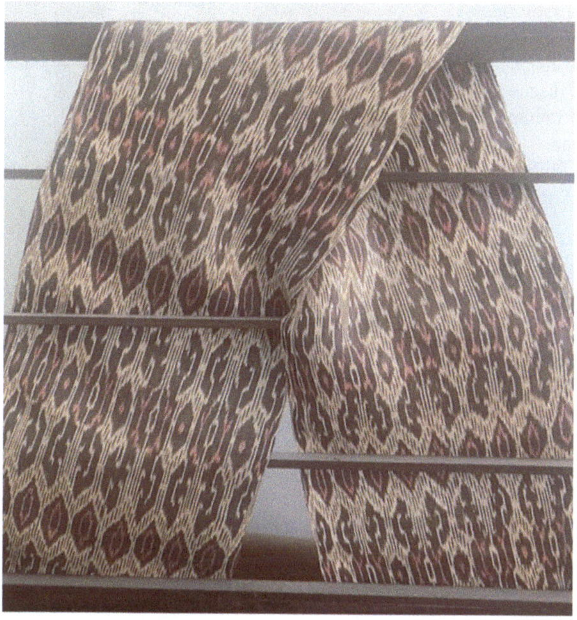

Fig. 1.8 T'nalak (or tenalak) is a traditional handwoven textile of the Tboli people from South Cotabato, Philippines. It is crafted using resist-dyed abaca fibers and holds deep cultural and spiritual significance. The weaving process is exclusively undertaken by Tboli women who receive the intricate designs in their dreams, a gift they attribute to Fu Dalu, the Tboli Goddess of Abaca. This unique practice makes each T'nalak piece not only a work of art but also a sacred expression of the weaver's connection to their heritage and spirituality. (Source: AGLalusin, ICropS, CAFS, the University of the Philippines Los Baños (UPLB))

offers potential solutions, and ongoing research to identify disease-resistant sources aims to sustain the industry.

Among these diseases, abaca bunchy top (Fig. 1.9), abaca mosaic, and bract mosaic virus are the most economically damaging, often coexisting in the same regions. For instance, in 1991, the Bicol region reported an average abaca bunchy top and mosaic disease incidence of 5.19%, resulting in fiber yield losses exceeding 800,000 kg, valued at approximately ₱18 million (Raymundo 2002). In Eastern Visayas, the incidence was 8.16%, with losses valued at around ₱8 million. Beyond reducing yield, these diseases also significantly degrade fiber quality. Bunchy top disease was first observed in Albay (Bicol region) between 1910 and 1911 (Ocfemia 1926) and later in Silang, Cavite, in 1915. Initially regarded as a minor concern, its increasing severity led to the abandonment of plantations in Cavite and Laguna by 1923. Today, the disease is widespread across abaca-growing regions and is highly destructive, causing severe stunting and significantly reducing productivity. Key symptoms include leaf crowding into a rosette-like formation and a transparent appearance of the main and secondary leaf veins (Ocfemia 1931b). The disease is caused by a virus closely related to the banana bunchy top virus (BBTV) and is

Fig. 1.9 Mature traditional abaca plant (**a**) which is very susceptible to the abaca bunchy top virus disease. (**b**). (Source: AG Lalusin, ICropS, CAFS, the University of the Philippines Los Baños (UPLB))

transmitted by the aphid *Petalonia nigronervosa* Coq. Genetic analysis of critical viral genes—replication initiation protein, coat protein, and movement protein—shows a 99% similarity between the abaca bunchy top virus (ABTV) and BBTV (Furuya et al. 2006).

Abaca mosaic disease was first documented in 1933 after an outbreak at the Odell Plantation in Tagum, Davao del Norte (Calinisan 1934) and has since caused significant economic losses. By 1941, Japanese planters in Davao estimated that around 50% of the abaca in eastern Davao was infected due to poor cultural practices. The disease spread rapidly due to several factors, including transmission by various aphid species, the abandonment of plantations during World War II, and the post-war expansion of abaca cultivation, which frequently relied on infected planting materials. The economic impact of abaca mosaic is evident in reduced fiber yields and income losses from poorer-quality fibers. Infected plants produce fibers with higher stretch percentages, a result of altered chemical composition. Additionally, studies show that abaca mosaic significantly reduces tensile strength, biomass yield, fiber output, plant height, stalk diameter, and farmer income. Effective disease management and the development of resistant varieties remain critical to mitigating these impacts and ensuring the industry's sustainability.

Abaca bract mosaic disease remains relatively understudied. The first confirmed case of natural infection with the banana bract mosaic virus in abaca was reported in the Philippines in 2000 (Sharman et al. 2000). Symptoms can appear at any growth stage, with infected plants showing young leaves marked by chlorotic stripes. The leaf lamina often develops spindle-shaped chlorotic streaks running parallel to the veins, though these may be less noticeable in newly infected younger leaves. In older leaves, raised leaf symptoms may also be observed.

Veins extending from the midrib form a pattern resembling continuous ripples. Petioles may develop greenish to yellowish streaks or spindle-shaped lesions, though these may be absent in older leaves already exhibiting leaf lamina symptoms. When dead leaf sheaths are removed from the pseudostems, distinct dark-colored mosaic patterns, stripes, or spindle-shaped streaks become apparent (Furuya et al. 2006). The defining characteristic of the disease is the dark reddish-brown mosaic pattern on the bracts of the inflorescence. In the absence of bracts, abaca bract mosaic symptoms can be easily mistaken for those of abaca mosaic.

Several factors have contributed to the widespread nature of these diseases, including a lack of resistant varieties. While some resistant germplasms have been identified, they often produce lower-quality fibers, limiting their commercial viability. Farmers often rely on traditional methods and fail to adopt improved techniques for disease management, such as crop rotation or the use of certified disease-free planting materials. During periods of economic expansion, new plantations were frequently established using infected seedlings, facilitating the spread of diseases to previously unaffected areas. Smallholder farmers, who dominate abaca cultivation in the Philippines, often lack the financial resources and technical knowledge to implement disease mitigation measures.

Efforts to combat these diseases focus on several key areas. Conventional breeding programs and advances in molecular biology are being utilized to develop varieties resistant to viral and pathogenic diseases. While progress has been made, balancing resistance with fiber quality remains a challenge. Extension services and research institutions are promoting integrated pest management (IPM) approaches, including biological control agents for aphid vectors and the use of certified disease-free planting materials. Farmers' cooperatives and agricultural extension programs aim to provide training on best practices, access to resistant planting materials, and financial support to enhance disease resilience in abaca farms. Innovations in biotechnology, such as the identification of resistance genes in wild relatives of abaca, offer hope for creating robust, high-yielding varieties that can withstand disease pressures.

The abaca industry remains a cornerstone of the Philippine economy, but it is under significant threat from persistent diseases and outdated farming systems. To sustain production and protect the livelihoods of farmers, a multi-faceted approach is essential. This includes continuous research on disease-resistant varieties, improved farming practices, and investments in education and infrastructure to enhance disease management. With these efforts, the abaca industry can secure its position as a global leader in fiber production, catering to both traditional and emerging markets.

1.3 Taxonomic Group and Botanical Description

Abaca (*Musa textilis* Nee), also known as Manila hemp, is a plant species native to the Philippines and belongs to the *Musaceae* family, which includes various cultivated banana species and their varieties (Table 1.1). The genus *Musa* consists of 30–50 species and numerous hybrids, the majority of which are triploids. Within this genus, *M. textilis* is classified under the *Australimusa* section. Morphologically, *M. textilis* shares similarities with edible banana cultivars, including a pseudostem structure, but the abaca plant is noticeably slimmer, has smaller leaves, and produces seeded fruits. Unlike bananas, abaca is valued not for its fruit but for its fibers, which are extracted from its leaf sheaths. The Philippines has over 400 identified abaca cultivars, yet only around 20 are commercially significant. These commercial varieties are primarily distinguished by their fiber quality, yield, and disease resistance. Despite its economic importance, further taxonomic research is needed to better understand the species and its full genetic diversity, which could support breeding and conservation efforts.

Spanish friar and botanist Manuel Blanco was the first to classify Philippine bananas, including wild varieties such as *butuhan* and *saging maching*, as well as abaca and the *Virgin* banana, all of which were categorized under *Musa troglodytarum* Linn. He listed abaca with the botanical variety *textoria* (Valmayor et al. 2002).

In 1915, Dr. Nicanor G. Teodoro reclassified Blanco's *Key to Musa Species in the Philippines* using collections from the College of Agriculture at the University of the Philippines Los Baños (UPLB) as reference materials. His *Key to Species and Varieties of the Genus Musa* became widely used for describing banana cultivars in the Philippines. However, it was later replaced by the taxonomic frameworks of Cheesman (1948), Simmonds and Shepherd (1955), and more recently, by the classification system adopted by curators of the national banana variety collections in Southeast Asia (Valmayor et al. 2002).

Table 1.1 Taxonomic classification of abaca (*Musa textilis* Nee)

Kingdom	*Plantae*—plantes, Planta, Vegetal, plants
Subkingdom	*Viridiplantae* - green plants
Infrakingdom	*Streptophyta*– land plants
Superdivision	*Embryophyta*
Division	*Tracheophyta*– vascular plants, tracheophytes
Subdivision	*Spermatophytina*– spermatophytes, seed plants, Phanérogames
Class	*Magnoliopsida*
Superorder	*Lilianae*– monocots, monocotyledons, monocotylédones
Order	*Zingiberales*
Family	*Musaceae*– banana
Genus	*Musa L.* – banana
Species	*Musa textilis* Née – Manila-hemp, abaca, abaca

Although there are many synonyms for *Musa textilis* L. Nee, (syns: *Musa trogloditarum* Linn. var. textoria Blanco, *Musa sylvestris* var. mindanauensis G. E. Rumphius, *Musa sylvestris* var.amboinensis G. E. Rumphius, *Musa mindanensis* G. E. Rumphius ex F. A. W. Miquel, *Musa sylvestris* L. A. Colla, *Musa abaca* G. (G. G.) S. Perrottet, *Musa sylvestris* Lamarie, *Musa textilis* Nee var. tashiroi Hayata, *Musa formosana* Hayata), there is little confusion and debate as to the taxonomic classification of abaca unlike its close relative, the banana. Abaca is a monocot belonging to the banana family *Musaceae* of the Order *Zingiberales,* Genus Musa, Section Australimusa (Copeland 1927; Valmayor et al. 2002; Szent-Ivany and Catley 1960). Its natural hybrids, Canton and Minay, *Musa balbisiana* Colla x Musa textilis Nee, are sources of fiber used to adulterate abaca fiber. Another source of adulteration is pacol, *Musa balbisiana* Colla, another indigenous species in the country.

Musa textilis Nee Section Australimusa has a basic chromosome number of 10 (n = 10 or 2n = 20). The following botanic description was that of Cheesman as modified by Brewbaker et al. (1956) and 1 here modified further to include variations described in various literatures. Indicated colors are based on the Royal Horticultural Society Color Guide (Tabora and Carlos 1978). Plant stooling freely, 7–51 suckers per hill; pseudostem or stalk 2.0–7.6 m high, glossy, color varies from greyed purple to blood red, nickel green, Agathia green, Scheele's green with either greyed yellow, green, Neyron rose, greyed purple streaks, tinge, coloration, or blotches and some with brown or jet-black base. The base circumference ranges from 12 to 63 cm and the top circumference ranges from 6 to 34 cm. The leaf sheaths range from 10 to 25 per stalk, compactly arranged, with the outermost ones being loose and completely detached from the pseudostem. Both the leaf sheaths and petioles lack a waxy coating. The leaf blades are oblong, tapering towards the apex, measuring 1.45–2.95 meters in length and 31–56 centimeters in width. They have a narrow truncate tip and a rounded, asymmetrical base. The midribs are green, occasionally exhibiting a purplish hue, similar to the lamina. The petioles, ranging from 30 to 62.75 centimeters in length, are relatively stout and stiff, positioning the leaves at a high angle. Their margins are well-developed and curved in, nearly enclosing the adaxial groove above, while remaining closely pressed against the pseudostem at the base without becoming scarious.

The inflorescence initially grows sub-horizontally, with a peduncle and rachis that are either glabrous or minutely puberulent. The sterile bract is green, reaching up to 50 cm in length with an acute tip. The basal flowers are female, with the number of female hands ranging from 3 to 9, while the upper hands are male (Fig. 1.10). Female flowers 6–12 per bract in two rows; ovary 5 cm long, green, glabrous; compound tepal 4 cm long, 1.2 cm wide, white with pale green lobes. The lateral lobes are oblong-lanceolate, measuring 5 mm in length, with filiform dorsal appendages of the same length. The central lobe also has a dorsal appendage, though slightly shorter. The free tepal is 3.5 cm long and 2 cm wide, white, boat-shaped, and corrugated in the upper third, with a minute apicula. Staminodes are about half the length of the styles.

Fig. 1.10 Inflorescences of traditional abaca (**a**), Dioscoro 1 or BANDALA (**b**), a second-generation backcross between abaca and pacol (wild banana) and developed fruits of traditional abaca (**c**). (Source: AG Lalusin, ICropS, CAFS, the University of the Philippines Los Baños (UPLB))

The male bud, when in advanced bloom, is oblong ovoid, ranging from 11 to 22 cm in length and 5–10 cm in width, with a blunt tip. The bracts are strongly imbricate, with the outermost being about two-thirds the length of the bud. Their coloration varies from greenish brown to purplish-brown, sometimes green at the rounded tip. The outer surface is smooth and polished, while the inner surface is much paler, with a white base. The bracts lift one at a time but persist for 2–3 days after the subtended flowers have fallen, so multiple bracts may remain open simultaneously. They are firm in texture, do not curl when fading, and are first strongly reflexed before becoming deciduous. Bract dimensions range from 6 to 16 cm in length and 4–9 cm in width.

Male flowers occur in clusters of 8–12 per bract in two rows. The compound tepal is 3.8 cm long, white or cream-colored, with a slightly more yellow upper part and lobes. The lobes measure about 3 mm in length, each with a 1 mm-long dorsal filiform appendage, and accessory teeth that are nearly as long as the lobes, sometimes also with appendages. The free tepal is 3.0 cm long and 1.2 cm wide, translucent white, acute, and barely apiculate. Stamens slightly extend beyond the flower at maturity.

The fruit grows in bunches, with individual fruits measuring 3.7–7.0 cm in length (including the pedicel) and 2.0–7.5 cm in diameter in the middle. They are oblong-ovate or ellipsoidal, with faintly angled surfaces at maturity. The base narrows into a stout pedicel of 1–1.5 cm in length, while the apex abruptly tapers into a truncate acumen about 7 mm long. The pericarp is 1 mm thick and turns yellow upon ripening. The pulp is scanty and pale buff in color.

Each fruit contains 36–160 seeds, which are sub-globose to turbinate, highly irregular in shape, and generally taller than they are wide, measuring approximately 2–3 mm across and 3–4 mm in height. The seed surface is smooth, with color variations, appearing either white or black.

1.3.1 Distribution

Abaca (*Musa textilis*) is widely grown across the Philippines, except for a few regions such as the Ilocos Region, Cagayan Valley, Region 3 (Central Luzon), Cavite, and Batangas. While these areas do not have commercial-scale cultivation, small-scale experimental plantations have been established in select locations, including Ilocos Norte, Rizal, Sultan Kudarat, and Tawi-Tawi, to assess the plant's adaptability to different environmental conditions.

Abaca thrives well in warm, humid climates with consistently high rainfall. Abaca requires substantial moisture to grow optimally. The Philippines receives an average annual rainfall ranging from 965 mm to 4064 mm, ensuring adequate water supply for the plant. Prolonged dry periods can be detrimental to abaca growth, as they may cause moisture stress, leading to reduced fiber yield and quality. Areas without a pronounced dry season provide the best environment for sustainable abaca cultivation. The ideal relative humidity for abaca cultivation ranges between 78% and 88%. These levels help maintain soil moisture and prevent excessive transpiration, allowing the plant to develop strong, high-quality fibers. The plant flourishes in areas where the average temperature remains below 27 °C, with a national mean of 26.6 °C. Extreme heat or sudden temperature fluctuations can negatively impact plant health and fiber quality.

The Philippines is classified into different climate types, with Types II, III, and IV being the most suitable for abaca farming. Climate Type II is characterized by a lack of a dry season and significant rainfall, particularly from November to January. This climate type ensures that abaca plants receive consistent water throughout the year, making it one of the most favorable environments for cultivation. Climate Type III does not have sharply distinct wet and dry seasons, it is relatively dry from November to April and wet during the rest of the year. This moderate variation still allows abaca to grow well, as long as irrigation is available during drier months. Climate Type IV, on the other hand, has evenly distributed rainfall throughout the year, providing stable moisture conditions that support continuous abaca growth and fiber development.

The combination of these climate conditions directly affects the productivity and quality of abaca fibers. In areas with optimal rainfall and humidity, the plant produces longer, stronger, and finer fibers, which are highly valued for export and industrial applications. Conversely, regions with prolonged dry seasons or inconsistent rainfall may experience lower fiber yields, weaker plant growth, and increased susceptibility to pests and diseases.

By cultivating abaca in suitable environments, farmers can maximize yield and fiber quality while ensuring sustainable production for both domestic and international markets.

1.3.2 Cultivation

Abaca productivity varies significantly based on environmental and soil conditions. The plant thrives best in fertile soils of volcanic or alluvial origin, which provide essential nutrients for healthy growth and high-quality fiber production. These soils are typically rich in organic matter, well-aerated, and capable of retaining moisture while allowing proper drainage. Ideal soil types include friable loam, very sandy loam, and silty loam, all of which support strong root development. For optimal growth, abaca requires at least 50 cm of topsoil underlain by another 50 cm of fine sandy loam. This layered soil structure ensures sufficient root penetration and prevents waterlogging. However, abaca struggles in areas with shallow soils, hardpan layers, or high-water ables, as these conditions restrict root expansion and impede water movement, ultimately reducing plant vigor and fiber yield.

In addition to soil quality, abaca flourishes in warm, humid environments with an average annual temperature of around 27 °C. While the plant can grow at elevations of up to 1200 meters above sea level, higher altitudes may lead to lower temperatures, which can slow growth and fiber development. Consistently high humidity also plays a crucial role in maintaining plant moisture levels and supporting strong fiber production. However, adverse conditions such as poorly drained soils, compacted subsoil layers, and low temperatures can hinder abaca's productivity. It tolerates a broad pH range (4.0–8.0) but prefers neutral to slightly alkaline soils. Abaca has a high potassium demand, necessitating substantial amounts of bases such as calcium and magnesium, and thrives in soils rich in organic matter (Raymundo 1974). Ensuring that abaca is cultivated in regions with optimal soil and climate conditions is key to maximizing fiber yield and maintaining sustainable production.

Suitable soils for abaca cultivation include Andepts like the Mayon, Guinobatan, Sariaya, Tupe, Isarog, and Magallanes series, as well as associated alluvium and colluviums; Udolf soils like the Annam, Ipa, Macolod, La Castellana, Miral, Ibaan, and Catbalogan series, along with associated well-drained alluvium and colluvium; and Udult soils like the Louisiana, Castilla, Guimbalaon, Tugbol, Kidapawan, Adtuyon series, and Paleudalf (Umingan clay loam) soil. However, Haplustult soil

(Maasin clay loam) deficient in phosphorus and Tropept soil (unmapped mountain soil) deficient in sulfur are unsuitable for robust abaca growth (Armecin and Ferraren 2001). Moisture is another critical factor for optimal abaca growth. The ideal soil moisture level ranges from 60% to 80% saturation, yet abaca cannot tolerate waterlogging (Hernais and Espino 1923). Abaca in the Philippines is produced by small farmers with average landholdings of about 2 hectares and by farmers living in or near forest zones. Abaca plantations remain productive until the 20th year but production peaks from the second year to the 15th year after which the production decreases to about 1/3 of peak production at the 20th year. Soil amelioration has been suggested prior to replanting old plantations.

An update of current knowledge about abaca by highlighting recent data on its characteristics, fiber, crop management, productivity, and market potential was reported by Araya-Gutiérrez et al. (2023). It was found that the strength and physical properties of abaca fiber are influenced by factors such as leaf position, plant maturity, and fiber extraction method. There are two main extraction methods: stripping and decortication. Productivity in abaca fiber extraction varies greatly depending on the method used. Traditional hand stripping, a labor-intensive process, produces approximately 20 kg of fiber per day. This highlights the need for more efficient extraction techniques to improve yield and meet increasing market demand. The study concludes that abaca has great market potential, but further research on its production and management is necessary to support its broader use.

A study on Indigenous Peoples (IPs) engaged in abaca farming in North Cotabato revealed that production relies on family labor, while postproduction tasks like harvesting, tuxying, and fiber stripping are manually performed under sharing agreements, taking about a month per hectare (Calica et al. 2024a). The resulting low fiber quality leads to reduced market prices and incomes. To supplement earnings, women engage in value-added activities like crafting abaca products. The findings highlight the need for interventions, including mechanization, training on fiber quality improvement, and capacity-building for value-added production, to enhance efficiency, incomes, and market opportunities for IP communities involved in abaca farming. The Indigenous Peoples (IPs) of North Cotabato were traditionally engaged in abaca fiber production using manual hand-stripping methods, yielding low-quality output and limited income (Calica et al. 2024b). A government project, in collaboration with PhilMech and the Kidapawan LGU, introduced mechanized processing systems, including a stripping machine, an all-weather dryer, and storage pallets. This innovation improved fiber quality, with 61% to 78% classified as excellent, compared to 44% with manual methods. Farmers achieved better financial outcomes and accepted higher processing fees due to reduced labor costs and faster processing times. However, hauling raw materials to processing centers remains labor-intensive, requiring solutions like a tramline system.

Aklan Province contributed 6% of national abaca production. However, increasing international demand remains unmet due to challenges such as improper practices, pests, diseases, and insufficient technology and government support. Abaca production in Aklan's top two municipalities was examined by surveying 408 farmers (Señeris 2024). Most respondents were male, aged 51–60, with over

21 years in abaca farming. Farmers averaged 928–890 plants/ha, far below the standard 2500 plants/ha. Challenges included natural calamities (35%) and lack of materials (34%). While 39.6% received support from PhilFIDA, this study highlights the need for stronger policies to boost production.

1.3.3 Domestication

Abaca (Musa textilis) is native to the Philippines and northern Indonesia, where it has been cultivated for centuries. The plant is highly valued for its strong, durable fibers, which are used to produce a wide range of products such as ropes, textiles, specialty papers, and more. Abaca was first domesticated by the Filipinos, and its cultivation remained largely confined to the Philippines until the early twentieth century (Spencer 1953). In 1925, abaca was introduced to Sumatra (Balifurnish 2012), where it began to be successfully cultivated. Prior to this, multiple attempts to introduce the plant to other regions were unsuccessful. For instance, India attempted cultivation in 1822, and later, there were efforts to grow abaca in German East Africa, the West Indies, and Florida, USA, but none of these efforts resulted in successful, widespread cultivation. Since the 1920s, however, abaca cultivation has expanded to other parts of Southeast Asia, as well as to regions such as Central America and Ecuador, where the plant has adapted well to local climates. Today, abaca is also grown in Equatorial Guinea and Kenya, further extending its reach beyond its native regions. Despite these expansions, the Philippines remains the dominant producer and exporter of abaca worldwide.

The plant's continued spread demonstrates its adaptability to various tropical climates and its importance as a natural fiber in global markets. Abaca fibers are highly sought after due to their strength, durability, and eco-friendly nature, making them ideal for products like cordage, textiles, and specialty papers. These qualities have allowed abaca to maintain its significance in both local and international industries.

1.3.4 Genetic Resources

The Philippines is widely regarded as the center of origin for abaca, from where it later spread southward to Borneo (Brewbaker and Umali 1956; Spencer 1953). According to the Vavilovian Crop Centers of Origin, abaca belongs to the Indo-Malayan Center, an area known for its rich biodiversity in plant species. The Philippine archipelago, in particular, falls within the region of the greatest *Musa* diversity, making it a crucial reservoir of genetic resources for the genus.

Several species within the *Musa* genus, including *Musa balbisiana Colla, Musa acuminata Colla,* and *Musa textilis Nee* (abaca), are indigenous to the Philippines. Their natural distribution overlaps, and because they are cross-compatible, various interspecific hybrids have emerged over time. This natural hybridization process has contributed to the vast genetic diversity of *Musa* in the country, further enriching its germplasm collection and offering opportunities for genetic improvement in breeding programs (Valmayor et al. 2002).

The collection and study of *Musa* germplasm in the Philippines date back to the Spanish era, when Spanish friar Manuel Blanco documented various banana species in his illustrated book *Flora de Filipinas*. During the American occupation (1911–1912), the Philippine Bureau of Agriculture launched an extensive banana collection program, initially assembling 276 accessions from 22 species. This collection, along with earlier descriptions by Blanco, laid the foundation for a comprehensive study by Dr. Eduardo Quisumbing, a professor at the University of the Philippines Los Baños (UPLB) in 1919. His research uncovered significant synonymy and duplication in the germplasm collection, leading to a reduction in the recorded accessions.

The outbreak of World War II (1941–1945) caused the loss of many plant and animal germplasm collections in the Philippines due to war-related destruction. However, by this time, the combined accessions of wild and cultivated *Musa* species at UPLB and the Bureau of Plant Industry had increased to 298. Following the war, efforts to rehabilitate the abaca industry were prioritized, especially in regions such as Davao and Bicol, where abaca mosaic disease had caused severe damage. Chemical control measures proved ineffective against the disease, prompting a shift toward long-term strategies, including breeding for resistance. Consequently, UPLB and the Bureau of Plant Industry launched breeding programs that utilized resistant wild Musa relatives to develop disease-resistant hybrids.

Today, the National Abaca Research Center (NARC) at Visayas State University in Baybay, Leyte, holds the largest abaca germplasm collection in the world, consisting of 518 cultivated species, 105 hybrids, and 150 wild relatives, totaling 773 accessions sourced from across the country. However, more than half of this collection was wiped out due to the abaca bunchy top virus disease. In addition to maintaining in-field accessions, abaca specimens are also preserved *in vitro* for conservation and research purposes. The NARC collection has been extensively studied for its fiber morphology, chemical composition, fiber quality, and physical properties, contributing to ongoing breeding and improvement programs. Notably, while *Musa textilis* has played a genetic role in some seedless hybrid banana varieties, it has not yet been incorporated into any formal breeding program for edible bananas (Fig. 1.11) (Vaughan 2011) .

Genetic diversity analysis using microsatellite markers of the abaca germplasm of the University of Philippines Los Banos (UPLB) found that the germplasm collection in Luzon Island has the highest diversity based on the Shannon diversity index (*H*) (Lalusin et al. 2007). The first comprehensive genetic diversity analysis of abaca (*Musa textilis* Née) germplasm in the Philippines using simple sequence repeat (SSR) markers was conducted by Yllano et al. (2020). This study examined 150 abaca accessions from Luzon, Visayas, and Mindanao, along with two banana cultivars, to assess genetic diversity and relationships. Out of 44 banana-based SSR primers tested, six were highly polymorphic, detecting 28 alleles with an average of

Fig. 1.11 Mature Pacol (wild banana) plant showing resistance to abaca bunchy top virus. (Source: AG Lalusin, ICropS, CAFS, the University of the Philippines Los Baños (UPLB))

4.55 alleles per locus and a polymorphism information content (PIC) mean of 0.56. Overall, the genetic diversity of abaca across the three major islands was high (I = 0.68), with Luzon exhibiting greater genetic variation than Visayas and Mindanao. Molecular variance analysis showed that 95% of the genetic variation occurred within populations, while only 5% was attributed to differences among populations. Cluster analysis using the Unweighted Pair Group Method with Arithmetic Mean (UPGMA) identified nine distinct clusters, forming two major groups and seven smaller groups, regardless of geographic origin. The use of SSR markers effectively distinguished abaca accessions and closely related species, identifying potential duplicate accessions. These findings provide strong genetic evidence of the extensive diversity within Philippine abaca germplasm. The molecular characterization of abaca using SSR markers is valuable for identifying superior genotypes and enhancing both *in situ* and *ex situ* conservation efforts. Additionally, these insights can support breeding programs aimed at improving abaca fiber quality, disease resistance, and overall genetic resource management.

1.3.5 Floral Biology and Pollination

Once the plant fully develops, the true stem emerges through the pseudostem, producing its inflorescence (Spencer 1953). The abaca inflorescence has a rachis with clusters (hands) of female and male flowers, maturing from base to tip. Clusters have two rows of flowers, covered with colorful bracts that change from light yellow-green to dark red-purple as they bloom. Female flowers are at the base and male flowers are at the tip of the inflorescence. Female flowers have a compound tepal, a free tepal, five degenerate stamens, and a well-developed inferior ovary. Male flowers are distinguished by a nonfunctional pistil, a reduced ovary, and large stamens.

Female flowers bloom between 6 PM and 7 AM, with bracts opening depending on the temperature. Typically, one bract opens per day, but on hot days, two may open. Female flowers bloom for 4–6 days, longer during cooler months (December and January) and shorter during hotter months (March to May) in Los Banos, Laguna. Blooming starts at the flower stalk base and continues downward, with all flowers in a cluster opening simultaneously. Peak receptivity of the stigma occurs between 11 AM and 12 Noon, and it may remain receptive for 2 days. Anther dehiscence in male flowers occurs early in the morning on sunny days and later in the afternoon on cloudy days, releasing pollen grains immediately upon corolla opening. Pollen grains lose viability after 24 hours under ordinary conditions (Javier and Oracion 1988).

Abaca's inflorescence structure prevents self-pollination, but cross-pollination within the same hill can occur. The first bracts contain female flowers, while inner bracts have only male flowers, making abaca strictly cross-pollinated (Torres and Garrido 1939). Insects like wasps, ants, and bees, attracted by the nectar, are common pollinators, along with bats and birds. The stigma remains receptive for about 2 days, aiding cross-fertilization. Once fertilized, the flower turns brown and shrivels within 24 hours. Male flower panicles can be stored in the shade for 7 days, with pollen remaining viable for fertilization (Torres and Garrido 1939; Torres et al. 1952).

The fruits of abaca, resembling small bananas with three ridges, mature about 150 days after pollination. Ripening starts with the first cluster, followed sequentially by others. Seeds are prepared for planting by washing off the pulp. They germinate in 10–52 days, with viability varying by variety. Air-drying seeds for 24 hours increases germination while sun-drying or soaking in hot water kills them. Seeds can remain viable for up to 155 days when buried.

1.4 Advances in Abaca Breeding in the Philippines

Advances in abaca breeding have been essential in enhancing the sustainability and profitability of abaca farming, which supports industries such as textiles, cordage, paper production, and specialty products. With increasing demand for abaca fiber, breeding efforts focus on improving key traits, including disease resistance, yield, fiber quality, and environmental resilience. Abaca is highly susceptible to diseases like the abaca bunchy top virus (ABTV) and mosaic diseases, which reduce yields and fiber quality. Breeding programs aim to develop resistant varieties through crossbreeding and genetic modification, helping reduce crop losses and improving plant health. Current average yields in the Philippines are around 850 kg per hectare, far below the potential 2000 kg per hectare. Breeding focuses on selecting plants with faster growth, higher fiber production, and better adaptation to local growing conditions, maximizing the output per hectare. High-quality fiber is critical for specialty paper and cordage. Breeding efforts focus on enhancing fiber length, strength, and uniformity, improving the commercial value of abaca and meeting industry demands. Abaca is vulnerable to climate change, including drought, heavy rains, and typhoons. Breeding programs aim to develop varieties with drought tolerance and flood resistance, ensuring long-term viability. New tissue culture and clonal propagation methods produce genetically uniform plants, improving planting efficiency and reducing establishment time for new plantations. Marker-assisted selection (MAS), genetic mapping and genomics technology allow breeders to select plants with desirable traits more quickly, speeding up the development of improved abaca varieties. Some breeding programs focus on preserving and incorporating indigenous abaca strains, which may have unique qualities or resistance traits, ensuring that local knowledge and cultural heritage are respected.

These advances are critical for maintaining abaca's role as a versatile, eco-friendly fiber source in global markets. As breeding techniques continue to evolve, abaca will remain an important agricultural commodity, meeting growing industry demands while supporting sustainable farming practices.

1.4.1 Conventional Abaca Breeding

The primary objective of the hybridization program at the time was a high-yielding and mosaic-resistant abaca. Two approaches were put in place, firstly inter-varietal crosses, where two varieties whose desirable traits may be found in their offspring. If the F_1 does not show the desired trait, the hybrid may be selfed and selection made in the F_2 generation or the F_1 may be backcrossed to the more desirable parent. Crossing commercial varieties was done before the start of CAIP but hybrids were lost either to disease or by project stoppage due to lack of funds. Secondly, the approach of inter-specific crosses was adopted because of the report that the search for resistance in abaca has proved futile. Pacol (*M. balbisiana*), primarily, was used in this approach which resulted in hybrids that 'failed to show symptoms of mosaic' (Umali et al. 1956).

Musa textilis can successfully cross with several *Musa* species, including *M. acuminata, M. balbisiana, M. lolodensis,* and *M. borneensis* (Brewbaker et al. 1956). However, their hybrids often exhibit partial or high sterility. Cross-pollination in abaca is facilitated by the stigma's receptivity, which lasts for approximately 2 days. Upon fertilization, the flower browns and shrivels within 24 hours. The male flower's heart or panicle can be collected and stored in the shade for up to 7 days, during which the pollen remains viable and effective for fertilization (Torres and Garrido 1939; Torres et al. 1952).

The native Musa species, including *M. acuminata*, *M. balbisiana*, and *M. textilis*, have overlapping distributions, leading to the existence of natural hybrids among them. A specific cultivar of *M. balbisiana* known as pacol produces low-quality fiber, which has been used as an adulterant in abaca production. In the Bicol region, natural hybrids between pacol and abaca are referred to as Canton and Minay (Valmayor et al. 1956).

The basic chromosome number for the *Eumusa* section, which includes edible bananas, is $n = 11$, while the *Australimusa* section, which includes abaca, has $n = 10$. The natural hybrid between diploid banana and abaca, known as Minay/Minary/Minray, has a chromosome count of $2n = 21$ (Pancho et al. 1959; Tabora and Carlos 1978), whereas Canton has $2n = 20$ (Valmayor et al. 1956). Canton is largely sterile, but Minay occasionally produces seeds. Crosses between *M. balbisiana* and *M. textilis* have resulted in hybrids with morphological traits and chromosome counts similar to Canton and Minay (Bernardo 1957). Additionally, hybrids between Minay and abaca, where abaca served as the male parent, have been successfully developed. These hybrids resemble Minay more than abaca (Brewbaker and Umali 1956). Artificial hybridization is generally more effective when abaca acts as the male parent. Crosses such as Libuton x Itom and Canorajan x Lagurhuan were also created (Oyardo 1974).

Between 1928 and 1931, 54 different crosses were attempted, with 29 successfully planted at the Guinobatan Abaca Experiment Station and 10 in Silang, Cavite (Torres and Garrido 1939). Additionally, 39 clones underwent disease-resistance screening. Heterosis was observed in F1 abaca hybrids, which produced more suckers than their parent plants. Crosses involving Maguindanao showed better adaptability and stronger root systems. In 1939, field testing was conducted on abaca hybrids, and they were named after their parent lines. For instance, hybrids included Linlay (*Linawaan x Laylay*), Linlib (*Linawaan x Libutanay*), and Linino (*Linawaan x Inosa*) (Labrador 1928). Other hybrids tested in the field included Itolaus (*Itom x Lausigon*), Itomag (*Itom x Maguindanao*), and Lausimag (*Lausigon x Maguindanao*) (Cruz and Balingkit 1974).

The abaca varietal improvement program at the University of the Philippines Los Baños (UPLB) began in the early 1950s through a collaboration between the university's College of Agriculture (UPCA) and the Bureau of Plant Industry (BPI). The program focused on collecting, classifying, and evaluating abaca varieties, establishing disease observation nurseries, clonal selection, and both intra- and inter-specific hybridization. A key goal was the development of disease-resistant abaca varieties, with the most significant breakthrough being the identification of Pacol as a source of resistance. Although hybridization efforts between Pacol and abaca were initiated, the project was discontinued in the 1960s. The abaca collection was subsequently maintained by the UPLB Forestry Abaca Gene Bank before being transferred to the UPLB Experiment Station in 1981.

Diaz (1997) developed F1 hybrids by crossing Mininonga with six abaca varieties and screening them for resistance to the bunchy top virus. Among the various crosses attempted, only the following successfully produced F1 seedlings: Malaniceron x Mininonga, Mininonga x Itolaus 39, Mininonga x Layahon, Mininonga x Putumag 22, Mininonga x Tinawagan Puti, and Sogmad Pula x Mininonga. These hybrids exhibited varying levels of resistance to bunchy top, with Malaniceron x Mininonga, Mininonga x Itolaus 39, and Mininonga x Layahon showing resistance, while Sogmad Pula x Mininonga demonstrated moderate resistance.

In 1999, the Institute of Plant Breeding at UPLB identified six abaca hybrid genetic stocks: Itolaus 39 x Magsarapong 2, Itolaus 39 x Magsarapong 3, Itolaus 39 x Magsarapong 4, Itolaus 39 x Magsarapong 7, Itolaus 39 x Magsarapong 8, and Tetraploid x Ilolaus 39. These genetic stocks were developed through conventional breeding involving Magsarapong, Tetraploid 1, and Itolaus 39—an abaca hybrid with moderate resistance to both mosaic and bunchy top viruses. Itolaus 39, a commercial hybrid of Itom and Lausigon, visually resembles Pacol. All identified stocks displayed moderate resistance to the mosaic virus and strong resistance to the bunchy top virus in greenhouse conditions, with high resistance observed in field evaluations. However, their fiber quality and recovery were lower than those of traditional abaca hybrids.

UPLB's abaca breeding program officially commenced in 1981. By 1986, the institute released six F1 hybrids between Pacol and abaca, which exhibited resistance to the bunchy top virus but had subpar fiber quality. Due to a lack of funding, the hybridization program was halted. Breeding efforts resumed in 2006 on a limited scale, leading to the evaluation of several BC1 crosses.

A total of 196 BC2 progenies were developed and evaluated for both resistance to the abaca bunchy top virus (AbBTV) and fiber quality. Among these, six promising clones were identified, exhibiting strong resistance to AbBTV along with desirable fiber characteristics. During the evaluation process, 1300 seedlings from four BC2 populations were manually inoculated with viruliferous aphids (*Pentalonia nigronervosa*). Each seedling was exposed to 10 aphids that had previously fed for 24 hours on abaca leaves infected with the bunchy-top disease. After 2 months, the majority of the inoculated seedlings showed no symptoms of the virus. A total of 196 seedlings were selected based on their resistance to bunchy top, overall vigor, and morphological similarity to true abaca. These included 12 seedlings from *BC1-19 x Abuab*, 132 from *BC1-20 x Abuab*, and 22 from *BC1-19 x Musa tex 51*. To further confirm their resistance, the selected seedlings underwent a second round of aphid inoculation. To screen for desirable morphological traits, the BC_2 abaca hybrids were selected and planted in the field with a 2.5 m × 2.5 m planting distance under coconut trees as partial shade. The abaca hybrids were used for the evaluation of morphological traits, fiber qualities and field resistance to bunchy top virus disease. These abaca populations were also used to screen the primers and to identify crosses with resistance to abaca bunchy top virus at the same time with good fiber qualities.

To verify whether progenies from the cross between *Musa textilis* Nee ($2n = 20$) and *Musa balbisiana* Colla ($2n = 22$) developed at the University of the Philippines Los Baños (UPLB) are true hybrids, a study was conducted to characterize abaca cultivars (Abuab and Inosa), a banana cultivar (Pacol), and their backcrosses (BC_1, BC_2, and BC_3) both morphologically and cytogenetically (Labrador et al. 2020). Agronomic traits and fiber quality were assessed through ANOVA and pairwise analysis, revealing significant differences in tensile strength, fiber quality, and fiber recovery percentage. However, no significant differences were observed in leaf sheath number and girth at the top. These findings confirmed that the recurrent parent genome (abaca) was successfully recovered through the third generation of backcrosses. Additionally, the study optimized cytogenetic techniques. The most suitable time for root tip collection was identified as 10:00 a.m. to 10:30 a.m. when a higher number of dividing cells were observed. A two-hour cold shock pre-treatment significantly improved chromosome size and the number of well-spread prometaphase cells, facilitating karyogram construction. Chromosome characteristics, including chromosome count and relative length, were analyzed and compared across plant samples. Comparative karyotyping confirmed a diploid chromosome number of $2n = 20$ for abaca cultivars and hybrids. However, the Inosa cultivar (*M. textilis* Nee) exhibited variability, with chromosome counts of either $2n = 20$ or $2n = 22$, differing from earlier reports that ranged from $2n = 17$ to $2n = 23$.

1.4.2 Breeding for Virus Resistance

Advances in abaca breeding have been ongoing to improve the quality, yield, and disease resistance of abaca plants. Abaca is susceptible to various diseases, including abaca bunchy top virus and abaca bract mosaic virus. Breeding programs aim to develop disease-resistant varieties through traditional breeding methods and biotechnology. Identifying and selecting disease-resistant plants and introducing these traits into commercial abaca varieties is a major goal. One of the primary objectives of abaca breeding is to enhance the quality of the fibers. This includes increasing fiber length, strength, and fineness. Researchers work on selecting plants with superior fiber characteristics to develop improved abaca varieties that meet the high-quality standards required by various industries. Increasing the yield of abaca fibers is crucial for the economic sustainability of abaca cultivation. Breeding programs aim to develop high-yielding abaca varieties through the selection of plants that produce more and better-quality fibers. Abaca is typically grown in tropical regions, and breeding programs work on developing varieties that can adapt to different environmental conditions. This includes developing abaca cultivars that can thrive in various soil types and climates, increasing the geographical range for abaca cultivation. Abaca plants typically take around 2–3 years to reach maturity before they can be harvested for fiber production. Breeding programs may target reducing the time to maturity, enabling more frequent harvests and higher overall fiber production. To ensure the genetic purity of cultivated abaca plants, clonal propagation techniques have been developed. These techniques involve reproducing abaca plants through tissue culture or vegetative propagation to maintain desirable traits and characteristics consistently. Advances in molecular genetics and genomics have enabled the identification of specific genes and markers associated with desired traits in abaca. Marker-assisted selection allows breeders to accelerate the development of improved varieties by selecting plants with the desired genetic traits. Abaca breeding programs may also focus on developing sustainable and environmentally friendly cultivation practices. This includes selecting varieties that require fewer inputs, such as fertilizers and pesticides, and have a lower environmental impact. Advances in abaca breeding are essential for the continued success of the abaca industry and for ensuring that abaca remains a valuable and sustainable natural fiber resource. These advancements not only benefit farmers and industries but also contribute to the conservation of abaca genetic diversity and the preservation of traditional abaca-based cultures in the Philippines and other abaca-growing regions.

Host-pathogen interactions typically lead to one of two host reactions: resistance or susceptibility. This reaction can be qualitative or quantitative, rooted in genetics. Resistance is defined as the host's ability to hinder pathogen development, encompassing pre-infection, post-infection, morphological, anatomical, and biochemical defenses. In contrast, a completely susceptible host allows unrestricted pathogen growth, while a resistant host employs various mechanisms to impede pathogen growth.

Resistance is diverse, reflecting different aspects of research specialization. Genetics and plant breeding relate to monogenic and polygenic resistance, while biochemical studies are associated with hypersensitive resistance. Terms like race-specific, race-nonspecific, differential, non-differential, vertical, horizontal, mature-plant, and seedling resistance describe various characteristics and dynamics of resistance. The durability and genetic basis of these resistance types vary.

Host-pathogen interactions are dynamic and complex, often involving multiple genetic interactions, making definitive conclusions challenging to reach. Developing pathogen-resistant plant varieties is more challenging than traits like yield or height enhancement. It necessitates a deep understanding of host-pathogen interactions and the influence of external factors on epidemics. Handling epidemics caused by more virulent pathogen genotypes, often referred to as new strains or races, is a continuing challenge driven by evolutionary forces. Breeding for resistance aims to manage this dynamic struggle. Successful breeding for resistance requires collaboration between plant breeders and plant pathologists. Deviations from this partnership can lead to varieties unable to keep up with pathogen evolution. In abaca breeding, collaborative efforts are essential for success.

Centers of origin and diversity of abaca as gene sources Leppik (1970) proposed a theory that had a far-reaching influence on breeding for resistance to plant pathogens. He postulated that the centers of origin of plant species are also their centers of diversity and that an array of different kinds of resistance and virulence genes would have emerged during the struggle between host plants and their pathogens in a gene-for-gene interaction as envisioned by Flor (1956). In this Darwinian 'survival of the fittest' setting, both host and pathogen are forced to develop various mechanisms to survive. As will become clearer later in this chapter, these mechanisms are different types of resistance in the host plant and of variable strains or races in the pathogen. The interplay of both mechanisms in evolutionary time will lead to a genetic equilibrium and co-existence. The Philippines, more specifically the Bicol Region, is recognized as the 'most probable center of origin of the species Musa textilis (Brewbaker et al. 1956; Umali and Brewbaker 1956). If one were to follow the tenets on gene centers expounded by Leppik (1970), a wealth of diverse germplasm harboring an array of resistance genes that would have emerged during the co-evolution between abaca and viruses based on the gene-for-gene hypothesis (Flor 1956; Person 1959) should be found. In this center of diversity, the mechanisms of indigenousness that lead to co-evolution and biodiversity creating an ecosystem equilibrium that allows host-pathogen co-existence, should be operating. This theory of indigenousness, which has been elucidated by Browning (1974), explains why epidemics do not occur in agroecosystems with diverse and heterogeneous plant populations. These populations over time develop polygenic, general, and other types of resistance as balancing mechanisms that negate the likelihood of epidemics. The role of the gene-for-gene theory in early attempts at breeding for resistance to abaca viruses was not apparent. There was a delayed awareness of this new theory. The hypothesis, which ensures that in the interplay of host genes and pathogen virulence genes, the host finds itself always one step behind, was proposed after Bernardo and Umali (1956), Umali and Brewbaker (1956) and other plant breeders have already instigated breeding programs designed to produce mosaic resistant varieties. In retrospect, as will be elucidated in later sections, the decision to utilize wild relatives in the Musa species is an implicit understanding of the center of diversity and B. Sources of genes for resistance to abaca virus's gene-for-gene theories although the genetics of host resistance to abaca viruses had yet to be understood while variability in abaca viruses was a peripheral issue. At the time, the techniques for evaluating host reaction to abaca viruses, likewise, appeared inadequate and inconsistent with the infection process. It was, likewise, not possible to 'inoculate' each offspring with viruliferous aphids due to the sheer number of offspring and the time needed for efficient insect transfer.

1.4.2.1 Sources of Resistance Genes

Abaca germplasm in the Philippines is quite diverse. Distinct abaca cultivars can be found in different regions although the possibility of the same genotypes being given different names exists. The nomenclature used in naming indigenous varieties suggests that many are strains from a common source, but which have undergone improvement by selection over time. Many popular varieties, such as Tinawagang Pula and Tangongon, are widely grown despite being susceptible to prevailing viruses. Some genotypes have been classified as resistant or highly resistant during the seedling and mature stages of plant growth based on various symptom expression criteria. In the course of a wide-ranging search for resistance against the virus disease of abaca during the early to middle part of the last century, there have been signs of resignation that this search will not be successful. Ocfemia (1931a) reported that their work 'to date has shown that all varieties of abaca used, including first-generation variety hybrids, are infected by the bunchy top in the same degree'. He added that 'there seems very little hope that resistant forms will be developed. Crossing wild species might give resistant forms but there is a danger that this may reduce the quality of the fiber'. Twenty-five years later, Umali and Brewbaker (1956), likewise, theorized that 'resistance to this disease (mosaic) does not appear to occur within the species M. textiles' but admitted that 'no thorough surveys have been made to substantiate this theory.' Due to the aforementioned uncertainty about the existence of resistance to viruses in abaca varieties, the search for resistance genes shifted to other Musa species and relatives. In a seemingly straightforward but remarkable article, Bernardo and Umali (1956) tested several Musa and related species, including putative hybrids and varietal crosses. The entries included: Musa species: Pacol, *M. balbisiana* Colla. Agotay, M. banksii F. V. V. M. Saguing maching, M. banksii F. V. M M. ornate Roxb. Putative hybrids Canton, *M. textilis* x *M. balbisiana*, Minay, *M. textilis* x *M. balbisiana* Red Alinsanay, *M. textilis* x *M. banksia* White Alinsanay, *M. textilis* x *M. banksia* Ensete species'Virgen' banana, *E. glaucum* (Roxb.) Cheesman Abaca (*M. textilis* Nee) Twenty-one varieties and four inter-varietal crosses in a result that provided proof of the validity of Leppik's center of origin and diversity theory and Flor's gene-for-gene hypothesis when applied to the Musa plant species, they found Pacol, Agotay and *M. ornata* completely resistant. Consequently, the hybrid crosses, Canton, Minay, and Alinsanay, showed a high to moderate degree of resistance indicating the possibility of the existence of 'strong' and 'weak' genes and different gene actions involving either one or both types. Additional experimental data were provided by Umali et al. (1956) from tests showing Pacol x abaca F_1 hybrids as completely resistant or immune to abaca mosaic and bunchy top, an indication of dominance gene action. The aforementioned interactions are common in many host-parasite systems with variable symptom expressions at different crop growth stages implying the effect of polygenes. The situation is quite apparent in the abaca bunchy top and mosaic virus disease systems. In 1970, two hybrids, P x CES 111-1 and P x CES 111-2 were released from a breeding program for resistance to bunchy top and mosaic viruses (Tabora 1970). Engle (1985), in screening 43 abaca cultivars and hybrids for resistance to bunchy top, identified Canton Bicol, Canton Ecogar, Canton Jamboree and Minsog as highly resistant while Baunan, Canton L. D., Layahon, Lunhan and Sogmad were moderately resistant. Subsequently, the Institute of Plant Breeding (1989) released six abaca hybrids that, after 3 years of tests, were found resistant to the bunchy top virus and moderately resistant to the Abaca mosaic virus under greenhouse conditions. All were rated as resistant to both viruses under field conditions. Some of the abovementioned varieties and resistant cultivars Malaniceron and Sogmad Pula together with moderately susceptible Putumag

22 and Tinawagan Puti were used by Diaz (1997) in generating hybrids that were subsequently evaluated for resistance to bunchy top utilizing the scale of Lomerio (1998). Villajuan-Abgona et al. (2004), using the quantitative approach proposed by Raymundo (1998), and Furuya et al. (2006), employing screen house and field tests, confirmed that resistance can indeed be found in many indigenous abaca varieties. Boguero et al. (2016), in a molecular screening of abaca accessions, arrived at the same conclusion.

1.4.2.2 Breeding for Resistance to Viruses

The abaca mosaic and abaca bunchy top viruses were so destructive during the last century that the initial program of abaca improvement in the country was focused on developing a clone that was resistant to the former (Umali and Brewbaker 1956). At the time, resistance to the Abaca bunchy top virus was a secondary concern. The pronouncements of early abaca researchers were prophetic. In 1931, Ocfemia wrote 'Whether varieties of abaca highly resistant to the disease may be developed eventually is not known. Crossing with wild species might give resistant forms but there is a danger that this may reduce the quality of the fiber`. Twenty-five years later, Umali and Brewbaker (1956) stated that 'Musa species collections are of considerable importance to the abaca breeder since it appears that resistance to mosaic, wilt and bunchy top may not be available within the species of Musa textilis.' They mentioned *M. ornate*, *M. banksii* and hybrids of two species, Canton, Minay and Alinsanay as potential resistance gene sources. Bernardo and Umali (1956) and Umali et al. (1956) proceeded to test these sources and found Pacol (*M. balbisiana*) as exhibiting complete resistance or immunity to both mosaic and bunchy-top diseases. The Philippines possesses a wealth of germplasm, within the *M. textilis* species, carrying genes for resistance to abaca viruses. These genes, when transferred to existing high-yielding but susceptible cultivars through classical or modern techniques, will circumvent problems, such as incompatibility or more specific plant characteristics, such as tensile strength, that might arise when utilizing wild relatives as gene donors. In retrospect, the search for virus-resistant genotypes was an important objective of many other abaca studies throughout the last century. Invariably, many of these studies were short-term undertakings and, in some cases, secondary in importance to the quest for better agronomic traits (Alcober 1986; Cruz and Balingit 1974; Engle et al. 1999; Moreno 2001; Tabora and Carlos 1978; Torres and Garrido 1939).

Techniques utilized by the Cooperative Abaca Improvement Program (CAIP), participated in by the UP College of Agriculture, Bureau of Plant Industry, and the Bureau of Agricultural Extension, coordinated abaca research in the Philippines starting in March 1954 (Umali and Brewbaker 1956). This was the first collaborative program on abaca breeding. The objective of the CAIP was to obtain a mosaic-resistant clone', mosaic being the main problem at the time. The program operated on the prevailing notion that resistance 'does not appear to occur within the species M. textilis.' Further, as previous research on abaca selection, evaluation and hybridization was limited, the following methods, based initially on experience with similar crops, were used.

Varietal Classification and Evaluation

This involved a systematic survey of available material as most of the information from experiment stations and other repositories was lost during World War II. The primary emphasis on varietal classification was on important agronomic and inflorescence characters. Evaluation of varieties was made largely on a regional basis in replicated trials with consideration of the growing conditions of respective regions. For instance, resistance to typhoons and dry seasons was the emphasis for the Bicol Region while, at the time, choosing varieties with resistance to mosaic was the priority for the Davao Region.

Observational and Disease Nurseries

Albay in the Bicol Region, the CAIP maintained the largest Philippine collection of abaca varieties, fittingly as the region is the most probable origin of the species. At the UP College of Agriculture, Musa species, including *M. ornate*, *M. balbisiana*, *M. banksii*, *M. textilis*, cultivated edible bananas of M. paradisiaca and *M. sapientum*, and suspected hybrids of two species, Canton, Minay and Alinsanay were tested for resistance to bunchy top and mosaic viruses.

Clonal Selections and Progeny Trials

This approach involved selection from seed-grown progenies on the assumption that clonal variants and mutants might arise from single-clone plantings. The assumption was based on the abaca plant being highly heterozygous. Seeds, harvested from selected hills of a variety, were considered selfed if the plantation was planted to a single variety. They were sown in seedboxes and later transplanted in nursery rows with the fast-growing and vigorous seedlings transplanted in the field.

1.4.2.3 Virus Variability and Resistance Breakdown

Abaca virus variability is not a current concern of growers and researchers alike as there has not been a case of 'breakdown of resistance', a phenomenon associated with the 'boom and bust cycle' (Brock 1967; Priestly 1978) and the gene-for-gene hypothesis (Flor 1956) as earlier described. Many of the varieties and hybrids with varying degrees of resistance, and the Bandala variety, presumed to be bunchy top resistant (Lalusin 2010, 2014), have not been extensively planted as to have elicited enough selection pressure to force a genetic change in the prevailing virus populations. Perhaps, in due time, such can happen even when abaca resistance to viruses is, as postulated, confirmed to be quantitative. Host-pathogen interaction involving sets of corresponding genes is possible (Person 1959). A co-evolutionary model based on multilocus gene-for-gene interaction has been proposed (Sasaki 2000) There have been accounts and circumstantial evidence alluding to the possible existence of variability in abaca viruses. Sharman et al. (2008) identified two isolates of a newly discovered Babuvirus associated with 'bunchy top' symptoms—one from abacá (*Musa textilis*) in the Philippines and the other from banana (*Musa* sp.) in Sarawak, Malaysia. They proposed naming this virus *Abaca bunchy top virus* (ABTV). Their research indicated that these two ABTV isolates might constitute distinct strains, as they were genetically more different from each other than the isolates belonging to the two geographic subgroups (Asian and South Pacific) of *Banana bunchy top virus* (BBTV). Although no distinct strain has, consequently, been categorically described, the genetic diversity of abaca viruses, particularly those of bunchy top and mosaic viruses, is a critical subject that should receive serious attention as host resistance becomes a major component of abaca virus disease management. The following accounts are indications of how pathogen populations in general and viruses, in particular, evolve and survive. More specifically, the mechanisms of evolution of the closely related Banana bunchy top virus are cited to underscore the risk abaca might encounter. McDonald and Linde (2002) emphasized that the evolutionary potential of pathogen populations is influenced by their genetic structure, with those having high potential being more likely to overcome host resistance. The key evolutionary forces shaping population genetic structure include mutation, population size and genetic drift, gene and genotype flow, as well as reproduction and mating systems. Galvani (2003) further noted that factors such as spatial structuring, within-host dynamics, polymorphism in host resistance, host longevity, and population size contribute to evolutionary changes affecting disease population dynamics. In plant viruses, the primary drivers of evolutionary change leading to molecular divergence and genetic structuring include mutation, recombination, re-assortment, and the migration of both viruses and their vectors across different geographical regions (Jeger et al. 2006). Harrison (2002) explained that genomic variation in plant viruses caused by mutations is further intensified by recombination, pseudo-recombination, and the acquisition of additional genomic components. Among DNA viruses, field isolates often exhibit unexpected levels of nucleotide diversity and frequent recombination, some of which occur between different species. In geminiviruses, recombination within the same region contributes significantly to their evolutionary divergence. Garcia-Arenal and McDonald (2003) evaluated the evolutionary potential of 35 plant viruses using a compound risk index based on pathogen population size, recombination rate, and the extent of gene and genotype flow. Regarding virus adaptation to quantitative plant resistance, Montarry et al. (2012) concluded that such adaptation is possible and may lead to resistance breakdown, supporting the view that quantitative resistance may not always be durable. Molecular characterization studies have reported re-assortment and recombination in the genome of *Banana bunchy top virus* (BBTV) (Hu et al. 2007; Hyder et al. 2011; Islam et al. 2010; Selvarajan et al.

2011). Additionally, Varsani et al. (2013) highlighted that high nucleotide substitution rates contribute to the rapid evolution of BBTV.

1.4.2.4 Genetics and Variability of Abaca Insect Vectors

Viruses Information on the genetics and variability of vectors of abaca viruses is as limited as that found on resistance of the abaca plant. Nonetheless, related studies on other crop species might provide insights that can be utilized in abaca. For example, Watanabe et al. (2013) investigated the localization, accumulation, and transmission efficiency of *Banana bunchy top virus* (BBTV) in four laboratory-established lineages of *Pentalonia* aphids, each derived from a different host plant: taro (*Colocasia esculenta*), *Heliconia* (*Heliconia* spp.), red ginger (*Alpinia purpurata*), and banana (*Musa* spp.). Mitochondrial sequencing identified three of these lineages as *P. caladii* van der Goot and one as *P. nigronervosa*. Additionally, microsatellite analysis distinguished the aphid lineages into four unique genotypes. The transmission test showed that all four lineages are competent vectors of the virus. In another study of the genetic diversity and population genetic structure of *P. nigronervosa* and *P. caladii* in Hawaii, Galambao (2011) found that genetic differentiation among and between populations of the two aphids exists and that this was correlated with host plant differences. Further, aphid populations were structured according to islands within host plant groups. This situation is similar to that found in the Philippines and should be investigated. The existence of aphid vector biotypes and their possible varying capabilities to transmit the banana bunchy top virus was mentioned by Rayapati (2013) in connection with the management of banana bunchy top diseases in Africa.

1.4.2.5 Gene Deployment and Durability of Resistance

Pathogen variability, breakdown of resistance and durability of resistance are events that are intertwined. Resistance either breaks down or remains effective indefinitely depending on the composition of the pathogen population. Wide-scale planting of a single hybrid that is uniform in morphological characteristics, usually derived from crosses between different genotypes, has been described as a monoculture. It is believed to be prone to a phenomenon called crop vulnerability when the population is defenseless as every plant is considered susceptible to a biotic agent such as a pathogen. However, it is not this crop uniformity that makes the population vulnerable but rather it is a situation called genetic vulnerability when host resistance is governed by a single gene and is therefore easily matched and overcome by a new virulence gene of a pathogen. When this happens, a destructive epidemic usually follows. This phenomenon has been called the ‘boom and bust cycle’ (Brock 1967; Priestly 1978) and operates according to the gene-for-gene hypothesis (Flor 1956) in many annual crops. In this ‘man-guided’ or ‘man-made’ evolution (Johnson 1961), the plant breeder and plant pathologist are the agents of change and have always been a step behind the pathogen adversary Abaca virus pathosystems may not have behaved as aforementioned due to the perennial nature of the abaca crop, the apparent quantitative type of host resistance prevailing, and the innate evolutionary capability of viruses. However, there are indications, as described above, of the possible emergence, albeit slowly, of virus forms with different virulent patterns at the slightest sign of pressure that is detrimental to their survival. This pressure, in the context of modern agriculture, can be the deployment of resistant cultivars. The durability of a resistance gene can be prolonged when pathogen variability and evolution are adequately understood allowing for gene deployment in time (cropping seasons) and/ or space (growing areas).

While a single gene can provide resistance to a specific pathogen population, its widespread use will eventually lead to resistance breakdown. To prolong resistance durability, combining multiple resistance genes, known as gene pyramiding, has proven to be an effective strategy. It is also known as gene stacking. Resistance gene pyramids might extend the durability of resistance due to a delay or low probability for the pathogen to respond with multiple virulence genes. Gene deployment to achieve durability of resistance against plant viruses has been reinforced by the integration of techniques from other fields of science. Durability is more than the case of multigenes, gene pyramids and species mixtures as above. It is now viewed from a perspective of evolutionary ecology, population genetics, landscape epidemiology (Fabre et al. 2012) and landscape genetics, an integration of population genetics and landscape ecology, whose focus is on assessing the degree to which landscapes facilitate the movement of organisms by relating gene flow patterns to landscape structure (Holderegger and Wagner 2008; Balkenhol et al. 2016). The infusion of the new field of landscape epidemiology, which deals with pathogen dynamics at the landscape scale, into the analysis further reinforces the approach (Plantegenest et al. 2007; Reisen 2010). It should be noted that abaca plantings are bound by landscape structures such as edges, shrubs, rolling terrains and other crop species.

1.4.2.6 Field Trials of Resistant Varieties Across Multiple Locations

Even as sufficient DNA markers that can expose variability in viruses from different geographical sources are developed, new varieties or hybrids that are products of resistance breeding programs must be tested in designed multilocation field experiments. Sources of inoculum must be provided in the experimental plot to avoid disease escape. The experiments should last for at least a year to allow for sufficient time for the host and pathogen to interact and for the host reaction to be fully expressed in terms of symptoms that typically describe the disease. Field tests in major abaca growing regions with clearly defined geographical and climatic differences that might have caused the emergence of distinct virus strains are an absolute necessity. A susceptible heck variety representative of each region must be included. Inferences on host resistance based on some laboratory assay tests or on greenhouse or screen house tests utilizing seedlings are seldom good indicators of the performance of a variety at a later growth stage. Under field conditions, this becomes a real concern especially as the host-parasite interaction will be influenced by climate change and cropping systems. The design, layout, and analysis proposed by Raymundo (1998) and described under the topic Screening for Quantitative Resistance are recommended as starting points for a suitable evaluation plan. Likewise, the following method employed by Niyongere et al. (2011) in screening Musa genotypes against the banana bunchy top virus can be modified to satisfy the requirements of abaca. Various banana types, including dessert bananas (AAA group), East African highland bananas (AAA, EHB), plantains (AAB), cooking bananas (ABB), a tetraploid hybrid, and wild diploid bananas (*Musa acuminata* and *Musa balbisiana*), were evaluated for resistance to *Banana bunchy top virus* (BBTV). Ten plants per genotype were arranged in a completely randomized design (CRD), with border rows consisting of BBTV-infected Yangambi Km 5 (AAA) plants. To facilitate disease transmission, colonies of *Pentalonia nigronervosa* collected from bunchy top-infected fields were introduced into the plots. After 28 months, eight genotypes remained free of symptoms. Molecular analysis indicated the presence of the virus in three of the visibly healthy cultivars which were considered tolerant and could act as reservoirs of the virus. 5. Genetic integrity of genotypes crossed with a wild relative. The genetic integrity of progenies from crosses between *M. textilis* and wild relatives, such as Pacol, *M. balbisiana* (Umali and Brewbaker 1956), is a lingering issue that should not diminish the gains of current abaca breeding programs. Additional cycles of backcrossing to the abaca recurrent parent should resolve this impasse quickly. This may not be needed as newer molecular techniques are used to confirm varietal identity (Collard and Mackill 2008; Ratnadewi 2001).

1.5 Molecular Breeding

Plant improvement, whether through natural selection or breeding efforts, has always depended on generating, evaluating, and selecting optimal allele combinations. Enhancing even the simplest traits often requires the manipulation of multiple genes. The advent of molecular markers now allows for the identification and mapping of valuable alleles within segregating populations. Once mapped, these markers facilitate the precise breakdown of complex traits into their genetic components, equipping breeders with advanced tools to manage these traits more effectively in breeding programs.

A more sustained program in abaca breeding at UP Los Baños, a continuation of intermittent projects as earlier mentioned (Bernardo and Umali 1956; Engle et al. 1999; Tabora 1970; Tabora and Carlos 1978; Umali and Brewbaker 1956) started in 2006 utilizing Pacol variety as the primary source of resistance genes. This program is the first to employ modern biotechnology techniques successfully in abaca breeding (Lalusin 2010, 2014). The use of DNA markers has improved the efficiency and precision of conventional plant breeding through marker-assisted selection (MAS) (Collard and McKill 2008). Marker-assisted backcrossing, one of its many applications, was the one used by Lalusin (2010, 2014) and Lalusin and Villavicencio (2015), in the development of a bunchy top-resistant variety, named BANDALA (Backcross Abaca with Native and Desirable Accessions to Lift the Abaca Industry), with the aid of microsatellite markers. Lalusin (2010) described a breeding process involving multiple crosses and backcrosses between the Pacol variety, which served as the source of resistance genes, and the abaca varieties Abuab and Musatex 51. The selection was based on resistance to *Bunchy top virus*, plant vigor, and morphological similarity to true abaca. To confirm resistance, seedlings were re-inoculated with the virus using viruliferous aphids. A total of 196 backcross two (BC2) abaca populations were chosen for further evaluation of morphological traits, fiber quality, and field resistance to *Bunchy top virus*. These populations were also screened with primers to identify crosses that combined resistance to *Abaca bunchy top virus* (ABTV) with desirable fiber qualities. From the 196 BC2 hybrids assessed, six promising clones exhibiting both bunchy top resistance and high fiber quality were ultimately selected.

Genetic marker systems have a wide range of applications in *Musa* improvement. They help enhance the heritability of traits that are difficult to select through indirect genotypic selection, simplify complex quantitative traits by breaking them down into Mendelian loci, and enable gene pyramiding for pest and disease resistance. Additionally, these markers facilitate the development of detailed genetic linkage maps (Faure et al. 1993) and support gene cloning based on mapping. Other key applications include accurate clone identification (Jarret et al. 1992, 1993), tracing evolutionary relationships among clones, detecting duplicates in germplasm collections, and monitoring somaclonal variation in micropropagated plants for commercial use (Damasco et al. 1996). PCR markers have also been developed to detect A and B genome sequences in *Musa* (Oriero et al. 2006; Pillay et al. 2000). Notably, three 10-mer RAPD primers have been found to produce distinct banding patterns that differentiate *M. acuminata* (A genome) from *M. balbisiana* (B genome). Various molecular techniques have been used to investigate the origin, genetic relationships, and diversity within *Musa* genomic groups. These include isozymes (Rivera 1983; Jarret and Litz 1986a, b), restriction fragment length polymorphisms (RFLP) (Jarret et al. 1992; Gawel and Jarret 1991a), random amplified polymorphic DNA (RAPD), repetitive elements, rDNA spacer length diversity (IGS) (Lanaud et al. 1992), sequence-tagged site markers (STS), and amplified fragment length polymorphism (AFLP) (Ude et al. 2002).

To date, no studies have been conducted to identify molecular markers associated with disease resistance or high yield in abaca and bananas. Currently, UPLB is developing molecular markers to accelerate the breeding of abaca varieties with high fiber yield and resistance to *Abaca bunchy top virus* (ABTV). The genetic background and relationships of various abaca accessions are being analyzed by fingerprinting BC_1 and BC_2 populations using specific primers (mMaCIR39, mMaCIR40, and mMaCIR45). This approach aims to streamline the screening process for ABTV resistance and aid in selecting clones with superior fiber quality (Fig. 1.5). Abaca accessions collected from FIDA Bicol, NARC in Leyte, Negros Occidental and

FIDA-Davao were also subjected to PCR analysis using the primer set and were found to be polymorphic. Fingerprinting analysis was carried out to establish the genetic relationship among the different accessions.

Barbosa et al. (2023) conducted whole-genome Illumina resequencing of abaca cultivars to identify genome-wide polymorphisms among abaca and other *Musa* species and assess their potential as genetic marker resources. Variant calling was performed using BCFtools, followed by genetic diversity and phylogenetic analyses. This study identified 20,590,381 high-quality single-nucleotide polymorphisms (SNPs) and DNA insertions/deletions (InDels) across 16 abaca cultivars. After filtering based on linkage disequilibrium (LD), 130,768 SNPs and 13,620 InDels were retained, contributing to gene diversity values of 0.396 ± 0.106 and 0.431 ± 0.111, respectively, across the cultivars. LD-pruned polymorphisms across abaca, *M. troglodytarum*, *M. acuminata* and *M. balbisiana* enabled genetic differentiation within abaca and across the four *Musa* spp. Phylogenetic analysis showed that the registered abaca varieties Abuab and Inosa had accumulated a significant number of mutations, prompting further research into the relationship between these mutations and their beneficial traits. This study was the first to develop marker resources for abaca using genome-wide polymorphisms, providing essential tools for varietal authentication and comparative genotyping with the more extensively studied *Musa* species.

1.6 Tissue Culture and Genetic Engineering

A standardized protocol for *in vitro* abaca culture has been developed, facilitating the large-scale production of disease-free abaca varieties through tissue culture. This advancement has been successfully transferred to the farmer level, allowing growers to access clean planting materials that help reduce the initial spread of diseases (Fig. 1.12). The use of tissue-cultured abaca has played a crucial role in improving

Fig. 1.12 Micropropagation of BANDALA seedlings through tissue culture. (Source: AG Lalusin, ICropS, CAFS, the University of the Philippines Los Baños (UPLB))

Fig. 1.13 (**a**, **b**) Potted out Dioscoro 1 or BANDALA seedlings in the nursery ready for field planting. (Source: AG Lalusin, ICropS, CAFS, the University of the Philippines Los Baños (UPLB))

plantation productivity by ensuring uniform, high-quality plants that are free from systemic infections at the time of planting.

However, despite the benefits of this technology, its long-term effectiveness is hindered by the persistence of viral inoculum in abaca plantations. Residual infections from previously affected plants, as well as alternative host plants and insect vectors, create a continuous cycle of disease transmission, making the complete eradication of viral diseases, such as *Abaca bunchy top virus* (ABTV), nearly impossible. Consequently, even tissue-cultured, disease-free planting materials can become infected within just a few growing seasons, limiting the sustainability of this approach as a long-term solution.

Given these challenges, developing new abaca cultivars with genetic resistance to pests and diseases is of paramount importance (Fig. 1.13). Resistant cultivars would provide a more durable and cost-effective strategy for disease management, reducing the reliance on continuous replanting with tissue-cultured materials and minimizing yield losses due to viral infections. Incorporating resistance traits into breeding programs would help ensure the long-term sustainability and resilience of abaca production, benefiting both farmers and the fiber industry.

Genetic engineering, particularly through plant transformation and gene cloning, is emerging as a valuable tool for crop improvement. However, the successful application of these techniques largely depends on the development of efficient and reproducible protocols for key processes such as establishing cell suspension cultures, inducing somatic embryogenesis, and facilitating protoplast fusion. Recent advancements have significantly improved the ability to regenerate plants from somatic embryoids, and notable progress has also been made in studying gene expression in abaca. These developments pave the way for further genetic enhancements, potentially leading to improved disease resistance, higher fiber quality, and overall better agronomic traits in abaca. The technique involves adventitious shoot induction and shoot multiplication in the medium supplemented with BA or 2,4-D and coconut water. Addition of 2,4-D also induced callus formation. Subsequent root induction in adventitious shoots was also obtained in MS medium supplemented with IAA, NAA and IBA (Aspuria 1984).

Buds and shoots were successfully induced from young floral sections and floral apices cultured in MS medium supplemented with kinetin (Ki) combined with either naphthalene acetic acid (NAA), indole-3-acetic acid (IAA), or indole-3-butyric acid (IBA). The initial formation of embryogenic (nodular) structures, which later developed into shoots, was first observed in explants grown in MS medium enriched with 2-isopentenyl adenine (2-ip) and Ki. Additionally, somatic embryos were generated from cell suspension cultures derived from globular structures formed on meristematic buds (scalps) or leaf sheaths. These cultures were maintained in a modified liquid MS medium supplemented with 2,4-dichlorophenoxyacetic acid (2,4-D), zeatin, and L-cysteine. The ability to produce somatic embryos through these methods is a critical step in the genetic engineering of abaca, as it facilitates the development of transgenic plants with improved traits such as disease resistance and enhanced fiber quality.

Transient expression of the GUS reporter gene has been detected in transformed meristematic globules of abaca (Aquino et al. 2005). In an effort to develop genetically engineered abaca with resistance to *Abaca bunchy top virus* (ABTV) and abaca mosaic disease, researchers from the Fiber Industry Development Authority (FIDA) are collaborating with scientists from the University of the Philippines Diliman (UPD) and Los Baños (UPLB). This joint research aims to enhance abaca's resistance to these viral diseases, contributing to more sustainable fiber production. The Fiber Industry Development Authority (FIDA) launched a project in partnership with the National Institute of Molecular Biology and Biotechnology at the University of the Philippines Diliman (UPD) and the Department of Horticulture at the University of the Philippines Los Baños (UPLB) to develop genetically modified abaca (FIDA 2008). This initiative aims to enhance abaca's resistance to diseases and improve its overall productivity through advanced biotechnological approaches. The project's three phases were (1) molecular cloning, characterization and development of gene constructs and biolistic transformation, (2) development and regeneration system for biolistic-mediated transformation of abaca, and (3) greenhouse characterization and evaluation of genetically modified abaca. The first year of the project saw the collection of Abaca bunchy top virus (ABTV)-infected abaca from five localities and extraction, amplification and cloning of ABTV genes. As part of phase 2, shoot tip cultures from 4-month-old suckers of two abaca cultivars were used or the induction of scalp and globule formation in establishing embryonic cell suspension cultures for transformation. The cultures are currently kept at the Tissue Culture Laboratory of the Department of Horticulture, UPLB. The scarcity of funds is suspected of as the reason for the project being temporarily shelved. The development of genetically modified or transgenic abaca most certainly would be in the future of abaca breeding. The proliferation of work in this field during the last several years on the closely related banana bunchy top disease is an indication of the momentum already attained and of the knowledge gained that abaca research can very well take advantage of. The work on transgenic bananas resistant to the Bunchy top virus of Borth et al. (2011) in Hawaii, Tripathi (2013) in sub-Saharan Africa, and Elayabalan et al. (2013) and Selvarjan et al. (2014) in India some of the current developments emerging from initiatives targeting varied environments.

1.7 Mutation Breeding

To enhance the genetic diversity of abaca, chemical mutation was induced using ethylmethane sulphonate (EMS). The Indonesian Sweetener and Fiber Crops Research Institute (ISFCRI) successfully applied this method to the abaca accessions Sangihe I and Tangongon (Purwati et al. 2018). A total of 26 abaca mutants were cultivated at the Cobanrondo experimental station in Malang, situated at an altitude of 1450 meters above sea level. Each mutant genotype was planted in separate clumps, with a spacing of three meters between them. Morphological character studies revealed that treating Sangihe I and Tangongon clones with 0.6% EMS significantly increased variability in the number of stalks, with a coefficient of variation (CV) of 91%. Additionally, all quantitative traits observed showed a positive and significant correlation with the weight of dry leaf sheaths, which serves as an indicator of fiber production potential. However, the chemical mutation did not enhance qualitative character diversity. Among the mutated clones, eight exhibited promising characteristics and potential for further development as high-yielding abaca varieties. These superior clones include Clone 3 (Sh 1.1.1-2), Clone 6 (Sh 4.1.2-2), Clone 11 (FK Sh 1.1-3-6), Clone 15 (Tg 3.1-2), Clone 17 (Tg 59.2-7), Clone 26 (Tg 70.2.3-7), Clone 13 (FK Sh 1.1.3-10), and Clone 18 (Tg 59.2-8). These findings highlight the potential of chemical mutation as a tool for improving abaca's productivity and fiber yield.

Gamma-ray irradiation using the Cobalt-60 (Co-60) method was applied to assess the sensitivity and response of *in vitro*-propagated abaca to *Banana bunchy top virus* (BBTV) and *Banana bract mosaic virus* (BBrMV). This study focused on the abaca cultivars Tinawagan Pula (TP) and Tangongon (TG) (Dizon et al. 2012). Both cultivars exhibited similar biological and physiological responses to gamma irradiation, with higher doses leading to increased mortality and reduced shoot proliferation. The lethal dose for 30% of the population (LD30) in terms of shoot proliferation was determined to be between 10 and 15 Gy. Post-irradiation analysis revealed that cv. TG demonstrated greater tolerance to gamma rays compared to cv. TP, based on shoot growth per subculture cycle. A total of 3210 and 3765 irradiated abaca plants, including non-irradiated controls, were exposed to BBTV and BBrMV through aphid transmission and mechanical inoculation, respectively. Three months after inoculation, PCR-based diagnostic tests on putative-resistant (PR) lines identified 43 TP (0.6%) and 9 TG (0.9%) plants that consistently tested negative for BBTV, while 36 TP (1.6%) and 15 TG (0.9%) plants initially showed no infection with BBrMV. Although further field trials and yield assessments are needed for final selection, this study established a foundation for mutation breeding as a strategy for developing virus-resistant abaca. With the identification of the optimal gamma irradiation dose (LD30) in combination with *in vitro* propagation, screening and generating PR lines for virus resistance in abaca can now be conducted more efficiently compared to conventional breeding methods.

1.8 Abaca Transcriptomics and Genomics

In Southeast Asia, something took place that led to the domestication of the *Musa* species 7000 years ago (D'hont et al. 2012). According to the molecular studies on *Musa* species, only the sections *Musa* and *Callimusa* of *Zingibrales* were kept, which was originally divided into four sections (Cheesman 1947). According to Häkkinen (2013), *Australlimusa* and *Callimusa* with chromosome number $n = 10$ were classified into *Callimusa*, whereas *Eumusa* and *Rhodoclamys* with chromosome number $n = 11$ were placed in the *Musa* section. Furthermore, it is somewhat challenging to evaluate the genomics of the *Musaceae* family because its known triploid or diploid cultivars are products of interspecific hybridization and polyploidization. Due to the polyploid nature of *Musa* spp., Martin (2016) used double haploid Pahang to improve the genome assembly of *Musa acuminata*. Pahang was chosen since it is one of the progenitors of domesticated cavendish. *M. Acuminata's* genome later became the first *Musa* to have its draft genome sequenced and the first monocot outside of the *Poales* order. Moreover, it was used as a model species for sequencing other *Musa* spp., including *M. itenerans* (Wu and Alexander 2016), *M. balbisiana Colla* (Niu et al. 2018) and *M. banksii* (Liu et al. 2019). The exploration of *M. itenerans*, provides insight into the complete genome duplication events that contributed to *M. itenerans*' divergence from related species (Wu and Alexander 2016). It was also mentioned that the discovery of disease-resistance components in this species' entire genome sequence supports the idea of using wild relatives for introgression for disease resistance and other agronomic traits (Wu and Alexander 2016). Despite being diploid of *Musa acuminata* (A genome), along with *Musa balbisiana* (B genome), both are typically triploid (Kumar et al. 2013) The double haploid Pahang belonged to *Musa aciminata* subspecies *malaccensis* and has a 523-megabase genome ($2n = 22$) (Dhon't et al. 2012). Through crossing between A- and B- genomes, several ploidies (diploid, triploid, and tetraploid) varieties were produced. However, triploid is the predominant genetic makeup of most cultivated banana cultivars up to date (Wang et al. 2016).

Banana genome hub (https://banana-genome-hub.southgreen.fr/) makes a total of eight *Musa* genomes available to the public, but none of them is a representation of the T genome, which includes abaca. Nonetheless, some papers described the utilization of the abaca's chloroplast DNA sequence to examine its evolutionary relationship with other Musa species (Liu et al. 2019; Niu et al. 2018; Shetty et al.

2016). It was found that *Musa balbisiana* Colla was closest to *Musa textilis Née* based on the phylogenetic study using their chloroplast genomes (Niu et al. 2018). Nonetheless, a draft genome for abaca might serve as the T genome representative and add to the growing collection of Musa genome sequences.

The University of the Philippines at Los Banos with funding from DA-Biotech PIU, conducted a research initiative to develop molecular markers for abaca breeding. This project aims to enhance the efficiency of selecting high-yielding abaca varieties with resistance to *Abaca bunchy top virus* (ABTV). Given the lack of studies identifying molecular markers linked to disease resistance or fiber yield in abaca, this research is crucial for improving breeding programs. To achieve this, researchers subjected abaca samples to PCR analysis using 12 banana primers. Among these, four primers (33%) produced scorable and reproducible banding patterns. Three primers—mMaCIR39, mMaCIR40, and mMaCIR45—were found to be polymorphic, meaning they can distinguish genetic variations among different abaca accessions. One primer, mMaCIR07, was monomorphic and did not show significant genetic diversity. Ongoing research is focused on fingerprinting BC1 and BC2 populations using the identified polymorphic markers to better understand the genetic background of abaca and its resistance traits. This will allow breeders to efficiently select superior abaca lines that combine high fiber yield and strong ABTV resistance. The integration of molecular marker-assisted selection (MAS) into abaca breeding programs is expected to significantly improve the industry. By accelerating the identification of disease-resistant and high-yielding cultivars, this approach will reduce reliance on traditional breeding methods, which are often time-consuming and labor-intensive. Ultimately, this research aims to develop improved abaca varieties that can withstand viral infections, enhance fiber production, and contribute to the long-term sustainability and productivity of abaca farming.

This research aimed to streamline the screening process for *Abaca bunchy top virus* (ABTV) resistance and the selection of clones with superior fiber quality. Currently, putative primers have been identified and are being used to assist in selecting high-fiber-yielding clones. Additionally, abaca accessions collected from various sources, including FIDA Bicol, NARC in Leyte, Negros Occidental, and FIDA Davao, were subjected to PCR analysis using the three previously identified polymorphic primers. Fingerprinting analysis was conducted to determine the genetic relationships among these different accessions. Further studies involve the characterization and development of microsatellite markers derived from RNA-Seq data, which will enhance the accuracy of genetic identification and selection in abaca breeding programs. These efforts aim to improve fiber yield and disease resistance, contributing to the long-term sustainability of abaca cultivation.

Characterization and development of microsatellite markers from RNA-Seq data of abaca (Musa textilis Nee cv. Abuab) was conducted by Palao (2019). The research aims to mine and characterize simple sequence repeats (SSRs) from RNA-Seq data, develop SSR markers for use in screening Musa textilis accessions, validate SSRs in silico and the laboratory, and create the abaca marker database (Abaca MarkerDB). Upon RNA extraction and sequencing, de novo assembly was conducted, instead of reference-guided assembly, to eliminate unique sequences. The in-silico approach of

SSR identification and characterization allowed for the development of robust and informative SSR markers. The developed markers in this study were concluded transferable to other Musa textilis varieties upon cross-amplification wherein 87.5% were identified as highly informative markers. The dendrogram of the abaca cultivars, Abuab, Pacol and its backcross progenies, showed that Abuab and BC_2 are included in the same group. Furthermore, an intuitive and user-friendly database named AbacaMarkerDB was designed to store all information about the SSRs of Musa textilis and the developed markers from this study. It is also hopeful that this database program can be useful for other crops despite minimal limitations. It is recommended to mine further the SSRs and DEGs or unigenes to produce more functional genic-SSRs and to find SSRs in unmerged backcross transcript data. It would also be useful if all other identified SSRs were characterized and validated. And later, update the database to include the name of the genes and other annotations useful for the completion of the program. This information is significant in terms of enabling the early detection of phenotypes of abaca cultivars.

In addition, Next Generation Sequencing (NGS) transcriptome analysis of abaca bunchy top virus resistance and differential expression of the EIF4G gene in Abuab, Pacol and BC_2 Hybrids was done by Muncada (2016). The study aims to discover abaca bunchy top antiviral resistance genes; compare and characterize the expression profile/transcriptome profile of BC_2, Pacol (Pc) and Abuab (Ab); validate differential gene expression using RT-qPCR; and propose a molecular mechanism of Abaca BC_2 hybrid resistance to abaca bunchy top virus. As preliminary research, a molecular analysis of the gene coding sequences of Musa acuminata was conducted in search of a candidate abaca bunchy top virus defense-related gene in Musa textilis Nee. As a result, a major grouping in Musa acuminata coding genes was identified as biotic defense response gene groups which include the transcription factors, defense-related genes, genes for cell wall biosynthesis and genes involved for secondary metabolism covering chromosomes 1 to 11. From these results, eIF4G genes were successfully isolated from Ab, Pc and BC_2 hybrid. This provided an initial step to understanding better its possible role in ABTV response. A full-length sequence of the eIF4G gene from these samples provided a more comprehensive structural and functional gene analysis such as SNPs analysis for identifying possible mutation events that may account for its role in establishing resistance to ABTV. Likewise, differential expression of eIF4G during early ABTV infection can further confirm its possible role in ABTV resistance. Abaca transcriptomes were assembled to validate the earlier results. A total of 349 differentially expressed genes (DEGs) and unique genes from each sample's assembled transcriptome were assayed. Highly expressed genes during early ABTV infection were identified resulting in the prediction that the 5 modes of innate immunity may be exhibited by BC_2-A2 hybrid. These PTI, ETI, UPS, UPR and chloroplast-dependent immunity can be potential candidate signaling pathways in the BC_2-A2 Abaca hybrid. BC_2-A2 hybrid has a sophisticated and robust response against ABTV such that a number of highly expressed genes unique in BC_2-A2 have been inherited from Pacol, the parent material found to be resistant to the bunchy top virus as confirmed from the mapping of Pacol consensus data set against banana and abaca bunchy top viruses. These evaluations provided preliminary insights into the system of molecular antiviral mechanisms in monocots. It is recommended that future studies on this matter through proteomics and metabolomics be conducted to fully understand the mechanism of innate immunity of abaca BC_2 hybrid against ABTV and its multiple antiviral resistance to Musa viruses. The study suggested that the validation by RT-qPCR of most of the DEGs identified in this study can further provide extensive application, especially of some of the interesting genes identified during the RNA-Seq conducted.

Moreover, genetic characterization of backcross progenies BC_1, BC_2 and BC_3 from *Musa textilis* Nee (Abuab) x *Musa balbisiana* Colla (Pacol) through transcriptomic analysis was conducted by Reamillo (2018). RNA samples were isolated from the target sample varieties and were subjected to next-generation sequencing. *De novo* assembly of the trimmed and cleaned sequence reads of each sample was performed. Common Musa acuminata transcripts between parents and backcrosses were identified. Based on the results, the backcrosses exhibited 69–77% of the Musa acuminata transcripts found in Abuab while exhibiting 68–78% of the Musa acuminata transcripts found in Pacol. This was further validated through BLAST wherein half of the contigs of the RNA samples were found to be Musa acuminata. These RNA transcriptomes were also subjected to BLAST against the Plant RefSeq to identify non-Musa contigs. Contigs with no hits (orphans) were also identified. These contigs were aligned to the 20 ABTV genomes in NCBI. Interestingly, only Abuab has an alignment, and the other 4 samples did not align at all with the ABTV sequences. This indicates that the Abuab sample obtained from the field was infected with ABTV. In developing the backcrosses, Pacol was used as the donor parent for the resistance genes. Since the backcrosses were found to be free of viral infection in the field, it could be possible that the backcrosses have acquired the resistant genes from the Pacol parent. However, this needs verification, not just in terms of the presence of the incorporated resistance genes in vivo but by its expression. Gene ontology (GO) annotations of proteins coded by the transcriptomes of BC_1, BC_2, and BC_3 were identified. According to results, most of the transcripts of BC_1 and BC_2 code for enzymes with ion binding, kinase, hydrolase and nuclei acid binding activities. On the other hand, it was identified in BC_3 that transcripts coding for transferase and oxidoreductase enzyme activities were defined in addition to ion binding, kinase, hydrolase and nuclei acid binding activities. The number of significantly expressed contigs/genes (DECs/DEGs) in each multi-group comparison involving the parents and backcrosses was also identified. A total of 9997 DEGs were identified between Abuab and Pacol wherein 9581 were defined as unique and 416 were common DEGs. Among the 5 comparisons of parents and backcrosses, 3661 DEGs were identified wherein 3482 were defined as unique and 179 were common. Likewise, the number of candidate genes for fiber quality and disease resistance among the parents and backcrosses in different sub-experiments were known. Between Abuab and Pacol, 193 genes were identified wherein 12 are the identified resistance genes analogues (RGAs). Among the 5 comparisons, 66 genes were identified and 6 were found to be RGAs. Among the genes mined are fiber quality genes such as genes for cell elongation, cell wall loosening, osmotic regulation, secondary cell wall synthesis, lipid transfer and metabolism, carbohydrate metabolism, and plant responses to auxin, ethylene and gibberellic acid, as well as putative disease resistance genes such as resistance gene analogs (RGAs) 1, 2, 3, and 4. With all the results gathered, it can be concluded that the totality of the expression patterns of BC1 was attributable to Abuab while BC2 and BC3 are attributable to Pacol.

Despite the increasing demand and economic significance of abaca (*Musa textilis* Nee) fiber, research utilizing RNA sequencing (RNA-seq) to identify genes linked to high fiber quality and *Abaca bunchy top virus* (AbBTV) resistance remains limited. Ereful et al. (2022a) addressed this gap by analyzing gene expression in three abaca varieties (Abuab, Inosa, and Tangongon), a wild banana variety (*Musa balbisiana* Colla, known as Pacol), and two developed backcrosses (Abuab × Pacol BC2 and BC3). These plants were grown at the Institute of Plant Breeding (IPB) in Laguna, Philippines, and RNA-seq was conducted on pseudostems collected from three-month-old suckers. Differential expression analysis using various models revealed that Abuab and BC3 exhibited the highest proportion (70%) of abaca-specific gene expression, highlighting their potential role in fiber production. Gene ontology (GO) enrichment analysis identified genes involved in cellulose synthase activity, callose synthase, β-glucosidase activity, and the glucan biosynthetic process, all of which contribute to fiber development. KEGG pathway analysis further revealed genes encoding enzymes associated with lignin biosynthesis, a key factor in fiber strength and quality. Additionally, genotypic differential expression (GDE) analysis comparing AbBTV-resistant and -susceptible groups identified candidate genes, including pathogenesis-related proteins and NBS-LRR, which are commonly associated with plant defense responses. However, since the genotypes studied were not infected with the virus, the role of these genes in disease resistance remains unconfirmed. Further research is required to validate their function, making them promising targets for future breeding and genetic improvement efforts in abaca.

The *Musa textilis* variety Abuab is known for its high fiber quality (FQ) but is susceptible to *Abaca bunchy top virus* (AbBTV), whereas the *Musa balbisiana* variety Pacol has low fiber quality but exhibits resistance to AbBTV. Their backcrosses (BC2 and BC3) successfully combine both desirable traits, offering improved fiber quality along with enhanced disease resistance (Ereful et al. 2022b). Analysis using RNA-seq showed that the regulatory divergence of Abuab and Pacol is largely explained by cis differences with 27.4% and 22.3% if we are to assess it using BC_2 and BC_3, respectively. Cis differences between the two genotypes are significantly reduced from BC_2 to BC_3 due to changes in genomic constitution. Trans, on the other hand, is robust to changes in allelic composition. These findings are primarily attributed to the loss of heterozygosity in BC3 compared to BC2. Further analysis revealed that both backcrosses demonstrated a genome-wide preference for the expression of Pacol-specific alleles over Abuab-specific alleles, despite Abuab having a broader genetic presence in the hybrids. Notably, the ratio of genotype-specific expressed transcripts was significantly disproportionate to the corresponding genetic composition, a phenomenon referred to as “genome-transcriptome incongruence.” Additionally, a preferential expression switch was observed, where certain genes that initially favored the Abuab (or Pacol) allele in BC2 shifted to expressing the Pacol (or Abuab) allele in the BC3 genome.

The entire genome of the abaca variety Abuab was sequenced using the Illumina NovaSeq 6000 and Pacific Biosciences Single-Molecule Real-Time Sequel platforms (Galvez et al. 2021). The genome size was estimated at 616 megabase pairs (Mbp) based on k-mer analysis. The assembly was conducted at 65× sequencing depth, successfully mapping 95.28% of the estimated genome size. BUSCO analysis indicated that 78.2% of the complete BUSCO genes were recovered, demonstrating the assembly's quality. A total of 33,277 gene structures were predicted, a number comparable to those identified in recently sequenced *Musa* species genomes. Additionally, 330 Mbp of repetitive elements were identified, making up 53.6% of the genome. The sequencing and assembly of the Abuab genome provide a valuable resource for gene discovery, aiding in crop improvement efforts and serving as an essential reference for genetic diversity studies within the *Musa* genus.

1.9 Breakthroughs in Abaca Breeding

1.9.1 The Dioscoro 1 or BANDALA Variety

An abaca hybrid developed by the Institute of Plant Breeding (IPB) and the Institute of Crop Science (ICropS), College of Agriculture and Food Science, University of the Philippines Los Baños (CAFS, UPLB) through the funding support of the Department of Agriculture Biotech Program, DOST, and DOST-PCAARRD is called Dioscoro 1 formerly named BANDALA which is an acronym for Backcross Abaca with Native and Desirable Accessions to Lift the Abaca Industry (Fig. 1.14). It is the result of the crossbreeding of Abuab (*M. textilis* Nee) and a wild banana cultivar 'Pacol' (*M. balbisiana* Colla) as parental lines to introduce virus-resistant genes from Pacol to the hybrid and simultaneously recover lines with good fiber quality from Abuab.

1.9.2 Dioscoro 1 or BANDALA for Textile

The Philippine Tropical Fabrics (PTF) are natural textile fibers that are spun, woven, knitted, and finished locally in the Philippines. The PTF Law, enacted on February 10, 2004, mandates the use of these fabrics for the uniforms of public officials and employees, as well as for other purposes. This legislation aims to support the local textile industry while fostering a sense of nationalism among government workers.

The law applies to various government entities, including government-owned or controlled corporations, local government units, state and local universities and colleges, and local water districts. Its primary goal is to encourage patriotism by promoting the use of locally produced textiles. However, exemptions are granted to

Fig. 1.14 Mature Dioscoro1 or BANDALA plant. Dioscoro 1 or BANDALA is a second-generation backcross between traditional abaca and a wild banana Pacol. BANDALA is very resistant to abaca bunchy top virus disease. (Source: AG Lalusin, ICropS, CAFS, the University of the Philippines Los Baños (UPLB))

certain professions that require specialized uniforms, such as firefighters, athletes, police officers, and military personnel who need protective gear like bulletproof vests.

A 2021 report highlighted challenges in implementing the Philippine Tropical Fabrics (PTF) Law, particularly in producing office uniforms. The main issues included the limited availability of PTF, the absence of local spinning technology, and a shortage of suppliers that meet the required standards.

To address these concerns, the PTF Technical Committee, led by the Civil Service Commission (CSC), conducted a review of the law's Implementing Rules and Regulations (IRR). In 2022, meetings were held with industry associations and private sector stakeholders to discuss improvements. As a result, the revised IRR introduced changes to the weight requirement for natural textile fibers, acknowledging their limited supply. The previous regulation mandated a minimum of 5% by weight for fibers such as abaca, banana, and pineapple, and 15% for silk. The updated standard now sets a uniform requirement of at least 5% for both natural textile fibers and silk.

The amended IRR also specifies that PTF must be sourced from accredited local suppliers or direct manufacturers and traders certified by the Department of Trade and Industry (DTI). These suppliers must operate within the Philippines and obtain a permit from the Department of Agriculture's Philippine Fiber Industry Development Authority (DA-PhilFIDA) for fiber transport. Additionally, the new IRR includes guidelines for planning, monitoring, and evaluating the implementation of the PTF Law, with oversight from the CSC, DA, Department of Science and Technology (DOST), and DTI.

On July 14, 2023, CSC Chairperson Karlo Nograles officially signed the amended PTF regulations. These updates aim to improve access to PTF for government employees while continuing to promote nationalism and support the local textile industry.

1.9.3 Testing for Philippine Tropical Fabrics

Maintaining fabric quality and ensuring adherence to textile standards are crucial, particularly for government office uniforms. To guarantee compliance, all Philippine Tropical Fabrics (PTF) must undergo testing and certification before being approved for official use. The Department of Science and Technology - Philippine Textile Research Institute (DOST-PTRI) leads this effort as the chair of the Technical Committee. The institute plays a key role in providing comprehensive textile testing services to verify whether fabrics meet the requirements outlined in the PTF Law. DOST-PTRI's textile testing laboratory specializes in identifying and quantifying natural textile fibers such as abaca, banana, pineapple, and silk in PTF and yarns. This is accomplished through optical micro-imaging analysis and fiber composition evaluation. Additionally, the laboratory conducts color fastness tests to determine fabric durability, measuring resistance to fading and staining caused by laundering, perspiration, and friction.

For natural textile fibers intended for textile applications (Fig. 1.15), the institute performs various assessments, including the analysis of essential components such as cellulose, lignin, residual gum, solvent extractives, and moisture content. Additionally, mechanical property tests evaluate factors like fabric cut, weight, pilling resistance, breaking strength, and dimensional stability. These efforts ensure that PTF maintains high-quality standards while also supporting the local textile industry.

To guarantee precision and reliability, all tests adhere to the Philippine National Standards (PNS) and the International Organization for Standardization (ISO) guidelines (PNS/ISO/IEC 17025-2017). Additionally, the laboratory follows internationally recognized testing methodologies from esteemed institutions such as the American Society for Testing and Materials (ASTM) and the American Association of Textile Chemists and Colorists (AATCC).

Quality assurance is an essential component of the Philippine Tropical Fabrics (PTF) procurement process. Beyond the usual procurement requirements, certification from DOST-PTRI is mandatory, along with an attestation from the textile producer verifying that their fabric is part of the certified PTF stock approved by

Fig. 1.15 Fibers extracted from the pseudostem of Dioscoro 1 or BANDALA. The fibers are suited for textile production and pulp and paper. (Source: AG Lalusin, ICropS, CAFS, the University of the Philippines Los Baños (UPLB))

DOST-PTRI. This ensures that only high-quality, compliant fabrics are used for government office uniforms.

1.9.4 What Are the Advantages of Dioscoro 1 or BANDALA?

Dioscoro 1 or BANDALA hybrid was developed using molecular markers in the breeding program to hasten the selection of progenies carrying the desired genes. The second backcross (BC_2) generation of the hybrid resulted in 87.5% abaca and 12.5% wild banana. With an 87.5% abaca genome, the hybrid's appearance and chemical composition are already the same as the traditional abaca varieties but with an added trait. It is resistant to the abaca bunchy top virus (AbBTV), a major abaca disease. Dioscoro 1 or BANDALA is envisioned to aid abaca farmers by reducing the losses due to diseases and helping to rehabilitate affected abaca farms.

1.9.4.1 Resistant to Abaca Bunchy Top Virus

Abaca bunchy top disease (AbBTD) was first reported in 1915 in Silang, Cavite (Ocfemia 1930). The disease is caused by *Banana bunchy top virus* (BBTV) and later became associated with *Abaca bunchy top virus* (ABTV). It is transmitted by the black aphid, *Pentalonia nigronervosa* Coq. AbBTD has severely impacted abaca production, particularly in the Bicol Region, Eastern Visayas, and Mindanao, where the crop is widely cultivated. It is considered the most destructive viral disease in abaca in the Philippines, causing fiber yield losses ranging from 13% to 77% (Bajet and Magnaye 2002). Infected abaca plants exhibit stunted growth, produce undersized suckers, and develop short, narrow, stiff, and up-curled leaves with yellow to brown margins.

Although some tolerant varieties have been identified within the abaca germplasm, they often possess inferior fiber quality. Conversely, traditional varieties known for their superior fiber quality are highly susceptible to virus diseases. Some improved varieties currently grown by farmers also remain vulnerable to AbBTD (Lalusin 2010).

A study titled *Characteristics of Selected Hybrids of Abaca (Musa textilis Nee) with Resistance to Bunchy Top* examined the incidence of bunchy top disease in the abaca hybrid Dioscoro 1 or BANDALA, comparing it to its parent varieties (Abuab and Pacol) and control varieties ('Tinawagan Pula' and Inosa) (Parac et al. 2021). Visual observations showed that Dioscoro 1 BANDALA did not exhibit symptoms of bunchy top disease—including dark green streaks on veins and midribs, marginal leaf chlorosis, narrow and stiff leaves, or leaf crowding at the apex during 6 months under screen house conditions. These findings were confirmed through enzyme-linked immunosorbent assay (ELISA) and polymerase chain reaction (PCR) analysis, which indicated that Dioscoro 1 or BANDALA remained free of ABTV symptoms under both field and screen house conditions. Similar results were observed in field trials conducted at Caraga State University in Ampayon, Butuan City. Over 5 years (2012–2017), Dioscoro 1 or BANDALA and BC2-2 remained free from symptoms of bunchy top disease despite natural exposure to infection (Table 1.2). These findings suggest that hybridization efforts, particularly involving Pacol as a resistance source, could be a promising approach for developing disease-resistant abaca varieties.

Meanwhile, Dioscoro 1 or BANDALA was rated as resistant to bunchy top disease based on its disease index (DI) value which measures the incidence and severity of disease after virus inoculation (Table 1.3).

1.9.4.2 Climate Resilience

Traditional abaca varieties were known to be sensitive to drought i.e., only a few weeks of dry soil will affect the fiber's final quality (Berglund et al. 2019). Moreover, they are vulnerable to sunlight and wind speed. It has been demonstrated that cutting the intensity of the sun by 50% gives a substantially larger yield of fiber than plants grown in less shade (Bande et al. 2013).

Through the project, Stability Analysis of BC_2 Abaca (*Musa textilis* Nee) Hybrids across Different Locations in the Philippines, researchers from IPB and ICropS, CAFS, and UPLB observed that Dioscoro 1 or BANDALA has a wide range of environmental adaptability as it was able to perform well even under unfavorable conditions (Parducho et al. 2020).

Table 1.2 Incidence of bunchy top disease in abaca hybrids, their parentals and susceptible control varieties

Abaca Hybrid/ Variety	Disease Incidence (%)					
	Screenhouse Evaluation			Field Evaluation		
	Visual Observation	Enzyme-linked immunosorbent assay	Polymerase chain reaction	Visual Observation	Enzyme-linked immunosorbent assay	Polymerase chain reaction
BC_2-2	7	7	7	0	0	0
Dioscoro 1 or BANDALA	0	0	0	0	0	0
Pacol	0	0	0	Not tested	Not tested	Not tested
Abuab	100	73	100	Not tested	Not tested	Not tested
Inosa	100	80	100	100	80	95
Tinawagan Pula	67	47	73	90	70	85

Source: AG Lalusin, ICropS, CAFS, the University of the Philippines Los Baños (UPLB)

Table 1.3 Response based on disease index of abaca hybrids, parentals, and control varieties to bunchy top disease

Hybrid/Variety	Disease Index (%)	Response[a]
Dioscoro 1 or BANDALA	0	Resistant
BC_2-2	4	Resistant
Pacol	0	Resistant
Abuab	72	Susceptible
Inosa	96	Susceptible
Tinawagan Pula	55	Moderately susceptible

Source: AG Lalusin, ICropS, CAFS, the University of the Philippines Los Baños (UPLB)
[a]Response based on DI value: 0–30%, resistant; 31–60%, moderately susceptible; and 61–100%, susceptible

In the project, they aimed to identify the stability of traditional abaca (Tinawagan Pula) and Dioscoro 1 or BANDALA by evaluating their performance in different environments.

Samples were tested in seven locations by using a randomized complete block design (RCBD) with three replications. The environmental characteristics of the seven locations were identified. Five locations (Los Baños, Laguna; Virac, Catanduanes; Baybay, Leyte; Mabini, Davao del Norte; and Kabacan, North Cotabato) are in an open environment; while two locations (Guinobatan, Albay, and Butuan, Agusan del Norte) are in a shaded environment.

According to the GGE (genotype main effect plus genotype by environment interaction) biplot analysis, the researchers have identified the most (North Cotabato, Leyte, Agusan Del Norte, and Albay) and the least (Davao del Norte, Laguna, and Catanduanes) favorable for growing environments for the three abaca genotypes, where Dioscoro 1 or BANDALA was able to thrive even in unfavorable environments (Fig. 1.16).

Further, the results of the project's nonparametric stability analyses also confirm the consistency of Dioscoro 1 or BANDALA's performance in a wide range of environments (Table 1.4). Parducho et al. (2020) cited Becker and Leon (1988) who noted that genotypes with consistent ranking or have few changes in rank in different nonparametric measures are considered more stable, further suggesting the stability of Dioscoro 1 or BANDALA. Dioscoro 1 or BANDALA can survive both in open and shaded conditions. It has a wide range of environmental adaptability and stability.

Eberhart and Russell's regression and deviation from the regression identified BC_{2-2} as a more stable genotype, showing that hybrids can perform better than Tinawagang Pula under unfavorable environments. However, Kang's rank sum and Nassar and Huehn's four stability measures identified Dioscoro 1 or BANDALA as the most stable among the three abaca genotypes (Table 1.4).

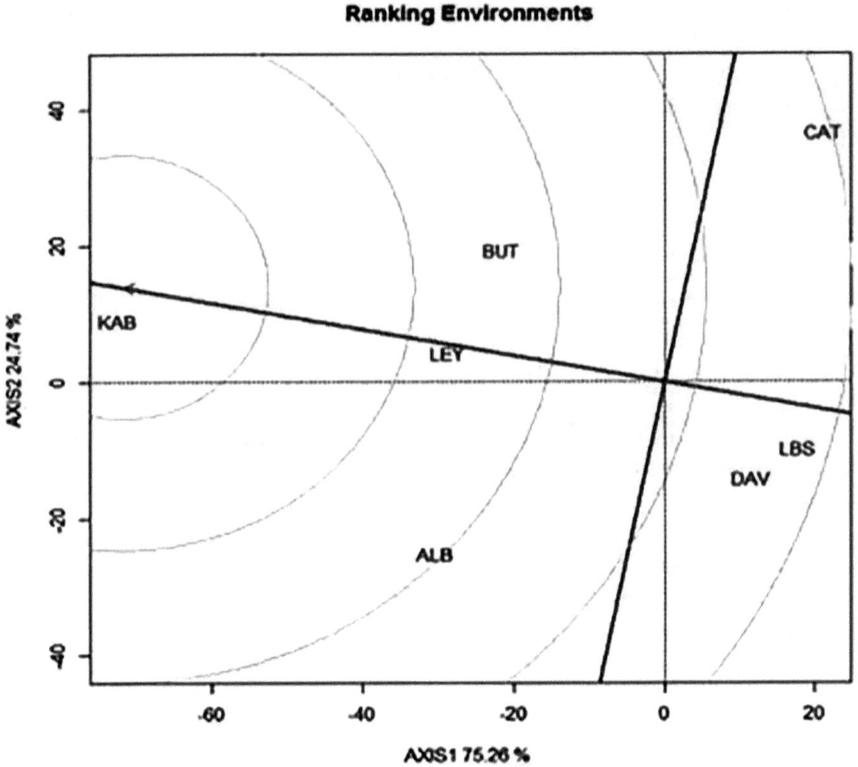

Fig. 1.16 GGE Biplot ranking of growing environments based on fiber yield. Note: KAB = N. Cotabato, BUT = Agusan del Norte, LEY = Leyte, ALB = Albay, DAV = Davao del Norte, LBS = Laguna, CAT = Catanduanes. (Source: AG Lalusin, ICropS, CAFS, the University of the Philippines Los Baños (UPLB))

Table 1.4 Stability analysis of the three abaca genotypes using parametric and nonparametric methods

	Stability measures						
Genotype	bij[a]	sd_{ij}	S1[b]	S2	S3	S6	Kang
BC_{2-2}	0.89	513	0.95	1	3.17	3.33	5
Dioscoro 1 or BANDALA	0.69	610.43	0	0.48	1	1	3
Tinawagan Pula	1.42	526.26	0.95	0.89	2.38	2.5	4

Source: AG Lalusin, ICropS, CAFS, the University of the Philippines Los Baños (UPLB)
[a]In this analysis, BC_{2-2} was identified as the most stable, with a regression coefficient of 0.69, which is not equal to 1 but has the closest value. Dioscoro 1 or BANDALA (0.69) did not deviate much from BC_{2-2}

1.9.4.3 Quality Fibers

Dioscoro 1 or BANDALA is found to have an average tensile strength of 43.35 kgf/gm and 28.91 kgf/mm2 based on the PhilFIDA and American Society for Testing and Materials (ASTM) protocol, respectively. Further, it is comparable with the three NSIC-released varieties (Abuab, Inosa, and Tangongon) in the Philippines based on their fiber morphology and chemical compositions (Lalusin 2022).

This is the output of the DOST-PCAARRD-funded project, Physicochemical Analysis and Fiber Tensile Strength Evaluation of Different Varieties of Abaca (*Musa textilis* Nee) in the Philippines, implemented by ICropS-CAFS, UPLB.

The project harvested fibers from 75 abaca accessions from different locations in the Philippines and was evaluated for their tensile strength and agronomic characteristics. They were also subjected to chemical analysis to compare Dioscoro 1 or BANDALA with traditional abaca varieties.

Based on the results using the PhilFIDA protocol (Fig. 1.17), approximately 73% of the accessions, including Dioscoro 1 or BANDALA, coincides with the reported range of PNS (2016) which is from 35 kgf/gm to 55 kgf/gm. Approximately 24% of the accessions have a lower tensile strength than the expected range while 3% have a higher tensile strength. Dioscoro 1 or BANDALA, with a tensile strength of 44.62 kgf/gm, ranks 13th among the 75 accessions.

Dioscoro 1 or BANDALA ranks 13th among the 75 accessions collected and is among the middle value in terms of tensile strength. The three NSIC-released varieties were also highlighted in their rankings: Tangongon (CARAGA) is among the top accessions; Abuab (Camarines Norte), Tangongon (Butuan), and Inosa (Caraga) are among the middle accessions, while Tangongon (Sarangani) is

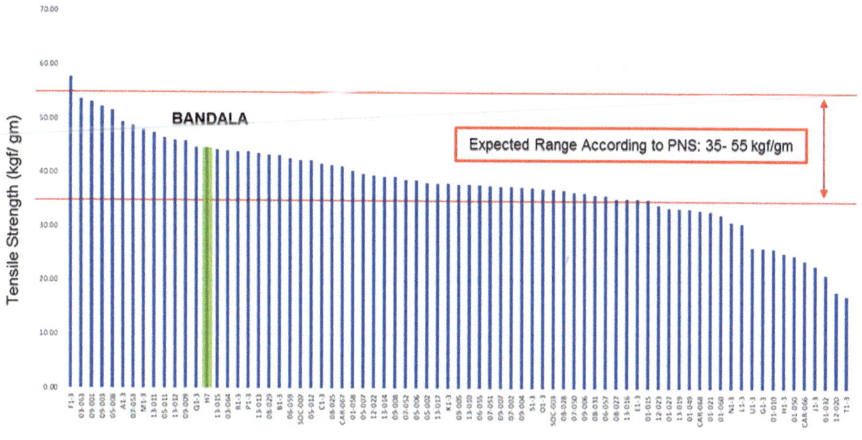

Fig. 1.17 Tensile strength values of 75 abaca accessions are based on the PhilFIDA protocol; diagonal red lines indicate the expected range according to the Philippine National Standard (2016). *Tested using Shimadzu AGX-V machine with 500 N load cell and set to 16 cm gauge length. (Source: AG Lalusin, ICropS, CAFS, the University of the Philippines Los Baños (UPLB))

among the bottom accessions. It can also be noted that 'Canton', a natural hybrid of abaca and wild banana, is also categorized as a bottom variety based on its tensile strength (Table 1.5).

The agronomic characteristics of the Dioscoro 1 or BANDALA as compared with the NSIC-released abaca varieties were also compared (Table 1.6). Pseudostem fresh weight (kg), pseudostem length (cm), and girth size (cm) are directly related to abaca yield i.e., higher values indicate higher yield. The number of leaf sheaths also indicates the volume of fibers that can be extracted from an abaca plant.

The chemical analysis conducted by BIOTECH, UPLB found that Dioscoro 1 or BANDALA was not significantly different from Tangongon (Caraga) and Tangongon (Sarangani) in terms of moisture content. In terms of ash content, Dioscoro 1 or BANDALA was found to be significantly different from the other NSIC-released varieties. While for the extractives, Dioscoro 1 or BANDALA was not significantly different from Tangongon (Butuan), Inosa (Caraga), and Inosa (Butuan).

In terms of alpha-cellulose content, Dioscoro 1 or BANDALA was found to be significantly different from the rest of the varieties tested. Moreover, for acid-insoluble lignin, Dioscoro 1 or BANDALA was not significantly different from Abuab (Camarines Norte) and Inosa (Butuan).

The cellulose content of natural fibers, a structural component of cell walls, is a primary factor that contributes to the strength of the fiber. High cellulose content means more hydrogen bonding and higher tensile strength of the fiber. Additionally, lignin correlated negatively with tensile strength and positively with failure strain.

These results show that Dioscoro 1 or BANDALA is already comparable with the three NSIC-released varieties in the Philippines based on their fiber chemical compositions (Table 1.7). Dioscoro 1 or BANDALA is abaca based on its fiber morphology and chemical compositions.

In the same project, the fiber morphology properties (slenderness, flexibility, fiber density) of selected abaca accessions, including Dioscoro 1 or BANDALA, were tested by the Forest Products Research and Development Institute (DOST-FPRDI), and computed through the Runkel ratio and Muhlstep ratio to predict their suitability for pulp and paper making. Results show that Dioscoro 1 or BANDALA has the highest value for slenderness (Table 1.8). The value for slenderness has no set limit, thus the higher value indicates higher suitability for pulp and paper making (Lalusin 2022). Dioscoro 1 or BANDALA is highly suitable for pulp and paper making.

1.9.5 Ways Forward

Studies on the quality of Dioscoro 1 or BANDALA fibers show their suitability for pulp, papermaking, fibercraft, textiles, and yarns. This untapped potential can widen opportunities both for abaca farmers and processors in the country. The utilization of abaca was further supported by DOST-FPRDI and the Philippine Textile Research Institute (DOST-PTRI) wherein Dioscoro 1 or BANDALA was used for papermaking and textile, respectively, through DOST-PCAARRD funded projects (FPRDI 2021).

Table 1.5 Abaca accessions with top, middle, and bottom tensile strength values based on PhilFIDA protocol

TOP			MIDDLE			BOTTOM		
Accession Name	Origin	Tensile Strength (kgf/gm)	Accession Name	Origin	Tensile Strength (kgf/gm)	Accession Name	Origin	Tensile Strength (kgf/gm)
Abuab Puti	Catanduanes	57.7096	Labo	Camarines Norte	38.6220	Irosin	Sorsogon	26.2813
Negro	Aurora	53.6425	Abuab[a]	Camarines Norte	38.0835	Linobloban	Catanduanes	26.1600
Putian	Zamboanga Del Norte	53.1639	Sarabiahon	CARAGA	38.0298	Wild Pula Caradgan	UPLB	26.0703
Bakayakan	Zamboanga Del Norte	52.2540	Tangongon[a]	Butuan	37.9718	Tagakan	Catanduanes	25.2574
Catandungan	Camarines Norte	51.6087	Lunhan	Zamboanga Del Norte	37.8830	Canton[b]	UPLB	24.7676
Laguis	Southern Leyte	49.3852	Inosa[a]	Caraga	37.8297	Natonin Hemp	CAR	23.8597
Kagisan	Negros Oriental	48.7146	Native	Panay Island	37.8221	Abuab Pula	Catanduanes	22.9113
Laguis	Butuan	47.9066	Native	Negros Oriental	37.6599	PSU	UPLB	21.1833
Tangongon[a]	CARAGA	47.4281	Kilayan	Zamboanga Del Norte	37.4947	Tangongon[a]	Sarangani	18.0753
Inisarog	Camarines Norte	46.4729	Jugbagon Native	Negros Oriental	37.4845	Bulusan	Sorsogon	17.3086

Source: AG Lalusin, ICropS, CAFS, the University of the Philippines Los Baños (UPLB)
[a]NSIC-released varieties;
[b]Natural hybrid of abaca and wild banana. Note: Dioscoro 1 or BANDALA has an average TS of 43.35 kgf/gm

Table 1.6 Agronomic characteristics of the BANDALA and NSIC-released abaca varieties Abuab, Inosa, and Tangongon

Characters	Abuab	Inosa	Tangongon	Dioscoro 1 or BANDALA
Number of stools	9a	8ab	8ab	9a
Plant height (cm)	273b	295ab	311a	290ab
Girth (cm)	53a	38b	47ab	52a
Stem fresh weight (kg)	25ab	24ab	29a	30.3a
Leaf-sheath number	22a	17b	19ab	22a
Fiber length (cm)	250ab	250ab	259ab	275a
% Fiber recovery	0.888ab	1.104a	0.321ab	0.961ab

Source: AG Lalusin, ICropS, CAFS, the University of the Philippines Los Baños (UPLB)
Note: In mean comparison, if the letter is the same, it means that the data is not significantly different. If the letters are different, the data is significantly different from each other

Table 1.7 Comparison of the chemical composition of NSIC-released varieties and Dioscoro1 or BANDALA using Tukey's HSD test

Varieties	% Moisture	% Ash	% Extractives	%Alpha Cellulose	%Acid Insoluble Lignin	% Acid Soluble Lignin
Dioscoro 1 or BANDALA	8.3600d	1.8950b	9.1950a	50.0850e	9.3100c	2.9250a
Tangongon (CARAGA)	8.3650d	1.5900d	7.9600b	58.6000cd	10.900ab	2.6250a
Abuab (Camarines Norte)	7.2900e	1.3700a	8.0300b	66.1700a	9.2700c	2.9100a
Inosa (CARAGA)	8.9700c	1.7750C	9.4950a	65.6150a	12.3000a	2.6250a

Source: AG Lalusin, ICropS, CAFS, the University of the Philippines Los Baños (UPLB)

Table 1.8 Computed values for predicting the suitability of samples for pulp and paper making on selected abaca accessions

Sample	Slenderness	Flexibility	Remarks	
			RR	MR
Dioscoro 1 or BANDALA	221.76410	0.67460	Very promising	Favorable
Tangongon (CARAGA)	162.05680	0.61530	Very promising	Favorable
Abuab (Camarines Norte)	183.41540	0.63260	Very promising	Favorable
Tangongon (Butuan)	172.94720	0.69640	Very promising	Favorable
Inosa (CARAGA)	183.14530	0.65590	Very promising	Favorable
Inosa (Butuan)	171.63880	0.60620	Very promising	Favorable
Tangongon (Sarangani)	219.97140	0.38500	Very poor	Not suitable

Source: AG Lalusin, ICropS, CAFS, the University of the Philippines Los Baños (UPLB)

1.9.5.1 Pulp and Paper Making

In the article, "DOST-FPRDI: Abaca Hybrid Yields Quality Fiber," Dioscoro or BANDALA possesses high-quality fiber fit for pulp and paper-making. It was mentioned that "the physical properties (basis weight, thickness, folding endurance, and burst, tensile and tear indices) of the Dioscoro 1 or BANDALA paper sheets we studied were comparable to those of commercial abaca" (DOST-FPRDI 2019).

1.9.5.2 Fibercraft, Textiles, and Yarns

Dioscoro 1 or BANDALA fiber was also used by DOST-PTRI in the project, "Community-Level Textile Raw Material Integration towards Self-sufficient Philippine Textiles Manufacturing of PROTex (Protective, Reengineered Occupational Technical Textiles)" where it served as one of the raw materials used in making repellent/antimicrobial natural textile Protective Outwears (POWs) (Fig. 1.18). The project was a part of the DOST-PCAARRD's Good Agri-Aqua Livelihood Initiatives towards National Goals or the GALING-PCAARRD Kontra

Fig. 1.18 For outdoor wear, the textile was made from 30% Dioscoro 1 or BANDALA fibers and 70% cotton fibers. (Source: AG Lalusin, ICropS, CAFS, UPLB)

COVID-19 Program (PCAARRD 2021). Aside from R & D projects, DOST-PTRI also featured Dioscoro 1 or BANDALA fabric in various exhibits such as the National Science and Technology Week exhibit and the Philippine Textile Stakeholders' Conference.

With pressing industry concerns, it is high time for the abaca industry to welcome the prospects and potentials of new breeding technologies. Dioscoro 1 or BANDALA is a product of many years of R&D toward varietal improvement. It has resistance against the bunchy top virus, resilience to environmental stress, and potential for value addition—all of which are expected to help in achieving the goals of the local industry. Research shows that Dioscoro 1 or BANDALA is already comparable with the three NSIC-released varieties in the Philippines and has promising results in the test for the suitability of fiber for pulp and papermaking. These prospects make Dioscoro 1 or BANDALA a desirable abaca hybrid. DOST-PCAARRD advocates for the recognition and acceptability of Dioscoro1 or BANDALA among the stakeholders and the abaca industry, as an additional or optional variety, and not a replacement for any traditional or naturally existing abaca variety. The Council, as outlined in its ISP for Abaca, will continue supporting Dioscoro 1 or BANDALA and other various S&T initiatives for the betterment of the Philippine abaca industry.

1.10 Conclusion and Prospects

The Philippine abaca industry is on track for continued expansion and remains a cornerstone of the country's agricultural economy, with its influence extending to both domestic and international markets. As the world places more emphasis on sustainability and environmental conservation, abaca, a natural and renewable resource, has found increasing demand. Its superior fiber properties, including strength, flexibility, and resistance to saltwater, make it an ideal material for various industrial uses. The versatility of abaca, combined with its biodegradability, makes it a preferred alternative to synthetic fibers and other plant-based materials that have a greater environmental impact. This trend aligns with global shifts toward greener, more sustainable products.

Currently, the global demand for abaca averages around 73,917 metric tons (MT) annually, with the Philippines meeting 85% of this demand. This level of production firmly establishes the country as the dominant global supplier. In the years to come, the Food and Agriculture Organization (FAO) has forecasted a rise in global consumption to over 85,000 MT. The increasing recognition of abaca's value, especially in industries where strength, durability, and environmental impact are key considerations, positions the industry for significant growth. Not only does the abaca industry provide a source of income for many Filipino farmers, but it also contributes to the broader goals of environmental sustainability and resource conservation.

The abaca pulp sector, which is a major component of the abaca industry, has shown particularly strong growth since 1989. The sector has experienced an annual growth rate of 6.6% in export volume and 7.9% in earnings, which demonstrates its rising significance in global markets. Abaca pulp is essential for specialty paper production, which is used in products such as currency notes, tea bags, sausage casings, and cigarette papers. As global economies improve and new markets emerge, this demand is projected to increase, further boosting the industry. Moreover, abaca pulp is considered an effective alternative to coniferous pulp, used in traditional paper production, offering a substitution ratio of 4:1. This makes abaca pulp a viable option for paper companies seeking to diversify their sources of raw materials and align with growing environmental standards.

In addition to its industrial applications, abaca is experiencing a surge in demand in the fibercraft sector. This segment has grown at an annual rate of 4.4%, driven by the increasing interest in handmade, eco-friendly products. Abaca is commonly used for crafting various decorative items, textiles, and accessories due to its natural look and sustainable qualities. As consumer preferences continue to lean toward environmentally friendly and ethical products, the market for abaca-based crafts is expected to expand. Abaca fabrics, which are used for fashion, home décor, and wrapping, are particularly popular in international markets, especially in Europe, North America, and Japan, where demand for eco-friendly textiles is on the rise. From 1989 to the present, exports of abaca fabrics have grown by 121.4% annually, reflecting their increasing popularity and the industry's ability to meet the evolving needs of the global market.

Though the abaca industry faces exciting prospects, certain challenges remain, particularly in the area of disease management. Abaca bunchy top disease (AbBTD), caused by viruses like Abaca bunchy top virus (ABTV), poses a significant threat to production, as infected plants suffer from reduced growth and yield. The concern about virus variability and pathogen evolution highlights the need for continuous research and the development of disease-resistant varieties. Insect vectors, such as aphids, play a critical role in spreading these diseases, and efforts are underway to understand the genetic variability of these insects to better manage the spread of the disease. To tackle this, strategies like gene deployment, mixed cropping systems, and the use of multiple resistance genes are being explored to ensure the long-term sustainability of the abaca industry. Multilocation field testing is essential to evaluate the effectiveness of various breeding approaches under diverse geographical and climatic conditions, ensuring that abaca varieties can thrive in different regions.

Breeding programs in the Philippines, spearheaded by institutions like the Institute of Plant Breeding at UPLB and the National Abaca Research Center (NARC) at Visayas State University, play a crucial role in the industry's sustainability. These institutions are focused on developing high-yielding, virus-resistant abaca varieties through both traditional and modern breeding techniques. One of their primary goals is to create abaca plants that not only offer higher fiber yields but also possess greater resistance to viruses, helping farmers maintain stable production levels. The success of these programs relies heavily on the expertise of dedicated researchers, technicians, and field workers who are committed to rehabilitating the abaca industry.

To support the growing demand for abaca, particularly as the industry continues to recover and expand, large-scale propagation and distribution of improved hybrids are essential. The Philippines currently has over 144,000 hectares dedicated to abaca production, but to meet the increasing market demands and support industry rehabilitation, around 230.4 million planting materials must be produced and distributed across abaca-growing areas. This requires extensive collaboration between government agencies, research institutions, and farmers, ensuring that the abaca industry continues to contribute to economic growth, sustainable agricultural practices, and environmental conservation in the Philippines. With ongoing research, breeding improvements, and strategic market expansions, the Philippine abaca industry is poised to not only recover but also thrive in the global marketplace for years to come.

Appendix 1.1 Research Institutes and Online Resources Available for Abaca Breeding

Institution	Area of specialization	Research activities	Website
Institute of Crops Science, College of Agriculture and Food Science, University of the Philippines Los Banos	Plant Breeding, Physiology, Production, Seed Technology, Tissue culture, genomics	The Institute of Crop Science (ICropS) is responsible for cereals, field legumes, root crops, fibers and forage industries. It shall also spearhead the expanding and more profitable fresh fruits and vegetables subsectors, as well as the promising ornamentals and herbal medicines businesses. It will continue to i.offer dynamic and relevant undergraduate and graduate curricular programs that shall promote science-based, modern and sustainable agriculture and develop competent and responsible crop scientists and professionals; (ii) generate sound scientific ideas, concepts, technologies, products and processes within the domains of crop science that are relevant to the needs and requirements of Philippine agriculture and end-user communities	https://cafs.uplb.edu.ph/icrops/

(continued)

Institution	Area of specialization	Research activities	Website
		(iii) effectively disseminate and transfer scientific knowledge and exploit the results of R&D to the local farming communities, agri-industries, consumers, export markets and other possible stakeholders; and (iv) contribute to the shaping of national policies and priorities in agriculture through the results of the ICropS R&D efforts.	
institute of Crop Science and Institute of Plant Breeding, College of Agriculture and Food Science, University of the Philippines Los Banos	Plant Breeding, tissue culture, genetic engineering, germplasm collection and conservation	Develop new and improved varieties of important agricultural crops; (b) Undertake investigations in plant breeding and allied disciplines related to crop improvement; (c) Systematically collect, introduce, preserve and maintain a germplasm bank of important and potentially useful agricultural crops; and (d) Assist other agencies in the multiplication of quality seeds and vegetative materials of recommended crop varieties	https://cafs.uplb.edu.ph/institute-of-plant-breeding/
National Abaca Research Center, Visayas State Univerisity, Baybay Leyte, Philippines	Development of high-yielding, pest and stress-resistant abaca varieties with high-quality fibers suited to specific end uses through conventional methods and biotechnology.	NARC developed abaca technologies that increase the income of small abaca farmers, generate employment for rural households, and increase value-added to the plant.	https://www.vsu.edu.ph/narc
Philippine Fiber Industry and Development Authority (PhilFIDA)	Promote good agricultural production practices, enforce fiber standards and other regulatory policies to maintain supply of good quality fibers and to	The Philippine Fiber Industry Development Authority or PhilFIDA is an agency of the Philippine Department of Agriculture responsible for promoting the	https://www.philfida.da.gov.ph

(continued)

Institution	Area of specialization	Research activities	Website
	ensure worldwide acceptability; Improve farm productivity	accelerated growth and development of the fiber industry in the Philippines, such as abaca, also known as Manila hemp and cotton	

Appendix 1.2 Genetic Resources for Abaca Breeding

Species	Important traits	Cultivation location
Musa balbisiana Colla (Pacol/Butuhan)	Resistance to abaca mosaic and abaca bunchy top virus	Growing wild all over the Philippines
Musa tex 51	Fiber Yield—2084 kg/ha per year	Cultivated in Bicol, Philippines
Abuab	Fiber Yield—1723 kg/ha per year	Cultivated in Bicol, Philippines
Tinawagan Puti	Fiber Yield—1607 kg/ha per year	Cultivated in Bicol, Philippines
Tinawagan Pula	Fiber Yield—1576 kg/ha per year	Cultivated in Bicol, Philippines
Linawaan	Fiber Yield—1320 kg/ha per year	Cultivated in Visayas, Philippines
Inosa	Fiber Yield—1270 kg/ha per year	Cultivated in Visayas, Philippines
Lalay	Fiber Yield—1090 kg/ha per year	Cultivated in Visayas, Philippines
Maguindanao	Fiber Yield—2100 kg/ha per year	Cultivated in Mindanao, Philippines
Bongolanon	Fiber Yield—1720 kg/ha per year	Cultivated in Mindanao, Philippines
Tangongon	Fiber Yield—1590 kg/ha per year	Cultivated in Mindanao, Philippines
Musa coccinea	*Callimusa*	Introduced
Musa exotica R. Valmayor, sp nov.	*Callimusa*	Introduced
Musa violascens	*Callimusa*	Introduced
Musa itenerans	*Eumusa*	Introduced
Musa acuminata Colla spp errans (Blanco) R. Valmayor, comb. nov.	*Eumusa*	Indigenous
Musa balbisiana	*Eumusa*	Indigenous
Musa alinsanaya	*Australimusa*	Indigenous
Musa textilis	*Australimusa*	Indigenous
Musa pekelii	*Australimusa*	Indigenous
Musa fehi	*Australimusa*	Introduced

(continued)

Species	Important traits	Cultivation location
Musa ornata	*Rhodochlamys*	Introduced
Musa velutina	*Rhodochlamys*	Introduced
Musa laterita	*Rhodochlamys*	Introduced
Canton (*Musa balbisiana x Musa textilis*)	Natural hybrid	Indigenous
Minay *(Musa balbisiana x Musa textilis)*	Natural hybrid	Indigenous
M x paradisiaca Linn	Cultivated banana	Indigenous
Musa acuminata variants and polyploids	Cultivated banana	Indigenous
Musa balbisiana variants and polyploids	Cultivated banana	Indigenous
*Ensete glaucum (*Roxb)	Ensete	Indigenous

References

Alcober ER (1986) Morphological characters and yield of abaca and related Musa clones in Baybay, Leyte, Philippines. Ann Trop Res (Philippines) 8(4):189–200

Aquino VM, Aspuria ET, Ruiz MAD, Santiago DS (2005) Genetic transformation of abaca by microprojectile bombardment. Proc 27th NAST Annual Scientific Mtg. Manila July 13–14, 2005

Araya-Gutiérrez D, Monge GG, Jiménez-Quesada K, Arias-Aguilar D, Cordero RQ (2023) Abaca: a general review on its characteristics, productivity, and market in the world. Rev Fac Nac Agron Medellín 76(1):10263–10273

Aspuria ET (1984) In vitro shoot and root formation and in vivo plantlet growth and development in abaca (*Musa textilis* Nee). MSc Thesis. College of Agriculture, UP Los Baños, pp. 167

Balifurnish (2012). www.balifurnish.com/indonesia/indonesia/. Accessed Nov 2012. Taken from http://www.abacaorganic.co.uk/. Accessed Aug 2012

Balkenhol N, Cushman SA, Storfer AT, Waits LP (2016) Landscape genetics, concepts, methods, applications. Wiley, New York, p 288. Banana bract mosaic potyvirus in the Philippines. Australasian Plt. Pathol

Bande MM, Grenz J, Asio VB, Sauerborn J (2013) Fiber yield and quality of Abaca (*Musa textilis* Var. Laylay) grown under different shade conditions, water and nutrient management. Ind Crop Prod 42:70–77

Barbosa CFC, Asunto JC, Koh RBL et al (2023) Genome-wide SNP and indel discovery in abaca (*Musa textilis* Née) and among other *Musa* spp. for abaca genetic resources management. Curr Issues Mol Biol 45:5776–5797. https://doi.org/10.3390/cimb45070365

Bernardo FA (1957) Plant characters, fiber and cytology of *Musa balbisiana* x *Musa textilis* F_1 hybrids. Phil Agric 41:117–156

Boguero APB, Parducho AML, Mendoza MR et al (2016) Molecular screening of abaca (*Musa textilis* Nee) accessions using microsatellite markers associated with resistance to bunchy top virus. Philipp J Crop Sci 41:13–19

Brewbaker JL, Umali DL (1956) Classification of Philippine Musae I. The genera of Musa L and Ensete Horan. Phil Agric 40:231–241

Brewbaker JL, Gorres DD, Umali DL (1956) Classification of Philippine Musae II. Canton, Minay, Putative hybrid forms of *Musa textilis* and *Musa balbisiana* Colla. Philipp Agric 40:242–257

Browning JA (1974) Relevance of knowledge about natural ecosystems to development of pest management programs for agro-ecosystems. Proc Am Phytopathol Soc 1:191–199

Calica GB, Sedillo KR, Galapon GMD, Macaranas RJP (2024a) Enhancing the Abaca fiber primary processing system in North Cotabato, Philippines. Int J Multidiscipl Res Publ 7(2):202–208

Calica GB, Galapon GMD, Macaranas RJP (2024b) Postproduction practices and marketing of Abaca in North Cotabato, Philippines. Int J Sci Res Manag 12(01):5727–5734. Website: https://ijsrm.net ISSN (e): 2321-3418. https://doi.org/10.18535/ijsrm/v12i01.em01

Calinisan MR (1934) Notes on the suspected "mosaic" of abaca in The Philippines. Philipp J Agric 5(4):255–256

CFC/UNIDO/FAO/FIDA (2004) Abaca Improvement of fiber extraction and identification of higher-yielding varieties. Final Technical Report CFC/FIGHF/09. Activities in Ecuador. Available from www.unido.org/fileadmin/import/48267_Activities_in_Ecuador.pdf. Accessed 6 June 2013

Cheesman EE (1947) Classification of bananas. I. The genus Ensete Horan and the genus *Musa* L. Kew Bull 2:97–117

Cheesman J (1948) Classification of bananas. III. Critical notes on the species (c) *Musa paradisiaca* Linn. and *Musa sapientum* Linn. Kew Bull 2:45–54

Copeland EB (1911) Abaca. Philippine Agric For 1(4):64–73

Copeland EB (1927) Nomenclature of the abaca plant. Philipp J Sci 33(2):141–153

Cruz OJ, Balingkit EV (1974) Performance of selected abaca varieties and promising hybrids in the Visayas. Phil J Plant Indus 39:53–107

D'hont A, Denoeud F, Aury JM et al (2012) The banana (*Musa Acuminata*) genome and the evolution of monocotyledonous plants. Nature 488(7410):213–217. https://doi.org/10.1038/Nature11241

Damasco OP, Graham GC, Henry RJ et al (1996) Random amplified polymorphic DNA (RAPD) detection of dwarf off-types in micropropagated Cavendish (*Musa* spp. AAA) bananas. Plant Cell Rep 16:118–123

Dempsey JM (1963) Long fiber development in South Vietnam and other Asian countries. US Department of Commerce, Washington DC

Diaz NT (1997) Resistance of abaca (*Musa textilis* Nee) F1 hybrids to abaca bunchy top virus. PhD Thesis, University of the Philippines Los Baños, College Laguna, Philippines, pp 73

Engle LM (1985) Abaca breeding. Annual Report No. 3 (January–December 31, 1985), University of the Philippines Los Baños, College, Laguna, pp 17

Engle LM, Peralta MTB, Dela Cueva FM et al (1999) Description of abaca (*Musa textilis* Nee) genetic stocks. Phil J Crop Sci 24:120

Ereful NC, Lalusin AG, Laurena AC (2022a) Assessing loss of regulatory divergence, genome-transcriptome incongruence, and preferential expression switching in abaca × banana backcrosses. Genes 13(8):1396. https://doi.org/10.3390/genes13081396. PMID: 36011307; PMCID: PMC9407414

Ereful NC, Lalusin AG, Laurena AC (2022b) RNA-Seq reveals differentially expressed genes associated with high fiber quality in abaca (*Musa textilis* Nee). Genes 13:519. https://doi.org/10.3390/genes13030519

Fabre F, Rousseau E, Mailleret L, Moury B (2012) Durable strategies to deploy resistance in agricultural landscapes. New Phytol 193:1064–1075. https://doi.org/10.1111/j.1469-8137.2011.04109.x

Faure S, Noyer JL, Horry JP et al (1993) A molecular marker-based linkage map of diploid bananas (*Musa acuminata*). Theor Appl Genet 87:517–526

Forest Products Research and Development Institute (2021) Dost-fprdi: abaca hybrid yields quality fiber. Retrieved from https://fprdi.dost.gov.ph/403-dost-fprdi-abaca-hybrid-yields-quality-fiber

Furuya N, Dizon TO, Natsuaki K (2006) Molecular characterization of banana bunchy top virus and cucumber mosaic virus from abaca (*Musa textilis* Nee). J Agric Sci Tokyo Univ Agric 51(2):92–101

Galambao MB (2011) The population genetics of *Pentalonia nigronervosa* and *P. caladii* (Hemiptera: Aphididae) in Hawaii. http://hdl.handle.net/10125/101701.scolarspace.manoa.hawaii.edu
Galvani AP (2003) Epidemiology meets evolutionary ecology. Trends Ecol Evol 18:132–139
Galvez LC, Koh RBL, Barbosa CFC et al (2021) Sequencing and de Novo assembly of Abaca (*Musa textilis* Née) var. Abuab Genome Genes 12(8):1202. https://doi.org/10.3390/genes12081202
Garcia-Arenal F, McDonald BA (2003) An analysis of the durability of resistance to plant viruses. Phytopathology 93:941–952
Gawel NJ, Jarret RL (1991a) Chloroplast DNA restriction fragment length polymorphisms (RFLPs) in *Musa* species. Theor Appl Genet 81(6):783–786
Häkkinen M (2013) Reappraisal of sectional taxonomy in *Musa (Musaceae)*. Taxon 62(4):809–813
Holderegger R, Wagner HH (2008) Landscape genetics. Bioscience 58:199–207. https://doi.org/10.1641/B580306
Hu J, Fu H, Lin C et al (2007) Re-assortment and concerted evolution in Banana bunchy top virus genomes. J Virol 81:1746–1761. https://doi.org/10.1128/JVI.01390-06
Hyder MZ, Shah SH, Hameed S, Naqvi SMS (2011) Evidence of recombination in the banana bunchy top virus genome. Infect Genet Evol 11:1293–1300. https://doi.org/10.1016/j.meegid.2011.04.015
Islam MN, Naqvi AR, Jan AT, Haq QMR (2010) Genetic diversity and possible evidence of recombination among Banana bunchy top virus (BBTV) isolates. Int Res J Microbiol 1:1–12
Jarret R, Litz R (1986a) Enzyme polymorphism in *Musa acuminata* Colla. J Hered 77:183–188
Jarret R, Litz R (1986b) Isozymes as genetic markers in bananas and plantains. Euphytica 35:539–547
Jarret R, Gawel N, Whittemore A, Sharrock S (1992) RFLP-based phylogeny of *Musa* species in Papua New Guinea. Theor Appl Genet 84:579–584
Jarret R, Vuylsteke DR, Pimenter RB (1993) Detecting genetic diversity in diploid bananas using PCR and primers from a highly repetitive DNA sequence. Euphytica 68:69–76
Javier DF, Oracion MZ (1988) Cytology and pollen viability of two abaca (*Musa textilis* Nee.) varieties found in Leyte (Philippines). Philipp. J Crop Sci 13(1):15
Kumar L, Lepoint P, Blomme G et al (2013) Building a collaborative, public-private R4D alliance for an integrated control of banana bunchy top disease in Sub-Saharan Africa. Workshop Report on 'Global Alliance for BBTV Control in Africa, Arusha, Tanzania, 2–3 Feb 2013, pp 62
Labrador AF (1928) The abaca project of La Carlota Experiment Station. Philipp Agric Rev 21:3–19
Labrador DA, Lalusin AG, Mendoza MRR, dela Viña CB (2020) Morphological characterization and karyotype analysis of abaca (*Musa textilis* Nee) and its hybrids with *Musa balbisiana* Colla. Philipp Agric Sci 103(1):13–28
Lalusin AG (2010) Abaca breeding for a more reliable Philippine abaca industry. Lecture No 5, Annual BSP-UP Professorial Chair Lectures, February 15–17, 2010 Bangko Sentral ng Pilipinas Malate, Manila
Lalusin AG (2014) Rehabilitation of abaca plantations through adoption of high-yielding and virus-resistant abaca hybrids. Paper presented during the Philippine Council for Agriculture, Aquatic and Natural Resources Research and Development (PCAARRD) Abaca and Rubber Program Review at the Prince Hotel, Butuan City, July 30, 2014
Lalusin AG (2022) Physicochemical analysis and fiber tensile strength evaluation of different varieties of abaca (*Musa textilis* Nee) in the Philippines. Institute of Crop Science, College of Agriculture and Food Science, University of the Philippines Los Baños, Philippines
Lalusin AG, Villavicencio MLH (2015) Abaca (*Musa textilis* Nee) breeding in The Philippines. In: Cruz VMV, Dierig DA (eds) Industrial crops, handbook of plant breeding, vol 9. Springer, New York

Lalusin AG, Castro SD, Laurena AC (2007) Genetic diversity and phylogenetic relationships of abaca (*Musa textilis* Nee) using microsatellite markers. 1st ASIAHORCS Joint Symposium, Nagoya University, Japan
Lanaud C, Tezenas Du Montcel H, Jolivot MP (1992) Variation of ribosomal gene spacer length among wild and cultivated bananas. Heredity 68:147–156
Leppik EE (1970) Gene centers of plants as sources of disease resistance. Annu Rev Phytopathol 8: 323–344
Liu J, Gao CW, Niu YF (2019) Complete chloroplast genome sequence of *Musa banksii* and its phylogenetic analysis. Mitochondrial DNA Part B: Resour 4(1):789–790. https://doi.org/10.1080/23802359.2019.1566794
Lomerio EO (1998) Protocol for screening abaca against Abaca mosaic virus (AMV) and Abaca bunchy-top virus (ABTV). Fiber Industry Development Authority Region 5, Bicol Univ Cpd, Legazpi City, pp. 8
Montarry J, Cartier E, Jacquemond M (2012) Virus adaptation to quantitative plant resistance: erosion or breakdown. J Evol Biol 25:2242–2252. https://doi.org/10.1111/j.1420-9101.2012.02600.x
Muncada GP (2016) Next generation sequencing transcriptome analysis of abaca bunchy top virus resistance and differential expression of EIF4G gene in abuab, pacol and BC_2 hybrid. PhD Thesis, UPLB
Niu YF, Gao CW, Liu J (2018) The complete chloroplast genome sequence of wild banana, *Musa balbisiana* variety 'Pisang Klutuk Wulung' (*Musaceae*). Mitochondrial DNA Part B: Resour 3(1):460–461. https://doi.org/10.1080/23802359.2018.146212
Niyongere C, Ateka E, Losenge T et al (2011) Screening Musa genotypes for banana bunchy top disease resistance in Burundi. Acta Hortic 897:439–447. https://doi.org/10.17660/ActaHortic.2011.897.60
Ocfemia GO (1926) Progress report on bunchy top of abaca or Manila hemp. Phytopathology 16: 894
Ocfemia GO (1930) Bunchy top of abaca or Manila hemp. I. A study of the cause of the disease and its method of transmission. Am J Bot 17:1–8
Ocfemia GO (1931a) The bunchy top of abaca and its control. Philipp Agric 20:328–340
Ocfemia GO (1931b) Save the abaca industry from ruin by bunchy-top. Philipp Agric 20:328–340
Ocfemia GO (1953) Note on the occurrence of the mosaic disease in the Abaca Experiment Station in Banao, Guinobatan, Albay. Philipp Agric 42:84–87
Ocfemia GO, Buha CG (1934) Bunchy-top of abaca or Manila hemp: further studies on the transmission of the disease and trial planting of abaca seedlings in bunchy-top devastated fields. Philipp Agric 22:567–581
Ocfemia GO, Celino MS, Garcia FJ (1947) Further studies on transmission of bunchy-top and mosaic of abaca (Manila hemp plant), separation of the two diseases, and mechanics of inoculation by *Pentalonia nigronervosa*. Philipp J Agric 31:87–97
Oriero CE, Odunola OA, Lokko Y, Ingelbrecht I (2006) Analysis of the B-genome derived simple sequence repeat (SSR) markers in *Musa* spp. Afr J Biotech 5(2):126–128
Oyardo EO (1974) Performance of 10 promising abaca hybrids in the Bicol region. Phil J Plant Indus 39:69–105
Palao DC (2019) Characterization and development of microsatellite markers from RNA-seq data of abaca (*Musa textilis* Nee) cv. Abuab, PhD, Thesis, UPLB
Pancho JV, Capinpin JM (1959) Cytotaxonomic study of the fiber-producing *Musa* in The Philippines I. Chromosome numbers in section Australimusa. Philipp Agric 43:397–403
Parac EP, Lalusin AG, Pangga IB, Sta Cruz FC (2021) Characteristics of selected hybrids of abaca (*Musa textilis* Nee) with resistance to bunchy top. Philipp Agric Sci 103(1):1–12
Parducho MAL, Rama RA, Lalusin AG (2020) Stability analysis of BC_2 Abaca (Musa textilis Nee) hybrids across different locations in The Philippines. Philipp J Crop Sci 45(3):44–49
Person C (1959) Gene-for-gene relationships in host: parasite systems. Can J Bot 37:1101–1130

Philippine Council of Agriculture, Aquatic, and Natural Resources Research and Development (2021) Development of PROTex protective apparels using fabric from local fibers takes off. Retrieved from https://www.pcaarrd.dost.gov.ph/home/portal/index.php/quick-information-dispatch/3826-development-of-protex-protective-apparels-using-fabric-from-local-fibers-takes-off
Philippine Fiber Industry Development Authority (2020) Philippine Abaca Industry Roadmap for 2021 to 2025. Retrieved from https://philfida.da.gov.ph/images/the-philippine-abaca-industry-roadmap/the-philippine-abaca-industy-roadmap-2021-2025.pdf
Philippine Statistics Authority (2024). https://www.psa.gov.ph/major-non-food-industrial-crops
Pillay M, Nwakanma DC, Tenkouano A (2000) Identification of RAPD markers linked to A and B genome sequences in *Musa*. Genome 43:763–767
Plantegenest M, Le May C, Fabre F (2007) Landscape epidemiology of plant diseases. J R Soc Interface 4(16):963–972. https://doi.org/10.1098/rsif.2007.1114
Ratnadewi D (2001) Identification of Musa textilis using RAPD markers. In: Adhi AK, Pambudy R, Burhanuddin (eds) Resource management: public-private partnership and knowledge sharing. Inter Symp-Workshop held on August 27–31, 2001, pp 75–80
Rayapati N (2013) Strategic research on aphid vector biotypes and transmission of Banana bunchy top virus for management of banana bunchy top diseases in Africa. In: Kumar L, Lepoint P, Blomme G et al (eds) Building a collaborative, public-private R4D alliance for an integrated control of banana bunchy top disease in Sub-Saharan Africa. Workshop report on 'Global Alliance for BBTVn Control in Africa', Arusha, Tanzania, Feb 2–3, 2013, p 62
Raymundo AD (1998) An alternative methodology of screening for resistance to bunchy-top and mosaic viruses of abaca. Paper presented during a technical training in connection with a UNIDO project on 'Screening for Resistance to Abaca Virus Diseases', July 24, 1998, Dep. Plant Pathol, Univ Philipp Los Los Baños, College, Laguna, Philippines
Raymundo AD (2002) Economic losses in abaca due to bunchy-top and mosaic virus diseases in the Bicol and Eastern Visayas region [Philippines]. J Tropical Plant Pathol (Philippines) 38(1–2): 31–34
Reamillo MCS (2018) Genetic characterization of backcross progenies BC_1, BC_2, and BC_3 from Musa textile is Nee (Abuab) x Musa Balbisiana Colla (Pacol) through transcriptomic analysis. PhD. Thesis, UPLB
Rivera R (1983) Protein and isozyme banding patterns among Philippine cooking bananas and their wild parents (*Musa* species). Paradisiaca 6:7–12
Sasaki A (2000) Host-parasite coevolution in a multilocus gene-for-gene system. Proc Roy Soc Lond B 267:2183–2188
Selvarajan R, Sheeba MM, Balasubramanian V et al (2011) Molecular characterization of geographically different banana bunchy top virus isolates in India. Indian J Virol 21:110–116. https://doi.org/10.1007/s13337-010-0020-1
Señeris GT (2024) Identifying key challenges in Abaca (Musa textilis Née) production: a study in two AklanMunicipalities, Philippines. Univ J Agric Res 12(1):24–34. https://doi.org/10.13189/ujar.2024.120103
Sharman M, Gambley CF, Oloteo EO (2000) First record of natural infection of abaca (*Musa textilis*) with banana bract mosaic potyvirus in The Philippines. Austr Plant Pathol l29:69
Sharman M, Thomas JE, Skabo S, Holton TA (2008) Abaca bunchy top virus, a new member of the genus Babuvirus (Family Nanoviridae). Arch Virol 153:135–147
Shetty SM, Shah MUM, Makale K et al (2016) Complete chloroplast genome sequence of *Musa balbisiana* corroborates structural heterogeneity of inverted repeats in wild progenitors of cultivated bananas and plantains. Plant Genome 9(2). https://doi.org/10.3835/plantgenome2015.09.0089
Simmonds NW, Shepherd K (1955) The taxonomy and origins of the cultivated bananas. J Linn Soc (Botany) 55:302–312
Spencer JE (1953) The abaca plant and its fiber. Manila Hemp Econ Bot 7(3):195–213
Tabora PC (1970) New abaca hybrids. Agric Los Baños 10(1):6

Tabora PC, Carlos JT Jr (1978) Cultivated species, varieties and relatives. In: Tabora PC, Carlos JT Jr (eds) The abaca. International Documentation on Abaca/UPLB Library, College Laguna, Los Baños

Torres JP, Garrido TG (1939) Progress report on the breeding of abaca (*Musa textilis* Nee). Philipp J Agric 10:211–230

Torres JP, Lanuza EA, Cruz PI (1952) Studies on abaca pollination and seed germination. Philipp J Agric 17:183–201

Ude G, Pillay M, Nwakanma D, Tenkouano A (2002) Analysis of genetic diversity and sectional relationship in Musa using AFLP markers. Theory Appl Genet 104:1239–1245

Umali DL, Brewbaker JL (1956) Abaca and its improvement. Philipp Agric 60:213–230

Umali DL, Bernardo FA, Ocfemia GO (1956) Reaction of Pacol-abaca F_1 hybrids to abaca mosaic and bunchy-top. Philipp Agric 60:47–50

Valmayor RV, Mendoza EM, Millare VE (1956) A cytological study of abaca and its relatives. Phillipp Agric 40:269–276

Valmayor RV, Espino RRC, Pascua OC (2002) The wild and cultivated bananas of The Philippines. PARFFI and BAR, Los Baños, Laguna, p 242

Varsani A, Stainton D, Thomas JE, Martin D (2013) Re-assortment, recombination and high nucleotide substitution rates: the drivers of rapid BBTV evolution. In: Kumar L, Lepoint P, Blomme G et al (eds) Building a collaborative,public-private R4D alliance for an integrated control of banana bunchy top disease in Sub-Saharan Africa. Workshop report on 'Global Alliance for BBTV Control in Africa', Arusha, Tanzania, Feb 2–3, 2013, p 62

Vaughan G (2011) *Musa textilis* Née. In: Brink M, Achigan-Dako EG (eds) Prota 16: Fibres/Plantes à fibres. [CD-Rom]. PROTA, Wageningen, Netherlands. Available from database.prota.org/.../Musa%20textilis_En.htm. Accessed Aug 2012

Villajuan-Abgona R, Oloteo EO, Bolivar AD et al (2004) Search for resistance to viral diseases in indigenous abaca varieties. In: Proc. international dissemination seminar on Abaca- improvement of fiber extraction and identification of higher yielding varieties held at the New World Renaissance Hotel, Makati City, Philippines, October 19, 2004

Wang Y, Xiao L, Guo S et al (2016) Fine mapping and whole genome resequencing identify the see coat color gene in *Brassica rapa*. PLoS One 11. https://doi.org/10.1371/journal.pone.0166464

Watanabe S, Greenwell AN, Bressnan A (2013) Localization, concentration, and transmission efficiency of Banana bunchy top virus in four asexual lineages of Pentalonia aphids. Viruses 5:758–776. https://doi.org/10.3390/v5020758

Wu X, Alexander LW (2016) Genetic diversity and population structure analysis of big leaf hydrangea using genotyping-by-sequencing. J Am Soc Hort Sci 144(40):257–263. https://doi.org/10.21273/JASHS04683-19

Yllano OB, Diaz MGQ, Lalusin AG, Laurena AC (2020) Genetic analyses of abaca (*Musa textilis* Née) germplasm from its primary center of origin, The Philippines, using simple sequence repeat (SSR) markers. Philipp Agric Sci 103(4):311–321

Chapter 2
Plant Breeding and Potential Biotechnology Uses of *Agave* as a Source of Fiber, with Emphasis on Maguey Tepeztate (*Agave marmorata* Roezl) From Central and South Mexico

Karen Y. Ruiz-Mondragón and Luis E. Eguiarte

Abstract The fibers extracted from Agave plants, commonly known as *ixtle*, *henequén*, and *sisal*, hold deep cultural significance in Mexico. These fibers have been a part of human heritage since pre-Hispanic times, at least since 2600 BC, used for clothing, ropes, nets, and other purposes, and are woven into Mexico's history and customs. However, the production and extraction of these fibers are declining in many communities due to competition with synthetic fibers, the loss of traditional manufacturing knowledge, and the absence of a market, along with the intense labor involved in extracting the fiber using conventional methods. In this chapter, we review all available information (from ecology, cytogenetics, genomics, evolution, tissue culture, crop strategies, and breeding) on *Agave marmorata* Roezl, a micro-endemic agave species from the Central-Southern states of Puebla and Oaxaca Mexico. *A. marmorata* is considered a multipurpose species due to its various uses, including as a construction material, live fence, culinary ingredient, fiber production, and its medicinal properties. Traditionally, its primary use has been to produce *pulque*, a non-distilled fermented drink. In recent years, there has been increasing interest in utilizing this species to create distilled mezcal, such as *pitzometl* and *tepextate*. However, the mezcal industry generates hundreds of tons of waste each year, mainly consisting of leaves and *bagazo* (the leftover fiber after fermentation). Consequently, it is crucial to develop strategies to reuse and repurpose this raw material. By rediscovering the ancestral uses of *A. marmorata*, we can diversify production methods, seek sustainable agricultural practices, and effectively manage waste from both pulque and mezcal production.

K. Y. Ruiz-Mondragón (✉) · L. E. Eguiarte
Laboratory of Molecular and Experimental Evolution, Department of Evolutionary Ecology, Institute of Ecology, University National Autonomous of Mexico, Mexico City, Mexico
e-mail: karen.ruiz@iecologia.unam.mx; fruns@unam.mx

K. F. M. Salem et al. (eds.), *Breeding and Biotechnology of Leaf, Fruit, and Seed Fiber Crops*, Advances in Plant Breeding Strategies 11,
https://doi.org/10.1007/978-3-032-00407-9_2

Keywords *Agave* · (*Agave marmorata* Roezl) · Agriculture · Breeding · Fibers · Management · Mezcal · Sustainable · Traditional uses

2.1 Introduction

The Agave genus holds significant historical, cultural, and ecological importance in the territories of Mexico and the southwestern United States of America, playing a key role in Mexico's economy. This genus is native to North and Central America as well as the Caribbean islands, with a few species also found in northern South America (Eguiarte et al. 2021; García-Mendoza 2002). The Agave genus is divided into two subgenera, mainly based on the morphology of its inflorescence. The species of the subgenus *Litteae* have a spiky inflorescence, while the subgenus *Agave* has a paniculate inflorescence (Gentry 1982). There are 287 accepted taxa (https://powo.science.kew.org/), with approximately 75% originating from Mexico (Eguiarte et al. 2021; García-Mendoza 2002; García-Mendoza et al. 2019; Gentry 1982). Mexico is considered the center of origin and diversification for the genus (Eguiarte et al. 2021; Gentry 1982). Gentry (1982) divided the distribution of Agave species in Mexico into areas of significant importance. The first is central Mexico, which includes the largest number of species coexisting in this region. Remarkably, the Tehuacán-Cuicatlán Valley, located in south-central Mexico, stands out for the richness of Agave species (García-Mendoza 2002), the Jalisco Plateau and the Chihuahuan Desert (Gentry 1982).

In Mexico, agave plants are commonly known as *maguey* (plural *magueys*). Their use dates back to pre-Hispanic times, and many archeological findings support this early and essential use. In central Mexico, the goddess of *maguey and pulque*, *Mayahuel*, was significant in the Nahuatl culture and cosmology, including the Mexicas and Aztecs (Nava-Cruz et al. 2015). In the sixteenth century, with the arrival of the Europeans, the Jesuit José Acosta called agave the tree of wonders (in Spanish, árbol de las Maravillas), describing a miraculous plant that provides water, wine, oil, honey, cloth, thread, needles, and much more. Also, in his second letter to the king of Spain, Hernán Cortés describes the magnificent market of Tenochtitlán. He writes about the various products available for sale, including "honey from bees and honey from plants, which those on other islands refer to as maguey. This plant is much larger than the dressing. From these plants, they also produce sugar and wine" (Blake 2012). Nowadays, more than 70 traditional uses are recognized for Agave species (Alducin-Martínez et al. 2023; Colunga and Zizumbo Villareal 2006), classified under 22 categories (ÁlvarezRíos et al. 2020), with alcoholic beverages being prominent (such as non-distilled pulque and distilled mezcal, tequila, bacanora, raicilla, among others). However, agave plants are also used in food, construction, and producing fibers.

Alcoholic beverages made from agave are an essential part of Mexico's history, tradition, and identity. Mexican laws, such as the Denomination of Origin (DO) NOM-251-SSA1-2009 for mezcal and NOM-006-SCFI-2012 and NOM-070-SCFI-2016 for tequila, specify that these beverages can only be produced in certain states of Mexico. *Tequila* is a globally recognized drink exclusively obtained from *Agave tequilana* Weber var. Azul. In contrast, mezcal can be produced from approximately 50 different species of agaves, as outlined in their Denomination of Origin (Colunga and Zizumbo Villareal 2006; Consejo Regulador del Mezcal 2018; Pérez-Zavala et al. 2020; SAGARPA 2017).

However, to produce these drinks, the Agave species must be allowed to grow and mature, a process that takes a minimum of 7–8 years in *Agave tequilana* Weber var. Azul, and up to 35 years in other species such as *Agave marmorata*. The only plants suitable for producing these drinks are those that are about to produce their only inflorescence after this lengthy growth period (they produce a single inflorescence and then die, as noted by Gentry 1982; Eguiarte et al. 2021). *Agave* plants are mainly composed of two aerial parts: their leaves (*pencas*) and the stem, commonly known as the *pineapple* (*piña*) or *head* (*cabeza*) (Nava-Cruz et al. 2015). When the plant starts to produce the inflorescence, it is removed, allowing all the resources (complex carbohydrates and water) to accumulate in the *piña*. Subsequently, the leaves or stalks of the stem are cut, or *jimados*, to cook the *piña* (with different methods depending on the region and kind of drink), allowing the polysaccharides to be hydrolyzed into glucose and fructose (Pérez-Zavala et al. 2020). The cooked stems are then ground (with different techniques), left to ferment for several days, and finally distilled (again with other methods) to produce a spirit drink (Pérez-Zavala et al. 2020).

Nowadays, alcoholic beverages play a crucial role in the Mexican economy. In 2020, the value of tequila production reached USD 1,870,186, with the sale of 374 million liters, equivalent to 1,519,000 tons of tequila agave plants. At the same time, the mezcal industry produced 8.1 million liters in 2021 for the official taxed market (www.gob.mx), with an unknown amount also produced for the informal, traditional market. However, throughout the entire production, process, hundreds of tons of *bagazo*, leaves, and fibers are generated as waste in the production of these distillates. *Bagazo* is the waste generated by crushing the stems (*heads*) after fermentation, and it is mainly made of agave fibers. This waste is sometimes used to make compost, but most of it is burned or left to naturally decompose (Pérez-Zavala et al. 2020).

In addition, agave leaves (*pencas*) are discarded during the *jima* before the cooking process, representing approximately 50% of the weight of the plant (Pérez-Zavala et al. 2020). These *pencas* could also be used for producing ethanol for biofuels, but not for human consumption (as they are rich in saponins and thus unpleasant). The tequila industry generated approximately 5,168,200 tons of leaf waste between 2015 and 2019, making waste management difficult and impacting the ecosystems (Álvarez-Chávez et al. 2021; Pérez-Zavala et al. 2020). Agave leaves are mainly composed of lignocellulose, consisting of strong and thick fibers traditionally used application for producing textiles, fibers, and, in some cases, gastronomy. However, most leaves are currently discarded. Therefore, it is necessary to create strategies that incorporate measures to reuse and take advantage of this raw material. Additionally, the leaves can be harvested every year during their maturation period (between 7–35 years, depending on the species). In *Agave americana*, it has been described that each plant yields 40–50 leaves per year, with a fiber content of 2.5–4.5% (Hulle et al. 2015).

The main objective of this chapter is to promote the sustainable use and management of waste, mainly leaves and *bagazo*, generated in the production of mezcal from *Agave marmorata.* These species have been minimally studied and have several gaps in the literature, but they show great potential for fiber production. We have carefully reviewed the most relevant aspects of the species, including published data on evolution, population dynamics, effects of climate change, cytogenetics, population genomics, and conservation. This comprehensive overview of information aims to provide insight into the potential uses of these species, considering the intrinsic characteristics of multipurpose plants.

*2.1.1 Botanical Classification and Distribution for Maguey Tepeztate (*Agave marmorata *Roezl)*

Agave marmorata was collected by Roezl in Tehuacán, Mexico (Gentry 1982). It is a species of plant in the *Asparagales* order, *Asparagaceae* family, and *Agave* genus, belonging to the Marmoratae group (Gentry 1982). It is a micro-endemic species found mainly in small areas of the Central-Southern states of Puebla and Oaxaca in Mexico, as shown in Fig. 2.1, within the Tehuacán—Cuicatlán Biosphere and in the Isthmus of Tehuantepec (Ornelas et al. 2002).

Agave marmorata has high population densities in the Tehuacán Cuicatlán Valley, Puebla, with 900–1100 plants per hectare. It has been identified that it grows much better with the protection of nurse plants (Godinez et al. 2008; Nobel 1998). The Zapotitlán Valley has even higher population density values, with 2620 individuals per hectare (Valiente-Banuet 2023).

Agave marmorata is characterized by large rosettes with bands that have a light green to grayish tone like marble and scabrous 30–50 leaves (100–135 cm), with teeth mostly 6–12 mm long, 2–5 cm apart (Gentry 1982). It is a monocarpic species, meaning it reproduces once and then dies, with occasional clonal reproduction occurring when it reaches approximately 60 cm in diameter (Jimenez-Valdés et al. 2010).

Agave marmorata is a large plant, reaching heights of 2 m and diameters of 2.5 m in mature, reproducing individuals with paniculate inflorescences that can reach up to 6–10 m high. The inflorescence contains 20–25 large diffuse decomposed umbels (Gentry 1982). It is a protandrous species with small flowers that are brilliant yellow, suggesting pollination by bats (Faegri and Van der Pijl 1971; Gentry 1982; Ornelas et al. 2002) because it generates a larger amount of pollen and nectar at night. The flowering period for this species is from May to the end of July (Ornelas et al. 2002), and it is one of the longest-lived agaves, taking up to 35 years to flower (Nogales 2017).

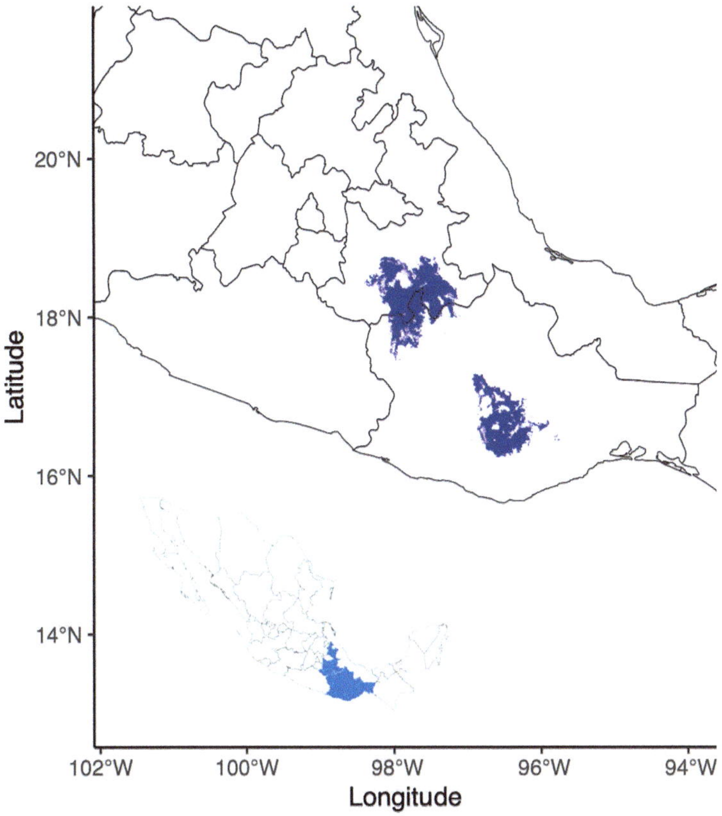

Fig. 2.1 *Agave marmorata* distribution. (Map created with R (package *raster* and *biomod*) using collection points from the GBIF.org (2 August 2023) database for *Agave marmorata*)

The fruits or dehiscent capsules generate an average of 224 tiny black seeds (Jimenez-Valdés et al. 2010). These seeds are nonphotoblastic, and their germination depends on the increase in soil water availability and the presence of trees and shrubs, such as nurse plants, which filter direct sunlight and provide protection to the plants (Arizaga et al. 2000; Jimenez-Valdés et al. 2010).

The name marmorata was given for the grayish marble-like leaves (Gentry 1982). It is commonly known by the names "*pixomel*", "*pizomel*", "*tepeztate,*" or "*maguey curandero,*" referring to its medicinal properties (Gentry 1982).

Kingdom	Plantae—plantes, Planta, Vegetal, plants
Subkingdom	Viridiplantae—green plants
Infrakingdom	Streptophyta—land plants
Superdivision	Embryophyta
Division	Tracheophyta—vascular plants, tracheophytes
Subdivision	Spermatophytina—spermatophytes, seed plants, phanérogames

Class	Magnoliopsida
Superorder	Lilianae—monocots, monocotyledons, monocotylédones
Order	Asparagales
Family	Asparagaceae
Genus	Agave L.—agave
Species	*Agave marmorata* Roezl.

2.1.2 Importance and Uses

Agave marmorata is mainly used to produce pulque and mezcal *tepeztate* or *pixomel*. However, the inflorescence must be cut to obtain this distilled drink, as the highest amount of carbohydrates is found in its stems or *heads*. This practice inhibits the sexual reproduction and propagation of its seeds from the harvested plants, leading to a loss of nectar and pollen resources for many animals that feed on them (Ornelas et al. 2002).

Agave marmorata has various interesting ecological and ethnobotanical traits. It is considered a multipurpose species (Aguilar Jiménez et al. 2018; Delgado-Lemus et al. 2014) since these plants are recognized as having the most significant number (14) of anthropogenic uses compared to other agaves (Delgado-Lemus et al. 2014). The entire plant is practically used; its uses range from purely ornamental, as the plants are beautiful (Fig. 2.2), to serving as a living fence, and forage, and having medicinal properties (analgesic and anti-inflammatory). Ecologically, *A. marmorata* helps retain rainwater and reduces soil erosion. The flowers, called *cacayas*, are usually cooked with eggs and are locally appreciated. The leaves and cuticles are consumed to wrap meat and are cooked in underground ovens (known as *mixiote*). The leaves also produce fiber (Aguilar Jiménez et al. 2018; Delgado-Lemus et al. 2014; García-Mendoza 2007). The dry inflorescences are used as decoration during religious celebrations (Casas et al. 2001; Lira et al. 2009), as construction material, as a tool for collecting fruits from trees, for obtaining biofuels and soap production (Delgado-Lemus et al. 2014).

The wide range of uses is mainly due to the enormous size and gigantism of *Agave marmorata* compared to other agaves grown in the region. For example, producing fibers can be very useful, although it is currently only practiced in some rural areas of the country. We believe it has strong potential but is not fully exploited at the industrial level. Additionally, this species is primarily used for the production of *pulque* and *mezcal*, so it is suggested that the leaves be harvested throughout their 35-year maturation period and the residue of the cooked *heads* and leaves be reused instead of discarded.

Physical, chemical, mechanical, and thermal analyses are still required to determine the specific characteristics and the amount of cellulose, hemicellulose, lignin, and water present in the fibers of *A. marmorata* leaves. However, we want to

Fig. 2.2 Different types of management in *Agave marmorata*. (**a**) wild populations in San Juan Raya, Puebla, México; (**b**) live fence in Zapotitlán, Puebla, México; (**c**) cultivated plants for mezcal and pulque production in Tehuacan Valley and (**d**) nurseries in San Juan Raya, Puebla, México. (Photo by: Ruiz-Mondragón Karen Y)

highlight the importance of rediscovering this species' ancestral uses to diversify production, seek sustainable agriculture strategies, and manage waste from pulque and mezcal production.

2.1.3 Fiber and Textile Properties

Natural fibers can be divided into animal or plant origin. Plant fibers can be classified as leaf fiber, seed fiber, and bast fiber, which are collected from the phloem of the plant (Samouh et al. 2021). Natural plant fibers are mainly composed of cellulose, hemicellulose, and lignin (Rosli et al. 2015), and they are considered a sustainable resource that can be found in nature (Madhu et al. 2017; Pathan and Veeresh Kumar 2023). Natural fibers have certain advantages over synthetic fibers since they are biodegradable, light, thermal, renewable, recyclable, non-toxic, low cost, and low density. They also have hardness, flexibility, and durability (Madhu et al. 2017; Pathan and Veeresh Kumar 2023).

In general, there are three main ways to extract natural fibers: (1) mechanical extraction, which is the oldest technique and is typically carried out in rural areas. In this method, the leaves are crushed and scraped with a knife (Hulle et al. 2015; Kolte et al. 2012; Mauersberger 1954; Steyn 2006; Young and Rowell 1986). More recently, a mechanical decorticator has been used to crush all the leaves and expose the fibers (Hulle et al. 2015). (2) Retting (enzymatic or natural) involves extracting fibers through a natural microbial process where non-fibrous matter is degraded without damaging the fiber (Ehrensing 2014; Hulle et al. 2015). (3) Chemical extraction involves the use of acids, alkalis, and enzymes as pretreatments of agave byproducts. Enzyme-extracted fibers are known to be more flexible and softer (Hulle et al. 2015; Kolte et al. 2012). During this process, fibers are extracted through hydrolysis of lignin and hemicellulose (Álvarez-Chávez et al. 2021). Hydrolysis with alkali allows the separation of lignin and cellulose and modifies the degree of polymerization and some structural properties (Perez-Pimienta et al. 2016; Thamae 2008). After extraction, the leaves are washed and dried. In 2013, the natural fiber industry was estimated to have generated 33 million tons, with the majority coming from cotton, followed by flax and sisal (made from *Agave sisalana*), generating approximately 45 billion euros (Ortega et al. 2019; Townsend and Sette 2015). There have been increases each year due to the growing demand in the market for obtaining sustainable products, creating a potential economic resource and new sources of employment. Using natural fibers as an alternative to cotton reduces the ecological footprint and plays an essential role in industrial and economic activity (Fowler et al. 2006; Gurunathan et al. 2015; Samouh et al. 2021).

Agaves are an important source for producing fibers and threads. They can produce the same amount of fiber as cotton fields but with a smaller water footprint since they require ten times less water (Davis et al. 2021). Agave plants use water efficiently due to CAM-type metabolism (Davis and Ortiz-Cano 2023). In addition, they have long, flexible, and resistant cellulosic fibers, a product of the anatomy of their leaves, as succulent plants require water storage and hydraulic insulation (Davis and Ortiz-Cano 2023; IPES 2016).

In Mexico, the extraction of fibers from agaves dates back to 2600 BC (Davis and Ortiz-Cano 2023; Turner and Miksicek 1984). For example, illustrations from Mayan codices such as the Dresden and TroCortesian (Irigoyen 1950) depict several people hunting, fishing, carrying bags, and using cords and ropes made from agave fibers (Colunga-GarcíaMarín et al. 1993). Knotting ropes was one of the earliest survival techniques used by humans. The first ropes were made from animal intestines, but this was a limited resource. Over time, people began using plants that coexisted with them. In pre-Hispanic times, various hard fibers were extracted from agaves, including *ixtle* (Ruschel Robinson 2013). Archaeological evidence shows that hard fibers from agaves were utilized even before the advent of agriculture, ceramics, and the weaving of soft fibers such as cotton, linen, silk, or wool (Ruschel Robinson 2013).

The production of fibers from Agave spread to the European and African continent after the conquest of America. The Spanish initially took ornamental plants, but eventually, agave fiber began to be used in Africa to produce carpets, rugs, and other textiles because of the durability of its fibers. The species with the most significant preference in obtaining fiber is *Agave fourcroydes*, known as *henequén* and processed mainly in the region of the Yucatan peninsula in Mexico, and *Agave sisalana*, known worldwide as sisal fibers.

In addition to Mexico, East Africa, India, and Brazil have been the most extensive regions for *Agave sisalana* cultivation for obtaining textiles (Gentry 1982; Raya et al. 2023). The first plant exports were made in the 1800s, mainly by the British and Germans, and they were sent to various botanical gardens throughout Europe with subsequent propagation in Eastern Africa (Brockway 1979; Davis and Ortiz-Cano 2023). In the 1960s, the fiber production market from *Agave sisalana* was led by Tanzania (Davis and Long 2014; Davis 2022), where a breeding program was carried out to generate a hybrid between *Agave amaniensis* × *Agave angustifolia* (H11648) (*A. amaniensis* is usually considered a synonym of *A. sisalana*), with the primary objective of producing a greater quantity of fibers in a wide variety of climatic conditions (Lock 1962). However, the most significant production is in Brazil today (Davis 2022; Raya et al. 2023). In Brazil, plant exports began in 1903, making it a leader in fiber production in the '60 s (FAO 2023; Medina 1954; Raya et al. 2023; Silva and de Macêdo Beltrão 1999).

In Mexico, the fibers extracted from agaves are known as *ixtle*, *henequén* or *sisal* and constitute the raw material for manufacturing fibers, sacks, ropes, carpets, and handicrafts (Garcia-Mendoza and Franco Martínez 2022). The production and extraction of these fibers are gradually disappearing in the communities that produce them, primarily due to competition with synthetic fibers, which are often cheaper and more readily available. Significant contributing factors include the effort and time required for traditional extraction methods, the high cost and scarcity of industrial machinery, the erosion of traditional manufacturing knowledge, and the lack of market demand (Garcia-Mendoza and Franco Martínez 2022). Additionally, the loss of indigenous languages plays a crucial role in this decline. Since fiber production is an ancestral and artisanal process, younger individuals in rural areas often no longer speak their native language. This linguistic loss disrupts the transmission of ancestral knowledge and results in diminished interest in fiber production (Ruschel Robinson 2013). Additionally, producers prefer to obtain more money from the production of mezcal or tequila, which has caused the loss of traditional knowledge (Garcia-Mendoza and Franco Martínez 2022).

However, recent attempts have been made to reestablish the market, and the potential use of other agave species for obtaining fibers has been explored. For instance, *Agave marmorata* and *Agave tequilana*, primarily used for producing mezcal and tequila, could also be utilized for fiber extraction. This could diversify the types of production and contribute to sustainable agriculture strategies (García-Méndez et al. 2020; Pathan and Veeresh Kumar 2023).

Although there is no information available on the properties of the fibers in *Agave marmorata*, we want to emphasize the importance of rediscovering the ancestral uses of this species. This can help diversify the types of production, seek sustainable agriculture strategies, and manage waste from pulque and mezcal production.

2.1.4 Domestication, Selection, and Early Improvements

Agave species have been significant since pre-Columbian times (Gentry 1982; Good-Avila et al. 2006). These plants were part of the diet in Central Mexico as far back as 9000 years ago (Callen 1995). Archaeological records indicate the presence of possibly domesticated agaves in the "Frightful" cave in Coahuila, dating back to 8000 BC (Bryant Jr 1974). Additionally, excavations conducted by McNeish in Tehuacán uncovered remnants of maguey plants that date to around 6500 BC (Spottedbird 2000).

The use of wild and cultivated plants in this region has been well documented with continuous cultivation as part of both traditional and modern commercial agriculture systems (Colunga-GarcíaMarín and Zizumbo-Villarreal 2006; Gil-Vega et al. 2006; Parker et al. 2010). In many parts of Mexico, wild agave plants are still harvested. Additionally, there are traditional agave plantations that undergo semi-domestication and domestication processes, with gradients ranging from wild to incipient management to fully domesticated populations (Blancas et al. 2010; Casas et al. 2007). Plant densities can vary from 20 to 3000 plants per hectare (Davis and Ortiz-Cano 2023; Martínez-Palacios et al. 1999). The domestication of *Agave* species is complex, mainly due to the long life cycle and the presence of sexual or clonal reproduction. Some *pulque* species have been considered domesticated (i.e., *Agave salmiana*) and *Agave fourcroydes* for fiber production.

Agave fourcroydes, commonly known as *henequén*, is the species most used in Mexico to produce fibers (Colunga-GarcíaMarín et al. 1999). This species has been used since the Mayan era in the Yucatan peninsula, Mexico (Colunga-GarcíaMarín and Zizumbo-Villarreal 2006; Colunga-García Marín et al. 1993). It was domesticated by the Mayans before the arrival of the Spanish conquest (Colunga-García Marín et al. 1993), with its possible wild ancestor being *Agave angustifolia* (Gentry 1982). Various domestication syndromes have been identified in this species compared to wild plants, such as more outstanding fiber production, reduced spines in leaves, larger leaves, larger plant size, higher sap content (with higher sugars content), higher production of vegetative propagules, and lower reproductive capacity (Álvarez-Rios et al. 2020; Colunga-GarcíaMarín et al. 2017; Figueredo et al. 2014, 2017, 2018; Leal-Díaz et al. 2015; Torres et al. 2015). The oldest traces of fiber production from this species date back to 2600 BC in the Maya era (Davis and Ortiz-Cano 2023; Turner and Miksicek 1984), with industrial and widespread fiber production continuing until the mid-1900s (Davis and Ortiz-Cano 2023; Evans 2007). However, the use of synthetic fibers and international competition with sisal fiber led to a drastic decline in the agave henequén fiber industry.

The use of Agave species for fiber has a long history among various indigenous groups, dating back approximately 3000 years ago. The tools and techniques for fiber extraction have remained largely unchanged and are similar across different regions of Mexico, depending on the specific species native to each area. For instance, *Agave marmorata* was used for fiber in the Tehuacán Valley of Puebla (Aguilar Jiménez et al. 2018; Delgado-Lemus et al. 2014; García-Mendoza 2007). In the Mezquital Valley of Hidalgo, *Agave americana*, *Agave mapisaga*, and *Agave salmiana* were identified as sources for weaving fabrics, combining *ixtle* with wool. Also, artisanal crafts made from these agave fibers include ayate, straw mats, carvings, headbands, necklaces, earrings, bags, and toys (Ruschel Robinson 2013). Moreover, *Agave lechuguilla* was an important resource for the Chichimecas. These agave species were primarily used to create sacks, ropes, cords, traps, fishing nets, sandals, clothing, bowstrings, and strings for tying arrowheads (Davis and Ortiz-Cano 2023).

2.2 Current Cultivation Practices and Challenges

Agave cultivation is common in Mexico and the southwest of the United States of America (Anderies et al. 2008). In Mexico, there is a cultivation gradient in the different agave species. This ranges from monocultures, mainly for the tequila species (*Agave tequilana* Weber var. Azul), as well as some mezcal species such as *Agave angustifolia* (Espadin mezcal). In these practices, the use of agrochemicals and vegetative propagation is encouraged to generate homogeneous plants (Gil-Vega et al. 2001) and there are polycultures where agaves are grown with different species, either with the milpa (corn, beans, and squash) or even with other species of agaves. They are also used as living fences to delimit land and retain soil in terraces using species such as *Agave marmorata*, *Agave salmiana* and *Agave americana*.

The use of agrochemicals in agave cultivation has become widespread due to the main pests that attack these plants, such as the insect *Scyphophorus acupunctatus*. This pest is considered the most harmful to agave species, affecting the tequila, mezcal, pulque, and fiber industry (Ramírez 1993; Siller-Jasso 1985). Other microbial pathogens include *Phytophthora* and *Erwinia*, which cause leaf damage and rot of the agave bud (Monja-Mio et al. 2019).

The biggest challenge in agave cultivation and breeding practice is mainly because most are monocarpic species that take 7–35 years to flower. In addition, there is genetic complexity present as some species are polyploid, leading to inefficient reproduction, sterility, and even seed recalcitrancy (Monja-Mio et al. 2019).

2.2.1 Current Cultivation Practices in Agave marmorata

As developed above, *A. marmorata* has been considered a multipurpose species. It has various forms of cultivation since it is used as a living fence to protect crop areas or cornfields from livestock, and in borders or terraces, it is also used to help retain soil and water (Delgado-Lemus et al. 2014). Recently, it has been grown in monocultures, but it is still rare. Due to its long maturation period of approximately

35 years and its large size, producers prefer to alternate crops with other agaves used in mezcal production (Ruiz-Mondragón et al. 2024).

2.2.2 Breeding Objectives in Agave marmorata

Recently, nurseries have been launched to produce seeds and plants, mainly for mezcal production. These plants were obtained and germinated from seeds from the region in Puebla, Mexico, and objectives have been proposed for breeding programs (Ruiz-Mondragón et al. 2024) as the genetic diversity in these nurseries has been quantified to be high in diversity and plasticity. However, these programs have not been promoted in regions with the highest production of this mezcal species, such as Oaxaca, Mexico.

2.3 Germplasm Biodiversity and Conservation

Agave germplasm has been found worldwide, either in the wild, mainly due to the transportation of plants during the conquest of America, or being cultivated to obtain fiber or for the production of spirits. Therefore, establishments have been started for germplasm conservation.

Botanical gardens play a crucial role in *the ex-situ* conservation of endemic plants. They also support research and promote the preservation and sustainable use of different species. In Mexico, a prime example is the Botanical Garden of the Institute of Biology at UNAM, which features 14 living collections comprising approximately 7000 plants and over 1500 biological species. Among these, there are 80 species of Agave located within a 4644 m^2 area, making it the largest collection of its kind (https://www.ib.unam.mx/).

In addition, the Botanical Garden FESC-UNAM, the Ethno Botanical Garden in Oaxaca, the Regional Botanical Garden Roger Orellana at CICY in Merida, Yucatan, the Helia Bravo Hollis Botanical Garden in Tehuacán, Puebla, Cadereyta Regional Botanical Garden, in Queretaro, UACH has established another collection in the State of Tlaxcala, and INIFAP (Instituto Nacional de Investigaciones Forestales Agrícolas y Pecuarias) has eight species in the Germplasm Bank by Maguey Mezcalero at Campo Experimental Valles Centrales de Oaxaca (Monja-Mio et al. 2019). Additionally, other places in the world, such as the Desert Botanical Garden in Phoenix, Arizona, USA, and the Royal Botanic Gardens at Kew in London, England, have living collections of agaves (Monja-Mio et al. 2019).

2.3.1 Cultivars Characterization and Phylogeny

Evolutionary and phylogenetic analyses have been carried out on Agaves, where two diversifications are mentioned in the *Agave sensu lato* group during the Miocene. Beginning to first diversify in the late Miocene 6.18 MA ago. This period was characterized by the emergence of arid biomes in America and the emergence of mountains in Mexico, such as the Trans-Mexican Neovolcanic Belt and the Sierra Madre Occidental (Jiménez-Barrón et al. 2020; Mastretta-Yanes et al. 2015). The second diversification, more recent at 3–2.5 million years ago, may be linked to the reproductive characteristics of the species and the influence and selection by pollinators (Eguiarte et al. 2006; Good-Avila et al. 2006; Rocha et al. 2006).

Using ITS markers, a Bayesian and ML tree was constructed, placing the species of interest, *Agave marmorata*, in group IV within *Agave sensu stricto*. It is the closest species phylogenetically to *Agave horrida*, *Agave pendula*, *Agave triangularis*, *A. antillarum*, *A. titanota*, *A. chiapensnsis*, *A. convallis*, *A. kerchoveu*, *A. colorara*, *A. attenuata*, *A. guiengola* and *A. ghiesberghtii* (Jiménez-Barrén et al. 2020). Within this group (IV), the tree credibility was estimated using BEAST, revealing a stem age of 1.41 million years, a crown age of 1.04 million years, a speciation rate of 6.15, an extinction rate (m) of 2.74, and a rate (r) of 3.41 (Jiménez-Barrén et al. 2020). Their diversification occurred during the Pliocene and Pleistocene, with the most significant number of diversification events described due to the periods of glaciations and interglacials, which caused expansion and contraction of the distribution areas of *Agave* species (Aguirre-Planter et al. 2020; Jiménez-Barrén et al. 2020; Ruiz-Mondragón et al. 2024; Scheinvar et al. 2017).

2.3.2 Genetic Resources Conservation Approaches: In Vitro, In Situ, Cryopreservation, and Gene Bank

Different approaches have been proposed for the conservation of this species. An *in-situ* conservation proposal was recently implemented in the Tehuacán- Cuicatlán Reserve. This area serves as a genetic reservoir of germplasm and a seed bank, thanks to the high genetic diversity found in wild localities in this region (Ruiz Mondragón et al. 2024).

Additionally, nurseries have been recently constructed for the production of seeds and plants, mainly for mezcal production, and to initiate some propagation programs. The genetic diversity in these nurseries is high and should be important in future improvement programs (Ruiz-Mondragón et al. 2024).

2.3.3 Cytogenetics

Agaves are characterized by having multiple ploidy levels (2x, 3x, 4x, 5x, 6x, and 8x) with a basic chromosome number of $n = 30$ (Monja-Mio et al. 2019). Additionally, these plants have asymmetric bimodal karyotypes, with five pairs of large chromosomes and twenty-five small ones, in both diploid and polyploid species (Castorena-Sánchez et al. 1991; Flores-Maya et al. 2015).

Agave marmorata is a diploid species with a chromosome number of $2n = 2x = 60$, consisting of five pairs of long chromosomes and 25 small pairs. The chromosomes were categorized into four subtelocentric, twelve metacentric, nine submetacentric, and five metacentric pairs (Flores-Maya et al. 2015).

2.4 Traditional Breeding

The cultivation of agaves was essential in pre-Hispanic times in Mexico and the southern United States. These species were cultivated before the arrival of Europeans and were a vital food source during periods of drought in Arid America 500–900 years ago (Evans 1992; Fish et al 1992; McClung de Tapia et al. 1992; Minnis and Plog 1976; Phippen 1999; Trombold 1985). Traditional agave growing systems were constructed as terraces in areas too dry to support corn. Agave phytoliths have also been found in the sediments trapped by these terraces (Trombold and Israde-Alcántara 2005).

Archaeological remains of agave cultivation dating back to 200 CE have been identified. This evidence includes charred agave fragments, ancestral ovens in pits with piles of rocks (Fish et al 1992; Van Buren et al. 1992), as well as tools used to scrape the stems of the agaves and utensils for obtaining sap and fiber (Parsons and Parsons 1990).

2.4.1 Traditional Breeding Methodologies

Since pre-Hispanic times, reproductive propagation in the Agave genus has primarily occurred through vegetative reproduction, especially through bulbils derived from the plant's rhizomes. In the case of *Agave fourcroydes* (*henequén*), traditional breeding and agricultural practices involve cutting the inflorescences shortly after they begin to develop in order to prolong the plant's lifespan. This practice limits the fertilization and dispersion of seeds. Additionally, it has been observed in this species that there is a low fertility rate and slow maturation of its seed (Piven et al. 2001), which is likely linked to domestication syndrome, polyploidy, or the inefficiency of pollinators (Piven et al. 2001).

In *Agave marmorata*, the management is very similar to *Agave fourcroydes* and other Agaves species since, due to its long reproduction time (up to 35 years), many of the plants are reproduced by bulbils derived from the mother plant and very rarely the germination of its seed is chosen.

2.5 Molecular Breeding

Due to the long reproductive life cycles (8–35 years) (Chen et al. 2014; Simpson et al. 2011), the different forms of reproduction (sexual or asexual), the large genome size (Huang et al. 2019), and polyploidy in the *Agave* genus (Monja-Mio et al. 2019; Simpson et al. 2011) biotechnology and genetic improvement have been a challenge (Davis 2022). The only case of genetic improvement and a breeding program was the production of the hybrid between *Agave amaniensis* × *Agave angustifolia* (H11648) carried out by the British in Tanzania to obtain a larger amount of fiber in different environmental conditions.

2.5.1 *Genomics*

Genomic diversity is essential for populations to adapt to different environmental and anthropogenic changes (Allendorf et al. 2013; Frankham et al. 2010; Govindaraj et al. 2015). It also gives us a complete overview of the evolutionary processes that occur in species at the population level. Limited information exists on population genomics and molecular breeding in *A. marmorata*. However, recent research on the genomic diversity and conservation of the species has been published using RADseq, identifying 29,101 SNPs across the species' distribution area with four different types of management (wild, nursery, live fence, and cultivated) (Ruiz-Mondragón et al. 2024).

Agave marmorata exhibits high levels of genomic diversity, with the highest multilocus heterozygosity (*MLH*) index found in wild localities ($MLH = 0.192$), followed by the live fence ($MLH = 0.192$), nursery populations ($MLH = 0.184$) and cultivated populations ($MLH = 0.170$). There was a significant difference between the wild and cultivated localities (*Wilcoxon test;* $p < 0.001$). For expected heterozygosity (H_E), greater diversity was observed in wild samples, with live fence samples presenting the highest genetic diversity ($H_E = 0.230$ and $H_E = 0.214$, respectively). In contrast, cultivated plants presented the lowest genetic diversity values ($H_E = 0.175$). The highest diversity at the population level was found in wild localities within the Tehuacán- Cuicatlán Reserve, suggesting this region is a germplasm reservoir for the species (Ruiz-Mondragón et al. 2024).

Furthermore, a positive inbreeding index at the species level of $F_{IS} = 0.11$ was reported, indicating possible crosses among relatives or self-pollination (Ruiz-Mondragón et al. 2024). Additionally, low genetic differentiation ($F_{ST} = 0.087$)

among populations and/or management categories may be attributed to intense gene flow, mainly related to its primary pollinators, nectarivorous bats. Nevertheless, Ruiz-Mondragón et al. (2024) detected two main genetic groups ($K = 2$), one in the state of Puebla and another in the state of Oaxaca, likely influenced by evolution, demographic history and geographical distance between populations (Ruiz-Mondragón et al. 2024).

2.6 Tissue Culture Applications

Commercial Agave plants are typically propagated through bulbils or axillary meristems from the inflorescences, a process that can take 7–35 years to reach reproductive age. Species that reproduce vegetatively can produce 3–4 stolons per year. Therefore, tissue culture propagation is one way to reproduce commercial plants more quickly and efficiently (Binh et al. 1990).

Tissue culture in Agave has been carried out with different objectives. One of them is to improve productivity and establish previously selected individuals with desired characteristics, such as a larger stem or *pineapple* size, greater sugar production, and a higher reproduction rate in the shortest time. Another use that has been given is to propagate or rescue varieties at risk of extinction and offer a means of protection against overexploitation of the natural environment because many agaves are extracted from wild populations (Martinez-Rodriguez et al. 2022; Monja-Mio et al. 2019).

2.6.1 Micropropagation in Agave marmorata

In vitro propagation is an effective tool for increasing production, multiplication and preserving plants that may be protected, such as *A. marmorata*. (Martinez-Rodriguez et al. 2022; Moraes et al. 2021; Slazak et al. 2014). Micropropagation techniques offer advantages, such as increasing seedling production in a relatively short time, improving agronomic traits, maintaining specific genotypes, and producing uniform plants (Bautista-Montes et al. 2022). This facilitates the management and future agronomic or industrial use of the plants. However, in agaves, they are generally propagated using the meristem of the leaves or by extracting clonal offspring, which produces a plant identical to the mother plant. This type of propagation reduces genetic variability and may increase disease susceptibility for instance (Martinez-Rodriguez et al. 2022).

There are reports of different types of micropropagation in *Agaves spp.* (see Bautista-Montes et al. 2022). Explant sources are used to induce direct or indirect morphogenesis. In direct morphogenesis, direct organogenesis can be developed, allowing the development of shoots without intermediate calluses, or direct embryogenesis, with the development of somatic embryos and subsequent development of the plant. In indirect morphogenesis, indirect organogenesis is induced by obtaining a callus or through indirect embryogenesis, which consists of the direct formation of somatic embryos from a small group of cells for the subsequent development of the plant (Bautista-Montes et al. 2022).

For micropropagation in *A. marmorata* from leaves or shoots, the best substrate identified is 100% Murashige and Skoog (MS), with 3% cane sugar, 100 mg L^{-1} myo-inositol, 0.7% agar, and 0.40 mg L^{-1} thiamine-HCl. This substrate should be supplemented with equal concentrations of 6-benzyl-adenine (BA) and indole-3-acetic acid (IAA) to promote the proliferation of new shoots (Aguilar Jiménez et al. 2018).

In organogenesis of stem tissues from *A. marmorata*, a culture medium was used containing 5.7 g of agar, MS (Murashige and Skoog) inorganic salts, 1 mg of thiamine-HCl, 100 mg of myo-inositol, 30 g of sucrose, 1 mg N6-Benzylaminopurine and 0.3 mg of indole acetic acid with a pH of 5.8. The medium was incubated for 70 days. A total of 228 stem tissue explants were obtained *in vitro* from 15 plants. Out of these 228 explants, 68 (29.82%) were viable (López Acevedo et al. 2018).

Somatic embryogenesis, which is the formation of an embryo from a group of cells, in *A. marmorata*, Alvarez-Aragón et al. (2020) reported a higher rate of embryo formation under light conditions with pretreatment of 10 mg/L BA in 0.1 mg/L of 2,4-dichloro phenoxy acetic (2,4-D) (Alvarez-Aragón et al. 2020). This technique has been considered an important tool for genetic regeneration and improvement (Copeland and McDonald 1995).

For *in vitro* micropropagation from seeds to generate clonal lines in *Agave marmorata*, Martinez-Rodriguez et al. (2022) used a culture medium with 25 mL with 15 g/L of bacteriological agar of solid modified Murashige and Skoog, added with growth regulator 6-benzyl aminopurine (BAP) cytokine, and incubated at 30 °C with 12 h. of photoperiod. These authors reported the formation of stable tissue, variation in multiplication capacity, adaptation to stress, continuous growth, and efficient multiplication of shoots of *A. marmorata*. In addition, variation was identified in the tissue multiplication capacity in the clonal lines, which may indicate genetic variability, and differences were found among clonal lines in bioactive phytochemical compounds, such as the presence of anthocyanins, previously reported with protective functions in wild agave plants (Martinez-Rodriguez et al. 2022).

2.7 Hybridization

Interspecific and intraspecific hybridization are common in the agave genus and have probably been influential in their evolution (Figueredo-Urbina et al. 2021; Gentry 1982; Granick 1944; Palomino et al. 2007; Pinkava and Baker 1985; Roll et al. 2021). This has made the phylogenetic and taxonomy of these plants complex since many may be the origin of natural hybridization.

This hybridization may play an important role mainly in ecological tolerance and its wide geographic distribution (Pinkava and Baker 1985; Stebbins 1980) due to the production of more combinations of adaptive genes (Pinkava and Baker 1985). In addition, some species of economic importance have been hypothesized as possible hybrids (*Agave sisalana*, *Agave fourcroydes*, and *Agave mapisaga*) (Gentry 1982; Pinkava and Baker 1985). Furthermore, thanks to its vegetative propagation mechanism, it favors the maintenance of hybrids over time (Roll Hodgson and Salywon 2021).

The production of hybrids has only been carried out between the species *A. angustifolia* × *A. amaniensis* × *A. amaniensis* (Lock 1962), generating the hybrid H11648 mainly carried out to obtain a more significant amount of fiber production.

2.7.1 Conventional Hybridization

In his classic study of agaves, Gentry discussed the hybridization between *Agave marmorata* and *Agave kerchovei* in 1982. He considers that *Agave peacockii*, from the state of Puebla, Mexico, was the product of this hybridization. *Agave peacockii* produces a raceme-shaped inflorescence with a morphology intermediate between the spicule of the *Littea* subgenus (to which *Agave kerchovei* belongs) and the umbrella of the *Agave* subgenus, which includes *A. marmorata*. Additionally, it has leaves resembling those of *A. kerchovei* and the characteristic band patterns on the leaves of *Agave marmorata* (Gentry 1982).

Valverde et al. (1996) supported the hypothesis of a hybrid origin for *Agave peacockii* through statistical analyses of the morphology of the three species. They measured were made leaf size, number of spines, and number of leaves. However, genetic and hybridization analyses are needed to confirm these ideas.

2.8 Ecology

Due to their ecological and physical adaptations, agave can grow in diverse ecosystems and temperatures. Many of these adaptations are thanks to its evolution towards a Crassulaceae Acid Metabolism (CAM) (Abraham et al. 2016; Andrade et al. 2007; Heyduk et al. 2016; Nobel 2010). This helps them have water use efficiency (WUE) since they open their stomata at night when temperatures are lowered and close during the day, resulting in high water use efficiency (WUE) (Garcia-Moya et al. 2011).

Another essential adaptation in these species was the development of a giant inflorescence with paniculate flowering (in some species more outstanding than 16 m), which is possibly the reflection of coevolution between their primary pollinators, the nectarivorous bats (Eguiarte et al. 2013, 2021), particularly the long-nosed bats *Leptonycteris spp.* These bat species are large compared to other bat species, fly for long distances, and transport larger pollen loads through distant populations of agaves (Fleming et al. 2009). Also, generating this enormous inflorescence attracts different pollinators, ensuring reproductive efficiency in agaves (Rocha et al. 2005; Schaffer and Schaffer 1977; Schaffer et al 1979).

In most species of agaves, after generating an enormous inflorescence, the plant dies. However, many species have an adaptation that allows them to reproduce through clonal propagation producing suckers via rhizomes at the base of the adult plant (Eguiarte et al. 2021; Gentry 1982). Sometimes, agaves also produce bulbs in the inflorescence, particularly if the inflorescence stem is damaged, which is the result of apomixis (Arizaga et al. 2000; Gentry 1982).

2.8.1 Reproduction, Pollinators and Seed Dispersal

Most Agave species exhibit sexual reproduction and are usually cross-pollinated, which favors genetic diversity. However, many species also reproduce asexually by forming rhizomes and bulbils (Alducin Martinez et al. 2023). Sexual reproduction occurs through pollination of the flowers in the large inflorescences (Gentry 1982; Monja-Mio et al. 2019).

Agaves are pollinated by a wide range of animals, from insects and birds to mammals, specifically nectarivorous bats (Arizaga et al. 2000; Gentry 1982; Howell 1979 Kuban et al. 1983; Ornelas et al. 2002; Schaffer and Schaffer 1977; Schaffer et al. 1979; Sutherland 1987; Slauson 2000). However, due to morphological, biochemical, and phenological issues, many species of agaves have been identified with a possible chiropterophily syndrome (Arizaga et al. 2000; Faegri and Van der Pijl 1971 Slauson 2000; Sutherland 1987). It is recognized that bats play a fundamental role in the pollination and reproduction of agaves and other columnar cacti and Bombacoideae trees (Bustamante et al. 2016; Fleming et al. 2009; Rojas-Martínez et al. 1999; Trejo-Salazar et al. 2015; Trejo-Salazar et al. 2021), with a focus on the long-nosed bat, *Leptonycteris yerbabuenae* (Glossophaginae), mainly due to its wide distribution in Mexico (Trejo-Salazar et al. 2021).

Nocturnal pollinators, such as nectar-feeding bats, *Choeronycteris mexicana* and *Leptonycteris nivalis*, as well as sphingid moths feeding on the pollen and nectar of *Agave marmorata* flowers have been observed (Ornelas et al. 2002; Rojas-Martínez and Valiente-Banuet 1996). Additionally, some diurnal species reported as floral visitors include *Cynanthus sordidus* (Gould) *C. latirostris Swainson*, *Amazilia violiceps* (Gould), *Lampornis clemenciae* (Lesson), *Eugenes fulgens* (Swainson), *Calothorax pulcher* Gould, *Melanerpes hypopolius* (Wagler), *Campylorhynchus jocosus* Sclater, *Mimus polyglottos* (Linnaeus), *Toxostoma curvirostre* (Swainson), *Icterus wagleri* Sclater, *I. spurius* (Linnaeus), *I. pustulatus* (Wagler), *I. parisorum* Bonaparte, *Carpodacus mexicanus* (Müller) and some insects, especially bees *Xylocopa* sp. (Ornelas et al. 2002). However, Ornelas et al. (2002) highlight the importance of birds as pollinators of *Agave marmorata*, suggesting that the production of pollen and nectar and receptivity of the stigma have promoted a multiple or generalist pollination syndrome.

Seed dispersal in *Agave marmorata* is mainly by wind, although ants (*Pogonomyrmex barbatus*) have been identified as main seed predators (Jimenez-Valdés et al. 2010).

2.8.2 Population Dynamics in Agave marmorata

Population dynamics is one of the fundamental areas of ecology and plays a central role in many approaches to preserving biodiversity (Hastings 2001). Jimenez Valdés et al. (2010) made models with projections of current conditions to calculate the finite rate of population growth in two populations of *Agave marmorata*: one unmanaged and the other managed. Both populations are located in the semi-arid valley of Zapotitlán Salinas, in the Biosphere Reserve of Tehuacán- Cuicatlán, Puebla, Mexico.

The researchers discovered that the unmanaged population had finite growth rates ($\lambda > 1$), with high and positive demographic processes related to greater fertility and population growth. In contrast, the managed population, where practices such as removing perennial plants and cutting flowering stalks for mezcal production were implemented, had negative effects on the population dynamics of this species (Jimenez Valdés et al. 2010). This is because cutting the inflorescence inhibits the only reproductive period in this species. Additionally, removing different plants decreases soil water and increases surface temperature (Jimenez Valdés et al. 2010).

2.9 Climate Change

A recent concern is how climate change will affect species distribution and interaction. These changes can affect ecological interactions by creating mismatches between species since losing a primary pollinator can cause chain effects in ecosystems (Brosi and Briggs 2013; Brosi et al. 2017; Gómez-Ruiz and Lacher Jr 2019). Addressing this possible problem, the potential distribution in the future (2050 and 2070) of different species of agaves and one of the most important pollinators, the nectarivorous bat *Leptonycteris nivalis*, has recently been modeled (Gómez-Ruiz and Lacher Jr 2019). The long-nosed bat (*L. nivalis*) closely interacts with agaves each spring, follows the paniculate agave flowers, and migrates up to 1200 km from central Mexico to the southwestern United States (Gómez-Ruiz and Lacher Jr 2019; Moreno-Valdez et al. 2000; Moreno-Valdez et al. 2004).

It was found that *Agave gentryi*, *A. horrida*, and *A. salmiana* reduced their distribution area by more than 80% in all scenarios and *A. parryi* and *A. palmeri* in three climate change scenarios. On the other hand, future models show a slight expansion of suitable areas of the *A. palmeri* species. However, when overlapping the areas of the presence of agaves and the bat *L. nivalis* in all future scenarios, there is at least a 75% decrease in the percentage of overlap in relation to the current distributions (Gómez-Ruiz and Lacher Jr 2019).

In *Agave marmorata*, various species distribution models (SDM) were analyzed. Two Shared Socio-economic Pathways climate change scenarios (SSP 2.45 and 5.85) and two General Circulation Models (GCMs: MPI-ESM1-2-HR and MIROC6) in two-time intervals: current vs. future 2041–2060, and current vs. future 2061–2080. Future range change was estimated by comparing the current period with the scenarios planned for the future (Ruiz-Mondragón et al. 2024).

For the time interval current vs. near future (2041–2060) with the lax model 2.45, and with the two GCM models (MPI-ESM1-2-HR and MIROC6), a gain in the distribution area was estimated (0.92–0.96%, respectively). In contrast, with the more pessimistic model, 5.85, the distribution area decreased (from −0.87 to −1, respectively). For current vs. future (2061–2080) with SSP 2.45 and General Circulation Models (GCMs: MPI-ESM1-2-HR and MIROC6), an increase in the distribution area was again estimated (0.93–0.99%, respectively), while for the model with SSP 5.85 a decrease (−1) was obtained (Ruiz-Mondragón et al. 2024).

In these models, differences found were related to the variation that exists between the General Circulation Models (GCMs) (Ureta et al. 2022) and mainly due to the different scenarios planned for Shared Socio-economic Pathways (SSPs) (carbonbrief.org; Riahi et al. 2017). These scenarios take into account different greenhouse gas emissions and CO_2 production, as well as the exploitation of fossil and energy resources (Ruiz-Mondragón et al. 2024; carbonbrief.org; Riahi et al. 2017). Therefore, the authors mention that in the best-case scenario, if the amount of CO_2 production can be reduced, the future of this species will be beneficial, and there is a possibility of an increase in the distribution area of *A. marmorata*. As mentioned throughout this chapter, *Agave* plants are worth considering due to the various types of products that can be obtained and because they can grow in places and future climates with high temperatures and low water availability due to their physiology.

It is important to highlight that high genetic diversity and plasticity have been estimated in this species, which are two critical aspects for its adaptation to climate change scenarios. However, genetic improvement programs and germplasm conservation *in situ* and *ex-situ* must be carried out to safeguard the species' adaptive potential.

2.10 Conclusions and Prospects

Considering the effects of climate change, the growing global population, higher consumption of raw materials, and the overproduction of a few species, it is important to find species that can withstand increased temperatures and drought. Agaves have played a key role during ancient periods of drought and climate change in the history of Arid America and Mesoamerica and possess valuable characteristics for obtaining various raw materials.

Specifically, *Agave marmorata* is a prominent species that has been used in multiple ways since ancient times. These species show excellent potential for obtaining fibers and other products, while also finding a way to use the waste derived from the mezcal industry. Additionally, they provide sources of work in rural areas, aiming to revalue and encourage the consumption of natural fibers. Furthermore, thanks to their CAM-type metabolism, they offer greater resilience and remarkable adaptation to climate change. They can withstand drought conditions, high and low temperatures, water and soil retention, and water use efficiency (WUE). These species stand out in the problems we face today, such as droughts, which are a consequence of climate change. However, sustainable use must be encouraged as this species has long life cycles, which could be counterproductive and lead to an extinction risk.

Exploring, rediscovering, and reassessing the ancestral practices and customs related to large multipurpose species like *Agave marmorata* in Mexico is of utmost importance. These remarkable plants, long cherished by indigenous communities, offer a wealth of knowledge and resources that can be vital for sustainable living. It is essential to inspire younger generations to delve into this rich heritage, encouraging them to understand and actively utilize the biological treasures that surround them. By fostering a deep connection to their land, cultural roots, and traditions, we can ensure that these invaluable ancestral techniques continue to thrive, passing down wisdom and skills even as older generations wane.

In the context of reducing waste produced by alcoholic beverage manufacturing, developing innovative techniques for sourcing and processing raw materials becomes imperative. This initiative must go hand in hand with the creation of sustainable and biodegradable products that respect the environment. To achieve this goal, a thorough analysis of *A. marmorata* and other Agave fibers is critical. This examination includes quantifying key components such as cellulose, lignocellulose, and hemicellulose derived from the vibrant green leaves and fibrous bagasse. Furthermore, assessing attributes like strength, color, durability, flexibility, and resistance is essential for exploring their potential applications as textile products on an industrial scale.

Appendix 2.1 Research Institutes and Online Resources Available for Agave Breeding

Institution name	Specialization and research activities	Address	Contact information and website
Institute of Ecology, Universidad Nacional Autónoma de México (UNAM).	Ecology, evolution, conservation of genetic resources and population genetics.	Circuito Exterior s/n Anexo al Jardín Botánico, Ciudad Universitaria. C.P. 04510. Ciudad de México, México.	https://www.ecologia.unam.mx/

(continued)

Institution name	Specialization and research activities	Address	Contact information and website
Centro de Investigación Científica de Yucatán, (CICY) Mexico	Genetics, ethnobotany, conservation	C. 43 No. 130-x 32 and 34 Col, Chuburna de Hidalgo, Pinzón II, 97205 Mérida, Yucatan.	https://www.cicy.mx/
Department of Plant Biology, Laboratory of Crop Physiology (LCroP), University of Campinas (Unicamp), Campinas, Brazil	Genetics, crop improvement and management practices	Cidade Universitária Zeferino Vaz – Barão Geraldo, Campinas – SP, Brazil	https://lcropbrazil.com/About/about.php
Institute of Biology and Botanic Garden, Universidad Nacional Autónoma de México (UNAM).	Ex-situ conservation, taxonomy and anthropogenic uses	Circuito Exterior Zona Deportiva S/N, Ciudad Universitaria, Coyoacán, C.P. 04510, Ciudad de México, CDMX	https://www.ib.unam.mx/
Instituto de Investigaciones en Ecosistemas y Sustentabilidad (IIES).	Ethnobotany and population genetics	Antigua Carretera a Pátzcuaro No. 8701 Col. Ex Hacienda San José de la Huerta C.P. 58190, Morelia, Michoacán, México.	https://www.iies.unam.mx/
Universidad Autónoma Chapingo.	Micropropagation, crop improvement and management practices	Km. 38.5, Carretera. México-Texcoco, Edo. México	https://chapingo.mx/
Centro de Investigación y de Estudios Avanzados del Instituto Politécnico Nacional (CINVESTAV)	Micropropagation and genetics	Libramiento Norte Carretera Irapuato León Kilómetro 9.6, Carr Panamericana Irapuato León, 36821 Irapuato, Guanajuato.	https://portal.cinvestav.mx/Unidad-Irapuato
Centro Universitario de Ciencias Biol.gicas y Agropecuarias, Universidad de Guadalajara	Biotechnology and Agriculture	Apartado Postal 1-139, Zapopan, Jalisco 45101, México.	
University of Illinois	Bioenergy crops	Energy Biosciences Institute and Department of Plant Biology University of Illinois at Urbana-Champaign, Urbana, IL, USA,	
University of California, Los Angeles UCLA	Physiological Ecology	Department of Ecology and Evolutionary Biology 621 Charles E. Young Drive, South Los Angeles, CA 90095-1606 USA	

References

Abraham PE, Yin H, Borland AM et al (2016) Transcript, protein and metabolite temporal dynamics in the CAM plant Agave. Nature Plants 2:1–10

Aguilar Jiménez D, Rodríguez De la O, Luis J (2018) Micropropagación y aclimatación de Maguey Pitzometl (*Agave marmorata* Roezl) en la Mixteca Poblana. Rev Colomb Biotecnol 20(2): 124–131. https://doi.org/10.15446/rev.colomb.biote.v20n2.77084

Aguirre-Planter E, Parra-Leyva JG, Ramírez-Barahona S et al (2020) Phylogeography and genetic diversity in a southern North American desert: *Agave kerchovei* from the Tehuacán-Cuicatlán Valley. Mexico Front Plant Sci 11:863

Alducin-Martínez C, Ruiz Mondragón KY, Jiménez-Barrón O, Aguirre- Planter E et al (2023) Uses, knowledge and extinction risk faced by agave species in Mexico. Plan Theory 12:124. https://doi.org/10.3390/plants12010124

Allendorf FW, Luikart GH, Aitken SN (2013) Conservation and the genetics of populations. Wiley

Alvarez-Aragón C, Martín A, Arzate-Fernández SY et al (2020) Regeneración de plantas de Agave marmorata Roezl, vía embriogénesis somática. Trop Subtrop Agroecosyst 23:36

Álvarez-Chávez J, Villamiel M, Santos-Zea L, Ramírez-Jiménez AK (2021) Agave by-products: an overview of their nutraceutical value, current applications, and processing methods. Polysaccharides 2:720–743. https://doi.org/10.3390/polysaccharides2030044

Anderies JM, Nelson BA, Kinzig AP (2008) Analyzing the impact of agave cultivation on famine risk in arid pre-hispanic Northern Mexico. Hum Ecol 36:409–422. https://doi.org/10.1007/s10745-008-9162-9

Andrade JL, de la Barrera E, Reyes GC et al (2007) El metabolismo acido de las crasulaceas: diversidad, fisiologia ambiental y productividad. Bol Soc Bot Mexico 81:37–50

Álvarez-Ríos GD, Pacheco-Torres F, Figueredo-Urbina CJ, & Casas A (2020) Management, morphological and genetic diversity of domesticated agaves in Michoacán, México. J Ethnobiol Ethnomedicine 16(1):3 https://doi.org/10.1186/s13002-020-0353-9

Arizaga S, Ezcurra E, Peters E et al (2000) Pollination ecology of *Agave macroacantha* (Agavaceae) in a Mexican tropical desert. II. The role of pollinators. Am J Bot 87:1011–1017

Bautista-Montes E, Hernández-Soriano L, Simpson J (2022) Advances in the micropropagation and genetic transformation of agave species. Plan Theory 11:1757. https://doi.org/10.3390/plants11131757

Binh LT, Muoi LT, Oanh HTK et al (1990) Rapid propagation of agave by in vitro tissue culture. Plant Cell Tissue Organ Cult 23:67–70. https://doi.org/10.1007/BF00116091

Blake M (2012) Azúcar, alcohol y la domesticación inicial del maíz. Arqueología mexicana las bebidas alcohólicas en el México antiguo raíces México volumen XIX num. 114 Instituto Nacional de Antropología e Historia, pp 34–39

Blancas J, Casas A, Rangel-Landa S et al (2010) Plant management in the Tehuacán Valley. Econ Bot 64:287–302

Brockway LH (1979) Science and colonial expansion: the role of the British Royal Botanic Gardens. Am Ethnol 6:449–465

Brosi BJ, Briggs HM (2013) Single pollinator species losses reduce floral fidelity and plant reproductive function. Proc Nat Acad Sci 110:13044–13048

Brosi BJ, Niezgoda K, Briggs HM (2017) Experimental species removals impact the architecture of pollination networks. Biol Lett 13:20170243

Bryant VM Jr (1974) Pollen analysis of prehistoric human feces from mammoth cave. In: Watson PJ (ed) Archaeology of the mammoth cave area. Academic Press Inc., New York, pp 203–249

Bustamante E, Búrquez A, Scheinvar E, Eguiarte LE (2016) Population genetic structure of a widespread bat-pollinated columnar cactus. PLoS One 11(3):e0152329. https://doi.org/10.1371/journal.pone.0152329

Casas A, Valiente-Banuet A, Viveros JL et al (2001) Plant resources of the Tehuacán-Cuicatlán Valley, Mexico. Econ Bot 55:129–166

Casas A, Otero-Arnaiz A, Pérez-Negrón E, Valiente-Banuet A (2007) In situ management and domestication of plants in Mesoamerica. Ann Bot 2007:1–15
Castorena-Sánchez I, Escobedo RM, Quiroz A (1991) New cytotaxonomical determinants recognized in six taxa of Agave in the sections Rigidae and Sisalanae. Can J Bot 69:127–1264
Chen Y, Chen X, Hu F et al (2014) Micropagation of Agave. Am Hort Sci 49:320–327
Colunga-García Marín P, Coello-Coello J, Espejo Peniche L, Fuente Moreno L (1993) Agave studies in Yucatan, Mexico. II. Nutritional value of the inflorescence peduncle and incipient domestication. Econ Bot 47(3):328–334
Colunga-García Marín P, Coello-Coello J, Eguiarte LE, Piñero D (1999) Isozymatic variation and phylogenetic relationships between henequén (*Agave fourcroydes*) and its wild ancestor A. angustifolia (Agavaceae). Am J Bot 86:115–123. https://doi.org/10.2307/2656960
Colunga-GarcíaMarín P, Zizumbo-Villarreal D (2006) Tequila and other Agave spirits from west-central Mexico: current germplasm diversity, conservation and origin. In: Hawksworth DL, Bull AT (eds) Plant conservation and biodiversity, Topics in biodiversity and conservation, vol 6. Springer, Dordrecht. https://doi.org/10.1007/978-1-4020-6444-9_6
Colunga-GarcíaMarín P, Zizumbo-Villarreal D, Torres I et al (2017) Los agaves y las prácticas mesoamericanas de aprovechamiento, manejo y domesticación. In: Casas A, Torres-Guevara J, Parra F (eds) Domesticación en el Continente Americano. Vol. 2. Investigación para el Manejo Sustentable de Recursos Genéticos en el Nuevo Mundo. UNAMUNALM, México, pp 273–308
Consejo Regulador del Mezcal (CRM) (2018) Informe Estadistico 2018. www.crm.org.mx
Copeland L, McDonald M (1995) Principles of seed science and technology, 3rd edn. Chapman & Hall, New York, pp 400–409
Davis SC (2022) Agave americana: characteristics and potential breeding priorities. Plants 11:2305. https://doi.org/10.3390/plants11172305
Davis SC, Long S (2014) Agave/Sisal. In: Cruz M, Dierig D (eds) Industrial crops: breeding for bioenergy & bioproducts. Springer, New York, pp 335–349
Davis SC, Ortiz-Cano HG (2023) Lessons from the history of Agave: ecological and cultural context for valuation of CAM. Ann Bot 132:819–833. https://doi.org/10.1093/aob/mcad072
Davis SC, Abatzaglou J, LeBauer D (2021) Expanded potential growing region and yield increase for Agave americana with future climate. Agronomy 11:2109. https://doi.org/10.3390/agronomy11112109
Delgado-Lemus A, Torres I, Blancas J, Casas A (2014) Vulnerability and risk management of Agave species in the Tehuacán Valley, México. J Ethnobiol Ethnomed 10:53. https://doi.org/10.1186/1746-4269-10-53
Eguiarte LE, Tambutti M, Silva-Montellano A et al (2006) Zur Naturgeschichte der Gattung Agave (Agavaceae): Taxonomie, Ökologie und Schutz (Teil 1). Avonia 24:8–19
Eguiarte LE, Aguirre-Planter E, Aguirre X et al (2013) From isozymes to genomics: population genetics and conservation of Agave in México. Bot Rev 79:483–506
Eguiarte LE, Jiménez Barrón OA, Aguirre-Planter E et al (2021) Evolutionary ecology of Agave: distribution patterns, phylogeny, and coevolution (an homage to Howard S. Gentry). Am J Bot 108(2):216–235. https://doi.org/10.1002/ajb2.1609
Ehrensing DT (2014) Feasibility of industrial Hemp product. Oregon State University Extension Services, Pacific Northwest. Available online: http://www.naihc.org/hempinformation.html
Evans ST (1992) Productivity of maguey terrace agriculture in central Mexico during the Aztec period. In: Killion TW (ed) Gardens of prehistory: the archaeology of settlement agriculture in greater Mesoamerica. University of Alabama Press, Tuscaloosa, pp 92–115
Evans S (2007) Bound in twine: the history and ecology of the henequen-wheat complex for Mexico and the American and Canadian Plains, 1880–1950. Texas A&M University Press, College Station
Faegri K, Van der Pijl L (1971) Principles of pollination ecology. Pergamon Press, Oxford, UK, p 291
FAO (2023) FAO statistics division. Available at: http://www.fao.org/faostat/en/

Figueredo CJ, Casas A, Colunga-GarcíaMarín P et al (2014) Morphological variation, management and domestication of 'maguey alto' (*Agave inaequidens*) and 'maguey manso' (*A. hookeri*) in Michoacán, México. J Ethnobiol Ethnomed 10(66). https://doi.org/10.1186/1746-4269-10-66

Figueredo-Urbina CJ, Casas A, Torres-García I (2017) Morphological and genetic divergence between *Agave inaequidens, A. cupreata* and the domesticated A. hookeri. Analysis of their evolutionary relationships. PLoS One 12(11). https://doi.org/10.1371/journal.pone.0187260

Figueredo-Urbina CJ, Casas A, Martínez-Díaz Y et al (2018) Domestication and saponins contents in a gradient of management intensity of agaves: *Agave cupreata, A. inaequidens* and *A. hookeri* in central Mexico. Genet Resour Crop Evol 65:1133–1146. https://doi.org/10.1007/s10722-017-0601-6

Figueredo-Urbina CJ, Álvarez-Ríos GD, García-Montes MA et al (2021) Morphological and genetic diversity of traditional varieties of agave in Hidalgo State, Mexico. PLoS One 16(7): e0254376. https://doi.org/10.1371/journal.pone.0254376

Fish SK, Fish PR, Madson JH (1992) Evidence for largescale agave cultivation in the Marana community. In: Fish SK, Fish PF, Madsen JH (eds) The Marana community in the Hohokam world, Anthropological Papers of the University of Arizona No 65. University of Arizona, Tucson

Fleming TH, Geiselman C, Kress WJ (2009) The evolution of bat pollination: a phylogenetic perspective. Ann Bot 104(6):1017–1043. https://doi.org/10.1093/aob/mcp197

Flores-Maya S, Vargas-Jurado MA, Suárez-Mota ME, Barrera-Escorcia H (2015) Análisis cariotípico de Agave marmorata y Agave peacockii (Agavaceae) ubicados en las terrazas aluviales del río Zapotitlán, Puebla, México. Polibotanica 40:109–125. https://doi.org/10.18387/polibotanica.40.7

Fowler PA, Hughes JM, Elias RM (2006) Biocomposites: technology, environmental credentials and market forces. J Sci Food Agric 86:1781–1789

Frankham R, Ballou JD, Briscoe DA (2010) Introduction to conservation genetics. Cambridge University Press, Cambridge, UK

García Mendoza AJ, Franco Martínez IS (2022) Los agaves, recursos fitogenéticos de los agroecosistemas tradicionales mexicanos de importancia para la agrobiodiversidad mundial. Universidad Nacional Autónoma de México. Instituto de Biología. Informe final Proyecto CONABIO No RG019//Proyecto Agrobiodiversidad mexicana, GEF 9380. Ciudad de México

García-Méndez RF, Cortés-Martínez CI, Almendárez-Camarillo A (2020) Thermochemical and tensile mechanical properties of fibers mechanically extracted from leaves of Agave angustifolia Haw. J Nat Fiber 19:3171–3185

García-Mendoza AJ (2002) Distribution of the genus Agave (Agavaceae) and its endemic species in Mexico. Cact Succ J 74(4):177–187

García-Mendoza AJ (2007) La biodiversidad de magueyes: base de La diversidad de mezcales. In: Foro tequilas y mezcales. GEA, AC, México

García-Mendoza AJ, Franco Martínez IS, Sandoval Gutiérrez D (2019) Cuatro especies nuevas de Agave (Aspaparagaceae, Agavoideae) del sur de México. Acta Bot Mex 126:e1461

Garcia-Moya E, Romero-Manzanares A, Nobel PS (2011) Highlights for *Agave* productivity. GCB Bioenergy 3:4–14. https://doi.org/10.1111/j.1757-1707.2010.01078.x

Gentry HS (1982) Agaves of continental North America. University of Arizona Press, Tucson

Gil-Vega K, Díaz C, Nava-Cedillo A, Simpson J (2006) ALFP analysis of Agave tequilana varieties. Plant Sci 170:904–909

Godinez AH, Jimenez M, Mendoza M et al (2008) Densidad, estructura poblacional, reproduccion y supervivencia de cuatro especies de plantas utiles en el Valle de Tehuacan Mexico. Rev Mex Biodivers 79:393–403

Gómez-Ruiz EP, Lacher TE Jr (2019) Climate change, range shifts, and the disruption of a pollinator-plant complex. Sci Rep 9:14048. https://doi.org/10.1038/s41598-019-50059-6

Good-Avila SV, Souza V, Gaut BS, Eguiarte LE (2006) Timing and rate of speciation in Agave (Agavaceae). Proc Natl Acad Sci USA 103:9124–9129. https://doi.org/10.1073/pnas.0603312103

Govindaraj M, Vetriventhan M, Srinivasan M (2015) Importance of genetic diversity assessment in crop plants and its recent advances: an overview of its analytical perspectives. Genet Res Int 2015:431487. https://doi.org/10.1155/2015/431487

Gonzalo D, Álvarez-Ríos Fernando, Pacheco-Torres Carmen Julia, Figueredo-Urbina Alejandro, Casas (2020) Management morphological and genetic diversity of domesticated agaves in Michoacán México Abstract J Ethnobiol Ethnomedicine 16(1) https://doi.org/10.1186/s13002-020-0353-9

Granick EB (1944) A karyosystematic study of the genus Agave. Am J Bot 31(5):283–298. https://doi.org/10.2307/2437546

Gurunathan T, Mohanty S, Nayak SK (2015) A review of the recent developments in biocomposites based on natural fibers and their application perspectives. Compos Part A Appl Sci Manuf 77:1–25

Heyduk K, McKain MR, Lalani F, Leebens-Mack J (2016) Evolution of a CAM anatomy predates the origins of Crassulacean acid metabolism in the Agavoideae (Asparagaceae). Mol Phylogenet Evol 105:102–113

Howell DJ (1979) Flock foraging in nectar-feeding bats: advantages to the bats and to the host plants. Am Nat 114:23–49

Huang X, Xiao M, Xi J et al (2019) De novo transcriptome assembly of Agave H11648 by Illumina sequencing and identification of cellulose synthase genes in Agave species. Genes 10:103

Hulle A, Kadole P, Katkar P (2015) Agave Americana leaf fibers. Fibers 3:64–75. https://doi.org/10.3390/fib3010064

IPES (2016) From uniformity to diversity: a paradigm shift from industrial agriculture to diversified agroecological systems. Report no 2. International Panel of Experts on Sustainable Food Sysystems, p 95. www.ipesfood.org. Last accessed 14 June 2023

Irigoyen R (1950) Los mayas y el henequ6n. Editorial Zamnfi, Mérida, Yucatán

Jiménez-Barrón O, García-Sandoval R, Magallón S et al (2020) Phylogeny, diversification rate and divergence time of *Agave sensu* lato (Asparagaceae), a group of recent origin in process of diversification. Front Plant Sci 11:536135

Jimenez-Valdés M, Godínez-Alvarez H, Caballero J, Lira R (2010) Population dynamics of Agave marmorata Roezl. under two contrasting management systems in central Mexico. Econ Bot 64(2):149–160

Kolte P, Daberao A, Miss Sharma A (2012) *Agave Americana*: the natural leaf fibre. Text Rev 7:1–5

Kuban JF, Lawley J, Neill RL (1983) The partitioning of flowering century plants by black-chinned and lucifer hummingbirds. Southwest Nat 28:143–148

Leal-Díaz AM, Santos-Zea L, Martínez-Escobedo HC et al (2015) Effect of *Agave americana* and *Agave salmiana* Ripeness on Saponin Content from Aguamiel (*Agave sap*). J Agric Food Chem 63:3924–3930. https://doi.org/10.1021/acs.jafc.5b00883

Lira R, Casas A, Rosas-López R, Paredes-Flores M et al (2009) Traditional knowledge and useful plant richness in the Tehuacán- Cuicatlán Valley, Mexico. Econ Bot 63:271–287

Lock GW (1962) Twenty-five years' Sisal research, 2nd (ed) edn. Longmans, Green & Co, Ltd, Harlow, UK, p 660

López Acevedo L, Merino Pérez YE, Enríquez J, del Valle JR et al (2018) Organogénesis in vitro EN TEJIDOS DE TALLO DE *Agave marmorata* Y *Agave angustifolia*. Rev Mexi Agroecosistemas 5(2):98–105

Madhu P, Sanjay MR, Senthamaraikannan P et al (2017) A review on synthesis and characterization of commercially available natural fibers: part II. J Nat Fiber 16:25–36

Martínez-Palacios A, Equiarte L, Furnier G (1999) Genetic diversity of the endangered endemic Agave victoriae-reginae (Agavaceae) in the Chihuahuan Desert. Am J Bot 86:1093–1096

Martinez-Rodriguez A, Beltran-Garcia C, Valdez-Salas B et al (2022) Micropropagation of seed-derived clonal lines of the endangered *Agave marmorata* roezl and their compatibility with endophytes. Biology 11:1423. https://doi.org/10.3390/biology11101423

Mastretta-Yanes A, Moreno-Letelier A, Piñero D et al (2015) Biodiversity in the Mexican highlands and the interaction of geology, geography and climate within the Trans-Mexican Volcanic Belt. J Biogeogr 42:1586–1600. https://doi.org/10.1111/jbi.12546

Mauersberger HR (1954) Mathew's textile fibres, their physical, microscopic and chemical properties, 6th (ed) edn. Wiley, New York

McClung de Tapia E, González Vázquez J, Zurita-Noguera J, Serra Puche MC (1992) Evidencia Para el uso del Agave sp. Durante el Periodo Formativo en el Sur de la Cuenca de México. Antropologí a y Técnica 5:99–114

Medina JC (1954) O sisal. Secretaria da Agricultura do Estado de São Paulo

Minnis PE, Plog SE (1976) a study of the site-specific distribution of *Agave parryi* in East Central Arizona. Kiva 41:299–308

Monja-Mio KM, Herrera-Alamillo MA, Sánchez-Teyer LF, Robert ML (2019) Breeding strategies to improve production of Agave (*Agave* spp.). In: Al-Khayri J, Jain S, Johnson D (eds) Advances in plant breeding strategies: industrial and food crops. Springer, Cham. https://doi.org/10.1007/978-3-030-23265-8_10

Moraes RM, Cerdeira AL, Lourenço MV (2021) Using micropropagation to develop medicinal plants into crops. Molecules 26:1752

Moreno-Valdez A, Grant WE, Honeycutt RL (2000) A simulation model of Mexican long-nosed bat (*Leptonycteris nivalis*) migration. Ecol Model 134:117–127

Moreno-Valdez A, Honeycutt RL, Grant WE (2004) Colony dynamics of *Leptonycteris nivalis* (Mexican long-nosed bat) related to flowering Agave in northern Mexico. J Mammal 85:453–459

Nava-Cruz NY, Medina-Morales MA, Martinez JL et al (2015) Agave biotechnology: an overview. Cr Rev Biotechnol 35(4):546–559. https://doi.org/10.3109/07388551.2014.923813

Nobel PS (1998) Los Incomparables Agaves y Cactos. Trillas, México

Nobel PS (2010) Desert wisdom/agaves and cacti: CO_2, water, climate change. iUniverse, New York

Nogales L (2017) Tepeztate, el maguey del monte. URL: https://mezcologia.mx/tepeztate/. Fecha de consulta: 19 de septiembre de, 2019

NORMA Oficial Mexicana NOM-006-SCFI (2012) Bebidas alcohólicas-Tequila-Especificaciones

NORMA Oficial Mexicana NOM-070-SCFI (2016) Bebidas alcohólicas-Mezcal-Especificaciones

NORMA Oficial Mexicana NOM-251-SSA1 (2009) Prácticas de higiene para el proceso de alimentos, bebidas o suplementos alimenticios, publicada en el Diario Oficial de la Federación el 1 de marzo de, 2010

Ornelas JF, Ordano M, Hernández A et al (2002) Nectar oasis produced by *Agave marmorata* Roezl. (Agavaceae) lead to spatial and temporal segregation among nectarivores in the Tehuacán Valley, México. J Arid Environ 52:37–51

Ortega Z, Castellano J, Suárez L et al (2019) Characterization of *Agave americana* L. plant as potential source of fibres for composites obtaining. SN Appl Sci 1:987. https://doi.org/10.1007/s42452-019-1022-2

Palomino G, Martínez J, Méndez I (2007) Variación inter e intraespecífica en especies de Agave por citometría de flujo y análisis de sus cromosomas. In: Colunga-García Marín P, Eguiarte L, Larqué A, Zizumbo-Villarreal D (eds) En lo ancestral hay futuro: del tequila, los mezcales y otros Agaves. CICY-CONACYT-CONABIO-INE, Mexico

Parker KC, Trapnell DW, Hamrick JL et al (2010) Inferring ancient Agave cultivation practices from contemporary genetic patterns. Mol Ecol 19(8):1622–1637. https://doi.org/10.1111/j.1365-294X.2010.04593.x

Parsons JR, Parsons MH (1990) Maguey utilization in highland central Mexico: an archaeological ethnography. Anthropological papers 82, Museum of Anthropology, University of Michigan, Ann Arbor

Pathan Y, Veeresh Kumar GB (2023) Potential of *Agave angustifolia* marginata for composite and textile applications—a new source of natural fibre. Ind Crop Prod 203:117213. https://doi.org/10.1016/j.indcrop.2023.117213

Perez-Pimienta JA, Flores-Gómez CA, Ruiz HA et al (2016) Evaluation of agave bagasse recalcitrance using AFEX™, autohydrolysis, and ionic liquid pretreatments. Bioresour Technol 211: 216–223

Pérez-Zavala ML, Hernández-Arzaba JC, Bideshi DK, Barboza-Corona JE (2020) Agave: a natural renewable resource with multiple applications. J Sci Food Agric 100:5324–5333. https://doi.org/10.1002/jsfa.10586

Phippen R (1999) Burned-rock features and middens: 4,800 years of roasting in Southeastern New Mexico. In: Whittlesey SM (ed) Sixty years of Mogollon archaeology. SRI, Tucson, pp 173–181

Pinkava DJ, Baker MA (1985) Chromosome and hybridization studies of *Agave*. Desert Plants 7: 93–100

Piven NM, Barredo-Pool FA, Borges-Argáez IC et al (2001) Reproductive biology of henequen (*Agave fourcroydes*) and its wild ancestor Agave Angustifolia (Agavaceae). i. Gametophyte development. Am J Bot 88(11):1966–1976

Ramírez CJL (1993) Plagas del henequén y su control. En: Barrera H, JA y R Díaz P (eds) Centro de Investigación Regional del Sureste. INIFAPSARH. Mérida Yucatán, pp 29–37

Raya FT, de Carvalho LM, José J et al (2023) Rescuing the Brazilian Agave breeding program: morphophysiological and molecular characterization of a new germplasm. Front Chem Eng 5: 1218668. https://doi.org/10.3389/fceng.2023.1218668

Riahi K, Van Vuuren DP, Kriegler E et al (2017) The shared socioeconomic pathways and their energy, land use, and greenhouse gas emissions implications: an overview. Glob Environ Change 42:153–168. https://doi.org/10.1016/j.gloenvcha.2016.05.009

Rocha M, Valera A, Eguiarte LE (2005) Reproductive ecology of five sympatric *Agave littaea* (Agavaceae) species in central Mexico. Am J Bot 92:1330–1341

Rocha M, Good-Ávila SV, Molina-Freaner F, Arita HT et al (2006) Pollination biology and adaptive radiation of Agavaceae, with special emphasis on the genus Agave. Aliso J Syst Evol Bot 22:329–344. https://doi.org/10.5642/aliso.20062201.27

Rojas-Martínez A, Valiente-Banuet A (1996) Análisis comparativo de la quiropterofauna del Valle de Tehuacán–Cuicatlán, Puebla–Oaxaca. Acta Zool Mex 67:1–23

Rojas-Martínez A, Valiente-Banuet A, ArizmendiMdel C et al (1999) Seasonal distribution of the long-nosed bat (Leptonycteris curasoae) in North America: does a generalized migration pattern really exist? J Biogeogr 26(5):1065–1077. https://doi.org/10.1046/j.1365-2699.1999.00354.x

Roll CM, Hodgson WC, Salywon AM (2021) Notes on a hybrid colony of *Agave parryi* var. huachucensis x *Agave palmeri* (Agavaceae) in southeastern Arizona. Phyton 25:1–7

Rosli NA, Ahmad I, Abdullah I et al (2015) Hydrophobic modification of cellulose isolated from *Agave angustifolia* fibre by graft copolymerisation using methyl methacrylate. Carbohydr Polym 125:69–75. https://doi.org/10.1016/j.carbpol.2015.03.002

Ruiz-Mondragón KY, Klimova A, Valiente-Banuet A et al (2024) Evolution history and the importance of genomic diversity facing climate change. The case of *Agave marmorata* Roezl., a microendemic agave used for mezcal production. Plants People Planet:1–17. https://doi.org/10.1002/ppp3.10532

Ruschel Robinson F (2013) El uso del ixtle en el Valle del Mezquital. Tesis de Maestria en estudios mesoamericanos, Facultad de filosofia y letras, Instituto de investigaciones filológicas, UNAM

Sagarpa S (2017) *Agave tequilero* y Mezcalero Mexicano. Planeación Agrícola Nacional 2017: 2030. Available at: www.gob.mx/cms/uploads/attachment/file/257066/Potencial-Agave_Tequilero_y_Mezcalero.pdf

Samouh Z, Cherkaoui O, Soulat D et al (2021) Identification of the physical and mechanical properties of Moroccan sisal yarns used as reinforcements for composite materials. Fibers 9: 13. https://doi.org/10.3390/fib9020013

Schaffer WM, Schaffer MV (1977) The reproductive biology of Agavaceae I.: pollen and nectar production in four Arizona agaves. Southwest Nat 22:157–168

Schaffer WM, Jensen DB, Hobbs DE et al (1979) Competition, foraging energetics, and the cost of sociality in three species of bees. Ecology 60:976–987

Scheinvar E, Gamez N, Catellanos-Morales G et al (2017) Neogene and Pleistocene history of *Agave lechuguilla* in the Chihuahuan desert. J Biogeogr 44:322–334. https://doi.org/10.1111/jbi.12851
Siller-Jasso MG (1985) Ciclo biológico en el laboratorio del Scyphophorus Picudo del Maguey acupunctatus Gyll (Coleóptera: Curculionidae) y algunas consideraciones sobre su impacto económico. Tesis profesional, Universidad Nacional Autónoma de México, México, p 91
Silva ORRF, de Macêdo Beltrão NE (1999) O agronegócio do sisal no brasil. Embrapa, Brasília, DF
Simpson J, Hernández AM, Juárez MJA et al (2011) Genomic resources and transcriptome mining in Agave tequilana. GCB Bioenergy 3:25–36
Slauson LA (2000) Pollination biology of two chiropterophilous agaves in Arizona. Am J Bot 87: 819–830
Slazak B, Sliwinska E, Saluga M et al (2014) Micropropagation of *Viola uliginosa* (Violaceae) for endangered species conservation and for somaclonal variation-enhanced cyclotide biosynthesis. Plant Cell Tissue Organ Cult 120:179–190
Spottedbird J (2000) Ethnobotanical information, Chihuahuan Desert Gardens, Centennial Museum, The University of Texas
Stebbins GL (1980) Polyploidy in plants: unsolved problems and prospects. In: Lewis WH (ed) Polyploidy, biological relevance. Plenum Press, New York, pp 495–520
Steyn HJH (2006) The evaluation of conventional retting *versus* solar baking of agave americana fibres in terms of textile properties. Master's thesis, University of Free State, Bloemfontein, South Africa
Sutherland SD (1987) Why hermaphroditic plants produce many more flowers than fruits: experimental tests with *Agave mckelveyana*. Evolution 41:750–759
Thamae T, Baillie C (2007) Influence of fiber extraction method, alkali and silane treatment on the interface of Agave Americana waste HDPE composites as possible roof ceilings in Lesotho. Compos Interfaces 14:821–836
Torres I, Blancas J, León A, Casas A (2015) TEK, local perceptions of risk, and diversity of management practices of *Agave inaequidens* in Michoacán, Mexico. J Ethnobiol Ethnomed 11: 61. https://doi.org/10.1186/s13002-015-0043-1
Townsend T, Sette J (2015) Natural fibers and the world economy. Discover Natural Fibers. http://www.dnfi.org/abaca/natural-fibre s-and-the-world econo my_1778/
Trejo-Salazar RE, Scheinvar E, Eguiarte LE (2015) Quién poliniza realmente los agaves? Diversidad de visitantes florales en 3 especies de Agave (Agavoideae: Asparagaceae). Rev Mex Biod 86:358–369. https://doi.org/10.1016/j.rmb.2015.04.007
Trejo-Salazar RE, Castellanos-Morales G, Hernández-Rosales D et al (2021) Discordance in maternal and paternal genetic markers in lesser long-nosed bat Leptonycteris yerbabuenae, a migratory bat: Recent expansion to the north and male phylopatry. Peer J 9:e12168. https://doi.org/10.7717/peerj.12168
Trombold CD (1985) A summary of the archaeology of the La Quemada Region. In: Foster MS, Weigand PC (eds) The archaeology of west and northwest Mesoamerica. Westview Press, Boulder, pp 327–352
Trombold CD, Israde-Alcántara I (2005) Paleoenvironment and plant cultivation on terraces at La Quemada, Zacatecas, Mexico: the pollen, phytolith and diatom evidence. J Archaeol Sci 32: 341–353
Turner BL, Miksicek CH (1984) Economic plant species associated with prehistoric agriculture in the Maya lowlands. Econ Bot 38:179–193. https://doi.org/10.1007/bf02858831
Ureta C, Ramírez-Barahona S, Calderón-Bustamante Ó, Cruz-Santiago P et al (2022) Evaluation of animal and plant diversity suggests Greenland's thaw hastens the biodiversity crisis. Commun Biol 5(1):985. https://doi.org/10.1038/s42003-022-03943-3
Valiente-Banuet A (2023) Mezcal boom and extinction debts. In: Jones RW, Ornelas-García CP, Pineda-López R, Alvarez F (eds) Mexican fauna in the Anthropocene. Springer. https://doi.org/10.1007/978-3-031-17277-9_14

Valverde PL, Vite F, Zavala-Hurtado JA (1996) A morphometric analysis of a putative hybrid between *Agave marmorata* Roezl and Agave kerchovei Lern.: Agave peacockii Croucher. Bot J Linnean Soc 122:155–161
Van Buren M, Skibo J, Sullivan A III (1992) The archaeology of an agave roasting location. In: Fish SK, Fish PF, Madsen JH (eds) The marana community in the hohokam world, Anthropological Papers of the University of Arizona No 65. University of Arizona, Tucson
Young RA, Rowell RM (1986) Cellulose structure, modification and hydrolysis. Wiley-Interscience Publication, New York

Chapter 3
American Century Plant (*Agave americana* L.) Breeding

Carlos A. Lecona-Guzmán, Federico A. Gutiérrez-Miceli, Sheila J. Reyes-Zambrano, Sugey Vásquez-Hernández, and José G. Joya-Dávila

Abstract The American century plant (*Agave americana* L.), endemic to the arid regions of Mexico, is notable for its important agronomic value and its ability to withstand extreme temperatures, attributed to its Crassulacean Acid Metabolism (CAM). This physiological system gives it unique adaptability to changing climatic conditions. Commonly known as *American agave* L., is classified as a monocotyledonous plant, within the *Asparagaceae* family. This species reproduces through seed germination or by transplanting basal shoots, with the latter being the preferred method for establishing plantations. In the countries where it is grown, it has multifaceted applications ranging from the textile industry (fibers) to the production of biofuels (lignocellulosic waste). Additionally, it is used in the pharmaceutical and food industries (fructans, saponins), highlighting its historical contribution to the production of traditional alcoholic beverages, known in Mexico as "spirit drinks". Over the centuries it has been a fundamental piece in agriculture, evolving from pre-Columbian to modern practices, facing phytosanitary and cultivation challenges. This chapter aims to provide a holistic view of *Agave americana* L., addressing crucial aspects including its origin, the transition from wild to cultivated status, taxonomic classification, and historical and contemporary uses. Additionally, it focuses on current practices, challenges, and advanced technologies for cultivation, with an emphasis on automation. Finally, the panorama of biotechnological techniques is explored as a key tool for the *in vitro* conservation of germplasm, utilizing techniques such as the isolation of cells and tissues. Likewise, it stands out in its use to obtain resistant and more productive plants, significantly contributing to the

C. A. Lecona-Guzmán · F. A. Gutiérrez-Miceli (✉) · S. J. Reyes-Zambrano · S. Vásquez-Hernández
Plant Tissue Culture Laboratory, National Technological Institute of Mexico/IT of Tuxtla Gutierrez, Graduate Studies and Research Division, Tuxtla Gutiérrez, Chiapas, Mexico
e-mail: carlos.lg@tuxtla.tecnm.mx; sheila.rz@tuxtla.tecnm.mx; D22270247@tuxtla.tecnm.mx

J. G. Joya-Dávila
Institute of Agricultural Sciences, Autonomous University of Baja California (ICA-UABC), Nuevo León, Baja California, Mexico
e-mail: jose.joya@uabc.edu.mx

K. F. M. Salem et al. (eds.), *Breeding and Biotechnology of Leaf, Fruit, and Seed Fiber Crops*, Advances in Plant Breeding Strategies 11,
https://doi.org/10.1007/978-3-032-00407-9_3

genetic improvement of this species, which is of both agronomic and industrial interest.

Keywords *Agave* spp. · Biodegradable fiber · Biotechnology · Fiber · Genetic variability · Somaclonal variation · Tissue culture

3.1 Introduction

Agaves are herbaceous plants with unique characteristics, that allow them to adapt to high temperatures, drought and even cold weather, characteristics that place them as species of utmost importance and relevance in agriculture due to the climate change events that occur today in a day of which in some parts of the planet the devastation due to water shortages is already beginning to be seen. Agaves are valuable resources for the food and pharmaceutical industries (such as steroid production for animals), the energy industry (as raw material for biofuels) and the ornamental industry.

Agave americana L. like many plants with Crassulacean acid metabolism (CAM) can survive long periods of drought; with as little as 200 mm of rain per year (Davis et al. 2017). It has been reported that *Agave americana* L. can be maintained with up to 200 mm of annual rainfall in regions where rainfall is not uniform throughout the year, the optimal average range for this plant is 600–800 mm (Gentry 1982).

The American Agave (*Agave americana* L.) stands out for being a versatile species with multiple uses in various industries and as a traditional food source for some communities in its place of origin.

It stands out among Agave species for its agronomic value and its ability to adapt to climate change. Specifically, its CAM enables it to survive prolonged periods of drought and extreme temperatures (−10 to 45 °C). This CAM process allows it to assimilate carbon dioxide (CO_2) during the night instead of during the day, thereby reducing water loss through evapotranspiration (Davis 2022; Davis and Ortiz-Cano 2023). *Agave americana* L. is a large succulent plant with slow growth, characterized by lanceolate leaves and a life cycle that ends with flowering (Davis 2022; Elbouzidi et al. 2023). The presence of the American century plant (*A. americana* L.) ranges from its wild state to its use as a decorative plant in gardens and urban environments, as well as in commercial crops (Fig. 3.1), mainly in climatic environments with low rainfall.

The cultivation of commercial crops has intensified, primarily for the production of alcoholic beverages, known in Mexico as "spirit drinks", the textile industry or fiber extraction and the manufacture of biofuels from lignocellulosic waste (Hulle et al. 2015; Lu and Wyman 2021). It is imperative to increase the productivity of *Agave Americana*, and in this context, the incorporation of new genotypes is essential to combat the damage caused by pests and phytopathogens.

Currently, the yield per hectare of *Agave americana* production is estimated at 16 Mg ha^{-1} year^{-1} and with the issue of climate change it may expand to higher latitudes (Davis et al. 2021). *Agave americana* L. reproduces only once during its

Fig. 3.1 *Agave americana* plants in different growing conditions. (**a**) Wild, (**b**) Urban environments. (Source: Photos by J.G. Joya-Davila)

entire life cycle, making it monocarpic. They can grow for several years before reaching their flowering stage. The cultivation period of Agaves can exceed 10 years before harvesting, so it is recommended to partially collect its leaves during this time (Irish and Irish 2000). There is low genetic variation in this crop due to its propagation method, which is through suckers (Reyes-Zambrano et al. 2021).

The objective of this chapter is to provide a general and comprehensive overview of *Agave americana* L., highlighting its origin, taxonomy, and both ancient and contemporary uses. In addition, it focuses on current practices and the challenges associated with its cultivation. It also discusses advanced technologies for monitoring and automation. Breeding strategies, both traditional and biotechnological, are explored, emphasizing genetic diversity, conservation of the local gene pool, and the application of biotechnological techniques, such as tissue culture, to expedite the selection of desirable traits. Furthermore, mutagenesis is highlighted as a tool that enables specific genetic changes to improve agronomic and industrial aspects. The importance of molecular markers and somaclonal variation in the genetic improvement process of this plant is also discussed.

3.1.1 Distribution and Botanical Classification

The *Agave* genus covers a variety of complex plants with high adaptive potential response to changing climatic conditions, thanks to their physiological characteristics. These characteristics allow them to store water and nutrients, ensuring the perpetuation of the species through the production of offspring and seeds. These plants play a crucial role in arid or semi-arid areas and are distributed in various geographical conditions around the world.

Mexico, as the center of origin for Agave, hosts a significant diversity of native genotypes, representing approximately 75% of the 200 known species in the country (Eguiarte et al. 2021). The geographical distribution of *A. americana* is primarily concentrated in the arid zones of Mesoamerica, but it has also spread to other continents such as Africa, Europe, and Asia (Gentry 1982; Davis 2022). *Agave americana* was first described by Carl Linnaeus in his work "Species Plantarum," published in 1753 (Davis and Ortiz-Cano 2023; Muriá 2012). This species plays a crucial ecological role both in its native habitats and in areas where it has been introduced. Its succulent leaves and high water storage capacity make it well adapted to arid environments, making it a valuable resource for wildlife. Additionally, the plant's flowers attract pollinators, contributing to local biodiversity.

Agave americana L. is native to northern Mexico and the southwestern United States, however, it is currently distributed throughout the world (Gentry 1982). During the colonization of America, the Spanish and Portuguese introduced *Agave americana* L. to the Canary Islands and the Azores, establishing cultivation in the Mediterranean region of Europe in the eighteenth century (Gentry 1982). Currently, it can be observed on the Mediterranean coast, as reported by CABI *A. americana* L. is considered an invasive species in countries such as New Zealand, Australia, Japan, South Africa, Namibia, Uganda, Kenya, Tanzania, Malawi, Greece, Ethiopia, Italy, Spain and Portugal (CABI 2022).

The taxonomic classification of *A. americana* is based on the work conducted by Gentry (1982). *A. Americana*, commonly known as "American Agave", "American century plant" or "Pitera", is a species that belongs to the *Agave* genus within the Asparagaceae family. This species is a monocotyledonous vascular plant taxonomically placed in the order Asparagales, as indicated by the information provided by the Integrated Taxonomic Information System (ITIS) (https://www.itis.gov). This order encompasses several notable plants, including species of high agronomic value such as *Asparagus officinalis* (Asparagus), *Iris germanica* (German Iris), *Ananas comosus* (Pineapple) and *Allium cepa* (Onion).

- Kingdom: Plantae (Plants)
  - Subkingdom: Viridiplantae (Green plants)
    - Infrakingdom: Streptophyta (Land plants)
    - Superdivision: Embryophyta
      - Division: Tracheophyta (vascular plants, tracheophytes)
        - Subdivision: Spermatophytina (Seed plants, phanérogames)
          - Class: Magnoliopsida
            - Superorder: Lilianae (Monocotyledons)
              - Order: Asparagales
                - Family: Asparagaceae
                  - Genus: *Agave* L. (Agave species)
                    - Species: *Agave americana* L. (American century plant)

The various varieties of wild *A. americana* L. share their habitat with cultivated agaves in cropping systems, such as live fences used to delimit properties. These agaves are also utilized as protective plants because of their rigid leaves and thorns, which serve to prevent the entry of unwanted animals or people. Additionally, they serve as decorative elements on farms catering to tourism and are used as food plants. As a cultivated plant, the most widely used varieties of *Agave americana* include var. oaxacensis Gentry, variety expansa (Jacobi) Gentry, var. marginata, protoamericana, var. sierrudita and var. americana (Mandujano et al. 2018; Mora et al. 2023). Morphologically, these varieties are distinguished by the size of the plant, the quantity, and dimensions of the leaves, as well as the size of teeth and spines. In its natural environment, the species is recognized for its erect, lanceolate to somewhat spatulate leaves, with a reflex or recurved arrangement, occasionally displaying greenish transverse bands, in addition, the flower, known as "quiote", is one of the most characteristic organs of this plant due to the notable height it reaches in its maturity stage. Botanically defined as an inflorescence that emerges from the center of the rosette, it can measure up to 12 m in height. It is important to note that this flower has both sexes in the same structure (Reynoso-Santos et al. 2012).

3.1.2 Importance and Uses

Throughout history, *Agave americana* has played essential roles in agriculture. In pre-Columbian civilizations, indigenous communities utilized its fibers for textile production and its sap for the crafting of fermented beverages, laying the groundwork for the current production of mezcal and tequila. In the contemporary context, the versatility of *A. americana* has significantly expanded, being considered a "zero waste"plant. Its applications span the textile, food, pharmaceutical, and biofuel industries. The fiber extracted from the American agave has multiple uses, commonly the fibers are used in the manufacture of ropes, hunting nets, footwear and clothing (de la Torre et al. 2018). Its resistance and durability make it a valuable resource for the manufacture of textile products, contributing to both artisanal tradition and modern industry. The fibers have also been used to make reinforced polymer composites, which provide thermal insulation, resistance, and biodegradability properties (Sathiamurthi et al. 2021).

In Mexico, *Agave americana* plays a significant role in the production of alcoholic beverages, also known as "spirits." Among these, pulque stands out, a pre-Hispanic drink made from various Agave species, including *A. americana.* In Oaxaca, it is known as "maguey arroqueño" and is used as raw material for mezcal production. In the Meseta Comiteca Tojolabal, in Chiapas, *A. americana* is used artisanally in the preparation of the beverage known as "Comiteco" (Mora et al. 2023). In the transformation of Comiteco liquor, historically both A. *americana* and *A. salmiana* were used; however, currently, the exclusive cultivation of *A. americana* plants is promoted (Reynoso-Santos et al. 2012). The residues generated during the Agave distillation process find significant utilization in the production of biogas (Santiago-Santiago et al. 2023).

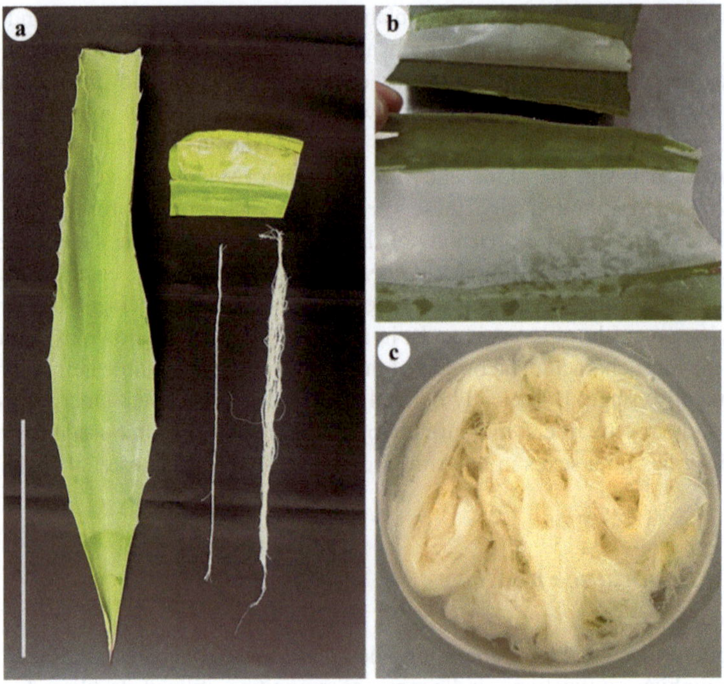

Fig. 3.2 Fibers of *Agave americana* L. (**a**) Leaf of *Agave americana* L. (**b** y **c**) Fibers. bar equals 30 cm. (Photos by J.G. Joya-Davila and C.A. Lecona-Guzmán)

The remarkable resistance of *Agave americana* L. fibers made it possible to use them on continents distant from the center of origin of the species, including Africa (until 1950), where its inhabitants used them to make heavy-duty ropes and cordage (Chaabouni et al. 2006). On the other hand, the fibers of this species have been used for the manufacture of paper, through a delignification pulping process (Fig. 3.2). The fibers of the sheets, after being pulped at 107 °C for 60 min with a 6% soda solution, generate optimal conditions for obtaining paper sheets with notable breaking length indexes, as well as resistance to the tear and burst indices. These paper sheets can find application in the manufacturing of various products, such as packaging, filters, and stationery (Jaouadi et al. 2011).

Additionally, the roots and leaves of *A. americana* stand out as sources of saponins, used in soap production, while the leaves can also be employed in paper manufacturing. On the other hand, the spines of the leaves have practical applications, being used as pins and needles (Zwane et al. 2011). Moreover, the extract from the leaves exhibits promising potential in the pharmaceutical industry because of its outstanding antioxidant activity (Hamissa et al. 2012).

3.1.3 Fiber and Textile Properties

The planet presents serious environmental problems, which is why the world scientific community is showing great interest in the use and exploitation of the potential of fibers from natural sources (Mylsamy and Rajendran 2011). There are different renewable resources from plants as well as an important and extensive resource of vegetable fibers some coconut fiber, jute, and banana among others which are found in a wide source for various natural fibers, coconut fiber, banana, and jute, which are found in abundance in various regions of the planet (Kolte et al. 2012). In this sense, there are other types of fibers from plants that are not yet used as textile fibers. A source of this type of strong fiber is "*Agave americana* L." The fibers of this plant are also known or named as "Pita fibers", which are cellulosic (Chattopadhyay and Khan 2012).

From the *A. americana* L. plant, the leaves are used for the extraction of fibers, however, only the lower leaves that are at angles greater than 45° concerning the vertical axis are harvested with sharp cutting tools, followed by the thorns. They are removed from the margins of the leaves as well as the spine found at the tip of the leaf (Mylsamy and Rajendran 2011).

The *Agave americana* L. is distinguished as the most weather-resistant species of agave. Its robust leaves can generate fibers, known as pita, ideal for the production of various products such as ropes, mats, rustic fabrics and long-lasting leather embroidery (Bezazi et al. 2014).

Three years of cultivation are required to begin harvesting the leaves for fiber production. For this purpose, only mature leaves with a minimum length of 1 m are used. Each plant can generate an average of 45 leaves annually with a fiber yield that varies between 2.5% and 4.5% depending on the variety (Hulle et al. 2015). Fiber extraction can be carried out by various methods, mechanically, it involves crushing or compressing the sheets so that only the fibers remain, in this method the fibers must be washed and dried immediately, to avoid hardening (Kolte et al. 2012).

If *A. americana* L. is cultivated for the use of fiber, it is necessary to homogenize the plantations through a selection of varieties that meet the requirements for obtaining fiber. Currently, the crop is very heterogeneous in the plantations there are plants with long leaves and straight, others have wide leaves with a very prominent lanceolate shape and others with less rigid and curved leaves. If the fiber extraction method is carried out with the methods used to obtain fiber from the sisal or henequen species, an important aspect would be the morphology of the leaf which must be long and straight, this could be taken as a starting point for the selection of elite cultivars for fiber extraction from *Agave americana* L. (Davis et al. 2019). If the use of the leaf is for another purpose such as obtaining sweeteners or bioethanol, the shape of the leaf is not important. Leaf shape is not relevant if *A. americana* L. is grown for bioethanol or sweetener production (Davis et al. 2011; Simpson et al. 2011).

Traditionally, field workers carried out decortication by hand, pulverizing the leaves, crushing them and removing pulp with a knife. Manual decortication requires more time as well as more labor. Currently, this practice is carried out more efficiently way using a decorticator machine. In the mechanical decortication phase, the leaves are shredded and macerated by a rotating disc with non-sharpened blades, resulting in only the fibers of the leaves (Kolte et al. 2012; Msahli et al. 2005). There are different models of decortication machines that are fed manually, and the process involves longitudinally splitting the leaf and inserting one half to extract the pulp, then repeating this procedure with the other half. In other decortication machines, the entire leaf can be introduced to extract the pulp in one step (Steyn 2006).

Once decortication is finished, the fibers are dried either directly in the sun or with the help of dryers. Typically, the process involves defibration, washing and drying. After completing the decortication and washing process, the fibers are dried using mechanical dryers or in the sun. Fiber extraction as well as the washing and drying processes must be carried out following a well-defined methodology and in a synchronized manner. Once the leaves are cut, the resin on the leaves hardens, causing the pulp to adhere to the fiber, complicating the cleaning of the obtained fiber. The efficiency of mechanical fiber extraction methods is low because they do not eliminate compounds that are part of the cell wall, mainly hydrocarbons, hemicellulose, waxes, and lignin between fibers (Kolte et al. 2012).

The chemical method focuses on enzymatic degradation and acidic or alkaline solutions. Likewise, fiber can be extracted by submerging the leaves in different media, such as seawater, or tap water, and even burying them under a layer of soil (Sathiamurthi et al. 2021). When choosing the extraction method, various factors must be considered, such as the type as well as the quality of the fiber required, allowing the process to be adapted to the specifications necessary for the final application of the fibers obtained. Properly extracted fiber has a shiny, whitish to yellowish appearance. In the dry state, it appears rigid and rough; however, when wet, the fiber becomes flexible and soft (Hulle et al. 2015). To achieve optimal mechanical properties, it is recommended to extract fibers in distilled water, subjecting the sheets to a hydrolysis treatment, followed by the fiber separation process through calendering and washing. To improve resistance, alkaline treatments with 2% NaOH are implemented, eliminating non-cellulosic compounds. However, it is highlighted that concentrations greater than 15% can cause fiber degradation. Finally, after these processes, fibers with outstanding characteristics are obtained, such as high hydrophobicity, tenacity, low density, and great extensibility (El Oudiani et al. 2011). The extraction method through degradation of the matrix in the soil is presented as an alternative in regions with scarcity of water resources. Higher fiber hardness is observed using this method compared to the water-based method. However, fibers derived from soil extraction may exceed the maximum diameters of Agave fibers used in textile applications, while only aqueous-based fibers seem to meet the maximum diameter requirements (Bezazi et al. 2014).

Regarding the structural composition of *A. americana* leaves, they are composed of cellulose strands inserted in a lignin matrix providing the characteristic resistance to textile products (Hulle et al. 2015). This organic matrix consists of molecules of lignin, pectin, hemicellulose, cellulose, and waxes. This composite structure is reinforced with cellulose microfibers, thanks to various types of hydrogen bonds or Van der Waals forces (Chaabouni et al. 2006). Agave fibers have an average chemical composition, with cellulose as the main component at 65.8%, followed by hemicellulose (12%), lignin (9.9%) and low percentages of pectin and waxes (0.8% and 0.3% respectively). In the particular case of *A. americana* L., it has a density of 1.36 g/cm^3 and can reach up to one and a half meters in length (Jaouadi et al. 2011). The lignin content is approximately 12%, varying according to the phenological stage of the plant, the total fresh mass of the leaves and the number of leaves in which it is distributed. This represents a lower value than the lignin content present in other plants, such as corn stover (13–14%) and hybrid poplar (24.9%). This characteristic is crucial for the industry, both in the manufacture of textiles and in the production of biofuels such as bioethanol (Lu and Wyman 2021).

The level of crystallinity of the fibers directly influences the chemical absorption capacity, with low values indicating amorphous regions more susceptible to water and chemicals. The outstanding crystallization and polymerization of *A. americana* fibers give them notable resistance making them ideal for application as a textile fiber (Bezazi et al. 2014).

A method of extracting fibers through a natural microbial process called retting has been widely studied. This method involves the elimination of non-micro matter found between the fibers of the woody portions of the plants, as well as the cellulose fibers. This method allows the retting method has been widely studied to extract fibers through a natural microbial treatment. Retting involves the removal of non-fibrous elements, an action that has the function of adhering between the fibers of the woody sections of the plants and the fibers without causing damage to the cellulose fiber. This procedure facilitates and favors the detachment of the individual fiber strands and the woody core. This process requires humidity and a warm temperature to activate the microorganisms.

Another method is the preferential decomposition procedure to split the fiber of the lignocellulosic biomass without causing damage to the cellulose fiber. This process consists of the microbial release of plant fibers from their environment (Mignoni 1999). This process can take at least 3 weeks. Retted microorganisms consume non-fibrous cementitious materials, mostly hemicellulose and pectin, which softens the sheet due to the degradation of intercellular adhesive substances with lower resistance. At the optimal point of fermentation, the fibers come off easily from the leaves. If the optimal point of fermentation is exceeded, the fibers lose quality because incomplete retting occurs which does not degrade the adhesive elements such as pectins, making fiber extraction more difficult. The leaves gradually soften due to the degradation of less resistant intercellular adhesive substances. When the rooting stage is not adequate, the compounds that serve as glue in the cell wall are not adequately removed, which can be pectins, causing difficulty in the fiber extraction stage. Although this process takes much longer, it is the most economical.

The most used traditional retting methods are two: the first by water and the second in the field or by dew. In the first, the leaves are submerged in a source of water, be it a tank, river, or pond, among others, in the second, the leaves are spread in an outdoor space, so that they are moistened by dew or rain for rooting. Water retting is the method that produces more uniform and higher quality fibers compared to the field method (Steyn 2006).

Another rapid retting method is enzymatic, in which commercially produced enzymes reduce the pectin contained in the fiber strands. This enzymatic procedure is faster than natural fermentation retting and produces softer fibers. The greatest advantage of this method is that it simplifies and reduces the costs of fiber extraction offering greater control, performance, and a shorter processing time. The enzyme solution can be reused multiple times making the procedure more profitable and environmentally friendly (Buschle-Diller et al. 1994; Steyn 2006).

Pectinases and xylanases are the enzymes used for the retting processes of plant fibers, which can separate or release portions of fibers from the leaves. The concentrations at which these enzymes can be used range from 1.5–5 g/L of water. The use of cellulase enzymes in any of the retting processes should be avoided as these enzymes degrade cellulose, resulting in lower resistance fibers (Buschle-Diller et al. 1994; Kashyap et al. 2001; Steyn 2006).

3.1.4 Fiber Evaluation

The fiber obtained from *Agave americana* L. is comprised of a series of cells known as final or last cells, which are connected by a serous lamina that produces strands of fibers. Each fiber is made up of one or more vascular bundles, consisting of transport tissues. The shape of each fiber is polygonal with a lumen inside. Fibers from the leaves of *A. americana* L contain phloem, xylem and clove cells dispersed throughout the leaf matrix. These cells are partially lignified and hard (Rolf Dahlgren and Clifford 1982). The diameter of the fibers varies depending on the number of final fibers in the leaf cross-section (Hatch 1993).

Using Scanning Electron Microscopy (SEM), the structure of *Agave americana* L. fibers has been analyzed, observing longitudinal stripes characteristic of long fibers of plant origin. The final fibers are held together by adherent and waxy compounds such as hemicellulose, pectin and/or lignin. The nature of this coating gives the *Agave americana* L. fibers greater resistance when exposed to external factors such as layering, chemical agents, alkalis and UV rays.

Agave americana L. fibers have oval and irregular sections with a helical structure of square-shaped needles. The characterization of these fibers has been carried out, measuring the length and diameter, the average length has been recorded at approximately 10.1 μm and the average diameter has been recorded at 3.1 μm. The average diameter of this fiber is smaller than that reported for other natural fibers such as sisal, alpha and flax. Compared to other natural fibers such as sisal, flax, and alpha the size of the *Agave american* L. fiber is much smaller. (Kadole and Hulle 2014; Msahli et al. 2006).

Regarding the length of *Agave americana* L. fibers, they are longer compared to sisal fibers with fiber threads reaching up to 2.8 m. The thickness of the fibers has been determined using the projection microscopy technique and has been recorded in a range of 100–150 μm, considered thick due to the large number of final cells that form the fiber bundle (Steyn 2006).

The fibers of *A. americana* L. have hydrophilic properties, this characteristic gives them the ability to absorb a greater amount of water than their own weight, and they have a moisture content in the range of 8–9% (Mylsamy and Rajendran 2011). The coloring of the fiber of *Agave americana* L. is varied, ranging from whitish to yellowish, this depends on the extraction technique and processing. Rooted fibers may present a light brown color due to the effect of the bacteria used.

The appearance of the fibers is semi-opaque and is believed to be due to the surface and section of the fiber being irregular in shape. Fibers with these characteristics of irregular cross sections separate light in all directions, giving them an opaque appearance. When the *Agave americana* L. fiber is dried it presents a rough, rigid texture and a hard surface, a typical characteristic of fibers that are extracted from leaves, however, when they are hydrated, they become soft, flexible, and slippery.

The *Agave americana* L. fiber has a breaking moment that is in a range of 2–4%, this means that this fiber can function when instantaneous forces are applied to the *Agave americana* L. fiber. The tensile value of the fiber suggests that the fiber is strong enough to be valued as a textile fiber. This characteristic is attributed to the polymerization and crystallization processes of cellulose. The tensile strength presented by *Agave americana* L., fiber can be applied to fabrics, carpets, rugs, and upholstery fabrics, among others.

Natural fibers present irregularities in growth which means that the tensile properties are not uniform in the fibers of *Agave americana* L. These irregularities are caused by factors such as the growth and size of the plant. Stronger fibers are obtained from the outer leaf sheaths, in contrast, the inner sheaths produce weaker fibers. The peripheral fibers have a lower tensile strength while the more internal fibers have a high fracture stress (Msahli et al. 2006).

It has been proven that the fibers of *Agave americana* L. maintain their stability in weak alkalis as well as in acids, which do not generate an alteration in the hardness and structure of the fibers of *Agave americana* L., so detergents can be applied to them dyes, bleaches of weakly acidic or basic nature. In strong bases or acids, they are unstable, these fibers distort or dissolve causing losses in tensile strength. Concentrated sodium hypochlorite causes whitening, however with continued exposure the fiber weakens and dissolves (Mylsamy and Rejendran 2011).

3.1.5 Applications of **Agave americana** *L. Fibers*

A low density, as well as a high tensile strength, are the characteristics of the fibers obtained from the leaves of *A. americana* L. For these characteristics, they have been used for many years in the manufacture of ropes and twine, which have been used in fishing activities and agriculture, as well as in the production of fish cords, mats, rugs, bags, padded curtains for furniture, fillings, baskets, hair, bracelets, items for decorating clothing and other woven objects. Papers have also been made from inferior quality *A. americana* L. fibers. Other applications for *A. americana* L. fibers are the production of paper, although it is of low quality, and they have been implemented in leather embroidery using a technique called piteado, *Agave americana* fibers may have potential applications for the production or manufacturing of furniture, fiberboard, bioplastic and/or geotextiles (Kolte et al. 2012; Msahil et al. 2006).

3.1.6 Fiber Industry

The use of Agave spp. fiber dates back to approximately 2600 BC in Mesoamerica (Fig. 3.3). During the Mayan era, henequen fiber was produced from this plant (Evans 2007). In the "Late Postclassic" period, textiles represented high-value commercial products for the Aztec civilization. Anthropological work in Mexico uncovered vestiges associated with tools such as obsidian blades and basalt scrapers for processing fiber as well as parts of textile works with maguey fiber. These findings suggest that the growth of the maguey fiber industry began in the Aztec city of Otumba (Nichols et al. 2000). The commercial use of Henequen (*Agave fourcroydes*) fibers began around 1830 as a measure to mitigate economic impacts, eventually becoming a prominent export product. During this period, the focus of commercial agave cultivation was on extracting fibers by processing the leaves (Cline 1987). Throughout the nineteenth century, fibers derived from Agave experienced an industrial increase (Fig. 3.3), reaching their maximum production due to their efficient use in the manufacture of ropes and twines for agricultural and fishing applications. Although their demand is reduced with the advent of synthetic fibers,

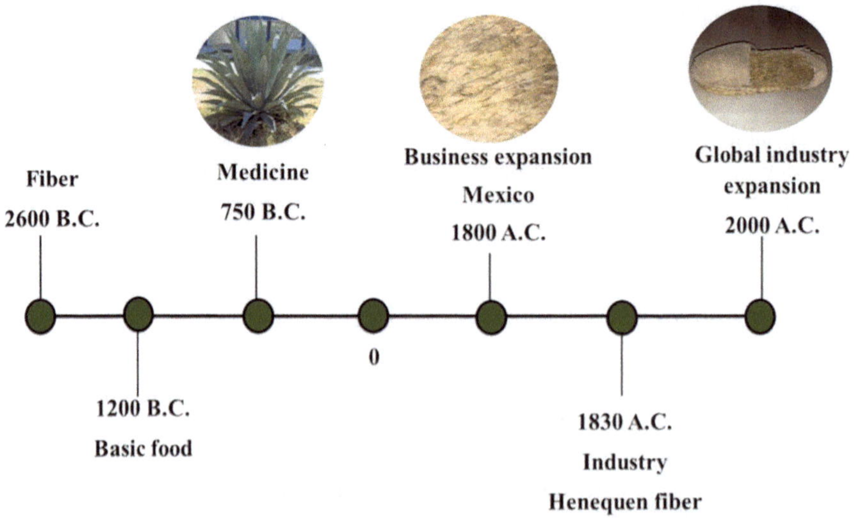

Fig. 3.3 *Agave americana* L. Timeline. Fiber and other uses. (Photos by J.G. Joya-Davila)

they still find applications in the creation of netting, carpets, rugs, geotextiles, and molded furniture (Davis and Ortiz-Cano 2023).

Regarding production, agave grown for fiber production is reported in countries such as Brazil, Madagascar, China, Indonesia, Kenya, Mexico, and South Africa, among others. A global production of around 300,000 tons is estimated at a value of 75 million dollars (FAO 2024). Brazil, Tanzania, Kenya, and Mexico stand out as the main producers of Agave fiber. However, it is relevant to note that, in terms of yield per hectare, China has occupied the leading position in the last 30 years (Davis 2022).

It is important to note that Agave has various applications in the industry. Its cultivation without generating unnecessary waste promotes sustainability in its production and processing, reducing the environmental impact since Agave is completely biodegradable. The organic waste derived from this crop is used to make compost or organic material.

Agave americana L. fibers are used by various industries to replace the aggressive chemical compounds and toxins used in the production of synthetic fibers. *American agave* fiber is used to strengthen asbestos sheets as well as in the manufacture of fiberglass. The fibers of *Agave americana* L. used in the industry can be classified into three types: low-quality fibers are used for making paper, medium-grade fiber is used in the cordage industry to make twine and rope, and high-quality fiber is used in the textile industry to make fine threads.

3.1.7 Ecological Utility of the **Agave americana** *Plant and fibers*

One of the properties of agave plants is that they have four times more cellulose than any other plant, effectively capturing CO_2 and requiring little irrigation. The production of agave fibers has a low environmental impact since no chemical inputs are needed for maintenance. When the leaves are harvested and processed, the only waste generated is organic waste which can be reused for the production of other products. The byproducts generated after processing the leaves are biodegradable and can be used for composting or as organic matter for soil nutrition, as well as for the production of biogas. Compared to synthetic fibers, agave fiber is one hundred percent natural (Kadole and Hulle 2014)

3.1.8 Selection, Domestication and Early Improvements

The initial advancements in *Agave americana* arose from observations and empirical knowledge passed down through generations. In the post-classic period of Mesoamerica, around 759 AD, Agave fibers had a transcendental role in the economy of the Aztec Empire. During this time, the Aztecs implemented cultivation techniques and established specialized workshops to produce ropes and cords (Davis and Ortiz-Cano 2023; Nichols et al. 2000). The semi-domestication strategies of *A. americana* are linked to its small-scale inclusion with other crops and its use as fencing for pastures (Davis et al. 2019).

The economic significance of *A. americana* L. has driven the expansion of its cultivation, both in Mexico (La Torre-Ruiz et al. 2016) and in the United States, where the possibility of cultivation in latitudes further north than those traditionally associated with *A. tequilana* is being evaluated. Furthermore, cultivation in drier areas with low rainfall has been explored (Davis et al. 2021). The establishment of *A. americana* crops has been carried out using offsets from selected plants in the field, as well as through the propagation of genotypes exhibiting outstanding agronomic characteristics. Additionally, tissue culture has enabled the large-scale multiplication of clones, utilizing various tissues for *in vitro* propagation, such as meristems and leaf sections. These processes have been conducted through both direct and indirect methods (Mora et al. 2023).

The application of plant tissue culture as a biotechnological tool emerges as a fundamental facilitator for the commercial improvement of agaves. Cellular elicitation and mutation induction are key processes that contribute to developing cell lines with stress resistance (Reyes-Zambrano et al. 2021). Additionally, molecular analyses play an essential role in confirming the induction of desirable genetic changes, allowing the determination of genetic diversity through polymorphism evaluation. This approach provides a solid foundation for the careful selection of the most suitable plant material for subsequent field establishment (Mandujano et al. 2018).

3.2 Current Cultivation Practices and Challenges

3.2.1 Current Cultivation Practices

Agave americana L. is a slow-growing perennial plant that flowers only once in its lifetime, producing a hermaphroditic flower. Occasionally, it develops a short woody stem at its base, presenting an inflorescence ranging from 6 to 12 m in height and succulent leaves measuring 1–2 m in length and 15–25 cm in width. These leaves are succulent, with a grayish-green or bluish-gray coloration (Elbouzidi et al. 2023). It primarily reproduces through seed germination and offsets. However, its rate of propagation in its natural environment is low due to its harvesting age, fluctuating between 10 and 12 years, which can be reduced to 6 or 8 years (Reyes-Zambrano et al. 2016).

Due to its reproductive system, plants die after reproduction, a reproductive strategy known as monocarpic or semelparous, which is uncommon in flowering plants. The reproductive effort and resource allocation to floral display ultimately led to the plant's death (Davis 2022).

In contemporary times, new digital tools, such as the use of sensors, offer an alternative to increase production and reduce costs in agriculture. These tools enable the monitoring of fertilizers, and pesticides, among other factors (Ammann et al.

2022), providing the opportunity to make informed decisions for prompt and intelligent actions based on the generated information (Carrillo-Riofrío et al. 2021). Among these technologies are vertical farming, digital agriculture, as well as artificial intelligence.

The use of these tools in agave production provides an opportunity not only for monitoring but also for automating certain aspects of the process, such as fertilization, irrigation, pest and disease detection, plant maturity, etc. However, they are expensive, and currently, there are no reports on the use of these technologies in agave cultivation. Nevertheless, it can be anticipated that in the coming years, they will be introduced to enhance the profitability of this crop.

3.2.2 Vertical Farming

The expression vertical agriculture was described by the American Gilbert Ellis Bailey in 1915, describing three systems: hydroponics, aquaponics, and systems with substrates (Bailey 1915). This technique has great technological production potential, however, it continues to be an expensive tool, having lower profitability, scalability and environmental sustainability compared to conventional practices (Van Gerrewey et al. 2021), its use is currently directed at species with a specific requirement of larger water, such as strawberry, lettuce, spinach, and tomato crops, among others. Therefore, to date, there are no reports of the implementation of this system in plants of the *Agave* genus.

3.2.3 Digital Farming

Digital agriculture, also known as smart agriculture, precision agriculture or agriculture 4.0, has been promoted to improve productivity (Eastwood et al. 2019). It encompasses high-performance methods, using computers, sensors, mobile devices, data storage and analytical models to aid agricultural development, improve quality, profitability, and productivity and reduce costs and risks (Queiroz et al. 2021). It provides a tool with great potential in agave crops, to carry out territorial monitoring, temperature, and precipitation, facilitating the detection of pests and some diseases, invasive flora, and fauna, among other factors that can harm the development of plantations.

3.2.4 Artificial Intelligence

Artificial intelligence is a field of computer science research that involves creating artificial systems capable of perceiving, reasoning, making decisions and solving problems (Jo et al. 2018). It incorporates technology into machines, equipment, and sensors in various agricultural production systems, allowing the automation of the process (Bassoi et al. 2019; Pivoto et al. 2018).

The application of artificial intelligence in agave production offers the opportunity to carry out not only monitoring but also automate some parts of the process, such as fertilization, irrigation, detection of pests, diseases, and plant maturity, among others. However, it is an expensive technology, and there are currently no reports on its use in agave crops, although it can be expected that in the coming years, it will be introduced as a tool to improve quality and production.

Specifically, in the case of fiber extraction, machinery has been developed that has been designed to crush leaves of various species of agave into particles for species of economic interest and greater commercial such as *A. sisalana* and *A. fourcoides* (Kumar et al. 2022; Murali and Morchhale 2014). However, the parameters for the design of these prototypes remain empirical (Pérez del Río et al. 2013). Designing a leaf shredder is a difficult task because the design parameters of the cutting elements must be selected, however, this could be improved by implementing new technologies such as artificial intelligence (AI). In this sense, genetic algorithms (GA) have been developed, which are adaptive tracking algorithms that at the time require evolutionary algorithms, which jointly generate high-quality solutions for complex engineering, design, and resolution problems (Aydin et al. 2020). Genetic algorithms are used as tools in process research, and scientific models, which provide precise formulations of a specific function (Mirjalili 2019).

In another context, AI has been used to classify the classes of 100% agave tequila, based on the concentration of volatile compounds, UV-Vis indices and RGB analysis; this technology has been called Artificial Neural Network. Using this technology, the maturation process can be demonstrated in wooden barrels in which there may be an increase in the concentration of congeneric compounds in tequila caused by oxidation reactions, and extraction of polyphenols and flavonoids, which impact the color that you can obtain the distillate (Warren-Vega et al. 2023).

3.2.5 Current Agricultural Challenges

In Mexico, the distribution of *Agave americana* L. has diminished due to the exploitation of its natural habitat, stemming from its use in the production of textile fibers and its employment in traditional medicine (Reyes-Zambrano et al. 2016). However, the major agricultural challenge facing *A. americana* is phytosanitary issues, such as wilt disease and dry bud rot bound with the fungus *Fusarium oxysporum* (López-Bautista et al. 2020).

Given the low availability of planting stock and the ongoing rise in the cost of agave piñas for cultivation, there has been an increase in the production and movement of offsets between states, heightening phytosanitary risks. This emphasizes the need to strengthen not only monitoring but also sampling, as outlined by the Campaign Against Regulated Agave Pests (Coria-Contreras et al. 2019; SENASICA 2017).

3.2.6 Breeding Objectives

Given the problems caused by the monoculture of agave, such as the reduction of crop diversity, soil erosion and phytopathological problems, which limit the possibilities of crossing and evolution of progeny, as well as the increased risk of extinction of the populations due to the poorly controlled extraction of specimens and seeds (Ruiz Mondragón et al. 2023), the need arises to reduce the maturity process of *A. americana*, which goes from 10 to 12 years, as well as its conservation using biotechnological tools (Raya et al. 2023).

3.3 Germplasm Biodiversity and Conservation

3.3.1 Germplasm Diversity

The diversity and complexity of the *Agave* genus, as well as its high demand for commercial purposes, has increased the production of agave species, among which *A. americana* stands out (Martinez-Rodriguez et al. 2019). Many studies show that in recent decades, Mexico has lost important species, which is why it is necessary to develop alternatives to exploit commercial species of interest without affecting natural resources, as well as generate conservation programs (Monja-Mio et al. 2019).

3.3.2 Cultivars Characterization and Phylogeny

A. americana is cultivated in arid and semi-arid areas, given its tolerance to drought, without harming its high biomass productivity (Davis et al. 2017; Liu et al. 2021), the excellent cold tolerance of *Agave americana* L. gives it a significantly wider advantage compared to other species of the same genus of growing in different regions. This tolerance could be given by the large amount of fibers present in the leaf tissues, thus maintaining rigidity, avoiding deformation, and conserving its capacity, to capture light to carry out the photosynthesis process (Monja-Mio et al. 2019).

It has flat, brown and black seeds that range from approximately 0.006–0.008 m, its root system is superficial, facilitating the absorption of rainwater, the flowers appear when the plant is around 10 years old between July and September, it has been described as that the optimal planting period is from April to June, given that the establishment of plants is favored by the rain cycle (Gentry 1982; Reynoso-Santos 2012).

A. americana presents differences with respect to its size, shape and number of leaves, and size of terminal and lateral spines, however, it can be recognized by combinations of characters, highlighting the shape of its leaves that range from lanceolate to a spatulate shape, being erect or recurved, glaucous, with the presence of greenish transverse bands in some cases, with a wavy to crenate margin, with small teeth (less than 1 cm), and a short terminal spine (up to 4 cm) (Gentry 1982). Different varieties of *A. americana* and subspecies have been reported, incorporating ssp. *protoamerican* Gentry, variety *expansa* (Jacobi) Gentry, and variety *oaxacensis* Gentry, variety *marginata* Trel., variety Trel *mediopicta.* and variety *striata* Tel (Davis 2022).

3.3.3 Genetic Resources Conservation Approaches

Agaves, including *A. americana*, have cultural, economic, and ecological importance (Cabrera-Toledo et al. 2020). However, traditional propagation has caused losses of genetic variability, as well as the induction of fungal diseases in plantations, reducing the number of plants (Martinez-Rodriguez et al. 2019). Biotechnology presents strategies for short, intermediate, and long-term storage, to conserve the germplasm of species of interest.

The preservation of genetic resources can be carried out in two ways: *in situ* and/or *ex situ*, being complementary to each other, ensuring the preservation of species in the short and long term, and the maintenance and sustainable use of species in their natural habitat is part of *in situ* conservation, while *ex situ* conservation involves different techniques and specialized infrastructure that favor the recovery and survival of specimens outside their habitats (Dávila et al. 2022).

3.3.3.1 *In Situ* Conservation

In situ conservation methods include spaces free of agricultural and livestock activities where agaves are grown, the planting of offspring obtained from adult plants (Reynoso-Santos2012), the implementation of nurseries, as well as avoiding the extraction of wild specimens. (Ruiz Mondragón et al. 2023). These populations managed *in situ* are part of the Agroforestry and Forestry Systems in Mexico (Cabrera-Toledo et al. 2020). In the state of Chiapas, Mexico, plantations have been established for the propagation of *A. americana*, to conserve the germplasm and satisfy agroindustrial demand (La Torre-Ruiz et al. 2016).

3.3.3.2 *Ex Situ* Conservation

Franco-Martínez (1995) describes the *ex situ* conservation strategies of Agavaceae, highlighting germplasm banks, seed banks, botanical gardens, and nurseries, one of the most important techniques is the so-called seed banks, however, the Seed collection, processing, storage, and recovery are key factors that can hinder their efficiency and effectiveness, with cryopreservation being the most feasible alternative for *ex situ* conservation (Liu et al. 2023).

3.3.3.3 Cryopreservation

Cryopreservation is considered a cost-effective as well as safe method for long-term storage of plant germplasm. This technique subjects living tissues to extremely low temperatures by subjecting them to liquid nitrogen (−196 °C) (Martínez et al. 2022), where they are They effectively stop biochemical and metabolic processes, as well as cell division, allowing the long-term preservation of unorthodox seeds (sensitive to dehydration), plants propagated *in vitro* or vegetatively, endangered plant species, plant cell lines, etc. .

The cryopreservation of species of *A. tequilana* Weber cultivar 'Chato' and *Agave peacockii* (Delgado-Aceves et al. 2022) has recently been reported, being the first cryopreservation studies of the *Agave* genus.

3.3.3.4 *In Vitro* Conservation

The use of biotechnological techniques in agave, such as the isolation of cells, tissues or complete organs of plants and their growth under *in vitro* conditions, is an alternative for safeguarding germplasm (Delgado-Aceves et al. 2022).

In vitro storage techniques for the *Agave* genus have been reported by Pérez Molphe Balch et al. (2012), using osmotic agents (sorbitol and mannitol), to reduce the growth rate, thus prolonging the subculture time, without harming the viability and ability to regenerate complete plants.

3.3.3.5 Gene Bank

Due to the physiological characteristics of *A. americana* that allow it to survive extreme conditions, they generate interest in conserving and studying the genetic conformation, to identify genes that help confer tolerance and resistance to abiotic and abiotic factors respectively, the genetic sequences of *A. americana* currently reported are found on the GeneBank platform (http://www.ncbi.nlm.nih.gov).

3.3.4 Cytogenetics

Agaves have a high number of chromosomes and a large genome, showing different levels of ploidy, almost 70% of agave species are polyploid (Figueredo et al. 2015). Recently, different types of molecular markers for DNA genetic fingerprinting have been used to carry out population-level studies in the *Agave* genus. (Trejo et al. 2020).

Different studies have described that the *Agave* genus includes species of different ploidy levels (2x, 3x, 4x, 5x, 6x and 8x) with a basic chromosome number of $n = 30$ made up of two groups of 25 small and 5 large chromosomes in a karyotype bimodal that increase in direct proportion to ploidy (Monja-Mio et al. 2019). *A. americana*, being monocarpic, presents evident genetic variation between wild plants and those available in nurseries (Davis 2022).

3.4 Traditional Breeding

The traditional propagation of *A. americana* is carried out through seeds or offsets. However, the use of seeds is infrequently employed due to the extended time required for their acquisition (10–12 years), despite the potential for obtaining a substantial quantity of seeds from a single flowering stalk. This method also demands careful attention to prevent issues of destruction. On the other hand, propagation through offsets is the most common, with the drawback being the low number of offsets produced annually (Álvarez-Duarte et al. 2018; Trejo et al. 2020). A promising alternative to address these challenges is the application of propagation and improvement techniques through biotechnology-derived methods in plant biotechnology.

Clonal propagation can be a viable alternative for obtaining quality plants and thus cover the raw material requirements for larger-scale agricultural production of *Agave americana* L. since it requires approximately 20 years or more for the plants to grow and reach their flowering stage (Chen et al. 2014; Monja-Mio et al. 2019). In order to apply this technique, one would first have to solve the problem of the wide variation that exists within the species and select desirable characteristics for the production of a specific product. These characteristics will depend on the valuable product that is to be obtained, for example, bioethanol, secondary metabolites, and fiber, among others (Huang et al. 2019; Simpson et al. 2011). However, long selection programs are required before selecting accessions that meet the desired characteristics to produce some valuable product.

Environmental conditions and the search for improvement strategies in agave cultivation are crucial. However, developing these programs proves challenging due to the plants' lengthy life cycles.

3.4.1 Improvement Strategies

Due to the great adaptive capacity of agave species, mainly *A. americana*, to develop optimally in various environmental conditions, the search for improvement strategies in agave crops is important, however, these programs are difficult to develop due to their long duration lifecycles.

Genetic improvement of *A. americana* L. has been limited due to many factors, the most important of which is the long life cycle of the species, which can last up to 10 years to reach a mature stage (Chen et al. 2014; Simpson et al. 2011), the size of its genome, which is very large (Huang et al. 2019) and the ploidy levels which show great variation (Monja-Mio et al. 2019; Simpson et al. 2011). Plant reproduction occurs sexually and/or asexually. Another reproduction alternative is microclonal propagation through which reproductive tissue can be obtained for agricultural production systems (Chen et al. 2014; Simpson et al. 2011). The study of transcriptomes is a tool that has proven to be very useful for studies in *Agave americana* L., specifically to study and identify the expression levels of genes that are related to cellulose synthase (Huang et al. 2019), however, there are infinite.

Opportunities to study transcripts and genes that have not yet been able to be identified with current techniques and methodologies (Yang et al. 2015). As molecular techniques and methodologies advance in the near future, opportunities will imminently arise for more efficient breeding programs focused on *Agave americana* L.

3.4.2 Traditional Breeding Methodologies

Traditional improvement methods in crops involve the crossing and selection of individuals exhibiting desirable traits to combine them. However, these techniques can take extended periods, up to 10 years in horticultural or herbaceous plants, while in woody or slow-growing species like agave, developing new varieties demands a greater investment of resources and time (Chaves 2021).

Among the prominent methods, hybridization stands out, a technique based on the crossbreeding of individuals with different genotypes (Llatas et al. 2021). In agave plantations, it is common to introduce other varieties from outside the locality, generating hybridization processes between varieties to obtain agaves with intermediate characteristics (Álvarez-Ríos et al. 2020; Figueredo-Urbina et al. 2021). Cross-pollination is another method that could provide an alternative for preserving genetic diversity in agaves of the same species and variety, under controlled conditions for hybrid production through sexual reproduction (Rodríguez-Garay et al. 2018). However, there are no reports of its application in *A. americana*, presenting an opportunity for further study.

3.4.3 Traditional Breeding Limitations

Currently, *A. americana* L. is a crop that has received interest due to its multiple previously described characteristics. However, this species is not commercially produced compared to other Agave species which are commercially cultivated in Central America and from South, Africa and Asia (Davis et al. 2011; Davis et al. 2014; Davis et al. 2019), experimental crops have not been generated as in Australia with the species *Agave rigida* var. Sisalana (Holtum et al. 2010; Yan et al. 2020). Davis et al. (2017) reported an experimental field trial of an *A. americana*

L. plantation in the USA. Therefore, there is no research or fieldwork focused on the production of *Agave americana* L. crops solely in the field of ornamental landscaping.

Added to the above is that, as in the entire *Agave* genus, harvesting plants before their sexual reproduction stage is a risk factor that prevents the production of pollen and, consequently, seeds in these plants, this has an impact on gene flow, that is, a low genetic variation of this genus of plants. Similarly, with this practice, sexual reproduction in this genus is very low, which represents a risk to the integrity of the agave populations (Félix-Valdez et al. 2016).

3.4.4 Role of Biotechnology

Due to its extensive application as an additive in animal feed, fertilization, and the extraction of fibers from the leaves of *A. americana* (Ortega et al. 2019; Thamae and Baillie 2007), coupled with the use of synthetic agrochemicals such as insecticides, bactericides, and fungicides to enhance crop yield, the safest and most cost-effective method for environmental control is the employment of resistant cultivars (Reyes-Zambrano et al. 2021). This involves the implementation of genetic improvement programs to generate superior genotypes/phenotypes, which necessitate the evaluation of both phenotypes and genotypes, as well as the selection and crossbreeding of superior phenotypes and the assessment of the obtained individual (Botero and Arias 2018; Dempewolf et al. 2017).

3.5 Molecular Breeding

3.5.1 Molecular Marker-Assisted Breeding

It is necessary to carry out alternative strategies which benefit and contribute to the improvement of Agave species. This family of plants has several species which are considered of high economic and ecological value. However, currently, there are very few programs or initiatives focused on the improvement and selection of outstanding agronomic characteristics, this is probably due to various factors, such as the life cycle of the plant, its morphology as well as its monocarpic nature. (Bautista-Montes et al. 2022). For this reason, breeding programs through crossing have been limited. The use of molecular biology tools, such as molecular genetic markers and genetic engineering, have been used for the improvement of many plant varieties. Molecular breeding approaches target specific regions of DNA and are therefore called marker-assisted breeding (MAB). MAB involves numerous modern plant breeding strategies, including marker-assisted selection (MAS), marker-assisted backcrossing (MABC), marker-assisted recurrent selection (MARS) and genome-wide selection. (GWS) or genomic selection (GS) (Bapela et al. 2023).

MAS allows you to speed up the playback process. It has been applied to various species of crops of agricultural interest. MAS involves the transfer of a DNA fragment associated with a specific phenotype, using molecular markers to identify the DNA fragment and define its location on the chromosome. Once the association between gene and phenotype is established, then the marked region can be transferred to successive generations without having to rely on generating the phenotype again. The classification of molecular markers was ordered in protein-based markers (such as isoenzymes) and DNA-based markers; the latter include restriction fragment length polymorphisms (RFLP), single nucleotide polymorphisms (SNP), amplified fragment length polymorphisms (AFLP) and simple sequence repeats (SSR) (Rauf et al. 2016).

3.5.2 Genomics

Genomics is a powerful tool to study and gain an understanding of the organization and function of biological systems, and it also allows tracking changes and molecular mechanisms under different conditions such as biotic stress, abiotic stress, and changes in plant physiology with those conditions (Yang and Guo 2018).

The complete genome of *Agave americana* L. is not yet reported, however, there are some sequences reported on the GeneBank platform (http://www.ncbi.nlm.nih.gov). In Mexico, in 2023, the Secretary of Agriculture and Rural Development (SADER) published the sequencing of the complete genome of *Agave tequilana* Weber var. blue, however, it has not yet been released.

3.5.3 Bioinformatics

Bioinformatics can help enhance genetic and genomic selection to determine the optimal combination of genotypes that will produce a desired phenotype (pathogen resistance, higher yield and quality), and can help accelerate the isolation of these new varieties. Bioinformatics is also essential for collecting and processing plant phenotypes and facilitating plant breeding (Mu et al. 2022).

Bioinformatics and genomics are tools that help us better understand the molecular mechanisms of plant tolerance to different stress conditions and can help generate new plant varieties in addition to improving the quality of crops. In the *Agave* genus, these technologies have not yet been used, although it would be of great help to improve this crop to improve the production of alcoholic beverages, inulin, fiber, or biofuel.

3.6 Tissue Culture Applications

3.6.1 Micropropagation

In *Agave americana* L., certain factors hinder mass multiplication through conventional methods, such as a long growth time, low rate of asexual reproduction, and limited obtaining of plants through sexual reproduction caused by a shortage of pollen, which affects the pollination of plants in the flowering stage, and low seed viability. *In vitro* culture can be employed to preserve and/or propagate plant species when methods based on seed germination are not viable (Chen et al. 2014; Lecona-Guzmán et al. 2017; Reyes-Zambrano et al. 2016).

Micropropagation offers several advantages directly applicable to the commercial production of Agave, including the mass production of seedlings in a short period, and pathogen-free micropropagated plants, thus reducing diseases from spreading other plants or crops and obtaining uniform plantations in terms of size and age. This results in homogeneous plantations that accelerate and improve the harvesting stage. *In vitro* micropropagation of *Agave americana* L. is an alternative that can overcome the limited availability of raw materials and lead to the development of protocols for the micropropagation of this species. Successful reports of *in vitro* micropropagation in this species are documented (Table 3.1).

In vitro micropropagation in *Agave americana* L. is a relatively recent method that began around 10 years ago. Among the published reports, the indirect organogenesis pathway has yielded the best results. Reyes-Zambrano et al. (2016), and Lecona-Guzmán et al. (2017), reported 74 and 71 shoots per explant, respectively using 2.4-D and BA as growth regulators (Fig. 3.4)

Table 3.1 *In vitro* Culture in *Agave americana* L

Explant	Hormone	Answer	Reference
Basal stem	Benzyladenine (BA)	Direct organogenesis 18.5 shoots/explant	Chen et al. (2014)
Hypocotyls obtained from seeds germinated *in vitro*	Glutamine, Asparagine and 2,4- Diclorofenoxyacetic acid (2,4-D)	Somatic embryogenesis	Oliva et al. (2014)
Meristem	BA combined with 2.26 μM 2,4-D	Indirect organogenesis 74 shoots/explant	Reyes-Zambrano et al. (2016)
Meristem	2,4-D and BA	Indirect organogenesis 71 shoots/explant	Lecona-Guzmán et al. (2017)
Meristem	2,4-D, Dicamba and Picloram	Direct and indirect somatic embryogenesis	Naziri et al. (2019)

Fig. 3.4 Indirect organogenesis of *A. americana* L. (**a**) Callus induced from meristems in MS medium supplemented with 0.5 mg/L 2–4, D and 9 mg/L of BAP, (**b**) Shoot formation at 90 days of induction, (**c**) Shoots after 120 days of induction, (**d**) Agave americana L. plantlets with roots 6-month-old. (Photos by S.J. Reyes-Zambrano and C.A. Lecona-Guzmán)

In the case of somatic embryogenesis, Oliva et al. (2014), reported the use of RITA bioreactors to increase the number of embryos using amino acids and 2.4-D. However, this report does not progress to the acclimatization stage. The most recent report on somatic embryogenesis is by Naziri et al. (2019), who achieved both direct and indirect embryogenesis using 2.4-D, dicamba, and picloram, reaching the rooting and acclimatization stage of plants. However, they did not specify the number of embryos obtained. These results suggest that obtaining shoots through the indirect organogenesis route in *Agave americana* L. depends on various factors such as the type of regulator used for induction, type of explant and culture medium used (Reyes-Zambrano et al. 2016). Currently, there are highly efficient protocols for the micropropagation of American agave, which can be used in genetic improvement programs of the species for the massive propagation of elite plants, which have important characteristics such as the production of alcohol, secondary metabolites, or processing of fibers (Lecona-Guzmán et al. 2017).

3.6.2 *Embryo Rescue*

Embryo rescue (ER) is one of the oldest and most successful techniques *in vitro* tissue culture for plant breeding. This technique is used to rescue and/or protect weak, immature/mature embryos and hybrid embryos, generated from interspecific and intergeneric crosses, to promote their development and growth until they become complete plants (Rogo et al. 2023).

With this technique, early rescue can be carried out (3–5 days after pollination) to cultivate immature seeds, flowers, or siliques (Smith 2013). These embryos can be dissected directly from the seed and grown in a suitable medium for germination and development of a complete plant. Embryo rescue allows the time for the next generation to be reduced since the rescued embryos become directly into seedlings (Lulsdorf et al. 2014).

In the Agave genus, this technique has not been developed because this crop only blooms once in its life and takes 10–12 years to bloom, and reaching flowering means the death of the plant. Therefore, to obtain plant material, other biotechnological techniques have been used, such as the cultivation of plant tissues.

3.6.3 In Vitro *Pollination*

In vitro pollination refers to the penetration of the pollen tube into the embryo sac by means other than the normal in situ process (Stewart 1981). In the *in vitro* pollination technique, plant reproductive cells (female and male) are separated and fused in a controlled environment to develop a zygotic embryo. This technique has been applied for *in vitro* pollination and flowering in several crops (Cardoso et al. 2017; Ewald and Ulrich 2012; Popielarska 2005). In *Agave americana* L. and no other species of this genus, this technique has been developed due to the problems that this crop has in its sexual reproduction.

3.6.4 *Synthetic Seeds*

Synthetic seeds are artificially encapsulated explants (embryos, propagules) derived from any actively meristematic tissue that can be manipulated and used like a real seed for transportation, storage, and planting. Eventually, it can develop into a complete plant, either *in vitro* or *ex vitro* (Nongdam 2016). This technology is promising as it offers several advantages, such as cost reduction, ease of handling, germplasm conservation, preservation, and multiplication of plants to maintain desirable and superior characteristics (Abbas et al. 2022).

To date, there are no reports on encapsulation for obtaining synthetic seeds of *A. americana* L. This may be due to the absence of an efficient protocol to produce somatic embryos. The only successful report on the encapsulation of synthetic seeds is for *Agave angustifolia* (HAW), where somatic embryos were encapsulated with 3% sodium alginate, in MS medium at 100%, 4.4 μmol of BAP, and immersed for 15 or 30 min in calcium chloride. This process led to regeneration and adaptation under ex vitro conditions, with the plants showing no morphological variations (Arzate-Fernández et al. 2016).

3.7 Genetic Engineering and Gene Editing

3.7.1 Genetic Transformation Methods

The development of *in vitro* regeneration protocols, either through organogenesis and/or somatic embryogenesis, represents the basis for the establishment of genetic transformation protocols. This technique can be a viable strategy to improve agronomic traits in plants such as greater productivity as well as tolerance or resistance to abiotic or biotic stress, gene introduction and the production of secondary metabolites of biotechnological interest.

Although in the *Agave americana* species, no methodologies have been developed for the development of genetic transformation protocols, in recent years transformation protocols have been developed in other Agave species. One of them is the patent for the genetic transformation of agave through bio-ballistics, *Agrobacterium tumefaciens* and *Agrobacterium rhizogenes* developed by Cabrera-Ponce (2022), in this protocol embryogenic calluses were bombarded with tungsten particles covered with plasmid DNA which contained marker genes. Subsequently, the embryogenic callus was placed in coculture with the bacteria for 48 h and then transferred to a selective medium to obtain transformed cells. The *PPT/Bar* and *hpt* genes (which confer resistance to phosphinothricin and hygromycin respectively) were used as selectable markers and the *uidA* (β-glucuronidase) gene was used as a reporter gene. Using these protocols, plants resistant to herbicides or antibiotics were obtained.

Flores-Benítez et al. (2007) reported genetic transformation in *Agave salmiana* through coculture with *Agrobacterium tumefaciens* and the bioballistics technique. The gene used as a reporter in both cases was *uidA* and the gene *nptII* (neomycin phosphotransferase II) and bar were used as selectable markers for *A. tumefaciens* or bioballistics-mediated transformation methods, respectively. The method that showed the most effectiveness was co-culture with *Agrobacterium tumefaciens* with which 32 transgenic plants were obtained, which were obtained from callus, the transformation efficiency was 2.7%. The transgenes were detected in 11-month-old plants. On the other hand, with the bioballistics protocol, a transgenic callus was obtained that revealed positive expression in the GUS gene assay after 14 months in a selective medium.

Transformed roots have also been obtained from explants of roots, leaves and stems of *Agave salmiana* through genetic transformation via *A. rhizogenes* A4, according to Rodríguez et al. (2007) *in vitro* plants of *A. salmiana* were inoculated with different concentrations of acetosyringone and bacteria, in this protocol leaf was the tissue in which the best response was observed, generating 63% of transformed roots when 200 μM acetosyringone and 1 × 10^9 bacteria mL were used. The *nptII* gene was used as a selectable marker and *uidA* as a reporter gene. In this protocol, a transformation rate of 80% of the tissues was obtained for the reporter gene and 60% for the selectable marker.

Gao et al. (2014) reported the obtaining of transgenic plants of Hybrid Agave 11,648 which presented greater resistance to *P. nicotianae* through a strategy of expression of hevein-like peptides in callus of Hybrid Agave 11,648. The medium for callus induction was SH, supplemented with 0.45 μM 2,4-D, 2.68 μM NAA and 13.2 μM BA. The shoot regeneration medium was SH, supplemented with 2.68 μM NAA and 6.66 μM BA. The effective infection time was 10 min and a concentration of 200 μM acetosyringone was used. The optimal time for callus pre-culture was 3 days and the optimal co-culture time was 4 days. Thirty-seven lines were obtained from 150 explants and the hevein-like gene was expressed in seven lines.

Gutiérrez-Aguilar et al. (2014) reported a methodology for genetic transformation for *A. tequilana* and *Agave desmettiana* via *A. tumefaciens*. The explants used were meristems and co-cultured with *Agrobacterium tumefaciens* strains LBA4404 and GV2260 using phosphinothricin (PPT/Bar) as a selective agent. These authors report that in *A. desmettiana* 2–20 shoots were obtained while in *A. tequilana* 1–2 shoots.

Although in the species *A. americana* there are no genetic transformation protocols mediated by *Agrobacterium* and/or bioballistics, there has been success in the different transformation protocols reported in other Agave species; however, these methodologies are complicated although they can be applied to any species of the Agave gender, require a lot of time as well as having an efficient and reproducible regeneration protocol.

3.7.2 Gene Editing Methodologies

Gene targeting technology in plants is based on donor-dependent homology-directed repair (HDR), which allows precise editing of the genome by introducing insertions, sequence replacements, and nucleotide substitutions (Atkins and Voytas 2020). However, the low editing efficiency achieved with HDR has limited its application in plants (Atkins and Voytas 2020; Huang et al. 2019). Deaminase- and reverse transcriptase-mediated primary editing technologies are alternative genome editing technologies; As they do not involve DNA double-strand break (DSB) formation and do not require donor DNA, these CRISPR-Cas-based tools induce precise sequence editing and are more efficient than HDR in plants. Following the development of the cytosine base editor (CBE) and the adenine base editor (ABE) in human cells, a dual base editor and precise DNA removal strategies derived from cell editing were developed for the first time in plant bases. Although CRISPR-Cas has demonstrated a great ability to improve crops, combining it with conventional breeding methods would benefit agricultural production even more. Several novel breeding approaches targeting reproduction-related genes using CRISPR-Cas have recently emerged. This technology has not been applied to the *Agave* genus, which could potentially enhance the biotechnological applications that can be given to the species of this genus, particularly in *Agave americana*, which can be used in the production of agaves, fiber and/or biofuel.

3.7.3 Nanotechnology

Nanotechnology is an important field that is based on the application and design of nanomaterials. Nanoparticles typically range from 1 to 100 nm. The main interest in nanoparticles is due to their different size, morphophonological area-volume ratio, and properties that are used in various fields, including food, medicine, and health (Rodríguez-Félix et al. 2021).

The applications of nanotechnology are promising in agriculture, since some nanomaterials can induce favorable effects in crops, such as better development, higher yield, and an increase in their nutritional content (Del Buono et al. 2021). In this sense, extracts from Agave Americana leaves have been used for the synthesis of silver nanoparticles (Ahmad et al. 2016), magnetic/cellulose nanocomposite (Majamo et al. 2024), and leaf extracts from other Agave species such as *A. potatorum* for the synthesis of gold nanoparticles (Moreno-Luna et al. 2019).

3.8 Mutation Breeding

The domestication of wild ancestors was what began genetic improvement in plants (Doebley 2004; Xiao et al. 2021). The main or primary source of genetic variations in plants is mutations (Jayaramachandran et al. 2020), described as hereditary changes in DNA that are eventually reflected in the phenotype (Singh et al. 2021).

It is a practical method for generating genetic diversity in different plant species (Nasri et al. 2021) and acquiring new biological functions (Mukai et al. 2017; Shelake et al. 2019). Tissue culture plays a crucial role in plant genetic improvement, facilitating the use of *in vitro* mutagenesis (Kashtwari et al. 2018).

3.8.1 Mutagenesis **In Vitro** *and Selection*

Mutations can be induced through chemical means (salicylic acid, 5-azacytidine, sodium azide, ethyl methanesulfonate, among others) (Masoabi et al. 2018) or physical methods (gamma rays, ultraviolet light). Both methods allow for the handling of large populations, increasing the efficiency of mutation induction and the subsequent recovery of mutant plants (Kashtwari et al. 2018).

In vitro mutagenesis techniques have been employed for the development of mutant plants with desirable traits, including disease resistance, early maturity, and tolerance to salt and drought (Dalvi et al. 2021; Patade et al. 2008). Unlike physical mutagens, chemical mutagens have more specific capabilities in generating mutations (Altindal 2019), with ethyl methanesulfonate being the most widely used chemical mutagen for this purpose (Corazón-Guivin et al. 2022; Krupa-Małkiewicz et al. 2017).

3.8.2 Ethyl Methanesulfonate (EMS)

Ethyl methanesulfonate (EMS) is the chemical mutagen that has been most used in plants because it causes metabolic and biological changes in plant cells. These changes can be reflected in an increase in the production of secondary metabolites, resistance to different types of biotic stress and/or tolerance to abiotic stress, as well as the generation of new varieties (Mallick et al. 2016; Uchida et al. 2011).

Ethyl methanesulfonate (EMS) has proven highly effective in generating point mutations (Nasri et al. 2021) with low frequencies of chromosomal aberrations (Masoabi et al. 2018) and has been employed to induce modifications in plants (Li et al. 2019b; Lu et al. 2015). Plants regenerated through mutations induced by EMS have exhibited alterations in their oil content and fatty acid composition across various crops (Lee et al. 2018). Additionally, an improvement in drought tolerance has been reported in some species (Sala et al. 1990; Witt et al. 2018).

EMS induces point mutations by changing cytosine to thymine, resulting in guanine/cytosine to adenine/thymine transitions (Krieg 1963; Li et al. 2019a, 2019b). It has also been reported to cause structural changes, insertions, and deletions of chromosomal segments (Li et al. 2019a, 2019b).

3.8.3 **In Vitro** *Mutagenesis in the Agave Genus*

Agaves serve as a source for various products in the food and pharmaceutical industries, as well as in biofuel production, among other applications (Rodríguez-Garay 2016). Their primary use is in the production of alcoholic beverages such as pulque, raicilla, tequila, mezcal, bacanora, and comiteco (Lecona-Guzmán et al. 2017; Martinez-Rodriguez et al. 2019; Nava-Cruz et al. 2015; Reyes-Zambrano et al. 2016).

Even though Agave is considered a crop of economic and agronomic importance, with limited natural regeneration, low genetic diversity, and recently a greater attraction for its antimicrobial, antioxidant, anti-inflammatory, antidiabetic, and anticancer attributes, among others, very little has been reported regarding genetic improvement of Agave species through the application of *in vitro* mutagenesis (Table 3.2)

Table 3.2 *In vitro* mutagenesis of *Agave* genus and other species

Especies	Agent mutagenic	Tissue	Answer	Reference
A. tequilana Weber	Co^{60} gamma rays	Vitroplants	Activation of its resistance mechanism against *Cercospora agavicola*	Ángeles-Espino et al. (2016)
Saccharum spp	Ethyl Methanesulfonate	Callus	Drought tolerance	Masoabi et al. (2018)
Aloe vera L.	Colchicine	Vitroplants with roots	Polyploid cells	Motero-Paredes et al. (2018)
A. americana L.	Ethyl Methanesulfonate	Callus	Dwarfism, different leaf shapes, increase in fructans and increase in PAL activity	Reyes-Zambrano et al. (2019)
Saccharum (cv. C-1051-73)	Sodium azide	Explants	Differences in the content of secondary metabolites (phenols, carotenoids and aldehydes) and chlorophyll	Gómez et al. (2019)
A. tequilana	Co^{60} gamma rays	Vitroplants	Somatic embryos	Espino et al. (2020)
Zingiber officinale Rosc.	Co^{60} gamma rays	Shoots	Resistance to yellow disease	Sharma et al. (2020)
Saccharum spp. hybrids	Ethyl Methanesulfonate	Callus	Agronomic improvements (early maturity, high sucrose content) and resistance to smut disease	Dalvi et al. (2021)
Musa ssp	Ethyl Methanesulfonate	Shoots	Cold resistance	Liu et al. (2024)

Díaz-Martínez et al. (2012) describes that the regeneration process for *Agave tequilana* weber var. Azul from different types of explants could be used to obtain information about the expression of genes that are obtained after mutagenesis processes, this information can be used for the silencing of genes that generate an undesirable response in plants obtained through mutagenic processes. Tejavathi et al. (2012) report that plants obtained from direct and indirect organogenesis in agave vera-cruz Miller, present genetic variation independently of the regeneration pathway. These plants were cultivated with arbuscular mycorrhizal fungi (AMF) and exposed to gamma radiation at 15Gy for 20 min at 2-min periods (physical mutagen) and ethyl methanesulfonate ranging from 0.1% to 1% for 1–5 h (chemical mutagen). They observed that a phenotypic variant which was obtained through indirect organogenesis presented a higher level of polymorphism compared to plants treated with mutagens. The plants that were obtained through direct organogenesis and that were subsequently treated with HMA were grouped in a separate subgroup. These results indicate genetic differences, which reflects that there is genetic variation in the plants that are treated with HMA, which suggests that different routes should be used. In vitro regeneration in combination with mutagenesis and symbiosis with AMF is a tool with which the genetic variability of agave plants can be increased.

On the other hand, Reyes-Zambrano et al. (2019) report that plants mutagenized with ethyl methanesulfonate cause changes or variations in morphological and morphometric parameters in shoots obtained *in vitro* from *Agave americana* L. (Fig. 3.5). In this work, dwarfism was reported in 60% of the regenerated shoots as well as different leaf shapes, which were characterized by lack of spines. In this

Fig. 3.5 *In vitro* mutagenesis of *Agave americana* L. callus obtained through chemical mutagenesis using Ethyl Methanesulfonate (EMS). (**a**) 15 mM exposed for 2 h, (**b**) 15 mM exposed for 4 h, (**c**) 30 mM exposed for 2 h, (**d**) 30 mM exposed for 4 h. (Photos by S.J. Reyes-Zambrano and C.A. Lecona-Guzmán)

same work, it is also reported that 30% of the shoots presented a higher fructan content compared to the control treatment shoots, as well as an increase in PAL activity.

Studies conducted by Espino et al. (2020) describe the generation of somatic embryos and mutant shoots regenerated from explants irradiated with Co^{60} gamma rays. They state that the radiation-induced mutations that altered competition in cells, reduced embryogenic callus and somatic embryo development. Meanwhile, Reyes-Zambrano et al. (2020) report that infection of *Agave americana* L. plants with *Fusarium oxysporum* significantly increases the activity of the chitinase enzyme 15 after infection compared to uninfected plants, while the enzymatic activity of β-1, 3 glucanase does not present statistically significant differences with the control plants. These results suggest that *Agave americana* L. plants respond to *Fusarium oxysporum* infection by activating chitinase-type resistance proteins (RP).

Recently, Reyes-Zambrano et al. (2021) reported the generation of somaclonal variants in *Agave americana* L., resistant *to Fusarium oxysporum*, as a result of mutation induction with EMS at a concentration of 200 mM over an exposure time of 2 h (Fig. 3.5)

3.8.4 Molecular Markers and Somaclonal Variation

Molecular markers consist of DNA segments associated with a part of the genome (Carrodeguas-Gonzalez and Zúñiga-Orozco 2021; Nadeem et al. 2018). They are employed in various fields such as taxonomy, genetic engineering, and plant breeding to detect DNA-level alterations or variations in plant material subjected to mutagenic treatment (Pardo et al. 2015; Kamińska et al. 2020).

It has been reported that molecular markers have been utilized in phylogenetic studies due to their ability to provide high levels of polymorphism (Abu-Ellail et al.2021), helping to determine genetic variation among plants (Bychappa et al. 2019).

The implementation of DNA markers contributes to the improvement of various crops of agronomic importance (Garrido-Cardenas et al. 2018). Notable among these are intersimple sequence repeats (ISSR), randomly amplified polymorphic DNA (RAPD), and direct amplification of minisatellite DNA (DAMD), which have been used in assessing clonal variability in many species (Bhattacharyya et al. 2018; Devi et al. 2013).

Pandian et al. (2018) determined genetic diversity in exotic and local millet genotypes using three different dominant markers (ISSR, RAPD, and DAMD), affirming their effectiveness and consistency. Villarreal-Fuentes et al. (2020) described the capability of SRAP molecular markers (Sequence-Related Amplified Polymorphisms) to assess the genetic diversity of eighty-eight soursop (*Annona muricata* L.) genotypes, demonstrating the utility and reliability of these markers for evaluating genetic diversity parameters. Recently, Gao et al. (2020) analyzed the genetic diversity of 12 phenotypic traits and 11 Sequence-Related Amplified

Polymorphism (SRAP) markers, reporting that the clustering of SRAP molecular markers showed the most accurate and reliable genetic relationship. Reyes-Zambrano et al. (2021) reported the use of RAPD, ISSR, and DAMD molecular markers to assess genetic variation caused by EMS-induced mutations in *A. americana* plants.

Somaclonal variation, as described by Larkin and Scowcroft (1981), refers to irregularities induced in plants, which can become genetic anomalies and can be inherited (Chawla 2011; Siddique and Yadav 2021). It is a phenomenon observed in cells and tissues *in vitro* (Pawełkowicz et al. 2021).

It is essential and of great importance for genetic improvement and the development of new cultivars, inducing morphological, and physiological changes, and resistance or tolerance to biotic and abiotic stress (Cao and Deng 2020).

Since somaclonal variation involves mutations, it is necessary to use techniques to detect mutations in crops, such as the use of molecular markers, due to their high precision (Bayati et al. 2021). It has been reported that EMS, as a chemical agent inducing mutations in crops, is an alternative for generating variability (Gómez et al. 2019).

3.9 Hybridization

3.9.1 Conventional Hybridization

Genetic variability is very important in the improvement of agricultural crops and can be generated through hybridization. After hybridization, the segregated progenies and advancing populations are managed, and the desired segregants are selected by various breeding methods, employing various selection procedures. These methods are pedigree, bulk, and single seed progeny and can be used alone or two methods can be combined.

In the *Agave Americana* species there are no developed hydrides, in the Agave genus there are very few individuals obtained by the hybridization technique, probably due to the long time that this genus takes to reach a flowering stage to carry out crosses between individuals.

In this sense, in the *Agave* genus, the hydride 1H11648 was reported. This hybrid, named after the date on which it was released (June 11, 1948), is the result of crossing *Agave angustifolia* x A. *amaniensis* Trel and Nowel plants and later backcrossed with *A. amaniensis*. The fiber content per leaf of this hybrid is greater than *Agave fourcroydes* and it is more productive in terms of the number of leaves per year obtained per plant than henequen, and the leaves lack marginal spines, making them easier to collect. Unfortunately, H11648 is badly affected by several fungal infections: Korogwe leaf spot, zebra leaf rot, and trunk rot. H11648 has since been the main hybrid planted on thousands of hectares in Tanzania, Kenya, and other parts of the world.

3.10 Conclusions and Prospects

Agave americana L. is a potential source for obtaining various products ranging from functional foods, fiber production, reinforced composite materials, paper manufacturing, textiles, and biofuels. However, the current state of research related to the cultivation of *Agave americana* L. While information has been reported to improve agronomic aspects such as nutrition, health, and phytochemical production, there is still a need to investigate how the morphological, morphometric, physiological, and biochemical aspects of plants are affected by biotic and abiotic factors interacting with the field cultivation. The most important biotic factors include infection with pathogenic fungi and bacteria, mainly Fusarium, nematodes, and herbivores. Abiotic factors include water scarcity, soil salinity and pH, and ambient temperature. It is necessary to employ omics techniques such as genomics, transcriptomics, proteomics, and metabolomics to generate information that can be used to enhance the profitability of cultivation.

Appendix 3.1 Research Institutes and Online Resources Available for American Century (*Agave americana* L.) Breeding

Institution	Area of specialization	Research activities	Website
National Technological of Mexico-Tuxtla Gutierrez Campus	Vegetal Biotechnology	Tissue culture, mutagenesis *in vitro*, micropropagation	https://www.tuxtla.tecnm.mx
Center for Research and Assistance in Technology and Design of the State of Jalisco A.C. (CIATEJ)	Biotechnology	Plant, Industrial, Medical, and Pharmaceutical Biotechnology, Environmental and Food Technology	https://www.ciatej.mx
National Institute of Research Forestry, Agricultural and Livestock	Agriculture and Forestry	Plant breeding, Production, Postharvest, Technology	https://www.gob.mx/inifap

Appendix 3.2 Genetic Resources for American Century (*Agave americana* L.) Breeding

Species	Important traits	Cultivation location
National Technological of Mexico-Tuxtla Gutierrez Campus	Resistance to *Fusarium oxysporum*	Tuxtla Gutiérrez, Chiapas, México

References

Aydin L, Artem HS, Oterkus (2020) Designing Engineering Structures using Stochastic Optimization Methods (1st ed.). CRC Press. https://doi.org/10.1201/9780429289576

Ahmad B, Shireen F, Bashir S et al (2016) Green synthesis, characterisation and biological evaluation of AgNPs using *Agave americana*, *Mentha spicata* and *Mangifera indica* aqueous leaves extract. IET Nanobiotechnol 10(5):281–287

Abbas MK, Mahood HE, Alhasan AS (2022) Production of synthetic seeds in vegetable crops: A review. IOP Conf. Ser. Earth Environ Sci 1060:012099. https://iopscience.iop.org/article/10.1088/1755-1315/1060/1/012099

Altindal N (2019) Molecular characterization of *Helianthus tuberosus* L. treated with ethyl methanesulfonate based on inter-simple sequence repeat markers. Inter. J Environ Sci Technol 16(9):5311–5318. https://doi.org/10.1007/s13762-019-02486-1

Abu-Ellail FF, El-Taib ABA, Hefny YAM (2021) Stability analysis and performance of promising sugarcane varieties for yield and quality traits. Electron J Plant Breed 12(3):623–636

Álvarez-Duarte MC, Garcia-Moya E, Suárez-Espinosa J et al (2018) Traditional knowledge, cultivation and use of maguey pulquero in municipalities of Puebla and Tlaxcala. Polibotanica 45:205–222. https://doi.org/10.18387/polibotanica.45.15

Álvarez-Ríos GD, Pacheco-Torres F, Figueredo-Urbina CJ, Casas A (2020) Management, morphological and genetic diversity of domesticated agaves in Michoacán, México. J Ethnobiol Ethnomed 16(3):1–17. https://doi.org/10.1186/s13002-020-0353-9

Ammann J, Umstätter C, El Benni N (2022) The adoption of precision agriculture enabling technologies in Swiss outdoor vegetable production: a Delphi study. Precision Agric 23: 1354–1374. https://doi.org/10.1007/s11119-022-09889-0

Ángeles-Espino A, Virgen-Calleros G, Valencia-Botín J et al (2016) Nondestructive evaluation of pathogenicity of *Cercospora agavicola* on agave azul tequilero plantlets irradiated with gamma rays Co60. Acta Universitaria 26(6):3–10

Arzate-Fernández AM, Piña-Escutia JL, Norman-Mondragón TH et al (2016) Regeneración de agave mezcalero (*Agave angustifolia* Haw.) a partir de embriones somáticos encapsulados. Rev Fitotec Mex 39(4):359–366

Atkins PA, Voytas DF (2020) Overcoming bottlenecks in plant gene editing. Curr Opin Plant Biol 54:79–84

Bailey GE (1915) Vertical farming. Lord Baltimore Press. In: Bailey GE (ed) Vertical farming. E. I. Du Pont de Nemours Powder Co., Wilmington

Bapela T, Shimelis H, Terefe T et al (2023) Breeding wheat for powdery Mildew resistance: genetic resources and methodologies—a review. Agronomy 13:1173. https://doi.org/10.3390/agronomy13041173

Bassoi LH, Inamasu RY, Bernardi AC et al (2019) Agricultura de precisão e agricultura digital. TECCOGS: Revista Digital de Tecnologias Cognitivas 20. https://doi.org/10.23925/1984-3585.2019i20p17-36

Bautista-Montes E, Hernández-Soriano L, Simpson J (2022) Advances in the micropropagation and genetic transformation of agave species. Plan Theory 11:1757. https://doi.org/10.3390/plants11131757

Bayati E, Gomarian M, Mirzaie-Nodousha H et al (2021) Producing a superior genotype from agria potato cultivar using somaclonal variation. Nexo Revista Científica 34(02):671–681. https://doi.org/10.5377/nexo.v34i02.11551

Bezazi A, Belaadi A, Bourchak M et al (2014) Novel extraction techniques, chemical and mechanical characterisation of *Agave americana* L. natural fibres. Compos B Eng 66(19): 194–203. https://doi.org/10.1016/j.compositesb.2014.05.014

Bhattacharyya P, Kumar V, Van Staden J (2018) *In vitro* encapsulation based short term storage and assessment of genetic homogeneity in regenerated *Ansellia africana* (*Leopard orchid*) using gene targeted molecular markers. PCTOC 133(2):299–310. https://doi.org/10.1007/s11240-018-1382-0

Botero K, Arias T (2018) Uso de las ciencias ómicas para el mejoramiento genético de cultivos. Rev Mexicana Cienc Agric 35(2):64–78

Buschle-Diller G, Zeronian SH, Pan N, Yoon MY (1994) Enzymatic hydrolysis of cotton, linen, ramie and viscose rayon fabrics. Text Res J 64:270–279

Bychappa M, Mishra MK, Jingade P, Huded AK (2019) Genomic alterations in coding region of tissue culture plants of *Coffea arabica* obtained through somatic embryogenesis revealed by molecular markers. PCTOC 139(1):91–103. https://doi.org/10.1007/s11240-019-01666-8

CABI (2022) Invasive species compendium. Centre for Agriculture and Bioscience International. Available online: https://www.cabi.org/. Accessed on 16 May 2022

Cabrera-Ponce JL (2022) Transformación Genética En El Género Agave Y Producción de Plantas Transgénicas Resistentes a Herbicidas. Available online: https://svt.cinvestav.mx/Portals/svt/descarga/Fichas/Agave%20tequilero-MXJL02000044A.pdf

Cabrera-Toledo D, Vargas-Ponce O, Ascencio-Ramírez S et al (2020) Morphological and genetic variation in monocultures, forestry systems and wild populations of *Agave maximiliana* of western Mexico: implications for its conservation. Front Plant Sci 11:817. https://doi.org/10.3389/fpls.2020.00817

Cardoso JC, Costa BT, de Almeida EJ (2017) Pollination and *in vitro* germination of seeds for interspecific hybridization of *Psidium guajava* and *Psidium cattleianum*. Euphytica 213:146. https://doi.org/10.1007/s10681-017-1929-x

Cao Z, Deng Z (2020) Morphological, cytological and molecular marker analyzes of 'Tapestry' caladium variants reveal diverse genetic changes and enable association of leaf coloration pattern loci with molecular markers. Plant Cell Tiss Organ Cult 143:363–375. https://doi.org/10.1007/s11240-020-01922-2

Carrillo-Riofrío FM, Segovia-Cáceres SM, Jijon-Paredes EM (2021) La Innovación en la Agricultura Digital. Domino de las Ciencias 7(3):1652–1658. https://doi.org/10.23857/dc.v7i3.2127

Carrodeguas-Gonzalez A, Zúñiga-Orozco A (2021) Aplicación de herramientas moleculares para el mejoramiento genético de pasturas. Pastos y Forrajes 44:e21

Chaabouni Y, Drean J, Msahli S, Sakli F (2006) Morphological characterization of individual fiber of *Agave americana* L. Text Res J 76(5):367–374

Chattopadhyay D, Khan J (2012) *Agave Americana*: a new source of textile fibre. Colourage 6:33–36

Chaves JP (2021) Biotecnología vegetal: mejoramiento de cultivos ante el cambio climático. Investiga TEC 14(42):3–5

Chawla H (2011) Introduction to plant biotechnology. CRC Press. https://doi.org/10.1201/9781315275369

Chen Y, Chen X, Hu F et al (2014) Micropropagation of *Agave americana*. Hort Sci 49(3): 320–327. https://doi.org/10.21273/HORTSCI.49.3.320

Cline HF (1987) El episodio del henequén en Yucatán. Secuencia. Revista de historia y ciencias sociales 08:186–186. https://doi.org/10.18234/secuencia.v0i08.185

Corazón-Guivin M, Arévalo-Rojas M, Acosta-Córdoba R et al (2022) Determinación de la DL_{50} de Metanosulfonato de Etilo (EMS) para la inducción de cambios morfológicos y fisiológicos en plántulas de Plukenetia volubilis. Rev Agrotec Amaz 2(1):209–e209. https://doi.org/10.51252/raa.v2i1.209

Coria-Contreras J, Mora-Aguilera G, Yáñez-Morales MDJ et al (2019) Epidemiología regional aplicada a la caracterización inductiva y pronóstico de la mancha gris del agave azul (*Cercospora agavicola*) en Jalisco, México. Rev Mex Fis 37(1):71–94. https://doi.org/10.18781/r.mex.fit.1809-4

Dalvi SG, Tawar PN, Suprasanna P et al (2021) EMS-based *In vitro* mutagenesis and mutant screening for smut resistance with agronomic traits in sugarcane. Sugar Tech 23(4):854–864. https://doi.org/10.1007/s12355-020-00931-x

Dávila P, Soto-Trejo F, Rodríguez-Arévalo I et al (2022) Wild plant conservation in Mexico in the 21st century. Bot Sci 100:175–197. https://doi.org/10.17129/botsci.3066

Davis S (2022) *Agave americana*: characteristics and potential breeding priorities. Plan Theory 11(17):2305. https://doi.org/10.3390/plants11172305

Davis S, Ortiz-Cano H (2023) Lessons from the history of *Agave*: ecological and cultural context for valuation of CAM. Ann Bot Mcad 72. https://doi.org/10.1093/aob/mcad072

Davis S, Dohleman F, Long S (2011) The global potential for Agave as a biofuel feedstock. GCB Bioenergy 3:68–78. https://doi.org/10.1111/j.1757-1707.2010.01077.x

Davis SC, LeBauer DS, Long SP (2014) Light to liquid fuel: theoretical and realized energy conversion efficiency of plants using Crassulacean Acid Metabolism (CAM) in arid conditions. J Exp Bot 65:3471–3478

Davis SC, Kuzmick ER, Niechayev N, Hunsaker DJ (2017) Productivity and water use efficiency of *Agave americana* in the first field trial as bioenergy feedstock on arid lands. GCB Bioenergy 9(2):314–325. https://doi.org/10.1111/gcbb.12324

Davis S, Simpson J, Gil-Vega K et al (2019) Undervalued potential of crassulacean acid metabolism for current and future agricultural production. J Exp Bot 70(22):6521–6537. https://doi.org/10.1093/jxb/erz223

Davis SC, Abatzoglou JT, LeBauer DS (2021) Expanded potential growing region and yield increase for *Agave americana* with future climate. Agronomy 11(11):2109. https://doi.org/10.3390/agronomy11112109

De la Torre LDL, Cummins I, Logan-Hines E (2018) *Agave americana* and *Furcraea andina*: key species to Andean cultures in Ecuador. Bot Sci 96(2):246–266. https://doi.org/10.17129/botsci.1813

Del Buono D, Di Michele A, Costantino F et al (2021) Biogenic ZnO nanoparticles synthesized using a novel plant extract: application to enhance physiological and biochemical traits in maize. Nano 11:1270. https://doi.org/10.3390/nano11051270

Delgado-Aceves L, Portillo L, Folgado R (2022) New approaches for micropropagation and cryopreservation of *Agave peacockii*, an endangered species. Plant Cell Tissue Organ Cult (PCTOC) 150(1):85–95. https://doi.org/10.1007/s11240-022-02246-z

Delgado-Lemus A, Torres I, Blancas J, Casas A (2014) Vulnerability and risk management of Agave species in the Tehuacán Valley, México. J Ethnobiol Ethnomed 10(1):1–15. https://doi.org/10.1186/1746-4269-10-53

Dempewolf H, Baute G, Anderson J et al (2017) Past and future use of wild relatives in crop breeding. Crop Sci 57(3):1070–1082. https://doi.org/10.2135/cropsci2016.10.0885

Díaz-Martínez M, Nava-Cedillo A, Guzmán-López JA, et al. (2012) Polymorphism and methylation patterns in Agave tequilana Weber var. 'Azul' plants propagated asexually by three different methods. Plant Sci 185–186, 321–330. https://doi.org/10.1016/j.plantsci.2012.01.004

Devi SP, Kumaria S, Rao SR, Tandon P (2013) *In vitro* propagation and assessment of clonal fidelity of *Nepenthes khasiana* Hook. f.: a medicinal insectivorous plant of India. Acta Physiol Plant 35(9):2813–2820. https://doi.org/10.1007/s11738-013-1314-x

Doebley J (2004) The genetics of maize evolution. Annu Rev Genet 38:37–59. https://doi.org/10.1146/annurev.genet.38.072902.092425

Eastwood C, Klerkx L, Ayre M, Dela Rue B (2019) Managing socio-ethical challenges in the development of smart farming: from a fragmented to a comprehensive approach for responsible research and innovation. J Agric Environ Ethics 32(5–6):741–768. https://doi.org/10.1007/s10806-017-9704-5

Eguiarte LE, Jiménez OA, Aguirre-Planter E et al (2021) Evolutionary ecology of *Agave*: distribution patterns, phylogeny, and coevolution (an homage to Howard S. Gentry). Am J Bot 108(2):216–235. https://doi.org/10.1002/ajb2.1609

El Oudiani A, Chaabouni Y, Msahli S et al (2011) Crystal transition from cellulose I to cellulose II in NaOH treated *Agave americana* L. fibre. Carbohydr Polym 86(3):1221–1229. https://doi.org/10.1016/j.carbpol.2011.06.037

Elbouzidi A, Baraich A, Bencheikh N et al (2023) *Agave americana* L. Asparagaceae. In: Ethnobotany of Northern Africa and Levant. Springer International Publishing, Cham, pp 1–10

Espino AA, Botín AJV, Serrano CR et al (2020) Indirect somatic embryogenesis on mutants of Agave tequilana weber cultivar blue induced with CO60 gamma rays. Trop Subtrop Agroecosyst 23(3):1–9

Evans S (2007) Bound in twine: the history and ecology of the henequen-wheat complex for Mexico and the American and Canadian Plains, 1880–1950. Texas A&M University Press, College Station

Ewald D, Ulrich K (2012) *In vitro* pollination in poplar of section *Populus*. Plant Cell Tissue Organ Cult 111:255–258. https://doi.org/10.1007/s11240-012-0189-7

Félix-Valdez LI, Vargas-Ponce O, Cabrera-Toledo D, Casas A et al (2016) Effects of traditional management for mescal production on the diversity and genetic structure of *Agave potatorum* (Asparagaceae) in central Mexico. Genet Resour Crop Evol 63:1255–1271. https://doi.org/10.1007/s10722-015-0315-6

Figueredo CJ, Casas A, González-Rodríguez A et al (2015) Genetic structure of coexisting wild and managed agave populations: implications for the evolution of plants under domestication. AoB Plants 7:plv114. https://doi.org/10.1093/aobpla/plv114

Figueredo-Urbina CJ, Álvarez-Ríos GD, García-Montes MA, Octavio-Aguilar P (2021) Morphological and genetic diversity of traditional varieties of agave in Hidalgo State, Mexico. Plos One 16(7):1–24. https://doi.org/10.1371/journal.pone.0254376

Flores-Benítez S, Jiménez-Bremont JF, Rosales-Mendoza S, Argüello-Astorga GR (2007) Genetic transformation of *Agave salmiana* by *Agrobacterium tumefaciens* and particle bombardment. PCTOC 91:215–224

Food Agriculture Organization (FAO) (2024) Future fibres: sisal. https://www.fao.org/economic/futurefibres/fibres/sisal/es/#:~:text=La%20producci%C3%B3n%20mundial%20de%20sisal,y%20Kenya%20(25%20000)

Franco-Martínez IS (1995) Conservación *in situ* y *ex situ* de las agaváceas y nolináceas mexicanas. Bot Sci 57:27–36

Gao J, Yang F, Zhang S et al (2014) Expression of a hevein-like gene in transgenic Agave hybrid No. 11648 enhances tolerance against zebra stripe disease. PCTOC 119:579–585

Gao S, Cong R, Gao L et al (2020) Genetic diversity analysis of phenotypic character and SRAP molecular markers in 45 tree peony cultivars. Rev Bras Bot 43(2):291–302. https://doi.org/10.1007/s40415-020-00596-6

Garrido-Cardenas JA, Mesa-Valle C, Manzano-Agugliaro F (2018) Trends in plant research using molecular markers. Planta 247(3):543–557. https://doi.org/10.1007/s00425-017-2829-y

Gentry HS (1982) Agaves of Continental North America. Arizona University Press, Tucson, p 670

Gómez D, Hernández L, Martínez J et al (2019) Sodium azide mutagenesis within temporary immersion bioreactors modifies sugarcane *In vitro* micropropagation rates and aldehyde, chlorophyll, carotenoid, and phenolic profiles. Acta Physiol Plant 41(7):1–7. https://doi.org/10.1007/s11738-019-2911-0

Gutiérrez-Aguilar PR, Gil-Vega KC, Simpson J (2014) Development of an *Agrobacterium tumefasciens* mediated transformation protocol for two Agave species by organogenesis. In: Gutiérrez-Mora A, Rodríguez-Garay B, Contreras-Ramos SM et al (eds) Sustainable and integral exploitation of Agave, p 32

Hamissa AMB, Seffen M, Aliakbarian B et al (2012) Phenolics extraction from *Agave americana* (L.) leaves using high-temperature, high-pressure reactor. Food Bioprod Proc 90(1):17–21. https://doi.org/10.1016/j.fbp.2010.11.008

Hatch KL (1993) Textile science. West Publishing Company, Eagan

Holtum JAM, Chambers D, Morgan T, Tan DKY (2010) Agave as a biofuel feedstock in Australia. GCB Bioenergy 3:58–67

Huang X, Xiao M, Xi J et al (2019) De novo transcriptome assembly of Agave H11648 by illumina sequencing and identification of cellulose synthase genes in Agave species. Genes 10:103. https://doi.org/10.3390/genes10020103

Hulle A, Kadole P, Katkar P (2015) *Agave americana* leaf fibers. Fibers 3(1):64–75. https://doi.org/10.3390/fib3010064

Irish M, Irish G (2000) Agaves, Yuccas, and related plants. Timber Press, Portland

Jaouadi M, Msahli S, Sakli F (2011) A study of the physical and mechanical properties of paper made from *Agave americana* L. Fibers Mater Manuf Process 26(4):567–572. https://doi.org/10.1080/10426910903124878

Jayaramachandran M, Saravanan S, Motilal A et al (2020) Genetic improvement of a neglected and underutilised oilseed crop: sesame (*Sesamum indicum* L.) through mutation breeding. Nucleus 63(3):293–302. https://doi.org/10.1007/s13237-020-00329-w

Jo Y, Cho H, Lee SY, Choi G et al (2018) Quantitative phase imaging and artificial intelligence: a review. IEEE J Sel Top Quantum Electron 25(1):1–14

Kadole P, Hulle A (2014) *Agave Americana* fibres: extraction, characterization, applications. LAP Lambert Academic Publishing, Saarbrucken

Kamińska M, Tretyn A, Trejgell A (2020) Genetic stability assessment of *Taraxacum pieninicum* plantlets after long-term slow growth storage using ISSR and SCoT markers. Biologia 75(4): 599–604. https://doi.org/10.2478/s11756-019-00377-x

Kashtwari M, Wani AA, Dhar MK, Jan S, Kamili AN (2018) Development of an efficient *In vitro* mutagenesis protocol for genetic improvement of saffron (*Crocus sativus* L.). Physiol Mol Biol Plants 24(5):951–962. https://doi.org/10.1007/s12298-018-0576-6

Kashyap DR, Vohra PK, Chopra S, Tewari R (2001) Application of pectinases in commercial sector: a review. Bioresour Technol Elsevier Sci 77:215–227

Kolte P, Daberao A, Miss Sharma A (2012) Agave Americana: the natural leaf fibre. Text Rev 7:1–5

Krieg DR (1963) Ethyl methanesulfonate-induced reversion of bacteriophage T4rII mutants. Genetics 48(4):561. https://doi.org/10.1093/genetics/48.4.561

Krupa-Małkiewicz M, Kosatka A, Smolik B, Sędzik M (2017) Induced mutations through EMS treatment and *In vitro* screening for salt tolerance plant of Petunia× atkinsiana D. Don Notulae Botanicae Horti Agrobotanici Cluj-Napoca 45(1):190–196. https://doi.org/10.15835/nbha45110578

Kumar NT, Shrivastava P, Jagadale M et al (2022) Fea technique for design and simulation of sisal decorticator raspador. J Krishi Vigyan 10(2):173–178

La Torre-Ruiz D, Ruiz-Valdiviezo VM, Rincón-Molina CI et al (2016) Effect of plant growth-promoting bacteria on the growth and fructan production of *Agave americana* L. Braz J Microbiol 47(3):587–596. https://doi.org/10.1016/j.bjm.2016.04.010

Lecona-Guzmán CA, Reyes-Zambrano S, Barredo-Pool FA et al (2017) *In vitro* propagation of *Agave americana* by indirect organogenesis. Hort Sci 52(7):996–999. https://doi.org/10.21273/HORTSCI10498-16

Larkin PJ, Scowcroft WR (1981) Somaclonal variation—a novel source of variability from cell cultures for plant improvement. Theor Appl Genet 60(4):197–214

Lee YH, Park W, Kim KS et al (2018) EMS-induced mutation of an endoplasmic reticulum oleate desaturase gene (FAD2-2) results in elevated oleic acid content in rapeseed (*Brassica napus* L.). Euphytica 214(2):1–12. https://doi.org/10.1007/s10681-017-2106-y

Li Y, Wang M, See DR et al (2019a) Ethyl-methanesulfonate mutagenesis generated diverse isolates of *Puccinia striiformis* f. sp. tritici, the wheat stripe rust pathogen. World J Micro Biotech 35(2):1–18. https://doi.org/10.1007/s11274-019-2600-6

Li S, Zhang N, Li J et al (2019b) A new male sterile line of "Duanye-13" radish (*Raphanus sativus* L.) produced by ethyl methanesulfonate mutagenesis. Genet Resour Crop Evol 66:981–987. https://doi.org/10.1007/s12355-017-0524-8

Liu D, Hu R, Zhang J et al (2021) Overexpression of an Agave phosphoenolpyruvate carboxylase improves plant growth and stress tolerance. Cells 10(3):582. https://doi.org/10.3390/cells10030582

Liu U, Gianella M, Dávila Aranda P et al (2023) Conserving useful plants for a sustainable future: species coverage, spatial distribution, and conservation status within the Millennium Seed Bank collection. Biodivers Conserv 32:2791–2839. https://doi.org/10.1007/s10531-023-02631-w

Llatas MNS, Ulloa WEV, Aguilar EEV et al (2021) Mejoramiento genético en plantas autógamas. Rebiol 41(1):136–153. https://doi.org/10.1071/FP13191

López-Bautista V, Mora-Aguilera G, Gutiérrez-Espinosa MA et al (2020) Morphological and molecular characterization of *Fusarium* spp. associated to the regional occurrence of wilt and dry bud rot in *Agave tequilana*. Rev Mex Fitopatol 38(1):79–106. https://doi.org/10.18781/r.mex.fit.1911-4

Lu ML, Wyman CE (2021) Elucidation of native California *Agave americana* and *Agave deserti* biofuel potential: compositional analysis. PLoS One 16(5):e0252201. https://doi.org/10.1371/journal.pone.0252201

Lu Y, Liu MY, Wang CS et al (2015) Effects of EMS mutagenesis on seed and seedling vigor and phenotypic variation of M2 population in *Chinese cabbage*. Plant Genet Resour 16:349–358

Lulsdorf MM, Ferrie A, Slater SM et al (2014) Methods and role of embryo rescue technique in alien gene transfer. In: Partab A, Kumar J (eds) Alien gene transfer in crop plants, vol 1. Springer, New York, pp 77–103

Majamo SL, Amibo TA, Tsegaw EZ (2024) Enhanced cellulose extraction from agave plant (*Agave americana* Species) for synthesis of magnetic/cellulose nanocomposite for defluoridation of water. Mater Today Communs 38:107683. https://doi.org/10.1016/j.mtcomm.2023.107683

Mallick M, Awasthi OP, Singh SK et al (2016) Physiological and biochemical changes in pre-bearing mutanst of Kinnow mandarin (*C. nobilis* Lour x *C. deliciosa* Tenora). Sci Hortic 199:178–185. https://doi.org/10.1016/j.scienta.2015.12.050

Mandujano A, Pons JL, Paredes R et al (2018) Genetic diversity of maguey (*Agave* spp.) in the mountains and plains of northern Guanajuato. Rev Mexicana Cienc Agrícola 9(3):511–523. https://doi.org/10.29312/remexca.v9i3.1212

Martínez MT, Suárez S, Moncaleán P et al (2022) Cryopreservation of holm oak embryogenic cultures for long-term conservation and assessment of polyploid stability. Plan Theory 11(9):1266. https://doi.org/10.3390/plants11091266

Martinez-Rodriguez A, Macedo-Raygoza G, Huerta-Robles AX et al (2019) Agave seed endophytes: ecology and impacts on root architecture, nutrient acquisition, and cold stress tolerance. In: Seed endophytes. Springer, Cham, pp 139–170. https://doi.org/10.1007/978-3-030-10504-4_8

Masoabi M, Lloyd J, Kossmann J et al (2018) Ethyl methanesulfonate mutagenesis and *In vitro* polyethylene glycol selection for drought tolerance in sugarcane (*Saccharum* spp.). Sugar Tech 20(1):50–59. https://doi.org/10.1007/s12355-017-0524-8

Mignoni G (1999) Cannabis as a Licit crop: recent developments in Europe. Kenaf Ecofibre, Rome

Mirjalili S (2019) Genetic algorithm. In: Mirjalili S (ed) Evolutionary algorithms and neural networks: theory and applications. Springer International Publishing, Cham, pp 43–55

Monja-Mio KM, Herrera-Alamillo MA, Sánchez-Teyer LF et al (2019) Breeding strategies to improve production of agave (Agave spp.). In: Al-Khayri J, Jain S, Johnson D (eds) Advances in plant breeding strategies: industrial and food crops, vol 6. Springer, Cham. https://doi.org/10.1007/978-3-030-23265-8_10

Mora AG, Mathis A, Camacho R (2023) Agave y derivados: Tendencias científicas, uso sostenible y patrimonio. Centro de Investigación y Asistencia en Tecnología y Diseño del Estado de Jalisco A.C. Av. Normalistas 800, Colinas de La Normal, 44270 Guadalajara

Moreno-Luna FB, Tovar-Corona A, Herrera-Perez JL et al (2019) Quick synthesis of gold nanoparticles at low temperature, by using *Agave potatorum* extracts. Mater Lett 235:254–257. https://doi.org/10.1016/j.matlet.2018.09.122

Motero-Paredes T, Vitoria Narvaez M, Vitoria Fernández E (2018) Induction of polyploidy with colchicine in Vitro plants of *Aloe vera* (L.). Rev Colomb Biotecnol 20(1):97–105

Msahli S, Ydrean J, Sakli F (2005) Evaluating the fineness of *agave Americana* L. fibers. Text Res J 75(7):540–543. https://doi.org/10.1177/0040517505053808

Msahli S, Sakli F, Drean JY (2006) Study of textile potential of fibres extracted from Tunisian *Agave Americana* L. Autex Res J 6(1):9–13. https://doi.org/10.1515/aut-2006-060102

Mu H, Wang B, Yuan F (2022) Bioinformatics in Plant Breeding and Research on Disease Resistance. Plan Theory 11:3118. https://doi.org/10.3390/plants11223118

Mukai T, Lajoie MJ, Englert M et al (2017) Rewriting the genetic code. Annu Rev Micro 71:557–577. https://doi.org/10.1146/annurev-micro-090816-093247

Murali S, Morchhale RK (2014) Sisal (*Agave sisalana*) fibre extraction for sustainable employment generation in India. In: Prakash Shukla J (ed) Technologies for sustainable rural development: having potential of socio-economic upliftment, vol 1, p 184

Muriá J (2012) A transition from indigenous to European technology in colonial Mexico: The case of tequila. Contrib Sci 8(1):93–98

Mylsamy K, Rajendran I (2011) Influence of fibre length on the wear behaviour of chopped *Agave americana* fibre reinforced epoxy composites. Tribol Lett 44:75–80. https://doi.org/10.1007/s11249-011-9829-z

Nadeem MA, Nawaz MA, Shahid MQ et al (2018) DNA molecular markers in plant breeding: current status and recent advancements in genomic selection and genome editing. Biotechnol Biotechnol Equip 32(2):261–285. https://doi.org/10.1080/13102818.2017.1400401

Nasri F, Zakizadeh H, Vafaee Y et al (2021) *In vitro* mutagenesis of *Chrysanthemum morifolium* cultivars using ethylmethanesulphonate (EMS) and mutation assessment by ISSR and IRAP markers. Plant Cell Tissue Organ Cult 149:657–673. https://doi.org/10.1007/s11240-021-02163-7

Nava-Cruz NY, Medina-Morales MA, Martinez JL et al (2015) Agave biotechnology: an overview. Crit Rev Biotechnol 35(4):546–559. https://doi.org/10.3109/07388551.2014.923813

Naziri M, Sh Sadat M, Howyzeh S (2019) The effect of different hormone combinations on direct and indirect somatic embryogenesis in *Agave Americana*. Iran J Plant Physiol 9(2):2739–2747. https://doi.org/10.30495/ijpp.2019.664579

Nichols DL, McLaughlin MJ, Benton M (2000) Production intensification and regional specialization: Maguey fibers and textiles in the Aztec city-state of Otumba. Anc Mesoam 11(2):267–291. https://doi.org/10.1017/S095653610011106X

Nongdam P (2016) Development of synthetic seed technology in plants and its applications: a review. Int J Curr Sci 19(4):86–101

Oliva VFG, Castro AC, Herrera I et al (2014) Response surface methodology of glutamine, asparagine and 2,4-Dichlorophenoxyacetic acid for *Agave americana* L. embryo number and their optimization in a RITA® automatic bioreactor system. Int J Biotechnol Wellness Ind 3(3): 88–94

Ortega Z, Castellano J, Suárez L et al (2019) Characterization of *Agave americana* L. plant as potential source of fibres for composites obtaining. SN Appl Sci 1:1–9. https://doi.org/10.1007/s42452-019-1022-2

Pandian S, Marichelvam K, Satish L et al (2018) SPAR markers-assisted assessment of genetic diversity and population structure in finger millet (*Eleusine Coracana* (L.) Gaertn) mini-core collection. J Crop Sci Biotechnol 21(5):469–481. https://doi.org/10.1007/s12892-018-0034-0

Pardo A, Hernández A, Méndez N et al (2015) Análisis genético, mediante marcadores RAPD, de microbulbos de ajo conservados e irradiados *In vitro*. Bioagro 27(3):143–150

Patade VY, Suprasanna P, Bapat VA (2008) Effects of salt stress in relation to osmotic adjustment on sugarcane (*Saccharum officinarum* L.) callus cultures. Plant Growth Regul 55(3):169–173. https://doi.org/10.1007/s10725-008-9270-y

Pawełkowicz ME, Skarzyńska A, Mróz T et al (2021) Molecular insight into somaclonal variation phenomena from transcriptome profiling of cucumber (*Cucumis sativus* L.) lines. Plant Cell Tissue Organ Cult 145(2):239–259. https://doi.org/10.1007/s11240-020-02005-y

Pérez del Río R, Caballero MC, Gómez LH et al (2013) Diseño y construcción de una desfibradora de hojas de *Agave angustifolia* Haw Design and construction of a leaf shredder angustifolia Haw agave. Rev Cie Téc Agr 22:5–14

Pérez Molphe Balch E, Esparza Araiza MJ, Pérez Reyes ME (2012) Conservación *in vitro* de germoplasma de *Agave* spp. bajo condiciones de crecimiento retardado. Rev Fit Mex 35(4): 279–287

Pivoto D, Waquil PD, Talamini E et al (2018) Scientific development of smart farming technologies and their application in Brazil. Inf Process Agric 5(1):21–32. https://doi.org/10.1016/j.inpa.2017.12.002
Popielarska M (2005) *In vitro* pollination of isolated ovules of sunflower (*Helianthus annuus* L.). Acta Biol Crac Ser Bot 47(1):85–92
Queiroz DMD, Coelho ALDF, Valente DSM et al (2021) Sensors applied to digital agriculture: a review. Cienc Agron 51. https://doi.org/10.5935/1806-6690.20200086
Rauf S, Al-Khayri JM, Zaharieva M et al (2016) Breeding strategies to enhance drought tolerance in crops. In: Al-Khayri J, Jain S, Johnson D (eds) Advances in plant breeding strategies: agronomic, abiotic and biotic stress traits. Springer, Cham. https://doi.org/10.1007/978-3-319-22518-0_11
Raya FT, Carvalho LM, José J et al (2023) Rescuing the Brazilian Agave breeding program: morphophysiological and molecular characterization of a new germplasm. Front Chem Eng 5:1218668. https://doi.org/10.3389/fceng.2023.1218668
Reyes-Zambrano SJ, Lecona-Guzmán CA, Barredo-Pool FA et al (2016) Plant growth regulators optimization for maximize shoots number in *Agave americana* L. by indirect organogenesis. Gayana Bot 73(1):124–131
Reyes-Zambrano SJ, Ramírez-Merchant ML, Arias-Castro C et al (2019) Morphometric and biochemical changes in *Agave americana* L. plantlets induced by ethyl methanesulfonate. Phyton 88(3):277. https://doi.org/10.32604/phyton.2019.06504
Reyes-Zambrano SJ, Lecona-Guzmán CA, Gutiérrez-Miceli FA et al (2020) Scanning electron microscopy and enzymatic analysis in *Agave americana* during *Fusarium oxysporum* infection. Rev Mex Fitopatol 38(3):408–419. https://doi.org/10.18781/r.mex.fit.2005-3
Reyes-Zambrano SJ, Lecona-Guzmán CA, Gutiérrez-Miceli FA et al (2021) Selection and analysis of polymorphisms in somaclonal variants of *Agave americana* resistant to *Fusarium oxysporum* via an ethyl methanesulphonate treatment. Phyton 90(6):1727. https://doi.org/10.32604/phyton.2021.016171
Reynoso-Santos R, Garcia A, López W et al (2012) Identificación taxonómica de agaves (*Agave* ssp.) utilizados para la elaboración del licor comiteco en Chiapas, México. Agro Productividad 5(4):9–17
Rodríguez G, Morales F, Gutiérrez R et al (2007) Generación de raíces transformadas de *Agave salmiana* Otto y su colonización por *Glomus intraradices*. Rev Fitotec 30:215–222
Rodríguez-Félix F, López-Cota AG, Moreno-Vásquez MJ et al (2021) Sustainable-green synthesis of silver nanoparticles using safflower (*Carthamus tinctorius* L.) waste extract and its antibacterial activity. Heliyon 7(4):e06923
Rodríguez-Garay B (2016) Somatic embryogenesis in Agave spp. In: Loyola-Vargas V, Ochoa-Alejo N (eds) Somatic embryogenesis: fundamental aspects and applications. Springer, Cham, pp 267–282. https://doi.org/10.1007/978-3-319-33705-0_16
Rodríguez-Garay B, López-Díaz S, Rodríguez-Domínguez JM et al (2018) Application of in casa pollination and embryo rescue techniques for breeding of agave species. Plant Cell Culture Protocols 1815:289–300. https://doi.org/10.1007/978-1-4939-8594-4_20
Rogo U, Fambrini M, Pugliesi C (2023) Embryo rescue in plant breeding. Plants 12(17):3106. https://doi.org/10.3390/plants12173106
Rolf Dahlgren RM, Clifford HT (1982) The monocotyledons: a comparative study. Academic, London, UK
Ruiz Mondragón KY, Klimova A, Aguirre-Planter E (2023) Differences in the genomic diversity, structure, and inbreeding patterns in wild and managed populations of *Agave potatorum* Zucc. used in the production of Tobalá mezcal in Southern Mexico. PLoS One 18(11):e0294534. https://doi.org/10.1371/journal.pone.0294534
Sala F, Baratta G, Todesco G et al (1990) Somaclonal variation and chemically induced mutagenesis for the production of salt and drought tolerance in tomato. In: I International symposium on in vitro culture and horticultural breeding, vol 280, pp 353–360. https://doi.org/10.17660/ActaHortic.1990.280.59

Santiago-Santiago AK, Arana-Coronado OA, Matus-Gardea JA et al (2023) Use of real options to evaluate the profitability of biogas production from stillage in the tequila industry. Agrociencia 57(7). https://doi.org/10.47163/agrociencia.v57i7.2991

Sathiamurthi P, Karthi KS, Sathishkumar TP et al (2021) Fiber extraction and mechanical properties of *Agave Americana*/Kenaf fiber reinforced hybrid epoxy composite. Mater Today Proc 46: 8594–8601. https://doi.org/10.1016/j.matpr.2021.03.571

SENASICA (2017) Servicio Nacional de Sanidad, Inocuidad y Calidad Agroalimentaria. Manual operativo de la campaña contra plagas reglamentadas del agave. Dirección de Protección Fitosanitaria. Clave: MOP-DPF-PRAV. http://www.sivea.org.mx/

Sharma V, Thakur M, Tomar M (2020) *In vitro* selection of gamma irradiated shoots of ginger (*Zingiber officinale* Rosc.) against *Fusarium oxysporum* f. sp. zingiberi and molecular analysis of the resistant plants. Plant Cell Tissue Organ Cult 143(2):319–330. https://doi.org/10.1007/s11240-020-01919-x

Shelake RM, Pramanik D, Kim JY (2019) Evolution of plant mutagenesis tools: a shifting paradigm from random to targeted genome editing. Plant Biotechnol Rep 13(5):423–445. https://doi.org/10.1007/s11816-019-00562-z

Siddique I, Yadav V (2021) Cytokinin influence on micropropagation system of Dianthus caryophyllus L. In: Siddique I (ed) Propagation and genetic manipulation of plants. Springer, Singapore, pp 33–41. https://doi.org/10.1007/978-981-15-7736-9_3

Simpson J, Martinez Hernandez A, Jazmín Abraham Juárez M et al (2011) Genomic resources and transcriptome mining in *Agave tequilana*. GCB Bioenergy 3(1):25–36. https://doi.org/10.1111/j.1757-1707.2010.01079.x

Singh H, Khar A, Verma P (2021) Induced mutagenesis for genetic improvement of Allium genetic resources: a comprehensive review. Genet Resour Crop Evol 68(7):2669–2690. https://doi.org/10.1007/s10722-021-01210-8

Smith RH (ed) (2013) Plant tissue culture: techniques and experiments. Elsevier, New Delhi

Stewart J (1981) *In vitro* fertilization and embryo rescue. Environ Exp Bot 21(3–4):301–315. https://doi.org/10.1016/0098-8472(81)90040-x

Steyn HJ (2006) The evaluation of conventional retting versus solar baking of *agave Americana* fibres in terms of textile properties. Doctoral dissertation, University of the Free State, South Africa

Thamae T, Baillie C (2007) Influence of fibre extraction method, alkali and silane treatment on the interface of *Agave americana* waste HDPE composites as possible roof ceilings in Lesotho. Compos Interf 14(7–9):821–836. https://doi.org/10.1163/156855407782106483

Tejavathi DH, Gayathramma K, Nijagunaiah R (2012) Genetic variability by in vitro mutagenesis and arbuscular mycorrhizal fungal symbiosis among micropropagated plants of Agave vera-cruz miller. Bioremediation, Biodiversity and Bioavailability 6:665–669

Trejo L, Reyes M, Cortés-Toto D et al (2020) Morphological diversity and genetic relationships in pulque production agaves in Tlaxcala, Mexico, by means of unsupervised learning and gene sequencing analysis. Front Plant Sci 11:524812. https://doi.org/10.3389/fpls.2020.524812

Uchida N, Sakamoto T, Kurata T et al (2011) Identification of EMS-induced causal mutations in a non-reference *Arabidopsis thaliana* accession by whole genome sequencing. Plant Cell Physiol 52:716–722. https://doi.org/10.1093/pcp/pcr029

Van Gerrewey T, Boon N, Geelen D (2021) Vertical farming: the only way is up? Agronomy 12(1): 2. https://doi.org/10.3390/agronomy12010002

Villarreal-Fuentes JM, Alia-Tejacal I, Hernández-Salvador MA et al (2020) *In situ* characterization of soursop (*Annona muricata* L.) in the Soconusco region, Chiapas, Mexico. Rev Chapingo Serie Hort 26(3):189–205. https://doi.org/10.5154/r.rchsh.2020.05.008

Warren-Vega WM, Contreras-Atrisco ZA, Ramírez-Quezada MF et al (2023) A novel approach of artificial intelligence for the study of the relation of physicochemical profile and color acquired by Tequila 100% agave in its maturation process. J Food Compost Anal 123:105533. https://doi.org/10.1016/j.jfca.2023.105533

Witt TW, Ulloa M, Pelletier MG et al (2018) Exploring ethyl methanesulfonate (EMS) treated cotton (*Gossypium hirsutum* L.) to improve drought tolerance. Euphytica 214(7):1–12. https://doi.org/10.1007/s10681-018-2206-3

Xiao Y, Jiang S, Cheng Q et al (2021) The genetic mechanism of heterosis utilization in maize improvement. Genome Biol 22(1):1–29. https://doi.org/10.1186/s13059-021-02370-7

Yan X, Corbin KR, Burton RA et al (2020) Agave: a promising feedstock for biofuels in the water-energy-food-environment (WEFE) nexus. J Clean Prod 261:121283. https://doi.org/10.1016/j.jclepro.2020.121283

Yang Y, Guo Y (2018) Elucidating the molecular mechanisms mediating plant salt-stress responses. New Phytol 217:523–539. https://doi.org/10.1111/nph.14920

Yang X, Cushman JC, Borland AM et al (2015) A roadmap for research on crassulacean acid metabolism (CAM) to enhance sustainable food and bioenergy production in a hotter, drier world. New Phytol 207(3):491–504. https://doi.org/10.1111/nph.13393

Zwane PE, Masarirambi MT, Magagula NT et al (2011) Exploitation of *Agave americana* L. plant for food security in Swaziland. Am J Food Nut 1(2):82–88. https://doi.org/10.5251/ajfn.2011.1.2.82.88

Chapter 4
Egyptian Clover (*Trifolium alexandrinum* L.) Improvement Over Decades from Traditional Selection to Next-Generation Breeding

Pummy Kumari, Dibyendu Seth, Neeraj Kharor, Subhash Chand, Satpal, and B. L. Sharma

Abstract Egyptian clover (*Trifolium alexandrinum* L.), commonly known as berseem, is a valuable winter multicut annual forage legume. It is highly prized for its high green fodder yield, fast regeneration potential, high nutritive value and suitability for livestock feed. Compared to alfalfa, another popular winter forage crop, berseem offers similar feed value without the risk of bloat. Although slightly less drought tolerant, berseem thrives in high moisture and alkaline soils. One of its key advantages is its early autumn planting window, which provides essential forage during the colder months and peaks in productivity as spring temperatures rise. Beyond direct grazing (though less common), berseem demonstrates exceptional versatility. It can be ensiled with oats, chaffed and mixed with straw for feed or used as a green manure crop to enrich the soil properties. Furthermore, berseem possesses a multi-cut nature that allows for 4–8 cuts, resulting in a prolonged supply of high-quality forage (with 20% crude protein, up to 65% *in-vitro* dry matter digestibility and high palatability) from November to May. These attributes collectively contribute to Berseem's popularity and productivity as a forage legume. This chapter aims to discuss the importance of Egyptian clover as a multicut *Rabi* fodder crop and its improvement using conventional plant breeding and advanced molecular approaches.

P. Kumari (✉) · N. Kharor · Satpal · B. L. Sharma
Forage Section, Department of Genetics and Plant Breeding, Chaudhary Charan Singh Haryana Agricultural University, Hisar, Haryana, India

D. Seth
Department of Genetics and Plant Breeding, Chaudhary Charan Singh Haryana Agricultural University, Hisar, Haryana, India

S. Chand
Division of Crop Improvement, ICAR-Indian Grassland and Fodder Research Institute, Jhansi, Uttar Pradesh, India

K. F. M. Salem et al. (eds.), *Breeding and Biotechnology of Leaf, Fruit, and Seed Fiber Crops*, Advances in Plant Breeding Strategies 11,
https://doi.org/10.1007/978-3-032-00407-9_4

Keywords Egyptian clover (*Trifolium alexandrinum* L.) · Multicut · Winter legume · Nutritional quality · Regeneration

4.1 Introduction

Egyptian clover (*Trifolium alexandrinum* L., 2n = 2x = 16), also known as "the king of fodders" is a crucial leguminous forage crop native to the Mediterranean and Middle Eastern regions. Belonging to the Fabaceae (*Leguminosae*) family, it is widely cultivated in northern India as a *rabi* season (winter) forage source (Kumari et al. 2023). This annual legume stands 30–80 cm tall with sparse hairs and an erect growth habit (Mohamed et al. 2024). A shallow taproot anchors the plant, while hollow stems branch from the base. Alternate leaves adorn the stem, each bearing leaflets 4–5 cm long and 2–3 cm broad. Inflorescences boast clusters of dense, elliptical yellowish-white flowers roughly 2 cm in diameter. Notably, cross-pollination by honey bees is essential for seed production. The fruit is a single-seeded pod containing a white to purplish-red seed.

Kingdom	Plantae—plantes, Planta, Vegetal, plants
Subkingdom	Viridiplantae—green plants
Infrakingdom	Streptophyta—land plants
Superdivision	Embryophyta
Division	Tracheophyta—vascular plants, tracheophytes
Subdivision	Spermatophytina—spermatophytes, seed plants, phanérogames
Class	Magnoliopsida
Superorder	Rosanae
Order	Fabales
Family	Fabaceae—peas, legumes
Genus	Trifolium L.—clover, trèfles
Species	*Trifolium alexandrinum* L.—Egyptian clover

Berseem reigns supreme as the main forage legume in Syria and Persia, where it serves as the primary green feed for horses, donkeys and camels. Its subtropical nature makes it a crucial winter crop that thrives in mild winters and exhibits strong regrowth after cutting (Chand et al. 2024). Notably, berseem has established itself as a leading Rabi (winter season) fodder crop across vast swathes of the North West Zone, Hill Zone and parts of Central and Eastern India, encompassing over two million hectares (Singh et al. 2019a, 2019b). However, its limitations under hot summer conditions restrict its widespread adoption. Interestingly, berseem can thrive as a summer annual in areas blessed with cool and moist summers. Berseem provides 4–7 cuttings, helps in the availability of green fodder for a longer duration and is critical for the sustainability of livestock productivity (Satpal et al. 2023). A key advantage of berseem lies in its ability to maintain a soft and succulent texture throughout its growth cycle. The incorporation of berseem clover in integrated farming systems has been reported to enhance soil fertility and provide sustainable forage solutions (Kumar et al. 2024; Seth et al. 2023). Berseem thrives well from the tropics till 35 °N, up to an elevation of 750 m from mean sea level (MSL) whereas 1500 m above MSL in the Himalayas. It exhibits some degree of frost tolerance, withstanding temperatures down to −6 °C, with some cultivars even tolerating as low as −15 °C (Dwivedi 2024). Rainfall requirements are modest with berseem performing well in areas receiving 550–750 mm rainfall annually. It performs better than Alfalfa under drought and short water logging periods indicating its moderate tolerance to such conditions (stresses) (Chand et al. 2023). Additionally, berseem also exhibits moderate tolerance to salinity and adapts to a wide range of soils, with a preference for fertile, loamy to clay soils with mild acidic to slightly alkaline pH (6.5–8). This adaptability and range of tolerance contribute significantly to its widespread cultivation. But in spite of all these advantages, there is a shortage of area-specific improved berseem cultivars apart from their low seed production potential as major bottlenecks. A field view of improved berseem varieties released from CCS HAU, Hisar is given in Fig. 4.1. The main objective of this chapter is to discuss possible breeding approaches and their use in combination with molecular and omic techniques to develop further resilient berseem varieties with high fodder and seed yield potential.

Fig. 4.1 (**a**, **b**) Improved varieties of berseem. (Source: Forage Section, CCS HAU, Hisar)

4.2 Crop Domestication

The timeline of Berseem's domestication remains unclear but evidence suggests that it originated in the eastern Mediterranean and India by 4000 BC. While its wild ancestor is likely extinct and its origin remains a topic of debate, Rouz et al. (2016) proposed berseem (*Trifolium berytheum* Boiss.) from the Middle East as a possible wild progenitor due to its full inter-fertility with cultivated berseem. Berseem's transformation from wild clover, possibly *Trifolium berytheum*, to a valuable crop, involved a combination of natural and human influence. Early cultivation likely involved unintentional selection. Farmers favored plants with desirable traits like palatability, high yield and fast regrowth after cutting (Roy et al. 2020). The well-adapted fodder plant cultivation provided better livestock forage and unknowingly shaped the early stages of berseem domestication. The selection became more deliberate with prolonged cultivation of berseem. Traditional practices like fallowing and crop rotation might have also inadvertently selected strains with beneficial nitrogen fixation and soil improvement properties. Over time, regional variations emerged as berseem spread to different areas probably due to adaptation to varying climates and farmer preferences. These variations became the foundation for further selection and improvement efforts. Archaeological evidence supports Berseem's established agricultural role by 2500 BC in ancient Egypt. Additionally, writings from Roman and Greek civilizations highlight its valuable qualities, suggesting potential selection for specific traits during this period.

T. salmoneum emerged as the most probable ancestor for Syrian Egyptian clover based on genetic similarity and the ability to cross-pollinate with *T. alexandrinum* (Badr et al. 2008). The close relationships observed between *T. salmoneum, T. berytheum* and cultivated Egyptian clover accessions, coupled with their reported ability to cross freely, suggest that these two wild species may be the primary progenitors of Egyptian clover. Analysis of the AFLP data revealed that accessions of *Trifolium apertum*, *Trifolium berytheum* and *Trifolium salmoneum* displayed the greatest genetic similarity to Egyptian clover accessions from Syria and Egypt. While all three wild species showed a close relationship, *T. apertum* was deemed a less likely ancestor due to reported cross-incompatibility and geographic distribution patterns. Selection under rain-fed irrigation conditions in Syria is hypothesized to have driven the initial domestication (Badr and El-Shazly 2012). Following domestication, the early cultivated forms likely spread to Palestine for rain-fed cultivation and Egypt for irrigated agriculture where these improved varieties subsequently achieved wider distribution.

Berseem exhibits significant variability, allowing for classification into cultivar groups based on branching behavior and productivity. Miskawi and Khadrawi cultivars stand out for their highly branched growth habit and superior yield potential (Ranjbar 2007). Berseem (*Trifolium alexandrinum* L.) boasts a remarkable geographic spread, having been introduced to new regions over centuries. Documented records show its arrival in Egypt by the sixth century (Mohamed et al. 2024), followed by India in the nineteenth century and wider adoption in the twentieth century across Pakistan, South Africa, the USA and Australia. Attempts to cultivate even extend to southern Europe and further northern areas like Brittany. This rapid spread, particularly in small-scale farming, highlights its value as a fodder crop.

4.3 Nutritional Profile and Health Benefits

Egyptian clover (*Trifolium alexandrinum* L.) is a multifaceted legume revered for its agricultural and environmental contributions. Thriving in mild winter environments, these versatile cool-season annuals emerge as a high-quality forage legume, boasting a rich nutrient profile, and offering a magnitude of benefits. As a powerhouse for livestock, berseem boasts an impressive 17% crude protein content along with substantial fiber and digestible nutrients making it a highly valuable winter forage source, especially when other food sources are scarce (Chand et al. 2024). Its green fodder contains 15–25% crude protein on a dry matter basis (DM basis) alongside essential minerals (11–19% on a dry matter basis) and carotene (Singh et al. 2019a, b). Berseem also offers excellent palatability and digestibility for ruminants. This non-bloating characteristic further enhances its value. Studies have shown that berseem can meet the nutritional demand of high-producing dairy cows (>30 L milk/day) providing sufficient dry matter intake. Early grazing, before foliage senescence, maximizes metabolizable protein content compared to other fresh forages. Berseem is also reported to support impressive growth rates in buffaloes (550 g/day) and milk production in dairy cows (up to 10 L/day). When combined with wheat straw, a common ruminant feed, berseem inclusion of up to 30% of the diet improved intake and digestibility (Enriquez-Hidalgo et al. 2020). Different quality parameters pertaining to green forage of berseem are shown in Table 4.1.

Berseem hay retains its nutritional quality, making it comparable to alfalfa hay when both are harvested at the same growth stage (Singh et al. 2020a, 2020b). Studies report organic matter digestibility ranging from 63% to 71% and crude protein digestibility from 63% to 75% in various ruminant species. Dry matter intake also falls within a favorable range (2.4–3.0% body weight) for growing heifers, buffalo, sheep and goats (Sharma and Murdia 1974). This high digestibility allows berseem hay, similar to alfalfa hay, to be used as a sole feed source for moderate

Table 4.1 Nutritive value of berseem green fodder (dry wt. Basis)

No.	Parameters	Nutritional value (%)
1.	Crude protein	17–22
2.	Essential minerals	11–19
3.	Organic matter digestibility	63–75
4.	Hemicellulose	7–10
5.	Cellulose	24–25
6.	Neutral detergent fiber	42–49
7.	Acid detergent fiber	35–38

Source: Forage Section, CCS HAU, Hisar (unpublished data)

weight gain in cattle, buffalo and sheep. Berseem hay can also effectively complement maize crop residues, much like alfalfa hay. The leguminous nature of berseem plays a crucial role in enhancing soil health. Through symbiosis with specific *Rhizobium* bacteria, berseem fixes atmospheric nitrogen into the soil, significantly reducing the dependence on chemical fertilizers. This fixed nitrogen becomes readily available for subsequent crops, improving soil fertility and increasing crop yields (Singh et al. 2022). Beyond these direct benefits, berseem offers significant soil conservation advantages. Its dense root system acts as a natural barrier, effectively preventing wind and water erosion, especially during the off-season when other crops are absent (Singh et al. 2021). This is beneficial, particularly in areas prone to adverse weather conditions. Additionally, incorporating berseem back into the soil as green manure further enriches the soil with organic matter, and nutrients and improves the soil structure thereby having a positive effect on the future crops to be grown on the same piece of land (Shabir et al. 2020).

Additionally, berseem can be intercropped with major *kharif* fodder crops, which not only provide livestock with additional food sources but also help suppress weeds and enhance soil fertility. Nasar et al. (2024) demonstrated that intercropping berseem with oats in a maize/soybean rotation boosted maize yield by 10% and enriched the soil, highlighting its contribution to sustainable and productive agricultural systems. Furthermore, berseem can be harvested at its peak and preserved as silage, a fermented and stored form of fodder. This allows farmers to manage fluctuations in forage availability, ensuring a reliable feed source during periods of scarcity or when fresh forage is limited. Finally, berseem's aesthetic appeal extends beyond the agricultural realm. Its attractive flowers and foliage create a haven for various wildlife species, including deer, rabbits and birds. This makes Berseem a valuable asset in wildlife habitat restoration projects, promoting biodiversity and a healthy ecological balance.

4.4 Current Cultivation Practices

Berseem is a valuable winter forage legume propagated solely through seeds. Sowing typically occurs in early autumn, either on a prepared seedbed or directly sown. Berseem can be grown as a monoculture or intercropped with other compatible crop species. For high green biomass production in the first cut, it is often mixed with *gobhia sarson, chinese cabbage, rye grass* or *winter cereals* like *oats*. In rice-wheat cropping systems, berseem is sown before or just after rice harvest and, hence thrives as winter and spring feed. Australian practices involve combining berseem with other legumes like Arrow leaf, Persian or Balansa clover. In certain areas, berseem is also intercropped along with vegetables (Sarson and Turnip). Berseem requires early sowing and weekly watering during initial establishment. Up to 10–15 irrigations might be necessary for optimal forage production. Studies in Morocco have shown irrigated berseem yielding 8–10 tons of dry matter per hectare (Bouazzama et al. 2017). Cultivar selection plays a role in the number and frequency of cutting. Highly productive varieties like Mescavi, HB 1 and HB 2 are released for cultivation from CCS HAU, Hisar, and Haryana which can be cut 4–6 times per season. Fahl berseem with its low branching habit is better adapted to drier regions. Cutting berseem at the appropriate stage is essential. Targeting a height where basal buds remain around 2–4 cm ensures proper re-growth and maximizes forage yield.

Berseem requires an initial cut at 50–60 days after planting followed by subsequent cuts every 30–40 days. The ideal cutting height for maximizing protein yield with lower fiber content is approximately 40 cm from the ground. Under irrigation, five to six cuts are achievable while dryland systems typically allow for one or two cuts during the cool season. Interestingly, berseem mixed with 20% ground maize creates high-quality silage and even a 5% molasses inclusion has proven successful. While berseem offers excellent forage, grazing is not ideal due to potential damage to vital growing buds. If intended for irrigated pastures, grazing should commence before the sward becomes too dense, maintaining a residual height of 5–6 cm with a 30–40-day rest period between grazing events. Haymaking with berseem is generally discouraged due to its succulent stems that impede proper drying. However, using the last spring cut, which is naturally drier or employing wilting techniques followed by rooftop drying can be attempted for hay production.

4.4.1 Current Agricultural Challenges and Breeding Objectives

Berseem cultivation despite its numerous advantages faces challenges that can be partially addressed by plant breeding advancements. Fungal diseases like crown rot and sclerotinia wilt significantly reduce yields while aphids, leafhoppers and cutworms inflict damage as common pests (Singh et al. 2019a, b). Chemical controls exist but raise concerns about environmental impact and potential insect resistance. Additionally, berseem's digestibility and protein content vary with cultivar and growing conditions. Breeding efforts target these issues and aim to develop varieties with enhanced disease and pest resistance as well as to improve forage quality and nutritional content. This often involves crossing existing berseem lines with germplasm sources containing desirable traits (*eg*. Resistance gene), followed by rigorous selection and testing to identify and propagate individuals with the desired traits. Furthermore, with climate change, berseem majorly faces stress due to drought, heat and salinity. Breeding for improved abiotic stress tolerance can enhance Berseem's adaptability to diverse growing conditions and bolster its resilience in the face of a changing climate.

Successful genetic improvement of crops through breeding programs relies heavily on understanding their natural mating behavior (self-pollination or cross-pollination). Berseem presents a unique challenge in this regard. While research suggests it's not entirely self-sterile (Roy et al. 2005a, 2005b), a good seed set often requires an external stimulus like tripping (movement of floral parts). Plant breeding, though time-consuming, offers a powerful tool to address this and other agricultural challenges pertaining to berseem cultivation. Utilizing diverse germplasm resources along with advanced techniques like marker-assisted selection and gene editing allows for targeted improvement. These techniques can introduce stress tolerance genes from other plant species into berseem thereby potentially enhancing its resilience (Hasan et al. 2021). However, introducing new traits can have unforeseen consequences. Thorough testing and evaluation are crucial before the widespread adoption of new berseem varieties. Balancing various breeding objectives *viz.,* stress resistance and quality improvement along with preserving existing traits such as nitrogen fixation capability requires careful consideration. Despite these challenges, plant breeding holds immense potential for berseem improvement. By focusing on specific traits like extended growth periods, increased biomass production, and resistance to diseases like root and stem rot, researchers can leverage advanced breeding techniques to develop improved berseem varieties. These new varieties can be more resilient and productive and better adapted to meet the evolving demands of farmers and consumers.

4.5 Biodiversity and Conservation

Among the 250–300 species within the clover genus *Trifolium* (Ellison et al. 2006), roughly 10% (20–30 species) hold value as forage plants in commercial agriculture, with an even greater number serving as local fodder sources in their native regions. Egyptian clover has been long cultivated in North Africa and Western Asia for fodder purposes. A detailed study by Zohary and Heller (1984), classified *Trifolium alexandrinum* into two varieties: *Var. alexandrinum* (known locally as Fahli) and *Var. serotinum* (known locally as Muscavi). The Fahli variety exhibits basal branching and yields one harvest per cultivation cycle, while the Muscavi variety exhibits apical branching and provides multiple harvests (4–6) within a single cycle. A third variety, Saidi, displays characteristics of both Fahli and Muscavi, producing basal and apical branches and yielding 2–3 harvests per cycle.

The genus *Trifolium* exhibits a variation in chromosome number and ploidy levels. The majority (80%) of *Trifolium* species studied have a diploid chromosome makeup with a base number (x) typically being 8. However, some species deviate from this pattern, possessing a base number as low as 5, 6 or 7 (Cleveland 1985). Interestingly, polyploidy, where an organism has more than two sets of chromosomes, is observed in around 16% of *Trifolium* species. These polyploid populations can have either uniform chromosome numbers or mixed ploidy levels.

There seems to be a correlation between life history strategy and chromosome number within *Trifolium.* Annual species generally have lower chromosome numbers, ranging from $2n = 10$ to 32 whereas perennial species tend to have higher chromosome numbers, spanning from $2n = 12$ to 130 (Taylor et al. 1977). This observation aligns with the fact that roughly one-third of *Trifolium* species are perennial and most of these perennials exhibit self-incompatibility, a breeding

Table 4.2 Ploidy levels of different closely related species of white clover

S. No.	Species	Ploidy	Annual/ perennial	References
1	*Trifolium cernuum*	2x	Annual	Kole (2011); Williams (2014)
2	*Trifolium glomeratum*	2x	Annual	
3	*Trifolium isthmocarpum*	2x	Annual	
4	*Trifolium nigrescens*	2x	Annual	
5	*Trifolium retusum*	2x	Annual	
6	*Trifolium suffocatum*	2x	Annual	
7	*Trifolium occidentale*	2x	Annual	
8	*Trifolium pallescens*	2x	Perennial	
9	*Trifolium parnassi*	2x	Perennial	
10	*Trifolium repens*	4x	Perennial	
11	*Trifolium thalii*	4x	Perennial	
12	*Trifolium montanum*	2x, 4x	Perennial	
13	*Trifolium ambiguum*	2x, 4x, 6x	Perennial	
14	*Trifolium uniflorum*	4x	Perennial	

system that prevents self-fertilization. This self-incompatibility likely promotes genetic diversity in these perennial populations. However, there are exceptions to this rule, with some perennials like *Trifolium fragiferum* existing in both self-incompatible and self-fertile forms (Evans 1976). Table 4.2 depicts the closely related species of white clover along with their ploidy levels. A list of cultivated Trifolium species is shown in Table 4.3.

India plays a significant role in conserving and managing *Trifolium* germplasm resources. The National Bureau of Plant Genetic Resources (NBPGR) acts as the nodal agency for activities related to plant genetic resources. This includes the exchange, quarantine, collection, conservation, evaluation and documentation of *Trifolium* species. NBPGR has successfully introduced and maintains over 500 *Trifolium* accessions from various countries under long-term storage conditions (Singh et al. 2019a, b). For the active management and utilization of germplasm wealth within the forage crop category, the ICAR-Indian Grassland and Fodder Research Institute (IGFRI) functions as the National Active Germplasm Site. IGFRI currently manages a collection of over 900 *Trifolium* accessions. IGFRI has meticulously cataloged information on various aspects of each accession to ensure effective utilization and maintenance of this conserved germplasm. This data includes qualitative and quantitative traits of the accession along with their origin. IGFRI also plays a crucial role in facilitating the utilization of Berseem germplasm by curators and scientists across the international plant genetic resources network. They have developed a comprehensive descriptor list specifically for Berseem. Significant progress has been made in generating new berseem germplasm lines with unique characteristics. These lines have been successfully registered at NBPGR, further enriching the available germplasm collection (data not shown in this passage but potentially detailed in Table 4.4).

Table 4.3 Cultivated *Trifolium* species along with their specific uses

S. No.	Species	Common name	Specific use	Ploidy	Lifespan	Reference
1	*T. ambiguum*	Kura/Caucasian	Used for revegetating in the Australian Alps	$2x = 16$	Perennial	Hely (1982)
2	*T. incarnatum*	Crimson	Winter pasture, in rotations, roadside stabilization	$2x = 14$	Annual	Knight (1985a)
3	*T. arvense*	Rabbit Foot	Adapted to infertile, dry, sandy soils	$2x = 14, 16$	Annual	Knight (1985b)
4	*T. lappaceum*	Lappa	With grass for spring grazing on calcareous soils	$2x = 16$	Annual	Knight (1985b)
5	*T. campestre*	Low -Large -hop	Grass for early spring grazing, hay	$2x = 16$	Annual	Bedard and Lapointe (1991)
6	*T. semipilosum*	Kenya White	commercial varieties available	$2x = 16$	Perennial	Ison and Parson (1992)
7	*T. subterraneum*	Subterranean	For grazing, particularly in temperate areas of Central and Northern Europe and America	$2X = 16$	Annual	Frame and Laidlaw (1998)
8	*T. cherleri*	Cupped	Grazing, cultivars released in Australia	$2x = 10$	Annual	Nichols et al. (2007)
9	*T. dubium*	Yellow suckling/ small hop/drooping	Grass for early spring grazing, hay	$4x = 28$	Annual	Jones (2012)
10	*T. fragiferum*	Strawberry	Grazing, particularly close, continuous grazing, and phase farming.	$2x = 16$	Perennial	Can et al. (2013)
11	*T. resupinatum*	Persian	For grazing, hay, silage. Some are water -logging tolerance	$2x = 14, 16; 4x = 32$	Annual	Hall et al. (2013)
12	*T. pratense*	Red Clover	In grass mixtures for hay, pasture and silage	$2x = 14$, $4x = 28$	Perennial	Moorby et al. (2016)
13	*T. repens*	White Clover	With grass for grazing, perennial pastures	$4x = 32$	Perennial	Chapman et al. (2016)
14	*T. hirtum*	Rose	Low-intensity grazing, especially in unproductive areas, soil stabilization.	$2x = 16$	Annual	Gornish and Ambrozio dos Santos (2016)
15	*T. hybridum*	Alsike/Swedish	Hay, pasture, particularly in cool moist areas	$2x = 16$	Perennial	Frame and Laidlaw (2019)
16	*T. alexandrinum*	Egyptian/Berseem	Green-chopped forage, pasture	$2x = 16$	Annual	Barik and Hedayetullah (2019)
17	*T. nigrescens*	Ball	Winter grazing	$2x = 16$	Annual	Dillard (2023)
18	*T. michelianum*	Balansa Clover	Grazing, silage, green manure. Some are water-logging tolerance. Cultivars released in Australia	$2x = 16$	Annual	Caudillo (2023)

Table 4.4 Registered berseem germplasm lines with unique characteristics at NBPGR, New Delhi, India

Crop name	Botanical name	National identity	Donor identity	INGR No	Year	Pedigree	Developing institute	Novel unique features
Egyptian Clover	*T. alexandrinum*	IC524012	PBL-123	INGR05017	2005	Spontaneous mutant	PAU, Ludhiana	Purple leaf and flower
Egyptian Clover	*T. alexandrinum*	IC0593646	IL-11-239	INGR12009	2012	*T. alexandrinum x T. Apertum*	ICAR-IGFRI, Jhansi	Stabilized Black (Dark tan) seeded against normal yellow seeded Berseem.
Egyptian Clover	*T. alexandrinum*	IC0593647	IL-11-238	INGR12010	2012	Penta-1	ICAR-IGFRI, Jhansi	Tetraploid pentafoliate.
Egyptian Clover	*T. alexandrinum*	IC0593648	190-P3	INGR12011	2012	Diploid Berseem	ICAR-IGFRI, Jhansi	Stabilized self-incompatible. Tetraploid.
Egyptian Clover	*T. alexandrinum*	EC329299	EC329299	INGR12012	2012	Saidi Ecotype of Berseem	ICAR-IGFRI, Jhansi	Self-compatible Berseem.
Egyptian Clover	*T. alexandrinum*	IC567683	Penta-1	INGR09045	2009	Spontaneous mutant of JHB 146	ICAR-IGFRI, Jhansi	Pentafoliate mutant
Egyptian Clover	*T. alexandrinum*	IC0613360	IL-13-106	INGR15026	2015	*T. alexandrinum x T. Apertum* cross	ICAR-IGFRI, Jhansi	Black seeded with pentafoliate leaf

Source: ICAR-NBPGR | MTS Database; All registered germplasm are conserved at NBPGR, New Delhi
PAU Punjab Agricultural University, *IGFRI* ICAR-Indian Grassland and Fodder Research Institute, Jhansi, Uttar Pradesh

4.5.1 Cultivars Characterization and Phylogeny

Understanding the intricate tapestry of diversity and evolutionary relationships among berseem cultivars is paramount for optimizing both cultivation practices and breeding programs. Cultivar characterization serves as a cornerstone in this endeavor, systematically evaluating and describing berseem genotypes to distinguish and document their unique traits. Germplasm characterization focuses on recording heritable traits that are consistently expressed across environments. This allows for efficient evaluation in a single location, ideally conducted upon acquisition and preferably near the collection site or under similar agro-climatic conditions. Standardized descriptors developed by organizations like UPOV, USDA or Biodiversity International guide the recording of these observations. A schematic flowchart of the various steps involved in berseem germplasm characterization and evaluation has been shown in Fig. 4.2. Modern approaches further enrich this process by incorporating highly reproducible molecular and biochemical markers (*viz.*, SSR, SNP, etc).

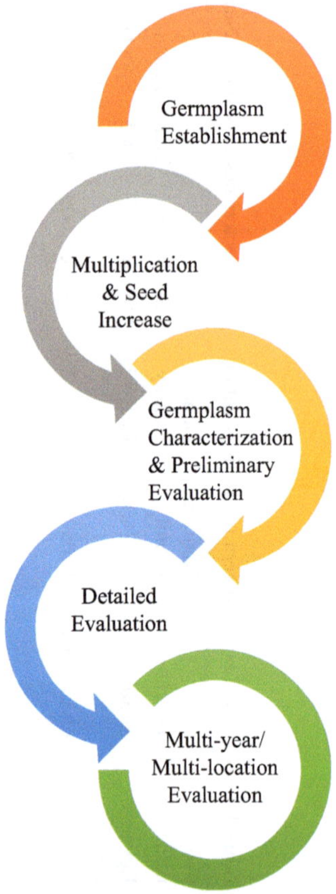

Fig. 4.2 Steps involved in germplasm evaluation and characterization. (Source: Created by authors)

Morphological characterization plays a vital role, meticulously evaluating physical attributes like plant height, stem branching patterns, leaf size and shape, flower color and seed morphology. This initial assessment provides valuable insights into cultivar diversity and potential adaptation to specific environmental conditions. Agronomic characterization on the other hand delves deeper, focusing on assessing performance parameters such as yield potential, forage quality, maturity period and response to various biotic and abiotic stresses like drought, salinity and diseases. Understanding these characteristics is crucial for recommending suitable cultivars for different regions and farming systems. Finally, molecular characterization takes the stage, utilizing advanced DNA-based techniques like microsatellite markers and SNPs. These powerful tools assess genetic diversity and identify unique molecular fingerprints for each cultivar (Hossain et al. 2020). This information proves invaluable for cultivar identification, purity testing and effective germplasm management, ultimately propelling berseem improvement strategies forward.

Cultivar characterization is vital in several aspects of berseem research and utilization. Firstly, detailed characterization data empowers informed decision-making. Farmers and researchers can leverage this information to select cultivars best aligned with their specific needs strategically. This selection considers environmental conditions, preferred farming practices and the intended end-use of the berseem crop (grazing or hay production). Secondly, cultivar characterization forms the cornerstone of successful breeding programs. The data provides a rich resource for identifying desirable traits within the berseem germplasm collection. Breeders can strategically select parental lines exhibiting complementary characteristics to target specific improvements in future berseem varieties. Finally, cultivar characterization is essential for effective berseem germplasm conservation efforts. Understanding the genetic diversity of the available germplasm collection is paramount. This knowledge allows researchers to develop targeted conservation strategies; ensuring valuable genetic resources are available for future generations.

Phylogenomics, a powerful tool in evolutionary biology sheds light on the evolutionary relationships between berseem cultivars and their kin within the *Trifolium* genus. This provides insights into the selection pressures that have shaped the current diversity of berseem, helping to identify wild relatives with desirable traits such as disease resistance or stress tolerance through phylogenetic analysis (Mohamed et al. 2024). Researchers can reconstruct the evolutionary history and branching patterns of these berseem cultivars by analyzing variations in DNA sequences, either from specific genes or entire genomes. Phylogenetic information can guide the development of core collections that represent the maximum genetic diversity of berseem with minimal redundancy for efficient germplasm management.

Traditionally, isozyme analysis has been used to assess genetic diversity and cultivar relationships by investigating variations in enzyme function. Modern techniques including DNA fingerprinting techniques like RAPD and AFLP, analyze variations in DNA fragments to establish unique genetic profiles (fingerprints) of cultivars, enabling inference of relatedness among them. The most precise method currently employed is DNA sequencing, which allows scientists to analyze complete genomes or specific gene sequences, providing the most detailed and accurate information on the genetic relationships between berseem cultivars.

The analysis of AFLP data using distance and parsimony methods showed that accessions of *T. apertum*, *T. berytheum*, and *T. salmoneum* are most similar to *T. alexandrinum* accessions from Syria and Egypt (Badr et al. 2008). Given their geographic range and cross-ability, *T. apertum* might be considered an unusual ancestor of the Egyptian clover. Meanwhile, the most likely ancestor of Syrian material for Egyptian clover has been identified as *T. salmoneum*. *T. alexandrinum* have also been found to have a tight relationship with *T. salmoneum* and *T. berytheum*. The ability of these species to cross freely indicates that *T. salmoneum* and *T. berytheum* may be regarded as the primary ancestors that had given rise to the Egyptian clover by selection in Syria under rain-fed irrigation (Zohary and Heller 1984). Following domestication, the early varieties of the domesticated crop might have been irrigated in Egypt and rain-fed in Palestine. It appears that the crop underwent genetic improvement in Egypt following cultivation, and Egyptian-developed cultivars were subsequently extensively dispersed.

4.6 Genetic Resources Conservation Approaches

4.6.1 Ex-situ *Conservation*

Recognizing the economic value of *Trifolium* species and the threat of genetic erosion, international efforts prioritized the *ex-situ* conservation of 21 *Trifolium* species back in the late twentieth century (Greene et al. 2014). This focus was further emphasized by the 1997 IUCN Red List of Threatened Plants, which identifies 39 *Trifolium* species, particularly the 18 found in the Mediterranean region, as critical for *ex-situ* conservation efforts (Walter and Gillett 1998). These efforts have resulted in the establishment of large *ex-situ* germplasm collections for *Trifolium*. Complementing these collections, researchers have conducted numerous eco-geographic surveys to identify and collect *Trifolium* germplasm in its natural habitat. Examples include surveys of legumes in southwest Turkey (Bennett et al. 1998) and annual *Trifolium* species in Morocco (Beale et al. 1993). Furthermore, extensive West Asia and North Africa collection missions have yielded over 3000 *Trifolium* accessions (Bounejemate et al. 1999). Figure 4.3 shows a number of accessions of different species of *Trifolium* held at various well-known gene banks.

Current collections of annual forage species are far from complete and often inadequate. While significant efforts have gone into collecting *Trifolium* germplasm for *ex-situ* conservation and breeding programs, the focus has primarily been on major cultivated species like *T. repens*, *T. pratense* and *T. subterraneum*. This focus has resulted in the neglect of potentially valuable yet less-common *Trifolium* species. Furthermore, Taylor et al. (1977) noted many older accessions lack sufficient passport data, including crucial details like collection site latitude and longitude. This lack of information significantly hinders the overall usefulness of the conserved germplasm.

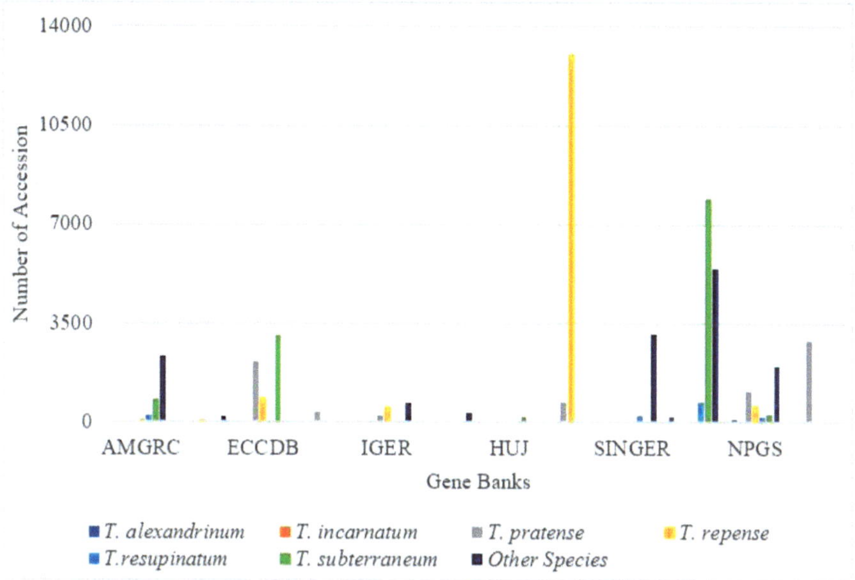

- AMGRC = Australia Medicago Genetic Resource Centre, South Australian Research and Development Institute, Adelaide, Australia
- ECCDB = European Central Crop Database.
- IGER = Institute of Grassland and Environmental Research, Aberystwyth, UK.
- HUJ = Department of Botany, Institute of Life Sciences Hebrew University of Jerusalem, Israel
- SINGER System-wide Information Network for Genetic Resources.
- NPGS= National Plant Germplasm System, Regional Plant Introduction Station, Washington, USA.

Fig. 4.3 Graph showing the number of accessions of different species of *Trifolium* held at various well-known gene banks

4.6.2 Cryopreservation and In-vitro *Conservation*

Cryopreservation offers a powerful tool for long-term storage of biological samples. This technique involves using liquid nitrogen (−196 °C) to achieve ultra-low temperatures, essentially halting biological activity and preventing tissue degradation. Cryopreservation allows for the long-term viability of biological materials by significantly slowing down metabolic processes. In the context of berseem (*Trifolium alexandrinum*), this technique can be applied for *in-vitro* conservation. Two potential methods for cryo-preserving berseem germplasm include somatic embryogenesis and organogenesis derived from callus cultures (Jacques et al. 2024).

4.6.3 DNA and Seed Banks

The emergence of improved crop varieties in the mid-twentieth century threatened the existence of landraces, prompting the establishment of gene banks for biodiversity preservation (Díez et al. 2018). These banks serve as a critical resource for future breeding programs by safeguarding genetic diversity. Gene banks employ long-term storage strategies, including DNA extraction and preservation from various plant sources, encompassing wild and cultivated species, along with endangered or even fossilized specimens. Locations of prominent gene banks dedicated to berseem plant genetic resource conservation in India are shown in Fig. 4.4. A prevalent method for

Fig. 4.4 Depicting *Trifolium alexandrinum* genetic resource conservation locations across India. (Source: ICAR-NBPGR, PGR and MAP) (ICAR-NBPGR | MTS Database (n.d.))

DNA storage involves dissolving it in a TE buffer and maintaining it at extremely low temperatures (−80 °C) or in an alcohol solution. For seed storage, the Svalbard Global Seed Vault project stands out as the world's largest secure facility. This vault, situated 1300 km north of the Arctic Circle within a permafrost zone, provides optimal conditions for seed preservation.

4.7 Traditional Breeding Approaches

One of the most important drawbacks in the genetic improvement of berseem is the lack of genetic variability (Malaviya et al. 2004; Roy et al. 2004). Breeders have employed various techniques to create new variations and select superior genotypes to overcome the limitations imposed by restricted genetic diversity. Mutation, polyploidization and inter-specific hybridization are conventional techniques that were widely employed to increase the variability in the existing gene pool of berseem. Selection can be practiced at various stages of plant growth, focusing on specific characteristics at each stage. Pedigree selection, mass selection and recurrent selection are commonly used methods. BL-1, BL-2, BL-22, BL-42, BL-180, HB1, HB 2 improved cultivars represent successful selection programs focusing on traits like yield, disease resistance and adaptation to local growing conditions (Roy et al. 2020).

The mutation breeding approach involves exposing germplasm to mutagenic agents such as radiation or chemicals to induce random mutations within the plant's genome. Mutation breeding has been elaborately discussed in the later section. Inter-specific hybridization of berseem with other *Trifolium* species also introduces new genetic variation. Induction of polyploidy breeding, a technique that alters a plant's chromosome number was a key milestone in berseem breeding. Early research focused on manipulating the Berseem genome to achieve larger leaves and stems. The use of colchicine treatment to induce autotetraploidy (doubling the chromosome number) combined with the selection at the tetraploid level resulted in the development of the first Berseem variety—'Pusa Giant'. This groundbreaking cultivar boasted increased fodder production, superior regeneration capacity and consistently higher yields throughout the season than existing diploid varieties released for general cultivation in India (Mehta and Swaminathan 1957). Another notable achievement in polyploidy breeding belongs to IGFRI, Jhansi. They successfully developed 'Bundel Berseem-3, an autotetraploid variety (Roy et al. 2020). This accomplishment involved colchicine treatment followed by a two-step selection process: recurrent single-plant selection and subsequent mass selection. 'Bundle Berseem-3' is recommended explicitly for cultivation due to its regional adaptation in the northeast zone, Bihar Orissa, West Bengal and Eastern Uttar Pradesh. Different genetic improvement programs utilizing breeding approaches like selection, polyploidy and mutation led to the development of varieties for different Berseem-growing regions of India (Table 4.5).

Table 4.5 List of Berseem varieties released/notified for different agro-climatic conditions in India

S. no.	Variety	GFY (q/ha)	Breeding method	Year of release/ notification	Parental Institutes	Area of adaptation
1	Mescavi	580–620	Selection from the introduced material from Egypt	1975	CCS HAU, Hisar	Punjab, Haryana, Delhi, Uttarakhand, UP and Madhya Pradesh
2	Pusa Giant	500–600	Selection after diploidization of C-10 cultivar (diploid)	1975	IARI, New Delhi	Pan India
3	BL 1	600–650	Selection from Mescavi cultivar	1978	PAU, Ludhiana	Punjab
4	Wardan	600–650	Selection from accession No. 526	1982	IGFRI, Jhansi	Pan India
5	JB 1	700–750	Selection from local material collected from Chinaware, MP	1981	JNKVV, Jabalpur	Central and northwestern zones
6	BL-10	800–900	Selection from indigenous material	1985	PAU, Ludhiana	Punjab, Haryana, Himachal Pradesh., Jammu
7	BL 22	480–520	Selection from an irradiated population of Mescavi cultivar	1988	PAU, Ludhiana	Sub-temperate hill regions of North India
8	BL 2	500–550	Selection from indigenous material	1989	PAU, Ludhiana	Punjab, Haryana, H.P., Jammu, Western UP, Uttarakhand
9	UPB 110	500–600	Composite selection from seven Berseem lines	1993	GBPUAT, Pantnagar	North-west, central and south India
10	BB 2	700–750	Selection from the indigenous material (25776–4-P6)	1997	IGFRI, Jhansi	North-West and Central zone
11	BB 3	500–550	Modified mass selection on tetraploid indigenous line JHB-89-3	2001	IGFRI, Jhansi	North-East and Eastern region
12	JB 5	550–580	Recurrent selection from the colchicine-treated material	2005	JNKVV, Jabalpur	Central zone
13	BL 42	600–700	Selection from the irradiated	2007	PAU, Ludhiana	North-West India

(continued)

Table 4.5 (continued)

S. no.	Variety	GFY (q/ha)	Breeding method	Year of release/ notification	Parental Institutes	Area of adaptation
			population of BL-2 cultivar			
14	BL 180	500–550	Selection from the irradiated population of BL-10 cultivar	2006	PAU, Ludhiana	Hill and North-West India
15	HB 1 (HFB 600)	600–650	Selection from germplasm no.6 (307011, 11-OP)	2006	CCS HAU, Hisar	North-West India
16	HB 2	650–700	Selection from the irradiated population of Mescavi cultivar.	2014	CCS HAU, Hisar	Haryana
17	JBSC-1	150–200	Selection from the material (EC 318954) introduced from Germany	2018	IGFRI, Jhansi	North-West and Central Zone
18	BL 43	550–580	Selection from the poly cross of five varieties	2019	PAU, Ludhiana	Punjab state
19	JB-05-09	650–680	Selection from the irradiated population of JB-1 cultivar	2019	JNKVV, Jabalpur	North-West India
20	BL 44	750–800	Selection from a poly cross of nine varieties	2021	PAU, Ludhiana	North West and North East India
21	BB-5	800–850	Phenotypic recurrent selection	2021	IGFRI, Jhansi	North West and North East India
22	BB-6	750–800	Phenotypic recurrent selection	2021	IGFRI, Jhansi	North West and North East India
23	BL-45	750–800	Selection from mutant line (BL-42)	2022	PAU, Ludhiana	Hill, North West and Central Zone
24	BB-7	750–820	Phenotypic recurrent selection	2022	IGFRI, Jhansi	Hill, North West and Central Zone
25	BB-8	750–820	Phenotypic recurrent selection	2022	IGFRI, Jhansi	Hill, North West and Central Zone

Source: Modification of Chand et al. (2024)

Traditional breeding approaches though significantly contributed to the improvement of the crop. Still, the process is time-consuming and labor-intensive and its effectiveness often relies on the availability of sufficient genetic variation within the gene pool. The future of berseem breeding likely lies in integrating traditional methods with modern advancements in molecular biology. Techniques like marker-assisted selection (MAS) can accelerate breeding cycles and target specific traits with greater precision. Additionally, exploring more comprehensive germplasm sources and utilizing tools like genetic engineering could further enhance breeding efforts.

4.7.1 Hybridization in Berseem Breeding

Interspecific hybridization, the crossing among different *Trifolium* species, has played a dual role in berseem breeding. Initially, it served as a tool to identify the closest relatives of *T. alexandrinum*. Successful hybridizations with *T. berytheum* and *T. salmoneum* (with a chromosome number of $2n = 16$) helped establish their likely ancestral relationships. The aim now has shifted from phylogeny analysis to leverage the wider adaptability and disease resistance traits such as against root rot caused by fungi like *Rhizoctonia solani* and *Fusarium semitactum*, stem rot caused by *Sclerotinia trifoliorum*, leaf blight by *Epicoccum sp.*, and powdery and downy mildew caused by *Oidium sp.* and *Peronospora trifoliorum* respectively (Bhaskar et al. 2002) that are potentially found in wild *Trifolium* species to improve cultivated berseem (Table 4.6). Interspecific incompatibility barriers, a common challenge across *Trifolium* species have hindered the incorporation of these beneficial traits into current cultivars. However, advancements in embryo rescue techniques are readily adopted by researchers, successfully employed to develop interspecific hybrids between cultivated berseem and *Trifolium apertum* (Malaviya et al. 2004), *T. constantinopolitanum* (Roy et al. 2004), *T. resupinatum* (Kaushal et al. 2005) and *T. vesiculosum* (Kaur et al. 2017). These hybrids have shown promising results with progeny exhibiting desirable characteristics. Drought-tolerant berseem lines derived from interspecific hybrids were also identified (Sevanayak et al. 2017).

Various biotic and abiotic resistance genes are known to be available in the secondary and tertiary gene pool of berseem. Table 4.6 lists a few species of *Trifolium* with their desirable characteristics which can be effectively utilized in crop breeding programs. Advanced molecular techniques can be employed to introduce these desirable stress tolerance genes into cultivated berseem varieties thereby significantly enhancing the resilience of berseem germplasm.

Table 4.6 Promising diploid *Trifolium* species along with their promising traits

Species	Gene pool	Chromosome number (2n = 2x)	Desirable characters	References
Trifolium alexandrinum ecotype Mescavi	Primary	2n = 16	Annual, multi-cut, highly productive, crude protein, high digestibility and palatability, basal branching	Malaviya et al. (2004)
Trifolium alexandrinum ecotype Fahli	Primary	2n = 16	Annual single-cut, self-compatible, stem branching	Singh et al. (2015)
Trifolium apertum	Secondary	2n = 16	Annual, profuse basal branching. Late flowering, resistance against root rot and stem rot, high protein content	Malaviya et al. (2004)
Trifolium resupinatum	Tertiary	2n = 16	Root rot and stem rot resistance, soil alkalinity tolerance	Kaushal et al. (2005)
Trifolium constantinopolitanum	Tertiary	2n = 16	Profuse basal branching, resistance against root rot and stem rot	Roy et al. (2004)
Trifolium vesiculosum	Tertiary	2n = 16	Lateness, disease resistance	Malaviya et al. (2004)

4.7.2 Beyond Tradition: Biotechnology's Role in Berseem Improvement

While conventional breeding methods have undeniably propelled berseem improvement, they face inherent limitations. Biotechnology offers exciting possibilities to create genetic variations beyond the reach of traditional techniques. These variations fall into three main categories:

1. **Somaclonal variation:** This approach involves culturing plant cells, during which genetic changes can occur. These changes can sometimes be beneficial, leading to the development of new and improved lines.
2. **Somatic hybridization**: This technique merges genetic material from different berseem plants, potentially combining desirable traits from each parent. However, compatibility challenges may arise when hybridizing certain lines.
3. **Genetic transformation**: This method directly inserts specific genes into the berseem genome. This targeted approach allows researchers to introduce traits not readily achievable through traditional breeding methods.

The methods employed for gene transfer in berseem are diverse. Delivery can be achieved chemically, electrically or through physical means like micro-projectile bombardment. These advancements aim to optimize berseem for intensive feeding systems. A key focus lies in reducing the levels of naturally occurring anti-nutritional compounds like hydrocyanic acid and saponins, which can negatively impact animal health (Samtiya et al. 2020). Molecular markers have become indispensable tools in berseem research. Researchers utilize various DNA markers (RFLP, AFLP, SSR, SNP) alongside protein and isozyme markers, each with specific advantages. These markers aid in:

- Characterizing germplasm collections: Markers help assess the genetic diversity within berseem germplasm collections.
- Cultivar identification: Markers can be used to definitively identify specific berseem cultivars.
- Hybrid detection: Markers can accurately identify hybrids resulting from crosses between different berseem lines.
- Genetic map construction: By analyzing marker inheritance patterns, researchers can construct genetic maps that reveal the chromosomal locations of genes.
- Identifying QTLs: Markers play a crucial role in pinpointing the locations of Quantitative Trait Loci (QTLs) associated with desirable traits like increased yield or enhanced nutritional content.

Further sections will delve deeper into the details of these various biotechnological methods. By harnessing the power of biotechnology, researchers can significantly improve berseem, leading to enhanced forage production and ultimately, improved animal nutrition.

4.8 Use of Biotechnology in Modern Berseem Breeding

Biotechnology offers a wide range of tools for berseem improvement. Molecular techniques allow researchers to understand the genetic makeup of berseem plants, directly manipulate genes and achieve in vitro plant regeneration from various plant parts. Techniques such as embryo rescue, micropropagation (including somatic embryogenesis and organogenesis), androgenic haploid production and induced mutation breeding can significantly enhance conventional breeding programs. These methods are time-saving and increase efficiency compared to traditional approaches. For instance, they can potentially reduce the time required to develop new berseem genotypes from 6 years (conventional methods) to 2 years, while simultaneously achieving greater genetic purity and lower costs (Ijaz and Haq 2021). Furthermore, biotechnology offers unique opportunities for creating variations in berseem that wouldn't be possible through conventional breeding. Examples include somaclonal variation, somatic hybridization and genetic transformation. These techniques can potentially improve berseem quality for intensive feeding systems by altering the balance of plant components and reducing the levels of undesirable compounds such as hydrocyanic acid and saponins, ultimately resulting in higher quality feed.

4.8.1 DNA Markers: Revolutionizing Berseem Improvement

DNA markers have emerged as powerful tools in crop improvement programs offering significant advantages over traditional breeding methods. They provide much-needed precision in selection, accelerate breeding efforts and enhance the evaluation and management of biodiversity and germplasm resources. Additionally, DNA markers hold promise for molecular identification, aiding in the intellectual property rights (IPR) protection of novel varieties and valuable germplasm. A key benefit of DNA markers lies in their ability to facilitate the rapid transfer of desirable genes between varieties. This allows breeders to efficiently incorporate beneficial traits from one cultivar into another. Furthermore, these markers enable the introgression of novel genes from wild relatives of berseem, potentially introducing valuable characteristics not present in cultivated varieties. Notably, DNA markers can effectively "tag" polygenic traits, which were previously challenging to analyze using traditional breeding methods. These complex traits, influenced by multiple genes, can now be more readily targeted for improvement (Mohamed et al. 2024).

Berseem research can leverage a variety of powerful molecular marker techniques. Some of the most prominent DNA marker types include:

- Restriction Fragment Length Polymorphism (RFLP): This established technique analyses variations in restriction enzyme cutting sites within DNA.
- Amplified Fragment Length Polymorphism (AFLP): This method utilizes selective PCR amplification to identify polymorphisms in restriction fragments.
- Random Amplified Polymorphic DNA (RAPD): RAPD markers employ short, random primers to amplify anonymous DNA segments, revealing polymorphisms in band patterns.
- Inter-Simple Sequence Repeat (ISSR): ISSR markers target regions between microsatellite repeats, offering a reliable and cost-effective approach.
- Simple Sequence Repeat (SSR) or Microsatellite Markers: These highly informative markers exploit polymorphism in short, tandem repeated DNA sequences.
- Single Nucleotide Polymorphism (SNP): SNPs, variations in single nucleotides within a DNA sequence, represent a highly specific and abundant marker type.

DNA and biochemical markers also play a role in berseem improvement. These markers, categorized as protein and isozyme markers, can provide valuable insights into genetic diversity. The applications of molecular techniques in berseem extend far beyond marker development. These techniques are instrumental in germplasm characterization, cultivar identification, and the detection of hybrids. Additionally, they play a crucial role in genetic mapping, facilitating the identification of Quantitative Trait Loci (QTLs) associated with desirable traits, and enabling the process of gene tagging (Aswini et al. 2023). The use of gene-associated sequences as genetic markers offers several significant advantages in the field of plant genetics, particularly in the study of legumes like berseem.

These markers leverage the power of bioinformatics and whole genome sequencing initiatives. This allows researchers to perform comparative genome analysis, enabling them to identify similarities and differences between berseem and other plant species. This comparative approach can provide valuable insights into the evolution, function and potential of genes within berseem. Utilizing gene-associated sequences facilitates the construction of function-oriented maps. These maps directly link specific genes to their functions, providing a powerful tool for understanding the genetic basis of various traits in Berseem (Kumaraswamy and Kashyap 2021). This knowledge can be crucial for targeted breeding programs aimed at improving specific characteristics like disease resistance, drought tolerance or forage quality. These markers also hold immense potential for identifying candidate genes for further investigation. By focusing on genes associated with specific traits of interest, researchers can delve deeper into the sequence-function relationships within Berseem. This allows them to understand the specific genetic mechanisms underlying these traits, paving the way for targeted manipulation and improvement. The availability of existing genetic maps, complete sequence information and markers from white clover (*Trifolium repens L.*) further enhances the utility of gene-associated sequences for berseem research. These resources provide researchers with a comprehensive set of tools for genome analysis across the entire *Trifolieae* tribe, which encompasses numerous important forage legumes like alfalfa (*Medicago sativa L.*). By utilizing linked markers identified through GWAS or other approaches, breeders can select desirable traits even before they are expressed in the plant, significantly accelerating the breeding process and improving selection efficiency.

Genomics holds immense potential to revolutionize berseem breeding by enabling the development of improved varieties with enhanced disease resistance, stress tolerance, forage quality and yield. Continued development of genomic resources, refinement of breeding strategies and capacity building will be essential to ensure the sustainable improvement of berseem and contribute to a more productive and resilient agricultural system. The foundation of successful genomics-assisted breeding lies in the development of comprehensive genomic resources. A high-quality reference genome sequence serves as a blueprint for understanding the organization and function of berseem genes (Zhou et al. 2014). This resource facilitates the identification and characterization of genes associated with desirable traits. Genetic markers that are DNA variations that can be linked to specific traits and used to track their inheritance during breeding are further used in breeding. Single nucleotide polymorphisms (SNPs) are the most widely used markers in berseem breeding due to their abundance and cost-effective genotyping methods (Tani et al. 2017).

Genome-wide association studies (GWAS), allow researchers to identify statistical associations between specific markers across the genome and complex traits like disease resistance, yield and forage quality. Genomic tools also facilitate the identification of candidate genes associated with specific traits. Functional characterization of these genes, such as their expression patterns and interaction with other genes, provides deeper insights into the underlying mechanisms of important traits. The genomic selection (GS) approach utilizes genome-wide marker information to predict the genetic merit of individuals for complex traits. This allows breeders to select superior individuals even in early generations without relying on phenotypic evaluation, leading to faster genetic gain. Despite the significant potential of genomics, several challenges remain which include

1. Limited genomic resources: While existing resources provide a good starting point, further development of high-density genetic maps, identification of functional markers and comprehensive gene expression data are necessary for in-depth understanding and efficient utilization of genomic information.
2. Integration with other breeding approaches: Integrating genomics with traditional breeding methods and advanced phenotyping techniques can accelerate genetic gain and address complex breeding objectives.
3. Capacity building and infrastructure development: Training breeders and researchers in utilizing genomic tools and establishing robust infrastructure for data analysis and management are crucial for the successful implementation of genomics in berseem breeding programs.

4.8.2 Genetic Engineering and Gene Editing

Agrobacterium-based transformation has emerged as a promising technique for introducing desirable traits into berseem (*Trifolium alexandrinum L.*). This method leverages soil bacteria, *Agrobacterium rhizogenes* as a vector to deliver genes of interest into berseem cells. Pioneering work by Tanaka et al. (2001) established this approach and employed innovative techniques to achieve successful berseem transformation. Their research revealed that the expression of the *rolB* gene within hairy roots (transformed roots induced by *Agrbacterium rhizogenes*) significantly influenced root growth potential. Maintaining an optimal level of *rolB* expression was critical for robust hairy root development. Additionally, their study suggested that endogenous indole-3-acetic acid (IAA), a plant hormone naturally produced by berseem, directly impacted root growth even without externally supplied auxins (plant growth hormones). This study highlights the potential of *Agrobacterium*-mediated transformation for berseem improvement by enabling the introduction of beneficial genes. However, further research is necessary to optimize the regeneration protocols, which will be crucial for developing transgenic berseem varieties with enhanced traits while maintaining normal plant morphology and growth.

A study by Abogadallah et al. (2011) explored the potential of genetic modification to improve berseem's tolerance to drought and salt stress, significant constraints on its cultivation. They introduced the *HARDY* gene from *Arabidopsis thaliana*, a well-known stress-resistant plant, into the berseem genome. This *HARDY* gene belongs to the AP2/ERF family of transcription factors, which play a critical role in a plant's response to environmental stresses. The study revealed that transgenic berseem lines L_2 and L_3, compared to wild-type berseem and displayed significant fresh weight increase under both drought and salt stress conditions. These increases ranged from 42% to 55% for drought stress and 38% to 95% for salt stress in L_2 and L_3 lines, respectively. Similar improvements were observed in dry weight.

Additionally, L_2 and L_3 exhibited thicker stems, potentially enhancing their resistance to lodging in the field. Line L_3 also possessed a higher number of xylem rows per vascular bundle, which could contribute to improved water transport and stress tolerance. Field trials simulating combined drought and salt stress conditions further supported these findings. Both L_2 and L_3 lines significantly outperformed wild-type berseem for fresh and dry weight with 31–46% increase, respectively. However, before widespread adoption, it is essential to acknowledge potential concerns and conduct further research regarding the long-term environmental and ecological implications of genetically modified crops.

4.8.3 Tissue Culture Applications

Biotechnology offers a promising avenue for the genetic improvement of berseem, complementing traditional breeding methods. Efficient plant regeneration through tissue culture is a critical prerequisite for utilizing these approaches, particularly those involving genetic transformation. Fortunately, advancements have been made in this area for berseem and closely related *Trifolium* species like *T. glomeratum*, *T. apertum*, and *T. resupinatum* (Mohamed et al. 2024). Researchers have successfully established protocols for in vitro plant regeneration from meristematic tissues and the development of regenerable callus cultures. Beyond establishing regeneration protocols, progress has been made in utilizing them for berseem transformation. Tanaka et al. (2001) demonstrated the successful regeneration of whole berseem plants from transformed shoots. Their approach involved a 2-month culturing period on Florilate media. Another valuable tool in the biotechnological toolbox is embryo rescue. This technique has proven effective in overcoming post-fertilization barriers encountered during interspecific crosses of berseem with species like *T. apertum*, *T. constantinopolitanum*, *T. resupinatum*, and *T. vesiculosum* (Kaur et al. 2017). Embryo rescue allows the development of hybrid embryos that might otherwise fail to mature naturally. Advancements in tissue culture regeneration and embryo rescue techniques are paving the way for the successful application of biotechnology in berseem improvement. These methods provide essential tools for genetic transformation and other biotechnological approaches, offering exciting possibilities for developing berseem varieties with improved production and productivity.

4.8.3.1 Micropropagation

In vitro propagation techniques show promise for efficient clonal multiplication of berseem (*Trifolium alexandrinum L.*) (Negi et al. 2024; Malik et al. 2023). Research has revealed that leaves are the most suitable explants for callus induction, with optimal results achieved using low concentrations of the growth regulator 2,4-D. Subsequent plant regeneration was achieved by manipulating the culture media composition. Benzyladenine (BA) effectively promoted shoot development, while indole-3-butyric acid (IBA) facilitated root formation in the regenerated plantlets. Sultan and Mohamed (1990) paved the way for utilizing *in vitro* propagation techniques for berseem, providing a valuable foundation for developing efficient and reliable methods for berseem micropropagation. Optimizing culture conditions and utilizing appropriate plant growth regulators are crucial for achieving this goal (Zayed 2013).

While research on berseem tissue culture primarily focuses on somatic embryogenesis, a method for plant regeneration through embryo formation, alternative approaches are also being explored. Somatic embryogenesis offers a robust regeneration method, but direct organogenesis, where shoots and roots develop directly from explants (plant tissues), presents a potentially more efficient option. Early research by Aly et al. (1994) established a protocol for somatic embryogenesis in berseem, outlining the steps required to produce plantlets through this method. However, Bhowal et al. (2011) investigated the feasibility of direct organogenesis in berseem. Their study identified the shoot tip as the most responsive explant for shoot development. By supplementing L2 media with specific plant growth regulators (BAP and IAA), they induced shoot formation within a remarkably short period of just 5 days. This method holds significant promise for berseem breeding programs. Direct organogenesis, particularly when focused on shoot regeneration, offers the potential to generate pure lines, which are invaluable tools for genetic improvement efforts.

Advancements in tissue culture techniques hold significant promise for improving berseem regeneration in laboratory environments. Research by Roy et al. (2005a, b) focused on optimizing callus formation, the initial step where undifferentiated cell masses develop from explants. Their study investigated the influence of various explants (plant tissue used to initiate cultures) and culture media combinations on callus formation success. The findings revealed significant interaction effects between the explant type, media composition and their combined influence on callus formation. Notably, explants derived from the petiolc (leaf stalk) cultured on a specific medium designated as "D" exhibited the highest potential for developing shoots and embryos. Following subculture, shoot formation was observed in 50% of the cultures, with many shoots successfully developing roots upon transfer to rooting media. These advancements hold tremendous potential for the future of berseem cultivation and utilization.

4.8.3.2 Embryo Rescue

Studies have demonstrated the potential of combining inter-specific hybridization with embryo rescue techniques. This approach offers a powerful tool for developing improved berseem varieties with enhanced traits not readily available within the existing gene pool. By overcoming cross-incompatibility barriers and providing optimal conditions for rescued embryos, researchers can create novel genetic combinations leading to berseem cultivars with superior agronomic performance. Kaushal et al. 2005 reported the utilization of embryo rescue and Kaushal et al.

(2006) identified an optimal medium formulation (EC3) containing kinetin and moderate sucrose levels for successful embryo germination and production of hybrids between *T. alexandrinum* and *T. resupinatum*, a species known for disease resistance and tolerance to alkaline soils. The hybrids developed displayed delayed flowering but possessed high pollen fertility (>90%), with some individuals even producing seeds through self-pollination. Similarly, Roy et al. (2004) used embryo rescue to generate hybrids between *T. alexandrinum* and *T. constantinopolitanum*, which exhibited intermediate morphological characteristics but had reduced pollen fertility (ranging from 55% to 65%). This finding highlights the importance of optimizing tissue culture conditions for maximizing embryo rescue success.

4.9 Mutation Breeding

Several researchers have explored mutation breeding using physical or chemical mutagens to introduce variation within the existing gene pool (Jindal et al. 2014). This approach involves exposing genotypes to mutagenic agents like physical or chemical, which can randomly alter the plant's DNA. While some mutations may be detrimental, others can create novel and beneficial traits. Mutation breeding has proven to be a valuable tool in overcoming the limited genetic diversity of Berseem. Researchers have developed cultivars with extended growing seasons by introducing variation and employing targeted selection, a key agronomic trait for Indian farmers facing summer forage shortages.

Conventional mutagenesis involves exposing berseem seeds to mutagenic agents such as ethyl methane sulfonate (EMS) or sodium azide (NaN_3). These agents can induce random changes in the DNA sequence of the seeds, leading to the creation of mutations in genes that control desired traits. Following mutagenesis, the M_1 (first mutant) generation seeds are grown, and individuals displaying beneficial mutations are selected for further evaluation in subsequent generations (M_2, M_3, etc.). The success of conventional mutagenesis in berseem relies on several factors, among which the selection of an appropriate mutagenic agent and dosage is crucial. High doses can increase mutation frequency but also lead to deleterious effects on plant growth and fertility. Conversely, low doses may result in fewer mutations reducing the possibilities of desirable mutants; 50% of individuals die, and maximum variation is achieved. Recent studies highlight the continued application of EMS (ethyl methanesulfonate) mutagenesis to improve berseem traits, especially for resilience against parasitic weeds. For instance, Pérez de la Vega et al. (2022) reported the partial resistance of legumes, including berseem, to *Orobanche crenata* through induced mutation breeding. Their research underscores the role of EMS mutagenesis in developing crop varieties with enhanced resistance against biotic stressors like Orobanche, contributing to higher and more reliable yields under parasitic weed pressures.

A significant success story emerged from this strategy: the induction of a longer-duration mutant in the 'Mescavi' variety through gamma-ray treatment (Sohoo et al.

1985). This mutant, later released as the 'BL-22' variety in 1988, was specifically targeted for cultivation in India's temperate and northwest zones. Similarly, 'BL-180,' released in 2006, was also developed for the northwest zone. Both cultivars share the extended growing season trait, allowing for an additional harvest of green fodder during the crucial summer fodder scarcity period. This extended duration is a highly valuable agronomic characteristic in the Indian context. Following the mutagenesis process, selection techniques play a vital role. Breeders meticulously screen the mutagenized population to identify plants exhibiting the desired mutations, such as the extended growing season observed in 'BL-22' and 'BL-180.' These desirable mutations are propagated to develop new, improved Berseem varieties like HB2.

Despite its potential, conventional mutagenesis has limitations. The induced mutations are random, making it challenging to target specific genes. Additionally, many mutations may be unwanted, requiring extensive screening to identify beneficial ones. The effectiveness of this approach depends on optimizing mutagenic treatments and implementing efficient selection strategies. Future research efforts should integrate conventional mutagenesis with newer techniques like molecular markers to facilitate the identification and characterization of desirable mutations, consequently accelerating berseem fodder improvement programs.

4.10 Conclusions and Prospects

Egyptian clover, being a major multicut winter season forage crop provides protein-rich nutritional fodder to livestock and contributes to the sustainability of the rice-wheat cropping system. However, the genetic improvement in this crop is hampered due to its narrow genetic base and lack of variability for fodder yield-related traits. Genetic variation in Egyptian clover can be created using mutagens, inter-specific hybrid formation and through the transfer of germplasm from its native location. Use of advanced tools like SSR markers, molecular linkage maps, and mapping of forage quality and biomass-contributing traits, there is enormous potential in Berseem to fortify the genomic resources. Future research on Berseem should focus on the exploration of its genetic resources. This includes investigating the genetic diversity within berseem germplasm collections and identifying novel genes associated with desirable traits. Additionally, research on the beneficial microorganisms associated with berseem, particularly those involved in nitrogen fixation and stress tolerance, could be valuable for the development of improved cultivars. Current research initiatives in Berseem focus on utilizing biotechnology tools to develop climate-resilient cultivars with improved water-use efficiency, nitrogen fixation, and feed quality. Techniques like DNA fingerprinting can authenticate berseem varieties' seeds and identify potential adulteration. Furthermore, molecular markers can be utilized to monitor the levels of anti-nutritional factors in berseem forage, allowing for harvesting and processing practices that optimize feed quality for livestock. These advancements are crucial for ensuring the sustainability of berseem production in a changing climate and contributing to a more climate-smart agricultural future.

Appendix 4.1 Research Institutes and Online Resources Available for Egyptian Clover Breeding

Institution	Area of specialization	Research activities	Website
ICAR – Indian Grassland and Fodder Research Institute, Jhansi, India	Exclusively works on grassland and fodder crops including berseem	Forage germplasm collection, characterization and utilization, pre-breeding, genetic improvement of forage crops, forage production, protection and conservation technologies	https://igfri.icar.gov.in/
Punjab Agricultural University, Ludhiana, India	Mutation and Polycross breeding	Germplasm collection, characterization and utilization, and mutation breeding	https://pau.edu/
Jawaharlal Nehru Krishi Vishwa Vidyalaya, Jabalpur, India	Mutation breeding	Germplasm collection, characterization and utilization, polycross and mutation breeding	https://jnkvv.org/
Chaudhary Charan Singh Haryana Agricultural University, Hisar, India	Mutation breeding	Germplasm collection, characterization and utilization, and mutation breeding	https://hau.ac.in/

References

Abogadallah GM, Nada RM, Malinowski R, Quick P (2011) Overexpression of HARDY, an AP2/ERF gene from Arabidopsis, improves drought and salt tolerance by reducing transpiration and sodium uptake in transgenic *Trifolium alexandrinum* L. Planta 233(6):1265–1276

Aly MAM, Khalil MFM, Moghieb REA et al (1994) Somatic embryogenesis in berseem (*Trifolium alexandrinum* L).The influence genetic background. Egypt J Genet Cytology 23:151–161

Aswini MS, Lenka B, Shaniware YA, Pandey P, Dash AP, Haokip SW (2023) The role of genetics and plant breeding for crop improvement: current progress and future prospects. Int J Plant Soil Sci 35(20):190–202

Badr A, El-Shazly H (2012) Molecular approaches to origin, ancestry and domestication history of crop plants: Barley and clover as examples. J Genet Eng Biotechnol 10(1):1–12

Badr A, El-Shazly HH, Watson LE (2008) Origin and ancestry of Egyptian clover (*Trifolium alexandrinum* L.) as revealed by AFLP markers. Genet Resour Crop Evol 55:21–31

Barik AK, Hedayetullah MD (2019) Berseem (Egyptian clover). In: Forage crops of the world. Apple Academic Press, Waretown, pp 213–230

Beale PE, Bounejemate M, Lahlou A et al (1993) Distribution of annual Trifolium species in Morocco. Aust J Agric Res 44:1303–1310

Bedard J, Lapointe G (1991) Responses of hayfield vegetation to spring grazing by greater snow geese. J Appl Ecol 28:187–193

Bennett SJ, Maxted N, Sabanci CO (1998) The ecogeography and collection of grain, forage and pasture legumes in south-west Turkey. Genet Resour Crop Evol 45:253–262

Bhaskar RB, Malavlya DR, Roy AK, Kaushal P (2002) Evaluation of exotic Trifolium accessions for disease incidence and resistance. In: Abstracts of national symposium on grassland and fodder research in the New Millennium held at IGFRI, Jhansi, pp 31–32

Bhowal M, Cherian KJ, Das L (2011) Direct organogenesis in fodder crop *Trifolium alexandruim* L. J Environ Res Develop 5(4):892–897
Bouazzama B, Mailhol JC, Bouaziz A et al (2017) Modelling of berseem (*Trifolium Alexandrinum* L.) growth and biomass yield under different levels of water stress in Tadla, Morocco. Revue Marocaine des Sciences Agronomiques et Vétérinaires 5(4):120
Bounejemate M, Gintzburger G, Christiansan S et al (1999) Forage and pasture legume genetic resources at ICARDA. In: Bennett SJ, Cocks PS (eds) Genetic resources of Mediterranean pasture and forage legumes. Kluwer, Dordrecht, pp 163–170
Can E, Arslan M, Sener O, Daghan H (2013) Response of strawberry clover (*Trifolium fragiferum* L.) to salinity stress. Res Crops 14(2):576–584
Caudillo MC (2023) Evaluating pollinator populations and lamb productivity from simple, diverse, and legume pastures in a dual-use pasture system
Chand S, Roy AK, Singh T et al (2023) Twenty-four years lucerne (Medicago sativa L.) breeder seed production in India: a retrospective study. Front Plant Sci 14:1259967. https://doi.org/10.3389/fpls.2023.1259967
Chand S, Roy AK, Kumar S et al (2024) Quality seed production scenario of Egyptian clover (*Trifolium alexandrinum*) in India: A 24-year retrospective analysis. Heliyon 10(15). https://doi.org/10.1016/j.heliyon.2024.e35735
Chapman DF, Lee JM, Rossi L et al (2016) White clover: the forgotten component of high-producing pastures? Anim Prod Sci 57(7):1269–1276
Cleveland RW (1985) Reproductive cycle and cytogenetics. Clover Sci Technol 25:71–110
Díez MJ, De la Rosa L, Martín I et al (2018) Plant genebanks: present situation and proposals for their improvement. The case of the Spanish network. Front Plant Sci 9:1794. https://doi.org/10.3389/fpls.2018.01794
Dillard SL (2023) Cool-season annual legumes and forbs for grazing and cover crop systems
Dwivedi S (2024) Fodder crops' potential: maximizing crop yield and income with berseem clover in crop diversification. Krishi Jagran. https://krishijagran.com/agripedia/fodder-crops-potential-maximizing-crop-yield-and-income-with-berseem-clover-in-crop-diversification/
Ellison NW, Liston A, Steiner JJ et al (2006) Molecular phylogenetics of the clover genus (*Trifolium* – Leguminosae). Mol Phylogenet Evol 39:688–705
Enriquez-Hidalgo D, Teixeira DL, Pinheiro Machado Filho LC et al (2020) Incorporating a fresh mixed annual ryegrass and berseem clover forage into the winter diet of dairy cows resulted in reduced milk yield, but reduced nitrogen excretion and reduced methane yield. Front Vet Sci 7: 576944
Evans AM (1976) Clovers. In: Simmonds NW (ed) Evolution of crop plants. Longman, London and New York, pp 175–179
Frame J, Laidlaw AS (1998) Managing white clover in mixed swards principles and practice. Pastos 28(1):5–33
Frame J, Laidlaw AS (2019) Prospects for temperate forage legumes. In: Grasslands. CRC Press, pp 3–28
Gornish ES, Ambrozio dos Santos P (2016) Invasive species cover, soil type, and grazing interact to predict long-term grassland restoration success. Restor Ecol 24(2):222–229
Greene SL, Kisha TJ, Yu LX, Parra-Quijano M (2014) Conserving plants in gene banks and nature: investigating complementarity with *Trifolium thompsonii* Morton. PLoS One 9(8):e105145
Hall EJ, Hughes SJ, Humphries AW, Corkrey R (2013) Habitat and plant diversity of *Trifolium tumens* (Steven ex M. Bieb.) collected in Azerbaijan and its characterisation and field evaluation in Tasmania, Australia. Crop Pasture Sci 64(4):374–387
Hasan N, Choudhary S, Naaz N et al (2021) Recent advancements in molecular marker-assisted selection and applications in plant breeding programmes. J Genet Eng Biotechnol 19(1):128
Hely FW (1982) Shoot and root responses to phosphorus by *Trifolium ambiguum* and Trifolium repens in a montane environment. N Z J Agric Res 25(1):77–85
Hossain M, Islam MM, Emon RM et al (2020) Microsatellite-based DNA fingerprinting and genetic analysis of some selected aus rice (Oryza sativa L.) genotypes. Ann Agric Crop Sci 5(3):1066

ICAR-NBPGR | MTS Database (n.d.). http://pgrinformatics.nbpgr.ernet.in/mts/mtsindex.aspx
Ijaz S, Haq IU (2021) Biotechnological applications for developing resistance against biotic and abiotic stresses and other quality traits in fodder crops. In: Sustainable winter fodder. CRC Press, pp 31–80
Ison RL, Parson EB (1992) Comparative growth and development of Kenya clover. Trop Grassl 26: 40–50
Jacques A, Duclos D, Danchin-Burge C, Mercat MJ, Tixier-Boichard M, Restoux G (2024) Assessing the potential of germplasm collections for the management of genetic diversity: the case of the French National Cryobank. Peer Commun J 4:1–24
Jindal Y, Dahiya GS, Phogat DS (2014) Evaluation of genetic variability in berseem after mutagenesis. Forage Research. Retrieved from https://forageresearch.in/wp-content/uploads/2014/03/201-204.pdf
Jones RA (2012) Virus diseases of annual pasture legumes: incidences, losses, epidemiology, and management. Crop Pasture Sci 63(5):399–418
Kaur A, Kaur KP, Kalia A et al (2017) Generation of interspecific hybrids between *Trifolium vesiculosum* and *T. alexandrinum* using embryo rescue. Euphytica 213:253. https://doi.org/10.1007/s10681-017-2042-x
Kaushal MP, Kumar S, Jindal SK (2005) Interspecific hybridization between berseem (*Trifolium alexandrinum*) and Indian clover (*Trifolium resupinatum*) using embryo rescue. Euphytica 141(1–2):111–116
Kaushal P, Tiwari A, Roy AK, Malaviya DR et al (2006) In vitro regeneration of Trifolium glomeratum. Biol Plant 50(4):693–696
Knight WE (1985a) Crimson clover. In: Clover science and technology, vol 25. Wiley, Hoboken, pp 491–502
Knight WE (1985b) Miscellaneous annual clovers. In: Clover science and technology, vol 25. Wiley, Hoboken, pp 547–562
Kole C (2011) Trifolium. In: Wild crop relatives: genomic and breeding resources. Springer, pp 249–272. https://doi.org/10.1007/978-3-642-14387-8_13
Kumar A, Dhanda SK, Pareek A, Bhardwaj S (2024) Enhancing production of berseem through integrated crop management practices in Haryana. Legume Res Int J 47(2):231–233
Kumaraswamy HH, Kashyap BK (2021) Genome mapping tools: current research and future prospects. In: Microbiomes and plant health, pp 125–202
Kumari P, Malik VK, Kumar V, Satpal S, Pahuja SK (2023) Nutritional enhancement of cultivated fodders using potential plant breeding strategies-a review. Forage Res 49(1):29–38
Malaviya DR, Roy AK, Kaushal P et al (2004) Evaluation of *Trifolium* species for defining multiple use gene pool for tropical Trifolium species. Ind J Genet 64(3):251–252
Malik P, Prajapati M, Bhoria S et al (2023) Efficient in vitro direct plant regeneration from mature cotyledon explants of berseem (Egyptian Clover, *Trifolium alexandrinum* L.). Ann Biol 39(2): 150–156
Mehta R, Swaminathan M (1957) Studies on induced polyploids in forage crops. 1. Survey of previous work. Ind J Genet Plant Breed 17:27–57
Mohamed HM, El-Assal SEDSM, Gamal El-Din AKY, El-Khishin DA (2024) Analysis of genetic diversity and population structure in some Egyptian berseem (*Trifolium alexandrinum*) accessions based on ISSR, SCoT and SRAP markers. Genet Resour Crop Evol 71:1–15
Moorby JM, Ellis NM, Davies DR (2016) Assessment of dietary ratios of red clover and corn silages on milk production and milk quality in dairy cows. J Dairy Sci 99(10):7982–7992
Nasar J, Ahmad M, Gitari H et al (2024) Maize/soybean intercropping increases nutrient uptake, crop yield and modifies soil physio-chemical characteristics and enzymatic activities in the subtropical humid region based in Southwest China. BMC Plant Biol 24(1):434
Negi S, Singh P, Trivedi VL et al (2024) The current trends and research progress globally in the plant tissue culture: 90 years of investigation. Plant Cell Tissue Organ Cult 157(3):73
Nichols PGH, Loi A, Nutt BJ et al (2007) New annual and short-lived perennial pasture legumes for Australian agriculture—15 years of revolution. Field Crop Res 104(1–3):10–23

Pérez de la Vega M, García García P, Gutierrez-Gonzalez JJ, Sáenz de Miera LE (2022) Tackling lentil biotic stresses in the genomic era. In: Genomic designing for biotic stress resistant pulse crops. Springer International Publishing, Cham, pp 253–308

Ranjbar GA (2007) Forage and hay yield performance of different berseem clover (*Trifolium alexandinum* L.) genotypes in Mazandaran conditions. Asian J Plant Sci 6(6):1006–1011

Rouz S, Jeddi F, Zouaghi M (2016) Study of genetic diversity of berseem accessions (*Trifolium alexandrinum* l. Spp) with RAPD markers

Roy AK, Malaviya DR, Kaushal P et al (2004) Interspecific hybridization of *Trifolium alexandrinum* × *T. constantinopolitanum* using embryo rescue. Plant Cell Rep 22(9):705–710

Roy AK, Malaviya DR, Kaushal P (2005a) Pollination behaviour among different breeding populations of Egyptian clover. Plant Breed 124(2):171–175

Roy AK, Tiwari A, Malaviya DR, Kaushal P, Kumar B (2005b) Regeneration efficiency in *Trifolium apertum* as influenced by explant media interaction. Cytologia 70(2):153–159

Roy AK, Agrawal RK, Chand S et al (2020) Database of forage crop varieties: 2020. ICAR-AICRP on Forage Crops & Utilization, ICAR-IGFRI, Jhansi, pp 1–362. ISBN-978-81-948917-2-7

Samtiya M, Aluko RE, Dhewa T (2020) Plant food anti-nutritional factors and their reduction strategies: an overview. Food Prod Process Nutr 2:1–14

Satpal AS, Kumari P, Panchta R (2023) Seed production of forage crop. In: Bhukar A, Kumar A, Singh S (eds) Compendium seed technology. Kalyani Publishers

Seth D, Ghosh P, Roy L (2023) Integrated farming systems a way for long-term farming viability: a review. Plant Arch 23(1):185–190

Sevanayak D, Singh T, Radhakrishna A (2017) Effect of drought stress on biomass and drought adaptive traits in berseem (*Trifolium alexandrinum* L.). In: National symposium—new directions in managing forage resources and livestock productivity in 21st century: challenges and opportunity, pp 4–17

Shabir MA, Khan HZ, Shabir K, Altaf F, Asad M, Nadeem M, Arshad U, Ibrahim M (2020) Egyptian clover green manuring improved grain nutritional contents, productivity and soil health of spring maize with different nitrogen rates. Commun Soil Sci Plant Anal 51(15): 1969–1978

Sharma VV, Murdia PC (1974) Utilization of berseem hay by ruminants. J Agric Sci 83(2):289–293

Singh T, Malaviya DR, Kaushal P (2015) Genetic analysis of some morphological traits in Egyptian clover (*Trifolium alexandrinum* L.)

Singh DK, Gupta S, Sahu NP, Srivastava·PP, Sardar P, Deo AD, Aklakur M (2019a) Chemical composition of berseem (*Trifolium alexandrinum*) leaf meal and leaf protein concentrate. J Entomol Zool Stud 7(4):1418–1421

Singh T, Radhakrishna A, Nayak DS, Malaviya DR (2019b) Genetic improvement of berseem (*Trifolium alexandrinum*) in India: Current status and prospects. Int J Curr Microbiol App Sci 8(1):3028–3036

Singh D, Choudhary A, Uikey V (2020a) Comparative analysis of exotic and notified berseem (*Trifolium alexandrinum* L.) varieties for fodder, quality and nutrients uptake. Forage Res 46(2): 168–175

Singh T, Radhakrishna A, Malaviya DR, Dheeravathu SN (2020b) Biomass accumulation, phenology and seed yield of *Trifolium alexandrinum* ecotypes evaluated in Central India. Trop Grassl Forrajes tropicales 8(1):28–34

Singh YP, Tomar SS, Singh S, Singh AK, Gupta RK (2021) Influence of tillage and relay cropping of berseem in mustard–pearl millet system on soil quality, productivity of crops and water. Agric Res 10(4):645–655

Singh G, Ward B, Levi A, Cutulle M (2022) Weed management by in situ cover crops and anaerobic soil disinfestation in plasticulture. Agronomy 12(11):2754

Sohoo MS, Beri SM, Bhardwaj BL (1985) Radiation induced late flowering and high yielding mutant in berseem (*Trifolium alexandrinum* L.). J Nucl Agric Biol 14(4):136–138

Sultan F, Mohamed M (1990) MSc thesis, Ain Shams University, Faculty of Agriculture, Department of Agronomyhttps://www.academia.edu/33725205/Regeneration_and_transformation_efficiencies_among_five_Egyptian_clover_cultivars_Trifolium_alexandrinum_

Tanaka N, Fujikawa Y, Aly MAM et al (2001) Proliferation and rol gene expression in hairy root lines of Egyptian clover. Plant Cell Tissue Organ Cult 66:175–182

Tani E, Abraham E, Chachalis D, Travlos I (2017) Molecular, genetic and agronomic approaches to utilizing pulses as cover crops and green manure into cropping systems. Int J Mol Sci 18(6): 1202

Taylor NL, Gibson PB, Knight WE (1977) Genetic vulnerability and germplasm resources of the true clovers. Crop Sci 17:632–634

Walter KS, Gillett HJ (1998) I997 IUCN red list of threatened plants. In: Compile by the world conservation monitoring centre. IUCN The World Conservation Union, Gland and Cambridge, UK, pp 1–862

Williams WM (2014) Trifolium interspecific hybridisation: widening the white clover gene pool. Crop Pasture Sci 65(11):1091–1106

Zayed EM (2013) Applications of biotechnology on Egyptian clover [(Berseem) (Trifolium alexandrinum L.)]

Zhou Y, O'Connell R, Maguire P et al (2014) High throughput secondary electron imaging of organic residues on a graphene surface. Sci Rep 4:7032. https://doi.org/10.1038/srep07032

Zohary M, Heller D (1984) The genus Trifolium. Israel Academy of Sciences and Humanities

Chapter 5
Maguey Espadin (*Agave angustifolia* Haw.) Breeding and Biotechnology

Ernestina Valadez-Moctezuma and Samir Samah

Abstract Agave plants are perennial evergreen xerophytes of varying diameters and heights. They consist of large, succulent leaves with spines on the edges and one at the tip, as well as a short stem called a "pineapple". Agave has held significant importance in Mexico since Mesoamerican times due to its divers derived products. Mexico is home to the greatest diversity of these plants, with the Agave genus comprising approximately 200 species and around 119 species are endemic to Mexico. Agave is distributed from the southern United States of America to South America. Today, several Agave species have been found to have multiple uses beyond those previously recognized, including general domestic use, alcoholic beverages, pharmaceuticals, medicinal, the food industry, textile, construction, agricultural, and ornamental purposes. Because Agave is not a plant used directly for human consumption, there have been limited initiatives for improving and selecting desirable characteristics in Agave. However, some successful results have been reported in plant mother selection, massive propagation, micropropagation and various biotechnological applications. Although there are numerous studies addressed to study the genetic variability of Agaves, not enough information has been generated that can be directly applied in breeding programs. The objective of this chapter is to review the current knowledge of the general biological, botanical, and agroecological characteristics, the products useful for man obtained from Agave; as well as its cultivation in the field, uses, applications, and technological advances, with an emphasis on *Agave angustifolia*.

Keywords Breeding · Century plant · Fiber · Maguey espadin (*Agave angustifolia* Haw.) · Mezcal · Micropropagation

E. Valadez-Moctezuma (✉)
Universidad Autónoma Chapingo, Departamento de Fitotecnia, Chapingo, State of Mexico, Mexico
e-mail: evaladezm@chapingo.mx

S. Samah
Biogenetix-Lab S. de R.L. de C.V, Ixtapaluca, State of Mexico, Mexico

K. F. M. Salem et al. (eds.), *Breeding and Biotechnology of Leaf, Fruit, and Seed Fiber Crops*, Advances in Plant Breeding Strategies 11,
https://doi.org/10.1007/978-3-032-00407-9_5

5.1 Introduction

The different plants of the genus Agave can grow and prosper in arid and semiarid environments since they developed characteristic morphological and physiological properties. Agave species are recognized to be obligate Crassulacean Acid Metabolism (CAM). This physiological condition allows these plants to produce some useful derivative products for humans, such as fiber and food. Moreover, drought and heat tolerance and water-use efficiency have allowed Agave plants to provide valuable products for humans for ages (Davis and Ortiz-Cano 2023). About 75% of the 200 known Agave species are distributed throughout Mexico, where they are recognized as native species. Agaves are used for the production of food, fiber, construction materials, fermented beverages and biofuel, Also, they are proven as antimicrobial and anti-inflammatory agents, and dietary supplements (Eguiarte et al. 2021). In Mexico, farmers harvest plants from the wild and traditional semi-domestication cultivation, from forest and grassland areas. Historically, Agave plants were used to generate several products as brooms, sandals and beverages, being the most popular traditional use that continues at present is to make fermented pulque (Escalante et al. 2016).

In this chapter, various aspects of Agave species are reviewed, with special attention to *A. angustifolia*. Topics addressed include the botanical and morphological description of Agaves, and their various uses, mainly the production of fibers and alcoholic beverages. Topics about the type of reproduction and transplant of Agave plants, and the recent applications of digital farming and artificial intelligence are described. Moreover, we review the accomplishments and prospects of Agave improvement and biotechnology. The chapter covers the efforts on the characterization and conservation of Agave species, the breeding strategies, including traditional breeding and its limitations, conventional hybridization and hybrid cultivars. Furthermore, we address topics about molecular breeding including genetic transformation and mutagenesis, molecular marker-assisted breeding and genomics. Finally, special attention was paid to *in vitro* cultivation, as it is the improvement method that has yielded the best results.

5.1.1 Botanical Classification and Distribution

The Agaves are endemic plants from America; are perennial plants that belong to the Asparagales order, and the Family Asparagaceae (Fig. 5.1). Eguiarte and Souza (2007), using molecular evidence, date the origin of the family at 25.8–11.7 million years ago, and at 10 million years for the genus, considering it a very recent genus with one of the most spectacular adaptive radiations. The number of species reported is variable, with some authors mentioning around 200, while others indicate 200–300 species. Agave is distributed from the southern United States of America to South America. Mexico is considered the principal center of origin and

Fig. 5.1 *Agave angustifolia* plant. (Source: Photograph by Valadez-Moctezuma E)

distribution of the genus (Fig. 5.2), with more than 75% in its territory, of which 119 species are endemic (García-Mendoza 2011). Some species have also been distributed through dry tropical regions and the Mediterranean basin (Davis et al. 2011).

Agave plants are hermaphrodites and monocotyledons, perennial evergreen xerophytes composed mainly of large succulent leaves with spines on the edges, and short stems called "pineapple". Its leaves are straight, narrow and rigid, light green to grayish in color. Its thorns are small, it has branched inflorescences up to 5 m high. The leaves are organized in rosettes at different levels and present one terminal spine (García-Mendoza 2011). Agaves can grow up to 2 m tall, and the roots penetrate around 30 cm into the ground. The life cycle in this genus varies among species, varying between 5–15 years to produce large flowering stalks of 2–15 m tall (Gentry 1982). The inflorescence, called panicle, generally self-pollinates and forms seeds, which are practically sterile. Agave plants are monocarpic, meaning that they bloom only once and then die. Due to their biology, Agaves survive in extreme conditions of salinity, drought, high temperatures, frosts, intense solar radiation, and scarce and nutritionally poor soils. They are found in habitats where few species survive, such as deserts, hillsides, coastal salt flats, and even humid tropical jungles as epiphytic plants (Escobar-Guzman et al. 2008). It reproduces by seed, bulbil and suckers.

Fig. 5.2 Distribution of Agaves species in Mexico. (Source: Edited by Valadez-Moctezuma E. from CONABIO (2006))

Specifically, Agave angustifolia Haw. is one of the Mexican species with wide morphological variation. Its distribution in Mexico is the widest compared to the other Agave species. In Mexico, *A. angustifolia* is recognized by different names according to the dialect of the region: chelem, doba-yej hamoc, juya cuu, yavi incoyo, amole, bacanora, maguey de campo, espadilla, zapupe, among others. In the state of Oaxaca, most of the mezcal drinks produced come from this Agave, which is grown for this purpose because it contains a high sugar content, matures in less time than other Agaves (4–5 years) and spreads easily through suckers. Due to this broad synonymy, it has been shown that some of the species, from which important products have been produced, are varieties of *A. angustifolia* but with different levels of ploidy, among them, *Agave fourcroydes* (2x and 5x); *Agave tequilana* (2x and 3x), *Agave Pacifica* (bacanora), among others (CONABIO 2006; Palomino et al. 2007).

Kingdom	Plantae
Subkingdom	Viridiplantae
Infrakingdom	Streptophyta
Superdivision	Embryophyta
Division	Tracheophyta
Subdivision	Spermatophytina
Class	Magnoliopsida
Superorder	Lilianae–monocotyledons
Order	Asparagales

Family	Asparagaceae
Genus	Agave L.
Species	*Agave angustifolia* Haw.

Agaves have crassulacean acid metabolism (CAM), which reduces water consumption and transpiration during the day, fixing CO_2, mainly at night with the help of the enzyme PEP carboxylase, and storing the product of the four-carbon reaction in vacuoles. Then, during the light period, CO_2 is assimilated in the chloroplasts by the C3 cycle. These plants assimilate carbohydrates through photosynthesis during the light day (Larcher 2003). At night, they open their stomata to release fewer amounts of water as is normal during the day. CAM metabolism allows plants to adapt and survive in semi-arid environments, with limited water or CO_2. This type of metabolism is found in more than 33 families of plants, with cacti (*Opuntia* spp., *Mammillaria* spp.) standing out. Among the adaptive advantages of CAM plants, are high efficiency in water use, better potential in net CO_2 absorption, resistance to drought and high biomass production, among others (Andrade et al. 2007; Niechayev et al. 2019). Despite the recent understanding of CAM, historical records suggest that the earliest American cultures also acknowledged the value of CAM plants. In particular, *Agave* plants have a valuable cultural heritage that provides products with commercial value (Davis and Ortiz-Cano 2023).

5.1.2 Presence of Agave in Mexico and Its Main Uses

The richness of the Agave genus in Mexico dates back around 8–10 million years in the territories of the Mexican Altiplano and has been used since approximately 8000 BC by its inhabitants. This natural region is located from the center of the country to the north; its eastern and western limits are marked by the two Sierra Madre, Oriental and Occidental (Fig. 5.2) (CONABIO 2006; Esparza-Ibarra et al. 2015). Numerous Agave species have been exploited in daily life from the pre-Columbian era until now. In Aztec cosmogony, the Maguey goddess was named "Diosa Mayahuel''for all the benefits that these plants offer (Nava-Cruz et al. 2014).

According to Pérez-Zavala et al. (2020), in México, the Agave plant has multiple uses with the main ones being: (1) Food and cooking: Agaves provides products for stews, candy, barbecue wrapping, mixiotes, white worms (*Acentrocneme hesperiaris*), red worms (*Hypopta agavis*), which are edible pulque bread, (2) Domestic: soap and natural detergents, shampoo, pots, chopsticks to extract edible worms, cellulose products for paper, cellulose and glucosides (Nava-Cruz et al. 2014), (3) Beverages: vinegar, syrup, aguamiel, honey, and spirit beverages such as mezcal, tequila, pulque, bacanora, and sotol (4) Nutraceuticals: Agave carbohydrates have been used as substitutes for sugars and fats in the food industry (Ortíz-Basurto et al. 2008), (5) Ornamental: accessories like earrings and rings, floral arches, Christmas decorations, gardens, crafts and basketry, (6) Agricultural: living fence, anti-erosion, leader plant of ecosystems, compost (fertilizer), (7) Forage: feed for domestic animals, (8) Building: fences, shacks, pens, tiles to cover the roof, collection channels for rainwater, (9) Natural Fibers: basketry, cordage, straw, brushes, polishing pads, woven clothes, (10) Industry Fibers: natural fibers are combined with cement, resins and other materials, to create materials mainly for construction, among others. These combined materials are more environmentally friendly, more manageable and lighter than traditional ones (Narváez-Zapata and Sánchez-Teyer 2009), (11) The pharmaceutical industry: includes chemicals, drugs and steroidal products (saponins), (12) Medical Importance: Agave has traditionally been used to treat some conditions like blows, internal injuries, and scurvy prevention. They used to treat diseases caused by many bacteria (*Streptococcus*, *Salmonella enterica*, *Escherichia coli*, *Pseudomonas aeruginosa*, *Shigella dysenteriae*, *Enterococcus faecalis*), as well as antifungal and antiparasitic diseases (Verastegui et al. 2008; Orestes et al. 2008); and oxidative stress (Ahumada-Santos et al. 2013). In addition, agaves have antihypertensive, anti-inflammatory and immunomodulatory effects (Duncan et al. 1999; Da Silva et al. 2002), (13) Metabolites: Agave plants contain various secondary metabolites viz. triterpenes, volatile coumarins, alkaloids, flavonoids, steroids, tannins, cardiotonic and reducing sugars, and (14) Biotechnology: among other uses the potential use in bioethanol production. Also, the CAM physiology condition allows the agave plant the ability to survive and produce in arid climates (Parascanu et al. 2021).

5.1.3 Main Agave Products Produced Biotechnologically

Given the architecture, physiology and biochemistry of Agave plants, biotechnological techniques have been applied to maximize their use. While many of its derivatives have been obtained through technical or traditional methods, the usefulness of omics techniques has quickly become apparent for these plants. Results obtained to date with genomic, transcriptomic, metagenomic (on Mezcal and Tequila fermentation), and metabolomic have allowed us to learn much more about the biology of Agave.

Biotechnological applications include the production of alcoholic beverages (both technical and traditional), industrial ethanol and bio-fuels based on the carbohydrates produced by Agave plants. Additionally, products used in pharmacy, chemistry and biomaterials are based on cellulose (Ruíz et al. 2011). Cellulase production involves microorganisms capable of degrading lignocellulosic material such as aerobic and anaerobic bacteria as well as several fungi like *Fusarium*, *Humicola*, *Penicillium*, and *Trichoderma*. Inulinase production involves fructanase to generate fructose syrups from inulin (Nava-Cruz et al. 2014). Three of these biotechnological developments involving *A. angustifolia* are described below.

5.1.3.1 Mezcal Production

Mezcal is a Mexican distilled alcoholic beverage produced from the Agave pineapples (stems). The production process of mezcal tequila is similar, with the difference that the mezcal production is manual (not industrialized) and different species of Agave can be used for its preparation. The word "mezcal" has its origin in the Nahuatl "mexcalli" or "cooked Agave", and is the name given to distilled alcoholic beverages obtained from the fermented carbohydrate (fructans) present in the cooked pineapples of various species of maguey. The production of Mezcal is a Mexican tradition and is carried out in rural areas, and with the main species used being *A. angustifolia* var. espadín (Aguierre-Dugua and Eguiarte 2013).

However, there are other Agave species, wild and cultivated, that are used depending on the region for the production of this drink. These include *Agave karwinskii* (cirial or cuishe), *A. potatorum* (tobalá), *A. marmorata* (tepeztate), *A. rhodacantha* (mexicano), *A. convallis* (jabalí o javalín), *A. americana* (coyote), *A. salmiana* (pulquero), *A. cantala* and *A. macrocantha* (mexicano o barril). Additionally, the use of *A. megalodanta* and *A. lyoboa* has also been reported, which correspond to recently discovered new species (species considered in danger of extinction), and which have also been used for the production of this drink (García-Mendoza et al. 2019). Some of these species are shown in Fig. 5.3. Sometimes, mixtures of species are made a create a homogeneous must resulting in a Mezcal known as "assembly", that acquires characteristics of each maguey used for its production.

The production of artisanal mezcal, mainly during the fermentation process is a crucial step. A complex and variable microbiota is present and involved in this process. Various environmental and plant-specific factors come into play too. These factors promote the growth of native microorganisms during fermentation, which in turn provide unique sensory characteristics to the final product. By developing spontaneously with native microorganisms, the production of compounds that give organoleptic qualities to the final product is favored. Having precise knowledge of the flora involved in mezcal production opens up opportunities for creating products

Fig. 5.3 Some Agaves species used for the preparation of the mezcal assembly: (1) Madrecuishe (*A. karwinskii* Zucc.), (2) Bicuishe (*A. karwinskii* Zucc.), (3) Coyote (*A. americana* L.) and (4) Pulquero (*A. salmiana* Otto ex Salm.). (Source: Photographs by Valadez-Moctezuma E)

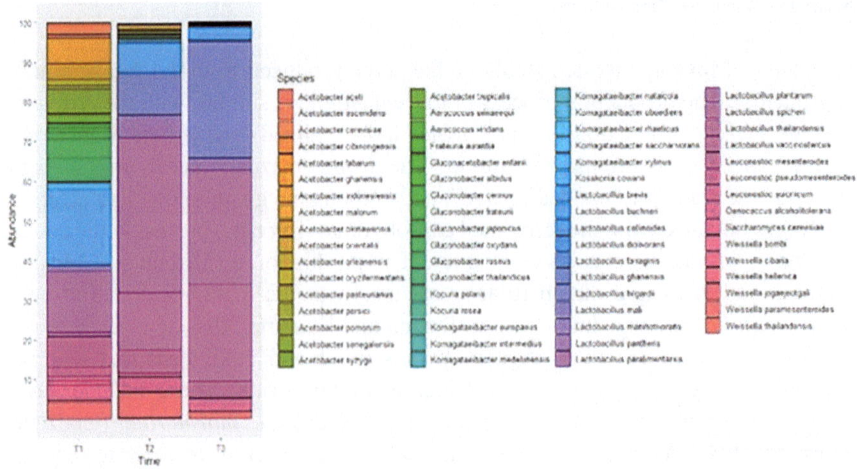

Fig. 5.4 Diversity of microorganisms identified with the MetaPhlAn program during must fermentation: Day 1 (freshly baked and crushed Agaves), T2 (fermented for 3 days) and T3 (fermented for 6 days). (Photographs by Valadez-Moctezuma E)

with more complex and enriched organoleptic characteristics. This is achieved through the combination of maguey species and native flora (Reyes-Carmona 2023; Torres-Velázquez et al. 2022).

In our laboratory, during the fermentation of assembled Agaves for mezcal production, we identified diverse microorganisms present in fresh must-processed mixtures of different Agave species through metagenomics analyses. Figure 5.4 shows the diversity of microorganisms identified at three different times (T1: 1 day, T2: 3 days and T3: 6 days) during the fermentation process. In this study, nine main genera and 66 microbial species were identified; whose presence was changed during the sampled times, except the Saccharomycetaceae family, which was present in all three times. The participation of different species such as Lactobacillus, Leuconostoc, Acetobacter and Saccharomyces provide different aromas to the final product, among other activities (Valadez-Moctezuma et al. unpublished data). This microbial richness is directly linked to the different Agaves involved in the assembly, their origin and whether they are cultivated or wild (Reyes-Carmona 2023).

The mezcal production process consists of five main steps: harvesting, cooking the pineapples, juice extraction, fermentation (spontaneous), distillation and sometimes maturation of the drink (Lappe-Oliveras et al. 2008) (Fig. 5.5). The biochemical changes that determine the primary sensory characteristics of mezcal occur during the fermentation of Agave juice, which continues for several days and is carried out spontaneously by various natural microorganisms. The production process varies depending on the agave growing region. The final result of production is influenced by several factors such as Agave species, cultivation practices and regional climatic conditions (Torres-Velázquez et al. 2022).

Fig. 5.5 Production process of artisanal mezcal: (**a**) cutting and pineapple selection, (**b**) pineapple cooking, (**c**) fermentation, (**d**) distillation, (**e**) alcohol determination, (**f**) bottled product. (Source: Photographs by Valadez-Moctezuma E)

The destination process of tequila and mezcal is very similar, with the main difference being that tequila is produced from the *A. tequilana* Weber blue variety, and is made in an industrialized way. On the other hand, mezcal production is generally manual. Morphological evidence suggests that *A. tequilana* is also a variety of *A. angustifolia*, and its production process has a Designation of Origin. In Mexico, a large industry of mezcal and tequila has spread since 2003, the United States, Spain and Germany, being the major consumers of tequila. For mezcal, the majority of the production is exported to the United States of America, China and Taiwan (Narvácz-Zapata and Sánchez-Teyer 2009; Nava-Cruz et al. 2014).

Tequila production reached 598.7 million liters in 2023 (Tequila Regulatory Council). It is estimated that tequila production will increase by approximately 27% over the next 10 years, while mezcal production is expected to increase by 8% (https://old.crt.org.mx). Figure 5.6 shows the production trend of mezcal from 2010 to 2023, showing a clear increase in demand and production. According to the Mexican Mezcal Quality Regulatory Council, mezcal production exceeded 12 million liters in 2023.

Due to the extensive semi-arid areas on the planet (Australia and Africa), there is an immediate land opportunity for Agave bioenergy crops to support the economy of its inhabitants (Davis et al. 2011). *A. tequilana* and *A. angustifolia* are mainly species used for the production of alcoholic beverages. However, *A. salmiana* is also of great interest for the production of "Pulque", which was used in religious rituals by pre-Columbian inhabitants, although its cultivation is not as extensive as that of

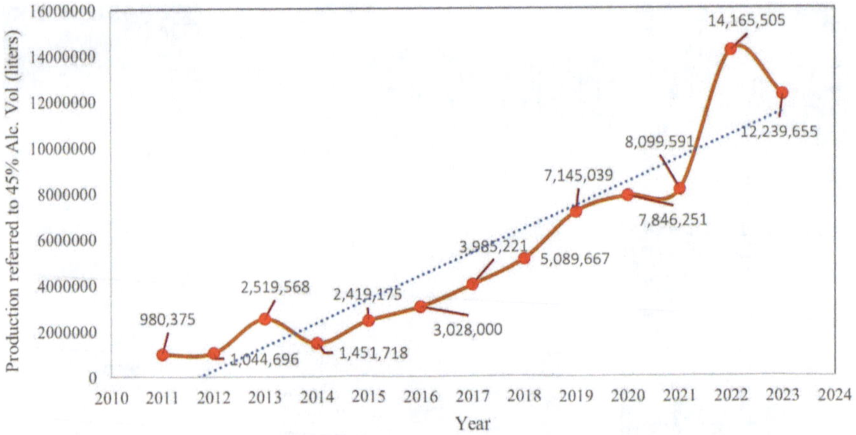

Fig. 5.6 Increase of mezcal production in Mexico during the last 13 years. (Source: https://comercam-dom.org.mx/)

A. tequilana and *A. angustifolia*. On the other hand, the drink Bacanora in the state of Sonora, Mexico is made from *A. pacifica*, which is a variety of *A. angustifolia* (Colunga-GarcíaMarín et al. 2007).

5.1.3.2 Fiber and Textile Properties

Sisal (*Agave sisalana* Perrine) and Henequen (*Agave fourcroydes* Lem.) are species that are mainly used to obtain fibers. Both species are native to Mexico. According to phylogenetic studies, it is assumed that *A. fourcroydes* is a clonal pentaploid variety of *A. angustifolia* (Castillo-Lara and Flores-Johnson 2020). These two types of fibers (Fig. 5.7) have been used since the pre-Hispanic era for the producing of threads, fabrics, sacks, bags, blankets, rugs, footwear, clothing (huipiles), belts, hammocks, sleeping mats (petals), hats and baskets (Esparza-Ibarra et al. 2015).

To obtain the fibers, the buds are used in the case of *A. lechuguilla*, or the lateral leaves in other types of Agave. Obtaining the fiber can be done by hand using special tools similar to knives (in the case of buds), or with shredding machines. Once the fiber is obtained, it is exposed to the sun to dry. The fiber obtained from *A. lechuguilla* is characterized by being quite hard, with high resistance and durability. It is highly resistant to chemical solvents, heat, acids and abrasive products. It is also resistant to water, even at high temperatures (in continuous exposure up to 150 °C). It has high liquid retention and a unique roughness on its surface due to the calcium oxalate crystals embedded in it, making it irreplaceable by synthetic substitutes. This fiber can be used in industry, commerce and domestic applications (Castillo et al. 2005).

Fig. 5.7 Henequen plant from which fibers of different sizes are obtained (**a** and **b**); **c**: this type of small fiber is used for constructions combined with appropriate materials. (Source: Castillo-Lara and Flores-Johnson (2020))

5.1.3.3 Fiber Industry

The fibers obtained from the Agave plant have been widely used for their high resistance and tension. They are biodegradable, recyclable, renewable and do not present any health risks. In their production, they promote sustainability. Evidence suggests that the Mayans cultivated henequen for the benefits they obtained from this plant. Some evidence can be found in various monoliths, leading to the belief that henequen was native and domesticated in Yucatan, Mexico (Castillo-Lara and Flores-Johnson 2020). The Mayans utilized henequen fibers in navigation, construction and trade. Yucatan remained significant due to the international demand for fibers during 1850–1950. However, this industry later declined with the introduction of synthetic fibers and the production of sisal fibers in other countries like Brazil and Tanzania (Colunga-GarcíaMarín and May-Pat 1993; Colunga-GarcíaMarín et al. 2007; Trejo-Torres 2018). However, Agave fibers are still used in some regions of Mexico, mainly as lightweight structural construction materials, crucial in rural areas. The use of Agave has led to increased cultivation in regions like Oaxaca and Puebla, Mexico, where species such as *A. angustifolia* Haw., *Angustiarum angustiarum* Trel, *A. americana* L. var. americana, and var. oaxacensis Gentry, and *Agave kerchovei* Lem, are also important for these purposes (Palma Cruz 2000).

In recent years, its usefulness has been investigated for its combination with other materials to produce sustainable combinations that are more friendly to the environment. The Yucatan Scientific Research Center (CICY), as part of its activity,

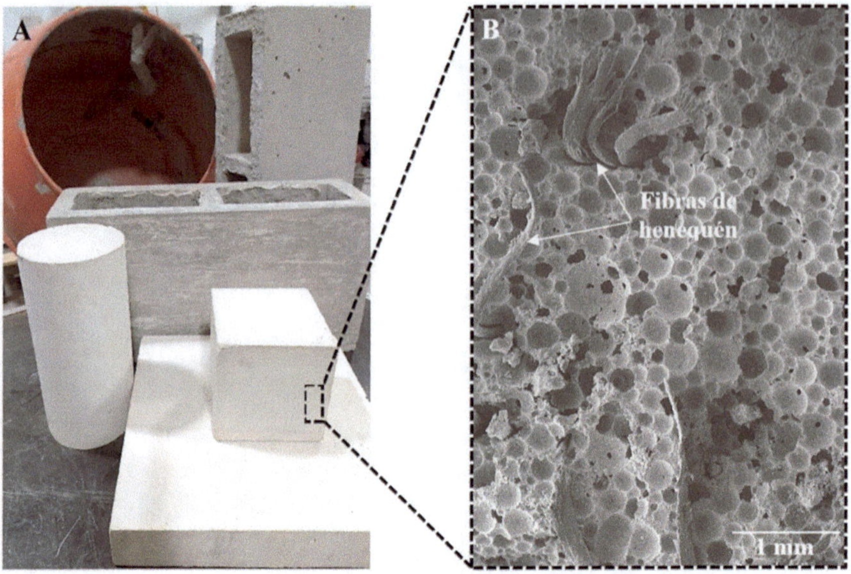

Fig. 5.8 (**a**) Foamed concrete materials for the construction industry manufactured in the CICY Materials Unit. (**b**) Cellular structure of foamed concrete with henequen fiber reinforcement observed through a scanning electron microscope, where the air cells and henequen fibers are observed. (Source: Castillo-Lara and Flores-Johnson (2020))

evaluates composite materials that are used in the construction, aerospace and automotive industries (e.g., carbon fiber materials and epoxy resin), among others. Figure 5.8 shows some examples of composite materials reported by the CICY.

To respect this, Fujiyama et al. (2014) evaluated the mechanical performance of sisal fiber-reinforced cement mortar. This item was produced from a mixture of cement, sand and water. The results showed a considerable increase in the impact resistance and considerably more significant when the longer fibers were used. On the other hand, innovative structural panels were elaborated based on the use of thin external layers of Sisal Fiber-Cement Composites combined with a core layer of Polypropylene Fiber-Reinforced Lightweight Concrete reported by Frazão et al. (2018).

5.2 Current Cultivation Practices and Challenges

5.2.1 Agave Cultivation

Due to the multiple uses of agave, there are issues of over-exploitation of various important species in Mexico, especially those found in the wild, as they are harvested. Currently, educational programs are being implemented for the population that uses these resources, to increase agro-ecological sustainability, protect the most vulnerable species, and raise awareness about not altering the environments where these plants develop.

Agave cultivation is multi-year; its leaves take 3–4 years to mature, and its maximum yield occurs between the 12th and 15th year. With the provision of sufficient nutrients to the soil and the application of adequate irrigation, yields of up to three tons per hectare can be achieved. To ensure a good crop, the following actions are carried out: Subsoil the soil 50–60 cm deep to ensure there is no compaction, and that root development is favored for nutrient absorption. Subsequently, harrowing is done with a disc implement to leave the soil more manageable aerated and conducive to crop development.

The layout of the plantation may vary depending on the soil needs (for example, the use of fertilizers or pesticides), its inclination, and the species to be cultivated. The Agave plants are distributed in rows, with a distance of 1 m between plants, and two meters between rows. This spacing works well for medium-sized Agaves, like *A, angustifolia*; but the distance between plants may need to be greater for larger types like *A. salmiana*. While it is possible to reduce the space between rows and plants, it is not recommended, as it can cause damage to the plants as they grow leading to wounds and infections. Figure 5.9, shows the Agave plantation and Fig. 5.10 shows the established cultivation of (*A. angustifolia*).

Fig. 5.9 Establishment of Agave cultivation. (Source: Photograph by Valadez-Moctezuma E)

Fig. 5.10 Cultivation of *A. angustifolia* already established. (Source: Photograph by Valadez-Moctezuma E)

The soil must drain well, as excess moisture is harmful to the Agave. Optimal growth temperatures are around 30 °C during the day and 15 °C at night, with susceptibility to frost when the temperature drops below 3 °C.

Agave requires minimal water, but it should be watered regularly. In winter the soil must be kept moist; in spring (during vegetative resumption) gradual watering is needed to keep the soil moist, not wet. Towards the end of summer, irrigation should be reduced.

The pineapple harvest is carried out through a process called "Jima" (a word in Nahuatl). This process involves pruning the Agave to cut the stalks and obtain the pineapple, which is a useful organ for the processing of fermented drinks. The Jima is a manual process, with techniques passed down from generation to generation where people learn to recognize a mature Agave plant.

For fiber extraction, the leaves are harvested by cutting them with great skill. The fleshy leaves are then taken to the mill where they are crushed, the fibers are extracted and then hung in the sun to dry.

5.2.2 *Reproduction of Agave and Transplantation*

The main method of reproduction for these plants is asexual through offsets from the rhizome or bulbils that develop in axillary meristems of the inflorescence, although they can also reproduce sexually through seeds. Plants obtained asexually are called clones, which are theoretically genetically similar. Analyzes carried out with molecular markers have shown that there is low genetic variability in these plants (Rodríguez-Garay et al. 2009).

The way to reproduce Agave plants varies, depending on the species, or the customs in the places where they are grown. For example, in *A. tequilana* var. blue propagation by suckers is more frequent, because, in addition to preserving its genetic characteristics, the development of the plants is faster and more vigorous than by bulbils (Rodríguez-Garay et al. 2009). During the twentieth century in large areas of East Africa, *A. sisalana* was mainly spread by bulbils. Under natural conditions, in *A. deserti* populations, only 1% of reproduction is by seed and 99% by offspring (Trejo-Torres 2018; Colunga-GarcíaMarín and May-Pat 1993).

The collection of suckers is normally obtained from a mother plant that is approximately 3 years old because the plant of that age is considered mature, and therefore, has a greater capacity for resistance to pests and diseases (among which *Fusarium oxysporum* is common). If the shoots are obtained from younger plants, there is a risk of damaging the rhizomes of the mother plant and causing wounds and infections that can provoke its death. When selecting offspring they must be completely healthy at the base; free of infections; have good weight and be smaller in size. It is considered that the younger and more vigorous the sapling, the greater the likelihood of success for the new crop.

5.2.3 Digital Farming

The evaluation of vegetation indexes that characterize plant growth, assess fruit ripeness or detect diseases are key factors to achieve high-quality and yield of fruit or vegetables. Both destructive and non-destructive techniques can used the accomplish these measurements (Gonçalves and Henriques 2015; Benelli et al. 2020). Among non-destructive techniques, few methods, such as digital farming, are used in Agave species. For Agave, a Convolutional Neural Network and Unmanned Aerial Systems were used to automate the detection and counting of Agave plants (Flores et al. 2021). Calvario et al. (2017) confirmed that High Spatial Resolution Satellite Images and images taken by Unmanned Aerial Vehicles (UAV) are useful for identifying specific areas or detecting production conditions. The proposed methodology by Calvario et al. (2017) for Agave crop monitoring is based on the combination of geomatics, photogrammetry, computer science and remote sensing through low-cost UAV. As a result, the precision detection value was 99%. In another study, the combination of geomatic technologies and field Spectroradiometry, associated with Spatial Scaling to Satellite Images were used to identify cultivation and conservation areas of *A. angustifolia* and *Agave cupreata* in Guerrero, Mexico. This combination also allows for the differentiation of the stages of development of each crop, as well as the calculation of the crop area (Olvera-Vargas et al. 2022).

5.2.4 Artificial Intelligence

Agriculture faces innumerable obstacles, for example, improper soil analysis, diseases, pest infestation, inadequate drainage, irrigation, low output, big data requirements and the inadequate use of new technologies by producers. These challenges lead to serious environmental threats and severe crop loss. Artificial Intelligence has grown to form an alternative approach to dealing with diverse agricultural issues. Some of the applications of artificial intelligence in crop management, weed and disease management and soil treatment were reviewed in the literature (Eli-Chukwu 2019; Sharma et al. 2022). Particularly, the process of shredding leaves of *A. angustifolia* is a completely traditional task named the "enriado" process (Campaña et al. 2020). Recently, a hybrid model established on artificial intelligence, that involves a genetic algorithm and a neural network, was developed by Pérez del Rio et al. (2022) to determine the relationship between the most efficient shredding method and the best design parameters selection for leaf shredder of *A. angustifolia* (Pérez del Rio et al. 2022).

For other Agave species, four artificial intelligence algorithms: vector support machine (VSM), climate space model (CSM), back propagation neural network (BPNN) and MaxEnt have been employed to model the distribution of *Agave durangensis*. Among these, the BPNN proved to be the most precise algorithm to predict areas of Agave repopulation (Flores and Sandoval 2022). Additionally, Warren-Vega et al. (2023) utilized an Artificial Neural Network to investigate the relationship between the physicochemical profile and the color of Tequila 100% during the maturation process.

5.2.5 Breeding Objectives

Genetic improvement in Agave has been minimally developed even for the two economically important species in Mexico (*A. tequilana* and *A. angustifolia*). This is due to the plant not being directly used for human consumption. The seeds produced by the Agave plant are typically infertile, they cannot be utilized for propagation. However, alternative methods such as cross-pollination have been attempted for genetic improvement. Some progress has been made using micro-propagation techniques, which have set the stage for genetic engineering experiments.

Another challenge to genetic improvement in Agave is the reluctance of the industry to adopt genetically improved or altered Agaves. For instance, in the production of tequila, only *A. tequilana* var. blue is used, with at least 50% of total reducing sugars coming from the pineapple. The remaining sugars may be sourced from other nearby varieties or cultivars. These conditions are intended not to alter the production process or the flavor of this product (NOM-006-SCFI-2012). In contrast, for mezcal production, Agave pineapples from different species can be utilized (referred to as assembly), resulting in diverse flavors and aromas in the final product. *A. angustifolia* is commonly used in the assemblies.

Characteristics of the Agave targeted for improvement include precocity, faster growth, spineless leaves, increased and superior fiber production, high pineapple weight yield, resistance to diseases, and sugar content, among others. However, these efforts have not yielded significant success or have shown low efficiency (García-Herrera et al. 2010).

5.3 Germplasm Biodiversity and Conservation

5.3.1 Germplasm Diversity and Phylogeny

The increase in mezcal production from *A. angustifolia* and its direct association with the increased removal of wild plants constitute important agricultural and environmental concerns (Cabrera-Toledo et al. 2022; Klimova et al. 2022). This problem is exemplified by the extension of tequila cultivation (*A. tequilana*), which promoted rapid expansion of plantations, soil erosion, reliance on vegetative propagation, massive destruction of natural vegetation, inhibition of gene flow and cross-pollination, and an augmented susceptibility to pathogens (Vargas-Ponce et al. 2009; Ruiz-Mondragon et al. 2022). A similar trend has been reported for other Agave species, showing a significant reduction in genetic diversity in cultivated agave plants (Aguirre-Dugua et al. 2013; Eguiarte et al. 2013; Lindsay et al. 2018). However, many Agave species still conserve relatively high genetic diversity due to the recent formation of cultivation areas and the current process of domestication. Nevertheless, there is a clear trend towards the reduction of genetic diversity (Eguiarte et al. 2021; Klimova et al. 2022, 2023; Trejo et al. 2018).

The use of molecular markers to estimate genetic diversity and genetic relationships between Agave populations/species can provide helpful information for crop breeding and resource conservation strategies. However, molecular markers-based research in populations of *A. angustifolia* are scarce. The use of genetic markers in Agave was focused mainly on determining the diversity between and within Agave populations, elucidating the relationship between cultivars and their wild relatives, estimating the genetic variation between mother plants and their seedlings or the genetic variation derived from the micropropagation of Agaves, in addition to the determination of the phylogenetic relationships between the different species of Agave. Therefore, molecular marker types such as inter-simple sequence repeats (ISSR) (Vargas-Ponce et al. 2009), randomly amplified polymorphic DNA (RAPD) (Rodríguez-Garay et al. 2009), inverse sequence tagged repeat (ISTR) (Torres-Morán et al. 2008), amplified fragment length polymorphism (AFLP) (Sánchez-Teyer et al. 2009; Rivera-Lugo et al. 2018) and microsatellites (simple sequence repeats, SSR) (Trejo et al. 2018) have been tested in *A. angustifolia*. Recently, based on 8616 SNPs obtained using the RAD-Seq of 94 cultivated and wild plants of *A. angustifolia* from Oaxaca and Puebla, Mexico, Klimova et al. (2023) found that genetic differentiation between the cultivated and wild genotypes was evident. However, they found several cultivated individuals with high genetic affinity with wild samples.

Taxonomically, *A. angustifolia* belongs to the "Rigidae" section along with 12 other species (Gentry 1982; Simpson et al. 2011). Initially, phylogenetic analysis for Agave species was carried out using unique or few molecular markers vis ITS, *rbc*L, *mat*K, *trn*H-*psb*A, etc. (Jimenez-Barron et al. 2020; Eguiarte et al. 2021). Recently, transcriptomic analyses were used to assess the relationships among several Agave species. A transcriptome profile similarity between *A. tequilana and Agave deserti* was presented by Gross et al. (2013). Also, the phylogenetic relationship between *A. deserti*, *A. sisalana* and *A. tequilana* was studied by Huang et al. (2018). Likewise, a phyllotranscriptomic reconstruction of ten *Agave* species created on 5863 orthologs displayed the clustering of the 10 species in two groups. One cluster was formed by *A. angustifolia*, *Agave* hybrid H11648, *A. deserti*, *A. tequilana*, *A. fourcroydes*, *A. salmiana*, and *A. sisalana*. Particularly, *A. fourcroydes* and *A. angustifolia* formed a unique subcluster that confirmed the ancestry hypothesis among the two species (Colunga-GarciaMarin et al. 1999).

5.3.2 Cytogenetics

Ploidy levels in *Agave* species range from diploid to octoploid with a basic chromosomic number of 30. The karyotype of Agave is conserved over numerous ploidy levels and it is arranged in a bimodal mode, with 25 small chromosome pairs and five large pairs.

In *A. angustifolia*, various ploidy levels have been informed viz. diploid (2x, 60 chromosomes), triploid (3x, 90 chromosomes), tetraploid (4x, 120 chromosomes) and hexaploid (6x, 180 chromosomes) (Palomino et al. 2007; Simpson et al. 2011). The diploid genome contains about 6–9.6 pg (Palomino et al. 2007, 2017). Based on the definition of 1 pg = 5980 Mbp, haploid (1Cx) genome size was assessed to vary between 2940 and 4704 Mbp of DNA for Agave (Moreno-Salazar et al. 2007; Palomino et al. 2007). *A. angustifolia* var. "letona" was described as tetraploid genotype and var. "chelem ki" hexaploids (Robert et al. 2008). Furthermore, *A. angustifolia* var. "Cimarron" and "Lineño" were considered diploids, both showed a bimodal karyotype that comprised ten large and 50 small chromosomes. *A. angustifolia* "Cimarron" had 42 metacentric chromosomes (m) + 4 submetacentric chromosome and (sm) + 6 subtelocentric chromosomes (st) + 8 telocentric chromosomes (t), while the karyotype observed in "Lineño" was 48 m + 2 sm + 2 st + 8 t (Palomino et al. 2017).

Gomez-Rodriguez et al. (2013) evaluate the physical mapping of the genomic regions 5S and 18S in Agave species. As a result, the 5S loci were placed in both arms of a small chromosome. Moreover, the 18S loci were associated with a secondary constriction of a large chromosome pair.

The most common polyploids in Agave may have been formed through non-reduced gametes. The formation of polyploids and new phenotypes arise that can contribute to the success of polyploids in nature or their selection and use in agriculture. Success, in polyploid evolution, is associated with changes in genome organization and regulation in gene expression. These variations give the polyploid a greater capacity for adaptation and response to extreme climatic conditions, increased tolerance to cold, resistance to pathogens and greater genetic plasticity to utilize new environments.

5.4 Traditional Breeding

5.4.1 Improvement Strategies

Genetic breeding of agaves depends on the selection of elite genotypes, and the efficiency and standardization of tissue culture processes to propagate massively high-yield plants. Various protocols for *in vitro* culture have been reported for several *Agave* species. The most challenging task is the selection of mother plants that would be the elite genotypes to initiate and maintain a breeding program. The selection of the mother plants from wild or cultivated genotypes is based on the combinations of some characteristics viz. comparative vigor of plants, and their physiological and health status (Monja-Mio et al. 2019).

5.4.1.1 Traditional Breeding Methodologies

Traditional improvement programs beyond crosses between and inter-species of Agave have been scarce. The most well-known and reported case is that of Hybrid H11648. This genotype was the result of the hybridization between *Agave amanuensis* and *Agave angustifolia* that then backcrossed to *A. amaniensis*. It was made available for cultivation during the early twentieth century in Tanzania. The main characteristic of this hybrid was to produce higher fiber yields in different climatic conditions and soil classes (Lock 1962). The H11648 hybrid is the preferred genotype to produce fiber worldwide, including in China and Brazil, where most of the agave fiber is manufactured, although its greatest weakness is susceptibility to pests and diseases (Huang et al. 2019).

Another case for improved Agave was for the production of fibers in Cuba. Different accessions were obtained from elite plants with few spines of *A. fourcroydes* obtained from henequen cultivation fields, and more genotypes were selected from Hybrid H11648. As a result, two genotypes were found to be the most outstanding, where one of them was characterized by greater production of vegetative mass and the other with greater production of fiber (Serrano and Gutiérrez 2009). Also, interspecies and intraspecies crosses enhanced the vegetative production and the viability of seeds in *A. americana* and *A. tequilana*. As a hypothesis, crosses between compatible Agaves genotypes could increase the level of fertility and expand the composition of germplasm (Escobar-Guzman et al. 2008). However, as mentioned earlier, there are restrictions for Tequila production using other varieties, or cultivars different from *A. tequilana* var. azul. These restrictions on these important commercial cultivars are a further obstacle to the application of improvement proposals.

5.4.2 Traditional Breeding Limitations

Although the agricultural and economic importance of some Agave species is remarkable, there are few programs to select and improve desirable traits. The lack of breeding initiatives in Agave is related principally to their biology and morphology, long lifecycles, extremely height-sized inflorescences (up to five meters), many agave species are polyploids, unproductive sexual reproduction, sterility and the recalcitrance of the majority of seeds (Monja-Mio et al. 2019). Therefore, it is important to implement complementary approaches that contribute to the improvement of Agave species. Attempts at breeding, germplasm diversification and conservation have been based mainly on the continuous selection of wild populations and the maintenance and improvement of ancient and recently selected local varieties. These activities are carried out with criteria that promote diversity related to a variety of flavors, duration of the life cycle and adaptation to multispecific and multivarietal agroecosystems.

5.5 Molecular Breeding

5.5.1 Molecular Marker-Assisted Breeding

Molecular breeding involves the discovery, improvement and utilization of molecular markers to increase the efficiency of breeding initiatives. Certainly, molecular Marker-Assisted Selection (MAS) is used as a complementary tool for traditional improvement strategies and has conferred prospects for plant breeding through indirect selection. Some traits that have been improved in cultivated plants (breeding for increasing yield, biotic and abiotic residence enhancement, etc.) have not been targeted for improvement in agave to date due principally to the limited genetics research and genomics information of agave Species.

Based on the literature available in our hands, there is no report on MAS breeding of any characteristic of any species of Agave. However, with the arrival of genetic markers, numerous researches were carried out to estimate the genetic variability between Agave genotypes and the diversity from their wild relatives. Molecular markers were used to characterize the germplasm of Agave species as an initiative to rescue the Brazilian Agave breeding program (Raya et al. 2023). Several research about the genotype characterization using and relationships between Agave species molecular markers have been detailed in the section "Germplasm Diversity and Phylogeny".

Furthermore, the availability of generic resources for Agave species could facilitate the integration of MAS into the breeding program. The first reference genome assembly of *A. angustifolia* (Taxonomy ID: 382117) was recently submitted by "Iridian Genomes" on Jan 11, 2024, at the Scaffold assembly level (Genome assembly ASM3558428v1; https://www.ncbi.nlm.nih.gov/datasets/genome/GCA_035584285.1/). The assembly consists of 713,698 contigs (327,581 Scaffolds) with a Guanine-Cytosine content of 37%. Nevertheless, no genome annotation is available so far. Moreover, there are only 41 Sequence Read Archive (SRA) resources, distributed between Transcriptomic, WGS, RAD-Seq, and Metagenomic strategies. Likewise, about 133 complete or partial genes were reported for *A. angustifolia* in GenBank (https://www.ncbi.nlm.nih.gov/gene/?term=Agave%20angustifolia).

Additionally, the first chloroplast genome of *A. angustifolia* was assembled by Qin et al. (2021). This cp genome accounts for 57,274 bp of which 37.8% is GC content. The Large Single Copy segment (LSC), the Inverted Repeat segments (IR) and the Small Single Copy segment (SSC) were 85,895 bp, 26,575 bp and 18,229 bp, respectively. Out of the 132 annotated genes: 86 were protein-coding genes, 38 tRNAs, and 8 rRNAs (Qin et al. 2021).

5.5.2 Genomics

Compared to model plant species, few genomics experiments have been executed on Agave species. Some of the causes are the large size of the genome (around four Gbp), the lack of an assembled genome, a huge amount of repetitive regions, the large life cycle (5–15 years) and the genetic flow among Agave populations intermediated by pollinators (Simpson et al. 2011; McKain et al. 2012).

Recently, transcriptome sequencing has been extensively experimented to study gene expression and compare different agricultural and biological conditions. For the Agave genus, researchers have directed to bioenergy production in *A. tequilana*, CAM photosynthesis in *A. americana* and response to abiotic stress in *A. sisalana* (Simpson et al. 2011; Abraham et al. 2016; Yin et al. 2018; Sarwar et al. 2019). Furthermore, genetic program variation during regeneration in *A. salmiana*, expression of cellulose synthase genes in the hybrid H11648, flavonoid production in *Agave lechuguilla*, fiber and cell wall biogenesis from *A. tequilana*, *A. sisalana,* H11648 and *A. fourcroydes,* and the identification of *Expansin* genes in *A. amanuensis* were searched through transcriptomic analyses (Avila de Dios et al. 2015; Cervantes-Pérez et al. 2018; Huang et al. 2019; Maceda-López et al. 2022; Morreeuw et al. 2021; Raya et al. 2021; Wang et al. 2023).

Particularly, an RNA-Seq experiment carried out on *A. angustifolia* revealed the expression behavior of cinnamyl alcohol dehydrogenase genes in other Agaves (Huang et al. 2022). Likewise, Valadez-Moctezuma et al. (2023) assembled the transcriptome of *A. angustifolia* using leaves, roots and pineapple as sources for mRNA, and discussed genes related to fructan and cellulose biosynthesis, and predicted molecular markers such as SSR and VNTR.

5.6 Tissue Culture Applications

5.6.1 Micropropagation

The technique of *in vitro* culture has been considered a viable alternative for the propagation of *A. angustifolia* and other Agave species, with somatic embryogenesis and massive propagation stand out (Arzate-Fernández and Mejía-Franco 2011). Considering the difficulties that genetic improvement represents in this plant, different biotechnological strategies and protocols have been developed for more than four decades, such as tissue culture, and even the genetic transformation of some agave genotypes. The development of various protocols continues to achieve these objectives; but also, to increase efficiency and costs.

In addition to the use of shoots from the rhizome, plant tissue and cell cultures have also been used for the asexual propagation of Agave. Specifically, micropropagation offers other alternatives to produce and improve the species. Several species have been micro-propagated to obtain new healthy plantlets on a large scale in a short time; besides, to massive and uniform production of seedlings for commercial plantations, inducing somaclonal variation to enhance agronomic characteristics through selection, maintaining specific genotypes, rejuvenated plants, the yield in foliar biomass through the selection and propagation, and for shortening their life cycle of the crops through the use of suckers (Bautista-Montes et al. 2022; Martínez and Arzate-Fernández 2022). Another benefit of applying this strategy is the opportunity to study molecular mechanisms in CAM plants through overexpression, gene silencing, molecular changes in new cells, metabolic pathways, etc., and complement/compare with the results obtained in the model plant *Arabidopsis thaliana*. This aspect became clear when knowing the genetic, physiological and anatomical mechanisms that occurred in the comparison of albino and variegated plants with respect to green plants in *A. angustifolia* (Hernández-Castellanos et al. 2020).

For the micropropagation of agave, different types of explants have been used, such as leaves meristems and axillary buds. Morphogenesis has been obtained directly or indirectly through any of these explants. The formation of organs (organogenesis) or embryos (embryogenesis) has also been achieved directly or indirectly, the latter involving the formation of callus. Another alternative consists of direct organogenesis employing the principal meristem from offsets and bulbils; these explants are free of pathogens, in addition, and allow obtaining the shoots in a short time. There is adequate literature regarding the *in vitro* cultivation of agave, where it is intended to adapt the operation of general media, such as Murashige and Skoog (MS) medium with a combination of hormone concentrations to achieve plant regeneration, multiplication or production of embryos of this plant. But also, to achieve the acclimatization process, since it is a critical step for the survival of the vitroplants. Some examples are mentioned below.

The different responses sought with in vitro regeneration have been achieved step by step through the evaluation of different concentrations of growth regulators like 2,4-D, IAA, IBA, BA, NAA, Zeatin, and KIN. But also, these responses can vary depending on the species, the explant and the genotype, among others. Higher proliferation of shoots has been reported when using BAP (up to 4 mg L^{-1}) and IAA (up to 1 mg L^{-1}), in culture medium to develop new adventitious shoots (organogenesis) of *A. angustifolia* stem tissue. In this research, it was concluded that the BAP had a synergistic effect to induce the formation of the highest number of shoots. The effect of organic nitrogen and sucrose on somatic embryogenesis has also been examined for *A. angustifolia.* As a result, the induction of embryogenic callus and the production of somatic embryos were positively correlated with the sucrose concentrations (Ríos-Ramírez et al. 2017; Reyes-Díaz et al. 2018).

In another study, it was noted that once *A. angustifolia* were regenerated, a critical stage is the acclimatization of the vitro plants because it is challenging to transfer them from a medium rich controlled in nutrients, growth regulators and saturated humidity, to a non-sterile substrate with uncontrolled relative humidity and temperature. Therefore, the challenge of survival of vitroplants is high; also, because the seedlings must develop structures such as stomata and functional roots from the new conditions in which they are found. After evaluating different types of substrates, it was reported that 100% of plant survival under shade conditions had a loamy sandy texture with a rich content of organic matter (Sánchez et al. 2020). Other evaluations were carried out in the acclimatization of vitro plants and it was reported that one of the main causes was the immaturity of the stomata too (Monja-Mio et al. 2015).

5.6.2 *Embryo Rescue*

Embryo rescue has been considered a viable tissue culture method for reproducing *A. angustifolia* and other agave species, with somatic embryogenesis standing out. Subsequently, a protocol was proposed that combines manual hybridization and subsequent embryo development using *in vitro* culture. With this protocol, it was possible to carry out classical genetic improvement in different species of agave including pollen, clonal and somatic cell selection, and phenotype and genotype selection for diverse genetic breeding purposes (Arzate-Fernández and Mejía-Franco 2011).

Broadly speaking, the protocol consists of the collection of panicles from selected mother plants, emasculation, and pollination occurring on mature flower stigma. Then, the pollinated flowers are protected with bags against undesired pollen. The resulting immature fruits are then ready to use for embryo rescue. Finally, plantlets germinated from these structures are obtained in Petri dishes with appropriate culture medium. If the purpose of the plant breeder is to obtain agave plants in a relatively short time, the most efficient technique is embryo rescue, since a greater number of plants are obtained in less time compared to propagation via seed (Rodríguez-Garay et al. 2018).

The procedure to follow for embryo rescue is as follows: select the panicle of interest and transport it to the laboratory, cut and sterilize the immature fruit from the panicle with ethanol 95%, open the sterilized fruit under a stereo microscope, and remove the immature seed. Recover the zygotic embryo and cultivate it with the adequate culture medium with incubation at 27 ± 2 °C for 7 days, then apply a 16 h photoperiod of fluorescent light. The recommended culture medium is MS basal medium with the addition of L2 Vitamins, 2 mg/L of 2,4-D and 0.3 mg/L of BA (Rodríguez-Garay et al. 2018).

On the other hand, it is possible to obtain agave embryos from mature seeds, but the maturation time will depend on the species of agave of interest (Rodríguez-Garay et al. 2018).

5.6.3 In Vitro Pollination

The pollination process of agave species is carried out manually, *ex situ*, to promote fertilization. The panicles are required to be cut and transported to the laboratory for *in home* fertilization. First, select the elite genotypes that will be used for hybridization. Then collect the inflorescence or the panicles, detach the anthers from the filament when the bud opens and place them over a clean paper. Allow the anthers to mature for 1–2 days after being collected, and eliminate the anther debris. Conserve the pollen in bags and/or vials in a desiccator at 4 °C. Pollen viability is estimated at 3–6 months, but it can be preserved for longer by covering it with oil and storing the vial at −20 °C (Rodríguez-Garay et al. 2018).

The flower female organs maturate 3–5 days after emasculation. The mature female organs are recognized by the visible nectar at the base of the style. The style should reach the same height as the filaments at the time of emasculation. Fertilize with a large quantity of pollen over the stigma with a fine brush. Depending on the environment, the breeder can protect the pollinated flowers with paper. Allow fertilization to take place (Rodríguez-Garay et al. 2018).

5.6.4 Synthetic Seeds

Agave angustifolia presents inefficient pollination, a single long-lasting flowering period (8–12 years), and low efficiency in seed production and germination. Furthermore, plants obtained by seed are slowly developed and show genetic variability and a low percentage of transplant regeneration compared to the propagation through suckers, bulbils or apical meristems cultures. Therefore, to ensure the conservation of the species, given its extensive use and to guarantee sufficient plant material to plant cultivation areas, other technologies must be applied such as obtaining synthetic seeds.

As a propagation and conservation strategy using synthetic seeds, somatic embryos of *A. angustifolia* were derived from the zygotic embryonic and each somatic embryo was summarized in sodium alginate in a calcium dichloride complex. As a result, encapsulated synthetic seeds with 3% sodium alginate allowed the germination of seed, and plantlet survival was achieved when supplemented with 4.4 μmol 6-benzylaminopurine (BAP). After 20 days of transplantation in greenhouse conditions, the first roots of the plants regenerated from synthetic seeds appeared with 80% of the synthetic seeds of *A. angustifolia* showing root formation within the capsules (Arzate-Fernández et al. 2016).

For *A. vera-cruz*, *in vitro* somatic embryos and shoots regenerated from embryogenic calli and organogenic were obtained and encapsulated in a sodium alginate/dihydrated calcium chloride complex. The conversion rate of synthetic seeds to seedlings ranged between 40% and 53% and about 75–80% of plantlets survived the transplantation (Tejavathi et al. 2006).

Recently, cryopreservation of somatic embryos was performed using the V-cryoplate technique, which consists of the encapsulation of plants on aluminum cryoplates, using two vitrification solutions for dehydration (Delgado-Aceves et al. 2021). As a result, the highest post-cryopreservation reactivation and plant conversion levels were accomplished with the PVS2 vitrification solution for 15 min at 0 ° C. These results represent feasible alternative techniques to produce and safely preservative of genetic material during an extended period, which can ensure the supply of plant material for commercial propagation, usefulness for genetic transformation programs and the production of elite genotypes of Agave species (Delgado-Aceves et al. 2021).

5.6.5 Other In Vitro Applications That Pertain to the Species

In vitro regeneration and propagation protocols for Agave, as with other plants, generally lead to mutations that cause somaclonal variation. This type of mutation is not always harmful to plants, but can sometimes positively affect important characteristics that lead to improvement. In *A. salmiana* and *A. marmorata*, somaclonal variations were determined through molecular genetic markers like ISSR and RAPD, in different explants of in vitro seedlings and meristematic zone of ex vitro seedlings through direct and indirect organogenesis (Arzate-Fernández et al. 2020). As a result, somaclonal variation was observed only in clones that went through a callus phase and consequent indirect organogenesis. On the contrary, the clones resulting from direct organogenesis had no genetic profile variation, as was shown with AFLP (González et al. 2003). Other Agave species namely, *A. tequilana*, *A. americana*, *A. cantala*, *A. fourcroydes*, *A. sisalana*, and *A. peacockii* have been manipulated for seed production and viability. Moreover, *A. crassispina*, *A. duranguensis*, *A. obscure*, *A. pigmaea*, *A. salmiana*, and *A. victoria-reginae* have been studied for axillary bud proliferation observing positive regeneration responses (Bautista-Montes et al. 2022).

Because the responses to in vitro culture are not the same for each agave species, there is extensive literature on the effect of hormone concentrations on explant development. Micropropagated shoots from *A. americana* were exposed to different concentrations of BA and NNA. Callus and embryo were induced by applying a combination of auxins. The results indicated that high concentrations of Dicamba had a positive effect on somatic embryo induction, but the number of embryos was affected (Naziri et al. 2019).

A novel protocol for the development of somatic embryogenesis of *Agave amica* L. was developed using various combinations for the generation and growth of callus, formation and maturation of embryos, and plant regeneration. For the germination of embryos, 10–15 mg L^{-1} goat colostrum was added to the MS medium. The embryos with leaves were regenerated within 90 days (Pourkhaloee et al. 2023).

5.7 Genetic Engineering and Gene Editing

5.7.1 Genetic Transformation of Agave Species

The exploitation of micropropagation techniques was a feasible strategy to generate genetic transformation procedures, to introduce genes and to ameliorate specific characteristics in *Agave* species. In recent years, some protocols have been tested to transform *Agave* plants with some encouraging results.

A protocol to transform genetically the Agave plants based on particle bombardment including *Agrobacterium tumefaciens* and *Agrobacterium rhizogenes* was patented (Cabrera-Ponce 2002). The embryogenic calli with a high regeneration capacity were bombarded with tungsten elements containing DNA of plasmid mixtures. For the case of transformation with *A. tumefaciens* and *A. rhizogenes*, the embryogenic calli are co-cultured with the bacteria for periods of 48 h and relocated to selective media to generate transformed cells. The genes *hpt* and *bar* were included as markers of selection and the *uid*A gene was incorporated as a reporter gene (Cabrera-Ponce 2002).

Flores-Benítez et al. (2007) applied two protocols to genetically manipulate *A. salmiana*, co-cultivation with particle bombardment and *A. tumefaciens*. The *uidA* gene was employed as a reporter while the *bar* and *npt*II genes were included as markers of selection. Eleven-month-old plants were used to extract DNA and to detect transgenes by PCR technique, and the GUS histochemical tests were carried out to verify the transgenic nature. Agrobacterium co-cultivation allowed the generation of 32 rooted plants with transgenic events and the transformation rate was about 2.7% (Flores-Benítez et al. 2007).

Parallelly, root manipulation mediated by different concentrations of *Agrobacterium rhizogenes* was carried out in leaves, pineapple, and roots of *A. salmiana* (Rodríguez-Hernández et al. 2007). The confirmation of the transfer of bacterial genes to roots was carried out by histochemical assays for the detection of GUS activity and PCR detection of the *npt*II and *rolB* genes. The leaves proved the biggest response with a transformed root production rate of 63%. A percentage of transformation of 60% of the tissues was resolved by the selectable markers (Rodríguez-Hernández et al. 2007).

Moreover, another protocol to genetically transform *A. desmettiana* and *A. tequilana*, using *A. tumefaciens* was established from direct organogenesis. For this method, the explants used were bulbil and apical meristems, co-cultivated with *A. tumefaciens* and the selective mediator was phosphinothricin. Consequently, *A. desmettiana* emitted more shoots per explant (up to 20 shoots) than *A. tequilana* (up to 2 shoots) (Gutiérrez-Aguilar et al. 2014).

A practical case of the use of genetic transformation in Agave genotypes was reported by Gao et al. (2014). The Agave Hybrid H11648 is susceptible to zebra disease (*Phytophthora nicotianae*). To overcome the large period of traditional breeding against *P. nicotianae* in Agave (more than 10 years), a genetic transformation protocol to overexpress *hevein* gene in the calli of H11648 was reported (Gao et al. 2014). As a result, the optimum experimental conditions were: 10 min for effective infection, 200 μM of acetosyringone, 3 days of sisal callus, and 4 days of *Agrobacterium* co-culture. Furthermore, the insertion, permanency and expression magnitude of the transgene (*hevein* gene) were determined by applying Southern blot and RT-PCR. As a result, the mycelial growth proportion of *P. nicotianae* was significantly inhibited by the leaf extracts from the transgenic H11648 lines. In total, 37 genotypes from 150 explants were generated and the *hevein* gene was detected in seven genotypes (Gao et al. 2014).

5.8 Mutation Breeding

5.8.1 In Vitro Mutagenesis and Selection

Breeding based on mutation induction could generate new desirable traits that are not found in nature or that have been lost during the evolution of a particular plant species. Genetic mutations can be induced by physical (X-rays, gamma rays, ionizing radiations, etc.) and/or chemical agents (sodium azide, ethyl methanesulphonate (EMS). Induced mutations can overcome the limitations of conventional breeding, for example, the long breeding cycle, loss of vigor, reduced fertility, complex genomes and a narrow gene pool. Mutagenesis has played a significant function in incorporating the desired genetic variability, which would have been difficult to achieve through traditional methods. Mutagenesis or breeding based on mutation has not been used in Agave due to the aforementioned industry and political restrictions. However, breeding progress may be accelerated by the induced mutations. In Agave, there are few reports for species treated with physical and chemical mutation agents (González-Oramas et al. 2007; Tejavathi et al. 2012; Ángeles-Espino et al. 2013, 2016, 2020; Valencia-Botín et al. 2020; Reyes-Zambrano et al. 2019, 2021); unfortunately, no study has addressed *A. angustifolia*.

González-Oramas et al. (2007) evaluated the radiosensitivity of shoots from Agave henequen to gamma radiations Co^{60} and determined the adequate dose of radiation for a breeding program. As a result, the percentage of mortality increased with the increase of irradiation dose. The mean lethal dose (LD_{50}) was estimated at 20 Gy, with which the explants increased their fresh mass by 49% compared to the control (González-Oramas et al. 2007). Moreover, LD_{50} of gamma rays Co^{60} to induce genetic changes within plantlets and *in vitro* callus of *Agave tequilana* was assessed by Ángeles-Espino et al. (2013). Calli of 6 weeks and plantlets of 12 weeks were irradiated. As a result, LD_{50} was estimated at 16 Gy for callus and 20–25 Gy for plantlets (Ángeles-Espino et al. (2013).

In embryogenic calluses of *A. tequilana* obtained by gamma irradiation, the 20 Gy doses triggered the embryogenic callus induction of 44% compared to the control (89%) after 28 days. Also, the development of somatic embryos and the transformation to plantlets were affected by low radiation amounts (Valencia-Botín et al. 2020).

Moreover, Reyes-Zambrano et al. (2019) evaluated the influence of ethyl methanesulfonate to induce changes in *in vitro* plantlets of *A. americana* explants produced from roots, meristems and leaves. About 60% of the plantlets presented morphologic transformations viz. enanism, variation in leaf forms, leaf spineless, and an increase of fructan content of 30%.

In more practical experiments, Ángeles-Espino et al. (2016) induced mutagenesis on *in vitro A. tequilana* plantlets irradiated with Gamma rays Co^{60} and selected potential resistant genotypes by evaluating the pathogenicity of *Cercospora agavicola*, a fungal of a high potential risk of Agave in the State of Jalisco, Mexico. The authors indicated that the seedling stage was not efficient enough to identify resistant genotypes and suggested that the selected mutants could be corroborated on lignified leaves in natural conditions. Likewise, the application of minimal amounts of gamma rays Co^{60} on *in vitro* explants of *A. tequilana* induced genetic changes and double irradiation increased the accumulation of glucose, fructose and sucrose in in stem, considering double irradiation at 20–20 Gy as optimal to increase carbohydrate accumulation (Ángeles-Espino et al. 2020).

Another practical example was presented by Reyes-Zambrano et al. (2021). Induced mutagenesis by EMS was analyzed to select *A. americana* somaclonal variants with the characteristic of being resistant to *Fusarium oxysporum*. The only EMS treatment that presented shoot formation was the callus exposed to 15 mM; the treatment duration of 2 h allowed the production of the largest amount of shoots per callus. To introduce the resistance to *F. oxysporum* in the somaclonal variants, different concentrations of the fungus as culture filtrate (CF) were applied. The selected CF produced necrosis and oxidation in about 71% of the callus. Nevertheless, they could generate shoots (3.5 shoots/callus). Molecular markers revealed a polymorphic profile of the somaclonal variants, indicating the induction of genetic changes produced by the mutagenesis EMS in the regenerated genotypes confers resistance to *F. oxysporum* (Reyes-Zambrano et al. 2021).

5.9 Hybridization

5.9.1 *Conventional Hybridization*

As mentioned previously, there are few reports on inter- and intraspecies hybridization of Agaves. The most documented program for the genetic improvement of Agaves was developed in Tanzania at the beginning of the twentieth century. The hybrid H11648 is the well-documented achievement of the selection and crossing of several diploid Agaves (Lock 1962).

5.9.2 *Hybrid Cultivars*

The hybrid H11648, named for the date of its release (11 June 1948), is a descendant of a cross among *Agave angustifolia* x *A. amaniensis*, and the generated genotype was backcrossed with *A. amaniensis*. The main characteristics of the H11648 include producing a higher yield of fiber per leaf and more amount of leaves per production cycle than sisal or henequen Agaves. Furthermore, the harvest of the leaves of this hybrid is facilitated due to the absence of marginal thorns in the periphery of the leaves. Unfortunately, H11648 is susceptible to some fungal diseases. Since its obtention, the H11648 has been the largest genotype cultivated in thousands of hectares in Tanzania and Kenya. Currently, H11648 is part of the germplasm to be studied and exploited, especially, in China and Brazil. Moreover, there is no known other cultivar where *Agave angustifolia* was involved in crosses (Gao et al. 2014).

5.10 Conclusions and Prospect

In this chapter, general and main attributes, cultivation, economic and industrial importance, and principal products of Agave species have been presented. This plant, like no other, has provided multiple benefits to humans since pre-Hispanic times in Mexico and other places, even when it is not directly related to human nutrition. The relevance of Agave focuses on its CAM metabolism, which allows for development with little water and growth in environments with adverse and semi-arid conditions. These two characteristics, regardless of their other benefits, designate these plants, as well as others, for example, of the genus Opuntia, as plants of the future; because they have the morpho-physiological characteristics to survive to a certain extent the climate change that is occurring in many areas of the world. These qualities compel us as users of them, to protect and conserve them using all possible alternatives.

The genomic and transcriptomic studies carried out on *Agave tequilana* and *Agave angustifolia*, mainly, offer data that could facilitate gene management for breedings in genetic and molecular research for these species; however, with the information already available, its usefulness can be extended to other related plants.

Few *Agave* species are employed as sources for fiber production (*A. angustifolia*, henequen, Hybrid H11648). The agave H11648 was the unique genotype that had the greatest well-documented agricultural and economic importance in some countries, principally Kenya, Tanzania, Mexico, and more recently Brazil and China. Conversely, the alcoholic beverages (tequila, mezcal and pulque) distilled mainly from *Agave tequilana* and *Agave angustifolia* have great prominence specifically in Mexico, where they have their respective designations of origin.

Although there are few reports on the development of traditional genetic breeding and the inclusion of molecular techniques in Agaves in general and specifically in *A. angustifolia*, the efforts for the selection of mother plants (elite genotypes), asexual propagation, and the *in vitro* regeneration of plants have been of great value in the production of these plants and the maintenance of the species. However, more efficient methods are required and it is required to create and conserve a more vast germplasm source. It is fundamental to rescue and safeguard *in vivo, in situ*, and *in vitro* the diversity of natural populations and elite lines, and generate new genotypes of great additional value. Moreover, the increase in research on the genomics of Agave plants in recent years could be of great contribution to a more effective improvement of Agave plants.

The development of socio-economic, ecological, cultural, sustainability and preservation programs is very important. In countries that grow Agave, promoting their protection and conservation is crucial, given the value they have as natural resources. These programs will generate significant economic income in the areas where they are grown, which will help improve the quality of life of its inhabitants.

Appendix 5.1 Research Institutes and Online Resources Available for Maguey Espadin (*Agave angustifolia* Haw.) Breeding

Institution name	Specialization and research activities	Address	Contact information and website
Universidad Autónoma Chapingo. Departamento de Fitotecnia	Biologist Professor-Research Area's Research: Molecular Biology, Biotecnology, Molecular Markers	Carretera Mexico-Texcoco, km 38.5 Chapingo, Esdad. de México CP. 56230	evaladezm@chapingo.mx

(continued)

Institution name	Specialization and research activities	Address	Contact information and website
CINVESTAV-Unidad Irapuato	Dra. June Kilpatrick Simpson Williamson	Km 9.6 Libramiento Norte Carretera Irapuato-León CP 36824 Irapuato, Gto. México	june.simpson@cinvestav.mx
Centro de Investigación en Alimentación y Desarrollo A.C.	M in Sc. Aldo Hiram Gutiérrez Saldaña. Ecology, Mycology, Biofertilizer, Bioremediation	Carretera Gustavo Enrique Astiazarán Rosas 46, La Victoria, 83304 Hermosillo, Sonora, México., Mexico	asaldana@ciad.mx
Centro de Investigación y Asistencia en Tecnología y Diseño del Estado de Jalisco, A.C.	Ph. D. José Manuel Rodríguez-Domínguez	Av. Normalistas No. 800, Colinas de la Normal. CP 44270, Guadalajara, Jalisco MEXICO	mrodriguez@ciatej.mx
Centro de Investigación Científica de Yucatá (CiCY)	Dr. Lorenzo Felipe Sánchez Teyer. Biochemical, Biotechnology	Centro de Investigación Científica de Yucatán, A.C. Calle 43 No. 130 x 32 y 34, Chuburná de Hidalgo; CP 97205, Mérida, Yucatán, México	santey@cicy.mx
Centro de Investigación Científica de Yucatá (CiCY)	Dr. Fulgencio Alatorre Plants Biotechnology Integrative Biology	Centro de Investigación Científica de Yucatán, A.C. Calle 43 No. 130 x 32 y 34, Chuburná de Hidalgo; CP 97205, Mérida, Yucatán, México	fulgencio.alatorre@.cicy.mx

Appendix 5.2 Genetic Resources for Maguey Espadin (*Agave angustifolia* Haw.) Breeding

Cultivar	Important traits	Cultivation location
Espadín	Plant height = 171 cm, Rosette diameter = 298 cm, Number of leaves = 147, Leaf length = 135 cm, Leaf width = 13 cm, Terminal spine length = 27 cm, Distance from the leaf	Puebla, Oaxaca, San Luis Postosi, Tamaulipas, Zacatecas, Michocan, Guerrero, Guanajuato, Durango

(continued)

Cultivar	Important traits	Cultivation location
	apex without teeth = 12 cm, Distance of the basal width of teeth = 5 mm, Number of teeth in 10 cm = 4, Minimum separation between teeth = 14 mm, Maximum separation between teeth = 31 mm, Tooth length = 5 cm.	
Lineño	Compared to var. espdain, the var. Lineño has greater plant height, leaf length, tooth length, minimum separation between teeth and number of teeth, and smaller plant diameter, leaf width, apical spine, leaf apex distance without teeth	Jalisco
Cimarron	Compared to var. espdain, the var. Cimarron has greater Plant height, Number of leaves Leaf length and Number of teeth, and smaller Terminal spine length, Minimum and maximum separation between teeth	Jalisco

References

Abraham P, Yin H, Borland A et al (2016) Transcript, protein, and metabolite temporal dynamics in the CAM plant Agave. Nat Plants 2:1–10. https://doi.org/10.1038/nplants.2016.178

Ahumada-Santos Y, Montes-Ávila J, Uribe-Beltrán M et al (2013) Chemical characterization, antioxidant and antibacterial activities of six Agave species from Sinaloa, Mexico. Ind Crop Prod 49:143–149. https://doi.org/10.1016/j.indcrop.2013.04.050

Andrade JL, Barrera EDL, Reyes García C et al (2007) El metabolismo ácido de las crasuláceas: diversidad, fisiología ambiental y productividad (Acid metabolism in Crassulaceae: diversity, environmental physiology and productivity). Bol Soc Bot Mex 81:37–50

Ángeles-Espino A, Valencia-Botín AJ, Virgen-Calleros G et al (2013) Lethal dosis (LD50) determination using Co^{60} on *Agave tequilana* var. Azul vitro plantlets Rev Fitotec Mex 36: 381–386

Ángeles-Espino A, Virgen-Calleros G, Valencia-Botín J et al (2016) Nondestructive evaluation of pathogenicity of *Cercospora agavicola* on Agave azul tequilero plantlets irradiated with gamma rays Co^{60}. Acta Univ 26:3–10

Ángeles-Espino A, Valencia-Botín AJ, Ramírez-Serrano C et al (2020) Molecular characterization of *Agave tequilana* mutants induced with gamma radiation Co^{60} and its effect in the fructooligosaccharides accumulation. Acta Univ 30:e2696. https://doi.org/10.15174/au.2020.2696

Arzate-Fernández AM, Mejía-Franco R (2011) Capacidad embriogénica de callos inducidos en ejes embrionarios cigóticos de *Agave angustifolia* Haw (Embryogenic capacity of induced calluses in zygotic embryonic axes of *Agave angustifolia* Haw.). Rev Fitotec Mex 34:101–106

Arzate-Fernández AM, Piña-Escutia JL, Norman-Mondragón TH et al (2016) Regeneration of Agave (*Agave angustifolia* HAW.) from encapsulated somatic embryos. Rev Fitotec Mex 39: 359–366

Arzate-Fernández AM, Martínez-Velasco I, Alvarez-Aragón SY et al (2020) Respuesta morfogenética de dos especies de Agave regeneradas in vitro (Morphogenetic response of two Agave species regenerated in vitro). Trop Subtrop Agroecosyst 23:1–17

Avila De Dios E, Gomez Vargas AD, Damian Santos ML et al (2015) New insights into plant glycoside hydrolase family 32 in Agave species. Front Plant Sci 6:594. https://doi.org/10.3389/fpls.2015.00594

Bautista-Montes E, Hernández-Soriano L, Simpson J (2022) Advances in the micropropagation and genetic transformation of *Agave* species. Plan Theory 11:1757. https://doi.org/10.3390/plants11131757

Benelli A, Cevoli C, Fabbri A (2020) In-field hyperspectral imaging: an overview on the ground-based applications in agriculture. J Agric Eng 51:129–139. https://doi.org/10.4081/jae.2020.1030

Cabrera-Ponce JL (2002) Transformación Genética en El Género Agave Y Producción de Plantas Transgénicas Resistentes a Herbicidas (Genetic Transformation in the Agave Genus and Production of Herbicide-Resistant Transgenic Plants). JL02000044. Available online: https://svt.cinvestav.mx/Portals/svt/descarga/Fichas/Agave%20tequilero-MXJL02000044A.pdf. Accessed on 20 June 2024

Cabrera-Toledo D, Mendoza-Galindo E, Larranaga N et al (2022) Genomic and morphological differentiation of spirit producing *Agave angustifolia* traditional landraces cultivated in Jalisco, Mexico. Plants 11:2274. https://doi.org/10.3390/plants11172274

Calvario G, Sierra B, Alarcón TE et al (2017) A multi-disciplinary approach to remote sensing through low-cost UAVs. Sensors 17:1411. https://doi.org/10.3390/s17061411

Campaña KO, Remache AP, Re MA (2020) Caracterización mecánica de fibras nuevas y centenarias de *Agave americana* (Mechanical characterization of new and centenary fibers of *Agave americana*). Rev Espacios 41:382–389

Castillo QD, Berlanga RCA, Cano PA (2005) Recolección, extracción y uso de la fibra de lechuguilla (*Agave lechuguilla* Torr) en el estado de Coahuila (Collection, extraction and use of lechuguilla fiber (Agave lechuguilla Torr) in the state of Coahuila). INIFAP, No 2, p 23

Castillo-Lara JF, Flores-Johnson EA (2020) La fibra de henequén (*Agave fourcroydes*) como una opción para materiales compuestos amigables con el medio ambiente (Henequen fiber (*Agave fourcroydes*) as an option for environmentally friendly composite materials). Ensayos desde el Herbario CICY 20220:99–105

Cervantes-Pérez SA, Espinal-Centeno A, Oropeza-Aburto A et al (2018) Transcriptional profiling of the CAM plant *Agave salmiana* reveals conservation of a genetic program for regeneration. Dev Biol 442:28–39. https://doi.org/10.1016/j.ydbio.2018.04.018

Colunga-GarcíaMarín P, May-Pat F (1993) Agave studies in Yucatan, Mexico I. Past and present germplasm diversity and uses. Economic Bot 47:312–327

Colunga-GarciaMarin P, Coello-Coello JN, Eguiarte LE et al (1999) Isozymatic variation and phylogenetic relationships between henequen (*Agave fourcroydes*) and its wild ancestor *A. angustifolia* (Agavaceae). Am J Bot 86:115–123. https://doi.org/10.2307/2656960

Colunga-GarcíaMarín P, Zizumbo-Villarreal D, MartínezTorres J (2007) Tradiciones en el aprovechamiento de los Agaves mexicanos: una aportación a la protección legal y conservación de su diversidad biológica y cultural (Traditions in the use of Mexican Agaves: a contribution to the legal protection and conservation of their biological and cultural diversity). In: Colunga-GarcíaMarín P, Larqué Saavedra A, Eguiarte L et al (eds) En lo ancestral hay futuro: del tequila, los mezcales y otros Agaves (In the ancestral there is a future: of tequila, mezcals and other Agaves). CICY-CONACYT-CONABIO-INE, pp 229–248

Comisión Nacional para el Conocimiento y Uso de la Biodiversidad (CONABIO) (2006) Mezcales y diversidad: Agave. 2ª edición. Conabio, México. https://bioteca.biodiversidad.gob.mx/janium-bin/detalle.pl?Id=20240627095659

Da Silva BP, De Sousa AC, Silva GM et al (2002) A new bioactive steroidal saponin from *Agave attenuate*. Z Naturforsch C J Biosci 57:423–428. https://doi.org/10.1515/znc-2002-5-603

Davis SC, Ortiz-Cano HG (2023) Lessons from the history of Agave: ecological and cultural context for valuation of CAM. Ann Bot 132:819–833. https://doi.org/10.1093/aob/mcad072
Davis SC, Dohleman FG, Logn SP (2011) The global potential for Agave as a biofuel feedstock. Bioenergy 3:68–78. https://doi.org/10.1111/j.757-1707.2010.01077.x
Delgado-Aceves L, González-Arnao MT, Santacruz-Ruvalcaba F et al (2021) Indirect somatic embryogenesis and cryopreservation of *Agave tequilana* Weber Cultivar 'Chato'. Plan Theory 10:249. https://doi.org/10.3390/plants10020249
Duncan AC, Jäger AK, van Staden J (1999) Screening of Zulu medicinal plants for angiotensin-converting enzyme (ACE) inhibitors. J Ethnopharmacol 68:63–70. https://doi.org/10.1016/s0378-8741(99)00097-5
Eguiarte LE, Souza V (2007) Historia natural del Agave y sus parientes: Evolución y Ecología (Natural history of Agave and its relatives: Evolution and Ecology). In: Colunga-García M, Larqué Saavedra A, Eguiart LE et al (eds) Lo ancestral hay futuro: del tequila, los mezcales y otros Agaves (The ancestral has a future: from tequila, mezcals and other Agaves). Centro de Investigación Científica de Yucatán, AC, pp 3–22
Eguiarte LE, Aguirre-Planter E, Aguirre X et al (2013) From isozymes to genomics: population genetics and conservation of Agave in México. Bot Rev 79:483–506
Eguiarte LE, Jiménez Barrón OA, Aguirre-Planter E et al (2021) Evolutionary ecology of Agave: distribution patterns, phylogeny, and coevolution (an homage to Howard S. Gentry). Am J Bot 108:216–235. https://doi.org/10.1002/ajb2.1609
Eli-Chukwu NC (2019) Applications of artificial intelligence in agriculture: a review. Eng Technol App Sci Res 9:4377–4383
Escalante A, Lopez Soto D, Velazquez Gutierrez J (2016) Pulque, a traditional Mexican alcoholic fermented beverage: historical, microbiological, and technical aspects. Front Microbiol 7:1026. https://doi.org/10.3389/fmicb.2016.01026
Escobar-Guzman RE, Zamudio HF, Vega GK et al (2008) Seed production and gametophyte formation in *Agave tequilana* and *Agave americana*. Botany 86:1343–1353
Esparza-Ibarra EL, Violante-González SJ, Monks JC et al (2015) Los Agaves mezcaleros del Altiplano Potosino y Zacatecano (The Mezcal Agaves of the Potosino and Zacatecan Altiplano). In: Pulido-Flore G, Monks S, López-Herrera M (eds) Estudios en Biodiversidad (Biodiversity Studies). University of Nebraska-Lincoln, pp 227–245
Flores EGJ, Sandoval S (2022) Artificial intelligence to model the potential distribution of *Agave durangensis*. In: IGARSS-2022, IEEE International Geoscience and Remote Sensing Symposium, Kuala Lumpur, pp 5828–5831
Flores D, González-Hernández I, Lozano R et al (2021) Automated agave detection and counting using a convolutional neural network and unmanned aerial systems. Drones 5. https://doi.org/10.3390/drones5010004
Flores-Benítez S, Jiménez-Bremont JF, Rosales-Mendoza S (2007) Genetic transformation of *Agave salmiana* by *Agrobacterium tumefaciens* and particle bombardment. Plant Cell Tissue Organ Cult 91:215–224. https://doi.org/10.1007/s11240-007-9287-3
Frazão C, Barros J, Toledo FR et al (2018) Development of sandwich panels combining Sisal Fiber-Cement Composites and Fiber-Reinforced Lightweight Concrete. Cem Concr Compos 86:206–223. https://doi.org/10.1016/j.cemconcomp.2017.11.008
Fujiyama R, Darwish F, Pereira MV (2014) Mechanical characterization of sisal reinforced cement mortar. Theor Appl Mech Lett 4:061002
Gao J, Yang F, Zhang S (2014) Expression of a hevein-like gene in transgenic Agave hybrid No. 11648 enhances tolerance against zebra stripe disease. Plant Cell Tissue Organ Cult 119: 579–585. https://doi.org/10.1007/s11240-014-0557-6
García-Herrera EJ, Méndez-Gallegos S, Talavera-Magaña D (2010) El género Agave spp. en México: principales usos de importancia socioeconómica y agroecológica (The genus Agave spp. in Mexico: main uses of socioeconomic and agroecological importance). Rev Salud Pub Nut 1:109–129

García-Mendoza AJ (2011) Flora del Valle de Tehuacán-Cuicatlán (Flora of the Tehuacan-Cuicatlan Valley). Universidad Autónoma de México, Mexico City

García-Mendoza AJ, Franco Martínez IS, Sandoval Gutiérrez D (2019) Cuatro especies nuevas de Agave (Asparagaceae, Agavoideae) del sur de México (Four new species of Agave (Asparagaceae, Agavoideae) from southern Mexico). Act Bot Mex 126:e1461. https://doi.org/10.21829/abm126.2019.1461

Gentry HS (1982) Agaves of Continental North America. University of Arizona Press, Tucson

Gomez-Rodriguez VM, Rodriguez-Garay B, Palomino G et al (2013) Physical mapping of 5s and 18s ribosomal DNA in three species of Agave (Asparagales, Asparagaceae). Comp Cytogenet 7: 191–203

Gonçalves J, Henriques R (2015) UAV photogrammetry for topographic monitoring of coastal areas. ISPRS J Photogramm Remote Sens 104:101–111. https://doi.org/10.1016/j.isprsjprs.2015.02.009

González G, Alemán S, Infante D (2003) Asexual genetic variability in *Agave fourcroydes* II: selection among individuals in a clonally propagated population. Plant Sci 165:595–601. https://doi.org/10.1016/S0168-9452(03)00227-9

González-Oramas G, Alemán-Garcia M, Garriga M et al (2007) Radio sensitivity to gamma rays (^{60}Co) in shoot tips of henequen. Biotech Veg 7:115–117

Gross S, Martin J, Simpson J et al (2013) De novo transcriptome assembly of drought tolerant CAM plants, *Agave deserti* and *Agave tequilana*. BMC Genomics 14:1–14. https://doi.org/10.1186/1471-2164-14-563

Gutiérrez-Aguilar PR, Gil-Vega KC, Simpson J (2014) Development of an *Agrobacterium tumefasciens* mediated transformation protocol for two Agave species by organogenesis. In: Gutiérrez-Mora A, Rodríguez-Garay B, Contreras-Ramos SM et al (eds) Sustainable and integral rxploitation of agave. Available online: http://www.ciatej.net.mx/Agave/1.7Agave.pdf. Accessed on 20 June 2024

Hernández-Castellanos S, Garruña-Hernández R, Us-Camas R et al (2020) *Agave angustifolia* albino plantlets lose stomatal physiology function by changing the development of the stomatal complex due to a molecular disruption. Mol Gen Genomics 295:787–805. https://doi.org/10.1007/s00438-019-01643-y

Huang X, Wang B, Xi J et al (2018) Transcriptome comparison reveals distinct selection patterns in domesticated and wild Agave species, the important CAM plants. Int J Genom 22:5716518. https://doi.org/10.1155/2018/5716518

Huang X, Xiao M, Xi J et al (2019) De Novo transcriptome assembly of Agave H11648 by Illumina sequencing and identification of cellulosesynthase genes in Agave species. Genes 10:103. https://doi.org/10.3390/genes10020103

Huang X, Xu B, Tan S et al (2022) Transcriptome sequencing of *Agave angustifolia* reveals conservation and diversification in the expression of cinnamyl alcohol dehydrogenase genes in Agave species. Agriculture 12:1003. https://doi.org/10.3390/agriculture12071003

Jimenez-Barron O, Garcia-Sandoval R, Magallon S et al (2020) Phylogeny, diversification rate, and divergence time of Agave sensu lato (Asparagaceae), a group of recent origin in the process of diversification. Front Plant Sci 11:536135. https://doi.org/10.3389/fpls.2020.536135

Klimova A, Ruiz Mondragón KY, Molina Freaner F et al (2022) Genomic analyses of wild and cultivated Bacanora Agave (*Agave angustifolia* var. pacifica) reveal inbreeding, few signs of cultivation history and shallow population structure. Plants 11:1426. https://doi.org/10.3390/plants11111426

Klimova A, Ruiz Mondragón KY, Aguirre-Planter E et al (2023) Genomic analysis unveils reduced genetic variability but increased proportion of heterozygotic genotypes of the intensively managed mezcal Agave, *Agave angustifolia*. Am J Bot 110:e16216. https://doi.org/10.1002/ajb2.16216

Lappe-Oliveras P, Moreno-Terrazas R, Arrizón-Gaviño J et al (2008) Yeast associated with the production of Mexican alcoholic non-distilled and distilled Agave beverages. FEMS Yeast Res 8:1037–1052. https://doi.org/10.1111/j.1567-1364.2008.00430.x

Larcher W (2003) Physiological plant ecology: ecophysiology and stress physiology of functional groups, 4th edn. Springer, New York. https://doi.org/10.1007/978-3-662-05214-3

Lindsay DL, Swift JF, Lance RF et al (2018) A comparison of patterns of genetic structure in two co-occurring Agave species (Asparagaceae) that differ in the patchiness of their geographical distributions and cultivation histories. Bot J Linn Soc 186:361–373

Lock GW (1962) Sisal: Twenty-Five Years' Sisal Research, 2nd edn. Longmans, Green & Co., Ltd., Harlow, UK

Maceda-López LF, Góngora-Castillo EB, Ibarra-Laclette E et al (2022) Transcriptome mining provides insights into cell wall metabolism and fiber lignification in *Agave tequilana* Weber. Plan Theory 11:1496. https://doi.org/10.3390/plants11111496

Martínez-Velasco I, Arzate-Fernández AM (2022) Analysis of the somaclonal variation in two *in vitro* regenerated Agave species. Trop Subtrop Agroecosyst 25:050. https://doi.org/10.56369/tsaes.3709

McKain MR, Wickett N, Zhang Y et al (2012) Phylogenomic analysis of transcriptome data elucidates co-occurrence of a paleopolyploid event and the origin of bimodal karyotypes in Agavoideae (Asparagaceae). Am J Bot 99:397–406. https://doi.org/10.3732/ajb.1100537

Monja-Mio KM, Barredo Pool F, Herrera Herrera G et al (2015) Development of the stomatal complex and leaf surface of *Agave angustifolia* Haw. 'Bacanora' plantlets during the in vitro to ex vitro transition process. Sci Horticul 189:32–40. https://doi.org/10.1016/j.scienta.2015.03.032

Monja-Mio KM, Herrera-Alamillo MA, Sánchez-Teyer LF et al (2019) Breeding Strategies to improve production of *Agave* (*Agave* spp.). In: Al-Khayri JM, Mohan JS, Johnson S (eds) Advances in plant breeding strategies: industrial and food crops, vol 6. Springer, Switzerland, pp 319–362

Moreno-Salazar SF, Esqueda M, Martínez J et al (2007) Tamaño del genoma y cariotipo en *Agave angustifolia* y *A. rhodacantha* de Sonora, México (Genome size and karyotype in Agave angustifolia and A. rhodacantha from Sonora, Mexico). Rev Fitot Mex 30:13–23

Morreeuw ZP, Escobedo-Fregoso C, Ríos-González LJ et al (2021) Transcriptome-based metabolic profiling of flavonoids in *Agave lechuguilla* waste biomass. Plant Sci 305:110748. https://doi.org/10.1016/j.plantsci.2020.110748

Narváez-Zapata JA, Sánchez-Teyer LF (2009) Agaves as a raw material: recent technologies and applications. Recent Pat Biotechnol 3:185–191. https://doi.org/10.2174/187220809789389144

Nava-Cruz NY, Medina-Morales MA, Martinez JL et al (2014) Agave biotechnology: an overview. Crit Rev Biotechnol 35:546–559

Naziri M, Sadat S, Soltani Howyzeh M (2019) The effect of different hormone combinations on direct and indirect somatic embryogenesis in Agave Americana. Iran J Plant Physiol 9:2739–2747

Niechayev NA, Pereira NP, Cushman JC (2019) Understanding trait diversity associated with crassulacean acid metabolism (CAM). Curr Opin Plant Biol 49:74–85. https://doi.org/10.1016/j.pbi.2019.06.004

Norma Oficial Mexicana NOM-006-SCFI-2012 (2012) Secretaría de Comercio y Fomento Industrial. Bebidas alcohólicas – Tequila – Especificaciones (Ministry of Commerce and Industrial Development. Alcoholic beverages – Tequila – Specifications). https://www.dof.gob.mx/nota_detalle.php?codigo=5282165&fecha=13/12/2012#gsc.tab=0. Accessed 14 June 2024

Olvera-Vargas LA, Pardo-Núñez J, Aguilar-Rivera N et al (2022) Detección de *Agave angustifolia* y *Agave cupreata* con técnicas geomáticas en Guerrero, México (Detection of *Agave angustifolia* and *Agave cupreata* with geomatic techniques in Guerrero, Mexico). Cienc Tecnol Agropecuaria 23:e2241. https://doi.org/10.21930/rcta.vol23_num2_art:2241

Orestes GJ, Meneses A, Simonet AM et al (2008) Saponinas esteroidales de la planta *Agave brittoniana* (Agavaceae) con actividad contra el parásito *Trichomona vaginalis*. Rev Biol Trop 56:1645–1652

Ortíz-Basurto RI, Pourcelly G, Doco T et al (2008) Analysis of the main components of the aguamiel produced by the maguey-pulquero (*Agave mapisaga*) throughout the harvest period. J Agric Food Chem 56:3682–3687. https://doi.org/10.1021/jf072767h

Palma Cruz FJ (2000) Agaves productores de fibras duras en el estado de Oaxaca, México (Agaves producing hard fibers in the state of Oaxaca, Mexico). Bol Soc Bot Mex 66:102. https://doi.org/10.17129/botsci.1615

Palomino G, Martínez J, Méndez I (2007) Variación inter e intraespecífica en especies de Agave por citometría de flujo y análisis de sus cromosomas (Inter and intraspecific variation in Agave species by flow cytometry and chromosome analysis). In: Colunga-García MP, Larqué S, Eguiarte LE et al (eds) En lo ancestral hay futuro: del tequila, los mezcales y otros Agaves (In the ancestral there is a future: of tequila, mezcals and other Agaves). CICY, CONACYT, CONABIO y SEMARNAT, Mérida, pp 41–64

Palomino G, Martínez J, Romero P et al (2017) Nuclear genome size and karyotype analysis of *Agave angustifolia* Haw. "Cimarron" and "Lineño" (Asparagales, Asparagaceae). Caryologia 70:93–101. https://doi.org/10.1080/00087114.2017.1279931

Parascanu MM, Sanchez N, Sandoval-Salas F et al (2021) Environmental and economic analysis of bioethanol production from sugarcane molasses and Agave juice. Environ Sci Pollut Res Int 28: 64374–64393. https://doi.org/10.1007/s11356-021-15471-4

Pérez del Rio R, Hidalgo M, Caballero M et al (2022) Multi-objective optimization through artificial intelligence for designing of an *Agave angustifolia* leaf shredder. Scientia Agropecuaria 13:291–299. https://doi.org/10.17268/sci.agropecu.2022.026

Pérez-Zavala ML, Hernández-Arzaba JC, Bideshi DK et al (2020) *Agave*: a natural renewable resource with multiple applications. J Sci Food Agric 100:5324–5333. https://doi.org/10.1002/jsfa.10586

Pourkhaloee A, Khosh-Khui M, Barba-Gonzalez R (2023) Somatic embryogenesis of tuberose (*Agave amica* L.) was improved by milk as a potential biostimulant in plant tissue culture. Int Hort Sci Technol 10:257–268. https://doi.org/10.22059/ijhst.2022.330660.504

Qin X, Yang X, Huang X et al (2021) The complete chloroplast genome of *Agave angustifolia*. Mitochondrial DNA Part B 6:3236–3237. https://doi.org/10.1080/23802359.2021.1941360

Raya FT, Marone MP, Carvalho LM et al (2021) Extreme physiology: biomass and transcriptional profiling of three abandoned Agave cultivars. Ind Crop Prod 172:114043. https://doi.org/10.1016/j.indcrop.2021.114043

Raya FT, de Carvalho LM, José J et al (2023) Rescuing the Brazilian Agave breeding program: morphophysiological and molecular characterization of a new germplasm. Front Chem Eng 5: 1218668. https://doi.org/10.3389/fceng.2023.1218668

Reyes-Carmona ED (2023) Diversidad microbiana en la fermentación de mezcal artesanal: un estudio metagenómico (Microbial diversity in artisanal mezcal fermentation: a metagenomic study). Dissertation, Universidad Autónoma Chapingo

Reyes-Díaz JI, Arzate-Fernández AM, Piña-Escutia JL (2018) Sucrose and organic nitrogen sources have an influence in Agave angustifolia somatic embryogenesis. Rev Mexi Cienc Agri 9:1508–1513. https://doi.org/10.29312/remexca.v9i7.1676

Reyes-Zambrano SJ, Ramírez-Merchant ML, Arias-Castro C et al (2019) Morphometric and biochemical changes in *Agave americana* L. plantlets induced by ethyl methanesulfonate. Phyton-Int J Exp Bot 88:277–284. https://doi.org/10.32604/phyton.2019.06504

Reyes-Zambrano SJ, Lecona-Guzmán CA, Gutiérrez-Miceli FA et al (2021) Selection and analysis of polymorphisms in somaclonal variants of *Agave americana* resistant to *Fusarium oxysporum* via an ethyl methanesulphonate treatment. Phyton-Int J Exp Bot 90:1727–1739. https://doi.org/10.32604/phyton.2021.016171

Ríos-Ramírez SC, Enríquez-del Valle JR, Rodríguez-Ortiz G et al (2017) Benzylaminopurine and indol-3-acetic acid concentrations in *in vitro* proliferation of *Agave angustifolia* adventitious shoots. Cienc Investig Agrar 44:285–294. https://doi.org/10.7764/rcia.v44i3.1810

Rivera-Lugo M, Garcia-Mendoza A, Simpson J et al (2018) Taxonomic implications of the morphological and genetic variation of cultivated and domesticated populations of the *Agave angustifolia* complex (Agavoideae, Asparagaceae) in Oaxaca, Mexico. Plant Syst Evol 304: 969–979. https://doi.org/10.1007/s00606-018-1525-0

Robert ML, Lim KY, Hanson L et al (2008) Wild and agronomically important Agave species (Asparagaceae) show proportional increases in chromosome number, genome size, and genetic markers with increasing ploidy. Bot J Linn Soc 158:215–222

Rodríguez-Garay B, Lomeli-Sencion J, Tapia-Campos E et al (2009) Morphological and molecular diversity of *Agave tequilana* Weber var. Azul and *Agave angustifolia* Haw. var Lineño. Ind Crop Prod 29:220–228. https://doi.org/10.1016/j.indcrop.2008.05.007

Rodríguez-Garay B, López-Díaz S, Rodríguez-Domínguez JM et al (2018) Application of *in Casa* Pollination and Embryo Rescue Techniques for Breeding of *Agave* species. Methods Mol Biol 1815:289–300. https://doi.org/10.1007/978-1-4939-8594-4_20

Rodríguez-Hernández G, Morales-Domínguez F, Gutiérrez-Campos R (2007) Generación de raíces transformadas de *Agave salmiana* Otto y su colonización por *Glomus intraradices* (Generation of transformed roots of *Agave salmiana* otto and its colonization by *Glomus intraradices*). Rev Fitotec Mex 30:215–222

Ruíz HA, Ruzene DS, Silva DP et al (2011) Development and characterization of an environmentally friendly process sequence (Autohydrolysis and organosolv) for wheat straw delignification. Appl Biochem Biotechnol 164:629–641. https://doi.org/10.1007/s12010-011-9163-9

Ruiz-Mondragon KY, Aguirre-Planter E, Gasca-Pineda J et al (2022) Conservation genomics of *Agave tequilana* Weber var. Azul: low genetic differentiation and heterozygote excess in the tequila Agave from Jalisco, Mexico. PeerJ 10:e14398. https://doi.org/10.7717/peerj.14398

Sánchez A, Coronel-Lara Z, Gutiérrez A et al (2020) Acclimatization and transplantation of *Agave angustifolia* Haw. vitroplants in wild conditions. Rev Mex Cienc Agric 11:1593–1605. https://doi.org/10.29312/remexca.v11i7.2403

Sánchez-Teyer F, Moreno-Salazar S, Esqueda M et al (2009) Genetic variability of wild *Agave angustifolia* populations based on AFLP: a basic study for conservation. J Arid Environ 73:611–616. https://doi.org/10.1016/j.jaridenv.2009.01.008

Sarwar MB, Ahmad Z, Rashid B et al (2019) De novo assembly of *Agave sisalana* transcriptome in response to drought stress provides insight into the tolerance mechanisms. Sci Rep 9:396–410. https://doi.org/10.1038/s41598-018-35891-6

Serrano V, Gutiérrez FO (2009) Mejoramiento genético de Agaves en Cuba (Genetic improvement of Agaves in Cuba). Temas Sci Tecnol 37:37–43

Sharma S, Verma K, Hardaha P (2022) Implementation of artificial intelligence in Agriculture. J Comp Cogn Eng 2:155–162. https://doi.org/10.47852/bonviewJCCE2202174

Simpson J, Martinez Hernandez A, Abraham Juarez JM et al (2011) Genomic resources and transcriptome mining in *Agave tequilana*. GCB Bioenergy 3:25–36. https://doi.org/10.1111/j.17571707.2010.01079.x

Tejavathi DH, Gayathramma K, Sowmya R (2006) Production of plantlets from encapsulated in vitro shoot buds and somatic embryos of agave vera-Cruz Mill. Plant Cell Biotech Mol Biol 7:183–186

Tejavathi DH, Gayathramma K, Nijagunaiah R (2012) Genetic variability by in vitro mutagenesis and arbuscular mycorrhizal fungal symbiosis among micropropagated plants of *Agave vera-cruz* miller. Bioremed Biodiv Bioavail 6:665–669

Torres-Morán MI, Almaraz-Abarca N, Velasco-Ramirez AP et al (2008) Taxonomic significance of ISTR to discriminate species in Agavaceae. Am J Agric Biol Sci 3:661–665. https://doi.org/10.3844/ajabssp.2008.661.665

Torres-Velázquez DS, Rojas-Contreras JA, Soto-Cruz NO et al (2022) Bacterial population monitoring during alcoholic fermentation of mezcal in Durango by DGGE. Int J Agric Nat Resour 49:112–122. https://doi.org/10.7764/ijanr.v49i2.2367

Trejo L, Limones V, Pena G et al (2018) Genetic variation and relationships among Agaves related to the production of Tequila and Mezcal in Jalisco. Ind Crop Prod 125:140–149. https://doi.org/10.1016/j.indcrop.2018.08.072

Trejo-Torres JC (2018) El Dr. Henry Perrine y su henequén verde, *Agave sisalana*: la fibra del mundo (Dr. Henry Perrine and his green henequen, Agave sisalana: the fiber of the world). Desde el Herbario CICY 10:227–232. http://www.cicy.mx/sitios/desde_herbario/

Valadez-Moctezuma E, Samah S, Mascorro-Gallardo JO et al (2023) The de novo transcriptome assembly of *Agave angustifolia* Haw, mining for carbohydrates and cellulose synthesis genes profiling, and molecular markers development. Plant Biotechnol Rep 17:889–904. https://doi.org/10.1007/s11816-023-00861-6

Valencia-Botín AJ, Ramírez-Serrano C, Virgen-Calleros G et al (2020) Indirect somatic embryogenesis on mutants of *Agave tequilana* weber cultivar blue induced with CO^{60} gamma rays. Trop Subtrop Agroecosys 23:1–9

Vargas-Ponce O, Zizumbo-Villarreal D, Martinez-Castillo J et al (2009) Diversity and structure of landraces of Agave grown for spirits under traditional agriculture: a comparison with wild populations of *A. angustifolia* (Agavaceae) and commercial plantations of *A. tequilana*. Am J Bot 96:448–457. https://doi.org/10.3732/ajb.0800176

Verastegui A, Verde-Star MJ, Garcia S et al (2008) Species of Agave with antimicrobial activity against selected pathogenic bacteria and fungi. World J Microbiol Biotechnol 24:1249–1252. https://doi.org/10.1007/s11274-007-9563-8

Wang X, Huang X, Chen L et al (2023) Transcriptome sequencing of *Agave amaniensis* reveals shoot-related expression patterns of Expansin A genes in Agave. Plan Theory 12:2020. https://doi.org/10.3390/plants12102020

Warren-Vega WM, Contreras-Atrisco ZA, Ramírez-Quezada MF et al (2023) A novel approach of artificial intelligence for the study of the relation of physicochemical profile and color acquired by Tequila 100% Agave in its maturation process. J Food Comp Anal 123:105533. https://doi.org/10.1016/j.jfca.2023.105533

Yin H, Guo HB, Weston DJ et al (2018) Diel rewiring and positive selection of ancient plant proteins enabled evolution of CAM photosynthesis in Agave. BMC Genomics 19:588. https://doi.org/10.1186/s12864-018-4964-7

Chapter 6
Recent Advances in Date Palm (*Phoenix dactylifera* L.) Biotechnology and Breeding

Plosha Khanum, Asif Ali Khan, Iqrar Ahmad Khan, Abdul Ghaffar, and Zulqurnain Khan

Abstract The date palm fruit (*Phoenix dactylifera* L.) is a valuable fruit crop that requires low inputs and has high nutritional value. Propagating date palms through suckers is slow and can be led to disease transfer. The extended juvenility phase and life cycle of date palms present challenges in breeding. Seed propagation is quicker, but determining the sex early is difficult, taking 7–8 years. To aid in date palm breeding, studies have been conducted to identify morphological, physiological, biochemical, and genetic markers for sex identification. Physical traits like plant height, leaf length, and leaf width have been considered, but they are influenced by the environment and may not be reliable for sex identification. It has been noted that morphological traits are not ideal for sex identification in date palms. Male plants exhibit higher stomatal conductance and transpiration rates compared to females. Biochemical analysis has shown variations in nitrogen, phosphorus, and potassium levels, with females having higher (NPK) concentrations due to increased energy needs for embryo growth. The *TPD1* gene is more expressed in males, while the *SHP1* gene, related to ovule development, is more expressed in females. Gene-based identification can enhance breeding efforts. Additionally, tissue culture techniques can introduce variation among plantlets, aiding breeders. These techniques can help establish date palm orchards with verified female plants early in the sex identification process, potentially, increasing yield. Early sex determination is crucial for a breeding program. This chapter focuses on the large-scale early sex determination of date palm plants using biotechnology techniques and enhancing yield through tissue culture and micropropagation.

P. Khanum · I. A. Khan · Z. Khan (✉)
Institute of Plant Breeding and Biotechnology, MNS-University of Agriculture, Multan, Pakistan
e-mail: plosha.khanum@mnsuam.edu.pk; zulqarnain.khan@mnsuam.edu.pk

A. A. Khan
Institute of Horticultural Sciences, University of Agriculture, Faisalabad, Pakistan

A. Ghaffar
Department of Agronomy, MNS-University of Agriculture, Multan, Pakistan
e-mail: abdul.ghaffar@mnsuam.edu.pk

K. F. M. Salem et al. (eds.), *Breeding and Biotechnology of Leaf, Fruit, and Seed Fiber Crops*, Advances in Plant Breeding Strategies 11,
https://doi.org/10.1007/978-3-032-00407-9_6

Keywords Date palm (*Phoenix dactylifera* L.) · Gene expression · Genetic marker · Micropropagation · *SHP1* · Tissue culture · *TPD1*

6.1 Introduction

The date palm (*Phoenix dactylifera* L.), a dioecious monocotyledonous fruit plant with 18 pairs of chromosomes, is a member of the Arecaceae family (Dransfield et al. 2008). One term for the date palm is called "tree of life." it grows well in harsh environments, and it is an important source of essential nutrients. Dates are relatively tolerant of water scarcity, high temperature and soil salinity (Yaish and Kumar 2015). Palm trees require a long warm climate with minimal rain, but need moisture around the roots, particularly from pollination through harvest. According to an old saying, *P. dactylifera* grows best with its head in the sun and its feet soaked in water. The genus *Phoenix* belonging to the Arecaceae or Palmaceae family has 14 species including the date palm, consisting of perennial monocotyledonous plants (Barrow 1998).

Palm trees can thrive in temperatures ranging from 12.7 to 27.5 °C on average, tolerating temperatures as high as 50 °C and brief spells of cold as low as −5 °C (Chao and Krueger 2007). The optimal temperature for date palm development throughout the maturing period of the fruit is between 21 and 27 °C. Dates are commonly produced in dry climates between 15 N° and 35 N° (Zaid and de Wet 2002). Palm trees have been remarkably adapted to long summers where water is scarce, and the temperature is high. They can also survive well in the desert due to their traits that allow them to use underground water to meet its requirements.

The date palm plant canopy provides a microclimate under which other sustenance crops can be cultivated. The actual date palm origin (*P. dactylifera* L.) was forgotten in ancient times. Moreover, the palm tree was probably cultured during 4000 BC Interestingly, it was used to build the sanctum of the moon god near Ur in southern Iraq, or Mesopotamia. (Chao and Krueger 2007). According to the Food and Agriculture Organization (FAO 2021), date production increased from approximately 1.8 million metric tons to approximately 6.9 million metric tons from 1962 to 2005. It was noted that 8400 tons of dates were being produced by 1.5 million date palms when the United Arab Emirates was founded in 1971. The production of date palms increased by 90 times in 2005, with an estimated 18 million date palms yielding 760,000 tons. The UAE now boasts the world's largest facility, producing 100,000 tissue-cultured plants yearly (Mohan 2006). Pakistan is among the top 10 high date producing countries (Fig. 6.1, FAOSTAT 2021).

Female and male plants are sheltered from the sun and the heat during the early days of the blooming process by a complex, fibrous covering (spathe). When the temperature in the shade rises above 18 °C, dates bloom, and when the temperature rises above 25 °C, dates bear fruit (Zaid and de Wet 2002). In female trees, its blossom which eventually, becomes the fruit, is held on tall, curving stalks also called the "fruit stalk." The blossom consists of many unbranched stalks arranged in spirals on the rachis. Flowers with three sepals and three petals are common in both

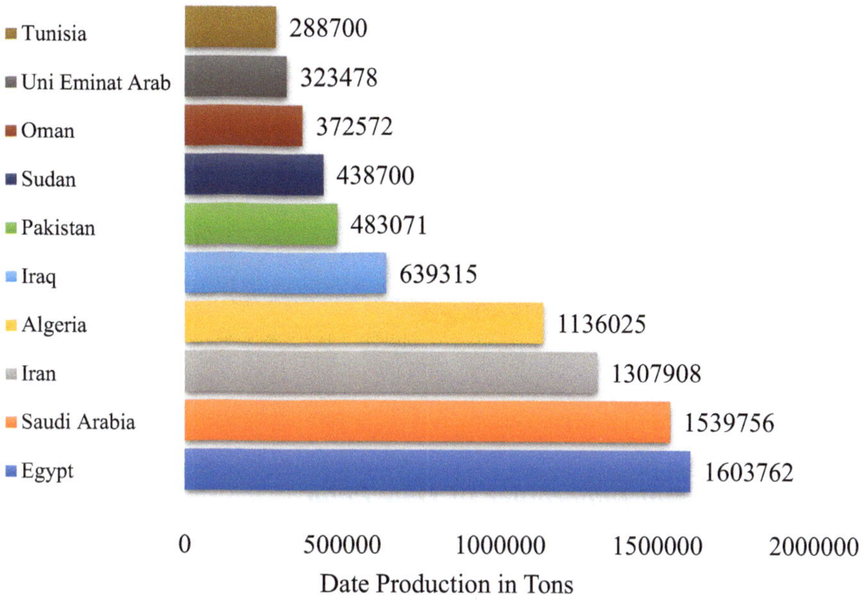

Fig. 6.1 Top dates producing countries in the world. (Source: FAOSTAT (2021))

male and female flowers. Male bloom is waxy white, whereas female blossom is often yellowish green. The average date palm yields 40 kg of fruit each year, but over 100 kg is achievable with diligent treatment. Dates can produce as little as 20 kg of fruit per year when grown with little input and maintenance. Female plants start bearing dates at the age of 4–6 years and attain peak output in 15–20 years (Nixon and Carpenter 1978). In addition, this plant is chosen for cultivation because of its good fruit quality and adaptability to harsh environments. Each cultivar is obtained by seed-based selection or through exceptional successor individual progeny.

Botanical research has identified 244 varieties in Morocco. The number of varieties was 250 in Tunisia, 370 (in Iraq), 400 (in Iran) (Zaid and de Wet 2002); 400 (in Sudan) (in Osman, 1984); 325 (in Pakistan), (Botes and Zaid 2002; Jamil et al. 2010); and even more varieties in other developing date palm nations. Globally, there are over 5000 different date palm varieties identified (Jaradat and Zaid 2004). The date palm produces basal suckers and has a large, cylindrical stem that can reach 10–20 m tall. A mature date palm‘s crown can produce 100–125 leaves on the petiole, each with 150 pinnae and acanthophylls. The various cultivars have different lengths of leaves, pinnae, and acanthophylls (Chao and Krueger 2007). This chapter aims to focus on the early sex determination of date palm plants on a large scale using biotechnology techniques in date palm breeding. It also aims to enhance the yield of date palms through tissue culture and micropropagation. Identifying morphophysiological, biochemical, and molecular parameters to differentiate the sex of date palms at an early stage will support date palm breeding (Fig. 6.2).

Fig. 6.2 https://en.wikipedia.org/wiki/Date_palm_Date_palm._2025

6.1.1 Historical Importance of Date Palm

Muslims regard the date palm to be the Prophet's holy fruit, and therefore, they prefer to eat it during Ramadan to break their fast. The date palm can help control hunger and provide comfort to the body after a day of fasting due to its nutritious composition. The Holy Prophet Hazrat Muhammad Peace Be Upon Him (PBUH) recommended the use and cultivation of date palms because they could cure several diseases. According to the Holy Prophet Muhammad (PBUH), "He will not be affected by magic or poison, who eats seven Ajwa dates every morning, on the day he eats them." The Prophet Muhammad (PBUH) also emphasized that "The family that has date palm fruit will not go hungry" (Al-Bukhari 2008).

In three major Abrahamic religions, Judaism, Christianity, and Islam, the date palm holds a unique sacred position. Christians use palm leaves to decorate their churches on palm Sunday, the Sunday before Easter. This tradition is still observed with rituals that date back to the early days of the Christian religion. Jewish women were given the name "Tamara," derived from the Hebrew Tamr, in hopes that their daughters would grow into beautiful, tall, intelligent, and fruitful adults, much like a date palm tree (Qudama 1985). The date palm is considered one of the heavenly blessings (Al-Quran, 55:68) and is mentioned nearly 26 times in the Holy Quran and 300 times in the Hadith of the Prophet Mohammad (Zaid et al. 2002).

Muslims regard the date palm to be the Prophet's holy fruit, and therefore, they prefer to eat it during Ramadan to break their fast. The date palm can help control hunger and provide comfort to the body after a day of fasting due to its nutritious composition. The Holy Prophet Hazrat Muhammad Peace Be Upon Him (PBUH) recommended the use and cultivation of date palms because they could cure several diseases. According to the Holy Prophet Muhammad (PBUH), "He will not be affected by magic or poison, who eats seven Ajwa dates every morning, on the day he eats them." The Prophet Muhammad (PBUH) also emphasized that "The family that has date palm fruit will not go hungry" (Al-Bukhari 2008).

In three major Abrahamic religions, Judaism, Christianity, and Islam, the date palm holds a unique sacred position. Christians use palm leaves to decorate their churches on Palm Sunday, the Sunday before Easter. This tradition is still observed with rituals that date back to the early days of the Christian religion. Jewish women were given the name "Tamara," derived from the Hebrew Tamr, in hopes that their daughters would grow into beautiful, tall, intelligent, and fruitful adults, much like a date palm tree (Qudama 1985). The date palm is considered one of the heavenly blessings (Al-Quran, 55:68) and is mentioned nearly 26 times in the Holy Quran and 300 times in the Hadith of the Prophet Mohammad (Zaid et al. 2002).

6.1.2 Medical Value of Date Palm

Date palms have numerous benefits, including nutritional, environmental, economic, and ornamental significance. Dates are a valuable source of macronutrients and micronutrients, making them affordable and readily available as cheap and plentiful (Al-Farsi and Lee 2008). The primary nutritional benefits of dates come from their carbohydrates which include soluble sugars and dietary fiber, as well as small amounts of fats or proteins (Ghnimi et al. 2017). Date palms have a variety of nutraceuticals benefits like antioxidant, antimutagenic, anti-inflammatory, antimicrobial, gastroprotective, anticancer, hepatoprotective, and immune stimulatory properties (Khalid et al. 2017). Date palm fruits have been used in old traditional medicines for many years in multiple cultures for their health benefits. They are used as a flavoring ingredient in dairy products and others besides their direct consumption in the human body. Dehydrated dates contain longer shelf life according to studies but their nutritional values is usually lower than fresh dates (Abdul-Hamid et al. 2016). Date palm types comprising Amber, Medjool, and Ajwa are high-value varieties and turn into considerably more income for farmers than ordinary varieties of date palm. Ajwa has a long history of health benefits dating back to 5000 BC and plays a significant role in Tib e Nabvi (PBUH), which is prophetic medicine based on the knowledge of the Prophet Mohammed (PBUH). Hadith and Islamic historical writing contain several mentions of Ajwa. Prophet Mohammed (PBUH) said that in a hadith "one who takes seven Ajwa dates in the morning will remain under the protection of Allah from magic/evil eyes and poison during the whole of the day." Only the Ajwa variety, also known as a Madinah-produced item, is native to the holy city of Madinah. When compared to other date palm types like Barni, Ajwa oil seed is of higher quality and can be utilized as a carrier for transversal enhancers in the pharmaceutical sector (Galeb et al. 2012).

Ajwa is highly beneficial for feeding mothers since it makes breast milk nutrient-rich. Additionally, numerous studies have demonstrated that children of women who frequently consume Ajwa are less susceptible to illnesses and infections. The nutritional content of Ajwa is higher than that of other date varieties. Both flesh and pits are rich in bioactive substances that help to improve human health and help to prevent, identify, and treat a variety of diseases (Al-Farsi and Lee 2008). Ajwa dates have potent anti-inflammatory, anti-cancer, anti-mutagenic, hepatoprotective, nephroprotective, and antioxidant properties that support a healthy immune system in humans (Sumaira et al. 2017). All these nutritional and medicinal values made Ajwa important among other date varieties. The specific origin of most of the date varieties, including Medjool, is still unclear, but they have been growing for thousands of years in the Americas, Africa, the Middle East, and other regions of the world. The world's oldest fruit to be cultivated is believed to be dates. There are other fascinating narratives of how the Medjool initially appeared in various nations, such as how it first reached America. When scientists traveled to Algeria to study dates and found that the conditions for Medjool cultivation were the same in the Coachella Valley of California. The natural sweetness of Medjool dates makes them a popular date variety among other varieties. They can be a healthy addition to the diet because they are a good source of fiber, antioxidants, and potassium.

The high potassium content of the Medjool lowers the risk of cardiovascular diseases by regulating blood pressure. High dietary fiber content in Medjool dates promotes healthy digestion and lowers low-density lipoprotein (LDL) cholesterol. Medjool dates help increase immunity, improve metabolism and encourage healthy bones (Salomón-Torres et al. 2019). The Amber is a sweet, dry date fruit variety of Saudi Arabia. It is the largest growing date of the Madina, grown for its large size. Although they are indigenous to the Middle East, amber dates are consumed and enjoyed worldwide, especially in India. It is one of the highest priced due to its nutritional value, which is also available in king size. Amber is a big source of additional nutrition like vitamin A, magnesium, potassium, vitamin B6, and niacin folate (Anonymous 2023). Therefore, due to high nutritional and commercial value, the Ajwa, Amber and Medjool varieties were selected for early sex determination to provide improved and cost-effective cultivars of date palm to farmers. The date palm has numerous benefits in terms of nutrition, ecology, commerce, and aesthetics. Dates are an inexpensive and abundant source of macronutrients and micronutrients (Al-Farsi and Lee 2008). According to Ghnimi et al. (2017), date fruits are mostly composed of carbohydrates like soluble sugars and dietary fiber, with trace amounts of lipids and proteins. Important nutraceuticals with an array of advantages, such as immune-stimulatory, gastro-protective, anti-cancer, anti-inflammatory, anti-microbial, anti-oxidant, and anti-mutagenic properties, are also present in dates (Khalid et al. 2017). Due to that reason, a lot of cultures have utilized these fruits in traditional medicine. Dates are utilized as a flavoring element in dairy products and other items besides being used directly. According to Abdul-Hamid et al. (2016), dehydrated dates have a longer shelf life than fresh dates, but they often have less nutritional value. This fruit is a natural folic acid and magnesium source that minimizes the danger of cardiovascular diseases and cancer, respectively (Abdullahi and Garko 2012). Date palm fruit can also be used to treat other conditions such as bronchial catarrh, sore throat, fever, and colds (Anjum et al. 2012). Date palm roots are used to cure toothache and diarrhea as minerals are present in the date palm, so it helps to strengthen the bones. Iron is present in significant amounts, so dates may be taken to cure anemic patients. Fluorine in date palms gave protection to teeth. Iron is present in significant amounts, so dates may be taken to cure anemic patients. Fluorine in date palms gave protection to teeth. Date palms, particularly their fruits, have been esteemed for their substantial nutritional and medicinal benefits for centuries. Rich in essential nutrients like fiber, potassium, magnesium, and copper, and dates provide a wholesome addition to any diet. Their high fiber content aids in digestion and helps alleviate constipation, while potassium contributes to regulating heart rate and blood pressure, promoting overall heart health. Dates are also a natural source of energy, offering a quick and sustained energy boost due to their high carbohydrate content. Additionally, the antioxidants in dates help reduce inflammation and protect against various diseases. Traditional medicine has long recognized the value of dates, using them to treat ailments such as respiratory and digestive issues. Furthermore, studies have shown that date palm extracts possess antiviral and antifungal properties, adding to their medicinal value. Despite their sweetness, dates have a low glycemic index, making them suitable for those looking to manage their blood sugar levels. In many cultures, dates hold religious significance and are often consumed during spiritual practices, highlighting their multifaceted importance in both health and tradition.

6.1.3 *Current Cultivation Practices and Challenges*

Current cultivation practices in agriculture are evolving to overcome the challenges of feeding a global population that grows rapidly and also minimizing the environmental impact. However, these practices also face significant challenges that need to be addressed for sustainable food production. Date palm cultivation is a significant agricultural activity in regions with arid and semi-arid climates. Major producers include Middle Eastern countries like Saudi Arabia, the United Arab Emirates, and Iraq; North African nations such as Egypt and Algeria; and South Asian countries, with Pakistan being one of the leading producers. These regions offer the hot, dry climate and well-drained soil that date palms require. Irrigation, although not essential, enhances growth, especially during flowering and fruiting stages. Regular pruning and pest control are critical to maintaining healthy palms. Harvesting, usually done by hand, depends on the variety and desired ripeness. However, challenges such as pests, diseases, and climate change necessitate adaptive agricultural practices. Economically, date palm cultivation generates employment, particularly during the harvest, and contributes to the export market, bolstering the economies of producing nations. Here's an overview of current cultivation practices and the challenges associated with them.

6.1.4 *Organic Farming*

Organic farming depends on the natural inputs and practices to maintain fertility of soil and manage pests and diseases without synthetic chemicals or genetically modified organisms (GMOs). Produces food with fewer chemical residues, supports biodiversity, and promotes soil health. Organic farming of date palms emphasizes sustainable and eco-friendly practices while avoiding synthetic chemicals like pesticides and fertilizers. The process begins with soil testing to ensure suitable conditions, as date palms thrive in well-drained sandy-loamy soils with a pH between 6.5 and 8.0. Soil fertility is enhanced by incorporating organic matter such as compost or well-rotted manure. Selecting high-yielding, pest-resistant varieties, like Medjool or Barhi, is crucial for successful cultivation. Date palm seedlings or offshoots are planted with adequate spacing (8–10 m) to allow for proper air circulation, and drip irrigation systems are employed to conserve water and provide consistent hydration, especially during establishment and fruiting stages.

Organic fertilizers, such as compost, bone meal, and seaweed extract, are applied twice a year to promote healthy growth, particularly during the growing and flowering seasons. Pest and disease management relies on natural solutions like neem oil, pheromone traps, and beneficial insects to control threats such as red palm weevils and fruit borers. Manual pollination is essential for date palms, as they are dioecious, requiring the transfer of pollen from male to female flowers during flowering. Weed control is achieved through organic mulching and manual removal to minimize competition for resources.

Dates are harvested at full maturity, depending on their type (soft, semi-dry, or dry), ensuring the best quality. Organic farming practices extend to post-harvest processes, which involve cleaning, sorting, and packaging without synthetic treatments. Farmers can further boost market value by obtaining organic certification, which involves maintaining meticulous records of farming activities to meet regulatory standards. Organic date palm farming not only produces healthier, chemical-free fruits but also improves soil health, promotes biodiversity, and caters to the growing demand for organic produce, often commanding premium prices in the market.

6.1.5 Vertical Farming and Controlled Environment Agriculture (CEA)

Involves growing crops indoors or vertically stacked in controlled environments using artificial lighting, temperature, and humidity control. It maximizes the efficiency of land use, reduces the use of water, and allows round production for the year in urban areas or regions with limited arable land.

6.1.6 Agroforestry

Agroforestry is the deliberate assimilation of trees and shrubs into crop and livestock farming systems to improve sustainability and diversify income sources. The significance of agroforestry lies in enhancing biodiversity, improving soil fertility, providing additional income streams from tree products, and mitigating climate change through carbon sequestration. Agroforestry is a sustainable land-use practice that integrates trees and shrubs with crops and/or livestock to create a mutually beneficial system. This approach combines agricultural and forestry techniques to optimize land productivity, enhance biodiversity, and promote environmental health. Trees in agroforestry systems provide shade, reduce wind erosion, and improve soil fertility through organic matter and nitrogen fixation. Meanwhile, crops benefit from moderated microclimates and reduced competition for water and nutrients. Livestock can graze beneath the trees, benefiting from shade and forage, while contributing manure to enrich soil quality. Agroforestry systems, such as alley cropping, silvopasture, and forest farming, are adaptable to diverse climates and landscapes. They offer economic benefits by diversifying income sources, improving crop resilience, and providing timber, fruits, and other forest products. Additionally, agroforestry helps combat climate change by sequestering carbon, conserving water, and preventing soil degradation. By harmonizing ecological and agricultural practices, agroforestry supports sustainable farming and rural livelihoods while safeguarding natural resources for future generations.

6.1.7 Challenges

Climate Change dangerous weather conditions, increasing the frequency of extreme events such as drought, stress and evolution in pest and disease dynamics. The challenges are adaptation of crops and practices to changing climatic conditions, maintaining productivity, and resilience. The causes of soil degradation are intensive farming practices, erosion, nutrient depletion, and loss of organic matter. The challenges are implementing sustainable soil management practices to improve soil health, fertility, and resilience. The water scarcity and quality issues is competition for water resources among agriculture, urban, and industrial sectors; and pollution of water bodies from agricultural runoff. The challenges are Efficient water use through technologies like drip irrigation, soil moisture sensors, and water recycling, as well as reducing nutrient runoff to protect water quality. The loss of biodiversity causes habitat destruction, monoculture farming practices, and chemical inputs affecting beneficial organisms. The challenges are promoting biodiversity-friendly farming practices, such as crop diversification, agroforestry, and habitat preservation.

Food security and equity issues are disparities in access to food due to poverty, conflicts, and inequitable distribution systems. The challenges are ensuring equitable access to nutritious food through sustainable agriculture practices and policies that support smallholder farmers and local food systems. Technological and knowledge gap issues are limited access to technology and knowledge among smallholder farmers, especially in developing countries. The challenges are bridging the digital divide and providing training and support for adoption of sustainable farming practices. Policy and Market Dynamics issues are the influence of subsidies, trade policies, and consumer preferences on agricultural practices. The challenges are aligning policies and market incentives to promote sustainable practices, incentivize biodiversity conservation, and support resilient food systems. Addressing these challenges requires collaboration among governments, farmers, researchers, and consumers to promote sustainable agriculture practices that ensure food security, protect the environment, and enhance resilience to future challenges.

6.2 Germplasm Biodiversity and Conservation

Germplasm biodiversity refers to the genetic diversity present within plants, animals, and microorganisms used in agriculture and related fields. It encompasses the genetic variation that enables species to adapt to changing environments, resist diseases, and improve productivity. Conservation of germplasm biodiversity is crucial for sustainable agriculture, food security, and biodiversity preservation.

6.2.1 Conservation Strategies

Ex situ conservation establishing and maintaining collections of seeds, tissues, and other genetic materials in gene banks and botanical gardens. Seed Vaults examples include the Svalbard Global Seed Vault, which stores duplicate samples of seeds from around the world for long-term storage. In situ conservation is on-farm conservation encouraging farmers to continue growing and conserving traditional varieties on their farms. Protected areas are established and reserves to conserve wild relatives and their habitats. International cooperation is a global initiative participation in international agreements and organizations such as the International Treaty on Plant Genetic Resources for Food and Agriculture (ITPGRFA) and the Convention on Biological Diversity (CBD). Community participation involves engaging local communities and indigenous peoples in germplasm conservation efforts, and respecting their knowledge and practices. Research and monitoring are genetic studies conducting genetic diversity assessments, trait characterization, and breeding programs to utilize genetic resources effectively. Monitoring includes regular monitoring of genetic erosion and conservation status of germplasm collections. In conclusion, conserving germplasm biodiversity is essential for ensuring food security, maintaining resilient agricultural systems, and preserving biodiversity. Efforts must continue to enhance conservation strategies, address challenges, and promote international cooperation to safeguard this valuable genetic heritage for future generations.

6.3 Date Palm Breeding and Challenges

The date palm has a long life cycle compared to other fruit crops; due to this reason, its genetics and morphology are difficult to understand. Date palms can be produced using several methods (Nixon and Carpenter 1978; Zaid and de Wet 2002). The most popular method that maintains the unique genetic makeup of donor varieties is vegetative propagation. Axillary buds on the trunk near the soil's surface give rise to offshoots in the date palm‘s early years. After 3–5 years of being attached to the parent palm, offshoots can be pulled off and planted since they have roots. At this age, the offspring will start to flourish and, in the case of female lines, a fruit. Its breeding is restricted (Carpenter 1979) because there is no easy way to differentiate between the sex of date palm plants at early stages. The utilization of random seedlings through sexual crossings is a further way of propagation. Approximately 50% of the seeds are male, however, they cannot be determined until the trees flower after a period of 4–5 years.

The breeding of date palms relies on genetic variation to create new, improved varieties, therefore seed propagation is essential to preserve that genetic diversity (Bendiab et al. 1993; Aberlenc-Bertossi et al. 2011). The plantation of productive females with the identification of sex allows the cost-effective palm tree cultivation. For the cultivation of a balanced ratio of female: to male plants in the date palm orchard, it is very necessary to differentiate the female and male plants at the initial stages (Sarkar et al. 2017). Furthermore, selecting and identifying superior seedling traits for increased production and improved physical and chemical attributes of fruits have economic importance. We have advanced our knowledge of the systems involved in determining a date palm‘s sex using traditional strategies. Sex identification of date plants from seedlings at an early stage is a challenging and hectic job that can be facilitated by identifying different morphological, physiological, biochemical, and molecular parameters. Different biochemical, physiological and molecular parameters are utilized in the early stages of sex identification in dioecious crops (Heikrujam et al. 2015). Morphologically, both sexes in date palms of the same variety differ in appearance. Male plants have a higher length of leaves and spin blades than female plants, but the environment highly influences these morphological characteristics and is unreliable (Ageez and Madboly 2011). It is exceedingly tough to predict the sex of date palm plants based on physical characteristics. Stomatal density is vital in regulating water usage efficiency because it influences leaf conductance, which is essential for photosynthesis and transpiration (Wang et al. 2007). Sexual dimorphism features, significantly bigger leaves, appear to help females and notably the sporophyte production by prolonging hydration (Moore et al. 2016). Females often have more extensive and denser leaves than males, which corresponds to higher photochemical production in females. Furthermore, greater considerable diversity among males within and across populations conflicts with the consistency of male features, which varies little between and among populations. Smaller stomata on plants with higher stomatal densities respond faster to dry conditions (Beerling and Woodward 1997; Wang et al. 2007). Because they produce fruits and flowers, females are thought to put out a more significant reproductive effort than males. The female genotypes exhibited higher concentrations of NPK compared with the males, providing further evidence that the females require more concentration of NPK to fulfill their needs during the reproductive process. Additionally, it was noted that compared to males, females are less able to continue growing at a lower concentration of NPK and are more susceptible to nutritional stress (Chen et al. 2014). Female plants maintain a higher N percentage than male plants in all genotypes. Female plants absorbed a higher percentage of P and K than the male plants of all genotypes Alghool and Benismail (2007) recorded the chemical aspects of date palm trees and observed that the female trees maintained higher NPK percentage than the males. Similarly, Ageez and Madboly (2011) also showed that the male date palm plants exhibit a lower percentage of NPK than the female plants. According to Chen et al. (2014) female *P. cathayana* would show increased flexibility to physiological changes if they were acclimated to N enrichment settings. They found that the females are more responsive to the changes in photosynthesis, growth, and N allocation with respect to the males. Consequently, females exhibit more seedling growth than males, which could be associated with greater phenotypic flexibility. The females maintained more biomass or shoot growth in nutrient-rich environments than the males (Hesse and Pannell 2011; Montesinos et al. 2012). The cost of reproduction as an evolutionary predictor of sexual dimorphism in life history features in dioecious species was explicitly studied because the greater cost of reproduction in females has driven the life history traits associated with sexual dimorphism in dioecious species (Obeso et al. 1998). Variations in material (evergreen vs deciduous species), growth phase (reproductive vs vegetative), and nutrient type provided could all have a role in this variance (compound vs single fertilizer). Retuerto et al. (2000) observed that gender specialization has been observed as an evolutionary response that can allow every gender to fulfill the distinct resource needs that are associated with reproduction, based on the theory that the varying costs of reproduction should take significantly different resource demands on the plants (Dawson and Bliss 1989; Dawson and Ehleringer 1993). The dioecious female plants are frequently studied to exert more hereditary struggle than that of male dioecious plants (Gehring and Linhart 1993; Sherman et al. 2019). Stress conditions like drought, and deficiency of nutrients those processes that are different in male and female plants (Li et al. 2020). The lack of nutrients that are stored in the stems of plants can have an extreme effect on the reproduction process in females than in males (Cipollini and Stiles 1991). It was shown that the female date palm plants maintained higher physiological activity than the male plants. These findings suggest that the differences in macronutrient acquisition for the plant metabolic processes could play a vibrant part in the date palm sex identification (male or female). Heterogeneous abundance protein profiling, especially in the expression of gender, can provide vital hints in dicotyledonous plant sex identification. Because of their greater engagement in reproduction, females are more inclined to make significant trade-offs with other aspects of their lives (Delph and Meagher 1995; Delph and Herlihy 2012; Barrett and Hough 2013). When the plants were well-watered and under moderate water stress, males had a higher abaxial stomatal density than females but a lower density when the plants were under high water stress. We can rule out the notion that increased stomatal density in highly water-stressed females results from lower leaf expansion because of the higher leaf water content and succulence in females compared to males. Females with a higher stomatal density may better manage their transpiration (Paoletti and Gellini 1993). Plants with a more stomatal density have smaller stomata, which respond to dryness quickly (Beerling and Woodward 1997; Wang et al. 2007). Because they produce fruits and flowers, females are thought to put out a more significant reproductive effort than males. Males predominate in xeric or low-nutrition locations, whereas females prefer mesic or high-nutrient environments (Dawson and Ehleringer 1993). For addressing sex-specific resource demands and decreasing intraspecific competition, the pattern of sexual segregation is advantageous (Eppley 2006; Li et al. 2007). Lengthy versatile advancement in diverse environments and differing reproductive outlays between the sexes lead to physiological specialization and stress sensitivity differences. Female plants are more susceptible to stressful situations than male plants (Xu et al. 2008; Zhao et al. 2009; Chen et al. 2010; Zhang et al. 2011). Female date palm trees were shown to have more activity in anabolic processes than male date palm trees, resulting in the accumulation of essential chemicals that aid in plant metabolism and date palm sex prediction (male or female). Biochemical examination of date palm leaves contributes a small information about the sex determination of immature plants (Qacif et al. 2007). Nitrogen and phosphorus are essential macronutrients present in date palm leaves, and their availability provides a tool for sex differentiation (Krouk et al. 2010; Warren 2011). Variations in nitrogen, phosphorus, and potassium concentration have been observed in the same date palm variety (Sewi cultivar) of the same age, and it has been reported that male date palm plants have lower nitrogen (0.54%) and females have higher Nitrogen recorded as 0.62%. Female trees showed a higher percentage of phosphorus (0.84%), while male trees recorded a significantly lower percentage (0.55%). As for the ratio of potassium, female trees recorded a higher percentage of potassium (0.84%), while male trees exhibited a significantly lower percentage (0.55%) (Ageez et al. 2011).

It has been noted in *populous cathayana* that under stress conditions if nitrogen, phosphorus, and potassium were present in limited quantity, male plants are more efficient than female plants (Sheng et al. 2014). The female plant showed a better growth response than male plants with the same nitrogen quantity. Sex-specific assortment pressures are created by different multiplicative provisions in females and males, resulting in sexual dimorphisms in physiological features (Barrett and Hough 2013). Because of their greater engagement in reproduction, females are more inclined to make significant trade-offs with other aspects of their lives (Delph and Meagher 1995; Delph and Herlihy 2012; Barrett and Hough 2013). Date palm breeding is a complex process aimed at improving desirable traits such as yield, fruit quality, pest and disease resistance, and tolerance to environmental stresses like salinity and drought. Traditional breeding methods rely on selecting and propagating superior female trees using offshoots, which ensures genetic uniformity but limits genetic diversity. Modern breeding techniques, including controlled pollination, genetic mapping, and biotechnological tools like tissue culture and molecular markers, are being increasingly employed to overcome these limitations. However, date palm breeding faces significant challenges. The long generation time of date palm often requiring several years to produce fruit slows the breeding process. Additionally, their dioecious nature (separate male and female plants) complicates the identification of superior male genotypes. Pests like red palm weevils and diseases such as Bayoud disease further threaten breeding efforts by limiting the availability of healthy germplasm. Climate change, with its associated impacts like water scarcity and extreme temperatures, also poses a challenge to breeding programs aimed at developing resilient varieties. Despite these hurdles, advancements in biotechnology and the conservation of genetic resources hold promise for accelerating breeding efforts and ensuring the sustainability of date palm cultivation.

6.4 Molecular Breeding

Proteomics is an influential tool for the identification of proteins and the analysis of gene functions. Therefore, protein expression may be analyzed to determine sex-related differences in dioecious plants (Barrett 2015). The protein reference map for date palm leaves identified various reported and novel proteins to differentiate the sex. A comparative proteomic study showed the role of ABC proteins in male and female sex determination by involving pollen development and male fertility (Sonia et al. 2013). Proteins were found to be possible molecular markers for determining sex in dioecious plants (Xiong et al. 2013). Modern biotechnology tools can assist date palm breeders in early sex detection. The *in-vitro* tissue regeneration environment can impact sex change, specifically through the revival of dormant organs (Komai et al. 2003). Date palm sex-specific DNA markers may also be detected using cell genetics methods. These markers are Inter Simple Sequence Repeat (ISSR) marker, Random Amplification of Polymorphic DNA (RAPD) marker, and Restriction Fragment Length Polymorphism (RFLP) marker. Even with unidentified markers, mapping data should yield predictions of the percentage of pairing and variance regions with X-linked motifs. In previous decades, there have been significant exertions to investigate the genetics of plant sex identification and to develop biomarker techniques that may be used to detect sex at an initial age (Mulcahy et al. 1992; Hormaza et al. 1994; Biffi et al. 1995). Gender-linked biomarkers are essential to make advances in empathetic sex chromosome progression and association in plants. The RAPD-PCR approach is the most widely used technology for determining sex in plants. In date palm, Younis et al. (2008) employed ISSR and RAPD markers to discover gender-DNA markers for the acquisition and recognition of outstanding male pollinators for breeding purposes, resulting in higher fruit yield enhanced quality features. There are many genetic markers to identify the sex, but the success is limited owing to the dioecious nature of date palms (Dhawan et al. 2013). The male plant has a heterogametic character that clearly shows that male has 'Y' chromosomes, which offers a faster approach for sex identification due to heterozygosity in the gender-linked region (Bekheet et al. 2011).

Sex-based DNA sequences offer a more competent and effective method for determining the sex of young plants than biochemical tests, which give little information in this regard (Qacif et al. 2007). The other variables and alleles that can modify the overall effect of sex-identifying genes have been observed and identified in vitis, spinacia, asparagus, and mercurialis (Durand et al. 1990). New methods for the characterization of genes and linkage map building have emerged because of the advancement of the polymerase chain reaction. Using 28,889 EST sequences from the date palm genome database, simple sequence repeats (SSRs) were discovered and gene-based markers, known as expressed sequence tag SSRs (EST-SSRs), were created (Zhao et al. 2013). Arbitrary primers were cast off for DNA template amplification in the random amplified polymorphic DNA (RAPD) method (Welsh and McClelland 1990). By employing marker-assisted selection, molecular breeding has the potential to accelerate fruit tree genetic advancement.

Molecular breeding is a cutting-edge strategy that leverages advancements in genetics and molecular biology to enhance traditional plant and crop breeding. Integrating tools like molecular markers, genetic mapping, and genome editing, enables the precise selection and incorporation of desirable traits into plant varieties. Unlike conventional breeding, which relies heavily on visual assessment and field trials to identify beneficial traits, molecular breeding uses DNA-based markers that are tightly linked to traits of interest. This allows breeders to identify superior genotypes at an early stage, saving time and resources. Key techniques in molecular breeding include marker-assisted selection (MAS), where specific DNA markers associated with traits like disease resistance, drought tolerance, or high yield are used to select parent plants for breeding. Another approach, marker-assisted backcrossing (MABC), focuses on transferring specific traits from one variety to another while maintaining the genetic identity of the recipient plant. Quantitative trait loci (QTL) mapping is another crucial aspect, used to locate regions on the genome that influence complex traits controlled by multiple genes, such as heat tolerance or grain size. Additionally, genomic selection predicts the performance of plants based on whole-genome data, allowing breeders to select individuals with optimal genetic potential for breeding programs.

Molecular breeding has been transformative in improving major food crops like rice, wheat, maize, and barley. For instance, molecular markers have been used to develop rice varieties resistant to bacterial blight and maize varieties tolerant to drought. It has also been applied in horticultural crops, such as tomatoes and grapes, to enhance flavor, shelf life, and resistance to pests and diseases.

Despite its potential, molecular breeding faces challenges. These include the high costs of developing molecular tools, the need for extensive and high-quality genomic data, and the requirement for advanced infrastructure and skilled personnel. Moreover, traits influenced by multiple genes (polygenic traits) pose a significant challenge due to their complex genetic architecture and interactions with environmental factors. However, recent advancements in genome sequencing, gene editing (e.g., CRISPR-Cas9), and bioinformatics are addressing these limitations, making molecular breeding more efficient and cost-effective.

In the context of global agriculture, molecular breeding offers a powerful solution to challenges such as climate change, food security, and the need for sustainable farming. By accelerating the development of crops with improved resilience, nutritional quality, and productivity, molecular breeding is poised to play a pivotal role in shaping the future of agriculture.

Gene-based sex determination in date palms is a reliable method. The genes that are utilized in the development of ovules must be expressed in females. Gene expression levels in every organism can be assessed using mRNA or protein. A constitutive, tissue/cell type-specific expression of any gene is possible or retort to exterior incentives. Understanding how an organism responds to environmental changes is now possible through the use of gene expression profiling. It results in a better understanding of both the biological processes that a gene controls. The degree to which its upstream regulatory section is expressed in any tissue may also be more easily observed. In plant biology, a number of expression profiling methods are used, and fresh innovations are always being developed. These processes are based on three concepts: sequencing, hybridization, and polymerase chain reaction (PCR) (Lohman et al. 2017). One of the most excellent tools for studying modest or medium-level gene expression is real-time polymerase chain reaction (RT-PCR) (Garson et al. 2005). After cDNA amplification and reverse transcription, as well as directly using genomic DNA, it evaluates gene expression. Real-time PCR can also be used to confirm microarray and Northern blotting results and to count the copies of a gene in transgenic plants (Li et al. 2002). There are some sex-linked genes (*GPAT3*, *CYP-703*, *LOG*) found in dioecious plants like white campion (Fechter et al. 2012) and papaya (Carvalho and Clark 2013). These genes also show their expression on dates to differentiate the sex of male and female plants. *CYP703* is a P450 family cytochrome that is only found in land plants. The chain hydroxylation of lauric acid, required for male organ development in higher plants, is a conserved and varied metabolic process (Morant 2007). The pollen-producing genes *CYP703* were encoded by a single gene in rice, Arabidopsis, maize, and other terrestrial plants. As *CYP703* is involved in pollen development, this gene will show expression in males and has no expression in female plants that will help differentiate sex at an early stage (Yang et al. 2014).

The Arabidopsis ovule has emerged as a model organism for researching the genomic bases of organ development (Grossniklaus and Schneitz 1998). *STK* caused abnormal transcription of *SHP1*, *SHP2*, and *SEP3* on the leaves of transgenic plants that showed homeotic conversions, indicating that *SHP1* and *SHP2* genes are involved in AG-independent carpel development, possibly promoting the observed phenotypes (Pinyopich et al. 2003). Many genes are involved in the formation and development of tissues like integuments or gametophytes. They have been discovered and can be used to identify sex at the initial phase. The study of genetics has shown that the specialization of tapetal cells requires Tapetum Determinant1 (*TPD1*) and Excess Microsporocytes1 (EMS1)/EXTRA Sporogenous Cells (EXS) (Zhao et al. 2002; Yang et al. 2003). *TPD1* is converted into a small cysteine-rich protein ligand for the leucine-rich reappearance receptor-like kinase *EMS1* (LRR-RLK). *TPD1* is secreted from generative cells to the plasma membrane of somatic cells, where active *TPD1–EMS1* signaling promotes periclinal cell division before regulating tapetal cell destiny. Furthermore, tapetal cells stop microsporocyte proliferation. Rather than developing into tapetal cells, the inner secondary parietal cells become microsporocytes when *TPD1* or EMS1/EXS are absent. This data suggests that the regulating route established by *TPD1* and EMS1/EXS is critical for tapetal cell specialization. We tried to make a case that identification of sex at early stages is important and possible for improving date palm production and quality. By evaluating an increase in fluorescent signals brought on by amplifying a specific gene using primers that are gene-specific, the target level is determined in real time (Livak and Schmittgen 2001). ABS forms heterodimers and tetramers with various MADS proteins, including the D-class floral homeotic proteins SEEDSTICK (*STK*), SHATTERPROOF 1 and 2 (*SHP1*, *SHP2*), mediated by SEPALLATA 3 (*SEP3*), according to protein interaction analysis Kaufmann et al. (2005). *ABS* and *STK* redundantly govern ovule inner integument production. *ABSSTK* lacks the whole inner integument layer and, as a result, the endothelium, resulting in poor seed laying (Mizzotti et al. 2012). The Bs-group MADS-box genes are required for controlling integument, seed coat, fertilization and endosperm development, and fruit growth in eudicot plants, such as *FBP24* in petunia and *ABS*, *STK*, and *GOA* in Arabidopsis (Nesi et al. 2002; De Folter et al. 2006; Prasad et al. 2010; Arora and Agarwal 2007; Mizzotti et al. 2012). Each locule in the anther has four layers of somatic cells encircling the germ cells (Goldberg 1993). The tapetum comes into intimate touch with the germ cells. It's the principal source of pollen expansion and pollen wall development precursors (Heslop-Harrison 1962; Mariani et al. 1990; Ariizumi and Toriyama 2011). One gene that contributes to the specialization of tapetal cells is *TPD1*. Another cell division in Arabidopsis thaliana is influenced by the repeat receptor-like kinase which is leucine-rich (LRR-RLK) EMS1 and the tiny protein *TPD1* (Zhao et al. 2002; Canales et al. 2002; Yang and Xie 2003). *TPD1* was used to identify the male plant at an early stage because this gene is exclusively expressed in male plants. The genes related to ovule development are only expressed in female plants. Hence molecular identification of sex supports date palm breeding.

6.5 Tissue Culture

The technique of propagating date palms through tissue culture from shoot tips using either embryogenesis or organogenesis was initially developed. While embryogenesis can be accomplished through the callus stage from a variety of meristematic tissues such as shoots, early leaves, stems, rachilla, and so on, organogenesis can be accomplished through the use of auxiliary buds and apical meristems. Different ideal circumstances are required for different cultivars since they react to tissue culture differently. But somaclonal variation, or off kinds, is one of the primary issues with TC propagation. Several typical signs are manifested by these somaclonal variants, such as diversity in leaves, variation in leaf structure and general plant growth patterns, trees that produce abnormal buds or do not form inflorescences, and trees that produce without seeds parthenocarpic fruits. Plant tissue culture refers to the regulated growth and multiplication of plant cells, tissues, and organs in media under controlled conditions (Ahloowalia et al. 2004). Techniques of tissue culture in plants are an essential part of agriculture's production improvement and crop management. Plant cells and tissue culture open new ways to study genetic and physiological plant processes and contribute by increasing genetic inequality to the reproduction of crops. The propagation of plants from various explants such as segments of nodes, tips of the shoot, and leave pieces has been observed previously (Zhang and Stolz 1991). A mass of undifferentiated parenchyma cells known as a callus is thought to have a large capacity for preserving cell proliferation, expansion, and development. The meristematic tissues' explant surfaces that have been cut are used to create the callus either naturally or by plant tissue culture. The kind of medium, the inclusion of growth regulators, and the incubation circumstances all affect callus induction (Al-Khayri and Ibraheem 2014).

The top genotype of female date palms has been subjected to *in vitro* callus culture and plantlet regeneration, which might be expanded to produce large-scale quality planting propagules for this commercially significant tree. The explant disinfection of Ajwa females was also supported by (Singh and Shekhawat 2009). Browning or darkening of media, contamination of either fungal, bacterial, or viral, tree's recalcitrance activity, and release of phenolic compounds are the key challenges that arose in culturing the fruit trees, which may lead to explant necrosis and phytotoxicity. In the culture establishment, a critical hindrance reported is phenolic excretions that cause medium browning (Krishna and Singh 2007). Moreover, the issue could be minimized by using younger explants for culturing. More immature plants release less or no phenolic compounds (Raj and Saudagar 2019).

The browning was minimized by placing explants in ddH_2O for 10 min. Explants were sterilized using $HgCl_2$ for 7–10 min or sometimes washed with 70% alcohol. Excellent consequences were attained on 10 min of sterilization. A few scientists conducted a substantial study on date palm tissue culture (Krikorian 1994; Tissert 1984; Al-Khayri and Al-Bahrany 2014). Fki et al. (2003) improved plant regeneration techniques by establishing embryogenic suspension culture.

Sudhersan and El-Nil (2004) used cell suspension culture to investigate the effects of sucrose on proteins, sugars, phenolics, and peroxidase activity. Zouine and El Hadrami (2004) developed an improved procedure for embryogenic suspension culture proliferation and somatic embryo germination. The delayed reaction was discovered on media with a slight dosage of 2,4-D, compared to other plant hormones for inducing callus culture, 2,4-D was a high-performing performer (Rathore et al. 2011) that is in line with our study concentration of 2, 4-D, 90 mg/L showed good response regarding callus induction was also validated as low concentration delayed callus induction or no callus induction.

Auxin in the media was required for callus culture induction, in vitro plantlet regeneration, and somatic embryogenesis (Gueye et al. 2009). The current findings revealed that date palm callogenesis necessitates a higher 2,4-D content in the culture medium (Kumar et al. 2010). Callus cultures kept on induction media for more extended periods (greater than 16–20 weeks) become hyper-hydrated, and the callus becomes glossy, watery and non-regenerative.

An increased level of hormones and water potential in the culture media caused hyperhydration (Rathore et al. 2011; Correll et al. 2000). Hyper-hydration is a biological condition that impairs the development and regeneration of in vitro cultures, resulting in difficult to handle cultures. The principle is to isolate many compact bonded tissues, and an appropriate nutrient media is used to cultivate these cells. For the survival and prosperity of such cells, nutrient media seem to be vital. Callus tissue and cell suspension provide a variety of sources for the isolation of individual cells.

Callus or suspension single cells can be cultivated on nutrients (Kramer and Amtmann 2012). According to reports, sucrose is crucial for embryo development and has a significant impact on embryogenesis (Veramendi and Navarro 1996). Higher sucrose concentrations in the culture medium aided in the maturity of the cultures and supported the callus regeneration. BAP, Kin, and 2-iP were discovered to have a synergistic effect on adventitious shoot bud development since cultures grew slowly when any one of these cytokinins was absent.

To sustain the regenerative capacity of cultures, it is recommended that media preparations, hormone concentration and media complements be optimized following tissue rejuvenation and the development of regenerative culture. Different scientists agree with our findings (Rathore et al. 2011) that hormonal content in culture media significantly impacts culture development, regeneration frequency, and plantlets recovered after sub-culture onto the hormone-free medium. The mineral needs for date palm tissue appear to differ between cultivars (Mazri and Meziani 2015). The MS and MS/2 basal preparations have been widely employed (Khierallah et al. 2017; Khan and Bi 2012; Al-Mayahi 2014). It was proved earlier that callus regeneration showed some variations (Naik and Chand 2011). Further studies would facilitate the development of genotype-dependent cultivation methods to improve the growth performance of re-calcitrant date palms to tissue culture response. This study has the potential to serve as a protocol for callus induction and plantlet regeneration to study plant transformation, soma-clones, as well as *in vitro* selection against several stress factors. This method might also be used for other genotypes of date palms to multiply desired genotypes within a short time.

6.5.1 Tissue Culture: A Tool for Advanced Plant Research and Improvement

Tissue culture is a laboratory technique that involves the in vitro propagation of plant cells, tissues, or organs under sterile and controlled conditions on a nutrient medium. This method has become a cornerstone of plant research, offering numerous applications in basic biology, crop improvement, and conservation. The process typically begins with the selection of explants, such as shoot tips, leaves, or callus tissues, which are sterilized and cultured on a growth medium containing essential nutrients, vitamins, and plant growth regulators (e.g., auxins and cytokinins). Depending on the goal, the cultured cells can develop into whole plants through processes such as organogenesis (formation of organs) or somatic embryogenesis (development of embryos from somatic cells).

One of the most significant advantages of tissue culture is its ability to produce genetically identical plants, known as clones, through micropropagation. This is particularly beneficial for propagating high-value crops, horticultural species, and medicinal plants that are difficult to reproduce through conventional means. Tissue culture also facilitates the production of disease-free plants by eliminating pathogens through meristem culture or thermotherapy. Additionally, somaclonal variation, which arises during tissue culture, serves as a valuable source of genetic diversity for plant breeding programs.

In crop improvement, tissue culture plays a pivotal role in the development of transgenic plants, where genetic modifications are introduced into the culture at the cellular level. It is also essential in protoplast culture and fusion, enabling the creation of hybrid plants that combine traits from two different species or genera. Furthermore, tissue culture is indispensable in the rapid propagation of elite varieties, speeding up the delivery of improved crops to farmers.

Tissue culture has profound applications in plant conservation, especially for endangered or threatened species. By storing plant tissues in cryopreservation or establishing in vitro gene banks, researchers can preserve genetic resources for future use. Moreover, tissue culture is widely employed in producing secondary metabolites, such as alkaloids and flavonoids, from medicinal plants under controlled conditions.

Despite its advantages, tissue culture faces challenges, including high costs, contamination risks, and the need for specialized infrastructure and expertise. Additionally, some plant species exhibit recalcitrance to in vitro culture, requiring customized protocols. However, advancements in biotechnology, automation, and cost-effective techniques are overcoming these limitations.

In conclusion, tissue culture has revolutionized plant science and agriculture, enabling precise and efficient manipulation of plant materials for research, breeding, and conservation. As technology continues to evolve, tissue culture will remain a vital tool for addressing global challenges in food security, biodiversity conservation, and sustainable agriculture.

6.5.2 Tissue Culture: A Transformative Approach in Plant Research and Agricultural Advancement

Tissue culture is a versatile and revolutionary technique in plant science, allowing the growth and regeneration of plants from small tissue segments under controlled laboratory conditions. This in vitro process is performed on a nutrient-rich medium supplemented with essential minerals, vitamins, amino acids, and plant growth regulators like auxins, cytokinins, and gibberellins. The process begins with the careful selection and sterilization of an explant (plant material), which is then cultured under sterile conditions. Depending on the plant species and objectives, various culture techniques—such as callus culture, organ culture, or somatic embryogenesis—are employed to achieve specific outcomes.

6.5.3 Applications of Tissue Culture

Micropropagation, or clonal propagation, is one of the most widely used applications of tissue culture. It facilitates the rapid multiplication of genetically identical, disease-free plants. This technique is particularly valuable for propagating crops like bananas, potatoes, orchids, and medicinal plants that are challenging to propagate through conventional means. Commercial industries rely on tissue culture to produce uniform, high-quality plants on a large scale, reducing the time required to bring crops to market.

Tissue culture serves as a vital tool for preserving plant genetic resources. Endangered or rare species can be conserved through cryopreservation, where plant tissues are stored at ultra-low temperatures for long-term preservation. This is especially critical for species with recalcitrant seeds that cannot be stored using traditional methods. In vitro gene banks maintain plant tissue cultures as a safeguard against biodiversity loss, ensuring the availability of germplasm for future research and breeding programs.

The elimination of viruses, bacteria, and fungi from plants is another critical application of tissue culture. By isolating meristematic tissue, which is usually free from systemic infections, scientists can produce healthy plants even from infected parent stock. This has been extensively applied in crops like sugarcane, strawberries, and citrus, ensuring improved yield and quality.

Tissue culture facilitates somatic embryogenesis, where somatic cells develop into embryos that can grow into complete plants. This has been a game-changer for crops like coffee, oil palm, and cereals, where traditional breeding methods are slow. Synthetic seed technology, derived from encapsulating somatic embryos, enables the storage, transport, and sowing of these embryos like conventional seeds, bridging the gap between laboratory and field applications.

Tissue culture is indispensable in genetic engineering, as it provides a platform for introducing desired genes into plant cells. Techniques such as Agrobacterium-mediated transformation and biolistic (gene gun) methods rely on tissue culture to regenerate transgenic plants from transformed cells. These approaches have led to the development of genetically modified (GM) crops with enhanced traits, such as pest resistance (Bt cotton), herbicide tolerance (Roundup Ready crops), and improved nutrition (Golden Rice).

The pharmaceutical and nutraceutical industries benefit immensely from tissue culture, which enables the production of valuable secondary metabolites, such as alkaloids, flavonoids, and terpenes, from medicinal plants. These compounds are produced in cell suspension cultures under controlled conditions, offering a sustainable alternative to wild harvesting.

6.5.4 Challenges in Tissue Culture

Despite its immense potential, tissue culture faces several challenges. Contamination by microbes is a persistent problem that can compromise entire experiments or commercial batches. Additionally, not all plant species are amenable to tissue culture; recalcitrant plants require extensive research to establish effective protocols. High costs associated with infrastructure, media preparation, and skilled labor can limit its accessibility in developing regions. Somaclonal variation, while sometimes beneficial, can lead to undesirable genetic changes in clonally propagated plants, requiring careful screening and selection.

Advances in tissue culture, combined with genomics and bioinformatics, are expected to overcome many current limitations. Automated systems and bioreactors are already streamlining large-scale production, making tissue culture more cost-effective. Integration with CRISPR-Cas9 gene editing technology is expanding the scope of genetic improvement, allowing for precise modifications in plant genomes. Additionally, tissue culture is increasingly being used in synthetic biology for creating novel plant traits and producing bioplastics, biofuels, and high-value compounds.

In summary, tissue culture has transformed plant science, offering unparalleled precision, efficiency, and versatility. Its applications extend beyond agriculture, contributing to conservation, industry, and fundamental research. As global challenges such as climate change, food security, and biodiversity loss intensify, tissue culture remains a cornerstone technology for developing sustainable and innovative solutions in agriculture and beyond.

Tissue culture, despite its wide range of applications and potential, faces several technical, economic, and biological challenges that can limit its effectiveness in research and commercial applications. Below are some of the key challenges: Contamination by bacteria, fungi, or other microorganisms is a major issue in tissue culture. Even minor lapses in aseptic handling can lead to the loss of entire cultures. Contaminants often thrive in nutrient-rich media and may remain undetected until they outcompete the plant tissue. Effective sterilization techniques and consistent monitoring are essential but can increase the cost and complexity of tissue culture operations.

Not all plant species respond well to tissue culture techniques. Some plants, particularly woody species or those with complex growth requirements, are recalcitrant to in vitro culture. This can manifest as poor callus formation, low regeneration rates, or failure to grow in culture. Developing species-specific protocols often require significant time and resources, limiting the adoption of tissue culture for certain crops.

While somaclonal variation (genetic changes during tissue culture) can be a source of diversity for breeding, it is often undesirable in clonal propagation. Such variations can lead to off-type plants with inconsistent traits, reducing the uniformity and quality of propagated crops. Screening and maintaining genetic stability in cultures are time-intensive and require advanced molecular tools for detection.

The infrastructure, equipment, and skilled labor required for tissue culture are expensive. Media preparation, sterile conditions, and labor-intensive protocols make tissue culture cost-prohibitive, especially for large-scale commercial operations in resource-limited regions. Moreover, the cost of maintaining specialized facilities often limits its accessibility for smallholder farmers and researchers.

Tissue culture techniques require expertise in handling sterile materials, preparing media, and monitoring cultures. A lack of adequately trained personnel can lead to errors, inefficiencies, and higher failure rates. In developing regions, the shortage of trained staff often hampers the adoption of tissue culture.

Plant tissue in culture is highly sensitive to environmental conditions, including temperature, humidity, and light. Suboptimal conditions can lead to issues such as vitrification (abnormal, waterlogged growth of plantlets), poor rooting, or dormancy in regenerated plants. Additionally, balancing plant growth regulators (e.g., auxins and cytokinins) in the medium is critical but varies between species and even cultivars, requiring extensive trial-and-error optimization.

Infected explants, especially those carrying latent pathogens such as viruses or endophytic bacteria, can introduce diseases into cultures. This is particularly challenging for crops prone to viral infections, like bananas and citrus. Eliminating such pathogens through techniques like meristem culture or thermotherapy adds complexity to tissue culture protocols. While tissue culture is highly effective at a laboratory scale, scaling up to industrial levels presents challenges. Large-scale production often requires bioreactors and automated systems, which are capital-intensive and demand sophisticated management. Maintaining consistency and quality in mass propagation remains a critical hurdle for commercial tissue culture. Certain applications of tissue culture, such as the production of genetically modified (GM) plants, face regulatory restrictions and public skepticism in some regions. Navigating the ethical and legal landscape can slow the adoption of tissue culture-derived technologies in agriculture.

Plants grown in vitro often struggle to acclimatize to external environments due to their sheltered growth conditions. Transitioning plants to natural conditions requires a gradual and controlled process to prevent stress-related losses. High mortality rates during acclimatization reduce the efficiency of tissue culture systems.

To address these challenges, advancements in automation, bioreactor technology, and synthetic media are being developed to reduce costs and improve scalability. Genetic tools, such as molecular markers and CRISPR-based diagnostics, are aiding in the detection of somaclonal variation and latent pathogens. Innovations in cryopreservation and protocol optimization are enhancing the viability of recalcitrant species. With ongoing research and technological improvements, tissue culture is becoming increasingly robust and accessible for a broader range of applications. Despite these challenges, tissue culture remains an indispensable tool in plant research, agriculture, and conservation, with significant potential for overcoming global food and biodiversity challenges.

6.6 Genetic Engineering and Gene Editing

Genetic engineering and gene editing are two closely related fields that have revolutionized biological research and have significant implications for medicine and agriculture. Genetic engineering involves altering an organism's genetic material (DNA) to achieve desired traits or outcomes. Recombinant DNA Technology involves inserting genes from one organism into another to transfer specific traits. For example, the insulin-specific gene can be inserted into bacteria for the production of human insulin. CRISPR-Cas9 is a powerful gene-editing tool that allows scientists to precisely edit DNA sequences within living cells. It has greatly simplified the process of making targeted changes to the genome. Transgenic organisms are organisms that have had foreign DNA inserted into their genome. They are used extensively in research and agriculture to study gene function or introduce desirable traits like pest resistance in crops.

Gene editing refers to the precise modification of DNA sequences within an organism's genome. Techniques like CRISPR-Cas9, TALENs (Transcription Activator-Like Effector Nucleases), and zinc finger nucleases enable scientists to target specific genes and make changes such as Knockout disabling a gene to study its function or to prevent the expression of a harmful trait. Knock-in introducing new genes or modifying existing ones to add desired functions or traits. Base Editing directly changes a single nucleotide base pair without cutting the DNA, which can be less disruptive and more precise than traditional gene editing methods.

The agriculture genetically engineered crops can be made more resistant to pests, diseases, and environmental stresses, potentially improving yield and reducing reliance on pesticides. The research area in which genetic engineering is crucial for understanding gene function and disease mechanisms, as well as for developing model organisms. The ethical concerns are issues such as the potential for unintended consequences, the ethical implications of modifying human embryos, and the equitable distribution of benefits and risks. The regulation there is ongoing debate and regulation concerning the use of gene editing technologies, particularly in agriculture, to ensure safety and ethical standards are met. Genetic engineering and gene editing are powerful tools with immense potential to advance science and improve our lives. However, their application raises important ethical, social, and regulatory challenges that must be carefully considered and addressed.

Gene editing is a transformative biotechnology that allows scientists to make precise modifications to an organism's DNA, offering unprecedented control over genetic material. The most widely used technique, CRISPR-Cas9, enables the targeted insertion, deletion, or alteration of specific genes with remarkable precision and efficiency. Other gene editing tools, such as TALENs and ZFNs, also provide valuable options, but CRISPR's simplicity and adaptability have made it the most popular. Gene editing has vast applications across various fields, especially in agriculture, medicine, and industrial biotechnology. In agriculture, it is used to develop crops with improved traits, such as drought tolerance, disease resistance, and enhanced nutritional content, while also creating livestock with desirable characteristics. In medicine, gene editing holds great promise for treating genetic disorders, such as sickle cell anemia, and developing personalized therapies for cancer and other diseases. Additionally, gene editing enables researchers to better understand gene function and study complex biological processes, accelerating advancements in functional genomics. Industrial biotechnology has benefited from gene editing by optimizing microorganisms for more efficient production of biofuels, enzymes, and pharmaceuticals. However, technology raises several challenges, including the potential for off-target effects, ethical concerns regarding human genetic modification, and regulatory hurdles, especially in agriculture. Despite these challenges, the potential of gene editing to revolutionize medicine, agriculture, and environmental conservation makes it one of the most promising and impactful tools in modern science, with continuous advancements expected to further expand its capabilities.

6.7 Mutation Breeding

Mutation breeding is a technique used in agriculture to introduce genetic variation into crops by inducing mutations in their DNA. Unlike genetic engineering, which involves precise manipulation of specific genes, mutation breeding relies on the random generation of mutations throughout the genome. These mutations can result in beneficial traits such as improved yield, disease resistance, or tolerance to environmental stress.

The techniques used in mutation breeding are radiation sources such as gamma rays, X-rays, or neutron beams are used to induce mutations by causing changes in the DNA sequence. This method is called irradiation breeding. Chemical mutagens like ethyl methane sulfonate (EMS) or nitrosoguanidine (NTG) can also induce mutations by altering DNA bases or causing DNA damage. Some biological agents like certain viruses or transposons (jumping genes) can induce mutations when inserted into the genome. The process of mutation breeding in Mutagen Treatment is seeds or plant tissues that are treated with the mutagen, either through irradiation, chemical application, or biological means. The selection process includes after treatment, plants are grown and screened for desirable traits such as improved yield, resistance to pests or diseases, better nutritional content, or other agronomic characteristics. The breeding plants showing desirable traits are selected as parents for further breeding. Crossbreeding with other varieties may be done to combine beneficial traits and develop new cultivars. The testing and release of selected mutant lines undergo rigorous testing to ensure they meet safety and regulatory standards. Once approved, new varieties are released to farmers for cultivation. An example of successful mutation breeding is dwarf wheat. The development of semi-dwarf wheat varieties through mutation breeding in the 1960s significantly contributed to the Green Revolution. The seedless citrus fruits include seedless varieties of citrus fruits such as oranges and grapefruits have been developed through mutation breeding.

Mutation breeding remains a valuable tool in crop improvement, offering a way to introduce genetic diversity and develop new varieties with desirable traits to meet the challenges of agriculture and food security.

6.8 Hybridization

Hybridization in the context of biology and agriculture refers to the process of crossing two genetically different individuals or varieties to produce offspring with desirable characteristics. This technique is widely used in plant and animal breeding to create hybrids that exhibit traits such as higher yield, disease resistance, or better adaptation to specific environmental conditions. The methods of hybridization are cross-pollination in which hybridization often involves controlled cross-pollination between two parent plants with desired traits. This can be achieved manually by transferring pollen from one plant to another or naturally with the assistance of insects or wind. The other is artificial insemination which occurs in animals. The hybridization may involve artificial insemination techniques to breed individuals from different breeds or species.

The types of hybrids are F1 Hybrids in which the first-generation offspring result from the cross between two purebred lines or varieties. F1 hybrids often exhibit hybrid vigor (heterosis), where they perform better than either of the parent lines. The other is backcross hybrids in which hybrids are produced by crossing an F1 hybrid back to one of its parent lines. This technique is used to introduce specific traits from one parent while retaining the desirable characteristics of the hybrid. Hybridization is a fundamental technique in modern agriculture and breeding programs, contributing to the development of new varieties that meet the evolving needs of food production, sustainability, and economic viability.

6.9 Conclusions and Prospects

In conclusion, the fields of genetic engineering, gene editing, mutation breeding, hybridization and tissue culture are at the forefront of advancing agriculture and biotechnology, offering innovative solutions to enhance food security, sustainability, and human health. Each approach has its unique strengths and applications. Their responsible application requires careful consideration of scientific, ethical, regulatory, and societal factors. Continued research, innovation, and collaboration will be key to harnessing these technologies effectively for sustainable development and addressing global food security challenges in the decades ahead. The findings of this chapter might be used on a broad scale to detect early sex in date palm to increased production, lowering production costs, and increasing yield per unit area. Micropropagation of selected plants could help to multiply desired plants on a large scale.

Appendix 6.1 Research Institutes and Online Resources Available for Date Palm (*Phoenix dactylifera* L.) Breeding

Institute name	Specialization research activities	Contact information and website
ICARDA (Science for resilient livelihood in dry areas)	Challenges and opportunities for growth	https://www.icarda.org/ icarda-comms@cgiar.org
Date Palm sub-station, Jhang	Main objective (selection and acclimatization of exotic and local varieties	https://www.Jhang.punjab.gov.pk
Shah Abdul Latif University, Khairpur	Plant Biotechnology (gene cloning, hydroponic cultivation)	info@salu.edu.pk
King Saud University, Saudi Arabia	Promoting date palm agriculture through education, research and extension	pps@ksu.edu.sa
ACSAD (Saudi Arabia)	Aims to improve the economic revenue of date palm trees by increasing production, improving fruit quality, raising the marketing and storing capacities, conserving superior seed lines and enhancing the research, development and extension capacities of national institutions.	email@acsad.org

Appendix 6.2 Genetic Resources of Date Palm

Cultivar/ genetic resources	Important traits	Cultivation location
Zahidi	Dry, semi-dry, oblong-ovait, thick skin, not very sweet, with smooth consistency	Paksistan, Iran, Iraq, UAE
Barhi	Soft fruit, ovait-rounded shape, medium thick skin, delicious taste, recommended for consumption fresh at the khalal stage	Iran, Pakistan, North Africa, Egypt
Deglect Nour	Soft with a unique taste, Oblong-ovait shape, medium thick skin, Semi dry	Mexico. United States, Algeria, North Africa
Hadrawi	Dry dates, sweet and fleshy	Saudi Arabia, Oman, Egypt, North Africa
Hallawy	Soft fruit, sweet taste, oblong with rounded apex, thin skin	Pakistan, Sudan, Middle East, North Africa
Hayani	Soft date, not too sweet, oblong shape	Libya, Tunisia, Algeria, Sudan

(continued)

Cultivar/ genetic resources	Important traits	Cultivation location
Kabkab	A soft, long shape, unripe date, can be consumed after cooking and drying	Iran, Iraq, Kuwait, Saudi Arabia, UAE
Lulu	Soft and sweet, thick fruit, less fibrous flesh	UAE, Qatar, Oman
Mazafati	Soft and flesh, cylinder shape, desirable taste	Pakistan, Iran
Khasab	Tough separating skin, thick fruit, round oval shape	Oman, Abu Dhabi, Saudi Arabia, Egypt

References

Abdul-Hamid NA, Mediani A, Maulidiani M, Abas F, Ismail IS, Shaari K, Lajis NH (2016) Discrimination and nitric oxide inhibitory activity correlation of Ajwa dates from different grades and origins. Molecules 21:1423

Abdullahi M, Garko M (2012) Medicinal value of date palm (*Phoenix dactylifera L.*). In: Proceedings of the Agricultural Society of Nigeria Conference Nsukka Nigeria, vol 10, pp 11–14

Aberlenc-Bertossi FA, Daher N, Chabrillange J, Tregear P, Mohamed N (2011) Sex determination in date palm: new perspectives on an old theme. In: Plant and animal genomes XIX conference, vol 11, pp 15–19

Ageez EA, Madboly E (2011) Identification of male-specific molecular markers in date palm sewi cultivar. Egypt J Genet Cytol 40:201–214

Ageez A, Madboly EA (2011) Identification of male specific molecular markers in date palm sewi cultivar. Egypt. J. Genet. Cytol 40:201-214

Al-Farsi MA, Lee CY (2008) Nutritional and functional properties of dates: a review. Crit Rev Food Sci Nutr 48:877–887

Al-Khayri JM, Ibraheem Y (2014) In vitro selection of abiotic stress tolerant date palm (Phoenix dactylifera L.): A review. Emir. J. Food Agric 26(11):921-933

Al-Mayahi AMW (2014) Effect of copper sulphate and cobalt chloride on growth of the In vitro culture tissues for date palm (Phoenix dactylifera L.) cv. Ashgar. Am. J. Agric. Biol. Sci 9:6

Alghool M, Benismail M (2007) Vegetative characters and fruit chemical analysis of date palms under rainfed conditions. The fourth symposium on date palm in Saudi Arabia, King Faisal University. Acta Hortic 736:136–145

Anjum FM, Bukhat SI, El-Ghorab AH, Khan MI, Nadeem M, Hussain S, Arshad MS (2012) Phytochemical characteristics of date palm (*Phoenix dactylifera*) fruit extracts. Pak J Food Sci 22:117–127

Al-Bukhari M (2008) Sahih Al-Bukhari. Book of Jihād. Muhammad Nasser. Beirut. Dār Ibn Kathīr

Anonymous (2023) Arabian Dates in Bulk. Available online with updates at https://arabiandatesinbulk.com/amber-dates/

Ahloowalia BS, Prakash J, Savangilkar VA, Savangilkar C (2004) Plant tissue culture. Low-cost option for tissue culture technology in developing countries. Proceeding of A Technical Meeting, Organized by The Joint FAO/IAEA 3-10

Ariizumi T, Toriyama K (2011) Genetic regulation of sporopollenin synthesis and pollen exine development. Annu Rev Plant Biol 62:437–460

Arora R, Agarwal P (2007) MADS-box gene family in rice: genome-wide identification, organization and expression profiling during reproductive development and stress. BMC Genomics 8:1–21

Barrett SCH (2015) Influences of clonality on plant sexual reproduction. Proc Natl Acad Sci 112: 8859–8866

Barrett S, Hough J (2013) Sexual dimorphism in flowering plants. J Exp Bot 64:67–82

Barrow SC (1998) A monograph of *Phoenix* L. (Palmae: *Coryphoideae*). Kew Bulletin, Springer, pp 513–575

Beerling DJ, Woodward FI (1997) Changes in land plant function over the Phanerozoic: Reconstructions based on the fossil record. Bot J Linn Soc 124:137–153

Bendiab K, Baaziz M, Brakez Z, Sedra M (1993) Correlation of isozyme polymorphism and beyond disease resistance in date palm cultivar and progeny. Euphytica 65:23–32

Biffi R, Restivo FM, Tassi F (1995) A restriction fragment length polymorphism probe for early diagnosis of gender in Asparagus officinalis L. Int. J. Hortic. Sci 30:1463-1464

Botes A, Zaid A (2002) The economic importance of date production and international trade

Canales CA, Bhatt M, Scott R, Dickinson H (2002) EXS, a putative LRR receptor kinase, regulates male germline cell number and tapetal identity and promotes seed development in Arabidopsis. Curr Biol 12:1718–1727

Carpenter JB (1979) Breeding of date palm in California. Date Grow Inst Rep 54:13–16

Carvalho AB, Clark AG (2013) Efficient identification of Y chromosome sequences in the human and drosophila genomes. Genome Res 23:1894–1907

Chao CT, Krueger RR (2007) The date palm (*Phoenix dactylifera* L.): overview of biology uses and cultivation. Int J Hortic Sci 42:1077–1082

Chen L, Zhang S, Zhao H, Korpelainen H, Li C (2010) Sex-related adaptive responses to interaction of drought and salinity in *Populus yunnanensis*. Plant Cell Environ 33:1767–1778

Chen J, Duan B, Wang M, Korpelainen H, Li C (2014) Intra and inter-sexual competition of *Populus cathayana* under different watering regimes. Funct Ecol 28:124–136

Cipollini ML, Stiles EW (1991) Costs of reproduction in *Nyssa sylvatica*: sexual dimorphism in reproductive frequency and nutrient flux. Oecologia 86:585–593

Correll MJ, Wu Y, Weathers PJ (2000) Controlling hyperhydration of carnations (Dianthus caryophyllus L.) grown in a mist reactor. Biotechnol. Bioeng 71:307-314

Dawson TE, Bliss LC (1989) Patterns of water use and the tissue water relations in the dioecious shrub, *Salix arctica*: the physiological basis for habitat partitioning between the sexes. Oecologia 79:332–343

Dawson TE, Ehleringer JR (1993) Gender-specific physiology carbon isotopes discrimination and habitat distribution in boxelder *Acer negundo*. Ecology 74:798–815

De Folter SA, Shchennikova V, Franken J, Busscher M, Baskar U, Grossniklaus R, Immink RG (2006) ABsister MADS-box gene involved in ovule and seed development in petunia and Arabidopsis. Plant J 47:934–946

Delph LFC, Herlihy CR (2012) Sexual fecundity and viability selection on flower size and number in a sexually dimorphic plant. Evolution 66:1154–1166

Delph LF, Meagher TR (1995) Sexual dimorphism masks life history trade-offs in the dioecious plant *Silene latifolia*. Ecology 76:775–785

Dhawan C, Kharb P, Sharma R, Uppa S, Aggarwal RK (2013) Development of male-specific SCAR marker in date palm (*Phoenix dactylifera* L.). Tree Genet Genomes 216:1143–1150

Dransfield JWJ, Baker M, Asmussen N, Lewis CE (2008) Genera Palmarum the evolution and classification of palms. Royal Botanic Gardens, London

Durand R, Durand G, Jacobs M (1990) Sexual determination and sexual differentiation. Crit Rev Plant Sci 9:295–316

Eppley SM (2006) Females make tough neighbors: sex-specific competitive effects in seedlings of a dioecious grass. Oecologia 146:549–554

Fechter IL, Hausmann M, Daum TS, Rosleff P, Viehöver B, Weisshaar R, Töpfer R (2012) Candidate genes within a 143 kb region of the flower sex locus in Vitis. Mol Gen Genomics 287:247–259

Food and Agriculture Organization of the United Nations (2021) FAOSTAT Database. Retrieved from http://www.fao.org/faostat/en/#data/QC

Galeb HAJ, Salimon EE, Eid NE et al (2012) The impact of single and double hydrogen bonds on crystallization and melting regimes of Ajwa and Barni lipids. Int Food Res J 48:657–666

Garson JPR, Grant U, Ayliffe RB et al (2005) Real-time PCR quantitation of hepatitis B virus DNA using automated sample preparation and murine cytomegalovirus internal control. J Virol Methods 126:207–213

Gehring JL, Linhart YB (1993) Sexual dimorphisms and response to low resources in the dioecious plant Silene latifolia (Caryophyllaceae). Int. J. Plant. Sci 154:152–162

Ghnimi SU, Syed K, Azharul A, Kamal-Eldin B (2017) Date fruit as food security crop in the arid lands seeking industrial valorization. Nutr Food Sci J 5:10–15

Goldberg LR (1993) The structure of phenotypic personality traits. Am. Psychol 48:26-34

Grossniklaus U, Schneitz K (1998) The molecular and genetic basis of ovule and megagametophyte development. Semin Cell Dev Biol 9:227–238

Gueye B, Saıd-Ahmed H, Morcillo F et al (2009) Callogenesis and rhizogenesis in date palm leaf segments. Are there similarities between the two auxin-induced pathways? Plant Cell Tissue Organ Cult 98:47–58

Heikrujam MK, Sharma M, Prasad T, Agrawal V (2015) Review on different mechanisms of sex determination and sex-linked molecular markers in dioecious crops: a current update. Euphytica 201:161–194

Heslop-Harrison J (1962) Evanescent and persistent modifications of chloroplast ultrastructure induced by an unnatural pyrimidine. Planta 58:237–256

Hesse E, Pannell JR (2011) Sexual dimorphism in a dioecious population of the wind-pollinated herb *Mercurialis annua*: the interactive effects of resource availability and competition. Ann Bot 107:1039–1045

Hormaza J, Dollo L (1994) Identification of a RAPD marker linked to sex determination in Pistacia vera using bulked segregant analysis. Theor. App. Genet 89:9-13

Jamil MSR, Nadeem MA, Hanif MA, Ali A, Akhter K (2010) Proximate composition and mineral profile of eight different date (*Phoenix dactylifera* L.) varieties from Pakistan. Afr J Biotechnol 9:3252–3259

Jaradat A, Zaid A (2004) Quality traits of date palm fruits in center of origin and center of diversity. J Food Agric Environ 2:208–217

Fki L, Masmoudi R, Drira N, Rival A (2003) An optimised protocol for plant regeneration from embryogenic suspension cultures of date palm, Phoenix dactylifera L., cv. Deglet Nour. Plant Cell Rep 21:517–524

Kaufmann K, Melzer R, Theißen G (2005) MIKC-type MADS-domain proteins: structural modularity, protein interactions and network evolution in land plants. Gene 347:183-198

Khierallah HS, Bader SM, Al-Khafaji MA (2017) NAA-induced direct organogenesis from female immature inflorescence explants of date palm. Mol. Biol 1637:17-25

Khalid S, Khalid N, Khan RS, Ahmed H, Ahmad A (2017) A review on chemistry and pharmacology of Ajwa date fruit and pit. Trends Food Sci Technol 63:60–69

Khan S, Bi TB (2012) Direct shoot regeneration system for date palm (Phoenix dactylifera L.) cv. Dhakki as a means of micropropagation. Pak. J. Bot 44:1965-1971

Komai F, Shikazono N, Tanaka A (2003) Sexual modification of female spinach seeds (*Spinacia oleracea* L.) by irradiation with ion particles. Plant Cell Rep 21:713–717

Kramer U Amtmann A (2012) Salt stress signals shape the plant root Carlos S Galvan-Ampudia and Christa Testerink. Plant Biol 14:296-302

Krikorian AD (1994) In vitro culture of plantation crops, Plant cell and tissue culture. J. Plant Biotechnol 497-537

Krishna H, Singh SK (2007) Biotechnological advances in mango (Mangifera indica L.) and their future implication in crop improvement a review. Biotechnol. Adv 25:223-243

Krouk GN, Crawford G, Coruzzi N, Tasy Y (2010) Nitrate signalling: adaptation to fluctuating environments. Curr Opin Plant Biol 13:265–272

Kumar N, Modi AR, Singh AS, Gajera BB, Patel AR, Patel MP, Subhash N (2010) Assessment of genetic fidelity of micropropagated date palm (Phoenix dactylifera L.) plants by RAPD and ISSR markers assay. Physiol. Mol. Biol. Plant 16:207-213

Li C, Xu G, Zang R, Korpelainen H, Berninger F (2007) Sex-related differences in leaf morphological and physiological responses in *Hippophae rhamnoides* along an altitudinal gradient. Tree Physiol 27:399–406

Li Z, Wu N, Liu T, Tang M, Chen H (2020) Gender-related responses of dioecious plant *Populus cathayana* to AMF drought and planting pattern. Sci Rep 10:11530

Li LC, Dahiya R (2002) MethPrimer: designing primers for methylation PCRs. Bioinformatics. 18: 1427-1431

Livak KJ, Schmittgen TD (2001) Analysis of relative gene expression data using real-time quantitative PCR and the 2−CT method. Methods 25:402–408

Lohman BK, Stutz WE, Bolnick DI (2017) Gene expression stasis and plasticity following migration into a foreign environment. Mol Ecol 26:4657–4670

Mazri MA, Meziani R (2015) Micropropagation of date palm: a review. Cell Dev. Biol 4:160

Mariani CMD, Beuckeleer J, Truettner J, Leemans Goldberg RB (1990) Induction of male sterility in plants by a chimaeric ribonuclease gene. Nature 347:737–741

Mizzotti CMA, Mendes E, Caporali A, Schnittger M, Kater L, Battaglia R, Colombo L (2012) The MADS-box genes seed stick and *Arabidopsis bsister* play a maternal role in fertilization and seed development. Plant J 70:409–420

Mohan JS (2006) Recent advances in date palm tissue culture and mutagenesis. Acta Hortic 736: 205–211

Montesinos DPP, Villar-Salvador P, García-Fayos M (2012) Genders in *Juniperus thurifera* have different functional responses to variations in nutrient availability. New Phytol 193:214–238

Moore JL, Kollar L, McLetchie D (2016) Does selection for gamete dispersal and capture lead to a sex difference in clump water-holding capacity? Am J Bot 103:1449–1457

Morant M (2007) *CYP703* is an ancient cytochrome P450 in land plants catalyzing in-chain hydroxylation of lauric acid to provide building blocks for sporopollenin synthesis in pollen. Plant Cell 19:1473–1487

Mulcahy D, Weeden N (1992) DNA probes for the Y-chromosome of Silene latifolia, a dioecious angiosperm. Sex. Plant Reprod 5:86-88

Naik SK, Chand PK (2011) Tissue culture-mediated biotechnological intervention in pomegranate: a review. Plant Cell Rep 30:707-721

Nesi NI, Debeaujon C, Jond AJ, Stewart GI, Jenkins M, Caboche L, Lepiniec R (2002) The transparent testa16 locus encodes the arabidopsis bsister mads domain protein and is required for proper development and pigmentation of the seed coat. Plant Cell 14:2463–2479

Nixon RW, Carpenter JB (1978) Growing dates in the United States. United States Department of Agriculture Science and Education Administration, Washington, D.C., pp 1–56

Obeso JR, Alvarez-Santullano NM, Retuerto RN (1998) Sex ratios, size distributions and sexual dimorphism in the dioecious tree *Ilex aquifolium* (Aquifoliaceae). Am J Bot 85:1602–1608

Osman AMA (1984) The performance of date palms in the Sudan. Acta Hortic 143:231-237

Paoletti E, Gellini R (1993) Stomatal density in beech and holm oak leaves collected over the last 200 years. Acta Ecol Sin 14:173–178

Pinyopich AGS, Ditta B, Savidge SJ, Liljegren E, Baumann E, Wisman M, Yanofsky F (2003) Assessing the redundancy of MADS-box genes during carpel and ovule development. Nature 424:85–88

Prasad K, Zhang X, Tobón E, Ambrose BA (2010) The Arabidopsis B-sister MADS-box protein, Gordita, represses fruit growth and contributes to integument development. Plant J 62:203–214

Qacif N, Baaziz M, Bendiab K (2007) Biochemical investigations on peroxidase contents of male and female inflorescences of date palm (*Phoenix dactylifera*.). Sci Hortic 114:298–301

Qudama A (1985) Dictionary of food medicinal plants. Darul-Nafaes Beirut

Raj S Saudagar P (2019) Plant cell culture as alternatives to produce secondary metabolites. Curr. Bioact. Compd 265-286

Retuerto RBF, Lema SR, Roiloa RT, Obeso JR (2000) Gender light and water effects in carbon isotope discrimination and growth rates in the dioecious tree *Ilex aquifolium*. Funct Ecol 14: 529–537

Salomón-Torres R, Ortiz-Uribe N, Valdez-Salas B, Rosas-González N, García-González C, Chávez D, Córdova-Guerrero I, Díaz-Rubio L, Haro-Vázquez MP, Mijangos-Montiel JL, Morales-Maza A, Mahadevan P, Krueger R (2019) Nutritional assessment, phytochemical composition and antioxidant analysis of the pulp and seed of Medjool date grown in Mexico. Peer J 7:e6821

Sarkar SJ, Banerjee S, Gantait S (2017) Sex-oriented research on dioecious crops of Indian subcontinent: an updated review. Biotechnol J 7:93–109

Sheng ZH, Jiang H, Zhao H, Korpelainen H (2014) Sexually different physiological responses of populous cathayana to Nitrogen and Phosphorus deficiencies. Tree Physiol 34:343–354

Sherman DA, Dahlgren JP, Ehrlén PA, García MB (2019) Sex and the cost of reproduction through the life course of an extremely long-lived herb. Oecologia 191:369–375

Sonia DSL, Laurent C, Pascal E, Salem C, Philippe V, David S, Amel T, Neila S, Ferid L, Thierry J (2013) The Date Palm (*Phoenix dactylifera* L.) leaf proteome: identification of a gender biomarker to screen male parents. Plant Omics 6:18–23

Tissert B (1984) Propagation of date palm by shoot tip cultures. Hortic. Sci. 19:230-231

Veramendi J, Navarro L (1996). Influence of physical conditions of nutrient medium and sucrose on somatic embryogenesis of date palm. Plant Cell Tissue Organ Cult 45:159–164

Wang HN, Ngwenyama Y, Liu JC, Walker S, Zhang S (2007) Stomatal development and patterning are regulated by environmentally responsive mitogen-activated protein kinases in Arabidopsis. Plant Cell 19:63–73

Warren CR (2011) How does P affect photosynthesis and metabolite profiles of *Eucalyptus globulus*? Tree Physiol 31:727–739

Welsh J, McClelland M (1990) Fingerprinting genomes using PCR with arbitrary primers. Nucleic Acids Res 18:7213–7218

Xiong E, Wu J, Shi X, Wang W (2013) Proteomic identification of differentially expressed proteins between male and female plants in *Pistacia chinensis*. PLoS One 8:64–76

Xu X, Yang F, Xiao X, Zhang P, Korpelainen S (2008) Sex-specific responses of *Populus cathayana* to drought and elevated temperatures. Plant Cell Environ 31:24–36

Yaish MW, Kumar PP (2015) Salt tolerance research in date palm tree (Phoenix dactylifera L.), past, present and future perspectives. Front. Plant Sci., 6:348

Yang SL, Xie LF (2003) *Tapetum Determinant1* is required for cell specialization in the Arabidopsis anther. Plant Cell 15:2792–2804

Yang SL, Xie LF (2003) Tapetum Determinant1 is required for cell specialization in the Arabidopsis anther. Plant Cell 15:2792-2804

Yang X, Wu DI, Shi J, He YI, Pinot F, Grausem B, Zhang D (2014) Rice CYP703A3, a cytochrome P450 hydroxylase, is essential for development of anther cuticle and pollen exine. J. Integr. Plant Biol 56:979-994

Younis RAA, Ismail OM, Soliman S (2008) Identification of sex-specific DNA markers for date palm (*Phoenix dactylifera* L.) using RAPD and ISSR techniques. Res J Agric Biol Sci 4:278–284

Zaid AP, de Wet F (2002) Climatic requirements of date palm. In: Zaid A (ed) Date palm cultivation, vol 156. Food and Agriculture Organization of the United Nations, Rome, Italy, pp 57–72

Zaid A, Arias-Jimenez EJ, Date Palm Cultivation (2002) FAO plant production and protection paper. FAO Rome pp: 156

Zhang S, Jiang H, Peng S, Korpelainen H, Li C (2011) Sex-related differences in morphological physiological, and ultrastructural responses of *Populus cathayana* to chilling. J Exp Bot 62: 675–686

Zhang B, Stoltz LP (1991) In vitro shoot formation and elongation of dwarf pomegranate. Int. J. Hortic. Sci 26:1084-1084

Zhao P, Liu F, Zhang B, Liu X, Wang B, Gong J, Liu H (2013) MAIGO2 is involved in abscisic acid-mediated response to abiotic stresses and Golgi-to-ER retrograde transport. Physiol. Plant 148:246-260

Zhao DZ, Wang GF, Speal B (2002) The *Excess Microsporocytes1* gene encodes a putative leucine-rich repeat receptor protein kinase that controls somatic and reproductive cell fates in the Arabidopsis anther. Genes Dev 16:2021–2031

Zouine J, El Hadrami I (2004) Somatic embryogenesis in Phoenix dactylifera L. Effect of exogenous supply of sucrose on proteins, sugars, phenolics and peroxidases activities during the embryogenic cell suspension culture. Biotechnol. J 3:114-118

Zhao HY, Duan B, Korpelainen H, Li C (2009) Sex-related adaptive response of *Populus cathayana* to photoperiod transitions. Plant Cell Environ 32:1401–1411

Chapter 7
Molecular Genetics and Breeding Strategies of Doum (*Hyphaene thebaica* L.)

Amal M. E. Abdel-Hamid, Eman Tawfik, and Khaled F. M. Salem

Abstract The doum palm (*Hyphaene thebaica* L.) is a flowering plant in the Magnoliopsida Superorder Lilianae—Monocots, Arecales order, and Arecaceae family. This perennial evergreen tree thrives in semi-arid regons and is one of 11 species in the genus Hyphaene, native to North Africa. It can be found in rainforests worldwide, though less commonly in arid environments, spanning the Sahel region of Africa from Mauritania to Sudan. Doum products are popular beverages with various contemporary and therapeutic uses, rich in monosaccharides, filaments, B-complex vitamins, and essential minerals like sodium, calcium, iron, potassium, magnesium, and phosphorus. Conventional plant breeding techniques are crucial for the continued development of the doum palm, with a focus on enhancing genetic diversity through hybridization and the introduction of new alleles. Environmentally friendly and therapeutic cultivars can be created to complement traditional methods. Challenges such as disease, economic impacts, and climate change affecting agricultural yields must be addressed to ensure successful cultivation. This chapter provides an overview of doum palm germplasm, origin, distribution, economic importance, health benefits, cultivation practices, genetic improvement,

A. M. E. Abdel-Hamid (✉)
Department of Biological and Geological Sciences, Faculty of Education, Ain Shams University, Cairo, Egypt
e-mail: amelmohamed@edu.asu.edu.eg

E. Tawfik
Botany and Microbiology Department, Faculty of Science, Helwan University, Helwan, Egypt
e-mail: emantawfik@science.helwan.edu.eg

K. F. M. Salem
Department of Biological Sciences, Shaqra University, Ad-Dawadimi, Saudi Arabia

Department of Plant Biotechnology, Genetic Engineering and Biotechnology Research Institute (GEBRI), University of Sadat City, Sadat City, Egypt
e-mail: khaled.salem@gebri.usc.edu.eg; khaledfathi@su.edu.sa

K. F. M. Salem et al. (eds.), *Breeding and Biotechnology of Leaf, Fruit, and Seed Fiber Crops*, Advances in Plant Breeding Strategies 11,
https://doi.org/10.1007/978-3-032-00407-9_7

diversity management, resource conservation, significance, applications, and traditional breeding. Advanced technologies like variation traits, molecular breeding, bioinformatics, tissue culture, molecular markers, embryo rescue, artificial seeds, and genomics are also discussed.

Keywords Biodiversity · Biotechnology · Breeding · Doum palm (*Hyphaene thebaica* L.) · Molecular breeding

7.1 Introduction

The doum palm (*Hyphaene thebaica*) is a desert palm indigenous to Egypt, sub-Saharan Africa, and western India. Referred to in Egypt as the doum palm or gingerbread palm, it attains a height of 6–9 m and generally features branched, branches with fan-shaped leaflets around 65–75 cm in length. It is categorized as one of the world's beneficial plants (Fletcher 1997). The trunk of the palm is utilized in construction and the production of numerous household things, while the leaves are employed for crafting mats, securing bundles, and producing writing paper.

The fruit is oblong, approximately the size of a yellow-orange apple features a red outer peel, possesses a thick, spongy, somewhat sweet, fibrous flesh reminiscent of gingerbread, and contains a huge pit. The fruit's skin is consumable and can be pulverized or sliced; the resulting powder is frequently desiccated and included in meals as a flavoring element (Orwa et al. 2009). The fruit pulp is utilized in culinary applications, with varied methods of preparation, and different types of edibility. The juvenile kernel is consumable, whereas the adult kernel is rigid and utilized solely as vegetable ivory (Doren 1997; Saber et al. 2023).

Prior research on doum has concentrated on the fruit due to its nutritional benefits since the fermented beverage produced by soaking dried fruit pulp in hot water is extensively used as a health tonic and is esteemed in the region for its numerous medical attributes. Investigations into the pulp of *H. thebaica* fruit have revealed the presence of trace minerals, proteins, and fatty acids, notably the necessary linoleic acid (Kamis et al. 2003). The aqueous extract of doum fruits exhibits antioxidant and anticancer properties, attributable to its high concentration of water-soluble phenolic compounds (Faten 2009; Hsu et al. 2006).

The doum palm is one of the most important plant families that provide humans with dietary fiber, carbohydrates and antihypertensive substances (Drouiche et al. 2024). The dom palm contains a phytochemical called lignin which has a pronounced anticancer effect as reported by Carter (1993); Abu-Farich et al. 2025; El-Sheikh et al. 2025).

East Africa is home to the common palm tree *Hyphaene compressa* (Kahn and Luxereau 2008; Stauffer et al. 2014). It belongs to the Coryphoideae subfamily of the Arecaceae family (Asmussen et al. 2006). The dominant genus *Hyphaene* in Africa consists of eight species, including *H. guineensis* Schumach. & Thonn, *H. compressa*, *H. coriacea* Gaertn., *H. reptans* Becc., *H. macrosperma* H. Wendl., *H. petersiana* Klotzsch ex Mart., *H. dichotoma* Furtado and *H. thebaica* L. *Mart.*

These species are sometimes collectively referred to as doum palm." Distributed across arid regions in Africa and India are *H. thebaica* (L.) Mart., *H. reptans* Becc., *H. petersiana* Klotzsch ex Mart., and *H. dichotoma* Furtado (Stauffer et al. 2014, 2018). Research has shown that the genus *Hyphaene* has a wide range of applications in Africa, including the production of non-timber items for food, medicine, construction materials, and textiles (Aboshora et al. 2014a; Kahn and Luxereau 2008).

Even though the genus has a significant economic function and contributes to the diversity of the palm family in Africa, it is still not well recognized and valued (Stauffer et al. 2018). The ongoing decline of the African doum palm population due to habitat destruction, drought, and overharvesting is concerning, as it is putting increased strain on the remaining subspecies of the doum palm. This may eventually lead to the loss of their gene pool (Stauffer et al. 2018).

People's lives in Kenya are significantly impacted by *H. compressa*, particularly for pastoralist groups who depend on it for construction materials, food, medicine, and revenue from the sale of woven goods (Amwatta 2004). In Kenya, *H. compressa* is mostly used for food. In certain arid and semi-arid regions of Kenya, there is an increasing interest in domesticating doum palms. The decrease in doum palm germplasm resources in these regions as a result of biotic stress and human intervention is the main cause of this (Omire et al. 2020b). Worldwide, it has been discovered that *H. compressa* is conserved in situ in six protected regions and five *ex situ* protected places (Cosiaux et al. 2017; GCI 2020). It is challenging to determine the Doum palm's conservation status in Africa (Johnson 1998). Furthermore, the Doum palm has not yet been included in the list of threatened species and is listed as a species of least concern on the IUCN Red List. This could be the cause of the area's meager conservation efforts. However, the biodiversity of Doum palms may be lost as a result of increased human activity, necessitating future conservation measures. Therefore, to speed up the process of domestication and genetic development, conservation and diversity research are required.

The competing demands of a growing human population and dwindling arable land have put traditional plant breeding under closer scrutiny. Conventional or classical plant breeding helped many domesticated plant species advance in the 1900s. Classical breeding relies on a thorough evaluation of the material under different conditions in comparison to known cultivars. The most cost-effective and environmentally friendly way to control the many pests and diseases (Atia et. al, 2021). Because drought is a serious environmental problem that impacts farming all over the world, plants that can withstand dry conditions are highly prized for their economic value (Ibrahim et al. 2024; Elansary et. al 2020).

Molecular markers eliminate the need for labor-intensive direct screening methods in controlled settings, such as greenhouses and fields, to expedite the selection process in crop breeding (Alqahtani et al. 2024; El Rabey et al. 2013; El-Absawy et al. 2015; Abdel-Hamid 2016a; Ibrahim et al. 2024; Abdel-Hamid and Mohamed 2014; Mattar et al. 2016; Salem 2009, 2015; Salem and Ibrahim 2024; Salem and Sallam 2016; Salem et al. 2022; 2024; Varshney et al. 2008). Molecular markers are particularly useful for identifying traits that are controlled by several genes (Abdel-Hamid 2016b). It is possible to create an improved cultivar by transferring multiple QTLs simultaneously or by combining QTLs for various agronomic traits once the appropriate molecular markers have been identified and the pertinent QTLs associated with an agronomic trait mapped (Abdel-Hamid and Salem 2021; Abu-Ellail et al. 2021; Abdel-Hamid 2020; Dita et al. 2006; Hassan et al. 2021; Salem 2004; Eltaher et al. 2022; Ghazy et al. 2020; Mourad et al. 2019). This chapter covers the origin, distribution, economic significance, health benefits, cultivation methods, diversity management, resource conservation, significance, applications, fiber properties, industry, genetics, selection, improvements, phylogenetics, and traditional breeding of doum palm germplasm. Additionally, cutting-edge technologies such as genetic engineering, gene editing, tissue culture, molecular markers, molecular breeding, variable traits, bioinformatics, embryo rescue, artificial seeds, and genomics are included.

7.1.1 Origin and Distribution

The Doum palm, *Hyphaene thebaica*, is a perennial and evergreen tree of considerable importance in semiarid regions, distinguished by its dichotomous stem. *Hyphaene thebaica* (Doum) is one of the 11 species in this class, originating from northern Africa. Extensive in the Sahel region of Africa, spanning from Mauritania and Senegal in the far West, via Kenya and Tanzania in the central area, to Egypt and Sudan in the East. It is a consumable forest product from Sudan. It is distributed across tropical rainforests globally and to a lesser extent in arid regions (FAO 2006). In Sudan, it is concentrated in the North, Focal, and West regions of the country. Abdel-Rahman et al. (2018). Additionally, it constitutes a local animal category in the Middle Eastern landmass, sub-Saharan Africa, and West India (Mansour 2021).

The conveyance of *H. thebaica* is restricted to valleys and slopes encompassing Rabigh, western Saudi Arabia. The low populaces of *H. thebaica* could be attributed to the rising aridity of this climate (Tarawneh and Chowdhury 2018). Anthropogenic impacts are probably going to prompt further decreases in the populace sizes of *H. thebaica* and other related species (Soultan et al. 2019). Doum palm (*Hyphaene thebaica* L.) is a flowering plant within the class Magnoliopsida, order Arecales and family Arecaceae. Doum palms are of the genus *Hyphaene* and *Hyphaene thebaica* species (https://www.itis.gov/servlet/SingleRpt/SingleRpt?search_topic=TSN&search_value=506725#null). The fruit morphology of the Doum palm (*H. compressa*) includes an orange, fibrous outer layer and a distinct, potent scent (Fig. 7.1). The flesh of the fruit blends seamlessly with its glossy skin (Omire et al. 2020a).

Fig. 7.1 Doum palm (*H. compressa*) fruit morphology. (**a**) Araka's ripe orange fruits. (**b**) Araka's immature fruits are maroon in color. (**c**) Kwali fruits are green when unripe. (**d**) Araka fruit morphology: from left to right, oblong-round (brown), oval (green), and reverse oval (orange). (**e**) Araka's unripe fruits are maroon. (**f**) Araka Nithi, Turkana, and Tana River fruits are displayed from left to right. (**g**) The orange mesocarp of an Araka palm fruit. (**h**) The hairy mesocarp of an Araka palm fruit and (**i**) A cross-section of an Araka palm fruit. (Source: Omire et al. (2020a))

Kingdom	Plantae—plantes, Planta, Vegetal, plants
Subkingdom	Viridiplantae—green plants
Infrakingdom	Streptophyta—land plants
Superdivision	Embryophyta
Division	Tracheophyta—vascular plants, tracheophytes
Subdivision	Spermatophytina—spermatophytes, seed plants, phanérogames
Class	Magnoliopsida Superorder Lilianae—monocots, monocotyledons
Order	Arecales
Family	Arecaceae
Genus	Hyphaene Gaertn.—hyphaene
Species	*Hyphaene thebaica* (L.) Mart.—Doum palm

7.1.2 Economic Importance and Health Benefits

Doum natural products are utilized as a well-known drink and for other restorative and modern advantages. Likewise, Doum organic products are plentiful in strands, cell reinforcements, B-complex nutrients, fundamental minerals such as sodium, calcium and iron in the epicarp and potassium, phosphorus and magnesium in the mesocarp, and have a significant substance of monosaccharides, for example, fructose and glucose (Aboshora et al. 2014b).

Doum natural product watery concentrate shows calming, antimicrobial, anticancer and pharmacological forthcoming because of its significant constituents like the phenolic, rejuvenating ointment and flavonoid intensifies (Mohamed 2009; Eldahshan et al. 2009; El-Beltagi 2018; El-Sheikh et al. 2025). Albeit the high worth of the drug properties of plant natural products, there is a need to investigate its hereditary variety on both morphological and sub-atomic levels. (Mansour 2021).

The powder of *H. thebaica* is utilized in certain contemporary products as a source of fiber and stabilizer (Elmoudnia et al. 2025). The edible portion of the fruit is intensely aromatic (thus the name Gingerbread Palm) and can be either crushed and ground or chipped. In addition to being incorporated with wheat and sugar to create Madeda drink. It is also utilized to produce nectars (Abdel-Rahman et al. 2018). Occasionally, it is desiccated and included as a flavor enhancer in culinary applications. In Turkana, Kenya, the powder is similarly utilized to prepare a mild cocktail by incorporating water and allowing it to rest. In several countries, the sap is employed to produce palm wine (Vogt 1995). Primarily, adolescents are consuming Doum natural goods as a snack.

Blend of processed wheat with Doum natural product powder expanding the rheological properties of the batter. As well as, upgraded absolute phenolic, complete flavonoid, and rising cancer prevention agent activity, and subsequently further developing the bread quality (Aboshora et al. 2015).

The water concentrate of Doum organic products has the option to diminish hyperlipidemia in nephrotic disorders and advisers for a decline in the risk of glomerulosclerosis and atherosclerosis and thusly the ordinary, protected and nontoxic *H. thebaica* natural product could be of colossal incentive for utilization as hypolipidemic medicines (Aboshora et al. 2014b).

The natural product is utilized to make a dark color frequently utilized in the tanning business. The organic product's seed is exceptionally strong and ivory-toned, which has been utilized economically for creating buttons and little carvings (El-Beltagi et al. 2018). Rugs, crates and composing paper are produced using leaves. Strips of Doum nuts are utilized to make molasses. Wood is sometimes utilized as kindling and for charcoal, yet frequently in development, offering help and rafters for ponies, posts for fencing and for building pontoons (Vogt 1995).

Trees and bushes play a critical part in keeping up with the regular biological system and in forestalling and battling desertification in the Sahel. They give a large number of valuable items to individuals because the wild-eatable plants are a great wellspring of supplements for the provincial populace.

7.2 Recent Cultivation Practices and Challenges

The investigation of bibliographic information connected with the assortment, drying, and capacity of *H. thebaica* uncovered fascinating outcomes. We found just a single report that was done by Ewansiha et al. (2021). It introduced all parts of gathering to the last planning of the example including the washing and drying stages. *H. thebaica* fills normally in the southwestern region of Egypt (Taha et al. 2020) and the Wina locale in Cameroon. It is collected in the period between November and February (Kolla et al. 2021). Harvests of this species were documented in the rural areas of Jambutu Yola (Ewansiha et al. 2021) and Kayauki in Adamawa and Katsina (Salisu and Saleh 2019) inside specific states in Nigeria. It also occupies the Tayba Garden in Elgazira Aba, Sudan (Aboshora et al. 2017).

Apparently, no bibliographic information referenced the rural acts of this species in the areas referenced previously. Most of the concentrates on *H. thebaica* zeroed in on the organic product, aside from the work revealed by Taha et al. (2020) who concentrated on the phytochemical content of the leaves and natural products. Salisu and Saleh (2019) concentrated on the phytochemical content of the stem.

Concerning assortment strategy, no bibliographic information portrayed this technique. The plant washing step was discretionary in the vast majority of the examinations, just a single report referred to washing the natural product with 70% ethanol, while another review referenced washing the new stems with refined water.

The drying of the plant material is perhaps the main move toward the planning of the concentrates. In this step, the plant grid is dried out to dispense with any gamble of microbial advancement subsequently expanding the capacity time of the plant material.

Three drying techniques were utilized including sun drying (Aboshora et al. 2014b), drying in obscurity (room temperature) (Aboshora et al. 2017; Bello et al. 2017), and in a stove (45 °C) (Taha et al. 2020) or (50 °C) (Gibril et al. 2020). The plant material was either squashed and straightforwardly extricated (Ewansiha et al. 2021) or squashed and put away for sometime later (Taha et al. 2020; Gibril et al. 2020; Kolla et al. 2021). The best techniques for saving plant material were either put away in a glass container or a plastic pack. These strategies permitted the examples to be kept airtight diminishing underlying changes brought about by oxidation peculiarities. One more review distributed by Kolla et al. (2021) referenced the utilization of cardboard boxes for the conservation of *H. thebaica* organic products (Saber et al. 2023).

7.3 Traditional Breeding

7.3.1 Improvement Strategies

The botanical species *Hyphaene thebaica*, also referred to as Doum palm or gingerbread tree, is indigenous to dry and semi-arid areas of Africa and the Middle East. Historically, it has been employed for diverse applications, encompassing sustenance, therapeutics, and substances. Although there is a lack of comprehensive contemporary breeding programs for *Hyphaene thebaica*, the utilization of traditional breeding and improvement tactics has contributed to the development of favorable characteristics in this particular plant species (Ghazal et al. 2016).

7.3.2 Selection of Desirable Traits

Selection of plants with preferred attributes, such as higher fruit quality, enhanced resistance to pests or diseases, and improved adaptation to certain environmental circumstances, is believed to have been a practice commonly observed among traditional farmers and communities (Janick and Moore 1996; Sansavini et al. 2004).

7.3.3 Propagation Techniques and Methods of Cultivation

Agriculturalists may have devised and transmitted propagation methods that safeguard characteristics. For instance, the practice of choosing seeds from trees that possess desirable fruit features or the practice of propagating by suckers in order to preserve advantageous traits in the progeny. The growth and development of *Hyphaene thebaica* may have been impacted by traditional agricultural practices such as intercropping and agroforestry. These approaches have the potential to enhance plant health and increase crop yield (Maussa et al. 1998; Shahin and Arafa 2007; Kahn 2008).

7.3.4 Genetic Diversity Management

Traditional agricultural societies frequently employ strategies such as seed exchange to effectively manage the genetic diversity of crops. This can aid in preserving a varied genetic reservoir, guaranteeing the population's ability to adapt to shifting environmental circumstances (Omire et al. 2022). Limited information exists regarding the genetic diversity of the doum palm, restricting access to its significant features and hence impeding its enhancement. Prior diversification research on *H. compressa* concentrated on examining the variety of its morphological characteristics (Shen et al. 2024). The research discovered five morphotypes, with accessions from the Kenyan Coastal region exhibiting the highest heterogeneity. Nevertheless, this approach is insufficient for accurately delineating the doum palm, as morphological markers can be influenced by environmental factors, are few in number, exhibit instability, develop slowly, and some manifest late in the plant's growth, complicating their assessment. Moreover, employing a singular marker such as morphology is insufficient for evaluating variety. In general, genetic markers surpass morphological indicators and may or may not align with the phenotypic manifestation of a genomic characteristic (Shen et al. 2024).

7.3.5 Ecological Adaptation

Throughout successive generations, farmers have likely chosen and nurtured plants that exhibit superior adaptability to certain regional circumstances, including soil composition, precipitation patterns, and temperature fluctuations. The acquisition of traditional knowledge pertaining to the plant's behavior, growth patterns, and reactions to various environmental factors would have occurred gradually over a period of time. The acquisition of observational knowledge holds significant value in facilitating informed decision-making regarding farming procedures. Plants may have been chosen by farmers and communities based on factors such as taste, nutritional composition, or therapeutic attributes. Over the course of time, this phenomenon may have exerted an influence on the frequency of particular cultivars exhibiting favored characteristics (Borlay et al. 2022).

7.3.6 Strategies for Conservation

Conservation measures for the preservation of plant genetic resources are frequently implemented by traditional cultures (Perrino and Perrino 2024). The sustainable utilization and preservation of *Hyphaene thebaica* may encompass several elements such as rituals, taboos, or specialized management techniques. Although contemporary molecular approaches offer greater precision and speed compared to traditional breeding methods, the latter have significantly contributed to the development of diversity and adaptation in *Hyphaene thebaica* throughout history. The incorporation of traditional knowledge into contemporary breeding methodologies has the potential to enhance the sustainable rearing and utilization of this significant botanical species (Anju et al. 2024).

7.3.7 Conventional Breeding Practices

The conventional practice of breeding *Hyphaene thebaica* entails the utilization of techniques implemented by indigenous populations to augment favorable characteristics and boost the overall efficacy of the species. It should be noted that within this particular context, traditional breeding may encompass a blend of human selection, seed management, and cultivation procedures (Hadrami and Hadrami 2009). The following are several facets of conventional breeding practices for *Hyphaene thebaica*.

7.3.7.1 The Process of Selecting and Collecting Seeds

Conventional breeders frequently choose seeds from parent plants that possess favorable traits like drought tolerance, abundant fruit production, and resistance to diseases.

Seeds are obtained from individuals who are in good health and strong in order to guarantee the transfer of advantageous characteristics to subsequent generations. Tarawneh and Chowdhury (2018) and Mansour et al. (2020).

7.3.7.2 Methods of Cultivation

Hyphaene thebaica may be cultivated by traditional farmers in particular sites or microenvironments that are conducive to its growth and development. Through the process of analyzing the performance of plants under various conditions, farmers can identify specific locations that facilitate enhanced development and adaptation (Kahn2008).

Targeted Extraction

Farmers engage in the deliberate selection of fruits or other plant parts based on their possession of desirable qualities. For instance, plants that possess larger or more flavorful fruits may be preferred for harvesting. Through selective harvesting, the reproduction and survival of plants with desirable features can be indirectly influenced (Shady et al. 2021).

Cultivar Upkeep

In certain instances, traditional communities may uphold certain cultivars or landraces that possess distinctive features, which have been selectively inherited across successive generations. Some regions may possess distinct cultivars that are well-suited to specific environmental circumstances, and these are preserved through cultural traditions and the careful handling of seeds (Dias et al. 2024).

7.3.7.3 Methods of Propagation

In order to enhance germination rates, traditional breeding practices sometimes use distinct propagation strategies, such as seed scarification or stratification. New individuals with identical features can be established by using cuttings or suckers from plants that possess desired attributes. Experimentation breeds: Traditional cultures frequently depend on intergenerational observations to ascertain and choose plants possessing optimal characteristics. The evaluation of plant performance by farmers often involves assessing many factors such as insect resistance, adaptability to environmental conditions, and general vigor (Taha et al. 2020).

7.3.7.4 Dissemination of Local Knowledge

Conventional breeding typically involves a collective effort within a community, where members of the community exchange local knowledge about the species. The accumulation of collective knowledge plays a significant role in the process of selecting and propagating plants that possess desirable features as seen by the community. It is noteworthy to acknowledge that although conventional breeding methods have demonstrated efficacy in facilitating crop adaptation to specific environmental circumstances, they may not encompass the same degree of regulated genetic manipulation as contemporary breeding technology. Integrating traditional knowledge into contemporary breeding programs might offer useful perspectives for enhancing *Hyphaene thebaica* and guaranteeing its long-term viability in various habitats (Smith et al. 2024).

7.3.7.5 Traditional Breeding Limitations

Traditional breeding techniques, although beneficial, have specific constraints when used for *Hyphaene thebaica.* Conventional breeding techniques are frequently laborious, necessitating multiple generations to attain substantial alterations in plant characteristics. This could impede the swift advancement of enhanced cultivars in reaction to evolving environmental or market circumstances. Its practices are predicated upon the indigenous genetic variability present within a given community. If the genetic variation in the natural population of *Hyphaene thebaica* is low, the ability to introduce new features may be restricted. The precision of modern biotechnology surpasses that of traditional breeding. The selection of traits is predicated upon observable qualities; however, the underlying genetic mechanisms may remain inadequately comprehended or regulated (Mansour 2021).

The conventional breeding methods employed in *Hyphaene thebaica* may not comprehensively account for the cultural and ethnobotanical dimensions that hold significance for indigenous populations. This may lead to the erosion of indigenous knowledge and customs linked to the plant. The efficacy of conventional breeding can be significantly influenced by environmental factors. The expression of features can be influenced by various factors, including climate, soil, and other ecological variables, which presents a challenge in isolating genetic determinants. Conventional breeding methods may encounter difficulties in obtaining a wide range of genetic resources. The natural distribution of *Hyphaene thebaica* may impose constraints on the accessibility of a wide range of genotypes for breeding. In order to address these constraints, it may be beneficial to consider a supplementary strategy that incorporates contemporary biotechnological methodologies, like molecular breeding and genetic engineering. The combined utilization of traditional knowledge and contemporary breeding practices has the potential to augment the efficacy of endeavors aimed at enhancing *Hyphaene thebaica*, benefiting both local communities and wider agricultural systems (Hatmaker et al. 2018; Khalil et al. 2020).

7.3.8 Molecular Breeding

The utilization of biotechnology is of paramount importance in the breeding of *Hyphaene thebaica*, since it provides sophisticated tools and methodologies that can effectively supplement conventional breeding approaches. Biotechnology enables a focused and expedited method to boost particular characteristics, overcome constraints, and augment the overall production and resilience of *Hyphaene thebaica*. Biotechnology plays a significant role in the breeding of *Hyphaene thebaica* through various means as described by (Kim and Sappington 2013): The utilization of biotechnological instruments facilitates the advancement of diagnostic techniques to monitor the robustness of *Hyphaene thebaica* plants. Early diagnosis of illnesses and pests is of utmost importance as it enables prompt response. The application of biotechnology can aid in the preservation and control of genetic resources in *Hyphaene thebaica* through the creation of effective gene banks, the safeguarding of germplasm, and the guarantee of a continuous supply of varied genetic material. The incorporation of biotechnological methodologies into the breeding initiatives for *Hyphaene thebaica* enables researchers and breeders to augment the accuracy, effectiveness, and expediency of cultivating enhanced cultivars. Additionally, it facilitates the resolution of certain obstacles associated with environmental strain, susceptibility to diseases, and market requirements, finally promoting the sustainable growth and utilization of *Hyphaene thebaica* (Khalil et al. 2020).

7.3.9 Genomics

7.3.9.1 Genetic Improvement

Appraisal of hereditary variety is a fundamental stage in plant-rearing projects, as it might aid the determination of appropriate lines and cultivars with higher variety and better execution. Joining phenotypic and sub-atomic assessments gives preferred data over their single evaluation. There are a few kinds of sub-atomic markers that contrast in their premise and target locales of the genome. The utilization of a multiple markers framework guarantees a decent inclusion of the genome and thus an improved outcome investigation could be obtained. One of the most well-known atomic marker frameworks utilized for the evaluation of plant hereditary variety is the entomb straight-forward succession rehashes, which depend on the enhancement of DNA districts flanked by microsatellites (Gupta et al. 1996; Salem 2004; Salem 2015; Mattar et al. 2016). ISSR markers have been utilized to evaluate hereditary variety in many plant species, for example, guava (Youssef and Ibrahim 2016), banana (Babu et al. 2018) and date palm (Purayil et al. 2018).

Broad endeavors are being proposed to foster new atomic devices to help in the sub-atomic assessment of the hereditary variety. One of the created sub-atomic markers is the beginning codon designated (SCoT) polymorphism (Collard and Mackill 2009), which is an original technique for producing plant DNA markers in light of the short-preserved locale flanking the ATG start codon in plant qualities. SCoT is applied in hereditary variety examination close by built segregant examination and quantitative attribute loci planning. Moreover, SCoT markers have been utilized in hereditary variety appraisal in a few plant varieties including Doum (Khalil et al. 2020), and Date palm (Al-Qurainy et al. 2015).

Khalil et al. (2020), evaluate the hereditary variety in 12 landraces of the doum palm filled in the Aswan professional flowerbed of "Egypt" utilizing a near examination of phenotypic qualities and sub-atomic markers for a first-time frame. This is the main report on the hereditary variety evaluation of doum palm (*H. thebaica*) in light of phenotypic and sub-atomic assessments. An assortment of 12 palm landraces was picked as a starter step to sort out their degree of inconstancy. In Egypt, the cultivation and creation area of doum is fundamentally moved in the southern areas (for example Qena, Luxor and Aswan Governorates). Sadly, a saw reduction in the quantity of plants was noticed. Likewise, doum palms were pronounced as a jeopardized species by the Service of Farming of the Sudan (Warrag et al. 2002). The high significance of its wholesome and therapeutic exercises makes a fundamental plan to concentrate more on doum by giving novel biotechnology methodologies and enhancement initiatives. Appraisal of hereditary variety is one of the most important assignments in rearing projects, consequently, legitimate genotypes with prevalent attributes can be chosen for additional get to the next development steps (Etminan et al. 2016). The aftereffects of the phenotypic assessment in the current review showed an elevated degree of fluctuation in most of the concentrated characteristics among the tried landraces (Khalil et al. 2020).

7.3.9.2 Molecular Marker-Assisted Selection

Molecular markers can enhance the efficiency of identifying and selecting plants with specific features. By immediately analyzing the genetic composition of the plant, breeders can circumvent the need for time-consuming and resource-intensive field tests (Mansour 2021).

7.3.9.3 Genetic Mapping and Characterization

Biotechnology tools facilitate the identification of the genes accountable for significant characteristics in *Hyphaene thebaica*. The comprehension of the genetic underpinnings of features offers valuable insights into the patterns of inheritance and enables the implementation of focused breeding endeavors (Hatmaker et al. 2018).

7.3.9.4 Marker-Assisted Backcrossing

The MAB approach integrates conventional breeding techniques with genetic markers in order to incorporate targeted genes derived from wild or unadopted material into high-quality plants. This phenomenon expedites the transmission of desirable characteristics while preserving the genetic heritage of the produced cultivar (Neiva et al. 2016).

7.3.9.5 Functional Genomics

The utilization of functional genomics methodologies, like transcriptomics and proteomics, facilitates the comprehension of gene functions and gene products. This body of information facilitates the identification of crucial biological pathways that underlie significant features (Mansour 2021; Abd El-Maksoud et al. 2020; Edris et al. 2019; Elansary et al. 2017).

7.3.9.6 Enhancing Traits Using Genetic Engineering

The field of genetic engineering enables the conscious alteration of a plant's genome in order to introduce or augment particular features (Abdel-Hamid et al. 2024). Potential areas of focus encompass raising resistance against pests, illnesses, or abiotic stressors, alongside augmenting nutritional composition or productivity. Precise genome editing skills are facilitated by tools such as CRISPR-Cas9, enabling the targeted change of certain genes. This approach can be utilized to introduce advantageous characteristics or alter preexisting ones in a regulated fashion (Omire et al. 2022).

There was limited availability of precise information regarding genetically enhanced features in *Hyphaene thebaica* (often known as doum palm or gingerbread tree). Nevertheless, there are some overarching perspectives on prospective characteristics that researchers may contemplate improving in palms or comparable crops described by (Tarawneh and Chowdhury 2018): (1) *Drought tolerance:* It will be beneficial to improve the resilience of *Hyphaene thebaica* to endure periods of water scarcity, particularly in dry and semi-arid areas. (2) *Resistance to pests and diseases:*

The enhancement of *Hyphaene thebaica's* overall health and productivity could be achieved through the development of cultivars that exhibit heightened resistance to pests and diseases. (3) *Enhanced Productivity and Fruit Characteristics:* The economic significance of the doum palm can be enhanced through the augmentation of fruit yield and the improvement of nutritional content or quality in its fruit. *Tolerance to stress:* The expansion of optimal growing circumstances could be facilitated by an increased capacity to withstand environmental challenges, such as salt or severe temperatures. (4) *Rapid Expansion:* The development of cultivars exhibiting accelerated growth rates could potentially yield advantages in terms of expedited establishment and enhanced productivity. (5) *The capacity to adapt to various soil types:* The cultivation potential of *Hyphaene thebaica* could be enhanced by increasing its adaptability to a broader spectrum of soil types. (6) *Accelerated Growth:* Enhancing the longevity of doum palm fruits is of paramount importance in terms of their storage, transportation, and market viability. (7) *Biofortification:* Biofortification is a process that aims to augment the nutritional composition of plants. In the case of *Hyphaene thebaica*, potential strategies may involve augmenting the concentrations of specific nutrients within the fruit. It is crucial to acknowledge that the advancement of improved characteristics in crops, such as genetic changes, frequently necessitates meticulous evaluation of environmental consequences, regulatory authorization, and public reception. Based on the most recent update, it is necessary to verify any particular progress or investigations pertaining to genetically modified characteristics in *Hyphaene thebaica* by consulting contemporary scientific literature or seeking information from agricultural research institutions (Aremu and Fadele 2011; Tarawneh and Chowdhury 2018).

7.3.9.7 Transgenic Cultivars

The available information regarding the development or commercialization of transgenic cultivars of *Hyphaene thebaica*, commonly referred to as the doum palm or gingerbread tree, was not extensively publicized. Transgenic cultivars are a term used to describe plants that have undergone genetic modification by introducing genes from a different species, enabling them to exhibit specific features.

It is crucial to acknowledge that the data regarding transgenic crops or genetically modified organisms may undergo modifications as time progresses. There may have been advancements in biotechnological research and genetic engineering since my previous update. For the most up-to-date information regarding transgenic cultivars of *Hyphaene thebaica*, it is advisable to consult recent scholarly literature, and agricultural databases, or establish communication with pertinent research institutions and authorities. Updates on the progress of transgenic cultivars can be obtained from researchers and organizations engaged in palm tree farming and genetic enhancement. It is important to take into consideration the local legislation and policies pertaining to the cultivation and release of genetically modified organisms, as these may exhibit variations across different countries and areas (Omire et al. 2022).

7.3.9.8 Bioinformatics

Biotechnology enables the utilization of bioinformatics tools to analyze genomic data on a big scale. This facilitates the identification of potential genes, the prediction of gene functions, and the development of strategies for specific characteristic enhancement (John and Anorue 2023).

7.4 Tissue Culture Applications

7.4.1 Tissue Culture and Micropropagation

Tissue culture, a biotechnological technique, facilitates the swift reproduction of superior plants. This approach has the potential to facilitate the large-scale cultivation of plants possessing desirable features, hence ensuring the accessibility of planting material for extensive production (Wijerathna-Yapa and Hiti-Bandaralage 2023). Figure 7.2 illustrates the recent applications for applying plant tissue culture techniques in doum plants.

7.4.2 Embryo Rescue

The utilization of embryo rescue is a prevalent method employed in the field of plant breeding and propagation to salvage embryos that may exhibit inadequate development or germination under typical circumstances. This method entails the formation and expansion of embryos by in vitro culture. The following is a proposed framework for the application of embryo rescue techniques to *Hyphaene thebaica*. Procedure for Embryo Rescue in *Hyphaene thebaica*: (1) *Embryo Collection:* Embryos

Fig. 7.2 Diagram for the recent purposes of *in vitro* and *in vivo* studies on doum (*Hyphaene thebaica* L.) (Saber et al. 2023)

can be obtained from *Hyphaene thebaica* fruits during their initial developmental stage. Typically, this is performed in aseptic circumstances to prevent contamination. (2) *Sterilization of the surface:* The embryos that have been gathered undergo surface sterilization in order to eradicate any bacteria that may be present on their surface. Bleach or ethanol are frequently employed as sterilization agents. (3) *The process of preparing a culture medium:* An appropriate nutrient medium is formulated by incorporating the essential macronutrients, micronutrients, vitamins, and plant growth regulators. The composition of the medium is modified in order to facilitate the growth of *Hyphaene thebaica* embryos. (4) *Embryo Inoculation:* Sterilized embryos are deposited onto the culture media that has been prepared in advance. The culture medium is responsible for supplying the essential nutrients and growth regulators required to facilitate the growth of embryos in an in vitro setting. (5) *Nurture and Surveillance:* The specimens are situated within a regulated setting that maintains ideal levels of temperature, humidity, and light. The monitoring of embryo growth and development is conducted on a regular basis. (6) *Subculturing:* Subculturing may be required in certain instances to move embryos to new media in order to ensure ongoing support for their growth and development (Saber et al. 2023).

7.4.3 Establishment and Adaptation

After the embryos transform into plantlets, the next step may entail stimulating the growth of roots and then adapting the plantlets to the surrounding environment (Saber et al. 2023).

7.4.4 Synthetic Seeds

Synthetic seeds are man-made structures designed to enclose somatic embryos or other propagules, serving as a mechanism for their preservation, conveyance, and reproductive process. The utilization of synthetic seeds presents several benefits, including the convenient storage and transportation of genetic material, safeguarding against infections, and the possibility of large-scale growth of plants possessing desirable characteristics.

In situations where conventional seed production encounters difficulties or where there is a necessity to safeguard particular genetic material, both embryo rescue and synthetic seed technologies can serve as valuable means for the conservation and propagation of *Hyphaene thebaica*. These strategies play a significant role in advancing the overarching objective of conserving, cultivating, and enhancing the genetic diversity of plant species (Rihan et al. 2017).

7.5 Atomic Rearing

The use of biotechnology is of fundamental significance in the reproducing of *Hyphaene thebaica*, since it gives modern devices and techniques that can enhance ordinary rearing methodologies. Biotechnology empowers an engaged and sped-up strategy to help specific qualities, conquer requirements, and expand the general creation and strength of *Hyphaene thebaica*. Biotechnology assumes a critical part in the rearing of *Hyphaene thebaica* through different means as depicted by Kim and Sappington (2013).

7.6 Conclusion and Prospects

Doum palm fruits are considered an alternative fruit rich in many active compounds, including polyphenolic flavonoids. These fruits are currently underutilized in society despite their wide application potential in culinary enterprises and domestic environments. The food sector should promote the successful manufacturing of Doum fruit powders. This includes creating new items such as confectionery, snacks, and beverages, as well as producing established powdered and liquid natural flavors to economically utilize these local raw ingredients. Local Doum palm fruits represent a potential source of functional food components, thanks to their significant antioxidant and antibacterial qualities, thus creating new commercial opportunities for both small and major food enterprises. This review proposes the following recommendations: Food companies should propose incorporating Doum fruits to enhance the nutritional quality of their products.

The promotion of new products using Doum palm fruits should be promoted among culinary entrepreneurs. Efforts should focus on the cultivation and domestication of Doum palm plants to prevent shortages and extinction. The government should provide modern technologies to producers of date palm fruits to improve product quality and enable effective competition with industrial goods.

References

Abdel-Hamid AME, Salem KFM (2021) Breeding strategies of garden pea (*Pisum sativum* L.). In: Al-Khayri JM, Jain SM, Johnson DV (eds) Advances in plant breeding strategies: vegetable crops. Springer, Cham. https://doi.org/10.1007/978-3-030-66969-0_9

Abdel-Rahman NA, Ismail IA, Elshafe'a EBB (2018) Utilization of four Sudanese wild fruits to produce nectars. Food Biol 7:01–07

Abdel-Hamid, A. M., & Mohamed, H. I. (2014). The effect of the exogenous gibberellic acid on two salt stressed barley cultivars. European Scientific Journal, 10(6)

Abu-Farich, B., Masalha, M., Hamarshi, H., El Ghouizi, A., Aboulghazi, A., El Ouassete, M., ... & Saad, B. (2025). In vitro evaluation of Hyphaene thebaica honey as a multitarget therapeutic product. Microbes & Immunity, 2(1), 78-91

Aboshora W, Lianfu Z, Dahir M et al (2014a) Physicochemical, nutritional and functional properties of the epicarp, flesh and pitted sample of doum fruit (*Hyphaene* Thebaica). J Food Nut Res 2(4):180–186

Atia, M. A., El-Khateeb, E. A., Abd El-Maksoud, R. M., Abou-Zeid, M. A., Salah, A., & Abdel-Hamid, A. M. (2021) Mining of leaf rust resistance genes content in Egyptian bread wheat collection. Plants, 10(7), 1378

Aboshora W, Lianfu Z, Dahir M et al (2014b) Effect of extraction method and solvent power on polyphenol and flavonoid levels in *Hyphaene thebaica* L. Mart Arecaceae doum fruit, and its antioxidant and antibacterial activities. Trop J Pharm Res 13:2057–2063

Aboshora W, Lianfu Z, Dahir M et al (2015) Influence of doum (*Hyphaene thebaica* L.) flour addition on dough mixing properties, bread quality and antioxidant potential. J Food Sci Technol 53(1):591–600

Aboshora W, Idriss SE, Omar KA et al (2017) Volatile flavor compounds of peel and pulp from doum (*Hyphaene thebaica* L) fruit. Am J Food Sci Nutri Res 4(5):165–169

Abu-Ellail FFB, Salem KFM, Saleh MM et al (2021) Molecular breeding strategies of beetroot (Beta vulgaris ssp. vulgaris var. conditiva Alefeld). In: Al-Khayri JM, Jain SM, Johnson DV (eds) Advances in plant breeding strategies: vegetable crops. Springer, Cham. https://doi.org/10.1007/978-3-030-66965-2_4

Abdel-Hamid, A.M.E. (2020) Microscopic description and molecular genetic variations of some taxa of asteraceae and its importance as grazing plants. Pakistan Journal of Botany, 52(4) pp 10. DOI: x.doi.org/10.30848/PJB2020-4(9).

Alqahtani MM, Saleh MM, Alwutayd KM et al (2024) Performance and genotypic variability in diverse date palm (Phoenix dactylifera L.) cultivars for fruit characteristics. Genet Resour Crop Evol. https://doi.org/10.1007/s10722-023-01719-0

Al-Qurainy F, Khan S, Nadeem M, Tarroum M (2015) SCoT marker for the assessment of genetic diversity in Saudi Arabian date palm cultivars. Pak J Bot 47(2):637–643

Amwatta CJM (2004) Diversity of use of doum (*Hyphaene compressa*) in Kenya. Palms 48:184–190

Anju TR, Theresa M, Mohanlal J (2024) Conservation approaches of plant genetic resources. In: Al-Khayri JM, Jain SM, Penna S (eds) Sustainable utilization and conservation of plant genetic diversity. Sustainable development and biodiversity, vol 35. Springer, Singapore. https://doi.org/10.1007/978-981-99-5245-8_19

Abdel-Hamid, A.M.E., Moursi, Y.S., Elatif, S.A., Razik, E.S.A., Salem, K.F.M. (2024). Genome Editing Technologies in Crops Improvement. In: Al-Khayri, J.M., Ingle, K.P., Jain, S.M., Penna, S. (eds) Plant Molecular Breeding in Genomics Era. Advances in Plant Breeding Strategies, vol 4. Springer, Cham. https://doi.org/10.1007/978-3-031-68598-9_11

Abdel-Hamid, A.M.E. (2016a): Molecular Genetic Markers for Salt Tolerance in Some Egyptian Barley. LAMBERT Academic Publication

Abdel-Hamid, A.M.E. (2016b) Characterization of four Salsola Species and their genetic relationship by AFLP. Pakistan Journal of Botany., 48(3): 1183-1187.

Aremu A, Fadele O (2011) Study of some properties of doum palm fruit (*H. thebaica* Mart.) in relation to moisture content. Afr J Agric Res 6:3597–3602. http://www.Academicjournals.org/AJAR

Asmussen CB, Dransfield J, Deickmann V et al (2006) A new subfamily classification of the palm family (Arecaceae): evidence from plastid DNA phylogeny. Bot J Linn Soc 151(1)15–38

Babu AG, Prabhuling G, Karani RS et al (2018) Genetic diversity analysis among banana cultivars through ISSR markers. J Pharm Phytochem 7(6)1576–1580

Bello BA, Khan SA, Khan JA et al (2017) Anticancer, antibacterial and pollutant degradation potential of silver nanoparticles from *Hyphaene thebaica*. Biochem Biophys Res Commun 490(3):889–894

Abd El-Maksoud R.M., Abdel-Hamid, A.M.E., Alnusairi, G.S.H. (2020): Differential Expression of Expressed Sequence Tags (ESTs) Regulated in Response to Salt Stress Conditions in Date Palm. Pak J Biol Sci 23(6):743-752

Borlay AJ, Mweu CM, Nyanjom SG, Omolo KM (2022) Physiological and biochemical evaluations for identification of drought tolerance potential in doum palms (*Hyphaene compressa*) under water deficit treatment. Univ J Agric Res 10(5):474–487

Elansary, H.O., Abdel-Hamid, A.M.E., Yessoufou, K. et al. (2020) Physiological and molecular characterization of water-stressed Chrysanthemum under robinin and chitosan treatment. Acta Physiol Plant, 42(3) pp 31. https://doi.org/10.1007/s11738-020-3021-8

Carter J (1993) Dietary fiber guide. Cereal Foods World 38(10):755–759

Collard BC, Mackill DJ (2009) Start codon targeted (SCoT) polymorphism: a simple, novel DNA marker technique for generating gene-targeted markers in plants. Plant Mol Biol Report 27:86–93

Cosiaux A, Gardiner L, Couvreur TL (2017) *Hyphaene compressa*, the IUCN red list of *threatened* species. IUCNRed List, England, UK

Dias N, Halvitigala A, Kumari Fonseka DLC (2024) Development of improved landraces in agriculture for rural development. In: Plant mutagenesis. Springer, Cham

Doren ET (1997) Vegetable ivory and other palm nuts/seeds as an art/craft medium. J Int Palm Soc 41(4):18–25

Drouiche F, Laouici H, Cheikh M (2024) Morphological, chemical, thermal and mechanical analysis of doum fibers as potential reinforcement of polymer composites. J Compos Mater 58(1)37–54

El Rabey H, Salem KFM, Mattar MZ (2013) The genetic diversity and relatedness of eight rice (*Oryza sativa* L.) cultivars as revealed by AFLP and SSRs markers. Life Sci J 10(1):1471–1479

El-Sheikh, D. M., & Ahmed, R. E. (2025). Anticancer, Anti-hepatitis A and Other Biological Activities of *Hyphaene Thebaica* and Its Application in Gluten-Free Biscuits. Food Technology Research Journal, 8(2), 72-91

El-Absawy EA, Nowar AA, Salem KFM et al (2015) Estimation of genetic diversity among some bread wheat (*Triticum aestivum* L.) genotypes using molecular markers and agronomic traits. Minufiya. J Agric Res 40(3):223–233

El-Beltagi HS, Mohamed HI, Yousef HN, Fawzi EM (2018) Biological activities of the doum palm (*Hyphaene thebaica* L) extract and its bioactive components. In: Shalaby E, Azzam GM (eds) Antioxidants in foods and its applications. Intechopen, pp 49–66. https://doi.org/10.5772/intechopen.74772

Eldahshan OA, Ayoub NA, Singab ANB, Al-Azizi MM (2009) Potential antioxidant phenolic metabolites from doum palm leaves. Afr J Pharm Pharmacol 3(4):158–164

Edris, S., Abo-Aba, S., Algandaby, M. M., Alshamrani, E. M. M., Atef, A., Ramadan, A. M., ... & Abdel-Hamed, A. M. E. (2019). Transcriptional dynamics of early developing tomato (Solanum lycopersicum L.) fruit based on RNA-seq analysis. Applied Ecology and Environmental Research, 17(6):13817-13831

Elmoudnia H, Millogo Y, Faria P et al (2025) Development of Doum palm fiber-based building insulation composites with citric acid/glycerol eco-friendly binder. J Compos Sci 9(2):67

Eltaher S, Sallam A, Emara HA et al (2022) Genome-wide association mapping revealed SNP alleles associated with spike traits in wheat. Agronomy 12:1469. https://doi.org/10.3390/agronomy12061469

Etminan A, Pour-Aboughadareh A, Mohammadi R et al (2016) Applicability of start codon targeted (SCoT) and inter-simple sequence repeat (ISSR) markers for genetic diversity analysis in durum wheat genotypes. Biotechnol Biotechnol Equip 30(6):1075–1081

Elansary H.O., Yessoufou K., Abdel-Hamid A.M.E., El-Esawi M.A., Ali H.M., Elshikh M.S. (2017): Seaweed extracts enhance *Salam turfgrass* performance during prolonged irrigation intervals and saline shock. Frontiers in Plant Science, 8:830.doi: 10.3389/fpls.2017.00830.

Ewansiha J, Ugo C, Kolawole D, Orji L (2021) Antibacterial activities of *Hyphaene thebaica* (doum palm) fruit extracts against intestinal microflora and potential constipation associated pathogens in Yola Metropolis. Nigeria Tanz J Sci 47(1):104–111

FAO (2006) Composition and characteristics of selected palm products. FAO Corporate Document Repository. Forestry Department. Tropical Palm. FAO, Rome

Faten MAE (2009) Antioxidant and anticancer activities of doum fruit extract (*Hyphaene thebaica*). Afr J Pure Appl Chem 3(10):197–201

Fletcher R (1997) Listing of useful plants of the world. Australian New Crops. http://www.newcrops.uq.edu.au/listing/hyphaenethebaica

GCI (2020) Botanic Gardens Conservation International, Richmond, UK. https://tools.bgci.org/plant_search.php

Ghazal MN, Barakat SA, Ali WAH, Riad RM (2016) Effect of supplementing doum (*Hyphaene thebaica*) to diets on reproductive and productive traits in rabbits. Egypt Poult Sci J 36 (III):711–723

Ghazy MI, Salem KFM, Sallam A (2020) Utilization of genetic diversity and marker-trait to improve drought tolerance in rice (*Oryza sativa* L.). Mol Biol Rep 48:157–170. https://doi.org/10.1007/s11033-020-06029-7

Gibril ME, Zhang N, Yi Y et al (2020) Physicochemical characterization and future beneficiation routes of wild fruit waste (*Hyphaene Thebaica* seed) as a source to extract mannan. J Clean Prod 267:121949

Gupta PK, Balyan HS, Sharma PC, Ramesh B (1996) Microsatellites in plants: a new class of molecular markers. Curr Sci 70:45–54

Hadrami IE, Hadrami AE (2009) Breeding date palm. In: Breeding plantation tree crops: tropical species. Springer New York, New York, NY, pp 191–216

Hassan MN, Mekkawy SA, Mahdy M et al (2021) Recent molecular and breeding strategies in lettuce (Lactuca spp.). Genet Resour Crop Evol 68:3055–3079. https://doi.org/10.1007/s10722-021-01246-w

Hatmaker EA, Staton ME, Dattilo AJ, Hadziabdic D, Rinehart TA, Schilling EE, Trigiano RN, Wadl PA (2018) Population structure and genetic diversity within the endangered species *Pityopsis ruthii* (Asteraceae). Front Plant Sci 9:943

Hsu B, Coupar IM, Ng K (2006) Antioxidant activity of hot water extracts from the fruit of the Doum palm, (*Hyphaene thebaica*). Food Chem 98(2):317–328

Ibrahim AA, Abd-Ellatif S, Abdel Razik ESS, Salem KFM (2024) Biodiversity of cereal crops and utilization in food and nutritional security. In: Al-Khayri JM, Jain SM, Penna S (eds) Sustainable utilization and conservation of plant genetic diversity. Sustainable development and biodiversity, vol 35. Springer, Singapore. https://doi.org/10.1007/978-981-99-5245-8_2

Janick J, Moore JN (1996) Fruit breeding, tree and tropical fruits, vol 1. John Wiley & Sons

John AO, Anorue DN (2023) Effect of doum palm mesocarp meal (*Hyphaene thebaica*) as partial replacement for maize on growth performance and haematological indices of weaned pigs. J Biotechol Bioinform Res 5(3):1–6

Johnson DV (1998) African and the western Indian Ocean region Africa. In: Non timber forest products, tropical palms. Food and Agriculture Organisation of the United Nations, New York. http://www.fao.org/3/i1590e/i1590e08.pdf

Kahn F, Luxereau A (2008) Doum palm habit and leaf collecting practices in Niger. Palms 52(1) 23–29

Kamis AB, Modu S, Zanna H, Oniyangi TA (2003) Preliminary biochemical and haematological effects of aqueous suspension of pulp of *Hyphaene thebaica* (l) mart in rats. Biokemistri 13(1): 1–7

Khalil OA, Ibrahim RA, Youssef M (2020) A comparative assessment of phenotypic and molecular diversity in Doum (*Hyphaene thebaica* L.). Mol Biol Rep 47(1):275–284

Kim KS, Sappington TW (2013) Microsatellite data analysis for population genetics. In: Kantartzi S (ed) Microsatellites. Methods in molecular biology, vol 1006. Humana Press, Totowa, NJ. https://doi.org/10.1007/978-1-62703-389-3_19

Kolla M, Laya A, Bayang J, Koubala B (2021) Effect of different drying methods and storage conditions on physical, nutritional, bioactive compounds and antioxidant properties of doum (*Hyphaene thebaica*) fruits. Heliyon 7(4):e06678

Mansour H (2021) Population genetics of *Hyphaene thebaica* Mart., in Rabigh province, Saudi Arabia: implications for conservation. GSC Biol Pharm Sci 17(03):153–159

Mansour H, Alsamadany H, Al-Hasawi ZM (2020) Genetic diversity and genetic structure of *Salvadora persica* L., rare plant species in Rabigh province, Saudi Arabia: implications for conservation. J Taibah Univ Sci 14(1):881–888

Mattar MZ, Salem KFM, El-Abed AB (2016) Identification of candidate microsatellite markers associated with agronomic traits in rice (Oryza sativa L.). J Bot 15(5):213–232

Maussa H, Margolis H, Dube P, Dongo JO (1998) Factors affecting the germination of doum palm (*Hyphaene thebaica*, mark) seeds from the semiarid zone of Niger, West Africa. For Ecol Manag 104:27–41

Mohamed AN (2009) Antioxidant and anticancer activities of doum fruit extract (*Hyphaene thebaica*). Afr J Pure Appl Chem 3(10):197–201

Mourad AMI, Alomari DZ, Alqudah AM et al (2019) Recent advances in wheat (*Triticum* spp.) breeding. In: Al-Khayri J, Jain S, Johnson D (eds) Advances in plant breeding strategies: cereals. Springer, Cham. https://doi.org/10.1007/978-3-030-23108-8_15

Neiva DS, Melo Júnior AF, Oliveira DA et al (2016) *Acrocomia emensis* (Arecaceae) genetic structure and diversity using SSR molecular markers. Genet Mol Res 15(1):1–11. https://doi.org/10.4238/gmr.15017785

Omire A, Budambula NL, Neondo J et al (2020a) Phenotypic diversity of doum palm (*Hyphaene compressa*), a semi-domesticated palm in the arid and semi-arid regions of Kenya. Scientifica 2020(1):4920830

Omire A, Neondo J, Budambula NLM et al (2020b) *Hyphaene compressa*, an important palm in the arid and semi-arid regions of Kenya. Ethnobot Res Appl 20:1–15

Omire A, Neondo J, Budambula NLM et al (2022) Genetic diversity and population structure of Doum palm (*Hyphaene compressa*) using genotyping by sequencing. Front Genet 13:762202. https://doi.org/10.3389/fgene.2022.762202

Orwa C, Mutua A, Kindt R et al (2009) Agroforestree database: a tree reference and selection guide. World Agroforestry Centre (ICRAF), Nairobi

Perrino EV, Perrino P (2024) Conservation of plant genetic resources in Italy. In: Sustainable utilization and conservation of plant genetic diversity. Springer Nature Singapore, Singapore, pp 895–971

Purayil FT, Robert GA, Gothandam KM et al (2018) Genetic variability in selected date palm (Phoenix dactylifera L) cultivars of United Arab Emirates using ISSR and DAMD markers. 3 Biotech 8:1–8

Rihan, H. Z., Kareem, F., El-Mahrouk, M. E., & Fuller, M. P. (2017). Artificial seeds (principle, aspects and applications). Agronomy, 7(4), 71.

Saber FR, Aly SH, Khallaf MA et al (2023) *Hyphaene thebaica* (Areceaeae) as a promising functional food: extraction, analytical techniques, bioactivity, food, and industrial applications. Food Anal Methods 16:1447–1467. https://doi.org/10.1007/s12161-022-02412-1

Saleh MM, Alnaddaf LM, Almuhammady AK et al (2021) Applications of plant-derived nanomaterials in mitigation of crop abiotic stress. In: Al-Khayri JM, Ansari MI, Singh AK (eds) Nanobiotechnology. Springer, Cham. https://doi.org/10.1007/978-3-030-73606-4_9

Salem KFM (2004) The inheritance and molecular mapping of genes for post-anthesis drought tolerance (PADT) in wheat. Martin-Luther-University, Halle-Wittenberg, p 124

Salem KFM (2009) Relationship between genetic diversity based on SSRs markers with heterosis and combining ability in diallel cross of bread wheat (Triticum aestivum L.). Minufiya J Agric Res 6(34):2159–2178

Salem KFM (2015) Allelic state at the microsatellite locus Xgwm261 marking the dwarfing gene Rht8 in Egyptian bread wheat (*Triticum aestivum* L.) genotypes released from 1947 to 2004. Genetika 47:741–750

Salem KFM, Ibrahim AA (2024) Plant biodiversity in the context of food security under climate change. In: Al-Khayri JM, Jain SM, Penna S (eds) Sustainable utilization and conservation of plant genetic diversity. Sustainable development and biodiversity, vol 35. Springer, Singapore. https://doi.org/10.1007/978-981-99-5245-8_1

Salem KFM, Sallam A (2016) Analysis of population structure and genetic diversity of Egyptian and exotic rice (Oryza sativa L.) genotypes using SSR markers. Comptes Rendus Biol 339:1–9. https://doi.org/10.1016/j.crvi.2015.11.003

Salem KFM, Röder MS, Börner A (2004) Drought tolerance in wheat germplasm- evaluation and molecular mapping. In: Institut für Pflanzengenetik und Kulturpflanzenforschung, IPK Tagung, 22–24/08/2004. Ag Ressourcengenetik und Reproduktion, Abteilung Genbank, Gatersleben

Salem KFM, Alloosh MT, Saleh MM et al (2021) Utilization of nanofertilizers in crop tolerance to abiotic stress. In: Al-Khayri JM, Ansari MI, Singh AK (eds) Nanobiotechnology. Springer, Cham. https://doi.org/10.1007/978-3-030-73606-4_11

Salem KFM, Alghuthaymi MA, Elabd AB et al (2022) Prediction of heterosis for agronomic traits in half-diallel cross of rice (*Oryza sativa* L.) under drought stress using microsatellite markers. Plants 11:1532. https://doi.org/10.3390/plants11121532

Salem KFM, Safhi FA, Alwutayd KM et al (2024) Analysis of genetic diversity, population structure and phylogenetic relationships of rice (Oryza sativa L.) cultivars using simple sequence repeat (SSR) markers. Genet Resour Crop Evol 71:2213–2227. https://doi.org/10.1007/s10722-023-01789-0

Salisu A, Saleh M (2019) Microwave-assisted mercerization of Hyphaene thebaica fiber for the adsorption of Pb (II) and Cu (II) ions from aqueous solutions. Niger Res J Eng Environ Sci 4(1): 251–261

Sansavini S, Donati F, Costa F, Tartarini S (2004) Advances in apple breeding for enhanced fruit quality and resistance to biotic stresses: new varieties for the European market. J Fruit Ornam Plant Res 12(Spec. ed. 2):13–52

Shady NH, Hassan HA, Elrehany MA et al (2021) *Hyphaene thebaica* (doum)-derived extract alleviates hyperglycemia in diabetic rats: a comprehensive in silico, in vitro and in vivo study. Food Funct 12(22):11303–11318

Shahin S, Arafa A (2007) Germination of doum palm (*Hyphaene thebaica* L. MART.) seeds as affected by some scarification treatments. J Prod Dev 12(2):453–462

Shen B, Li W, Zheng Y et al (2024) Morphological and molecular response mechanisms of the root system of different Hemarthria compressa species to submergence stress. Front Plant Sci 15: 1342814

Smith G, Chowenga M, Karsters J (2024) Local knowledge matters: understanding the decision-making processes of communities under climate change in Suriname. Front Clim 5:1294271. https://doi.org/10.3389/fclim.2023.1294271

Soultan A, Wikelski M, Safi K (2019) Risk of biodiversity collapse under climate change in the Afro-Arabian region. Sci Rep 955(9):1–12

Stauffer FW, Ouattara D, Stork AL (2014) Palmae (arecaceae)/15. Trop Afr Flower Plants 8:326–354

Stauffer FW, Roguet D, Christe C et al (2018) A multidisciplinary study of the doum palms ('Hyphaene'Gaertn.): origin of the project, current advances and future perspectives. Saussurea 47:97–115

Taha GA, Abdel-Farid IB, Elgebaly HA et al (2020) Metabolomic profiling and antioxidant, anticancer and antimicrobial activities of Hyphaene thebaica. PRO 8(3):266

Tarawneh QY, Chowdhury S (2018) Trends of climate change in Saudi Arabia: implications on water resources. Climate 6(8):1–19

Varshney RK, Salem KFM, Röder MS et al (2008) Assessment and comparison of genetic diversity in barley germplasm collection using gSSR, eSSR and SNP markers. Plant Genet Res 6:167–174

Vogt K (1995) A field worker's guide to the identification, propagation and uses of: common trees and shrubs of dryland Sudan. SOS Sahel International (UK), London. p 100, 101, 125. Printed by: Antony Rowe Ltd. Chippenham

Warrag EI, Elsheiksh EA, Elfeel AA (2002) Forest genetic resources conservation in Sudan. For Genet Resour 30:48–51

Wijerathna-Yapa A, Hiti-Bandaralage J (2023) Tissue culture: a sustainable approach to explore plant stresses. Life (Basel).https://doi.org/10.3390/life13030780

Youssef M, Ibrahim R (2016) Molecular markers associated with high vitamin-C content in guava. J Agric Chem Biotech 7(3)49–55

Chapter 8
Maguey Cenizo (*Agave durangensis* Gentry) Breeding

Norma Almaraz-Abarca and José Antonio Ávila-Reyes

Abstract Maguey Cenizo (*Agave durangensis* Gentry) is a species of Asparagaceae, with a natural distribution limited to northern-central Mexico. It serves as the raw material for a regional mescal industry in this country. *Agave durangensis* has been little studied for aspects that may contribute to its genetic improvement. However, the ecological and economic importance of the species has prompted producers and government research institutions to conduct studies on various biological aspects of the plant, which can support the development of breeding and genetic improvement programs. The results of these efforts have yielded a robust understanding of the genetic and chemical variation as well as the potential of certain biotechnological tools to achieve this goal. This knowledge will likely expedite the development of varieties capable of thriving under a wide range of biotic and abiotic stress conditions in the near future. This is important because *A. durangensis*, with its phenotypical attributes and reproductive strategy, has the potential not only to sustain a successful distilled beverage industry but also the fiber and bioproduct industries. The existing knowledge and future discoveries must also contribute to the conservation of this resource, which is a valuable element of Mexico's natural heritage. This chapter provides an overview of the current molecular, phytochemical, breeding and biotechnological knowledge on *A. durangensis* as well as the progress in breeding strategies, genetic lines improvement, and conservation approaches necessary to address the challenges posed by the global current climate change and human population growth on fiber, construction materials, food, and beverage supply.

Keywords Breeding · Fiber · Genetic improvement · Genetic variability · Improvement programs · Maguey cenizo (*Agave durangensis* Gentry) · Propagation strategies

---

N. Almaraz-Abarca (✉) · J. A. Ávila-Reyes
Centro Interdisciplinario de Investigación para el Desarrollo Integral Regional unidad Durango (CIIDIR Durango), Instituto Politécnico Nacional, Durango, Mexico
e-mail: nalmaraz@ipn.mx; jaavila@ipn.mx

K. F. M. Salem et al. (eds.), *Breeding and Biotechnology of Leaf, Fruit, and Seed Fiber Crops*, Advances in Plant Breeding Strategies 11,
https://doi.org/10.1007/978-3-032-00407-9_8

8.1 Introduction

The genus *Agave sensu lato* (*Agave sensu stricto* clade, *Manfreda*, *Polianthes*, and *Prochnianthes*) hosts over 250 species (Jiménez-Barron et al. 2020). It originated in the Miocene, between 7.9 and 9.8 million years ago (Good-Avila et al. 2006; Scheinvar et al. 2017) in the American continent. The genus belongs to the family of monocots Asparagaceae (Agavaceae) and is endemic to this continent (García-Mendoza and Galván 1995). The highest richness and diversity of the genus is found in Mexico; this is why this country is considered the center of origin of *Agave* (García-Mendoza 1995).

Agave, along with *Opuntia*, is an emblematic genus of Mexico. Its species are found throughout the entire territory of this country, mainly in the arid and semiarid ecosystems, which cover more than half of its territory (Perez-Aguilar et al. 2021). One of the most economically important species of *Agave* is *A. durangensis*, commonly named Agave Cenizo. The species has a natural geographical distribution limited to northern-central Mexico. This chapter aims to provide an overview of the advances in the molecular, phytochemical, breeding and biotechnological knowledge of *A. durangensis*, which despite its limited natural distribution holds significant ecological importance and both present and potential economic value.

8.1.1 Importance and Uses

The diversity, distribution and abundance of *Agave* have led to its extensive and varied use by human societies settled in Mexican territory since pre-Hispanic times. Traditional uses include construction materials, textile fiber sources, livestock feed, materials for fermented beverages (pulque) and distilled beverages (tequila and mescal) (Ávila-Reyes et al. 2010; Zizumbo-Villarreal et al. 2009), as well as human food (the flowers of many *Agave* species are edible) (Barriada-Bernal et al. 2014). Recently, species of the genus have been suggested as sources of sugars, to replace sucrose (Yargatti and Muley 2022), bioactive metabolites (Nava-Cruz et al. 2015), for the treatment of different health disorders (Herrera-Ruiz et al. 2022), and feed additives (Jiménez-Sánchez et al. 2020). Also as sources of fuel (Duran et al. 2021), surfactant substances for removing toxic elements, such as arsenic (Gonzalez-Valdez et al. 2013), fluoride (Majamo et al. 2024), and chromium (Rathnayaka et al. 2024) from water, as well as fiber for the preparation of biocomposites for non-structural applications in the automotive industry (Arumugam et al. 2024).

While only a few species of *Agave* have been studied for their carbohydrate content and composition, published results indicate that they are significant sources for biofuel production. For example, Barragán-Trinidad and Buitrón (2024) highlighted the potential of *A. tequilana* var. Azul as a reliable substitute for biofuel production after anaerobic digestion by ruminal microorganisms. *Agave* species synthesize and accumulate pentoses, which helps address the issue of low yeast consumption of this carbohydrate type, ultimately reducing the cost of using commercial enzymes. *Agave lechuguilla* has also been revealed as an important element in promising biofuel production systems, as its juice is a low-cost substrate for *Clostridium acetobutylicum* to produce hydrogen and butanol (Oliva-Rodríguez et al. 2024).

Given the complex phytochemical composition of the genus (Almaraz-Abarca et al. 2013a; Morreeuw et al. 2021; Soto-Castro et al. 2021), it is expected that the current range of uses will expand in the coming years. The wide diversity of actual and potential uses of diverse species within the genus also enhances their acceptance as introduced species in different countries around the world. Currently, several species of *Agave* are cultivated in various European countries (Evdokimova et al. 2021; Verloove et al. 2019), Japan (Onozuka and Osawa 2022), and some African countries (El Mokni and Verloove 2021).

8.1.2 Biological Aspects and Distribution of Maguey Cenizo (Agave durangensis Gentry)

Agave durangensis Gentry (Fig. 8.1) is one of the less studied species of the genus, perhaps due to its geographical distribution in a small area in north-central Mexico (Fig. 8.2). The species belongs to the subgenus *Agave*, a group proposed by Gentry (1982). Subgenus *Agave* is characterized by having panicle-shaped inflorescences (Fig. 8.3). The division of the genus into the *Agave* subgenus and the *Littaea* subgenus (spike-shaped inflorescence) is morphological and has not been supported by molecular analysis (Eguiarte et al. 2021).

Fig. 8.1 Maguey Cenizo (*Agave durangensis* Gentry) from the population El Registro, Durango, Mexico. (Photo by Norma Almaraz-Abarca)

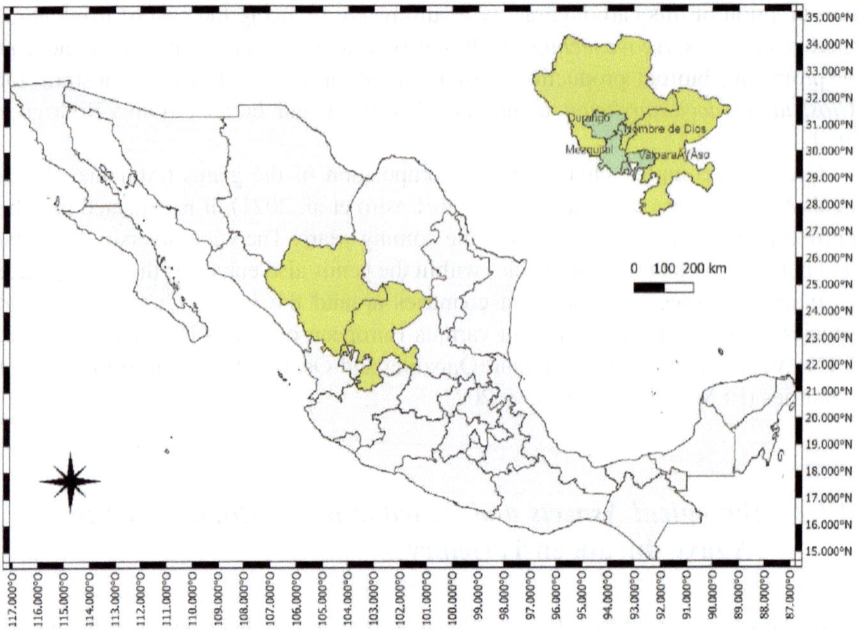

Fig. 8.2 Natural distribution (in dark green) of Maguey Cenizo (*Agave durangensis* Gentry) in Mexico. (Figure courtesy of Liliana Wallander Compeán, Interdisciplinary Research Center for Integral Regional Development, Durango, National Polytechnic Institute, Mexico)

Fig. 8.3 Panicle-shaped inflorescences of Maguey Cenizo (*Agave durangensis* Gentry) in a plantation in Durango, Mexico. (Photo by Norma Almaraz-Abarca)

The taxonomic classification of the species is as follows:

Kingdom	Plantae
Subkingdom	Viridiplantae (green plants)
Infrakingdom	Streptophyta (land plants)
Superdivision	Embryophyta
Division	Trcheophyta (vascular plants)
Subdivision	Spermatophytina (seed plants)
Class	Magnoliopsida
Order	Asparagales
Family	Asparagaceae
Genus	*Agave*
Species	*Agave durangensis* Gentry

Agave durangensis was described by Gentry and Gilly in 1951 (Gentry 1982) from an individual collected in Sierra El Registro, near Durango City, Mexico; approximately 29 Km away. This population is the type population. The description provided by Gentry (1982) of the type specimen is as follows: Medium to large plants, between 80 and 120 cm tall and 120 and 180 cm wide, individual or cespitose; gray to glaucous in color; broad leaves measuring 40–90 × 14–22 cm, lanceolate, with prominent teeth 1–2 cm long, by 1–2 cm, with a 4–6 cm long spine, that is grooved at the top. The inflorescence reaches 7–8 m in height; flowers are 60–80 mm long, yellow, with outer tepals 10–12 mm long and the inner ones smaller (Fig. 8.4); capsules are oblong 4.5–6 × 1.6–1.8 cm in size (Fig. 8.5); and small seeds (Fig. 8.6). Pollination of *Agave* species is often done by bats. However, the flowers of *A. durangensis* produce both nectar and pollen, attracting too many insects, like *Apis mellifera*, which may play a significant role in the pollination of the species (Fig. 8.7). Recently, Mayagoitia-Toulet and Zamora-Gutiérrez (2024) reported 16 species having as an interaction center the flowers of *A. durangensis*, among these species are bats (*Choeronycteris mexicana*), birds (hummingbirds (*Cynanathus* sp.), Mexican woodpecker (*Dryibates scalaris*), calandria (*Icterus* sp.), bees, wasps, moths (*Manduca quinquematulata*), butterflies, mosquitoes, and flies. This interaction of species around the flowers of *A. durangensis* gives an idea of the ecological importance of this species of *Agave*.

Agave durangensis is considered a highly variable species (González et al. 2009). A morphological study based on nine vegetative attributes (plant height, plant width, leaf length, leaf medial width, number of teeth/10 cm, terminal spine length, plant height/plant width, plant height/leaf length, and leaf length/leaf medium width) revealed high intra- and interpopulation variation in three natural populations of *A. durangensis*, which were distinguished by these characteristics (Almaraz-Abarca et al. 2013b), confirming the species high morphological variation. Figure 8.8 displays some of the variations in the shape and arrangement of the lateral spines of the leaves. In contrast, distinguishing between populations based on morphological characters has not been possible for other *Agave* species, such as *A. salmiana* and *A. mapisaga* (Alfaro-Rojas et al. 2007).

Fig. 8.4 Maguey Cenizo (*Agave durangensis* Gentry) flowers. (Photo by José Antonio Ávila-Reyes)

Fig. 8.5 Maguey Cenizo (*Agave durangensis* Gentry) capsules and seeds. Capsules show viable seeds (black) and unviable seeds (white). (Photo by José Antonio Ávila-Reyes)

Fig. 8.6 Viable Maguey Cenizo (*Agave durangensis* Gentry) seeds. (Photo by José Antonio Ávila-Reyes)

Fig. 8.7 Maguey Cenizo (*Agave durangensis* Gentry) flowers visited by honey bees (*Apis mellifera*). (Photo by José Antonio Ávila-Reyes)

Fig. 8.8 Some of the variations that can be found in the shape and disposition of the leaf lateral spines of Maguey Cenizo (*Agave durangensis* Gentry). (Photos by Norma Almaraz-Abarca)

The high morphological variability of *A. durangensis* makes it difficult to establish its specific limits. In order to contribute to defining these limits, chemotaxonomic (Almaraz-Abarca et al. 2009) and genetic variability studies have been carried out (Almaraz-Abarca et al. 2013b). However, further research on the topic is needed to provide new insights into evolutionary, phylogenetic, and breeding approaches.

González et al. (2009) established that the species is relatively abundant in its natural distribution area, indicating no conservation problems. However, *A. durangensis* supports a mescal industry, which is supplied mainly with plants gathered from natural populations according to traditional methods (Ávila-Reyes et al. 2010). This practice may pose a risk to the species. To address this, studies utilizing genetic algorithms and geographic information systems (Escobar and Sandoval 2022; Ocaña-Nava et al. 2007), as well as studies studies of site quality (Loera-Gallegos et al. 2018) have been conducted to model the species potential distribution. However, further research on breeding strategies and stress tolerance is necessary to conserve natural populations, preserve valuable alleles, establish plantations for the mescal industry, and expand cultivation areas.

Rosas et al. (2014) stated that the natural current distribution area of *A. durangensis* is around 17,000 hectares; however, Ríos-Saucedo et al. (2018) estimated that within three million hectares there are varying levels of probability for the presence of the species. The area has been protected by the Denomination of Origin Mezcal since 1994 (Almaraz-Abarca 2011) and the distilled beverage must be produced in accordance with the Mexican Official Standard NOM-070-SCFI-2016.

Maguey Cenizo (*Agave durangensis* Gentry) is a monocarpic species of the genus *Agave* (not all species in the genus are monocarpic). Some authors have reported that *A. durangensis* is propagated by rhizomes (López et al. 2021); however, field observations made by the authors of the current paper indicate that the main reproductive strategy of the species is by seeds. In the wild populations of *A. durangensis*, only between 3% and 6% of the 3 or 4-year-old plants produce between one and six rhizomes (Orea et al. 2006; Almaraz-Abarca 2011). These values are very low compared to those registered for other species of *Agave*. For instance, *A. asperrima*, a species whose main reproductive strategy is by rhizomes, has populations, in which more than 50% of the plants produce more than ten rhizomes per plant (Almaraz-Abarca 2011). In contrast to *A. durangensis*, other species of *Agave*, such as *A. fourcroydes*, propagate asexually through the production of rhizomes, without producing viable seeds (Infante et al. 2003).

8.2 Maguey Cenizo (*Agave durangensis* Gentry) Phenolic Variation

Plants produce a wide variety of compounds, known as secondary metabolites that were once thought to have no direct relationship with their growth and development. However, it is now understood that these metabolites play crucial physiological and ecological roles in plants. This is why they are now referred to as "specialized metabolites" (Corso et al. 2020). These compounds act as protective systems against different types of abiotic stress (Cuong et al. 2020; Farooq et al. 2021; Ghanem et al. 2021; Linić et al. 2019), defend against herbivores (Singh et al. 2021) and pathogens (Modesto et al. 2021; Yao et al. 2019), and also have allelopathic functions (Bi et al. 2022). They play a role in determining early patterns of water entry, which triggers protein synthesis (Siddiqui and Khan 2010) in seeds, affecting germination. This aspect is particularly important for *A. durangensis*, as its reproduction primarily relies on seeds.

Among specialized metabolites, phenolic compounds have been extensively studied due to their widespread presence in the plant kingdom (Taiz and Zeiger 2003), their chemical diversity (see Table 8.1), and their various physiological and ecological functions in plants (Mierziak et al. 2014; Pagare et al. 2015; Zhang et al. 2018). Additionally, phenolic compounds have significant biological activities that benefit human health (Hussain et al. 2019; Montané et al. 2020).

Phenolic acids and flavonoids are two types of phenolic compounds that are widely synthesized and accumulated in different groups of plants. The most common phenolic acids found in plants are derivatives of benzoic acid and derivatives of cinnamic acid. The basic structure of flavonoids contains 15 carbons (C6-C3-C6)

Table 8.1 Diversity of phenolic compounds reported for several species of *Agave*

Species	Compound	Tissue	References
A. americana L.	Kaempferol-3-*O*-glucoside Kaempferol-3-*O*-rutinoside	Flowers	Subramanian and Nair (1970)
	5,7-Dihydroxy-6,5′-dimethoxy-3′ 4′-Metilenedioxiflavanone	Somatic tissue	Parmar et al. (1992)
	Ellagic acid hexoside Kaempferol-*O*-glucuronide Kaempferol-*O*-rutinoside Kaempferol-*O*-hexoside Quercetin-*O*-rutinoside Quercetin-*O*-hexoside Quercetin-*O*-rhamnoside Isorhamnetin-*O*-hexoside Isorhamnetin-*O*-rutinoside	Leaves	Maazoun et al. (2019)
	Gallic acid Psidic acid Vanillic acid glucoside Coumaroyl quinic acid Hydrocinnamic acid glucoside Eeucomic acid Caffeic acid ethyl ester derivative Dihydroxybenzoic acid pentosylrhamnoside Caffeic acid derivative Rutin Kaempferol-rutinoside Kaempferol-glucosyl-rutinoside Kaempferol-rhamnosyl-rutinoside	Leaves	Khadiri et al. (2024)
A. angustifolia Haw.	Thirty-two phenolic compounds, among them glycosylated flavonoid derivatives and anthocyanins	Leaves	Bermúdez-Bazán et al. (2024)
A. asperrima Jacobi	Three kaempferol glycosides	Leaves	Almaraz-Abarca et al. (2009)
A. barbadensis Trel.	7-Hydroxy-3-(4-methoxybenzyl)-chroman 7-Hydroxy-3-(4-methoxybenzyl)-chroman-4-one 5,7-Dihydroxy-3-(4-methoxybenzyl)-chroman-4-one	Rhizomes	Tinto et al. (2005)
A. durangensis Gentry	Twelve kaempferol glycoside derivatives (among them kaempferol-3-*O*-[glucosyl(1–2) glucoside] and 7-*O*-methylkaempferol-3-*O*-[rhamnosyl(1–2) glucoside]) One myricetin-3-*O*-rhamnoside	Pollen	Almaraz-Abarca et al. (2009)
	Twenty nine kaempferol glycosides (among them kaempferol-3,7-*O*-diglucoside, kaempferol-3-*O*-[6-acetylglucoside]-7-*O*-glucoside, kaempferol-3-*O*-[rhamnosyl(1–6) glucoside], and kaempferol-3-*O*-rhamnoside) Three quercetin glycosides (among them quercetin-3-*O*-arabinoside)	Leaves	Almaraz-Abarca et al. (2009)

(continued)

Table 8.1 (continued)

Species	Compound	Tissue	References
A. lechuguilla Torr	Five phenolic acids Seven flavonoids (among them four quercetin-3-*O*-glycosides and one kaempferol-3-*O*-glycoside) One flavone Seven dihydro flavonoids	Leaves	Almaraz-Abarca et al. (2013c)
	Three flavones Epicatechin gallate Twelve anthocyanins/anthocyanidins Four dihydro flavonoids Thirty-two flavonols (isorhamnetin and seven glycoside derivatives, kaempferol and nine glycoside derivatives, myricetin and four glycoside derivatives, as well as quercetin and 8 glycoside derivatives)	Central part (cogollo)	Morreeuw et al. (2021)
A. marmorata Roezl	Catechin Gallic acid Caffeic acid Sinapic acid Isorhamnetin rutoside Isorhamnetin-3-*O*-diglucoside Isorhamnetin-glucuronide Isorhamnetin-3-glucoside Myricetin-3-*O*-glucoside Myricetin-3-*O*-rutinoside Rutin Quercetin-glucuronide Quercetin-3-*O*-diglucoside Quercetin-3-*O*-glucoside Kaempferol-diglucoside Kaempferol rutinoside	In vitro obtained-shoots	Martinez-Rodriguez et al. (2022)
A. shrevei Gentry	Four kaempferol derivatives One quercetin derivative One flavone	Leaves	Delgado-Alvarado et al. (2021)
A. sisalana Perrine ex. Engelm.	5,7-Dihydroxy flavanone Kaempferol-3-rutinoside-4′-glucoside Kaempferol-3-(2^G-rhamnosylrutinoside) 7-*O*-methyleucomol 3′-Desoxysappanone (±)-3,9-Dihydroeucomin Dihydro-bonducellin 7-Hydroxy-3-(4-hydroxy benzyl) chromane 5,7-Dihydroxy-3-(4′-hydroxy-benzyl)-4-chromanone 5,7-Dihydroxy-3-(3′-hydroxy-4′--methoxybenzyl)-4 chromanone	Leaves	Chen et al. (2009)
	Catechin Kaempferol Isorhamnetin Chrysin	Roots	dos Santos Mazo et al. (2024)

(continued)

Table 8.1 (continued)

Species	Compound	Tissue	References
	Gallic acid Psidic acid Benzyl diglucoside Gallic acid caffeoyl-glucose Dihydroxybenzoic acid-rutinoside Hydroxycinnamic acid-glucoside Eucomic acid Benzyl-diglucoside Dihydroxybenzoic acid-pentosylrhamnoside Gallic acid caffeoylglucose Gallic acid Rutin Kaempferol-rutinoside Kaempferol-glucosylrutinoside Kaempferol-glucoside Isorhamnetin-rutinoside	Leaves	Khadiri et al. (2024)
A. tequilana Weber	5,7-Dihydroxy-3-(4-methoxybenzyl)-chroman-4-one 7-Hydroxy-3-(4-hydroxy benzyl)-chroman-4-one 4′-Dimethyl-3,9-dihydropunctatin	Leaves	Morales-Serna et al. (2010)
A. striata Zucc.	Three phenolic acids Thirteen flavonols (among them ten quercetin-3-*O*-glycosides, one kaempferol-3-*O*-glycosides, and one herbacetin-3-*O*-glycoside) One flavone Four dihydorflavonoids	Leaves	Almaraz-Abarca et al. (2013c)
A. victoriae-reginae T. Moore	Sixteen phenolic acids Eleven flavonols (among them seven quercetin-3-*O*-glycosides, one kaempferol-3-*O*-glycoside, kaempferol (the aglycone form), and one isorhamnetin derivative) Five flavones Eight dihydro flavonoids	Leaves	Almaraz-Abarca et al. (2013c)

Source: Prepared by Norma Almaraz-Abarca and José Antonio Ávila-Reyes

distributed in three rings: A, B and C (Fig. 8.9). Rings A and B are aromatic, while ring C is a heterocycle. Differences in the degree of oxidation in the C ring, as well as the presence and positioning of different hydroxyl, methoxyl, and/or glycosyl groups in their basic structure, are the characteristics that distinguish the different types of flavonoids. Among the most commonly found in plants are flavones and flavonols, both characterized by having a double bond between carbons 2 and 3 and a carbonyl group in ring C. The difference between flavones and flavonols is the absence of a hydroxyl group on carbon 3 in the former and its presence in the latter. Another important group is that of dihydroxyflavonoids, characterized by not having a double bond between carbons 2 and 3 of ring C.

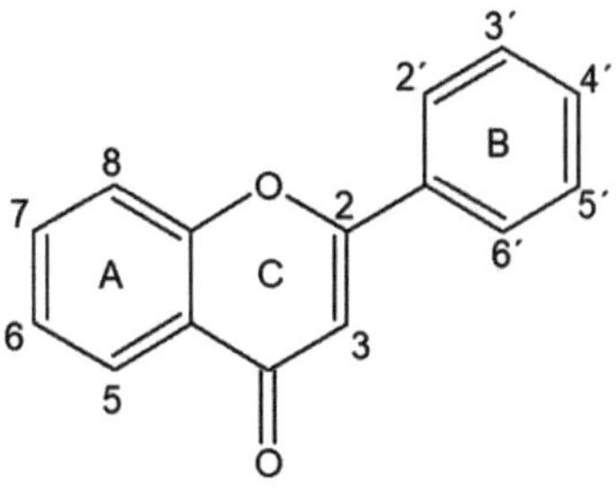

Fig. 8.9 General basic structure of flavonoids. (Figure courtesy of Eli Amanda Delgado-Alvarado, Interdisciplinary Research Center for Integral Regional Development, Durango, National Polytechnic Institute, Mexico)

The various phenolic compounds are synthesized from the shikimate pathway, which produces aromatic amino acids such as phenylalanine, tyrosine and tryptophan. The most common precursor is phenylalanine, which loses its amino group by the action of the enzyme phenylalanine ammonium lyase (PAL) to create cinnamic acid. Cinnamic acid serves as the starting point for the synthesis of several types of phenylpropanoids and more complex phenolic compounds (Corso et al. 2020).

In the biosynthesis pathway of phenolics, each step involving hydroxylation, methylation, glycosylation, or acylation of simple compounds to form more complex compounds is catalyzed by a specific enzyme, in a defined sequential order, which appears to be unique to each plant species (Heller and Forkman 1994). This specificity adds value to the profiles of these compounds as chemical markers. The extensive research on the vast diversity of phenolic compounds found in the plant kingdom suggests a corresponding high enzymatic diversity.

The concentration of individual compounds in plant phenolic profiles can be modulated by factors such as the genotype (Yang et al. 2018), geographical origin (del Valle et al. 2015), and growth conditions (Hernández-Pacheco et al. 2022). Both quantitatively and qualitatively, individual phenolics vary among different tissues (Medina-Medrano et al. 2015) and at various stages of plant development (Cobaleda-Velasco et al. 2017), highlighting their significance as chemo markers in the quality control of plant-based foods and their derivatives.

Maguey Cenizo (*A. durangensis* Gentry) is not only morphologically variable but also chemically diverse. A study focused on determining the phenolic composition of the species (Almaraz-Abarca et al. 2009) revealed that the chemical profile of 23 individuals of one natural population varied between three and nine compounds, while individuals from another wild population had between three and four compounds. The variation in foliar phenolic profiles allowed the differentiation between the two natural populations and also identified three different chemotypes within one of the wild populations.

Considering that phenotypic variation is a result of genetic variability and the plasticity of response to environmental factors (Liu et al. 2017), the high morphological and chemical variation of *A. durangensis* reflects a high genetic variability.

The phenolic variation among different species of *Agave* is also high (Almaraz-Abarca et al. 2013a). The approaches carried out for several species of the genus have been taxonomic and for exploring the impact of phenolic compounds on human health. Table 8.1 shows the different phenolic compounds reported for several species of *Agave*, including *A. durangensis*.

Flavonoid profiles have been revealed by several authors as chemical attributes that exhibit a species-specific trend in various groups of plants, such as the genus *Physalis* (Medina-Medrano et al. 2015) and the genus *Hypericum* (Bruňáková et al. 2021). In the genus *Agave*, specific taxonomic potential was uncovered for *Agave victoriae-reginae*, *A. striata*, and *A. lechuguilla*, as these taxonomic entities were distinguished from each other based on their unique profiles. Additionally, the distinct phenolic profiles support the proposal made by González-Elizondo et al. (2011) to consider *A. victoriae-reginae* as a complex of species and to separate *A. pintilla* as an independent species. For *A. durangensis*, further analysis of natural populations is necessary to determine their phenolic composition as specific chemomarkers. Nevertheless, the high phenolic diversity reported for the species (Almaraz-Abarca et al. 2009) indicates that *A. durangensis* is a valuable source of bioactive phenolics.

8.3 Maguey Cenizo (*Agave durangensis* Gentry) Genetic Variability

The seed reproduction of *A. durangensis* is one explanation for the high genetic variability registered for the species. However, physical barriers that impede gene flow, along with the capacity for producing interspecific hybrids or genetic introgression, as established by Morales-Nieto et al. (2021), can also significantly contribute to this variability.

Currently, numerous molecular markers have been developed to estimate the genetic variability of various *Agave* species. Some markers are based on unique genome regions, while others rely on repeated regions, like microsatellites or SSR (Simple Sequence Repeats), and ISTR (inverse sequence tagged repeats). ISTR detects polymorphisms between regions that separate retrotransposons, mobile genome elements that are abundant, active, and major contributors to genetic variation in plants (Orozco-Arias et al. 2022; Paszkowski 2015). Retrotransposons replicate through RNA and are delimited by specific sequences known as LTRs (long terminal repeats) (Ramakrishnan et al. 2023). These LTR sequences have been used to design primers for detecting ISTR through PCR (polymerase chain reaction). Rhode (1996) designed universal primers for detecting these markers, which have proven to be useful in estimating genetic variability in various plant groups, such as potato (Lightbourn and Veilleux 2003) and Agavaceae species (Torres-Morán et al. 2008). While ISTR markers were utilized to estimate the genetic variability of *A. durangensis* (Almaraz-Abarca et al. 2013b), other types of molecular markers have been preferred for similar studies in different *Agave* species (Table 8.2). Figure 8.10 illustrates the variability observed in the ISTR amplification profiles among 17 individuals from a natural population of *A. durangensis*.

The study conducted by Almaraz-Abarca et al. (2013b) revealed a distinct ISTR amplification profile among the three natural populations of *A. durangensis* analyzed. These profiles were found to be different from the profile of *A. asperrima*,

Table 8.2 Molecular markers used with different aims in some species of *Agave*

Species	Molecular marker	Main contribution	References
A. americana *A. americana* var. *marginata* *A. angustifolia* var. *marginata*	SCoT* ISSR	Both types of markers are useful for the authentication of the species analyzed	El-Hawary et al. (2020)
A. americana *A. angustifolia* *A. asperrima* *A atrovirens* *A. dasyliroides* *A. filifera* *A. mapisaga* *A salmiana salmiana* *A. salmiana* Calvillito *A. salmiana* Cenizo *A. salmiana crassispina* *A. salmiana* Verde *A. stricta* *A. tequlana* Azul *A. tequilana* Chato *A. tequilana* Moraleño *A. tequilana* Siguin *A. weberi*	Six primers for the cytochrome P450 gene	Discrimination of the species consistently with the taxonomy Useful fingerprinting markers	González-Pérez et al. (2024)
A. angustifolia	AFLP	Speciation process under way. Some genetic variability is generated during asexual reproduction	Sánchez-Teyer et al. (2009)
A. angustifolia *A. tequilana* *A. rhodacantha*	SNP	Genetic groups, congruent with ethnobotanical landraces were distinguished No evidence to genetically delimitate the species	Cabrera-Toledo et al. (2022)
A. angustifolia *A. cocui* *A. tequilana*	ISSR	A closer genetic relationship between A. angustifolia and A. cocui than between A. angustifolia and A. tequilana	Dávila et al. (2007)
A. cupreata	ISSR	Species not genetically eroded, retaining their historical pattern of diversity inside populations	Aguirre-Dugua and Eguiarte (2013)

(continued)

Table 8.2 (continued)

Species	Molecular marker	Main contribution	References
A. desmetiana	SCoT ISSR	Both types of markers are useful for the authentication of the species analyzed The highest similarity was found between A. angustifolia var. marginata and A. desmetiana	El-Hawary et al. (2020)
A. durangensis	ISTR	High interpopulation genetic variability. A current advanced diversification process is observed in one population.	Almaraz-Abarca et al. (2013b)
	SSR	Moderate to high heterozygosity in four wild populations Transferability of SSR loci between populations	Torres-Morán et al. (2024)
	An inverted region, containing the genes *trnE*, *trnY* and *trnD*, in the chloroplast genome	Discrimination of the species from other species of the genus. A monophyletic group for the species, on the base of the evolutionary tree of the genus.	González-Trillo et al. (2024)
A. ferox	SCoT ISSR	Both types of markers are useful for the authentication of the species analyzed	El-Hawary et al. (2020)
A. fourcroydes	AFLP	Genetic variability through asexual reproduction	Infante et al. (2003)
	AFLP ISTR	Comparison of molecular markers to quantify genetic diversity. ISTR revealed more variability	Demey et al. (2004)
	AFLP MSAP (methylation levels)	Epigenetic changes after embryogenesis induction were revealed. Differentiation among two clonal lines before somatic embryogenesis induction	Monja-Mio et al. (2018)
A. inaequidens Koch	Microsatellites	High genetic diversity. No differences among wild, sylvicultural-managed, and cultivated populations. Domestication in an incipient stage	Figueredo et al. (2015)
A. kerchovei Lem.	Three non-coding chloroplast regions	High chloroplast genetic variability and genetic structure, low intrapopulation variability	Aguirre-Planter et al. (2020)
A. pygmae	SCoT ISSR	Both types of markers are useful for the authentication of the species analyzed	El-Hawary et al. (2020)

(continued)

Table 8.2 (continued)

Species	Molecular marker	Main contribution	References
A. potatorum	SSR	Intermediate to high interpopulation genetic diversity. Slight effects of management on genetic diversity	Félix-Valdez et al. (2016)
A. salmiana *A. marmorata*	RAPD	Micropropagation pathway (direct or indirect) determines the genetic fidelity of the regenerated plants	Martínez-Velasco and Arzate-Fernández (2022)
A. sisalana Perrine	ISSR	Higher genetic intrapopulation variability than genetic interpopulation variability. Populations are divided into two groups according to their geographical location.	Bispo dos Santos et al. (2015)
A. striata subsp. *striata* *A. striata* subsp. *falcata*	ISSR	Correlation between genetic variability and environmental variables	Trejo et al. (2016)
A. tequilana (nine varieties)	AFLP	Discrimination among varieties. Wide gene pool for the species	Gil-Vega et al. (2006)
A. tequilana	AFLP	Molecular differences between rhizome offsets and inflorescence bulbils	Abraham-Juárez et al. (2009)
	ISTR	Asexual genetic variability observed	Torres-Morán et al. (2010)
	AFLP	Genetic somaclonal variability. Existence of asexual variability. Wide genetic differences among asexually propagated individuals without correlation with phenotypic variation in response to Fusarium solani	Chávez-Sánchez et al. (2022)
A. tequilana and *A. angustifolia*	RAPD	Specific discrimination	Rodríguez-Garay et al. (2009)

Source: Prepared by Norma Almaraz-Abarca and José Antonio Ávila-Reyes
**SCoT* Start Codon Targeted

which was included in the study for comparative purposes. Additionally, the results revealed the high interpopulation level of genetic variability with one of the populations showing signs of an advanced diversification process. The significant intra and interpopulation genetic variability observed in this study has led the authors to consider the possibility that *A. durangensis* may be a complex of taxa rather than a clearly defined species. However, further research involving a large number of

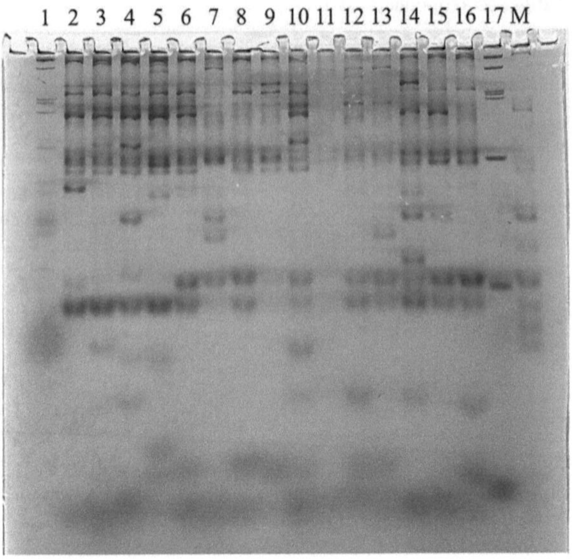

Fig. 8.10 Variability in the ISTR amplification profiles (acrylamide gel) among 17 individuals of a wild population of Maguey Cenizo (*Agave durangensis*) from Durango, Mexico. The letter M indicates molecular marker. (Photo by Norma Almaraz-Abarca)

natural populations and a variety of markers (not limited to molecular) is needed. Additionally, studies on the reproductive biology of the taxon are necessary to either support or refute this proposal.

8.4 Maguey Cenizo (*Agave durangensis* Gentry) Fiber

The use of agave fibers in Mexico for the manufacturing of fabrics and ropes dates back to pre-Hispanic times and continues today among several original ethnic groups, who use and preserve their traditional knowledge. Currently, the use of fibers from certain species of the *Agave* has significantly diversified. Plant-based fibers are now being explored as an alternative for the production or reinforcement of building materials. The thermal properties, noise isolation capabilities, humidity modulation, low cost and eco-friendliness of the production process, make plant fibers attractive for the development and innovation of construction materials (Saini et al. 2024) to meet the needs of the growing human population.

Research on the mechanical properties of sisal (*Agave sisalana*) fibers has revealed their potential as an additive to enhancing the strength and durability of 3D printing construction materials (Zavaleta et al. 2024). Fibers from this species, when combined with other natural materials, have been found feasible for the production of reinforced epoxy composites for use in construction and automotive materials (Natarajan et al. 2024).

Agave americana and *A. tequilana* are two other species of the *Agave* that have been studied as sources of fiber for various purposes. *Agave americana* cellulose, when combined with magnetic components, has been identified as a suitable material for the production of nanocomposites for fluoride removal from water (Majamo et al. 2024). Research on the tensile resistance and elasticity of the fibers from the flower stems of *A. americana* revealed these structures as a source of durable natural fiber (Lalaymia et al. 2024). Additionally, the potential of *A. tequilana* bagasse fiber as a material for constructing lithium-sulfur batteries was reported by Arvizu-Rodríguez et al. (2024).

The high morphological, chemical and genetic diversity of *A. durangensis* presents an opportunity to select alleles for establishing plantations with desirable characteristics, such as plant size, soluble carbohydrate content, or bioactive compound content, to support mescal production and pharmaceutical and fiber manufacturing. While the potential of *A. durangensis* as a fiber source has been minimally explored, Kumar and Ram (2021) have already noted its potential due to the large amount of discarded leaves during the mescal production process, estimated at 1200 tons annually (Contreras-Hernández et al. 2018). The size of *A. durangensis* leaves suggests that this species could be a valuable fiber source (Fig. 8.11); however, further studies on the length, diameter, color, thickness, linear density (mass/unit length), and strength, among other attributes of its fibers are necessary, similar to those conducted on other *Agave* species, like *A. americana* (Hulle et al. 2015), *A. salmiana* (Reyes-Samilpa et al. 2023), and *A. cantala* (Sumarago et al. 2024).

Fig. 8.11 Leaf fiber of Maguey Cenizo (*Agave durangensis*). (Photo courtesy of Vicente Hernández Vargas, Interdisciplinary Research Center for Integral Regional Development, Durango, National Polytechnic Institute, Mexico)

Fig. 8.12 Bagasse fiber of Maguey Cenizo (*Agave durangensis* Gentry) as a waste material from the mescal industry. (Photo courtesy of Rene Torres Ricario, Interdisciplinary Research Center for Integral Regional Development, Durango, National Polytechnic Institute, Mexico)

A study on the chemical composition of *A. durangensis* bagasse fiber showed that its holocellulose and lignin content are suitable for the elaboration of particleboards, whether combined or not with wood particles (Moreno-Anguiano et al. 2021). Figure 8.12 shows the bagasse fiber of *A. durangensis* as a waste material from the mescal industry. Another study explored the potential of *A. durangensis* bagasse (the waste generated after boiling and crushing the agave heads in the production process of mescal) in the manufacturing of medium-density fiber boards, which have several non-structural applications. The physical and mechanical characterization (Fig. 8.13) of those boards indicated that the fiber of *A. durangensis* bagasse has the potential to be used as a complement in the manufacturing of wood-based medium-density fiber panels (Moreno-Anguiano et al. 2022).

The lignocellulosic residue of bagasse from the leaves of *A. durangensis* also has potential for the production of polyhydroxybutyrate (PHB) when used as a substrate for *Bacillus cereus* strain 4 N. PHB is a material with the potential to replace polypropylene and polyethylene, as revealed by Martínez-Herrera et al. (2021).

Our research group determined the potential of the lignocellulosic fiber of *A. durangensis* (from the waste of the mescal industry) to obtain high-quality activated carbon to elaborate a nanocomposite combined with chitosan, with the

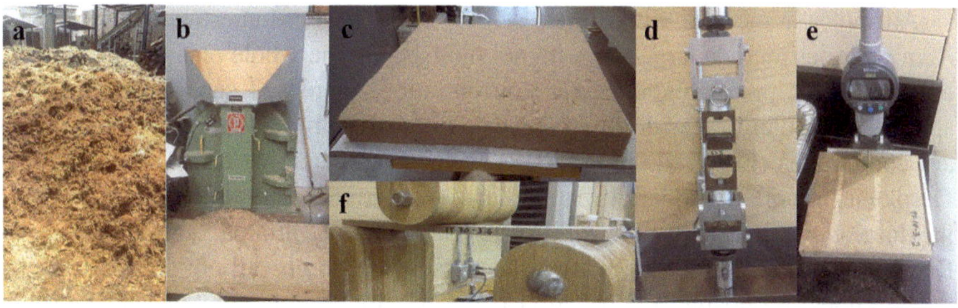

Fig. 8.13 Board preparation from Maguey Cenizo (*Agave durangensis* Gentry) bagasse fiber. (**a**) Fiber collection, (**b**) Ground fiber, (**c**) Bagasse fiber panel, (**d**) Internal cohesion test, (**e**) Thickness swelling measurement and (**f**) Bending test. (Photos courtesy of Oswaldo Moreno-Anguiano, Institutional Doctorate Program in Agricultural and Forestry Sciences, University Juárez of Durango State, Mexico)

capacity to remove oxytetracycline and enrofloxacin from water (Molina-Amaya et al. 2024). The nanocomposite obtained represents a feasible economic alternative to remove these two antibiotics, which are considered difficult to remove emerging contaminants by some authors (Lin et al. 2024; Zhang et al. 2024).

The results of the studies mentioned above give an idea of the diversity of uses that the leaf fiber and bagasse fiber of *A. durangensis* can have and should encourage further explorations on the topic.

8.5 Domestication

Despite having accompanied various American human groups in their technological and cultural development for about 12,000 years (Álvarez-Ríos et al. 2020), *Agave* species have not undergone an intense domestication process like other plant species, such as corn. The species of *Agave* are still rustic and it is difficult to perceive domestication syndrome traits. The long time it takes to reach maturity and fructify, which occurs after 4–25 years, depending on the species (González et al. 2009), as well as the monocarpic condition of many species in the genus and their morphological attributes (very tall inflorescences and leaves with jagged edges and terminal spines arranged in a rosette) have hindered controlled pollination. Some species in the genus are cultivated but have not been domesticated. Figueredo et al. (2014) consider *A. tequilana*, *A. fourcroydes*, *A. angustifolia*, and *A. rhodacantha* Trel. as domesticated, based on phenotypic divergence from their putative ancestors. More recently, cultivated varieties of *A. americana*, *A. salmiana*, and *A. mapisaga* were incorporated into the domesticated status because they are larger plants, bear fewer lateral denticles on the leaves than their wild forms, characteristics considered part of the domestication syndrome (Figueredo-Urbina et al. 2021).

Attributes associated with human selection during the domestication process of *Agave* species include larger size, higher biomass (of the whole plant or useful parts like leaves or pine), higher sugar concentration, increased sap production, less spikiness (smaller or less dense of spines), and lower production of caustic secondary compounds (Figueredo et al. 2018) such as saponins that cause skin irritation. Despite its importance as a raw material in mescal production, studies contributing to the domestication process of *A. durangensis* are currently limited. One such study by Delgado-Alvarado et al. (2019) focused on selecting high carbohydrate-producing genetic lines, proposing a correlation between carbohydrate content and morphological features as a tool to identify outstanding plants for mescal production. Another study by Pérez-González et al. (2021) explored the brix degrees concentration of seedlings of *Agave durangensis* and two other species of the genus, associating foliar morphology with accumulated carbohydrates. More recently, Luna-Solis et al. (2024) revealed the techno-functional properties of the foliar agavines of *A. durangensis*, proposing them for food, cosmetic, and pharmaceutical applications. A third study by Ramírez-Malagón et al. (2008) developed reliable in vitro propagation protocols for *A. durangensis* and five other *Agave* species. Despite these important contributions, more studies are needed to develop the domestication and improvement of *A. durangensis*.

Other species in the genus have been studied for their ecological, physiological, and genetic aspects, leading to the development of biotechnological tools for selecting desirable genetic lines. For example, Delgado-Aceves et al. (2021) reported in vitro techniques to induce indirect regeneration of somatic embryos to obtain and safely cryopreserve material for long-term commercial propagation, elite lines production, and genetic transformation programs of *A. tequilana*. The study by Gomez-Vargas et al. (2022) identified candidate genes of the regulatory mechanism of fructan metabolism in *A. tequilana*.

Recently, Vásquez-Hernández et al. (2023) reported that few studies are using in vitro mutagenesis for the genetic improvement of *Agave* species. Among the few that have received attention are *A. americana* L., *A. angustifolia* Haw., *A. marmorata* Roezl., *A. sisalana* Perr., *A. tequilana*, *A. guiengola* Gentry, and *A. salmiana*. These represent less than 3% of the species in *Agave sensu lato*.

8.6 Current Cultivation Practices and Challenges

The current cultivation of *A. durangensis* is still in its early stage because until a few years ago the supply of plants for the mescal industry mainly came from natural populations. However, the growing demand for this beverage (Dévora-Rodríguez et al. 2021) has prompted producers and governmental research institutions to conduct studies supporting breeding programs to intensify and expand plant production areas. This supply raw material while also conserving natural populations of the species. Information regarding the institutions involved in this effort is provided in Appendix 8.1.

Fig. 8.14 Harvest of Maguey Cenizo (*Agave durangensis* Gentry) plants. The process involves removing the leaves (a process called "jimado"). (Photo by José Antonio Ávila-Reyes)

In the monocarpic species of *Agave*, prior to flowering, an intense carbohydrate accumulation (fructans) occurs in the "piña" of the plants, which is important to producing mescal. To harvest the piñas, plants are previously removed from their leaves, a process called "jimado" (Fig. 8.14).

While any species of *Agave* can be used to produce mescal, larger species accumulating higher levels of fructan are preferred. Producers select those species that provide an appropriate aroma and flavor to the final product, although the accumulation of carbohydrates and organic compounds that contribute to aroma and flavor can be influenced by climate and soil composition (Vera-Guzmán et al. 2018). In Durango, Mexico, *A. durangensis* is the preferred species for mescal production. *Agave durangensis* takes approximately 8 years to develop its inflorescence or scape (though some individuals may take up to 16 years). The scape must be removed and the plants left for several days to several months, depending on the empirical criteria established by each producer (Valenzuela-Ruiz et al. 2003), to achieve higher levels of fructan accumulation. The interruption of the reproductive cycle caused by scape removal and the long life cycle of *Agave* species used for alcoholic beverages are two aspects that conservation and breeding programs must take into account. Additionally, since sexual reproduction is the main perpetuation strategy for *A. durangensis*, studies on its reproductive biology are essential.

Around 20 years ago, the establishment of nurseries in the natural distribution area of *A. durangensis* began to produce plants for establishing plantations. According to Barraza-Soto et al. (2014), plants from nurseries take a shorter period (5 years) to develop the floral scape, reducing time and costs for producers.

8.6.1 Conventional Cultivation Practices

Nurseries have been established to produce plants of *A. durangensis* (Fig. 8.15) using seeds collected from natural populations. The seeds are sown in bags or seedbeds, using conventional nursery methods. After a year and a half, seedlings are ready to be transplanted into the field for plantation establishment (Barraza-Soto et al. 2014). Plantations are typically established in lands with organic-rich soils and medium to sandy textures, with plants spaced 1.5–3 m apart (Fig. 8.16). The average plantation density is 2000 plants per hectare (Valenzuela-Ruiz et al. 2003).

Fig. 8.15 Plantlets (**a**) and young plants (**b**) of Maguey Cenizo (*Agave durangensis* Gentry) growing in a nursery of Durango, Mexico. (Photos courtesy of Ricardo Martínez, Interdisciplinary Research Center for Integral Regional Development, Durango, National Polytechnic Institute, Mexico)

Fig. 8.16 Maguey Cenizo (*Agave durangensis* Gentry) plantation in Durango, Mexico. (Photo by Norma Almaraz-Abarca)

In sexual propagation, understanding the germination behavior and seed quality is important. The effect of different temperatures on the germination of *A. durangensis* seeds has been studied. Orea et al. (2006) found that 25 °C was the optimal temperature for germination, leading to faster growth of seedlings compared to those germinating at higher or lower temperatures.

The quality of the seeds of *A. durangensis* was explored by Barriada-Bernal et al. (2013) across three natural populations. Interpopulation variations in morphometric, germination, and biochemical indicators of seed vigor were revealed, with seeds from the population, Veracruz identified as having the highest vigor. This study also revealed population-specific morphological attributes and population-specific physiological, chemical, and biochemical seed vigor indicators. Additionally, the chemical and biochemical markers showed a high correlation with germination markers, suggesting that evaluating chemical and biochemical markers as indicators of seed vigor can help in selecting seed lots with high germination performance. The results of the study by Barriada-Bernal et al. (2013) suggested that the variability of the natural populations of *A. durangensis* is a valuable source of alleles for the genetic improvement of the species.

The germination potential of *A. durangensis* seeds has also been evaluated in the field under biotic (mainly *Dodonea viscosa* cover) and abiotic (rock cover) conditions (Candia-Acosta et al. 2019). In field conditions, the maximum germination recorded for *A. durangensis* seeds was less than 10%, under rock cover (Candia-Acosta et al. 2019). This value, compared to that obtained under nursery conditions (between 99% and 100%) reported by Barriada-Bernal et al. (2013) highlights the

Fig. 8.17 Preparation of growth trays for young seedlings from germinated seeds of Maguey Cenizo (*Agave durangensis* Gentry). (Photo courtesy of Ricardo Martínez, Interdisciplinary Research Center for Integral Regional Development, Durango, National Polytechnic Institute, Mexico)

importance of establishing nurseries to meet the demand for plants for the mescal industry and conservation programs of *A. durangensis*. Figure 8.17 shows young seedlings from germinated seeds of Maguey Cenizo (*Agave durangensis* Gentry).

The impact of fertilization on the development of *A. durangensis* seedlings was explored by Orea et al. (2006). These authors found that the nutritional requirements of *A. durangensis* seedlings begin 12 weeks after sowing, enhancing plant growth.

To enhance growth and yield, Hernández-Vargas et al. (2015) tested the effect of different doses of three types of manure (cattle, sheep, and vermicompost) on the growth and yield of *A. durangensis* plants, suggesting a vermicompost-based fertilization regimen to stimulate biomass accumulation and plant growth.

Rosales et al. (2014) developed a successful and cost-effective technique for producing *A. durangensis* plants under protected conditions by implementation a mulch and drip irrigation system.

The results of the studies commented on above have supported the establishment of *A. durangensis* plantations in different regions within its natural distribution area. Currently, there is a trade in plants produced in nurseries that aid in establishing plantations providing producers with agricultural information to maintain these plantations.

Results of our research group revealed the high capacity of *A. durangensis* to tolerate saline stress during germination and seedling growth (Gallegos-Hernández et al. 2024), which makes it a species of ecological importance for forestation and reforestation programs in ecosystems with saline soils. This is significant as soil salinity has become a major agricultural problem (Hapani and Marjadi 2015) affecting over 800 million hectares of arable land globally (Acosta-Motos et al. 2017). This limits the establishment of traditional crops, particularly in arid and semiarid regions, where saline soils are predominant and the lack of precipitation is insufficient for leaching (Soni et al. 2017).

8.6.2 Alternative Cultivation Practices

The challenges posed on agricultural production by current climate change and the increasing human population have encouraged governments, and scientific and technological institutions to develop innovative agricultural practices. The significance of adopting alternative agricultural practices for *A. durangensis* and for all cultivated species of *Agave*, in general, is a topic that needs further exploration. Some alternative agricultural practices are agroforestry, vertical farming, and digital agricultural techniques.

8.6.2.1 Agroforestry

Agroforestry was recognized as a recent innovative technology by Konfo et al. (2024). This type of farming has been used in Mexico since pre-Hispanic times (Torres-García et al. 2019). Agroforestry can be understood as the simultaneous cultivation of broad cover species (trees) and horticultural or limited cover species in the same crop area, the former playing different roles in the system, such as living fences and sources of different products. Agroforestry contributes to the conservation of agrobiodiversity, traditional agricultural knowledge, and the reduction of pollution due to the reduced use of agrochemicals, as well as the efficient use of crop land, with a consequent economic improvement. Thirty-one species and four variants of *Agave* have been reported in agroforestry systems in Mexico (Torres-García et al. 2019). *Agave durangensis* was not mentioned; however, given the benefits of this system, establishing agroforestry systems for this species should be encouraged. Recently, Rosales-Serna et al. (2023) evaluated the production of *A. durangensis*, a common bean variety and an oat variety, under an agroforestry system, with three fertilization conditions. The authors concluded that the agroforestry system is a viable option to enhance production, diversify crops, and meet immediate nutritional needs.

8.6.2.2 Vertical Farming

Vertical farming is another plant cultivation technology, designed and developed mainly for the establishment of urban agriculture. The advantages and types of this kind of agriculture have been commented on by several authors (Al-Kodmany 2018;

Silva et al. 2023). While vertical farming is suitable for horticultural production, it may not be appropriate for large plant species, with extensive canopies, such as several species of *Agave*, including *A. durangensis*. Cultivating these species under this system may not be feasible.

8.6.2.3 Digitalized Agriculture

The digitalization of agriculture is a practice increasingly used around the world. For instance, the digital systematization of irrigation and fertilizer supply in crop fields as well as the systematization of protected agriculture systems. The benefits of modernizing agriculture with digital technologies are undeniable, as highlighted by various authors (Konfo et al. 2024; Zolkin et al. 2024).

The cultivation of *A. durangensis* stands to gain from technological advancements in agriculture. However, several challenges must be addressed for this to occur. These challenges include (1) the need for more scientific knowledge about the reproductive biology and technical management of breeding and production, (2) the cost associated with modernizing production, and (3) resistance to change from producers, who based on their traditional knowledge, have established successful plantations.

8.6.3 Artificial Intelligence

Tools of artificial intelligence have an important potential for the cultivation, breeding and improvement of *Agave* species. These tools have been utilized to evaluate algorithms predicting reforestation areas for *A. durangensis* in Durango, Mexico as well as supporting population density estimations and early disease detection for the species (Escobar-Flores et al. 2022; Escobar and Sandoval 2022). The results obtained by these studies have implications for selecting plantation sites and managing them effectively.

8.7 Biotechnology Strategies for Genetic Improvement

To overcome the natural obstacles to genetic improvement using conventional methods, such as the life cycle and long inflorescence of *Agave* species, which make pollination control difficult, it is necessary to develop alternative strategies for genetic manipulation more shortly and simply.

The capacity of reprogramming gene expression, through which plants can regenerate tissues, organs, and whole individuals, named totipotency or pluripotency, is essential for their survival and perpetuation, and it has been used for different purposes by humans. Plant gene expression reprogramming involves different mechanisms. Cell reprogramming can be modulated by transcription factors, which promote changes in the expression of a considerable number of genes, inducing callus formation and organogenesis in wounding plant tissues (Ikeuchi et al. 2017). Epigenetic changes, such as histone acetylation, have also been revealed as inductors of plant cell reprogramming by modulating gene expression (Rymen et al. 2019). In vitro culture of plant cells and tissues is a proper technique for reprogramming gene expression and genetic transformation.

8.7.1 *In Vitro Plant Cell and Tissue Culture*

8.7.1.1 Micropropagation

In vitro plant cell and tissue culture, particularly micropropagation, is an option for (a) the maintenance of elite individuals bearing highlighting agronomic and industrial attributes, (b) the massive production of plants, and (c) plant genetic improvement, thus in vitro plant cell and tissue culture can support micropropagation and breeding programs.

Since the beginning of the twentieth century, the technique of plant cell culture started its development, taking advantage of the different totipotency expressions of plant cells under different environmental stimuli. Protocols for producing agave vitro plants, either directly (through direct organogenesis) or indirectly (through callus intermediary), were developed in the 1970s. Robert et al. (1987) mentioned that in 1977 Groenewald et al. achieved callus induction and plant regeneration from seed fragments of a species of *Agave*. Thereafter, work continues to date either to develop protocols for previously unconsidered species or to improve existing protocols for species of commercial interest. Some species of the genus *Agave* for which in vitro protocols have been developed to produce vitro plants are *A. fourcroydes* (Robert et al. 1987), *A. cantala* (Binh et al. 1990), *A. parrasana* Berger (Santacruz-Ruvalcaba et al. 1999), and *A. tequilana* (Valenzuela-Sánchez et al. 2006). Table 8.3 shows the species of *Agave*, for which micropropagation protocols have been recently developed or improved.

For developing tissue culture-based strategies supporting conservation and massive propagation programs for a particular plant species, it is necessary to evaluate, because the success depends on them: (1) the genotype of the donor plant, (2) the type of explant (plant part or type of tissue), (3) explant density, (4) growth medium composition and growth regulators balance, and (5) incubation conditions and duration of incubation (Ahmad et al. 2021; Monja-Mio et al. 2021). Detailed studies to reveal the impact of all these aspects on micropropagation have been done for some species, such as Chamomile (*Matricaria chamomilla* L.) (Ahmad et al. 2021). The data shown in Table 8.3 indicate that different types of explants can be used directly and indirectly to reach the micropropagation of several species of *Agave*; however, the achievements are very variable. This suggests that despite the important contributions already made, different aspects, that determine success in

Table 8.3 Species of *Agave* for which micropropagation protocols have been newly developed or improved

Species	Type of explant	Micropropagation via	Achievements	Reference
A. americana	Segments of the basal part of shoots extracted from rhizomes	Direct and indirect somatic embryogenesis	87.5 shoots/explant by direct embryogenesis	Naziri et al. (2019)
A. angustifolia	Shoots from clonal lines obtained using a previous protocol	Direct organogenesis	6.23 shoots/explants	Monja-Mio et al. (2021)
	Sproutings from the stem of seedlings	Direct organogenesis	729 shoots/explant 100% ex vitro survival	Sánchez et al. (2020)
A. guiengola Gentry	Axillary sprouting from stem segments	Direct organogenesis	43 shoots/explant 100% ex vitro survival	Chávez-Ortiz et al. (2021)
A. fourcroydes	Vitro plants previously obtained through micropropagation of adventitious shoots	Direct somatic embryogenesis	Description of the thin layer technique as a more efficient method of micropropagation for the species	Monja-Mio et al. (2022)
A. marmorata Roezl.	Sygotic embryonic axis of seeds Meristematic zone of seedlings germinated in vitro Meristematic zone of nursery generated seedlings	Callogenesis Direct organogenesis	26.7 shoots/explant (indirect organogenesis) 100% ex vitro survival	Arzate-Fernández et al. (2020)
	Seedlings from clonal lines	Direct organogenesis	83.4 shoots/explant 100% ex vitro survival	Martinez-Rodriguez et al. (2022)
	Stems of 1-year-old plants	Direct organogenesis	21.40 shoots/explant 99% ex vitro survival	Mancilla-Álvarez et al. (2024a)
A. maximiliana Baker	Apical meristems	Direct organogenesis	26.93 shoots/explant 100% ex vitro survival	Santacruz-Ruvalcaba et al. (2022)
A. peacokii	Rhizomatous shoots	Direct organogenesis	87 shoots/explant 100% ex vitro survival	Delgado-Aceves et al. (2022)

A. potatorum Zucc.	Individual shoots	Direct organogenesis	9.87 shoots/explant 98%–100% ex vitro survival	Ramírez-Mosqueda et al. (2022)
	Stems (about 1-year old)	Direct organogenesis	30.73 shoots/explant 98.67% ex vitro survival	Correa-Hernández et al. (2022)
	Stems proximal area of cotyledons Proximal area of leaves of in vitro germinated seedlings	Direct and indirect organogenesis	81 shoots/explant (indirect organogenesis) 50 shoots/explant (direct organogenesis) 71% ex vitro survival	Téllez et al. (2023)
	Two shoots joined, previously obtained through micropropagation	Direct organogenesis under different conditions of temporary immersion conditions	17.33 shoots/explant 98–100 ex vitro survival	Mancilla-Álvarez et al. (2024b)
A. salmiana	Sygotic embryonic axis of seeds Meristematic zone of seedlings germinated in vitro Meristematic zone of nursery germinated seedling	Callogenesis and direct organogenesis	23.8 shoots/explant by direct organogenesis	Arzate-Fernández et al. (2020)
A. tequilana cultivar Chato	Young leaves of rhizomatous shoots in vitro cultivated	Indirect somatic embryogenesis	83% post-cryoconservation regrowth 70% plant conversion rate	Delgado-Aceves et al. (2021)
A. tequilana cv. azul	Segmented stems of plantlets	Direct organogenesis	26.7 shoots/explant	Vázquez-Martínez et al. (2022)

Source: Prepared by Norma Almaraz-Abarca and José Antonio Ávila-Reyes micropropagation, still need to be explored for the species of the genus. The review of the publications cited in Table 8.3 indicates that each species of *Agave* has particular requirements for developing in vitro plants depending on the type of explant and micropropagation strategy.

The survival of in vitro plants in ex-situ conditions is a bottleneck in the micropropagation protocols of many species. However, the data shown in Table 8.3 indicate that this is not the case for several species of *Agave*, such as *A. angustifolia*, *A. guiengola*, *A. marmorata*, *A. maximiliana*, *A. peacokii*, and *A. potatorum*, for which the protocols used for the in vitro plant production have ensured the production of enough robust individuals to survive under ex vitro conditions and/or that the ex vitro preparation practices have worked well with those species.

The requirements for *A. durangensis* began to be explored about 15 years ago. Table 8.4 shows the achievements of the research concerning this species.

The data presented in Table 8.4 suggest that *A. durangensis* is a recalcitrant species to micropropagate and further studies are needed to develop successful in vitro protocols to support conservation and massive propagation programs for the species. Currently, our research group is exploring the requirements for producing in vitro plants of *A. durangensis* through both direct and indirect organogenesis.

Table 8.4 Micropropagation protocols developed for *Agave durangensis*

Type of explant	Micropropagation via	In vitro conditions	Achievements	Reference
Offsets previously multiplied in vitro	Direct organogenesis	Shoot generation: Semisolid MS culture medium with modified concentrations of N + 0.049 μM IBA + 4.44 μM BA	100% explants generated 5.9 shoots/explant The authors only comment that in vitro plants were easily hardened and established in soil	Ramírez-Malagón et al. (2008)
Shoots in vitro cultivated	Direct organogenesis	Shoot generation: Semisolid 100% MS culture medium + 0.3 mg/L IBA + 3 mg/L kinetin Root development: 100% MS culture medium + 0.3 mg/L IBA + 0.1 mg/L kinetin	4.33 shoots/explant 100% root induction	González and Núñez (2015)
Foliar parts of in vitro-generated seedlings	Direct organogenesis	Shoot generation: 100% MS culture medium + 4.0 mg/L BAP, without auxins.	1.4 shoots/explant after 75 days of incubation. The authors gave no information about root development	Dévora-Rodríguez et al. (2021)

Source: Prepared by Norma Almaraz-Abarca and José Antonio Ávila-Reyes

Fig. 8.18 *Agave durangensis* callus (**a**) and rooted shoot (**b**) obtained under different growth regulators balance. (Photo courtesy of Heberto Salas Ayala, Interdisciplinary Research Center for Integral Regional Development, Durango, National Polytechnic Institute, Mexico)

Additionally, given the richness of bioactive phenolic compounds in the species (Almaraz-Abarca et al. 2009), we are exploring the effect of different in vitro growth conditions on the synthesis and accumulation of these compounds from friable callus to in vitro plants (Fig. 8.18).

Micropropagation can lead to molecular and/or epigenetic variability (Duta-Cornescu et al. 2023), which is reflected in phenotypic differences between in vitro plants and the mother plant (somaclonal variation). Depending on the objective of micropropagation, somaclonal variation can be either advantageous or disadvantageous. This topic has been extensively covered by Duta-Cornescu et al. (2023) in the context of medicinal plants. We believe that somaclonal variation offers an opportunity for selecting genotypes with desired characteristics and can represent a very useful biotechnological tool for the genetic improvement of plants with characteristics and life cycles such as those of *Agave* species.

Both, genetic and epigenetic variability have been corroborated by in vitro plants of some *Agave* species, such as *A. tequilana* by Díaz-Martínez et al. (2012). They proposed a possible correlation between methylation patterns, expression of transposable element-associated genes, and somaclonal variation in the species. Epigenetic changes, specifically changes in DNA methylation, modulated by nutrient and exogenous cytokinin levels in the culture media for long periods, were found to be related to phenotypic variation in *A. angustifolia* (Duarte-Aké and De-la-Peña 2021). The importance of the type of organogenesis (direct or indirect) in generating somaclonal variation was revealed by Martínez-Velasco and Arzate-Fernández (2022) for *A. salmiana* and *A. marmorata*, with indirect organogenesis producing the greatest somaclonal variation. The importance of genotype in determining methylation/demethylation events was revealed by Monja-Mio et al. (2018) for *A. fourcroydes* clonal lines. This information commented above gives an idea of the diversity of factors that determine somaclonal variability in some *Agave* species, and as in other aspects, the exploration of these factors for a broader range of species of the genus is required. For *A. durangensis*, investigations focused on the subject would provide valuable information to reveal the conditions for maintaining genetic fidelity in vitro and the conditions under which selection for specific desirable phenotypic traits could be generated.

8.7.1.2 Embryo Rescue and Culture

Embryo culture is a methodology derived from the in vitro culture of cells and tissues. Embryo rescue and culture involve the aseptic removal of developing embryos directly from reproductive structures and their subsequent cultivation in vitro to develop into seedlings (Li et al. 2015).

Many factors determine the success of in vitro embryo culture, which are practically the same as those mentioned in Sect. 8.7.1.1 of this chapter, for micropropagation.

The technique has important implications for plant improvement, contributing to reducing the frequency of abortive events, improving the survival of hybrids, and facilitating the selection of genetic variants.

To overcome the obstacles imposed by the natural morphology, structure, and life cycle of *Agave* species in the practice of controlled pollination, Rodríguez-Garay et al. (2018) developed a hybridization technique called "in casa pollination and embryo rescue". In this practice, manual pollination is carried out using pollen from the desired origin (plants from different populations and even different *Agave* species), in flowers kept in the panicles of the inflorescences cut from the whole plants. The cut panicles or inflorescences are left to mature inside a greenhouse or laboratory. Subsequently, when the fertilized ovules begin to develop into embryos, they are aseptically recovered and cultured in vitro. Details of the technique can be found in the report by Rodríguez-Garay et al. (2018).

The in-casa pollination and embryo rescue methodology requires procedures to identify hybrid formation. According to our experience, any of the molecular markers described in Table 8.2 of this chapter can be useful for this purpose. Many of these markers generate species-specific amplification profiles, which would assist in identifying hybrid individuals.

Among the *Agave* species for which the technique has been successfully applied is *Agave salmiana* subsp. salmiana. For this species, an in vitro protocol of plant regeneration was developed by Ángeles-Vázquez et al. (2023) using indirect somatic embryogenesis through callus formation and zygotic embryos obtained by "in casa pollination" as explants.

Much work still needs to be done to develop embryo rescue and culture protocols for most *Agave* species, as these protocols can be species-specific, as are their micropropagation protocols in many cases. We consider that the technique can be very relevant for the improvement of *A. durangensis*, to select genetic variants produced by hybridization between individuals from genetically variable populations. This is because the technique makes pollination control much easier and the in vitro procedure can produce a high number of seedlings in a shorter time.

8.7.2 In Vitro Genetic Modification

In vitro culture is a valuable tool for plant genetic modification. For species with life cycles and characteristics like *Agave* species, in vitro genetic transformation could be a useful alternative to accelerate the generation of new genotypes. However, Mexico, as the center of origin of *Agave* (due to its rich and diversity of species) (García-Mendoza 1995) and the high endemism level in the country (García-Mendoza et al. 2019), as well as the sexual reproductive mechanism of several species of *Agave*, the development of genetically modified plants of the genus must be carried out with the greatest caution to protect this natural resource, which is part of the natural heritage of Mexico. This aspect is very cautiously taken care of by the country's government, due, moreover, to the fact that patenting of genetically modified organisms limits and/or makes more expensive the use of resources that have been naturally and ancestrally used by many communities and ethnic groups in the country.

To speed up the breeding process, changing the genetic composition of plants to obtain and select desirable phenotypic attributes, in vitro genetic transformation is a suitable alternative. Currently, transformation protocols have been developed for only a few species of *Agave*. For instance, Flores-Benítez et al. (2007) compared two transformation methods, co-cultivation with *Agrobacterium tumefaciens* and particle bombardment, to integrate *nptII* and *bar* genes in *Agave salmiana*, finding that the transformation via co-cultivation of explants with *A. tumefaciens* was the most effective method of transformation. Additionally, Gao et al. (2014) achieved the integration and expression of a heaven-like gene (forming part of a family of plant antimicrobial peptides) with antifungal activity from seeds of *Pharbitis nil* (Convolvulaceae) in the *Agave* hybrid No. 11648 (H.11648) [(*A. amaniensis* × *A. angustifolia* Haw.) × *A. amaniensis*] for enhancing tolerance against zebra stripe disease (caused by *Phytophthora nicotianae*).

Given that, as previously reviewed in the current paper, the requirements to produce micro-plants can be highly variable from species to species, likely, the requirements to insert, modify, and express genes are also highly variable among the species of *Agave*. This suggests that a lot of effort should be made to develop successful specific protocols.

8.7.3 Genome and Epigenome Editing

Regulation of gene expression is an option for the generation of transgenic plants for plant breeding. Genome editing and epigenome editing have been recently developed for the modulation of gene expression. These technologies, like genetic transformation and somaclonal variation, are assisted by cell and tissue culture, but unlike these last ones, the insertion of foreign genetic material is not involved and the genetic modifications are targeted and exact, nonrandom as those caused by physical and chemical mutagenic agents.

Genome editing uses site-specific nucleases for inducing, mainly in the promoter region of genes, double-stranded breaks in a predetermined site in DNA, in order to permanently insert, delete or substitute sequences, after inducing endogenous DNA repair systems (Miglani et al. 2020). Epigenome editing induces changes at specific sites of the epigenome, by using an effector domain, which is part of a protein that can covalently attach or remove a selected epigenetic mark on DNA or histone at a targeted locus (Kungulovski and Jeltsch 2016).

To apply genome and epigenome editing in plant breeding, it is necessary to have extensively explored the genome and epigenome of the plant species of interest. These aspects have been little explored for most *Agave* species. However, relevant knowledge has been obtained for *A. fourcroydes* and *A. angustifolia* (de-la-Peña et al. 2012; Díaz-Martínez et al. 2012). Other contributions have been published recently. For instance, the determination of the size, organization, and sequence of the chloroplast genome of *A. durangensis* was revealed by González-Trillo et al. (2024), whereas the molecular mechanisms underlying the chloroplast biogenesis of *A. angustifolia* was unveiled by Andrade-Marcial et al. (2024). These advances suggest that successful genome and epigenome editing protocols for these species will likely be developed soon. However, for *A. durangensis*, extensive and intensive research on genetic variability, molecular markers, epigenetic variation, gene expression modulation and micropropagation protocols is still necessary before genome and epigenome editing methods can be developed.

8.7.4 Hybridization

Natural interspecific and subgeneric hybridization in *Agave* is common (Figueredo-Urbina et al. 2021) and has been documented approximately 57 years ago (Gentry 1967). The ecological and evolutionary implications of hybridization have been fundamental in modeling the current composition of the genus. Despite the abundance of natural hybrids in this group of plants, only a few have achieved formal taxonomic and horticultural recognition. Among these hybrids are *A.* × *arizonica* Gentry & J.H. Weber (*A. chrysantha* Peebles × *A. toumeyana* var. *Bella* Breitung), *A.* × *ajoensis* W.C. Hodgs. (*A. simplex* (Gentry) Salywon & W.C. Hodgs. × *A. schottii* var. *Schottii* Engelm.), and *Agave parryi* var. *Huachucensis* × *Agave palmeri* (Roll et al. 2021).

An example of a hybrid obtained expressly by cross-pollination between *A. angustifolia* and *A. amaniensis*, and backcrossed with *A. amanuensis*, in Tanzania, resulted in plants that produced longer fibers than *A. sisalana*. This is the hybrid H11648, selected for its long fibers and spine-free leaf margins. However, hybrid 11648 was very susceptible to fungal diseases and pest attacks (Kimaro et al. 1994). Even so, the hybrid is the only one highly produced and marketed in China, where studies to determine both the genomic and CAM pathway evolution of the hybrid across many lineages have been carried out (Yang et al. 2023), highlighting the activity of different LTR retrotransposon lineages in the conformation of its genome. The contribution of differential accumulation of LTR retrotransposons found in the genome of 11648 led Ramos et al. (2023) to propose that this has made an important contribution to the bimodal karyotype of the genus *Agave*.

Another interesting type of hybrid, involving *Agave* species, is generated by artificial intergeneric crosses between *Agave* and *Manfreda* (the latter currently considered a subgenus of *Agave*), called Mangaves. These hybrids have shorter flowering times than *Agave* and many have lost the monocarpic condition. Only one natural Mangave has been reported and was found in Mexico (*Manfreda variegata* × *Agave mitis*). Mangaves have been maintained and propagated by micropropagation (Walker 2023).

Roll et al. (2021) reported that the persistence of *Agave* hybrids in natural populations is largely maintained by vegetative reproduction through the production of rhizomes, as the germination rate for hybrid seeds is low (6% for *Agave parryi* var. *Huachucensis* × *Agave palmeri*). This suggests that for species, whose reproductive mechanism is essentially by seeds, like *A. durangensis*, the development of in vitro protocols, like embryo rescue and culture, would be very useful for the rescue of hybrid embryos, genetic improvement and development of varieties or genetic lines with outstanding attributes.

Biotechnological tools have also contributed to solving the fungal susceptibility problem of the hybrid H11648 (Gao et al. 2014), as discussed above in in this same Sect. 8.7.4 of this chapter.

The results discussed above, although still few, allow us to glimpse a promising panorama for improvement with biotechnological tools for *Agave* species and suggest a broad diversification scenario of potential characteristics to select from artificial and natural hybrids of the genus.

8.7.5 Germplasm Biodiversity and Conservation

8.7.5.1 Ex-situ Conservation: Botanical Gardens and Nurseries

One of the goals of natural resource conservation is to systematically perpetuate genetic variability. Plant conservation can be carried out in situ or ex-situ, with ex-situ conservation often taking place in botanical gardens. Mexico has a rich history of botanical gardens dating to pre-Hispanic times. Several botanical gardens, in the country serve medicinal, religious, and aesthetic purposes, with notable examples being those created by King Nezahualcoyotl in Texcoco and Aztec king Moctezuma Ilhuicamina in Oaxtepec (Almaraz et al. 2005). Currently, the Botanical Garden of the Biology Institute of the National Autonomous University of Mexico stands out as one of the most important in the country, boasting an expansive area of 12.7 hectares and hosting over 1600 species, including a significant collection of Agavaceae species (https://www.ib.unam.mx/ib/jb/).

Fig. 8.19 Maguey Cenizo (*Agave durangensis* Gentry) experimental plantation in Durango, Mexico. (Photo by Norma Almaraz-Abarca)

The Interdisciplinary Research Center for Integral Regional Development, Durango, National Polytechnic Institute established a nursery for *A. durangensis* in 2006, in Durango, Mexico, which is still maintained to date. Young individuals are preserved in this nursery, which provides plants to maintain experimental plantations (Fig. 8.19). Germplasm is collected from plants in the flowering-fruiting stage in these plantations. The nursery and experimental plantations also support research on germination, phytochemistry, genetic variability, resistance to abiotic stress, as well as in vitro tissue culture and epigenetic variation.

8.7.5.2 In Situ Conservation

It is important to note that *A. durangensis* was not considered severely endangered, at high risk of extinction, or threatened species, based on several parameters, such as the risk level established by the International Union for the Conservation of Nature (IUCN) Red List, as well as the available knowledge on endemism in Mexico, general uses, pollinators, floral visitors, and/or reproduction, as discussed by Alducín-Martínez et al. (2023). However, explorations carried out by our research group indicate that conservation efforts should be addressed, including the creation of preservation areas and seed collection to maintain the natural genetic variability of the species and the establishment of quality plantations.

Other urgent conservation actions should include designating areas for in situ preservation, and collecting seeds from natural populations in order to create a germplasm bank, in which strategies for long periods of conservation can be developed for *Agave durangensis.*

8.7.5.3 Cryopreservation

Cryopreservation represents a valuable alternative for the long-term protection and conservation of plants, mainly for species with short-lived or sterile seeds. Although, our observation suggests that *A. durangensis* produces many viable seeds, with a high germination potential even after a couple of years, cryopreservation of pollen, cell cultures, hybrid embryos, or other tissues would still be beneficial for breeding and improving the species.

During cryopreservation, plant tissues are immersed in a cryoprotective solution, dehydrated, and then placed in liquid nitrogen (Śliwińska et al. 2024). After a variable preservation period, the tissue is thawed and stimulated to resume growth. In many cases, this stimulation occurs under in vitro conditions, making micropropagation protocols necessary.

Cryopreservation can induce genome changes, as informed by Santana et al. (2024) for *Hylocereus costaricensis*. In this species of Cactaceae, the authors reported that the liquid nitrogen used in seeds for 30–60 days induced endoreplication, increasing the ploidy level of seedlings. Presumably, endoreplication was a consequence of the triggering of oxidative stimuli. This finding indicates that the genetic stability of cryopreserved tissues must be constantly monitored. But it also suggests that the cryopreservation method itself represents a mechanism for generating genetic variability, which can be useful in breeding.

Some promising results on the application of the technique to *Agave* species come from the work of Delgado-Aceves et al. (2021) and (2022). The first authors reported 83% post-cryopreservation regrowth and a 70% plant conversion rate for *A. tequilana* Weber cultivar 'Chato', through indirect somatic embryogenesis and cryopreservation of intermediate callus. The latter reported 96% post-cryopreservation regrowth for *A. peacockii*, from shoot-tips of in vitro propagated plants.

The results of Díaz and Garay (2008), informing about the relevance of the use of olive oil for cryopreservation for 6 months of the pollen of *A. tequilana* and *A. angustifolia*, are also promising. This should encourage further exploration and development of cryopreservation protocols for other *Agave* species.

8.8 Germoplasm Collection of Maguey Cenizo (*Agave durangensis* Gentry)

To the best of our knowledge, there is no register of any improved variety or transgenic cultivar of *A. durangensis* in the Catálogo Nacional de Variedades Vegetales of Mexico (Rojas et al. 2022) or in any other country. However, important efforts are being carried out at the Interdisciplinary Research Center for Integral Regional Development, Durango, National Polytechnic Institute (México) to maintain a germplasm collection and select high carbohydrate-producing genetic lines for the development of *A. durangensis* varieties. Nevertheless, a series of obstacles must be overcome to achieve this objective, including the development of appropriate biotechnological strategies (micropropagation, rescue and cultivation of hybrid embryos, genetic transformation) to facilitate the development, maintenance and propagation of these genetic lines in a timely manner.

As discussed in previous sections, the potential of *A. durangensis* as a source of fiber with diverse applications is very important, so simultaneously, the selection of genetic lines with desirable fiber characteristics should not be ignored. The high intra and interpopulation genetic and chemical variability present in wild populations of *A. durangensis* offers a valuable opportunity for selecting mother plants.

Other institutions involved in conducting valuable efforts to maintain germplasm collection or developing biotechnological tools for *A. durangensis* are listed in Appendix 8.2.

8.9 Conclusion and Prospects

Maguey Cenizo (*Agave durangensis* Gentry) is a biologically interesting and economically important species of *Agave*. It is endemism and the current intensity of exploitation makes it urgent to carry out studies and develop in situ and ex situ conservation approaches. The progress of biotechnological breeding strategies, which at present are very limited for the species, is also urgent to assist in the maintenance of, on the one hand, the high genetic variability of their natural populations (sources of valuable alleles), and on the other hand, the genetic fidelity of elite specimens for some characteristic of commercial interest.

The high genetic variability of the natural populations of *A. durangensis* offers the opportunity for developing varieties with many diverse attributes to broaden the spectrum of uses, beyond its use as raw material for the production of mescal. The results discussed here on the composition of the bagasse of this species suggest that can be used as a source of fiber with very diverse applications.

Agave durangensis also has a significant potential to mitigate some impacts of global climate change, as indicated by the research results on its tolerance to salt stress conditions and the results from the use of artificial intelligence, which suggest that the species has an important capacity for adaptation to different stressed environments. However, to achieve this, it is necessary the involvement of more government and private research institutions and producers, as well, as the development of public policies for evolving robust strategies, which can combine and increase the positive impact of traditional and biotechnological breeding approaches, always taking care of the conservation of natural populations and considering the economic and cultural benefit of the human communities that take advantage of the resource.

Appendix 8.1 Research Institutes and Online Resources Available for Maguey Cenizo (*Agave durangensis* Gentry) Breeding

Institute	Area of specialization and research activities	Contact information
Universidad Autónoma de Coahuila	Analysis of agroindustrial residues and lignocellulosic biomass	Blvd. Venustiano Carranza e Ing. José Cárdenas V. S/N, Saltillo, Coahuila 25280, México. https://www.uadec.mx
Centro para la Investigación en Materiales Avanzados S. C.	Analysis of agroindustrial residues and lignocellulosic biomass	Alianza Norte 202, Autopista Monterrey-Aeropuerto km 10, Apodaca, Nuevo León 66628, México https://cimav.edu.mx/
Centro Interdisciplinario de Investigación para el Desarrollo Integral Regional unidad Durango, Instituto Politécnico Nacional	Plant genetic resources, plant diversity, high-quality seed production, morphological, phytochemical, and molecular characterization, plant breeding using conventional and biotechnological approaches, resistance to abiotic stresses, in situ and ex-situ genetic resources conservation	Sigma 119, Durango, Dgo., 34220, México https://www.ciidirdurango.ipn.mx
Instituto Tecnológico Nacional de México, Instituto Tecnológico de Durango	Analysis of agroindustrial residues and lignocellulosic biomass	Felipe Pescador 1803 Ote, Durango, Dgo., 34080, Mexico http://www.itdurango.edu.mx
Universidad Autónóma Metropolitana, Iztapalapa	Analysis of agroindustrial residues and lignocellulosic biomass	PA 55–535, Iztapalapa, CDMX, 09340, México http://www.iztapalapa.uam.mx/
Comisión Nacional Forestal	Conservation programs, forest cover estimation	Periférico Poniente 5360, Zapopan, Jalisco, 45,019, México https://www.gob.mx/conafor
Instituto Tecnológico Nacional de México, Instituto Tecnológico del Valle del Guadian	Crop management, plant breeding	Carr. Durango-México Km. 22.5, Montemorelos, Dgo., 34371, México http://www.itvalledelguadiana.edu.mx/
University of Guanajuato	Plant development, organogenesis, in vitro propagation	Lascuráin de Retana 5, Guanajuato, Gto., 36000, México webugto@ugto.mx
Universidad Juárez del Estado de Durango	Plant micropropagation, plant conservation	Av. Veterinarias S/N Circuito, Durango, Dgo., 34120, México https://www.ujed.mx/

(continued)

Institute	Area of specialization and research activities	Contact information
Universidad Michoacana de San Nicolás de Hidalgo	Analysis of agroindustrial residues and lignocellulosic biomass	Francisco J Mújica S/N, Morelia, Michoacán, 58040, México https://www.umich.mx/contacto.html
Instituto Nacional de Investigaciones Forestales, Agrícolas y Pecuarias (INIFAP, Valle del Guadiana)	Plant breeding, crop management, plant breeding for resistance to biotic and abiotic stresses; conservation of genetic resources in situ and ex-situ	Valle del Guadiana, Durango, Dgo., 34170, México www.inifap.gob.mx

Appendix 8.2 Genetic Resources for Maguey Cenizo (*Agave durangensis* Gentry) Breeding

Collection type	Important traits/uses	Location
Germplasm collection, nursery propagation, experimental plantations	High carbohydrate-producing lines/Conservation, research, and education	Centro Interdisciplinario de Investigación para el Desarrollo Integral Regional unidad Durango, Instituto Politécnico Nacional Sigma 119, Durango, Dgo., 34220, México https://www.ciidirdurango.ipn.mx
Nursery propagation, experimental plantations	Resistance to biotic and abiotic stress lines/Conservation, research	Instituto Nacional de Investigaciones Forestales, Agrícolas y Pecuarias (INIFAP, Valle del Guadiana) Valle del Guadiana, Durango, Dgo., 34170, México www.inifap.gob.mx
Botanical Garden	Conservation and education	Jardín Botánico, Universidad Nacional Autónoma de México (UNAM), Ciudad de México, México. https://www.ib.unam.mx/ib/jb/
Private Company (Production of agave plants El Álamo)	Conservation, propagation, and sale	Durango-El Mezquital highway, Durango. Dgo., México

References

Abraham-Juárez MJ, Ramírez-Malagón R, Gil-Vega KC et al (2009) AFLP analysis of genetic variability in three reproductive forms of *Agave tequilana*. Rev Fitotec Mex 32:171–175

Acosta-Motos JR, Ortuño MF, Bernal-Vicente A et al (2017) Plant responses to salt stress: adaptive mechanisms. Agronomy 7:18. https://doi.org/10.3390/agronomy7010018

Aguirre-Dugua X, Eguiarte LE (2013) Genetic diversity, conservation and sustainable use of wild *Agave cupreata* and *Agave potatorum* extracted for mezcal production in Mexico. J Arid Environ 90:36e44. https://doi.org/10.1016/j.jaridenv.2012.10.018

Aguirre-Planter E, Parra-Leyva JG, Ramírez-Barahona S et al (2020) Phylogeography and genetic diversity in a southern north American Desert: *Agave kerchovei* from the Tehuacán-Cuicatlán Valley. Mexico Front Plant Sci 11:863. https://doi.org/10.3389/fpls.2020.00863

Ahmad A, Tahir ul Qamar M, Shoukat A et al (2021) The effects of genotypes and media composition on callogenesis, regeneration and cell suspension culture of chamomile (*Matricaria chamomilla* L.). PeerJ 9:e11464. https://doi.org/10.7717/peerj.11464

Alducin-Martínez C, Mondragón KYR, Jiménez-Barrón O et al (2023) Uses, knowledge and extinction risk faced by *Agave* species in Mexico. Plants 12:124. https://doi.org/10.3390/plants12010124

Alfaro-Rojas G, Legaria-Solano JP, Rodríguez-Pérez JE (2007) Diversidad genética en poblaciones de agaves pulqueros (*Agave* spp.) del nororiente del Estado de México. Rev Fitotec Mex 30:1–12. https://doi.org/10.35196/rfm.2007.1.1

Al-Kodmany K (2018) The vertical farm: a review of developments and implications for the vertical city. Buildings 8:24. https://doi.org/10.3390/buildings8020024

Almaraz AN, González EM, Naranjo JN et al (2005) Los jardines botánicos y la conservación de cactáceas. Vid Supra 1:18–22

Almaraz-Abarca N (2011) Agave durangensis, 1st edn. IPN-CONACYT-COCYTED, Distrito Federal

Almaraz-Abarca N, Delgado-Alvarado EA, Hernández-Vargas V et al (2009) Profiling of phenolic compounds of somatic and reproductive tissues of *Agave durangensis* Gentry (Agavaceae). Am J Appl Sci 6:1076–1085

Almaraz-Abarca N, Delgado-Alvarado EA, Ávila-Reyes JA et al (2013a) The phenols of the genus *Agave* (Agavaceae). J Biomater Nanobiotechnol 4:33624. https://doi.org/10.4236/jbnb.2013.43A002

Almaraz-Abarca N, Delgado-Alvarado EA, Torres-Morán MI et al (2013b) Genetic variability in natural populations of *Agave durangensis* (Agavaceae) revealed by morphological and molecular traits. Southwest Nat 58:314–324. https://doi.org/10.1894/0038-4909-58.3.314

Almaraz-Abarca N, González-Elizondo MS, Campos MG et al (2013c) Variability of the foliar phenol profiles of the *Agave victoriae-reginae* complex (Agavaceae). Bot Sci 91:295–306. https://doi.org/10.17129/botsci.9

Álvarez-Ríos GD, Pacheco-Torres F, Figueredo-Urbina CJ et al (2020) Management, morphological and genetic diversity of domesticated agaves in Michoacán. México J Ethnobiol Ethnomed 16:3. https://doi.org/10.1186/s13002-020-0353-9

Andrade-Marcial M, Pacheco-Arjona R, Hernández-Castellano S et al (2024) Transcriptome analysis reveals molecular mechanisms underlying chloroplast biogenesis in albino *Agave angustifolia* plantlets. Physiol Plant 176:e14289. https://doi.org/10.1111/ppl.14289

Ángeles-Vázquez BV, Alvarez-Cervantes J, Tovar-Jiménez X et al (2023) Plant regeneration from indirect somatic embryogenesis of *Agave salmiana* Otto ex Salm-Dyck subsp. salmiana using zygotic embryo obtained by *in-casa* pollination as explants. Polibotánica 56:171–182. https://doi.org/10.18387/polibotanica.56.9

Arumugam C, Arumugam GS, Ganesan A et al (2024) Novel short *Agave americana* based biocomposite and nanobiocomposites for automotive applications. Int J Lightweight Mater Manuf 7:556–571. https://doi.org/10.1016/j.ijlmm.2024.03.003

Arvizu-Rodríguez LE, Olvera-Sosa M, Arcibar-Orozco JA et al (2024) Carbons derived from *Agave tequilana* fibers as efficient sulfur supports for high-performance lithium-sulfur batteries. Energ Technol 12:2300743. https://doi.org/10.1002/ente.202300743

Arzate-Fernández AM, Martínez-Velasco I, Alvarez-Aragón C et al (2020) Respuesta morfogenética de dos especies de *Agave* regeneradas *in vitro*. Trop Subtrop Agroecosyst 23: 47. https://doi.org/10.56369/tsaes.3201

Ávila-Reyes JA, Almaraz-Abarca N, Delgado-Alvarado EA et al (2010) Phenol profile and antioxidant capacity of mescal aged in oak wood barrels. Food Res Int 43:296–300. https://doi.org/10.1016/j.foodres.2009.10.002

Barragán-Trinidad M, Buitrón G (2024) Pretreatment of agave bagasse with ruminal fluid to improve methane recovery. Waste Manag 175:52–61. https://doi.org/10.1016/j.wasman.2023.12.041

Barraza-Soto S, Domínguez-Calleros PA, Montiel-Antuna E et al (2014) La producción de mezcal en el municipio de Durango, México. Ra Ximhai 10:65–74

Barriada-Bernal G, Almaraz-Abarca N, Gallardo-Velázquez T et al (2013) Seed vigor variation of *Agave durangensis* Gentry (Agavaceae). Am J Plant Sci 4:2227–2239. https://doi.org/10.4236/ajps.2013.411276

Barriada-Bernal LG, Almaraz-Abarca N, Delgado-Alvarado EA et al (2014) Flavonoid composition and antioxidant capacity of the edible flowers of *Agave durangensis* (Agavaceae). CyTA J Food 12:105–114. https://doi.org/10.1080/19476337.2013.801037

Bermúdez-Bazán M, Estarrón-Espinosa M, Castillo-Herrera GA et al (2024) *Agave angustifolia* Haw. leaves as a potential source of bioactive compounds: extraction optimization and extract characterization. Molecules 29:1137. https://doi.org/10.3390/molecules29051137

Bi B, Tong Q, Wan C et al (2022) *Pinus sylvestris* var. *mongolica* mediates interspecific belowground chemical interactions through root exudates. For Ecol Manag 511:120158. https://doi.org/10.1016/j.foreco.2022.120158

Binh LT, Muoi LT, Oanh HTK et al (1990) Rapid propagation of agave by in vitro tissue culture. Plant Cell Tissue Organ Cult 23:67–70. https://doi.org/10.1007/BF00116091

Bispo dos Santos SL, Rodrigues Passos A, de Oliveira Domingos Queiroz SR et al (2015) Genetic variability in populations of *Agave sisalana* Perrine detected by inter simple sequence repeats. Biosci J Uberlândia 31:1624–1633. https://doi.org/10.14393/BJ-v31n6a2015-26420

Bruňáková K, Bálintová M, Henzelyová J et al (2021) Phytochemical profiling of several *Hypericum* species identified using genetic markers. Phytochemistry 187:112742. https://doi.org/10.1016/j.phytochem.2021.112742

Cabrera-Toledo D, Mendoza-Galindo E, Larranaga N et al (2022) Genomic and morphological differentiation of spirit producing *Agave angustifolia* traditional landraces cultivated in Jalisco. Mexico Plants 11:2274. https://doi.org/10.3390/plants11172274

Candia-Acosta JA, Díaz-Vasquez MA, Torres Herrera SI et al (2019) Germinación de semillas de *Agave durangensis* bajo diferentes coberturas en Durango, México. Revista Mexicana de Agroecosistemas 6:18–27

Chávez-Ortiz LI, Morales-Domínguez JF, Rodríguez-Sahagún A et al (2021) In vitro propagation of *Agave guiengola* Gentry using semisolid medium and temporary immersion bioreactors. Phyton Int J Exp Bot 90:1003–1013. https://doi.org/10.32604/phyton.2021.012862

Chávez-Sánchez C, Mancilla-Margalli NA, Montero-Cortés MI et al (2022) Asexually propagated *Agave tequilana* var. azul exhibits variation in genetic markers and defense responses to *Fusarium solani*. AoB Plants 14:1–10. https://doi.org/10.1093/aobpla/plac027

Chen PY, Kuo YC, Chen CH et al (2009) Isolation and immunomodulatory effect of homoisoflavones and flavones from *Agave sisalana* Perrine ex Engelm. Molecules 14:1789–1795. https://doi.org/10.3390/molecules14051789

Cobaleda-Velasco M, Alanis-Bañuelos RE, Almaraz-Abarca N et al (2017) Phenolic profiles and antioxidant properties of *Physalis angulata* L. as quality indicators. J Pharm Pharmacogn Res 5: 114–128

Contreras-Hernández MG, Ochoa-Martínez LA, Rutiaga-Quiñones JG et al (2018) Effect of ultrasound pre-treatment on the physicochemical composition of *Agave durangensis* leaves and potential enzyme production. Bioresour Technol 249:439–446. https://doi.org/10.1016/j.biortech.2017.10.009

Correa-Hernández L, Baltazar-Bernal O, Sánchez-Páez R et al (2022) In vitro multiplication of agave tobala (*Agave potatorum* Zucc.) using Ebband-flow bioreactor. S Afr J Bot 147:670–677. https://doi.org/10.1016/j.sajb.2022.03.009

Corso M, Perreau F, Mouille G et al (2020) Specialized phenolic compounds in seeds: structures, functions, and regulations. Plant Sci 296:110471. https://doi.org/10.1016/j.plantsci.2020.110471

Cuong DM, Kwon SJ, Van Nguyen B et al (2020) Effect of salinity stress on phenylpropanoid genes expression and related gene expression in wheat sprout. Agronomy 10:390. https://doi.org/10.3390/agronomy10030390

Dávila M, Castillo MA, Laurentin H (2007) Uso de marcadores moleculares ISSR para inferir las relaciones genéticas y la variabilidad intraespecífico en *Agave*. Rev Fac Agron (MARACAY) 33:93–111

del Valle JC, Buide ML, Casimiro-Soriguer I et al (2015) On flavonoid accumulation in different plant parts: variation patterns among individuals and populations in the shore campion (*Silene littorea*). Front Plant Sci 6:939. https://doi.org/10.3389/fpls.2015.00939

de-la-Peña C, Nic-Can G, Ojeda G et al (2012) KNOX1 is expressed and epigenetically regulated during in vitro conditions in *Agave* spp. BMC Plant Biol 12:203. http://www.biomedcentral.com/1471-2229/12/203

Delgado-Aceves L, González-Arnao MT, Santacruz-Ruvalcaba F et al (2021) Indirect somatic embryogenesis and cryopreservation of *Agave tequilana* Weber cultivar 'Chato'. Plants 10:249. https://doi.org/10.3390/plants10020249

Delgado-Aceves L, Portillo L, Folgado R et al (2022) New approaches for micropropagation and cryopreservation of *Agave peacockii*, an endangered species. Plant Cell Tissue Organ Cult 150: 85–95. https://doi.org/10.1007/s11240-022-02246-z

Delgado-Alvarado EA, Ávila-Reyes JA, Almaraz-Abarca N et al (2019) Genetic studies on morphological variability and carbohydrate content of *Agave durangensis* as potential tools for conservation and propagation programs. J Anim Plant Sci 29:1308–1317

Delgado-Alvarado EA, Ávila-Reyes JA, Torres-Ricario R et al (2021) Caracterización fitoquímica de *Agave shrevei* Gentry. e-CUCBA 8:56–59. https://doi.org/10.32870/ecucba.vi16.200

Demey JR, Gamez E, Molina S et al (2004) Comparative study of the discriminating capacity of AFLP and ISTR markers for genetic analysis of *Agave fourcroydes*. Plant Mol Biol Report 22: 29–35. https://doi.org/10.1007/BF02773346

Dévora-Rodríguez DG, Pulido-Díaz C, Chávez-Simental JA et al (2021) Reguladores de crecimiento en el desarrollo vegetativo de vitroplantas de *Agave durangensis* Gentry. Ecosis Recur Agropec 8:e2720. https://doi.org/10.19136/era.a8n1.2720

Díaz SL, Garay BR (2008) Simple methods for *in vitro* pollen germination and pollen preservation of selected species of the genus *Agave*. e-Gnosis 6:2

Díaz-Martínez M, Nava-Cedillo A, Guzmán-Lópeza JA et al (2012) Polymorphism and methylation patterns in *Agave tequilana* Weber var. 'Azul' plants propagated asexually by three different methods. Plant Sci 185–186:321–330. https://doi.org/10.1016/j.plantsci.2012.01.004

dos Santos Mazo G, Rodrigues FJA, Silvério da Costa LT et al (2024) Development of an antioxidant, anti-aging, and photoprotective phytocosmetic from discarded *Agave sisalana* Perrine roots. Cosmetics 11:104. https://doi.org/10.3390/cosmetics11030104

Duarte-Aké F, De-la-Peña C (2021) High cytokinin concentration and nutrient starvation trigger DNA methylation changes in somaclonal variants of *Agave angustifolia* Haw. Ind Crop Prod 172:114046. https://doi.org/10.1016/j.indcrop.2021.114046

Duran GMD, Weber B, Jiménez GJ et al (2021) The application of solid biofuels as a source of process energy in Mexico: case studies using agave and coffee waste. Biofuels Bioprod Biorefin 15:1233–1244. https://doi.org/10.1002/bbb.2230

Duta-Cornescu G, Constantin N, Pojoga DM et al (2023) Somaclonal variation-advantage or disadvantage in micropropagation of the medicinal plants. Int J Mol Sci 24:838. https://doi.org/10.3390/ijms24010838

Eguiarte LE, Barrón JOA, Aguirre-Planter E et al (2021) Evolutionary ecology of *Agave*: distribution patterns, phylogeny, and coevolution (an homage to Howard S. Gentry). Am J Bot 108: 216–235. https://doi.org/10.1002/ajb2.1609

El Mokni R, Verloove F (2021) Additional species of *Agave* (Agavoideae/Agavaceae, Asparagaceae *sensu lat.*) introduced and naturalising in Tunisia and North Africa. Bradleya 39:221–235. https://doi.org/10.25223/brad.n39.2021.a23

El-Hawary SS, El-Kammar HA, El-Dine RS (2020) Antimicrobial activity and genetic variability between *Agave* cultivars using ISSR and SCoT analyses. Egypt J Chem 63:3817–3822. https://doi.org/10.21608/ejchem.2020.20732.2241

Escobar FJG, Sandoval S (2022) Artificial intelligence to model the potential distribution of *Agave durangensis*. International Geoscience and Remote Sensing Symposium, Kuala Lumpur, pp 5828–5831. https://doi.org/10.1109/IGARSS46834.2022.9883767

Escobar-Flores JG, Sandoval S, Gamiz-Romero E (2022) Unmanned aerial vehicle images in the machine learning for agave detection. Environ Sci Pollut Res 29:61662–61673. https://doi.org/10.1007/s11356-022-18985-7

Evdokimova OL, Alves CS, Whiffen RMK et al (2021) Cytocompatible cellulose nanofibers from invasive plant species *Agave americana* L. and *Ricinus communis* L.: a renewable green source of highly crystalline nanocellulose. J Zhejiang Univ-Sci B 22:450–461. https://doi.org/10.1631/jzus.B2000683

Farooq M, Ahmad R, Shahzad M et al (2021) Differential variations in total flavonoid content and antioxidant enzymes activities in pea under different salt and drought stresses. Sci Hortic 287: 110258. https://doi.org/10.1016/j.scienta.2021.110258

Félix-Valdez LI, Vargas-Ponce O, Cabrera-Toledo D et al (2016) Effects of traditional management for mescal production on the diversity and genetic structure of *Agave potatorum* (Asparagaceae) in Central Mexico. Genet Resour Crop Evol 63:1255–1271. https://doi.org/10.1007/s10722-015-0315-6

Figueredo CJ, Casas A, Colunga-García Marín P et al (2014) Morphological variation, management and domestication of 'maguey alto' (*Agave inaequidens*) and 'maguey manso' (*A. hookeri*) in Michoacán, México. J Ethnobiol Ethnomed 10:66. http://www.ethnobiomed.com/content/10/1/66

Figueredo CJ, Casas A, González-Rodríguez A et al (2015) Genetic structure of coexisting wild and managed agave populations: implications for the evolution of plants under domestication. AoB Plants 7:plv114. https://doi.org/10.1093/aobpla/plv114

Figueredo UCJ, Casas A, Martínez-Díaz Y et al (2018) Domestication and saponins contents in a gradient of management intensity of agaves: *Agave cupreata*, *A. inaequidens* and *A. hookeri* in Central Mexico. Genet Resour Crop Evol 65:1133–1146. https://doi.org/10.1007/s10722-017-0601-6

Figueredo-Urbina CJ, Álvarez-Ríos GD, García-Montes MA et al (2021) Morphological and genetic diversity of traditional varieties of agave in Hidalgo state. Mexico PLoS ONE 16: e0254376. https://doi.org/10.1371/journal.pone.0254376

Flores-Benítez S, Jiménez-Bremont JF, Rosales-Mendoza S et al (2007) Genetic transformation of *Agave salmiana* by *agrobacterium tumefaciens* and particle bombardment. Plant Cell Tissue Organ Cult 91:215–224. https://doi.org/10.1007/s11240-007-9287-3

Gallegos-Hernández G, Almaraz-Abarca N, Delgado-Alvarado EA et al (2024) Variation of the chemical and biochemical responses to salinity during germination and early growth of seedlings of two populations of *Agave durangensis* Gentry. Plant Sci Today 11:193–204. https://doi.org/10.14719/pst.2963

Gao J, Yang F, Zhang S et al (2014) Expression of a hevein-like gene in transgenic *Agave* hybrid no. 11648 enhances tolerance against zebra stripe disease. Plant Cell Tissue Organ Cult 119: 579–585. https://doi.org/10.1007/s11240-014-0557-6

García-Mendoza A (1995) Riqueza y endemismos de la familia Agavaceae en México. In: Linares E, Dávila P, Ching F et al (eds) Conservación de plantas en peligro de extinción: diferentes enfoques, 1st edn. Universidad Nacional Autónoma de México, Distrito Federal, pp 51–75
García-Mendoza A, Galván R (1995) Riqueza de las familias Agavaceae y Nolinaceae en México. Bol Soc Bot México 56:7–24. https://doi.org/10.17129/botsci.1461
García-Mendoza S, Franco MIS, Sandoval GD (2019) Cuatro especies nuevas de *Agave* (Asparagaceae, Agavoideae) del sur de México. Acta Bot Mex 126:e1461. https://doi.org/10.21829/abm126.2019.1461
Gentry HS (1967) Putative hybrids in *Agave*. J Hered 58:32–36. https://doi.org/10.1093/oxfordjournals.jhered.a107536
Gentry HS (1982) Agaves of continental North America, 1st edn. The University of Arizona Press, Tucson
Ghanem A-MFM, Mohamed E, Kasem AMMA et al (2021) Differential salt tolerance strategies in three halophytes from the same ecological habitat: augmentation of antioxidant enzymes and compounds. Plants 10:1100. https://doi.org/10.3390/plants10061100
Gil-Vega K, Díaz C, Nava-Cedillo A et al (2006) AFLP analysis of *Agave tequilana* varieties. Plant Sci 170:904–909. https://doi.org/10.1016/j.plantsci.2005.12.014
Gomez-Vargas AD, Hernández-Martínez KM, López-Rosas ME et al (2022) Evidence for light and tissue-specific regulation of genes involved in fructan metabolism in *Agave tequilana*. Plants 11: 2153. https://doi.org/10.3390/plants11162153
González DBM, Núñez PHG (2015) Micropropagación de *Agave durangensis* en un sistema de inmersión temporal (SIT). Jóvenes en la Ciencia Revista de Divulgación Científica 1:8–13
González EM, Galván VR, López EIL et al (2009) Agaves, magueyes, lechuguillas y noas del estado de Durango y sus alrededores, 1st edn. IPN-Conabio-Cocyted, Distrito Federal
González-Elizondo MS, González-Elizondo M, López-Enriquez IL et al (2011) El Complejo *Agave victoriae-reginae* (Agavaceae). Acta Bot Mex 95:65–94. https://doi.org/10.21829/abm95.2011.268
González-Pérez E, Flores-Benítez S, Valera-Montero LL et al (2024) Exploring genetic profile of agaves (*Agave* spp.) from Mexican arid and semiarid land using molecular markers for cytochrome P450 gene. Pak J Agric Sci 61:27–32
González-Trillo AC, Reyes-López MA, Almaraz-Abarca N et al (2024) Characterization of the complete chloroplast genome sequence of *Agave durangensis* (Asparagales: Asparagaceae: Agavoideae). Mitochondrial DNA B 9:536–540. https://doi.org/10.1080/23802359.2024.2338546
Gonzalez-Valdez LS, Almaraz-Abarca N, Proal-Nájera JB et al (2013) Surfactant properties of the saponins of *Agave durangensis*, application on arsenic removal. Int J Eng Appl Sci 4:87–94
Good-Avila SV, Souza V, Gaut BS et al (2006) Timing and rate of speciation in *Agave* (Agavaceae). Proc Natl Acad Sci 103:9124–9129. https://doi.org/10.1073/pnas.0603312103
Hapani P, Marjadi D (2015) Salt tolerance and biochemical responses as a stress indicator in plants to salinity: a review. CIBTech J Biotechnol 4:23–46
Heller W, Forkman G (1994) Biosynthesis of flavonoids. In: Harborne JB (ed) The flavonoids advances in research since 1986, 1st edn. Chapman and Hall, London, pp 500–535
Hernández-Pacheco CE, Almaraz-Abarca N, Rojas-López M et al (2022) Salinity generates varying chemical and biochemical responses in *Physalis ixocarpa* (Solanaceae) during different times of exposure. Electron J Biotechnol 59:25–35. https://doi.org/10.1016/j.ejbt.2022.06.002
Hernández-Vargas V, Orea-Lara G, Gómez-Ortiz S (2015) Crecimiento de las plantas de agave mezcalero con aplicación de abonos. Vid Supra 7:50–53
Herrera-Ruiz M, Jiménez-Ferrer E, González-Cortazar M et al (2022) Potential use of *Agave* genus in neuroinflammation management. Plants 11:2208. https://doi.org/10.3390/plants11172208
Hulle A, Kadole P, Katkar P (2015) *Agave americana* leaf fibers. Fibers 3:64–75. https://doi.org/10.3390/fib3010064

Hussain G, Huang J, Rasul A et al (2019) Putative roles of plant-derived tannins in neurodegenerative and neuropsychiatry disorders: an updated review. Molecules 24:2213. https://doi.org/10.3390/molecules24122213

Ikeuchi M, Iwase A, Rymen B et al (2017) Wounding triggers callus formation via dynamic hormonal and transcriptional changes. Plant Physiol 175:1158–1174. https://doi.org/10.1104/pp.17.01035

Infante D, González G, Peraza-Echeverría L et al (2003) Asexual genetic variability in *Agave fourcroydes*. Plant Sci 164:223–230. https://doi.org/10.1016/S0168-9452(02)00404-1

Jiménez-Barron O, García-Sandoval R, Magallón S et al (2020) Phylogeny, diversification rate, and divergence time of *Agave sensu lato* (Asparagaceae), a group of recent origin in the process of diversification. Front Plant Sci 11:536135. https://doi.org/10.3389/fpls.2020.536135

Jiménez-Sánchez DE, Calderón-Santoyo M, Herman-Lara E et al (2020) Use of native agave fructans as stabilizers on physicochemical properties of spray-dried pineapple juice. Dry Technol 38:293–303. https://doi.org/10.1080/07373937.2019.1565575

Khadiri M, Boubaker H, Askarne L et al (2024) Exploration of the antifungal potential of aqueous extracts from two agave species against major postharvest diseases in apples. Postharvest Biol Technol 214:112992. https://doi.org/10.1016/j.postharvbio.2024.112992

Kimaro DN, Msamya BM, Takamura Y (1994) Review of sisal production and research in Tanzania. Afr Stud Monogr 15:227–242

Konfo TRC, Chabi ABP, Amoussouga GA et al (2024) Recent climate-smart innovations in agrifood to enhance producer incomes through sustainable solutions. J Agric Food Res 15:100985. https://doi.org/10.1016/j.jafr.2024.100985

Kumar A, Ram C (2021) *Agave* biomass: a potential resource for production of value-added products. Environ Sustain 4:245–259. https://doi.org/10.1007/s42398-021-00172-y

Kungulovski G, Jeltsch A (2016) Epigenome editing: state of the art, concepts, and perspectives. Trends Genet 32:101–113. https://doi.org/10.1016/j.tig.2015.12.001

Lalaymia I, Belaadi A, Boumaaza M et al (2024) Weibull statistic and artificial neural network analysis of the mechanical performances of fibers from the flower *Agave* plant for eco-friendly green composites. J Nat Fibers 21:2305228. https://doi.org/10.1080/15440478.2024.2305228

Li J, Wang X, Wang X, Wang Y (2015) Embryo rescue technique and its applications for seedless breeding in grape. Plant Cell Tissue Organ Cult 120:861–880. https://doi.org/10.1007/s11240-014-0656-4

Lightbourn G, Veilleux R (2003) Retrotransposon based markers to characterize somatic hybrids and variation induced by protoplast fusion of monoploid potato. Acta Hortic 619:35–43. https://doi.org/10.17660/ActaHortic.2003.619.4

Lin X, Fang H, Wang L et al (2024) Photocatalytic degradation of sulfamethoxazole and enrofloxacin in water using electrospun composite photocatalytic membrane. Water 16:218. https://doi.org/10.3390/w16020218

Linić I, Šamec D, Grúz J et al (2019) Involvement of phenolic acids in short-term adaptation to salinity stress is species-specific among Brassicaceae. Plants 8:155. https://doi.org/10.3390/plants8060155

Liu W, Strong DR, Pennings SC et al (2017) Provenance-by-environment interaction of reproductive traits in the invasion of *Spartina alterniflora* in China. Ecology 98:1591–1599. https://doi.org/10.1002/ecy.1815

Loera-Gallegos HM, Corral-Rivas JJ, Montiel-Antuna E et al (2018) Calidad de sitio para *Agave durangensis* Gentry en la Sierra de Registrillo, Durango. Rev Mex Cienc Agric 9:1171–1180. https://doi.org/10.29312/remexca.v9i6.1582

López SPM, Hernández RA, Méndez GJ et al (2021) Mejores prácticas de manejo y ecuaciones alométricas de biomasa de *Agave durangensis* Gentry en el estado Durango, Proyecto: 2017-4-292674. CONACYT-CONAFOR-UJED-INIFAP, Zapopan

Luna-Solis HA, Ochoa-Martínez LA, González-Herrera SM et al (2024) Extraction of agavins from *Agave durangensis* leaves: Structural and techno functional characterization. Food Biosci 57:103563. https://doi.org/10.1016/j.fbio.2023.103563

Maazoun AM, Hamdi SH, Belhadj F et al (2019) Phytochemical profile and insecticidal activity of *Agave americana* leaf extract towards *Sitophilus oryzae* (L.) (Coleoptera: Curculionidae). Environ Sci Pollut Res 26:19468–19480. https://doi.org/10.1007/s11356-019-05316-6
Majamo SL, Amibo TA, Tegaw EZ (2024) Enhanced cellulose extraction from agave plant (*Agave americana Species*) for synthesis of magnetic/cellulose nanocomposite for fluoridation of water. Mater Today Commun 38:107683. https://doi.org/10.1016/j.mtcomm.2023.107683
Mancilla-Álvarez E, Spinoso-Castillo JL, Muñoz-Márquez TRA et al (2024a) Temporary immersion bioreactor as an efficient method for in vitro propagation of *Agave marmorata*. S Afr J Bot 169:6–11. https://doi.org/10.1016/j.sajb.2024.04.011
Mancilla-Álvarez E, Spinoso-Castillo JL, Schettino-Salomón SS et al (2024b) Temporary immersion systems induce photomixotrophism during in vitro propagation of agave Tobalá. 3 Biotech 14:74. https://doi.org/10.1007/s13205-024-03928-5
Martínez-Herrera RE, Alemán-Huerta ME, Flores-Rodríguez P et al (2021) Utilization of *Agave durangensis* leaves by *Bacillus cereus* 4N for polyhydroxybutyrate (PHB) biosynthesis. Int J Biol Macromol 175:199–208. https://doi.org/10.1016/j.ijbiomac.2021.01.167
Martinez-Rodriguez A, Beltran-Garcia C, Valdez-Salas B et al (2022) Micropropagation of seed-derived clonal lines of the endangered *Agave marmorata* Roezl and their compatibility with endophytes. Biology 11:1423. https://doi.org/10.3390/biology11101423
Martínez-Velasco I, Arzate-Fernández AM (2022) Analysis of the somaclonal variation in two *in vitro* regenerated agave species. Trop Subtrop Agroecosyst 25:50. https://doi.org/10.56369/tsaes.3709
Mayagoitia-Toulet DJ, Zamora-Gutiérrez V (2024) Flores de agave, centros de interacción del semidesierto. Therya ixmana 3:101–103
Medina-Medrano JR, Almaraz-Abarca N, González-Elizondo MS et al (2015) Phenolic constituents and antioxidant properties of five wild species of *Physalis* (Solanaceae). Bot Stud 56:24. https://doi.org/10.1186/s40529-015-0101-y
Mierziak J, Kostyn K, Kulma A (2014) Flavonoids as important molecules of plant interactions with the environment. Molecules 19:16240–16265. https://doi.org/10.3390/molecules191016240
Miglani GS, Kaur A, Kaur L (2020) Plant gene expression control using genome- and epigenome-editing technologies. J Crop Improv 34:1–63. https://doi.org/10.1080/15427528.2019.1678541
Modesto I, Sterck L, Arbona V et al (2021) Insights into the mechanisms implicated in *Pinus pinaster* resistance to pinewood nematode. Front Plant Sci 12:690857. https://doi.org/10.3389/fpls.2021.690857
Molina-Amaya JM, Ávila-Reyes JA, Delgado-Alvarado EA et al (2024) Potential of chitosan and carbon crystals biocomposites for removing enrofloxacin and oxytetracycline from water. Rev Int Contam Ambie 40:313–329
Monja-Mio KM, Quiroz-Moreno A, Herrera-Herrera G et al (2018) Analysis of two clonal lines (embryogenic and non-embryogenic) of *Agave fourcroydes* using AFLP and MSAP. Am J Plant Sci 9:745–762. https://doi.org/10.4236/ajps.2018.94059
Monja-Mio KM, Olvera-Casanova D, Herrera-Alamillo MA et al (2021) Comparison of conventional and temporary immersion systems on micropropagation (multiplication phase) of *Agave angustifolia* Haw. 'Bacanora'. 3 Biotech 11:77. https://doi.org/10.1007/s13205-020-02604-8
Monja-Mio KM, Rescalvo-Morales A, Robert ML (2022) Use of thin cell layer technique for induction of somatic embryogenesis of *Agave fourcroydes*. Methods Mol Biol 2527:59–67. https://doi.org/10.1007/978-1-0716-2485-2_5
Montané X, Kowalczyk O, Reig-Vano B et al (2020) Current perspectives of the applications of polyphenols and flavonoids in cancer therapy. Molecules 25:3342. https://doi.org/10.3390/molecules25153342
Morales-Nieto CR, Siqueiros-Candia M, Álvarez-Holguín A et al (2021) Diversidad, estructura genética e hibridación en poblaciones de *Pinus arizonica* y *P. durangensis*. Madera Bosques 27: 1–14. https://doi.org/10.21829/myb.2021.2722170

Morales-Serna JA, Jiménez A, Estrada-Reyes R et al (2010) Homoisoflavanones from *Agave tequilana* Weber. Molecules 15:3295–3301. https://doi.org/10.3390/molecules15053295

Moreno-Anguiano O, Carrillo-Parra A, Rutiaga-Quiñones JG et al (2021) Chemical composition of *Luffa aegyptiaca* Mill., *Agave durangensis* Gentry and *Pennisetum* sp. PeerJ 9:e10626. https://doi.org/10.7717/peerj.10626

Moreno-Anguiano O, Cloutier A, Rutiaga-Quiñones JG et al (2022) Use of *Agave durangensis* bagasse fibers in the production of wood-based medium density fiberboard (MDF). Forest 13: 271. https://doi.org/10.3390/f13020271

Morreeuw ZP, Escobedo-Fregoso C, Ríos-González LJ et al (2021) Transcriptome-based metabolic profiling of flavonoids in *Agave lechuguilla* waste biomass. Plant Sci 305:110748. https://doi.org/10.1016/j.plantsci.2020.110748

Natarajan P, Mohanraj M, Kumar M et al (2024) Experimental investigation on mechanical and tribological analysis of pineapple leaf (*Ananas comosus*) and sisal (*Agave sisalana*) fibers reinforced hybrid epoxy composites. Int Polym Process 39. https://doi.org/10.1515/ipp-2023-4433

Nava-Cruz NY, Medina-Morales MA, Martinez JL et al (2015) *Agave* biotechnology: an overview. Crit Rev Biotechnol 35:546–559. https://doi.org/10.3109/07388551.2014.923813

Naziri M, Sadat S, Howyzeh MS (2019) The effect of different hormone combinations on direct and indirect somatic embryogenesis in *Agave americana*. Iran J Plant Physiol 9:2739–2747

Ocaña-Nava D, García-Mendoza A, Larson J (2007) Modelación supervisada de la distribución de los magueyes mezcaleros en México. In: Colunga-GarcíaMarín P, Eguiarte L, Larqué SA et al (eds) En lo ancestral hay futuro: del tequila, los mezcales y otros agaves, 1st edn. CICY-CONACYT-CONABIO-INE, Distrito Federal, pp 153–171

Oliva-Rodríguez A, Morales-Martínez TK, Rodríguez-De la Garza JA et al (2024) Optimization of hydrogen and butanol production from agave guishe juice using *Clostridium acetobutylicum* ATCC824. Int J Hydrogen Energy 108:206–214. https://doi.org/10.1016/j.ijhydene.2024.06.121

Onozuka M, Osawa T (2022) Utilization potential of alien plants in nature-based tourism sites: a case study on *Agave americana* (century plant) in the Ogasawara Islands. Ecol Econ 195: 107362. https://doi.org/10.1016/j.ecolecon.2022.107362

Orea LG, Cifuentes-Díaz de León A, Gómez OS et al (2006) Germinación de semillas (*Agave durangensis*) a diferentes temperaturas y efecto de la fertilización en el desarrollo de las plántulas. Vid Supra 1:51–56

Orozco-Arias S, Candamil-Cortes MS, Jaimes PA et al (2022) Automatic curation of LTR retrotransposon libraries from plant genomes through machine learning. J Integr Bioinform 19:20210036. https://doi.org/10.1515/jib-2021-0036

Pagare S, Bhatia M, Tripathi N et al (2015) Secondary metabolites of plants and their role: overview. Curr Trends Biotechnol Pharm 9:293–304

Parmar VS, Jha HN, Gupta AK et al (1992) Agamanona, a flavanona from *Agave americana*. Phytochemistry 31:2567–2568. https://doi.org/10.1016/0031-9422(92)83333-T

Paszkowski J (2015) Controlled activation of retrotransposition for plant breeding. Curr Opin Biotechnol 32:200–206. https://doi.org/10.1016/j.copbio.2015.01.003

Perez-Aguilar LY, Plata-Rocha W, Monjardin-Armenta SA et al (2021) The identification and classification of arid zones through multicriteria evaluation and geographic information systems—case study: arid regions of Northwest Mexico. ISPRS Int J Geo Inf 10:720. https://doi.org/10.3390/ijgi10110720

Pérez-González KA, Ríos-Saucedo JC, Luján-Olguín R et al (2021) Evaluación de la concentración de grados brix (bx) en plántulasde tres especies de agave mezcalero: efecto de la fertilización. Agrofaz 3:16–25

Ramakrishnan M, Papolu PK, Mullasseri S et al (2023) The role of LTR retrotransposons in plant genetic engineering: how to control their transposition in the genome. Plant Cell Rep 42:3–15. https://doi.org/10.1007/s00299-022-02945-z

Ramírez-Malagón R, Borodanenko A, Pérez-Moreno L et al (2008) In vitro propagation of three *Agave* species used for liquor distillation and three for landscape. Plant Cell Tissue Organ Cult 94:201–207. https://doi.org/10.1007/s11240-008-9405-x

Ramírez-Mosqueda MA, Cárcamo-Corona RG, Aguilar-Jiménez D et al (2022) Micropropagation of agave (*Agave potatorum* Zucc.) through direct organogenesis. Agrociencia 56:1133–1143. https://doi.org/10.47163/agrociencia.v56i6.2823

Ramos LC, Báez M, Fuchs J et al (2023) Differential repeat accumulation in the bimodal karyotype of *Agave* L. Genes 14:491. https://doi.org/10.3390/genes14020491

Rathnayaka RMH, Priyantha N, Gunathilake WSS (2024) Removal of trivalent and hexavalent chromium from aqueous solution using fiber of *Agave americana* plant and its modified forms. Colloids Surf C: Env Aspects 2:100029. https://doi.org/10.1016/j.colsuc.2024.100029

Reyes-Samilpa A, Reyes-Agüero JA, van't Hooft A et al (2023) Fibers of *Agave salmiana* cultivar *Xa'mni*: physical characterization and comparison between leaf maturity and growth environment. Emerg Mater 6:543–549. https://doi.org/10.1007/s42247-022-00439-y

Rhode W (1996) Inverse sequence-tagged repeat (ISTR) analysis, a novel and universal PCR-based technique for genome analysis in the plant and animal kingdom. J Genet Breed 50:249–261

Ríos-Saucedo RS, Reyes-Rodríguez CM, Rosales-Serna R et al (2018) Actualización preliminar de inventarios para áreas productoras de maguey cenizo en diferentes municipios de Durango. Cienc Innov 1:323–332

Robert ML, Herrera JL, Contreras F et al (1987) In vitro propagation of *Agave fourcroydes* Lem. (Henequen). Plant Cell Tissue Organ Cult 8:37–48. https://doi.org/10.1007/BF00040731

Rodríguez-Garay B, Lomelí-Sención JA, Tapia-Campos E et al (2009) Morphological and molecular diversity of *Agave tequilana* Weber var. Azul and *Agave angustifolia* Haw. var. Lineño Ind Crops Prod 29:220–228. https://doi.org/10.1016/j.indcrop.2008.05.007

Rodríguez-Garay B, López-Díaz S, Rodríguez-Domínguez JM et al (2018) Application of *in casa p*ollination and embryo rescue techniques for breeding of *agave s*pecies. In: Loyola-Vargas VM, Ochoa-Alejo N (eds) Plant cell culture protocols, methods in molecular biology, vol 1815. Humana Press, New York, pp 289–300. https://doi.org/10.1007/978-1-4939-8594-4_20

Rojas SAL, Martínez NB, Pacheco AMA (2022) Catálogo Nacional de Variedades Vegetales. Secretaría de Agricultura y Desarrollo Rural- Servicio Nacional de Inspección y Certificación de Semillas, México

Roll CM, Hodgson WC, Salywon AM (2021) Notes on a hybrid colony of *Agave parryi* var. *Huachucensis x Agave palmeri* (Agavaceae) in southeastern Arizona. Phyton 25:1–7

Rosales MS, Hernández AM, Sarmiento LH et al (2014) Producción de plantas de *Agave durangensis* en sistema de acolchado y riego por goteo. Folleto técnico 76. INIFAP, México

Rosales-Serna R, García-Rodríguez JL, Sigala-Rodríguez JA, Santana-Espinoza S (2023) Desarrollo y rendimiento de cultivos anuales en sistemas agroforestales: alternativa productiva para el Altiplano Semiárido Mexicano. Agrofaz 5:59–65

Rosas MI, Colmenero RA, Naranjo JN et al (2014) El mezcal de Durango, México: un análisis histórico, biológico y económico para su producción sustentable. Ingeniantes 1:75–83

Rymen B, Kawamura A, Lambolez A et al (2019) Histone acetylation orchestrates wound-induced transcriptional activation and cellular reprogramming in *Arabidopsis*. Commun Biol 2:404. https://doi.org/10.1038/s42003-019-0646-5

Saini K, Matsagar VA, Kodur VR (2024) Recent advances in the use of natural fibers in civil engineering structures. Constr Build Mater 411:134364. https://doi.org/10.1016/j.conbuildmat.2023.134364

Sánchez A, Coronel-Lara Z, Gutiérrez A et al (2020) Acclimatization and transplantation of *Agave angustifolia* Haw. vitro plants in wild conditions. Rev Mex Cienc Agric 11:1593–1605. https://doi.org/10.29312/remexca.v11i7.2403

Sánchez-Teyer F, Moreno-Salazar S, Esqueda M et al (2009) Genetic variability of wild *Agave angustifolia* populations based on AFLP: a basic study for conservation. J Arid Environ 73:611–616. https://doi.org/10.1016/j.jaridenv.2009.01.008

Santacruz-Ruvalcaba F, Gutiérrez-Pulido H, Rodríguez-Garay B (1999) Efficient *in vitro* propagation of *Agave parrasana* Berger. Plant Cell Tissue Organ Cult 56:163–167. https://doi.org/10.1023/A:1006232911778

Santacruz-Ruvalcaba F, Castañeda-Nava JJ, Villanueva-Gónzalez JP et al (2022) Micropropagación de *Agave maximiliana* Baker por proliferación de yemas axilares. Polibotánica 54:139–151. https://doi.org/10.18387/polibotanica.54.9

Santana BID, Paiva R, Valquíria dos Reis M et al (2024) Seed cryopreservation without vitrification (PVS2) induces oxidative stimuli to promote endoreplication in red pitaya seedlings. Plant Cell Tissue Organ Cult 156:11. https://doi.org/10.1007/s11240-023-02642-z

Scheinvar E, Gamez N, Catellanos-Morales G et al (2017) Neogene and Pleistocene history of *Agave lechuguilla* in the Chihuahuan desert. J Biogeogr 44:322–334. https://doi.org/10.1111/jbi.12851

Siddiqui ZS, Khan MA (2010) The role of seed coat phenolics on water uptake and early proteins synthesis during germination of dimorphic seeds of *Halopyrum mucronatum* (L.) Staph. Pak J Bot 42:227–238

Silva BC, Canovas BA, Garcia de Almeida GL et al (2023) Brazilian agribusiness: investment in vertical farming as a competitive advantage. In: Harmony of knowledge exploring interdisciplinary synergies, vol 101. https://doi.org/10.56238/sevened2023.006-101

Singh S, Kaur I, Kariyat R (2021) The multifunctional roles of polyphenols in plant-herbivore interactions. Int J Mol Sci 22:1442. https://doi.org/10.3390/ijms22031442

Śliwińska AA, Tomiczak K, Obrębski M, Wileńska B, Kiełkiewicz RM, Podwyszyńska M, Zieleźnicka JM, Mikuła A, Sykłowska-Baranek K (2024) An effective cryobanking approach preserving the genetic, physiological and phytochemical stability of *Polyscias filicifolia* Bailey hairy roots. Ind Crop Prod 208:117845. https://doi.org/10.1016/j.indcrop.2023.117845

Soni A, Dhakar S, Kumar N (2017) Mechanisms and strategies for improving salinity tolerance in fruit crops. Int J Curr Microbiol Appl Sci 6:1917–1924. https://doi.org/10.20546/ijcmas.2017.608.226

Soto-Castro D, Perez-Herrera A, Garcia-Sanchez E et al (2021) Identification and quantification of bioactive compounds in *Agave potatorum* Zucc. leaves at different stages of development and a preliminary biological assay. Waste Biomass Valori 12:4537–4547. https://doi.org/10.1007/s12649-020-01329-2

Subramanian SS, Nair AGR (1970) Chlorogenein and kaempferol glycosides from the flowers of *Agave amaricana*. Phytochemistry 9:2582. https://doi.org/10.1016/S0031-9422(00)85782-9

Sumarago EC, de la Cerna MFM, Leyson AKB et al (2024) Production and characterization of nanocellulose from maguey (*Agave cantala*) fiber. Polymers 16:1312. https://doi.org/10.3390/polym16101312

Taiz L, Zeiger E (2003) Plant physiology online, 3rd edn. Sinauer Associates

Téllez TAG, Jiménez RJA, González CO et al (2023) Cultivo in vitro de *Agave potatorum*, especie amenazada endémica de México. Bot Sci 101:883–894. https://doi.org/10.17129/botsci.3279

Tinto WF, Simmons-Boyce JL, McLean S et al (2005) Constituents of *Agave americana* and *Agave barbadensis*. Fitoterapia 76:594–597. https://doi.org/10.1016/j.fitote.2005.04.013

Torres-García I, Rendón-Sandoval FJ, Blancas J et al (2019) The genus *Agave* in agroforestry systems of Mexico. Bot Sci 97:263–290. https://doi.org/10.17129/botsci.2202

Torres-Morán MI, Almaraz-Abarca N, Velasco-Ramírez AP et al (2008) Taxonomic significance of ISTR to discriminate species in Agavaceae. Am J Agric Biol Sci 3:661–665

Torres-Morán MI, Escoto-Delgadillo M, Molina-Moret S et al (2010) Assessment of genetic fidelity among *Agave tequilana* plants propagated asexually via rhizomes versus in vitro culture. Plant Cell Tissue Organ Cult 103:403–409. https://doi.org/10.1007/s11240-010-9777-6

Torres-Morán MI, Hormaza JI, Larranaga N (2024) DNA analyses reveal high heterozygosity values and low differentiation among populations of mezcal-producing *Agave durangensis*. Genet Resour Crop Evol. https://doi.org/10.1007/s10722-023-01850-y

Trejo L, Alvarado-Cárdenas LO, Scheinvar E et al (2016) Population genetic analysis and bioclimatic modeling in *Agave striata* in the Chihuahuan Desert indicate higher genetic variation and lower differentiation in drier and more variable environments. Am J Bot 103:1020–1029. https://doi.org/10.3732/ajb.1500446

Valenzuela-Ruiz JF, Velasco-González OH, Márquez-Linares MA (2003) Desarrollo sustentable del agave mezcalero en Durango. SEP-IPN-SAGDR, Distrito Federal

Valenzuela-Sánchez KK, Juárez-Hernández RE, Cruz-Hernández A et al (2006) Plant regeneration of *Agave tequilana* by indirect organogénesis. In Vitro Cell Dev Biol-Plant 42:336–340. https://doi.org/10.1079/IVP2006788

Vásquez-Hernández S, Montes-Molina JA, Gutiérrez-Miceli FA et al (2023) In vitro mutagenesis for the improvement of *Agave* genus. Phyton-Int J Exp Bot 92:2065–2078. https://doi.org/10.32604/phyton.2023.028784

Vázquez-Martínez O, Núñez-Palenius HG, Pérez-Molphe BEM et al (2022) In vitro-propagation of *Agave tequilana* Weber cv. azul in a temporary immersion system. Phyton-Int J Exp Bot 91:83–96. https://doi.org/10.32604/phyton.2022.017281

Vera-Guzmán AM, Guzmán-Gerónimo RI, López MG et al (2018) Volatile compounds profile in mezcal spirits as influenced by agave species and production processes. Beverages 4:9. https://doi.org/10.3390/beverages4010009

Verloove P, Thiede J, Rodríguez AM et al (2019) A synopsis of feral *Agave* and *Furcraea* (Agavaceae, Asparagaceae s. lat.) in the Canary Islands (Spain). Plant Ecol Evol 152:470–498. https://doi.org/10.5091/plecevo.2019.1634

Walker C (2023) Mangaves. Cactus Succul Rev 37:5–10

Yang Q-Q, Gan R-Y, Ge Y-Y et al (2018) Polyphenols in common beans (*Phaseolus vulgaris* L.): chemistry, analysis, and factors affecting composition. Compr Rev Food Sci Food Saf 17:1518–1539. https://doi.org/10.1111/1541-4337.12391

Yang Z, Yang Q, Liu Q, Li X et al (2023) A chromosome-level genome assembly of *Agave hybrid* no.11648 provides insights into the CAM photosynthesis. Hortic Res 269. https://doi.org/10.1093/hr/uhad269

Yao Q, Peng Z, Tong H et al (2019) Tomato plant flavonoids increase whitefly resistance and reduce spread of tomato yellow leaf curl virus. J Econ Entomol 112:2790–2796. https://doi.org/10.1093/jee/toz199

Yargatti R, Muley A (2022) *Agave* syrup as a replacement for sucrose: an exploratory review. Funct Foods Health Dis 12:590–600

Zavaleta D, Ñañez R, Silva G et al (2024) Additive construction using enhanced earthen-based composites: improvement of the mechanical strength and water durability using chitosan and agave fibers. Constr Build Mater 411:134159. https://doi.org/10.1016/j.conbuildmat.2023.134159

Zhang TJ, Zheng J, Yu ZC et al (2018) Functional characteristics of phenolic compounds accumulated in young leaves of two subtropical forest tree species of different successional stages. Tree Physiol 38:1486–1501. https://doi.org/10.1093/treephys/tpy030

Zhang X, Wang Z, Zhang J et al (2024) Polar organic chemical integrative sampler to predict changes in bioavailability of oxytetracycline and sulfamethoxazole from soil. Water Air Soil Pollut 235:65. https://doi.org/10.1007/s11270-024-06886-z

Zizumbo-Villarreal D, González-Zozaya F, Olay-Barrientos A et al (2009) Archaeological evidence of the cultural importance of *Agave* spp. in Pre-Hispanic Colima, Mexico. Econ Bot 63:288–302. https://doi.org/10.1007/s12231-009-9092-5

Zolkin AL, Matvienko EV, Shevchenko OP et al (2024) Digital technologies and automation in breeding and agriculture. E3S Web Conf 474:01017. https://doi.org/10.1051/e3sconf/202447401017

Chapter 9
Pineapple (*Ananas comosus* L.) Breeding: Current Status and Future Prospects

Luli Wang, Xiaohong Dai, Dongliang Li, and Jingjing Chen

Abstract As a tropical fruit, pineapple (*Ananas comosus* (L.) Merr.) is widely cultivated in tropical areas, including Asia, Africa, and the Americas. The fruit is mainly cultivated for the fruit market, with pineapple leaves serving as a by-product that can be utilized for leaf fiber extraction or as feed. Pineapple leaf fiber is a high-quality fiber, with natural antibacterial, bactericidal, and deodorizing properties. However, the fiber industry remains underdeveloped. Pineapple originated in South America and has since spread to other regions. Pineapple ranks as one of the top two tropical fruits worldwide, coming after bananas. It significantly contributes to the improvement of tropical fruit accessibility while generating substantial economic gains for nations situated in tropical and subtropical climates. Large-scale cultivated pineapple varieties are primarily the result of clone selection and cross-breeding for fresh consumption or processing, with less focus on breeding for leaf fiber. Pineapple breeding is constrained by limited germplasm resources and self-incompatibility. Modern biotechnology encompasses a variety of advanced techniques such as molecular marker-assisted breeding, functional genomic studies, mutation breeding, plant tissue culture, and transgenic breeding, all of which are being utilized to improve the efficiency of plant breeding programs. This chapter provides an overview of the origin, distribution, germplasm biodiversity and conservation, fruit and leaf fiber utilization, cultivation practices and recent advancements in traditional breeding methods and the integration of modern biotechnology to improve conventional breeding techniques for pineapple.

Keywords Biotechnology · Domestication · Germplasm · Leaf fiber · Pineapple (*Ananas comosus* L.) · Traditional breeding

L. Wang · X. Dai · D. Li · J. Chen (✉)
Key Laboratory of Tropical Fruit Biology, Ministry of Agriculture and Rural Affairs, South Subtropical Crops Research Institute, Chinese Academy of Tropical Agricultural Sciences, Zhanjiang, China
e-mail: luli_wang@catas.cn; daixiaohong@catas.cn; lidongliang@catas.cn; chenjingjing@catas.cn

K. F. M. Salem et al. (eds.), *Breeding and Biotechnology of Leaf, Fruit, and Seed Fiber Crops*, Advances in Plant Breeding Strategies 11,
https://doi.org/10.1007/978-3-032-00407-9_9

9.1 Introduction

Pineapple (*Ananas comosus* L. Merr.), is now widely cultivated in most tropical and subtropical regions worldwide ranking second in global production of tropical fruits. Pineapple is classified under the genus Ananas within the Bromeliaceae family *Ananas* Mill genus. The family includes 56 genera and 2600 species (https://www.itis.gov/servlet/SingleRpt/SingleRpt?search_topic=TSN&search_value=42335#null). The pineapple has a chromosome number of $2n = 2x = 50$ and a genome size of approximately 526 Mbp (Ming et al. 2015). It is cultivated worldwide for food or ornamental purposes due to its fruit. The main function of pineapple fruit is to supply the fresh fruit market. Pineapple fruits are rich in nutrients, including various vitamins, sugars, polyphenols, organic acids, and minerals. It is one of the most widely accepted tropical fruits in the market. Additionally, pineapple leaves are fiber content in pineapple leaves, which is valuable for industrial applications such as weaving or making clothes. Pineapple has its origins in South America and has since been spread widely in the region of tropical and subtropical through continuous expansion and domestication.

Currently, pineapple cultivars are mainly divided into five categories, including Smooth Cayenne, Spanish, Pernambuco, Queen, and Perolera (Li et al. 2022). Queen and Smooth Cayenne varieties dominate the world pineapple industry, and now pineapple cultivation worldwide covers 1,046,712 hectares (Table 9.1). About 90 countries worldwide have cultivated pineapples (Li et al. 2022). According to 2021 country-level data, the five leading pineapple-producing nations were Costa Rica, Indonesia, the Philippines, Brazil, and China (Table 9.2). The list of the top 20 producers is largely occupied by Asian and American nations. Among them, pineapple yield is the highest in the Americas region, and the highest yield per hectare is in Indonesia (129.2117 t/ha). This chapter introduces the utilization and industry status of pineapple leaf fiber (PALF), domestication, selection and improvement of cultivars for fresh food and processing, while also examining traditional and modern biological approaches to enhance pineapple production.

Table 9.1 Regional statistics on pineapple crop acreage, annual yield volumes, and agricultural productivity in Africa, the Americas, Asia, and Oceania

Continent	Harvested area (ha)	Production (mt)	Yield(hg/ha)
Africa	366,480	5,251,254.32	143,289
America	253,410	10,221,506.77	403,358
Asia	422,150	13,062,025.04	309,416
Oceania	4671	113,079.56	242,079
World (Total)	1,046,712	28,647,865.69	273,694

* = average data, mt = metric tons, ha = hectares; Source: FAOSTAT (2021)

Table 9.2 Global hierarchy of top 20 pineapple-producing nations: Production quantity ranking with associated yield per hectare metrics (2021)

Country	Production (t)	Yield (t/ha)
Costa Rica	2,938,334.00	73.4584
Indonesia	2,886,416.84	129.2117
Philippines	2,860,202.36	42.6150
Brazil	2,317,554.00	36.4458
China	2,301,836.00	30.5526
China, mainland	1,899,000.00	28.1154
Thailand	1,800,558.00	24.1718
India	1,799,000.00	16.9717
Nigeria	1,541,979.75	8.2095
Mexico	1,271,520.75	49.8206
Colombia	927,050.01	92.7050
Viet Nam	726,129.64	18.4835
Ghana	668,093.22	60.7080
Angola	663,263.00	66.3263
Peru	588,175.20	58.8175
Venezuela (Bolivarian Republic of)	478,521.32	47.8521
Dominican Republic	443,226.26	58.0048
Benin	406,220.00	60.3546
China, Taiwan Province of	402,836.00	51.6655
United Republic of Tanzania	372,178.85	24.2685

Source: FAOSTAT (2021)

9.1.1 Importance for Human Nutrition and Uses

Pineapple fruit has enormous health benefits for humanity. As an excellent source of essential micronutrients, it provides substantial amounts of ascorbic acid (vitamin C), and thiamine (vitamin B1), along with vital minerals like potassium and manganese. Beyond these core components, its nutritional profile extends to retinoids (vitamin A), phylloquinone (vitamin K), as well as bioactive compounds including choline, betaine, calcium phosphate, zinc selenide, and trace concentrations of volatile phytochemicals.

Pineapple juice harbors over thirty distinct nutritive and physiologically active components, comprising carbohydrates, phenolic compounds, and organic acids with citric acid and L-malic acid as dominant forms. Furthermore, this tropical fruit possesses diverse phytochemical constituents such as flavonoids and tannins, recognized for their capacity to alleviate oxidative stress linked to premature aging processes.

Pineapple is rich in vitamin C which enhances immune system function. Vitamin C is the main water-soluble antioxidant in the human body, protecting the body from attack and damage. Although pineapple has relatively fewer calories, a cup of pineapple juice contains up to 131% of the daily antioxidant vitamin C value. Vitamin C is typically used to support immune system function.

Pineapple is also rich in bromelain. Pineapple protease was previously extracted from pineapple juice. Now it is obtained from mature plant stems salvaged during field cleaning. The yield of bromelain in pineapple stem juice is 2.15%. This enzyme, like papain in papaya, is used for meat tenderization and cold protection of beer. The proteolytic properties enable its incorporation into gelatin formulations to enhance hydrosolubility, while also finding industrial applications in latex emulsion stabilization and hide curing applications. The contemporary medical practice utilizes this cysteine protease as a postprandial digestive aid and postoperative anti-inflammatory management agent.

PALF has good antibacterial functions and can be spun pure or mixed with fibers such as cotton, wool, silk, soybean protein fiber, polyester, acrylic, polypropylene, etc. In China, the Philippines, Thailand and other countries, PALF has also been used to make fiber textile products.

9.1.2 Fiber and Textile Properties

Pineapple leaves contain abundant fibers that possess similar properties and quality to hemp fibers. In the main pineapple-producing areas, the leaves are often extracted as a by-product for fiber extraction or used as animal feed. The fiber quality varies among different varieties, with *A. comosus* var. *erectifolius* pineapple being particularly prized for its lack of thorns on its leaf edges and upright, which is the result of people's long-term evolutionary selection of wild pineapple owing to this trait makes it easier to obtain fiber (Fig. 9.1). The fiber content can reach up to 6%, and in Central America, people have been using these fibers to weave hammocks and fishing nets.

The appearance of PALF is similar to hemp fiber, existing in the form of bundle fibers, each containing about 10–20 single fibers. After alkali treatment, the rough substance on the surface dissolves, leaving a smooth fiber with some distortion or bending. The crystallinity and orientation of PALF are higher than flax fiber and yellow horse fiber, contributing to its higher density, strength and stiffness, with lower elongation at break (Neto et al. 2013). The fiber fineness of pineapple leaves ranges from 2.00–3.33 tex, similar to jute but inferior to ramie and flax. The length of PALF is comparable to leaf fiber length, but the single fiber length is only 3–8 mm, necessitating the use of semi-degumming and other processes to create process fiber suitable for spinning. The strength of PALF is 30–40 cN/dtex (centimeter Newton/decimetre tex, breaking strength unit) with an elongation at break of 3–8%. PALF contains more hydrophilic groups, and there are a lot of gaps and holes between single fibers, in addition, PALF hemicellulose and pectin content is high, so its moisture absorption and moisture release are fast (Sun 2019).

PALF is strong and naturally possesses antibacterial, bactericidal, and deodorizing properties (Fig. 9.1). It can be spun alone or blended with other fibers to make cloth, clothing, fishing nets, sewing thread, crafts, and paper. Additionally,

Fig. 9.1 Samples of Pineapple and fiber. (**a**, **b**): Pineapple plant sample; (**c**, **d**): Pineapple leaf sample; (**e**, **f**): Fiber extracted from pineapple leaves; (**g**): Products produced from pineapple fiber. (Figures constructed by our lab)

considering the degradation of synthetic polymers in the environment has numerical negative effects, pineapple fibers can also be used to prepare natural polymers (Baran et al. 2018).

9.1.3 Fiber Industry

Countries like the Philippines, India, Japan and China have been involved in the development and utilization of PALF in textiles. The process of extracting PALF involves extraction, drying, degumming and carding (Sun 2019). Traditional methods for extracting PALF include extraction using hemp water, biochemical, extraction and mechanical extraction (Joshi and Patel 2021; Kengkhetkit and Amornsakchai 2012). The water bubble extraction method includes soaking pineapple leaf tissue in water for 7–10 days, manually collecting the fibers, washing them, and drying them. Enzyme extraction involves soaking the leaves in enzymes or alkaline reagents, scraping, cleaning, and drying them. Mechanical extraction uses physical methods to break down the epidermis and mesophyll tissue of pineapple leaves to extract fibers.

Freshly extracted leaf fibers contain gum and impurities, that need to be removed by washing. In order to prevent the mildew and blackening of the fiber, it is necessary to ensure that the moisture content of the fiber is not more than 13%, so it is necessary to dry the wet fiber in time, and the drying can be natural drying and equipment drying. Natural drying generally takes 4–6 hours, but the degree of drying is not easy to determine. Equipment drying is mainly the use of fiber dryer drying.

The degumming of PALF mainly includes chemical degumming, microbial degumming, biochemical combined degumming, mechanical degumming, and lye boiling. Chemical degumming is easy to pollute the environment, microbial degumming has not been used in industrial production, and lye boiling is the most important means of PALF degumming (Sun 2019). After extraction, the fibers need to be combed to remove impurities, cut long fibers, and eliminate short fibers and hemp grains. The main processing types of leaf fiber combing include flat combing and silk round combing.

PALF has various uses in the textile industry for product development, clothing making, and weaving. Meanwhile, due to the superior mechanical and physical properties of PALF, they can be used to produce Polymer matrix composites. Recent advancements focus on developing eco-friendly composites using biopolymer matrices combined with natural fibers like PALF as sustainable alternatives to conventional materials (Alam et al. 2021; Mohanty et al. 2000). Therefore, incorporating PALF reinforcement significantly enhances material performance while promoting environmental sustainability. To overcome the processing bottleneck of the utilization of PALF, Tian et al. (2024) optimized a dual-enzyme system (laccase/xylanase) for bio-degumming of mildly alkaline-treated PALF. Current research demonstrates PALF's potential in creating high-performance biodegradable composites, including natural rubber-PALF biocomposites functioning as plant-derived leather substitutes (Duangsuwan et al. 2023). In fiber-reinforced polymer materials, epoxy resins, polyester polymers, and vinyl ester compounds serve as common matrix components (Xiao et al. 2024). While PALF textiles have achieved industrial production, there is still no professional, complete PALF textile production line. Degumming, spinning, printing and dyeing processes still rely on other hemp production lines to complete, leading to poor technology and equipment matching. The industrialization of PALF textiles still has a long way to go.

9.1.4 Domestication, Selection and Early Improvements

The earliest scientific documentation of wild pineapple species was established by Baker (1889), who named it *Acanthostachys ananassoides*. However, historical records indicate initial documentation of diminutive wild pineapple varieties as early as 1647, with Father Gumila's 1741 account further confirming their proliferation in the Orinoco Basin, pineapples were already a major tropical fruit crop. Even before Columbus arrived in America in 1493, several local varieties of pineapple had differentiated and distributed in tropical Americas (Collins 1949, 1960). The hypothesis was proposed in 1919 that pineapples originated in southern South America and gradually migrated northward (Collins 1960; Pickersgill 1976; Purseglove 1972). European cultivation commenced during the sixteenth century, followed by subsequent expansion to Asian and African territories through the 1700s (Collins 1949). This is also the most widely accepted hypothesis currently. However, the hypothesis of the origin and migration of early pineapple is based on incomplete polymorphism and distribution data. Since 1981, the analysis of the morphological diversity of pineapples distributed north of the Amazon has provided more support for the hypothesis of southern origin.

Before Europeans arrived in South America, it was the early domestication stage of South American pineapples. Such as 'Spanish' (syn. Roja Espanola), 'Perola', 'Smooth Cayenne', 'Queen', 'Singapore Spanish' and 'Perolera', among others, in existence when Columbus arrived in the Caribbean (Levitt 2014; Mendiola et al. 1951). The second domestication occurred during the sixth to the ninth century, which was the basis for modern breeding, as collected and selected, for preexisting clones by Europeans. The third stage is modern commercial breeding, and the earliest recorded breeding activities were in 1913 and 1915, with some crossing between 'Cayenne' and 'Queen'. Filipino breeders have also used 'Red Spanish' and 'Smooth Cayenne' as parents, producing 20,000 seedlings annually, but the project has not been continued (Wortmann and Kerns 1964). The breeding program implemented in the 1950s was relatively complete, supported by the Pineapple Research Institute (PRI), and two commercial varieties, 'MD-2' (73–114) and 'CO-2' (73–50), were cultivated. The project produced 591,000 seeds in 1958 and 160,000 seeds comprised 115 crosses in 1964 (Smith 1965). By 1970, the program was producing 400,000 seeds per year (Williams 1970). In the period from the 1960s to the 1970s, a large number of hybrid combinations were produced (Williams and Fleisch 1993). This provides a large amount of material basis for pineapple breeding.

Crop domestication refers to the process of artificial selection of crop growth and yield phenotypes during human breeding and cultivation of specific crops. In order to obtain crop phenotypes that meet people's expectations, the domestication process may lead to degrees of reproductive isolation of crops and even the generation of new species. However, the physiological, morphological, and reproductive isolation phenotypic changes resulting from the domestication process of pineapples were not significantly different. Compared with wild species, the domestication phenotype of domesticated pineapples mainly focuses on the enlargement of pineapple fruits, the increase in fruitless numbers, the shortening and widening of leaf types to improve photosynthetic efficiency, the lengthening of reproductive periods to produce larger fruits, and the decrease in seed number due to lower sexual reproduction fertility (Leal and Coppens d'Eeckenbrugge 1996). There are several main domesticated pineapple varieties, *A. comosus* var. *comosus*, *A. comosus* var. *erectifolius* and *A. comosus* var. *bracteatus*. These pineapple varieties are mainly used for consumption, fiber extraction, and decorative purposes.

9.2 Current Cultivation Practices and Challenges

9.2.1 Current Cultivation Practices

Pineapple leaves are thick and have well-developed aerial roots, which reduce transpiration and help absorb moisture from the air, giving them a certain degree of drought resistance. Pineapple production areas are mainly found in regions with high temperatures and abundant rainfall, so pineapple cultivation typically does not require additional watering and relies mainly on rainfall. When selecting areas for pineapple planting, it is recommended to choose mountainous or gently sloping terrain with loose and permeable e soil and a pH value between 4.5 and 6.5. Pineapples require a large amount of nitrogen, phosphorus, potassium and various trace elements for growth. Increasing plastic film coverage can ensure the soil temperature and moisture conditions in the soil, which is beneficial for accelerating the growth rate of pineapples, enabling them to bear fruit earlier. During the process of pineapple cultivation, methods such as ethephon and calcium carbide are mainly used to promote flowering, among which ethephon is mainly used for promoting flowering. At the same time, some pineapple varieties are not very sensitive to ethylene, and the ethylene content should be increased to have a better-promoting effect on flowering. Plant growth regulators can be used to increase fruit size and improve quality. To prepare a fruit preservation solution, dissolve 1 g of gibberellin crystal dry powder with a 6% content in a 75% alcohol solution to make 100 mL. Removing the crown buds of young pineapple fruits manually and adding 1–2 drops of medication to the top of the fruit can significantly enhance fruit size and quality. Pineapples should not be grown in paddy fields or poorly drained soil types as these conditions can lead to issues such as pineapple mold and heart rot due to high humidity and alkaline soil.

9.2.2 Current Agricultural Challenges

With the reliance on manual labor, the cost of production is high. This method limits the growth of the pineapple industry. Mechanizing pineapple production is crucial for increasing economic benefits and promoting high-quality industrial development. Currently, there is limited research on integrating pineapple agricultural machinery and agronomy, and a scalable model has not been established. Exploring suitable mechanized pineapple cultivation models to integrate agricultural machinery and agronomy is essential for advancing pineapple production mechanization technology.

Various factors impact pineapple yield, including planting methods, climate, soil types, and cultivation techniques. For example, the traditional fertilization methods for pineapple mainly include sprinkling and furrow application. With the prevalence of continuous pineapple cultivation in recent years and the problems caused by seasonal drought, traditional fertilization can no longer meet the needs of pineapple industry development. Promoting new integrated water and fertilizer technology will play a certain promoting role in the development of the pineapple industry.

Pineapple disease is a significant factor that affects the yield and quality of pineapple, including wilt disease, black heart disease, and pineapple mealworms. Wilt disease can cause the root system of pineapple plants to stop growing and necrosis, and the leaves of the plants cannot undergo photosynthesis, which hinders the growth of the leaves and leads to chlorosis, withering, resulting in a decrease in pineapple yield and quality. The pineapple powder scale is an important factor in the infection of pineapple wilt disease, and it is the main vector for spreading wilt disease. The occurrence of "black heart disease" in pineapple is not only influenced by weather factors but also by cultivation practices such as excessive use of chemical fertilizers, expanding agents and ripening agents. The planting technique that focuses on reducing chemical fertilizer use, increasing organic fertilizer application, minimizing expanding and ripening agents, and reducing planting density can effectively enhance pineapple quality and efficiency.

In summary, during the breeding process of pineapple varieties, it is necessary to ensure that pineapple varieties have strong stress resistance and disease resistance and to increase the application of genetic engineering breeding technology. Planters should familiarize themselves with the key points of pineapple cultivation techniques, such as water and fertilizer management, and disease and pest control.

9.2.3 Breeding Objectives

Because the PALF industry is underdeveloped, there are limited varieties of specially cultivated leaf fiber. The main focus is on cultivating fresh fruit or processing pineapple. The goals of pineapple breeding include developing high-yielding and disease-resistant cultivars. These improved varieties demonstrate enhanced phytopathological resistance, specifically against soil-borne pathogens causing heart rot and root rot syndromes. However, breeding objectives vary from country to country based on geography, climate, and market demand (Li et al. 2022). Consumer-oriented fresh market selections prioritize mid-range fruit weights (1.0–2.0 kg) with uniform cylindrical morphology and sucrose levels exceeding existing commercial varieties. When it comes to processing, larger-fruits weighing ≥2 kg, cylindrical in shape, containing elevated brix values coupled with improved juice yield are ideal for canning purposes (Kumar 2006; Li et al. 2022).

9.3 Germplasm Biodiversity and Conservation

9.3.1 Germplasm Diversity and Phylogeny

Understanding the diversity of pineapple is crucial for understanding the evolution and domestication history of the fruit, as well as for new breeding of pineapple varieties. Currently, all pineapple germplasm has been reclassified into the genus Ananas Millers, which encompasses two taxonomically validated species: *A. comosus* L. Merr. and *A. macrodontes* Morren (Coppens d'Eeckenbrugge and Leal 2003). *A. comosus* is a diploid plant with 50 chromosomes, while *A. macrodontes* are tetraploids with 100 chromosomes (Brown and Gilmartin 1986; Brown et al. 1997; Marchant 1967). *A. comosus* complex comprises five recognized infraspecific taxa:: comosus, microstachys, parguazensis, erectifolius, and bracteatus (Ferreira and Cabral 1993; Scherer et al. 2015). Research on the phenotype and genetic diversity of pineapple germplasm has shown that *A. comosus* var. *microstachys* and *A. comosus* var. *parguazensis* are wild varieties, while *A. comosus* var. *comosus* (the edible pineapple), *A. comosus* var. *erectifolius* (used for fiber), and *A. comosus* var. bracteatus (an ornamental pineapple) are domesticated varieties (Coppens d'Eeckenbrugge and Leal 2003; Coppens d'Eeckenbrugge and Govaerts 2015). The main traits involved in pineapple domestication are fruit quality, yield, and regulation of flowering time (Wee and Thongtham 1991; Duval and Coppens d'Eeckenbrugge 1993). Varieties like 'Smooth Cayenne' are popular commercial choices due to their high yields, robust growth, and ease of flowering control. 'Perola' is favored in Brazil and some African countries for its sweet flavor, and white flesh. Thornless varieties like 'Primavera', 'Manzana', 'Monte Lirio', and 'Perolera' are grown in specific regions for easier planting and harvesting. *A. comosus* var. erectifolius pineapples have the highest leaf fiber content and best quality in the pineapple genus, originating from wild pineapple domesticated under various conditions.

With the advancement of genomics and molecular marker technologies, DNA-based molecular markers such as RAPD, RFLP, AFLP, SSR, and SNP have been extensively applied to assess genetic diversity in pineapple (Ruas et al. 1995; Kato et al. 2004; Wang et al. 2017). Ruas et al. (1995) employed random amplified polymorphic DNA (RAPD) markers to identify and differentiate four cultivated pineapple varieties: 'Perola', 'Smooth Cayenne', 'Primavera', and 'Perolera'. Duval et al. (2001) utilized restriction fragment length polymorphism (RFLP) markers to analyze 301 Ananas accessions, which revealed that the genetic diversity of cultivated pineapples is lower compared to their wild counterparts. These studies highlight the effectiveness of molecular markers in evaluating genetic variation within pineapple germplasm..

Kato et al. (2005) used amplified fragment length polymorphism (AFLP) markers to distinguish 148 accessions of *A comosus* and found high genotype diversity in cultivated pineapple species. Shoda et al. (2012) were able to distinguish 31 cultivated varieties in Japan through 18 SSR markers. Feng et al. (2013) used 18 newly developed SSR markers to define the genetic diversity of 48 pineapple germplasms. Zhou et al. (2015) confirmed that 96 of the 213 SNP markers obtained by mining public databases can be used to distinguish Ananas accessions within the USDA pineapple germplasm collection located in Hawaii. This above indicates that molecular markers serve as effective tools for evaluating pineapple germplasm diversity. With the utilization and popularization of genomics and metagenomics technologies, the excavation and utilization of pineapple germplasm genetic resources will be greatly accelerated.

9.3.2 Genetic Resources Conservation Approaches

Currently, the primary storage facilities for pineapple germplasm worldwide are the United States Department of Agriculture (USDA) - Agricultural Research Service in Hilo, Hawaii. They maintain more than 180 pineapple cultivars and wild accessions.

Other institutions also store pineapple germplasm, such as EMBRAPA/CNPMF in Cruz das Almas, Brazil, and CIRAD-FLHOR in Martinique. CIRAD has collected over 600 pineapple germplasm materials (Harry et al. 2005), while CNPMF in Brazil collected 500 germplasm materials (Ferreira and Cabral 1993). In China, the South Subtropical Tropical Crop Research Institute (SSCRI, CATAS) of the Chinese Academy of Tropical Agricultural Sciences, FTRI, and the Tropical Crops Genetic Resource of Chinese Academy of Tropical Agricultural Sciences (TCGR, CATAS) collect and preserve various pineapple germplasm resources. Additionally, 130 pineapple germplasm resources are preserved in the National Tropical Fruit Field Gene Bank (SSCRI) (Li et al. 2022). Due to some degree of self-incompatibility, pineapple germplasm is primarily stored through vegetative propagation by crows, slides, feeders, or in vitro culture. However, during pineapple tissue culture, there is a certain accumulation of somatic mutations. This presents challenges for germplasm identification but also provides a material foundation for clone selection.

9.4 Traditional Breeding

9.4.1 Improvement Strategies, Methods and Limitations

Although hybrid breeding is limited in pineapple, there are also some pineapple varieties obtained through this method. The 'MD-2' variety, selected by Hawaiian Pineapple Research Institute using 'Smooth Cayenne' as a parental hybrid, has high sugar and ascorbic acid content. Including 'BRS Vitória' and 'BRS Imperial', the hybrid strains developed by the Embrapa Research Center in Brazil exhibit strong resistance to Fusarium and high yield. Malaysia's 'Josapine', which was bred by crossbreeding 'Spain' and 'Smooth Cayenne' as parents, has good fresh eating characteristics and strong resistance to black heart disease. Australia's 'Aus Jubilee', 'Aus Carnival', and 'Aus Festival' are all pineapple varieties selected through hybridization. Pineapple cross-breeding in Taiwan has bred the 'Tainung' series of pineapple varieties, such as 'Tainung 16', 'Tainung 17'and 'Tainung 21′. Currently, the new pineapple varieties bred in the Chinese Mainland mainly include 'Yuecui', 'Yuetong', 'Yuetian' and 'Renong 8′ (Li et al. 2022).

In addition to traditional hybrid breeding, mutation is also an important way to improve pineapple varieties. Mutation by ethyl methyl sulfonate (EMS) is currently the main reported mutagenic method, and pineapple stems, buds, and other materials can be used as mutagenic materials. Peng (2017) obtained a cold-resistant mutant line through EMS treatment. From the clone selection perspective, the excellent variant variety 'N67-10' selected by Japan-based 'Smooth Cayenne' not only has a high and stable yield, but also has advantages such as high sugar, low acid, aroma, and juiciness (Ikemiya et al. 1984). In addition, some spiky variants have been found in 'Cayenne'. For example, pineapple varieties such as 'Tainung 1', 'Tainung 2', and 'Tainung 3' were selected from 'Smooth Cayenne'. For trait improvement varieties,

'Jinxiang' pineapple is an early maturing pineapple variety selected from the main domestic pineapple variety 'Comte de Paris', and has good stress resistance and better cold resistance than 'Comte de Paris' (Li et al. 2022).

Early pineapple breeding selection and genetic analysis were achieved through the assistance of DNA markers. Genetic diversity can be distinguished between wild and cultivated species using various types of DNA markers (Ruas et al. 2001; Duval et al. 2001, 2003; Kato et al. 2004; Paz et al. 2005). Both wild and cultivated pineapples exhibit high heterozygosity, which is maintained through self-incompatibility and vegetative propagation. This characteristic provides abundant genetic resources for developing DNA markers. The construction of the first pineapple genetic map utilized molecular markers from 46 individuals in the F1 generation of *A. comosus* and *A. bracteatus* (Wai et al. 2014; Carlier et al. 2004). The genetic map for var. comosus includes 156 DNA markers involving (33 RAPDs, 115 AFLPs, 8 ISSRs) across 30 linkage groups. There are more molecular markers in var. bracteatus (60 RAPDs, 264 AFLPs, 11 ISSRs), and 50 linkage groups. A new genetic map from 142 F2 populations contains 574 DNA molecular markers (454 AFLP, 79 RAPD, and 41 ISSR), and 46 linkage groups in total (Carlier et al. 2007; Botella and Smith 2008). These genetic maps have been continuously improved in future research (Carlier et al. 2006), with the number of linkage groups decreasing to 28, approaching the chromosome number level of $n = 25$ (Sousa et al. 2013).

Most cultivated pineapple varieties, experience self-incompatibility, while some self-compatible varieties see a rapid decline in their offspring after a few generations. This poses a challenge to traditional hybrid breeding methods in improvement. Pineapples are typically reproduced through vegetative propagation, lacking sexual reproduction and homologous recombination. This can lead to the accumulation of harmful mutations in pineapples during the long-term breeding process, which is not conducive to the selection and improvement of pineapple germplasm. At present, it is not clear whether the self-incompatibility of pineapples is naturally generated spontaneously or obtained through domestication, and the relevant mechanism has not been elucidated.

9.4.2 Traditional Breeding Methodologies

9.4.2.1 Clone Selection

Due to the extensive reliance on vegetative propagation within many cultivated pineapple varieties, a significant amount of mutations and adverse variations have accumulated, fostering opportunities for the development of novel pineapple varieties. There have been instances of clone selection through mutation. For example, approximately 20% of plants within the 'Smooth Cayenne' cultivar are mutated (Smith et al. 2005). In Taiwan, 'Tainung No. 1/2/3', was developed through clonal selection from 'Smooth Cayenne', through clone selection. A mutant line called

PQM-, characterized by late ripening was developed from the Queen group in India (Prakash et al. 2009). In Australia, two variations 'magligo' and 'Alexander' were selected from 'Queen', while other varieties were chosen from 'Spain' and Pinar respectively (Pérez et al. 2009, 2011, 2012). 'Jinxiang' is a variety with early maturing traits selected from Japan through the bud variation bud of 'Comte de Paris', and it also exhibits stronger stress and cold resistance compared to its parent plants (Wang et al. 2016).

9.4.2.2 Conventional Hybridization

Hybridization is the primary method of modern pineapple breeding, utilizing existing varieties, such as 'Smooth Cayenne', 'Monte Lirio', 'Queen' and 'Singapore Canning' as parents. The sequencing results of the pineapple genome, have shown that cultivated pineapples have relatively high gene heterozygosity. For example, pineapple varieties F153, MD2 and CB5, have heterozygosity rates of 1.89%, 1.98%, and 2.93%, respectively (Ming et al. 2015). Currently, there are various breeding around the world, such as the Japanese variety 'Okinawa 17', which is a hybrid developed from a second-generation cross involving 50% 'Cream Pineapple', 25% HI101, and 25% 'McGregor' (Ogata et al. 2016). Furthermore, varieties such as 'Soft Touch', 'Gold Barrel', and 'Julio Star', have also been bred through hybridization in the Japanese region. Varieties generated in Taiwan, such as 'Tainung 4', were bred by crossing between 'Singapore Spanish' and 'Smooth Cayenne'. 'Tainung 17', generated by 'Smooth Cayenne' and 'Rough' (syn. 'Queen') cross. 'Tainung 21' is composed of 5/8 'Smooth Cayenne', 1/4 'Queen', and 1/8 'Singapore Spanish'. 'Tainung 22', was selected by cross Since 1991, Australia has employed PRI cultivars, including PRI 73–50, PRI 71–92, PRI 53–116, and PRI 59–656, as well as pre-Columbian cultivars like 'Smooth Cayenne' and 'Queen', as parental lines (Sanewski and DeFaveri 2017). At the same time, breeding activities in the region have also utilized varieties such as 'Bracteatus', 'Monte Irio', 'PRI 62–85', 'PRI 75–970', and 'MD - 2' to develop 'Aus Festival' (Smooth Cayenne × PRI 73–50), 'Aus Carnival' (PRI 73–50 × PRI 53–116), and 'Aus Jubilee' ('Smooth Cayenne' × 'PRI 73–50'), among other varieties. Some commercial companies are also involved in pineapple hybrid breeding activities, such as Dole has developed the 'Dole-14' variety through a selection process involving 'MD-2' and the PRI variety 'P-1972' as parent lines (Sanewski and DeFaveri 2017). 'Yue Crispy', the first hybrid pineapple variety, came into existence in 1957 as a result of a cross between 'Smooth Cayenne' and 'Yellow Mauritius'. It secured a variety of rights in 2015 (Liu et al. 2006). Despite some progress in hybrid breeding, it is still limited by the availability of parental materials. Variety selection, primarily focuses on phenotype screening, with less emphasis on genetic analysis of genes (Gonzales and Vriesenga 2005; Young and Gonzales 2009; Young 2016).

9.4.3 Enhanced Traits and Hybrid Cultivars

Cultivating pineapple varieties with excellent characteristics is typically achieved through hybridization. To develop cultivars resistant to the fungal pathogen fusariosis, researchers primarily employed outcross combinations involving 'Perolera', 'Primavera', 'Perola', and 'Smooth Cayenne', resulting in the generation of 60,537 seedlings for selection (Cabral et al. 1993). The hybrid population includes 'Perolera' × 'Smooth Cayenne' and 'Primavera' × 'Smooth Cayenne' (Cabral and De Matos 2009; Ventura et al. 2009). Furthermore, some individuals have expanded the population through crossing to screen for resistance to the fungal pathogen Fusarium variety. Breeders have also developed some varieties resistant to Fusarium, such as 'Vitória' created by crossing between 'Primavera' and 'Smooth Cayenne'(Ventura et al. 2007). Additionally, since 1998, molecular markers that can be associated with resistance locality have been developed (Rohrbach et al. 1998). When cultivating disease-resistant varieties, it is important to note that domesticated '*A. comosus* var. *braceatus*' is essentially a larger and more edible version of wild pineapple (Kilian et al. 2016; Zhou et al. 2015). The process has also led to an enhancement in fruit size, along with modifications such as more refined leaf edges, lower fruit fiber concentrations, reduced peduncle length, heightened self-incompatibility, and improved flavor profiles.

9.4.4 Role of Biotechnology

Clone selection is often random, and traditional cross-breeding requires a breeding cycle of at least 15 years (Ogata et al. 2016), so it is important to establish other breeding techniques. With the development of modern biotechnology, there are plenty of opportunities to improve pineapple breeding. The techniques include genetic engineering, development of molecular markers, assessment of genetic diversity, development of DNA markers associated with traits of interest and their use in molecular assisted breeding (MAS), construction of genetic linkage maps, and genome, transcriptome analysis (Ogata et al. 2016). The use of next-generation sequencing platforms to fully explore genetic variations and create DNA markers, developing genetic markers that can be used to screen materials for agronomic traits and other traits of interest, particularly those related to fruit, will be very valuable in pineapple. This approach will reduce breeding timelines and enhance the identification of promising candidates, thereby lowering overall breeding duration and expenses. However, few DNA markers associated with simple inherited traits have been found in pineapple, and few phenotypes are controlled by quantitative trait loci (QTL) (Ogata et al. 2016). Moreover, pineapple transformation provides a precise method for making targeted modifications to the plant genome, serving as an exceptional strategy for enhancing genetic improvement. Both direct gene transfer and *Agrobacterium*-mediated transformation are viable approaches (Ko et al. 2008).

Ko et al. (2006, 2013) successfully obtained an anti-dark heart strain with polyphenol oxidase (*PPO*) using a particle bombardment transformation system, demonstrating that down-regulating *PPO* expression in pineapple could reduce the incidence of black heart disease.

9.5 Tissue Culture Applications

The number of sprouts per pineapple plant determines its reproductive coefficient to be only around 10–20. At the same time, long-term asexual reproduction can lead to variations, product degradation, and low quality in seed offspring. Tissue culture offers an efficient method for propagating pineapple seedlings, enabling the acceleration of pineapple reproduction and the generation of a significant number of plants within a brief timeframe. Compared with conventional asexual reproduction methods, this method boasts benefits such as rapid propagation speed, high yield, and independence from seasonal and climatic constraints. The tissue culture initiation from pineapple crows could take about 3–4 months (Fig. 9.2) (Nagai and Rosal 1995). Multiplication also needs another few months to ensure a steady growing number of in vitro growth shots. The shoot initiation and replication media used in the tissue culture process are the same in most research systems (Ko et al. 2006; Nagai and Rosal 1995; Soneji et al. 2002). Moreover, prolonged use of high concentrations of cytokinin in the culture medium during the tissue culture process can increase the probability of variation in pineapple (Bairu et al. 2011). Therefore,

Fig. 9.2 Phenotype of pineapple tissue culture process. (**a**–**d**): Pineapple callus tissue induced differentiation into seedlings; (**e**–**f**): Root induction and seedling formation. (Figures constructed by our lab)

the use of regulators should be strictly controlled during the pineapple tissue culture process. Callus and embryogenic cells/cell clusters can be derived from leaf bases and subsequently utilized for transformation purposes or to generate new plants, completing the pineapple breeding process (Espinosa et al. 2002; Firoozabady et al. 2006; Gangopadhyay et al. 2009; Ko et al. 2005, 2006; Ma et al. 2012; Nan et al. 1996; Nan and Nagai 1998; Yabor et al. 2006a, b). At present, the tissue culture technology of pineapple is relatively mature, but due to certain variations in the process of tissue culture, tissue culture seedlings have not been used in the production of pineapple seedlings.

9.6 Molecular Breeding

9.6.1 Molecular Marker-Assisted Breeding

With the release of genetic maps of pineapple developing by markers such as random amplification of polymorphic DNA (RAPD), AFLP (amplified fragment length polymorphism), and simple sequence repeat (SSR), a solid foundation has been provided for exploring pineapple genetic resources and improving breeding efforts (Carlier et al. 2004, 2006, 2012; Sousa et al. 2013). Especially with the widespread use of whole genome sequencing technology, the discovery of a large number of SNP loci as markers has greatly improved Marker-Assisted Breeding (Myers et al. 2000; Schmid et al. 2003; Newcomb et al. 2006). Due to the publication of the reference genome for *A. comosus* var. comosus F153, an F1 segregating population was constructed using F153 × CB5. Comparison was conducted using Bowtie2 followed by SNP identification using GATK. An ultra-high density genetic map of F153 was generated, including 296,896 SNPs in 25 linkage groups. This high-density genetic map was utilized to align the scaffolding and contiguities derived from the genomic scaffolding into 25 chromosomal groups (Ming et al. 2015).

The P site is the locus that determines the development of pineapple leaf spines. This site greatly facilitates the planting and harvesting of pineapple by controlling the presence of leaf thorns. Recognizing the agricultural importance of spine traits in pineapple production, Urasaki et al. (2015) implemented RAD-seq methodology and detected five CAP and SSR markers linked to the P site. These markers align with the previously identified SSR_ AJ845056 located in the unified linkage group, confirming the linkage of these DNA markers to the P site.

9.6.2 Functional Genomics

Currently, the genomes of some pineapple varieties have been sequenced and annotated, such as ‘F153’, ‘MD-2’, ‘GL1’ and ‘CB5’ , ‘BL’, ‘YLL’ and ‘LY’ varieties (Ming et al. 2015; Redwan et al. 2016; Chen et al. 2019; Feng et al. 2024). Due to its self-incompatibility property, most pineapple cultivars such as 'F153' are heterozygous. To assemble the complex genome of 'F153', sequencing technologies including Illumina HiSeq2000, Moleculo, Roche 454, as well as PacBio P6-C4 for third-gen sequencing, were employed (Ming et al. 2015). These sequences were assembled with contigs averaging 2 kb and scaffolds at 13 kb, and it is an unsatisfactory result. Owing to the high heterozygosity, the assembly of 'F153' remained poor. To address this issue, an F1 hybrid population based on 'F153' and 'CB5' was used. Because the F1 offspring incorporate both parental genetic types, it was used to improve the quality of pineapple genome assembly (v1 assembly). Finally, by combining the construction of the BAC library (v2 assembly) and the sequencing data of the third-generation genome, the assembly level of the pineapple genome was continuously improved. Finally, by combining v2 assembly with third-generation PacBio sequences sequencing data, this assembly of the pineapple genome has been improved compared with the initial Illumina-only assembly. The completed genome assembly covered 382 Mb, representing approximately 72.6% of the projected 526 Mb genome size (Luo 2012; Myers et al. 2000).

The genome of the 'MD-2' pineapple variety was first completed using third-generation long-read sequencing technology. However, due to the high probability of errors in the PacBio sequencing method, Illumina's method was combined to improve the accuracy of sequencing. Compared to other plant genomes, the pineapple genome performs better in terms of gene coverage and continuity, resulting in the final assembly version named ACMD2. This high-quality pineapple genome has proven to be a valuable resource for fruit crop breeders.

The release of the CB5 genome employed a combination of technologies including PacBio SMRT sequencing for long-range sequencing, Hi-C scaffolding for chromosomal organization, and Illumina short-read sequencing for high-throughput data generation (Chen et al. 2019). Sequencing the raw data, resulted in 513 Mb assembled into 25 chromosomes with 29,412 genes. By comparing the assembled CB5 with MD2 and F153 for genomic analysis, the relevant characteristics of functional genes such as sugar content, color development, fruit ripening, and fiber development were analyzed. Gene re-sequencing was performed on 89 *Ananas* accessions. Among these varieties, 67 accessions belonging to *A. comosus* var. *comosus*, nine accessions of var. *bracteatus*, two accessions belonging to var. *erectifolius*, nine accessions belonging to the wild var. *microstachys*, and two accessions coming from *Pitcainia gracilis* and *P. punicea* as outgroups.

Wild pineapple 'GL1' belonging to *Ananas comosus* var. *bracteatus* a distinct pineapple variety. *Ananas comosus* var. *bracteatus* f. tricolor (GL1) is categorized as a red pineapple accession and exhibits white leaf margins that transition to red during flowering in spring and autumn (Feng et al. 2022). The GL1 genome was assembled to a size of 461 Mb, with (1) a contig N50 value of 2.97 Mb and a Benchmarking Universal Single-Copy Ortholog (BUSCO) score of 97.3%, indicating high assembly quality. Based on the assembly results, GL1 is organized into 25 pseudochromosomes. 'BL', 'YLL' and 'LY' is telomere-to-telomere assembly (Feng et al. 2024). Those three varieties are wild pineapple variety ('YLL'), fiber pineapple variety ('LY'), and globally cultivated edible pineapple variety ('BL'), respectively. The LY variety, characterized by its approximately 6% dry fiber content, is utilized for fiber extraction (Leal and Amaya. 1991; Feng et al. 2024). For 'BL', 'YLL', and 'LY', they generated HiFi sequencing data (~65X) and Hi-C data (~118X), yielding genome assemblies with BUSCO completeness scores of 99.3% (reference genome) and 99.2% (annotation). The results demonstrate a high level of completeness in the three assembled genomes. Their heterozygosity rates were 1.45% ('BL'), 1.45% ('YLL'), and 2.23% ('LY'), respectively, aslo indicating their highly heterozygous of pineapple genomes.

Pineapple diverged from the cereals (Poaceae) about 100 million years ago (Mya), suggesting that pineapple is an important evolutionary reference for investigating the genome evolution in the Poaceae genome (Givnish et al. 2014; Magallón et al. 2015). The release of the pineapple genome provides important evidence for determining some important events in the evolutionary process of monocotyledonous plants (Ming et al. 2015). In angiosperms, many have experienced recurrent WGDs (Tang et al. 2008). Contrasting with most sequenced cereal crops, pineapple demonstrates simpler genomic architecture. Examination of collinear regions identified 388 conserved chromosomal segments and 4891 duplicated gene pairs attributable to ancient whole genome duplication events (Liang et al. 2018). Additionally, 64% of the gene regions in the 25 linkage groups are in a collinearity state. The results of collinearity analysis indicate that about 35% of the pineapple genome has a different number of duplicated segments, indicating there were perhaps two WGDs happened in the pineapple (Paterson et al. 2012; Tang et al. 2010).

9.6.3 Analysis of Transcriptome

Through gene annotation, 27,024 protein-encoded genes can be annotated in the 'F153' variant genome and 27,017 in the 'MD-2' variant genome(Ming et al. 2015; Redwan et al. 2016). Moyle et al. (2005, 2006) generated more than 5600 ESTs associated with fruit maturation and root-nematode interactions, and the data was deposited in an online resource PineappleDB (http://genet.imb.uq.edu.au/Pineapple/). Ong et al. (2012) sequenced transcriptional materials extracted from mature pineapple fruits using next-generation sequencing and assembled 28,728 unique transcripts. Koia et al. (2012) constructed a pineapple-specific microarray to profile transcriptional dynamics during fruit maturation. Currently, transcriptome data for roots, leaves, and flowers have been obtained in the 'F153' variety, and for fruits in the 'MD-2' variety (Wai et al. 2016a, b). RNA-seq mapping revealed 22,057 transcriptionally active genes, though 4967 remained undetected across these tissues (Wai et al. 2016a). A total of 17,522 genes were screened for subsequent differential expression analysis. The expression data of these differential genes can be downloaded from (http://proteomics.ysu.edu/publication/data/Pineapple/). In ripening fruits, a substantial proportion of genes—such as those producing proteases, glycosyl hydrolases, and related catalytic proteins—exhibited increased expression (Wai et al. 2016b). For CAM-related carbon fixation genes that have been identified, beta-carbon anhydrase 2 and alpha-carbonic anhydrase 1 were upregulated in leaf tissue (Ming et al. 2015).

9.7 Genetic Engineering

9.7.1 *Transformation Methods*

Transgenic technology is highly significant for improving the quality, growth and yield traits of pineapple enhancing its economic value. In pineapple, most transformations are achieved through *Agrobacterium*-mediated biological methods. Genetically modified seedlings can be produced from leaf bases or stem segments through organogenesis and then selected by using hygromycin (Wang et al. 2009). Firoozabady et al. (2006) employed embryogenic cell cultures and chloroform-based selection techniques. Sripaoraya et al. (2001) and Ko et al. (2005, 2006) utilized particle bombardment for pineapple gene transformation and demonstrated that transgenic materials transformed through biological methods often exhibit elevated transgene copy numbers compared to physical methods. Many common *A. tumefaciens* strains employed in pineapple genetic engineering include LBA4404, EHA101, EHA105, AGL0, AT2260, and GV3101. However, some studies have shown that GV3101 (=C58pMP90) has higher efficiency in GUS transformation compared to EHA101 and LBA4404. Optimal bacterial cell densities for transformation span 8 × 10^7 to 10^9 cells/mL (Firoozabady et al. (2006). To get higher transformation efficiency of pineapple, acetosyringone (AS) is usually added to the transformation system during the infection and co-culture stages, with working concentrations commonly maintained between 100 and 300 μM (Firoozabady 2011; Firoozabady et al. 2006; Wang et al. 2009; Gangopadhyay et al. 2009; Mhatre 2013). To increase the contact area between *Agrobacterium* and the transformation material, a vacuum is usually used to enable *Agrobacterium* to better penetrate the gaps in plant tissue (Ko et al. 2005; Firoozabady et al. 2006; Wang et al. 2009; Gangopadhyay et al. 2009; Firoozabady 2011). Finally, dark culture for 3 days at 24–25°C is the most commonly used co-culture condition. Scientists found that depositing the pineapple callus and *A. tumefacien* under 25°C for 24 h resulted in the highest GUS expression in pineapple calli. However, the expression of transformed genes cannot be detected after co-culturing for a longer time (Fullner and Nester 1996). Moreover, GUS transformation experiments have shown that the best co-cultivation temperatures for *Phaseolus acutifolius* and *Nicotiana tabacum* were 22°C and 20°C respectively. It can be seen that the *Agrobacterium* transformation system of pineapple is not static and needs to be selected based on the materials used for transformation and the selected *Agrobacterium* strain (Dillen et al. 1997; Su et al. 2012).

9.7.2 *Enhanced Traits and Transgenic Cultivars*

Controlling the flowering period of pineapples is an important aspect of commercial cultivation. Allowing pineapples to bloom naturally or early, can lead to a decrease in pineapple yield and inconsistent flowering periods, which can also increase the harvesting cost of pineapples (Paull et al. 2017). In pineapple plants, the flowering phenotype is regulated by ethylene signaling, making it possible to regulate the flowering stage by regulating ethylene synthesis (Burg and Burg 1966; Min and Bartholomew 1996). Aaminocyclopropane synthase (ACS) is a key gene in the ethylene synthesis pathway, so inducing silencing of this gene to regulate flowering has been applied in practice (Trusov and Botella 2006). After constructing a gene silencing vector, A*grobacterium tumefaciens* mediated transformation was carried out using electromagnetic cells/issues as the material (Firoozabady et al. 2006). Polyphenol oxidation (PPO) can oxidize phenols in fruit pulp, leading to the development of pineapple black heart disease. This disease usually occurs when the fruit is stored at low temperatures (<10° C) and in low light conditions (Stewart et al. 2001). Silencing *PPO* gene vectors through gene bombardment and *Agrobacterium* mediated biotransformation in leaf-induced callus tissue has been successful (Ko et al. 2005, 2006). Transgenic plants regenerated from calli showed that approximately 38% of transgenic lines exhibited no blackheart syndrome (Ko et al. 2013).

Some attempts have also been made to improve the resistance of pineapples through genetic modification. For example, utilizing a gene gun transformation method to transfer the bialaphos resistance (bar) gene into pineapple plants can achieve basta resistance (Sripaoraya et al. 2001). Oomycete pathogens like *Phytophthora* and *Pythium* induce core and root necrosis in pineapple crops (Green and Nelson 2015). Espinosa et al. (2002) integrated genes expressing chitinase and pathogenesis-related AP24 proteins into pine-apple calli through *Agrobacterium*-mediated transformation. To engineer nematode-resistant cultivars, researchers introduced a truncated rice cystatin gene (OcIΔd86) regulated by the sugarcane ubiquitin 9 (scubi9) promoter, transfected into leaf base and stem explants via *Agrobacterium* co-cultivation. However, phenotypic assessment of transgenic lines revealed subthreshold transgene expression levels incapable of achieving enough nematode resistance (Wang et al. 2009).

There are also relevant studies on improving the quality of pineapple fruit through genetically modified methods. For example, a protein from soybean that can store iron, ferritin, was cloned and transformed into leaf primordia through *Agrobacterium tumefaciens*. Regenerated plants were screened using kanamycin selection. Experimental results showed that the transgenic plants had the characteristic of increasing storing levels of both iron and zinc (Mhatre 2013). Furthermore, Del Monte company increased the accumulation of lycopene by overexpressing a gene derived from oranges and inhibiting other genes, resulting in the production of red flesh 'Plinkglow' pineapple.

9.8 Mutation Breeding

The mutation breeding in pineapple mainly includes physical mutation, chemical mutation, and mutation during the cell engineering process. Various types of mutagenesis have been reported, but no mature and widely promoted varieties have been cultivated through mutagenesis methods. Using pineapple's vegetative organs, tissues, and callus as explants, pineapple mutated tissues produced through cell engineering typically exhibit chimeric forms (Li et al. 2022). Somatic embryos can reduce the generation of chimeras, and the occurrence and regeneration system for pineapple through somatic embryogenesis have been successfully established in pineapple (Sripaoraya et al. 2003; Firoozabady and Moy 2004; Yapo et al. 2011). For example, Liang et al. (2009) and He et al. (2007, 2009) have successfully induced somatic embryos from calli from the leaf base of pineapple and pineapple slices.

Pineapple breeding through physical radiation induction has been attempted by different research institutions and scholars around the world. The Institute of Agricultural Research and Development of Malaysia used an apical meristem of the pineapple variety 'Jose' as an explant material to induce 16 mutant to 'bacterial black heart rot' mutant strains (Chan 2008; Ibrahim et al. 2009). Marz (1964) reported the introduction of self-fertile mutants by X-raying pollen during meiosis. In Kerala, India, researchers obtained mutant strains with growth inhibition (Anonymous 1964). In the Philippines, researchers used gamma rays to obtain spiny or spin-free materials by irradiating 'Queen' (Lapade et al. 1995). Researchers in Ghana researched the 'Smooth Cayenne' and 'Sugar Loaf gamma Pineapple' varieties with drought tolerance and heat tolerance properties obtained through ray-induced mutagenesis (Lokko and Amoatey 2001; Osei-Kofi et al. 1997).

Chemical mutagenesis is currently the most common method of mutation used in plant breeding. Ethyl methanesulfonate (EMS) used to treat plant tissues, can introduce point mutations in the genome. Current reports, also provide examples of EMS-induced mutations. Huang et al. (1995) treated pineapple calluses with EMS mutations, followed by selective media containing hydroxyproline. Similarly, Peng (2017) obtained cold tolerance mutant strains through EMS mutagenesis.

9.9 Conclusions and Prospects

Over the past 12,000 years, humans have domesticated hundreds of crops for various purposes, such as food, clothing and medicine. Most domesticated crops no longer exhibit self-incompatibility, and the main breeding method is through seed propagation. The classic concept of species domestication selects ideal germplasm from wild sources, but this is not always the case. The mechanisms of sexual and asexual reproduction in pineapples are not fully understood. Currently, there are two domestication models, one through sexual reproduction and the other through asexual reproduction. The sexual reproduction model involves the random pairing of wild plants for propagation through seeds, while the asexual reproduction model suggests that pineapple domestication is mainly achieved through accumulating physiological mutations during asexual reproduction. This process involves selecting phenotypic variations such as fruit quality, yield, or resistance phenotype.

The commercial cultivation of pineapple varieties typically involves hybrid breeding of existing varieties such as 'Smooth Cayenne', 'Monte Lirio', 'Queen' and 'Singapore Canning'. Therefore, making good use of existing resources, conducting germplasm evaluation, selecting suitable hybrid combinations, and utilizing hybrid breeding methods are of great value for breeding pineapple varieties. However, the limitations of self-incompatibility in cultivated varieties have limited hybrid breeding in the Baltic region, and the longer cycle of hybrid breeding poses challenges. Therefore, using other breeding methods such as chemical and physical mutagenesis to select new varieties is significant for pineapple breeding. Additionally, combining the accumulation of physiological mutations during the asexual reproduction process of pineapple can greatly accelerate this breeding process. However, there is currently no large-scale commercial cultivation of varieties through this method.

With the release of the pineapple genome, the relationships, diversity, and historical origins of this fruit have been unveiled. The annotation of the genome provides valuable gene structure information opening the doors for in-depth research on biological mechanisms and gene functions. Compared with other crops, research on the molecular mechanism analysis of the physiological traits and phenotypes related to pineapple is still very limited. Therefore, exploring the functional genes of pineapple and applying them to breeding practice has significant potential. At the same time, by developing molecular markers through the genome and combining them with genetic maps related to the pineapple, important trait localization and pineapple germplasm identification and evaluation can be carried out. In addition, based on genomic and molecular marker information, hybrid breeding design and variety selection can be promoted. However, there are a limited number of genomes currently sequenced. These results limit the application of markers for polymorphism and identification of genetic relationships between pineapples. While resequenced accessions cover a wide range of diversity in *Ananas*, key subspecies, cultivars, and wild germplasm remain unsequenced.

With the maturity of pineapple tissue culture and asexual reproduction technology system, improving pineapple traits and quality through transgenic technology is also an important direction for future development. Currently, pineapple transgenic technology primarily relies on *Agrobacterium* mediated biological transformation methods, with some utilizing gene gun methods for transformation. Progress has been made in scientific research through transgenic regulation of pineapple flowering, enhancing pineapple resistance, and improving pineapple quality, validating the regulatory role of relevant genes in shaping pineapple traits. This lays the foundation for future application of gene functions in developing new pineapple varieties.

Appendix 9.1 Research Institutes and Online Resources Available for Pineapple (*Ananas comosus* L.)

Group	Variety name	Characteristics	Site of cultivation
Smooth Cayenne	Smooth Cayenne	The fruit is large, oblong and cylindrical in shape, with an average fruit weight of 1.8–4.5 kg. When the fruit is ripe, the skin is orange yellow, the flesh is light yellow, and the fruit is low in fiber and juicy. Soluble solids ranging from 14% to 16%, up to over 20%, with an acid content of 0.5% to 0.6%, and a slightly lighter aroma (Li et al. 2008).	Venezuela and French Guiana discovered it for the first time and continuously improved it around the world
	Sarawak	The plant is relatively tall, with less leaf opening and more buds. The fruit is larger and conical in shape, with a thicker flesh texture and a stronger aroma (Li et al. 2008).	Malaysia; Indonesia; Kalimantan Island
	Hilo	The fruit is small, cylindrical, weighing 1.0–1.5 kg. The fruit is ripe before the skin turns yellow with deep eyes, and the flesh is yellow and transparent. The fruit is not resistant to storage and transportation and is suitable for canning (Li et al. 2008).	Hawaii, USA
	St. Michael	The fruit weighs 2.25–2.75 kg, with a small crown and fruit core, low acidity and sweetness, and a flat taste when fully ripe (Li et al. 2008).	The Azores
	Giant Kew	The fruit is large, with an average weight of 2.75 kg, often exceeding 4.5 kg, occasionally	India

(continued)

Group	Variety name	Characteristics	Site of cultivation
		reaching 10 kg, and a large fruit core (Li et al. 2008).	
	Charlotte Rothschild	The fruit is small and weighs 0.8–2.0 kg. The crown of the fruit is conical, and at maturity, the fruit is orange yellow, fragrant, and juicy (Li et al. 2008).	Guyana
	Pattawia	The skin is dark green, orange or yellow when mature. The flesh is lemon yellow, sweet and slightly sour, with a fruit weight of 2.0–2.5 kg and Brix 12°. Commonly used in canned or concentrated juice production (Li et al. 2008).	Thailand
	Nanglae Pineapple	The leaves are dark green with purple spots and have thorns. The fruit is short and round, with shallow eyes, dark green and yellow skin, low fiber flesh, Brix 16° – 20°, and an acid content of 0.3% (Li et al. 2008).	Thailand
	Cayenne Lisse	The fruit weighs 1.0–1.5 kg, with a very small crown and many buds, without bracts (Li et al. 2008).	Martinique Island and Ivory Coast
Queen	Queen (Alexandria, Buitenzorg, Common Rough, McGregor, Moris, Malacca, Victoria)	The fruit is small, tubular or slightly conical in shape, with an average fruit weight of 0.5–1.0 kg, with a few weighing up to 2.5 kg and maturing early. When the fruit is ripe, the fruit's skin and flesh are golden yellow, with a rich aroma. The Brix is 14°–18°, the soluble solids are 12–15%, the acid content is 0.47%, and the fiber is low,	South Africa; Queensland; Australia; and the Philippines

(continued)

Group	Variety name	Characteristics	Site of cultivation
		mainly used for fresh food (Li et al. 2008).	
	Mauritius (European Pine, Malacca Queen, Red Ceylon, Red Malacca)	The leaves are dark green with a wide red stripe in the center and red thorns on the edges. The fruit weighs 1.36–2.25 kg, with yellow skin, a small fruit core, and extremely sweet flesh (Li et al. 2008).	Malaysia; Sri Lanka; India
	Ripley (Ripley Queen)	The plant is short, with slightly dark red leaves arranged tightly. The fruit weighs 1.36–2.7 kg, with a light copper skin and a pale yellow flesh. It is fiber-free and very sweet (Li et al. 2008).	Queensland, Australia
	Natal Queen	More buds are sucked, with a fruit weight of 0.75–0.9 kg (Li et al. 2008).	South Africa; El Salvador; United States
	James Queen (Z)	It is a variant of the national Queen, tetraploid, with a large fruit and a square fruit base (Li et al. 2008).	South Africa
	MacGregor	It is a variant of National Queen, with scattered and robust plants, wide leaves, large but few suckers, cylindrical fruits, medium to large, firm flesh, and a fragrance resembling Queen (Li et al. 2008).	South Africa; Queensland, Australia
	Alexandria	The variety selected from Ripley Queen is more robust, with larger buds and fruits, cone-shaped fruits, tender flesh, and a Ripley Queen aroma (Li et al. 2008).	Queensland, Australia
	Vazhakulam	Locally known as 'Kannara' is suited for table purposes. The average fruit weight is 1.3–1.6 kg. The fruit has	Indian

(continued)

Group	Variety name	Characteristics	Site of cultivation
		a slightly conical shape with deep eyes. The flesh is crisp and golden yellow. Its juice is sweet with Brix 14°–16° and acidity is 0.50–0.70% (Joy and Anjana 2013).	
	Comte de Paris	It's a cultivar in China, with short, densely spiny plants. The fruit is cylindrical and small with an average fruit weight of 0.5–1.0 kg, and deep eyes. When the fruit is ripe, the fruit's skin and flesh are golden yellow with a rich aroma. The Brix is 14°–18°, the soluble solids are 12–15%, the acid content is 0.47%, and the fiber is low. Mainly used for fresh food.	China
	Yellow Mauritius	Queen's cultivated species in China. Similar to Comte de Paris.	China
Spanish	Spanish	The main commercial variety in South America. The plant is large, with broad and long leaves, thin and soft, dark green leaves, and sharp and hard red thorns on the leaf edges. The fruit is medium, slightly round, with a weight of 0.9–2.7 kg and deep eyes. When the fruit is ripe, the skin is orange-red, and the flesh is light yellow to golden yellow with a large fruit core. The flesh has more fiber, Brix 10°–12°, a low acid content, and a strong aroma. The ripened fruit is hard and easy to harvest and transport, used for canned food and fruit juice (Li et al. 2008).	Malaysia; Singapore; Florida; Honduras; Costa Rica; Puerto Rico; Cuba; Mexico; etc.

(continued)

Group	Variety name	Characteristics	Site of cultivation
	Red Spanish (Espanola Roja)	The fruit is medium-sized, cylindrical, orange, and weighs 1.2–2.0 kg. The flesh is nearly white, fragrant and sweet, with Brix 12°, but low acidity. Resistant to high temperature, drought, browning, white rot, wilt, and phytophthora, but not resistant to high manganese soil and nematode diseases, and highly sensitive to South American lepidopters (Li et al. 2008).	Caribbean basin
	Singapore Spanish (Singapore Canning, Ruby, Red Pine, Nanas Merah, Nangka Gandol Betek)	The fruit is small, with few, large and shallow eyes, about 1.0 kg, and cylindrical. The skin of ripened fruit is dark purple to copper orange, the flesh is golden, and the Brix is 10°–12°. The fruit juice has good color and quality, mainly used for canned food processing. The leaf thorns vary greatly with different strains, from more thorns to completely spineless. The plant is robust, with a large number of stem and suction buds, ranging from 2 to 6, with multiple crowns. It is resistant to phytophthora and exhibits severe verticillium wilt when the soil manganese content is high. It is prone to fruit cracking and is sensitive to nematode disease (Li et al. 2008).	Malaysia
	Masmerah	It is similar to 'Red Singapore', but with no thorns and upright leaves. The cylindrical fruit with translucent	Malaysia

(continued)

Group	Variety name	Characteristics	Site of cultivation
		golden flesh is suitable for canning. The maturity period is 22 weeks (Li et al. 2008).	
	Red Singapore (Red Jamaica, Singapore Spanish, Singapore Queen, Singapore Common)	The fruit is cylindrical, with dark yellow skin, slightly red skin, deep eyes, and a weight of 1.6–2.25 kg. The fruit has a small fruit and fibrous flesh, making it suitable for canning. The plant is resistant to pests and diseases (Li et al. 2008).	Honduras; Costa Rica; Puerto Rico; Mexico in Central America
	Cabezona (Bull Head, Pina de Ague)	It is a famous variant of RedSpanish, triploid, with a plant height of over 1 m. The fruit weighs 1.8–2.75 kg, sometimes reaching 8.0 kg or more. The ripened fruit is yellow with less fiber, only serving fresh fruits (Li et al. 2008).	The semi-acidic region of Rajas in Puerto Rico, El Salvador, Philippines
	Green Selangor	It is a mutant line of Singapore Spanish, retaining its main quality. Due to the absence of anthocyanins, the leaves and inflorescences are all green. The bracts are light yellow, and the thorns are only at the leaf tips. The flesh is golden, often with white spots, and this variety is used for making canned goods (Li et al. 2008).	Malaysia
	Cumanesa	A variant of 'Red Spanish', with an oval-shaped fruit, a large crown, and an orange-yellow skin. The fruit weight of 0.9–1.70 kg, and yellow-white flesh (Li et al. 2008).	The states of Sucre and Ramonagas in Venezuela, as well as Colombia and El Salvador
	Valera (Negrita, Andina)	Plants are small to medium-sized, with	

(continued)

Group	Variety name	Characteristics	Site of cultivation
		spiny leaves and purple-green color. The fruit is conical and columnar, weighing 1.5–2.5 kg, with white flesh and purple skin (Li et al. 2008).	The states of Lara, Merida, and Trujillo in Venezuela
Pernambuco	Perola (Pernambuco, Branco de Perambuco, Eleuthera, Abacaxi, Jupi, Abakka, Abachi, White Abacaxi of Pernambuco)	The fruit is small to medium-sized, with a weight of 0.9–5.0 kg. The fruit is conical and green. The flesh is white or light yellow, sweet and juicy, with a narrow fruit core, Brix 13°–16° and high vitamin C content, considered the most delicious pineapple (Li et al. 2008).	Brazil; Bahamas; Florida; Philippines; etc
	Sugarloaf (Pan deAzucar)	The fruit is nearly conical, sometimes a bit round, and weighs 0.68–1.36 kg. The flesh is white to yellow, sweet and juicy (Li et al. 2008).	Mexico; Venezuela; Puerto Rico; Cuba; Philippines; etc
Perolera	Perolera (Tachirense, Ca-pachera, Motilona, Lebrja)	The plant is medium in size, with 35–40 reddish-green leaves and short thorns. The fruit is ovoid, small, with a small crown. The sepals are prominent, and the fruit has deep eyes. The fruit weighs 0.45–0.70 kg and has yellow skin and flesh, less fiber, and a small fruit core (Li et al. 2008).	Colombia; Venezuela; Brazil; Ivory Coast
	Manzana (Bumanguesa)	It may be a mutant line of 'Perolera'. The leaves are bright green and the fruit shape is round to cylindrical. The fruit has large and shallow eyes, dark yellow flesh, a thin fruit core, and red or purple skin. In tropical highlands, it shows an attractive dark red to	The Andes Mountains in Colombia and Venezuela

(continued)

Group	Variety name	Characteristics	Site of cultivation
		bright red color. This variety is resistant to root-knot nematodes (Li et al. 2008).	
	Monte Lirio (Cambray)	The leaves are smooth and have no thorns at the end, and the fruit is medium-sized and round. The flesh is sweet and white, and the aroma is pleasant (Li et al. 2008).	From Mexico to Ecuador, it is cultivated throughout Central America
	Maipure	A strain with tall, red spotted and smooth leaves, the fruit with 170–190 fruitlets (Li et al. 2008).	Venezuela in the Amazon River Basin
Hybridization and other categories	Amritha	It was bred by crossing 'Kew' and 'Ripley Queen' at the Pineapple Research Center in India. Each fruit exceeds 2 kg in weight, featuring a compact crown and a golden-yellow hue with a desirable cylindrical form. It boasts the additional benefits of appealing flesh color, a pleasant aroma, high TSS and total sugars, along with low acidity (Li et al. 2008).	India
	Aus-Jubilee	It has higher sweetness, slightly yellower fruit color than regular varieties, a beautiful appearance, twice the vitamin C content of regular commercial varieties, and low acidity. Mature flesh can remain intact and can grow and bear fruit annually (Li et al. 2008).	Queensland, Australi7
	Josephine	It is a hybrid of 'Johor' ('Singapore Spanish') and 'Sarawak', a fresh fruit successfully cultivated by MARDI, with a growth cycle of 1 year.	Malaysia

(continued)

Group	Variety name	Characteristics	Site of cultivation
		The plant is robust with prickly leaf tips and 2–3 buds. The fruit weighs 1.1–1.5 kg and is cylindrical, with a deep purple color. The ripened fruit is orange-red, and the flesh boasts a rich yellow hue and a captivating fragrance. The Brix is between 17° and 22°. The fruit exhibits an extended shelf life and demonstrates resistance to black heart disease or browning resulting from low temperatures (Li et al. 2008).	
	P.R.1–67	It is a hybrid variety of 'Smooth Cayenne' and 'Red Spanish', with an average fruit weight of 2.5 kg and high yield (79 t/hm^2). It is slightly sour than 'Red Spanish', and suitable for fresh consumption and juice making. It has high resistance to gum disease and mustard wilt disease (Li et al. 2008).	Puerto Rico; Venezuela
	RL41	It is a hybrid breed of CIRAD-FLHOR developed by 'Smooth Cayenne' and 'Manzana'. The fruit weighs 1.4–2.0 kg, is orange to red, and has a regular cylindrical shape. The fruit has a large and shallow eye, firm flesh, thin fruit core, less fiber in the flesh, and Brix 15°–18°. The crown is shallow and erect, with dense plants with smooth and erect leaves and long flower stalks. Compared to 'Smooth Cayenne', it has a shorter growth cycle, higher vitamin C	

(continued)

Group	Variety name	Characteristics	Site of cultivation
		content, and is sensitive to small fruit heart rot (Li et al. 2008).	
	MD-2/CO-2	Two similar hybrid varieties cultivated from PRI hybrid lines 73–114 and 73–50 were named by the Del Monte company. The fruit is medium to large, cylindrical, 1.3–2.5 kg in weight, with large and shallow eyes, bright orange-yellow skin, slightly higher sugar content than 'Smooth Cayenne', low acidity and fiber content, Brix 15°–17°, and high vitamin C content. Compared to the 'Smooth Cayenne', it is more resistant to internal browning, but sensitive to phytophthora, highly resistant to parasites, and has strong storage capacity (Li et al. 2008).	The world's main pineapple-producing areas: are Hawaii; Costa Rica; Ecuador and the Philippines
	Imperial	A Fusarium wilt-resistant variety developed in Brazil in 1998 by -cross-breeding 'Perolera' with 'Smooth Cayenne'. The fruit weighs 1.67 kg, has a yellow flesh, Brix 15.8°, and an elegant aroma. The plant is 49 cm tall, spineless, and dark green (Li et al. 2008).	Brazil
	Brecheche	The plant should not exceed 1 m, with olive green leaves. The leaf is narrow and spineless. The fruit is small, with a small crown without thorns, weighing 0.7–1.0 kg. It has yellow skin and flesh, white fiber, and a rich aroma (Li et al. 2008).	Southern Venezuela

(continued)

Group	Variety name	Characteristics	Site of cultivation
	Caicara	The tall of the plant is 1.5 m tall. The number of leaves does not exceed 30, with long, spiny, and light green leaves. The fruit is large, cylindrical and conical, with deep eyes, a large crown and thorns, weighing 1.8–2.5 kg. It has yellow skin and white flesh, a large heart, juicy flesh, and less fiber (Li et al. 2008).	Bolivarian State of Venezuela
	Cabezona (Cabazoni)	It has a sturdy, spiny, yellow or orange fruit, and weighs 2.25–4.5 kg. The fruit is succulent, white, and sweet (Li et al. 2008).	South America
	Congo Red	The plant has a long bright red flowering period, and the fruit is prone to rot and burst when hot and dry. The fruit weighs over 2.25 kg and is waxy, with a fragrant yellow flesh (Li et al. 2008).	
	Panare	The plant is medium in size, with 35–40 reddish-green leaves and short thorns. The fruit is ovoid, and small, with a small crown, prominent sepals, and deep eyes. The fruit weighs 0.45–0.70 kg, with yellow skin and flesh, slightly fragrant, less fiber, and a small fruit core (Li et al. 2008).	Indian tribes; Bolivar state of Venezuela
	Black Antigua	The leaves have thorns, the fruit is small to medium in size, and the skin is dark green. When the fruit ripens, it turns yellow to orange. The flesh is golden, and the fruit is sturdy and delicious (Li et al. 2008).	South America
	Phuket	The fruit is extremely sweet, weight 1.3–1.5 kg, with over	Thailand

(continued)

Group	Variety name	Characteristics	Site of cultivation
		Brix 15°. It can be used for fresh food or making concentrated juices (Li et al. 2008).	
	Pulae	It is a miniature variety hybrid between 'Phuket' and 'Nang Lare'. The fruit has sweet and crispy golden flesh, resembling honey, is rich in fiber, and has high vitamin C content. It also has an over Brix 15° (Li et al. 2008).	Thailand
	Sr-Ratcha	Sweet and slightly sour, with yellow flesh, 1.5–1.8 kg, and Brix 12°–15 °. Used for fresh food or making concentrated juices (Li et al. 2008).	Thailand
	N67–10	A vegetatively derived cultivar from 'Hawaiian Smooth Cayenne' was introduced in 1985. This plant is robust, producing limited slips and suckers. It reaches maturity by late August in Okinawa, with an average fruit weight of approximately 1.4 kg. The flesh has a yellowish-white color. This cultivar is well-suited for processing applications (Ogata et al. 2016).	Japan
	Soft Touch	A vegetatively selected hybrid of 'Hawaiian Smooth Cayenne' and I-43-880 was introduced in 1999. The plant exhibits moderate vigor, with an average of 6.6 slips and 0.4 suckers per plant. The fruit typically weighs around 0.85 kg, and featuring creamy white flesh. The juice has	Japan

(continued)

Group	Variety name	Characteristics	Site of cultivation
		a Brix level of 16.0% and acidity of 0.60%, and the cultivar is distinguished by its appealing peach-like aroma (Ogata et al. 2016).	
	Gold Barrel	This cultivar was developed through cross-breeding between 'Cream Pineapple' and 'McGregor ST-1' and was introduced in 2009. The plant demonstrates moderate vigor and produces a limited number of slips and suckers. Its harvest period can be effectively managed due to its sensitivity to ethephon. The fruit typically weighs approximately 1.4 kg, with yellow flesh. The juice has a Brix level of 16.0% and acidity of 0.5%, and the cultivar is noted for its outstanding eating quality (Ogata et al. 2016).	Japan
	Julio Star	This cultivar resulted from the hybridization of 'N67–10' and 'Cream Pineapple', released in 2009. It matures in late July in Okinawa. The plant exhibits moderate vigor, producing approximately five slips and a few suckers. The fruit typically weighs around 1.2 kg, with pale yellow flesh. The juice has a Brix level of 16.0% and acidity of 0.6%. This cultivar is notable for its resistance to fruitlet core rot and its extended shelf life (Ogata et al. 2016).	Japan
	Okinou P17	It was bred by cross-breeding from 'Yugafu' × 'Summer Gold' and was released	Japan

(continued)

Group	Variety name	Characteristics	Site of cultivation
		in 2015, and a dominant characteristic is its high Brix (around 19%). Plant vigor is moderate and plants produce 3.2 slips and 3.0 suckers on average. The fruit typically weighs around 1.1 kg and features yellowish-white flesh. The juice boasts a Brix level of 19.0% and acidity of 0.6%. This cultivar stands out for its outstanding eating quality, resistance to fruitlet core rot, and resilience against typhoon winds (Ogata et al. 2016).	
	Vitória	The 'Vitória' pineapple is a hybrid variety developed at Embrapa Cassava & Tropical Fruits through the cross-breeding of 'Primavera' and 'Smooth Cayenne'. This cultivar is notable for its resistance to fusariosis and features spineless light green foliage. The fruit has an intermediate weight of 1.5 kg, a cylindrical shape, and golden skin. When fully ripe, it achieves a high sugar content of 15.8° Brix. The 'Vitória' pineapple was officially recommended for commercial cultivation in 2006 (Cabral and De Matos 2009).	Brazil
	Gigante de Tarauacá	The 'Gigante de Tarauacá,' also known as 'Arroba Tarauacá,' is a pineapple cultivar originating from the Tarauacá region in Acre State, located in Northern Brazil. This variety produces fruits that typically	Brazil

(continued)

Group	Variety name	Characteristics	Site of cultivation
		measure 24.2 cm in length, and 18.2 cm in diameter, and weigh an average of 4.4 kg (Scherer et al. 2015).	
	Tainung No. 16 (sweet honey)	The fruit exhibits a conical or elliptical form, with a yellow to vividly golden exterior and a yellow interior when fully ripe. On average, each fruit weighs 1.23 kg and contains 17.5% soluble solids (Li et al. 2022).	Taiwan, China; Guangdong China; Hainan, China; Guangxi China etc
	Tainung No. 17 (gold diamond)	The fruit features a cylindrical shape, with its exterior displaying a yellow hue and its interior a consistent yellow tone upon ripening. It achieves an average weight of 1.45 kg. Additionally, the fruit boasts 15.8% soluble solids alongside 0.28% titratable acid (Li et al. 2022).	Taiwan, China; Guangdong China; Hainan, China
	Tainung 4 (Easy Peeler)	'Tainung 4' (Easy Peeler) was selected which resembled its parent Queen. The average fruit weight is 1.2 kg. The fruit eyes are easy to peel. The flesh is soft and crispy, the sugar content is 19.5° Brix and the acidity is 0.43%. The aroma is strong, and it is durable for storage and transportation (Fitchet 1989).	Taiwan, China; Guangdong China; Hainan, China
	Yue crispy	It was bred by crossing'Smooth Cayenne' and 'Yellow Mauritius', released in 2005. The fruit is long and conical in shape, with yellow skin and yellow flesh at maturity. The average fruit weight is	Guangdong, China

(continued)

Group	Variety name	Characteristics	Site of cultivation
		1.5 kg, and the soluble solid content is 15.2–23.0%. The quality of fresh fruits surpassed that of 'Comte de Paris,' while processing characteristics were superior to those of 'Cayenne'.	
	Renong 8	Bred by the South Subtropical Tropical Crop Research Institute of the Chinese Academy of Tropical Agricultural Sciences, with 'MD-2' (female parent) and 'Perola' (male parent), released in 2021. This variety has strong growth potential and compact plant type. The leaves are relatively upright, with no thorns at the edges, and the color of the leaves is strong green. The average fruit weight is 1.46 kg. The fruit is cylindrical, with shallow eyes. The flesh is yellow with the soluble solid content 24.30%.,the titratable acid content 0.35%, and the Vc content 9.89 mg;100 g^{-1}.	Guangdong, China
	Yuetian	Obtained through hybrid breeding by the Fruit Tree Research Institute of Guangdong Academy of Agricultural Sciences, released in 2020. The plant is smaller and the fruit stalk is short. The ripen fruit skin is yellow, and the fruit is medium size, upright, short tube shaped with shallow eyes. The flesh is orange yellow with a crisp texture, less fiber, sweet and juicy taste, and a soluble solid content of 21 ~ 25%. It has good	Guangdong, China

(continued)

Group	Variety name	Characteristics	Site of cultivation
		cold and drought resistance and is suitable for fresh consumption and processing(Li et al. 2022).	

Note: These institutions have the most significant role at a global level. Nonetheless, there are a significant number of universities and international institutes that also contribute to pineapple breeding activities in the world

Appendix 9.2 Genetic Resources of Pineapple A: Most Popular Common Pineapple Varieties in the World, Their Characteristics and Site of Cultivation

Country	Institution name	Specialization and research activities	Website
French	French Agricultural Research Centre for International Development (CIRAD), Pôle de Recherche Agro-environnementa e de la Martinique (PRAM)	Genetic resources and Breeding	Website: http://collections.antilles.inra.fr/BRCPortal/initHome.do
America	University of Hawaii	Pineapple genetic resources and breeding	Website: https://manoa.hawaii.edu/
Brazil	Embrapa Cassava & Tropical Fruits	The research unit conducts research on cassava, citrus, banana, pineapple, mango, papaya, passion fruit, Barbados cherry, hog plum, yellow mombin, and others.	Website:https://www.embrapa.br/en/busca-geral/-/busca/Embrapa+Cassava+%26+Fruits?buscaPortal=Embrapa+Cassava+%26+Fruits
Malaysian	Malaysian Agriculture Research and Development Institute (MARDI)	Pineapple genetic resources and breeding	Website: https://www.mardi.gov.my/ (link is external)
Japan	Okinawa Prefectural Agricultural Research Center (OPARC)	Pineapple genetic resources and Systematic cross-breeding programs	820 Makabe, Itoman, Okinawa 901-0336, Japan.
India	Pineapple Research Centre, Kerala Agricultural University	Pineapple genetic resources and breeding	Website: https://www.kau.in/
China	Chiayi Agricultura Experimental Station, Taiwan Agricultural Research Institute(TARI)	Research for horticulture crops in breeding and biotechnology	Website: https://www.tari.gov.tw/english/content/index

(continued)

Country	Institution name	Specialization and research activities	Website
China	South China Agricultural University (SCAU)	The construction of the genetic transformation system and somatic embryogenesis of pineapple were studied	Website: https://www.scau.edu.cn/
China	South Subtropical Crop Research Institute of the Chinese Academy of Tropical Agricultural Sciences (SSCRI, CATAS)	Collection and preservation of germplasm resources, and breeding	Website: https://www.catassscri.cn/
China	Institute of Fruit Tree Research of Guangdong Academy of Agricultural Sciences (IFTR, GAAS)	Collection and preservation of germplasm resources, and breeding	Website: http://www.gdaas.cn/gdfruit/

Note: Most popular pineapple varieties in the world, their characteristics and site of cultivation

References

Alam MA, Sapuan SM, Ya HH et al (2021) Application of biocomposites in automotive components: a review. In: Biocomposite and synthetic composites for automotive applications. Woodhead Publishing, pp 1–17

Anonymous (1964) Pineapple Research Station, Kannara, Kerala Agric. Res Indian Counc Agric Res 4:141–142

Bairu MW, Aremu AO, Van Staten J (2011) Somaclonal variation in plants: causes and detection methods. Plant Growth Regul 63:147–173

Baker JG (1889) Handbook of the Bromeliaceae. George Bell & Sons, London, p 22

Baran A, Vrábel P, Olcák D, Chodák I (2018) Solid state 13C-NMR study of a plasticized PLA/PHB polymer blend. J Appl Polym Sci 135:1–7

Botella JR, Smith M (2008) Genomics of pineapple, crowning the king of tropical fruits. In: Moore PH, Ming R (eds) Genomics of tropical crop plants. Plant genetics and genomics: crops and models, vol 1. Springer, New York, pp 441–451

Brown G, Gilmartin A (1986) Chromosomes of the Bromeliaceae. Selbyana 9:88–93

Brown GK, Palací CA, Luther HE (1997) Chromosome numbers in Bromeliaceae. Selbyana 18:85–88

Burg SP, Burg EA (1966) Auxin-induced ethylene formation: its relation to flowering in the pineapple. Science 152:1269–1269

Cabral JRS, De Matos AP (2009) Imperial, a new pineapple cultivar resistant to fusariosis. Acta Hortic 822:47–50

Cabral JRS, de Matos AP, de Cunha GAP (1993) Selection of pineapple cultivars resistant to fusariose. Acta Hortic 334:53–58

Carlier JD, Reis A, Duval MF et al (2004) Genetic maps of RAPD, AFLP and ISSR markers in *Ananas bracteatus* and A. comosus using the pseudo-testcross strategy. Plant Breed 123(2): 186–192

Carlier JD, Nacheva D, d'Eeckenbrugge GC, Leitão JM (2006) Genetic mapping of DNA markers in pineapple. Acta Hortic (ISHS) 702:79–86

Carlier JD, d'Eeckenbrugge GC, Leitão JM (2007) Pineapple. In: Kole C (ed) Genome mapping and molecular breeding in plants, Fruits and nuts, vol 4. Springer, Heidelberg, pp 331–342
Carlier JD, Sousa NH, Santo TE et al (2012) A genetic map of pineapple (*Ananas comosus* (L.) Merr.) including SCAR, CAPS, SSR and EST-SSR markers. Mol Breeding 29(1):245–260
Chan YK (2008) Breeding of seed and vegetatively propagated tropical fruits using papaya and pineapple as examples. Acta Hortic 787:69–76
Chen LY, VanBuren R, Paris M et al (2019) The *bracteatus* pineapple genome and domestication of clonally propagated crops. Nat Genet 51:1549–1558
Collins JL (1949) History, taxonomy and culture of the pineapple. Econ Bot 3:335–359
Collins JL (1960) The pineapple-botany, cultivation and utilization. Interscience Publishers Inc., New York, p 294
Coppens d'Eeckenbrugge G, Govaerts R (2015) Synonymies in *Ananas* (Bromeliaceae). Phytotaxa 239:273–279
Coppens d'Eeckenbrugge G, Leal F (2003) Morphology, anatomy and taxonomy. In: Bartholomew DP, Paull RE, Rohrbach KG (eds) The pineapple: botany, production and uses. CABI, Wallingford, Oxford, pp 13–32
Dillen W, De Clercq J, Kapila J et al (1997) The effect of temperature on *Agrobacterium tumefaciens*-mediated gene transfer to plants. Plant J 12:1459–1463
Duangsuwan S, Amornsakchai T, Phinyocheep P et al (2023) Achieving high-performance green composites from pineapple leaf fiber–poly(butylene succinate) through both fiber alignment and matrix orientation across the thickness. ACS Omega 8(39):13
Duval MF, Coppens d'Eeckenbrugge G (1993) Genetic variability in the genus *Ananas*. Acta Hortic 334:27–32
Duval MF, Noyer JL, Perrier X et al (2001) Molecular diversity in pineapple assessed by RFLP markers. Theor Appl Genet 102:83–90
Duval MF, Buso GC, Ferreira FR et al (2003) Relationships in *Ananas* and other related genera using chloroplast DNA restriction site variation. Genome 46:990–1004
Espinosa P, Lorenzo J, Iglesias A et al (2002) Production of pineapple transgenic plants assisted by temporary immersion bioreactors. Plant Cell Rep 21:136–140
Feng S, Tong H, Chen Y et al (2013) Development of pineapple microsatellite markers and germplasm genetic diversity analysis. Biomed Res Int 2013:317912
Feng L, Wang J, Mao M et al (2022) The highly continuous reference genome of a leaf-chimeric red pineapple (*Ananas comosus* var bracteatus f. tricolor) provides insights into elaboration of leaf color. G3 (Bethesda) 12(2):jkab452
Feng J, Zhang W, Chen C et al (2024) Telomere-to-telomere assembly, manually curated annotation, and comparative analysis. J Integr Plant Biol 66(10):2208-2225
Ferreira F, Cabral J (1993) Pineapple germplasm in Brazil. Acta Hortic 334:23–26
Firoozabady E (2011) Organogenic transformation and regeneration. US Patent Number US8049067 B2
Firoozabady E, Moy Y (2004) Regeneration of pineapple plants via somatic embryogenesis anorganogenesis. In Vitro Cell Dev Biol Plant 40(1):67–74
Firoozabady E, Heckert M, Gutterson N (2006) Transformation and regeneration of pineapple. Plant Cell Tiss Org Cult 84:1–16
Fitchet M (1989) Observations on pineapple improvement in Taiwan. Subtropica 10(11):10–12
Food and Agriculture Organization of the United Nations (2021) FAOSTAT: Food and Agriculture Organization Corporate Statistical Database. https://www.fao.org/faostat/en/#data
Fullner KJ, Nester EW (1996) Temperature affects the T-DNA transfer machineries of *Agrobacterium tumefaciens*. J Bacteriol 178:1498–1504
Gangopadhyay G, Roy SK, Gangopadhyay SB, Mukherjee KK (2009) *Agrobacterium*-mediated genetic transformation of pineapple var. Queen using a novel encapsulation-based antibiotic selection technique. Plant Cell Tiss Org Cult 97:295–302
Givnish TJ, Barfuss MH, Van Ee B et al (2014) Adaptive radiation, correlated and contingent evolution, and net species diversification in Bromeliaceae. Mol Phylogenet Evol 71:55–78

Gonzales J, Vriesenga J (2005) Pineapple plant named ‘P-1972’. US patent Application Number 20050283863

Green J, Nelson S (2015) Heart and root rots of pineapple. Plant Disease PD-106 published by College of Tropical Agriculture and Human Resources, University of Hawaii at Manoa, Cooperative Extension Services

Harry JP, Lenoir H, Perrier X et al (2005) The CIRAD pineapple germplasm database. Acta Hortic 666:73–76

He YH, Luo J, Wu HT et al (2007) Somatic embryogenesis from leaf base callus of *Ananas comosus*. J Fruit S 24(1):59–63. (In Chinese with English abstract)

He YH, Fang SQ, Lu M et al (2009) Establishment of embryogenic cell suspension line in pineapple. Acta Hortic Sinica 36(S):1938

Huang JH, Kong DQ, Huang F (1995) Callus induction and antimetabolite mutant selection for pineapple. Chin J Trop Crops 16(S):1–6. (In Chinese with English abstract)

Ibrahim R, Hamzah A, Jam ZJ et al (2009) Gamma irradiation-induced mutation for the improvement of Josapine pineapple against bacterial heart rot disease and improved fruit quality. An International Joint FAO/IAEA Symposium

Ikemiya H, Onaha A, Nakasone H (1984) Excellent strain N67-10 in pineapple. Kyushu Agric Res 46:268

Joshi S, Patel S (2021) Review on mechanical and thermal properties of pineapple leaf fiber (PALF) reinforced composite. J Nat Fibers 19:10157–10178

Joy PP, Anjana R (2013) Pineapple varieties (Tech. Rep.). Pineapple Research Station, Kerala Agricultural University

Kato CY, Nagai C, Moore PH et al (2004) Intra-specific DNA polymorphism in pineapple (*Ananas comosus* (L.) Merr.) assessed by AFLP markers. Genet Resour Crop Evol 51:815–825

Kato CY, Nagai C, Moore PH et al (2005) Intra-specific DNA polymorphism in pineapple (*Ananas comosus* (L.) Merr.) assessed by AFLP markers. Genet Resour Crop Evol 51:815–825

Kengkhetkit N, Amornsakchai T (2012) Utilisation of pineapple leaf waste for plastic reinforcement: 1. A novel extraction method for short pineapple leaf fiber. Ind Crop Prod 40(11):55–61

Kilian A, Sanewski G, Ko L (2016) The application of DArT-seq technology to pineapple. Acta Hortic 1111:181–188

Ko L, Hardy V, Jabin-Décor M et al (2005) The introduction of transgenes to control blackheart in pineapple: biolistics vs *Agrobacterium* transformation. In: Contributing to a Sustainable Future. International Association for Plant Tissue Culture and Biotechnology (IAPTC) Conference (Australian Branch- 7th National Meeting), 21–24

Ko HL, Campbell PR, Jabin-Décor MP et al (2006) The introduction of transgenes to control blackheart in pineapple (*Ananas comosus* L.) cv. Smooth Cayenne by microprojectile bombardment. Euphytica 150:387–395

Ko L, Sanewski G, Smith M (2008) Pineapple. In: Kole C, Hall T (eds) Compendium of transgenic crop plants. Transgenic tropical and subtropical fruit and nuts, vol 5. Wiley-Blackwell, Oxford, UK, pp 109–130

Ko L, Eccleston K, O’Hare T et al (2013) Field evaluation of transgenic pineapple (*Ananas comosus* (L.) Merr.) cv. ‘Smooth Cayenne’ for resistance to blackheart under subtropical conditions. Sci Hortic 159:103–108

Koia JH, Moyle RL, Botella JR (2012) Microarray analysis of gene expression profiles in ripening pineapple fruits. BMC Plant Biol 12:240

Kumar N (2006) Breeding of horticultural crops: principles and practices. New India Publishing Agency, Delhi, pp 118–120

Lapade AG, Veluz AMS, Santos IS (1995) Genetic improvement of the Queen variety of pineapple through induced mutation and in vitro culture techniques. In: Conference on the international symposium IAEA and food agric Org of the UN IAEA. Proceedings induced mutations and molecular techniques for improvement, Vienna

Leal F, Amaya L (1991) The curagua (Ananas lucidus, Bromelia-ceae) crop in Venezuela. Econ. Bot 45:216–224

Leal F, Coppens d'Eeckenbrugge G (1996) Pineapple. In: Janick J, Moore JN (eds) Fruit breeding. I. Tree and tropical fruits. Wiley, New York, pp 515–557
Levitt R (2014) A noble present of fruit: a transatlantic history of pineapple cultivation. Gard Hist 42(1):106–119
Li YL, Zeng XH, Sun GM (2008) Foreign pineapple variety resources. World Agri 2008(1):55–58
Li DL, Jing MM, Dai XH et al (2022) Current status of pineapple breeding, industrial development, and genetics in China. Euphytica 218:85
Liang XL, Chen XL, Cheah K et al (2009) Establishment of direct somatic embryogenesis system through transverse thin cell layer culture in pineapple (*Ananas comosus* L. Merrill). Acta Hortic Sinica 36(11):1597–1602. (In Chinese with English abstract)
Liang PP, Chen XQ, Zhang XT, Tang HB (2018) Comparative genomics of pineapple and other angiosperm genomes. In: Min R (ed) Genetics and genomics of pineapple, vol 22. Springer, Urbana, pp 131–153
Liu Y, Zhong Y, Meng XC et al (2006) A new variety for Pineapple'Yuecui'. Acta Hortic Sinica 33: 214. (In Chinese with English abstract)
Lokko Y, Amoatey H (2001) Improvement of pineapple using in vitro and mutation breeding techniques. In: M.S.R, Asad S, Zafar Y et al. (eds) In vitro techniques for selection of radiation-induced mutations adapted to adverse environmental conditions, pp 5–29.
Luo R (2012) Soapdenovo2: an empirically improved memory-efficient short-read de novo assembler. GigaScience 1(1):1–6
Ma J, He Y, Wu C et al (2012) Effective *Agrobacterium*-mediated transformation of pineapple with CYC1A1 by kanamycin selection technique. Afr J Biotechnol 11:2555–2562
Magallón S, Gómez-Acevedo S, Sánchez-Reyes LL, Hernández-Hernández T (2015) A meta-calibrated time-tree documents the early rise of flowering plant phylogenetic diversity. New Phytol 207(2):437–453
Marchant C (1967) Chromosome evolution in the Bromeliaceae. Kew Bull 21:161–168
Marz GS (1964) Study of self-incompatibility in the Pineapple. Agric Res Pretoria 2–6:561–562
Mendiola NB, Capinpin JM, Mercado TM (1951) Pineapple breeding in The Philippines 1922–1941. Philippine J Agr 16:51–84
Mhatre M (2013) Chapter 33. *Agrobacterium*-mediated genetic transformation of pineapple (*Ananas comosus* L. Merr.). In protocols for micropropagation of selected economically—important horticultural plants. Methods Mol Biol 994:435–453
Min XJ, Bartholomew DP (1996) Effect of plant growth regulators on ethylene production, 1-aminocyclopropane-1-carboxylic acid oxidase activity, and initiation of inflorescence development of pineapple. J Plant Growth Regul 15:121–128
Ming R, VanBuren R, Wai CM et al (2015) The pineapple genome and the evolution of CAM, photosynthesis. Nat Genet 45:1435–1442
Mohanty AK, Misra M, Hinrichsen G (2000) Biofibres, biodegradable polymers and biocomposites: an overview. Macromol Mater Eng 276−277(1):1–24
Moyle R, Fairbairn DJ, Ripi J et al (2005) Developing pineapple fruit has a small transcriptome dominated by metallothionein. J Exp Bot 56:101–112
Moyle R, Ripi J, Fairbairn DJ et al (2006) The pineapple EST sequencing and microarray project. Acta Hortic 702:47–50
Myers EW, Sutton GG, Delcher AL et al (2000) A wholegenome assembly of Drosophila. Science 287(5461):2196–2204
Nagai C, Rosal L (1995) Pineapple micropropagation. Hawaii Sugar Planters' Association Annual Report. pp 34–35
Nan GL, Nagai C, Moore PH, Sun SSM (1996) Tissue culture and genetic transformation studies on pineapple. In Vitro Cell Dev Biol Plant 32:102A
Nan GL, Nagai C (1998) Genetic transformation of pineapple (Ananas comosus [L.] Merr.) via particle bombardment. In Vitro Cell Dev Biol Anim 34:1052

Neto ARS, Araujo MAM, Souza FVD, et al (2013) Characterization and comparative evaluation of thermal, structural, chemical, mechanical and morphological properties of six pineapple leaf fiber varieties for use in composites. Ind Crop Prod 43(2):529–537

Newcomb RD, Crowhurst RN, Gleave AP et al (2006) Analyses of expressed sequence tags from apple. Plant Physiol 141(1):147–166

Ogata T, Yamanaka S, Shoda M et al (2016) Current status of tropical fruit breeding and genetics for three tropical fruit species cultivated in Japan: pineapple, mango and papaya. Breed Sci 66: 69–81

Ong WD, Voo LYC, Kumar VS (2012) De novo assembly, characterization and functional annotation of pineapple fruit transcriptome through massively parallel sequencing. PLoS ONE 7:e46937

Osei-Kofi F, Amoatey HM, Lokko Y (1997) Improvement of pineapple using biotechnology and mutation breeding techniques. In: Report of 2 FAO/IAEA Resarch coordination meeting on invitro techniques for selection of randiation-induced mutant adapted to adverse environmental conditions, pp 23–27

Paull RE, Bartholomew DP, Chen CC (2017) Pineapple breeding and production practices. In: Lobo MG, Paull RE (eds) Handbook of pineapple technology. Production, postharvest science, processing and nutrition. Wiley Blackwell, West Sussex, pp 16–38

Paterson AH, Wendel JF, Gundlach H et al (2012) Repeated polyploidization of Gossypium genomes and the evolution of spinnable cotton fibres. Nature 492(7429):423–427

Paz EY, Gil K, Rebolledo L, et al (2005) AFLP characterization of the Mexican pineapple germplasm collection. J Am Soc Hortic Sci 130:575–579

Peng XN (2017) Optimization of Bromeliaceae culture in vitro and its cold-tolerant mutants induction. Dissertation, Agricultural University Of Hunan (In Chinese with English abstract)

Pérez G, Yanes E, Isidrón M et al (2009) Phenotypic and AFLP characterization of two new pineapple somaclones derived from in vitro culture. Plant Cell Tissue Organ Cult 96(1):113-116

Pérez G, Mbogholi A, Sagarra F et al (2011) Morphological and physiological characterization of two new pineapple somaclones derived from in vitro culture. In Vitro Cell Dev Biol Plant 47(3): 428–433

Pérez G, Yanez E, Mbogholi A et al (2012) New pineapple somaclonal variants: P3R5 and Dwarf. Am J Plant Sci 3(1):1–11

Pickersgill B (1976) Pineapple. In: Simmonds NW (ed) Evolution of crop plants. Longman, London, pp 14–18

Prakash J, Bhattacharyya S, Chattopadhyay K et al (2009) PQM-1: a new developed superior clone of pineapple for northeastern India as evident through phenotype, fruit quality and DNA polymorphism. Sci Hortic 120(2):228–291

Purseglove JW (1972) Pineapple. In: Tropical crops, Monocotyledons. Longman, London, pp 75–91

Redwan RM, Saidin A, Kumar SV (2016) The draft genome of MD-2 pineapple using hybrid error correction of long reads. DNA Res 23(5):427–439

Rohrbach KG, Christopher D, Hu J et al (1998) Management of a multiple goal pineapple genetic engineering program. In: III International Pineapple Symposium, vol 529. pp 111–114

Ruas PM, Ruas CF, Fairbanks DJ et al (1995) Genetic relationship among four varieties of pineapple, *Ananas comosus*, revealed by random amplified polymorphic DNA (RAPD) analysis. Braz J Genet 18:413–416

Ruas CD, Ruas PM, Cabral JRS (2001) Assessment of genetic relatedness of the genera *Ananas* and Pseudananas confirmed by RAPD markers. Euphytica 119:245–252

Sanewski G, DeFaveri J (2017) The Australia fresh market pineapple breeding program. International Symposium on GA3 Tropical Fruit (Guava, Wax Apple, Pineapple, Sugar Apple) 1166:6

Scherer RF, Olkoski D, Souza FVD et al (2015) Gigante de Tarauacá: A triploid pineapple from Brazilian Amazonia. Sci Hortic 181:1–3

Schmid KJ, Sorensen TR, Stracke R et al (2003) Large-scale identification and analysis of genome-wide single-nucleotide polymorphisms for mapping in *Arabidopsis thaliana*. Genome Res 13(6A):1250–1257

Shoda M, Urasaki N, Sakiyama S et al (2012) DNA profiling of pineapple cultivars in Japan discriminated by SSR markers. Breed Sci 62:352–359

Smith JB (1965) Multiple objective breeding: new dimensions in pineapple improvement. PRI News 13(1):121–117

Smith MK, Ko HL, Sanewski GM et al (2005) *Ananas comosus* pineapple. In: Litz RE (ed) Biotechnology of fruit and nut crops. CABI, Wallingford, pp 158–172

Soneji JR, Rao MN, Mhatre M (2002) Somaclonal variation in micropropagated dominant axillary buds of pineapple (*Ananas comosus* L. Merr.). J Hortic Sci Biotech 77:28–32

Sousa N, Carlier J, Santo T, Leitão J (2013) An integrated genetic map of pineapple (*Ananas comosus* (L.) Merr.). Sci Hortic 157:113–118

Sripaoraya S, Marchant R, Power JB et al (2001) Herbicide-tolerant transgenic pineapple (*Ananas comosus)* produced by microprojectile bombardment. Ann Bot 88:597–603

Sripaoraya S, Marchant R, Power JB, Davey M (2003) Plant regeneration by somatic embryogenesis and organogenesis in commercial pineapple (*Ananas comosus* L.). In Vitro Cell Dev Biol Plant 39(5):450–454

Stewart RJ, Sawyer BJB, Bucheli CJ, Robinson SP (2001) Polyphenol oxidase is induced by chilling and wounding in pineapple. Aust J Plant Physiol 28:181–191

Su G, Park S, Lee S, Murai N (2012) Low co-cultivation temperature at 20°C resulted in the reproducible maximum increase in both the fresh weight yield and stable expression of GUS activity after *Agrobacterium tumefaciens*-mediated transformation of tobacco leaf discs. Am J Plant Sci 3:537–545

Sun GM (2019) Fruit science and practice in China: Pineapple. Shaanxi Science and Technology Press, Shanxi, pp 219–233. (In Chinese)

Tang H, Bowers JE, Wang X et al (2008) Synteny and collinearity in plant genomes. Science 320(5875):486–488

Tang H, Bowers JE, Wang X, Paterson AH (2010) Angiosperm genome comparisons reveal early polyploidy in the monocot lineage. Proc Natl Acad Sci U S A 107(1):472–477

Tian YH, Teng LQ, Yang JQ et al (2024) Characterization of pineapple leaf fiber following enzymatic degumming. Cellulose Chem Technol 58(1–2):125–132

Trusov Y, Botella JR (2006) Silencing of the ACC synthase gene *ACACS2* causes delayed flowering in pineapple [*Ananas comosus* (L.) Merr.]. J Exp Bot 7:3953–3960

Ventura J, Costa H, Cabral J, de Matos A (2007) 'Vitoria': new pineapple cultivar resistant to fusariosis. In: VI International Pineapple Symposium, vol 822. pp 51–56

Ventura JA, Costa H, Caetano LCS (2009) Vitoria pineapple: scab resistant cultivar. Rev Bras Frutic 31:4

Wai CM, Vanburen R, Wang ML, et al (2014) Development of pineapple (Ananas comosus L.) genetic map with restriction site associated DNA (RAD) markers. XII Int PAG Conf, P:452

Wai CM, Powell B, Ming R, et al (2016a) Analysis of alternative splicing landscape in pineapple (Ananas comosus). Trop Plant Biol 9(3):150–160

Wai CM, Powell B, Ming R, et al (2016b) Genome-wide identiffcation and analysis of genes encoding proteolytic enzymes in pineapple. Trop Plant Biol 9(3):161–175

Wang ML, Uruu G, Xiong L et al (2009) Production of transgenic pineapple (*Ananas comosus* (L) Merr) plants via adventitious bud regeneration. In Vitro Cell Dev Biol Plant 45(2):112–121

Wang XM, Ren H, Liu YQ, et al (2016) Breeding report of a new pineapple variety 'Jinxiang Boluo.' J Fruit Sci 33(04):509–512 (In Chinese with English abstract)

Wang JS, He JH, Chen HR et al (2017) Genetic diversity in various accessions of pineapple [*Ananas comosus* (L.) Merr.] using ISSR and SSR markers. Biochem Genet 55:347–366

Wee YC, Thongtham MLC (1991) *Ananas comosus* (L.) Merr. In: Verheij EWM, Coronel RE (eds) Plant resources of South-East Asia No. 2 edible fruits and nuts. Pudoc, Wageningen, pp 66–71

Williams DDF (1970) Production, propagation, and testing of new varieties. PRI News 18:9

Williams DF, Fleisch H (1993) Historical review of pineapple breeding in Hawaii. Acta Hortic 334: 67–76

Wortmann S, Kerns K (1964) The plant breeding program. PRI Res Rep 64:180

Urasaki N, Goeku S, Kaneshima R et al (2015) Leaf margin phenotype-specific restriction-site-associated DNA-derived markers for pineapple (Ananas comosus L.). Breeding Science 65.3: 276-284

Xiao HY, Mohamed THS, Muhammad IN et al (2024) A short review on the recent progress and properties of pineapple leaf fiber reinforced composite. E3S Web Conf 477:00096

Yabor L, Arzola M, Aragón C et al (2006a) Biochemical side effects of genetic transformation of pineapple. Plant Cell Tiss Org Cult 86:63–67

Yabor L, Espinosa P, Arencibia AD, Lorenzo JC (2006b) Pineapple [*Ananas comosus* (L.) Merr.]. In *Agrobacterium* protocols, vol 2. Methods Mol Biol 344:219–226

Yapo ES, Kouakou TH, Kone M et al (2011) Regeneration of pineapple (*Ananas comosus* L.) plant through somatic embryogenesis. J Plant Biochem Biot 20(2):196–204

Young RA (2016) Pineapple production on Dole farms in Latin America. Acta Hortic 111:227–230

Young R, Gonzales J (2009) Pineapple plant named 'Dole 14'. United States Patent Application 20090328260

Zhou L, Matsumoto T, Tan HW et al (2015) Developing single nucleotide polymorphism markers for the identification of pineapple (*Ananas comosus*) germplasm. Hortic Res 2:15056

Chapter 10
Sago Palm (*Metroxylon sagu* Rottb.) Breeding

Teuku Tajuddin, Devit Purwoko, Albertus Fajar Irawan, Imam Civi Cartealy, Imron Riyadi, Yusuf Sigit Ahmad Fauzan, Karyanti, Dwi Asmono, and Mochamad Hasjim Bintoro

Abstract Sago palm (*Metroxylon sagu* Rottb.) has garnered significant attention as a starch crop due to the large amounts of starch accumulated in its trunk. Sago palms are a crucial source of food and carbohydrates. To this day, many people in Maluku, North Maluku, Papua and West Papua, continue to consume sago as a staple food. Apart from serving as an alternative carbohydrate source to rice in local food diversification efforts and supporting food security, sago also boasts health benefits. The growth and development of sago palms are heavily influenced by genetic and environmental factors. The plant exhibits remarkable genetic diversity as it is widely distributed across the archipelago, with each sago production area housing diverse accessions. Sago belongs to the Arecaceae (Palmae) Family and completes its life cycle after flowering. The highest starch production occurs as the sago palm enters the flowering phase. Sago starch can be utilized as an industrial raw material and

T. Tajuddin (✉) · D. Purwoko · Y. S. A. Fauzan
Research Center for Applied Botany, Research Organization for Life Sciences and Environment, National Research and Innovation Agency (BRIN), Science and Technology Park of B.J. Habibie, South Tangerang City, Banten Province, Indonesia
e-mail: teuku.tajuddin@brin.go.id; devit.purwoko@brin.go.id; yusuf.sigit.ahmad.fauzan@brin.go.id

A. F. Irawan
Faculty of Animal and Agricultural Science, Diponegoro University, Jl Prof. Jacub Rais. Diponegoro University Campus, Tembalang, Semarang, Central Java Province, Indonesia
e-mail: albertusfajarirawan@lecturer.undip.ac.id

I. C. Cartealy
Research Center for Computing, Research Organization for Electronics and Informatics, National Research and Innovation Agency (BRIN), Science and Technology Park of Soekarno, Cibinong, West Java Province, Indonesia
e-mail: imam.civi.cartealy@brin.go.id

I. Riyadi
Indonesian Oil Palm Research Institute, Bogor Unit, Bogor, West Java Province, Indonesia
e-mail: imron.riyadi@iribb.org

K. F. M. Salem et al. (eds.), *Breeding and Biotechnology of Leaf, Fruit, and Seed Fiber Crops*, Advances in Plant Breeding Strategies 11,
https://doi.org/10.1007/978-3-032-00407-9_10

biofuel. Vegetative propagation, through methods such as suckers, in vitro and ex vitro techniques, is the most common way to propagate sago palms. When devising new breeding programs, it is essential to consider current scientific advancements, like tissue culture, mutagenesis, functional genomics, molecular markers, and genetic engineering, as they have the potential to improve breeding efficiency. Information on genetic diversity is necessary for germplasm conservation. Documenting the genetic diversity of organisms is vital for preserving biological diversity and the existence of specific organisms, such as sago palms. This chapter delves into the potential of sago palm and the research progress supporting the breeding program for this plant.

Keywords Advance breeding · Biotechnology · Fiber · Food security · Vegetative propagation · Marker-assisted breeding · Sago palm (*Metroxylon sagu* Rottb.)

10.1 Introduction

Indonesia is known as the center of diversity for the sago palm (*Metroxylon sagu* Rottb). Sago is native to the Provinces of Papua and West Papua, characterized by a large number of accessions found and dominated in several places in these two provinces. From this center, the palm spread to Maluku, Sulawesi, Mindanao and western Indonesia, such as Sumatra, Java and Kalimantan (Fathnoer et al. 2020; Yamamoto et al. 2020a). Based on their kinship, sago can be divided into three groups, namely the Papua, West Papua and Maluku groups, the Sulawesi and Mindanao groups, and the Sumatra, Kalimantan and Java groups. In Southeast Asia, sago palms grow mostly in Indonesia. Bintoro et al. (2017a, b) stated that the distribution of sago ranges from 180° East Longitude (Countries in the South Pacific) to India (90° East Longitude) and from 10° North Latitude in Mindanao to 10° South Latitude, encompassing several islands in Indonesia. The sago palm in Indonesia is worth considering because of its potential.

Karyanti
Research Center for Estate Crops, Research Organization for Agriculture and Food, National Research and Innovation Agency (BRIN), Science and Technology Park of Soekarno, Cibinong, West Java Province, Indonesia
e-mail: karyanti@brin.go.id

D. Asmono
Department of Research and Development, PT Sampoerna Agro Tbk, Sampoerna Strategic Square, North Tower, Jakarta, Indonesia
e-mail: dwi.asmono@sampoernaagro.com

M. H. Bintoro
Department of Agronomy and Horticulture, Faculty of Agriculture, IPB University, IPB Dramaga Campus, Bogor, West Java Province, Indonesia
e-mail: bintoro@apps.ipb.ac.id

Sago palms in Indonesia have the widest growth area in the world. According to Flach (1997), there are around 2.306 million hectares of sago globally, with approximately 1 million hectares in Papua New Guinea, 1.25 million hectares in Indonesia, 45,000 hectares in Malaysia, and 3000 hectares each in Thailand and the Philippines. The remaining 5000 hectares are spread across various countries, including Brunei Darussalam, and Pacific island countries such as the Marshal Islands, Solomon Islands, Vanuatu and New Caledonia. Djoefrie et al. (2014) reported that the Accelerated Development of Papua and West Papua unit had conducted aerial photography in Papua and West Papua Provinces, revealing that the sago area in Papua is 4,749,325 hectares and in West Papua 510,213 hectares. Most sago in Papua and West Papua is found on the south coast. The area of sago in Indonesia is 5,579,637 hectares spread across various islands, namely in 15 provinces and 60 districts and cities, representing 85% of the world's sago area.

Arif (2019) revealed that sago has long been known in Indonesia. This is characterized by the name sago in several regional languages. For example, *kirai* in Sundanese, *ambulung* or *kersulu* in Javanese, *lapia* in Ambon and Seram. In other countries, sago is also known as *rumbia* or *balau* in Malaysia, *lumbia* in the Philippines, *thagu bin* in Myanmar, *saku* in Cambodia, and *sakhu* in Thailand (Ruddle et al. 1978). Instead of food, sago palm has a huge potential as an energy bioresource and raw material for various industries. Its utilization has diversified into various products and materials, showing potential and prospects for both food and non-food industries. Research has shown that the starch produced by sago palms can be used as raw materials for beer and glutamate fermentation, and also biodegradable plastics (Okazaki et al. 2005), acetone–butanol–ethanol fermentation (Gumira et al. 1996), ethanol production (Hung et al. 2018), alcohol (Tjahjono et al. 2024), cyclodextrin (Fridiano 2021), glucose syrup (Sarungallo and Murtiningrum 2005), and biodegradable plastic manufacturing (Amni et al. 2019; Dewi et al. 2023). Other parts of the sago palm can be used as pulp and material for making paper (Maurila et al. 2024). Based on this research, sago starch can be utilized by various industries such as the food industry, animal feed industry, pharmaceuticals, plastics industry and paper industry.

Sago palm is propagated generatively by seed and vegetatively by suckers. Due to the small number of seeds found in the fruit bunch and the low rate of seed germination ability, the production of seedlings from seed is not recommended. Germination studies have shown a low germination rate of only 3.50–6.43% (Limbongan et al. 2005). Sago palm propagation by seed resulted in high genotypic variation. Suckers are the most commonly used planting material when developing sago palms on plantations. As a result of constant suckering, the palm multiplies and gathers around the mother palm. Other vegetative propagation methods using in vitro and ex vitro cultures have been performed to avoid these problems and provide vigorous and quality planting materials. In vitro propagation of sago palm has been conducted by various laboratories (Hisajima et al. 1991; Sinta and Riyadi 2008; Sumaryono et al. 2007; Tahardi et al. 2002; Tajuddin et al. 2015). Additionally, callus-derived somatic embryogenesis can generate a plentiful amount of superior planting materials (Riyadi et al. 2005; Sumaryono et al. 2012).

Phenotypically, plant identification has limitations due to variations in subjective analysis of raw materials and the confusion of plant characteristics. Genetic characterization based on PCR is a method for identifying and verifying the validity of raw materials for sago palms. DNA markers are highly accurate since they give polymorphism information and a unique genetic composition for each species, regardless of age or physiological variables such as environmental factors.

In order to assist future sago palm breeding (Abbas et al. 2010) and genetic resource conservation information (Kjaer et al. 2004), an understanding of the genetic variability of sago palm is required. It is necessary to create robust genetic markers to back up these remarkable achievements. Sequence analysis using bioinformatics can be carried out using two approaches, namely web-based and programming language algorithm-based. Web-based bioinformatics analysis can be used for annotations, genome mapping, and advanced sequence analysis, such, as identifying homologous genes and interactions between related genes. Web-based bioinformatic analysis can be conducted online through programs available with public access on the web. This method is economical and can be used as preliminary research before real experiments are carried out (Narita et al. 2012). However, when analyzing large sequence data, a programming language algorithm is supported by a supercomputer to run simultaneously.

This chapter describes the sago palm, its enormous potential, various products produced from this plant, the biodiversity covering morphological and genotypic diversity, several propagation methods, cultivation, omics and bioinformatics, and research progress supporting the breeding program of this plant.

10.1.1 Fiber and Textile Properties

Sago dreg is a type of waste generated during the processing of sago to produce starch. Sago dregs in the form of fiber have not been widely used and are often left to pile up near sago processing plants. Some are even thrown into rivers, raising concerns about it polluting the surrounding environment and waterways.

According to Numberi et al. (2023), sago dregs contain 1.70% crude fiber, 10.4% water, 1.85% crude protein, 3.64% ash, 0.01% fat, and 82.4% carbohydrates. From their experiment, Wina et al. (1986) estimated that sago fiber contains starch ranging from 2.60% to 4.03%. Furthermore, Latuconsina (2015) reported that sago contains 65.7% carbohydrates, 21% lignin, and 20% cellulose. Fresh sago fiber is light brown. However, after 4–5 days, the color changes to dark brown. The color changed due to an oxidative reaction of phenolic compounds (Fig. 10.1).

Scientists have conducted research and innovation to reduce sago dregs while obtaining useful products. Sago dregs have a high nutritional content in the form of carbohydrates, making them a potential energy source for feed. Sunardi et al. (2021) found that sago frond fiber components contain 16.41% water-soluble extractives, 27.12% hemicellulose, 35.53% cellulose, and 18.25% lignin. High lignin content can be successfully reduced using chemical methods, such as the ammonia process (Sangadji et al. 2019; Syadik et al. 2022), or biological methods, like microbial fermentation.

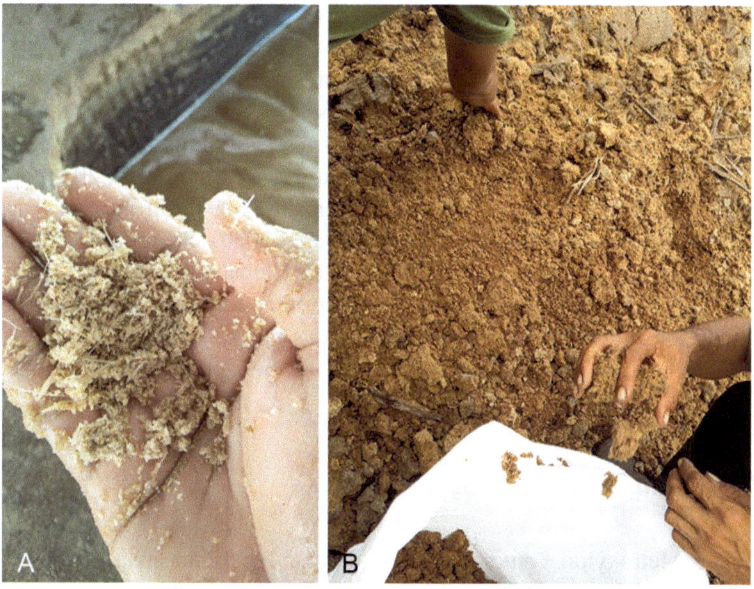

Fig. 10.1 Sago fiber produced by a modern sago mill, (**Panel A**) Fresh sago fiber, and (**Panel B**) 5-day-old dried sago fiber. (Source: AF Irawan)

The high cellulose content in sago dregs can be an alternative source of cellulose for various applications such as paper making, additives in medicines, bioplastics, and textiles (Kim et al. 2006).

In addition to its benefits, sago also plays an important role in the textile industry. Sago is used to help hold fibers together and make spinning machines work better, in a process called sizing. Sago can easily bind fibers together, which makes them more manageable during textile production. If we are cautious, new fabrics or clothes that have been sized can contain sago residues that disappear after washing (Karim et al. 2008).

10.1.1.1 Fiber Industry

Because there is a lot of residual waste from the sago industry each year, people are trying to find ways to turn the waste into something useful. There are a few applications related to the utilization of sago wastes. Sago waste can also be utilized to produce fermentable sugars, nanomaterials, as well as ceramics (Amin et al. 2019). The sago palm leaves and trunks are used for construction materials and make roofs (Anggraeni and Ngantung 2018). The middle part of the leaves and the outer part of the stem are used to make mats and baskets by weaving them together (Pasolon 2015). In the Visayas Region of the Philippines, people prefer sago as the best material for making shingles for roofs on small houses or huts. Many shingle makers in the region like to use sago leaves instead of nipa leaves because sago leaves are stronger, especially when used in coastal areas. Sago shingles have a higher price compared to nipa shingles (Quevedo et al. 2005).

Maurila et al. (2024) characterize sago fiber waste for its suitability as a raw material for paper production. The resulting paper has a grammage of 62 g/m^2, a thickness of 0.1294 mm, a water absorption capacity of 627.27 g/m^2, and other relevant properties. Other products that can be produced from sago dregs include activated carbon, used as an adsorbent to reduce household liquid waste (Che et al. 2023), and iodine substances in hospital liquid waste (Rahmi et al. 2018). Furthermore, mixing the sago dregs with polyester resins, can result in a new material, known as bio-composite (Syahputra and Yulianto 2020). Sago dregs and bark are also beneficial for making charcoal and briquettes (Jamaludin et al. 2022), for fuel or other energy sources. Sago fiber has been used in concrete mixes as an alternative material to substitute sand (Ginting et al. 2022), as an additional material for concrete brick (Umar et al. 2020), and in particle boards for soundproof walls (Pratama 2022).

10.1.2 Botany of Sago Palm

The name Metroxylon comes from two words, Metro meaning pith and xylon meaning xylem (Flach 1997), while sago is starch. *Metroxylon sagu* is a plant that stores starch resulting from photosynthesis in its stem.

The taxonomic hierarchy of the sago palm is as follows (ITIS 2024).

Kingdom	Plantae
Subkingdom	Viridiplantae
Infrakingdom	Streptophyta
Superdivision	Embryophyta
Division	Tracheophyta
Subdivision	Spermatophytina
Class	Magnoliopsida
Superorder	Lilianae
Order	Arecales
Family	Arecaceae
Genus	*Metroxylon* Rottb.
Species	*Metroxylon sagu* Rottb.

The sago palm generally grows in freshwater swamps or peat swamp areas, areas on the riverbanks, close to water sources and swamp forests. Sago is known for growing in clusters. Optimal growth occurs in clusters consisting of 1–8 sago stems, with each sago base producing 5–7 young stems. In natural or clay environments, sago trees thrive, growing wider with more offspring at various levels. Sago's palms have a tree crown, comprised of leaves, branches, twigs, flowers and fruit, formed from midribs with finned leaves (Ehara et al. 2000). Adult sago palms can reach heights of 8–17 m, with some trees reaching 30 m, depending on the type and growing conditions.

The sago palm is an arborescent, monocotyledonous palm, often with a more or less thorny columnar trunk. The trunk consists of a hard outer layer of bark and an inner layer in the form of a pith containing fibers and starch (Nurmiati et al. 2022). The hard outer bark is approximately 3–5 cm thick and is often used as a building material in the Maluku area. Young sago trees have thinner bark than mature sago palms (Fathnoer et al. 2020). The trunk serves as storage for sago starches during the vegetative growth period, with heavier and longer trunks containing more starch. Sago palm trunks can weigh up to 1.2 tons (Yater et al. 2019). The weight of sago bark is around 17–25% while the weight of the pith is around 75–83% of the weight of the stem. At 3–5 years old, it contains little starch, but at the age of 11 years, the sago pith contains 15–20% sago starch. The average stem diameter is 35–50 cm, with the widest diameter reaching 80–90 cm. Typically, the lower part of a sago tree trunk is a larger diameter than the upper part and contains more starch. The cylindrical trunk serves as a food source for the sago itself. At the harvest age of 10–12 years, the starch content can reach 150–300 kg of wet starch (Konuma 2018). The roots of the sago palm form a fibrous root system, spreading and thickening as the tree grows.

The leaves of the sago palm are elongated (lanceolate) and slightly wider with a stem in the middle, having leaf stalks and a unique leaf texture. When young, the leaves or fronds of a sago palm are layered, falling off as they mature and independently attaching to stem segments. Sago trees typically have 18 leaf stalks with lengths reaching 6–7 m. The leaf stalks vary in size and length, with an average length of 60–180 cm and a width of 5 cm. As the leaf stalks age, they fall off the trunk. Initially light green, sago leaves darken to dark green and then reddish brown when ripe (Ellen 2006). Sago's palms have around 1000 stomata per/mm^2 leaf, facilitating photosynthesis by absorbing carbon dioxide and converting it to carbohydrates (Omori et al. 2000).

Sago palm flowers are numerous lateral inflorescences that usually appear at the end of the tree trunk, brownish-red in color and densely clustered with many branches. Thc branches of sago palm flowers consist of primary, secondary and tertiary branches, each containing a pair of male and female flowers that are ready to open and bloom. The initial flowering phase is marked by the emergence of the shortest leaves, or what are usually called flag leaves. It is estimated that these flowers appear when the tree is 10–15 years old, depending on the type and after flowering (Ellen 2006). The flowerings appear once in the life cycle of a palm, and then the palm dries out and starts to die. Sago palm flowers are also cross-pollinated plants, so these plants grow solitary and rarely form fruit. When the flowers begin to appear, the fruit formation phase begins, which lasts about 2 years. The fruit of the sago palm is identical with a scaly texture and a round shape with a yellowish-brown color (Karim et al. 2008).

10.1.3 Harvesting and Starch Extraction

10.1.3.1 Harvesting

Determination of sago readiness for harvest is determined by observing the morphological characteristics of sago palms. Sago palm trees that are ready for harvest are categorized into three maturity phases, namely the early maturity phase, the mature phase and the post-maturity phase (Astuti et al. 2014; Ruamba and Sumule 2020).

The early maturity phase of the sago palm is marked by the young fronds or shoots and the opening and widening of some of the older fronds, with the thorns on the sago fronds starting to disappear. In this phase, the starch content in the sago trunk is still low, because all the segments of the sago trunk are not yet filled. This is characterized by a pith that is still liquid or wet (Astuti et al. 2014).

The mature phase is characterized by the largest and highest stem size in one cluster and the number of leaves on the shoot/crown being between 3–4 fronds, and no appearance of flowers yet, with the shoot part appearing swollen. Some of the leaf midribs begin to dry and decrease, and then the top of the midrib releases shoots that have not yet opened. The top of the tree is flat compared to younger sago trees. This phase is the right phase for harvesting sago palm because the sago is already full. All sago segments from the bottom segment to the first segment of the sago midrib are filled (Ruamba and Sumule 2020). If sago is harvested in this phase, the pith of the sago palm is hard and dry, indicating dense starch content. According to Ruddle et al. (1978), to ensure that sago contains enough starch, tests can be carried out by making a hole in the sago trunk approximately 1 m above the ground, then taking out the pith, chewing and squeezing it. If the juice is turbid, it means the starch content is sufficient and the sago palm is ready to be harvested.

The third stage is the post-maturity phase, marked by the sago flowers having bloomed, then the sago fronds begin to dry until they disappear, and then the flowers will slowly produce fruit. In this phase, the sago starch is reduced until it is used up (Ruamba and Sumule 2020).

Harvesting is done by clearing the land to create an entrance to the cluster and cleaning the trunk before cutting to make the process of cutting and transporting the harvested products easier. The trunk is cut as close to the root as possible (Fig. 10.2), using an axe or chainsaw. The stems are cleaned of the midribs and some of the tree tops, leaving only 6–15 meter-long sago logs. These logs are cut into 1–2 m pieces to facilitate transportation, with each log weighing approximately 120 kg, having a diameter of 45 cm, and a bark thickness of around 3.1 cm. The second harvest is carried out 2 years later.

Fig. 10.2 The process of sago harvesting. (**a**) Sago palm at about harvesting stage, (**b**) At the start of cutting the palm with a chainsaw, (**c**) Cutting of the palm, (**d**) Measuring the length of trunk of about 42 inches for log section, (**e**) Sectioning the trunk, (**f**) Log counting by plantation team, (**g**) Making nose on the log for the rope to group, (**h**) Rolling log to canal via harvesting rail, (**i**) Chaining the logs by rope through the nose, (**j**) Group of logs is pulled by diesel boat to mill, and (**k**) Sago logs are gathered on mill pond awaiting to be processed. (Source: AF Irawan)

10.1.3.2 Starch Extraction

Sago starch is obtained by processing the trunks of the sago tree. This process involves extraction to separate the sago dregs and starch found in the trunk. From one sago tree, approximately 17–25% sago starch and 75–83% sago dregs can be obtained (Latuconsina 2014). Additionally, Wina et al. (1986) discovered the presence of free sugars in sago starch samples, sucrose at 37%, glucose at 26% and fructose at 12%.

In general, the sago extraction process is divided into two methods, namely traditional methods and modern (manufactured) methods. Traditional methods are commonly used by farmers, while modern methods are typically operated by companies.

Traditional Extraction

The traditional extraction method is carried out by farmers or local people in the form of small-scale processing plants. Farmers cut down the sago palm tree and process it all at once in one place in the cultivating area, making the sago extraction process easier because there is no need to transport heavy sago trunks (Karim et al. 2008).

Extraction of sago starch goes through several steps, starting with selecting mature sago palm trees, and cutting them into several pieces or logs with a length of 40–70 cm. The woody outer layer of the logs is then removed, revealing the pithy interior. After chopping the pith into several pieces, they are placed on a sieve and then squeezed by hand. During this process, water is poured over it to remove the starch. The starch liquid drips into a container through the filtered sieve. Starch particles settle to the bottom of the container while the water overflows into the outer vessel. The water is re-filtered several times to ensure that all starch is extracted. The sago starch is separated from the fiber, dried and made into flour. This process is repeated several times until the flour is completely smooth. The packaged sago flour is then ready to be distributed to consumers (Ruddle et al. 1978).

Industrial Extraction

Extraction using modern methods is an advancement of processing techniques used by traditional farmers. This company collects sago stems that have been cut into short pieces measuring 1–1.2 m from their gardens or purchases them from farmers' gardens. Modified technology is employed in large-scale factories, to extract starch, with the entire process being mechanized. The process begins with cutting the sago palm tree, removing the bark and then feeding it into a slicer machine to extract the pith from the sago logs. The extracted pith is then grated into fine pieces using a conveyor belt machine (Fig. 10.3). The sago starch extraction process produces a starch slurry, which is then filtered using a centrifugal filter machine to separate crude fiber (dregs), sago starch, and washing and filtering wastewater. To obtain pure starch, it is further separated and purified using a sieve equipped with a nozzle separator and a cyclone separator. The starch is then dried using a rotary vacuum dryer with a hot air dryer. All these processes are mechanized and carried out using machines.

10.1.4 Source for Human Energy and Food Security

To strengthen the food security initiative and fulfill the country's dietary requirements, it's crucial to cultivate alternative crops beyond rice for carbohydrates and to enhance food variety. The sago palm is one of the plants that has caught the attention of the Indonesian government as a potential alternative non-rice food plant, with

Fig. 10.3 Industrial extraction for sago starch. (**a**) Log debarking, (**b**) Log splitting, (**c**) Placing the pith on a running belt to be crushed, (**d**) Pith entering chopping machine-generated by diesel, (**e**) Crushed pith is mixed with water and run to the sieved batch, (**f**) Starch and fiber milk passed through the sieve is stirred to separate fiber, (**g**) Separation of fiber and starch by sucking the milk from the previous batch, (**h**) Starch sedimentation on batch, (**i**) Wet starch stayed on batch after releasing the water out, (**j**) Wet starch, (**k**) Wet starch are being dried above concrete-floored on oven house, (**l**) Burned sago barks are used for the heat to dry wet starch, and (**m**) Dry starch produced by wet extraction process. (Source: AF Irawan)

environmentally friendly functions. Sago palm was included in a program to accelerate its development (Directorate General of Estates Crop 2013). In 100 g of sago starch, there are 285 calories, 77 g of water, 0.2 g of protein, 30 mg of calcium, 71 g of carbohydrate, 0.7 mg of Fe and 0.3 g of fiber (Johnson 2010).

Sago has the potential to support national food security and can be utilized for environmental management, control, and preservation, as and as an alternative food for Indonesian people as a rice substitute. This supports Indonesian Presidential Decree No. 22 of 2009 concerning policies for diversifying food consumption based on local resources as a basis for strengthening food security to improve the quality of human resources and preserve natural resources. The potential of these natural resources can be mobilized to achieve food security through food diversification efforts. A sustainable food system will support food security, through optimal use of natural and human resources, be acceptable and easily accessible, environmentally friendly, and meet nutritional needs that are adequate, safe, healthy and available for now and in the future.

The sustainable management and use of sago palm align with the national movement to control food inflation, which the government is intensifying to prepare for future crises. Many areas in Indonesia have the potential to become hubs for developing national food security. Sago, as the largest starch producer, can ensure starch production year-round. Each trunk can produce around 200 kg of wet sago flour per year, or 25–30 tonnes per hectare.

Additionally, sago has health potential seen from its superiority compared to rice in terms of nutrition and health, such as (1) it has a low glycemic index (GI), 28 for sago noodles, sago macaroni, 48 for plate sago and 53 for sago porridge, so it does not rapidly increase blood glucose levels (Haliza et al. 2006; Hariyanto 2014); (2) increases immunity, (3) reduces the risk of colon and lung cancer, (4) reduces obesity, (5) makes defecation easier (Flach 1993), and (6) and its benefit for non-gluten diets for autism patients (Alamri 2020; Metaragakusuma et al. 2016) improved immunity, and reduced risk of heart disease (Rana et al. 2022).

Sago has the potential to not only maintain food diversification and security but also improve the economy of local communities, especially during the current global food crisis. Therefore, sago palm is used to strengthen food security, especially in the eastern region of Indonesia. The Ministry of Agriculture of Indonesia has prepared a more detailed concept regarding sago-based food security. Now, various food industries derived from sago starch are increasing and improving, with demand for products at home and abroad continuing to grow.

10.2 Biodiversity

Biodiversity encompasses several levels of diversity: morphology, genetics, and ecosystem diversity. Evaluating morphological diversity with traditional methods includes assessing observable traits such as leaf shape, size, and color, as well as characteristics of the trunk and inflorescence. This investigation also includes estimating sago starch output based on physical traits. The production of sago starch is closely related to biomass production, influenced significantly by factors of the harvest stage, indigenous varieties, environmental circumstances, and farming methods. It is essential to thoroughly observe sago palm production and utilize characteristics like leaf area, total leaf marks, functional leaves, and trunk height to measure starch output and identify the best time for harvesting (Riyanto et al. 2018). Morphological evaluations offer valuable insights into the observable diversity among sago palm populations. The morphological diversity of sago palm seedlings exhibited high variations, with the similarity coefficient ranging from 10% to 69%.

The color of old sago palm leaves is generally consistent, but when they are young, especially at the sago palm bud stage, the color can vary depending on the genetic characteristics. According to Firmansyah (2022), there are many variations

Table 10.1 The color of sago leaf buds in Mentawai Island (West Sumatra) and Lingga Island (Riau Archipelago)

Accession	Mentawai	Lingga	Colour
Saikot	183A		Dark red
Limu	N199B		Dark greyish yellowish brown
Betaet	183A		Dark red
Ukra	187A		Dark red
Siri Langga	143A		Strong yellow green
Sibeotcuo	133A		Moderate olive green
Nerekeh		180B	Moderate red
Panggak Laut		180B	Moderate red
Musai		179A	Moderate red
Pekaka		180B	Moderate red
Keton		180B	Moderate red
Teluk		179A	Moderate red

Source: Al Manar et al. (2023) and Firmansyah (2022)
Note: The color is identified using the tool from the Royal Horticulture Society Color Chart (2015)

in the color of sago palm leaf buds in Mentawai Island, West Sumatra, including dark red, dark greyish-yellowish brown, strong yellow-green, and moderate olive green. In contrast, Al Manar et al. (2023) found that the color of sago leaf bud in Lingga Island, Riau Archipelago Province is mostly moderate red (Table 10.1).

The colors of young sago leaf buds in Timika District are diverse, ranging from strong green yellow, moderate green orange, light red yellowish, orange-red greyish, strong yellow greyish, and strong red, to light orange (Table 10.2). In Sorong District, the majority of leaf bud colors were light olive, followed by moderate yellow-green, strong yellow-green and moderate red. In Mappi District, the predominant color of the leaf bud is pale green, with strong yellow-green and dark yellowish green also present (Table 10.2).

The sago found in Mappi Regency, South Papua consists of two types: thorny sago and non-thorny sago. The average production of dry starch per sago in Mappi Regency was 477.44 kg/stem (Table 10.3). The highest starch yield was from heim sago with long thorns found in Taragai village, producing 1197.20 kg/stem per tree. In comparison, *katage* produced 624.35 kg, *mani* produced 718.15 kg, *wiyage* produced 320.33 kg, and *arare* produced 232.65 kg per tree. *Katage* and *mani* had very high production, while *wiyage* and *arare* had high production in Assue District. The production of sago starch was also high, with 288.53 kg of dry starch/tree (*homang*) and 261.30 kg of dry starch per tree (*sigare*).

There are three types of sago accession in Mamuju Regency, West Sulawesi, namely *kasimpo*, *kapas*, and *ruwi*. All accessions were vegetative accessions or had only experienced one generative (flowering) phase. The highest dry starch production was observed in the *kapas* accession, which was 436.61 kg per stalk, while the *kasimpo* accession was 243.14 kg per stalk. The lowest dry starch production was in the *ruwi* accession, which was 81.67 kg per stem. The average production of sago dry starch is 304.38 kg per stem. The *kapas* accession had the

Table 10.2 The color of sago leaf buds of the Center, Southwest and South Papua

Accession	Mimika	Sorong	Mappi	Color
Omoroko	144A			Strong green yellow
Durumu	137A			Moderate green orange
Oko	179D			Light red yellowish
Taina	177C			Orange red greyish
Mbapare	153C			Strong yellow greyish
Omiya-1	N144C			Strong red
Omiya-2	183A			Light orange
Ikimina	152B			Moderate yellow green
Warumo		148B		Light olive
Wasulagi		152A		Light olive
Wasnan		152A		Moderate yellow green
Wayuluh		152B		Light olive
Wagelik		138B		Moderate red
Wanegles		152A		Strong yellow green
Wawun		N34A		Light olive
Wafaban		143A		Strong yellow greyish
Moor-Tarau		152A		Light olive
Ghanu-Bhai Bande			N144C	Strong yellow green
Komru-Heim			N138C	Pale green
Taragai-Heim			N138C	Pale green
Wiyage-Nan Duyemoo			N138C	Pale green
Katago-Yemoo			N138C	Pale green
Mani-Naa Duyemoo			141C	Strong yellowish green
Arage-Cham			143C	Strong yellow green
Sigare-Bakhan			N134A	Dark yellowish green
Homang-Naa Du Yemoo			N134A	Dark yellowish green

Source: IPB University team (2021a); IPB University team (2021b)
Note: The color is identified using the tool from the Royal Horticulture Society Color Chart (2015)

highest dry starch production, which reached 717.90 kg per stem (Table 10.4). Several variables affect the production of starch, such as height, diameter, moisture content, and yield of sago stems. According to Ehara (2009), there is a direct relationship between sago starch production and the weight of the stem. The accumulation of starch initiates at the lower stem base and peaks in the middle to upper two-thirds of the stem, excluding the midrib, as noted by Schuiling (2009).

The yield percentage per accession from lowest to highest is *ruwi* 11.05%, *kasimpo* 21.10%, and *kapas* 23.89%. Starch yield was calculated to determine dry starch production (kg per stem).

Accession of sago spread across Indonesia has several types, including thorny sago and non-thorny sago. All sago accessions found in several provinces have many variations, from the lowest to the highest yield and from the lowest to the highest dry starch production. The highest production yield was identified in South

Table 10.3 Production per village in Mappi Regency, South Papua

Accession	Yield (%)	Dry starch production (kg per stem)
moor[1]	15.56	171.15
ghanu[2]	30.53	376.00
komru[2]	27.45	584.80
taragai[2]	27.52	1197.20
wiyage[2]	30.94	320.35
katage[2]	39.01	624.25
mani[2]	32.32	718.15
arare[1]	34.81	232.65
sigare[2]	22.03	261.30
homang[2]	28.16	288.55
Average	28.83	477.44
Standard deviation	6.54	312.73
Coefficient of variability (%)	22.70	65.50

Source: IPB University team (2021a)
Description: 1 = non-thorny sago, 2 = thorny sago

Table 10.4 Production per accession in Mamuju Regency, West Sulawesi

Accession	Sample	Yield (%)	Dry starch production (kg per stem)
	1	18.80	183.64
	2	16.94	106.51
kasimpo	3	20.44	144.68
	4	24.46	391.93
	5	25.80	388.93
Average		21.10	243.14
	1	19.16	258.27
	2	24.61	425.76
kapas	3	27.37	717.90
	4	24.41	344.52
Average		23.89	436.61
ruwi	1	11.05	81.670
Average		21.20	304.38
Standard deviation		4.920	192.39
Coefficient of variability (%)		23.21	63.210

Source: Abhal (2022)

Papua, Mappi Regency, of 1197 kg per stem. High production yields are also discovered in Papua Province with the accession of *he bang* having a dry starch weight of 447 kg per stem.

Various studies that have been carried out showed that the genetic resources of sago (GRS) in Indonesia, in general, and Papua in particular, have high diversity. Analysis of the sago genome based on molecular markers of the chloroplast genome (cpDNA) and nuclear genome (Wx Gene) shows that sago palms in Indonesia are diverse (Abbas 2018). The high genetic diversity of sago palms detected in the results of genome analysis has relevance to the morphological variation uncovered by numerous sago palm researchers, highlighting the importance of genetic diversity identified through molecular markers in the chloroplast and nuclear genomes. This suggests that sago palms in Indonesia exhibit significant diversity, although genetic markers indicate a lower level of diversity compared to morphological markers. Matanubun et al. (2005) revealed that the variety of sago palm in Papua is very large based on morphological characters, namely that in total there are 96 varieties found from eight locations, that are *Waropen*, *Salawati*, *Wasior*, *Inanwatan*, *Onggari*, *Sentani*, *Kaureh*, and *Windesi* in Papua. Bintoro (2011) reported that in Papua, there are 36 types of sago with thorns and 17 types of sago without thorns, and Yamamoto et al. (2005) found that around Sentani there are 15 varieties of sago. Furthermore, Ehara et al. (2000) found 11 varieties in Southeast Sulawesi, North Sulawesi and North Ambon based on morphological characteristics.

Based on the number of haplotypes found in several sampling locations for sago palms, Papua is the center of sago palm diversity because in Papua the highest number of haplotypes is found and there are natural populations. Hedrick (2011) revealed that large amounts of biological diversity exist in natural populations. Based on this statement, it can be said that the sago palms in Papua are a natural population (not a migrant population). If we talk about sources of diversity, the islands of Papua, Sulawesi and Kalimantan are sources of diversity for sago palms because, on these seven islands, specific haplotypes are found. Vendramin et al. (1999) revealed that the large number of haplotypes reflects the high variation or diversity in a population. Mengoni et al. (2003) documented that differences in chloroplast haplotypes in each population reflect differences in genetic entities (sources of variation).

In order to promote the implementation of germplasm conservation, particularly sago palm, information on the genetic diversity of germplasm is crucial. Preserving the sago palm germplasm is necessary to prevent its degradation. To demonstrate the persistence of richness and the existence of particular organism types, the diversity of an organism must be recorded. Plant breeders look through a large number of accessions to locate new sources of genes from diverse gene pools that can identify and study their potential for genetic improvement. Lack of genetic variety significantly hinders breeding programs. The more diverse germplasm features that are present during selection, the more likely that something new and beneficial will be discovered. Thus, both in situ and out of situ collections are crucial.

10.3 Propagation of Sago Palm

Sago palm can be propagated using two methods: the generative, which involves seeds, and the vegetative, which involves offshoots known as suckers. Other methods have also been developed for propagating sago seedlings by some researchers, which were known as in vitro and ex vitro cultures.

10.3.1 Generative Propagation

Seeds are seldom used as planting material since sago palm is harvested before fruiting. Moreover, according to the farmers, seedlings are slow in establishment by about 1 to 2 years compared to suckers. During fruiting, sago palm could produce over 2000 fruits per palm (Jong 1995). However, a poor germinability of sago palm seeds has been reported previously (Alang and Krishnapillay 1986; Flach 1984; Jaman 1985; Johnson and Raymond 1956; van Kraalingen 1984). In a later study, Ehara et al. 1998 reported that it could be increased to 20–27% when the air temperature increased from 25 to 30 °C. Prior to the experiment, the removal of the husk (exocarp and mesocarp) and sarcotesta was performed. At 30 °C, they recorded about 40% of germination rate. An earlier study performed by Rembon et al. (2008) on the sago wild type of local varieties named *Manno Kecil* and *Manno Besar* growing in mineral soil, in Jayapura, Papua, found that fertile and well-developed seeds displayed a higher germination rate than in the previous study. Physiological maturity of well-mature seeds was indicated by their positions which had been unbeared from the stalks and fallen to the ground. Using the same treatment method by removing husk and sarcotesta prior to the experiment as described by Ehara et al. (1998), germination rates of varieties of *Manno Kecil* and *Manno Besar* were 70.1% and 77.3%, respectively. Exocarp and mesocarp present on fruits are known as germination inhibitors (Ehara et al. 2001). After gaining success on the sago palm seed germination rate, a lot of sago palm research has recently been conducted using seeds as an experimental material (Fig. 10.4a) such as to study the photosynthetic activities in relation to groundwater level (Azhar et al. 2020a, b), effects of soil types and water regimes (Rembon et al. 2010), and individual exclusion of macro- and micronutrients during early establishment (Irawan et al. 2011a).

Reports on the growth and starch yield of sago palm growing on a natural sago habitat in a remote area where there is almost no human intervention in the propagation or cultivation are still few. An investigation on starch yield and growth of sago palm on a swampy sago palm habitat in Kaureh Sub-District, Jayapura, Papua Province was conducted (Irawan et al. 2013). On this site, a generative propagation was supposed to naturally take place. Fallen and germinated sago palm seeds, after their exocarps and mesocarps were naturally decomposed, were found fully and partially immersed in the soil surface (Fig. 10.4b, c). The process of germination might have been supported by high humidity and temperature around the sites. Mature fruits or seeds might have spread to the areas following the water current. Once they are successfully established, they will form a sago habitat. This kind of process should have taken place for years.

Most of the sago palms found on this site were mixed with indigenous tree species at varying densities. The study found that the sampled palms displayed various starch content on their trunks ranging from 17.8–63.4% at flowering and flower

Fig. 10.4 (**A**) Sago palm at fruiting stage found on *Manno Besar* local variety in Sentani Papua, (**B**) fallen fruits, (**C**) several germinated seeds collected from the ground, and (**D**) several well-developed sago palm seedlings grown at a controlled room temperature at Makino Botanical Garden, Kochi, Japan, prepared for sago research. (Source: AF Irawan)

initiation stages with an average of 47.5%, calculated on a dry basis. When expressed as a percentage of dry starch per fresh trunk, the range was 2.3–32.7% with an average of about 19%. These results are higher than those reported by Jong (2018) in the natural sago habitat in South Sorong, West Papua where the average was about 10–11% and the highest percentage was about 17%. The trunk shape was slender, with an average length and diameter of 9.7 m and 34 cm, respectively, although it grew on mineral soil. The average trunk diameter was much lower than that found in semi-cultivated gardens growing on peat in Tebing Tinggi Island, Riau Province which was around 42 cm (Monda et al. 2022). The size of the diameter tended to reduce as the groundwater level was lower than the soil surface (Monda et al. 2022). Competition for light was supposed to take place. Lower trunk diameter from the height of 2–6 m compared to 7 m above on 6 sample palms, indicating that the palm suffered from light at the young trunk growth stage (Irawan et al. 2013).

10.3.2 Vegetative Propagation

10.3.2.1 Propagation by Suckers

Sago palm is commonly propagated by suckers. Both smallholders (Shimoda and Power 1986) and commercial plantations (Jong 2006) use suckers as an effective material to expand their cultivation fields instead of seeds. Once a sago cluster is established, it normally comprises the main palm and suckers generated from the basal rhizome of the main palm with various stages of development. However, up to the present, there is almost no research studying the exact age of the suckers that appeared on the cluster and their relationship with growth and duration from their emergence to harvest. All the reports determine the suckers' age by estimation which is usually judged from the sucker weight following the estimation method by experienced sago farmers (Irawan et al. 2009a, b). These data are very important to understand the growth pattern of suckers on clusters. Questions remain about whether the position of the suckers on the cluster, the size of their frond base diameters, and even their generation orders affect the duration of their harvesting times since the first emergence. However, such data requires a very long research observation.

Several constraints facing the establishment of sago palm on a large scale are supplies of planting material, high mortality rate during nursery and after transplanting, and a huge gap in the harvest interval commonly experienced from the same cluster. Besides sucker, sago seeds may provide huge potential as a planting material. However, the study of sago breeding is very limited. The selection of the sago variety by far is based on regionally based local varieties, such as in Maluku (Yamamoto et al. 2020a), Kendari, Southeast Sulawesi (Yamamoto et al. 2010), and Papua (Yamamoto et al. 2020b). Apart from that, the survival rate after transplanting is one of the important keys to creating high cluster density which may lead to keeping starch productivity stable. Nevertheless, there is almost no evidence that well-cultivated sago plantations can produce above 10 tons of dry starch per hectare per year consistently. A lot of factors are involved in the irregular production of harvestable palm from the same cluster which may need support from the growth of the suckers generated. A comprehensive and detailed study on the differentiation and appearance of sucker buds on each node of leaf position during the rosette stage was reported by Nabeya et al. (2015) using a stereomicroscope. The sucker buds started to differentiate from the third relative basipetal leaf's position (rbL 3) from the growing point (rbL 0), and they began to be more visible after rbL 6. This study indicates that suckers should be gradually produced as long as newly emerged fronds are produced during the rosette stage. However, questions remain whether large numbers of suckers are still consistently produced when the palm is at the trunk growth stage or the number of suckers may be reduced. In addition, the practices of sucker extraction and harvest from the cluster may have effects on the production of newly generated suckers, or their production is also related to the status of soil nutrition around the cluster. To find out the answer, further studies are required.

One of the main issues in sago palm establishment is the availability of planting material. Normally, suckers are used for planting material for establishment. Based on the experiences of the establishment of a sago palm plantation during 2011–2014 in one of the commercial sago plantations cultivated on peat, the size of the expansion area depends on the number of planting materials. Since sago planting uses an 8 × 8 m square distance, it needs 156 suckers per hectare. In oil palm plantations such as those that belonged to Indonesian state-owned enterprises, the expansion of cultivated area could attain 2745 ha in a year from 2018–2019 (Directorate General of Estate Crops 2021). However, in the case of sago, the number of ready-stock suckers is very limited. Unlike oil palm seedlings which are ready to be ordered once the cultivated field is ready to plant, the preparation of sago palm suckers as planting material needs a lot of consideration. They have to be purchased from nearby islands as suckers provided by nearby local sago farmers are very limited. A sufficient number of suckers are normally gathered for several days or even weeks before they are ready to be sent. Those conditions reduce the freshness of the suckers which later affects their survival rates (Jong 1995). Based on the authors' experience on sago establishment on deep peat from 2011–2014 in Tebing Tinggi Island, Riau Province, Indonesia, around 5000–8000 suckers could be provided in a month by local sago farmers (Figs. 10.5 and 10.6) and farmers from nearby islands. By only relying on these material supplies, the size of the expansion area for sago could not even meet 1000 ha per year. Another source of sucker-derived planting material may come from the existing clusters. However, exploiting them has consequences for the reduction of the number of followers in the cluster which may lead to low production of harvestable palm. Considering the low survival rate of suckers during nursery and after transplanting, opening a cultivated area of 500 ha per year for sago establishment may be more reasonable.

Vegetative propagation by sucker is considered a common case and it can be performed massively. On the other case, they are also required for the next harvested palms. Other massive propagation can also be done by tissue culture by utilizing the shoot of the sucker as a source of explant (Sinta et al. 2008; Sumaryono et al. 2012).

Fig. 10.5 (**a**) Farmers place frond-trimmed suckers on a raft, and (**b**) the growth performance of suckers at about 3 months during the nursery. (Source: AF Irawan)

Fig. 10.6 Small-size suckers used for polybag-type nurseries. (**a**) Extracted suckers prepared for nursery, (**b**) planting trimmed suckers on 5 kg-size polybags, (**c**) growth of survived suckers at about 5 months after planting under the shade, (**d**) sucker at about 5 months after planting which is ready to be transplanted. (Source: AF Irawan)

This method may reduce the number of suckers generated and this is a dilemmatic situation facing the establishment of sago palm. Instead of using suckers, utilizing the shoot of the harvested palm as a source of explant may be considered appropriate. However, the success rates of plantlets obtained from one shoot need to be further studied.

10.3.2.2 In Vitro Propagation

Research and development of sago palm in vitro culture was carried out at the Cell Culture and Plant Micropropagation Laboratory, Indonesian Oil Palm Research Institute Bogor Unit (IOPRI BU) located in Bogor, West Java Province. The in vitro culture method developed was somatic embryogenesis (SE) that aimed to produce clonal superior sago seedlings which have the advantage of being more homogeneous, efficient and on a large scale. The superior characteristics referred to were high starch content, high production, early maturation time, and other superior characteristics.

Preparation of Explants

Explants of sucker were selected at the age of 3–5 months and the shoot base diameter of 5–10 cm. The suckers were then cut off from the mother plants using an axe or pruning knife. The selected suckers were collected in sacks and then taken to a cleaner place to be tidied up and cleaned. The cleaning process was carried out by cutting the outer fronds until 2–3 innermost fronds were left. Then the sucker shoots were cut off and leaving a height of 20–30 cm. They were cut clean and trimmed close to the coleoptile. The cutting suckers were then washed clean or wiped with cotton soaked in alcohol. Subsequently, the sucker was neatly packed and transported to the laboratory in Bogor. The time needed for the process of taking the sucker to the culture process at the Bogor Laboratory for a maximum of 5 days, so that the sucker is still fresh and the tip meristem is still viable to be used as an explant. When arriving at the laboratory, the packaging was immediately opened and cleaned using cotton wool soaked in 70% alcohol. The sucker was then ready to be used as a source of explant which was cleaned by washing with detergent. After the explant was dry, it was transferred to a dry container and taken to the laminar airflow cabinet (LAF).

Callus Induction

In the LAF cabinet, the sucker was peeling the last frond into the deepest. Afterward, it was sterilized with 0.5% fungicide by shaking in an Erlenmeyer flask for 10 minutes. After removing the fungicide solution, add a 20% Clorox or Na hypochlorite solution and shake for 10 minutes. Followed by rinsing with sterile distilled water three times, and then draining on sterile filter paper in a petri dish.

The meristem tip was cut off from the clean and sterile sucker, which was used as an explant (Fig. 10.7a). The meristem tip was then cut into four pieces, cultured on solid media in a petri dish (Fig. 10.7b), and sealed with parafilm. The explants were then cultured in a dark culture room for 8–14 months until the callus appeared. If no callus appeared, the subculture was prepared every 6–8 weeks on new media with the same composition. The medium used for callus induction was modified MS, added with 2,4-D with a concentration of 50–200 ppm, 0.1–2.0 ppm kinetin and 10% coconut water.

The callus induction stage was the most critical because it's difficult to obtain an embryogenic callus that is capable of being initiated into a somatic embryo. Not all calluses can develop into somatic embryos. Therefore, it is important to understand the accurate explant dissection techniques and mix appropriate media for embryogenic callus induction.

The explant response would be observed after 2 weeks of culture, showing swelling and changing color to yellowish white (Fig. 10.7c). Callus appeared as early as 8 months after culture and more often appeared at >12 months after culture. Callus initiation was observed from the edge of the explant where the cut or incision was made. There were only a few primary calluses that have just appeared and there

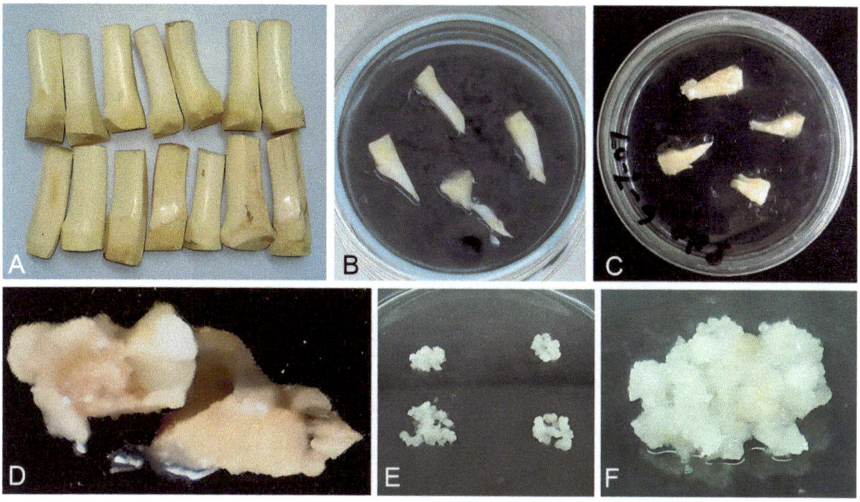

Fig. 10.7 The process of somatic embryogenesis at the callus induction stage. (**a**) Sucker that has been peeled to the deepest and sterilized was ready to be cut into explants. (**b**) Explants cultured in solid media for callus induction. (**c**) Explant growth response at 2 weeks after culture. (**d**) Initiation of callus at 8–14 months after culture. (**e**) Primary callus subculture on fresh media, and (**f**) Primary callus growth after subculturing. (Source: I Riyadi)

are also a lot of them in aggregate form (Fig. 10.7d). After the callus begins to grow and increase its mass, subculture is carried out on the new petri dish with the same media composition, however with a slight reduction in the concentration of auxin (2,4-D) by 0.20–0.25 times with a constant concentration of cytokinin (kinetin) (Fig. 10.7e). After more than 2 weeks, the callus increased in number with more oval or rounded structure and fresh white or yellowish white (Fig. 10.7f). This phenomenon agreed with the research conducted by Tahardi et al. (2002) and Riyadi et al. (2005) who presented the results of callus induction using sucker explants.

Embryogenic Callus Proliferation

Callus proliferation uses material or explant in the form of a secondary callus that has been subcultured at least once. This process can be done on solid media in jar bottles (Fig. 10.8a, b) or liquid media using the Temporary Immersion System (TIS) (Fig. 10.8c, d). The media utilized for callus proliferation is modified (Murashige and Skoog) media (MS) plus 2,4-D with a concentration of 10–30 ppm and kinetin or BAP with a concentration of 0.1–2.0 ppm. The callus is cultured in a dark room similar to callus induction conditions. The duration for the callus proliferation stage is typically 4–6 weeks, after which it can be subcultured for further callus proliferation or used as an explant for somatic embryo induction (Fig. 10.8e).

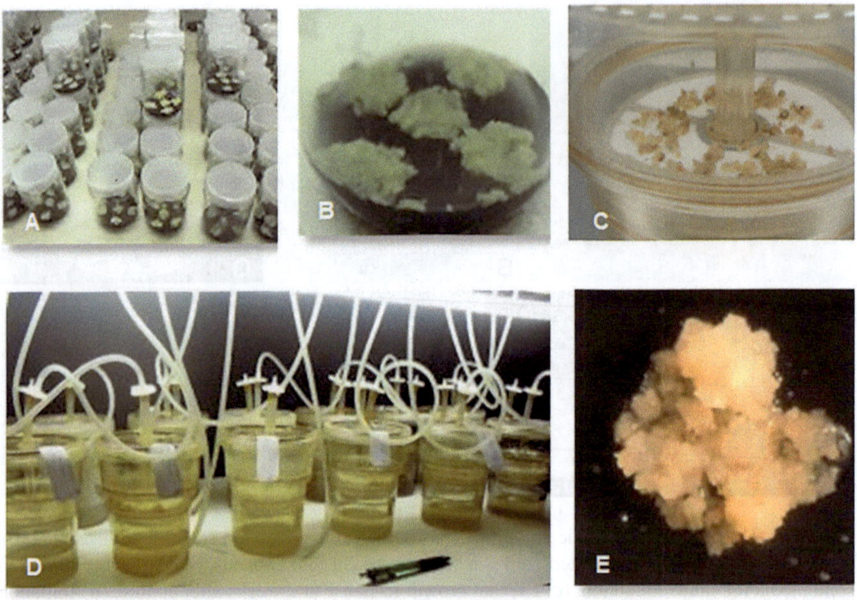

Fig. 10.8 (**a**) Callus proliferation on solid media in jar bottles. (**b**) The results of callus proliferation on solid media 4 weeks after culture. (**c**) Callus proliferation in the liquid media of the TIS system (temporary immersion system). (**d**) Callus culture in TIS liquid media for more mass production. (**e**) Embryogenic callus resulting from proliferation. (Source: I. Riyadi)

Two weeks after subculture, the amount of callus begins to increase and reaches its maximum biomass 6 weeks after culture (Fig. 10.8b, c). Compared with solid media, callus proliferation in TIS liquid media is more efficient and has better proliferation rates and quality of callus produced (Fig. 10.8c). The rate of callus proliferation on solid media increases by 3–5 times, while in TIS culture it increases by 7–10 times. The increase in callus biomass is faster and higher compared to oil palm (*Elaeis guineensis*) using the TIS culture system (Sumaryono et al. 2007). Usually, embryogenic callus proliferation is carried out at least 3 times in a culture cycle, sometimes up to 6–7 culture cycles.

Induction and Proliferation of Somatic Embryos

Somatic embryo induction was accomplished when there was a sufficient stock of embryogenic calluses. A callus that had reached the embryogenic or pro-embryoid phase could be easily and quickly induced to become a somatic embryo. Therefore, embryogenic callus was selected to be used as explants in the somatic embryo induction and initiation process (Fig. 10.9a).

The media used in somatic embryo induction and proliferation is MS media with the addition of 2,4-D with a concentration of 0.05–0.1 ppm plus kinetin or BAP at a concentration of 0.5–2.0 ppm. Explants were cultured in a well-lit room under a fluorescent lamp or LED with a light intensity of 20μmol photons/m^2sec and

Fig. 10.9 Induction and proliferation stages of somatic embryos. (**a**) Clump of embryogenic or pro-embryoid callus as somatic embryo induction explant. (**b**) Induction and proliferation of somatic embryos on solid media 2 weeks after culture. (**c**) The somatic embryo initiation in solid media. (**d**) The induction and proliferation of somatic embryos in the TIS of the Rita model. (**e**) somatic embryo performance in the TIS of the Nalgen model. (Source: I. Riyadi)

12 hours of light exposure at a temperature of 26 ± 1 °C for 4–6 weeks per culture cycle. This method was consistent with the findings of research conducted by Riyadi et al. (2016).

Somatic embryos began to emerge 2–4 weeks after culture (Fig. 10.9b, d), and were visible 4–6 weeks after culture (Fig. 10.9c, e). Though sometimes in the initial culture somatic embryos were not visible, there was a rapid increase in biomass, especially in TIS cultures (Riyadi et al. 2018). Likewise, this yield was faster and higher compared to oil palm in TIS culture (Sumaryono et al. 2007). After subculturing, somatic embryos continued to develop while there was still an increase in biomass. The embryos produced were young or in the globular phase (Riyadi et al. 2005). These globular embryos could be further proliferated by subculturing in fresh MS medium, resulting in a two- to threefold increase in biomass after 4–6 weeks of culture. Somatic embryo proliferation was achieved two to three times in a culture cycle.

Maturation of Somatic Embryos

The maturation stage aimed to encourage and accelerate the maturation process of young phase somatic embryos so that they could improve into further developmental phases starting from elongated-scutellar and coleoptile (mature phase embryos). The final goal of the maturation stage was to have as many embryos as possible in each of the adult or coleoptile development phases to produce more sprouts/germinants and plantlets.

The explant used in the maturation stage culture was a somatic embryo at the young or globular phase (Fig. 10.10a). The maturation stage culture could be achieved on solid media in jar bottles or TIS liquid culture media. The media used was modified Murashige and Skoog (MS) media supplemented with ABA at a concentration of 0.01–0.1 ppm and kinetin or BAP at a concentration of 0.5–2.0 ppm. The explants were cultured in light culture under a fluorescent lamp or LED with a light intensity of 30μmol photons/m^{2}sec. and 12 hours of light exposure with a temperature of 26 ± 1 °C for 4–6 weeks per culture cycle. Normally only 2 culture cycles are needed to result in an adult embryo (all coleoptile) (Riyadi 2010).

The somatic embryo maturation process occurred at 3 weeks after culture (Fig. 10.10) and was visible at 4–6 weeks after culture (Fig. 10.10b, c). This same phenomenon was also observed by Riyadi and Sumaryono (2009) in their

Fig. 10.10 The process of development of somatic embryogenesis of superior sago plants in the somatic embryo maturation phases. (**a**) Clump globular phase of somatic embryo as an explant for somatic embryo maturation culture. (**b**) Results of somatic embryo maturation on solid media. (**c**) Example of somatic embryo development phases, from 1. pro-embryo (2–3 weeks), 2. globular (3–4 weeks), 3. elongated (5–6 weeks), 4. scutellar (7–8 weeks) to 5. coleoptillar phases 8–14 weeks). (Source: I Riyadi)

experiment. After 4 or 6 weeks when the maximum culture cycle was reached, the mature embryos that had formed were immediately selected and subcultured to an advanced stage media. The remaining young embryos were subcultured to the maturation media again. The rate of somatic embryo maturation reached 70–100%, depending on the genetic characteristics of the sago palms. TIS culture produced a higher percentage of maturation rate compared to solid media. On the other hand, the increase in biomass at the maturation stage was 2 times lower compared to the proliferation stage.

Somatic Embryos Germination

The germination stage culture required mature somatic embryos that were coleoptillar (Fig. 10.11a), in sufficient quantities to obtain more germinants which were used as explants. At this stage, nearly no additional biomass was observed. The germination stage culture process could be accomplished on solid media in jar bottles or liquid media on the TIS system.

The media used in germination culture was MS media containing kinetin or BAP with a concentration of 0.5–2.0 ppm and GA3 with a concentration of 0.1–0.5 ppm

Fig. 10.11 (**a**) Initiation of somatic embryo germination on solid media. (**b**) The result of germination (germinant) grew into shoots on solid media. (**c**) Initiation of somatic embryo germination in liquid media of TIS. (**d**) The results of germination (germinant) grew into shoots in liquid media of the Temporary Immersion System. (Source: I Riyadi)

(Riyadi 2010). The culture was placed in a light culture room under a fluorescent lamp or LED with a light intensity of 30μmol photons/m^2 sec. and 12 hours of light exposure at a temperature of 26 ± 1 °C for 4–6 weeks per culture cycle. Similar to the maturation stage, the germination stage culture was completed within a maximum of 2 culture cycles.

In the germination process, the plumules of potential shoots arose at 2–4 weeks after culture (Fig. 10.11b, c). The number of germinants increased 5–6 weeks after culture (Riyadi et al. 2005). During harvesting at 6 weeks of culture, the percentage of germinants obtained in TIS culture was higher (Fig. 10.11d) compared to solid media (Fig. 10.11b). Additionally, the level of homogeneity in TIS culture was much higher compared to solid media (Fig. 10.11b and d), as revealed by Riyadi et al. (2018).

The percentages of germination of sago somatic embryos were around 40–100% depending on the genetic characteristics of sago palms. TIS culture produced a higher germination percentage, reaching 60–100% (Fig. 10.11d), than solid media, which only reached 30–80% (Fig. 10.11b).

Enlargement of Germinants into Plantlets

The germinant enlargement stage was the final stage in the tissue culture activities of sago palms. The explants used were germinants derived from the culture of the germination stage (Fig. 10.12a). This stage only utilized solid media in jar bottles (Fig. 10.12a, b), followed by static liquid media in culture tubes (Fig. 10.12c).

The media used was MS media containing kinetin or BAP with a concentration of 0.5–2.0 ppm and GA_3 with a concentration of 0.5–2.0 ppm. After the explant was cultured in a germinant growth medium, it was then placed in a light culture room under a fluorescent lamp or LED with a light intensity of 30μmol photons/m^{2}s, and 12 hours of light exposure with a temperature of 26 ± 1 °C for 4–6 weeks per culture cycle. The germinant enlargement stage required 1–2 culture cycles to complete. The subsequent stage was subculturing into a culture tube containing liquid media to stabilize the rooting system until they were ready for acclimatization (Fig. 10.12c).

The growth of sprouts was observed after 2–3 weeks of culture, showing elongating shoots and the emergence of leaves. The height of the shoot reached 5–10 cm in one culture cycle in 6 weeks (Fig. 10.12b). Subsequently, the plantlets were then subcultured into new media for the growth of shoots and root systems using liquid MS media in a culture tube. The media contained NAA at a concentration of 1–5 ppm, IBA at a concentration of 1–3 ppm and BAP or kinetin at a concentration of 1–2 ppm. At this stage, the root system was formed after 5–6 weeks of culture (Riyadi 2010).

If the root system was not perfected, the sprout was subcultured into fresh media with the same composition until the root system was developed and ready for acclimatization. If the acclimatization process was required at a distant location, the plantlets were prepared, arranged and packed with safety and moisturizer to ensure they remained fresh and undamaged during transportation.

Fig. 10.12 The stage of enlarging the germinant into plantlets. (**a**) The germinants as initial explant. (**b**) The growth of germinants into shoots. (**c**) Plantlets with shoot and root systems in liquid media. (Source: I Riyadi)

Acclimatization

A higher success rate for acclimatization could be achieved when perfect plantlets were presented. The acclimatization process for sago plantlets was more complicated than for plantlets of other palms, such as coconut, oil palm, date palm and other palms. Plantlets that were acclimatized had to meet certain requirements: a minimum of 1 open leaf, a height of at least 10 cm, and roots. Before acclimatization, the plantlets were removed from the culture tube, cleaned of any remaining media and dried (Fig. 10.13a). They were then plantlets were planted in acclimatization media in a glass pod containing a mixture of soil, organic material (cocopeat/manure) and sand in a ratio 2:1:1 v/v (Fig. 10.13b).

After 10–12 weeks, the plantlets were transferred to new media in polybags with the same composition. They were placed under a parent with a 50–60% light absorption rate or pre-nursery (PN) for 6–8 weeks (Fig. 10.13c). Subsequently, the plantlets were moved to a fully open area, without shading or main nursery (MN).

Fig. 10.13 The process of sago plantlet acclimatization. (**a**) Plantlets after cleaning and sterilizing. (**b**) Plantlets grew in a hood and were ready to be transferred into polybags. (**c**) Seedlings are ready to be planted or transported. (Source: I. Riyadi)

After 6–8 weeks, the plantlets were stronger and larger, ready to be used as planting materials in the field.

Field Test of SE Sago Palm

The field test was conducted to observe and track the growth, development and genetic characteristics of sago palms produced from SE culture. The tests took place in an experimental plantation in Bogor, where 30 seedlings were planted in two rows with a spacing of 9 m × 9 m in a rectangular pattern. The sago seedlings tested included two types: the thorny type from Papua and Maluku, and the thornless type from Riau and Kalimantan. One year after planting, sucker shoots had densely grown on all main trees.

Fig. 10.14 (**a**) Ten years old sago palms from SE culture. (**b**) Harvesting starch of sago. (Source: I Riyadi)

In another block, field tests have also been carried out on thornless sago from Parung Bogor which was planted about 5–6 years earlier. At around 10 years old, the trees produced flowers and fruits (Fig. 10.14a). In order to examine the stem's characteristics and starch content, harvesting was done by cutting down the sago palm tree and making logs of the stem 80–100 cm long. The log was split with a saw into four sections to make it easier to extract the starch from the trunk (Fig. 10.14b). Based on observations of the harvested sago stems and pith, this sago is considered good and has a large pith that can be harvested (predicted >600 kg starch). As for the organoleptic testing, the sago flour or starch showed a sweet and delicious taste.

10.3.2.3 Ex Vitro Propagation

The ex vitro was developed based on the method of in vitro culture, but it is carried out outside of a tube. Ex vitro propagation is a vegetative propagation method that involves selecting seedling shoots, sterilizing the injured parts of the stem tuber, inducing new roots by soaking in a growth regulator solution, planting in a medium with organic material, incubation in a controlled environment and regular watering. One advantage of the ex vitro technique is its ability to use small, young suckers, resulting in seedlings with high viability, reaching up to 100% when planted in the field.

Ex vitro technology for sago palms was developed to produce high-quality sago seeds for rehabilitating sago plantations and forests using suckers weight between 200–500 g and standing around 75 cm tall. This method has been applied through technology transfer in sago seed production activities in collaboration with PT

Fig. 10.15 (**a**) Shading net for ex vitro culture. (**b**) incubator facilities covered with UV plastic. (Source: YSA Fauzan)

National Sago Prima, Indonesia. Seedling production took place in Unit I in Kepau Baru Village, and Unit II in Teluk Kepau Village, both located in Tebing Tinggi District, Meranti Islands Regency, Riau Province, Indonesia. Unit I produced 47,928 seedlings, while Unit II produced 67,361 seeds. Supporting facilities for seedling production, in ex vitro technology include a shading net (Fig. 10.15a), an ultraviolet (UV) plastic incubator (Fig. 10.15b), a water treatment unit, a warehouse for production facilities and a management office.

The incubator room provides controlled environmental conditions for growing plant explants. The frame can be made of iron, wood or bamboo, with the size tailored to the planned seed production capacity. For sago seeding, an incubator measuring 740 cm long, 425 cm wide and 170 cm high is used. The incubator room is covered with UV plastic, ensuring a light intensity of 14% and air-tight seals. Additionally, the incubator room features two side doors for easy access when placing or removing seedlings.

Sterilization and Root Induction

The planting medium that will be used in propagating sago seeds can be mineral soil or organic soil (peat). The choice of planting medium depends on the availability of materials at the seed production location. In regions outside of Java, such as the Riau Islands, Kalimantan and Papua, peat is often used as a planting medium. Peat has a high organic material content, crumbly texture and high porosity. Sago suckers which are used as explants, grow on the clump of stems of the parent palm and must meet the following criteria: healthy, approximately ±6.3 cm in diameter at the base of the stem, have shoots in the form of reddish spear leaves measuring 5–45 cm in height, weight between 300–500 g and come from a grove that has produced at least a parent palm with a stem height of over 5 m.

After cleaning the explants, they are taken to the processing room for pre-sterilization treatment. Pre-sterilization involves soaking the explants in a

Fig. 10.16 (**a**) Sterilization process of sucker explants. (**b**) The sucker explants are soaked in root induction solution. (**c**) planting explants in a polybag. (Source: YSA Fauzan)

solution of Metroxylon A formula (Tajuddin et al. 2012) in plastic boxes. Each plastic box can hold 50 explants with a soaking solution volume of 90 l. The explants are arranged alternately between the tuber and the shoot in the box, soaked for 10 minutes and then rinsed to remove any remaining dirt.

The next step is sterilization to eliminate fungi and bacteria by soaking the explants in a mixture of Metroxylon B and C formula (Tajuddin et al. 2012) solutions for 30 minutes, followed by drying the explants (Fig. 10.16a).

The sterile explants are then transferred to boxes containing a mixture of Metroxylon D1, D2 and D3 formulas (Tajuddin et al. 2012) for root induction. Arrange the explants in an upright position neatly and carefully so as not to damage the roots. Fill each box with 50 explants and add a mixture of Metroxylon D1, D2 and D3 formulas (Tajuddin et al. 2012) until the bases are completely submerged. Soaking the explants was carried out for 17 hours in the processing room (Fig. 10.16b).

The explants that had been soaked in the root induction formula were then planted in polybags of 30 × 30 cm (Fig. 10.16c). A pesticide acaricide was sprayed on each polybag at a dose of 2 g per plant in order to prevent pest attacks.

Explants Incubation

Incubation is an important stage so that explants remain in optimal conditions for growth. Explant incubation was conducted under controlled microclimate conditions (temperature, humidity, light intensity), for 8 weeks. To maintain temperature, humidity and intensity of sunlight, misting was carried out periodically. In order for the explants to grow well, regular maintenance is carried out by watering them every day in the morning. Apart from watering, to prevent fungal and bacterial attacks, it was sprayed using the Metroxylon B formula (Fig. 10.17). Control of explants in the incubator was performed regularly (Fig. 10.18). Disease-infected leaf sheaths were immediately separated and removed from the incubator to be handled separately so they do not spread.

Fig. 10.17 Incubation of sucker explants. (**a**) Sucker explants. (**b**) incubator facility. (Source: YSA Fauzan)

Fig. 10.18 Controlling sucker explants that have become seedlings in the incubator. (Source: YSA Fauzan)

Acclimatization and Maturation

Pre-nursery Stage

Seedling acclimatization after incubation is divided into two stages, namely pre-nursery and nursery. In the pre-nursery stage, explants that have successfully grown into seedlings (after the incubation stage) are marked by the emergence of new leaves or respiratory roots from within the media. The sago seeds are slowly removed from the incubator so that they can adapt to the external environmental

Fig. 10.19 Sago seedlings, (**a**) the pre-nursery stage, (**b**) the nursery stage and (**c**) the adaptation stage. (Source: YSA Fauzan)

conditions. At this stage, the seeds are placed under double shading with a light intensity of 70% (Fig. 10.19a). The goal of pre-nursery is to reduce stress due to excessive water evapotranspiration in plants and planting media.

Maintenance is carried out by watering and fertilizing (a mixture of urea and NPK fertilizer) during the 10-days pre-nursery period at a dose of 3 g/plant. Apart from that, preventing pest and disease attacks is done by spraying insecticides at the recommended doses. Pre-nursery lasts for 8 weeks so that the seedlings grow better and stronger.

Nursery Stage

Nursery is an advanced stage of pre-nursery, where the seedlings are expected to adapt to more extreme temperatures, humidity and light intensity before being placed outside the shading house. This stage aims to increase the growth ability of plant organs such as roots, stems and leaves so that they can grow more optimally under single shading (Fig. 10.19b).

Adaptation Stage (Open Area)

Adaptation is the stage of a seedling's adjustment to the external environment. Seed adaptation is carried out in conditions without shade so that the seedlings can adapt to natural conditions in the land (Fig. 10.19c) in order to increase the survival rate. Maintenance and pest prevention are also carried out at this stage.

Distribution and Planting

Good seedlings must meet two criteria (BSN 2018), namely physical and physiological requirements. Physical criteria consist of seedlings with single and straight stems, healthy, and 50% woody. Meanwhile, physiological criteria include the firmness and compactness of the roots and media.

Distribution and planting of seedlings on land is the final goal of seedling production. Distribution to the planting area requires good handling (Fig. 10.20).

Fig. 10.20 Sago seeds ready for planting. (**a**) Quality of sago seeds ready for planting. (**b**) Transportation of sago seeds using a pompous boat and, (**c**) Sago seeds that have grown on the land. (Source: YSA Fauzan)

Several things must be considered when handling seeds, especially when loading and unloading them from transportation equipment and transporting them to the planting area.

10.4 Cultivation

The optimal temperature for sago palm growth is around 24.5–29.0 °C with an average of 26 °C and a minimum of 15 °C (Flach 1984). Bintoro et al. (2010) added that although the minimum temperature for sago growth is 15 °C, when the temperature drops below 13 °C in the seedling phase sago palms cannot grow and the seedling death rate increases. New leaf growth is slower than normal and stops growing below 17 °C (Schuiling 2009). Minimum air temperature is an important factor that can limit the growth of sago palms (Okazaki et al. 2015) because the photosynthetic enzyme Ribulose-1.5-bisphosphate carboxylase/oxygenase (Rubisco) is very sensitive to air temperature. Rubisco activity decreases by 80% at air temperatures below 25 °C (Yamori and Caemmerer 2009). The maximum air temperature for the morphological growth of sago seedlings is 35 °C. At this temperature, sago leaves can develop normally but the color of the leaves changes to pale green (Irawan et al. 2011b).

Relative humidity for sago palms is 90% (Bintoro et al. 2010). If sago palms grow at temperatures below 20 °C and relative humidity lower than 70%, starch production decreases by around 25%. (Bintoro et al. 2010). The optimal rainfall for sago palms is 2000 mm per year (Okazaki et al. 2015). According to Schmidt and Ferguson's classification (Schmidt and Ferguson 1952), sago grows optimally in climate types A and B with rainfall between 2500 and 3500 mm and 142–209 days of rainfall per year (Turukay 1986). Under sub-optimal conditions, sago takes longer to reach maturity. Under these conditions, new leaf growth also slows, delaying maturation and starch production.

The optimal sunlight for sago palms is 900 J/cm^2/day. Sago palms require full sunlight in connection with the formation of starch in their trunks (Flach and Schuiling 1989). To obtain an optimum electron transport rate in the light reaction, sago palms require a light intensity of 700–800μmol (Azhar et al. 2018). Excess light intensity has an impact on reducing the efficiency of light absorption and is followed by a decrease in the electron transport rate due to excess energy.

Elevation is related to air temperature. The atmospheric environment that influences sago palms consists of the macro and microenvironments. The distribution of sago palm is between 10° North Latitude and 10° South Latitude and is generally found in wet lowlands. According to Flach (1983), sago palms can grow well at altitudes up to 500 m above sea level. The most suitable habitat for sago palms is below 400 m above sea level. If sago is planted above 400 m above sea level, it will grow slowly and produce low production (Djoefrie 1999). According to Schuiling and Flach (1985), sago growing above 600 m above sea level has stunted growth, slender stems and a plant height of less than 6 m. Naturally, sago grows on riverbanks, lakeshores, swamps, peatlands and damp lands.

Another issue that has recently been faced with the sago palm establishment is low starch productivity. This may be due to a low number of harvestable palms per hectare per year. Two major factors affecting this issue are low cluster density and a huge gap in production obtained from the same cluster due to irregular and long harvest intervals. Flach (1977) stated that the sago harvesting pattern in a cluster could be regulated by arranging the number of successive suckers generated at different ages. Theoretically, the next harvestable palms would replace the previously harvested palms at two-year intervals. However, such a practice is not easy to implement and up to the present, there is a lack of evidence.

To study the composition of palms at various development stages, an observation was carried out on 30 random clusters growing on deep peat which were established in 2012 at one of the cultivated blocks in a commercial sago plantation of PT National Sago Prima. The census for different categories of trunk length and sucker size was done in 2021 or about 9 years after transplanting. The harvestable palm commonly produces 7 logs per palm or the trunk length is estimated to be around 7 m. The estimated year of harvest is calculated by using the assumption that 5 years are required for the rosette stage and 7 years are needed from the stages of just after trunk formation to flower initiation. The total years are around 12 years. This assumption is based on Jong et al. (2006) who mentioned that sago may require over 10 years from planting to harvesting. The results found that the harvest interval varies considerably. Some clusters show that the next harvested palms are obtained 1 year after the previous ones. The others show that the next ones are over 4 or even 7 years later. However, a longer period of the next harvest occurs thereafter meaning that the issue of irregular harvest interval is present in sago palm establishment and it may lead to low starch productivity. Yanagidate et al. (2009) investigated the number of harvestable palms per hectare per year in cultivated sago palm gardens growing on mineral soil in Kendari, Southeast Sulawesi Province. They found that 22 palm ha^{-1} year^{-1} can be harvested on average. However, the number showed a downtrend thereafter. This study found that an average of 27.2% harvestable palm

Fig. 10.21 (**a**) A 9-year-old cluster composed of a mother palm forming a trunk at ±3.5 m high and suckers but none of its suckers produced a trunk. (**b**) A-9-year-old cluster composed of two trunking palms (a mother palm at ±3.4 m high and its young trunk growing sucker at ±1 m high) and suckers at the rosette stage. (**c**) A 9-year-old cluster composed of two trunking palms (a mother palm and its sucker) at about the same height (± 3.2 m) and suckers at the rosette stage. (**d**) A 9-year-old cluster composed of three trunking palms (a mother palm at ±1.7 m and two suckers at ±1.5 (elder) and 0.5 m (younger)) and suckers at the rosette stage. (Source: AF Irawan)

can be harvested per year by using the data from the years 2023–2028. Assuming there are 100 cluster ha^{-1}, then it is about 27 harvestable palm ha^{-1} year^{-1}. This result is slightly higher than the above study reported.

The huge gap of growth between the mother palm and its suckers generated can be seen on Fig. 10.21. A 9-year-old cluster composed of a mother palm forming a trunk at ±3.5 m high and suckers but none of its suckers formed a visible trunk (Fig. 10.21a). The other clusters were composed of two (Fig. 10.21b, c) and three (Fig. 10.21d) palms with visible trunk while the other suckers were also still at rosette stage. From the visual growth of the palms shown in those figures, it may take years for the rosette stage-suckers to form the trunk or to attain the young trunk growth stage. Although the number of suckers was not recorded in that figure, the density of suckers produced by the cluster also varied considerably.

A similar study was conducted on a semi-cultivated sago plantation in Lingga Island, Province of Riau Island, in May 2017. The observation included 42 random clusters from three different locations where a large number of palms had been previously harvested. These clusters had been established since ancient times and farmers used the suckers to expand their sago cultivation areas. Eight categories of trunk length on mature palms and two categories of suckers are listed. Similar to the previous report, although suckers are produced from the rhizome since the establishment, the number of suckers on the cluster and mature palm is very low. The growth of mature palms with visible trunks cannot keep up with the incoming harvestable palms and it creates a huge gap in production. Further studies on factors affecting the growth and newly-produced suckers on a cluster are required.

Figure 10.22 shows the composition of harvested clusters in a semi-cultivated garden. Similar to the condition of clusters described on commercial plantations, the harvesting gap between the visible trunks and the remaining suckers may be over 4 (Fig. 10.22a) or even 7 (Fig. 10.22b, c) years. After the palm is harvested, the remaining suckers are still at the rosette stage (Fig. 10.22d). Moreover, compared to commercial plantations, the average number of suckers with a frond base diameter over 15 cm in a cluster produced from these clusters (3.5) is lower than that produced from a cluster in a commercial plantation (6.7).

10.5 Molecular Marker-Assisted Breeding

Genetic variability within a species is the raw material for breeding programs. In the context of sago palm, understanding and preserving germplasm diversity is essential for developing cultivars that can withstand environmental challenges, resist diseases, and adapt to changing climatic conditions. The broader the genetic base, the greater the potential for breeding programs to produce improved varieties. To access information on the genetic diversity of sago palms, it is necessary to develop genetic markers.

Molecular tools have made it easier to understand the genetics behind the attributes that plant breeders seek, as well as to develop technology and genetics. The development of molecular markers has expedited and enhanced the capacity of plant breeding to yield superior plants. DNA markers help to identify germplasm, transfer single genes, and select desired features from genomic regions linked to the expression of specific traits. To apply selection using efficient and effective markers to enhance polygenic traits, new technology and innovation are required.

The development of genetic markers in plant breeding is a technique for searching for and utilizing molecular markers (DNA markers) in plant breeding activities. The development of genetic markers aims to identify, isolate and map genes related to required traits in plant breeding activities. Through the use of genetic markers, it is hoped that the speed and efficiency of plant breeding activities can be increased. Genetic marker development activities in plant breeding are usually pre-breeding activities. Plant breeding assisted with genetic markers can obtain plant genetic information that is not influenced by climate and environmental conditions with high precision (Guler and Imamoglu 2023).

The use of biotechnology to enhance the quality of sago palms has grown significantly. To improve the quality of sago palms, biotechnology facilities can be used for sago palm breeding. Because sago palms only flower once in their lifespan and only begin to produce flowers when they are around 10 years old, conventional

Fig. 10.22 Gap of growth among suckers at various development stages and a few numbers of suckers on more than 20-years-old harvested clusters growing in the semi-cultivated garden. Clusters are composed of one (**a**–**c**) visible trunk and none and (**d**) visible trunk while other suckers are at the rosette stage. (Source: AF Irawan)

sago palm breeding takes a very lengthy period. This biotechnology method for sago palm breeding has a lot of potential. There is currently a shortage of early-stage research on using biotechnology to improve the quality of sago palms. Propagations in vitro have started, and they might provide a superior variety (Sumaryono et al. 2013). Because of breeding with a biotechnological approach, new types of sago palms have not yet been reported.

The genetic diversity of sago palms has been shown by partial genome analysis utilizing a variety of molecular markers. Numerous studies that used molecular markers also revealed genetic variety. This indicates that sago palms have a high degree of genetic diversity, the capacity to increase quality and productivity, and the ability to respond to climate change. In light of the impending scarcity of sources of carbohydrates and in order to be more cost-effective, time- and money-efficient, a more advanced biotechnological method for producing improved kinds of sago is needed.

10.6 Genomics and Transcriptomics

The genetic information publicly available on sago palm is limited. For example, searching the NCBI short read archive (SRA) database, which is a public repository of raw second or third-generation sequencing data, on *Metroxylon sagu* in January 2024 returns 24 accessions.

Both genomics and transcriptomics studies of sago palm are still in their early phase. There are only a few publications available on genomics studies of sago palm using second and third-generation sequencing technology. Lim et al. (2021, 2022) studied the genomics of sago palms by sequencing the whole genome of sago. The group performed functional genomics analysis and identified 33,242 coding sequences (CDS), successfully annotating 32,041 of them. However, Yan et al. (2022) reported a higher number of annotated genes. They performed transcriptome sequencing and reported 56,600 annotated genes.

Since starch production is important in sago palm, Yan et al. (2022) further inspected the sago palm's annotated genes involved in carbohydrate-related metabolic pathways. As many as 15 metabolic pathways have contributed to carbohydrate production in sago palm including glycolysis/gluconeogenesis, citrate cycle (TCA cycle), pentose phosphate pathway, pentose and glucuronate interconversions, fructose and mannose metabolism, galactose metabolism, ascorbate and aldarate metabolism, starch and sucrose metabolism, amino sugar and nucleotide sugar metabolism, pyruvate metabolism, glyoxylate and dicarboxylate metabolism, propanoate metabolism, butanoate metabolism, C5-branched dibasic acid metabolism, and inositol phosphate metabolism.

Transcriptomics studies can reveal important genes contributing to different expressions or phenotypes in plants. Yan et al. (2022) began to study the expression pattern between non-trunking and trunking sago palms and described the raw sequencing data and assembled transcripts. This study is promising, as it will identify a gene or genes that might affect the trunk development of sago palms. Non-trunking sago palm is well-known phenomenon amongst sago breeders or farmers where sago palms planted on deep-peat permanently flooded soil usually fail to develop their trunks, thus yielding little or no starch (Flach 1997). The plants remain in the rosette stage even after 4 years of being planted.

10.7 Bioinformatics

Marker-assisted selection breeding is a frequently used method in plant breeding that heavily relies on the availability of genetic diversity information. For economically important crops, plant breeders can often find this information publicly available in various publications or databases. Unfortunately, genetic information may not be readily accessible for underutilized or neglected crops due to the lack of attention given to them.

Unlike major staple crops that have received significant attention from the research community, sago palm is largely overlooked. This is because its distribution is limited to the tropical Southeast Asian archipelago, spanning from Thailand, Malaysia, Indonesia, the Philippines, to Papua New Guinea. In some regions, such as the eastern part of Indonesia, residents consume starch from sago palm as the main source of carbohydrates. Thus, the breeding of sago palm in these areas becomes important for its sustainable development and assurance of continuity of supplying the sago starch for consumption, since most sago palm harvests are still coming from wild stands. According to official statistics of the Agriculture Ministry of Indonesia, in 2018, commercial plantation of sago palm was only 12,588 ha (4%) of the total 311,954 ha of sago coverage in Indonesia (Directorate General of Estate Crops 2021).

The limitation of genetic information on sago palm hinders the research development and its sustainable conservation. However, efforts are being made to explore the genetic diversity of sago palm. Many scientists studied sago palm genetic diversity using various genetic markers such as SSR, RAPD, or AFLP (Abbas 2018; Nisar and Hussain 2022; Zulaeha et al. 2023). With this approach, they amplified various loci by PCR to detect the polymorphisms of the loci and studied their diversity.

Recently, others have begun to take advantage of the latest developments in sequencing technology by using second- or third-generation sequencing technology to sequence the whole genome or transcriptome of sago palm (Lim et al. 2020; Lim et al. 2022; Yan et al. 2022) or develop genetic markers to explore the genetic diversity of sago palm (Purwoko et al. 2019). Unlike Sanger sequencing, second and third-generation sequencing is designed to perform sequencing in parallel, thus, in theory, it can perform whole genome sequencing in a single run. These two technologies produce different sizes of reads (or sequences of DNA), while second-generation produces short sequences of 100–600 bp, third-generation produces longer sequences, on average of 10–100 kbp. Thus, it is also called short-read sequencing for second-generation sequencing and long-read sequencing for third-generation sequencing. Each technology has its limitations; for short-read sequencing, its inability to sequence long stretches of DNA causes trouble for assembling a highly complex and repetitive sequence longer than the read size. For long-read sequencing, it is the low nucleotide-level accuracy rate, which is only 85%, much lower than that of second-generation sequencing which can reach as high as 99.9% (Athanasopoulou et al. 2021; Fox et al. 2014). Thus, depending on the challenge that needs to be resolved, one may need to choose a specific technology or combine both technologies. For example, for variant-sensitive applications, second-generation sequencing is still preferred by many, whereas for de novo whole genome sequencing, third-generation sequencing is a more attractive option.

However, a major factor hampering the application of these technologies in sequencing the genomes of many species is that the computation step requires both computational resources and specialists to analyze the generated data. For many biologists studying minor (underutilized or neglected) crops such as sago palm, this factor alone can be a hurdle hard to solve. Analyzing second- or third-generation sequencing data requires computational skills and knowledge since the data must be processed using a series of tools linked together to form a workflow (Goecks et al. 2010).

Nowadays, many tools or frameworks are designed to ease the use of bioinformatics tools. One of many frameworks available is Galaxy (Goeks et al. 2010), which provides two main public galaxy servers, https://usegalaxy.org and https://usegalaxy.org.au. Galaxy framework is quite popular among biologists since its ease of use: there are tons of bioinformatics tools available in the framework, ready to use, a drag and drop system to create workflows, and the published shareable workflows among its users. Other frameworks are also available for specific technologies, such as epi2me (https://labs.epi2me.io), which is a framework specifically designed for analyzing the third-generation Oxford Nanopore (ONT) sequencing data. Other frameworks are designed to handle package, dependency, and environment management required for a specific tool, such as Conda (https://docs.conda.io/en/latest/), so that it helps to ease the installation process of bioinformatics tools for inexperienced users.

10.8 Conclusions and Prospects

Molecular markers, including DNA sequences linked to particular traits, are crucial in Marker-Assisted Selection (MAS). Unlike traditional breeding methods, MAS enables breeders to pinpoint and choose plants with desired traits at the molecular level. In the case of sago palm, this leads to a more accurate and effective breeding process, saving both time and resources. For example, the discovery of molecular markers associated with disease resistance or increased yield can greatly accelerate the development of resilient and high-yielding sago palm varieties.

The integration of Genomic Selection (GS) takes sago palm breeding to a predictive and highly accurate level. By analyzing large sets of genomic data, breeders can predict the performance of specific genotypes, enabling the selection of superior individuals for further breeding. This not only expedites the development of desirable traits but also minimizes the need for extensive field testing, making the breeding process more efficient.

In germplasm conservation, NGS (next-generation sequencing) technology is used to identify and analyze genetic variations in germplasm collections. This can be done by DNA sequence analysis, which allows researchers to identify genetic variations present in germplasm populations. Bioinformatics is also used to build integrated databases and information systems, allowing researchers to access and share information about conserved germplasm.

In the utilization of germplasm, NGS can be used to identify and utilize genetic variation in germplasm collections for the development of plant varieties that are more productive, resistant to difficult environmental conditions, and have good nutritional quality. The NGS method, through the utilization of an integration platform between genomics, transcriptomics, proteomics, metabolomics, and genomics can be used in the management and development of germplasm to produce superior clones/varieties.

Germplasm diversity is the lifeblood of successful sago palm breeding programs. Preserving, understanding, and strategically utilizing the rich genetic resources within sago palm populations not only ensures the resilience of this vital crop but also paves the way for sustainable agriculture and food security in the face of evolving environmental and agricultural challenges. As we continue to explore the intricate tapestry of germplasm diversity in sago palm, we unlock the potential for a brighter and more sustainable future for this essential crop. Molecular breeding is a game-changer in sago palm breeding programs, offering precision, efficiency, and the potential for transformative improvements. As researchers continue to unlock the genetic secrets of sago palm, the application of molecular techniques will play a pivotal role in shaping the future of this vital crop, ensuring its continued contribution to global agriculture and livelihoods.

Appendix 10.1 Research Institutes and Online Resources Available for Sago Palm (*Metroxylon* spp.) Breeding

Institution name	Specialization and research activities	Address	Contact information and website
Department of Agronomy and Horticulture, IPB University	Specializes in tropical agriculture and biosciences, Tropical Agriculture: plant cultivation and natural resource management, biosciences, biotechnology, genetics and molecular biology as applied to agriculture.	Jl. Meranti, Kampus IPB Dramaga, Bogor 16680, Jawa Barat, Indonesia	https://repository.ipb.ac.id/
Postgraduate, IPB University	Specializes in tropical agriculture and biosciences, Tropical Agriculture: plant cultivation, animal	Jl. Raya Darmaga, Gedung Sekolah Pascasarjana IPB, Kampus IPB Darmaga Bogor Jawa	https://repository.ipb.ac.id/

(continued)

Institution name	Specialization and research activities	Address	Contact information and website
	husbandry and natural resource management, biosciences, biotechnology, genetics and molecular biology as applied to agriculture.	Barat, Indonesia, 16680.	
IPB Press	Publications and areas of focus mentioned on their website and online store specialize in publishing books related to Agriculture, Forestry, Management, Business, Animal Husbandry, Fisheries and Veterinary Medicine	Jl. Taman Kencana No. 3 16151 Kota Bogor West Java	https://www.ipbpress.com
Indonesian Journal of Agronomy	Research related to the field of agronomy. This generally encompasses: Crop Production, Plant Breeding, Weed Science and Crop Physiology	Jl. Meranti, Kampus IPB Dramaga Bogor 16680 West Java, Indonesia	https://journal.ipb.ac.id/index.php/jurnalagronomi/index
Kalpataru	Thematic Scientific Journal, which presents results of recent investigations in the field of Archaeology, environment and culture.	Pusat Arkeologi Nasional Jalan Raya Condet Pejaten No. 4, Pasar Minggu, Jakarta Selatan 12510 Indonesia	http://www.setjen.kemdikbud.go.id/arkenas/
University of Nusa Bangsa	Specializes in tropical agriculture and biosciences, Tropical Agriculture: plant cultivation and natural resource management as applied to agriculture.	Jalan KH. Sholeh Iskandar KM. 4, Tanah Sereal, Cibadak, Tanah Sereal, Bogor, Jawa Barat 16166, Indonesia	https://unb.ac.id/
Wageningen University	Public research university in Wageningen, Netherlands, specializing in life sciences with a focus on agriculture, technical and engineering subjects. Areas of	Droevendaalsesteeg 2, 6708 Pb, Wageningen, Belanda	https://www.wur.nl/

(continued)

Institution name	Specialization and research activities	Address	Contact information and website
	specialization include: Agriculture, Food Science, Environmental Sciences, Animal Sciences, Plant Sciences and Social Sciences		
Faculty of Agriculture, IPB University	Specializes in tropical agriculture and biosciences, Tropical Agriculture: plant cultivation, animal husbandry and natural resource management, biosciences, biotechnology, genetics and molecular biology as applied to agriculture.	Jl. Raya Dramaga Kampus IPB Dramaga Bogor 16680 West Java, Indonesia.	https://ipb.ac.id/
University of Papua	Biotechnology and Plantations; Plant Science; Tissue Culture; Molecular Ecology. Research activity: Sago palm, Sago palm collection and breeding	Perumahan Dosen Unipa, Jl. Amban Pantai No 15, Amban, Manokwari, West Papua Province, Post code: 98314	barahimabas@gmail.com; b.abbas@unipa.ac.id https://barahimaabbas.wordpress.com/
Indonesian Oil Palm Research Institute, Bogor Unit	Plant *In Vitro* Culture	Jl. Taman Kencana No 1 Bogor West Jawa – Indonesia 16128	admin@iopri.org; admin@iribb.org; imronriyadibogor@gmail.com; www.iopri.co.id
PT Nasional Sagu Prima	Sago palm collection, breeding and production	lk. Buntal, Kec. Tebing Tinggi Tim., Kabupaten Kepulauan Meranti, Riau 28753	By persons
Diponegoro University	Agro-physiology of sago palm, Sago palm vegetative propagation by suckers, Agronomy, Sago palm applied science on large cultivation.	Faculty of Animal and Agricultural Sciences, Department of Agriculture, Study Program of Agroekoteknologi. Jl. Prof. Jacub Rais. Kampus Universitas Diponegoro, Tembalang. Semarang, Indonesia 50275.	albertusfajarirawan@lecturer.undip.ac.id https://agroekoteknologi.fpp.undip.ac.id/en/homepageen/

(continued)

Institution name	Specialization and research activities	Address	Contact information and website
PT Sampoerna Agro Tbk.	Oil palm breeder, Sago palm management on a large scale.	Department of Research and Development. PT Sampoerna Agro Tbk. Sampoerna Strategic Square, North Tower, 28/f Jl. Jendral Sudirman Kav. 45 DKI Jakarta, Indonesia 12930	dwi.asmono@sampoernaagro.com https://sampoernaagro.com/home

Appendix 10.2 Genetic Resources of Sago Palm (*Metroxylon* spp.)

Cultivar	Important traits	Cultivation location
Fablen	Production of sago-type dry starch was 169.81 kg stem^{-1} and Rendemen 15.64%.	Kabupaten Sorong, Papua Barat
Fafion	Production of sago-type dry starch was 310.00 kg stem^{-1} and Rendemen 18.31%.	
Failik	Production of sago-type dry starch was 165.61 kg stem^{-1} and Rendemen 21.27%.	
Fakattao	Production of sago-type dry starch was 185.29 kg stem^{-1} and Rendemen 20.60%.	
Fakreit	Production of sago-type dry starch was 288.17 kg stem^{-1} and Rendemen 21.17%.	
Falia	Production of sago-type dry starch was 259.06 kg stem^{-1} and Rendemen 20.52%.	
Fanke	Production of sago-type dry starch was 129.65 kg stem^{-1} and Rendemen 17.46%.	
Fanomik	Production of sago-type dry starch was 414.32 kg stem^{-1} and Rendemen 19.63%.	
Fasai	Production of sago-type dry starch was 161.97 kg stem^{-1} and Rendemen 12.48%.	
Fasampe	Production of sago-type dry starch was 244.53 kg stem^{-1} and Rendemen 17.97%.	
Fasinan	Production of sago-type dry starch was 287.80 kg stem^{-1} and Rendemen 17.42%.	
Fasongka	Production of sago-type dry starch was 224.15 kg stem^{-1} and Rendemen 15.90%.	
Moor	Production of sago-type dry starch was 171.15 kg stem^{-1} and Rendemen 15.56%.	Kabupaten Mappi, Papua Selatan

(continued)

Cultivar	Important traits	Cultivation location
Ghanu	Production of sago-type dry starch was 376.00 kg stem^{-1} and Rendemen 30.53%.	
Komru	Production of sago-type dry starch was 584.80 kg stem^{-1} and Rendemen 27.45%.	
Taragai	Production of sago-type dry starch was 1197.20 kg stem^{-1} and Rendemen 27.52%.	
Wiyage	Production of sago-type dry starch was 320.35 kg stem^{-1} and Rendemen 30.94%.	
Katage	Production of sago-type dry starch was 624.25 kg stem^{-1} and Rendemen 39.01%.	
Mani	Production of sago-type dry starch was 718.15 kg stem^{-1} and Rendemen 32.32%.	
Arare	Production of sago-type dry starch was 232.65 kg stem^{-1} and Rendemen 34.81%.	
Sigare	Production of sago-type dry starch was 261.30 kg stem^{-1} and Rendemen 22.03%.	
Homang	Production of sago-type dry starch was 288.55 kg stem^{-1} and Rendemen 28.16%.	
Taina	Production of sago-type dry starch was 250.57 kg stem^{-1} and Rendemen 22.28%.	Kabupaten Mimika, Papua
Oko	Production of sago-type dry starch was 384.10 kg stem^{-1} and Rendemen 18.25%.	
Omiya 1	Production of sago-type dry starch was 402.09 kg stem^{-1} and Rendemen 22.18%.	
Omiya 2	Production of sago-type dry starch was 143.87 kg stem^{-1} and Rendemen 15.82%.	
Durumu	Production of sago-type dry starch was 312.14 kg stem^{-1} and Rendemen 14.54%.	
Mbapare	Production of sago-type dry starch was 239.18 kg stem^{-1} and Rendemen 18.73%.	
Omoroko	Production of sago-type dry starch was 169.33 kg stem^{-1} and Rendemen 15.35%.	
Ikimina	Production of sago-type dry starch was 220.49 kg stem^{-1} and Rendemen 19.76%.	
Waruwo	Production of sago-type dry starch was 289.21 kg stem^{-1} and Rendemen 40.53%.	Kota Sorong, Papua Barat Daya
Wasulagi	Production of sago-type dry starch was 248.54 kg stem^{-1} and Rendemen 43.80%.	
Wasnan	Production of sago-type dry starch was 361.51 kg stem^{-1} and Rendemen 43.38%.	
Wayuluk	Production of sago-type dry starch was 211.84 kg stem^{-1} and Rendemen 39.51%.	
Wagelik	Production of sago-type dry starch was 215.72 kg stem^{-1} and Rendemen 25.52%.	
Wanegles	Production of sago-type dry starch was 128.54 kg stem^{-1} and Rendemen 49.01%.	

(continued)

Cultivar	Important traits	Cultivation location
Wawun	Production of sago-type dry starch was 110.69 kg stem^{-1} and Rendemen 37.24%.	
Wafabala	Production of sago-type dry starch was 149.30 kg stem^{-1} and Rendemen 47.29%.	
Saikoat	Production of sago-type dry starch was 123.16 kg stem^{-1} and Rendemen 19.04%.	Kabupaten Kepulauan Mentawai, Kepulauan Riau
Limu	Production of sago-type dry starch was 216.48 kg stem^{-1} and Rendemen 24.93%.	
Betaet	Production of sago-type dry starch was 383.87 kg stem^{-1} and Rendemen 31.55%.	
Ukra	Production of sago-type dry starch was 544.37 kg stem^{-1} and Rendemen 32.24%.	
Sirilanggai	Production of sago-type dry starch was 322.55 kg stem^{-1} and Rendemen 23.35%.	
Sibeotcun	Production of sago-type dry starch was 285.62 kg stem^{-1} and Rendemen 24.27%.	
Kapa	Production of sago-type dry starch was 477.00 kg stem^{-1} and Rendemen 22.22%.	Kabupaten Luwu Utara, Sulawesi Selatan
Kasimpo	Production of sago-type dry starch was 359.61 kg stem^{-1} and Rendemen 18.45%.	
Ute	Production of sago-type dry starch was 140.35 kg stem^{-1} and Rendemen 17.46%.	
Use	Production of sago-type dry starch was 280.38 kg stem^{-1} and Rendemen 16.30%.	
Sabbe	Production of sago-type dry starch was 221.78 kg stem^{-1} and Rendemen 21.02%.	
Kiduri	Production of sago-type dry starch was 179.66 kg stem^{-1} and Rendemen 18.40%.	
Kasimpo	Production of sago-type dry starch was 243.14 kg stem^{-1} and Rendemen 21.10%.	Kabupaten Mamuju, Sulawesi Barat
Kapas	Production of sago-type dry starch was 436.61 kg stem^{-1} and Rendemen 23.89%.	
Ruwi	Production of sago-type dry starch was 304.38 kg stem^{-1} and Rendemen 21.20%.	
Lameuru	Production of sago-type dry starch was 445.00 kg stem^{-1} and Height 10.30 m.	Kabupaten Konawe Selatan, Sulawesi Tenggara
Arono	Production of sago-type dry starch was 314.36 kg stem^{--1} and Height 7.25 m.	
Aopa 2	Production of sago-type dry starch was 237.86 kg stem^{-1} and Height 8.00 m.	
Aopa 1	Production of sago-type dry starch was 236.54 kg stem^{-1} and Height 8.40 m.	
Lamoen 1	Production of sago-type dry starch was 297.93 kg stem^{-1} and Height 8.25 m.	
Lamoen 2	Production of sago-type dry starch was 228.19 kg stem^{-1} and Height 5.93 m.	
Sandei	Production of sago-type dry starch was 280.59 kg stem^{-1} and Height 9.60 m.	

(continued)

Cultivar	Important traits	Cultivation location
Mataiwoi 1	Production of sago-type dry starch was 195.05 kg stem^{-1} and Height 9.55 m.	
Mataiwoi 2	Production of sago-type dry starch was 243.24 kg stem^{-1} and Height 6.35 m.	
Landabaru 2	Production of sago-type dry starch was 294.23 kg stem^{-1} and Height 7.30 m.	
Landabaru 3	Production of sago-type dry starch was 261.06 kg stem^{-1} and Height 7.70 m.	
Watundeho	Production of sago-type dry starch was 318.62 kg stem^{-1} and Height 9.90 m.	
Unggulino	Production of sago-type dry starch was 432.46 kg stem^{-1} and Height 13.88 m.	
Mahang	Production of sago-type dry starch was 102.11 kg stem^{-1} and Rendemen 22.31%.	Kalimantan Selatan
Buntal	Production of sago-type dry starch was 178.00 kg stem^{-1} and Rendemen 24.56%.	
Salak	Production of sago-type dry starch was 91.04 kg stem^{-1} and Rendemen 14.29%.	
Madang	Production of sago-type dry starch was 157.84 kg stem^{-1} and Rendemen 18.38%.	
Gandut	Production of sago-type dry starch was 127.82 kg stem^{-1} and Rendemen 19.13%.	
Tada timur 1	Production of sago-type dry starch was 386.10 kg stem^{-1} and Rendemen 19.02%.	Kabupaten Parigi Moutong, Sulawesi Tengah
Sigenti	Production of sago-type dry starch was 429.19 kg stem^{-1} and Rendemen 20.80%.	
Tada timur 2	Production of sago-type dry starch was 329.33 kg stem^{-1} and Rendemen 19.29%.	
Khatulistiwa	Production of sago-type dry starch was 367.70 kg stem^{-1} and Rendemen 23.45%.	
Sigenti selatan 1	Production of sago-type dry starch was 68.07 kg stem^{-1} and Rendemen 8.13%.	
Sigenti selatan 2	Production of sago-type dry starch was 56.52 kg stem^{-1} and Rendemen 12.89%.	
Sigenti selatan 3	Production of sago-type dry starch was 320.05 kg stem^{-1} and Rendemen 18.08%.	
Sigenti selatan 4	Production of sago-type dry starch was 264.87 kg stem^{-1} and Rendemen 16.67%.	
Sigenti utara 1	Production of sago-type dry starch was 209.23 kg stem^{-1} and Rendemen 25.24%.	
Sigenti utara 2	Production of sago-type dry starch was 128.54 kg stem^{-1} and Rendemen 23.93%.	
Maninili utara	Production of sago-type dry starch was 122.99 kg stem^{-1} and Rendemen 23.31%.	
Maninili	Production of sago-type dry starch was 316.87 kg stem^{-1} and Rendemen 23.16%.	
Siaga	Production of sago-type dry starch was 112.75 kg stem^{-1} and Rendemen 8.37%.	

(continued)

Cultivar	Important traits	Cultivation location
Sinei induk	Production of sago-type dry starch was 330.39 kg $stem^{-1}$ and Rendemen 20.16%.	
Nerekeh	Production of sago-type dry starch was 156.68 kg $stem^{-1}$ and Rendemen 15.99%.	Kepulauan Riau, Pulau Lingga
Panggak laut	Production of sago-type dry starch was 95.08 kg $stem^{-1}$ and Rendemen 14.81%.	
Musai	Production of sago-type dry starch was 238.66 kg $stem^{-1}$ and Rendemen 16.76%.	
Pekaka	Production of sago-type dry starch was 210.38 kg $stem^{-1}$ and Rendemen 22.10%.	
Keton	Production of sago-type dry starch was 227.40 kg $stem^{-1}$ and Rendemen 15.20%.	
Teluk	Production of sago-type dry starch was 288.64 kg $stem^{-1}$ and Rendemen 23.17%.	
Yebha	Spineless, high production, white starch	Unipa experimental garden
Phara	Spiny, high production, white starch	Unipa experimental garden
Osukhulu	Spineless, moderate production, cream color of starch	Unipa experimental garden
Local sago	Unthorns stalk, on mineral soil	Parung, Bogor
Local sago	Unthorns stalk, Thorns stalk, tolerants on peat	Meranti, Riau
Local sago	Unthorns stalk, Thorns stalk, high starch rendement	Konawe, Southern Sulawesi
Yeba, Para, Rondo, Hapholo	Unthorns stalk, Thorns stalk, high starch rendement, early harvesting	Koya, Papua
Mbov and Alitir	Unthorns stalk, Thorns stalk, high starch rendement,	Merauke, Papua
Yeba and Para	Unthorns stalk, Thorns stalk, high starch rendement,	Manokwari, Papua
Yeba and Para	Unthorns stalk, Thorns stalk, high starch rendement,	Sorong, Papua
Local sago	Unthorns stalk	Banjarmasin, South Kalimantan
Molat and Tuni	Unthorns stalk, Thorns stalk, high starch rendement,	Seram, Maluku

References

Abbas B (2018) Sago palm genetic resource diversity in Indonesia. In: Ehara H, Toyoda Y, Johnson D (eds) Sago palm. Springer, Singapore. https://doi.org/10.1007/978-981-10-5269-9_5

Abbas B, Renwarin Y, Bintoro MH et al (2010) Genetic diversity of sago palm in Indonesia based on chloroplast DNA (cpDNA) markers. J Biol Div 11(3):112–117

Abhal NEI (2022) Identification of various sago (*Metroxylon* spp) accessions in Mamuju subdistrict, Mamuju Regency, West Sulawesi Province. Thesis, IPB (Institut Pertanian Bogor) University, Bogor

Al Manar P, Zuhud EAM, Andarwulan N et al (2023) Morphological characteristics and potential of sago (*Metroxylon* spp.) in Lingga Regency, Riau Islands, Indonesia. J Manajemen Hutan Tropika 29(1):11–21

Alamri ES (2020) Efficacy of gluten- and casein-free diets on autism spectrum disorders in children. Saudi Med J 41:1041–1046

Alang ZC, Krishnapillay B (1986) Studies on the growth and development of embryos of the sago palms (*Metroxylon* spp.) in vivo and in vitro. In: Stanton R (ed) Sago-'85 proc. third int. sago symposium, Tokyo, pp 121–129

Amin N, Sabli N, Izhar S et al (2019) Sago wastes and its application. Pertanika J Sci Technol 27(4): 1841–1862

Amni C, Ismet, Aprilia S et al (2019) Study on biodegradable plastic from sago with addition of glycerol and sorbitol. IOP Conf. Series: Earth and Environmental Science 365(2019):012052. IOP Publishing. https://doi.org/10.1088/1755-1315/365/1/012052

Anggraeni D, Ngantung JM (2018) Study of compressive strength of concrete using a mixture of sago pollen as an additive to fine aggregate. Portal Sipil 7(2):9–30

Arif A (2019) Papuan sago for the world. Gramedia & ANJ, Jakarta

Astuti M, Hafiza, Yuningsih E et al (2014) Guidelines for better sago (*Metroxylon* spp) cultivation. Directorate General of Plantations, the Ministry of Agriculture of Indonesia, Indonesia

Athanasopoulou K, Boti MA, Adamopoulos PG et al (2021) Third-generation sequencing: the spearhead towards the radical transformation of modern genomics. Life (Basel) 12:30. https://doi.org/10.3390/life12010030

Azhar A, Makihara D, Naito H et al (2018) Photosynthesis of sago palm (*Metroxylon sagu* Rottb.) seedling at different air temperatures. Agriculture 8:4. https://doi.org/10.3390/agriculture8010004

Azhar A, Makihara D, Naito H et al (2020a) Evaluating sago palm (*Metroxylon sagu* Rottb.) photosynthetic performance in waterlogged conditions: utilizing pulse-amplitude-modulated (PAM) fluorometry as waterlogging stress indicator. J Saudi Soc Agric Sci 19(1):37–42. https://doi.org/10.1016/j.jssas.2018.05.004

Azhar A, Makihara D, Naito H et al (2020b) Sago palm (*Metroxylon sagu* Rottb.) response to drought condition in terms of leaf gas exchange and chlorophyll a fluorescence. Plant Prod Sci 24(1):65–72. https://doi.org/10.1080/1343943X.2020.1794914

Bintoro MH (2011) Progress of sago research in Indonesia. Proceedings of the 10th international sago symposium sago for food security, bio-energy, and industry from research to market, 29–30 Oct 2011, pp 16–34, IPB (Institut Pertanian Bogor) International Convention Center, Bogor

Bintoro MH, Purwanto MYJ, Amarillis S (2010) Sago palm in peatlands. IPB, Institutu Pertanian Bogor, University Press, Bogor

Bintoro MH, Pratama AJ, Ahmad F et al (2017a) Sago forest arrangement. Research Institutions and Community Service. IPB (Institut Pertanian Bogor) University, Bogor

Bintoro MH, Supijatno, Trikoesoemaningtyas et al (2017b) Identification of sago palm (*Metroxylon* spp) in Mimika District, Papua Province. Digreat Publishing, Bogor

BSN (2018) Forest plant seeds. SNI 8420. National Standardization Agency of Indonesia, Indonesia

Che SN, Basirun AA, Hameed Sultan NS et al (2023) Modified nanocellulose-based adsorbent from sago waste for diclofenac removal. Sustainability 15(7):5650. https://doi.org/10.3390/su15075650

Dewi R, Sylvia N, Riza M (2023) Characterization of degradable plastics from sago and breadfruit starch-based with addition of zinc oxide (ZnO) catalyst and polyvinyl alcohol (PVA). J Kimia Sains dan Aplikasi 26(11):427–436. https://doi.org/10.14710/jksa.26.11.427-436

Directorate General of Estate Crops (2021) Statistical of national leading estate crops commodity 2019–2021. Ministry of Agriculture of Indonesia, Indonesia

Directorate General of Estates Crop (2013) Increasing production, productivity and quality of annual crops- Technical guidelines for sago palm development in 2024. Ministry of Agriculture of Indonesia, Indonesia

Djoefrie MHB (1999) Empowerment of sago plants as producers of alternative food ingredients and potential agro-industrial raw materials in the context of national food security. Scientific oration by permanent professor of plantation crop science, Faculty of Agriculture. IPB, Institut Pertanian Bogor, University, 11 Sept 1999
Djoefrie MHB, Herodian S, Ngadiono et al (2014) Sago for the welfare of the people of Papua. Papua and West Papua Development Acceleration Unit, Jakarta
Ehara H (2009) Potency of sago palm as carbohydrate resource for strengthening food security program. Indonesian J Agron 37(3):209–219. https://doi.org/10.24831/jai.v37i3.1255
Ehara H, Komada C, Morita O (1998) Germination characteristics of sago palm seeds and spine emergence in seedlings produced from spineless palm seeds. Principles 42:212–217
Ehara H, Susanto S, Mizota C et al (2000) Sago palm (*Metroxylon sagu*, Arecaceae) production in the eastern archipelago of Indonesia: variation in morphological characters and pith. Econ Bot 54(2):197–206
Ehara H, Morita O, Komada C et al (2001) Effect of physical treatment and presence of the pericarp and sarcotesta on seed germination in sago palm (*Metroxylon sagu* Rottb.). Seed Sci Tech 29: 83–90
Ellen R (2006) Local knowledge and management of sago palm (*Metroxylon sagu* Rottboell) diversity in south central Seram, Maluku, eastern Indonesia. J Ethnobiol 26(2):258–298
Fathnoer V, Bintoro MH, Lubis I (2020) Assessment of morphological attributes of sago palm accessions of Aimas, Sorong, West Papua, Indonesia. J Trop Crop Sci 7(1):7–13
Firmansyah ANA (2022) Identification of various sago accessions (*Metroxylon* spp) in Siberut National Park, Mentawai Islands, West Sumatra. Dissertation, IPB (Institut Pertanian Bogor) University, Bogor
Flach M (1977) Yield potential of the sago palm and its realization. In: Tan K (ed). Sago-'76. Proc. of the 1st int. sago symposium, Kuala Lumpur, Malaysia, 5–7 July 1976. Kemajuan Kanji Sdn. Bhd, Kuala Lumpur, pp 157–177
Flach M (1983) The sago palm. Domestication, exploitation, and product. FAO Plant Production and Protection, Rome, p 85
Flach M (1984) The sago palm. FAO Plant Production and Protection Paper 47 AGPC/MIS/8O, FAO, Rome, p 85
Flach M (1993) Problem and prospect of present sago palm development. Agricultural University, Department of Tropical Crop Science
Flach M (1997) Sago palm. *Metroxylon sagu* Rottb. Promoting the conservation and use of underutilized and neglected crops. Institute of Plant Genetics and Crop Plant Research, Gatersleben/International Plant Genetic Resources Institute, Rome
Flach M, Schuiling DL (1989) Revival of an ancient starch crop. A review of the agronomy of the sago palm. Agr Syst 7:259–281
Fox EJ, Reid-Bayliss KS, Emond MJ et al (2014) Accuracy of next generation sequencing platforms. Next Gener Seq Appl 1:1000106. https://doi.org/10.4172/jngsa.1000106
Fridiano A (2021) Design process of cyclodextrin production from sago (*Metroxylon* spp.) starch. Dissertation. IPB (Institut Pertanian Bogor) University, Bogor
Ginting A, Adi P, Santosa B et al (2022) The effect of adding sago fiber on the compressive strength of non-sand concrete. Potensi 24(1):36–42
Goecks J, Nekrutenko A, Taylor J et al (2010) Galaxy: a comprehensive approach for supporting accessible, reproducible, and transparent computational research in the life sciences. Genome Biol 11:R86. https://doi.org/10.1186/gb-2010-11-8-r86
Grabherr MG, Haas BJ, Yassour M et al (2011) Full-length transcriptome assembly from RNA-Seq data without a reference genome. Nat Biotechnol 29:644–652. https://doi.org/10.1038/nbt.1883
Guler BA, Imamoglu E (2023) Molecular marker technologies in food plant genetic diversity studies: An overview. Foods Raw Mater 11(2):282–292. https://doi.org/10.21603/2308-4057-2023-2-575
Gumira SED, Mangunwidjaya D, Darmoka A et al (1996) Production of acetone-butanol-ethanol from hydrolyzed substrates of sago starch and tapioca piles resulting from enzymatic hydrolysis. In: Potential of wetlands in agribusiness development efforts in wetland areas. Proceedings of the national sago symposium III. Pekan Baru, Indonesia, 27–28 Feb 1996

Haliza W, Purwani EY, YulianiS (2006) Evaluation of enzymatically resistant starch and glycemic index of sago noodle. J Food Technol Ind 17(2):149–152
Hariyanto B (2014) Development of sago-based food product technology to support food availability. Technology Assessment and Application Agency, Serpong
Hedrick PW (2011) Genetics of populations, 4th edn. Jones & Bartlett Learning, Sudbury MA 01776, Massachusetts, p 675
Hisajima S, Jong FS, Arai Y et al (1991) Propagation and breeding of sago palm (*Metroxylon sagu* Rottb.) plant in vitro: 1. Embryo culture and induction of multiple shoots from sago embryos in vitro. Jap J Trop Agric 35(4):259–267
Hung HC, Adeni DSA, Johnny Q et al (2018) Production of bioethanol from sago hampas via simultaneous saccharification and fermentation (SSF). Nusantara Biosci 10(4):240–245. https://doi.org/10.13057/nusbiosci/n100407
IPB (Institut Pertanian Bogor) University Team (2021a) Study of regional land suitability in Mappi Regency, Papua Province in collaboration with the Mappi Regency Regional Development Planning Agency and IPB University. Internal Report, Bogor
IPB (Institut Pertanian Bogor) University Team (2021b) Feasibility study of developing sago/tapioca starch industrial production in Konawe Raya, Southeast Sulawesi. Faculty of Agricultural Technology, IPB University, Internal Report. Bogor, Indonesia
Irawan AF, Yamamoto Y, Yoshida T et al (2009a) Characteristics of sago palm (*Metroxylon sagu* Rottb.) suckers from various mother palms at different growth stages in Tebing Tinggi Island, Riau, Indonesia. Trop Agr Dev 53(1):1–6. https://doi.org/10.11248/jsta.53.1
Irawan AF, Yamamoto Y, Yoshida T et al (2009b) Characteristics of suckers from sago palm (*Metroxylon sagu* Rottb.) grown in different soil types in Tebing Tinggi Island, Riau, Indonesia. Trop Agr Dev 53(4):103–111. https://doi.org/10.11248/jsta.53.103
Irawan AF, Yamamoto Y, Yoshida T et al (2011a) Comparison of early growth of sago palm (*Metroxylon sagu* Rottb.) seedlings in a culture solution with individual exclusion of macro- and micronutrients. Trop Agr Dev 55(1):1–10. https://doi.org/10.11248/jsta.55.1
Irawan AF, Yamamoto Y, Yoshida T et al (2011b) Effect of root and leaf trimming in sago palm (*Metroxylon sagu* Rottb.) suckers prior to planting on the survival rate in nursery. Trop Agr Dev 55(1):51–54. https://doi.org/10.11248/jsta.55.51
Irawan AF, Nicoll DA, Jamaluddin S (2013) Investigation on trunk characters and starch yield potencies of some existing local sago palm (*Metroxylon sagu* Rottb.) varieties in a swampy sago forest of Kaureh District, Papua. In: Pranamuda H, Toyoda Y, Osok RM, Chan MYK, Haska N (eds) Proc. of the 11th int. sago symp, Indonesian Society of Sago Palm Development and Utilization, Jakarta, pp 107–118
ITIS (2024) *Metroxylon sagu* Rottb., Taxonomic Serial No. 506732. Integrated Taxonomic Information System
Jaman OH (1985) The study of sago seed germination. In: Proc. 22nd research officers' conference, Kuching. Department of Agriculture, Sarawak, pp 69–78
Johnson DV (2010) Tropical palms. Food and Agricultural Organization of United Nation, p 242
Johnson RM, Raymond WD (1956) Sources of starch in colonial territories. In: The sago palm : Colonial Plant Anim Prod 6:20–32
Jong FS (1995) Research for the development of sago palm (*Metroxylon sagu* Rottb.) cultivation in Sarawak, Malaysia. PhD, Dissertation of Agricultural University, Wageningen, p 139
Jong FS (2006) Technical recommendations for the establishment of a commercial sago palm (*Metroxylon sagu* Rottb.) Plantation. Jpn J Trop Agr 50(5):224–228. https://doi.org/10.11248/jsta1957.50.224
Jong FS (2018) An overview of sago industry development, 1980s–2015. In: Ehara H, Toyoda Y, Johnson DV (eds) Sago palm: multiple contributions to food security and sustainable livelihoods. Springer Nature Singapore Press, pp 75–89. https://doi.org/10.1007/978-981-10-5269-9_6
Jong FS, Watanabe A, Hirabayashi D et al (2006) Growth performance of sago palm (*Metroxylon sagu* Rottb.) in peat of different depth and soil water table. Sago Palm 14(2):59–64. https://doi.org/10.57418/sagopalm.14.2_59

Karim AA, Pei-Lang Tie A, Manan DMA et al (2008) Starch from the sago (*Metroxylon sagu*) palm tree-properties, prospect, and challenges as a new industrial source for food and other uses. Comp Rev Food Sci Food Safety 7(3):215–228. https://doi.org/10.1111/j.1541-4337.2008.00042.x

Kim J, Yun S, Ounaies Z (2006) Discovery of cellulose as a smart material. Macromolecules 39(12):4202–4206

Kjær A, Barfod AS, Asmussen CB et al (2004) Investigation of genetic and morphological variation in the sago palm (*Metroxylon sagu*; Arecaceae) in Papua New Guinea. Ann Bot 94(1):109–117. https://doi.org/10.1093/aob/mch112

Konuma H (2018) Status and outlook of global food security and the role of underutilized food resources: sago palm. In: Ehara H, Toyoda Y, Johnson DV (eds) Sago palm: multiple contributions to food security and sustainable livelihoods. Springer Nature Singapore Press, pp 3–16. https://doi.org/10.1007/978-981-10-5269-9_1

Latuconsina MH (2014) Light bricks with a mixture of sago dregs waste. Thesis, Gadjah Mada University, Indonesia

Latuconsina MH (2015) Light bricks mixed with sago pulp waste. Gadjah Mada University, Indonesia

Lim LWK, Chung HH, Hussain H (2020) Complete chloroplast genome sequencing of sago palm (*Metroxylon sagu* Rottb.): molecular structures, comparative analysis and evolutionary significance. Gene Rep 19:100662. https://doi.org/10.1016/j.genrep.2020.100662

Lim LWK, Chung HH, Hussain H et al (2021) Genome survey of sago palm (*Metroxylon sagu* Rottbol). Plant Gene 28:100341. https://doi.org/10.1016/j.plgene.2021.100341

Lim LWK, Lau MML, Chung HH et al (2022) First high-quality genome assembly data of sago palm (*Metroxylon sagu* Rottboll). Data Brief 40:107800. https://doi.org/10.1016/j.dib.2022.107800

Limbongan J, Hanafiah A, Nggobe M (2005) Development of Papuan sago. Papua Agricultural Technology Research Center, Agricultural Research and Development Agency, Department Agriculture, Indonesia

Jamaludin, Mangera Y, Witdarko Y (2022) Utilization of solid waste from processing sago into briquettes as alternative energy in the Dwitrap farming group, Tambat Village, Merauke Regency. Musamus Devotion J 4(2):58–68

Matanubun H, Santoso B, Nauw M et al (2005) Feasibility study of the natural sago forest for the establishment of the commercial sago palm plantation at Kaureh District, Jayapura, Papua, Indonesia. Abstracts of the eight international sago symposium in Jayapura, Indonesia, Japan Society for the Promotion of Science

Maurila TD, Rahmayanti HD, Ardiani S et al (2024) Study of pulp and paper making characteristic produces from sago fiber waste. AIP Conf Proc 3070:040007. https://doi.org/10.1063/5.0199176

Mengoni A, Gonelli C, Brocchini C et al (2003) Chloroplast genetic diversity and biogeography in the serpentine endemic Ni-hyperaccumulator Alyssum bertolonii. New Phytol 157:349–356

Metaragakusuma AP, Katsuya O, Bai H (2016) An overview of the traditional use of sago for sago-based food industry in Indonesia. KnE Life Sci 3:119–124

Monda Y, Kuwahara S, Aoki Y et al (2022) Productivity of sago palms on smallholder plantations after rewetting in previously drained peatland: a case study on Tebing Tinggi Island Riau Province, Indonesia. Tropics 31(1):11–32. https://doi.org/10.3759/tropics.MS21-15

Nabeya K, Nakamura S, Goto Y (2015) Position and development of differentiated lateral buds in sago palm (*Metroxylon sagu* Rottb.). Plant Prod Sci 18(4):435–442. https://doi.org/10.1626/pps.18.435

Narita V, Arum AL, Isnaeni S et al (2012) Web-based bioinformatics analysis for exploration of the chitosanase enzyme based on sequence similarity. J Al-Azhar Indones 1:197–203

Nisar M, Hussain H (2022) Assessment of the genetic variations of sago palm *Metroxylon sagu* in three regions of Sarawak, Malaysia using amplified fragment length polymorphism (AFLP) markers. Chem Biol Technol Agric 9(46):1–13. https://doi.org/10.1186/s40538-022-00315-1

Numberi JJ, Joni, Ekayuliana A (2023) Analysis of the physicochemical properties of sago palm dregs as a source of 65% bioethanol fuel. G-Tech 7(3):1169–1179

Nurmiati, Santoso P, Periadnadi et al (2022) The existence and characterization of potential fermentative bacteria from fresh sago pith (*Metroxylon sagu* Rottb.). Int J Prog Sci Technol 36(1):461–467. https://doi.org/10.52155/ijpsat.v36.1.4894

Okazaki M, Tadenuma S, Ohmi M (2005) Diverse utilization and industrial development of sago biomass. Proceeding Abstract p 2. In: Karafir YP, Fere VE, Toyoda Y (eds) The eight international sago symposium, 4–6 Aug 2005. The Japan Society for Promotion of Science, Manokwari

Okazaki M, Toyota K, Kimura SD (2015) Sago project in Leyte. Tokyo University of Agriculture and Technology, Tokyo

Omori K, Yamamoto Y, Nitta Y et al (2000) Stomatal density of sago palm (*Metroxylon sagu* Rottb.) with special reference to positional differences in leaflets and leaves, and change by palm age. Sago Palm 8:2–8

Pasolon YB (2015) Environment, growth and biomass production of sago palm (*Metroxylon sagu* Rottb.): a case study from Halmahera, Papua and Kendari. Int J Sustain Trop Agric Sci 2(1): 97–104

Pratama YE (2022) Feasibility of sago fiber particle board (45°) rice bran matrix in accordance with SNI Standard 03-2105-2006 based on mechanical and physical properties (in Indonesian). Proceeding of Dissemination 2021/2022, Bandung, Feb 2022, Faculty of Industrial Technology, Bandung National Institute of Technology, Bandung

Purwoko D, Cartealy IC, Tajuddin T et al (2019) SSR identification and marker development for sago palm based on NGS genome data. Breed Sci 69:1–10. https://doi.org/10.1270/jsbbs.18061

Quevedo MA, Loreto AB, Mariscal AM et al (2005) Distribution and traditional uses of sago palms (*Metroxylon sagu* Rottb) in the Eastern and Central Visayas regions of The Philippines. Sago Palm 13(2):17–25. https://doi.org/10.57418/sagopalm.13.2_17

Rahmi, Fachrudin S, Nurmalasari (2018) Utilization of sago fiber waste (*Metroxylon sago*) as iodin adsorbent. J Rekayasa Kimia dan Lingkungan 13(1):70–77. https://doi.org/10.23955/rkl.v13i1.10072

Rana A, Samtiya M, Dhewa T et al (2022) Health benefits of polyphenols: a concise review. J Food Biochem 46(10):e14264. https://doi.org/10.1111/jfbc.14264

Rembon FS, Pasolon YB, Yamamoto Y (2008) Characteristics of seed and germination of wild-type sago "Manno" (*Metroxylon sago* Rottb.) collected from sago palm field around Lake Sentani near Jayapura, Indonesia. Sago Palm 16:1–6. https://doi.org/10.57418/sagopalm.16.2_79

Rembon FS, Yamamoto Y, Pasolon YB (2010) Effects of soil types and water regimes on the early growth of sago palm (*Metroxylon sagu* Rottb.) seedlings. Sago Palm 18:28–40

Riyadi I (2010) Effect of kinetin and BAP on the growth and development of somatic embryos in sago palm (*Metroxylon sagu* Rottb.). AgroBiogen J 6(2):101–106

Riyadi I, Sumaryono (2009) The effect of soaking interval and length on the growth and maturation of somatic embryos in sago palm (*Metroxylon sagu* Rottb.). Menara Perkebunan 77(2):94–103

Riyadi I, Tahardi JS, Sumaryono (2005) The development of somatic embryos of sago palm (*Metroxylon sagu* Rottb.) on solid media. Menara Perkebunan 73(2):35–43

Riyadi I, Efendi D, Purwoko BS et al (2016) Indirect somatic embryogenesis in sago palm (*Metroxylon sago* Rottb.) using suspension culture, instant immersion and solid media. AgroBiogen J 12(1):37–44

Riyadi I, Efendi D, Purwoko BS et al (2018) Effect of TDZ on induction of somatic embryos in sago palm (*Metroxylon sagu* Rottb.) in three different culture methods. Menara Perkebunan 86(1): 11–20

Riyanto R, Widodo I, Abbas B (2018) Morphology, growth, and genetic variations of sago palm (*Metroxylon sagu*) seedlings derived from seeds. Biodiversitas 19:682–688

Robertson G, Schein J, Chiu R et al (2010) De novo assembly and analysis of RNA-seq data. Nat Methods 7:909–912. https://doi.org/10.1038/nmeth.1517

Ruamba A, Sumule AI (2020) Characteristics of maturity of several types of sago (*Metroxylon sago* sp) ready for harvest according to the understanding of the people of Randawaya village, Ampimoi Bay, Yapen Islands, Papua Province. J Agrotek 8(1):7–13
Ruddle K, Johnson D, Townsend PK et al (1978) Palm sago—a tropical starch from marginal lands. Australian National University Press, Canberra
Sangadji I, Patty CW, Salamena JF (2019) Crude fiber content of sago by product fermented with white root fungi (*Pleurotus ostreatus*) and urea. Agrinimal J Ilmu Ternak dan Tanaman 7(1): 20–25. https://doi.org/10.30598/ajitt.2019.7.1.20-25
Sarungallo ZL, Murtiningrum (2005) Production and characterization of glucose syrup of Papua sago starch. Abstracts of the eight international sago symposium in Jayapura, Indonesia, Japan Society for the Promotion of Science
Schmidt FH, Ferguson JHA (1952) Rainfall types based on wet and dry period ratios for Indonesia with Western New Guinea. Kementerian Perhubungan dan Djawatan Meteorologi dan Geofisik, Jakarta
Schuiling DL (2009) Growth and development of true sago palm (*Metroxylon sagu* Rottboll) with special reference to accumulation of starch in the trunk: a study on morphology, genetic variation and ecophysiology, and their implications for cultivation. PhD Thesis, Wageningen University
Schuiling DL, Flach M (1985) Guidelines for the cultivation of sago palm. Department Tropical Crop Science. Agricultural University, Wageningen
Shimoda H, Power AP (1986) Investigation into development and utilization of sago palm forest in the East Sepik region, Papua New Guinea. In: Yamada N, Kainuma K (eds) Tokyo: Sago Palm Research Fund and Tropical Agricultural Research Association of Japan. Cambridge University Press, Tokyo, pp 94–104
Sinta MM, Riyadi I, Sumaryono (2008) Somatic embryogenesis of sago palm (*Metroxylon sagu* Rottb.) from different origins in Indonesia. Int. Biotech. Conf. Estate Crops 2017. IOP Publishing, p 6. https://doi.org/10.1088/1755-1315/183/1/012008
Sumaryono, Riyadi I, Pauline DK et al (2007) Growth and development of embryogenic callus and somatic embryos of oil palm (*Elaeis guineesis* Jacq.) in the temporary immersion system. Menara Perkebunan 75(1):1–11
Sumaryono, Muslihatin W, Ratnadewi D (2012) Effect of carbohydrate source on growth and performance on in vitro sago palm (*Metroxylon sagu* Rottb.) plantlets. Hayati J Biosci 19(2): 88–92. https://doi.org/10.4308/hjb.19.2.88
Sumaryono, Riyadi I, Abbas B (2013) In vitro propagation for large scale production of sago planting material. Proceedings of the 11th sago palm symposium
Sunardi, Noviyanti N, Istikowati WT et al (2021) Analysis of fiber component of sago (*Metroxylon sago*) fronds and the study of cellulose morphology after oxidation using ammonium persulphate. J Sains dan Terapan Kimia 15(1):48–63. https://doi.org/10.20527/jstk.v15i1.9724
Syadik F, Satria, Youlandari (2022) The protein and crude fiber content of sago (*Metroxylon sago*) dregs through chemical method as alternative ruminant feed. J Sains dan Teknologi Peternakan 3(2):49–54
Syahputra A, Yulianto D (2020) Utilization of sago tree for fiber waste making biocomposite. J Ren Ener Mech 3(1):22–31. https://doi.org/10.25299/rem.2020.vol3(01).4263
Tahardi JS, Sianipar NF, Riyadi I (2002) Somatic embryogenesis in sago palm (*Metroxylon sagu* Rottb.). In: Kaimuna K, Okazaki M, Toyoda Y, Cecil JE (eds) New frontiers of sago palm studies. Universal Academic Press, Inc., Tokyo, pp 75–81
Tajuddin T, Minaldi, Erwinda et al (2012) Ex vitro propagation of sago palm (*Metroxylon sagu* Rottb.). Patent No. IDP000031658, Indonesia
Tajuddin T, Karyanti, Sukarnih T et al (2015) Inducing the in vitro rooting of sago palm (*Metroxylon sagu* Rottb.) using auxin. Indones J Biotechnol Biosci 2(2):73–80. Corpus ID: 64642123
Tjahjono AE, Rustiaty B, Mulawati I et al (2024) Enhanced ethanol production from sago pith using very high gravity fermentation by Saccharomyces cerevisiae adapted in media containing high ethanol concentration. AIP Conf Proc 2957(1):030020. https://doi.org/10.1063/5.0184585

Turukay B (1986) The role of the sago palm in the development of integrated farm system in Maluku Province of Indonesia. In: Yamada N, Kainuma K (eds) Tokyo: sago palm research fund and tropical agricultural research association of Japan. Cambridge University Press, Tokyo, pp 7–15

Umar MZ, Hasan LOA, Arsyad M (2020) The use of sago fiber waste *Metroxylon sagu* rottb. as added material to concrete bricks. Proc USN Kolaka-ADRI International Conference on Sustainable Coastal-Community Development 1:25–29. https://doi.org/10.31327/icusn-adri.v1.1056

Van Kraalingen DWG (1984) Some observations on sago palm growth in Sepik River Basin, Papua New Guinea. Report of Department of Minerals and Energy, Konedobu, Papua New Guinea, pp. 69

Vendramin GG, Degen B, Petit RJ et al (1999) High level of variation at Abies alba chloroplast microsatellite loci Europe. Mol Ecol 8:1117–1126

Wina E, Evans AJ, Lowry JB (1986) The composition of pith from the sago palms *Metroxylon sagu* and *Arenga pinnata*. J Sci Food Agric 37:352–358

Xie Y, Wu G, Tang J et al (2014) SOAPdenovo-trans: de novo transcriptome assembly with short RNA-Seq reads. Bioinformatics 30:1660–1666. https://doi.org/10.1093/bioinformatics/btu077

Yamamoto Y, Yoshida T, Miyazaki A et al (2005) Biodiversity and productivity of several sago palm varieties in Indonesia. Eighth international sago symposium in Jayapura, Indonesia. Japan Society for the Promotion of Science, pp 35–40

Yamamoto Y, Rembon FS, Omori K et al (2010) Growth characteristics and starch productivity of three varieties of sago palm (*Metroxylon sagu* Rottb.) in Southeast Sulawesi, Indonesia. Trop Agr Dev 54(1):1–8. https://doi.org/10.11248/jsta.54.1

Yamamoto Y, Yoshida T, Yanagidate I et al (2020a) Studies on growth characteristics and starch productivity of the sago palm (*Metroxylon sagu* Rottb.) folk varieties in Seram and Ambon Islands, Maluku, Indonesia. Trop Agr Dev 64(3):125–134. https://doi.org/10.11248/jsta.64.125

Yamamoto Y, Yanagidate I, Yoshida T et al (2020b) Characteristics of growth and starch productivity of wild sago folk variety "Manno" grown near Jayapura, Papua, Indonesia. Trop Agr Dev 64(3):135–145. https://doi.org/10.11248/jsta.64.135

Yamori W, Caemmerer SV (2009) Effect of rubisco activase deficiency on the temperature response of CO2 assimilation rate and rubisco activation state: insights from transgenic tobacco with reduced amount of rubisco activase. Plant Physiol 151:2072–2082

Yan WJ, Hussain H, Chung HH et al (2022) Transcriptome dataset of sago palm in peat soil. Data Brief 41:107908. https://doi.org/10.1016/j.dib.2022.107908

Yanagidate I, Rembon FS, Yoshida T et al (2009) Studies on trunk density and prediction of starch productivity of sago palm (*Metroxylon sagu* Rottb.): A case study of a cultivated sago palm garden near Kendari, Southeast Sulawesi Province, Indonesia. Sago Palm 17(1):1–8. https://doi.org/10.57418/sagopalm.17.1_1

Yater T, Hw T, Meliala C et al (2019) A comparative study of phenotypes and starch production in sago palm (*Metroxylon sagu*) growing naturally in temporarily inundated and non-inundated areas of South Sorong, Indonesia. Biodiversitas 20(4):1121–1126. https://doi.org/10.13057/biodiv/d200425

Zulaeha S, Fadhullah H, Cartealy IC et al (2023) Assessment of genetic diversity among sago palms (*Metroxylon sagu* Rottb.) in Bengkulu, Indonesia using simple sequence repeats. Genet Resour Crop Evol 70(8):1–13. https://doi.org/10.1007/s10722-023-01819-x

Chapter 11
Sisal (*Agave sisalana* Perrine) Breeding

Muhammad Shahzad, Muhammad Ahsan Qureshi, Sohail Kamaran, Hafiz Amjad Ali Rana, Muhammad Iftikhar, and Gulfam Hassan

Abstract This chapter highlights the significance of natural fiber crops within the Asparagaceae family. With the earth's population surpassing eight billion in 2022 and continuing to grow rapidly, there is a pressing need to boost agricultural productivity to meet the increasing demand for food and fiber all while minimizing ecological and environmental impacts. Given the current challenges posed by climate change, it is crucial to cultivate eco-friendly crops without compromising yield and quality. Sisal (*Agave sisalana* Perrine) stands out as a key natural fiber crop originating from Mexico and Central America offering great potential to bridge the gap between fiber demand and supply. However, existing sisal varieties suffer from low yield and fiber quality, prompting plant breeders to prioritize these traits. Additionally, climate change is also affecting the overall yield and fiber quality of Sisal. There is a dire need for time to address those threats being faced by the Sisal

M. Shahzad (✉)
Department of Agriculture Extension Education, Faculty of Agriculture & Environment, The Islamia University of Bahawalpur, Bahawalpur, Punjab, Pakistan

M. A. Qureshi
Institute of Horticultural Sciences, University of Agriculture, Faisalabad, Faisalabad, Pakistan

S. Kamaran
Department of Plant Breeding and Genetics, Faculty of Agriculture, University of Agriculture Faisalabad, Faisalabad, Pakistan
e-mail: skamran@gudgk.edu.pk

H. A. A. Rana
Department of Outreach and Continuing Education, Faculty of Social Sciences and Humanities, Muhammad Nawaz Shareef University of Agriculture, Multan, Pakistan
e-mail: amjad.ali@mnsuam.edu.pk

M. Iftikhar · G. Hassan
Institute of Agricultural Extension, Education and Rural Development, Faculty of Social Sciences, University of Agriculture Faisalabad, Faisalabad, Pakistan
e-mail: muhammad.iftikhar@uaf.edu.pk; gulfam.hassan_stu@uaf.edu.pk

K. F. M. Salem et al. (eds.), *Breeding and Biotechnology of Leaf, Fruit, and Seed Fiber Crops*, Advances in Plant Breeding Strategies 11,
https://doi.org/10.1007/978-3-032-00407-9_11

crop in order to sustain the future of Sisal as a potential fiber crop. Genetic variability serves as a fundamental prerequisite for successful breeding programs, with conventional now benefiting from advanced genomic tools like sequencing technologies and Marker-Assisted Selection (MAS) enabling targeted and trait-specific breeding efforts. Marker-Assisted Selection (MAS) is a modern breeding technique that utilizes molecular markers to aid in the selection of plants with specific traits. In the context of sisal breeding, MAS can significantly enhance the efficiency and precision of the breeding process. Researchers have identified DNA markers associated with traits of interest in sisal, including fiber quality, disease resistance, and stress tolerance. Breeding is a basic driver in agriculture to modify and improve crops for better yield and quality for the prosperity of humans.

Keywords Sisal (*Agave sisalana* Perrine) · Breeding · Biotechnology · Fiber · Genetic diversity · Marker Assisted Selection (MAS)

11.1 Introduction

This chapter highlights the importance of natural fiber crops from the Asparagaceae family in the context of agricultural sustainability. With the global population surpassing eight billion in 2022, there is an urgent need to increase agricultural productivity to meet the growing demands for food and fiber while concurrently mitigating ecological and environmental ramifications. In light of prevailing climate change scenarios, there exists a critical need to cultivate crops that are ecologically sound yet do not compromise on yield or quality (Partheepan et al. 2024). The historical lineage of sisal (*Agave sisalana* Perrine) showed its record in the regions of Mexico and Central America. The indigenous residents of these locations were among the pioneer observers and users of sisal fibers. They were using sisal fiber in the manufacturing of bags, baskets and some garments. Historical documentation verifies its cultivation before Columbus discovered America (Lock 1962; Mande 1980). The spread of sisal around the world started during colonial times when European travelers and traders observed its economic benefits. Sisals introduction to warm and hot regions of Africa, Asia and the Caribbean demonstrated its adaptability to diverse climates (Caujapé-Castells et al. 2010).

Sisal (*Agave sisalana*) traces its origins to Mexico, where it was initially identified in the port town of Sisal. Initially cultivated solely as a fiber crop in Mexico, sisal cultivation spread to Tanzania and is now widely grown in arid and semi-arid regions worldwide (Borland et al. 2014; Ramankutty et al. 2008; Vorosmarty et al. 2010). Over time, sisal cultivation has expanded, leading to the development of approximately 275 varieties, including the sisal hybrid known as "Hybrid 11648" (Peregrine 1969).

Historically, sisal was primarily utilized as a natural fiber in tropical and subtropical regions. However, advancements in cultivation and processing techniques have broadened its application. Sisal production witnessed significant growth in African countries and other regions during the early decades of the twentieth century. For instance, Tanzania alone produced nearly 230,000 metric tons of sisal in 1960, while Brazil accounted for 40% of global sisal fiber production in 2008 (Sabea 2001). Currently, sisal cultivation is widespread in various African countries, including Tanzania, Madagascar, South Africa, Kenya, Angola, Mozambique, Ethiopia, and Jamaica. Additionally, countries such as Brazil, China, and India also cultivate sisal.

Agave sisalana stands out as a significant producer of robust natural fibers renowned for their quality. These fibers, extracted from the leaves of the plant, hold a notable position in global plant fiber production, accounting for approximately 2% of the total (Nava-Cruz et al. 2015). Traditionally, sisal fiber finds its application in the manufacturing of cords and ropes, leveraging its inherent strength and durability. However, its utility extends far beyond traditional uses, with widespread adoption across various industrial sectors.

In recent years, sisal fiber has seen increasing utilization in reinforcing building materials, furniture, panels, and automobile upholstery, reflecting its versatility and adaptability to diverse applications. Notably, sisal fiber presents several advantages over synthetic alternatives, including lower density (making it lighter), reduced production costs, and its biodegradable and recyclable nature. These inherent qualities align with the global trend towards favoring sustainable natural resources with minimal environmental impact (Pinos-Rodríguez et al. 2009).

The versatility, durability, and eco-friendliness of sisal fiber make it an attractive choice for industries seeking sustainable materials. Its utilization not only meets functional requirements but also aligns with environmental stewardship goals, contributing to the growing preference for eco-conscious solutions. As such, sisal fiber represents a valuable resource in the pursuit of sustainable development and responsible resource management (Ahmad et al. 2022).

The extraction of sisal fiber has evolved from traditional retting methods to modern decortication machines, facilitating more efficient fiber extraction. Moreover, valuable products are derived from sisal waste materials, enhancing their economic viability (Mohankumar et al. 2021).

Belonging to the Asparagaceae family, sisal garnered commercial interest in the 1880s following the invention of the machine grain binder and the increasing demand for low-cost twine. Commercial cultivation began in regions like the Bahamas and Tanganyika (now Tanzania). Sisal fiber found diverse applications across marine, agricultural, shipping, and industrial sectors. Additionally, sisal fiber was utilized in the production of various goods, including matting, rugs, millinery, and brushes.

The widespread cultivation and utilization of sisal underscore its economic and industrial significance. Its resilience in arid and semi-arid environments makes it a valuable crop for regions facing water scarcity or unfavorable climatic conditions. Furthermore, ongoing research and technological advancements continue to enhance sisal cultivation practices, contributing to its sustainability and economic viability (Fig. 11.1).

Fig. 11.1 Sisal seedlings planted in two sowing seasons. (Photo by M. Shahzad)

11.1.1 Botany of Sisal

The sisal plant grows to a height of approximately 90 cm (3 feet) with a diameter of about 38 cm (15 inches). Its lance-shaped leaves form a dense rosette, exhibiting a fleshy and rigid texture, ranging in color from gray to dark green. Each leaf measures 60–180 cm (2–6 feet) in length and 10–18 cm (4–7 inches) in width, culminating in a sharp spine. After 4 to 8 years, the mature plant produces a central flower stalk reaching about 6 m (20 feet) in height. Clusters of yellow flowers, with an unpleasant odor, appear at the top of branches growing from the flower stalk. As the flowers wither, bulbils develop at the upper angle between the stem and flower stalk, falling to the ground to take root. Similar to other Agave species, the old plant dies after flowering. Sisal (*Agave sisalana* Perrine) is a perennial succulent plant belonging to the Asparagaceae family. It typically exhibits a rosette growth habit, characterized by a dense cluster of thick, sword-shaped leaves arising from a central stem. The leaves are succulent and rigid, with serrated edges and a pointed tip, often reaching lengths of 1–2 m and widths of 10–15 cm. The leaves are arranged in a spiral pattern, forming a symmetrical rosette structure. The color of sisal leaves can vary, ranging from bluish-green to grayish-green, depending on environmental conditions and genetic factors. The leaves are covered with a thick cuticle to minimize water loss and protect against harsh environmental conditions.

Sisal plants are monocarpic, meaning they flower only once in their lifetime, typically after several years of growth. The flowering stalk, known as an inflorescence, emerges from the center of the rosette and can reach heights of up to 6 m. The inflorescence consists of numerous small, tubular flowers arranged in dense clusters along the stalk. Sisal flowers are typically greenish-yellow in color and attract pollinators such as bees and hummingbird.

After flowering and seed production, sisal plants generally undergo senescence, with the original rosette dying back. However, sisal plants produce prolific suckers or offshoots from the base, allowing for vegetative propagation and the continuation of the plant's life cycle. Overall, sisal is a robust and drought-tolerant plant with distinctive succulent leaves, well-adapted to arid and semi-arid environments. Its botanical features make it a valuable source of fiber and contribute to its widespread cultivation in various regions worldwide.

11.1.2 Sisal Cultivation

Sisal is a perennial plant that can be grown in any season of the year in various geographical locations across the sub-continent ranging from arid to semi-arid and humid to sub-humid areas (Weink 1969). Sisal can thrive in rain-fed areas and on undulating lands with steep slopes. It can be cultivated in all types of soils except saline and waterlogged. Propagation of sisal is done through bulbils produced in flower stalks and suckers that originate around the base of the plant. These bulbils and suckers are grown in nurseries until they reach a height of 50–70 cm before being transplanted into the field. The ideal plant to plant distance is 1 m and the row to row distance should be maintained at 2 m for optimal cultivation of sisal. Sisal cultivation does not require a significant amount of inorganic fertilizer and the pH of the soil should be maintained within a range of 5.5–6.5.

Sisal plants can thrive in poor soil with proper drainage, preferring warm and moist climates. Young plants propagated from bulbils or rhizomes of mature plants, spend their initial 12–18 months in nurseries before being transferred to the field at the onset of the rainy season. Maturation occurs in about 3–5 years, depending on the climate, yielding satisfactory fiber for 7 or 8 years. During the productive period, approximately 300 leaves are produced, with outer leaves being cut off close to the stalk as they reach full length. The initial harvest involves around 70 leaves, followed by an annual production averaging about 25.

The extraction of sisal fiber involves machine decortication, where the leaves are crushed between rollers. The resulting pulp is removed, and the fiber undergoes washing and drying through mechanical or natural means. Typically creamy white and ranging from 100 to 125 cm (40 to 50 inches) in length, sisal fiber strands are coarse and inflexible. Sisal has many distinguished characteristics such as its ability to absorb a wide range of light and a great amount of carbon dioxide. It is environmentally sustainable, bio-degradable, input efficient and produces its own nutrients. Sisal production is also helpful in mitigating climate change because it is a good source of renewable bio-energy (Bessou et al. 2011).

Pakistan has great potential to grow many fiber crops like cotton, banana, jute, sisal and hemp. Among all these fiber crops, sisal has the unique ability to grow in marginal and wastelands, accounting for almost 50 million hectares of the total area in Pakistan. Sisal can be successfully grown in these areas because it requires less maintenance and a small amount of input compared to other fiber crops. A sisal plant can produce fiber for up to 8 years on average. It can survive in harsh climates and severe drought conditions which often occur in arid areas of Pakistan. The plant can shrink its extensive roots to minimize water loss and can produce fine roots very quickly for efficient water uptake. Sisal plants require almost no fertilizer and other chemicals, making them marvelous plants that can be inter-cultivated and inter-grazed in a variety of soil conditions.

During the nineteenth century, the diverse adaptability and utility of sisal made it a valuable commodity and its cultivation began on a larger scale due to the increased need for strong fibers. By the early 1900s, sisal became the most cultivated and used fiber crop worldwide. Its ability to grow in dry climates made it popular for farming. However, after the 1920s, its cultivation declined and synthetic fibers replaced natural fibers, posing threats to the environment. Due to the growing population, the demand for fiber is increasing so it is essential to grow more natural fiber crops to reduce reliance on synthetic fiber. Currently, lack of awareness and better varieties is a significant obstacle hurdle in sisal cultivation. Therefore, plant breeders should develop early maturing, better fiber-yielding and input-responsive varieties to meet the growing demand for fiber from a multiplying population without disturbing the environment. Agricultural extension services can play a role in disseminating awareness among farmers by highlighting the economic importance of sisal.

Sisal (*Agave sisalana*) is a valuable fiber crop known for its resilience and economic significance. However, the cultivation of sisal is often hindered by various diseases, collectively referred to as sisal perrine diseases (Singh and Singh 2018). Sisal perineal diseases encompass a group of fungal, bacterial, and viral pathogens that adversely affect the health and productivity of sisal plants (Singh and Singh 2018). These diseases pose significant challenges to sisal cultivation worldwide, leading to yield losses and economic setbacks for growers (Acosta-Gallegos et al. 2007). Despite their importance, comprehensive studies on sisal perrine diseases are relatively limited, necessitating a thorough examination of the current state of knowledge and future research directions.

Fusarium wilt (*Fusarium oxysporum* f. sp. sisalana) is among the most devastating fungal diseases affecting sisal (Lang 1985). This soil-borne pathogen infects the vascular system of sisal plants, causing wilting, yellowing, and ultimately death (Lang 1985). Anthracnose (*Colletotrichum gloeosporioides*) is another prominent fungal disease, characterized by dark lesions on leaves and stems, leading to defoliation and reduced fiber yield (Lang 1985). Additionally, leaf spot diseases caused by various fungal pathogens contribute to foliage necrosis and reduced photosynthetic efficiency in sisal plants (Nweke et al. 2013). Bacterial soft rot (*Erwinia* spp.) is a common bacterial disease affecting sisal, particularly during periods of high humidity (Neumann and Metz 1971). This disease presents as water-soaked lesions on leaves and stems, progressing to tissue decay and plant collapse (Neumann and Metz 1971). Bacterial leaf streak (*Xanthomonas campestris* pv. dieffenbachiae) is another bacterial pathogen causing yellow streaks and necrotic lesions on sisal leaves, impairing photosynthesis and fiber production (Neumann and Metz 1971).

Fig. 11.2 Sisal plantation in the field. (Photo by M. Shahzad)

Sisal mosaic virus (SiMV) and Sisal yellow mottle virus (SiYMV) are the primary viral pathogens affecting sisal (Singh and Singh 2018). These viruses cause mosaic symptoms, chlorosis, and stunting in infected plants, ultimately reducing fiber quality and yield (Singh and Singh 2018).

Integrated disease management (IDM) approaches combining cultural, chemical, and biological control methods are essential for mitigating the impact of sisal perrine diseases (Acosta-Gallegos et al. 2007). Measures such as crop rotation, sanitation, and the use of resistant cultivars can help reduce disease incidence and severity (Acosta-Gallegos et al. 2007). Additionally, the application of fungicides, bactericides, and virucides may be necessary to manage disease outbreaks effectively (Nweke et al. 2013) (Figs. 11.2 and 11.3).

11.2 Economic Importance of Sisal Fiber

Derived from the leaves of the Agave sisalana plant, sisal fibers have assumed pivotal significance across various industries owing to their versatile utility. The economic assessment of sisal fibers is predicated upon their diverse applications within different domains. Boosting remarkable strength and durability, sisal emerges as an ideal material for the manufacturing of ropes and yarns. Particularly, the maritime sector extensively relies on sisal ropes due to their robust performance in challenging environments (Gentry 2004; Nobel 1994; Weber 2003).

Sisal exhibits remarkable versatility, finding applications in a spectrum of industrial products, spanning heavy-duty uses like composite material in the automotive sector and reinforcement in construction to more delicate purposes such as paper, newsprint, and telephone cable paper. The sisal fiber is instrumental in the

Fig. 11.3 Sisal plants of different ages. (Photo by M. Shahzad)

production of wood-free thinning paper, Bible paper, and floor coverings. Beyond industrial realms, sisal also serves domestic needs, acting as household fuel and contributing to building materials. Notably, sisal floor coverings have garnered substantial consumer attraction in the United Kingdom due to their stain-resistant properties. The burning of petroleum products and their increasing demand have forced policymakers to shift towards a sustainable paradigm in order to keep the environment safe from all hazards (Leadm 2006). This sustainable paradigm shift generally demands changes in many sectors like construction, automotive and marine. (Han and Choi 2010).

Keeping in view this scenario, sisal is considered a model fiber and bio-energy crop due to many characteristics such as its durability, input efficiency, sustainability, affinity for dyeing, environmental conservatism, flexibility in fiber and resistance of plant under harsh climatic conditions. All these qualities of the sisal crop have caught the attention of many countries like China, Brazil, Mexico and South Africa to increase per hectare yield and establish local industry to export many valuable products to earn foreign exchange (Salum and Hodes 2001).

The demand for sisal is increasing throughout the world due to its multidimensional uses in different products. Sisal fiber can be utilized as an alternative to synthetic fiber in the manufacturing of various products. Due to its recyclable nature, it is widely believed that this fiber has the potential to replace synthetic fiber in the near future to minimize health and environmental hazards (Anandjiwala 2006). Many automobile corporations are using sisal fiber in dashboard making and ceiling of cars. Almost 12,000 tons of sisal fiber were used only in Brazil during the year 2013 by automobile corporations for making bonnets, insulation parts, panels, dashboards and seat cushions (Vilance et al. 2014).

11.3 Sisal Fiber Extraction

Sisal fiber is extracted from mature leaves using hammering and retting methods. Hammering is a modern mechanical method widely used for commercial fiber extraction. In this process, a sisal decortication machine is used to extract fibers from the leaves. The machine contains a raspador which is a rotating drum with blades spaced 10 cm apart. Sisal leaves are fed into one side of the raspador and high-quality fiber is obtained quickly from the other side (Cooper et al. 1981; Saxena and Roy 1995). During this process, all the waste material is gathered on one side leaving only the fiber strands.

The other procedure of fiber extraction is called the retting method which involves a biodegradation process and normally takes 15–21 days to separate fibers from leaves. In this method sisal leaves are dumped under soil and decomposition takes place. It is time time-consuming method and yields poor-quality fiber. In order to remove contamination, the obtained fiber is dried in the sun or washed with hot water. Drying is one of the most crucial elements to get good quality fiber (Svetlitchnyi et al. 2013). The artificial drying method is more suitable because it gives fine-quality fiber as compared to sun drying.

Fiber extraction involves the process of decortication, where leaves undergo crushing, beating, and brushing by a rotating wheel with blunt knives, leaving only the fibers. In East Africa, leaves are sent to a central decortication plant, employing water to wash away waste parts. The extracted fiber is then dried, brushed, and baled for export, with proper drying crucial for maintaining quality. While artificial drying generally yields better grades than sun drying, it may not always be practical in less industrialized sisal-producing countries. In the arid climate of northeast Brazil, smallholders use portable raspadors without water for fiber extraction.

The quality of sisal fiber is determined by its length, strength and color. After drying, the fiber makes up about 4% of the total leaf weight. The dried fiber is creamy white with a diameter of 0.2–0.4 mm and a length of 80–120 cm. It is categorized into three types based on durability and length: longer (80–90 cm), medium (70–80 cm) and shorter (60–70 cm). Sisal fiber is rigid and possesses properties such as strength, durability, dye affinity stretchability and resistance to decay in salt water (Yu 2005).

11.4 Uses of Sisal Fiber

For many years, the sisal crop has been grown for its fiber which forms 3–4% of the plant. The fiber has been used for manufacturing twine, ropes, and other hand-woven crafts (TSGA 1992). Sisal fiber, extracted from the leaves of the Agave sisalana plant, serves a multitude of purposes across diverse industries. Primarily renowned for its exceptional tensile strength and durability, sisal finds traditional application in cordage and ropes. Its robust nature makes it indispensable in agriculture, where sisal twines are extensively used for bundling crops and securing fences. In the marine industry, sisal's resistance to saltwater degradation is harnessed for critical tasks like mooring lines and fishing nets. Beyond its utilitarian roles, sisal contributes to the textile industry by being woven into carpets, mats, and rugs, providing a rustic aesthetic. Craftsmen and artists incorporate sisal into handmade products, while the automotive industry utilizes fibers for reinforcing components like door panels and carpets. Sisal's stiff and durable qualities also make it ideal for brush manufacturing, offering effective bristles for various applications. Erosion control blankets and geotextiles woven from sisal fibers provide sustainable solutions in construction and environmental conservation. From packaging materials to erosion prevention, sisal's versatility and eco-friendly attributes position it as a timeless and indispensable material in modern applications. Sisal fiber has been traditionally used for rope and carpet making due to the least technological development but with time sisal fiber is becoming popular in the light of the blooming sisal industry. Sisal fiber is classified into three major categories based on durability and strength (Yu 2005).

The first category includes fine quality and high-grade fiber used for manufacturing carpets, dashboards, marine suiting, fiber boats, fiberglass, handicrafts, fiber guns and cabin interiors. The second category includes medium-graded fiber used in marine rope making, binder twins, marine suiting and hand-woven products. The third category or low-quality sisal fiber is used in the paper industry because it contains cellulose and semi-cellulose in bulk amounts (Evans 1990; Parsons and Parsons 1990; Plunket and Urunuela 2012). Due to its stain-resisting quality, sisal fiber is also used as a floor covering in many developing countries worldwide (Morley 2011).

The process of fiber extraction is performed by a decorticator (a machine that has a rotating wheel with sharp blades connected to an axle) and demands extensive water and energy consumption. Sisal leaves are entered into one side of the machine and crushed and beaten by a rotating wheel leaving the waste material on the other side. The obtained fiber accounts for only 5% of the total weight of sisal leaf, leaving 95% as waste, which is washed away using water into a nearby disposal site. The majority of sisal factories in Kenya and Tanzania usually separate solid material from sisal waste before disposing of it to make many valuable products (Shamte 2001). In many cases, waste material after fiber extraction is disposed of without utilizing it but nowadays, means are being explored to use it for making fertilizers and energy generation. The juice of plants is used in the preparation of pharmaceutical products like inulin and hecogenin while crushed parenchyma tissues can be used as animal feed. The residue left after the biodegradation process contains a large amount of nutrients called bio-slurry and can be used as a bio-fertilizer (Nijaguna 2002). The nutrient composition of bio-slurry versus sisal waste is given in Table 11.1.

The juice extracted after acidic hydrolysis may be used for fructose extraction which is an alternative to honey and table sugar. Recent technological developments in the sisal industry, have led to the creation of valuable products such as tiles and bio-fuel as well as exploration into the production of various pharmaceutical

Table 11.1 Nutrient composition of bio-slurry vs. sisal waste

Nutrients	Composition of bio-slurry (kg/t)	Composition of sisal waste (kg/t)
Nitrogen (N)	4.0	6.0
Calcium (Ca)	17.5	25.0
Magnesium (Mg)	1.1	1.6
Potassium (K)	0.6	0.8
Phosphorus(P)	0.7	1
Sulphur(S)	1.75	2.5

products (Corradini et al. 2009). The waste material which makes up more than 95% of the total weight of sisal leaves is rich in nutrients and can be used as a natural fertilizer in the field.

Furthermore, research indicates the potential of sisal in pharmaceutical applications, cattle feed production, decorative panels, handbags, fashion accessories, and geotextiles. The strong fibers of sisal help these products last longer and maintain their strength. Sisal fibers are utilized in the production of twine, netting and other agricultural materials. The use of sisal as a raw material is environmentally friendly due to its low or nonexistent environmental. The impact makes it a smart choice for farming practices. Sisal fibers are commonly used to make bags and sacks for packing agricultural materials such as coffee, cocoa, and grains. The strength and durability of the fibers help to keep the material safe during transportation. Sisal geotextiles are important for stopping soil from washing away and keeping the ground stable. The uses of the sisal plant have continued to expand. The diversified uses of sisal are illustrated in Fig. 11.4.

11.5 Production and Management of Sisal: Opportunities and Prospects

Pakistan experiences extreme climatic conditions with dry, hot summers and cold winters. The geographical landscape is diverse, featuring high mountains in the north, central, northwest, and southwest regions, plateaus in the center, and plains, deserts, and an extensive coastline in the southwest. Each region exhibits distinct climatic variations, ranging from very cold to extremely hot, with some areas maintaining moderate temperatures throughout the year. Despite this diversity, historical data indicates lower precipitation compared to neighboring regions. The country relies heavily on its rich river water system to meet agricultural water needs, emphasizing the vital role of agriculture in ensuring food security and poverty reduction. However, the agricultural sector faces challenges from temperature increases affecting cropping seasons, heightened irrigation demands, increased evapotranspiration, and growing heat stress on crops.

Due to its distinct climatic conditions, Pakistan is highly susceptible to climate change, with a vulnerability index surpassing that of many other countries globally.

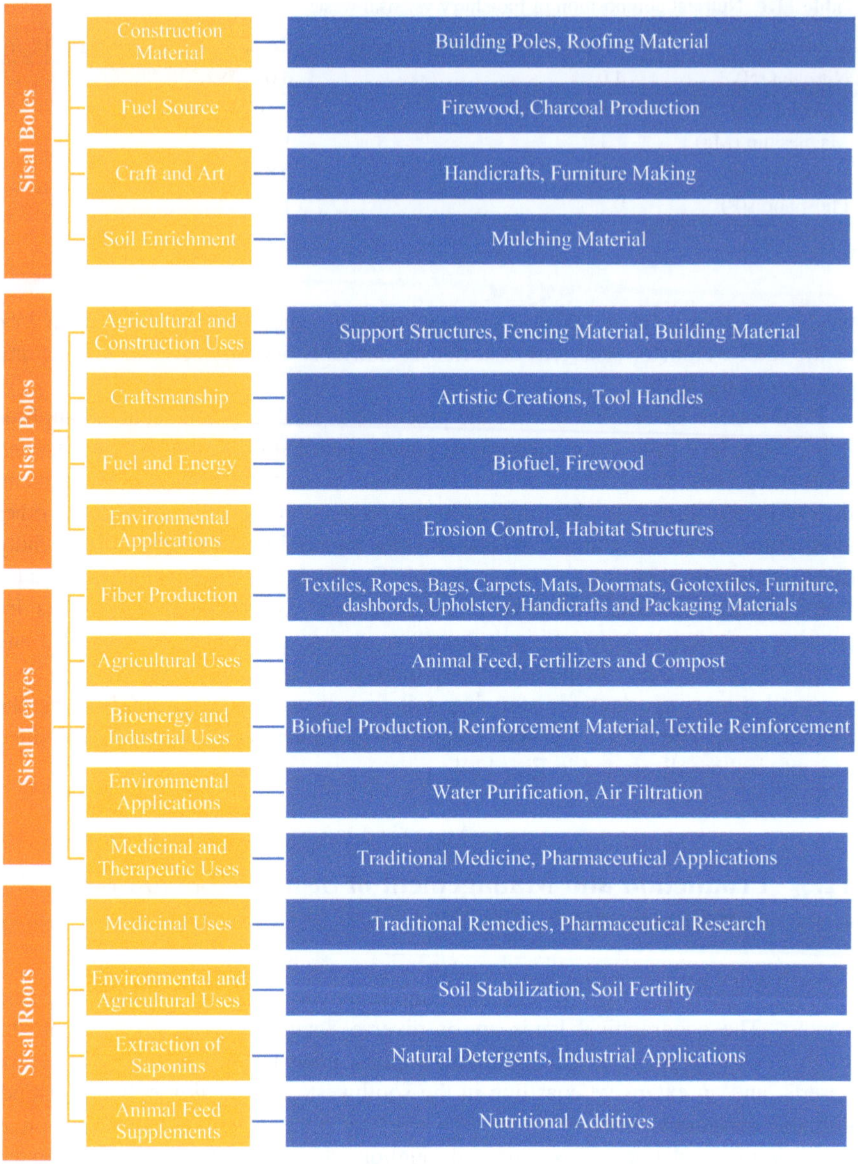

Fig. 11.4 Sketch diagram of sisal plant showing its diversified uses. (Photo by M. Shahzad)

Recent climatic changes, including elevated temperatures, altered precipitation patterns, weather shifts, floods and earthquakes have posed significant challenges. Despite not being a major contributor to emissions causing climate change, Pakistan's vulnerability necessitates urgent adaptation measures. Since 2010, the agricultural sector has endured three massive floods, severely impacting agriculture and the economy. The floods of 2010, 2011, 2014 and 2022 resulted in extensive destruction of crops, livestock, forestry, and fisheries, along with damage to critical infrastructure and livelihoods. Massive losses in production, standing crops, and livestock were recorded, emphasizing the vulnerability of the agricultural sector to climate-induced disasters.

In particular, the floods of 2011 in Baluchistan and Sindh caused substantial losses, affecting rural populations heavily dependent on agriculture. Standing crops, livestock, and infrastructure were ravaged, resulting in a predicted total loss of US $1840.3 million, with the agricultural sector bearing the brunt. The floods in 2014 further aggravated the situation, affecting millions of people, causing deaths, damaging houses, and disrupting agricultural activities. The recovery and resilience-building costs were estimated in the hundreds of millions of dollars, underlining the severe impact on agriculture. In light of these challenges, there is a growing interest in adaptation strategies to mitigate the impact of climate change.

Crop cultivation in different parts of the country has limited crop production due to bad weather. A sizeable area of Pakistan is natural deserts, unused mounds and flat areas that can be a potential area for sisal cultivation. Sisal is advantageous because it can grow on poor land, deserts and even near the sea without replacing food crops. Some areas that are not suitable for other crops can also grow sisal, like areas along roads and dry forests. In the first 2–3 years of sisal farming, intercropping would be encouraged, yielding extra income. This provides income to farmers during this gap period and makes it a useful and viable choice for farming in areas where crop production is not practiced due to factors like salinity, drought and poor drainage. Sisal production in these areas may bring new livelihood opportunities for unemployed households both in plant production and the sisal fiber extraction process.

Along with fiber extraction, sisal could be used in bioenergy production thus reducing the usage of non-renewable energy sources. Therefore, government and non-government sectors must provide technical and financial assistance to small farmers to adopt sisal as a cash crop. Furthermore, biomass left over after fiber extraction amounting to 96% can be used as bio fertilizers. Its production in poor and marginal areas will especially help the poor who rely directly or indirectly on farming. Adopting sisal as a major crop will help in boosting livelihood opportunities and strengthen the country's economy. High-quality fiber production can be exported to meet international demands, bringing revenue and economic prosperity. Pakistan's climatic conditions are diverse, with dry, hot summers and cold winters. The geography encompasses high mountains in the north, plateaus in the center, and plains, deserts, and a lengthy coastline in the southwest. Each region experiences distinct climatic variations, with some areas maintaining moderate temperatures year-round. Despite this diversity, historical data indicates lower precipitation compared to neighboring regions. The country relies heavily on its river water system for agriculture, emphasizing its crucial role in ensuring food security and poverty reduction. However, the agricultural sector faces challenges from temperature increases affecting cropping seasons, heightened irrigation demands, increased evapotranspiration, and growing heat stress on crops.

Pakistan is highly susceptible to climate change, with a vulnerability index surpassing that of many other countries globally. Recent climatic changes, including elevated temperatures, altered precipitation patterns, weather shifts, floods, and earthquakes, have posed significant challenges. Despite not being a major emitter of greenhouse gases, Pakistan's vulnerability necessitates urgent adaptation measures. Since 2010, the agricultural sector has endured three massive floods, severely impacting agriculture and the economy. The floods resulted in extensive destruction of crops, livestock, forestry, fisheries, critical infrastructure, and livelihoods, leading to massive losses in production, standing crops, and livestock.

Crop cultivation in different parts of Pakistan faces limitations due to adverse weather conditions. A significant portion of the country comprises natural deserts, unused mounds, and flat areas, which have the potential for sisal cultivation. Sisal is well-suited for cultivation on poor land, deserts, and coastal areas without displacing food crops. Intercropping during the initial years of sisal farming can provide additional income to farmers, making it a viable option in areas where traditional crop production is constrained by factors such as salinity, drought, and poor drainage.

Adopting sisal as a major crop has the potential to boost livelihood opportunities and strengthen the economy of Pakistan. High-quality fiber production can be exported to meet international demands, generating revenue and contributing to economic prosperity (Fig. 11.5).

11.6 Sisal Breeding

Production of quality fiber is mandatory to meet international fiber quality standards in order to compete in markets. Sisal fiber could be improved by breeding for fiber length, strength and fineness. The use of advanced breeding techniques to modify and enhance already adapted sisal varieties would result in better fiber yield and quality (Abdurakhmonov et al. 2014). Sisal fiber with additional features can be used in delicate fabrics, special strings and expensive items. Sisal breeding can help enhance fiber yield per capita. A better crop yield makes sisal farming more profitable due to its steady supply to industries. Sisal faces issues with bugs and diseases that can impact its quality and yield. This problem can be addressed by breeding varieties resistant to various diseases and by implementing smart farming practices (Bolek 2003). Efficient sisal varieties are resistant to environmental and biological stresses and are necessary to maintain a steady supply chain. This balance is crucial for a sustainable livelihood framework in different industrial zones.

Efforts have been made to enhance the qualities of Agave sisalana through breeding techniques, aiming to develop superior varieties. The initial breeding endeavors began in 1931 to cultivate longer fiber Agaves. Five principal cultivated species were utilized, namely *A. sisalana*, *A. fourcroydes, A. cantala, A. angustifolia*, and *A. amaniensis*. Various breeding methods were employed, with specific hybridization proving more effective for better yield and fiber quality.

Fig. 11.5 (**a**, **b**) Sisal field. (Photo by M. Shahzad)

Notably, cross-pollination between *A. angustifolia* and *A. amaniensis* yielded promising results. Subsequent back-crossing and self-pollination efforts led to the development of a sisal variety known as Hybrid 11648 during the 1940s and 1950s, demonstrating superiority over conventional *A. sisalana* varieties. Hybrid 11648, when planted at low densities, exhibited fewer leaves but with higher fiber content, making harvesting easier (Monja-Mio et al. 2019; Raya et al. 2023).

Currently, the predominant sisal cultivars include Hybrid 11648 and *A. sisalana*. Ongoing research activities focus on refining cultivars beyond the existing ones, emphasizing specific hybridization, and clonal selection, including selfing and backcrossing, as well as variety evaluation and mutation breeding. The Ministry of Agriculture, Livestock Development, and Cooperatives (MALDC, 1990) spearheaded initiatives related to Agave species and hybrid collection, aimed at enhancing the sisal industry. The inter-specific hybridization research program aims to incorporate disease resistance, high yield with superior fiber quality, and adaptability to diverse agroecological zones into sisal hybrids (Cruz-Magalhães et al. 2019).

Mutation breeding aims to induce spine-free leaf margins, increase leaf numbers in Agave sisalana, and enhance disease resistance, particularly against "Korogwe leaf spot," zebra leaf rot, and bole rot diseases commonly encountered in Hybrid 11648. Various Agave species and hybrids are meticulously maintained at the sisal research station to serve as a germplasm repository for future sisal breeding endeavors. This collection is expected to address future challenges in the sisal industry by providing breeding parents for the development of alternative cultivars tailored to specific needs (Souza et al., 2018).

Sisal breeding programs are designed to cope with climate change, diseases and pests. The outcomes of these programs are new types that can better handle changing weather conditions. This involves qualities like surviving dry spells, handling high temperatures and adapting well to different soil types. Sisal varieties that are suitable for the climate help improve farming better and healthier by lowering how much it harms nature. Strong plants can better handle big storms, making them last for a long time in places that grow sisal. As people like natural fibers more and less plastic sisal is a good choice for sustainable development. New advances in genetics, chemistry and biotechnology have paved the way for improving new types of sisal plants. Collaboration with researchers, schools and industry professionals worldwide enhances our collective knowledge of breeding sisal. This collaboration provides more resources for task-strengthening breeding projects with better ideas (Chen et al. 2017).

In recent years, significant advancements have been made in breeding sisal (*A. sisalana* Perrine) to enhance its productivity, fiber quality, and resilience to environmental stressors, paving the way for its sustainable cultivation in diverse agro-climatic conditions. Sisal, a member of the Asparagaceae family, is a crucial natural fiber crop known for its versatility and economic importance. Traditional breeding approaches have been complemented with modern genomic techniques, ushering in a new era of precision breeding aimed at addressing the evolving challenges faced by the sisal industry. Genetic variability is essential for successful breeding programs, enabling breeders to leverage desirable traits such as high fiber yield, improved fiber quality, disease resistance, and drought tolerance.

Advancements in genomic technologies, including next-generation sequencing (NGS) and marker-assisted selection (MAS), have revolutionized sisal breeding by enabling the identification and utilization of molecular markers linked to target traits. These tools facilitate the rapid and accurate selection of superior genotypes, accelerating the breeding process and enhancing efficiency (Yang et al. 2015). Recent studies have identified genomic regions associated with fiber quality traits, providing valuable insights into the genetic basis of fiber production in sisal (Sun et al. 2018). Additionally, transcriptomic analyses have elucidated the molecular mechanisms underlying key physiological processes, offering potential targets for genetic manipulation to improve agronomic performance and stress tolerance in sisal.

Furthermore, advances in tissue culture and genetic transformation techniques have opened new avenues for the introgression of desirable traits into elite sisal cultivars. Tissue culture-based approaches, including somatic embryogenesis and micropropagation, offer efficient methods for clonal propagation and the rapid multiplication of superior genotypes, ensuring the maintenance of desirable traits and genetic purity (Dong et al. 2016). Moreover, genetic transformation techniques enable the targeted introduction of genes conferring traits such as herbicide resistance, pest resistance, and enhanced fiber quality, expanding the genetic resources available for sisal improvement.

The integration of multidisciplinary approaches, including genomics, transcriptomics, metabolomics, and biotechnology, holds promise for accelerating sisal breeding efforts and developing tailored solutions to address specific challenges faced by the sisal industry. Collaborative research initiatives involving academia, industry, and government agencies are essential for fostering innovation, sharing resources, and translating research findings into practical breeding applications. Moreover, initiatives aimed at enhancing germplasm conservation, characterization, and utilization are critical for preserving genetic diversity and ensuring the long-term sustainability of sisal cultivation (Zhang et al. 2021).

11.6.1 Genetic Basis of Sisal

The sisal plant, scientifically known as Agave sisalana Perrine, is a tropical species characterized by herbaceous consistency and a prominent floral scape that can reach a height of 12 feet or more (Silva and Beltrão 1999). Its primary use revolves around extracting fibers from its leaves, which significantly contributes to the global production of hard fibers (Seagri 2014). Understanding the genetic behavior of sisal, population is crucial due to its economic importance, especially in generating income for semi-arid areas. Population genetics is a valuable tool for studying genetic diversity, describing variability, and examining mechanisms for maintaining this diversity (Nei 1987). Molecular classification of samples is essential for selecting individuals that adequately represent genetic diversity and establishing differentiation standards.

The study of genetic variability in sisal populations greatly benefits from molecular genetic techniques. Various molecular markers, such as isoenzymes, microsatellites, RAPD, AFLP, and ISSR, have been used to study the genetic variability of different Agave species. These molecular approaches provide a comprehensive understanding of the genetic makeup of sisal populations, aiding in the selection of individuals that contribute to genetic diversity and providing insights for differentiation standards (Colunga-Garcíamarín et al. 1999; Demey et al. 2004;

Eguiarte et al. 2003; Martínez-Palacios et al. 1999; Navarro-Quezada et al. 2003; Silvamontellano; Teyer-Sánchez et al. 2009; Vargas-Ponce et al. 2009).

The genetic structure of *A. sisalana* is very complex and there is limited knowledge available about it. Different sisal genotypes show significant genotypic differences, making genetic improvement possible. Genetic variations are essential for a successful breeding program aiming to develop new varieties suitable for cultivation in various environmental conditions, with improved fiber yield and quality. Identifying important genes involved in fiber growth, quality and resistance to pests or diseases has equipped scientists to breed new varieties. Advanced breeding strategies and tools like next-generation sequencing help in studying the sisal genome comprehensively. Molecular markers precisely identify traits of interest for targeted trait improvement programs. Furthermore, preserving genetic resources is very important to maintaining genetic variability in sisal. The genetic base will be broadened because of the huge gene pool available for study and improvement in future experiments. With advancements in molecular genetic knowledge and sequencing techniques, plant scientists understand how genes control the fiber traits of sisal, promising improvements in quality and yield for better economic benefits.

11.6.2 Genetic Diversity in Sisal

Research on the genetic diversity of sisal collections has been somewhat limited, but several important studies have provided valuable insights. For example, Navarro-Quezada et al. (2003) conducted an in-depth analysis of genetic differentiation among sisal populations distributed across multiple clusters. Their findings revealed a pattern of genetic isolation by distance, suggesting that geographic separation has contributed to the genetic divergence observed in these populations. This phenomenon indicates that populations further apart geographically are more genetically distinct from each other, likely due to limited gene flow between them.

Similarly, Moreira and Vieira (1999) explored genetic divergence using agronomic traits collected from seven different sisal accessions. Despite the relatively small sample size, they were able to classify these accessions into five distinct groups based on their agronomic characteristics. This study highlighted the significant genetic variability present within these accessions, which could be valuable for breeding purposes. Of particular interest were accessions 11648 and IAC 069, which showed the highest level of genetic divergence among the groups studied. This high divergence makes these accessions prime candidates for inclusion in future breeding programs.

The potential for achieving heterosis, or hybrid vigor, is especially promising with these genotypes. Heterosis occurs when crossbreeding leads to offspring that exhibit superior qualities compared to their parents. Genotype 11648, in particular, stands out due to its genetic background as an interspecific hybrid, as described by Doughty (1938). The unique genetic makeup of this accession could contribute significantly to the development of new sisal varieties with improved traits, making it a valuable resource for breeding programs aimed at enhancing the genetic diversity and agronomic performance of sisalHereditary variation in sisal (Agave sisalana) is a diverse peculiarity enveloping different qualities pivotal for plant improvement. The hereditary information of sisal shows varietal differences in fiber quality, which is a positive sign for improvement. Genetic variability is good for the development of new strains that can cope with environmental stress without compromising yield and fiber quality. The versatility of sisal to various ecological zones is a strong signal of its natural adaptability in diverse conditions. By utilizing and understanding the hereditary variations inside sisal populations, researchers can breed better varieties that can withstand changing environments.

Research on the genetic diversity of sisal (*A. sisalana*) collections has provided valuable insights into the potential for future breeding advancements. The genetic diversity present within sisal populations serves as a foundation for plant improvement efforts. Variability in genetic traits, particularly those related to fiber quality, offers opportunities for the development of improved sisal varieties capable of withstanding various environmental stressors while maintaining high yields and fiber quality. Sisal's adaptability to diverse ecological zones further underscores its potential for genetic enhancement. By leveraging and comprehensively understanding the genetic variations inherent in sisal populations, researchers can strategically breed new varieties that exhibit enhanced resilience and productivity under changing environmental conditions.

Furthermore, advances in breeding techniques offer additional avenues for improving sisal cultivars. Marker-assisted selection (MAS) is a modern breeding approach that utilizes molecular markers to identify and select plants with desired traits more efficiently (Kumar et al. 2015). In the context of sisal breeding, MAS holds promise for accelerating the breeding process by enabling the targeted selection of plants with superior fiber quality, disease resistance, and stress tolerance. By integrating MAS into sisal breeding programs, breeders can enhance the efficiency and precision of trait selection, thereby expediting the development of improved sisal varieties.

Moreover, emerging genomic technologies, such as next-generation sequencing (NGS), offer unprecedented opportunities for genetic analysis and manipulation in sisal breeding. NGS technologies enable the rapid and cost-effective sequencing of entire genomes, facilitating the identification of genetic markers associated with desirable traits and the characterization of genetic diversity within sisal populations (Huang et al. 2020). By leveraging NGS data, researchers can gain deeper insights into the genetic basis of important traits in sisal and develop more targeted breeding strategies aimed at enhancing desired traits.

Queiroz et al. (2015) conducted a comprehensive study to assess the genetic diversity of 140 sisal populations using Inter-Simple Sequence Repeat (ISSR) molecular markers. This research was carried out in six counties within the state of Bahia, Brazil, providing a broad geographic representation of sisal diversity in the region. The researchers employed a total of 18 ISSR primers to amplify DNA samples from the sisal populations. Of these, nine primers were found to be particularly effective, leading to the detection of 143 polymorphic loci, which are variations in DNA sequences that can be used to measure genetic diversity.

The results of this study revealed that the average percentage of polymorphism within the populations was 64%, indicating a relatively high level of genetic variability. Polymorphism percentage is a key indicator of the diversity present within a population, as it reflects the proportion of loci that are variable among the individuals studied.

In addition to polymorphism, the study also calculated the average heterozygosity (He) and the Shannon-Wiener index (I), which are both measures of genetic diversity. The average heterozygosity was found to be 0.180, while the Shannon-Wiener index was 0.279. Heterozygosity measures the likelihood that two alleles randomly selected from a population will be different, and a higher value generally indicates greater genetic diversity. The Shannon-Wiener index, on the other hand, is a more comprehensive measure that takes into account both the richness and evenness of alleles within the population.

The study further explored the distribution of genetic variation within and between the sisal populations. It was revealed that approximately 73% of the molecular variance could be attributed to differences within the populations themselves, while the remaining variance was due to differences between populations. This suggests that most of the genetic diversity in these sisal populations is found within individual populations rather than between different populations.

The researchers also conducted a clustering analysis that grouped the populations into two distinct clusters based on their geographical distribution. This indicates that there is a spatial structure to the genetic variability observed in the sisal populations, with populations from different regions exhibiting distinct genetic characteristics. This spatial structuring is often the result of limited gene flow between populations that are geographically separated, leading to the development of unique genetic profiles in different areas.

The calculated GST value, a measure of genetic differentiation between populations, was 0.235. This value indicates a moderate level of population differentiation, meaning that while there is some genetic differentiation between the populations, it is not so pronounced as to suggest strong isolation. Instead, the moderate GST value suggests that there is sufficient genetic variation within the sisal populations to prevent significant divergence among them. This level of differentiation may be beneficial for the long-term sustainability of sisal as it indicates that there is a reservoir of genetic diversity that can be drawn upon in breeding programs to develop new varieties with desirable traits. These findings underscore the potential for utilizing the genetic structure of sisal for the establishment of gene banks aimed at conserving genetic diversity both in situ (in the natural environment) and ex situ (in controlled environments). This conservation strategy could ensure the preservation of individuals with desirable commercial traits. Furthermore, the presence of genetic variability among sisal genotypes highlights the importance of employing ISSR molecular markers for analyzing the genetic divergence between different genotypes, thereby aiding improvement efforts in sisal breeding programs. Overall, this study provides valuable insights into the genetic diversity and structure of sisal populations, offering important implications for sisal conservation and breeding strategies.

In conclusion, the genetic diversity inherent in sisal populations provides a valuable resource for breeding programs aimed at developing improved varieties with enhanced resilience and productivity. Through the integration of advanced breeding techniques such as MAS and NGS, researchers can accelerate the breeding process and develop sisal cultivars tailored to meet the demands of diverse environmental conditions and end-user requirements. Continued research and investment in sisal breeding are essential to harnessing the full potential of this versatile fiber crop and ensuring its continued contribution to global agricultural sustainability and economic prosperity.

11.6.3 Elite Mother Plant Selection Through Clonal Methods

When initiating a cloning program sourced from natural populations, the challenging aspect lies in identifying the most suitable individuals to serve as mother plants. Selection is based on comparative variability or morphology keeping in mind factors such as the age of the plant, physiological and biochemical traits as well as environmental conditions (e.g., rainfall and temperature) and ecological conditions (e.g., soil and topography). In the case of certain agaves, selection often needs to occur in remote and sometimes inaccessible locations where the population size may be minimal. In some instances, only a single individual of certain mezcal agaves has been discovered after multiple expeditions. Selection criteria should focus on the performance of mature individuals exhibiting the desired traits, as young plants haven't had the opportunity to showcase their characteristics. However, using mature plants as starting material for tissue culture is impractical due to their tough and fibrous nature. Their tissues are prone to phenolization, and susceptibility to microorganism infestation, and, most importantly, their meristematic tissues exhibit low responsiveness. Consequently, the preferred approach involves utilizing young rhizomes attached to robust mature mother plants. Regardless of the scenario, it is essential to use the best, largest, and healthiest plants to initiate lines for subsequent second and third rounds of selection.

Alternatively, when feasible, starting from seeds cultivated in plots designed for selection purposes is an option. Vitroplants, derived from various clonal lines produced in the first generation, must be planted under consistent conditions for a genuine first round of selection. While not immediate solutions, these strategies prove to be efficient and shorter in duration compared to crossbreeding programs (Arce-Montoya et al. 2006, 2007).

Starting with plantation materials proves to be a more expeditious and straightforward approach compared to initiating from wild populations. The advantage lies in the abundance of plants thriving under relatively uniform conditions. However, in such instances, the selection process necessitates the direct involvement of agronomists capable of discerning the optimal plants within their land parcels. It is crucial to bear in mind that genetic enhancement follows a recurrent selection process aimed at attaining specific traits.

11.6.4 Micropropagation

Micropropagation involves initiating and cultivating new plants from cells or tissues of selected mother plants cultured in vitro. Under favorable conditions, thousands of genetically identical or very similar clonal plants can be produced within a short period. Another advantage of micropropagation is its ability to eliminate pathogenic microorganisms and propagate healthy lines, which can enhance crop productivity. The use of plant tissue culture (PTC) techniques, specifically the culture of meristematic tissues to generate new plants, is a crucial aspect of micropropagation. In some cases, somatic embryogenesis is also utilized (Siahsar et al. 2011).

Numerous reports describe PTC/micropropagation protocols for various Agave species, showing different methods for culturing agaves. While some protocols are more efficient, others may be lengthier or more costly. Many reports fail to address potential issues such as microbial contamination or morphological abnormalities, making them less suitable for large-scale micropropagation. These protocols encompass a range of Agave species (e.g., *A. tequilana, A. fourcroydes, A. angustifolia*, and *A. sisalana*) using diverse explant tissues (e.g., bulbils, rhizomes, seeds, and roots) and various culture media (e.g., MS, SH, LS), as well as different concentrations of auxins, cytokinins, sugars, and vitamins., Tailored for specific species, these protocols can serve as a basis for large-scale applications or in vitro germplasm conservation.

11.6.4.1 Large-Scale Propagation of Agaves

Producing hundreds or even thousands of plants is achievable in a well-equipped laboratory using standard techniques. However, scaling up to tens and hundreds of thousands of plants requires meticulous procedures, skilled personnel, specialized laboratory facilities to prevent contamination, and a well-designed production system to address bottlenecks and minimize costs and losses. Plant quality and health are critical considerations when generating thousands of plants for plantations.

Large-scale propagation is primarily a commercial endeavor undertaken by industrial companies in specialized labs, known as bio-factories. These facilities need well-defined protocols for each species, including Standardized Operational Procedures (SOP) that detail manipulations, sterilization and disinfection processes, culture media preparation, plant quality aspects, and cost-related factors like reagents, working hours, cabin usage, autoclave utilization, medium volume per production lot, and plant production rates on a weekly, monthly, and annual basis.

11.6.5 Rescuing Endangered Species and Safeguarding Germplasm

Numerous agave species thrive exclusively in the mountains and arid regions of Mesoamerica and North America, with Mexico being the primary host to 75% of all recognized Agave species (Gentry 2004; Martinez-Palacios et al. 1999). Due to their extended life cycles, sluggish growth, limited reproductive rates, and significant economic value as raw materials for both industrial products and ornamental purposes, these species are particularly vulnerable to genetic erosion resulting from overexploitation. Numerous studies indicate a concerning decline in species and diversity within Mexico over the past two to three decades.

To counteract this trend, it is imperative to develop alternative methods for commercially exploiting valuable species without jeopardizing natural resources. Implementation of conservation programs is crucial, encompassing in situ efforts through national laws that prohibit the exploitation of endangered species, as well as ex situ approaches involving botanic gardens and nurseries and in vitro techniques such as slow growth or cryopreservation methods.

11.7 Recent Advancements in Genomic Research

Despite the societal and economic significance of Agave species and their hybrids, there is a notable shortage of studies, particularly at the molecular level. This shortage is primarily attributed to several factors, including the absence of fundamental genetic knowledge, the substantial size of genomes ranging from 2940 to 4704 Mbp, and extended life cycles lasting between 5 and 12 years. Despite these challenges, the Agave genus stands out for its substantial genetic diversity, with ploidy levels varying from 2n to 8n ($n = 30$) even within species. Frequent hybridization in wild cultivars further contributes to the proliferation of variation in ploidy levels and the intricacy of hybrid genomes (Davis and Long 2015; Simpson et al. 2011).

While DNA-seq data for Agave tequilana exists in the NCBI database, the absence of a published reference genome sequence remains a notable gap in current research. Additionally, a mere six studies have explored agave transcriptomics, with none delving into the molecular intricacies of cell wall biosynthesis (Abraham et al. 2016; Cervantes-Pérez et al. 2018; Gross et al. 2013; Huang et al. 2018, 2019; McKain et al. 2012; Sarwar et al. 2019; Simpson et al. 2011). The dearth of molecular investigations in this realm underscores the untapped potential and the need for a comprehensive exploration of the genetic and molecular facets of Agave species and hybrids.

In the recent past, there has been an expanded focus on advancing genomic research to enhance the understanding of sisal fiber development and quality. High-throughput sequencing innovations, for example, next-generation sequencing (NGS), have likely assumed an urgent part in disentangling the hereditary premise of sisal fiber qualities (Cheng et al. 2023). These innovations empower specialists to investigate the whole sisal genome all the more productively, distinguishing key qualities, administrative components, and atomic pathways related to fiber advancement. Headways in practical genomics, including transcriptomics and proteomics, could offer experiences into the powerful cycles of basic fiber arrangement in sisal. By concentrating on quality articulation examples and protein collaborations during various phases of fiber improvement, specialists can acquire a more far-reaching comprehension of the sub-atomic components included. Marker-assisted selection (MAS) and genomic selection have likely been used to speed up sisal breeding programs (Prasad et al. 2017). The identification of molecular markers linked to fiber traits allows for more exact and sped-up breeding processes, selecting plants with desirable fiber characteristics at earlier stages. Collaborations between genomics researchers, plant scientists and agronomists have likely strengthened to integration of genomic findings into practical applications.

11.7.1 Identification of Key Areas for Fiber Quality and Disease Resistance

The identification of key factors related to fiber quality and disease resistance in sisal (*Agave sisalana*) involves extensive genomic and molecular studies. Several studies have been conducted to investigate the genetic basis of disease resistance in sisal and highlight areas for improvement. Monja-Mio et al. (2019) explored breeding strategies to enhance *Agave* spp., including sisal, to improve disease resistance. Their findings emphasized the importance of incorporating disease-resistance traits into sisal cultivars to mitigate the impact of pathogens. Additionally, a study by Trejo-Torres et al. (2018) examined the genetic diversity and population structure of sisal, providing insights into potential sources of disease resistance within the species. Understanding the genetic basis of resistance can inform breeding efforts aimed at developing resistant varieties. Furthermore, genetic mapping studies, have identified genomic regions associated with disease resistance in sisal. These findings offer valuable markers for marker-assisted selection (MAS) in breeding programs, enabling the targeted introgression of resistance alleles. In conclusion, disease resistance in sisal is a complex trait influenced by genetic factors. Continued research efforts to elucidate the genetic basis of resistance and incorporate resistance traits into breeding programs hold promise for developing disease-resistant sisal cultivars and ensuring sustainable production in the face of disease pressures.

Scientists plan to pinpoint explicit genes that assume vital parts in deciding the attributes of sisal fiber and its resistance to diseases. Entire genome sequencing of sisal gives a fundamental comprehension of its hereditary makeup. Bioinformatics procedures are utilized to elucidate the genome structure and its potential to modify or improve varieties. Transcriptomics includes contemplating the transcriptome, the total arrangement of RNA involved in the genome and RNA sequencing (RNA-Seq) recognizes qualities that are effectively translated during fiber improvement and in light of microbe challenges (Chen et al. 2014).QTL mapping involves identifying regions of the genome associated with specific traits, such as fiber quality or disease resistance. Molecular markers linked to these QTLs can be used for marker-assisted selection in breeding programs. The discovery of CRISPR-Cas9 technology has revolutionized molecular biology and may be used to improve fiber quality traits of sisal (Zhang and Showalter 2020). By consolidating these methodologies, specialists can recognize key genes associated with sisal fiber improvement and safeguard against diseases. This information is important for breeding programs intending to foster sisal improvement.

11.8 Traditional Breeding Methods

Conventional breeding techniques employed for sisal (*Agave sisalana*) include selection and controlled fertilization for desired traits. Sisal is primarily propagated through seeds, with breeders aiming to improve fiber quality, yield, disease resistance, and other agronomic important attributes (Monja-Mio et al. 2019). Here are the key stages and strategies involved in conventional sisal breeding:

1. Selection of Parental Plants: Breeders identify plants with favorable traits, such as superior fiber quality, disease resistance and adaptability to diverse environmental conditions. Parental plants are chosen based on their genetic makeup and observable characteristics.
2. Fertilization: Sisal plants are cross-pollinated, with the emasculation of buds (removing male reproductive organs) often performed to prevent self-fertilization. Pollen from selected plants with desired traits is then applied to the emasculated flower.
3. Seed formation: After successful fertilization, seeds begin to develop in flowers. Seeds are then harvested and genetic variability is checked.
4. Nursery establishment: Sisal seeds are germinated to establish nurseries for raising seedlings. Seedlings are cared for until they are ready to be transplanted into the field.
5. Field evaluation Assessment: Seedlings are transferred to the field and their phenotypic traits such as fiber quality, yield, and disease resistance, are monitored.

11.9 Marker-Assisted Selection (MAS)

Marker-Assisted Selection (MAS) is a modern breeding technique that utilizes molecular markers to aid in selecting plants with specific traits. In sisal (*A. sisalana*) breeding, MAS can enhance the efficiency and precision of the process. Researchers have identified DNA markers associated with traits of interest in sisal, such as fiber quality, disease resistance, and stress tolerance. Single Nucleotide Polymorphisms (SNPs) and Simple Sequence Repeats (SSRs) are commonly used markers. These markers are then mapped to specific locations on the sisal genome, crucial for understanding the genetic basis of the targeted traits. Association studies establish correlations between specific markers and desired traits, leading to marker-trait associations. DNA-based tests are developed to quickly identify specific markers in individual plants, serving as proxies for trait expression without time-consuming phenotypic evaluations. Breeders use MAS information to selectively breed plants with desired traits, choosing parents with favorable marker profiles for the next generation. MAS enables early identification of desirable traits, streamlining the breeding process (Hospital 2003).

MAS accelerates the breeding cycle by allowing breeders to make early selections based on genetic information instead of waiting for phenotypic expression, which can be time-consuming. MAS is especially useful for traits that are hard to measure directly or controlled by multiple genes, allowing for more precise targeting of complex traits. MAS is often used alongside traditional breeding methods, providing a molecular-level understanding of genetic makeup (Bilyeu et al. 2006; Bernardo and Yu 2007). This enhances the precision of trait selection, reducing the risk of selecting plants with undesirable traits and informing decisions on advancing in the breeding program (Flint-Garcia et al. 2003). By simplifying the breeding process and reducing the time for trait evaluation, MAS contributes to the development of improved sisal varieties.

11.10 The Role of Breeding in Sustainable Agriculture

Plant breeding is a crucial sector that has significantly contributed to improving sustainable crop production in the past and will likely play an even larger role in the future. Breeding is a fundamental driver in agriculture, aimed at modifying and enhancing crops for increased yield and quality to benefit humans. The ongoing enhancement improvement of breeding programs is essential to maintaining varietal purity, yield and quality. As of 2022, Earth's population has exceeded eight billion and continues to grow rapidly. The main challenge facing humanity is the provision of food, not only in developed nations but also in underdeveloped regions. Agriculture, a major source of pollution, poses a threat to the environment if not managed properly (Adegbeye et al. 2020). Therefore, there is an urgent need to boost the productivity of the agricultural sector to meet the increasing demand for food and fiber while minimizing ecological and environmental impacts. In this context, sustainability involves the efficient long-term utilization of resources and a reduction in production-related environmental and landscape effects. To tackle this challenge, a focus on improving cultivation practices and plant materials is crucial. Plant breeding plays a key role in providing farmers with new varieties and hybrids. While genotypes have inherent yield potentials, various limiting factors affect their productivity. Instead of relying solely on intensive technologies, active pest and disease management, and heavy use of inorganic fertilizers to maximize yields, a shift towards sustainability involves enhancing abiotic and biotic stress tolerance in genotypes. The objective is not just to maximize yield but to develop plant materials that can sustain food and fiber production in organic and integrated farm management systems (Le Campion et al. 2019; Mojid and Mainuddin 2021; Ullah et al. 2019).

11.11 Conclusions and Prospects

The production of high-quality sisal fiber is essential to meet international standards and compete effectively in the global market. Improving sisal fiber can be achieved through advanced breeding techniques that focus on enhancing fiber length, strength and fineness. By using smart agricultural practices and developing biotic and abiotic stress-tolerant varieties, yield and quality can be ensured for sustainable supplies to industries. The genetic basis of sisal, especially Agave sisalana, is complex and requires comprehensive molecular studies to unravel its complexity. Molecular markers such as isozymes, microsatellites, RAPD, AFLP, and ISSR provide valuable insights into the genetic variation of sisal populations. Understanding genetic diversity within sisal populations is essential for successful breeding programs that produce varieties suitable for diverse environmental conditions, ensuring better fiber yields and quality.

Micropropagation is a technique that initiates plant growth from cells or tissues grown in the laboratory, providing a fast and efficient way to generate genetically identical clones. Large-scale production of clones can be achieved in well-equipped laboratories using precise protocols. Conservation of endangered sisal species and protection of germplasm is essential to counter genetic erosion resulting from overexploitation. In situ and ex situ conservation efforts, including laboratory techniques such as slow growth and cryopreservation, play a critical role in preserving valuable species without endangering natural resources.

Recent advances in genomic research have been limited for Agave species, mainly due to challenges such as genome size and extended life cycles. High-throughput sequencing technologies, such as next-generation sequencing, offer promising ways to unravel the genetic basis of sisal fiber traits. Collaboration between genomics researchers, botanists, and agronomists is crucial for translating lab research to the field. Identifying key areas of fiber quality and disease resistance involves comprehensive genetic and molecular studies. Whole genome sequencing, transcriptomes and molecular markers help identify specific genes responsible for sisal fiber traits and disease resistance. Marker-assisted selection (MAS) represents a contemporary breeding methodology that harnesses molecular markers to facilitate the selection of plants harboring desired traits. Within the sphere of sisal (Agave sisalana) breeding, MAS emerges as a potent tool capable of augmenting the efficacy and precision of the breeding endeavor. Researchers have successfully pinpointed DNA markers correlating with pivotal traits in sisal, including fiber quality, disease resistance, and stress tolerance. Among the repertoire of molecular markers, Single

Nucleotide Polymorphisms (SNPs) and Simple Sequence Repeats (SSRs) feature prominently. These discerned markers are subsequently mapped to precise loci on the sisal genome, furnishing indispensable insights into the genetic underpinnings of the targeted traits. Marker-assisted selection (MAS) accelerates the breeding program by using DNA markers associated with desired traits, enabling early and accurate trait selection.

Traditional breeding methods, which include controlled fertilization and selection of parent plants based on desirable traits, remain an integral part of sisal breeding. Marker-assisted selection (MAS) complements traditional methods, providing a molecular-level understanding of genetic makeup for more efficient trait selection. The role of breeding in sustainable agriculture is crucial, as it addresses the challenges of producing food and fiber while minimizing environmental impacts. Plant breeding contributes to the development of varieties capable of resisting biotic and abiotic stresses, and promoting sustainable agricultural practices.

Appendix 11.1 Research Institutes and Online Resources Available for Sisal (*Agave sisalana* Perrine) Breeding

Institution	Area of specialization	Research activities	Website
International Center for Agricultural Research in the Dry Areas (ICARDA	Agricultural research in dry areas	Crop improvement, including genetic enhancement, to address challenges in dryland agriculture. ICARDA focuses on developing resilient crop varieties and sustainable agricultural practices suitable for arid and semi-arid regions	ICARDA
International Plant Genetic Resources Institute (IPGRI)	Conservation and use of plant genetic resources for food and agriculture	Conservation, characterization, and utilization of plant genetic resources, including sisal germplasm collections. IPGRI facilitates access to genetic materials and information for plant breeding and crop improvement	Bioversity International
National Agricultural Research Centers (NARCs)	Agricultural research and development tailored to local conditions	Conduct research on crop improvement, including sisal breeding, adapted to specific agro-climatic zones. NARCs collaborate with national and international partners to address agricultural challenges and enhance productivity and resilience	Varies by country. Examples include: Pakistan Agricul tural Research Council (PARC) Kenya Agricul tural and Live stock Research Organization (KALRO)

(continued)

Institution	Area of specialization	Research activities	Website
International Sisal Institute (ISI)	Promotion and advancement of sisal production and utilization	Facilitates research and collaboration among stakeholders in the sisal industry. Provides resources and information on sisal cultivation, processing, and utilization to support sustainable production practices and market development	International Sisal Institute
Consultative Group on International Agricultural Research (CGIAR)	Global agricultural research for development	Conducts research on various aspects of agriculture, including crop improvement, natural resource management, and climate change adaptation. CGIAR collaborates with partners worldwide to develop innovative solutions for food security, poverty alleviation, and environmental sustainability	CGIAR
Online Genetic Resources Information Network (OGRIN)	Online platform for genetic resources information	Provides access to genetic data, germplasm catalogs, and breeding resources for various crops, including sisal. OGRIN facilitates the exchange of genetic resources information and supports research and conservation efforts	OGRIN
International Union for Conservation of Nature (IUCN)	Conservation of biodiversity and natural resources.	Provides information and resources on plant conservation, including endangered and economically important plant species like sisal. IUCN supports conservation initiatives, research, and policy advocacy to promote the sustainable use and management of natural resources.	IUCN

Appendix 11.2 Genetic Resources for Sisal (*Agave sisalana* Perrine) Breeding

Species	Important traits	Cultivation location
Agave angustifolia	High fiber content, drought tolerance	Native to Mexico, Central America, and the Caribbean
Agave fourcroydes	High-quality fiber, resistance to pests and diseases	Primarily grown in Mexico, particularly in the Yucatan Peninsula
Agave tequilana	Sugar content for tequila production, drought tolerance	Native to Mexico, primarily cultivated in the Jalisco region
Agave sisalana (Sisal)	High fiber yield, tolerance to arid conditions	Widely cultivated in tropical and subtropical regions worldwide, including Brazil, Tanzania, Kenya, and Madagascar
Agave americana	Ornamental value, drought tolerance	Native to Mexico, but cultivated as an ornamental plant in various regions globally

References

Abdurakhmonov IY, Buriev ZT, Saha S (2014) Phytochrome RNAi enhances major fiber quality and agronomic traits of the cotton *Gossypium hirsutum* L. Nat Commun 5(1):1–10

Abraham PE, Yin H, Borland AM et al (2016) Transcript, protein and metabolite temporal dynamics in the CAM plant Agave. Nat Plants 2(12):16178

Acosta-Gallegos JA, Rosales-Serna R, Gallegos-Pérez JO, Hernández-Delgado S (2007) Current status of sisal (Agave sisalana) production and research in Mexico. Ind Crop Prod 26(3): 292–298

Adegbeye M, Reddy PRK, Obaisi A et al (2020) Sustainable agriculture options for production, greenhouse gasses and pollution alleviation, and nutrient recycling in emerging and transitional nations-An overview. J Clean Prod 242:118319

Ahmad J, Majdi A, Deifalla AF et al (2022) Concrete reinforced with sisal fibers (SSF): overview of mechanical and physical properties. Crystals 12(7):952

Bernardo R, Yu J (2007) Prospects for genomewide selection for quantitative traits in maize. Crop Sci 47:1082–1090

Bessou C, Ferchaud F, Gabrielle B, Mary B (2011) Biofuels, greenhouse gases and climate change. In: Eric L, Marjolaine H, Mireille N, Philippe D (eds) Sustainable agriculture, vol 2. Springer, Dordrecht, pp 365–468

Bolek Y (2003) Status of genome mapping and use in cotton improvement. KSÜ Fen ve Mühendislik Dergisi 6(2):72–79

Caujapé-Castells J, Tye A, Crawford DJ, Santos-Guerra A et al (2010) Conservation of oceanic island floras: present and future global challenges. Perspect Plant Ecol Evol Syst 12:107–129

Cervantes-Pérez SA, Espinal-Centeno A, Oropeza-Aburto A et al (2018) Transcriptional profiling of the CAM plant *Agave salmiana* reveals conservation of a genetic program for regeneration. Dev Biol 442(1):28–39

Chen Y, Liu YS, Zeng JG (2014) Progresses on plant genome sequencing profile. Life Sci Res 18: 66–74

Chen X, Lu X, Shu N et al (2017) Targeted plant breeding - current and future views on mutagenesis in cotton (*Gossypium hirsutum* L.) using the CRISPR/Cas9 system. Sci Rep 7:44304

Cheng C, Fei Z, Xiao P (2023) Methods to improve the accuracy of next-generation sequencing. Front Bioeng Biotechnol 11:982111

Crawley MJ (2007) The R book. Wiley, Manchester, UK

Cruz-Magalhães V, Andrade JP, Figueiredo YF et al (2019) Sisal bole rot: an important but neglected disease. In: Plant diseases – current threats and management trends. IntechOpen

Davis SC, Long SP (2015) Sisal/Agave. In: Cruz VMV, Dierig DA (eds) Industrial crops. Handbook of plant breeding, vol 9. Springer, New York. https://doi.org/10.1007/978-1-4939-1447-0_15

Demey JR, Gamez E, Molina S, Infante D (2004) Comparative study of the discriminating capacity of AFLP and ISTR markers for genetic analysis of *Agave fourcroydes*. Plant Mol Biol Rep 22: 29–35

Eguiarte LEF, Dugua XA, Munive MR, Díaz CT, Montellano AS, Vázquez AV (2003) Diversidad Genetica en dos Especies Mezcaleras. Proyecto Conabio. México, DF 38–33

Flint-Garcia SA, Darrah LL, McMullen MD, Hibbard BE (2003) Phenotypic versus marker-assisted selection for stalk strength and second-generation European corn borer resistance in maize. Theor Appl Genet 107:1331–1336

Gentry HS (2004) *Agave sisalana*. In: Gentry HS (ed) Agaves of continental North America. The University of Arizona Press, Tucson, pp 628–631

Gross SM, Martin JA, Simpson J, Abraham-Juarez MJ, Wang Z, Visel A (2013) De novo transcriptome assembly of drought tolerant CAM plants, *Agave deserti* and *Agave tequilana*. BMC Genomics 14(1):1–14

Hospital F (2003) Marker-assisted breeding. In: Newbury HJ (ed) Plant molecular breeding. Blackwell Publishing and CRC Press, Oxford and Boca Raton, pp 30–59

Huang X, Wang B, Xi J, Zhang Y et al (2018) Transcriptome comparison reveals distinct selection patterns in domesticated and wild Agave species, the important CAM plants. Int J Genom:1–12

Huang X, Xiao M, Xi J, He C, Zheng J, Chen et al (2019) De novo transcriptome assembly of agave H11648 by illumina sequencing and identification of cellulose synthase genes in Agave species. Genes 10(2):103

Lang PLM (1985) Variability of Mexican isolates of *Colletotrichum gloeosporioides* Penz. from agave hosts. Phytopathology 75(12):1457

Le Campion A, Oury FX, Heumez E, Rolland B (2019) Conventional versus organic farming systems: dissecting comparisons to improve cereal organic breeding strategies. Org Agri 10:63–74

Lock GW (1962) Sisal. Wiley, New York

Mande SM (1980) An economic analysis of crop diversification in sisal estates in Morogoro and Kilosa Districts. M.Sc. thesis, University of Dar es Salaam

Martínez-Palacios A, Eguiarte LE, Furnier YG (1999) Genetic diversity of the endangered endemic *Agave victoriae*-reginae (Agavaceae) in the Chihuahuan desert. Am J Bot 86:1093–1098

Mbogoni JDJ, Ndondi PM, Kips PHA (1989) Soil conditions and agricultural production potential for hybrid sisal and selected fruit crops of Bomb wera estate.

McKain MR, Wickett N, Zhang Y, Ayyampalayam S, McCombie WR, Chase MW et al (2012) Phylogenomic analysis of transcriptome data elucidates co-occurrence of a paleopolyploid event and the origin of bimodal karyotypes in Agavoideae (Asparagaceae). Am J Bot 99(2):397–406

Mohankumar D, Amarnath V, Bhuvaneswari V et al (2021) Extraction of plant based natural fibers–a mini-review. In: IOP conference series: materials science and engineering. 1145(1-012023). IOP Publishing.

Mojid MA, Mainuddin M (2021) Water-saving agricultural technologies: regional hydrology outcomes and knowledge gaps in the Eastern gangetic plains—a review. Water 13:636

Monja-Mio KM, Herrera-Alamillo MA, Sánchez-Teyer LF, Robert ML (2019) Breeding strategies to improve production of Agave (*Agave* spp.). In: Advances in plant breeding strategies: industrial and food crops, vol 6, pp 319–362. Springer Cham

Navarro-Quezada A, González-Chauvet R, Molina-Freaner F, Eguiarte LE (2003) Genetic differentiation in the *Agave deserti* (Agavaceae) complex of the Sonoran desert. Hered Cardiff 90: 220–227

Nei M (1987) Molecular evolutionary genetics. Columbia University Press, New York

Neumann H, Metz R (1971) *Erwinia carotovora* subsp. carotovora, the causal agent of sisal soft rot. East Afr Agric Forestry J 37(2):196–201

Nobel PS (1994) Remarkable agaves and cacti. Oxford University Press, Inc., New York

Nweke IA, Echendu TN, Iwuagwu MA (2013) The use of fungicides for controlling leaf spot disease of sisal (Agave sisalana) in Nigeria. I J Agric Forestry 3(5):265–269

Partheepan K, Musthafa MM, Bhavan T (2024) Mitigating climate change and its impact on food security by integrating ecosystem services to agroecosystems in Sri Lanka. YSF Thematic Publication, Colmbo, Srilanka p 377

Peregrine WTH (1969) Investigations on chemical control of zebra disease in Agave hybrid no 11648. Ann Appl Biol 63:45–51

Prasad TS, Mohanty AK, Kumar M et al (2017) Integrating transcriptomic and proteomic data for accurate assembly and annotation of genomes. Genome Res 27(1):133–144

Ramankutty N, Evan ATC, Monfreda, Foley JA (2008) Farming the planet: 1. Geographic distribution of global agricultural lands in the year 2000. Springer Cham

Raya FT, Quintanilha-Peixoto G, de Oliveira AB, Marone MP, dos Reis WJV, JGO M et al (2023) Molecular epidemiology of sisal bole rot disease suggests a potential phytosanitary crisis in Brazilian production areas. Front Chem Eng 5:1174689

Sabea H (2001) Reviving the dead: entangled histories in the privatization of the Tanzanian sisal industry. Africa 71:286–313

Sarwar MB, Ahmad Z, Rashid B, Hassan S, Gregersen PL, Leyva MD et al (2019) De novo assembly of Agave sisalana transcriptome in response to drought stress provides insight into the tolerance mechanisms. Sci Rep 9(1):396

Seagri (2009) Cultura do Sisal. Disponível em: http://www.bahia.ba.gov.br/seagri. Acesso outubro de 2009

Siahsar B, Rahimi M, Tavassoli A (2011) Plant tissue culture: current status and opportunities. J Agric Environ Sci 10(7):94–101

Simpson J, Martínez Hernández A, Jazmín Abraham Juárez M et al (2011) Genomic resources and transcriptome mining in Agave tequilana. GCB Bioenergys 3(1):25–36

Singh AN, Singh AP (2018) Review on various diseases of sisal and its management. Int J Chem Stud 6(5):1122–1125

Ullah H, Santiago-Arenas R, Ferdous Z et al (2019) Improving water use efficiency, nitrogen use efficiency, and radiation use efficiency in field crops under drought stress: a review. Adv Agron 156:109–157

Vargas-Ponce O, Zizumbo-Villarreal D, Martínez-Castillo J et al (2009) Diversity and structure of landraces of Agave grown for spirits under traditional agriculture: a comparison with wild populations of *Agave angustifolia* (agavaceae) and commercial plantations of Agave tequilana. Am J Bot Saint Louis 96:448–457

Vorosmarty CJ, McIntyre PB, Gessner MO et al (2010) Global threats to human water security and river biodiversity. Nature 467:555–561

Weber E (2003) Invasive plant species of the world: a reference guide to environmental weeds. CAB International, Wallingford

Weink JF (1969). Long Agave fibres. In: Ferwerda FP, Wit F (eds) Outlines of perennial crop breeding in the tropics. H. Veenman and Zonen, N.V. Wageningen. The Netherlands, pp 1–21

Zhang Y, Showalter AM (2020) CRISPR/Cas9 genome editing technology: a valuable tool for understanding plant cell wall biosynthesis and function. Front Plant Sci 11:589517

Chapter 12
Recent Advances in Sugar Palm (*Arenga pinnata* Wumb. Merr.) Biotechnology and Breeding

Asmah Awal, Illy Kamaaina Ramle, and Nazatul Asikin Muda

Abstract Sugar palm (*Arenga pinnata* Merr) is a new promising crop as a source of sugar and biofuel worldwide. *Arenga pinnata* is a multipurpose palm with high economic value; it is a good candidate for crop integration or agroforestry because almost all plant parts (stem, leaf, fruit, palm blossom, fiber, root) are useful for human life. The species is well distributed throughout Southeast Asia. They are either partly domesticated or grow wild in the jungle. Generally, the production is at a small scale and run by rural farmers for income generation. Exploitation of the usefulness of *Arenga pinnata* has been an essential income-generating venture among rural households in many Southeast Asian countries. Although considered a minor forest species, the *Arenga pinnata* tree provides some important products that have great export potential and are of economic importance to Malaysia. This palm tree is a multipurpose species, where its sap, black fibers, trunk, leaves and many others can be used to produce various types of food and beverages, timber commodities, fibers, biopolymers and biocomposites. It is currently used as a source of renewable energy or bio-fuel, in the form of bio-ethanol, which is produced through the fermentation of the sugar palm sap. This chapter will focus on the botanical distribution and description of sugar palm as the multipurpose usage of the plant can be widely used in human living. Despite the challenges of sugar palm breeding conservation, plant biotechnology techniques have been used in regenerating the sugar palm tree and producing the synthetic seed. The plant tissue culture method was widely used and can solve the problems and challenges of sugar palm breeding.

Keywords Sugar palm (*Arenga pinnata* Wumb. Merr.) · Biotechnology · Breeding · Multipurpose · Plant tissue culture

---

A. Awal (✉) · I. K. Ramle · N. A. Muda
Faculty of Plantation and Agrotechnology, UiTM Cawangan Melaka, Merlimau, Melaka, Malaysia
e-mail: asmah138@uitm.edu.my; nazatulasikin@uitm.edu.my

K. F. M. Salem et al. (eds.), *Breeding and Biotechnology of Leaf, Fruit, and Seed Fiber Crops*, Advances in Plant Breeding Strategies 11,
https://doi.org/10.1007/978-3-032-00407-9_12

12.1 Introduction

12.1.1 A Brief History of Sugar Palm

One of the great versatility of sugar palm (*Arenga pinnata* Wumb. Merr.) is, that it is one of the oldest cultivated plants. The history of sugar palm cultivation and the commercialization of its fibers in Malaysia is recorded to date back to the Malacca Sultanate era in 1416. In 1786, a British settlement was established in Penang and an industry was formed based on sugar palm seeds which were brought together with cloves and nutmegs from Indonesia (Lawrence Long Ltd 2008). Realizing the potential of sugar palm trees at the time, the British East India Company started to commercially cultivate them in the 1800s to make rope from sugar palm fibers due to their high durability characteristics (Othman and Harun 1992). In the nineteenth century, sugar palm cultivation was widely cultivated in Malacca, and seawater-resistant fibers were the primary product (Mogea et al. 1991). Around the same century on the island of Singapore, a large-scale sugar palm tree plantation was planned by the Dutch East India Company. However, after the loss of Napoleon, the Dutch handed over Singapore to the English who then handed it back to the Dutch Colonies. In East Kalimantan, the Japanese have figured out how to plant sugar palm trees, but the project was later abandoned after their loss in the Second World War (Smits 2011). To date, particularly sophisticated cultivation methods have been developed in South East Asia. Long ago, aside from being cultivated for their strong fibers, sugar palm trees were also cultivated for their sugary sap as a source of sugar and alcoholic drinks. The tapping of sugar palms for sugary juice is a very old tradition. As a source of sugar, it was probably cultivated long before sugar cane. In traditional belief among the Minahassa (the indigenous tribe of North Sulawesi), consumption of toddy (alcoholic drink) dates back to the first human being, Toar and Luimiuut, on Earth (Mogea et al. 1991).

Arenga pinnata also known as the areng palm is a versatile and economically important palm species native to Southeast Asia. This palm has a rich history and plays a significant role in the cultural, economic and ecological landscapes of the regions where it grows. The sugar palm tree is native to tropical regions including Indonesia, Malaysia, Thailand, the Philippines, and Papua New Guinea. Due to its resilience to different ecological conditions, it can be found in a variety of environments, such as lowland woods, upland areas, and mountainous regions. The sugar palm tree is a tall, thin palm tree that can grow up to 20 meters in height. Its trunk is smooth and greyish-brown, with prominent rings formed by old leaf scars. The enormous, feather-like leaves of the palm can reach a length of 6 meters, giving them a remarkable contrast to the tropical surroundings. This species is further distinguished by clusters of small, spherical fruits and branched inflorescences carrying clusters of flowers.

The sugar palm has been used for generations by indigenous cultures and has profound cultural roots. Each of the palm sections has its own essential functions. The major resource used to make palm sugar, palm wine (toddy), and vinegar is the sap that is collected from the inflorescences. The sap of sugar palm is also called toddy. These goods have cultural significance and are essential to regional cuisines and customs. Other than that, sugar palm leaves and fibers are used in traditional crafts such as mats, baskets, thatch roofs, and other woven goods. In addition to boosting the regional economy, these handicrafts protect cultural heritage.

In the rural communities where sugar palm is grown, the crop is vital to the local economy. The manufacturing of palm sugar, palm wine, and other goods obtained from palm trees sustains a large number of families and makes a substantial contribution to regional trade and economy. Besides, the habitats for sugar palms sustain a variety of ecosystems and aid in the preservation of biodiversity. Tropical landscapes require sugar palms because of their deep root system, which both prevents soil erosion and improves soil health.

The sugar palm has attracted attention lately due to its contemporary uses that go beyond customary ones. Research is being done on its sap as a possible source of bioethanol and other bio-based products that could support sustainable energy projects. In order to maintain the sugar palm's long-term viability, efforts are also being made to preserve and sustainably manage its genetic variety. To sum up, the sugar palm is an exceptional species of palm with strong cultural ties and major ecological and economic significance in Southeast Asia. Its adaptability, historical applications, and contemporary uses demonstrate its worth as an important resource for both local communities and society at large. In order to maintain the sugar palm's legacy and preserve lives in tropical areas, conservation and sustainable use of the plant are imperative.

12.2 Origin and Classification

Sugar Palm (*Arenga pinnata* Wumb. Merr.) is a terminal flowering plant in the class Magnoliopsida, order Arecales, and family Arecaceae. Many edible plants, including economically important palms such as coconut (*Cocos nucifera*) and oil palm (*Elaeis guineensis*), belong to the Arecaceae family. Sugar Palm (*Arenga pinnata* Wumb. Merr.) is the most economically significant species in its genus. *Arenga* is one of about 20 genera in the Arecaceae family, which includes important economic palms such as coconut (*Cocos nucifera*) and oil palm (*Elaeis guineensis*) (Fig. 12.1 and Table 12.1).

12.3 Geographical Distribution

Sugar palm is native to the regions of South East Asia countries including Indonesia, Malaysia, Myanmar, Philippines and Thailand and is widely distributed throughout India, Sri Lanka, Southern China, Southeast Asia, New Guinea and Guam (Mogea et al. 1991; Duke 2000; Mogea 2003). *Arenga pinnata*, also referred to as the sugar

Fig. 12.1 Sugar Palm Trees in Plantation. Location: Kampung Peruang, Benta, Pahang, Malaysia. (Photted by Dr. Nazatul Asiki)

Table 12.1 Taxonomy classification of sugar palm

Kingdom: Plantae (plants)
Subkingdom: Viridiplantae
Infrakingdom: Streptophyta
Superdivision: Embryophyta
Division: Tracheophyta
Subdivision: Spermatophytina
Class: Magnoliopsida
Superorder: Lilianae (monocotyledons)
Order: Arecales
Family: Arecaceae
Genus: Angera Labill. (arenga palm)
Species: *Arenga pinnata* (Wurmb) Merr. (sugar palm)

Source: Lim (2012), Witt and Luke (2017)

palm or areng palm, is a tropical Asian plant native to Southeast Asia and portions of the Indian subcontinent. It is a member of the Arecaceae family. This medium-sized palm tree has a robust, solitary trunk that grows to a height of 20–30 meters. Its big, pinnate leaves have many leaflets that are distributed regularly along the rachis. According to scientific classification, the species belongs to the order Arecales (palms), class Liliopsida (monocotyledons), phylum Tracheophyta (vascular plants), kingdom Plantae, genus Arenga, and species *A. pinnata*. Its natural distribution extends throughout Southeast Asia, including Myanmar, Thailand, Laos, Vietnam, Cambodia, Malaysia, Indonesia, and the Philippines, from India, Sri Lanka, and Bangladesh (Haagen 2015). Typically found in lowland rainforests, *A. pinnata* can be found near rivers or in swampy places, demonstrating its adaptability to humid tropical climates (Chen et al. 2021). *A. pinnata* is an important ecological component of its ecosystem, offering food and shelter to a variety of wildlife species. Additionally, across its distribution, this palm is extremely important to local populations economically. The species is prized for its many applications; (a) sap utilization: the sap, known as toddy, is tapped and fermented to produce sugar, vinegar, and alcoholic beverages. (b) edible parts: the young inflorescence and inner part of the stem are edible and commonly used in local cuisine. (c) fiber production: the leaves yield fibers used for making ropes, mats and thatch. (d) wood utilization: the trunk wood is employed in construction and furniture making (Dai et al. 2019).

The cultivation of *A. pinnata* goes beyond its native environment. The palm is grown in tropical gardens and parks, especially in areas with a humid climate and well-drained soil. It is a sought-after addition to gardens and landscapes because of its usefulness and visual charm. The adaptable palm species *A. pinnata* has cultural, ecological, and commercial value throughout tropical Asia. Its use in food, drink, crafts, and building materials emphasizes its worth and significance both locally and globally (Li et al. 2018).

Sugar palm is abundantly distributed throughout Indonesia. It inhabits almost the entire archipelago of Indonesia. Data from the Ministry of Agriculture in 2013 showed that sugar palm was extensively dispersed over an area of over 70,000 hectares in 14 Indonesian regions. These provinces included Banten, South Kalimantan, West Sumatra, North Sumatra, Bengkulu, North Maluku, North Sulawesi, North Sulawesi, Southeast Sulawesi, Central Java, West Java, and Papua.

In Peninsular Malaysia, sugar palm is found in many states, especially in the mountainous and lowland rainforest regions. The species is often found in regions with humid tropical climates, such as Pahang, Johor, Perak, Selangor, and Kelantan (Kamarudin and Jamaludin 2012). Sugar palms are commonly grown in these areas for their sap, which is used to make palm sugar and other customary goods. Sugar palm is also widely distributed in both wild and cultivated environments in the Borneo states of Sarawak and Sabah. It can be found in these regions' foothills, riverine areas, and rainforests (Suhaimi et al. 2016). Sabah and Sarawak's sugar palm plantations provide palm sugar and other byproducts that are used in regional cuisines and customs (Fig. 12.2).

Fig. 12.2 Distribution of sugar palm around the Asia-Pacific. (Source: Haagen (2015))

12.3.1 Current Cultivation Practices of Sugar Palm

A series of specialized techniques are employed in the cultivation of sugar palms (*Arenga pinnata*) in order to maximize growth, productivity, and sustainability; these techniques are especially important for extracting sap that is used to make traditional products like palm sugar. Practices for growing sugar palms differ depending on regional customs, environmental factors, and intended results; they combine traditional knowledge with contemporary farming methods.

12.3.1.1 Choosing Planting Sites

Sugar palmsgrow best in tropical and subtropical areas, where they usually like well-drained soils that retain moisture well and receive enough sunlight. Lowland locations with enough rainfall to sustain good development are carefully selected as cultivation sites (Suhaimi et al. 2016). The species is frequently included in agro-forestry systems, where it enhances the growth of other crops like oil palm or rubber.

12.3.1.2 Methods of Propagation

Suckers or seeds can be used to grow sugar palms. Mature fruit seeds are gathered, germinated in nurseries, and then released into the field. In order to preserve genetic integrity, suckers—shoots that emerge from the base of adult palms—can also be utilized for vegetative propagation (Sreekala et al. 2000).

12.3.1.3 Planting and Spacing

For maximum growth and ease of management, appropriate spacing is essential when creating sugar palm plantations. In order to ensure proper canopy development and avoid overcrowding, which can lower productivity, seedlings or suckers are usually planted at a recommended spacing of 5–7 meters apart (Ishak et al. 2019).

12.3.1.4 Management of Mature Palm

To maintain productivity, mature sugar palm plants need to receive regular maintenance. This involves routine trimming to get rid of outdated fronds and make it easier to reach the inflorescence so sap may be collected. To guarantee the best nutrient availability for growth and sap production, palm trees can also be treated with organic fertilizers regularly (Kamarudin and Jamaludin 2012).

12.3.1.5 Techniques for Collecting Sap

The main goal of sugar palm cultivation is to gather sap from the inflorescence, which is then processed to make palm sugar and other goods. In order to reach the flower stalks without endangering the tree, competent climbers must ascend the palms using either conventional techniques or specialized equipment (Suhaimi et al. 2016).

12.3.1.6 Sap Processing

To avoid fermentation, the harvested sap is treated right away after collection. It is reduced to a boil in order to concentrate the sugar and create palm sugar, also known as jaggery, a traditional sweetener that is extensively utilized in regional cuisines. Palm wine (toddy) and vinegar are two other items made from sugar palm sap (Kamarudin and Jamaludin 2012).

12.4 Botanical Description

12.4.1 Stem/Trunk

The sugar palm tree stands solitary, tall, stout and unbranched (Fig. 12.3). They can grow up to 8.0–15.0 m in height with a stem width of 0.5 m. The sugar palm trunk is marked with rather distinct bases of old leaf sheaths' remains and long dark brown fibers topped by a dense crown of leaves.

12.4.2 Leaf/Leaflets

The leaves grow in an ascending manner, pinnate and can reach up to 9.1 m long, and 3.1 m wide. Every single individual palm holds around 100–160 leaflets on each

Fig. 12.3 Solitary sugar palm tree in plantation. (Photted by Dr. Nazatul Asikin)

side. The leaflets of sugar palm are erect, induplicate, pinnate, and dark green with a pale, white waxy coating on their lower surface (Fig. 12.4). They are also tip lobed, variously toothed and usually with jags, 'fishtail' tips. A singular leaflet can measure up to 1.0–1.5 m long, 6.0–8.0 cm wide. The maturity of the sugar palm tree is indicated by the simultaneous appearances of two short leaves at the top of the stem.

12.4.3 *Inflorescence*

Sugar palm trees flower all year round in the lowland tropics. It takes about 7–10 years for the trees to flower at lower elevations of <700 m. At higher elevations of >1000 m, flowering will take more than 15 years. The flowers of sugar palms are naturally insects pollinated. The palm's first flowering appears at the age of 10–12 years but sometimes occurs as early as 5–6 years. The first inflorescence arises from a node near the top meristem, while subsequent inflorescences arise from the lower-located nodes in descending order from the first flowering node

Fig. 12.4 Sugar palm's leaf or leaflets. (Photted by Dr. Asmah Awal)

downwards. Each node bears only one inflorescence. Female and male flowers occur in different inflorescences (Duke 2000). Usually, 6–8 inflorescences at the top of the stem are female, the others at the lower nodes are male (Mogea et al. 1991).

12.4.3.1 Female Inflorescence

Female flowers (Fig. 12.5) are scentless with 3 unequal green imbricated orbicular sepals, measuring one-third length of petals, persistent; petals coriaceous, 1.5–2.5 cm long, light-green, ovate, or triangular, valvate, persistent with sepals as cupule at base of the fruit; staminodes absent, or if present, sometimes producing nectar (Duke 2000).

Fig. 12.5 Sugar palm's female inflorescence. (Photted by Dr. Nazatul Asikin)

12.4.3.2 Male Inflorescence

Male inflorescences of sugar palm are scentless, usually solitary, sometimes paired and rarely in threes (Fig. 12.6). The male flowers consist of 3 unequal green imbricated, persistent sepals that measure one-fourth length of the petals. The apex is broadly acute and thin-margined. Each flower also consists of 3–4 navicular and valvate petals which measure 2.5 cm long, red-brown or red-purple on the outside, and yellow on the inside. Stamens are yellow, numerous, with elongated apiculate anthers, borne on short filaments and with no rudimentary ovary (Duke 2000).

12.4.3.3 Bisexual Inflorescence

Sometimes, sugar palm trees bear bisexual flowers in which both stamens and pistils are present in a single flower or hermaphrodite. Stamens are usually abortive (Duke 2000).

12.4.4 Fruits

Sugar palm fruits are produced in colossal bunchs the size of a golf ball each (Chan 2000) (Fig. 12.7). The ovoid fruits of sugar palm are 5.0–7.0 cm in diameter, green when immature, turning yellow and gradually black with a fleshy white mesocarp that is irritating to the skin. Each fruit contains two to three grey-brown seeds with a length of 2.5–3.5 cm and 2.0–2.5 cm wide (Janick and Paull 2008). The outer husk of the fruit contains stinging crystals of calcium oxalate (raphides) which cause an intense burning sensation when it touches the skin. This part of the fruit is also poisonous (Florido and De Mesa 2003).

Fig. 12.6 Sugar palm's male inflorescence. (Photted by Dr. Asmah Awal)

Fig. 12.7 Fresh fruits of sugar palm. (Photted by Dr. Nazatul Asikin)

12.5 Ecology and Climate Requirements

12.5.1 Ecology

Sugar palm is a rainforest palm in its natural state, and thrives on rich, moist soil in partial shade, but is adaptable to drier areas and full sun such as in tropical rain forests and dry forests. In its native range, sugar palm is found growing in some forested areas but never far from settled areas as in ravines, along streams, on slopes and in areas under semi-cultivation. It is also occasionally found in virgin forests since its fruits are scattered by fruit bats, wild pigs, civet cats and probably small mammals (Florido and De Mesa 2003). Sugar palm trees also usually grow adjacent to human settlements where antropic propagation plays a major role (Orwa et al. 2009). Generally, sugar palm trees can grow and survive in most soil conditions and improve the ecosystems of forest floors (Duke 2000).

12.5.2 Climate Requirements

Sugar palm trees grow best in hot, wet and humid climates from sea level to 1400.0 m elevation. It is also tolerant of temperatures below freezing but is damaged at −2 °C. Specimens have been known to recover from the temperature as low as −4 °C (Janick and Paull 2008).

12.6 Horticulture

12.6.1 Propagation

Sugar palm is naturally propagated from mature black seeds which germinate in 3–12 months (Fig. 12.8).

12.6.2 Diseases, Pest and Weeds

Sugar palm is rarely troubled by pests or disease. The stem-rotting fungus *Ganoderma pseudoferrum* has been reported to affect sugar palm (Fig. 12.9a). In Southeast Asia, the rhinoceros beetle (*Orcytes rhinoceros*) (Fig. 12.9b) has been known to feed on the foliage (Henderson 2009; Lim 2012).

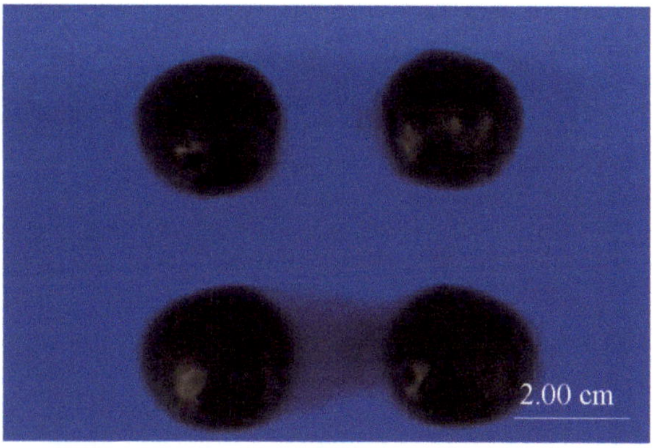

Fig. 12.8 Matured Seeds of Sugar Palm. (Photted by Dr. Nazatul Asikin)

Fig. 12.9 (**a**) *Ganoderma pseudoferrum* fruiting body on sugar palm trunk. (**b**) Adult rhinoceros beetles (*Orcytes rhinoceros*) observed on leaf surfaces. (Photted by Dr. Nazatul Asikin)

12.7 Economic Potential

Sugar palm trees have been around for making a variety of by-products for hundreds of years. In Malaysia, the production of sugar blocks locally known as *gula enau* or *kabung* is the most popular usage of sugar palm as the main material of the sugar block is from the palm sap. Tapping palm sap activity is still going on among the locals as income can be generated by making the traditional sugar blocks and it can be processed to become crystalized and brown sugar. The production of brown sugar can function as an alternative to commercialized sugarcane granular sugar. Other than that, bioethanol can also be produced by the process of fermentation of palm sap. Bioethanol can form varieties of other products such as chemical products, solvents, pharmaceuticals, medicines, and beverages and it can also be used for the production of biofuel. The next important part after palm sap is its fruit. It can be processed for making pickles, juices, desserts, canned foods and also cooked for making traditional sugary syrup.

Nevertheless, the most important part after its palm sugar and its fruit is its fiber. Its black fiber or locally known as *ijuk* fiber is used for making ropes, brooms, brushes, paint brushes, septic tank base filters, roofing, fishing tools and handicrafts. There are three main advantages of sugar palm fiber; it has high tensile strength and can withstand a longer life to degrade; is not affected by heat and moisture compared to coir fiber, it is durable and has good resistance to seawater. Traditionally, sugar palm fibers were used for making ropes for ship cordages and were proven to have good properties in seawater. Not like other natural fibers (kenaf, banana pseudo-stem, bagasse, pineapple leaf, oil palm fibers), it does not require secondary processes such as water retting or mechanical decorticating process to yield the fibers. It is due to the fibers that were originally wrapped along the base of the palm leaf ribs.

The fiber from sugar palms (*Arenga pinnata*) is used in a wide variety of conventional and contemporary applications that take advantage of this natural material's special qualities. For generations, Southeast Asian indigenous cultures have used sugar palm fiber, which is mostly obtained from the palm's leaves and other parts, for a variety of applications.

The fiber from sugar palms has long been used for roofing and thatching. The long, strong fibers are used to thatch roofs, offering efficient insulation and weather protection, by being weaved into mats or panels. Sugar palm thatched roofs are renowned for their sturdiness and resilience to pests and dampness (Sreekala et al. 2000).

Fibre from sugar palms is also preferred for basketry and handicrafts. Expert weavers create beautiful patterns and motifs that are well-liked in regional markets and artisanal communities by weaving strong and flexible fibers into baskets, mats, caps, and bags (Gnanaharan and Sundari 2014). Owing to its inherent strength and durability, sugar palm fiber is utilized not only in handicrafts but also in the production of ropes and cordage. Ropes for binding, tying, and lashing different objects are made by twisting and braiding the fibers (Kumar and Satyanarayana 2007).

The prevention of erosion and stabilization of soil are two further significant uses for sugar palm fiber. In reforestation and restoration operations, for instance, mats or blankets composed of sugar palm fibers are placed on sloping terrain to stop soil erosion and encourage vegetation development (Nurhayati et al. 2012).

Sugar palm fiber has become more well-known as a sustainable packaging material in recent years. Sugar palm fiber is used to make biodegradable packaging items that are an environmentally beneficial substitute for traditional plastics, helping to lessen the impact of plastic waste on the environment (Khalina et al. 2019).

Bio-composite materials for the manufacturing, automotive, and construction sectors are also made from sugar palm fiber. Sugar palm fiber and resin composite panels are used in lightweight structural applications because of their advantageous mechanical qualities (Sreekala et al. 2000).

Moreover, the textile and clothing sectors may find a use for sugar palm fiber. Textiles for apparel, accessories, and home furnishings are made by blending sugar palm fibers with other natural or synthetic fibers (Khalina et al. 2019).

In conclusion, sugar palm fiber's many uses demonstrate its adaptability, toughness, and environmentally beneficial qualities. Sugar palm fiber has a wide range of applications, from handicrafts and traditional thatching to cutting-edge erosion control, packaging, and bio-composites. These applications help to provide sustainable solutions that support regional economies and communities.

Sugar palm trees have always been appreciated by local communities thanks to their multipurpose uses and ecological benefits (Dalibard 1999). Generally, sugar palm trees are cultivated for their delectable sugary sap and sweetened meat (endosperms) for food purposes, while the black fibers are used mostly for industrial purposes. Nowadays, the sugar palm tree has risen to a certain level in engineering. Different uses of sugar palm parts are summarized in Table 12.2.

12.7.1 Economic Importance and Health Benefits of Sugar Palm

The sugar palm (*Arenga pinnata*) is widely cultivated and used in tropical countries due to its numerous health advantages and economic significance. This variety of palm is prized for its many uses in both traditional medicine and many industries.

12.7.2 Economic Importance

Making palm sugar from the sap of the sugar palm is one of the main economic advantages of the plant. The sap, called toddy, is extracted from the inflorescences and then boiled and crystallized to create palm sugar. A common traditional sweetener in baking, cooking, and drinks is palm sugar. Its demand is fueled by its distinct flavor profile and nutritional attributes, which draw customers from both domestic and foreign markets (Nair 2014). Additionally, sugar palm sap is fermented to create palm wine, often known as toddy, which is a well-liked alcoholic beverage throughout Southeast Asia. The manufacturing of palm wine boosts regional economies and small enterprises, giving rural communities access to a source of income (Nair 2014).

Sugar palm leaves and fibers are used in crafts such as thatch manufacture, mat-making, and basket weaving. According to Rushdan et al. (2019), these handicrafts support rural economies and give communities that grow sugar palms new sources of income. Traditional house materials, tools, and furniture are all made from the durable wood of sugar palm trunks. The production of sugar palm supports local economies and livelihoods, which promotes the timber sector (Nair 2014). Activities related to the production and processing of sugar palms give rural populations job opportunities. Through the selling of goods made from palm oil, the industry promotes trade and commerce, aiding in the economic development of tropical areas (Rushdan et al. 2019).

Table 12.2 Uses of different parts of sugar palm trees

Parts	Uses
Sap	Fresh beverage; sugar/palm sugar
	Source of biofuel (bioethanol); palm wine/arrack/ alcohol;
	vinegar.
	Sweetener and flavor stimulant for cakes, pastries, pancakes,
	toast, sandwiches, cereals, beverages, fruit juices, dessert
	toppings.
Fibers	Fibers of leaf bases as rigging, brushes, covers of underwater
	telegraph cables, road construction, rope, mattress stuffing,
	sieves, brooms, roof material;
	Fibers of trunk pith as ship cordage;
	The hairs found on the base of the leaf sheaths as tinder for
	igniting fire;
	Fibers of roots are used for making capes in Sulawesi and hats in
	Indo-China.
Root	Folk medicine as tea to cure kidney stones; Insect repellent; Erosion control for natural conservation (root system penetrates up 3.0 meters deep and 10.0 meters wide in soil helps to stabilize soil in place); Improve soil macro conditions, soil porosity, and trapping rainwater
Stem	Stem core as sago starch and fibers; Pith of young stems eaten in soups or fried and pickled preserve; the starch is also applied as biopolymer or plasticizer; The very hard outer part of the trunk is used for barrels, flooring and furniture, posts for pepper vines, boards, tool handles and musical instruments like drums.
Leaf	Young leas as salads, cigar paper; Mature leaf as material for thatching houses, wrappers, rough brooms and woven baskets; Old woody leaf bases and long leaves can be used for fuel
Petioles	Material for basketry, marquetry work on tables, stands, screens, boxes and light pieces for furniture; walking sticks, butts of blow-pipes
Fruits	Edible endosperms of unripe fruits as ko lang-Kaling (sweetmeat); Pulps as natural cosmetic bases proclaimed to have an anti-aging effect
Male inflorescence	Tapped to collect sap
Flowers	Source of nectar for honey production
Seed	Black seed as toy beetle
Pith of leaf's rachis	Drinking cup
Terminal bud or 'cabbage'	Salads, raw or cooked
Saw-milled waste	Edible mushroom media

Source: Redhead (1989), Sastra et al. (2006), Ishak et al. (2013), Djarwanto and Suprapti (2016), Mogea et al. (1991), Yanti and Ali (2017), Sahari et al. (2013), Johnson (1996), Soeseno (2000), Smits (2011)

12.7.3 Health Benefits

Sugar palm offers various health benefits attributed to its nutritional composition and tradition. The sugar palm is rich in vitamins (especially B vitamins), minerals (including potassium, iron, and zinc), and antioxidants including sugar palm sap and products such as palm sugar. Because of its higher nutritious content and lower glycemic index, palm sugar is regarded as a better substitute for refined sugar (Kale and Rajasree 2016). The sugar palm tree has several uses in traditional medicine. While several palm preparations are used to cure diseases like coughs, fevers, and digestive difficulties, the sap is said to offer cooling and purifying properties (Kale and Rajasree 2016). Because of its slow-release carbs, palm sugar offers a sustained energy source that can be used to sustain energy levels without quickly raising blood sugar levels (Nair 2014). Sugar palm sap and products include antioxidants that lower the incidence of chronic illnesses including diabetes and heart disease and fight oxidative stress (Kale and Rajasree 2016). By encouraging regular bowel movements and maintaining the balance of the gut flora, the natural fibers included in sugar palm products enhance digestive health (Kale and Rajasree 2016).

A significant crop with many uses in the economy and health, sugar palm has several advantages. Its use and cultivation support the preservation of cultural history, the availability of wholesome goods in domestic and international markets, and sustainable livelihoods. The economic and health-promoting potential of sugar palm is further enhanced by ongoing research and innovation in product development and farming.

12.8 Challenges

12.8.1 Breeding

Despite the known importance of sugar palms economically in tropical and subtropical regions, very few farmers have actively propagated or domesticated the species (Martini et al. 2012). Its development and vegetative propagation potential are neglected as there have been quite limitations reported on the species. The limitations of the sugar palm include its slow fruit development, slow seed germination, non-uniform seedlings, and long seed dormancy period due to impermeable seed coat (Asikin and Pusitaningtyas 2000). The rapid propagation of and its propagation from the mature specimen (seed) is impossible as the fruit takes 2 years to mature and it takes approximately 3–12 months for the seeds to germinate (Janick and Paull 2008). Germinated seeds usually produce non-homogenous, degradable, low viability and poor field growth seedlings (Putih et al. 2003; Soeseno 2000;). In open storage, the seed's behavior is short-lived (Devi et al. 2014).

Nursery germination of sugar palm seedlings (Fig. 12.10) is poor due to specific temperature and storage requirements to maintain seed viability. The population of

Fig. 12.10 Sugar Palm Seedling Raised in Nursery. (Photted by Dr. Nazatul Asikin)

sugar palms is maintained traditionally by the harvesting of natural seedlings nurtured as nursery seedlings. Continuous intensive harvest of sugar palm fruits for domestic use eventually reduces the source of seed for natural regeneration (Martini et al. 2012). Rapid propagation of sugar palms from offshoots is impossible due to the limited resources as the offshoot production is limited to a certain period of their life cycle (Zaid and De Wet 1999). While some degree of regional selection has been applied to sugar palms, mostly towards reducing the juvenile period, no rigorous breeding or cultivar evaluation program has ever been conducted (Janick and Paull 2008). Without systematic enrichment planting and conservation efforts of the seedlings by humans, it is feared that the number of sugar palm trees will continuously decline and might lead to extinction.

12.8.2 Current Agricultural Challenges

The cultivation of sugar palm (*Arenga pinnata*) presents several agricultural challenges that impact its productivity, sustainability, and economic viability. Addressing these challenges requires innovative solutions and improved farming practices to ensure the long-term success of sugar palm cultivation. Here is an in-depth exploration of the current agricultural challenges faced by sugar palm growers:

Sugar palm has a notably long gestation period, typically taking 5–7 years to reach maturity and begin producing sap (Kong 2015). This slow growth rate delays the onset of economic returns for farmers and necessitates significant upfront investment in plantation management without immediate profitability.

Another significant challenge is the alternate bearing pattern exhibited by sugar palm, which it produces fruit in irregular cycles rather than annually (Nair 2014). This inconsistency in flowering and fruiting leads to unpredictable harvests and complicates production planning for farmers, impacting overall productivity.

Moreover, sugar palm populations often exhibit limited genetic diversity due to predominant clonal propagation through suckers (Kong 2015). This genetic uniformity limits the availability of diverse genetic resources for breeding programs aimed at improving important traits like yield, disease resistance, and environmental adaptability.

Pest and disease management remains a concern despite the general resilience of sugar palms. Pathogens such as Phytophthora palmivora, causing stem rot, and pests like bagworms can still pose significant threats to palm health and yield (Nair 2014).

Sustainable cultivation practices face environmental challenges related to land use, water management, and agrochemical usage. The expansion of sugar palm plantations may encroach on natural ecosystems, leading to biodiversity loss and environmental degradation if not managed responsibly (Rushdan et al. 2019).

Market constraints also affect sugar palm farmers, with issues such as price fluctuations and limited market access impacting economic viability (Nair 2014). Farmers may struggle to secure fair prices for their products and diversify market opportunities beyond traditional uses of sugar palm.

Furthermore, climate variability and change pose risks to sugar palm cultivation, altering rainfall patterns, temperature regimes, and pest dynamics. Extreme weather events like droughts or floods can negatively impact palm health and production, necessitating adaptive strategies to mitigate risks (Kong 2015).

In regions where research and extension services are limited, the dissemination of best practices, innovative technologies, and pest management strategies to sugar palm farmers may be hindered (Nair 2014). This lack of support can impede the adoption of sustainable and efficient farming methods.

In conclusion, addressing thc agricultural challenges faced by sugar palm growers requires collaborative efforts and targeted interventions. Investments in research and development, improved agronomic practices, capacity-building initiatives, and policy support are essential to enhance the resilience, sustainability, and productivity of sugar palm farming systems, ensuring sustainable livelihoods for communities dependent on this valuable palm species.

12.8.2.1 Challenges of Sugar Palm Conventional Breeding

The biological traits and reproductive habits of the sugar palm (*Arenga pinnata*) present considerable problems to conventional breeding. These difficulties may affect the efficacy and efficiency of conventional breeding initiatives meant to enhance desired sugar palm features (Barcelos et al. 2019). First of all, the sugar palm requires several years to mature and yield viable seeds due to its lengthy reproductive cycle. The breeding procedure is considerably prolonged by this prolonged generation time, which makes it time-consuming to assess and choose improved genotypes (Barcelos et al. 2019).

Second, breeding programs are complicated by the sugar palm's alternating bearing habit. Because of this behavior, the trees' fruiting patterns are erratic as they bear fruit in cycles as opposed to yearly. As a result, obtaining seed material for assessment and selection regularly can be difficult (Barcelos et al. 2019).

The low genetic diversity found in natural populations of sugar palms presents another difficulty. The species has limited genetic variety since it primarily reproduces clonally by suckers. This restriction makes it more difficult to access a variety of genetic resources for breeding, which may impede the creation of novel cultivars with enhanced features (Barcelos et al. 2019).

In addition, controlling pollination and avoiding unintentional cross-pollination during breeding studies are challenging due to the intricate pollination systems of sugar palms, which depend on wind and insects. Breeding outcomes' accuracy and dependability are impacted by this complexity (Barcelos et al. 2019).

Moreover, breeding programs face logistical difficulties due to the huge size and area needs of sugar palm trees. Large populations of sugar palm trees might be difficult to manage and assess in field trials due to resource constraints (Barcelos et al. 2019).

In sugar palm populations, there is also restricted access to regulated breeding material, such as particular male and female parent trees with desirable features. Breeding efforts become more challenging when trying to find and keep suitable parents for a long time (Barcelos et al. 2019).

Furthermore, sugar palms frequently display biannual flowering, which happens every 2 years. The continuity of breeding programs is impacted by this erratic flowering pattern, which throws off breeding dates and produces an unpredictable seed supply (Barcelos et al. 2019).

We need creative methods and tactics to get beyond these obstacles. The production of improved sugar palm varieties with increased features can be accelerated by combining traditional breeding procedures with cutting-edge molecular approaches including genetic transformation and marker-assisted selection (Barcelos et al. 2019).

Future breeding initiatives aiming at guaranteeing the sustainability and productivity of sugar palm farming depend on the preservation and use of a variety of genetic resources through germplasm collections and ex situ conservation activities (Barcelos et al. 2019).

The commercial breeding of sugar palm plants is hindered by various constraints. Dormancy of the seeds, seeds that do not store well, and seedlings with low field development are some of the challenges associated with building a successful conventional sugar palm breeding program. Before being cultivated, wild seedlings are usually collected and raised in nurseries to maintain a constant population of sugar palms. The amount of fruit collected for human consumption has significantly decreased the source of seeds because natural regeneration has a significant impact on the population of sugar palms. Another cause for concern is the high manpower expenses involved in processing items made from sugar palm.

12.8.3 Labour

The establishment of sugar palm farms on a commercial scale may have also been prevented by the concern of high labor. The harvesting of sugar palm products particularly fruits and sap requires farmers and tappers to work hard enough and they normally cannot afford more energy to further process them. There is also concern regarding the processing of sugar palm fruits. Softening the fruit flesh of sugar palm by soaking before de-pulping requires caution. Decomposing fruits develop a fluid causing intense itching, burning sensation and irritation when it comes into contact with the skin (Schmidt 2007).

12.9 Mutation Breeding

In sugar palm (*Arenga pinnata*) breeding, genetic mutations are purposefully introduced to introduce novel and desirable features, to enhance the productivity, disease resistance, and adaptability of this economically significant palm species. Through targeted genetic alterations, this method—which is frequently employed in crop development programs—has been utilized to improve a number of sugar palm features. Genetic alterations in sugar palm are initiated by the use of mutation induction techniques. In order to cause random mutations in the DNA sequence, mutagenic agents such as chemicals (ethyl methane sulfonate, sodium azide) or ionizing radiation (gamma rays, X-rays) are applied to sugar palm seeds, seedlings, or tissue cultures (Priyadarshan and Mohan Jain 2019). The population becomes genetically diverse as a result of these mutations, offering a pool of fresh genetic resources for improving traits.

Mutant populations are screened after mutagenesis to find individuals displaying desired characteristics like increased nutritional quality, disease resistance, increased yield, or tolerance to environmental stress. Promising mutants with desired features are found by applying sophisticated screening methods, such as phenotypic analysis and molecular markers (Ahloowalia et al. 2004). According to Priyadarshan and Mohan Jain (2019), sugar palms have shown notable success in strengthening their resistance to environmental challenges including salt and drought through mutation breeding. This includes improved disease resistance against infections such as *Phytophthora palmivora*, which causes stem rot. Superiority-displaying mutants are chosen for additional analysis and propagation. To evaluate their performance under various agroclimatic situations, field experiments and evaluations are conducted on selected mutant lines from mutation breeding programs. Commercial release and integration into sugar palm breeding programs are contemplated for mutants exhibiting exceptional qualities concerning sap production, sugar content, or fruit attributes (Ahloowalia et al. 2004).

Other palm species, such as the oil palm (Elaeis guineensis), have benefited from the effective application of mutation breeding techniques to create superior variations. This strategy has the potential to improve breeding initiatives and encourage environmentally friendly farming methods for sugar palms (Priyadarshan and Mohan Jain 2019). Mutation breeding is therefore a useful technique for improving sugar palm crops, providing chances to create new varieties with improved features for both commercial viability and sustainable cultivation. The possibility for further advancing sugar palm breeding programs and meeting the changing needs of growers and industries relying on this significant palm species exists in the ongoing research and implementation of mutant breeding techniques.

12.9.1 Biotechnology Breeding

By providing powerful tools to change the genetic makeup of the sugar palm (*Arenga pinnata*) and enhance desired features like yield, disease resistance, and stress tolerance, biotechnology breeding techniques are revolutionizing the improvement of this plant. These cutting-edge techniques enhance conventional breeding strategies and hasten the creation of better sugar palm cultivars that are compatible with a wide range of agroclimatic circumstances.

Tissue culture and micropropagation are two essential biotechnology tools utilized in sugar palm breeding. Tissue culture facilitates the rapid replication of superior variations while maintaining genetic purity by enabling the mass creation of homogeneous and disease-free plantlets from chosen elite genotypes (Rao et al. 2019). This method is essential for increasing the output of premium planting supplies. In sugar palm, somatic embryogenesis is another biotechnological technique used. It avoids the requirement for seeds by stimulating the creation of embryos from somatic cells. This technique facilitates the effective propagation of superior genetic material by enabling the regeneration of plants from cells exhibiting desired features and the clonal proliferation of elite sugar palm varieties (Rao et al. 2019).

The induction of particular genes into sugar palms to give desired features is mostly dependent on genetic transformation. Genes encoding characteristics like improved production or resistance to disease can be inserted into sugar palms through genetic transformation through biolistic or Agrobacterium-mediated techniques (Singh et al. 2018). With the use of this technology, sugar palm cultivars can now inherit desirable characteristics from a variety of genetic sources.

Another biotechnology approach used in sugar palm breeding programs is marker-assisted selection (MAS). MAS makes it possible to identify plants with desired characteristics by using particular DNA markers associated with those qualities. This streamlines trait introgression into breeding lines and speeds up the breeding process by enabling early selection of individuals bearing target genes without waiting for phenotypic expression (Rao et al. 2019).

Utilizing omics technologies, such as transcriptomics, proteomics, and genomics, it is possible to understand the genetic and molecular underpinnings of significant sugar palm features. These methods contribute to the identification of candidate genes for trait improvement by offering insights into the genetic diversity and regulatory processes underpinning important traits (Singh et al. 2018).

The analysis and conservation of genetic diversity in sugar palm germplasm collections are further aided by biotechnology methods. Characterizing genetic resources, evaluating genetic diversity, and identifying desirable features for breeding programs are all accomplished through the use of molecular markers and DNA profiling techniques (Rao et al. 2019).

Modern genome editing techniques like CRISPR-Cas9 have the potential to modify sugar palm genes precisely and effectively. By precisely targeting and changing particular genes, this ground-breaking technique enables unparalleled accuracy in trait improvement, including increased production and disease resistance (Singh et al. 2018).

To sum up, the application of biotechnology in breeding techniques is revolutionizing the improvement of sugar palms by speeding up the creation of superior variants with improved features. Sugar palm cultivation could undergo a revolution if biotechnology tools and traditional breeding practices are integrated. This would provide sustainable output and resilience to changing environmental problems.

12.9.2 Plant Tissue Culture of Sugar Palm

Arenga pinnata, also referred to as the sugar palm, has been studied and propagated due in large part to plant tissue culture techniques. The goal of tissue culture efforts for this palm species, which is a member of the Arecaceae family, is to efficiently reproduce plants while preserving genetic resources (Li et al. 2018). The selection of appropriate explants—plant portions used to commence cultures—is usually the first step in the process of cultivating *Arenga pinnata* plant tissue culture. Young leaf tissues, branch tips, or early embryos are frequently utilized as explants for palm species. Using disinfectants like alcohol or bleach, these explants are thoroughly sanitized on their surfaces to remove any impurities, guaranteeing a sterile foundation for the creation of a culture (Dai et al. 2019).

After being sterilized, the explants are put on a nutrient medium that is balanced in terms of the number of vital nutrients and plant growth regulators like auxins and cytokinins. According to Chen et al. (2021), this media gives the explants the support they need to grow and develop, which triggers the beginning of shoots from the explant tissues.

Using particular combinations of plant growth regulators and culture procedures, it is possible to produce several shoots from a single explant under optimal conditions. These formed shoots can then be moved to a rooting media that has been enhanced with auxins to promote the growth of roots. In order for rooted plantlets to develop into independent plants, they must first become acclimated to ex vitro conditions (environment outside the culture vessel) (Li et al. 2018).

There are numerous advantages to using tissue culture methods for *Arenga pinnata*. First of all, tissue culture makes it possible for plants to multiply quickly, producing a huge number of uniform, disease-free plantlets. Furthermore, it contributes to the preservation of A. pinnata's valuable germplasm and genetic diversity, which is crucial for species conservation initiatives in areas where the palm is endangered or vulnerable (Dai et al. 2019). Additionally, tissue culture makes it easier to research the genetics, physiology, and biochemistry of *Arenga pinnata*, which provides important information for crop productivity and development. The tissue culture of palms such as *A. pinnata* has its benefits, but it can also have drawbacks such as genotype dependency, long culture times, and the requirement for specialized facilities and knowledge (Chen et al. 2021).

Thus, *Arenga pinnata* study, conservation, and propagation have found a useful ally in plant tissue culture. This method benefits many industries, including biotechnology, horticulture, and agriculture, by facilitating the sustainable management and use of this economically and ecologically significant palm species (Li et al. 2018).

12.9.2.1 Embryo Culture of Sugar Palm

It has been established that embryo culture, one of the earliest kinds of tissue culture methods, offers plant breeders the most benefits. Through the process of overcoming seed dormancy, embryo culture allows plants to abbreviate their breeding cycle. In order to assess the relationship between the age of the embryos and the various nutrient medium compositions and their impact on the rate of in vitro germination and plant development within a certain time frame, a sugar palm embryo culture procedure was established. Explants were generated from fresh fruit embryos removed between 15 and 30 months following anthesis. Explants from 15-month-old fruits exhibited the highest rate of germination (90%), whereas explants from 30-month-old fruits supported a 72% rate of germination, according to the study's conclusion. The altered nutrient medium composition did not affect the culture's growth or development, however, anomalies in the haustorium and cotyledonary petiole formation were noted.

Another study described a well-established technique for producing sugar palm seedlings aseptically using embryo culture. Given that sugar palm seeds take a long time to germinate when planted in nurseries, the plant tissue culture method has been shown to encourage aseptic seedlings to regenerate quickly. From surface-sterilized immature fruits, immature embryos were removed as explants, and these embryos were then injected into both standard MS medium and MS medium that had been altered with varying doses of 6-benzylaminopurine (BAP). After 8 weeks of observation, the results revealed that 60% of explants grown on MS media supplemented with lower doses of BAP induced fast regeneration in the form of radicles and an enclosed sheath. It took 32 weeks for the development of accidental shoots to become apparent. After 24 weeks, explants grown on free-hormone MS media showed the highest proportion of shoot regeneration, with 90% of the shoots emerging. After 32 weeks, aseptic seedlings were successfully developed.

Embryo culture in sugar palm (*Arenga pinnata*) is a specialized technique within plant tissue culture that involves the isolation, culture, and development of embryos under controlled conditions. This method is instrumental in the propagation, genetic improvement, and conservation of sugar palm germplasm.

Embryo culture begins with the isolation of embryos from mature seeds of sugar palms. The embryos are carefully dissected and surface sterilized to remove contaminants, ensuring aseptic conditions for subsequent culture (Kamarudin et al. 2012). The isolated embryos are then placed onto nutrient-rich culture media supplemented with specific growth regulators and nutrients tailored to promote embryo development. In sugar palm embryo culture, the initiation of embryo development is stimulated by providing optimal growth conditions, including the appropriate balance of plant hormones such as auxins and cytokinins (Yaacob et al. 2015). These hormones play crucial roles in regulating cell division, differentiation, and organogenesis within the developing embryos. Under controlled environmental conditions, the embryos undergo morphological changes and progress through developmental stages.

Successful embryo culture in sugar palm may encounter challenges related to genotype-specific responses, contamination issues and embryo recalcitrance (Yaacob et al. 2015). Optimization of culture protocols, including media composition, growth regulator concentrations and culture conditions is essential for enhancing the efficiency and success rate of embryo culture techniques. The purpose of continued research in sugar palm embryo culture is to optimize protocols for mass propagation of elite varieties, conservation of genetic diversity and development of novel genetic resources. Advances in biotechnology, including genomic tools and gene editing techniques will further expand the applications of embryo culture in sugar palm breeding and genetic improvement.

12.9.2.2 Organogenesis of Sugar Palm

In plant tissue culture, the ability to produce adventitious roots and shoots through organogenesis is thought to be of the utmost significance. Explants of basal stem segments removed from aseptic seedlings and immature zygotic embryos were used to directly observe the organogenesis response in sugar palms. After each explant was injected with varying dosages of plant growth regulators (PGRs) on MS media, the explants' capacity for organogenesis was assessed. The examined explants showed a response after 8 weeks of culture; however, the PGR compositions in the culture media had a significant impact on the number of shoots and roots that were established. Basal stem explants cultivated on MS + 1.0 mg/L kinetin (Kin) + 2.0 mg/L Naphthaleneacetic acid (NAA) showed the best organogenesis response, with 90% regeneration rate, nine adventitious shoots, and twenty-five roots formed. Plantlet development was optimized through frequent sub-culturing of the regenerated shoots on basal MS medium (MS0). A 70% optimal rate of root regeneration was reported in immature zygotic embryo explants seeded on MS + 2.0 mg/L BAP + 2.0 mg/L NAA at this time. Also, it was reported that the best adventitious shoot regeneration (10 shoots) in culture was promoted by the embryo explants inoculated on MS + 2.0 mg/L BAP + 1.0 mg/L gibberellin acid (GA_3) + 1.0 mg/L silver nitrate ($AgNO_3$). MS + 3.0 mg/L Indole-3-butyric acid

(IBA) encouraged a large number of culture rootings. Before being transferred to a greenhouse, all plantlets were given a late acclimatization period on the planting medium.

A crucial step in the biotechnology of the sugar palm (*Arenga pinnata*) is called organogenesis, which is the act of growing new organs, like roots or shoots, from explant tissues in a controlled culture environment. This method is essential for producing elite sugar palm plantlets, genetic transformation, and tissue culture propagation. We explore the scientific nuances and real-world uses of organogenesis in sugar palm biotechnology in this essay. Organogenesis is the process by which specialized organs form and differentiate from undifferentiated tissues; this process is usually generated in vitro through the use of tissue culture techniques. In sugar palm, the process of organogenesis is started by adjusting the amounts of growth regulators and nutrients in the culture media to encourage the development of particular organs from explants, like shoots or roots (Kamarudin et al. 2012).

The process of selecting and preparing appropriate explants, such as stem tips, leaf segments, or inflorescence tissues, is the first step in the organogenesis of sugar palms. In order to encourage organ formation, these explants are cultivated on nutrient-rich media supplemented with certain growth regulators, like auxins and cytokinins (Yaacob et al. 2015). Depending on the intended result, auxins encourage root development whereas cytokinins stimulate the creation of shoots. The explants go through a number of biological changes during organogenesis, such as cell division, elongation, and differentiation, which result in the development of new organs. The development of the target organ type is largely dependent on the presence of particular growth regulators and ideal culture conditions (Kamarudin et al. 2012). It is possible to subculture and propagate the shoots or roots that emerge from the explants further to create whole plantlets.

Organogenesis has several important applications and benefits in sugar palm biotechnology including clonal propagation where organogenesis allows for the rapid multiplication of elite sugar palm genotypes, resulting in the production of uniform plantlets with desirable traits. Next, organogenesis plays a role in genetic transformation which facilitates genetic engineering experiments by providing a platform for introducing and expressing foreign genes in sugar palm tissues. Other than that, it can enable the regeneration of whole plants from small tissue explants, offering a reliable method for propagation and conservation of valuable germplasm.

Organogenesis in sugar palms may encounter difficulties because of genotype recalcitrance, culture contamination, and genotype responsiveness, despite its promise (Yaacob et al. 2015). Improving the effectiveness and repeatability of organogenesis procedures requires optimizing tissue culture protocols, which includes choosing the right explants, growth regulators, and culture conditions.

To sum up, organogenesis is an essential process in sugar palm biotechnology that has several uses in genetic enhancement, conservation, and propagation. Researchers and breeders can create better sugar palm cultivars more quickly by utilizing the concepts of organogenesis to create cultivars that are more suited to particular industrial and agricultural applications. Sugar palm biotechnology will grow as a result of ongoing study and advancements in tissue culture techniques, which will help tropical countries' sustainable agricultural and economic growth.

12.9.2.3 Callogenesis and Somatic Embryogenesis of Sugar Palm

For plant tissue culture, inducing callogenesis and somatic embryogenesis is another crucial strategy. To create genetically identical replicas of plants with desired traits, callus cultures are helpful. In the meantime, somatic cells divide during somatic embryogenesis to create entire embryos that have zygotic-like functions. Somatic embryogenesis is a method of producing embryos that can be used to research plant cell development. Using basal stem explants from aseptic seedlings is one of the usual micropropagation procedures for achieving callogenesis in sugar palms. The sugar palm explants cultivated on MS media supplemented with 0.1–0.5 mg/L 4-Dichlorophenoxyacetic acid (2,4-D) in combinations of equal concentrations of Kin ultimately showed the optimal rate of callogenesis. It took 4–8 weeks for the explants to regenerate a friable, translucent, whitish primary callus induction.

A sugar palm's somatic embryogenesis response and callogenesis can also be induced by using immature zygotic embryos as explants. To induce callus, the chosen explants were cultivated on MS medium supplemented with various concentrations and a combination of 2,4-D + NAA. The callogenesis response was monitored while all cultures were incubated at a temperature of 25 ± 2 °C in total darkness. The addition of $AgNO_3$ encouraged the formation of heart-shaped and torpedo somatic embryos, high concentrations of sucrose and the addition of casein hydrolysate (CH) to the current culture medium composition to improve the growth and development of somatic embryos at the globular stage. Cotyledonary embryos multiplied when established somatic embryos were subcultured on maturation media. Regeneration of clonal roots was seen when the cultures were moved to a half-strength basal MS medium for 4 months. A useful tropical plant with many uses in industry and agriculture is the sugar palm (*Arenga pinnata*). The application of biotechnological methods, such as callogenesis and somatic embryogenesis, is very promising for this significant palm species' genetic improvement, conservation, and propagation. In light of sugar palm biotechnology, this essay examines the intricate scientific details and useful applications of callogenesis and somatic embryogenesis.

The technique of starting callus production from explants, or plant tissues, in a controlled laboratory setting, is known as callogenesis. To produce callus in sugar palms, a variety of explants can be used, such as leaf segments, stem tips, or inflorescence tissues (Yaacob et al. 2015). Usually, the induction of callus is accomplished by adjusting the doses of growth regulators, especially auxins like 2,4-Dichlorophenoxyacetic acid (2,4-D), which promote uncontrolled cell proliferation and growth.

In tissue culture, the development of callus tissue is an essential stage that yields a source of manipulable totipotent cells for use in somatic embryogenesis and other applications. In order to produce somatic embryos, callogenesis facilitates the quick proliferation of plant material and the creation of embryogenic cultures.

The technique of inducing embryos to develop from somatic (non-reproductive) cells within the callus tissue is known as somatic embryogenesis. According to Kamarudin et al. (2018), the process of inducing somatic embryogenesis in sugar palms entails moving embryogenic callus cultures to particular growth media that are ideal for embryo development. Growth regulators and nutrient compositions are precisely manipulated to bring about this transformation.

Somatic embryos that resemble zygotic embryos are formed when the callus tissue goes through organogenesis and differentiation during somatic embryogenesis. Under carefully regulated conditions, these somatic embryos can be further developed and germinate into whole plantlets, allowing for the quick clonal replication of particular genotypes of sugar palms.

Callogenesis and somatic embryogenesis offer several important applications and benefits for sugar palm biotechnology. The benefits of using tissue culture techniques include mass propagation which the techniques allow for the rapid multiplication of elite sugar palm genotypes, facilitating large-scale production of uniform plantlets for commercial cultivation. Additionally, the tissue culture method aids in genetic improvement, as somatic embryogenesis offers a platform for genetic transformation studies intended to introduce desired qualities into sugar palm, such as increased yield and disease resistance. Other than that, sugar palm germplasm can be preserved since tissue culture techniques allow the genetic variety to be preserved and rare or endangered sugar palm variants to be preserved, supporting efforts to conserve biodiversity.

Although there is great promise for callogenesis and somatic embryogenesis, obstacles including genotype specificity, culture contamination, and genotype recalcitrance may reduce their effectiveness and usefulness (Kamarudin et al. 2018). To improve the effectiveness, repeatability, and scalability of these methods, tissue culture protocols—including hormone concentrations, culture fluid compositions, and culture conditions must be optimized.

To sum up, callogenesis and somatic embryogenesis are advanced biotechnological instruments that have the potential to transform sugar palm farming and enhance genetic diversity. Researchers and breeders can expedite the creation of enhanced sugar palm cultivars suited to certain industrial and agricultural requirements by utilizing these strategies. The complete potential of sugar palm biotechnology will be further unlocked through ongoing research and development in tissue culture techniques, opening the door for sustainable agriculture and regional economic growth in tropical areas.

12.9.2.4 Synthetic Seed Establishment of Sugar Palm

A promising method for maintaining plant species and propagating germplasm is synthetic seed. The primary goal of the establishment of synthetic seeds is to transform micro-propagules, primarily somatic embryos, into artificial seeds that may be grown in greenhouses or fields as needed. The term "artificial seeds" refers to the concept of mechanically encapsulating somatic embryos, which are intended to be handled, stored, and planted similarly to natural seeds. Individual sugar palm somatic embryos at the cotyledonary stage were removed from the aseptic culture and encapsulated in a matrix solution made up of 1.0–4.0% sodium alginate, 30.0 g/

L sucrose, and various PGR concentrations. The matrix solution was autoclaved at 121 °C for 20 minutes prior to the encapsulation process.

To create artificial beads, each somatic embryo was put into the prepared solutions and then pipetted into sterile calcium chloride solutions. After being cultured on basal MS media for germination, the synthetic beads containing somatic embryos were rinsed and desiccated on distilled paper. Eight weeks of dark culture at 25 ± 2 °C resulted in an 80% germination rate for the synthetic seeds, encouraging the growth of adventitious shoots and roots. The exhibited synseed method would be an excellent way to encourage the quick growth of sugar palm seedlings meant for large-scale production, both aseptic and clonal, while methodically retaining its genetic resources for later use. In order to preserve and propagate sugar palms (*Arenga pinnata*), synthetic seed technology has shown great potential. It provides effective mass propagation of genetically homogeneous plants from certain elite genotypes. This novel method allows for regulated germination, storage, and transportation by encasing somatic embryos or embryogenic cells in protective coatings.

The generation of somatic embryos from explants, such as leaf segments or embryogenic cells generated from tissue culture, is the first step in the production of synthetic seeds in sugar palms. Then, a gel matrix comprising growth regulators, nutrition, and protective agents is used to encapsulate these somatic embryos (Rao et al. 2019). throughout encapsulation, alginate or other gel-forming materials are frequently utilized to give the embryos mechanical support and protect them throughout germination and storage (Huang et al. 2010).

In order to maintain ideal circumstances for germination and plantlet growth, the gel-coated embryos are additionally encased in a semipermeable membrane that controls gas and water exchange (Fayaz et al. 2011). Under carefully managed in vitro or ex vitro settings, sugar palm synthetic seeds can germinate, allowing the encapsulated embryos to develop into plantlets under the protective shell (Singh et al. 2014).

The following are some benefits of using synthetic seed technologies for sugar palm conservation and propagation:

(a) Mass propagation: it enables rapid multiplication of selected elite genotypes, ensuring uniformity and genetic purity.
(b) Storage and transportation: synthetic seeds provide a convenient method for long-term storage and transportation of valuable germplasm, reducing the reliance on traditional seed storage facilities.
(c) Disease-free propagation: production under sterile conditions minimizes the risk of pathogen transmission, ensuring the production of disease-free plants.
(d) Conservation of genetic resources: synthetic seeds contribute to the conservation of genetic diversity within sugar palm germplasm collections by preserving elite genotypes for future breeding and research.

In order to increase the quality and viability of synthetic seeds, ongoing research focuses on optimizing sugar palm synthetic seed production processes, including refining gel composition, encapsulating techniques, and germination conditions (Rao et al. 2019).

Synthetic seed technology has the potential for large-scale sugar palm plantation establishment in practical applications. It offers an economical and effective substitute for traditional seedling production methods. Through faster breeding programs and better germplasm management, sugar palm farming can be improved by utilizing the advantages of synthetic seed technology.

12.9.2.5 In Vitro Pollination of Sugar Palm

A complex method in agricultural biotechnology, in vitro pollination of sugar palms (*Arenga pinnata*) allows for exact control over the breeding process and seed output. Given the difficulties in generating specific genetic crossings and the uncertainty in pollen availability that comes with natural pollination, this technique is especially helpful for the development of sugar palms. Through the judicious use of in vitro pollination, researchers and breeders can create better cultivars, quicken the breeding cycle, and increase genetic variety.

The meticulous selection of parent plants based on desirable features like yield, disease resistance, or quality of palm products is the first step in the in vitro pollination process. To ensure maximum maturity and viability, mature pollen grains are harvested from male parent palms that have been identified. After being gathered, the pollen is treated under carefully monitored circumstances to get rid of impurities and keep it viable for storage.

Simultaneously, a selection of female parent palms' pistillate flowers are ready for pollination. We closely inspect these flowers to make sure they are open to pollen, which is necessary for fertilization to occur successfully. To ensure accuracy and prevent contamination, the pollination process is carried out in a sterile setting within the controlled atmosphere of a tissue culture laboratory.

The stigma of the female flowers is directly touched with pollen grains using specialized instruments like micropipettes or fine brushes. By applying pollen directly, the fertilization process is started and the pollen reaches its intended destination. Breeders can create targeted crosses between parent plants in the regulated environment of in vitro pollination, which makes it easier to create hybrid seeds with predefined genetic features.

To create the best conditions possible for the development of the embryo, the pollinated flowers or ovaries may occasionally be encased in gel matrices or culture media after pollination. This method of encapsulation supports the growth and development of embryos in vitro, increasing the efficiency of seed output.

In vitro pollination has several advantages for sugar palm farming. (a) Precision and Control: Breeders may accomplish specific genetic crossings and create cultivars with desired features thanks to the exact control that in vitro pollination provides over the breeding process. (b) Effectiveness: Compared to natural pollination methods, this technology boosts seed production rates, improving the scalability of sugar palm farming and breeding programs. (c) Genetic Improvement: In vitro pollination helps to create genetically varied and hardy sugar palm cultivars that are suited to a range of environmental circumstances by encouraging controlled genetic crosses and (d) Accelerated Breeding Cycles: By shortening the time needed to obtain hybrid seeds and assess progeny, in vitro pollination speeds up the development and deployment of cultivars.

In vitro pollination plays a critical role in research and practical applications that promote sustainable agriculture and advance sugar palm breeding programs. It makes it easier to create new cultivars with enhanced qualities, which eventually raises production and makes sugar palm farming more profitable.

In summary, in vitro pollination is a revolutionary method that highlights the convergence of conventional wisdom with state-of-the-art biotechnology in sugar palm breeding. The whole genetic potential of sugar palms may be accessed by researchers and breeders through the use of controlled pollination, opening the door to improved agricultural resilience and sustainability.

12.10 Conclusions and Prospects

An important tropical crop with major cultural and economic significance across many locations is the sugar palm (*Arenga pinnata*). Because tissue culture techniques offer accurate approaches for genetic improvement, conservation, and multiplication of this essential palm species, they have revolutionized the cultivation of sugar palms. Tissue culture techniques that are revolutionizing sugar palm technologies include callogenesis, somatic embryogenesis, organogenesis, and embryo culture.

Plant cells, tissues, or organs are first manipulated under carefully monitored laboratory conditions in tissue culture. Callogenesis, or the development of callus from explants like leaf segments or shoot tips, starts the synthesis of undifferentiated cells that constitute the basis for later processes. These callus-derived cells are stimulated to develop into somatic embryos through a process known as somatic embryogenesis, which produces embryos that can germinate into plantlets. Organogenesis allows for the regeneration of entire plants by controlling the differentiation of callus into particular organs, such as shoots or roots. To create plantlets under controlled circumstances, embryos from mature seeds are isolated and cultured in an embryo culture system.

Tissue culture in sugar palms has a wide range of useful uses. First, tissue culture ensures uniformity and scalability of plant production by facilitating the fast reproduction of elite sugar palm genotypes by clonal propagation. Second, by using tissue culture and genetic transformation studies to introduce desired qualities like disease resistance or increased yield into sugar palm tissues, genetic improvement is accomplished. Thirdly, impediments to seed growth and germination can be overcome by rescuing hybrid embryos from controlled crosses through tissue culture procedures like embryo culture. Finally, by maintaining genetic variety and uncommon sugar palm types, tissue culture serves a critical role in germplasm conservation and aids in the preservation of biodiversity.

Although sugar palm tissue culture has many potential uses, there are drawbacks, including genotype-specific reactions, contamination problems, and genotype recalcitrance. To improve the effectiveness and repeatability of tissue culture procedures, culture protocols—including growth regulator concentrations, medium formulations, and culture conditions—need to be optimized.

The potential for sugar palm tissue culture is promising and diverse going forward. The optimization of tissue culture procedures to improve effectiveness, scalability, and application across various sugar palm genotypes will be the main focus of future research and innovation. Furthermore, the use of molecular markers and genomic technologies will expedite the genetic advancement and trait selection in sugar palm. Another interesting approach is the use of tissue culture to develop sugar palm cultivars with improved nutritional content for higher yield and nutritional value. Furthermore, developing sugar palm cultivars with improved environmental resilience—such as pest and drought resistance—will be essential to the success of sustainable agriculture.

To sum up, tissue culture techniques play a crucial role in the advancement of sugar palm biotechnology by providing accurate means of propagation, genetic enhancement, and conservation. The future of sugar palm production shows potential for sustainable agriculture and economic development in tropical regions with continued research and cooperation.

Appendix 12.1 Research Institutes and Online Resources Available for Sugar Palm *Breeding*

Institution	Area of specialization	Research activities	Website
Forest Research Institute Malaysia (FRIM)	Forestry research, sustainable resource use	Genetic improvement of sugar palm, disease resistance studies, sustainable cultivation practices	www.frim.gov.my
Universiti Putra Malaysia (UPM)	Agricultural sciences, biotechnology	Development of high-yield sugar palm varieties, studies on sap extraction techniques	www.upm.edu.my
Malaysian Agricultural Research and Development Institute (MARDI)	Crop research, food security	Studies on sugar palm adaptability, propagation techniques, and breeding for resilience	www.mardi.gov.my
Sabah Forestry Department	Tropical forest management	Studies on sugar palm cultivation in forested areas, integrating agroforestry techniques	www.forest.sabah.gov.my
Indonesian Palm Crops Research Institute (IPCRI/Balit Palma)	Breeding and genetic improvement of palm crops	Studies on sugar palm sap yield, genetic diversity, and propagation systems	balitpalma.litbang.pertanian.go.id

(continued)

Institution	Area of specialization	Research activities	Website
Indonesian Agency for Agricultural Research and Development (IAARD)	Agricultural innovation and crop development	Research on high-yield sugar palm varieties, disease resistance, and agroforestry systems	www.litbang.pertanian.go.id
Bogor Agricultural University (IPB University)	Agricultural sciences and biodiversity	Advanced research on sugar palm physiology, breeding, and bioproduct potential	www.ipb.ac.id
Center for International Forestry Research (CIFOR)	Sustainable forestry and agroforestry systems	Research on integrating sugar palm cultivation into agroforestry for sustainability and rural livelihoods	www.cifor.org
Universitas Gadjah Mada (UGM)	Agriculture and environmental sciences	Studies on sugar palm sap processing, bioproducts, and sustainable cultivation	www.ugm.ac.id

Appendix 12.2 Genetic Resources for Sugar Palm Breeding

Species	Important traits	Cultivation location
Arenga pinnata	High sap yield for sugar and bioethanol production	Tropical regions in Malaysia, including Sabah and Sarawak
Arenga undulatifolia	Drought tolerance and adaptability to lowlands	Southeast Asian regions, potentially in experimental trials in Malaysia
Arenga tremula	High fiber quality and disease resistance	Tropical forests and select plantations in Malaysia
Arenga microcarpa	High adaptability to rocky or hilly terrains	Native to Indonesia, cultivated in remote highland areas
Arenga westerhoutii	High biomass production, the potential for fiber extraction	Found in tropical forests in Southeast Asia, including Indonesia and Malaysia

References

Abdullah WG, Rianse U, Muhidin W, Widayati E, Mihrad S, Taridala SAA, Rianse IS, Baka WK (2018) Farmer's motivation in Aren sugar processing business. IOP Conf Ser Earth Environ Sci 122:012007

Ahloowalia BS, Maluszynski M, Nichterlein K (2004) Global impact of mutation-derived varieties. Euphytica 135(2):187–204

Anwar R, Fatah FA, Othman N, Nawawi H, Tajuddin Z (2014) Socioeconomic study of Arenga pinnata smallholders in Malaysia. In: Proceedings of the international symposium on research of arts, design and humanities (ISRADH 2014)

Arjchariyaartong W (2007) The Competitiveness of the Sugar Industry in Thailand. Ph.D Thesis. Analyse, Planung und Organisation der land wirtschaftlichen Produktion. University of Hohenheim, Stuttgart, Germany

Asikin D, Puspitaningtyas DM (2000) Study on *in vitro* and *in vivo* seed germination of *arenga pinnata* (Wurmb) Merr. Seminar Hasil Penelitian dan Pengembangan Bioteknologi, Cibinong, Bogor, Indonesia. Puslitbang Bioteknologi

Chan E (2000) Periplus nature guides: tropical plants. Periplus Editions (HK) Ltd., Singapore

Dalibard C (1999) Overall view on the tradition of tapping palm trees and prospects for animal production. Livest Res Rural Dev 11(1):1–37

Devi MG, Purwito A, Husni A (2014) Globular embryo induction of sugar palm (*Arenga pinnata* (Wurmb) Merr.). Int J Bio Sc, Bio Chem, Bio Inf 4(1):60–66

Djarwanto D, Suprapti S (2016) Utilization of aren (*Arenga pinnata* Merr.) sawmilling waste for edible mushroom cultivation media. Indonesian J Forest Res 3(1):9–18

Duke JA (2000) Handbook of nuts: herbal reference library, vol 4. CRC Press

Fayaz A, Jaskani MJ, Hafeez A, Khan SA (2011) Plant biotechnology and its application in Pakistan. In: Plant biotechnology and molecular markers. Springer, Dordrecht, pp 103–116

Florido HB, De Mesa PB (2003) Sugar Palm (*Arenga pinnata* Wurmb Merr.). Res Inform Ser Ecos 15(2):2–6

Gnanaharan R, Sundari C (2014) Palm fibre for handicrafts. VNR Publishing House, Sivakasi

Haagen A (2015) Arenga pinnata (Wurmb) Merr. In: Promising source of bioethanol and low glycemic index sugar. https://doi.org/10.13140/RG.2.2.36701.36325

Henderson A (2009) Palms of southern Asia. Princeton University Press, Princeton

Huang Y, Zhao W, Zhang H, Chen W, Liu D (2010) Advances and prospects in plant embryogenic somatic cells and synthetic seed. In: Frontiers in plant science. Springer, Berlin/Heidelberg, pp 77–85

Ishak MR, Sapuan SM, Leman Z, Rahman MZA, Anwar UMK, Siregar JP (2013) Sugar palm (*Arenga pinnata*): its fibres, polymers and composites. Carbohydr Polym 91(2):699–710

Ishak AM, Aziz NA, Zakaria Z, Yusof MAM (2019) Sugarcane Arenga pinnata (Wurmb) Merr: an underutilized crop for integrated farming systems in Malaysia. In: Zakaria NA, Yusof MAM, Ishak AM (eds) Sabah's agricultural development: issues and challenges. Springer, Singapore, pp 103–117

Janick J, Paull RE (eds) (2008) The encyclopedia of fruit and nuts. CABI

Johnson DV (1996) Palms: their conservation and sustained utilization: status survey and conservation action plan, vol 31. IUCN, Gland Switzerland and Cambridge. UK. 116 + viii pp

Kale V, Rajasree SR (2016) Ethnomedicinal uses, phytochemistry and pharmacology of Arenga pinnata: a review. J Ethnopharmacol 191:181–200

Kamarudin N, Jamaludin N (2012) Screening of local sugar palm (Arenga pinnata) accessions for the production of coconut sugar. Agric Agric Sci Proc 2:1–7

Khalina A, Juliana AH, Suriani MJ, Paridah MT (2019) Potential of sugar palm (Arenga pinnata) fibre reinforced composites for packaging applications. Bioresources 14(2):2551–2566

Kong L (2015) Sugar palm (Arenga pinnata): a promising multipurpose tree for the tropics. Agrofor Syst 89(4):695–703

Kumar V, Satyanarayana KG (2007) Natural fiber reinforcements and their composites: development in production and properties. Springer, Berlin

Lawrence Long Ltd. United Kingdom. Retrieved 27 April 2008 from http://www.lawrencelong.co.uk/arenga.html

Lim TK (2012) Edible medicinal and non-medicinal plants, vol 1. Springer, Dordrecht, pp 285–292

Martini E, Roshetko JM, Van Noordwijk M, Rahmanulloh A, Mulyoutami E, Joshi L, Budidarsono S (2012) Sugar palm (*Arenga pinnata* Wurmb Merr.) for livelihoods and biodiversity conservation in the orang Utan habitat of Batang Toru, North Sumatra, Indonesia: mixed prospects for domestication. Agrofor Syst 86(3):401–417

Mogea JP (2003) Four new species of Arenga (Palmae) from Indonesia. Reinwardtia 12(2): 181–189

Mogea J, Seibert B, Smits W (1991) Multipurpose palms: the sugar palm (*Arenga pinnata* Wurmb Merr.). Agrofor Syst 13(2):111–129
Nair KPP (2014) Agronomy and economy of black pepper and cardamom: the "King" and "Queen" of spices. Elsevier
Nurhayati E, Darmawan W, Aini FN (2012) Potential of sugarcane and sugar palm fiber for erosion control in Tanah Laut Regency, South Kalimantan. J Trop Soils 17(1):75–81
Orwa C, Mutua A, Kindt R, Jamnadass R, Simons A (2009) Agroforestree database: a tree species reference and selection guide Version 4.0. World Agroforestry Centre ICRAF, Nairobi, KE
Othman AR, Haron NH (1992) Potensi industri kecil tanaman enau. FRIM Report:7–18
Priyadarshan PM, Mohan Jain S (eds) (2019) Breeding plantation tree crops: tropical species. Springer
Putih R, Satria B, Thaib R (2003) Upaya perbanyakan vegetatif enau (*Arenga pinnata* (Wurmb) Merr.) melalui regenerasi tunas secara *in vitro*. Stigma 11:208–212
Rao SR, Kshemkalyani S, Mohan Jain S, Priyadarshan PM (2019) Biotechnology in plantations: application in tree improvement—an overview. In: Biotechnology for plantation tree crops. Springer, Singapore, pp 1–18
Redhead J (1989) Utilization of tropical foods: trees. FAO Food Nutr 47(3):52. FAO, Rome, Italy
Rianse IS, Abdullah WG, Midi LO, Hartono S, Suryantini A, Jamhari WW (2016) Financial, economic and environmental feasibility analysis of palm sugar domestic industry in Kolaka Indonesia. Int J Econ Manag Syst 1:104
Rushdan I, Othman N, Arof A, Roslan A (2019) Biochemical and antioxidant properties of sugar palm (*Arenga pinnata*) fibre: a potential food industry by-product. Food Biosci 27:82–88
Sahari J, Sapuan SM, Zainudin ES, Maleque MA (2013) Thermo-mechanical behaviors of thermoplastic starch derived from sugar palm tree (*Arenga pinnata*). Carbohydr Polym 92(2): 1711–1716
Sastra HY, Siregar JP, Sapuan S, Hamdan MM (2006) Tensile properties of *Arenga pinnata* fiber-reinforced epoxy composites. Polym-Plast Technol Eng 45(1):149–155
Schmidt L (2007) Seed processing. In: Tropical forest seed. Springer, Berlin Heidelberg, pp 67–142
Singh BD, Singh AK, Ansari MW (2014) Plant tissue culture: an introductory text. Springer, New Delhi
Smits W (2011) Masarang Foundation
Soeseno S (2000) Bertanam Aren. Penebar Swadaya, Jakarta (ID)
Sreekala MS, Kumaran MG, Thomas S (2000) Oil palm fibre reinforced phenol formaldehyde composites: influence of fibre surface modifications on the mechanical performance. Appl Compos Mater 7(5–6):295–329
Suhaimi AA, Bakar BH, Hassan N, Hamzah MA (2016) Agro-economic potential of sugarcane (Arenga pinnata) and its contributions towards the income of the rural community in the Highlands of Sabah, Malaysia. Int J Agric Ext 4(2):117–122
Widodo TW, Elita R, Asari A (2009) Sugar palm (Arenga pinnata Merr) plantation for bio ethanol production, sustainable development and environmental conservation. In: Presentation paper for research workshop on sustainable biofuel development in Indonesia, 4–5 February 2009. Indonesian Center for Agricultural Engineering Research and Development (ICAERD)
Witt A, Luke Q (2017) Guide to the naturalized and invasive plants of Eastern Africa. CABI
Yaacob JS, Meon S, Zakaria M (2015) Tissue culture techniques for the micropropagation of Arenga pinnata (Wurmb) Merr. J Plant Biotechnol 42(3):201–207
Yanti M, Ali S (2017) Cosmeceutical effects of galactomannan fraction from *Arenga pinnata* fruits *in vitro*. Pharm Res 9(1):39
Zaid A, De Wet PF (1999) Chapter I botanical and systematic description of date palm. In: FAO plant production and protection papers, p 1

Chapter 13
Advances in Breeding and Biotechnology Approaches for Genetic Enhancement in Yucca (*Yucca* spp.)

Meghdad Kamali Moghaddam, Eman H. Afifi, and Khaled F. M. Salem

Abstract In recent years, various techniques for extracting lignocellulose fibers from yucca leaves have been reviewed. These fibers are obtained from the leaves of Yucca, a genus of perennial shrubs and trees in the Agavoideae subfamily native to the Americas. Yucca plants have evergreen rosettes with sword-shaped leaves and large clusters of white or whitish flowers. They can be found in wet rainforests, subtropical habitats, and hot deserts. The leaves of yucca plants have a prickly texture and grow in a rosette pattern. They produce tall stalks of white or yellow flowers that are used in soap-making. Yucca plants are known for their resilience in drought conditions and their attractive appearance. Some species of yucca produce edible fruits or leaves with medicinal uses. Throughout Central America, yucca fiber has been used for various purposes, such as sandal production, tying, weaving, making textiles with feathers and fur, creating mats, and crafting baskets. Yucca is a valuable resource for indigenous communities due to its widespread availability and accessibility. In addition to reviewing fiber extraction techniques, the physical and chemical properties of these yucca fibers have been evaluated in this chapter. This chapter helps in understanding the characteristics and potential applications of the fibers derived from yucca leaves, Yucca species, the fiber extraction method, germplasm biodiversity and conservation, the objectives and stages of beetroot

M. K. Moghaddam (✉)
Department of Textile Engineering, University of Bonab, Bonab, Iran
e-mail: kamali@ubonab.ac.ir

E. H. Afifi
Fruit Breeding, Ornamental Plants and Woody Tress Department, Horticulture Research Institute (HRI), Agriculture Research Centre (ARC), Giza, Egypt

K. F. M. Salem
Department of Biological Sciences, Shaqra University, Ad-Dawadimi, Saudi Arabia

Department of Plant Biotechnology, Genetic Engineering and Biotechnology Research Institute (GEBRI), University of Sadat City, Sadat City, Egypt
e-mail: khaled.salem@gebri.usc.edu.eg; khaledfathi@su.edu.sa

K. F. M. Salem et al. (eds.), *Breeding and Biotechnology of Leaf, Fruit, and Seed Fiber Crops*, Advances in Plant Breeding Strategies 11,
https://doi.org/10.1007/978-3-032-00407-9_13

breeding programs, and conventional breeding methods. Additionally, it addresses modern methods of plant breeding, including marker-assisted breeding and genetic engineering.

Keywords Agavoideae · Germplasm Biodiversity · Ligno cellulose Fiber · Molecular Breeding · Tissue Culture

13.1 Introduction

The Yucca plant is an evergreen shrub native to North and South America, found in a variety of environments from humid rainforests and moist subtropical ecosystems to hot and deserts. It has long, sharp, spiky leaves that grow in a rosette pattern and can reach several feet in length. In the summer, the plant produces tall stalks of white or yellow flowers, which can be used to make soap. Yucca plants are known for their resilience and ability to withstand drought and other challenging growing conditions, as well as their aesthetic appeal. While all yucca varieties share large blooms and sword-shaped leaves, each type has a unique native region, preferred growing climate, and distinct characteristics. Some species of yucca produce edible fruit or leaves that are used for medicinal purposes. They are popular as ornamental plants due to their distinctive appearance and low maintenance needs (Table 13.1).

Yucca plants require well-draining soil and plenty of sunlight. They are susceptible to root rot and sensitive to overwatering. Yuccas also benefit from fertilizing with a balanced, slow-release fertilizer. Pruning can help the plant maintain its shape and remove dead or damaged leaves. Yuccas are low-maintenance plants that can survive in a wide range of environments, making them a popular choice for both indoor and outdoor landscaping.

As the plant ages, its lower leaves fall off, and the stem becomes bare. Yucca leaves are translucent and waxy when young, but they become tougher as they mature. Flowering stems appear once a year in tropical regions. The yucca moth (family Prodoxidae), an insect, pollinates yuccas through a very specific process in which it transfers pollen from one plant's stamens to another while simultaneously laying an egg in the flower. The moth larva then consumes some of the developing seeds, but not all of them.

Table 13.1 Taxonomic rank of yucca

Scientific classification	
Kingdom:	Plantae
Division:	Magnoliophyta
Class:	Liliopsida
Order:	Asparagales
Family:	Agavaceae
Subfamily	Agavoideae
Genus:	*Yucca*

Yuccas are frequently planted in gardens as ornamental plants. Some yuccas also have edible parts, including fruits, seeds, flowers, flowering stems, and occasionally roots, although consumption of these parts is rare, with references to yucca as food usually stemming from confusion with the botanically unrelated but similarly spelled yucca. Moreover, yuccas are recognized as a wonderful source of medicine and have numerous advantages over other plants, from their roots to their fruit.

Early reports of the species were confused with Manihot esculenta. Consequently, Linnaeus mistakenly derived the generic name from the Taíno word for the latter, yucca. The Aztecs, who had been living in Mexico before the Spanish arrival, called the local yucca species (*Yucca gigantea*) iczotlin Nahuatl, which gave the Spanish izote. Izote is also used for *Yucca filifera*.

Yucca is native to both South and North America and has been introduced to various regions including Argentina Northeast, Argentina Northwest, Bahamas, Bangladesh, Bolivia, the Cayman Islands, Cuba, the Dominican Republic, the East Aegean Islands, Ecuador, El Salvador, France, the Free State, Galápagos Islands, Great Britain, Haiti, Hungary, Italy, Jamaica, Kentucky, KwaZulu-Natal, Leeward Islands, the Marianas, Massachusetts, Mauritius, New Caledonia, New South Wales, New York, Pakistan, Poland, Portugal, Puerto Rico, Queensland, Romania, Réunion, Southwest Caribbean, Spain, Trinidad-Tobago, Turkey, European Turkey, Uruguay, and the Windward Islands.

The competing demands of a growing human population and dwindling arable land have put traditional plant breeding under closer scrutiny. Conventional or classical plant breeding helped many domesticated plant species advance in the 1900s (Aboshama et al. 2015; Emara et al. 2011, 2013). Classical breeding relies on a thorough evaluation of the material under different conditions in comparison to known cultivars. The most cost-effective and environmentally friendly way to control the many pests and diseases (Lorenzen et al. 2010). Because drought is a serious environmental problem that impacts farming all over the world, plants that can withstand dry conditions are highly prized for their economic value (Ibrahim et al. 2023b; Börner et al. 2009, 2010; Elabd et al. 2020; Saleh et al. 2021a, b; Salem and Börner 2004; Salem et al. 2004a, 2021a, b, d, e).

Molecular markers eliminate the need for labor-intensive direct screening methods in controlled settings, such as greenhouses and fields, to expedite the selection process in crop breeding (Alqahtani et al. 2024; El Rabey et al. 2013; Elabd and Salem 2009; El-Absawy et al. 2012, 2015; El-Zanaty et al. 2011, 2013; Esmail et al. 2009; Ibrahim et al. 2023a, 2024; Ismail et al. 2013; Mattar et al. 2016; Salem 1997, 2009, 2015; Salem and El-Zanaty 2015; Salem and Ibrahim 2024; Salem and Mattar 2014a, b; Salem and Sallam 2016; Salem et al. 2007a, b, 2008a, b, 2010, 2015a, b, 2021c, f, 2022, 2024; Varshney et al. 2008). Molecular markers are particularly useful for identifying traits that are controlled by several genes. It is possible to create an improved cultivar by transferring multiple QTLs simultaneously or by combining QTLs for various agronomic traits once the appropriate molecular markers have been identified and the pertinent QTLs associated with an agronomic trait mapped (Abdel-Hamid and Salem 2021; Abu-Ellail et al. 2021; Aldahak et al. 2021; Hassan et al. 2021; Salem 2004; Eltaher et al. 2017, 2018, 2021,

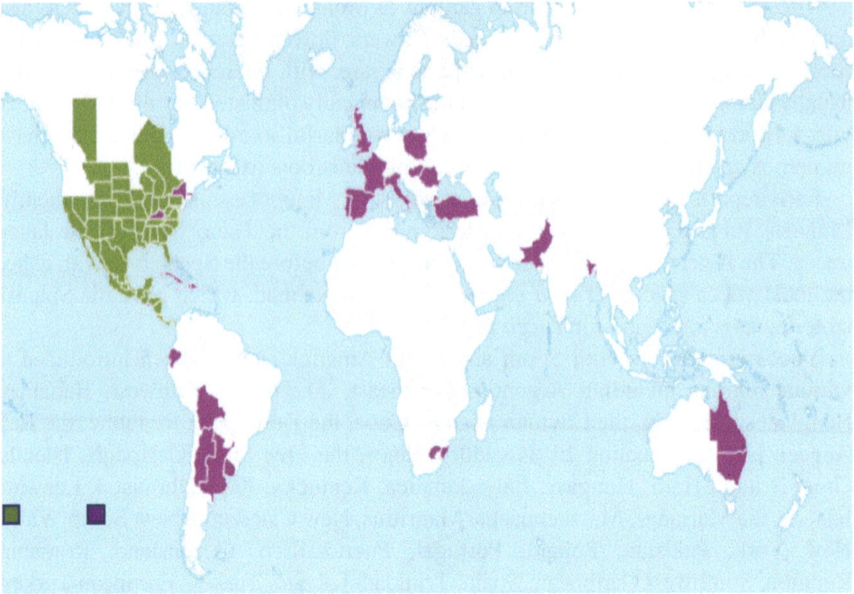

Fig. 13.1 Yucca plant distribution in the worldwide. (Source: Powo (2023))

2022; Ghazy et al. 2020; Mourad et al. 2019; Salem and Börner 2003; Salem et al. 2003, 2004b, 2007b, 2023). This chapter aims to provide insight into the characteristics and potential applications of fibers derived from yucca leaves, Yucca species, fiber extraction methods, germplasm biodiversity and conservation, objectives and stages of beetroot breeding programs, and conventional breeding methods. Additionally, it discusses modern methods of plant breeding, such as marker-assisted breeding and genetic engineering (Fig. 13.1).

13.2 Yucca Species

The International Union for Conservation of Nature's (IUCN) Red List of Threatened Species™ is the most comprehensive source of information on the global extinction risk status of animal, fungus, and plant species. In the IUCN Red List of Threatened Species (2020), various types of yucca are categorized, and presented in Table 13.2. Detailed information can be found on the web page listed in Appendix 13.1.

Yucca is a flowering plant belonging to the class Magnoliopsida, order Asparagales, and family Asparagaceae. Many species are also classified in the Chenopodiaceae family (https://www.itis.gov). Yucca (*Yucca* spp.) is part of the genus Yucca.

Table 13.2 Yucca species and population trend (2022)

	Yucca	Common Name(s):	Population Trend
1	Aloifolia	Spanish Bayonet, Aloe Yucca	Unknown
2	Angustissima	Narrow-leaf yucca	Stable
3	Arkansana	Arkansas yucca	Stable
4	Baccata	Banana yucca	Stable
5	Baileyi	Bailey's Yucca	Stable
6	Brevifolia	Western Joshua Tree	Decreasing
7	Campestris	Plains yucca	Decreasing
8	Capensis	Cape Region Yucca	Decreasing
9	Carnerosana	Palma Samandoca	Stable
10	Cernua	Nodding Yucca	Decreasing
11	Coahuilensis	Coahuila Texas Yucca	Decreasing
12	Coonstrica		Stable
13	Decipiens	Chinese Palm	Decreasing
14	Declinata		Unknown
15	Elata	Soapweed	Decreasing
16	Endlichiana	Pitilla	Decreasing
17	Faxoniana		Unknown
18	Filamentosa		Unknown
19	Filifera	Izote	Stable
20	Flaccida		Stable
21	Glauca	Great Plains Yucca	Stable
22	Gloriosa	Spanish Dagger	Decreasing
23	Grandiflora	Sahuiliqui	Decreasing
24	Guatemalensis		Decreasing
25	Harrimaniae	Harriman's Yucca	Unknown
26	Intermedia		Stable
27	Iouisianesis	Gulf Coast Yucca	Stable
28	Jaegeriana	Eastern Jashua Tree	Decreasing
29	Jaliscensis		Decreasing
30	Kanabensis	Kanab Yucca	Unknown
31	Lacandonica	Quim	Decreasing
32	Linearifolia		Decreasing
33	Madrensis	Mountain Yucca	Unknown
34	Mixtecana		Stable
35	Nana		Stable
36	Necopina	Glen Rose Yucca	Decreasing
37	Neomexicana	New Mexico Yucca	Stable
38	Pallida	Pale Leaf Yucca	Decreasing
39	Periculosa		Decreasing
40	Pinicola	PlamaDatil	Stable
41	Potosina		Decreasing
42	Queretaroensis		Decreasing

(continued)

Table 13.2 (continued)

	Yucca	Common Name(s):	Population Trend
43	Reverchonii	San Angelo Yucca	Decreasing
44	Rigida	Yucca Azul	Decreasing
45	Rostrata	Big Bend Yucca	Decreasing
46	Rupicola	Twisted leaf Yucca	Stable
47	Schidigera	Moave Yucca	Decreasing
48	Schotti	Amole	Stable
49	Thompsoniana	Trans Pecos Yucca	Stable
50	Toftiae		Decreasing
51	Torreyi		Stable
52	Treculeana	Spanish Bayonet	Decreasing
53	Utahensis		Decreasing
54	Valida	Palma	Stable

Kingdom Plantae – plantes, Planta, Vegetal, plants
Subkingdom Viridiplantae – green plants
Infrakingdom Streptophyta – land plants
Superdivision Embryophyta
Division Tracheophyta – vascular plants, tracheophytes
Subdivision Spermatophytina– spermatophytes, seed plants, phanérogames
Class Magnoliopsida
Superorder Lilianae – monocots, monocotyledons, monocotylédones
Order Asparagales
Family Asparagaceae
Genus *Yucca* L. – yucca species, yucca

13.3 Yucca Fiber

Yucca, an ancient plant fiber, has been used in various ways throughout Central America. The early Pueblo, Hohokam, and Cave cultures all found major applications for yucca fiber and split leaves in the production of sandals. The fiber was commonly used as a tying material by all cultures, woven into nets and fabrics, and served as the foundation for feather and fur cloth. The leaves were extensively used in creating mats and baskets. Yucca was favored in industries where Indians used fiber due to its high accessibility, with the species closest to hand being the one most commonly used (Bell and Castetter 1941).

13.3.1 Fiber Extraction Method

Bell and King (1944) found that the arrangement of the vascular system in relation to the accompanying fibers is a defining feature of the genus, with no discernible variation among the different species such as *Y. baccata*, *Y. elata*, *Y. torreyi*, *Y. carnerosana*, *Y. glauca*, and *Y. thompsoniana* (Bell and King 1944). Yucca possesses a fibrous group that encloses a vascular bundle on both ends. The leaves of the different Yucca species display fibrous clusters spread throughout the internal tissue. These clusters can include xylem and phloem groups, covered with fibers on both ends, or they can solely consist of fibers without any accompanying vascular tissue. Among these bundle types, the former is highly distinctive to the genus and thus serves as a valuable means of identification.

According to the authors, the fiber obtained from *Y. glauca* leaves was glossy white in appearance and was the same length as the leaf. However, the fiber from *Y. aloifolia* L. is of fair quality, the leaves are too short and difficult to obtain to be useful for fiber production. The sisal hemp machine used in 1983 to mechanically extract *Y. filamentosa* L. fiber produced very dark, yellowish-colored, rough, and slightly brittle fiber that was weaker than commercial cordage fibers and considerably weaker than Manila and common hemp (Bell and Castetter 1941). Coarse fiber can be obtained from the leaves of *Y. Treculeana* Carr., and fine fiber may be extracted from *Y. Schottii* jaliscensis Trel.

Yucca fiber has relatively high tenacity in general, but it is brittle and cannot easily be softened. The filaments often have a glossy white appearance and are composed of irregular bundles, most of which are large. These bundles split into finer fibers when rubbed vigorously between the fingertips, but they often retain a substantial bit of rigidity. The fibers often have strong walls and a clear, hollow core. The ends are rounded at the extremities and routinely become slim (Bell and Castetter 1941).

Stevenson believed that the yucca fiber prepared by Indians relied on chewing. Upon gathering central or freshly grown leaves, each leaf is neatly wrapped around a 4-inch span and secured using a fragment of another leaf. These folded leaves undergo boiling along with cedar ashes. Once the folds have been cooked and cooled, the outer layer is carefully peeled off by young individuals. The leaf is then chewed from one end and gradually drawn into the mouth before being chewed from the other end. Following the chewing process, the fibers are separated, straightened, and arranged on the ground. They are then bundled, tied together with strings, and suspended within an interior chamber. Osborne (1965) thought that chewing was not a part of the process for making cordage because mature leaves taste revolting and weedy and contain a significant amount of saponin, a potent laxative. In general, fibers collected from the inner (younger) leaves of yucca species display a finer nature when seen on a larger scale than fibers obtained from the outer, more mature leaves. Prior to leaf opening, the fibers harvested from the new growth were particularly soft, white, and delicate; however, it is important to note that these fibers lacked several characteristics (Osborne 1965). Osborne (1965) discovered that yucca fiber may be extracted either mechanically (with a Hummer's scraper) or by boiling and soaking the leaves for a predetermined amount of time (Osborne 1965). According to Bell's 1941 ethnobiological research in the American Southwest (Bell and Castetter 1941), before the leaves were put into hot water barrels and heated to boiling point to prepare, they were crushed between two cylinders to eliminate non-fibrous tissues. The crushed bulk was then supported by hurdles in order to maintain the fiber's straightness and separation. The bundles were then dropped into a boiling bath of water and ashes that was used to create an alkaline solution (Bell and Castetter 1941).

Recently, researchers have directed their focus toward extracting fibers from yucca leaves and investigating their mechanical and chemical properties. This shift in attention is a result of advancements in testing apparatuses and procedures.

Azanaw et al. (2019) conducted a study on the extraction of cellulose fibers from *Yucca Elephantine* plant leaves (Azanaw et al. 2019). They employed two methods: water retting (for 26 days) and chemical retting using sodium hydroxide at concentrations of 3, 5, and 10%. The process involved collecting raw yucca leaves, immersing them in water, and subjecting them to the chemical treatment at a temperature of 100 °C for 2 hours, with a mass-to-volume ratio of 20:1. The results of the study indicated that the extracted fibers possessed a relatively good tensile strength. In the water retting method, the fibers exhibited a lower tensile strength than those in the chemical retting process using 3% sodium hydroxide (Azanaw et al. 2019). Extraction of fibers from *Yucca Gloriosa* leaves by mechanical method and using blades has been studied by different researchers (Soltani et al. 2020; Taban et al. 2020; Samaei et al. 2020, 2021). Moghaddam and colleagues, in their study, employed a chemical approach to derive cellulose fibers from yucca leaves. They utilized varying concentrations of sodium hydroxide, as well as different temperatures and durations (Moghaddam and Karimi 2022). Their findings revealed that bleaching allows for the production of stronger and whiter fibers (Moghaddam and Karimi 2020). Additionally, achieving cellulose microfibers necessitates subjecting the plant's leaves to recurrent alkaline treatment and alkaline oxidation (Moghaddam and Torabi 2022). Belaadi et al. (2022) conducted an extraction of cellulose fiber from *Yucca Treeleana* L. by immersing the leaves in water for 10–20 days at temperatures ranging from 25 to 35 °C (Belaadi et al. 2022). Nascimento et al. used the pull-out method to obtain lignocellulose fibers from *Yucca aloifolia L.* leaves (Nascimento et al. 2021).

13.3.2 Fiber Morphology

The Yucca fiber is composed of individual fibers that are connected and coated with waxy and gum substances such as lignin, pectin, and others (Nascimento et al. 2021; Moghaddam and Karimi 2022). By subjecting the fiber to boiling in an alkaline solution or immersing it in a water and microbial solution, the non-cellulosic components, namely pectin and hemicelluloses, are eliminated (Moghaddam 2022,

2024). This process leads to an enhancement of the fiber's fineness (Moghaddam and Karimi 2020, 2022). The study by Azanaw found that the extracted fiber maintains a consistent and even diameter throughout its length. Furthermore, the fibers extracted using the chemical process are whiter and have a smoother surface compared to fibers obtained using the water-retting method (Azanaw et al. 2019). Ekunsanmi and Tripathi also demonstrated that fibers extracted chemically have a whiter appearance than fibers extracted microbially (Ekunsanmi and Tripathi 2019). The microbial retting process results in fibers that have a slight greenish hue due to leftover plant materials.

The scanning electron microscope image of fibers shows that the mechanically extracted fibers are covered with binding components (Fig. 13.2a, b). Additionally,

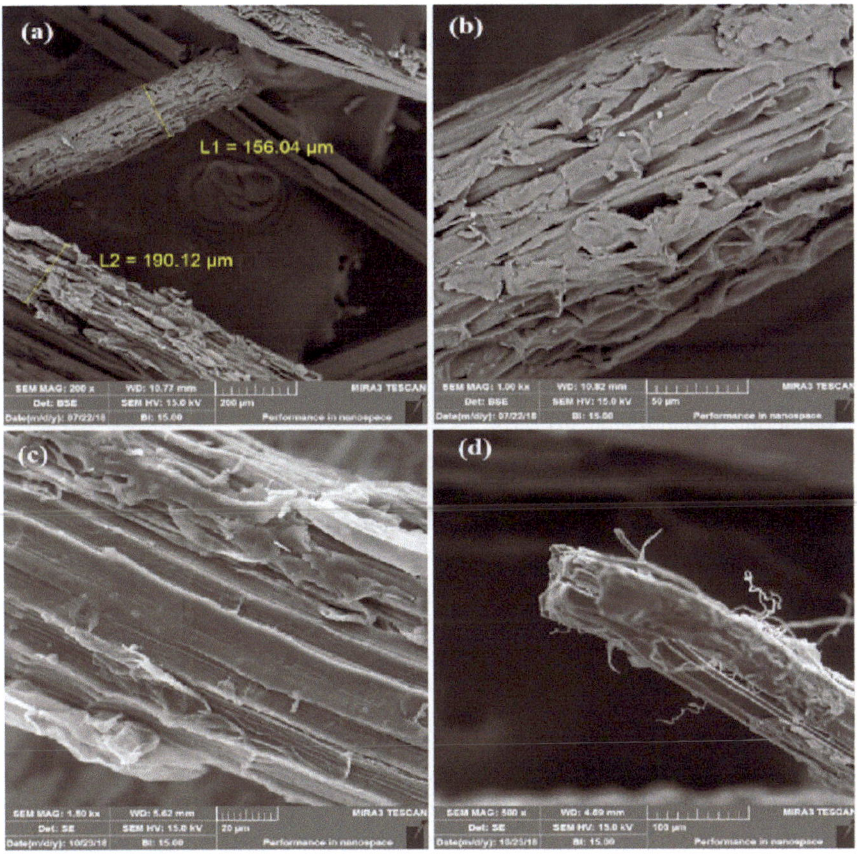

Fig. 13.2 Scanning electron microscopic image of the surface of yucca fiber: (**a**, **b**) mechanically extracted fibers and (**c**, **d**) alkaline treated fiber. (Source: Taban et al. (2020))

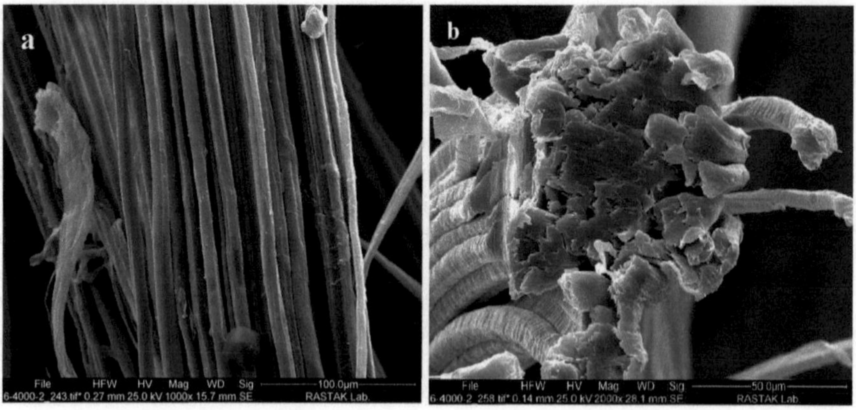

Fig. 13.3 Scanning electron microscopy of the bleached yucca fibers: (**a**) longitudinal section and (**b**) cross-section. (Source: Moghaddam and Karimi (2020))

the alkaline treatment of fibers led to a cleaner and softer surface of the fiber (Fig. 13.2c, d) according to Taban et al. (2020).

Performing more intensive chemical treatments, such as increasing alkali concentration, temperature, and treatment time, results in greater separation of fibers from each other. As depicted in Fig. 13.3a, fibers treated with alkali and bleached, with non-cellulosic materials removed, demonstrate high fineness and a smooth surface. The irregular and non-uniform cross-sectional shape of the fiber is visible in Fig. 13.3b (Moghaddam and Torabi). Belaadi et al. discovered that Yucca Treeleana L fibers possess an elliptical shape, which is a crucial factor for composite applications (Beladdi et al. 2022).

The researcher discovered that the diameter of fibers isolated using a chemical solution was smaller compared to those obtained through water retting. Additionally, as the concentration of alkali increased, the extracted fiber diameter decreased (see Table 13.3). This decrease in diameter can be attributed to the removal of non-cellulosic components present in the fibers (Azanaw et al. 2019).

13.3.3 Fiber Composition

Yucca fiber is a lignocellulosic fiber composed mainly of cellulose, lignin, and hemicellulose. The percentage of each compound in the fiber depends on the retting method used (Table 13.4). Separating fiber from plant leaves through water retting results in fiber with a higher percentage of cellulose and hemicellulose compared to the chemical method using a low alkaline concentration (Azanaw et al. 2019). Unlike fibers obtained from water retting, chemically extracted fibers have fewer impurities on the surface and smaller diameters. The mean fiber diameter is related to the impurities removal percentage during alkali treatment. By oxidative treatment of

Table 13.3 Retting methods for yucca fiber extraction

Retting method		Retting Conditions	Fiber Diameter (μm)	References
Water retted	Untreated river water,	Room temperature, 26 Days	242	Azanaw et al. (2019)
	Unknown water	Room temperature, 10–20 Days	205	Belaadi et al. (2022)
Chemical retted	3% NaOH	Boiling Temp., 2 h, L:R: 20:1	397	Azanaw et al. (2019)
	5% NaOH	Boiling Temp., 2 h, L:R: 20:1	214	Azanaw et al. (2019)
	10% NaOH	Boiling Temp., 2 h, L:R: 20:1	135	Azanaw et al. (2019)
	4% NaOH	4 h, 80 °C, L:R: 30:1	12–65	Moghaddam and Karimi (2022)
	3% NaOH	2 h, 100 °C, L: R: 30:1	12–65	Moghaddam and Karimi (2022)
Mechanical extraction and chemical treatment	5% NaOH	3 h, 25 °C	72–168	Taban et al. (2020)
Chemical retted and oxidative treatment	3% NaOH (twice time); 6%H_2O_2	2 h, 100 °C; 1 h, 100 °C	7.22	Moghaddam and Torabi (2022)
Chemical retted and oxidative treatment	4% H_2O_2, 4% Sodium silicate, 0.5% NaOH	1 h, 90 °C	6–15	Moghaddam and Karimi (2020)
Pull-out	–	–	60–147	Nascimento et al. (2021)

yucca fiber with hydrogen peroxide, researchers found that more non-cellulosic material was removed from the fibers and the physical and chemical properties of the treated fibers were increased (Moghaddam and Karimi 2020).

Of course, as in the chemical method, factors such as temperature, time, type of substances and concentration of chemical materials are important. In the water retting method, however, water temperature, type of water, and duration of treatment are key factors to consider (Azanaw et al. 2019; Moghaddam and Karimi 2022). It has been observed that increasing chemical treatment factors lead to the removal of lignin, hemicellulose, and other non-cellulosic components from the fiber. However, it should be noted that higher alkali concentration results in a decrease in cellulose percentage.

In general, lignocellulose fiber extracted from yucca leaves has a cellulose content similar to that of other natural cellulose fibers (Table 13.5). Increasing the cellulose content in the fiber enhances its mechanical, thermal, moisture absorption and chemical properties. Therefore, lignocellulosic fibers with a high percentage of cellulose are always more suitable for industrial applications.

Table 13.4 Yucca fiber composition

Retting method		Retting Conditions	Cellulose	Hemicellulose	Lignin	References
Water retted	Untreated river water,	Room temperature, 26 Days	64.2	22	11	Azanaw et al. (2019)
	Unknown water	Room temperature, 10–20 Days	82.3	12.1	5.6	Belaadi et al. (2022)
Chemical retted	3% NaOH	Boiling Temp., 2 h, L:R: 20: 1	66	23	13	Azanaw et al. (2019)
	5% NaOH	Boiling Temp., 2 h, L:R: 20: 1	59	17	10	Azanaw et al. (2019)
	10% NaOH	Boiling Temp., 2 h, L:R: 20: 1	57	14.5	5.5	Azanaw et al. (2019)
Mechanical and chemical treatment	5% NaOH	3 h, 25 °C	66.4	19.7	6.7	Taban et al. (2020)
Pull-out	–	–	52.5	20.5	10	Nascimento et al. (2021)

Table 13.5 Cellulose fiber composition

Fiber	Cellulose	Hemicellulose	Lignin
Yucca	57–82	14–23	5–13
Hemp	68	15	10
Jute	61–71	14–20	12–13
Kenaf	72	20.3	9
Ramie	68–76	13–16	0.6–0.7
Sisal	65	12	9.9
Flax	71	18.6–20.6	22
Bamboo	26–43	30	21–31

Source: Ngo (2018)

13.3.4 *Thermal Analysis of Fiber*

The thermal degradation of lignocellulose fibers is influenced by both non-cellulosic components and the crystalline arrangement of cellulose (Moghaddam et al. 2024; Sonia and Dasan 2013). The breakdown of lignocellulose fibers occurs in two stages: amorphous phase degradation, which involves the degradation of components like lignin and hemicelluloses, and crystalline phase degradation, which involves the degradation of cellulose (Dorez et al. 2014). The crystalline phase,

Table 13.6 Thermal analysis of yucca fiber

Fiber extraction conditions	Peak 1	Peak 2	Peak 3	Char at 600 °C	References
3% NaOH, 100 °C, 2 h, one time	49.6	332.8	445.8	5.92	Moghaddam and Torabi (2022)
3% NaOH, 100 °C, 2 h, twice time	39.1	310.1	410.1	1.49	Moghaddam and Torabi (2022)
Cellulose microfiber	39.4	333.4	422.9	0.07	Moghaddam and Torabi (2022)
4% NaOH, 80 °C, 4 h	–	322.12	–	43.3	Moghaddam and Karimi (2022)
4% NaOH, 100 °C, 2 h	–	317.6	–	33.7	Moghaddam and Karimi (2022)
3% NaOH, 100 °C, 2 h	110	330	–	25	Azanaw et al. (2019)
5% NaOH, 100 °C, 2 h	–	340	–	21	Azanaw et al. (2019)
10% NaOH, 100 °C, 2 h	–	350	–	21	Azanaw et al. (2019)
Water retted, Room temperature, 10–20 Days	120	293.5	360.6	–	Belaadi et al. (2022)
Mechanical retted	55–110	260–315	340	26.5	Taban et al. (2020)
Mechanical retted, 5% NaOH, 3 h, 25 °C	55–110	260–335	364	6.5	Taban et al. (2020)
Pull-out	<100	294.7	357.7	18	Nascimento et al. (2021)

characterized by the ordered arrangement of cellulose chains, exhibits greater thermal stability compared to the amorphous phase. Therefore, the presence of a well-organized crystalline structure in the fiber contributes to its higher thermal stability.

Lignin, one of the non-cellulosic components, plays a role in the production of char during the thermal degradation of lignocellulose fibers. It is responsible for the formation of residual carbonaceous material. Consequently, raw lignocellulose fibers generally have a higher ash content due to the presence of more lignin. However, during treatment processes such as chemical extraction, lignin is often removed, resulting in a lower ash level in the treated fiber (Moghaddam and Torabi 2022). Table 13.6 shows the thermal characteristics of the yucca fiber. When chemically removing fibers with NaOH (sodium hydroxide), the weight loss of the fiber is proportional to the concentration of NaOH applied. As the NaOH concentration increases, the weight loss decreases, and the beginning degradation temperature (the temperature at which degradation starts) increases. This phenomenon may be attributed to the elimination of non-cellulosic components by higher concentrations of NaOH during the treatment process (Azanaw et al. 2019; Moghaddam and Karimi 2020; Taban et al. 2020).

Based on the information provided, Thermogravimetric Analysis (TGA) determined that Yucca fiber is thermally stable up to 260 °C. This indicates that it meets the requirements for thermal stability as a natural cellulosic material, making it suitable for use in thermoplastic polymeric composites.

13.3.5 Mechanical Properties of Fiber

The retting methods such as chemical, mechanical, and water retting can affect the physical properties of the retted fibers. Table 13.7 shows the physical properties of yucca fibers. Research conducted by Azanaw et al. (2019) indicates that the concentration of alkaline substances, specifically sodium hydroxide, can influence the tensile strength of extracted fibers. Increasing the concentration of sodium hydroxide during the extraction process leads to a decrease in the strength of the obtained fiber. Chemical retting at an optimal concentration can effectively remove non-cellulosic impurities from fiber surfaces, enhance fiber crystallinity and improve tensile strength. This is primarily achieved through high alkali penetration in the amorphous region of the cellulose structure. However, at high alkali concentration levels, the decrease in strength can be attributed to the partial degradation of lignin and hemicellulose within the crystalline structure of cellulose. These components are responsible for binding cellulose chains together, and their degradation can weaken the overall strength of the fibers (Moghaddam and Karimi 2022). Comparing the strength of fibers extracted through chemical retting and water retting, it is found that water-retted fibers exhibit higher strength compared to fibers extracted using a chemical solution with a high concentration of alkaline substances (Azanaw et al. 2019). However, Taban et al. found that the action of mechanically extracted fibers with alkaline substances increases the strength of the fibers (Taban et al. 2020).

The overall findings indicate that various techniques for extracting yucca fibers result in commendable tensile strength, modulus, and elongation. When compared to alternative plant fibers sourced from leaves like sisal or stem fibers like jute, hemp, and kenaf, yucca fibers exhibit favorable mechanical characteristics (Azanaw et al. 2019; Nascimento et al. 2021; Taban et al. 2020).

Table 13.7 Mechanical properties of the yucca fiber

Fiber extraction conditions	Tenacity (MPa)	Elongation (%)	Modulus (GPa)	References
3% NaOH, 100 °C, 2 h	350–480	–	–	Moghaddam and Karimi (2022)
Water retted, Room temperature, 10–20 Days	222	2.47	16.85	Belaadi et al. (2022)
Mechanical retted	573	7.3	10.4	Taban et al. (2020)
Mechanical retted, 5% NaOH, 3 h, 25 °C	617	6.4	10.9	Taban et al. (2020)
Pull-out	801	2	39	Nascimento et al. (2021)

13.3.6 *Acoustic Behavior of Fiber*

Taban et al. (2020) focused on creating a sound insulation layer using Yucca fibers. They prepared a composite layer by mixing Yucca Gloriosa fibers with a 5% polyvinyl alcohol binder. The process involved pressing the leaves on a wood stick, stripping the fibers by hand, and soaking them in a 5% alkaline solution at room temperature for 3 hours. The researchers found that a layer made with fiber thicknesses ranging from 20 to 40 mm had an absorption coefficient of 0.75–0.9 at frequencies of 2000 Hz.

Soltani et al. (2020) utilized water retting and sharp wood to extract fibers from non-fibrous Yucca Gloriosa leaves. They demonstrated that a composite layer made of Yucca fibers and a polyvinyl alcohol binder effectively insulated sound. Layers with thicknesses of 15 and 30 mm had the highest sound absorption coefficient at frequencies of 4000 and 2500, respectively.

In a study by Samei et al. (2021), it was found that fibrous structures exhibit strong absorption in high-frequency bands, while their performance in low-frequency bands is primarily determined by thickness and flow resistance. The researchers noted that the peak sound absorption coefficient for 100% Yucca fiber samples occurred at 2500 Hz. This peak shifted towards lower frequency bands as the kenaf component content in the kenaf/yucca fiber composite increased. Increasing the kenaf component content was found to enhance sample absorption in low-frequency bands. Samples containing 70% kenaf and 30% yucca showed superior absorption performance at both low and high frequencies. The average fiber diameter for yucca and kenaf fibers in the study was 130 μm and 80 μm, respectively.

13.4 Germplasm Biodiversity and Conservation

13.4.1 *Germplasm Diversity*

Biological and plant geneticresources are crucial food and agricultural assets, as both food and industrial production rely on genetic variation among different strains and wild varieties (Leon-Lobos et al. 2025). Humans utilize and cultivate these genetic resources to adapt to changing environmental conditions and bridge the gap in food and industrial needs.

Plants have a limited genetic base, making them susceptible to environmental threats. Genetic resources are currently being lost at an alarming rate. As these resources diminish, humans lose the ability to adjust to new social, economic, and environmental challenges.

According to Arce-Montoya et al. (2006), clonal lines may have a higher internal coefficient of variation compared to plants grown from seeds, but they still exhibit significant heterogeneity. Some clonal lines show superior performance, indicating that cloning and selection can produce fast-growing individual lines for plantations without depleting natural resources.

13.4.2 Cultivars Characterization and Phylogeny

The description and classification of species depend on the relationship between Yuccas and yucca moths, as confirmed by Olle Pellmyr et al. (2005). Yuccas and yucca moths (Lep., Prodoxidae) are widely known for their obligatory mutualistic pollination. Yuccas are home to several non-pollinating yucca moths in addition to pollinators. Here, morphological and molecular data are used to revise the genus Prodoxus, the non-pollinating sister group of the pollinators, analyze their evolutionary relationships and investigate the evolution of host tissue specialization. Twenty-two species have been identified, including *Prodoxusgypsicolor* sp.n., *P. sonorensis* sp.n., *P. carnerosanellus* sp.n., *P. tamaulipellus* sp.n., *P. weethumpi* sp.n., *P. tehuacanensis* sp.n., *P. californicus* sp.n., *P. mapimiensis* sp.n., and *P. atascosanellus* sp.n. Descriptions of Prodoxus y-inversus Riley, *P. coloradensis* Riley, and *P. sordidus* Riley have been rewritten. Prodoxus and the genus Agavenema are synonyms. According to phylogenetic analysis, stalk-feeding is the most primitive behavior within the group, leaf-mining has one genesis from a stalk-feeding progenitor and fruit-feeding has three distinct origins. While it is common for species with dissimilar eating patterns to live in the same host, the findings indicate that ecological specialization and diversification may have only happened in one or perhaps two hosts. As explained by Olle Pellmyr et al. (2008) yucca moths (Tegeticula and Parategeticula; Lepidoptera, Prodoxidae) are the exclusive pollinators of yuccas, in an obligatory association. The number of described species has increased from four to twenty as a result of recent revisionary work conducted in recent years, which revealed significantly more species diversity than previously thought. Drawing from field surveys conducted in Mexico and analysis of specimens, we present the following five new species: *T. californica*, *Pellmyr sp.* Nov. and *Tehuacana T. tambasi*, *Pellmyr & Balcázar-Lara sp.* nov. *T. baja, Pellmyr & Balcázar-Lara sp.* nov. *Balcázar-Lara* sp. nov. and *P. ecdysiastica*, *Pellmyr & Balcázar, Lara & Pellmyr sp.* nov. Trevuleanella tegeticula It is determined that Pellmyr is a junior synonym of *T. mexicana Bastida.* All adult species of the *T. yuccasella* complex have a diagnostic key available. The two pollinator taxa' monophyly was supported by a phylogeny based on a 2104-bp stretch of mitochondrial DNA (mtDNA) in the cytochrome oxidase I and II areas, and the 17 species of the *T. yuccasella* complex that have been identified were strongly supported. A recent and highly diversified collection of 11 taxa has most associations well supported, but others are less so and in one case conflict with a whole-genome data set (amplified fragment length polymorphism, or AFLP). In one of the traditional models of coevolution, the latest mtDNA-based analyses, in conjunction with previously released AFLP data, offer a solid phylogenetic framework for future research on life-history evolution and host interactions.

13.4.3 Genetic Resources Conservation Approaches: In Vitro, In Situ, Cryopreservation and Gene Bank

Plant genetic diversity conservation is essential for the future of agriculture, food security, and nutrition, as well as for ensuring that future generations can adapt to global climate change (Joshi 2019). Twenty-four strategies are being considered for the conservation of unconventional plant species in Nepal. In-situ strategies include protected areas, Ramsar sites, World Heritage Sites, community forests, and legal protection for non-orthodox plant species. Ex-situ strategies involve field gene banks, botanical gardens, city parks, government farms, religious sites, in-vitro cultures, and evolutionary plant breeding (Abulaila et al. 2025; Abdullaev et al. 2025; Akparov et al. 2025; Aladele et al. 2025; Al Lawati 2025; Almarri et al. 2025; Arab et al. 2025; Békefi et al. 2025; Breidy 2025; Cañar et al. 2025; de Galarreta et al. 2025; Desheva et al. 2025; Díez et al. 2025; Galea-Gragera and Llera Cid 2025; Ghidan and Abdulmajid 2025; Gudaye et al. 2025; Harzli et al. 2025; Hidayatun and Kurniawan 2025; Ihsan and Prihatini 2025; Jatoi and Siddiqui 2025; Laghetti et al. 2025; Mallor and Luna 2025; Monteros-Altamirano et al. 2025; Osterc et al. 2025; Popović et al. 2025; Rocha et al. 2025; Wang et al. 2025). On-farm strategies include community field gene banks, school field gene banks, household field gene banks, community/public orchards, field gene banks at the village level, geographical indication, and participatory landrace enhancement. The most effective method is to establish field gene banks, which have been implemented in approximately 20 research stations. This approach should be expanded to include government farms and agriculture offices to safeguard the regional APGRs found in their respective regions.

According to Dang, biodiversity is not only reflected in species composition, ecosystem diversity, and genetic resources, but also plays a crucial role in providing biodiversity services (provisioning, regulatory, cultural, and support services) that significantly contribute to the national economy by ensuring food security, preserving animal and plant genetic resources, and providing materials for construction, fuel, and pharmaceutical resources. Biodiversity for food and agriculture has undergone significant changes due to various factors, including changes in land and water use, pollution, over-exploitation, climate change, natural disasters, and other threats. There is a lack of uniform laws and regulations protecting biodiversity; community involvement in biodiversity conservation is inadequate, which results in weak law enforcement; deforestation and the illegal wildlife trade pose serious threats to biodiversity; overall investments in biodiversity are insufficient, leading to a shortage of financial, human, and technological resources and a lack of effective intersectoral coordination mechanisms to respond to overlap in functions among relevant ministries and agencies.

To meet national and CBD targets, the following priority activities are recommended: (1) Enhance state management of biodiversity by defining the roles and responsibilities of the Ministry of Agriculture and Rural Development (MARD) and MONRE in managing biodiversity; (2) Encourage more intimate and coordinated collaboration between important agencies and stakeholders in conservation; (3) Enforce laws and regulations pertaining to biodiversity conservation; and (4) Raise funding for biodiversity conservation, especially related to food and agriculture. These funds should be allocated toward creating an inventory of biodiversity, establishing a monitoring system to track changes in biodiversity, developing and managing a database system to share, exchange, and manage data, enhancing staff capabilities, promoting oversight of law enforcement, and ultimately increasing state budget allocations for biodiversity conservation. Ensuring the maintenance of a national network of protected areas (forest, marsh, and terrestrial areas); and ensuring that vital ecosystems are preserved. Important ecoregions should prioritize protecting Protected Areas. Encourage the management and protection of biodiversity at the three levels of ecosystem, species, and genetic diversity. Maintaining and developing genetic resources through the completion of an inventory, the gathering of data on biodiversity resources, and the integration of indigenous knowledge across the country; Developing risk management and risk control of alien species, with a special focus on genetically modified organisms (GMOs), their use, and any potential impacts on the environment, biodiversity, and human health; Analyzing and evaluating the role of biodiversity in response to climate change and proposing appropriate solutions; Encourage the incorporation of biodiversity conservation into development strategies, plans, and initiatives at the federal, provincial, and ministry levels; Boost funding for biodiversity conservation and ensure the public budget for conservation is managed well and Preserve and encourage assistance.

13.5 Cytogenetics

Mohammed and Najem (2020) investigated the cytogenetic effects of Yucca elephantine leaf crude saponins on mitosis using Allium cepa L. root tips as a plant test system in vivo. The root tips of *A. cepa* were exposed to four different concentrations of crude saponins (0.00, 6.25, 12.5, 25, or 50 mg/ml) for four hours. Cytogenetic diagnoses considered in this investigation included chromosomal aberrations, mitotic index, and phase index.

The findings revealed that saponins decreased the mitotic index (MI) to less than 50%. Specifically, the mitotic index increased to 25.48%, 17.98%, and 18.81% when treated with 12.5, 25, or 50 mg/ml, respectively. This reduction was considered hazardous and sublethal. At dosages of 25 and 50 mg/L, saponins caused 1.30–8.30% of chromosomal abnormalities, including micronuclei and nuclear lesions (chromatin lesions) in prophase and interphase. They also arrested chromosomes at metaphase (c-metaphase) up to 100% at 6.25 mg/ml (Mathew et al. 2024).

As methods for assessing chromosomal number and structure continue to evolve, the field of cytogenetics has seen significant advancements. The detection of chromosome alterations has progressed from bands to fluorescence in situ hybridization (FISH) to chromosomal microarrays, similar to the evolution of single nucleotide variation detection from Sanger sequencing to next-generation sequencing. Emerging technologies like genome sequencing and optical genome mapping have greatly improved clinical laboratory testing in cytogenetics.

In cytogenetic diagnostics, techniques such as banding and focused evaluations like FISH remain essential. These approaches allow for a quick turnaround time, especially for diseases with known etiologies involving identified cytogenetic abnormalities. Furthermore, laboratories can now utilize more flexible, higher-throughput methods to improve resolution in more complex cases.

13.6 Traditional Breeding

13.6.1 Improvement Strategies

Improvement can occur either through natural adaptation to changing environmental conditions, as discussed by Rickelle et al. (2024). The chaparral yucca, also known as Hesperoyucca Whipplei (HW), is a native yucca species found in the southwestern states of the United States, specifically Arizona and California. This plant is a fascinating subject for investigating the mechanical characteristics of its physical structures due to its adaptation to dry, arid environments. Research has shown that chaparral yucca possesses unique physical and structural traits that enable it to resist environmental stresses and external influences. For example, the fibrous structure of the yucca's leaves and stems provides both rigidity and flexibility, allowing them to withstand physical pressures like wind and bend without breaking. In order to determine the mechanical properties of HW and understand its unique physical and structural features, tests were conducted for this research. Different plant parts, such as complete leaf samples, fiber bundles, braided fibers, and single fibers, underwent tensile tests. Additionally, the outer bark of stalks was tested using 3-point bending. Analyzing the mechanical characteristics of HW can provide insights into how the plant adapts to its environment and can lead to advancements in materials science and engineering.

Additionally, improvement can occur through hybridization between different varieties as explained by Karolina et al. (2021). In plants, hybridization can result in significant phenotypic and genotypic changes that impact hybrid physiology, ecology, and overall fitness levels. *Yucca gloriosa*, a recent homoploid hybrid species within the Asparagaceae family, originated from a cross between Yucca filamentosa and *Yucca aloifolia*. Previous studies identified a solitary *Y. aloifolia* genotype as the maternal parent of *Y. gloriosa* through analysis of nuclear-encoded microsatellites and approximately 11 kb of the chloroplast genome. To assess the hybrid origins of *Y. gloriosa*, chloroplast genomes from 41 accessions of all three species were reconstructed using whole genome resequencing. Additionally, mRNA-seq was used to examine transposon transcription and re-sequencing data to annotate transposon abundance in the three species. Haplotype analysis and chloroplast phylogeny suggest that multiple hybridization events, with both parental species acting as the maternal donor, contributed to the formation of *Y. gloriosa*. The three species exhibited significant differences in transposon abundance at the superfamily level with the hybrid often showing similarities to one or both parents or being intermediate in transposable element (TE) superfamily abundance. *Y. gloriosa* displayed higher abundance in Copia LTR transposons compared to either parent in a single instance. Limited transcriptional activity of transposons was observed in the expression patterns across the three species, indicating that either *Y. gloriosa* did not activate transposons upon hybridization, or any activeated transposons were quickly silenced. The identification and quantification of transposon families, along with expression data, provide opportunities for further exploration of the relationship between epigenetics and the significant phenotypic variation observed in this homoploid hybrid system.

13.6.2 Traditional Breeding Methodologies

Traditional breeding is considered one of the most effective ways to address many of the challenges facing plant production, especially in light of climate change population growth and increased pollution, which have led to the emergence of new diseases. Among the traditional breeding methods, hybridization between different varieties stands out. This process can transfer one or more traits from the parents, but it is time-consuming and unpredictable in terms of the traits that will be inherited.

Another breeding method is mutations, whether they occur naturally or are induced in the field or laboratory. This approach is quicker than hybridization, but like hybridization, it is difficult to control which trait will be introduced. Selection and evaluation are also important traditional breeding methods. Some strains are chosen and evaluated under different climatic conditions than those in which they were originally grown.

In conclusion, traditional breeding methods play a crucial role in improving plant production and addressing the challenges posed by changing environmental conditions and population growth.

13.6.3 Traditional Breeding Limitations

Some restrictions can hinder the hybridization process, even if the percentage of its occurrence is low (Immaculate et al. 2019). Hybrids can have either positive or negative effects. One advantageous quality of hybrids is heterosis, which can arise through epistasis, dominance, or over-dominance. The negative effects include embryo abortion, infertility, and halted pollen tube growth. Chromosome doubling, the application of hormones like 2, 4-Dichlorophenoxyacetic acid (2, 4-D), and embryo rescue have been used to address these issues, respectively, regarding sterility, pollen tube growth arrest, and embryo abortion (Karolina et al. 2021).

Additionally, hybridization can disrupt the epigenetic regulation of transposable elements, leading to their overabundance. The variations in ploidy that occur in hybrid plant species often complicate our understanding of the processes that maintain genomic integrity after hybridization. Homoploid hybrid species, which have the same number of chromosomes as their parents, offer a unique opportunity to study the genomic effects of hybridization when ploidy is not altered.

13.6.4 *Role of Biotechnology*

Technology plays a major role in meeting the increasing demand for food, addressing numerous challenges in agriculture, and providing explanations for various uncertainties. According to Saurav et al. (2023), the objective of sustainable agriculture is to produce enough food and fiber to meet current needs while conserving and enhancing natural resources for future generations. Achieving agricultural sustainability involves striking a balance between social responsibility, environmental stewardship, and economic viability. This can be particularly difficult when dealing with biotic and abiotic stressors such as diseases, pests, climate change, soil degradation, and water scarcity. The main barrier to sustainable agriculture is the abundance of pests and illnesses that can drastically reduce crop yields and quality. To solve these problems, biotechnology can be utilized to produce crops that are resistant to illnesses and pests. Another barrier to sustainable agriculture is a lack of nutrients in the soil, which can lower crop yields and plant health. More nutrient-dense and productive crops could be developed with the help of biotechnology. On the other hand, it is imperative to guarantee that these technologies are developed responsibly and that the advantages they bring about are shared fairly among communities and regions (Khatik 2023).

The production of food is greatly impacted by several difficulties that agriculture is currently grappling with. These issues include deforestation, soil degradation, heavy metal and hydrocarbon contamination, and one potential solution is plant biotechnology. Biotechnology is a cutting-edge method that helps preserve and safeguard the environment by, for example, using fewer chemical pesticides and herbicides. Biotechnology includes, among other things, genetic engineering, plant tissue culture, marker-assisted technology and nanotechnology.

13.7 Molecular Breeding

Ranranet et al. (2023) discovered that Yucca, a member of the class B flavin-dependent monooxygenases, plays a role in regulating plant growth and stress response by catalyzing a crucial step in the synthesis of endogenous auxin. A systematic analysis of the Yucca gene family and its stress response can benefit breeding applications and help dissect regulatory mechanisms.

In a study by Pantla et al. (2024), it was also found that molecules can enhance the speed and efficiency of breeding processes. To ensure sustainable production and global food security, future research should focus on improving trait mapping studies, refining marker technologies, and promoting the use of Marker-Assisted Selection (MAS) in crop breeding.

Molecular markers and MAS are valuable tools for accelerating crop breeding programs, leading to increased efficiency, precision, and genetic gains. However, challenges such as marker validation, cost-effectiveness, and trait complexity still need to be addressed.

13.7.1 Molecular Marker-Assisted Breeding

More accurate identification of novel cultivars can be achieved by using molecular markers in addition to traditional techniques. Molecular markers are abundant, unaffected by changes in tissue or environment, and enable cultivar identification at an early stage of plant development, making them a valuable addition to the morphological and physiological characterization of cultivars (Elshafei et al. 2011; Maria et al. 2020). Two sister species of yucca found on the Baja California Peninsula, *Y. valida* and *Y. capensis*, have coevolved with Tegeticulabaja, the pollinator. The peninsula is home to both endemic yucca species, which now have allopatric populations. Some plants in the southern region of the Magdalena flatland, a geographically isolated area within Yucca valida's distribution, have been hypothesized to have hybrid origins based on their morphological traits. We analyzed the genomes and climate of the two yucca species and the hypothetical hybrid populations. Samples were taken from 35 different locations, and we genotyped 3423 single nucleotide polymorphisms in 120 individuals. Geographic cline analysis and Bayesian testing were conducted on the genomic data. Species distribution models were projected across various periods to assess changes in the distributional range. Additionally, a statistical test was performed to determine niche divergence between the parental species and the putative hybrid populations using climate data from the occurrence sites. Bayesian models supported the concept of post-divergence gene flow between the yucca species while structure analysis of hybrid populations' genomes indicated mixed ancestry. Our species distribution models revealed that the geographic ranges of the parental species largely overlapped during the last glacial maximum, potentially facilitating genetic mixing between the species. Ultimately, we discovered that the majority of the evaluated environmental axes between the parents and hybrid populations are divergent, suggesting that the hybrid populations' climatic niche is changing in comparison to that of their progenitors. Our findings demonstrate that the populations in the southern Magdalena flatland are the outcome of the genetic admixture of two different species. Because they were born in a different environment than their parents, hybrid individuals with this unique genetic mix show signs of ecological divergence. This divergence creates barriers to reproduction both temporally and spatially.

13.7.2 Genomics

Rocio et al. (2022) stated that insect-pollinated plants, pollinator behavior during pollen transfer and plant dispersal affect gene flow through pollen. The study focused on assessing gene flow facilitated by TegeticulaBaja, the yucca-moth that serves as the obligate pollinator of the perennial species Yucca valida. The researchers utilized nuclear microsatellites to genotype 405 individuals, including 105 trees and progenies from 15 plants (300 seedlings). They estimated genetic diversity and population structure across the geographical distribution of the plant. Additionally, the study examined the outcrossing rate, pollen donors, dispersion distance, genetic autocorrelation index, as well as diversity and differentiation indexes. Results indicated that *Y. valid* primarily exhibits an outcrossing mating system. Within a limited reproductive neighborhood (1.4 ha) genetically linked individuals showed that pollen carries genes over short distances (42 m). The offspring of every mother plant have at least 2.8 pollen donors. The interaction between yucca life-history features may sustain high levels of genetic diversity and slight differentiation among *Y. valida* populations, despite observed inbreeding. To comprehend the distribution of genetic variation within and between populations, it is crucial to understand the spatial scale at which gene flow functions.

13.7.3 Bioinformatics

Yucca schidigera plants produce saponins as secondary metabolites (Guadalupe et al. 2024). These metabolites have useful qualities for use in food, medicine, and cosmetics. Therefore, developing effective techniques for removing these metabolites is a crucial area of research. In order to extract high saponin extracts from *Yucca schidigera*, it is suggested to employ two technologically aided methods: ultrasound-assisted extraction (YE-US) and deep eutectic solvent extraction (YE-DES). These methods were compared to the traditional sequential extraction method (YE-AB). The efficiency of each method was evaluated based on extraction time, solvent volume, and extract yield. YE-US was found to have the fastest extraction time, and highest yield, and used less solvent overall compared to YE-AB and YE-DES. While there were no significant differences, the saponin content extracted was generally higher for YE-US and YE-AB than for YE-DES. Comparing YE-US to YE-AB and YE-DES, a statistically significant increase in antioxidant activity was observed. Although functional property results varied among samples, they showed promise for use as stabilizers, antibacterials, and surfactants. The functional qualities of the studied procedures suggest they could be used in industry, contribute to the development of more environmentally friendly processes, and improve the extraction of metabolites from yucca plants.

13.8 Tissue Culture Applications

By 2050, the world's population is predicted to gradually increase from 6.7 to 9 billion. By 2030, agricultural production must rise by 50% to meet the growing demand for food. However, increasing food production at a lower cost is becoming increasingly crucial. Improved associated activities and a wider variety of agriculture are therefore necessary. It is imperative to implement innovative technologies that provide optimal results for production and agriculture to meet present and future demands. One biotechnological approach to addressing the issue of food accessibility in developing countries that are grappling with a rapidly expanding population on limited land is plant tissue culture (Khatik 2023). Both biotic and abiotic factors affect crop productivity, making it necessary to identify genes that affect important characteristics such as yield, quality, and resistance (Gangadhara et al. 2024). Tissue culture has expanded the range of breeding options, yet conventional breeding techniques, like hybridization, have proved essential in overcoming these issues. To enhance crop genetics, plant breeders utilize various methods, with the plant's morphological, physiological, and genetic characteristics being key factors in determining its success. Plant tissue culture stands out among these techniques as it can be used for somatic embryogenesis, inducing polyploidy, genetic alteration, and producing disease-free plants. Through accelerating breeding cycles, conserving germplasm, and generation variation, this strategy significantly contributes to agricultural development. Plant tissue culture can be used to develop plants with improved genotype, phenotype, and totipotency potential. This method has various applications, such as creating transgenic plants with enhanced agronomic and disease resistance, functional genomics, crop improvement, developing virus-free plant material, and germplasm conservation of vegetative propagative plant material to protect endangered plant species. Genome editing in plant tissue culture plays a significant role in crop improvement by creating plants with superior traits. The Omics method enables a deeper knowledge of complex gene functions in somatic embryogenesis and organogenesis, which will enhance the protocols for refractory species' regeneration. Using the metabolomics approach in tissue culture will help with better natural product extraction and characterization. This will result in exceptional research with fresh discoveries (Elshafei et al. 2011). Few plants are produced when Yucca is propagated using cuttings and offsets. Compared to conventional propagation techniques, micropropagation has significant production and marketing benefits. Therefore, in order to facilitate the release of the employed Yucca cultivars at large-scale production, the current research was carried out to design a quick technique for rooting, acclimation, and in vitro multiplication.

13.8.1 Micropropagation

Yucca valida, closely linked to *Yucca schidigera*, is a species harvested from the wild for steroidal extracts, and represents a significant potential source of steroidal saponins (Arce-Montoya et al. 2006). Due to their long life spans before harvesting and modest growth, neither species has been domesticated. Amicropropagationtechnique has been developed to produce isogenic or clonal lines for use in plantations. In an experimental plantation, seventeen clonal lines were grown, assessed, and compared to the performance of plants grown from seeds over 26 months.

The primary and secondary shoot numbers (2.02 and 3.21 per explant, respectively) indicated that the "Green" cultivar had the highest average mean value, as per the data (Elshafei et al. 2011). The highest mean value for shoot multiplication from primary and secondary cultures was observed in medium protocol B, which consisted of MS medium, 0.2 mg/L of NAA, and 4 mg/L of BAP (1.63 and 2.96 per explant, respectively). The lateral bud explant derived from propagules from the primary and secondary culture proved to be the most effective for multiplying shoots.

It can be challenging to find variegated leaf cultivars when producing true-to-type plants on a large scale (Alexandre et al. 2021). Shoot tips from variegated cultivars of three different species Yucca gloriosa 'Variegata,' Phormiumtenax 'Jessie,' and Cordyline australis 'Pink Passion' were introduced in vitro for propagation. This was done to determine the potential for micropropagation of ornamental variegated monocotyledons in an industrial and commercial context, assess the stability of their variegated phenotypes, and understand when and showoff-type plants may appear. The proliferation rates and variegation stability were monitored through individual plant observation over multiple subcultures. Samples were periodically preserved and examined by staining histological sections of the shoot tips during the in vitro procedure. The leaf tissues were analyzed using confocal and light microscopy. After three or four subcultures, the multiplication rates for Yucca, Phormium, and Cordyline stabilized at approximately 2.5–3. After six multiplication cycles, approximately 10% of Yucca, 25% of Phormium, and 60% of Cordyline were off-type plants. Histology investigation revealed that in a PGR-free medium, thc shoot apical meristem (SAM) completely or partially repressed axillary bud meristems (AxM) in Yucca, Cordyline and Phormium. For Yucca, only AxM was detected in the presence of BAP or BAP plus NPA; for Phormium, both AxM and some adventitious bud meristems (AdM) were primarily formed; and for Cordyline, both AxM and AdM were developed. The three variegated cultivars were found to be periclinal chimeras, as observed in the leaf section.

The callus tissue of Yucca gloriosa L. buds can be used to regenerate whole plants under the right culture conditions (Sergi et al. 1983). It was discovered that morphogenesis was aided by increasing the content of 2,4-dichlorophenoxyacetic acid in the nutritional medium to 1 mg/L. According to White, when a mixture of vitamins was used with 1 mg/L 2,4-D, vigorous rhizogenesis took place. Platelet regeneration from callus tissue was achieved by mixing vitamins and 1 mg/L of 2,4-D at 25 °C, as per Staba. The ability of the callus from *Yucca gloriosa* L. buds to morphogenically expand was improved by raising the cultivation temperature to 28 °C.

13.8.2 Embryo Rescue

One of the most established and effective in vitro tissue culture methods for plant species is the embryo rescue (ER) procedure. ER refers to a group of techniques that help transform a lethal or underdeveloped embryo into a plant capable of survival. Important alleles of agricultural significance from wild species, such as resistance or tolerance to abiotic and biotic stressors or morphological traits in crops, can be introduced through intraspecific, interspecific, or intergeneric crosses. However, effective hybridization is often hindered by pre- and post-zygotic reproductive barriers. Pre-zygotic barriers can occur in the stigma, style, or ovary and are characterized by incompatibility reactions that hinder pollen germination, pollen tube growth, or entry into the ovule. Several ways are used to get around these obstacles, such as placenta pollination, in vitro ovule pollination, mixed pollen from different species, and cut-style or graft-on-style techniques. Conversely, post-zygotic barriers function at various tissues and stages, starting with the early stages of embryonic development and continuing through the offspring's growth and reproduction. Embryo abortion is the outcome of several crossings between distinct genera when endosperm development fails. In these situations, ER methods are required to save these hybrids. In addition to making successful crosses easier, ER has the potential to produce haploids, doubled haploids, and ploidy levels that can be altered for chromosome engineering through monosomic and disomic addition and replacement lines. Moreover, ER can be utilized to propagate uncommon plants and reduce the reproductive cycle. It has also been employed extensively to investigate the stages of embryonic development, particularly in mutants that are embryo-lethal. The most popular ER technique involves taking immature embryos and cultivating them straight on culture media. In certain instances, the effective growth of early embryos from the zygote stage to maturity is made possible by the in vitro culture of ovules, ovaries, or placentas (Ugo et al. 2023).

13.8.3 Synthetic Seeds

One of the most vital resources for plant tissue culture researchers and breeders is synthetic seed technology. It offers significant benefits such as large scale, rapid multiplication, low labor and cost propagation of various easily infected vegetatively propagated, hybrid, and seed-free plants. Transgenic plants, somatic and cytoplasmic hybrids, sterile and unstable genotypes, and the cryopreservation of desirable elite genotypes can all be multiplied using this technique. Some horticultural crops' seeds are difficult to propagate due to their small size and heterozygosity, which is especially evident in most cross-pollinated crops like orchids where endosperm content is minimal. Some seeds that require the association of mycorrhizal fungi to germinate, like orchids, do not form seeds at all. Synthetic seeds can be produced to facilitate vegetative propagation of certain crop species. These could include branch buds, enfolded cell aggregates, somatic embryos, or any other type of tissue that could be used as planting material or that could develop into a plant in situ or outside while maintaining storage potential. An artificial seed with an artificial seed coat, an artificial liquid for nourishment, and an artificial seed housing an embryo created through somatic embryogenesis. Toshio Murashige coined the term "synthetic seed" for the first time in 1977. Somatic carrot embryos were used by Kitto and Janick in 1982 to generate the first artificial seeds. Since plants are unable to replicate or generate seeds, this technique is required to reproduce hybrids. Additionally, it permits the creation of transgenic plants, elite genotypes of disease-free plants, conservation of resistant species, and the manufacture of male- or female-sterile plants (Sanjoy 2024).

13.9 Genetic Engineering and Gene Editing

Crops can be genetically modified to increase their resistance to pests, diseases, flooding, rising water levels, and high temperatures (Elshafei et al. 2011). Plant biotechnology has had a significant impact on crop enhancements and sustainable agriculture (Khatik 2023). By genetically modifying crops to enhance their resilience to diseases, salinity, insects, flooding, rising water levels, and temperatures, plant biotechnology has influenced crop improvements and sustainable agriculture in India. Researchers can effectively scale up the production of metabolites and gain valuable insights into the biosynthesis and regulation of bioactive substances through the use of omics technology (Shagun et al. 2023).

13.9.1 Genetic Transformation Methodologies

Expanding the use of new artificial promoters can reduce the occurrence of homology-based gene silencing and enhance transgenic stability. Bidirectional synthetic promoters, which allow for the expression of up to eight genes through a single regulatory element, are a particularly intriguing class of promoters. Complex segregation patterns and random integration of multiple transgenes are much less common when several genes are introduced and expressed under the control of a single transgenic event. These issues are frequently observed throughout the traditional, multistep transformation process of transgenic crop development, which is facilitated by genetic constructs with a single gene. Since the homology and complexity of previously evaluated DNA promoter segments are low, their abundance can be increased by using orthogonal expression systems composed of trans factors and synthetic promoters that are either not found in nature or originate from distinct species (Piotr and Malgorzata 2024).

13.9.2 Gene Editing Methodologies

With the advent of gene editing tools, strategies that have reduced the complexity and time required for genome sequencing have substantially increased the ability to make modifications in genomes, particularly plant genomes, in a site-specific manner. These technologies offer a useful starting point for research on metabolic engineering, gene function, and enhancing corp traits. Approximately years ago, the first genome editing techniques were developed using Mega nucleases. These were later followed by zinc finger nucleases, transcriptional activator-like effector nucleases, and, most recently, the widely used technique, clustered regulatory interspaced short palindromic repeat (CRISPR)-associated protein (Cas) system (CRISPR/Cas). The fundamental concept behind CRISPR/Cas is that an endonuclease protein called Cas cleaves one or both strands of DNA, and then the DNA repairs itself through cellular mechanisms. Its widespread adoption and revolutionary impact on precise genome alteration in various organisms can be attributed to its user-friendly design, increased targeting flexibility, and cost-effectiveness. Gene editing has been carried out in numerous plant species, including major food crops like potatoes, wheat, rice, and maize, using the CRISPR/Cas system (Nawara et al. 2016, 2017; Saleh et al. 2020a, b). This overview outlines the several methods, including the most recent ones, being used to insert DNA fragments or make alterations as small as a single base pair mutation. Methods for knocking out or modulating through activation and repression are described at the level of gene expression. Prerequisites for CRISPR/Cas-mediated editing are covered, along with the current challenges in the field.

To improve plant characteristics and obtain a thorough grasp of gene activities, plants require precision-targeted gene control. In eukaryotes, transcriptional and epigenetic mechanisms can be used to control gene expression. Programmable gene regulation tools have been made possible in large part during the past 10 years by developments in gene-targeting technology and a deeper comprehension of the principles behind epigenetic gene regulation (Yuejing et al. 2024).

13.9.3 Enhanced Traits

Genetic changes occur in plants in nature without intervention to adapt to climate changes, as discussed by Rickelle et al. (2024). The chaparral yucca, also known as Hesperoyucca Whipplei (HW), is a native yucca species found in the southwestern states of the United States, specifically Arizona and California. This plant is a fascinating subject for investigating the mechanical characteristics of its physical structures because it has adapted to thrive in dry, arid environments. Research shows that chaparral yucca possesses unique physical and structural traits that enable it to resist stressors from the environment and external influences. For example, the fibrous structure of the yucca's leaves and stems provides both rigidity and flexibility, allowing them to withstand physical pressures like wind and bend without breaking. Analyzing the mechanical characteristics of HW can provide valuable insights into how the plant adapts to its surroundings and may inspire the development of novel materials.

13.9.4 Transgenic Cultivars

Modern agriculture has entered a new age thanks to advancements in bioinformatics, which seamlessly combine computational techniques with crop genomes. Deciphering the complex genetic information of crops is made possible by bioinformatics tools and techniques, which provide specialized solutions for increased crop productivity, disease resistance, and environmental sustainability. High-throughput sequencing technologies are used in the first part of this revolution, producing enormous amounts of genetic data. In order to collect and analyze these data sets and effectively decipher the intricate genetic codes of crops, bioinformatics techniques are essential. Precision breeding initiatives are made possible by this genomic information, enabling researchers to pinpoint desirable features and hasten the production of crops with increased resilience and productivity. Moreover, bioinformatics makes it easier to identify biological markers linked to important agronomic features. With the use of this information, marker-assisted breeding can be used to select desired traits in crops more quickly, saving time and money compared to traditional breeding approaches. As a result, crop types that are more productive and resistant to diseases and pests are developed, enhancing the security of food supplies worldwide. Crop genomics driven by bioinformatics not only improves crop performance but also tackles environmental issues. Researchers can create crops that use less water, fertilizer, and pesticides by figuring out the genetic basis of stress tolerance and resource efficiency. This minimizes the ecological footprint of agriculture, encourages sustainable methods, and helps farmers both financially and environmentally (Prajval et al. 2024).

13.9.5 Nanotechnology

The urgent need to produce enough food to meet the rising demand worldwide is driving the application of nanotechnology in agriculture. Nanoparticles, due to their extremely small size, that bridge the gap between atomic or molecular structures and macroscopic materials, are ideal for various agricultural applications. By optimization of resource management, environmental sustainability, and nutrient usage, nanotechnology has the potential to revolutionize traditional agricultural methods. Numerous areas, such as nutrient delivery, pest control, soil fertility enhancement, precision farming, water management, post-harvest preservation, environmental sustainability, smart delivery systems, genetic modification, and Nano fertilizers

(NFs), are affected by nanotechnology in agriculture. In comparison to conventional fertilizers, NFs, in particular, have drawn attention for their capacity to increase crop development and improve nutrient delivery while limiting environmental damage and cutting costs. These nutrients in nano size greatly increase the bioavailability of nutrients to plants, which in turn increases agricultural output and growth. Nevertheless, the use of nanomaterials in agriculture also prompts questions because of their possible effects on soil microbial diversity, which is essential to preserving the health of the soil. (Singh et al. 2024). Abd-Elsalam (2024) agrees regarding the potential of agricultural nanotechnology in improving global agriculture productivity and addressing security, biodiversity, and global warming. Biotechnological advancements, especially for crops, are being developed to increase agricultural output and manage the disease, as indicated by current trends in publications and patents. In the present issue, we highly recommend, as a precondition for substantial and in-depth research, the use of biosynthesized nanoparticles from a range of sources, including plants, agricultural waste, and microbes. Agronomic applications for nanomaterials are diverse and include smart delivery systems for regulated agrochemical release, Nano pesticides, Nano herbicides, Nanosensors, and fertilizers. Furthermore, genetic modification and plant breeding also use nanotools. To reduce the influence of nanoparticles on plant and soil biota, a detailed analysis of the physicochemical soil qualities of the agricultural fields where they will be applied is necessary. Last but not least, and perhaps most significantly, we urge that when creating regulatory policies to address biosafety concerns, risk assessment, legislation, nanotoxicity, and biosafety be given top priority. Comprehensive efforts based on identified knowledge deficits must be made to advance and develop futuristic work starting now.

The yucca plant has been utilized in the field of nanotechnology to develop medical materials for treating cancer through a novel green nanotechnology approach that synthesizes biocompatible gold nanoparticles without the need for external chemical reducing agents. This is achieved by leveraging an electron-rich cocktail of phytochemicals from *Yucca filamentosa* L. (Velaphi 2023). In this study, we also evaluate the anti-cancer potential of these nanoparticles.

To create *Y. filamentosa* phytochemicals encapsulated gold nanoparticles (YF-AuNPs), an extract of *Yucca filamentosa* L. containing a variety of phytochemicals was synthesized. Additionally, to enhance in vitro and in vivo stability, gum Arabic stabilized YF-AuNPs, or GAYF-AuNPs, were produced. The anticancer activities of these nanoparticles were assessed using the PC-3 prostate cancer cell line and the MDAMB-231 breast cancer cell line. The targeting capabilities of the gold nanoparticles were also examined using pro-tumor macrophage cell lines.

YF-AuNPs and GAYF-AuNPs, two novel nanomedicine agents, were extensively characterized. The results indicated that the AuNPs were spherical and monodisperse, with moderate zeta potentials (-19 and -20 mV, respectively), suggesting stability in both in vitro and in vivo environments. The core sizes of YF-AuNPs (14 ± 5 nm) and GAYF-AuNPs (10 ± 5 nm) allow for optimal penetration into tumor cells through receptor-mediated endocytosis and the enhanced permeability and retention (EPR) effect.

Notably, YF-AuNPs exhibited significant anticancer activity against breast tumors (MDAMB-231) and prostate cancers (PC-3) by inducing early and late apoptotic phases. Moreover, YF-AuNPs resulted in elevated levels of pro-tumor cytokines (IL-6 and IL-10) and reduced levels of anti-tumor cytokines (TNF-α and IL-12), demonstrating the immunomodulatory potential of YF-AuNPs.

In conclusion, these functionalized nano-Ayurvedic medicinal agents derived from *Y. filamentosa* phytochemicals showed specific toxicity to cancer cells while preserving normal cells. This study is the first to demonstrate the effectiveness of YF-AuNPs in targeting pro-tumor macrophage cell lines, suggesting an immunomodulatory approach for cancer treatment. By combining immunomodulatory potential with selective efficacy, this research presents a promising new therapeutic approach for human cancer using herbal and nano-Ayurvedic remedies.

13.10 Mutation Breeding

Since mutations are heritable changes in an organism's genetic makeup, they play a significant role in both species' variety and evolution. Both spontaneous and artificial mutations can occur, and those that are more adapted to their surroundings, whether natural or artificial have a selection advantage. The introduction of wild species into domestication and agriculture has mostly been attributed to mutation. In plant breeding, purposeful mutation has proven to be an extremely effective tactic. Currently, over 210 plant species have over 3220 officially released mutant cultivars. For the majority of breeding objectives, if not all of them, mutant traits can be created. These traits include yield, quality, stature, disease resistance, pest resistance, tolerance to abiotic stresses, postharvest degradation, and unique end-user characteristics (Souleymane et al. 2015).

13.10.1 Conventional Mutagenesis

The process of causing mutations in an organism's DNA is known as mutagenesis. Producing crops with improved quality, disease and pest resistance, stress tolerance, and higher yields is essential for food security and sustainable agriculture can be accomplished through induced mutagenesis. Mutations in seeds and other plant propagules can be caused by physical, biological, or chemical mutagens. This is followed by the selection of better agronomic features in the first generation of plants. The second and third generations demonstrate the phenotypic stability of agronomic traits produced by the mutagenesis process, with subsequent generations undergoing additional analysis. Only mutant lines displaying the necessary traits are selected for further development. Worldwide, induced mutagenesis procedures are used to produce new crop varieties, developed around the world utilizing induced mutagenesis techniques such as irradiation or chemical mutagens (David and Ebenezer 2024).

13.10.2 *In Vitro Mutagenesis and Selection*

There are different types of physical, chemical, and biological mutagens. Currently, physical mutagen treatments are the most commonly used method to induce mutations in crop production and plant breeding (Souleymane et al. 2015). Mutations, which are genetic alterations, can be inherited by future generations. Natural mutations occur spontaneously and play a crucial role in driving evolution.

Random mutagenesis in plants, whether conducted in vivo or in vitro, involves the use of physical and chemical mutagens to increase the frequency of mutations, thereby accelerating the selection of important agronomic traits. The majority of physical and chemical mutagenesis techniques have been successfully applied in both in vivo and in vitro experiments.

The process of mutation and the mechanisms of repair operate at the cellular level, meaning that applying a mutagen in either setting does not affect the organism. The type of mutations expected to be induced by a specific mutagen remains the same whether it is applied in vitro or in vivo. In fact, both in vitro and in vivo random mutagenesis can lead to the creation of identical mutants with the same desired trait in a particular plant species.

Therefore, the GMO Panel has concluded that there is no need to differentiate between plants obtained through in vitro or in vivo methods (Mullins et al. 2021).

13.11 Hybridization

Varieties of crops have emerged through natural selection. Plant breeding occurs through intentional cross-breeding and other methods, which serve as the foundation for most modern plant breeding programs. In plant breeding, crossing is used to create new combinations of traits specific to each variety. Each variety has its own set of advantages and disadvantages. When two are crossed, it should be possible to choose from the offspring those lines that combine the advantages of both parents while excluding those that combine the disadvantages. Selecting the correct parents is a difficult and artistic decision. Depending on the kind of flower, there are variations in the science of relatively basic crossing techniques. Techniques are used to remove a flower's anthers before they have an opportunity to burst and disperse pollen, which could lead to self-pollination. Genetic incompatibility that makes crossover between specific varietal types problematic can occur even within a species. For each crop, the safest way to extend crossing must be found through trial and error. Choosing desired features and developing breeding plans that go beyond the technicalities of genetics and genomics constitute the "art" of plant breeding. This requires an imaginative and intuitive approach. There are aspects of both science and art in breeding, and it is undoubtedly a changing field. Plant breeders employ a complex dance of science and creativity in their art, fusing technical expertise with instinct, cultural sensitivity, and a profound love for the usefulness and beauty of plants. This combination leads to the creation of crop types that are not only able to flourish in a variety of settings but also have a deep emotional connection with people (Kwon-Ndung 2024).

Plant breeders have been relying on the genetic variety that arises from sexual crossings between members of the same species for decades. However, there is not enough diversity within species populations, which is why hybridization procedures are used to take advantage of beneficial features of interest in distantly related or even unrelated plants. There are two types of hybridization: somatic and sexual. Pollination, either naturally occurring or induced, is the process by which two genomes from distinct parental taxa are combined through sexual hybridization, often known as wide or distant hybridization. The process of somatic hybridization, which combines somatic cells rather than gametes, is heavily reliant on the capacity to produce viable protoplasts and ultimately differentiate them into whole plants in vitro (Immaculate et al. 2019).

13.11.1 Conventional Hybridization

The relationship between *Yucca filamentosa* and Tegeticulayuccasella, the yucca moth (Lepidoptera: Prodoxidae), the species-specific pollinator, is frequently cited as a prime example of a coevolved plant-pollinator mutualism (Massey and Hamrick 1999). Initially, based on behavioral observations, it was thought that moth dispersal was limited and self-fertilization rates for the self-compatible *Y. filamentosa* would be high. However, an analysis of its mating system revealed that *Y. filamentosa* was primarily outcrossed.

Ten polymorphic allozyme loci were examined to investigate the breeding structure of a wild *Y. filamentosa* population in order to gain a better understanding of efficient breeding patterns in the *Y. filamentosa* population. The analysis showed that *Y. filamentosa* is primarily outcrossed, with many sired fruits each pollinated by a distinct batch of pollen. There appears to be some linked mating inside fruits, as evidenced by the effective number of pollen contributors per fruit, ranging from 1.56 to 3.13.

Paternity analyses revealed that pollen within the sample population traveled distances ranging from 6 m to 293 m (mean = 118 m), and at least 10% of the offspring were fathered by pollen originating outside of the community. These findings are discussed about the mutualism between yucca and yucca moths.

13.11.2 Somatic Cell Hybridization

Somatic hybridization is the process of fusing isolated protoplasts from somatic cells to create hybrid plants. This method completely bypasses sexual reproduction and allows for the merging of genomes from two desired parents, regardless of their taxonomic relationships. The nuclear, mitochondrial, and plastid genomes of the parents are combined in hybrid cells formed by the fusion of two unrelated protoplasts. The use of somatic hybridization in crop development is constantly evolving. However, it is important to note that a major limitation of somatic hybridization is genetic incompatibility after protoplast fusion (Swapnil et al. 2020).

13.11.3 Hybrid Cultivars

In agriculture, hybridization has long been used to increase crop yields. Essentially, it can result in hybrid vigor, characterized by larger, more biomass, fertile, and disease-resistant hybrid plants than their parent plants. Due to the mixing of distant parental genomes during hybridization, the hybrid offspring experience a genomic shock that causes a variety of genomic rearrangements, including transpositions, changes in genome size, chromosomal rearrangements, and other impacts on the chromatin. It has recently been discovered that hybridization can change the somatic mutation rates in plants in addition to causing chromosomal rearrangements (Tufail et al. 2018).

Following their formation, hybrids have been tested using a variety of hybrid identification techniques, including fluorescence in situ hybridization, cytogenetic analysis, and the use of morphological and molecular markers (Immaculate et al. 2019). The application of hybridization techniques in plant improvement is still a crucial strategy for overcoming species barriers and utilizing significant traits in unrelated crop plants that would not have been possible with traditional plant breeding methods.

13.12 Conclusions and Prospects

Yucca is a perennial shrub and tree genus in the Agavoideae subfamily, with 40–50 species. Its evergreen rosettes, sword-shaped leaves, and large terminal panicles of white or whitish flowers make it endemic to the Americas. Yuccas have evolved to various climatic and biological conditions, thriving in diverse environments. Due to the global expansion and diversity of this plant, researchers have focused on extracting fibers from its leaves in recent decades. The fibers were extracted from plant leaves using mechanical, chemical, and natural processes, and their physical and chemical properties were studied. Research findings indicate that the resulting fiber contains a high percentage of cellulose, and a high purity percentage can be achieved through purification operations such as alkali treatment and bleaching. Under the extraction conditions, yucca fibers exhibited good fineness and strength, as well as good crystallinity, thermal resistance, and mechanical qualities when compared to various plant fibers, including stem fibers like flax, hemp and leaf fibers such as sisal. Yucca fibers can be used in the production of yarn, rope, composite structural reinforcement, and sound insulation. It is anticipated that yucca fibers will play a significant role in the textile, paper, composite, and construction industries in the near future.

Appendix 13.1 Short description of Yucca species

	*Yucca aloifolia* (Aloe yucca, Spanish Bayonet) Spanish bayonet is an attractive yucca with sword-like leaves and large, white, occasionally purple flowers, making it a popular decorative shrub.
	*Yucca brevifolia* (Joshua tree) The Mojave Desert in the southwest of the US is home to the fast-growing Joshua tree, a dense, dark green tree with dense branches reaching 49 feet. It can live for hundreds of years in harsh conditions. Grown outdoors, it thrives in low humidity and warm temperatures.
	*Yucca constricta* (Buckley's yucca) Yucca Buckley is an evergreen perennial shrub that thrives in hot deserts and sandy soils. It forms stemless clumps with distinctive pale blue to dark green leaves. It is suitable for poor soil and can produce tall spikes reaching 5 feet. Grow in full light and provide sufficient space for cultivation.
	*Yucca baccata* (Banana yucca, datil) The evergreen yucca banana plant, found in deserts of the US and Mexico, has thin, clustered leaves ranging from 1 to 3 feet and a height of 3 feet. Its flower stalks are shorter than other species. Identifying the plant involves observing its white flowers and fruit, which resemble bananas and turn red when ripe.
	*Yucca decipiens* (Palma China) Yucca decipiens is a large, branching member of the Asparagaceae, native to north-central Mexico from Durango to San Luís Potosí. It is evergreen, up to 20 feet (6 m) tall, growing at elevations of 1500–2000 m in the mountains.
	*Yucca elata* (Soaptree yucca) Soaptree yucca, also known as Yucca elata, is a drought-tolerant, tall, tree-like plant in the southwestern US and northern Mexico. Its spiky leaves and drought tolerance make it a popular ornamental plant in gardens and landscapes. Additionally, soaptree yucca is crucial for ecosystems as a food source for butterflies.
	*Yucca filamentosa* (Spoonleaf yucca or Filament yucca) Adam's needle is a tall, magnificent yucca with white flowers and cottony filaments that differentiate it from other species. It can grow up to 8 feet tall.

(continued)

	*Yucca flaccida* (Flaccid leaf yucca) The leaves of a weak-leaf yucca have a propensity to droop or curl. This plant species can be recognized by its sword-shaped leaves with pointed ends, like the majority of yuccas. When this yucca flowers, it produces clusters of bell-shaped, white flowers that are at the ends of long stems.
	*Yucca filifera* (Palma Chuna yucca) Yucca filifera, a fast-growing yucca native to northern Mexico, forms a branched tree with spear-shaped, upwards-facing leaves and curled filaments. Its tall, heavily branched yucca has straight, ensiform leaves growing in rosette-shaped bunches, with an inflorescence of white flowers hanging over.
	*Yucca glauca* (Great Plains yucca) Soap yucca, a drought-resistant desert plant, produces green-white flowers on a 3-foot-tall stem, requires sunlight and is also known as Spanish bayonet yucca.
	*Yucca gloriosa* (Moundlily yucca, Adam's needle, Spanish Dagger) The Spanish dagger, a native yucca plant from southern America, has slender, dark green leaves used for rope and fabric. Growing to two to three meters, it produces white blooms with purple-red hues. Although it can be grown indoors, it thrives outdoors in full sun.
	*Yucca grandiflora* Yucca grandiflora, native to the Sierra Madre Occidental in Chihuahua and Sonora, is an arborescent species found in slopes and ridges at elevations 2500–4500 ft. It grows among oaks and in desert scrub and thorn scrub areas. It can grow up to twenty feet tall and occasionally has branches at the base or top.
	*Yucca harrimaniae* (Dwarf Yucca; Narrow-leaf Yucca; Yucca nana) *Yucca harrimaniae*, native to the southwestern US, grows on plateaus, hillsides, and rocky outcrops. It prefers sandy-loam soil and full sun. Leaves and roots are used for ropes, mats, baskets, paint brushes, and soap production.
	*Yucca intermedia* (Intermediate Yucca) *Yucca intermedia* McKelvey is a small plant in the Asparagaceae family, forming clumps of rosettes in dry steppes, juniper-pinyon woodlands, and desert grassland areas of New Mexico. Its leaf blade is linear, plano-convex, and slightly flexible, with entire margins. Inflorescences are mostly green, racemose, and glabrous, arising within rosettes.

(continued)

	*Yucca jaliscensis* (Jalisco Yucca, Jalisco Soapwort, Izote Yucca)
	Yucca jaliscensis is a tall tree-like Yucca species native to southwestern Mexico. It can grow up to 10 meters in height and 6 meters in spread but is typically found at less than 7.5 meters. The leaves are stiff to flexible, variably blue to green, strap-like, spineless, and can be up to 1 meter in length and 8 cm in width. Old leaves typically self-clean from the trunk over time.
Yucca kanabensis	Kanab yucca; *Yucca angustissima* subsp. Narroleaf yucca
	*Yucca angustissima* is an evergreen shrub with spear-shaped leaves 20–80 cm long and 4–20 mm wide. It can form colonies or solitary rosettes. Harvested for local food and materials, Yucca's roots are rich in saponins, and it is grown as an ornamental in gardens.
Yucca lacandonica	Tropical yucca
	*Yucca lacandonica*, native to Belize and southern Mexico, is an epiphytic plant species in the yucca genus. It grows in tree branches, with a trunk up to 3 meters long and partially horizontal. Its leaves are long and narrowly lanceolate, tapering at both ends.
Yucca madrensis	Soco yucca; Sierra Madre Yucca
	*Yucca madrensis* is an evergreen, succulent perennial plant with thin, open rosette leaves ranging from 50 to 100 cm long and 20–35 mm wide. It is harvested for local food and soap production, with unusually long and arching leaves.
Yucca nana	Dwarf yucca
	*Yucca nana* is a dwarf yucca native to Utah and Colorado, growing to one foot tall and wide, producing creamy white flowers. It is gaining popularity in gardens.
Yucca pallida	Pale yucca
	Pale leaf yucca, also known as yucca pallida, is a flowering plant in arid southwestern and northern Mexico. Its pale green leaves produce white flowers in spring and summer, making it an ornamental plant and essential for ecosystems. The yucca butterfly feeds on its nectar, pollinating the plant.

(continued)

Yucca periculosa	Izote
	*Yucca periculosa* is a tree-like evergreen perennial plant with a much-branched main trunk, the branches starting about 100–200 cm from the base of the trunk. The plant can grow up to 10 meters tall, with a rosette of long, sword-shaped leaves at the apex of each stem.
Yucca recurvifolia	Curve-leaf yucca; *Y. gloriosarecurvifolia. Y. pendula. Y. recurva.*
	*Yucca recurvifolia* is an ornamental evergreen shrub in the southeastern US, known as curve-leaf or pendulous yucca. It is a cultivar of Yucca gloriosa and belongs to the Agavaceae family. The plant has long, thin leaves that are semi-flexible and can lean over.
Yucca rigida	Blue yucca, Izote Azul, Mexican blue yucca, Rigid Yucca
	*Yucca rigida*, also known as Blue yucca, is an elegant desert plant with an agave-like form. It grows to a 12-foot-tall tree with minimal maintenance, making it suitable for garden settings. Yucca rigida is exceptionally hardy and tough, able to withstand droughts, cold temperatures, and hot temperatures.
Yucca rostrata	Big Bend yucca
	*Yucca rostrata* is a slow-growing tree-like yucca with upright stems and narrow gray-blue foliage. Its jagged, transparent yellowish leaves have small, thin blades. Large white flowers emerge from its foliage. *Yucca rostrata* is cold-resistant and should not be confused with Mexican blue yucca (*Yucca rigida*), which has stiffer leaves and is less cold-hardy.
Yucca rupicola	Texas yucca, or Twist-leaf yucca
	*Twisted yucca*, also known as Texas yucca, stone yucca, or twisted leaf yucca, has sword-shaped leaves with light-green to dark green color. It requires care due to sharp, serrated leaves with sharp tips. These leaves can grow up to 2 feet long and curl as they mature.
Yucca schidigera	Mojave yucca
	The Mojave yucca, native to the desert, has sharp, dangerous leaves and grows up to 16 feet (5 meters). Its medium-sized tree has clustered green crowns and yellowish or bluish-green foliage.

(continued)

	*Yucca schottii (*Hoary yucca or Mountain yucca*)* Yucca × schottii, a mountain yucca, is a tall species with thin trunks and branching occasionally above ground. Its leaves are lanceolate, broad, and tapering, with a hard spine at the tip. The flowers are white, and the fruit is fleshy and egg-shaped.
Yucca standleyi	
	*Yucca thompsoniana (*Beaked yucca, Soyate and Palmita*)* Thompson's yucca, native to Texas, Chihuahua, and Coahuila, is an Asparagaceae plant with a trunk up to 1 m tall. It flowers before the trunk, continuing after stem growth. Leaves are narrow, glaucous, and have a panicle inflorescence. Flowers are white and appear in late March to early May. Fruit is a dry, egg-shaped capsule.
	*Yucca thornberi* (*Yucca baccata* var. brevifolia; Fleshy-fruit Yucca) *Yucca baccata* ssp. thornberi forms thicket-like clumps with 6–25 stems and rosettes of leaves. The stems are 65–150 cm long and covered with old dead leaves. They are rarely procumbent and branch at 130 cm above ground. The leaves are yellow-green, straight, and concavoconvex, with a leaf margin without fibers or splitting into fine long fibers.
	*Yucca torreyi* (Torrey yucca; Spanish Bayonet, Spanish Dagger, Old Shag, Palma) The Torrey yucca, named after Columbia University botanist John Torrey, is a bell-shaped, 2–3 inch long, creamy-white or purple-tinged yucca. Native Americans ate the pulpy fruits raw or roasted and used the coarse fibers of the long leaves for ropes, mats, sandals, baskets, and cloth. The species was designated as a new variety in 1859.
	*Yucca treculiana* (Texas bayonetted, Trecul's yucca Spanish dagger) *Yucca treculiana* is a large, tree-like flowering plant in the Asparagaceae family, native to Texas, New Mexico, and Coahuila. Its leaves can reach up to 128 cm, and its cream-colored flowers can be purple-tinged. The fruit is fleshy and succulent, measuring up to 19 cm.
	*Yucca valida* (Datilillo) *Yucca valida* is an evergreen, tree-like shrub with large stems branching from the base and near the trunk. It can grow up to 15 feet tall and has yellow-green leaves with white margins. As the plant matures, the stem grows erect, and dead leaves persist until five to six feet from the crown.
	*Yucca yucatana* (Yucatan yucca) *Yucatan yucca*, a native succulent plant in Mexico, is a perennial shrub and tree in the agave family. Mayan healers use aloe yucca leaves for arthritis pain relief, using them as a cleansing paste rub.

References

Abdel-Hamid AME, Salem KFM (2021) Breeding strategies of garden pea (*Pisum sativum* L.). In: Al-Khayri JM, Jain SM, Johnson DV (eds) Advances in plant breeding strategies: vegetable crops. Springer, Cham. https://doi.org/10.1007/978-3-030-66969-0_9

Abd-Elsalam KA (2024) Agricultural nanotechnology. J Plants 13(4):489. https://doi.org/10.3390/plants13040489

Abdullaev FK, Rustamov NSU, Yakubov MD (2025) Uzbekistan Genebank. In: Al-Khayri JM, Salem KFM, Jain SM (eds) Plant gene banks. Springer, Singapore. https://doi.org/10.1007/978-981-99-4236-7_57-1

Aboshama HM, Salem KFM, Hassan HMF, Tahoun MH (2015) Estimating heterosis and combining ability for in vivo and in vitro traits using diallel cross on tomato (*Solanum lycopersicum* L.). Minufiya. J Agric Res 40(3):757–770

Abu-Ellail FFB, Salem KFM, Saleh MM et al (2021) Molecular breeding strategies of beetroot (*Beta vulgaris* ssp. vulgaris var. conditiva Alefeld). In: Al-Khayri JM, Jain SM, Johnson DV (eds) Advances in plant breeding strategies: vegetable crops. Springer, Cham. https://doi.org/10.1007/978-3-030-66965-2_4

Abulaila K, Al Sane KO, Tahabsom Z (2025) Jordan Seed Bank. In: Al-Khayri JM, Salem KFM, Jain SM (eds) Plant gene banks. Springer, Singapore. https://doi.org/10.1007/978-981-99-4236-7_45-1

Akparov ZI, Mammadova SA, Kalantarova NS, Mirzaliyeva IA, Mammadov AT (2025) National Genebank of Azerbaijan. In: Al-Khayri JM, Salem KFM, Jain SM (eds) Plant gene banks. Springer, Singapore. https://doi.org/10.1007/978-981-99-4236-7_31-1

Al Lawati AH (2025) Plant genetic resource conservation of Oman. In: Al-Khayri JM, Salem KFM, Jain SM (eds) Plant gene banks. Springer, Singapore. https://doi.org/10.1007/978-981-99-4236-7_50-1

Aladele SE et al (2025) Plant genetic resources conservation and utilization in Nigeria. In: Al-Khayri JM, Salem KFM, Jain SM (eds) Plant gene banks. Springer, Singapore. https://doi.org/10.1007/978-981-99-4236-7_17-1

Aldahak L, Salem KFM, Al-Salim SHF, Al-Khayri JM (2021) Advances in chicory (Cichorium intybus L.) breeding strategies. In: Al-Khayri JM, Jain SM, Johnson DV (eds) Advances plant breeding strategies: vegetable crops. Springer, Cham. https://doi.org/10.1007/978-3-030-66969-0_1

Alexandre R, Latifa H, Laurence HO, Agnès G (2021) Effects of the in vitro behavior of micropropagated plants on the stability of variegation in *Yucca gloriosa*, Phormiumtenax, and *Cordyline australis* cultivars. Sci Hortic 287:20. https://doi.org/10.1016/j.scienta.2021.110115

Almarri NB, Ahmad S, Alsubaie GTG, Elshal MH (2025) Saudi Arabia Gene Bank. In: Al-Khayri JM, Salem KFM, Jain SM (eds) Plant gene banks. Springer, Singapore. https://doi.org/10.1007/978-981-99-4236-7_54-1

Alqahtani MM, Saleh MM, Alwutayd KM et al (2024) Performance and genotypic variability in diverse date palm (Phoenix dactylifera L.) cultivars for fruit characteristics. Genet Resour Crop Evol 71:1759–1772. https://doi.org/10.1007/s10722-023-01719-0

Arab S et al (2025) National Gene Bank of Egypt. In: Al-Khayri JM, Salem KFM, Jain SM (eds) Plant gene banks. Springer, Singapore. https://doi.org/10.1007/978-981-99-4236-7_5-1

Arce-Montoya M, Rodríguez-Álvarez M, Hernández-González JA, Robert ML (2006) Micropropagation and field performance of Yucca valida. Plant Cell Rep 25:777–783. https://doi.org/10.1007/s00299-006-0144-3

Azanaw A, Haile A, Gideon RK (2019) Extraction and characterization of fibers from *Yucca elephantine* plant. Cellulose 26:795–804

Békefi Z, Szilágyi S, Desiderio F, Demku T, Varga J (2025) Prunus Gene Bank in Hungarian University of Agriculture and Life Sciences. In: Al-Khayri JM, Salem KFM, Jain SM (eds) Plant gene banks. Springer, Singapore. https://doi.org/10.1007/978-981-99-4236-7_71-1

Belaadi A, Amroune S, Seki Y et al (2022) Extraction and characterization of a new lignocellulosic fiber from Yucca treculeana L. leaf as potential reinforcement for industrial biocomposites. J Nat Fiber 19(15):12235–12250

Bell WH, Castetter EF (1941) The utilization of yucca, sotol, and beargrass by the aborigines in the American Southwest. 5:1–74

Bell WH, King CJ (1944) Methods for the identification of the leaf fibers of mescal (Agave), yucca (Yucca), beargrass (Nolina) and sotol (Dasylirion). Am Antiq 10(2):150–160

Borner A, S. Landjeva, K.F.M. Salem, U. Lohwasser (2009) Pflanzengenetische Ressourcen als Grundlage für die Züchtung klimatoleranter Sorten. Proceedings of a German Conference 'Weizenforum = Wheat Forum' Norddeutsches Weizenforum Schriftenreihe des Inst. Für Pflanzenbau und –züchtung der CAU Kiel, Germany, 63, 23–27

Börner A, Landjeva S, Salem KFM, Lohwasser U (2010) Plant genetic resources- a prerequisite for drought tolerance breeding in cereals. Proc Jahrestagung der Vereinigung der Pflanzenzüchter und Saatgutkaufleute Österreichs, 24–26 November 2009, Raumberg-Gumpenstein, pp 11–13

Breidy J (2025) Lebanon Gene Bank. In: Al-Khayri JM, Salem KFM, Jain SM (eds) Plant gene banks. Springer, Singapore. https://doi.org/10.1007/978-981-99-4236-7_48-1

Cañar DY et al (2025) Colombia Gene Bank. In: Al-Khayri JM, Salem KFM, Jain SM (eds) Plant gene banks. Springer, Singapore. https://doi.org/10.1007/978-981-99-4236-7_95-1

David A, Ebenezer OO (2024) Mutagenesis in crop improvement: methods and applications. J Crop Improv 38(1):1–23. https://doi.org/10.1080/15427528.2024.2336257

de Galarreta JIR, Barandalla L, Ritter E, Velázquez ME, Rios D (2025) Spanish Potato Germplasm Bank. In: Al-Khayri JM, Salem KFM, Jain SM (eds) Plant gene banks. Springer, Singapore. https://doi.org/10.1007/978-981-99-4236-7_114-1

Desheva G, Valchinova E, Deshev M (2025) Bulgarian gene bank. In: Al-Khayri JM, Salem KFM, Jain SM (eds) Plant gene banks. Springer, Singapore. https://doi.org/10.1007/978-981-99-4236-7_66-1

Díez MJ, Valcárcel JV, Solbes E, Muñoz E (2025) Forty years conserving Spanish biodiversity in the Gene Bank of the Universitat Politècnica de València, Spain. In: Al-Khayri JM, Salem KFM, Jain SM (eds) Plant gene banks. Springer, Singapore. https://doi.org/10.1007/978-981-99-4236-7_86-1

Dorez G, Ferry L, Sonnier R et al (2014) Effect of cellulose, hemicellulose and lignin contents on pyrolysis and combustion of natural fibers. J Anal Appl Pyrol 107:323–331

Ekunsanmi T, Tripathi S (2019) A comparison of the tensile strengths of yucca fiber extracted by microbial and chemical methods. J Life Sci 13:1–4

El Rabey H, Salem KFM, Mattar MZ (2013) The genetic diversity and relatedness of eight rice (Oryza sativa L.) cultivars as revealed by AFLP and SSRs markers. Life Sci J 10(1):1471–1479

Elabd AB, Salem KFM (2009) Detection of genetic diversity in rice (Oryza sativa L.) Varieties using microsatellite DNA-Markers and Quantitative Traits. 6th Egyptian Plant Society of Breeding Conference, 4–7 May 2009, Ismailia, Egypt

Elabd AB, Saleh MM, Aldahak L, Salem KFM (2020) Identification of superior rice (Oryza sativa l.) genotypes under irrigated conditions based on multivariate analysis. 7th International Scientific Conference of Genetic and Environment from 8th –9th December 2020, Baghdad, Iraq

El-Absawy EA, Salem KFM, Abdellatif KF, H. ElShamy (2012) Estimation of genetic diversity among some Egyptian barley (Hordeum vulgare L.) cultivars using RAPD markers and agronomic traits. 1st International Conference On Biotechnology Applications In Agriculture, Moshtohor-Hurghada, 18–22 February 2012, Faculty of Agriculture, Benha University, Egypt

El-Absawy EA, Nowar AA, Salem KFM et al (2015) Estimation of genetic diversity among some bread wheat (Triticum aestivum L.) genotypes using molecular markers and agronomic traits. Minufiya J Agric Res 40(3):223–233

Elshafei AA, Esmaiel NM, Al-Doss AA, Barakat MN (2011) An assessment of the cultural capabilities of clonal propagation and molecular characterization of *Yucca elephantipes* cultivars. J Med Plants Res 5(13):2896–2905

Eltaher A, Sallam V, Belamkar HA et al (2017) Genome-Wide Association Mapping for Drought Tolerance Traits at Seedling Stage Using Nebraska Winter Wheat. International Conference On Productive crop design (genome to phenome), April 2017, USA

Eltaher S, Sallam A, Belamkar V et al (2018) Genetic diversity and population structure of F3:6 Nebraska winter wheat genotypes using genotyping-by-sequencing. Front Genet 9:76. https://doi.org/10.3389/fgene.2018.00076

Eltaher S, Baenziger S, Belamkar V et al (2021) GWAS revealed effect of genotype x environment interactions for grain yield of Nebraska winter wheat. BMC Genomics 22:1–14

Eltaher S, Sallam A, Emara HA et al (2022) Genome-wide association mapping revealed SNP alleles associated with spike traits in wheat. Agronomy 12:1469. https://doi.org/10.3390/agronomy12061469

El-Zanaty AM, Salem KFM, Esmail RM (2011) Detection of genetic diversity in Egyptian cotton (Gossypium barbadense L.) varieties using RAPD markers and morphological traits. J Am Sci 7(12):1107–1115

El-Zanaty AM, Salem KFM, Ismail M, Esmail RM (2013) Impact of plant breeding on genetic diversity of the Egyptian cotton (Gossypium barbadense L.) varieties as revealed by RAPD markers. J Agric Chem Biotechn Mansoura Univ 4:217–227

Emara HA, Nower AA, Salem KFM, Eltaher S (2011) Estimating heterosis and combining ability for in vivo and in vitro traits using diallel cross in Maize (Zea mays L). J Prod Dev 16:101–120

Emara HA, Salem KFM, Abdel-Halim AA, Abd El-Kowy RNF (2013) An in vivo and in vitro traits analysis of a diallel cross in rice (Oryza sativa L.). Egyptian J Plant Breed 17(2):591–610

Esmail RM, Salem KFM, El-Zanaty AM (2009) Genetic diversity studies of Egyptian Cotton (Gossypium barbadense L.) varieties using RAPD markers and agronomic traits. 6th Egyptian Plant Society of Breeding Conference, 4–7 May 2009, Ismalia, Egypt

Galea-Gragera FA, Llera Cid F (2025) Spain Germplasm Bank of Annual Forage Legumes Seeds at the Scientific and Technological Research Center of Extremadura (CICYTEX). In: Al-Khayri JM, Salem KFM, Jain SM (eds) Plant gene banks. Springer, Singapore. https://doi.org/10.1007/978-981-99-4236-7_102-1

Gangadhara D, Deepak CA, Santhoshini E, Shobha I (2024) Plant tissue culture and its role in plant breeding. In: Vennela M, Varma D, Kumari S et al (eds) Genetics and plant breeding vistas: emerging trends in genetics and plant breeding. Stella International Publication

Ghazy MI, Salem KFM, Sallam A (2020) Utilization of genetic diversity and marker-trait to improve drought tolerance in rice (Oryza sativa L.). Mol Biol Rep 48:157–170. https://doi.org/10.1007/s11033-020-06029-7

Ghidan WF, Abdulmajid DA (2025) Egyptian Rice Gene Bank. In: Al-Khayri JM, Salem KFM, Jain SM (eds) Plant gene banks. Springer, Singapore. https://doi.org/10.1007/978-981-99-4236-7_106-1

Guadalupe JGC, Jaime LM, Rosalba PM (2024) *Yucca schidigera* saponin rich extracts: evaluation of extraction methods and functional properties. Sustain Chem Pharm 38:101470. https://doi.org/10.1016/j.scp.2024.101470

Gudaye TY, Lakew BF, Selamo MM, Tessema ES (2025) Ethiopian Crop Gene Bank. In: Al-Khayri JM, Salem KFM, Jain SM (eds) Plant gene banks. Springer, Singapore. https://doi.org/10.1007/978-981-99-4236-7_7-1

Harzli I, Mostafa K, Abdulla MF, Mehmood M (2025) Sustaining Turkish genetic resources. In: Al-Khayri JM, Salem KFM, Jain SM (eds) Plant gene banks. Springer, Singapore. https://doi.org/10.1007/978-981-99-4236-7_88-1

Hassan MN, Mekkawy SA, Mahdy M, Salem KFM et al (2021) Recent molecular and breeding strategies in lettuce (Lactuca spp.). Genet Resour Crop Evol 68:3055–3079. https://doi.org/10.1007/s10722-021-01246-w

Hidayatun N, Kurniawan H (2025) Conservation of agricultural genetic resources in Indonesia. In: Al-Khayri JM, Salem KFM, Jain SM (eds) Plant gene banks. Springer, Singapore. https://doi.org/10.1007/978-981-99-4236-7_40-1

Ibrahim AA, Salem KFM, Abdein MA, Ramadan SM (2023a) Advances in summer squash (Cucurbita pepo L.) molecular breeding strategies. In: Singh S, Sharma D, Sharma SK, Singh R (eds) Smart plant breeding for vegetable crops in post-genomics era. Springer, Singapore. https://doi.org/10.1007/978-981-19-5367-5_8

Ibrahim AA, Salem KFM, Ramadan S (2023b) Allium sativum and stressful conditions. In: Abdel Latef AAH (ed) Book medicinal plant responses to stressful conditions. CRC Press

Ibrahim AA, Abd-Ellatif S, Abdel Razik ESS, Salem KFM (2024) Biodiversity of cereal crops and utilization in food and nutritional security. In: Al-Khayri JM, Jain SM, Penna S (eds) Sustainable utilization and conservation of plant genetic diversity. Sustainable development and biodiversity, vol 35. Springer, Singapore. https://doi.org/10.1007/978-981-99-5245-8_2

Ihsan F, Prihatini R (2025) Indonesian tropical fruit Gene Bank. In: Al-Khayri JM, Salem KFM, Jain SM (eds) Plant gene banks. Springer, Singapore. https://doi.org/10.1007/978-981-99-4236-7_41-1

Immaculate MM, Joshua KM, Johnstone ON (2019) Plant hybridization as an alternative technique in plant breeding improvement. Asian J Res Crop Sci 4(1):1–11

Ismail M, Salem KFM, El-Zanaty AM, Esmail RM (2013) Association mapping of fiber and agronomic characters in Egyptian cotton (Gossypium barbadense L.) genotypes. J Agric Chem Biotech Mansoura Univ 4:205–216

Jatoi SA, Siddiqui SU (2025) National Gene Bank of Pakistan. In: Al-Khayri JM, Salem KFM, Jain SM (eds) Plant gene banks. Springer, Singapore. https://doi.org/10.1007/978-981-99-4236-7_51-1

Joshi (2019) Twenty-four approaches for conservation of non-orthodox agricultural plant genetic resources in Nepal. J Nepal Agric Res Council 5(1):22–33. https://doi.org/10.3126/jnarc.v5i1.23801. License CC BY-NC

Karolina H, Edward VM, Jane G et al (2021) Hybridization history and repetitive element content in the genome of a homoploid hybrid, *Yucca gloriosa* (Asparagaceae). J Front Plant Sci 11. https://doi.org/10.3389/fpls.2020.573767. License CC BY 4.0

Khatik N (2023) Role of plant biotechnology in agricultural sustainability and improvement of crops. Univ J Agric Res 11(6):929–936. https://doi.org/10.13189/ujar.2023.110601. License CC BY 4.0

Kwon-Ndung E (2024) THE art and science of plant breeding. Conference: science and humanities: bridging the divide for human development at: Federal University of Lafia

Laghetti G, Piergiovanni AR, De Paola D, Direnzo P, Bucci G, Sonnante G (2025) Italy IBBR Gene Bank. In: Al-Khayri JM, Salem KFM, Jain SM (eds) Plant gene banks. Springer, Singapore. https://doi.org/10.1007/978-981-99-4236-7_73-1

León-Lobos P et al (2025) Plant gene banks in Chile. In: Al-Khayri JM, Salem KFM, Jain SM (eds) Plant gene banks. Springer, Singapore. https://doi.org/10.1007/978-981-99-4236-7_94-1

Mallor C, Luna E (2025) The genetic diversity of the Spanish vegetable Genebank BGHZ-CITA: conservation and utilization. In: Al-Khayri JM, Salem KFM, Jain SM (eds) Plant gene banks. Springer, Singapore. https://doi.org/10.1007/978-981-99-4236-7_109-1

Maria CA, Rafael BB, Jaime GP (2020) Hybridization between yuccas from Baja California: genomic and environmental patterns. Front Plant Sci 11. https://doi.org/10.3389/fpls.2020.00685. License CC BY

Massey LK, Hamrick JL (1999) Breeding structure of a yucca filamentosa (agavaceae) population. J Eval 53(4):1293–1298. https://doi.org/10.1111/j.1558-5646.1999.tb04542.x

Mathew TM, Melanie B, Ying-Chen CH et al (2024) Clinical cytogenetics: current practices and beyond. J Appl Lab Med 9(1):61–75. https://doi.org/10.1093/jalm/jfad086

Meghdad, Kamali Moghaddam (2024) Morphologies and properties of lignocellulose fiber extracted from Typha leaves with potential for composite applications The J TextE Inst 115(2) 284-293 https://doi.org/10.1080/00405000.2023.2200316

Mattar MZ, Salem KFM, El-Abed AB (2016) Identification of candidate microsatellite markers associated with agronomic traits in rice (Oryza sativa L.). J Bot 15(5):213–232

Moghaddam MK, Karimi E (2020) The effect of oxidative bleaching treatment on Yucca fiber for potential composite application. Cellulose 27:9383–9396

Moghaddam MK, Gheshlagh FG, Moezzi M (2024) Extraction and characterization of cellulose microfibers from cornhusk for application as reinforcing agent in biocomposite International J Biol Macromol 264130669-https://doi.org/10.1016/j.ijbiomac.2024.130669

Moghaddam MK, Karimi E (2022) Structural and physical characteristics of the yucca fiber. J Ind Text 5(55):8018s–8034s

Moghaddam MK, Torabi T (2022) Cellulose microfibers isolated from Yucca leaves: structural, chemical, and thermal properties. J Nat Fibers 19(16):13120–13130

Meghdad, Kamali Moghaddam (2022) Structural and Thermal Properties of Cellulose Microfiber Isolated from Typha Australis by Sequential Alkali-Oxidative Treatment J Nat Fibers 19(15) 10526–10538 https://doi.org/10.1080/15440478.2021.1994094

Mohammed RK, Najem AA (2020) Cytogenetic effects of the crude saponins of *Yucca elephantipes* L. leavesonthe *Allium cepa* L. root tips mitosis. Iraqi J Agri Sci 51(2):542–549

Monteros-Altamirano Á et al (2025) Plant genetic resources in Ecuador. In: Al-Khayri JM, Salem KFM, Jain SM (eds) Plant gene banks. Springer, Singapore. https://doi.org/10.1007/978-981-99-4236-7_99-1

Mourad AMI, Alomari DZ, Alqudah AM et al (2019) Recent advances in wheat (Triticum spp.) breeding. In: Al-Khayri J, Jain S, Johnson D (eds) Advances in plant breeding strategies: cereals. Springer, Cham. https://doi.org/10.1007/978-3-030-23108-8_15

Mullins E, Jean-Louis B, Tamas D et al (2021) In vivo and in vitro random mutagenesis techniques in plants. EFSA J 19(11):6611. https://doi.org/10.2903/j.efsa.2021.6611. License CC BY-ND 4.0

Nascimento HMD, Santos AD, Duarte VA, Bittencourt PRS, Radovanovic E, Favaro SL (2021) Characterization of natural cellulosic fibers from Yucca aloifolia L. leaf as potential reinforcement of polymer composites. Cellulose 28:1–17

Nawara HM, Mattar MZ, Salem KFM, Eissa OA (2016) Estimating heterosis and combining ability for agronomic and tissue culture traits using diallel cross on bread wheat (Triticum aestivum L.). Egypt J Bot 2016:177–200

Nawara HM, Mattar MZ, Salem KFM, Eissa OA (2017) Diallel study on some in vitro traits of bread wheat (Triticum aestivum L.) under salt stress. Inter J Agric Environ Res 3(1):1988–2006

Ngo T-D (2018) Natural fibers for sustainable bio-composites. Natural and artificial fiber-reinforced composites as renewable sources. IntechOpen, pp 102–126

Olle P, Manuel B, Zar L, David MA, Kari AS, James LM (2005) Phylogeny and life history evolution of Prodoxus yucca moths (Lepidoptera: Prodoxidae). Systematic Entomology

Olle P, Manuel B, Kari AS, David MA, Rik JL (2008) Phylogeny of the pollinating yucca moths, with revision of Mexican species (Tegeticulaand Parategeticula; Lepidoptera, Prodoxidae). Zool J Linnean Soc 152:297–314

Osborne CM (1965) The preparation of yucca fibers: an experimental study. Memoirs Soc Am Archaeol 19:45–50

Osterc G, Gostinčar K, Hudina M, Usenik V, Luthar Z (2025) Slovene Fruit Gene Bank. In: Al-Khayri JM, Salem KFM, Jain SM (eds) Plant gene banks. Springer, Singapore. https://doi.org/10.1007/978-981-99-4236-7_84-1

Pantla ST, Nilesh T, Vineeth RA (2024) Applications of molecular marker implementation for enhanced oilseed breeding through marker-assisted selection (Mas). J Exp Agric Int 46(5): 865–876. https://doi.org/10.9734/jeai/2024/v46i52441

Piotr S, Malgorzata M (2024) Plant synthetic promoters. J Appl Sci 14(11):4877. https://doi.org/10.3390/app14114877. License CC BY 4.0

Popović V et al (2025) Economic seed production and storage of fiber crops in gene banks in Serbia. In: Al-Khayri JM, Salem KFM, Jain SM (eds) Plant gene banks. Springer, Singapore. https://doi.org/10.1007/978-981-99-4236-7_108-1

Powo (2023) Yucca L., https://powo.science.kew.org/taxon/urn:lsid:ipni.org:names:30003895-2#source-KB

Prajval V, Krishnamoorthi A, Riya T, Ruchitha T, Manish K, Anushi, Abhishek S, Karthik C (2024) From code to crop: how bioinformatics is transforming crop genomics in modern agriculture and bettering environment. J Adv Biol Biotechnol 27(4):27–49. https://doi.org/10.9734/JABB/2024/v27i4737

Ranran W, Jingbin C, Yun L et al (2023) Genome-wide identification, expression analysis, and potential roles under abiotic stress of the yucca gene family in mungbean (*Vigna radiata* L.). Int J Mol Sci 24:1603. https://doi.org/10.3390/ijms24021603

Rickelle S, Kyle R, Gustavo VS, Daryl M, Mariappan J (2024) Mechanical characterization of yucca plant for potential biomimetic applications. ASME 2023 international mechanical engineering congress and exposition. https://doi.org/10.1115/IMECE2023-113300

Rocha F et al (2025) Portugal Gene Bank. In: Al-Khayri JM, Salem KFM, Jain SM (eds) Plant gene banks. Springer, Singapore. https://doi.org/10.1007/978-981-99-4236-7_79-1

Rocio A, Maria CA, Rafael BB, Fernando RP (2022) Pollen dispersal and genetic diversity of Yucca valida (Asparagaceae), a plant involved in an obligate pollination mutualism. Biol J Linn Soc. https://doi.org/10.1093/biolinnean/blac031/6565171

Saleh MM, Ali WA, Alwan MH et al (2020a) Multi-environmental evaluation of wheat tetraploid genotypes for agronomic traits under rainfed conditions in Syria. J Aridland Agric 6:1–5

Saleh MM, Salem KFM, Elabd AB (2020b) Definition of selection criterion using correlation and path coefficient analysis in rice (Oryza sativa L.) genotypes. Bull Natl Res Cent 44(1):1–6

Saleh MM, Alnaddaf LM, Almuhammady AK et al (2021a) Applications of plant-derived nanomaterials in mitigation of crop abiotic stress. In: Al-Khayri JM, Ansari MI, Singh AK (eds) Nanobiotechnology. Springer, Cham. https://doi.org/10.1007/978-3-030-73606-4_9

Saleh MM, Mahmoud AS, Abbas HS et al (2021b) Nanotechnological approaches for efficient delivery of plant ingredients. In: Faizan M, Hayat S, Yu F (eds) Sustainable agriculture reviews, vol 53. Springer, Cham. https://doi.org/10.1007/978-3-030-86876-5_11

Salem KFM (1997) Breeding studies on rice (Oryza sativa L.). M Sc Thesis, Faculty of Agriculture, Minufiya University, Shibin El-Kom, Egypt

Salem KFM (2004) The inheritance and molecular mapping of genes for post-anthesis drought tolerance (PADT) in wheat. Martin-Luther-University, Halle-Wittenberg, p 124

Salem KFM (2009) Relationship between genetic diversity based on SSRs markers with heterosis and combining ability in diallel cross of bread wheat (Triticum aestivum L.). Minufiya. J Agric Res 6(34):2159–2178

Salem KFM (2015) Allelic state at the microsatellite locus Xgwm261 marking the dwarfing gene Rht8 in Egyptian bread wheat (Triticum aestivum L.) genotypes released from 1947 to 2004. Genetika 47:741–750

Salem KFM, Börner A (2003) Evaluation of wheat accessions from Gatersleben Genebank for post-anthesis drought tolerance in the field. In: Gesellschaft für Pflanzenzüchtung, GPZ-Tagung, Vortragsveranstaltung zum Thema: "Abiotischer Streß—Herausforderung für die Pflanzenzüchtung", 11–12 Juni 2003, Groß Lüsewitz, Germany

Salem KFM, Börner A (2004) Post-anthesis drought tolerance in genus Triticum. In: Botanikertagung, 5–10 September 2004, Braunschweig, Germany

Salem KFM, El-Zanaty AM (2015) Assessing genetic diversity of old and modern Egyptian rice (Oryza sativa L.) using microsatellite markers. Minufiya J Agric Res 40(6):1507–1518

Salem KFM, Ibrahim AA (2024) Plant biodiversity in the context of food security under climate change. In: Al-Khayri JM, Jain SM, Penna S (eds) Sustainable utilization and conservation of plant genetic diversity. Sustainable development and biodiversity, vol 35. Springer, Singapore. https://doi.org/10.1007/978-981-99-5245-8_1

Salem KFM, Mattar MZ (2014a) Genetic diversity in old and modern Egyptian bread wheat (*Triticum aestivum* L.) varieties revealed by simple sequence repeats. Egypt J Genet Cytol 43:143–156

Salem KFM, Mattar MZ (2014b) Identification of microsatellite alleles for salt tolerance at seedling stage in wheat (Triticum aestivum L.). Life Sci J 11(12s):1064–1073

Salem KFM, Sallam A (2016) Analysis of population structure and genetic diversity of Egyptian and exotic rice (Oryza sativa L.) genotypes using SSR markers. Comptes Rendus Biol 339:1–9. https://doi.org/10.1016/j.crvi.2015.11.003

Salem KFM, Röder MS, Börner A (2003) Mapping of quantitative trait loci (QTLs) for post-anthesis drought tolerance in bread wheat (*Triticum aestivum* L.). In: Gesellschaft für Pflanzenzüchtung, GPZ-Tagung, 11. Vortragsveranstaltung "Harnessing Genetic Diversity: Genomics and Allele Mining"., 16–17.09.2003, Institut für Pflanzengenetik und Kulturpflanzenforschung (IPK), Gatersleben, Germany

Salem KFM, Röder MS, Börner A (2004a) Drought tolerance in wheat germplasm- evaluation and molecular mapping. In: Institut für Pflanzengenetik und Kulturpflanzenforschung, IPK Tagung, 22–24/08/2004, Ag Ressourcengenetik und Reproduktion, Abteilung Genbank, Gatersleben, Germany

Salem KFM, Röder MS, Börner A (2004b) Molecular mapping of quantitative trait loci (QTLs) determining post-anthesis drought tolerance in hexaploid wheat (*Triticum aestivum* L.). In: 7 Gesellschaft für Pflanzenzüchtung, GPZ-Tagung, Vortragsveranstaltung zum Thema: " Klimatische und edaphische Sortenanpassung und Züchtung für Nachwachsende Rohstoffe". 3–5 March 2004, Halle/Saale, Germany. Vort Pflanzenzüchtung 64, pp 21–24

Salem KFM, Röder MS, Börner A (2007a) Evaluation of some barley varieties for the presence of thermostable alleles of ßamylase. 8th African Crop Science Society Conference, El-Minia, Egypt, 27–31 October 2007, pp 643–648

Salem KFM, Röder MS, Börner A (2007b) Identification and mapping quantitative trait loci for stem reserve mobilisation in wheat (Triticum aestivum L.). Cereal Res Common 35:1367–1374

Salem KFM, Varshney RK, Salem KFM, Röder MS, Börner A (2008a) Detection of genetic diversity in Egyptian barley genotypes using barley expressed sequence Tag-SSR markers. In: 10th international barley genetics symposium (IBGS), Alexandria, Egypt 5–10 April 2008. The International Center for Agricultural Research in the dry areas (ICARDA) and the Bibliotheca Alexandrina, Alexandria, Egypt

Salem KFM, El-Zanaty AM, Esmail RM (2008b) Assessing wheat (Triticum aestivum L.) genetic diversity using morphological characters and microsatellite markers. World J Agric Sci 4:538–544

Salem KFM, Varshney RK, Röder MS, Börner A (2010) EST-SSR based estimates on functional genetic variation in a barley (Hordeum vulgare L.) collection from Egypt. Genet Resour Crop Evol 57(4):515–521

Salem KFM, Röder MS, Börner A (2015) Assessing genetic diversity of Egyptian hexaploid wheat (Triticum aestivum L.) using microsatellite markers. Genet Resour Crop Evol 62:377–385

Salem KFM, Yaseen YM, El-Abd AB, Ayoub A (2015a) Inheritance of some yield components and grain quality traits in rice (Oryza sativa L.). J Product Dev 20:161–177

Salem KFM, Yaseen YM, El-Abd AB, Ayoub A (2015b) Breeding studies on some agronomic and quality traits in rice (Oryza sativa L.). J Product Dev 20(2):211–226

Salem KFM, Alloosh MT, Saleh MM et al (2021a) Utilization of nanofertilizers in crop tolerance to abiotic stress. In: Al-Khayri JM, Ansari MI, Singh AK (eds) Nanobiotechnology. Springer, Cham. https://doi.org/10.1007/978-3-030-73606-4_11

Salem KFM, Elabsawy A, Aldahak L et al (2021b) Evaluation of stem reserve mobilization in Egyptian bread wheat (Triticum aestivum L.) genotypes and F1 hybrids under post-anthesis chemical desiccation stress. J Genet Environ Resour Conserv 9(1):176–182

Salem KFM, El-Absawy EA, Khidr YA, El-Shamy H (2021c) Assessment genetic diversity among some Egyptian bread wheat (Triticum aestivum L.) cultivars using RAPD markers. J Genet Environ Resour Conserv 9(1):168–175

Salem KFM, Saleh MM, Abu-Ellail FFB et al (2021d) Role of quantum dots, polymeric NPS and dendrimers in emphasizing crops tolerate biotic and abiotic stresses. In: Faizan M, Hayat S, Yu F (eds) Sustainable agriculture reviews, vol 53. Springer, Cham. https://doi.org/10.1007/978-3-030-86876-5_1

Salem KFM, Saleh MM, Abu-Ellail FFB et al (2021e) The role of salicylic acid in crops to tolerate abiotic stresses. In: Hayat S, Siddiqui H, Damalas CA (eds) Salicylic acid- a versatile plant growth regulator. Springer, Cham. https://doi.org/10.1007/978-3-030-79229-9_7

Salem KFM, Saleh MM, Aldahak L, Elabd AB (2021f) Assessment phenotypic diversity of rice (Oryza sativa L.) genotypes by multivariate analysis. J Arid Agric 7:52–59. https://doi.org/10.25081/jaa.2021.v7.6516

Salem KFM, Alghuthaymi MA, Elabd AB et al (2022) Prediction of heterosis for agronomic traits in half-diallel cross of rice (Oryza sativa L.) under drought stress using microsatellite markers. Plants 11:1532. https://doi.org/10.3390/plants11121532

Salem KFM, Tawfik E, Mahdy M et al (2023) Advances in lettuce (Lactuca spp.) molecular breeding strategies. In: Singh S, Sharma D, Sharma SK, Singh R (eds) Smart plant breeding for vegetable crops in post-genomics era. Springer, Singapore. https://doi.org/10.1007/978-981-19-5367-5_11

Salem KFM, Safhi FA, Alwutayd KM et al (2024) Analysis of genetic diversity, population structure and phylogenetic relationships of rice (Oryza sativa L.) cultivars using simple sequence repeat (SSR) markers. Genet Resour Crop Evol 71:2213–2227. https://doi.org/10.1007/s10722-023-01789-0

Samaei SE, Mahabadi HA, Mousavi SM, Khavanin A, Faridan M (2020) Effect of Alkali treatment on diameter and tensile properties of Yucca Gloriosa Fiber using response surface methodology. J Nat Fibers 17:2429–2442

Samaei SE, Berardi U, Soltani P, Taban E (2021) Experimental and modeling investigation of the acoustic behavior of sustainable kenaf/yucca composites. Appl Acoust 183:108332

Sanjoy KB (2024) Prospects and limitations of synthetic seeds. Res J Environ Sci 6(3):6–16. https://doi.org/10.5281/zenodo.11401084

Saurav D, Manjit KR, Dinesh P, Piyush KM (2023) Role of biotechnology in creating sustainable agriculture. J PLOS Sustain Transform 2(7). https://doi.org/10.1371/journal.pstr.0000069. License CC BY 4.0

Sergi VD, Mzevinar KG, Mzia NM (1983) Regeneration of plants from callus tissue of Yucca gloriosa buds. J Zeitschrift für Pflanzenphysiologie 111(2):179–182. https://doi.org/10.1016/S0044-328X(83)80044-0

Shagun S, Pankaj K, Rohit S, Ashish RW (2023) In vitro propagation and omics breakthroughs for understanding specialized metabolite production in high-value Himalayan Fritillaria species. J Indus Crops Prod 205:117541. https://doi.org/10.1016/j.indcrop.2023.117541

Singh A, Upadhyay P, Rami E (2024) Nanotechnology interventions for sustainable plant nutrition and biosensing. J Soil Sci Plant Nutr 24:1775–1798. https://doi.org/10.1007/s42729-024-01772-3

Soltani P, Taban E, Faridan M, Samaei SE, Amininasab S (2020) Experimental and computational investigation of sound absorption performance of sustainable porous material: Yucca Gloriosa fiber. Appl Acoust 157:106999

Sonia A, Dasan KP (2013) Chemical, morphology and thermal evaluation of cellulose microfibers obtained from Hibiscus sabdariffa. Carbohydr Polym 92:688–674

Souleymane B, Brian PF, Stephan N, Abdelbagi MA, Pierre JLL, Bradley JT (2015) Plant breeding: current progress and future assessment. In: Plant breeding reviews -edition: 39-chapter: 2. Wiley Blackwell -Editors: Jules Janick

Swapnil, Digvijay S, Jenny PE, Priyanka K (2020) Somatic hybridization: an effective tool for cropimprovement. J Food Sci Rep 1(7):21–23. ISSN 2582-5437

Taban E, Mirzaei R, Faridan M, Samaei E, Salimi F, Tajpoor A, Ghalenoei M (2020) Morphological, acoustical, mechanical and thermal properties of sustainable green Yucca (Y. gloriosa) fibers: an exploratory investigation. J Environ Health Sci Eng 18:883–896

Tufail B, Ratnesh CM, Md Mohidul H, Tapan KM, Hanhong B (2018) Effect of hybridization on somatic mutations and genomic rearrangements in plants. Int J Mol Sci 19:3758. https://doi.org/10.3390/ijms19123758

Ugo R, Marco F, Claudio P (2023) Embryo rescue in plant breeding. J Plants 12(17):3106. https://doi.org/10.3390/plants12173106. License CC BY 4.0

Varshney RK, Salem KFM, Röder MS et al (2008) Assessment and comparison of genetic diversity in barley germplasm collection using gSSR, eSSR and SNP markers. Plant Genet Res 6:167–174

Velaphi T, Ananya J, Alice RK, Katti KK, Katti KV (2023) Green nanotechnology of *Yucca filamentosa*- phytochemicals-functionalized gold nanoparticles-antitumor efficacy against prostate and breast cancers. J Nanotechnol Sci Appl 16:19–40. https://doi.org/10.2147/NSA.S437812. License CC BY-NC 3.0

Wang J et al (2025) China National Gene Bank (CNGB). In: Al-Khayri JM, Salem KFM, Jain SM (eds) Plant gene banks. Springer, Singapore. https://doi.org/10.1007/978-981-99-4236-7_103-1

Yuejing C, Yu Z, Ming W (2024) Targeted gene regulation through epigenome editing in plants. J Curr Opin Plant Biol 80:102552

Part II
Fruit Fiber Crops

Chapter 14
Breeding and Biotechnology Approaches for Genetic Enhancement in Citrus (*Citrus* spp.)

Venus E. Hasan and Khaled F. M. Salem

Abstract Citrus, the world's most important fruit crop, is consumed globally due to its rich content of phytochemicals like limonoids, synephrine, hesperidin flavonoid, polyphenols, and pectin, as well as essential nutrients like calcium, potassium, thiamine, niacin, and magnesium. Citrus also acts as a natural antioxidant, boosting the immune system and aiding in cholesterol reduction, kidney stone prevention, ulcer prevention, cancer prevention, and arteriosclerosis prevention. Studies have shown that citrus fiber has high apparent viscosity and water-holding capacity, with pectin making up to 42% of its composition. The size of citrus fiber particles affects the performance of different models, altering the apparent viscosity of their solutions. Citrus fibers swell upon hydration, and their functional characteristics can be changed by an alkali mixture. Chemo-mechanical treatment damages the citrus fiber's crystalline area, creating a rough exterior and porous interior structure, increasing its soluble dietary fiber content and exposing more hydrogen bonds to water. Citrus fiber produced through this process may have applications as culinary additives. The three genera of citrus are *Fortunella, Poncirus*, and *Citrus* ssp., with most citrus species being a result of hybridization processes. Abiotic and biotic stressors have significantly impacted citrus productivity, emphasizing the need for reliable breeding programs to manage diseases and meet increasing demands for fruit quality and quantity. Due to challenges like lengthy juvenile phases, sterility, self-incompatibility, parthenocarpy, and polyembryony in traditional breeding methods, genetic engineering technologies offer alternative strategies. Genomics-assisted breeding (GAB) using genome-wide association studies to identify

V. E. Hasan (✉)
Department of Biotechnology, General Commission for Scientific Agricultural Research, Latakia, Syria

K. F. M. Salem
Department of Biological Sciences, Shaqra University, Ad-Dawadimi, Saudi Arabia

Department of Plant Biotechnology, Genetic Engineering and Biotechnology Research Institute (GEBRI), University of Sadat City, Sadat City, Egypt
e-mail: khaled.salem@gebri.usc.edu.eg; khaledfathi@su.edu.sa

K. F. M. Salem et al. (eds.), *Breeding and Biotechnology of Leaf, Fruit, and Seed Fiber Crops*, Advances in Plant Breeding Strategies 11,
https://doi.org/10.1007/978-3-032-00407-9_14

quantitative trait loci can help overcome these challenges. Virus-free plants can be produced through shoot tip grafting and micro budding. This chapter covers the history, taxonomy, cultivation methods, and challenges associated with citrus (*Citrus* spp.), highlighting different production techniques and genetic resources. It also discusses advancements in citrus breeding, such as hybridization protocols for new cultivars, genomic techniques for identifying critical genes, and molecular markers for marker-assisted selection.

Keywords Citrus (*Citrus* spp.) · Biotechnology · Conservation · Tissue culture · Traditional breeding · Mutation · Fiber

14.1 Introduction

14.1.1 Background, Objectives and Content Description

Citrus plants are large shrubs or small trees that can grow to a height of 5–15 meters. They have smooth margins, alternately placed evergreen leaves, and spiky branches. The flowers are either single or in small corymbs, each about 2–4 cm in diameter, with many stamens and five (occasionally four) white petals. They often emit a strong perfume. The fruit of a citrus plant is a hesperidium, a specialized berry that is either globose or elongated, measuring 4–30 cm in length and 4–20 cm in diameter, and has a leathery rind on the outside. The flavedo, or zest, is the outermost layer of the pericarp. The mesocarp (middle layer), consists of the white, spongy "albedo" in citrus fruits. The endocarp is the deepest layer of the pericarp. The segments, also known as "liths," each contain a locule filled with pulp or juice vesicles. String-like hairs that extend from the endocarp provide feeding to the growing fruit.

Because most citrus fruits are high in juice, their scent is partly attributed to the rind's flavonoids and limonoids, which are terpenes. Their distinct acidic flavor comes from the juice's high concentration of citric acid. Numerous species in the genus are grown for their fruit, which can be consumed raw, pressed to make juice, or preserved in pickles and marmalades, making them significant from a commercial standpoint.

Fresh citrus fruits are typically consumed, including oranges, tangerines, grapefruits, and clementines. Usually peeled, they are simple to divide into sections. More often than not, grapefruits are cut in half and spooned out of the skin. For this reason, there are specialized spoons known as grapefruit spoons with serrated points. Juices from oranges and grapefruits are also highly favored for breakfast. Lemons and limes, which are more acidic varieties of citrus, are usually not consumed by themselves. Meyer lemons are both sweet and tart, and they can be eaten with their skin. Popular drinks like limeade and lemonade are made by diluting the fruit juices and then adding sugar. Lime and lemon juice can also be added to cooked food or used as a garnish. Their juice is a prominent element in vegetable salad and cooked meat, among other uses (Korlapati et al. 2014).

Citrus fruits, which include oranges, lemons, grapefruits, and limes, are members of the Rutaceae family. Citrus is a genus that originated in Southeast Asia and has been cultivated for thousands of years. It is prized for its distinct flavor profile and high nutritional content (Ali et al. 2020). Citrus fruit agriculture has expanded over the centuries, adjusting to various temperatures and growth environments (Vigouroux et al. 2011). Citrus fruit harvests are important to the world's agricultural industry. According to Duarte et al. (2016), they are one of the most extensively cultivated fruit crops, with significant production centers located in the USA, Brazil, China, and India. In addition to being consumed raw, citrus fruits are also processed into jams, juices, and other goods, supporting a multibillion-dollar business (Farris 2010). Millions of people work in the citrus sector, which is also a major source of revenue for many rural areas (Karamura et al. 1998). Citrus fruits are valuable additions to diets globally because of their nutritional qualities, which include a high concentration of vitamin C (Jain et al. 2023). This chapter explores the history, taxonomy, cultivation methods, and issues associated with citrus (*Citrus* spp.). It highlights genetic resources and various production techniques. Additionally, it delves into recent advancements in citrus breeding, including the use of molecular markers for marker-assisted selection, genomic methods for identifying key genes linked to important traits, and the creation of hybridization techniques for producing new citrus varieties.

14.1.2 Botanical Classification and Distribution

Citrus belongs to the Rutaceae family, order Sapindales, and class Magnoliopsida of flowering plants. Citrus species, including lime, pummelo, and grapefruit, are direct descendants of this genus (https://www.itis.gov).

- Kingdom: Plantae
  - Subkingdom: Viridiplantae
    - Infrakingdom: Streptophyta
      - Division: Superior Embryophyta
        - Division: Tracheophyta (tracheophytes and vascular plants)
          - Subdivision: Phanérogames, Seed Plants, and Spermatophytes
            - Class: Magnoliopsida
              - Superorder: Rosanae
                - Order: Sapindales
                  - Family: Rutaceae
                    - Genus: *Citrus*

Direct Descendants: Citrus X aurantifolia (Christm.) Swingle- Key lime, Citrus X aurantium L.– sour orange, *Citrus hystrix* DC sour lemon, *Citrus jambhiri* Lush. - sour lemon, *Citrus limetta* Risso - bitter orange, Citrus X limon (L.) Burm. f.– lemon, Citrus X Limonia Osbeck- mandarin lime, *Citrus macroptera* Montrouz species, Citrus maxima (Burm. f.) Merr. - Pummelo, shaddock pomelo, *Citrus medica* L. – Citron, Citrus X nobilis Lour – Tangor, Citrus X paradisi Macfad- grapefruit, paradise citrus, *Citrus reticulata* Blanco – Tangerine, Citrus X sinensis (L.) Osbeck- sweet orange, Orange tachibana (*Citrus tachibana* (Makino) Tanaka), Citrus X tangelo J.W. Ingram and H.E. Moore Tangelo. Figures 14.1, 14.2, 14.3, 14.4, and 14.5 depict some species of Citrus.

Fig. 14.1 Some species of Citrus. (**a**–**d**) *Citrus sinensis* var. Washington navel. (Photoed by Dr. V. E. Hasan)

Fig. 14.2 Some species of Citrus. (**a**–**d**) *Citrus deliciosa Tenore*). (Photoed by Dr. V. E. Hasan)

Fig. 14.3 Some species of Citrus. (**a**) *Citrus Paradis* and (**b**, **c**) *Citrus unshiu*. (Photoed by Dr. V. E. Hasan)

Fig. 14.4 Some species of Citrus. (**a**) *Citrus meyeri* and (**b**) *Citrus sinensis* var. sanquinelli orange. (Photoed by Dr. V. E. Hasan)

14.1.3 Importance and Uses

Important micronutrients, such as vitamins C and E, carotenoids, and flavonoids, are found in the human diet and are required for maintaining good health. These chemicals can be found in almost all plant material and have multiple dietary sources (Di Majo et al. 2005). Antioxidant nutraceuticals or phytochemicals, as well as functional food additives, contribute to the nutritional value of food. Edible fruits and vegetables contain phytochemicals that, when consumed, may favorably alter human metabolism and fend off degenerative and chronic illnesses (Tripoli et al. 2007). Consuming more fruits and vegetables can protect against degenerative diseases like atherosclerosis and cancer (Keys 1995). Epidemiological studies have demonstrated a negative correlation between the consumption of citrus-based dietary flavonoids and cardiovascular diseases (Di Majo et al. 2005; Hertog et al. 1993).

For a long time, citrus fruits have been valued for their healthful, antioxidant-rich qualities and for being the main source of important phytochemical elements. Scientific research has shown that oranges, in particular, have a high vitamin and mineral content that offers numerous health benefits. Additionally, it is now known that citrus fruits contain other physiologically active, non-nutritional substances like soluble and insoluble dietary fibers and phytochemical antioxidants which may reduce the risk of cancer and various chronic illnesses such as obesity, arthritis, and coronary heart disease (Crowell 1999).

As an antioxidant, a mature orange with good color intensity evenly dispersed across the surface is considered to be of high quality. These oranges need to be solid, devoid of faults, rot, and other flaws, and have a texture and shape that are fairly smooth and typical of the kind. Citrus flavonoids' biological activity and positive health effects have been documented (Tripoli et al. 2007). These pigments, and anthocyanin, contribute to the color of flowers and fruits.

Fig. 14.5 Some species of Citrus. (**a**) *Citrus reticulata* var. Fremont mandarin, (**b**) *Citrus sinensis* var. jaffa orange and (**c**, **d**) Hybrid. Kara mandarin(*Citrus nobilis* × *Citrus unshiu*). (Photoed by Dr. V. E. Hasan)

Furthermore, they can be found in dietary fruits and vegetables and use a variety of strategies, such as metal chelation, to exert their antioxidant activity (Bombardelli and Morazzoni 1993). Due to their capacity to both inhibit and contribute hydrogen atoms to the hydroxyl radical, flavonoids have been shown in studies to be efficient radical scavengers of the hydroxyl radical (Cillard and Cillard 1988; Darmon et al. 1990; Di Majo et al. 2005; Tripoli et al. 2007). In addition to being a great source of vitamin C, oranges also include dietary fiber, folate, carotenoids, and flavonoids, which are bioactive substances that protect against degenerative diseases and cancer (Ejaz et al. 2006). Eating meals high in vitamin C strengthens the body's defenses against infections and helps the blood rid itself of dangerous, pro-inflammatory free radicals. Numerous phytochemicals, including narigenin and hesperetin, are found in sweet oranges. Naringenin functions as an immune system modulator, antioxidant, scavenger of free radicals, and anti-inflammatory in humans.

Anti-inflammation: Because citrus flavonoids contain regulatory enzymes (phosphodiesterase, phospholipase, lipoxygenase, and cyclooxygenase) that regulate the formation of biological mediators, which activate specialized cells involved in inflammation and endothelial cells, they have anti-inflammatory properties. These enzymes' inhibition by flavonoids may be linked to their suppression of the inflammatory and immunological responses (Tripoli et al. 2007). In fact, phosphodiesterases and kinases necessary for cell signal transduction and activating can be inhibited by citrus flavonoids. Additionally, they have an impact on the activating of T and B lymphocytes, among other immune response-related cells (Manthey et al. 2001), and preventing atherosclerosis (Hertog et al. 1993).

According to Tripoli et al. (2007), hesperidin derived from citrus cultures may be effective as a mild anti-inflammatory agent in therapeutic settings. It can also give flavonoids that have the same effect (Da Silva et al. 1994). Hesperidin also inhibits the over-expression of cyclooxygenase-2, inducible nitric oxide synthase (iNOS), prostaglandin E2 overproduction, and nitric oxide (NO) generated by lipopolysaccharide (LPS), according to studies conducted on mouse macrophage cells (Sakata et al. 2003).

Anti-Cancer and Anti-Arteriosclerosis: Citrus flavonoids have unique cytotoxic, antiproliferative, and apoptotic properties that can help prevent cancer (Elangovan et al. 1994; Hirano et al. 1994). They are antimutagenic, can absorb UV light, and protect DNA from oxidative damage (Stapleton and Walbot 1994). These flavonoids fight free radicals in the cell nucleus and shield DNA from harm. This has been demonstrated in mice exposed to x-ray radiation (Shimoi et al. 1994). By directly interacting with the chemicals that cause cancers, such as chromosomal aberrations induced by bleomycin, flavonoids can also preserve DNA (Heo et al. 1994). Studies have shown that citrus flavonoids prevent rat malignant cells from proliferating and forming tumors in the liver and heart tissue of syngenetic rats (Bracke et al. 1989). The ability of citrus flavonoids to impede cell mobility is the basis for their action (Bracke et al. 1989).

Nutrients such as potassium, folic acid, calcium, phosphorus, iodine, iron, chlorine, manganese, zinc, pectin, beta-carotene, fiber, and amino acids are also abundant in oranges. A single orange is believed to contain about 170 phytonutrients and over 60 flavonoids with anti-inflammatory, anti-tumor, antioxidant, and blood clot-inhibiting properties. According to Cha et al. (2001), each of these characteristics supports overall health.

Anti-Obesity: Sweet oranges are low in calories, cholesterol, and saturated fats, but they are high in dietary fiber and pectin, which are particularly beneficial for obese people. As a bulk laxative, pectin shields the mucous membrane from toxins and also acts as a barrier against cancer-causing agents in the colon by binding to them. Pectin has also been demonstrated to lower blood cholesterol levels by binding to bile acids in the colon and preventing re-absorption of the lipids (Walton et al. 1945). The alkaloid synephrine, found in orange peels, lowers the production of cholesterol. Oranges include antioxidants that fight oxidized blood LDL (low-density lipoprotein).

14.1.4 Fiber and Textile Properties

More than 110 million tons of citrus trash are produced annually by the citrus processing industry worldwide (Mahato et al. 2020). Forty to sixty percent of the fruit mass is made up of citrus trash, which includes seeds, peels, and pulp leftovers. Without additional treatment, citrus waste from the processing of citrus fruit poses a threat to the environment. Citrus pomace is commonly used in the pectin extraction business to maximize the usage of citrus waste (Song et al. 2021a).

The dietary fiber that is removed from citrus waste is called citrus fiber. Pectin which has been isolated from insoluble dietary fiber (IDF) makes up the majority of it (Hua et al. 2019). Research has indicated that citrus fruit has superior physicochemical features, such as water holding capacity (WHC) and water swelling capacity (WSC), and higher total dietary fiber (TDF) than grain fruit. Citrus fruit has also been shown to have outstanding physiological and functional qualities. IDF, SDF, pectin, and fructooligosaccharides make up citrus fruit. These elements specifically promote the development and activity of the gut microbiota, particularly the bacteria lactobacillus and bento. As a result, dietary fiber intake promotes the growth of gut microorganisms (Liu et al. 2022a, b). Citrus fruit's water-soluble dietary fiber can absorb cholesterol and lower blood cholesterol levels (Figuerola et al. 2005).

The composition and amount of dietary fiber vary amongst citrus fruit varieties and are associated with the method and source of dietary fiber extraction.

Citrus fruit's physicochemical characteristics have been thoroughly investigated. Citrus fruit's cellulose side chain contains a large number of hydroxyl groups that, when combined with H2O, can form more stable hydrogen bonds and produce superior WHC. Citrus fruit may hold more water and have WSC because of its flexible surface structure and tiny space barrier (Mudgil and Barak 2013).

14.1.5 Fiber industry

High-quality foods are influenced by various factors, including appearance, flavor, texture, and nutrition (Bourne 1992). The International Organization for

Standardization defines texture as the rheological and structural characteristics of a food product that can be perceived through mechanical, tactile, and sometimes visual and auditory senses. A customer's evaluation of texture begins when they open, handle, and cook a food product and continues until they eat it. Customer preferences for texture are influenced by factors at every stage of the process, from the initial sight of the product to consumption, during which the food may undergo several changes. Creating food products with desirable textures can be challenging and involve many dynamic factors (Borwankar 1992). It is essential to understand how fibrous materials function and why they behave in specific ways, as fibrous plant cell walls often significantly impact the quality of the foods they are used in (Van Buren 1979). Rheological characteristics and composition play a crucial role in determining food texture (Bourne 1992). Therefore, in addition to understanding the rheology, function, and composition of components, a food product engineer must comprehend quality control, process control, and the design of process equipment (Ofoli et al. 1987; Saravacos 1970). One way to improve the functionality of fibrous materials, such as food processing wastes, is to increase their internal surface area, making them more suitable for enhancing the nutritional value and quality of food by improving their apparent viscosity and water-binding ability (Turbak et al. 1983). Ruan et al. (1996) developed highly refinable cellulose, a product similar to microfilmed cellulose, using agricultural leftovers and food industry wastes. Additionally, citrus fruit has a high internal surface area, water retention capacity, and apparent viscosity (Lundberg 2005). Due to its excellent water-holding capacity, yield-increasing properties, and moisture retention abilities, citrus fruit is commonly used in various food products such as baked goods, meats, dairy products, sauces, and dressings. Moreover, the neutral color, flavor, and odor of processed citrus fruit make it suitable for uses that other fibrous materials may not be suitable for (Garau et al. 2007; Grigelmo-Miguel and Martin-Belloso 1999). One particular challenge with citrus fruit is drying it without compromising its functionality upon rehydration (Braddock 1999). While progress has been made in addressing the issue of drying citrus fruit without losing its functionality, further research is needed to fully understand the mechanisms involved. A deeper understanding of this topic could lead to more diverse uses of plant cell walls and contribute to the production of healthier food.

Analyzing the composition, images, length/width, and rheological characteristics of citrus fibers in this study can provide insight into their structure, shape, and functional properties. Adding citrus fruit to food products can enhance their nutritional value and quality while maintaining or improving their flavor profile. The findings of the rheological investigation can support this goal.

The heat stability of citrus fruits across a range of pH values, for example, is a significant discovery presented in this chapter. Making food products that are more heat stable can benefit from this. Pectin accounts for about 42% of the composition of citrus fruits, which is significant because of the fruit's high insoluble content and effect on water retention. Understanding the chemical makeup of citrus fruits will help you better understand their moisture. Citrus fruit size has a significant influence on apparent viscosity and the flow models that explain its flow characteristics. For example, the apparent viscosity and yield stress rose with increasing particle size. Particle size also affected the shape of the rheology curves, and as the fiber length shrank, rheograms became more linear. The citrus fruits with the longest fruit segments were well-represented by the Herschel–Bulkley model. However, the optimal fitting flow model differed for the shorter fiber lengths. With a strong correlation coefficient for every particle size examined in this project, the Casson model appeared to function effectively. Since the water-holding capacity did not appear to change with fiber length, but the flow behavior did, it was assumed that opposing forces were countering each other. Shorter fiber lengths result in both an increase in surface area that may store more water and a loss of matrix structure, which also helps with water retention (Lundberga et al. 2014).

14.1.6 Domestication, Selection and Early Improvements

Three genera, Citrus, Poncirus (trifoliate orange), and Fortunella (kumquat)—of the six genera that make up the Rutaceae family are valuable commercially. Citrus is a genus of evergreen Old World fruits that originated in southeast Asia, which encompasses southern China, eastern India, and Burma (Soost and Roose 1996).

Citrus (*Citrus medica*), lemon (*Citrus limon*), lime (*Citrus aurantifolia*), pummelo (*Citrus grandis*), sour orange (*Citrus aurantium*), sweet orange (*Citrus sinensis*), mandarin (*Citrus reticulata*), and grapefruit (*Citrus paradisi*) are the main horticultural categories. Although erroneous, these "species" labels provide a handy framework for thinking about this group. All but grapefruit were domesticated before 1000 BCE.

Due to outcrossing, apomixis, and selection, the taxonomy of citrus plants is problematic and illustrates philosophical disagreements regarding the definition of connections in an old group of cultivated plants (Roose et al. 1995). Thus, Walter T. Swingle identified 16 species, Toshio Tanaka proposed 162 species, and R.W. Scora has identified three main species based on biochemical analysis: *C. medica* (citron), *C. reticulata*, as well as *C. grandis* (pummelo). Many so-called species had hybrid origins, as supported by molecular data.

In order to distinguish between the development of Citrus species and their close relatives, the Japanese botanist Tyozaburo Tanaka proposed a line of demarcation, which is today known as the Tanaka Line, extending from the northeastern border of India to the island of Hainan. The majority of mandarins, Poncirus, Fortunella, and lemons, limes, citrons, and pummelo, as well as sweet and sour oranges, are found north of this line. While pummelo is cleistogamous, most citrus is cross-pollinated, and many of them include apomixes, which are linked to nucellar polyembryony and allow for seed propagation for cloning.

As their names suggest, sexual hybridization between domesticated species has produced a wide variety of new citrus varieties. These include Citradia (Poncirus × sour orange), Citrangequat (citrange × kumquat), Tangelos (tangerine × grapefruit), Tangor (tangerine × orange), and Orangelo (orange × grapefruit). The grapefruit, sometimes known as the forbidden fruit, originated through interspecific hybridization and initially showed up in Barbados in the eighteenth century. It is the result of a natural cross between the apomictic sweet orange and the non-apomictic pummelo, known as a Shaddock in Barbados (Gmitter 1995). The grapefruit can be multiplied by seed propagation and inherits the apomictic feature from the sweet orange. Every grapefruit appears to be a single clone; later variations, like seedlessness and pink flesh and fruit, are the result of somatic mutation selection.

The citron, whose name is C, was the first citrus to arrive in the West. Medica points to the media (now Iran and Persia) as the source. Citron, according to De Candolle, was first mentioned in Iran around 300 BCE and is found in the Himalayas of India. Although Postgate (1987) does not support Goor and Nurock's (1968) theory that it was mentioned in Nippur, South Babylon, as early as 4000 BCE, it is possible that it reached the West before this date. Egyptian literature does not mention it till the second century CE.

According to Leviticus (23:40), the Hebrew word for the citron is etrog. Biblical allusions to the fruit of the goodly tree would place it around 1200 BCE. Jews quickly began to revere the tree, and on Succoth (the Feast of Tabernacles), citron branches still bearing the style were given as presents. The Romans connected the citron with winter, and they can be seen in mosaics honoring the changing of the seasons (D. Parrish, pers. commun.). Various citrons were used by baroque painter Bimbi (1648–1729) in his depictions of fruit varieties (Consiglio Nazionale delle Ricerche 1982).

Through the Arabs, the sweet and sour orange made its way to Europe in the eleventh century. The Arabic name haranj, the Persian name harang, and the Spanish term naranja are where the name orange originates. Lime and pumello arrived in Europe in the thirteenth century, whereas lemons and pumello arrived in the twelfth. The largest citrus wave to hit Europe was in the fifteenth century when traders traveled overland from southeast Asia.

Portuguese sailors created a second wave in the sixteenth century by bringing citrus that was better suited for commercial use. The nobles and the wealthy, who appreciated the fragrant bloom and exquisite fruit, started growing citrus in containers when protected culture (orangeries) was developed in Europe during the Renaissance. Originating in Vietnam and China, mandarins were prized fruits in China and Japan throughout the twelfth century, but they weren't introduced to Europe until 1805 (Roose et al. 1995). While propagation from polyembryonic (nucellar) citrus seeds is true-to-type, grafting is preferred due to the lengthy juvenile stage and juvenile characteristics like thorniness and thick albedo. Numerous citrus varieties have served as rootstocks. Nonetheless, nucellar clones have been utilized to eradicate viruses from citrus since seedlings are virus-free. In order to create virus-free clones, meristem grafting and thermotherapy have been used recently.

Changes in citrus fruit characteristics such as size, shape, flesh and peel color, acidity, peel-easy (especially for mandarins), and seedlessness are linked to selection.

Radiology scans may cause seedlessness. The popularity of orange juice in the US has led to the growth of a massive business for concentrate and so-called "fresh" chilled juice. A large portion of citrus processing has been relocated to Brazil, where Sao Paulo's environment makes citrus growers quite happy. Consumers choose easy-peel citrus, which is indicative of the future of the fresh fruit market.

14.2 Current Cultivation Methods and Challenges

14.2.1 Current Cultivation Methods

In the nursery, tiny seedlings are kept under close supervision until they are ready to be planted. It is a managed facility created to generate seedlings cultivated in ideal conditions until planting time. The nursery's primary purpose is to create an environment where seedlings can grow rapidly into healthy plants prepared for field planting.

When the seedlings are small, the nursery needs to shelter them from wind, rain, and harsh sunlight (30% shade is ideal, or 70% of full sunlight). Depending on the requirements of the plants, it must also provide adequate water and nutrients as well as protection from weeds, pests, and diseases. This will make it easier to produce high-quality seedlings that meet the demands of consumers, including farmers. There isn't a set design for creating a citrus nursery. Every nursery will have a different design that satisfies minimal standards based on specific demands, resources, and requirements.

The potential success or failure of an orchard is largely determined by the superiority of the planting material (genotype, compatibility between the rootstock and scion, healthy planting material, and growing media quality), as well as the efficiency with which the crop was propagated and grown at the nursery.

Due to the risk of flooding during heavy downpours, it is important to avoid low-lying areas or depressions where rain may accumulate at the nursery site. Many of the nursery's agronomic practices could be affected by this. It is crucial to ensure there is never any standing water in the nursery. This can be prevented by improving drainage, such as digging small channels to allow water to flow away. Ground nurseries and polybag nurseries are the two main types. For commercial production, it is advisable to use a polybag nursery, where planting material is grown in polybags.

When it comes to agronomic procedures such as seeding, mulching, trimming, sorting, budding, managing water and fertilizer, gathering data, and organizing planting materials, polybag nurseries are more organized than ground nurseries. Large quantities of planting materials can be produced in a relatively small area. Additionally, plants grown in polybags show early establishment, a higher survival rate, and easier transportation of planting materials because the roots are preserved during transplantation.

Seedlings raised in 17.5 cm × 25 cm polybags produced more leaves, longer tap roots, larger and taller seedlings, and significantly greater total dry matter per seedling compared to those raised in 12.5 cm × 15 cm polybags. When transplanting seedlings, be sure to select them carefully. Choose robust, thriving seedlings with well-developed roots, and avoid seedlings with curled roots. Transplant the seedlings when they are 4–5 weeks old.

Bamboo, wooden poles, and galvanized steel tubing can be used to construct the crossbars and poles. In addition to installing nursery poles or posts and a high-quality water supply or irrigation system to water the nursery site, shade netting and palm fronds are typically used. Side netting and shade nets should also be installed for the roof.

Relevance of vegetative propagation: Budded materials are used to construct commercial citrus plantations. This is because using the seeds of the chosen citrus would result in a plant that is unable to withstand the majority of diseases prevalent in that area, and it will be challenging to treat virus-caused diseases. It will also take some time for the new plant to produce fruit and generate returns on investment. The yield is typically lower than that of budded material, and the trees will grow very tall, which can make harvesting and other cultural practices challenging. Additionally, the new plant may not fully inherit all of the horticultural traits of the parent material. Vegetative propagation is the method of choice because it shortens the immaturity period from 7 to 3 years; it addresses environmental issues such as drought, soil acidity, and diseases like tristeza; it yields material that closely resembles the parent in terms of quality and yield; it lowers the height of the plant.

Rootstock: The term refers to the part of the grafted or budded planting material that is in the soil. The rootstock significantly influences the scion's performance in terms of yield, disease tolerance, juice quality (brix, total acidity, volume, and color), plant vigor and size, and plant spacing, more so than the scion influences the rootstock. Because of its well-known effects on fruit quality and size, tree size, crop load, and disease tolerance, choosing a citrus rootstock is essential for citrus plantations. It is believed that intrinsic differences across rootstocks that affect plant-water relations are the primary cause of differences in vigor, crop load, and fruit quality among citrus rootstocks. Citrus fruit's varying fruit growth and sugar accumulation are connected to these variations. The genetic composition of various rootstocks affects plant water relations characteristics such as root dispersion and the effectiveness of nutrient and water uptake.

Different rootstocks have diverse effects on the performance of the scion, impacting factors such as yield, plant height, and canopy size. Rootstocks also influence the canopy, scion shape, brix, acidity, disease resistance, and juice content. Additionally, rootstocks vary physiologically in terms of nutrient intake, drought tolerance, low pH tolerance, and salinity.

Rough lemon and Cleopatra mandarin are two examples of the recommended rootstock types from which fruits should be harvested for seed extraction. To prevent the seeds from being destroyed, cut the fruit halfway rather than all the way through. When harvesting rough lemons, it is important to remove the seeds just as the fruit is ready to turn orange because over-ripening drastically lowers the number of seeds present. To get rid of the mucilaginous coating, wash the seeds. Seeds should be submerged in water to eliminate any off-types and undeveloped seeds; discard any that float. If seeds won't be sown right away, they can be kept in polybags at a low temperature. To prevent infections caused by seeds, soak seeds in water at 50 °C for 10 minutes. For sufficient seed survival and germination, proper seed storage is necessary. You can either plant the seeds directly into the potting media in a polybag or you can plant them in a wooden box with loamy, well-drained soil. The rootstock seed germination box should promote drainage with a minimum depth of 15 cm.

After germination, the seedlings are transferred into polybags filled with potting media. This approach is recommended because it allows the nurseryman to select high-quality, uniform, and vigorous material for establishment. The seedlings are then allowed to grow to the three-leaf stage after being placed in a pre-nursery box. The nurseryman can also remove sick seedlings and avoid damaging the polybags by thinning out the seedlings to one per polybag when they are being raised in boxes or beds.

Scion: The scion is the part of a budded citrus plant from which the farmer will harvest the fruits. Like the rootstock, the gardener must be familiar with the horticultural characteristics of the scion, such as its maturation duration, potential yield, and juice quality. Essentially, scion selection is an economic process in which specific cultivars are chosen either in response to or in anticipation of market demands. Improved planting materials are the most crucial input for citrus production because they impact both potential yields and overall productivity per unit area.

Because errors made at this early stage won't be noticeable until the trees start to bear fruit, the nurseryman must certify the legitimacy of the cultivars chosen. The selected scion will determine the plant's maturity group and citrus variety (sweet orange, grapefruit, mandarin, or tangor). Instead of producing true-to-type materials, trees cultivated from seeds or seedlings create trees with widely varying yields, growth patterns, and levels of pest and disease tolerance. Grafting procedures can be used to eliminate such differences in plants. The interactions between the scion and the rootstock are crucial from a citrus business standpoint. A hybrid plant with different development patterns from what would have occurred if the parts had been grown separately can be produced by grafting or budding two different plants (genotypes) together. Some of these effects are detrimental and should be avoided, but others are crucial for gardening. Compatibility with the rootstock, maturity dates, potential yield, and the quality of the fruit juice produced are all important considerations when selecting a scion. Numerous rootstock/scion incompatibility systems can be found in citrus plantations. During budding, two separate plant parts with disparate genetic makeup are joined. These two different materials grow at different rates. The composite tree diminishes if the stock develops faster (Swingle citrumelo as rootstocks), but a complete union happens when the scion and stock develop at similar rates. These two conditions do not affect the crop's yield.

On the other hand, if the scion grows faster than the stock (referred to as Rotuma in Cleopatra Mandarin), the shoot may place additional strain on the stock, making it more likely to break. The following steps should be followed when choosing scion or budwood: Choose disease-free, highly productive mature trees (about 10 years old) with desired traits for scion or budwood. Choose mature budwood with swollen buds from the growth of the preceding season. Using the proper instruments, cut budwood that is 10–15 cm long (secateurs and knives). To prevent desiccation, keep budwood fresh by wrapping it in clean, damp newsprint or a damp towel, cloth, or jute bag and storing it in an icy Styrofoam box with moist sawdust (sprinkle water on the budwood before covering).

Note: When collecting multiple types from the orchard at the same time, it is easy for the budwood of the varieties to get mixed. Label batches right away to prevent this, and make sure you have accurate records for every batch. Take caution when storing budwood for longer than 5 days after harvesting or cutting it down. Avoid cutting budwood that has broken buds, picking budwood from trees older than 8 years, and taking budwood from trees that have gummosis infection, dieback of trees, or greening. Successful budding requires the use of specialized tools. These consist of an ice chest or Styrofoam box, a basin or bucket, secateurs for cutting budwood, a sharpening stone, a well-sharpened budding knife, and budding tape.

Budding: It's critical to make sure that the rootstock and scion materials used align with the citrus grower's target's ultimate goal. This is because rootstocks are selected based on a number of criteria, including yield and fruit quality, salinity, drought tolerance, resistance to the predominant virus disease in that particular region, and fruiting time or season. Because vegetatively propagated materials are true to type, mistakes in their selection cannot usually be fixed. For instance, although being a highly common rootstock type in Brazil, sour orange is unsuitable for usage in Ghana due to the country's high incidence of tristeza virus illness. Again, choosing the best rootstock for the job is crucial since rootstocks have a big impact on the scion's fruit yield and quality, and different rootstocks have different rates of water and nutrient uptake. The amount of heat units needed by the scion to develop its fruits to maturity varies. As a result, we have late maturing and early mid-season types. If the farmer doesn't choose the right variety of budwood, they risk losing money and receiving less yield. For the reasons mentioned above and others, it is not advised to buy citrus planting supplies from colleges that are not accredited. Suggested are reputable organizations where one can go for information about citrus.

Grafting and budding are accomplished using a variety of techniques, including (1) A "T" or a "T" shield inverted, (2) Chip's shield-building steps. To reduce the risk of disease infection in the budwood or sprouting scion, plant the bud rootstocks 20–30 cm above the ground. Make a "T"-shaped or inverted-T-shaped cut in the bark. Carefully remove the cambium's bark. Shape the bud accordingly and place it in the gap between the bark and the cambium. Hold it in place and secure it with grafting tape.

Two weeks after budding, check the plants to see if the buds have been taken. A successful take is indicated by a green bud, whereas a failure necessitates another blossoming effort at a different rootstock position. If the bud is green when plucked, cut the entire rootstock 2–3 cm above the bud union to promote sprouting and best growth.

Regular pruning is necessary for any branches that appear on the rootstock beneath the bud. Label each budling with the variety and the date, then group the plants. Write the variety, day budder, budwood supply, and budling count on the label of every batch of seedlings. Avoid: Using unclean tools, Touching exposed cut surfaces, and Soiling buds.

14.2.2 Artificial Intelligence

Globally, artificial intelligence (AI) has revolutionized various sectors, such as finance, healthcare, transportation, and industry. The aim of AI is to develop intelligent systems that facilitate quick and accurate decision-making. By utilizing intricate algorithms and data analytics, AI systems offer valuable insights to assist in decision-making processes. As technology advances, the impact of AI is extending to numerous industries, paving the way for increased productivity, enhanced decision-making, and potentially groundbreaking discoveries. The growing utilization of AI underscores its significance as a powerful tool to enhance human potential and drive innovation across a variety of industries (Meshram et al. 2021; Singh and Kaur 2022; Zhang and Lu 2021).

With major advancements and contributions to the sector, artificial intelligence is vital to agriculture. By independently carrying out intricate activities like irrigation, phytosanitary treatment, and gathering crucial data on vegetation and water stress levels, intelligent drones have significantly changed farming operations (Hafeez et al. 2022; Rejeb et al. 2022). These unmanned aerial aircraft allow more precise agricultural treatments and provide farmers with insightful information.

With their specific designs, agricultural robots have reduced resource waste by greatly increasing the productivity and efficiency of jobs like planting, weed identification, and crop harvesting, especially in orchards (Ren et al. 2020). For example, the Terra-Sentia robot was created by Atefi et al. (2021) for field-based agricultural study, allowing for quick and efficient data collection on plant attributes like biomass, canopy cover, and height. The autonomous navigation system of the TerraSentia robot moves around agricultural areas with efficiency, gathering vital information for breeding and crop management projects.

In a similar vein, Grimstad and From's (2017) Thorvald robot is dedicated to vegetable farming and helps with a range of agricultural duties, including planting, weeding, and crop monitoring. Its sophisticated cameras and sensors provide precise plant-weed separation, and its modular construction permits customization to fit particular farming demands. Field traversal is streamlined by the robot's autonomous navigation technology, which lowers manual labor costs and boosts output. The Thorvald robot is one example of how robotic technologies have the ability to transform agricultural processes through improved precision and less resource waste (Grimstad and From 2017).

Citrus yield prediction is a significant area of focus that has garnered attention from researchers. Many studies have concentrated on utilizing fruit identification and counting within citrus trees to estimate yields. This process involves capturing images of trees and using object detection software to identify fruits, requiring extensive image data collection and categorization (data labeling). Deep learning models trained on annotated datasets are then employed to predict fruit numbers in new photos (Li et al. 2021; Zheng et al. 2021; Zhang et al. 2022). Pre-trained deep-learning object detection models, such as RCNN, Faster R-CNN, Retina-Net, and YOLO, can enhance citrus yield prediction by effectively identifying and categorizing fruits. The development and training of these models using large-scale datasets have enabled accurate fruit identification (Bharati and Pramanik 2020; Fan et al. 2020; Jiang et al. 2022).

However, these techniques have limitations, such as the reliance on ripe fruits for prediction, which may not be practical for growers seeking pre-harvest estimates. Additionally, applying these techniques in large-scale orchards can be challenging and labor-intensive. An alternative approach involves compiling datasets with key variables related to yield and employing supervised learning methods in machine learning to predict yields accurately. By training algorithms on collected data, farmers can make informed decisions and plans in agriculture. The agricultural sector, particularly citrus farming, provides valuable data on yield, including irrigation water usage, fertilization techniques, phytosanitary treatment products, and climate data (Albrigo et al. 2019; Dutta et al. 2023).

Additional information about orchard tree counts, sizes, flora traits, and water stress levels can be obtained using satellite or drone photography. In their 2017 study, Rodríguez et al. used field data from 964 orange, mandarin, and lemon plantations in Argentina to forecast citrus yields. Temperature, rainfall, species, variety, tree age, spacing, irrigation, and orchard location were all included in the dataset. Root Mean Square Error (RMSE) values for lemon, mandarin, and orange were determined following the testing of several machine learning regression tree models and parameter optimization, and the results showed encouraging outcomes. Citrus yield forecasting using climate data was investigated by Al-Shanableh (2022), who found important insights. Moussaid et al. (2021) demonstrated enhanced prediction accuracy with spectrum information from satellite photographs by developing a machine-learning model to anticipate orchard output prior to harvest using historical data on climate, irrigation, fertilizer, and field information.

A novel breeding strategy known as mimic breeding, integrating MAS and GS, has shown promise in selecting new hybrids by mimicking parental combinations. This approach, akin to complimentary breeding, focuses on identifying candidate individuals with desired qualities, leading to the creation of unique and high-end hybrids. NARO's "CitrusBreeding2.0" initiative combines GAB and citrus ancestry to enhance breeding efficiency and predict promising cultivars. By increasing genetic diversity and utilizing simulation-based breeding, breeders can optimize cross-combinations and predict elite hybrids more accurately, ultimately improving breeding efficiency and cultivar development.

14.2.3 Breeding Objectives

Fruit breeding initiatives, which encompass a wide range of particular aims addressing abiotic and biotic stress resistance, tree design, precocity, and productivity, all aim to produce high-quality fruit (Kader 2002). Depending on the fruit species, fruit quality varies and is comprised of many characteristics (Callahan 2003). For instance, the degree of sweetness alone determines the taste of some fruits, while a balance of acidity and sweetness is necessary for others (Kader 2002). For these species, crisp texture is considerably more acceptable than soft, and vice versa. But all commercial cultivar releases need to have mouthwateringly appetizing fruits. The quality of the fruit determines whether a tree succeeds or fails in the commercial market, regardless of its resistance to disease, insects, or great productivity (Ray 2002).

The simple definition of fruit quality is: Whatever the consumer chooses, as it is challenging to define a precise description because it fluctuates depending on the consumer's taste or needs (Barritt 2001; Kupferman 2002). Because people differ in their inclinations and conceptions of what constitutes quality, breeders must offer a wide range of substitute forms in order to satisfy consumer demand. Factors including color, size, nutrition, shelf life, processing compatibility, texture, taste, and sweetness can also be considered factors of quality (Kader 2002; Kumar et al. 2023). However, because fruit crops are perennial, breeding for quality enhancement is complicated by several factors, such as the vast size of the plant, the length of the juvenile period, and abiotic stress, like fruit drops because of natural conditions. Furthermore, the quality of fruit is a quantitatively inherited multiple genes character, which complicates the breeding effort for fruit crop quality development. A character or trait is any attribute that demonstrates heritable variation in an individual. Quality features are characteristics that specify a certain facet of produce quality. Every crop has a unique set of quality attributes that are frequently quite distinct from one another. Fruits are an important part of a healthy diet because they can supply all the necessary components, such as fiber, vitamins, minerals, proteins, fats, and carbohydrates, that are typically lacking in agronomical crops. This makes fruits an excellent way to treat nutritional disorders.

In contrast to agronomical crops, relatively few efforts have been made to improve fruit trees. Up to now, efforts have been undertaken consistently to increase the quality attributes of staple crops. Fruit breeding goals are mostly concentrated on tree design, precociousness, yield, resistance to biotic and abiotic stressors, and physiological diseases (Ray 2002). Fruit quality maintenance has received very little attention. Fruit crop breeding is hampered by several factors, such as market preference for native or local varieties over improved ones.

The main breeding program obstacle is the vast size and perennial behavior of the fruit trees, which makes it challenging to carry out tasks like emasculation, bagging, tagging, data collection, harvesting, etc. Most fruits are multiplicated in asexual ways, which typically provide fruit and blossoms after 5 years of age (Arias et al. 2012). In contrast, fruit trees derived from seedlings require more than 10 years to bear fruit, thus breeders need a long time to receive the results. Furthermore, even after a protracted wait, there is still a possibility of receiving undesirable outcomes, which makes the breeder less enthusiastic. For instance, just three types of apples were ultimately released for sale after a 26-year apple breeding program in Dresden-Pillnitz, Germany, screened 52,000 seedlings (Myles 2013). Fruit and flower drops, pest and disease infestations, along natural disasters, frequently cause enormous losses in breeding and commerce (Arias et al. 2012).

Enhancing fruit quality requires a solid understanding of genetic inheritance and variation. Fruit trees are heterozygous crops, excellent traits are inherited quantitatively and governed by several genes (Kenis et al. 2008). Biotechnological methods offer an accurate and dependable, means of resolving certain issues in traditional breeding. When handling labor-intensive crops, molecular techniques such as marker-assisted selection, candidate genes, genomics (Murovec and Bohanec 2012), transgenic, and cis-genics have proven to be beneficial in terms of the time, patience, and effort needed. The quantity and quality of fruits are negatively correlated. The overall output of crop plants is impacted when quality is improved by breeding; for example, if quality rises, yield falls, and vice versa.

Origin of Fruit Quality Characteristics: (1) A developed variety is the most favored one. (2) A line of germplasm. (3) an artificial or spontaneous mutant. (4) A wild relative. (5) A mutation.

Fruit Quality Characteristics: Taste, sweetness, aroma, flavor, and texture; appearance (size, shape, and color); and maintaining quality (softening, ripening control).

Breeding Strategies for High-Quality Fruit: The process of bringing wild species under human cultivation is known as domestication, and it began when man began to cultivate superior plants for use (Simmonds 1979).

Introducing a plant's genotype, or a combination of genotypes, into previously uncultivated areas is known as plant introduction. New cultivars of a crop already cultivated nearby, wild relatives of the crop species, or an entirely new crop species could all be introduced. The majority of materials are imported from other continents or nations (Singh et al. 2005).

The foundation of every crop development is selection. Selection is the process of removing attractive plant kinds from the population. It is one of the two essential steps—the other being—of any breeding effort. Two selections and the production of variation. The process of selection is carried out by two entities: man (artificial selection) and nature itself (natural selection). While they may occasionally work well together, their goals are generally at odds with one another because of the differences between the two circumstances (domestication and nature).

The most important step in hybridization is germplasm conservation, especially when using wild species in breeding programs, as it involves crossing parents and selecting progenies. These parents are selected after screening of the natural populations and wild relatives being conserved (Sharma et al. 2015). Different forms of hybridization have developed as a result of polyploidization, interspecific hybridization, or a mix of the two (Jalikop 2015).

The goals are to establish biotic and abiotic resistant varieties and to create genetic variation through the transfer of one or more qualitative features. Changes in DNA sequence that are not caused by genetic separation or recombination are known as mutations (Van 1998).

In plant evolution and diversification, polyploidy is defined as having more than two complete sets of chromosomes per cell nucleus (Ray 2002). True allopolyploids are formed through hybridization between distantly related species; in this instance, the divergent chromosome.

Polyploids: Aspects morphological and cytological, delayed blossoming and sluggish development, decreased dry weight but increased flesh weight, growth in size but a drop in the number of fruits, flowers, and leaves, Larger cell size, Greater stomatal gourd cell size (reduction in stomata count/unit area), ploidy variation among different species.

Utilizing polyploidy. Enhancement of quality: such as Seedlessness. Use directly as a new variable or kind. Gene transfer between different species. Expanding allopolyploids' genetic base is now in place.

A technique that allows the manipulation of cellular genomes by protoplast fusion is known as somatic hybridization, and it may lead to the development of novel varieties" (Loredana et al. 2012). Accordingly, a novel arrangement of the nuclear, chloroplast, and mitochondrial genomes may facilitate the production of novel cultivars with fruit-quality features (Singh and Rajam 2009). In the citrus group, somatic hybridization has been attempted to develop interspecific/inter-generic crosses to overcome issues such as pollen or ovule sterility, sexual incompatibility, and polyembryony (Loredana et al. 2012). Hybrids were created by protoplast fusion of sweet orange (Citrus sinensis L. Osbeck) and lemon by (Singh and Rajam 2009).

Molecular Techniques: Compared to annual crops, there is a very small amount of work about molecular biology in the case of perennial crops" (Pena and Seguin 2001). Some of the fruit tree breeding issues have been resolved through the use of molecular techniques. It would be very useful to be able to predict qualities like color, taste, texture, and nutrients by identifying biomarkers in seedlings. Conventional breeding requires more time, labor, money, and other inputs for germplasm preservation and evaluation, which is costly (Myles 2013). The marker-assisted selection (MAS) and genomics approaches are important fruit crops (Kumar et al. 2006). Furthermore, it has been demonstrated that cutting-edge genetic transformation technologies decrease a tree's juvenile stages and increase a tree's tolerance to biotic stressors and phytoremediation in perennial trees (Pena and Seguin 2001).

14.2.4 Quantitative Traits Loci (QTLs) and Marker-Assisted Selection

Gelderman is credited with coining the term QTL (Geldermann 1975). When it comes to using the markers in the identification and selection process, those that are closely associated with the studied characteristic are informative. Using molecular markers SSR and SNPs to identify the desired attribute, are the most favored. This information is provided by QTLs and has been used in the breeding of fruit quality. The close relationship with other quality measures is beneficial when they have desirable qualities for breeders. It gets harder to distinguish them in the offspring, though, if one is unacceptable in hybrids. In order to find out, several crossings between cultivars possessing good and bad traits are conducted, and the population data from these crosses yields QTL, which may be utilized as markers in MAS.

(Longhi et al. 2012) carried out such analyses for fruit texture, finding QTLs using SSR and SNP markers. The most widely used QTL technique is interval mapping.

Candidate Genes Approach: Using a reverse breeding strategy to move from the gene to the desired characteristic, this approach is based on the idea that a molecular polymorphism is linked to phenotypic (Pflieger et al. 2001). Since fruit quality is closely correlated with the metabolites it contains, the candidate gene approach to understanding the signaling mechanism and biosynthetic route may be helpful. Fruits provide good flavor, color, and antioxidants such as flavonoids, phenolics, and other active substances. Specific or widespread metabolic processes are used to produce secondary metabolites. Genes that are regulatory and structural encode the enzymes involved in these pathways. (Nookaraju et al. 2010). In fruit crops such as citrus, the biosynthesis pathways of pigments have been thoroughly investigated (Kato et al. 2004). For instance, citrus fruit taste development is influenced by candidate genes (Kato et al. 2004).

The study of a specific fruit crop's full genetic makeup is included in the genomics approach. Dr. T. created the phrase genomics. Roderick. It can be applied to hybrid development as well as the selection of better parents. A sort of marker-assisted selection known as genomic selection (GS) occurs when genetic markers present across the genome manifest phenotypic variance (Myles 2013). In fruit trees such as citrus, genomic analysis related to produce quality at developmental stages has been evaluated (Terol et al. 2008; Zhu et al. 2013).

In genetically modified fruit crops, desired genes are introduced (Dias and Ortiz 2014), primarily in order to create elite kinds with better quality attributes. Through protoplast transformation and agrobacterium-mediated transformation, the fruit quality gene was transmitted, as a result of the recovery of transgenic citrus plants (Guo et al. 2005).

14.3 Biodiversity and Conservation of Germplasm

14.3.1 Diversity of Germplasm

Citrus and its close relatives are part of a total of 28 genera within the family Rutaceae and the tribe Citreae of the subfamily Aurantioideae (Swingle and Reece 1967). Before the use of molecular data, citrus classification was based on morphology or isozymes. Currently, citrus is categorized into two main groups: Swingle (Swingle and Reece 1967) and Tanaka (1977). There are 162 species within the genus Citrus according to the Tanaka system, but only 16 species are due to the Swingle system.

According to Barrett and Rhodes (1976) and Swingle and Scora (1975), there are only three fundamental, authentic species of citrus in the subgenus Citrus: citron (C. medica L.), mandarin (C. reticulata Blanco), and pummelo (C. maxima L. Osbeck). Some of the other domesticated citrus species in the subgenus Citrus are thought to be hybrids of these closely related genera, subgenus species, or real species. In recent years, this notion has been supported by data derived from molecular markers (Federici et al. 1998; Nicolosi et al. 2000).

14.3.2 Cultivars Characterization and Phylogeny

In the citrus genus, both nature and human selection have played significant roles in shaping the current variation. Citrus taxonomy is influenced by various factors, such as the frequent occurrence of bud mutations, sexual compatibility with related genera and species within the genus, and asexual reproduction through nucellar embryony, a characteristic of several citrus species. Consequently, there are variations in the number and diversity of species among different classification schemes.

14.3.3 Genetic Resources Conservation Approaches

In order to supply genetic resources for breeding and research initiatives, citrus germplasm is preserved in gene banks at various locations across the globe (Abulaila et al. 2025; Abdullaev et al. 2025; Akparov et al. 2025; Aladele et al. 2025; Al-Khayri et al. 2025; Al Lawati 2025; Almarri et al. 2025; Arab et al. 2025; Békefi et al. 2025; Breidy 2025; Cañar et al. 2025; de Galarreta et al. 2025; Desheva et al. 2025; Díez et al. 2025; Galea-Gragera and Llera Cid 2025; Ghidan and Abdulmajid 2025; Gudaye et al. 2025; Harzli et al. 2025; Hidayatun and Kurniawan 2025; Ihsan and Prihatini 2025; Jatoi and Siddiqui 2025; Laghetti et al. 2025; León-Lobos et al. 2025; Mallor and Luna 2025; Monteros-Altamirano et al. 2025; Osterc et al. 2025; Popović et al. 2025; Rocha et al. 2025; Wang et al. 2025). The importance of genebank collections is especially clear in light of the threat posed by diseases and climate change to citrus-growing regions. There are 132 citrus accessions in the collection of the Russian Research Institute of Floriculture and Subtropical Crops (RRIFSC) in Sochi, Russia; these accessions represent 50 taxa. Numerous accessions in the RRIFSC collection were chosen especially for their cold tolerance to support citrus production in southern Russia, which is one of the world's most northern citrus-growing regions.

Data collected from tree responses during the region's severe winters has led to a better understanding of the cold tolerance of various RRIFSC accessions. The RRIFSC boasts an impressive collection of cold-tolerant lemon accessions. The National Clonal Germplasm Repository for Citrus and Dates (NCGRCD) of the Agricultural Research Service (ARS) under the United States Department of Agriculture (USDA), along with the Aurantioideae collections at the University of California Citrus Variety Collection in Riverside, California, collectively house 1328 accessions of citrus cultivars and wild relatives. At these facilities, accessions undergo treatments to eliminate known diseases and are screened for graft-transmissible pathogens.

In contrast to Sochi, the California collection primarily focuses on desert-adapted grapefruit and lemon cultivars. Huanglongbing, a disease currently threatening the US citrus industry, is being studied for genes that provide resistance or tolerance to citrus diseases among the collection of citrus and its relatives. Cooperative exchange efforts could enhance the diversity of the two gene banks, providing breeding and research groups worldwide with more valuable genetic resources.

Certain citrus cultivars are grown as grafted trees in orchard environments after being propagated clonally. Diverse citrus genetic resource representatives are frequently kept in gene banks as grafted scions on rootstocks in fields, greenhouses, or greenhouse environments. Collections can be lost in weather-related or pathogen-related disasters even with extensive protection. Although they can be costly, off-site security backups are essential to preventing loss (Volk 2011).

Maintaining a secondary planting is doable with sufficient resources, but off-site preservation of materials can prove to be more cost-effective. Citrus seeds might be difficult to preserve, but it is doable. For most species, it is effective to lower the moisture content of seeds or excised embryos before storing them for an extended period at liquid nitrogen temperatures (−196 °C). Citrus sinensis L. embryonic axes are one example. Could be quickly chilled to liquid nitrogen temperatures after being dried (Santos and Stushnoff 2003). When the varied alleles found in wild populations of citrus species are the targeted conservation aim, citrus seed conservation becomes even more important.

Hamilton et al. (2009) and Ashmore et al. (2011), for instance, have reported on the use of desiccation-based cryopreservation techniques for the seeds of four wild citrus species found in Australia: *Citrus inodora, Citrus australasica, Citrus garrawayae*, and *Citrus australis*. These species are fundamentally orthodox, and it is advised to store them in fluid nitrogen since storing these oil-rich seeds at −20 ° C and low moisture contents (5% or less) might cause substantial degradation. Somatic development in *C. inodora* and *C. microproliferation. C. australasica, C. garrawayi* and C. Microcuttings have also been shown to produce inodora (Ashmore et al. 2011). These offer more choices for medium- and long-term ex situ conservation.

The particular allelic mix of the scions is very important for cultivars. Cryopreservation techniques based on vegetative propagules can be used to back up these materials, which can then be kept in secondary sites (Engelmann 2010).

For several years, somatic embryos and shoot tips generated from in vitro plants have been preserved using cryopreservation techniques. Citrus nucellar cell suspensions were used as a model system in some of the initial studies on plant cryoprotectant solutions (Sakai et al. 1990). *Citrus aurantium* L. and Poncirus trifoliata in vitro plants were effectively kept alive by encasing the apices of the removed shoot tips in alginate beads before being desiccated by air, sucrose preculture, and liquid nitrogen (Al-Ababneh et al. 2002; Gonzales-Arnao et al. 1998; Wang et al. 2002). Poncirus trifoliata in vitro stock cultures were also utilized as shoot tip source materials for vitrification and quick freezing after the application of cryoprotectant solutions in cryopreservation techniques (Wang et al. 2003). In these instances, the recuperating shoot tips were cultivated in vitro with optimal growth media by being placed directly on a semi-solid medium. Citrus can also be effectively and reliably derived from somatic embryos.

Gonzalez-Arnao et al. (2008) discuss cryopreservation. Somatic embryo methods offer a consistently high survival rate, but they require several optimized culture procedures.

Vitrification freezing techniques have recently been established as a cryopreservation approach for citrus shoot tips removed directly from screen houses or greenhouse stock plants (Volk et al. 2012). It uses micrografting for recovery following LN exposure and does not require in vitro source materials. Therefore, since the rootstocks are cultivated in vitro from seeds, just their media needs to be tuned. Carrizo citrange (*Poncirus trifoliata*) was chosen as the rootstock to which recovered branch tips from several citrus species would be micrografted because of its high degree of compatibility (Edriss and Burger 1984; Volk et al. 2012). Its prominent trifoliate leaves make it easy to discern visually, adding to its advantages of widespread availability.

The majority of citrus sanitation programs across the globe employ the internationally renowned in vitro quarantine technique, which is also advised by the FAO for the interchange of citrus germplasm (Frison and Taher 1991). It has been used in Spain to clean 330 locally chosen accessions and import over 290 accessions from various citrus locations. To reduce phytosanitary hazards, it would be ideal for this tried-and-true method to be used in every nation that collects citrus.

14.3.4 Cytogenetics

The term "molecular cytogenetic map" refers to the organization of DNA sequences on chromosomes in the genome to identify homologous chromosomes and chromosomal rearrangements, as well as to examine genome homology across various species (Meng et al. 2020; Zhao et al. 2021).

Fluorescence in situ hybridization, or FISH, is the main method used to create molecular cytogenetic maps (Jiang 2019). For many years, bacterial artificial chromosomes (BAC) and repetitive DNA sequences have been employed as FISH probes in plants (Deng et al. 2018, 2020). As a standard for FISH in Solanum, Zea, Cucumis, Populus, Brassica, and Saccharum, oligonucleotide (oligo) probes have been created. These probes, which target repeated DNA sequences, particular chromosomal regions, or entire chromosomes, can be simple-sequence repeats or a pool of single-copy DNA sequences (Bi et al. 2020; Xin et al. 2020; Yu et al. 2021; Zhao et al. 2021). Furthermore, Chromoycin A3 (CMA) staining was used to identify the chromosomes in citrus and cuscuta (Ibiapino et al. 2022). According to CMA bands, citrus chromosomes can be divided into seven groups: A, B, C, D, E, F, and G (Carvalho et al. 2005). First, BAC clones, 45S rDNA, and FISH were used to generate a basic cytogenetic map of C. sinensis (Lan et al. 2016). FISH was used to map RSs in blood orange (Deng et al. 2019). Nevertheless, it is still difficult to identify every chromosome because there aren't enough markers for citrus.

Three C genome sequences were recently published, providing a chance to investigate oligo-based probes for FISH mapping on each chromosome. However, technical problems with genome sequencing and assembly result in gaps in the genome, particularly in areas rich in RSs (Liu et al. 2022). Version 1 (v1) of the sweet orange genome had only bp-level sequencing lengths, whereas versions 2 and 3 (v3) had kb-level sequencing lengths (Wang et al. 2021). Gapless telomere-to-telomere (T2T) sequence assemblies can be utilized to create whole genomes, even though the majority of plant genome sequence assemblies are still fragmented (Belser et al. 2021; Liu et al. 2020; Navratilova et al. 2022). Using ultra-long Oxford Nanopore Technologies (ONT) reads that range from 100 kb to 1 Mb, centromere assembly has been accomplished in both rice and Arabidopsis (Naish et al. 2021; Song et al. 2021b). Even with ultra-long read sequencing, it is still difficult to assemble high tandem repetitive sequences (HTRSs), such as 45S rDNA regions (Navratilova et al. 2022).

High genomic collinearity between C. sinensis and C. maxima was found when chromosomal painting probes (CPPs) were used on the two species (Chenqiao et al. 2019). On certain homologous chromosomes, each CPP produced modest noise signals; yet, on other pairs, they created strong and unique FISH signals.

Song et al. effectively investigated molecular probes for C. sinensis chromosomal identification in 2023. Every chromosome was successfully recognized by the 30 oligo probes they created, which also offered comprehensive details on chromosomal orientation and diversity regions. Whole chromosomes in a number of plant species have been seen using CPPs (He et al. 2020; Xin et al. 2020; Yu et al. 2021). When it came to viewing certain chromosomal areas, the OPs developed by Song et al. (2021b) outperformed CPPs. Four RSs (45S rDNA, 5S rDNA, CsCen2, and CL-2) are also helpful markers for identifying sweet orange chromosomes. Additionally, they presented a technique that effectively identified chromosomes 1–9 in a single experiment by utilizing oligo-FISH with CMA labeling. This approach was found to be more effective than the earlier one, which involved several rounds of sequential FISH with CPPs in different plant species (He et al. 2020; Yu et al. 2021; Zhao et al. 2021).

CsCen7 was studied as a centromeric marker in C. sinensis in addition to markers for chromosomal identification. In eukaryotes, the centromere is essential for chromatid segregation during mitosis and meiosis (Hofstatter et al. 2022). Citrus' studies on gap-free genome assembly, epigenetics, and centromeric genetics have profited immensely from the centromeric repeat CsCen7. FISH signals of the 181 bp HTRS known as CL at the ends of chromosomes indicate otherwise, despite bioinformatics analysis's prediction that it is a centromeric repeat in C. sinensis (Melters et al. 2013). Significant heterozygosity was shown in the sweet orange's high-resolution molecular cytogenetic map, with two Chr B carrying 45S rDNA locations and two Chr C exhibiting homology (Lan et al. 2016; Silva et al. 2010). The map showed that whereas Chr B with 45S rDNA and Chr B are not similar, Chr CS and Chr Bf with 45S rDNA locations are. One pair of homologous chromosomes was found to have Chr CL and rDNA fragile spots. These results highlight the necessity of a thorough molecular cytogenetic map that uses both oligo and RS FISH signals to detect homologous chromosomes in heterozygous plants.

In addition to helping with genome viewing and quality evaluation, the high-resolution molecular cytogenetic map also helps with genome assembly, especially when assembling RSs. While 30 oligo sequences were correctly constructed on the sweet orange genome v3, HTRSs were not properly assembled, according to a study on citrus genome assembly accuracy using the molecular cytogenetic map. For instance, genome v3 did not display any 45S rDNA on Chr 6, despite 45S rDNA FISH revealing a substantial 45S rDNA location at the end of Chr 6.

According to reports, it is still difficult to completely create HTRSs with multiple copies of repetitive DNA elements in plant genomes (Liu et al. 2022; Navratilova et al. 2022). For upcoming research on citrus chromosomes and related species, the high-resolution molecular cytogenetic atlas will be helpful. According to recent research, CRISPR technology can be used to remodel plant chromosomes, such as those of Arabidopsis thaliana (Rönspies et al. 2022a, b). Chromosome-targeted rearrangements, including translocations, for genetic enhancement in citrus are made possible by the molecular cytogenetic map. According to Huang et al. (2004), C used to exclusively break its longest chromosomal peak. In the future, our molecular cytogenetic map will make it easier to identify and gather individual citrus chromosomes or chromosome segments of any size. Plant diversification and taxonomy placements can be inferred from chromosome pan- and core-karyotypes (Cheng et al. 2022). We may create citrus pan- and core-karyotypes, investigate chromosome collinearity and the evolution of the citrus family, and look into other translocations and rearrangements in citrus plants by using the high-resolution molecular cytogenetic map of sweet orange as a guide.

14.4 Traditional Breeding

14.4.1 Improvement Strategies

The selection of spontaneous bud mutations from commercial orchards produced most citrus varieties. Fruit quality (size, color, and seedlessness) and prolonging the harvest season are the main selection objectives. New cultivars are frequently introduced, particularly for sweet oranges, particularly those in the navel group, satsumas, and clementines (17 new cultivars protected in Spain over the past 10 years). Seedless Mandarin mutants were recently introduced in Israel (Mor) and the United States (Tango).

One important requirement for the fruit business is seedlessness. According to Yamamoto et al. (1995), sterility can be divided into three categories: self-incompatibility, male sterility, and female infertility.

Some male-sterile and self-incompatible accessions are unable to produce seedless fruits because they do not have parthenocarpy. Citrus germplasm seems to possess a significant level of parthenocarpy, which is an essential trait for the generation of seedless fruit. Citrus ('Navel' orange or 'Satsuma' mandarin) parthenocarpy is an autonomic parthenocarpy, meaning that it produces fruit without seeds and pollination (Vardi et al. 1988). Fruits without seeds may be produced by male sterile or self-incompatible accessions if cross-pollination is prevented. However, those accessions may produce seedy fruits in mixed plantings, particularly if pollinating insects are present. In areas such as the Mediterranean Basin, where the main seedless mandarins are self-incompatible clementines, selection is required for both male and female sterility.

The primary goal of rootstock breeding is to produce plants that are tolerant to soil conditions and diseases, with tolerance to CTV, Phytophthora sp., and nematodes being common requirements. Tolerance to salinity, iron chlorosis in calcareous soils, and tolerance to malsecco reduces the nursery survival of vulnerable rootstocks. One of the biggest challenges in the Mediterranean Basin is breeding rootstocks that would enable the management of limited water supplies. Blight tolerance in Florida has emerged as a critical goal. Many features of the rootstock/scion interaction can be altered by the genotype of the rootstock. In all citrus-producing regions, so rootstock with CTV tolerance must be bred.

14.4.2 Traditional Breeding Methodologies

Depending on the goal of genetic improvement, several breeding techniques are used for citrus, focusing on either the scion or rootstock. Rootstock breeding offers greater strategic possibilities compared to scion breeding. In the past, traditional farmers selected main rootstocks like Rangpur lime, Volkamer lemon, trifoliate orange, sour orange, and *C. macrophylla* based on popular true species or hybrids. However, new alternatives have emerged through systematic breeding efforts, replacing many of these traditional rootstocks (Caruso et al. 2020). Novel tetraploid rootstock candidates have been created through somatic hybridization by protoplast fusion, with somatic hybrids serving as breeding parents for new tetraploid candidates (Grosser and Gmitter 1990a, b, 2011).

Rootstock selection is not influenced by the taste of the fruit, allowing for greater genetic and taxonomic diversity in rootstock improvement strategies compared to scion improvement. This diversity can be drawn from genebank collections, regardless of whether sexual hybridizations occur at diploid, tetraploid, or somatic levels. Gametophytic self-incompatibility, which is typical of pummelo, is present in a number of Mandarin and Mandarin hybrid cultivars (Soost 1968).

Some male-sterile and self-incompatible accessions are unable to produce seedless fruits because they do not have parthenocarpy. Citrus germplasm seems to possess a significant level of parthenocarpy, which is an essential trait for the generation of seedless fruit. Citrus ('Navel' orange or 'Satsuma' mandarin) parthenocarpy is an autonomic parthenocarpy, meaning that it produces fruit without seeds and pollination (Vardi et al. 1988). Fruits without seeds may be produced by male sterile or self-incompatible accessions if cross-pollination is prevented. However, those accessions may produce seedy fruits in mixed plantings, particularly if pollinating insects are present. In areas such as the Mediterranean Basin, where the main seedless mandarins are self-incompatible clementines, selection is required for both male and female sterility.

Furthermore, all of the aforementioned, aside from clementines, reproduce from apomictic seeds, maintaining the phenotype and genotype of their mother. Thus, according to Wu et al. (2014 2018, 2021), these types represent a remarkably small portion of the genetic diversity found within the species. As is typical of pummelo, gametophytic self-incompatibility is present in a number of mandarin and mandarin hybrid cultivars (Soost 1968).

Because they lack parthenocarpy, several male-sterile and self-incompatible accessions are unable to produce fruit without seeds. Hence, parthenocarpy is an essential trait for the growth of fruit without seeds, and citrus germplasm seems to possess a lot of this trait. Citrus (such as "Navel" oranges and "Satsuma" mandarins) have autonomic parthenocarpy, meaning that pollination is not necessary for the production of seedless fruit (Vardi et al. 1988). Male sterile or self-incompatible accessions can produce fruit without seeds if cross-pollination is prevented. However, if pollinating insects are available, certain accessions may produce seedy fruits in mixed plantings. In areas like the Mediterranean Basin, where self-incompatible clementines are the main seedless mandarins, selection is required for both male and female sterility.

There are ongoing debates worldwide regarding modifications to national and international laws that regulate cultivar-type identification restrictions (Stover et al. 2020). If these arbitrary parameters are relaxed, future scion breeding will have greater potential for sexual hybridization. With $2n = 2x = 18$, citrus and related taxa are typically diploid. *Fortunella hindsii* Swing, a closely related genus, has at least one wild tetraploid variety; however, polyploidy is rare in wild or unselected citrus. Over the years, numerous Citrus and Poncirus accessions have been shown to have spontaneous tetraploid forms. These forms can also be created artificially through methods like somatic hybridization and colchicine treatment. Sometimes, spontaneous crossings between diploids can result in triploids; this could be the source of "natural" triploids like large-fruited acid limes. To create seedless offspring, some breeding initiatives employ a technique known as the creation of triploids by controlled diploid-by-tetraploid crosses (Grosser and Gmitter 2011; Soost and Roose 1996).

Novel genome-based genetic enhancement techniques are now being researched and are probably going to gain popularity in the future. These tactics, which employ CRISPR technology, include gene or genome editing, marker-assisted selection, and genome selection based on genome-wide association studies (GWAS) (Germana et al. 2020; Mattia et al. 2022; Minamikawa et al. 2017; Shimizu 2020).

14.4.3 Traditional Breeding Limitations

By utilizing vegetative propagation, citrus breeders focus on inciting polymorphism in one cycle, from selected clones. So Selection is linked to the genotypes obtained through hybridizations, induced mutagenesis, or somatic hybridization and transformation, developed in citrus to address challenges with traditional breeding.

Spontaneous mutations are identified in orchards. Optimizing the transfer of the parents' key commercial features to the progeny is necessary to increase the efficacy of such tactics. This requires a deep understanding of the inheritance of the chosen traits as well as systematic handling of heterozygosity. It is specially selected by using a marker for all the features that cannot be determined in the juvenile stage.

Numerous apomictic citrus genotypes have seeds with multiple embryos that are generated from nucellar cells or maternal somatic tissue. The formation of nucellar embryos seems to require fertilization (Kepiro and Roose 2007; Soost 1987).

However, the genotypes of C. maxima and C. medica are solely nonapomictic monoembryonic. Partial apomixis is influenced by genetics and environmental variables (Khan and Roose 1988). Although it hinders the development of sexual seedlings and self-fertilization progenies, and thus breeding programs, nucellar embryony provides a straightforward and economical method for clonal rootstock propagation by seed.

14.4.4 Role of Biotechnology

Citrus breeding techniques combine the latest technological advancements with the advantages of vegetative propagation. Since triploids are often sterile in both male and female sexes, selecting triploid lines is the traditional method for creating seedless cultivars. As a result, most trees in a triploid offspring under field evaluation should be seedless, allowing for the effective selection of other traits. Furthermore, the typical reduction in fruit size seen in seedless mutants of seedy cultivars should be counterbalanced by the increased fruit size linked to triploidy (Ollitrault et al. 2008).

14.5 Molecular Breeding

14.5.1 Molecular Marker-Assisted Breeding

With a small number of genetic markers and alleles that helped describe and distinguish amongst citrus accessions, citrus plants were genotyped utilizing antiquated protein-based techniques. With the advent of DNA-based methods like restriction fragment length polymorphisms (RFLPs) and amplified fragment length polymorphisms (AFLPs), the number of genetic markers accessible for genotyping accessions increased and the cost per data point gradually decreased. Newer, more potent, and more reasonably priced marker systems, like more widely distributed single nucleotide polymorphisms (SNPs) and expressed sequence tags-simple sequence repeats (EST-SSRs), were developed as a result of the integration of gene expression data with genomic information made possible by improved DNA and RNA sequencing technologies. These developments made it possible to genotype citrus germplasm more thoroughly and discriminatively as well as to characterize it at a much deeper molecular level. The most advanced technology is whole genome sequencing and assembly. The swift development of sequencing technology has made it feasible to use this platform to precisely describe and distinguish even mutants from a single genetic resource or cultivar.

Numerous citrus genotyping techniques, such as RAPD and DaRT markers, simple sequence repeats (SSRs), genotyping by sequencing (GBS), and SNP arrays, have been used in various projects over the years (Barkely et al. 2006). High-density SNP arrays, like those developed by Hiraoka (2020) (~58 K SNP features), are ideal for comparing genotypes within and between collections due to the polymorphism they offer. It is important to note that citrus SNP arrays have a limited genetic diversity range compared to global germplasm banks, limiting their applicability. To create more comprehensive genotype collections in the future, new arrays must incorporate more taxonomic variation from the citrus germplasm pool. The increasing number of citrus genome sequences being generated makes it possible to develop a more robust array platform.

Genebanks that want to use molecular characterization tools to scan the genotypic diversity and understand the relatedness among accessions within their collection must carefully weigh the costs of using these new platforms against the absolute necessity for such information in order to make the most impactful management decisions. They must also make prudent use of their limited financial resources. Seven SSR primers were used to select sexually tolerant individuals since citrus seeds are polyembryonic Fig. 14.6 and generate nucellar embryos with genotypes comparable to the seed parent. Consequently, hybrids may be successfully identified using four primers (Hasan et al. 2021a, b, 2022).

14.5.2 Genomics

Citrus genealogy research has provided some important insights. Several well-known cultivars, such as Kishu, Kunenbo, Yuzu, and pummelo, have been crossed multiple times over a few generations to create the diverse range of cultivars we have today. This indicates that the genetic diversity of these original cultivars has decreased, a fact confirmed by a study using SSR markers (Shimizu et al. 2016). Fruit characteristics, such as size, shape, color, rind thickness, SSC, acidity, and peelability, have undergone significant changes in just one generation (Food and Agriculture Organization 2016). This suggests that creating an elite hybrid does not

Fig. 14.6 Polyembryonic in Citrus. (Photoed by Dr. V.E. Hasan)

always require multiple generations (Caruso et al. 2020). The detailed genealogies reveal that several unique cultivars are descendants of cultivars that were previously not used in breeding programs. Utilizing these exotic, unfamiliar cultivars as breeding parents could lead to the development of new cultivars with distinct characteristics.

The ancient cultivar Kishu (*C. kinokuni* hort. ex Tanaka) has been recorded as far back as 740 A.D. in Japan, according to Shimizu et al. (2018). It was discovered that Kishu served as the indirect parent of 24 cultivars, including Jabara, Mochiyu, Kabosu, Henka mikan, Kizu, Konkoji, Hyoukan, Hassaku, Asahikan, Kaikoukan, Yamabuki, Iyo, USSR tangelo, Shunkokan, Kawabata, Unzoki, Kabuchi, Keraji, Yuge hyoukan, Ujutkitsu, Kawachi bankan, Sanbokan, Andokan, and Yuukunibu, through Naruto (*C. medioglobosa* hort. ex Tanaka) or Kunenbo (*C. nobilis* Lour. var. kunep Tanaka).

According to Shimizu et al. (2016), Kunenbo is a descendant of Kishu, resulting from a cross with an unknown seed father. This was confirmed using whole-genome sequencing analysis (Shimizu et al. 2017). Increasing the effectiveness of choosing elite hybrids with premium quality, genomics-assisted breeding utilizing MAS and GS is the key to resolving the breeding trilemma. However, the problem still lies in striking a balance between rapid breeding and diversity growth.

High-quality and distinctive cultivars, like Hassaku, Sudachi, Kabosu, Satsuma, and Hyuganatsu, were chosen from a single cross.

The disclosed lineage suggests that, although it is unlikely in a short amount of time, selecting an elite hybrid of good quality from a single cross can be accomplished by assessing numerous seedlings over an adequate amount of time. Increasing diversity is a conflicting goal in citrus breeding to increase quality. Numerous families have noted—most notably Kaikoukan and Tatachibana with their progeny, that selecting a superior elite hybrid that significantly deviates from the parent in terms of fruit characteristics can be accomplished with just one generation of crosses. With the advent of genome-wide prediction approaches using GS, citrus breeding efficiency can now be increased by quickly choosing a large number of seedlings. Breeding new cultivars with superior quality and distinguishing characteristics in a single generation is conceivable by combining genomics-assisted breeding (GAB) approaches to select thousands to tens of thousands of seedlings, thereby avoiding the breeding trilemma. The genomic breeding strategy lowers the cost of breeding while increasing the virtual breeding size and improving the possibility of selecting the best seedling.

When performing GAB, cost is still another crucial consideration. The cost of genotyping every seedling is too great to be disregarded, even if the cost of genotyping the entire genome has been decreasing. All things considered, the cost of breeding will be lowered in part by the two-step selection of seedlings prior to grafting. MAS uses many DNA markers to pick a big group of seedlings in two steps. First, it selects the primary candidate seedlings and then it removes the majority of the hopeless individuals. SSR or SNP markers derived from GWAS are appropriate for low-cost selection of a sizable population in this first selection. The few surviving promising individuals' genome-wide genotypes are acquired for the second selection, which then uses GS fruit trait prediction to identify elite individuals. Compared to annual crops, the total expense of breeding citrus to keep and assess a large number of individuals throughout time is substantially higher. As a result, compared to the expense of doing phenotypic selection on a large number of seedlings, executing GAB in two steps will reduce overall costs (Shimizu et al. 2017).

14.5.3 Bioinformatics

Data related to genebank materials must be documented. Spreadsheets, local databases, and shared database structures can all be used to store information. Passport information is best kept in a database that is somewhat accessible to the general public since it should be accessible to the user community. The location, amount, propagation history, pathogen test findings, etc. of each inventory item stored are crucial pieces of information for administering a genebank. There may occasionally be a hybrid system in place, with information kept on a central server that is accessible to the gene bank staff and the general public, but not shared with them. In any event, adhering to sound data management procedures is essential to guarantee system security, backup, and integrity. GRINGlobal is one tool for managing gene banks (USDA 2022). Genesys compiles data for ten global citrus collections. Documentation of germplasm collections may benefit from the inclusion of other pertinent databases. Among them is the United States' National Center for Biotechnology Information (NCBI). The National Institutes of Health is a widely acknowledged and utilized source of genomic and transcriptome data that has a wealth of knowledge on citrus accessions. Such data is maintained by citrus-specific databases, which also contain information on breeding and metabolomics. Each database has a set of tools designed especially to enable the research community to query the databases for details on particular genes, species, cultivars, etc. These include, for example, the Citrus Genome Database at Washington State University in Pullman, Washington, USA; the Citrus Pan-genome2breeding Database at Huazhong Agricultural University in Wuhan, Hubei, China; and the Citrus Genomic Variation Database at Southwest University/Chinese Academy of Agricultural Sciences in Beibei, Chongqing, China.

14.6 Tissue Culture Applications

14.6.1 Micropropagation

The process of grafting tiny shoot tips from sick plants onto immature rootstock seedlings is known as shoot tip grafting in vitro (STG) (Navarro 1992; Navarro and Juarez 2007; Navarro et al. 1975). Since the procedure enables the recuperation of matured citrus trees devoid of all diseases, the little shoot tips are typically clear of pathogens. Because of these benefits, most citrus-growing nations use shoot tip grafting (STG) to restore viable plants for commercial replication.

A different citrus tissue culture method (Navarro 1991) was created for the safe introduction of citrus and has shown to be highly effective in keeping citrus pests and illnesses out. Budsticks imported from other countries are first disinfected, cleaned, and then micropropagation to cause lateral buds to sprout and the production of flushes, from which shoot tips are removed and micrografted in vitro. Essentially, the only material imported is a little shoot tip, which is often pest- and disease-free.

Comparing this tissue culture method to the conventional quarantine methods reveals several advantages. The quarantine time is shortened by the early introduction of pest and disease elimination that may have been present in the original material. Citrus research stations may function as quarantine stations since tissue culture allows test tubes to take the place of greenhouses in remote locations. Many of these stations use shoot tip grafting (STG) to clean up local crops, and the staff and facilities needed are typically on hand. As a result, many nations can easily set up the tissue culture process to ensure the secure importation of citrus vegetative reproduction parts. This approach has been approved by law in several nations and is advised for the exchange of citrus germplasm (Frison and Taher 1991). It has been applied effectively in Spain to import more than 280 genotypes from different citrus-growing regions. Some of the introductions are being widely utilized in breeding programs, and others are part of the Citrus Germplasm Bank of IVIA in Valencia.

In order to propagate elite genotypes or create individuals that cannot be recuperated through conventional methods, somatic hybridization, genetic transformation, and the recuperate of haploid and tetraploid plants are all using shoot tip grafting (STG) more and more (Navarro and Juarez 2007). The difficulty of rooting transgenic shoots is one of the main obstacles to citrus transformation. Therefore, using shoot tip grafting (STG) to restore plants from changed shoots or buds has become standard practice, significantly improving the effectiveness of the genetic transformation process. Figure 14.7 shows the stages of shoot tip grafting in citrus.

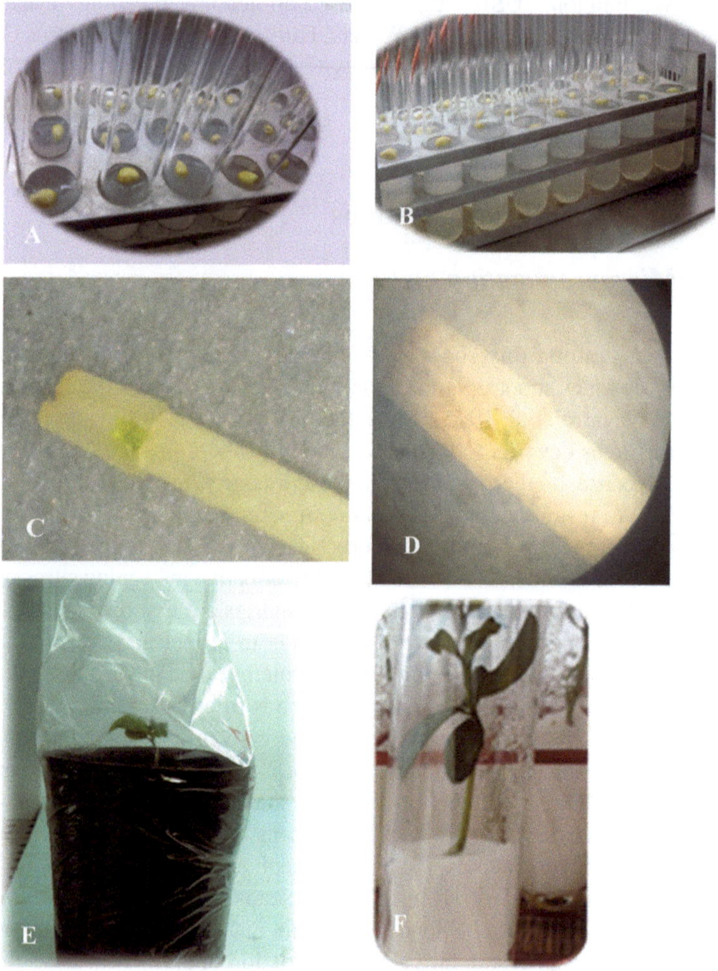

Fig. 14.7 Stages of shoot tip grafting in Citrus, (**a**, **b**) Seed culturing in Murashige and Skoog (MS) medium, (**c**, **d**) Grafting the growing shoot onto the rootstock, (**e**) Adaption in Lab and (**f**) Success grafting. (Photoed by Dr. V. E. Hasan)

14.6.2 Embryo Rescue

To create triploid citrus, several techniques have been established (Aleza et al. 2010; Navarro et al. 2003; Ollitrault et al. 2008). One of these techniques utilizes flow cytometry and embryo rescue to select triploids in 2x × 2x crossings, taking advantage of natural polyploidization occurrences like 2n gametes. While Chen et al. (2008) reported first meiotic division restitution (FDR) in sweet oranges, also Luro et al. (2004) postulated second meiotic division restitution (SDR) for diploid megagametophyte formation in clementine. These tetraploid plants result from somatic hybridization or apomictic seedlings, which occur when the chromosomal stock of nucellar cells naturally doubles (Grosser et al. 2000). Shoot tips grafted in vitro treated by Colchicine and used to produce tetraploid monoembryonic lines (Juárez et al. 2004; Navarro and Juárez 2007). The maternal tetraploid Clementine can now be involved in 4x × 2x crossings thanks to these tetraploids.

In vitro embryo rescue is successful in the case of embryos that are produced by interploid or diploid x diploid hybridization which contains malformed or underdeveloped seeds because of uneven ploidy ratio between the embryo and the endosperm. Maheshwari and Ranga Swamy (1958) demonstrated that fully formed embryos taken from apomictic cultivars can be cultured in vitro to produce zygotic plantlets. By extracting the seeds after 3–4 from anthesis, Nowadays, triploid breeding programs utilize embryo rescue, typically with undeveloped seeds from mature fruit (Navarro et al. 2003, 2004; Ollitrault et al. 2007; Starrantino and Recupero 1982).

14.6.3 In Vitro Pollination

Ovule cultures of apomictic genotypes are typically used to generate embryogenic callus lines (Ollitrault et al. 1995; Pérez et al. 1998, 1999; Rangan et al. 1969). Plants regenerated from such a callus show little to no somaclonal diversity (Kobayashi 1987; Olivares-Fuster et al. 2000; Starrantino and Russo 1983). On the other hand, it is difficult to induce genetically homogenous embryogenic callus from the ovules of parthenogenesis cultivars, and the regenerated plants differ phenotypically from the parent plants (Navarro et al. 1985). So many experiments have been conducted by the culture of style and stigma culture (Carimi et al. 1995, 1999) or the culture of the anther (Germanà 2003).

14.6.4 Other In Vitro Applications Pertaining to the Species

Trans-grafting, a technique that combines genetic engineering and traditional grafting methods, is crucial in citrus biotechnology breeding. This procedure involves grafting a non-genetically modified scion onto a transgenic rootstock (Haroldsen et al. 2012a, b). In trans-grafted plants, transgene products can move from the transgenic rootstock to the non-transgenic scion through the phloem, allowing the latter to acquire beneficial traits without altering the genetics of the end products, such as fruits (Song et al. 2015). Citrus crops grafted onto transgenic rootstocks that overexpress a microbial peptide showed lower infection rates from HLB bacterial disease compared to non-grafted plants (Bergey et al. 2015). De Francesco et al. (2020) demonstrated tolerance to citrus psoriasis virus in non-transgenic scions after transgrafting a transgenic sweet orange interstock that overexpressed a hairpin CP-mRNA.

14.7 Genetic Engineering and Gene Editing

14.7.1 Genetic Transformation Methodologies

In some Citrus varieties, such as orange (*C. sinensis*), lemon (*C. limon*), grapefruit (*C. paradisi*), and mandarin groupings like satsumas and clementines, specific features can be introduced into genotypes without altering their elite genetic background through genetic transformation. Genetic modification should be considered the most promising method of improvement due to the complex and heterozygous genetic structure of these varieties.

Genetic transformation methods in citrus are based on somatic embryogenesis internodes from greenhouse-grown plants (Duan et al. 2007; Fleming et al. 2000; Hidaka et al. 1990; Vardi et al. 1990), and more commonly, organogenesis (Kaneyoshi et al. 1994; Moore et al. 1992; Peña et al. 1995). For commercially important citrus species, effective transformation technologies rely on selecting appropriate source plant material in optimal ontological states, identifying competent cells for transformation in citrus explants, using suitable marker genes, establishing ideal infection and co-cultivation conditions, and utilizing proper culture media.

The Agrobacterium strain is crucial for transforming citrus and rapidly producing whole transgenic plants by grafting transgenic shoots onto rootstocks at first in vitro and later in the greenhouse. Successfully transforming mature plant material beyond the juvenile stage was a significant advancement in utilizing genetic transformation for citrus breeding (Cervera et al. 1998a, 2008).

Transgenic and control plants flowered after 14–18 months in the greenhouse, confirming their mature status. This method significantly reduces the time required for flowers to bloom and fruit to develop by several years. Various genotypes have been included in the process of transforming mature explants, including multiple cultivars of sweet orange, sour orange, Mexican limes, Fino lemon, Cleopatra mandarin, *C. macrophylla*, and clementine Clemenules (Peña et al. 2008).

14.7.2 Gene Editing Methodologies

One way to overcome issues with Agrobacterium-mediated transformation in certain genotypes is to directly introduce DNA into protoplasts through electroporation (Niedz et al. 2003) or PEG-mediated DNA uptake (Fleming et al. 2000). The PEG-mediated transformation of W serves as a clear example. Compared to the traditional epicotyl-mediated Agrobacterium method, using protoplasts in the Murcott tangor significantly increases transformation efficiency (Dutt et al. 2018).

Recently, the use of protoplasts has been reconsidered for their potential in gene editing, with or without the use of DNA molecules, thereby reducing the need for integrating foreign DNA (Huang et al. 2020).

The most challenging citrus species to infect and transform are the juvenile tissues from mandarin hybrids, particularly the epicotyls (Cervera et al. 1998b).

14.7.3 Enhanced Traits

Leaf disks from germinating seedlings are a valuable source of starting material for Agrobacterium-mediated transformation because they are readily available, rapidly accessible, and carry a lower risk of contamination than epicotyls. Mature leaf transformation allows the introduction of new traits without sacrificing clonal fidelity (Sandal et al. 2007).

Citrus leaf disks were used as explants by Moore et al. (1992), who compared their organogenic potential with stem segments. They discovered that using stems as explants was significantly more effective in producing shoots than using leaf segments. This is likely because organogenesis occurs more efficiently from stems (Almeida et al. 2003).

The success of traditional breeding programs and transformation techniques based on juvenile tissue explants is significantly hindered by long juvenile phases, which can often exceed 5 years. Traditional breeding methods can take up to 25–30 years to fully develop and commercially distribute new varieties (Caruso et al. 2020). The identification of new genes associated with early flowering in citrus is crucial for transgenesis or traditional breeding of desirable traits. Some strategies to shorten the lengthy juvenile stages of citrus trees are based on our knowledge of Arabidopsis thaliana flowering pathways. The constitutive overexpression of Arabidopsis LEAFY (LFY) or APETALA1 (AP1) genes in juvenile seedlings was the first method successfully used to reduce generation times in citrus plants (Peña et al. 2001). In transgenic citrange rootstocks, the independent expression of both genes led to early flowering and fruiting. These genes play a role in flower induction. However, LFY caused anomalies in vegetative growth, making AP1 more effective than LFY.

The citrus orthologue of Flowering Locus T (FT), CiFT, was ectopically overexpressed in P. trifoliata (Endo et al. 2005), resulting in a reduction of the juvenile period. However, once again, the vegetative growth and architecture of the plant were abnormal. Velázquez et al. (2016) developed the Citrus Leaf Blotch Virus (CLBV), which carries the AtFT or CiFT genes, as a viral vector-based tool to induce early flowering. Blooming began 4–6 months after the vector was injected into different genotypes, with no additional phenotypic anomalies. Additionally, Soares et al. (2020) described a unique method to induce premature flowering in Carrizo citrange rootstocks by overexpressing the orthologue of *Citrus clementina* (CcFT3) under the control of an AtSUC2 phloem-specific promoter. Plants with normal morphology resulted from this technique, flowering 16 months after transformation. Furthermore, faster flowering was achieved by grafting juvenile scions. To mitigate the pleiotropic effects of constitutive overexpression of early flowering genes, other strategies have been employed. Sinn et al. (2021), for instance, created transgenic grapefruits that expressed P. trifoliata FT1 (PtFT1) through a translational fusion using an antibody against a single-chain variable fragment. Transgenic FT chimeras with premature flowering were produced by reducing FT activity.

Recently, citrus genes that regulate various desirable traits have been cloned and identified as novel candidates for intragenesis and cis strategies. These include CsSAMT1 (salicylic acid carboxyl methyltransferase 1), which provides tolerance to HLB; CsMADS5, a fruit ripening-associated transcription factor that positively regulates carotenoid biosynthesis in citrus; a CsMYB96 transcription factor that enhances citrus fruit resistance against fungal pathogens; CiMADS43, a MADS-Box gene involved in citrus flowering and leaf development; and CiNPR4, an NPR1-like gene that increased resistance of transgenic citrus plants to HLB (Khan et al. 2021; Li et al. 2020; Lu et al. 2021; Peng et al. 2021; Zhang et al. 2021; Zou et al. 2021).

14.7.4 Transgenic Cultivars

The biology of citrus tree reproduction is complex. In citrus, the presence of apomixis inhibits the development of active zygotic embryos, leading to the formation of adventitious embryos directly from maternal nucellar cells. Citrus trees also have long juvenile stages. In subtropical regions, the flowering phase typically begins after at least 5 years, with several more years needed to reach full maturity. Another challenge hindering traditional breeding in citrus genetic improvement is the intricate taxonomic relationships among cultivar groups (Gmitter and Talon 2008). Genetic transformation, which involves introducing specific traits into established genotypes without altering their elite genetic background, shows promise for enhancing citrus genetics. Biotechnological advancements have facilitated the rapid production of new varieties and the improvement of existing cultivars' germplasm (Peña and Navarro 2000).

After acclimating to a specific photoperiod, elongated epicotyl stems are used as explants for DNA-coated gold or tungsten particle bombardment. The decision on whether to use circular or linear DNA (plasmid) is determined by the strategy. For example, linearizing the DNA is necessary if homologous recombination is to be induced (Cabrera-Ponce et al. 1997; Tomas et al. 1995). If not, a circular DNA molecule can be effectively converted into a relaxed or supercoiled state. Citrus stems can generally withstand the high-speed particle impact, with tests conducted on citrus tissue microparticle bombardment using pressures ranging from 900 to 2000 psi.

On the other hand, transforming apical stem tissues increases transformation efficiency. Following bombardment, both the middle and ends of the stems produce visible photosynthetic tissue, although in smaller amounts. Within the next 2 months, dissection of the embryos is possible for future heterografting. Regeneration is technically challenging for plants with radicular systems, so it is recommended to graft the plantlet onto a robust rootstock. Wu et al. (2019) utilized a biolistic approach based on mannose selection and phosphomannose isomerase (PMI) to achieve increased transformation efficiency and decreased escape rates. For information on selection markers, refer to the section "Selection Markers, New Reporter Genes, and Marker-Free Systems for Citrus Transformation."

14.7.5 Nanotechnology

The initial paper by Sanford et al. (1987) on the biolistic-mediated transformation of plant cells has been modified using several techniques. The incorporation of transgenes in citrus species was established by Yao et al.'s (1996) work, which reported on the transformation of tangelo using nondifferentiated embryonic callus. Bespalhok Filho et al. (2003) conducted epicotyl bombardment using GUS as the reporter gene, noting expression in the Carrizo citrange meristematic cambial tissue. Plant regeneration was a challenge in every instance because citrus explants are reluctant to form roots and, as a result, yield a plant that is similar to those grown in greenhouses or orchards. Wu et al. (2016) have described the epicotyl transition in Carrizo citrange. Using this technique, microprojectiles coated with the target DNA were fired onto the apical region of epicotyl explants. After the DNA reaches the cell, it is anticipated that it will go via the nucleus and get permanently integrated into the genome. Tissue culture was used to develop callus formation in order to produce regenerating plantlets, which are then grafted to create a whole plant. Figure 14.1 shows the steps of the biolistic transformation protocol. After being adjusted to a photoperiod, elongated epicotyl stems are utilized as explants for DNA-coated gold or tungsten particle bombardment. The strategy will determine whether to use circular or linear DNA (plasmid). For instance, linearizing the DNA is necessary if homologous recombination is to be induced (Tomas et al. 1995; Cabrera-Ponce et al. 1997). If not, a circular DNA molecule can be effectively changed into a relaxed or supercoiled state. Citrus stems often withstand the accelerated particle impact, although tests using pressures ranging from 900 to 2000 psi have been conducted on citrus tissue microparticle bombardment. On the other hand, transformation efficiency is increased by transforming apical stem tissues. Following bombardment, both the middle and ends of the stems produce visible photosynthetic tissue, but in smaller amounts. The embryos can be removed for additional heterografting after 2 months. It is technically difficult for plants with radicular systems to regenerate. Grafting the plantlet onto a robust rootstock is therefore advised.

A novel method called pollen magnetofection has been employed to create transgenic seeds devoid of in vitro plant regeneration. It involves creating nanoparticles, functionalizing them with DNA or RNA that encodes the desired characteristic, and creating magnetic nanoparticles (MNPs), which are subsequently incorporated into pollen by applying a magnetic field. Given that MNP-DNA/RNA complexes interact through electrostatic attractions, this goal is technically efficient. To be more precise, 1–10 million pollen particles can be combined with the loaded MNPs and spread out over a magnetic plate so that the particles can penetrate the pollen through their own pores. To fertilize emasculated blooms, pollen must be applied right away. Taking into account that pollen viability is expressed in hours, treated pollen has to be utilized right away. The desired transgene may be enclosed in seeds produced by magnetofected pollen pollination. Foreign DNA is inherited in a Mendelian fashion and integrated into the plant genome.

Citrus could benefit from the application of pollen magnetofection, which would make the creation of genetically modified plants easier. Furthermore, gene editing may also benefit from this technology. Although this method of transforming citrus is easily introduced, there are a few things to keep in mind (Xoconostle-Cázares, personal communication). Plants need to be maintained and sourced from certified orchards.

In a biosafety greenhouse with an 18:6 light/dark cycle, complete watering and fertilizing, it is possible to artificially or naturally induce flowering. In 3 months, 20–100 flowers can be produced. *C. aurantifolia*, grown in tropical orchards and sometimes known as Mexican lime, can yield flowers seven to eight times a year. On the other hand, C. sinensis, the sweet orange, can flower up to three times in a single growing season. A standard magnetofection experiment can pollinate up to 100 flowers. Mature fruits would be ready for harvesting after about 3 months and should have a smooth surface. Mature converted seeds can be recovered by collecting mature fruits; these seeds can germinate at 30 °C in a seedling nursery. After leaving the nursery, plantlets can be moved to individual pots for further examination. In a standard experiment, an average of five transgenic plantlets expressing the transgene is expected. Fruit quality may have an impact on the efficiency of transformation. It is important to keep transgenic plants in a safe environment. (Marcelino-Perez et al. 2021; Zhang et al. 2019a, b).

14.8 Mutation Breeding

14.8.1 Conventional Mutagenesis

The majority of citrus varieties grown today originated from the earliest citrus breeding technique, which involved the selection of spontaneous mutations.

Research conducted in Spain, Morocco, and Corsica has led to the development of clementines with longer maturation times and larger, more vibrant fruits. The diversification of oranges, lemons, grapefruits, and Satsuma mandarins in Japan can also be attributed to spontaneous mutations over a longer period Some examples include pink grapefruits, the navel orange, acidless sweet oranges and lemons, where mutations in bud meristems have resulted in chimera formation.

Since 1935, induced mutagenesis has been commonly used to enhance cultivars, with gamma irradiation being the most frequently employed method. Gamma irradiation can cause various mutations, such as point mutations, chromosomal breaks, and rearrangements. This method is particularly interesting for seedless selection due to the potential for sterility from translocations and inversions (Hearn 1984). According to Roose and Williams (2007), budwood should be exposed to 30–50 Gy to induce mutations. For example, the Star Ruby red flush grapefruit was attained by irradiating Hudson grapefruit seeds (Hensz 1971), showcasing the success of irradiation programs in creating new cultivars.

The "Mor" mandarin low-seeded was selected by the budwood mutation of Murcott (Vardi et al. 1993). Like that, seedless mandarin Tango occurred by mutation of Afourer (Nadorcott) budwood (Roose and Williams 2007).

14.8.2 Enhanced Traits and Improved Cultivars

To improve sweet oranges in Florida, little seed content, late harvest time, and other fruit quality characteristics have been used for somaclonal variation (Grosser et al. 2007).

Due to the narrow genetic base of grapefruit cultivars, hybridization is preferred to increase genetic diversity. The grapefruit market cultivars with significant commercial value were not produced through traditional breeding methods. Many of these cultivars can be traced back to a series of spontaneous bud sports or artificially induced mutations of the original white grapefruit, resulting in cultivars ranging in color from white to peach, red, and pink.

It is believed that a deep-golden-fleshed grapefruit found in Texas is a somatic mutation of a white grapefruit with a Duncan-like appearance. A grapefruit family tree illustrates the cultivars that were created through sexual crosses, induced mutations, and spontaneous bud sports (Louzada and Ramadugu. 2021).

14.9 Hybridization

14.9.1 Conventional Hybridization

Two cross hybridizations were conducted between Sour orange and Trifoliate: (Sour orange ♀ × Trifoliate ♂) and (Trifoliate ♀ × Sour orange ♂), as well as between

Cleopatra mandarin as the seed parent ♀ and sour orange as the Pollen Parent ♂. In the second season, Trifoliate was the pollen parent ♂ and each Sour orange and Cleopatra mandarin was the seed parent ♀. This research was carried out over two seasons: the first season was 2018–2019 and the second was 2019–2020 at the Agricultural Scientific Research Center in Lattakia, Syria. The aim was to study the tolerance of the resulting hybrids to cold and calcareous stress, which are challenges faced by farmers in this region.

The flowers of the pollen parent (♂) were isolated and bagged before opening to prevent foreign pollen from contaminating the pollen parent's pollen. As for the seed parent (♀), the petals and stamens were removed from the buds at the balloon stage, and only large and healthy ovaries were selected to complete the hybridization process (emasculation). The flowers of the pollen parent were collected and placed in Petri dishes the day before starting the hybridization process, allowing the pollen to be ready for spreading when the time for hybridization arrived. The hybridization process was then carried out early the next morning by shaking the pollen parent flowers, ensuring that the pollen grains landed on the stigma of the seed parent flowers. The pollinated flowers were covered with paper bags and tied to prevent any foreign pollen from contaminating them (Hasan et al. 2022).

14.9.2 Somatic Cell Hybridization

It is common for malformed embryos to be created in protoplast fusion research. These include among other things, several fasciated cotyledons, embryos that only generate shoots, embryos that germinate but lack a strong vascular link between the shoot and the root, and aberrant shoot growth. The inability of these embryos to develop into greenhouse-ready plants lowers the effectiveness of somatic hybrid recovery and results in the loss of potentially useful genotypes. In order to recover plants that are established in the greenhouse, IVIA regularly grafts in vitro shoots created by these embryos.

Citrus somatic hybrids are typically created by fusing protoplasts isolated from embryogenic callus or suspension cultures of one parent with leaf-derived protoplasts of the other parent. (Ollitrault et al. 1996). The embryogenic parent increases the ability for plant regeneration from the fusion products. Using a transgenic citrus plant that expresses the Green Fluorescent Protein as one of the donors in somatic fusion assays, the processes of protoplast fusion, regeneration, and hybrid plant selection have been seen (Olivares-Fuster et al. 2002). Citrus protoplast fusion can occur electrically (Saito et al. 1991; Ling and Iwamasa 1994; Hidaka et al. 1995; Ollitrault et al. 1996) or chemically (Ohgawara et al. 1985; Grosser and Gmitter 1990, b). In order to facilitate membrane fusion, Olivares-Fuster et al. (2005) have recently presented an electrochemical protoplast fusion approach that combines chemical protoplast aggregation with a DC pulse. Plant regeneration occurs after protoplast fusion without the need for selection. But while some hybrid plants are produced in the bulk of the documented trials, most of the regenerated plants are allotetraploid somatic hybrids (Grosser et al. 2000).

It has been shown that in the event of a fusion between leaf and callus protoplasts, mitochondrial DNA was inherited from the embryonic callus parent (Moreira et al. 2000; Cabasson et al. 2001), even though chloroplasts have monoparental inheritance with random segregation (Ollitrault et al. 2001; Liu et al. 2004). These findings point to a potential connection between mitochondria and the capacity for regeneration (Grosser et al. 1996). There have also been a few documented instances of possible mitochondrial recombination (Moreira et al. 2000; Liu et al. 2004).

14.9.3 Hybrid Cultivars

Sweet orange is one of the natural hybrids or secondary species that has been extensively studied. Agrobacterium-mediated methodology (De Oliveira et al. 2009; Fávero et al. 2012), protoplast transformation (Fleming et al. 2000), and embryogenic suspension cells (Dutt and Grosser 2010) are the main foundations for the numerous protocols that have been published. Additionally, an effective transformation technique based on Agrobacterium co-cultivation was developed for sour oranges. Rough lemon is another naturally occurring hybrid that has been genetically modified by Savita et al. (2011) and Ali et al. (2012). Other crops include Femminello Siracusano lemon (*C. limon* L. Burm.), Mexican lime (De Oliveira et al. 2015), and grapefruit, which have undergone effective transformation and regeneration (Jia et al. 2017). Numerous biotechnological advancements for various citrus species and types have been documented so far. Currently, mature materials from several mandarin genotypes, as well as sweet orange, sour orange, and lime, have been effectively transformed and regenerated. For the transformation of mature tissue, a selection of stem pieces from the initial flushes of generated adult buds was chosen. In this sense, 14 months after being moved to the greenhouse, transgenic sweet orange plants that had been regenerated from mature tissues bloomed and produced fruits (Cervera et al. 2005).

The importance of citrus for human health lies in its ability to treat arteriosclerosis, prevent cancer, kidney stones, and stomach ulcers, as well as reduce cholesterol levels, and high blood pressure, and strengthen the immune system. These health benefits are a result of vitamins, especially vitamin C, and phytochemical compounds such as limonoids, synephrine, hesperidin flavonoids, polyphenols, and pectin. A single orange is said to contain about 170 phytonutrients and over 60 flavonoids with anti-tumor, anti-inflammatory, blood clot inhibiting, and antioxidant properties.

The composition, images, length/width, and rheological properties of citrus fibers in this study provide a much better understanding of the structure and functional properties of citrus fibers. The rheological study can be used to improve how citrus fiber can be applied in food products to enhance their quality and nutrition while maintaining or improving their taste profile. For instance, an important finding reported in this paper is the heat stability of citrus fibers across different pH values, which can help create more heat-stable food products. Regarding composition, pectin makes up approximately 42% of the citrus fiber composition, which is significant due to the fiber's largely insoluble nature and its impact on water-holding capacity. Understanding the chemical composition can help in better understanding the hydration of citrus fibers.

The functional properties of citrus fiber, such as water swelling and holding capacities, were significantly improved by the chemo-mechanical treatment. This treatment also greatly decreased the crystallinity value of citrus fiber and helped it form rougher surfaces and porous structures. The alkaline treatment broke some hydrogen bonds within cellulose or hemicellulose chains, loosening the microstructure of citrus fiber and exposing more hydrophilic groups to water. Subsequent high-speed homogenization further destroyed the cellulose's crystalline region, broke more chains of cellulose and hemicellulose in citrus fiber, increased the content of soluble dietary fiber (SDF), and further loosened the microstructure of citrus fiber. During the processing, the pectin chains inhibited the irreversible re-aggregation of broken cellulose chains when dried. The chemo-mechanically modified citrus fiber could potentially be added to jams and beverages for texture modification. Our present research provides a theoretical basis for further research and application of citrus fiber.

Citrus improvement requires a continuous effort for success. Emerging biotechnologies are providing the research community with new tools that can increase the speed and efficiency of the process. Public perception of transgenic citrus, however, is a real concern and should be taken into consideration. Even though it could be argued that pathogen-resistant transgenic citrus could improve sustainability by reducing pesticide applications, they still bring other concerns. Some of these concerns include the effect of the modified plants on the environment, the risk of transgene dissemination by pollen, potential damage to local production and small growers who cannot adopt the new technology, and potential risks to human health due to the consumption of transgenic citrus fruits, among others. The development of virus-free planting material is the best viable option to deal with viral diseases. It can be achieved through shoot tip grafting and microbudding techniques. Drip irrigation and fertigation systems in citriculture are another tool to save water and losses of fertilizers. It also facilitates better yield with quality production.

14.10 Conclusion and Prospects

The importance of citrus for human health is significant, as it includes the treatment of arteriosclerosis, prevention of cancer, kidney stones, stomach ulcers, and reduction in cholesterol levels, high blood pressure, and strengthening of the immune system. These health benefits are a result of vitamins, especially vitamin C, as well as phytochemical compounds like limonoids, synephrine, hesperidin flavonoids, polyphenols, and pectin. A single orange is said to have about 170 phytonutrients and over 60 flavonoids with anti-tumor, anti-inflammatory, blood clot inhibiting, and antioxidant properties.

The composition, images, length/width, and rheological properties of citrus fibers in this study provide a much better understanding of the structure and functional properties of citrus fibers. The rheological study can be used to improve the application of citrus fiber in food products to enhance their quality and nutrition while maintaining or improving their taste profile. An important finding reported in this paper is the heat stability of citrus fibers across different pH values, which can be beneficial in creating more heat-stable food products. In terms of composition, pectin makes up approximately 42% of citrus fiber, which is significant due to the fiber's largely insoluble nature and its impact on water-holding capacity. Understanding the chemical composition can help in better hydrating citrus fibers.

The functional properties of citrus fiber, such as water swelling and holding capacities, were significantly improved by the chemo-mechanical treatment. This treatment also greatly decreased the crystallinity value of citrus fiber, resulting in a rougher surface and porous structures. The alkaline treatment broke hydrogen bonds within cellulose or hemicellulose chains, loosening the microstructure of citrus fiber and exposing more hydrophilic groups to water. Subsequent high-speed homogenization further destroyed the crystalline region of cellulose, breaking more chains of cellulose and hemicellulose in citrus fiber, increasing the content of soluble dietary fiber, and further loosening the microstructure of citrus fiber. During the treatment process, pectin chains prevented the irreversible re-aggregation of broken cellulose chains when dried. Chemo-mechanically modified citrus fiber could potentially be added to jams and beverages for texture modification. Our current research provides a theoretical basis for further research and application of citrus fiber.

Citrus improvement requires a continuous effort for success. Emerging biotechnologies are providing the research community with new tools that can increase the speed and efficiency of the process. Public perception of transgenic citrus, however, is a genuine concern and should be taken into consideration. While it could be argued that pathogen-resistant transgenic citrus could improve sustainability by reducing pesticide applications, they also bring other concerns. Some of these concerns include the effect of the modified plants on the environment, the risk of transgene dissemination by pollen, potential damage to local production and small growers who may not be able to adopt the new technology, and potential risks to human health due to the consumption of transgenic citrus fruits, among others.

The best viable option for dealing with viral diseases is the development of virus-free planting material. This can be achieved through techniques such as shoot tip grafting and microbudding. In citriculture, implementing a drip irrigation and fertigation system is another tool to save water and reduce losses of fertilizers. Additionally, it can help improve yield and produce higher-quality crops.

Appendix 14.1 Research Institutes and Online Resources Available for Citrus Breeding

Institution name	Specialization and research activities	Address	Contact information and website
NIFTS: Division of Citrus Research	we are developing new citrus cultivars with excellent cultivability and market competitiveness as well as breeding materials	NIFTS: Division of Citrus Research, Japan	https://www.naro.go.jp
AES - Citrus Research Center	–	University of California, the Agricultural Experiment Station (AES) has three branches: one each at the Berkeley, Davis, and Riverside campuses, USA	https://cnas.ucr.edu
Central Tissue Culture Laboratory	This lab is open for all the PG students and Scientists of the Institute for work related to tissue culture of various agriculturally important crops	Indian Agricultural Research Institute, India	https://www.iari.res.in
IBBR - Division of Palermo	The IBBR Division of Palermo was founded in 1970 as the Center for Genetic Improvement of Citrus (CMGA). In 1999 it was renamed Research Institute of Citrus.	IBBR - Division of Palermo, Italy	https://www.ibbr.cnr.it
Citrus Research International (CRI)	A research and technical services organization based in South Africa that focuses primarily on citrus	Citrus Research International (CRI), South Africa	https://www.citrusres.com
Citrus - Plant Breeding Program - UF/IFAS - University of Florida	The UF/IFAS citrus-breeding program began in the mid-1980s to develop scions and rootstocks that resist disease and withstand cold.	University of Florida, USA	https://programs.ifas.ufl.edu

References

Abdullaev FK, Rustamov NSU, Yakubov MD (2025) Uzbekistan genebank. In: Al-Khayri JM, Salem KFM, Jain SM (eds) Plant gene banks. Springer, Singapore. https://doi.org/10.1007/978-981-99-4236-7_57-1

Abulaila K, Al Sane KO, Tahabsom Z (2025) Jordan seed bank. In: Al-Khayri JM, Salem KFM, Jain SM (eds) Plant gene banks. Springer, Singapore. https://doi.org/10.1007/978-981-99-4236-7_45-1

Akosah DK, Adjei-Nsiah S, Brentu FC (2021) Response of late Valencia sweet orange (*Citrus Sinensis* (L.) Osbeck) to fertilization on acrisols of the semi-deciduous forest agroecological zone of Ghana. Commun Soil Sci Plant Anal 52(11):1275–1285

Akparov ZI, Mammadova SA, Kalantarova NS, Mirzaliyeva IA, Mammadov AT (2025) National Genebank of Azerbaijan. In: Al-Khayri JM, Salem KFM, Jain SM (eds) Plant gene banks. Springer, Singapore. https://doi.org/10.1007/978-981-99-4236-7_31-1

Al Lawati AH (2025) Plant genetic resource conservation of Oman. In: Al-Khayri JM, Salem KFM, Jain SM (eds) Plant gene banks. Springer, Singapore. https://doi.org/10.1007/978-981-99-4236-7_50-1

Al-Ababneh NS, Karam, Shibli RA (2002) Cryopreservation of sour orange (*Citrus aurantium* L.) shoot tips. In Vitro Cell Dev Biol-Plant 38:602–607

Aladele SE et al (2025) Plant genetic resources conservation and utilization in Nigeria. In: Al-Khayri JM, Salem KFM, Jain SM (eds) Plant gene banks. Springer, Singapore. https://doi.org/10.1007/978-981-99-4236-7_17-1

Albrigo LG et al (2019) Citrus. CABI

Aleza P, Juárez J, Cuenca J et al (2010) Recovery of citrus triploid hybrids by embryo rescue and fl ow cytometry from 2x ´ 2x sexual hybridisation and its application to extensive breeding programs. Plant Cell Rep 29:1023–1034

Ali S, Mannan A, El Oirdi M et al (2012) Agrobacterium mediated transformation of rough lemon (*Citrus jambhiri* Lush) with yeast HAL2 gene. BMC Res Notes 5:285. https://doi.org/10.1186/1756-0500-5-285

Ali MM, Hashim N, Abd Aziz S, Lasekan O (2020) Pineapple (*Ananas comosus*): a comprehensive review of nutritional values, volatile compounds, health benefits, and potential food products. Food Res Int 137:109675

Al-Khayri JM, Salem KFM, Jain SM (2025) Plant gene banks. Springer Singapore. https://doi.org/10.1007/978-981-99-4236-7

Almarri NB, Ahmad S, Alsubaie GTG, Elshal MH (2025) Saudi Arabia gene bank. In: Al-Khayri JM, Salem KFM, Jain SM (eds) Plant gene banks. Springer, Singapore. https://doi.org/10.1007/978-981-99-4236-7_54-1

Almeida WAB, Mourão Filho FAA, Mendes BMJ et al (2003) Agrobacterium-mediated transformation of Citrus sinensis and Citrus limonia epicotyl segments. Sci Agric 60:23–29. https://doi.org/10.1590/S0103-90162003000100005

Al-Shanableh F (2022) Prediction of the annual yield of citrus growth in the Guzelyurt district using fuzzy inference systems. In: Intelligent and Fuzzy systems: digital acceleration and the new normal-proceedings of the INFUS 2022 conference, vol 1. Springer, pp 168–176

Arab S et al (2025) National Gene Bank of Egypt. In: Al-Khayri JM, Salem KFM, Jain SM (eds) Plant gene banks. Springer, Singapore. https://doi.org/10.1007/978-981-99-4236-7_5-1

Arias RS, Borrone JW, Tondo CL et al (2012) Genomics of tropical fruit tree crops. In: Schnell R, Priyadarshan P (eds) Genomics of tree crops. Springer, New York. https://doi.org/10.1007/978-1-4614-0920-5_7

Ashmore SE, Hamilton KN, Offord CA (2011) Conservation technologies for safeguarding and restoring threatened flora: case studies from eastern Australia. In Vitro Cell Dev Biol Plant 47:99–109

Atefi A, Ge Y, Pitla S, Schnable J (2021) Robotic technologies for high-throughput plant phenotyping: contemporary reviews and future perspectives. Front Plant Sci 12:611940

Barkley NA, Roose ML, Robert R, Krueger Claire T (2006) Federici Assessing genetic diversity and population structure in a citrus germplasm collection utilizing simple sequence repeat markers (SSRs). Theor Appl Genet 112:1519–1531. https://doi.org/10.1007/s00122-006-0255-9

Barrett HC, Rhodes AM (1976) A numerical taxonomic study of affinity relationships in cultivated Citrus and its close relatives. Syst Bot 1:105–136

Barritt BH (2001) Apple quality for consumers. Int Dwarf Fruit Tree Assoc 34:54–46

Békefi Z, Szilágyi S, Desiderio F, Demku T, Varga J (2025) Prunus Gene Bank in Hungarian University of Agriculture and Life Sciences. In: Al-Khayri JM, Salem KFM, Jain SM (eds) Plant gene banks. Springer, Singapore. https://doi.org/10.1007/978-981-99-4236-7_71-1

Belser C, Baurens FC, Noel B et al (2021) Telomere-to-telomere gapless chromosomes of banana using nanopore sequencing. Commun Biol 4:1047

Bergey D, Dutt M, Barthe G et al (2015) Genetically modified Citrus rootstocks expressing a synthetic cationic antimicrobial peptide slows down bacterial infection in the non-transgenic scion. In Vitro Cell Dev Biol 51:499

Bespalhok Filho JC, Kobayashi AK, Pereira LFP et al (2003) Transient gene expression of beta-glucuronidase in citrus thin epicotyl transversal sections using particle bombardment. Braz Arch Biol Technol 46:1–6

Bharati P, Pramanik A (2020) Deep learning techniques—R-CNN to mask R-CNN: a survey. Comput Intell Pattern Recognit Proc CIPR 2019:657–668

Bi Y, Zhao Q, Yan W et al (2020) Flexible chromosome painting based on multiplex PCR of oligonucleotides and its application for comparative chromosome analyses in Cucumis. Plant J 102:178–186

Bombardelli E, Morazzoni P (1993) The flavonoids: new perspectives in biological activities and therapeutics. Chimica Oggi 11:25–28

Borwankar RP (1992) Food texture and rheology: a tutorial review. J Food Eng 16(1–2):1–16

Bourne MC (1992) Food texture and viscosity. Academic, New York

Bracke ME, Vyncke B, Van Larebeke NA et al (1989) The flavonoid tangeretin inhibits invasion of MO_4 mouse cells into embryonic chick heart in vitro. Clin Exp Metastasis 7:283–300

Braddock RJ (1999) Handbook of citrus by-products and processing technology. Wiley, New York

Breidy J (2025) Lebanon gene bank. In: Al-Khayri JM, Salem KFM, Jain SM (eds) Plant gene banks. Springer, Singapore. https://doi.org/10.1007/978-981-99-4236-7_48-1

Cabasson C, Luro F, Ollitrault P et al (2001) Non Random inheritance of mitochondrial genomes in Citrus hybrids produced by protoplast fusion. Plant Cell Rep 20:604–609

Cabrera-Ponce JL, Lopez L, Assad-Garcia N et al (1997) An efficient particle bombardment system for the genetic transformation of asparagus (*Asparagus officinalis* L.). Plant Cell Rep 16:255–260. https://doi.org/10.1007/bf01088276

Callahan MA (2003) Breeding for fruit quality. Acta Hortic 622:255–302. https://doi.org/10.17660/ActaHortic.2003.622.27

Cañar DY et al (2025) Colombia gene bank. In: Al-Khayri JM, Salem KFM, Jain SM (eds) Plant gene banks. Springer, Singapore. https://doi.org/10.1007/978-981-99-4236-7_95-1

Carimi F, Pasquale F, De Crescimanno FG (1995) Somatic embryogenesis in Citrus from styles culture. Plant Sci Limerick 105:81–86

Carimi F, Pasquale F, De Crescimanno FG (1999) Somatic embryogenesis and plant regeneration from pistil thin cell layers of Citrus. Plant Cell Rep 18:935–940

Caruso M, Smith MW, Froelicher Y et al (2020) Traditional breeding. In: Talon M, Caruso M, Gmitter FG (eds) The genus citrus. Woodhead Publishing, pp 129–148. https://doi.org/10.1016/B978-0-12-812163-4.00007-3

Carvalho R, Filho WS, Brasileiro-Vidal AC et al (2005) The relationships among lemons, limes and citron: a chromosomal comparison. Cytogenet Genome Res 109:276–282

Cervera M, Juárez J, Navarro A et al (1998a) Genetic transformation and regeneration of mature tissues of woody fruit plants bypassing the juvenile stage. Transgenic Res 7:51–59

Cervera M, López MM, Navarro L et al (1998b) Virulence and super virulence of Agrobacterium tume faciens in woody fruit plants. Physiol Mol Plant Pathol 52:67–78. https://doi.org/10.1006/pmpp.1997.0135

Cervera M, Juárez J, Navarro L et al (2005) Genetic transformation of mature citrus plants. Methods Mol Biol 286:177–188. https://doi.org/10.1385/1-59259-827-7:177

Cervera M, Navarro A, Navarro L et al (2008) Production of transgenic adult plants from clementine mandarin by enhancing cell competence for transformation and regeneration. Tree Physiol 28:55–66

Cha JY, Cho YS, Kim I et al (2001) Effect of hesperetin, a citrus flavonoid on the liver triacylglycerol content and phosphatidate phosphohydrolase activity in orotic acid-fed rats. Plant Foods Hum Nutr 56:349–358

Chen C, Lyon MT, O'Malley D et al (2008) Origin and Frequency of 2n Gametes in Citrus sinensis × Poncirus trifoliata and Their Reciprocal Crosses. Plant Sci 174:1–8

Cheng M, Li X, Cui H et al (2022) Fish-based "pan" and "core" karyotypes reveal genetic diversification of *Roegneria ciliaris*. J Genet Genomics 50(2):135–138

Chenqiao Z, Zheng X, Huang Y et al (2019) Genome sequencing and CRISPR/cas9 gene editing of an early flowering mini citrus (*Fortunella hindsii*). Plant Biotechnol J 17:2199–2210

Cillard J, Cillard P (1988) Compose's phe'noliques et radicaux libres. STP Pharma 4:592–596

Consiglioi Nazionale Delle Ricerche (1982) Agrumi, frutta e uve nella Firenze di Bartolomeo Bimbi pittore mediceo. Edizione Fuori Commercio, F. & B. Parretti Grafiche, Florence, Italy

Crowell PL (1999) Prevention and therapy of cancer by dietary monoterpenes. J Nutr 129(3):775S–778S. PMID 10082788

Da Silva EJA, Oliveiraand AS, Lapa AJ (1994) Pharmacological evaluation of the anti-inflammatory activity of a citrus bioflavonoid, hesperidin, and the isoflavonoids, duartin and claussequinone, in rats and mice. J Pharm Pharmacol 46:118–122

Darmon N, Ferrandiz V, Canal MT, Mitjavilla S (1990) Activite antiradicallaire de flavonoides vis-a-vis de l'anion superoxide et Du radical hydroxyle. Liaison-Grupe Polyphenols Bull 15:158–162

De Francesco A, Simeone M, Gómez C et al (2020) Transgenic Sweet Orange expressing hairpin CP-mRNA in the interstock confers tolerance to *Citrus psorosis* virus in the non-transgenic scion. Transgenic Res 29:215–228. https://doi.org/10.1007/s11248-020-00191-1

de Galarreta JIR, Barandalla L, Ritter E, Velázquez ME, Rios D (2025) Spanish Potato Germplasm Bank. In: Al-Khayri JM, Salem KFM, Jain SM (eds) Plant gene banks. Springer, Singapore. https://doi.org/10.1007/978-981-99-4236-7_114-1

De Oliveira MLP, Febres VJ, Costa MGC et al (2009) High-efficiency Agrobacterium-mediated transformation of citrus via sonication and vacuum infiltration. Plant Cell Rep 28:387–395. https://doi.org/10.1007/s00299-008-0646-2

De Oliveira MLP, Moore G, Thomson JG et al (2015) Agrobacterium-mediated transformation of Mexican lime (*Citrus aurantifolia* Swingle) using optimized systems for epicotyls and cotyledons. Adv Biosci Biotechnol 6:657–668. https://doi.org/10.4236/abb.2015.611069

Deng H, Cai Z, Xiang S et al (2019) Karyotype analysis of diploid and spontaneously occurring tetraploid blood orange [*Citrus sinensis* (L.) osbeck] using multicolor fish with repetitive DNA sequences as probes. Front Plant Sci 10:311

Deng H, Tang G, Xu N et al (2020) Integrated karyotypes of diploid and tetraploid carrizo citrange (*Citrus sinensis* L. Osbeck x *Poncirus trifoliata* L. Raf.) as determined by sequential multicolor fluorescence in situ hybridization with tandemly repeated DNA sequences. Front Plant Sci 11: 569

Desheva G, Valchinova E, Deshev M (2025) Bulgarian gene bank. In: Al-Khayri JM, Salem KFM, Jain SM (eds) Plant gene banks. Springer, Singapore. https://doi.org/10.1007/978-981-99-4236-7_66-1

Di Majo D, Giammanco M, La Guardia M et al (2005) Flavanones in Citrus fruit: structure antioxidant activity relationships. Food Res Int 38:1161–1166

Dias JS, Ortiz R (2014) Advances in transgenic vegetable and fruit breeding. Agric Sci 05(14): 1448–1467

Díez MJ, Valcárcel JV, Solbes E, Muñoz E (2025) Forty years conserving Spanish biodiversity in the gene bank of the Universitat Politècnica de València, Spain. In: Al-Khayri JM, Salem KFM, Jain SM (eds) Plant gene banks. Springer, Singapore. https://doi.org/10.1007/978-981-99-4236-7_86-1

Dong G, Shen J, Zhang Q et al (2018) Development and applications of chromosome-specific cytogenetic bac-fish probes in s. Spontaneum. Front Plant Sci 9:218

Duan YX, Guo WW, Meng HJ et al (2007) High effi cient transgenic plant regeneration from embryogenic calluses of Citrus sinensis. Biol Plant 51:212–216

Duarte A, Fernandes MJ, Bernardes JP, Miguel MG (2016) Citrus as a component of the Mediterranean diet. J Spatial Organiz Dynam 4(IV):289304

Dutt M, Lee DH, Grosser JW (2010) Bifunctional selection-reporter systems for genetic transformation of citrus: mannose- and kanamycin-based systems. Vitr Cell Dev Biol Plant 46:467–476. https://doi.org/10.1007/s11627-010-9300-0

Dutt M, Erpen L, Grosser JW (2018) Genetic transformation of the 'W Murcott' tangor: comparison between different techniques. Sci Hortic 242:90–94. https://doi.org/10.1016/j.scienta.2018.07.026

Dutta SK, Gurung G, Yadav A, Laha R, Mishra VK (2023) Factors associated with citrus fruit abscission and management strategies developed so far: a review. New Zealand J Crop Hortic Sci 51(4):467–488

Edriss MH, Burger DW (1984) Micro-grafting shoot-tip culture of Citrus on three trifoliate rootstocks. Sci Hortic 23:255–259

Ejaz S, Ejaz A, Matsuda K, Chae WL (2006) Limonoids as cancer chemopreventive agents. J Sci Food Agric 86:339–345

Elangovan V, Sekar N, Govindasamy S (1994) Chemopreventive potential of dietary bioflavonoids against 20-methylcholanthrene-induced tumorigenesis. Cancer Lett 87:107–113

Endo T, Shimada T, Fuji H et al (2005) Ectopic expression of an FT homolog from Citrus confers an early flowering phenotype on trifoliate orange (Poncirus trifoliata L. Raf.). Transgenic Res 14: 703–712. https://doi.org/10.1007/s11248-005-6632-3

Engelmann F (2010) Use of biotechnologies for the conservation of plant biodiversity. In Vitro Cell Dev Biol-Plant 47:5–16

Fan J, Huo T, Li X (2020) A review of one-stage detection algorithms in autonomous driving. In: Proceedings of the 2020 4th CAA international conference on vehicular control and intelligence (CVCI). IEEE, pp 210–214

Farris AL (2010) The natural aversion: the fda's reluctance to define a leading food-industry marketing claim, and the pressing need for a workable rule. Food Drug LJ 65:403

Fávero P, de Alves Mourão Filho FA, Stipp LCL et al (2012) Genetic transformation of three sweet orange cultivars from explants of adult plants. Acta Physiol Plant 34:471–477. https://doi.org/10.1007/s11738-011-0843-4

Federici CT, Fang DQ, Scora RW, Roose ML (1998) Phylogenetic relationships within the genus Citrus (Rutaceae) and related genera as revealed by RFLP and RAPD analysis. Theor Appl Genet 94:812–822

Figuerola F, Hurtado ML, Estévez AM et al (2005) Fibre concentrates from Apple Pomace and Citrus peel as potential fibre sources for food enrichment. Food Chem 3:395–401

Fleming GH, Olivares-Fuster O, Fatta Del-Bosco S et al (2000) An alternative method for the genetic transformation of sweet orange. In Vitr Cell Dev Biol 36:450–455

Food and Agriculture Organization (2016) Intergovernmental Group on Citrus Fruits. Citrus fruit statistics 2015. Food and Agriculture Organization of the United Nations, Rome, Italy

Frison E, Taher MM (1991) FAO/IBPGR Technical guidelines for the safe movement of citrus germplasm. Food and Agriculture Organization of the United Nations, Rome/International Board for Plant Genetic Resources, Rome, p 50

Galea-Gragera FA, Llera Cid F (2025) Spain Germplasm Bank of Annual Forage Legumes Seeds at the Scientific and Technological Research Center of Extremadura (CICYTEX). In: Al-Khayri JM, Salem KFM, Jain SM (eds) Plant gene banks. Springer, Singapore. https://doi.org/10.1007/978-981-99-4236-7_102-1

Garau MC, Simal S, Rossello C et al (2007) Effect of air-drying temperature on physico-chemical properties of dietary fibre and antioxidant capacity of orange (Citrus aurantium v. Canoneta) by-products. Food Chem 104(3):1014–1024

Geldermann H (1975) Investigations on inheritance of quantitative characters in animals by gene markers. I. Methods. Theor Appl Genet 46:319–330

Germanà MA (2003) Somatic embryogenesis and plant regeneration from anther culture of Citrus aurantium and *C. reticulata*. Biologia-Bratislava 58:843–850

Germana MA, Aleza P, Grosser JW et al (2020) Citrus biotechnology. In: Talon M, Caruso M, Gmitter FG Jr (eds) The genus Citrus. Woodhead Publishing/Elsevier, Oxford, Sawston, Cambridge, UK, pp 171–192. https://doi.org/10.1016/B978-0-12-812163-4.00003-6

Ghidan WF, Abdulmajid DA (2025) Egyptian Rice Gene Bank. In: Al-Khayri JM, Salem KFM, Jain SM (eds) Plant gene banks. Springer, Singapore. https://doi.org/10.1007/978-981-99-4236-7_106-1

Gmitter FG Jr (1995) Origin, evolution, and breeding of the grapefruit. Hortic Rev 13:345–363

Gmitter FG, Talon M (2008) Citrus genomics. Int J Plant Genom 2008:528361. https://doi.org/10.1155/2008/528361

Gonzalez-Arnao MT, Engelmann F, Urra C et al (1998) Cryopreservation of Citrus apices using the encapsulation-dehydration technique. CryoLetters 19:177–182

Gonzalez-Arnao MT, Panta A, Roca WM et al (2008) Development and large scale application of cryopreservation techniques for shoot and somatic embryo cultures of tropical crops. Plant Cell Tissue Organ Cult 92:1–13

Goor A, Nurock M (1968) The fruits of the holy land. Israel University Press, Jerusalem

Grigelmo-Miguel N, Martin-Belloso O (1999) Characterization of dietary fiber from orange juice extraction. Food Res Int 31(5):355–361

Grimstad L, From PJ (2017) The Thorvald II agricultural robotic system. Robotics 6(4)24

Grosser JW, Gmitter FG Jr (1990a) Protoplast fusion and citrus improvement. Plant Breed Rev 8: 339–374

Grosser JW, Gmitter FG Jr (1990b) Somatic hybridization of Citrus with wild relatives for germplasm enhancement and cultivar development. HortScience 25(2)147–151

Grosser JW, Gmitter FG Jr (2011) Protoplast fusion for production of tetraploids and triploids: applications for scion and rootstock breeding in citrus. Plant Cell Tissue Organ Cult 104:343–357. https://doi.org/10.1007/s11240-010-9823-4

Grosser JW, Gmitter FG Jr et al (1996) Further evidence of a cybridization requirement for plant regeneration from citrus leaf protoplasts following somatic fusion. Plant Cell Rep 15:672–676

Grosser JW, Ollitrault P, Olivares-Fuster O (2000) Somatic hybridization in Citrus: an effective tool to facilitate cultivar improvement. In Vitro Cell Dev Biol Plant 36:434–449

Grosser JW, Deng XX, Goodrich RM (2007) Somaclonal variation in sweet orange: practical applications for variety improvement and possible causes. In: Khan IA (ed) Citrus genetics, breeding and biotechnology. CAB International, Wallingford, pp 219–233

Gudaye TY, Lakew BF, Selamo MM, Tessema ES (2025) Ethiopian Crop Gene Bank. In: Al-Khayri JM, Salem KFM, Jain SM (eds) Plant gene banks. Springer, Singapore. https://doi.org/10.1007/978-981-99-4236-7_7-1

Guo W, Duan Y, Olivares-Fuster O et al (2005) Protoplast transformation and regeneration of transgenic valencia sweet orange plants containing a juice quality-related pectin methylesterase gene. Plant Cell Rep 24(8):482–486

Hafeez A et al (2022) Implementation of drone technology for farm monitoring & pesticide spraying: a review. Inf Proc Agric 10:192

Hamilton KN, Ashmore SE, Pritchard HW (2009) Thermal analysis and cryopreservation of seeds of Australian wild *Citrus* species (Rutaceae*): Citrus australasica, C. inodora and C. garrawayi.* CryoLetters 30:268–279

Haroldsen VM, Chi-Ham CL, Bennett AB (2012a) Transgene mobilization and regulatory uncertainty for non-GE fruit products of transgenic rootstocks. J Biotechnol 161:349–353. https://doi.org/10.1016/j.jbiotec.2012.06.017

Haroldsen VM, Szczerba MW, Aktas H (2012b) Mobility of transgenic nucleic acids and proteins within grafted rootstocks for agricultural improvement. Front Plant Sci Plant 3:Article 39. https://doi.org/10.3389/fpls.2012.00039

Harzli I, Mostafa K, Abdulla MF, Mehmood M (2025) Sustaining Turkish genetic resources. In: Al-Khayri JM, Salem KFM, Jain SM (eds) Plant gene banks. Springer, Singapore. https://doi.org/10.1007/978-981-99-4236-7_88-1

Hasan V, Elkhateeb A, Khojah H (2021a) Response of some Citrus rootstocks to abiotec stresses. Doctorate Thesis, Department of Horticulture, Faculty of Agriculture, Tishreen University, Syria

Hasan V, Elkhateeb A, Khojah H (2021b) Identification of F1 hybrids from crosses of Citrus rootstocks tolerant to shilling stress using molecular markers SSR.Al-Muthanna j. For Agric Sci 8:35–45. https://doi.org/10.52113/mjas04/8.4/4

Hasan V, Elkhateeb A, Khojah H (2022) Selection of tolerant of some citrus hybrids F_1 to Calcareous stress and identification sexuall individuals by SSR marker. Am J Plant Sci 13: 60–82. https://doi.org/10.4236/ajps.2022.131005

He L, Zhao H, He J et al (2020) Extraordinarily conserved chromosomal synteny of citrus species revealed by chromosome-specific painting. Plant J 103:2225–2235

Hearn CJ (1984) Development of seedless orange, Citrus sinensis, cultivar Pineapple and grapefruit, Citrus paradisi, cultivars through seed irradiation. J Am Soc Hortic Sci 109:270–273

Hensz RA (1971) 'Star Ruby', a new deep-red-fl eshed grapefruit variety with distinct tree characteristics. J Rio Grande Valley Hort Soc 25:54–58

Heo HY, Lee SJ, Kwon CH et al (1994) Anticlastogenic effects of galangin against bleomycin induced chromosomal aberrations in mouse spleen lymphocytes. Mutat Res – Fundamental and Molecular Mechanisms of Mutagenesis 311:225–229

Hertog MG, Hollman PCH, Katan MB, Kromhout D (1993) Dietary antioxidant flavonoids and risk of coronary heart disease. Lancet 342:1007–1011

Hidaka T, Omura M, Ugari M et al (1990) Agrobacterium-mediated transformation and regeneration of Citrus spp. from suspension cells. Japan J Breed 40:199–207

Hidaka T, Moriguchi T, Motomura T et al (1995) Development of a new electrode chamber and its efficiency in protoplast fusion in Citrus. Breed Sci 45:237–239

Hidayatun N, Kurniawan H (2025) Conservation of agricultural genetic resources in Indonesia. In: Al-Khayri JM, Salem KFM, Jain SM (eds) Plant gene banks. Springer, Singapore. https://doi.org/10.1007/978-981-99-4236-7_40-1

Hirano T, Gotoh M, Oka K (1994) Natural flavonoids and lignans are potent cytostatic agents against human leukemia HL-60 cells. Life Sci 55:1061–1069

Hiraoka Y (2020) Application of high-density SNP genotyping array in citrus germplasm characterization and genetic dissection of traits. PhD Dissertation, University of California Riverside

Hofstatter PG, Thangavel G, Lux T et al (2022) Repeat-based holocentromeres influence genome architecture and karyotype evolution. Cell 185:3153–3168 e3118

Hua M, Lu J, Qu D (2019) Structure, physicochemical properties and adsorption function of insoluble dietary fiber from ginseng residue: a potential functional ingredient. Food Chem 286:522–529.

Huang D, Wu W, Zhou Y et al (2004) Microdissection and molecular manipulation of single chromosomes in woody fruit trees with small chromosomes using pomelo (citrus grandis) as a model. I. Construction of single chromosomal DNA libraries. Theor Appl Genet 108:1366–1370

Huang X, Wang Y, Xu J et al (2020) Development of multiplex genome editing toolkits for citrus with high efficacy in biallelic and homozygous mutations. Plant Mol Biol 104:297–307. https://doi.org/10.1007/s11103-020-01043-6

Ibiapino A, Baez M, Garcia MA et al (2022) Karyotype asymmetry in *Cuscuta* L. subgenus *Pachystigma* reflects its repeat DNA composition. Chromosom Res 30:91–107

Ihsan F, Prihatini R (2025) Indonesian tropical fruit gene bank. In: Al-Khayri JM, Salem KFM, Jain SM (eds) Plant gene banks. Springer, Singapore. https://doi.org/10.1007/978-981-99-4236-7_41-1

Jain S, Singh H, Kore DS et al (2023) A comprehensive review on postharvest physiological disorders in citrus fruit crops (Citrus spp.). Int J Environ Climate Change 13(10):2863–2873

Jalikop SH (2015) Allied species utilization in fruit improvement-way forward. Souvenir of lead papers and abstracts of National meet on distant hybridization in horticultural crop improvement. pp 25–35
Jatoi SA, Siddiqui SU (2025) National Gene Bank of Pakistan. In: Al-Khayri JM, Salem KFM, Jain SM (eds) Plant gene banks. Springer, Singapore. https://doi.org/10.1007/978-981-99-4236-7_51-1
Jia H, Zhang Y, Orbovic V et al (2017) Genome editing of the disease susceptibility gene CsLOB1 in citrus confers resistance to citrus canker. Plant Biotechnol J 15:817–823. https://doi.org/10.1111/pbi.12677
Jiang J (2019) Fluorescence in situ hybridization in plants: recent developments and future applications. Chromosom Res 27:153–165
Jiang Z, Zhang M, Huang Y et al (2022) Comparison and characterization of the structure and physicochemical properties of three citrus fibers: effect of ball milling treatment. 11:2665. https://doi.org/10.3390/foods11172665. https://www.mdpi.com/journal/foods
Juárez J, Aleza P, Olivares-Fuster O et al (2004) Recovery of tetraploid clementine plants (*Citrus clementina* hort. ex Tan .) by in vitro colchicine treatment of shoot tips. Proc Int Soc Citric 1: 151–154
Kader AA (2002) Fruits in the global market. In: Knee M (ed) Fruit quality and its biological basis. Sheffield Academic Press, Sheffield, UK, pp 1–16
Kaneyoshi J, Kobayashi S, Nakamura Y et al (1994) A simple and effi cient gene transfer system of trifoliate orange. Plant Cell Rep 13:541–545
Karamura E, Frison E, Karamura DA, Sharrock S (1998) Banana production systems in eastern and southern Africa. Bananas Food Secur 401:412.;17:36. 10
Kato M, Ikoma Y, Matsumoto H, Sugiura M, Hyodo H, Yano M (2004) Accumulation of carotenoids and expression of carotenoid biosynthetic genes during maturation in citrus fruit. Plant Physiol 134(2):824–837
Kenis K, Keulemans J, Davey MW (2008) Identification and stability of qtls for fruit quality traits in Apple. Tree Genet Genomes 4(4):647–661
Kepiro JL, Roose ML (2007) Nucellar embryony. In: Khan IA (ed) Citrus genetics, breeding and biotechnology. CAB International, Wallingford, pp 141–150
Keys A (1995) Mediterranean diet and public health: personal reflections. Am J Clin Nutr 61:1321–1323
Khan IA, Roose ML (1988) Frequency and characteristics of nucellar and zygotic seedlings in three cultivars of trifoliate orange. J Am Soc Hortic Sci 113:105–110
Khan M, Hu J, Dahro B et al (2021) ERF 108 from Poncirus trifoliata (L.) Raf. functions in cold tolerance by modulating raffinose synthesis through transcriptional regulation of PtrRafS. Plant J 108:705–724. https://doi.org/10.1111/tpj.15465
Kobayashi S (1987) Uniformity of plants regenerated from orange (Citrus sinensis Osb.) protoplasts. Theor Appl Genet 74:10–14
Korlapati S, Sushil SN, Jeyakumar P, et al (2014) AESA based IPM package for Citrus. pp 53
Kumar GBS, Srinivas L, Ganapathi TR et al (2006) Hepatitis B surface antigen expression in transgenic tobacco (Nicotiana tabacum) plants using four different expression cassettes. Plant Cell Tissue Organ Cult 84(3):315–323
Kumar M, Kumar R, Singh VP et al (2023) Breeding approaches for quality improvement in fruit crops: strategies and achievements. Int J Environ Climate Change 13(5):75–93. Article no. IJECC.97844 ISSN: 2581-8627
Kupferman E (2002) Critical aspects of harvest and quality management. Postharvest Information Network, 14–18. Available: www.postharvest.tfrec.wsu.edu/EMK2002A.pdf
Laghetti G, Piergiovanni AR, De Paola D, Direnzo P, Bucci G, Sonnante G (2025) Italy IBBR Gene Bank. In: Al-Khayri JM, Salem KFM, Jain SM (eds) Plant gene banks. Springer, Singapore. https://doi.org/10.1007/978-981-99-4236-7_73-1
Lan H, Chen CL, Miao Y et al (2016) Fragile sites of 'valencia' sweet orange (citrus sinensis) chromosomes are related with active 45s rdna. PLoS One 11:e0151512

León-Lobos P et al (2025) Plant gene banks in Chile. In: Al-Khayri JM, Salem KFM, Jain SM (eds) Plant gene banks. Springer, Singapore. https://doi.org/10.1007/978-981-99-4236-7_94-1

Li Q, Qin X, Qi J et al (2020) CsPrx25, a class III peroxidase in Citrus sinensis, confers resistance to citrus bacterial canker through the maintenance of ROS homeostasis and cell wall lignification. Hortic Res 7:15–17. https://doi.org/10.1038/s41438-020-00415-9

Li G et al (2021) Lemon-YOLO: an efficient object detection method for lemons in the natural environment. IET Image Process 15(9):1998–2009

Ling JT, Iwamasa M (1994) Somatic hybridization between Citrus reticulata and Citropsis gabunensis through electrofusion. Plant Cell Rep 13:493–497

Liu JH, Xu XY, Deng XX (2004) Advances on Citrus somatic hybrids and inheritance of their nuclear and cytoplasmic components. J Agric Biotech 12:237–246

Liu J, Seetharam AS, Chougule K et al (2020) Gapless assembly of maize chromosomes using long-read technologies. Genome Biol 21:121

Liu H, Wang X, Liu S et al (2022) Citrus pan-genome to breeding database (cpbd): a comprehensive genome database for citrus breeding. Mol Plant 15:1503–1505

Liu Z, Zhao J, Sun R et al (2022a) Lactobacillus Plantarum 23-1 improves intestinal inflammation and barrier function through the Tlr4/Nf-Kb signaling pathway in Obese Mice. Food Funct 13: 5971–5986

Liu Z, Zhou X, Wang W et al (2022b) Lactobacillus Paracasei 24 attenuates lipid accumulation in high-fat diet-induced Obese Mice by regulating the Gut Microbiota. J Agric Food Chem 15: 4631–4643. [CrossRef] [PubMed]

Longhi S, Moretto M, Viola R et al (2012) Comprehensive QTL mapping survey dissects the complex fruit texture physiology in Apple (Malus X domesticaborkh.). J Exp Bot 63(3): 1107–1121

Loredana A, Nicasio T, Sergio B et al (2012) Genetic improvement of Citrus fruits:new somatic hybrids from Citrus sinensis (L.) Osb. And Citrus limon (L.) Burm. F. Fd. Theatr Res Int 48: 284–290

Louzada ES, Ramadugu C (2021) Grapefruit: history, use, and breeding. 31(3). https://doi.org/10.21273/HORTTECH04679-20. https://creativecommons.org/licenses/by-nc-nd/4.0/

Lu S, Ye J, Zhu K et al (2021) A fruit ripening-associated transcription factor CsMADS5 positively regulates carotenoid biosynthesis in citrus. J Exp Bot 72:3028–3043. https://doi.org/10.1093/JXB/ERAB045

Lundberg BM (2005) Using highly expanded citrus fiber to improve the quality and nutritional properties of foods. Cereal Foods World 50(5):248–252

Lundberg B, Pana X, Whitec A et al (2014) Rheology and composition of citrus fiber. J Food Eng. wwwelseviercom/locate/jfoodeng 125:97–104

Luro F, Maddy F, Jacquemond C et al (2004) Identifi cation and evaluation of diplogyny in clementine (*Citrus clementina*) for use in breeding. Acta Hortic 663:841–847

Mahato N, Sharma K, Sinha M et al (2020) Bio-Sorbents, industrially important chemicals and novel materials from Citrus processing waste as a sustainable and renewable bioresource: a review. J Adv Res 23:61–82.

Maheshwari P, Ranga Swamy NS (1958) Polyembryonny and in vitro culture of embryos of Citrus and Mangifera. Indian J Hortic 15:275–282

Mallor C, Luna E (2025) The genetic diversity of the Spanish Vegetable Genebank BGHZ-CITA: conservation and utilization. In: Al-Khayri JM, Salem KFM, Jain SM (eds) Plant gene banks. Springer, Singapore. https://doi.org/10.1007/978-981-99-4236-7_109-1

Manthey JA, Guthrie N, Grohmann K (2001) Biological properties of citrus flavonoids pertaining to cancer and inflammation. Curr Med Chem 8:135–153

Marcelino-Perez G, Ruiz-Medrano R, Gallardo-Hernandez S et al (2021) Adsorption of recombinant human b-Defensin 2 and two mutants on mesoporous silica nanoparticles and its effect against Clavibacter michiganensis sbp. michiganensis. Nano 11:2144

Mattia MR, Du D, Yu Q et al (2022) Genomewide association study of healthful flavonoids among diverse mandarin accessions. Plants 11(3):317. https://doi.org/10.3390/plants11030317

Melters DP, Bradnam KR, Young HA et al (2013) Comparative analysis of tandem repeats from hundreds of species reveals unique insights into centromere evolution. Genome Biol 14:R10
Meng Z, Han J, Lin Y et al (2020) Characterization of a *Saccharum spontaneum* with a basic chromosome number of x = 10 provides new insights on genome evolution in genus saccharum. Theor Appl Genet 133:133–199
Meshram V et al (2021) Machine learning in agriculture domain: a state-of-art survey. Artif Intell Life Sci 1:100010
Minamikawa MF, Nonaka K, Kaminuma E, Kajiya-Kanegae H, Onogi A, Goto S, Yoshioka T, Imai A, Hamada H, Hayashi T et al (2017) Genome-wide association study and genomic prediction in citrus: potential of genomics-assisted breeding for fruit quality traits. Sci Rep 7: 4721.
Monteros-Altamirano Á et al (2025) Plant genetic resources in Ecuador. In: Al-Khayri JM, Salem KFM, Jain SM (eds) Plant gene banks. Springer, Singapore. https://doi.org/10.1007/978-981-99-4236-7_99-1
Moore GA, Jacono CC, Neidigh JL et al (1992) Agrobacteruim mediated transformation of Citrus stem segments and regeneration of transgenic plants. Plant Cell Rep 11:238–242
Moreira CD, Chase CD, Gmitter Jr et al (2000) Transmission of organelle genomes in citrus somatic hybrids. Plant Cell Tissue Organ Cult 61:165–168
Moussaid A, Fkihi SE, Zennayi Y (2021) Tree crowns segmentation and classification in overlapping orchards based on satellite images and unsupervised learning algorithms. J Imaging 7(11)241
Mudgil D, Barak S (2013) Composition properties and health benefits of indigestible carbohydrate polymers as dietary fiber: a review. Int J Biol Macromol 61:1–6.
Murovec J, Bohanec B (2012) Haploids and doubled haploids in plant breeding. Plant Breed 4(1): 364–369. https://doi.org/10.5772/29982
Myles S (2013) Improving fruit and wine: what does genomics have to offer? Trends Genet 29(4): 190–196
Naish M, Alonge M, Wlodzimierz P et al (2021) The genetic and epigenetic landscape of the Arabidopsis centromeres. Science 374:eabi7489
Navarro L (1991) Citrus shoot-tip grafting and its applications: a review. Proc Int Soc Citricult 1: 452–456
Navarro L (1992) Citrus shoot tip grafting in vitro. In: Bajaj JPS (ed) Biotechnology in agriculture and forestry, High-Tech and Micropropagation, vol 18. Springer, Berlin, pp 327–338
Navarro L, Juarez J (2007) Shoot-tip grafting in vitro: impact in the citrus industry and research applications. In: Khan IA (ed) Citrus genetics, breeding and biotechnology. CAB International, Wallingford, pp 353–364
Navarro L, Roistacher CN, Murashige T (1975) Improvement of shoot tip grafting in vitro for virus-free citrus. J Am Soc Hortic Sci 100:471–479
Navarro L, Ortiz JM, Juarez J (1985) Aberrant citrus plants obtained by somatic embryogenesis of nucelli cultured in vitro. HortScience 20(2):214–215
Navarro L, Juárez J, Aleza P et al (2003) Recovery of triploid seedless mandarin hybrids from 2n x 2n and 2n x 4n crosses by embryo rescue and fl ow cytometry. In: Vasil IK (ed) Plant biotechnology 2002 and beyond. Kluwer Academic, Dordrecht, pp 541–544
Navarro L, Olivares Fuster O, Juarez J et al (2004) Applications of biotechnology to citrus improvement in Spain. Acta Hortic 632:221–234
Navratilova P, Toegelova H, Tulpova Z et al (2022) Prospects of telomere-to-telomere assembly in barley: analysis of sequence gaps in the morexv3 reference genome. Plant Biotechnol J 20: 1373–1386
Nicolosi E, Deng ZN, Gentile A et al (2000) Citrus phylogeny and genetic origin of important species as investigated by molecular markers. Theor Appl Genet 100:1155–1166
Niedz P, Mckendree WL, Shatters RC (2003) Electroporation of embryogenic protoplasts of sweet orange (*Citrus sinensis* (L.) Osbeck) and regeneration of transformed plants. In Vitro Cell Dev Biol - Plant 39:P586–P494

Nookaraju A, Upadhyaya CP, Pandey SK et al (2010) Molecular approaches for enhancing sweetness in fruits and vegetables. Sci Hortic 127(1)1–15

Ofoli RY, Morgan RG, Steffe JF (1987) A generalized rheological model for inelastic fluid foods. J. Texture Stud. 18, 213–230.

Ohgawara T, Kobayashi S, Ohgawara E et al (1985) Somatic hybrid plants obtained by protoplast fusion between Citrus sinensis and Poncirus trifolata. Theor Appl Genet 71:1–4

Olivares-Fuster O, Asins MJ, Duran-Vila N (2000) Cryopreserved callus, a source of protoplasts for citrus improvement. J Hort Sci Biotech 75:635–640

Olivares-Fuster O, Peña L, Duran-Vila N et al (2002) Green fluorescent protein as visual marker in somatic hybridization. Ann Bot 89:491–497

Olivares-Fuster O, Duran-Vila N, Navarro L (2005) Electrochemical protoplast fusion in citrus. Plant Cell Rep 24:112–119

Ollitrault P, Dambier D, Cabasson C et al (1995) optimized management of citrus embryogenic calli for breeding programmes. Fruits 49:394–397

Ollitrault P, Dambier D, Sudahono S et al (1996) Somatic hybridisation in Citrus: some new hybrid and alloplasmic plants. Proc Int Soc Citricult 2:907–912

Ollitrault P, Dambier D, Froelicher Y et al (2001) La diversité des agrumes ; structuration et exploitation par hybridation somatique. Comptes rendus de l'Académie d'Agriculture 86:197–221

Ollitrault P, Froelicher Y, Dambier D et al (2007) Seedlessness and ploidy manipulation. In: Khan IA (ed) Citrus genetics, breeding and biotechnology. CAB International, Wallingford, pp 197–218

Ollitrault P, Dambier D, Luro F et al (2008) Ploidy manipulation for breeding seedless triploid citrus. Plant Breed Rev 30:323–352

Osterc G, Gostinčar K, Hudina M, Usenik V, Luthar Z (2025) Slovene Fruit Gene Bank. In: Al-Khayri JM, Salem KFM, Jain SM (eds) Plant gene banks. Springer, Singapore. https://doi.org/10.1007/978-981-99-4236-7_84-1

Peña L, Navarro L (2000) Transgenic Citrus. In: Bajaj YPS (ed) Transgenic trees. Springer Berlin Heidelberg, Berlin, pp 39–54. https://doi.org/10.1007/978-3-642-59609-4_4

Pena L, Seguin A (2001) Recent advances in the genetic transformation of trees. Trends Biotechnol 19(12):500–506

Peña L, Cervera M, Juárez J et al (1995) High effi ciency Agrobacterium-mediated transformation and regeneration of citrus. Plant Sci 104:183–191

Peña L, Martín-Trillo M, Juárez J et al (2001) Constitutive expression of Arabidopsis LEAFY or APETALA1 genes in citrus reduces their generation time. Nat Biotechnol 19:263–267. https://doi.org/10.1038/85719

Peña L, Cervera M, Fagoaga C et al (2008) Citrus. In: Kole C, Hall TC (eds) Compendium of transgenic crop plants, Fruits & Nuts (Tropical & Sub-tropical), vol 5. Willey-Blackwell

Peng A, Zou X, He Y, Chen S, Liu X, Zhang J et al (2021) Overexpressing a NPR1-like gene from Citrus paradisi enhanced huanglongbing resistance in C. sinensis. Plant Cell Rep 40:529–541. https://doi.org/10.1007/s00299-020-02648-3

Pérez RM, Galiana AM, Navarro L et al (1998) Embryogenesis in vitro of several Citrus species and cultivars. J Hortic Sci Biotech 73:796–802

Pérez RM, Mas O, Navarro L et al (1999) Production and cryoconservation of embryogenic cultures of mandarin and mandarin hybrids. Plant Cell Tissue Organ Cult 55:71–74

Pflieger S, Lefebvre V, Causse M (2001) The candidate gene approach in plant genetics:a review. Mol Breed 7(4):275–291

Popović V et al (2025) Economic seed production and storage of fiber crops in gene banks in Serbia. In: Al-Khayri JM, Salem KFM, Jain SM (eds) Plant gene banks. Springer, Singapore. https://doi.org/10.1007/978-981-99-4236-7_108-1

Postgate JN (1987) Notes on fruit in the cuneiform sources. Bul Sumerian Agr III:115–144

Rangan TS, Murashige T, Bitters WP (1969) In vitro studies of zygotic and nucellar embryogenesis in citrus. In: Chapmn H (ed) Proc. First Int. Citrus Symp., Univ. California, Riverside, vol 1, pp 225–229

Ray PK (2002) Breeding tropical and subtropical fruits. Springer, Berlin, Heidelberg. ISBN, 978-3-540-42855-8, pp 56–67

Rejeb A et al (2022) Drones in agriculture: a review and bibliometric analysis. Comput Electron Agric 198:107017

Ren G et al (2020) Agricultural robotics research applicable to poultry production: a review. Comput Electron Agric 169:105216

Rocha F et al (2025) Portugal Gene Bank. In: Al-Khayri JM, Salem KFM, Jain SM (eds) Plant gene banks. Springer, Singapore. https://doi.org/10.1007/978-981-99-4236-7_79-1

Rönspies M, Schindele P, Wetzel R, Puchta H (2022a) Crispr-cas9-mediated chromosome 622 engineering in Arabidopsis thaliana. Nat Protoc 17:1332–1358. 623

Rönspies M, Schmidt C, Schindele P, Lieberman-Lazarovich M, Houben A, Puchta H (2022b) Massive crossover suppression by crispr–cas-mediated plant chromosome engineering. Nat Plants 8:1153

Roose ML, Williams E (2007) Mutation breeding. In: Khan IA (ed) Citrus genetics, breeding and biotechnology. CAB International, Wallingford, pp 345–352

Roose ML, Soost RK, Cameron JW (1995) Citrus (Rutaceae). In: Smartt J, Simmonds NW (eds) Evolution of crop plants, 2nd edn. Longman Scientific & Technical, Essex, pp 443–449

Ruan R, Lun J, Zhang J et al (1996) Structure-function relationships of highly refined cellulose made from agricultural fibrous residues. Appl Eng Agric ASAE Publ 12(4)465–468

Saito W, Ohgawara T, Shimizu J et al (1991) Acid citrus somatic hybrids between Sudachi (*Citrus sudachi* Hortex Shirai) and lime (*C. aurantifolia* Swing) produced by electrofusion. Plant Sci 77:125–130

Sakai A, Kobayashi S, Oiyama I (1990) Cryopreservation of nucellar cells of navel orange (*Citrus sinensis* Osb. var. braziliensis Tanaka) by vitrification. Plant Cell Rep 9:30–33

Sakata K, Hirose Y, Qiao Z, Tanaka T, Mori H (2003) Inhibition of inducible isoforms of cyclooxygenase and nitric oxide synthase by flavonoid hesperidin in mouse macrophage cell line. Cancer Lett 199:139–145

Sandal I, Saini U, Lacroix B et al (2007) Agrobacterium-mediated genetic transformation of tea leaf explants: effects of counteracting bactericidity of leaf polyphenols without loss of bacterial virulence. Plant Cell Rep 26:169–176. https://doi.org/10.1007/s00299-006-0211-9

Sanford JC, Klein TM, Wolf ED et al (1987) Delivery of substances into cells and tissues using a particle bombardment process. Part Sci Technol 5:27–37. https://doi.org/10.1080/02726358708904533

Santos IRI, Stushnoff C (2003) Desiccation and freezing tolerance of embryonic axes from *Citrus sinensis* [L.] Osb. pretreated with sucrose. CryoLetters 24:281–292

Saravacos GD (1970) Effect of temperature on viscosity of fruit juices and purees. J Food Sci 35(2): 122–125

Savita Singh B, Virk GS, Nagpal AK (2011) An efficient plant regeneration protocol from callus cultures of Citrus jambhiri Lush. Physiol Mol Biol Plants 17:161–169. https://doi.org/10.1007/s12298-011-0055-9

Scora RW (1975) On the history and origin of Citrus. Bull Torr Bot Club 102:369–375

Sharma TVRS, Singh S, Singh A K, et al (2015) A note on possibility of utilizing wild relatives of crop plants present in Andaman & Nicobar Islands for creating new variability. Souvenir of lead papers and abstracts of National meet on Distant Hybridization in Horticultural Crop Improvement. pp 1–1

Shimizu T (2020) Genomic breeding. In: Talon M, Caruso M, Gmitter F Jr (eds) The genus Citrus. Woodhead Publishing/Elsevier, Oxford, Sawston, Cambridge, UK, pp 149–169

Shimizu T, Kitajima A, Nonaka K et al (2016) Hybrid origins of citrus varieties inferred from DNA marker analysis of nuclear and organelle genomes. PLoS One 11:e0166969. [CrossRef]

Shimizu T, Tanizawa Y, Mochizuki T et al (2017) Draft sequencing of the heterozygous diploid genome of Satsuma (*Citrus unshiu* Marc.) using ahybrid assembly approach. Front Genet 8: 180.

Shimizu T, Kitajima A, Nonaka K et al (2018) Amodel for the domestication and diversification processes of modern citrus cultivars in Japan. In: Proceedings of the IV international symposium on citrus biotechnology, Canelones, Uruguay, 16–18 April 2018, p 7–14.

Shimoi K, Masuda S, Furogori M, Esaki S, Kinae N (1994) Radioprotective affect of antioxidative flavonoids in c-ray irradiated mice. Carcinogenesis 15:2669–2672

Silva A, Marques A, Santos K et al (2010) The evolution of cma bands in citrus and related genera. Chromosom Res 18:503–514

Simmonds NW (1979) Principles of crop improvement. Longman, London

Singh P, Kaur A (2022) A systematic review of artificial intelligence in agriculture. In: Deep learning for sustainable agriculture, pp 57–80

Singh S, Rajam MV (2009) Citrus biotechnology: achievements, limitations and Future Directions. Physiol Mol Biol Plants 15(1):871–884

Singh RV, Chand D, Tyagi V, Singh AK, Singh SP, Singh S, Binda PC (2005) Priorities for introduction of fruit crops in India. Indian J Plant Genet Resour 18(1):67–68

Sinn JP, Held JB, Vosburg C et al (2021) Flowering Locus T chimeric protein induces floral precocity in edible citrus. Plant Biotechnol J 19:215–217. https://doi.org/10.1111/pbi.13463

Soares JM, Weber KC, Qiu W et al (2020) The vascular targeted citrus FLOWERING LOCUS T3 gene promotes non-inductive early flowering in transgenic Carrizo rootstocks and grafted juvenile scions. Sci Rep 10:21404. https://doi.org/10.1038/s41598-020-78417-9

Song GQ, Walworth AE, Loescher WH (2015) Grafting of genetically engineered plants. J Am Soc Hortic Sci 140:203–213. https://doi.org/10.21273/jashs.140.3.203

Song LW, Qi JR, Liao JS et al (2021a) Enzymatic and enzyme-physical modification of citrusfiber by Xylanase and planetary ball milling treatment. Food Hydrocoll 121:107015.

Song JM, Xie WZ, Wang S et al (2021b) Two gap-free reference genomes and a global view of the centromere architecture in rice. Mol Plant S1674–2052(1621):00230–00236

Soost RK (1968) The incompatibility gene system in citrus. In: Chapmn H (ed) Proc. First Int. Citrus Symp., Univ. California, Riverside, vol 1, pp 189–190

Soost RK (1987) Breeding citrus-genetics and nucellar embryony. In: Abbott AJ, Atkin RK (eds) Improving vegetatively propagated crops. Academic Press Limited, London, pp 83–110

Soost RK, Roose ML (1996) Citrus. In: Janick J, Moore JN (eds) Fruit breeding, Tree and Tropical Fruits, vol 1. Wiley, New York, pp 257–323

Stapleton AE, Walbot V (1994) Flavonoids can protect maize DNA from the induction of ultraviolet radiation damage. Plant Physiol 105:881–889

Starrantino A, Recupero G (1982) Citrus hybrids obtained in vitro from 2x females X 4x males. Proc Int Soc Citricult 1:31–32

Starrantino A, Russo F (1983) Reproduction of seedless orange cultivars from undeveloped ovules raised 'in vitro'. Acta Hortic 131:253–258

Stover E, Gmitter FG Jr, Grosser J et al (2020) Rationale for reconsidering current regulations restricting the use of hybrids in orange juice. Hortic Res 7:38. https://doi.org/10.1038/s41438-020-0277-5

Swingle WT, Reece PC (1967) The botany of Citrus and its wild relatives. In: Reuther W, Webber HJ, Batchelor LD (eds) The Citrus industry, vol 1. University of California Press, Berkeley, pp 190–430

Tanaka T (1977) Fundamental discussion of Citrus classification. Stud Citrologia 14:1–6

Terol J, Conesa A, Colmenero JM et al (2008) The areas of flavor, health, color and ripening. BMC Genomics 9:351

Tomas DT, Ross MC, Songstad DD (1995) Direct DNA transfer into intact plant cells via microprojectile bombardment. In: Gamborg OL, Phillips GC (eds) Plant cell, tissue and organ culture (Editorial Springer Lab Manual). Springer, Berlin, pp 197–213

Tripoli E, La Guardia M, Giammanco S, Di Majo D, Giammanco M (2007) Citrus flavonoids: molecular structure, biological activity and nutritional properties: a review. Food Chem 104: 466–479

Turbak AF, Snyder FW, Sandberg KR (1983) Microfibrillated cellulose, a new cellulose product: properties, uses, and commercial potential. J Appl Polym Sci Appl Polym Symp 37(2):815–827

USDA (2022) GRIN-Global National Plant Germplasm System. Citrus descriptors. Accessed 25 Oct 2022. https://npgsweb.ars-grin.gov/gringlobal/cropdetail?type=descriptor&id=112

Van Buren JP (1979) The chemistry of texture in fruits and vegetables. J Texture Stud 10(1):1–23

Van Harten AM (1998) Mutation breeding: theory and practical applications. Cambridge Univ. Press, Cambridge

Vardi A, Frydman Shani A, Weinbaum SA (1988) Assessment of parthenocarpic tendency in citrus using irradiated marker pollen. Proc Int Soc Citricult 1:225–230

Vardi A, Bleichman S, Aviv D (1990) Genetic transformation of citrus protoplasts and regeneration of transgenic plants. Plant Sci 69:199–206

Vardi A, Spiegel-Roy P, Elchanati A (1993) Mandarin tree named Mor. US patent PP8, 378

Velázquez K, Agüero J, Vives MC et al (2016) Precocious flowering of juvenile citrus induced by a viral vector based on Citrus leaf blotch virus: a new tool for genetics and breeding. Plant Biotechnol J 14:1976–1985. https://doi.org/10.1111/pbi.12555

Vigouroux Y, Barnaud A, Scarcelli N, Thuillet AC (2011) Biodiversity, evolution and adaptation of cultivated crops. C R Biol 334(5–6):450–457

Volk GM (2011) Citrus conservation at the National Center for Genetic Resources Preservation. Citrograph 2:42–43

Volk GM, Bonnart R, Krueger R et al (2012) Cryopreservation of Citrus shoot tips using micrografting for recovery. CryoLetters 33:418–426

Walton BS, Bartholomew ET, Ramsey RC (1945) Analysis of the organic acids of orange juice. Plant Physiol 20(1):3–18

Wang QC, Batuman O, Li P et al (2002) Cryopreservation of in vitro-grown shoot tips of 'Troyer' citrange [*Poncirus trifoliata* (L.) Raf. X *Citrus sinensis* (L.) Osbeck] by encapsulation-dehydration. Plant Cell Rep 20:901–906

Wang Q, Li P, Batuman O (2003) Effect of benzyladenine on recovery of cryopreserved shoot tips of grapevine and citrus cultured in vitro. CryoLetters 24:293–302

Wang L, Huang Y, Liu Z et al (2021) Somatic variations led to the selection of acidic and acidless orange cultivars. Nat Plants 7:954–965

Wang J et al (2025) China National Gene Bank (CNGB). In: Al-Khayri JM, Salem KFM, Jain SM (eds) Plant gene banks. Springer, Singapore. https://doi.org/10.1007/978-981-99-4236-7_103-1

Wu GA, Prochnik S, Jenkins J et al (2014) Sequencing of diverse mandarin, pummelo and orange genomes reveals complex history of admixture during citrus domestication. Nature Biotechnol 32(7)656–662. https://doi.org/10.1038/nbt.2906

Wu H, Acanda Y, Jia H, Wang N et al (2016) Biolistic transformation of Carrizo citrange (*Citrus sinensis* Osb. _*Poncirus trifoliata* L. Raf.). Plant Cell Rep 35:1955–1962. https://doi.org/10.1007/s00299-016-2010-2

Wu GA, Terol J, Ibanez V et al (2018) Genomics of the origin and evolution of Citrus. Nature 554(7692):311–316. https://doi.org/10.1038/nature25447

Wu H, Acanda Y, Canton M et al (2019) Efficient biolistic transformation of immature citrus rootstocks using phosphomannose isomerase selection. Plants 8:390. https://doi.org/10.3390/plants8100390

Wu GA, Sugimoto C, Kinjo H et al (2021) Diversification of mandarin citrus by hybrid speciation and apomixis. Nat Commun 12:4377. https://doi.org/10.1038/s41467-021-24653-0

Xin H, Zhang T, Wu Y et al (2020) An extraordinarily stable karyotype of the woody populus species revealed by chromosome painting. Plant J 101:253–264

Yamamoto M, Matsumoto R, Yamada Y (1995) Relationship between sterility and seedlessness in citrus. J Japan Soc Hort Sci 64:23–29

Yao JL, Wu JH, Gleave AP et al (1996) Transformation of citrus embryogenic cells using particle bombardment and production of transgenic embryos. Plant Sci 113:175–183. https://doi.org/10.1016/0168-9452(95)04292-X

Yu F, Zhao X, Chai J et al (2021) Chromosome-specific painting unveils chromosomal fusions and distinct allopolyploid species in the saccharum complex. New Phytol 233:1953–1965

Zhang C, Lu Y (2021) Study on artificial intelligence: the state of the art and future prospects. J Ind Inf Integr 23:100224

Zhang H, Yin L, Wang M, Yuan X, Liu X (2019a) Factors affecting the accuracy of genomic selection for agricultural economic traits in maize, cattle, and pig populations. Front Genet 10: 189.

Zhang R, Meng Z, Abid MA et al (2019b) Novel pollen magnetofection system for transformation of cotton plant with magnetic nanoparticles as gene carriers. In: Zhang B (ed) Transgenic cotton: methods and protocols. Humana Press, New York, pp 56–74. https://doi.org/10.1007/978-1-4939-6911-1

Zhang M, Wang J, Luo Q et al (2021) CsMYB96 enhances citrus fruit resistance against fungal pathogen by activating salicylic acid biosynthesis and facilitating defense metabolite accumulation. J Plant Physiol 264:153472. https://doi.org/10.1016/j.jplph.2021.153472

Zhang W, Wang J, Liu Y et al (2022) Deep-learning-based in-field citrus fruit detection and tracking. Hortic Res 9:uhac003

Zhao Q, Meng Y, Wang P et al (2021) Reconstruction of ancestral karyotype illuminates chromosome evolution in the genus cucumis. Plant J 107:1243–1259

Zheng Z, Xiong J, Lin H et al (2021) A method of green citrus detection in natural environments using adeep convolutional neural network. Front Plant Sci 12:705737

Zhu C, Feng C, Li X, Xu C, Sun C, Chen K (2013) Analysis of expressed sequence tags from Chinese bayberry fruit (*Myrica rubrasieb* and zucc.) at different ripening stages and their association with Fruit Quality Development. Int J Mol Sci 14(2):3110–3123

Zou X, Zhao K, Liu Y et al (2021) Overexpression of salicylic acid carboxyl methyltransferase (CsSAMT1) enhances tolerance to huanglongbing disease in Wanjincheng Orange (*Citrus sinensis* (L.) Osbeck). Int J Mol Sci 22:2803. https://doi.org/10.3390/ijms22062803

Chapter 15
Exploring Innovations in Plant Breeding and Biotechnological Approaches for the Sustainable Cultivation of Coconut (*Cocos nucifera* L.)

H. Mary Henrietta

Abstract A significant tropical crop with enormous economic, cultural, and ecological significance is the coconut (*Cocos nucifera* L.). Its many uses, from the production of food and oil to building materials and cosmetics, highlight the necessity of ongoing advancements in the field of coconut agriculture. This study examines the various strategies used to increase coconut production, including hybridization, tissue culture, genetic engineering, and conventional breeding. Coconut improvement has historically been greatly aided by traditional breeding, which has made it possible to select for desired qualities including disease resistance, high production, and increased fruit quality. The lengthy reproductive cycle of coconut trees and the imprecise control over trait inheritance, however, make this approach difficult. The ability to rapidly produce premium coconut variants in a controlled environment is what has revolutionized the propagation of coconuts thanks to tissue culture techniques. Planting material free of diseases and shorter generation times are two benefits of this strategy. A promising method for precisely manipulating coconut traits is genetic engineering. Although worries over public acceptance and regulatory obstacles remain, transgenic techniques have been investigated to enhance features like pest resistance and higher oil content. Conversely, hybridization makes use of the inherent genetic diversity found in each species to produce hybrids with superior qualities. To sum up, many techniques for improving coconuts are being developed to fulfill the needs of the contemporary world. The integration of tissue culture, genetic engineering, hybridization, and conventional breeding provides a versatile toolkit for optimizing coconut yield, safeguarding the long-term viability of this emblematic tropical commodity against new threats. Breeders have fascinating opportunities as well as obstacles due to the distinctive characteristics of coconuts. The entire potential of coconut improvement can be realized by incorporating

H. M. Henrietta (✉)
Department of Mathematics, Saveetha Engineering College (Autonomous), Chennai, Tamil Nadu, India
e-mail: henriettamaths@gmail.com

K. F. M. Salem et al. (eds.), *Breeding and Biotechnology of Leaf, Fruit, and Seed Fiber Crops*, Advances in Plant Breeding Strategies 11,
https://doi.org/10.1007/978-3-032-00407-9_15

cutting-edge technologies, implementing effective breeding techniques, and encouraging international cooperation. This will guarantee coconut's continuous assistance to food security and rural lives in tropical regions. This chapter covers the history, taxonomy, cultivation methods, and difficulties associated with *Cocos nucifera* L. It focuses on genetic resources and different production methods.

Keywords Sustainable agriculture · Genomic-assisted breeding · Coconuts (*Cocos nucifera* L.) · Biotechnological tools · Coconut breeding · Tissue culture · Hybridization · Mutation breeding

15.1 Introduction

A crop with a great proportion of genetic diversity is coconut. The coconut palm (*Cocos nucifera* L.), a division of the palm tree brood Aceraceae, is the only extant species in the genus Cocos. This enormous hand may stretch up to 30 meters (100 feet) in height. The pinnate leaves can grow to a length of 4–6 meters (13–20 feet), while the pinnae measure 60–90 centimeters (2–3 feet). The stem is smooth when old leaves fall off neatly. In healthy soil, a tall coconut palm tree can develop up to 75 fruits a year. However, it usually yields no more than 30. Coconut palm bears fruit in 6–10 years and reach their optimum productivity in 15–20 years when given proper care and growing circumstances. Since ancient times, Asian cultures have been cultivating true-to-type dwarf genres of Pacific coconuts. The coconut is a flexible and all-purpose fruit. Coconut water is a liquid that can be used to make coconut cream and milk or just drank on its own. You can consume the flesh dry or unprocessed. The coconut shell itself can be utilized to model bowls, cups, and several utensils, while the fibers from the shell can be utilized to make rope, mats, and supplementary textiles. Coconut oil is widely used in cooking, but it may also be made into cosmetics and additional by-products. Coconut fiber can be used to make rope, mats, baskets, and other handicrafts. It is a durable material.

Additionally, the fiber is utilized to create coconut matting, a kind of flooring composed of interwoven coconut fibers. Because it is soft, long-lasting, and low maintenance, coconut matting is a common option for flooring in tropical regions. A significant sector in many developing nations is the production of coconut fiber. Each year, the business brings in billions of dollars and employs millions of people. Because of the equally rising market for coconut fiber by-products in both domestic and international markets, the business is expanding quickly. The domestication of the coconut palm dates back thousands of years, with its origins thought to be in Southeast Asia.

Though the plant was soon found to have many additional applications, it is most likely that the earliest coconut palms were farmed for food and water. Coconut palms with huge fruit sizes, excellent yields, and resistance to pests and diseases were eventually chosen by humanity. Due to these early advancements, the modern coconut palm-an important crop in many regions of the world—was developed.

Table 15.1 Countries worldwide production of coconut

Country/Region	2022	2021	2020
Indonesia	17,190,328	17,100,000	16,800,000
Philipinnes	14,931,158	14,717,294	14,490,923
India	13,317,000	14,301,000	14,006,000
Brazil	2,744,418	2,465,180	2,434,095
Sri Lanka	2,204,150	2,496,00	2,233,600
Vietnam	1,930,182	1,866,700	1,720,661
Papua New Guinea	1,258,149	1,180,000	1,180,000
Myanmar	1,217,442	1,220,000	1,252,215
Mexico	1,119,847	1,074,400	1,074,400
Thailand	679,232	651,866	618,246

Source Food and Agriculture Organization Corporate Statistical Database (FAOSTAT)

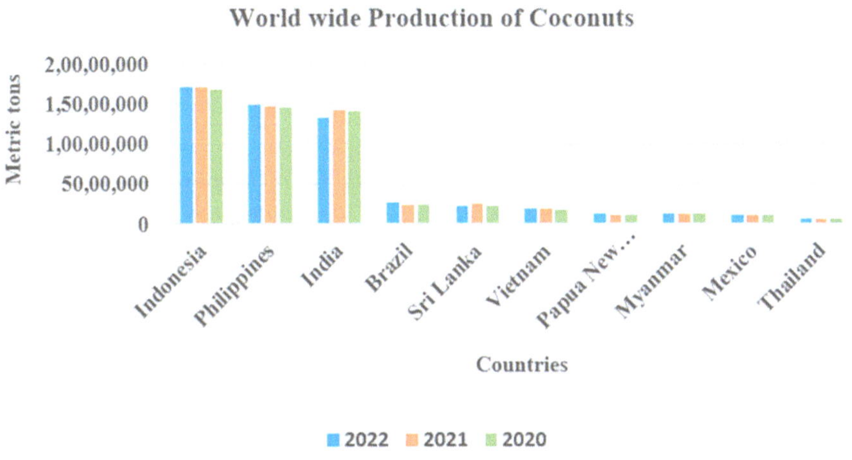

Fig. 15.1 Top Coconut Producers Worldwide in 2023 (FAOSTAT)

In accordance with the graph for consecutive in the past 3 years, Indonesia is the world's top producer of coconuts produced in 2022 (FAOSTAT). With 16.7 million tonnes produced, the Philippines comes in second. With 13.8 million tonnes produced, India comes in third. Other significant coconut-producing nations are Brazil, Sri Lanka and Vietnam (Table 15.1 and Fig. 15.1). The graph also demonstrates that, during the previous 5 years, coconut cultivation has increased in the majority of countries. Indonesia has experienced the most increase. India, the Philippines, and Sri Lanka are among the other nations where coconut agriculture has increased significantly Discuss the coir extraction and trends in industries like building materials, absorbent and toxic materials.

Many in tropical regions rely on the coconut palm (*Cocos nucifera* L.) for food, shelter, and a source of money. It is a highly adaptable and widespread plant. Because of its unique biological characteristics and wide range of uses, breeding efforts looking to increase production, resilience, and adaptability face both new problems and opportunities. In addition to highlighting the obstacles preventing further advancements in coconut breeding, this article delves into the essential elements that make coconuts special. This chapter addresses the history, taxonomy, cultivation techniques, and challenges related to *Cocos nucifera* L. It emphasizes genetic resources and various production techniques.

15.1.1 Origin, Classification, Diversity and Uniqueness

Coconut (*Cocos nucifera* L.) is a flowering plant within the class *Magnoliopsida,* order *Arecales* and family *Arecaceae.* Many edible plants are also in the *Arecaceae* (https://www.itis.gov). Coconut (*Cocos nucifera* L.) are of the genus *Cocos* and species.

Kingdom *Plantae – plantes, Planta, Vegetal, plants*
Subkingdom *Viridiplantae – green plants*
Infrakingdom *Streptophyta – land plants*
Superdivision *Embryophyta*
Division *Tracheophyta – vascular plants, tracheophytes*
Subdivision *Spermatophytina– spermatophytes, seed plants, phanérogames*
Class *Magnoliopsida*
Superorder *Lilianae – monocots, monocotyledons, monocotyledons*
Order *Arecales*
Family *Arecaceae*
Genus *Cocos L. – coconut palm*
Species *Cocos nucifera* L.*– niu, ololani, cocotier, coconut palm, Palma*

(a) Botanical features: The height, inflorescence structure, nut shape, and copra quality of coconuts show a remarkable variation. A valuable genetic pool is provided by this natural variety for breeding and selection (Fig. 15.2). (b) Perennial habit and long generation time: Unlike annual crops, coconuts have a protracted juvenile period (5–7 years) and a long lifetime (60–80 years), which contribute to their prolonged and resource-intensive breeding cycles. This calls for quick generation advancement methods and effective breeding procedures. (c) Outcrossing and cross-pollination: Designing breeding programs and selecting traits is made more difficult by wind-mediated pollination with significant heterozygosity. In order to maintain genetic integrity and promote hybrid development, controlled pollination methods are essential and (d) A wide range of applications and products: Coconuts can be used to make an extensive range of goods, including fiber, lumber, coconut water, oil, and copra. Multiple qualities must be accommodated in breeding systems, frequently with contradictory selection. Stelte (2023) discussed the coir extraction and trends in industries like building materials, absorbent and toxic materials.

Fig. 15.2 (**a**) Coconut varieties in Dharmapuri district, Tamil Nadu. (**b**) Coconuts varieties in Chennai, Tamil Nadu. (Internal Source)

15.2 Current Cultivation: Practices and Challenges

An overview of the methods used currently in coconut agriculture, as well as the difficulties and goals the sector faces in breeding. Henrietta et al. (2022) gave a demonstration of coconut farming, coconut varietals, and coconut product variants.

15.2.1 Current Cultivation Practices

15.2.1.1 Conventional Farming

Traditionally, coconuts are farmed in wide-open fields, with farmers sowing and tending to the trees by hand. This farming method can be effortful and ineffective due to its difficulty in controlling diseases and pests.

15.2.1.2 Vertical Farming

This innovative farming technique involves cultivating crops in layers that are piled vertically. Because it uses less area and water than traditional farming, this kind of cultivation may be more effective. It is very early in the development of vertical farming; thus it is unclear if it will be profitable to produce coconuts using this method.

15.2.1.3 Digital Farming

This refers to the application of technology to raise agricultural productivity and efficiency. This can involve the use of software to control irrigation, drones to spray pesticides, and sensors to track the health of the crops. While digital farming can help farmers save costs and boost yields, its implementation can be costly.

15.2.1.4 Artificial Intelligence

With its potential to completely transform agriculture, artificial intelligence (AI) is a quickly evolving technology. AI can be used to enhance pest and disease management, create novel crop varieties, and improve fertilization and irrigation. Since artificial intelligence is still in its infancy, it is unclear how technology gets adopted in the coconut industry. The approach, which uses the DenseNet-121 architecture to classify images of coconut leaves, offers a novel approach to developing disease recognition models (Anitha and Santhi 2023). Since everything in the modern world is getting mechanized, farmers can save money by using automatic recognition. There are 526 photos in the hand-picked dataset (both healthy and sick). The suggested approach uses the DenseNet121 architecture, which offers an accuracy of roughly 99% and outperforms existing techniques by removing vanishing gradients and improving feature propagation by lowering the feature count.

15.2.2 Current Agricultural Challenges

There are challenges that the fruit crop must take into account.

15.2.2.1 Pests and Diseases

Root rot, mealybugs, rhinoceros beetles, and other diseases are among the pests and diseases that coconut palms are vulnerable to. Coconut palms are susceptible to consequential damage from several pests and diseases, which can result in lower yields or even tree mortality.

15.2.2.2 Climate Change

The production of coconuts is seriously threatened by changes in the climate. Droughts and floods may result from changes in rainfall patterns brought on by rising sea levels and temperatures. Because of these alterations, coconut trees may become stressed and more vulnerable to illnesses and pests.

15.2.2.3 Ineffective Production Techniques

Low yields are a possible result of using inefficient traditional methods of coconut cultivation. For instance, planting coconut palms too near to one another might result in resource rivalry and overcrowding.

15.2.3 Breeding Objectives

The coconut genome is still mostly unknown, despite the species' notable morphological variety (Arumugam et al. 2022). The methods and strategies used in breeding to increase coconut production (Batugal et al. 2009). It might address subjects including hybridization, selection techniques, genetic improvement, and trait enhancement in coconut plants. In order to meet the requirement of enhanced production, standard, and livability in the coconut business, the article may also discuss possibilities, problems, and forthcoming directions for coconut breeding. (a) High yields: Breeders of coconut palms are attempting to create coconut palm cultivars with high yields. Selecting characteristics like big fruit size, early bearing, and disease and insect tolerance will help achieve this. (b) Quality enhancement: Breeders of coconuts are also attempting to create coconut palm cultivars with better quality. This can be achieved by choosing characteristics like long shelf life, high oil content, and sweet coconut water and (c) Pest and disease resistance: Breeders of coconut palms are also attempting to create coconut palm cultivars that are resistant to pests and illnesses. This is crucial because diseases and pests can seriously harm coconut palms and lower their production.

15.3 Germplasm Biodiversity and Conservation

The morphological, physiological, and biochemical characteristics of coconuts vary greatly, making them a very diversified crop. Coconut varieties are primarily categorized as tall or dwarf. The most popular kinds are the tall varieties, which are cultivated for their nuts and utilized to make copra, coconut milk, and coconut water. Although dwarf varieties are less widespread, they are nonetheless planted for their smaller, more oil-rich nuts and their aesthetic appeal. The genetic makeup of coconut germplasm can be better understood through the use of SSR markers, which help to preserve and manage the genetic resources of coconuts in the Pacific (Rajesh et al. 2008b).

15.3.1 Cultivars Characterization and Phylogeny in Coconut

Many techniques, such as morphological studies, molecular markers, and cytogenetics, have been used to study the phylogeny of coconut cultivars. These studies' findings imply that there was just one wild progenitor from whom all coconut cultivars descended and that natural selection and human activity have since led to their diversification. Baudouin and Lebrun (2002b) uses Amplified Fragment Length Polymorphism (AFLP) markers to investigate the genetic diversity of coconut germplasm in the Pacific region. The study likely looks into the linkages, population structure, and genetic variation between various populations or accessions of coconuts in the Pacific. The AFLP markers offer important information on genetic diversity and possible uses for conservation and breeding initiatives involving coconuts. The goal of Mauro-Herrera (2007) is to use AFLP to investigate the genetic diversity of coconut germplasm in the Indian Ocean region. The study most likely looks at the linkages, population structure, and genetic variation between several populations or accessions of coconut in the Indian Ocean region. Insights into the genetic configuration of coconut germplasm are obtained through the use of AFLP markers, which support efforts to conserve and breed this significant crop. The analysis in Ashburner (1997) attends to the fruit components of populations of coconut palms in the South Pacific. The study probably looks at morphological features, genetic diversity, or other aspects of coconut palm fruits in this region. The study's main conclusions show how diversified the coconut germplasm is throughout the Indian Ocean. The primary cause of the diversity is the distinction in coconut germplasm between Southeast Asian and Indian Ocean regions. There are now two large genetic clusters known to exist: one in Southeast Asia and one in the Indian Ocean. Two sub-clusters, one in the western part of the Indian Ocean and one in the eastern Indian Ocean, were further separated from the main Indian Ocean cluster. A prime resource for conservation and breeding is the coconut germplasm found in the Indian Ocean.

15.3.2 Genetic Resources Conservation in Coconut

The genetic amenities of the coconut can be preserved in a variety of ways. In coconut breeding projects, beneficial features can be identified and selected through molecular markers. Propagating coconut plants in vitro is an economical and successful method. Coconuts can be genetically engineered to have greater disease and insect resistance. The functionality and uses of the specialized program in analyzing genetic diversity, population distribution, and relatedness among various coconut accessions may also be covered in the article (Baudouin and Lebrun 2002a). Through conventional breeding or genetic engineering, coconuts can be bred to have higher stress tolerance. Linda and Abdullah (2023) improvements in the cultivation of coconuts employing multi-omics, marker-assisted selection, and conventional breeding, with a focus on dwarf cultivars. Genetic engineering or traditional breeding can be used to raise the caliber of coconut goods. These methods consist of (a) In vitro conservation is the process of cultivating coconut cells or embryos in a regulated setting. This technique is applied to uncommon or endangered cultivar propagation as well as small-scale germplasm conservation.

(b) Growing coconut trees in their native environment is known as "in situ conservation." Large volumes of germplasm are preserved using this technique, which also preserves the genetic variety of coconut populations. (c) Coconut embryos or cells are frozen in liquid nitrogen during the cryopreservation process. Germplasm can be preserved with this technique for extended periods. (d) Germplasm is kept in gene banks in several formats, such as cryopreserved, in vitro, and in situ. Germplasm is stored in gene banks for use in research and future projects like cultivating new cultivars.

15.3.3 Cytogenetics

The study of an organism's chromosomes is known as cytogenetics. The diploid chromosomal number of coconuts is $2x = 32$. The coconut is a useful model organism for cytogenetic research because of its large and easily studied chromosomes. Utilizing cytogenetic research, scientists have been able to comprehend the evolution of the coconut, pinpoint the genes responsible for key features, and create novel breeding strategies. Additionally, cytogenetics has been applied to create techniques for disease detection and management in coconuts.

15.4 Traditional Breeding

The productivity, quality, and disease resistance of coconut palms have all been improved by the application of breeding techniques in coconut studies. Although traditional breeding techniques have been employed for millennia, they have several drawbacks. These days, biotechnology is employed to get around these restrictions and quicken the breeding process. In coconut breeding, there are two primary improvement tactics employed: (a) Conventional breeding: To create a hybrid with the necessary characteristics, two distinct coconut trees are crossed in this process. The long process of traditional breeding can result in a new variety that is inferior to its parents after many years of breeding. (b) Biotechnology: This entails introducing new genes into coconut trees through genetic engineering methods. The breeding process can be sped up and genes not present in naturally occurring coconut palm populations can be introduced with the help of biotechnology. (Oropeza et al. 1999) offers a thorough synopsis of the most modern developments in coconut biotechnology. The information is divided into three sections: applications of coconut biotechnology, basic concepts of coconut biotechnology, and prospects for coconut biotechnology in the future. Additionally, it addresses a variety of subjects.

15.4.1 Traditional Breeding Methodologies

Coconut quality and yield have increased as a result of the effectiveness of traditional breeding techniques. These techniques can, however, be labor- and time-intensive. Growing interest has been seen in improving coconuts through the use of contemporary breeding methods like genetic engineering in recent years. With these methods, trees may be produced with even more desired characteristics, like resilience to pests and diseases and tolerance to drought. Using conventional breeding techniques, creating a new coconut variety can take several generations. The correct trees must be chosen, their seeds must be collected, and planting them takes a lot of time and work. Not every breeding program succeeds in creating new coconut kinds. Traditional breeding techniques have been effective in raising the quantity and quality of coconuts despite these obstacles. To produce even superior coconut types in the future, these methods—along with contemporary breeding techniques—are probably going to be adopted. A variety of techniques have been employed in traditional coconut breeding, such as (a) Mass selection: In this method, the best coconut palms are chosen based on each desirable characteristic, and their seeds are then planted. Since it can take several years to develop a fresh variety that is better than its parents, this process is slow and ineffective. (b) Hybridization: To create a hybrid with the required features, two distinct coconut palms are crossed. Although hybridization is a more effective breeding technique than mass selection, it might still take some time to complete. (c) Mutation breeding: In order to cause mutations, coconut palms are exposed to mutagenic substances, such as radiation or chemicals. New features, including tolerance to illnesses or pests, can occasionally be produced by mutations. (d) Clone breeding: This technique entails grafting desired tree cuttings onto a rootstock. At that point, the grafted tree and the original tree will share the same genetic makeup. To grow trees with desirable qualities, including an abundant yield or resistance to pests and diseases, clone breeding is frequently utilized.

15.4.2 Traditional Breeding Limitations

Traditional breeding faces several constraints, including (a) The procedure is moderate: it can take several years to create a new coconut type that is better than its parents. This is due to the lengthy growth period of coconuts, which might take several years for them to mature and bear fruit. (b) It's not always effective: Not every cross between several coconut palms works, and not every hybrid has the necessary characteristics. This is because several genes frequently regulate the desired qualities of a coconut tree, such as high production, disease resistance, and drought tolerance. Because of this, selecting these qualities using conventional breeding techniques is challenging. (c) New gene introduction is not always feasible: genes not present in coconut palm natural populations cannot be introduced through traditional breeding. (d) Labor-intensive: The process of selecting the appropriate trees, gathering their seeds, and planting them takes a great deal of time and work when using traditional breeding methods. Small-scale farmers may find this to be a significant obstacle because they lack the funding for conventional breeding initiatives.

Traditional breeding has been successful in producing some enhanced coconut cultivars despite these drawbacks. By introducing new genes through genetic engineering, coconut trees can acquire new characteristics like resilience to pests and diseases or tolerance to drought. However, the application of contemporary breeding methods, including genetic engineering, is probably going to speed up the production of new coconut kinds. With the application of these contemporary breeding techniques, coconut plantations could be improved by creating new coconut kinds that are more productive, disease- and pest-resistant, and climate-adaptable. One method for locating trees with desired traits, even if they are not readily apparent—is marker-assisted selection. Enabling breeders to concentrate on trees that have the best chance of succeeding, helps expedite the breeding process.

15.4.3 Role of Biotechnology

Nowadays, biotechnology is being utilized to accelerate the breeding process and get around the drawbacks of traditional breeding. Biotechnology can be used for (a) Genetic engineering: In order to give coconut trees desired characteristics, including higher yield or resistance to pests or diseases, genes from other organisms are inserted into the plant. It is possible to introduce genes using biotechnology that are absent from coconut palm populations in the wild. The genes can be inserted into the coconut palm genome by genetic engineering methods. (b) Quicken the breeding process: Breeders may pick for desired traits more quickly thanks to biotechnology, which speeds up the breeding process. Molecular markers can be used to identify plants that have the desired genes. (c) Increase breeding efficiency: Breeders can select numerous desirable features at once by using biotechnology, which increases breeding efficiency. The application of marker-assisted selection (MAS) can do this. The DNA sequences are useful for tracking and identifying genes. They can be applied to select desirable features, identify parents and offspring, and evaluate genetic diversity and (d) Tissue culture: This entails cultivating plants in a lab environment from cells or tissues. This can be used to propagate uncommon or endangered kinds or to swiftly create vast quantities of plants.

15.4.4 Biotechnology in Coconut Development

The coconut sector could transform thanks to biotechnology. It can aid in raising yields, boosting pest and disease resistance, and creating new kinds with desired characteristics. But certain issues must be resolved, like the price of biotechnology and the requirement for regulatory clearance. A new and promising tool for the development of coconuts is biotechnology. It has the potential to greatly increase the sustainability and productivity of the coconut sector with more study and development. Here are some instances of the application of biotechnology in the growth of coconuts: in India, scientists have employed biotechnology to create a coconut variety immune to the deadly Ganoderma butt rot disease. (a) Tissue culture has been employed by researchers in the Philippines to propagate uncommon and endangered coconut types. Researchers in Indonesia have selected and identified high-yielding, drought-resistant coconut types using molecular markers. Added advantages of applying biotechnology to the development of coconuts: (b) Higher yields: By creating cultivars that can withstand harsher growing conditions or that are more resistant to pests and diseases, biotechnology can be employed to increase yields. (c) Better quality: By increasing their nutritional value or lowering the concentrations of dangerous pollutants, biotechnology can enhance the quality of coconut goods like oil and milk. (d) Sustainability: By creating coconut cultivars that use less water or fertilizer or that are less likely to harm the environment, biotechnology can help make the coconut business more sustainable. All things considered; biotechnology could have a big impact on how the coconut sector develops. In addition to helping to keep the industry safe from illnesses and pests, it can help to increase yields, quality, and sustainability.

15.5 Molecular Breeding

Coconut varieties are improved by molecular breeding, which employs molecular biology methods. A significant tropical crop, coconut is used for a variety of purposes, including food, cosmetics, and the manufacture of biofuel. Because of its lengthy breeding cycle and complicated genetics, traditional coconut breeding techniques can be laborious and slow. Using molecular breeding, better coconut varieties with desired features can be developed more quickly and precisely.

15.5.1 Molecular Marker-Assisted Breeding in Coconut Study: Genomics and Bioinformatics

Enhancing the productivity, quality, and disease resistance of coconut palms can be accomplished through the use of breeder technology. Although there are several drawbacks to traditional breeding techniques, biotechnology is now being utilized to get around these issues and speed up the breeding process. A potent technique used in coconut breeding efforts to hasten the production of superior coconut types is molecular marker-assisted breeding or MAB. Researchers can find and use molecular markers to improve the accuracy and efficiency of coconut breeding by utilizing genomics and bioinformatics. This note highlights recent citations that demonstrate improvements in the area and gives an overview of the use of genomics and bioinformatics in molecular marker-assisted coconut breeding.

The application of molecular markers for marker-assisted breeding (MAB) in coconuts was assessed by Shasany (2021). A total of one thousand coconuts were screened using one thousand molecular markers, such as SSR, AFLP, and SNP markers. Individuals with desired qualities, like a high yield, a high oil content, and resilience to illnesses and pests, were identified using the markers. New coconut cultivars were subsequently created using the identified individuals. According to the study's findings, MAB is a viable approach for raising coconut output and quality. With the help of technology, it is possible to identify people who possess desired qualities and create new coconut cultivars that are more nutritional, prolific, and pest- and disease-resistant. A promising technique to raise the yield and quality is MAB.

15.5.2 Bioinformatics in Molecular Marker-Assisted Breeding

Single nucleotide polymorphisms (SNPs) and simple sequence repeats (SSRs) are two examples of molecular markers that can be developed and identified with the use of bioinformatics techniques and resources. These markers act as reference points genetically for the desired features. Single nucleotide polymorphism (SNP) markers were employed in this work to evaluate the genetic diversity of coconut germplasm in the Indian Ocean (Rajesh et al. 2021). A total of 1248 SNPs were found and 134 accessions from 17 different nations were genotyped using them. The findings demonstrated that, with a mean genetic distance of 0.32, the coconut germplasm in the Indian Ocean is quite diversified. The primary cause of the diversity was the distinction in coconut germplasm between Southeast Asian and Indian Ocean regions. Two significant genetic clusters were also found by the investigation, one in the Indian Ocean and one in Southeast Asia.

The western and eastern Indian Ocean sub-clusters comprise the Indian Ocean cluster, which was further subdivided into two smaller groups. The study concluded that the Indian Ocean's coconut germplasm is a great resource for conservation and breeding. Mathew (2021) used comparative genomics research and whole-genome sequencing (WGS) to create simple sequence repeat (SSR) markers in coconuts. The WGS data of coconuts led to the identification of 10,000 SSR markers. Thereafter, a genomic map of coconut was created using the SSR markers. QTLs for significant agronomic variables, like yield, oil content, and drought tolerance, were found using the genetic map.

Subsequently, the QTLs were employed to create marker-assisted selection (MAS) instruments aimed at enhancing coconut yield and quality. The study concluded that comparative genomics and WGS are effective methods for creating SSR markers in coconuts. To increase coconut yield and quality, SSR markers can be utilized to create genetic maps, pinpoint QTLs, and create MAS tools.

15.5.3 Benefits of Breeding Coconuts with Marker-Assisted Breeding (MAB)

(a) By lowering the time and costs associated with selecting desired features, MAB can aid in increasing the effectiveness of breeding operations. (b) By recognizing people who possess the required attributes even when they do not exhibit them externally, MAB can aid in increasing the accuracy of selection. (c) MAB can aid in boosting the genetic diversity of breeding populations, which may facilitate the emergence of novel cultivars with increased pest and disease resistance. (d) MAB is an effective method that may be utilized to raise coconut yield and quality. Although the technology is still in its infancy, it has the potential to completely transform the breeding process for coconuts and turn them into a more lucrative and sustainable crop.

15.5.4 Genomics in Molecular Marker-Assisted Breeding

Whole-genome sequencing has made it possible to generate extensive genomic resources for coconuts. Identification of genes and genomic areas linked to significant agronomic features is facilitated by the use of reference genomes. In Tanaka (2021), a reference genome for coconuts (*Cocos nucifera* L.) at the chromosomal scale is presented. Given that the coconut genome was sequenced, assembled, and annotated, it probably offers significant new insights into the chromosomal structure and organization. Examine how breeding programs benefit from the reference genome, how to comprehend the biology of coconuts, and which genes are associated with important traits.

15.5.5 Genomic and Bioinformatics

In the molecular marker-assisted coconut breeding process, genomics and bioinformatics are essential components. The generation of markers, QTL mapping, and marker-trait association studies have been made easier by the availability of a reference genome for coconuts and the use of genomic and bioinformatics tools. These developments aid in the creation of improved coconut cultivars with increased resilience, quality, and yield. QTL mapping, or quantitative trait loci mapping, is the process of locating genetic areas linked to desired traits. This method supports marker-assisted selection (MAS) in breeding programs and helps choose candidate genes for additional study (Singh et al. 2020). The author mapped quantitative trait loci (QTLs) for coconut oil content and yield components using simple sequence repeats (SSRs) markers.

Two coconut cultivars, Malayan Yellow Dwarf (MYD) and Laguna Tall (LAGT) were crossed, and 100 progenies were genotyped using 278 SSR markers. The findings indicated that QTLs for yield components were found on chromosomes 1, 2, 3, 4, 5, 6, 7, 8, 9, and 10. These yield components included the number of nuts per bunch, nut weight, and copra yield. The following chromosomes had QTLs for oil content: 1, 2, 3, 4, 5, 6, 7, 8, 9, and 10. A subset of fifty progenies from the same cross were used to validate the QTLs for yield components and oil content. The findings demonstrated that there was consistency between the two mapping populations' QTLs for yield components and oil content.

15.6 Tissue Culture Applications

Crops can be propagated from a single cell or tissue using a technique called tissue culture. This can be used to swiftly and effectively grow a huge number of plants as well as plants with desirable traits, including early bearing or disease resistance. One useful technique for raising the sustainability and output of coconut crops is tissue culture. There are several possible uses for tissue culture techniques, which have been employed in coconut farms for several years. Within the coconut industry, two of the most significant uses of in vitro technology are tissue culture and micropropagation. The technique of cultivating cells, tissues, or organs in a regulated setting is known as tissue culture. Using a sort of tissue culture called micropropagation, many plants can be grown from a tiny number of cells. The hydrophilic and hydrophobic properties of the fatty acid and its by-products are particularly important for use in the production of surfactants and foam stabilizers for detergents, cosmetics, shampoos, and pharmaceuticals, as well as corrosion inhibitors, emulsifiers, and wetting agents, and as plasticizers for PVC products. A sizable portion of all coconut oil is used in the oleochemicals industry (Blake 1990). (a) Mass propagation: The cost of planting materials can be decreased by mass propagating coconut plants through tissue culture. This can be especially helpful in developing nations, where the price of planting supplies can be a significant obstacle to the production of coconuts. (b) Better varieties: Better coconut plant types, such as those with a higher yield or disease resistance, can be created using tissue culture. Farmers' lives may be enhanced and coconut plantations' productivity may rise as a result. (c) Conservation: Rare or endangered coconut kinds can be preserved through tissue culture. Plant samples can be taken and grown in vitro to do this. If the wild population decreases or vanishes, this guarantees that the plants will not be lost.

The development of tissue culture technology holds great promise for revolutionizing the production of coconuts. Tissue culture can assist boost the output of coconut plantations and enhance farmers' lives by mass-propagating improved types of coconut plants. Coconut plantlets free of pests and diseases and tailored to certain settings can be created using tissue culture. This is significant because a variety of diseases, including yellowing, wilting, and crown rot, can affect coconut plants. This is significant since the placement of coconut plants affects their needs for temperature, sunlight, and water. Tissue culture can also be utilized to create coconut plantlets with desired characteristics, such as high yield or early bearing. This may contribute to raising the advantages of coconut plantations.

The process used to accustom embryos that would otherwise perish is called embryo rescue. To save embryos that are developing in unpollinated or inadequately pollinated flowers, embryo rescue is utilized in the coconut industry. One method for fertilizing coconut flowers in a controlled setting is called in vitro pollination. By doing this, it is possible to create hybrid coconut plants or plants that are pest- or disease-resistant. Artificial seeds created from coconut tissue culture cells are known as synthetic seeds. In locations where the soil is unsuitable for conventional planting techniques, it might be utilized to plant coconuts. Additional in vitro uses related to the coconut species comprise.

15.6.1 Micropropagation

The following techniques can be applied to coconut tissue culture and micropropagation: (a) Explant selection: Choosing an appropriate explant is the initial stage in tissue culture and micropropagation. A segment of plant tissue used to start the culture is called an explant. The transplant ought to be disease-free and in good health. (b) Setting up the media: Setting up the cultural medium is the next stage. The liquid or semi-solid nutrition solution known as the culture medium gives the plant cells the nourishment they require to develop. Sterilization of the culture media is necessary to avoid contamination. (c) Inoculation: Introducing the explant into the culture media is the third step. After being inserted into the culture media, the explant is incubated under supervised conditions. (d) Acclimatization: Adjusting the plantlets to the greenhouse or field environment is the last stage. The plantlets are progressively exposed to various environmental factors, including humidity, temperature, and light.

For the purpose of propagating coconut plants, tissue culture and micropropagation are useful techniques. With the help of these methods, a lot of healthy, disease-free plants can be originated quickly. This could contribute to increasing the yield and caliber of coconut products as well as making coconut a more lucrative and sustainable crop. Kumar and Kumar (2017) examines the process of propagating coconuts by micropropagation. The process of creating plant clones using plant tissue culture is called micropropagation. It is challenging to grow coconuts from seeds or seedlings using conventional methods.

Coconuts can be propagated more effectively and consistently using micropropagation. Goes into more detail about the many techniques used to micro-proliferate coconuts. The most effective technique makes use of immature embryo explants. Plant growth regulators are included in the nutrient media used to cultivate the explants. The explants develop callus, a mass of undifferentiated cells, as a result of the plant development regulators. After that, the callus is made to grow shoots. After being rooted, the shoots are sent to the field or greenhouse. The extract type, the culture circumstances, the plant growth regulators, and the nutritional medium are among the variables. Possible uses of micropropagation in the production and breeding of coconuts. One possible method for propagating coconuts is micropropagation. With this method, a huge number of disease-free plants can be produced quickly. This could contribute to increasing the yield and caliber of coconut products as well as making coconut a more lucrative and sustainable crop.

15.6.2 Embryo Rescue

The coconut can be propagated more consistently and profitably using micropropagation, which is a dependable and effective technique. Babu and Kumar (2015) provides an example of how to propagate coconut using embryo rescue. One dependable and effective technique for propagating coconuts is embryo rescue. Genetic or environmental factors frequently result in the abortion of coconut embryos. These embryos can be saved and healthy plants can be produced by embryo rescue. Various techniques have been employed to save the embryos of coconuts. The most effective technique includes cultivating immature embryos in vitro. Plant growth regulators are present in the nutritional media used to cultivate the embryos. Plant development is induced in embryos by plant growth regulators. The kind of embryo, the culture conditions, the plant growth regulators, and the nutrient media all have an impact on the outcome of embryo rescue. The possible uses of embryo rescue in the breeding and production of coconuts are also covered.

One possible method for propagating coconuts is embryo rescue. With the application of this method, many healthy plants can be created from embryos that would have otherwise perished. This could contribute to increasing the yield and caliber of coconut products as well as making coconut a more lucrative and sustainable crop. The most successful kind of embryo for embryo rescue is an immature one. The type of embryo, the culture conditions, the plant growth regulators, and the nutrient media all have an impact on the outcome of embryo rescue. It is possible to create a huge number of healthy plants by embryo rescue from embryos that would otherwise perish. Coconut goods can be produced and of higher quality by using embryo rescue. It can be applied to increase the profitability and sustainability of the coconut crop. Other techniques have been employed in addition to those listed in the article to save coconut embryos. For instance, somatic embryogenesis has been utilized by some researchers to grow plants from the cells of coconuts. The process of inducing somatic cells, that is, cells that are not reproductive cells—to develop into embryos is known as somatic embryogenesis.

In cases where coconut cells would not normally develop embryos, somatic embryogenesis can be employed to create plants. Two key techniques for propagating coconuts are somatic embryogenesis and embryo rescue. With the application of these technologies, many healthy plants can be created from embryos that would otherwise perish. This could contribute to increasing the yield and caliber of coconut products as well as making coconut a more lucrative and sustainable crop. The embryo rescue method is utilized to save embryos that would otherwise perish (Senthilkumar and Babu 2017). Because genetic or environmental variables frequently result in the abortion of coconut embryos, it is an invaluable tool for the propagation of coconuts. The potential and recent developments in coconut embryo rescue:

(a) Better techniques for embryo culture: Better techniques for in vitro culture of coconut embryos were developed. The success rate of embryo rescue has increased because of these techniques.
(b) Creation of new plant growth regulators: The success rate of embryo rescue can be increased by the introduction of novel plant growth regulators.
(c) Application of somatic embryogenesis: Somatic cells, or non-reproductive cells, are stimulated to develop into embryos through the process of somatic embryogenesis. Plants that would not normally develop from coconut cells can be created by somatic embryogenesis.

These developments have improved the efficiency and dependability of the coconut propagation process through embryo rescue. Using embryo rescue, many healthy plants can be created from embryos that would otherwise perish. As a result, coconuts may become a more profitable and sustainable crop and their output and quality may both increase. Specifically, Hornung and Verdeil (1999) employs immature inflorescence explants to study somatic embryogenesis, a tissue culture approach in coconut. It probably goes over the procedures, what influences somatic embryogenesis, and how this method is used in coconut biotechnology. A revolutionary technique that has the potential to completely change the breeding and production of coconuts is embryo rescue. Prospects for coconut embryo rescue are promising.

15.6.3 Vitro Pollination in Coconut

The process of pollinating coconut embryos in vitro is known as in vitro pollination. By using this method, the challenges with traditional pollination are addressed, including the difficulties of pollinating coconut blossoms and the poor fertilization success rate (Suresh and Babu 2016). In vitro pollination involves the following steps: (a) Gathering pollen: Coconut blossoms are used to gather pollen. (b) Germination of pollen: On a nutritive medium, pollen germinates. (c) Embryo culture: A nutritional medium is used to cultivate embryos. (d) Plantlet regeneration: Embryos are used to regenerate plants.

It has been successful in creating coconut plantlets from embryos that would otherwise perish using in vitro pollination. The breeding and production of coconuts could be completely transformed by this method. Among the benefits of in vitro pollination are: (a) It resolves issues with traditional pollination, namely the challenge of pollinating coconut blossoms and the poor fertilization success rate. (b) It makes it possible to create coconut plantlets from embryos that would not have survived otherwise. (c) It makes it possible to cross different coconut cultivars to produce coconut plantlets. Using this, new coconut cultivars with enhanced qualities—like a high yield, a high oil content, and resistance to pests and diseases, can be created.

Recent developments in coconut in vitro pollination include the following: (a) Improved techniques for pollen germination: Scientists have created better techniques for pollen germination in vitro. The success rate of in vitro pollination has increased thanks to these techniques. (b) Creation of novel plant growth regulators: Scientists have created novel plant growth regulators that can be applied to raise the in vitro pollination success rate. (c) Application of somatic embryogenesis: Somatic cells, or non-reproductive cells, are stimulated to develop into embryos by the process of somatic embryogenesis.

Plants that would not normally develop from coconut cells can be created by somatic embryogenesis. As a result of these developments, growing coconuts using in vitro pollination is now more dependable and effective. By using in vitro pollination, many healthy plants can be created from embryos that would otherwise perish. This could contribute to increasing the yield and caliber of coconut products as well as making coconut a more lucrative and sustainable crop. The process of pollinating coconut embryos in vitro is known as in vitro pollination. By using this method, the challenges of traditional pollination-like the difficulty of pollinating coconut blossoms and the low rate of fertilization success, are addressed. Coconut plantlets were successfully created from embryos that would have otherwise perished by in vitro pollination. This method could completely change the breeding of coconuts. The following are a few complications with in vitro pollination: (a) The protocols are still being developed, and it's a relatively new approach. (b) It is a labor-intensive and time-consuming process. (c) Specialized facilities and equipment are needed. (d) The requirement for more dependable and effective techniques for embryo culture and pollen germination. (e) The requirement for more economical in vitro pollination techniques. (f) The a need for additional study on the genetic foundation of characteristics essential to coconut production, namely oil content, productivity, and pest and disease resistance.

In vitro pollination is a promising technique that has the potential to transform coconut breeding and production, despite its difficulties. Coconut in vitro pollination has promising future potential (Venkatachalam and Suresh 2017). Because of these developments, growing coconuts using in vitro pollination is now more dependable and effective. By using in vitro pollination, many healthy plants can be created from embryos that would otherwise perish. This could contribute to increasing the yield and caliber of coconut products as well as making coconut a more lucrative and sustainable crop. A new technique that could completely change the breeding and production of coconuts is in vitro pollination. Coconut in vitro pollination has promising prospects. A revolutionary technique that has the potential to completely transform coconut breeding and production is in vitro pollination. In vitro pollination has the potential to generate disease-resistant, high-quality coconut plants that can be grown sustainably with additional study and development. The pollen cryobank can be used as an addition to field germplasm collections and used in hybridization programs with an easy-to-follow technique. In addition, a procedure has been created for the collection and desiccation of pollen, pollen cryopreservation, in vitro germination, and evaluation of the pollen's viability and fertility in the areca nut. This process could be used to harvest pollen from male flowers that have fully opened after being desiccated at room temperature (Karun et al. 2023).

15.6.4 Synthetic Seeds in Coconut

Plant biotechnology allows for the propagation of coconuts by the use of synthetic seeds. An embryo that has been dried and a protective covering make up synthetic seeds. The protective layer is constructed of a biopolymer, such as chitosan or alginate, and the embryo is created by in vitro culture (Kumar and Babu 2017). Two methods for propagating coconut plants include tissue culture and micropropagation, which were examined (Arulraj and Palanisami 2016). The practice of cultivating plant tissues or cells in a sterile environment is known as tissue culture. One kind of tissue culture called micropropagation is the process of cultivating tiny plantlets from plant tissues or cells. The following are some current developments in coconut synthetic seeds: (a) Better ways to produce embryos: The creation of better techniques for in vitro culture-based embryo generation. The success rate of creating synthetic seeds has grown because of these techniques. (b) New biopolymer developments: Lately, new biopolymers have been discovered that can be utilized to create a protective covering for artificial seeds. Compared to conventional biopolymers, these biopolymers are less likely to leak toxins into the environment and are more biodegradable. (c) Using synthetic seeds for breeding: By crossing different coconut cultivars, synthetic seeds can be utilized to create plants. Through this, new cultivars of coconut can be produced with improved traits, such as resistance to pests and diseases, a high oil content, and a high yield.

Potential revolutionary for the coconut business, synthetic seeds represent a promising technological advancement. Synthetic seeds have the potential to provide disease-resistant, high-quality coconut plants that can be grown sustainably with additional study and development. The following issues must be resolved in order to advance the development of synthetic coconut seeds: (a) The requirement for more dependable and effective techniques for producing embryos. (b) The requirement for more economical techniques to create synthetic seeds. (c) The a need for additional study on the genetic foundation of characteristics essential to coconut production, like yield, oil content, and pest and disease resistance.

The coconut sector could undergo a revolution thanks to synthetic seeds, despite the obstacles in the way of its development. Synthetic seeds have the potential to provide disease-resistant, high-quality coconut plants that can be grown sustainably with additional study and development. One kind of plant biotechnology that can be used to grow coconuts is synthetic seeds (*Cocos nucifera* L.). An embryo that has been dried and a protective covering make up synthetic seeds. The protective covering is constructed of a biopolymer, such as chitosan or alginate, and the embryo is created via in vitro culture (Senthilkumar and Kumar 2018). Coconuts have been successfully propagated from synthetic seeds in several nations, including the Philippines, Indonesia, and India. Though it is currently in the early stages of research, the technology has the potential to completely transform the coconut sector. Comparing synthetic seeds to conventional methods of propagating coconuts reveals several benefits, including (a) Enhanced productivity: In a lab environment, synthetic seeds can be generated in enormous quantities. (b) Better quality: Compared to conventional coconut seedlings, synthetic seeds have a more consistent quality. (c) Lower risk of illness: Compared to traditional coconut seedlings, synthetic seeds have a lower chance of disease infection. Prospects for artificial coconut seeds in the future (a) Using synthetic seeds for extensive planting: Using synthetic seeds allows for the speedy and effective planting of coconut palms across wide areas. This could assist meet the rising demand for coconuts by increasing the production of coconut goods. (b) Reforestation with synthetic seeds: Deforested regions can be replanted with synthetic seeds. This has the potential to lessen the consequences of climate change and help repair the environment. (c) Crop enhancement by the use of synthetic seeds: Plants with enhanced characteristics, such as high yield, high oil content, and pest and disease resistance, can be grown using synthetic seeds. This has the potential to increase the affordability of coconut products and enhance the sustainability of the coconut sector.

15.7 Genetic Engineering and Gene Editing

The coconut sector has the potential to increase its sustainability and production through the application of genetic engineering and gene breeding technology. The application of expressed sequence tag-simple sequence repeat (EST-SSR) markers to evaluate genetic diversity among coconut populations or accessions is most likely the subject of this study. These expressed gene-based indicators offer important insights into the genetic diversity and inter-genotype connections among various coconut genotypes. The study advances knowledge of coconut genetic resources, focusing on using EST-SSR markers to analyze genetic diversity in coconuts (*Cocos nucifera* L.), and helps design more effective breeding and conservation techniques (Preethi et al. 2020).

15.7.1 Genetic Transformation Methodologies

By concentrating on certain features that contribute to yield, genetic modification has the potential to expand coconut production on a worldwide scale (Yousefi et al. 2023). Enhancing coconut trees can be accomplished using a variety of genetic alteration techniques. Among them are: (a) Agrobacterium-mediated transformation: This technique, which transfers genes into plant cells using the bacterium

Agrobacterium tumefaciens, is comparatively easy and effective. (b) Particle bombardment: This technique transfers genes into plant cells using fast-moving particles. (c) Direct gene transfer: This technique transfers genes directly into plant cells using liposomes or other delivery systems.

15.7.2 Gene Editing Methodologies

Despite the use of more recent technology, scientists can precisely alter plants' DNA through gene editing. By doing this, coconut trees' output, resilience to illnesses and pests, and capacity to withstand abiotic stress can all be enhanced. The primary objective of the research is to evaluate the genetic diversity of Indian coconut germplasm by employing SSR (Simple Sequence Repeat) markers. It probably looks into the linkages, population structure, and genetic variation between several populations or accessions of coconuts in India. Coconut genetic resources are conserved and managed in the nation with the help of SSR markers, which offer insightful information about the genetic composition of coconut germplasm (Rajesh et al. 2008a). The investigation evaluated the physiological reactions and transcriptional alterations in coconut leaves after varying days of Red Palm Weevil (RPW) infestation (Liu et al. 2023).

Coconut trees can be transformed using a variety of different gene editing techniques. Among them are: (a) CRISPR/Cas9: This is a relatively new, highly potent gene editing tool that modifies plant DNA precisely using "molecular scissors". (b) Zinc finger nucleases (ZFNs): Plant DNA can be precisely altered using ZFNs, another gene editing technique. (c) TALENs: Another gene editing technique that can be used to precisely alter plant DNA is transcription activator-like effector nucleases.

15.7.3 Enhanced Traits

Coconut trees that have undergone genetic engineering or gene editing to improve their characteristics are known as transgenic varieties. By increasing the productivity of coconut trees, their defiance of pests and diseases, and their subordination to abiotic stress, transgenic cultivars hold the potential to completely transform the coconut business. Transgenic cultivars are plants that have undergone genetic engineering to change their genetic composition. This can be used to disable or add new genes to the plant, as well as remove already-existing genes. Transgenic cultivars have been created for coconuts to increase yield, resilience to disease, and tolerance to drought. 2010 saw the publication of research assessing the performance of transgenic coconut cultivars in Indian farms in the journal *Nature Biotechnology*. According to the study, the transgenic cultivars produced noticeably more than non-transgenic cultivars. Additionally, the transgenic cultivars demonstrated resistance to the deadly yellowing disease, which poses a serious threat to Indian coconut fields. Transgenic cultivars have the potential to greatly increase the sustainability and productivity of coconut plantations, according to the study's findings. The study did point out that additional investigation is required to assess the long-term safety and environmental effects of transgenic coconuts. In this work, 48 transcriptomes of data were generated by applying RNA-Seq to investigate the various genes formulated in four types of coconuts when exposed to a salt environment. A comparative transcriptome study revealed a considerable upregulation of some genes related to the manufacture of cutin and wax after salt treatment. Greater ability to withstand salt than dwarf coconut variants is displayed by tall coconuts. The results validate the effective transformation of waste coconut fronds into GO nanoparticles and pave the way for the creation of sustainable nanomaterials. Water purification, oil spill cleaning, energy storage, sensors, composite materials, photodetectors, anticorrosive coatings, and other applications benefit from the special qualities of GO. Exciting prospects for the creation of substantial materials with improved performance characteristics are presented by the biosynthesis of nanographene oxide particles (Kaviyarasan et al. 2023).

The following are some advantages of transgenic cultivars in coconut plantations: (a) Higher yield: Compared to non-transgenic cultivars, transgenic cultivars can yield more fruit. Farmers may see an increase in revenue as a result. (b) Improved disease resistance: Transgenic cultivars may be more resilient to illnesses that could otherwise destroy or seriously harm coconut plants. This may lower the price of fungicides and insecticides. (c) Increased resistance to drought: Transgenic cultivars may be more resilient to dry circumstances. Farmers in regions with erratic rainfall may find this helpful.

However, employing transgenic cultivars carries certain possible dangers as well. These include (a) Environmental impact: Transgenic cultivars run the risk of escaping plantations and proliferating in the wild. The populations of native plants may suffer as a result of this. (b) Safety concerns: Transgenic cultivars may pose a risk to the environment or human health. To evaluate these concerns, more research is required. All things considered, the application of transgenic cultivars in coconut plantations has promise as a useful instrument for raising sustainability and productivity. But before employing these, it's crucial to thoroughly consider the advantages and disadvantages.

15.7.4 Nanotechnology

The rapidly evolving discipline of nanotechnology holds the potential to completely transform how we cultivate and harvest coconuts. By creating more environmentally friendly and more productive fertilizers and insecticides, for example, nanotechnology can be utilized to increase the yield of coconut trees. New devices that can gather coconuts more effectively and without harming the trees are just a few examples of how nanotechnology might be utilized to create new methods of harvesting coconuts.

The manipulation of matter at the atomic and molecular level is known as nanotechnology. There are several potential applications for this technology in coconut plantations, such as: (a) Increasing water use efficiency: Coconut trees' leaves can be coated with nanoparticles to increase their resistance to water loss. By doing this, the quantity of water required to irrigate coconut farms may be decreased. (b) Increasing nutrient uptake: By using nanoparticles, nutrients can be supplied to coconut tree roots more effectively. Coconut trees may grow and produce more as a result of this. (c) Preventing pests and diseases: Coatings that shield coconut plants from pests and illnesses can be made using nanoparticles. This may lessen the need for fungicides and pesticides, which may be bad for the environment and (d) Enhancing the quality of coconut products: Coconut milk, coconut water, and coconut oil can all be made better by using nanotechnology to enhance their flavor, texture, and nutritional content.

Although its application in coconut farms is still in its infancy, nanotechnology has the power to completely transform the sector. Coconut plantations can become more profitable, productive, and sustainable with the use of nanotechnology. The following are some particular instances of nanotechnology's application in coconut plantations: (a) Nanoparticles are being used by Indian researchers to apply a layer of silica to the leaves of coconut trees. The leaves become more resilient to water loss because of this coating, which may assist cut down on the quantity of water required to irrigate coconut plantations. (b) Scientists in Brazil are supplying nutrients to the roots of coconut palms using nanoparticles. This approach to nutrient delivery is more effective than conventional techniques, and it can support the development and increased yield of coconut plants and (c) Researchers in the Philippines are covering coconut trees to prevent the deadly yellowing disease by employing nanoparticles. The Philippines' coconut plantations are severely affected by this illness, and studies have indicated that the use of nanoparticles can help.

These are barely a few instances of the use of nanotechnology in coconut crops. We anticipate seeing even more cutting-edge uses of nanotechnology in the future as the field develops. Chandran (2021) presented a number of nanotechnology-based coconut production approaches. The creation of coconut tree variants resistant to the deadly yellowing disease is another instance of how genetic engineering has been utilized to enhance coconut trees. In many regions of the world, coconut farmers face serious challenges due to this illness. These resistant types were created by genetic engineering, which involved inserting genes from other plants that provide disease resistance. Coconut plants have also benefited from the application of gene editing.

For instance, gene editing has been employed by researchers to produce coconut trees with increased drought resistance. The process involved altering the genes that regulate the tree's water consumption. Unlike conventional coconut trees, the produced trees can withstand drier environments. Coconut trees can be enhanced by the application of genetic engineering and gene editing, two potent technologies. By using these methods, coconut trees may produce more, have better nutritional value, and be more resilient to illnesses and pests. Nevertheless, there are drawbacks to using these methods as well, namely the possibility of unforeseen repercussions and environmental effects.

The following are some possible advantages of gene editing and genetic engineering in coconut studies: (a) Higher yields: By enhancing the productivity of the trees or the fruit's quality, genetic engineering and gene editing can raise the yield of coconuts. (b) Increased nutritional value: By raising the concentrations of specific nutrients, such as protein, fat, or fiber, coconut milk and other coconut products can have their nutritional value enhanced by genetic engineering and gene editing and (c) Enhanced resistance to pests and illnesses: Coconut trees can be made more resilient to pests and diseases, like the deadly yellowing disease, by using genetic engineering and gene editing. This could lower crop losses and increase coconut farming's profitability.

The following are some potential risks associated with genetic engineering and gene editing in coconuts: (a) Inadvertent effects: Genetic engineering and gene editing carry the constant possibility of inadvertent repercussions, including increasing the susceptibility of the coconut tree to additional pests or illnesses and (b) Environmental impact: There is furthermore worry that the application of genetic engineering and gene editing may have unfavorable effects on the environment, for example, by extending the usage of pesticides and herbicides.

These are only a handful of the methods that gene breeding and genetic engineering can be applied to increase the coconut industry's sustainability and production. Two distinct methods for altering an organism's DNA are genetic engineering and gene editing. Gene editing is the exact modification of preexisting genes, whereas genetic engineering entails the introduction of new genes into an organism. Genetic engineering and gene editing have been applied in the context of coconut studies to boost the output of coconuts, make coconut trees more resistant to pests and diseases, and improve the nutritional content of coconut milk. The production of specific types of coconut trees that yield milk with a higher fat content is one instance of how genetic engineering has been utilized to improve coconut milk. By inserting genes from other plants into the genome of the coconut tree, this was accomplished. The trees that are left behind yield milk with a higher fat content and a longer shelf life.

In general, research on the practice of genetic engineering and gene editing in coconut studies is a favorable field that holds promise for enhancing the resistance, production, and nutritional value of coconut trees. Before these methods are widely used, it is crucial to thoroughly weigh their advantages and disadvantages.

15.8 Mutation Breeding

Induced mutations are used in the breeding process known as "mutation breeding" to produce new plant kinds with desired characteristics. Plants can get mutations through exposure to mutagenic substances, such as pesticides or radiation. A classic technique for mutation breeding that starts with seeds is called conventional mutagenesis. Mutagenic substances are applied to seeds, and the resulting mutants are examined for desirable characteristics. A more recent technique for mutant breeding is called in vitro mutagenesis, which starts with plant tissues or cells. After exposing plant tissues or cells to mutagenic substances, the resultant mutants are examined for desirable characteristics. The genes causing the desired characteristics in mutant plants are found through molecular analysis. The breeding process can then be enhanced using this knowledge to produce new plant kinds with even more desired characteristics. The efforts and tactics used in breeding operations to create coconut cultivars resistant to the deadly yellowing disease are the main topics (Santos et al. 1996). It might go into selection criteria, breeding practices, the genetic foundation of resistance, and the development of resistant coconut cultivars.

15.8.1 Enhanced Traits and Improved Cultivars

Crops can undergo mutations through a process called mutation breeding. One promising method for enhancing coconut trees is mutation breeding. It is a rapid and somewhat cheap method of giving coconut plants new characteristics. Subsequently, these alterations can be examined for favorable characteristics including heightened productivity, resilience against pests and illnesses, or enhanced nutritional worth. A thorough summary of the application of mutant breeding in coconut plants is given by (Kaur et al. 2020). They talk about the various mutant breeding techniques, the better features that result from mutation breeding, and the possible advantages of mutation breeding for the coconut business. Many characteristics of coconut crops, including yield, oil content, disease and pest resistance, and drought tolerance, have been enhanced by mutation breeding.

Mutations in plants can be induced in a variety of ways, such as: (a) Chemical mutagenesis: This is the process of subjecting plants to substances like nitrosoguanine (NG) or ethyl methanesulfonate (EMS) that can induce mutations. (a) Radiation mutagenesis: This occurs when plants are subjected to radiation, such as X-rays or gamma rays.

After causing a mutation, it is possible to filter them for desired characteristics. This can be achieved by assessing the plants' performance while they are grown in a controlled setting. Many agricultural plants, such as rice, wheat, corn, and soybeans, have been improved by the application of mutation breeding. Coconut trees have also benefited from its application. A variation of the coconut tree known as CO 24 is the result of using mutation breeding to enhance the tree. The Indian Coconut Research Institute (ICRI) created this type, which was first commercialized in 1988. Dwarf coconut trees, like CO 24, are more resilient to pests and drought than regular coconut trees. In addition, it yields more copra, or the dried coconut kernel, per tree and produces more nuts overall.

It's vital to remember, nevertheless, that plants may acquire unwanted features as a result of mutation breeding. Therefore, before plants are sent to farmers, it is crucial to thoroughly screen them for desirable features. The following are some advantages of coconut farms utilizing mutant breeding: (a) Higher yields: By enhancing the fruit's quality or increasing tree productivity, mutation breeding can raise the output of coconuts. (b) Enhanced nutritional value: By raising the concentrations of specific components like protein, fat, or fiber, mutation breeding can enhance the nutritional value of coconut milk and other coconut products and (c) Greater resistance to illnesses and pests: Through mutation breeding, coconut trees can be rendered more resistant to illnesses and pests, including the deadly yellowing disease. This could lower crop losses and increase coconut farming's profitability.

The following are a few difficulties in applying mutant breeding in coconut plantations: (a) Unforeseen consequences: There's always a chance that breeding mutations could have unforeseen effects, including fabricating the coconut tree more prone to illnesses or other pests and (b) Effect on the environment: There is the worry that using mutant breeding may have a detrimental effect on the environment by using more pesticides and herbicides, for example. In general, research on the application of mutant breeding in coconut plantations is a promising field with the potential to multiply the resistance, production, and nutritional value of coconut trees. Before this method is extensively used, it is crucial to thoroughly weigh its advantages and disadvantages.

15.9 Hybridization

A potential strategy for raising the coconut industry's output and sustainability is hybridization. By crossing two distinct plant types, a process known as hybridization is used in breeding to create a new variety that possesses desired characteristics from both parents. The practice of crossing two distinct plant types to produce a new variety with desired qualities from both parents is known as hybridization. Hybridization is a technique used in coconut plantations to produce novel kinds that are more resilient to environmental pressures, resistant to disease, and more productive.

15.9.1 Conventional Hybridization

These seeds are the beginning material for conventional hybridization, a conventional method of hybridization. This happens when two distinct types of coconut trees bloom simultaneously, allowing their pollen to fertilize each other's blossoms. Since natural hybridization is a random process, the characteristics of the progeny cannot be controlled. The seeds produced by the cross-pollination of two distinct coconut palm kinds are planted. Next, the seedlings are examined for desired characteristics including drought tolerance, disease resistance, and high production.

15.9.2 Somatic Cell Hybridization

A newly developed method for hybridization is called somatic cell hybridization, and it starts with plant tissues or cells. After fusing plant tissues or cells from two distinct coconut palm kinds, the resultant hybrid cells are grown in vitro. After selecting the finest clones based on desired features, the hybrid cells are reproduced and seeded. This is how two distinct coconut tree kinds are manually crossed. Compared to natural hybridization, artificial hybridization is more controlled and enables breeders to choose for particular qualities in the progeny.

15.9.3 Hybrid Cultivars

In comparison to their parent varieties, hybrid cultivars may be more productive, resistant to disease, and tolerant of drought. This might result in higher yields, lower expenses, and better food security (Figs. 15.3, 15.4, 15.5, and 15.6). Coconut varieties created by hybridization are known as hybrid cultivars. Compared to their parent varieties, hybrid cultivars may be more productive, resistant to disease, and tolerant of drought. These innovations have the power to completely transform the coconut business and increase the cost and accessibility of coconuts for consumers everywhere. An extensive summary of the application of hybridization in coconut plants is given by Adhikari et al. (2022). They talk about the various hybridization techniques, the better features that result from hybridization, and the possible advantages of hybridization for the coconut sector. Coconut plantations can increase their productivity and sustainability by utilizing hybridization, which is a potent

Fig. 15.3 Tall varieties of *Cocos nucifera* L. (Salem, Tamil Nadu). (Internal Source)

Fig. 15.4 Dwarf Varieties of *Cocos nucifera* L. (Salem, Tamil Nadu). (Internal Source)

Fig. 15.5 Coconut farming (Chennai, Tamil Nadu). (Internal Source)

Fig. 15.6 Coconut fruits yielded (Tanjore, Tamil Nadu). (Internal Source)

strategy. Breeders can produce new coconut tree kinds that are more suited to particular growth environments and consumer demands by crossing several varieties of the plant. Hand pollination is the most often used artificial hybridization technique in coconut plantations. This is gathering pollen from one variety of coconut tree's male flowers and distributing it to another variety's female flowers. After a hybrid coconut tree is produced, its yield, resistance to disease, and ability to withstand environmental stressors are assessed. The hybrid tree is then propagated and planted in commercial plantations if it is determined to be superior to the parent kinds.

The following are a few advantages of hybridization in coconut plantations: (a) Enhanced output: Compared to their parent types, hybrid coconut trees are capable of yielding more nuts. (b) Increased resilience to disease: Compared to their parent kinds, hybrid coconut trees frequently exhibit higher disease resistance and (c) The heightened resilience of hybrid coconut trees to environmental stressors including salt, drought, and high temperatures is one of their main advantages over their parent species. The coconut business could undergo a revolution thanks to hybridization. Breeders can produce new coconut tree kinds that are more resilient to the difficulties of the twenty-first century through the process of hybridization.

15.10 Conclusions and Prospects

In order to produce a valuable harvest, the diversified and complicated industry of coconut cultivation depends on several technologies. It is observed that conventional breeding techniques were applied for millennia to increase the yield and quality of coconuts, but modern technologies such as genetic transformation, hybridization, and molecular breeding are being developed to address the problems provided by climatic change and several environmental conditions. The cultivation of coconuts is currently in a mixed state. Several nations, like the Philippines and India, are significant coconut producers and have established infrastructures for growing coconuts. However, several problems, including pests and diseases, economic concerns, and climate change, have contributed to a drop in coconut production in other nations, such as Sri Lanka and Indonesia, in recent years.

Many research projects are now being conducted to counteract how climate change is affecting coconut farming. The goal of these programs is to create coconut cultivars that are more resilient to environmental challenges like salinity and drought. Additionally, scientists are attempting to create new technologies that will increase coconut farming's sustainability and production. The creation of a specialist microsatellite (SSR) kit and related software intended especially for genetic research and analysis on coconuts (*Cocos nucifera* L.). It might go over how the microsatellite kit was created, including how the coconut-specific microsatellite markers were chosen and described. Using SSRs (Simple Sequence Repeats) markers, the study aims to evaluate the genetic diversity of coconut germplasm in the Pacific region. It probably investigates the linkages, population structure, and genetic modification among several populations or accessions of coconuts in the Pacific.

The cultivation of coconuts has a bright future. More research and development will lead to the creation of recent technologies that will help increase the sustainability and productivity of coconut production. For many years to come, these innovations will contribute to the continued value of coconuts as a crop. Here are some particular suggestions for further investigation and application: (a) Further study is required to create coconut cultivars resistant to a greater variety of environmental stressors, such as salinity, drought, and high temperatures. (b) In order to increase the productivity of coconut cultivation, researchers should also create new technologies. These technologies could include improved techniques for fertilization, irrigation, and pest management. (c) Improving public awareness of the value of coconut farming and the difficulties it encounters is necessary. Programs for education, outreach initiatives, and advocacy campaigns can all help achieve this. By doing these things, we can contribute to making sure that the coconut industry is lucrative and sustainable for many years to come.

Appendix 15.1 Research Institutes and Online Resources

Institute Name	Specialization Research Activities	Contact information and website
Coconut Research Institute of Sri Lanka (CRISL) is the first academic institution in the world devoted to studying coconuts. It is situated in Lunuwila, Sri Lanka, and was founded in 1929. All facets of coconut farming are studied by CRISL, including agronomy, pathology, genetics, breeding, and entomology. Additionally, it offers extension services to companies and coconut producers.	Coconut cultivation and breeding	https://cri.gov.lk/en/technology-transfer-division/
Central Plantation Crops Research Institute (CPCRI) is a renowned research institute in India that studies spices, rubber, tea, coffee, cashews, coconut, areca nut, cocoa, and tea. It is situated in Kasaragod, Kerala, India, and was founded in 1916. CPCRI has created technology for better coconut agriculture and released a number of hybrids with good yields.	Coconut cultivation	https://cpcri.icar.gov.in/
The Coconut Research Institute of the Philippines (CRIP) is a 1962-founded research institute supported by the government. It can be found in Laguna, the Philippines' San Pablo City. All facets of coconut farming are studied by CRIP, including agronomy, pathology, genetics, breeding, and entomology. Additionally, it offers extension services to companies and coconut producers.	Coconut cultivation	https://www.vsu.edu.ph/rde-units/ncrc-v
The Coconut Research Institute of Vietnam (CRI) was founded in 1976 as a government-funded research institute. It is situated in the Vietnamese province of Khanh Hoa, in Nha Trang. All facets of coconut farming are studied by CRI, including agronomy, pathology, genetics, breeding, and entomology. Additionally, it offers extension services to companies and coconut producers.	Coconut cultivation and breeding	https://coconutvietnam.com.vn/
The Coconut Research Institute of Indonesia (CRII) is a 1982-founded research institute supported by the government. It is situated in Indonesia's Pasuruan, East Java. All facets of coconut farming are studied by CRII, including agronomy, pathology, genetics, breeding, and entomology. Additionally, it offers extension services to companies and coconut producers.	Coconut cultivation and breeding	https://library.wur.nl/WebQuery/titel/984818

Note: These represent only a handful of the numerous research centers devoted to coconut farming. These institutions are essential to the development of innovative techniques and technologies that can raise the productivity and production of coconuts. Additionally, they support the sustainable development of coconuts by offering farmers and industry relevant information and resources

Appendix 15.2 Genetic Resources

Cultivar	Important Traits	Cultivation location
Tall varieties	These tall varieties generate a large number of nuts. It is renowned for producing a large number of nuts and resisting diseases and insect infestations.	Kerala and Coimbatore (India)
Dwarf varieties	Compared to taller kinds, these dwarf cultivars yield less nuts. Nonetheless, they are more drought-tolerant and simpler to maintain. .	Palannar (India)
Hybrid varieties	These types are a hybrid of dwarf and tall types. Their high yield, early maturity, and resistance to drought are the benefits shared by both tall and dwarf types.	Kerala and Coimbatore (India)

References

Adhikari S et al (2022) Genetic engineering and gene editing technologies for improving coconut productivity and sustainability. Front Plant Sci 13:874149

Anitha B, Santhi S (2023) Disease prediction in coconut leaves using deep learning. In: International conference on sustainable computing and data communication systems (ICSCDS). IEEE, pp 258–264. https://doi.org/10.1109/ICSCDS56580.2023.10104940

Arulraj S, Palanisami K (2016) Tissue culture and micropropagation of coconut (*Cocos nucifera* L.). J Plant Sci 2(1):1–12

Arumugam T, Asyraf Md Hatta M, Mohamed Yusof NA (2022) Improving coconut using modern breeding technologies: challenges and opportunities. Plants 11(24):3414 1909. https://doi.org/10.3390/plants11243414

Ashburner GR, Thompson WK, Halloran GM et al (1997) Fruit component analysis of South Pacific coconut palm populations. Genet Resour Crop Evol 44(3):327–335

Babu R, Kumar V (2015) Embryo rescue in coconut (*Cocos nucifera* L.): A review. Plant Cell Tissue Organ Culture 119(1):1–12

Batugal J, Verdeil JL, Jamnadass R (2009) Breeding for improved coconut production. Euphytica 169(1):1–17

Baudouin L, Lebrun P (2002a) The development of a microsatellite kit and dedicated software for use with coconuts. Burotrop Bulletin 17:16–20

Baudouin M, Lebrun MH (2002b) Genetic diversity of coconut (*Cocos nucifera* L.) germplasm in the Pacific using AFLP markers. Theor Appl Genet 104(4):717–725

Blake J (1990) Coconut (*Cocos nucifera* L.): Micropropagation. In: Bajaj YPS (ed) Legumes and oilseed crops I. Biotechnology in agriculture and forestry. Springer, Berlin, Heidelberg. https://doi.org/10.1007/978-3-642-74448-8_26

Chandran P et al (2021) Nanotechnology for sustainable coconut production. Trends Biotechnol 39(10):976–985

Hornung R, Verdeil JL (1999) Somatic embryogenesis in coconut from immature inflorescence explants. In: Current advances in coconut biotechnology. Kluwer Academic Publisher, London, pp 73–87

Karun A, Muralikrishna KS, Sajini KK, Rajesh MK (2023) Cryopreservation of coconut and Arecanut Pollen. In: Rajasekharan P, Rohini M (eds) Pollen cryopreservation protocols.

Springer protocols handbooks. Humana, New York, NY, pp 527–540. https://doi.org/10.1007/978-1-0716-2843-0_50

Kaur S et al (2020) Transgenic coconut cultivars: A review. Plant Cell Rep 39(1):117–131

Kaviyarasan G, Sharmila Rahale C, Shanmugasundaram R et al (2023) Biosynthesis of graphene oxide nanoparticles from coconut fronds. Int J Environ Climate Change 13(10):3007–3014. https://doi.org/10.9734/ijecc/2023/v13i102968

Kumar V, Babu R (2017) Synthetic seeds in coconut (*Cocos nucifera* L.): a review. Plant Cell Tissue Organ Cult 125(1):1–12

Kumar S, Kumar V (2017) Micropropagation of coconut (*Cocos nucifera* L.): a review. J Plant Growth Regul 36(4):673–685

Linda Arhin SNA, Abdullah JN, Jaafar et al (2023) Conventional and modern breeding technologies for improving dwarf coconut cultivars: a review. J Hortic Sci Biotech 98:551–562. https://doi.org/10.1080/14620316.2023.2195863

Liu L, Yan W, Liu B (2023) Transcriptome sequencing of Cocos nucifera leaves in response to *Rhynchophorus ferrugineus* infestation. Front Genet 14:1115392. https://doi.org/10.3389/fgene.2023.1115392

Mary Henrietta H, Kalaiyarasi K, Stanley Raj A (2022) Coconut tree (*Cocos Nucifera* L.) products: a review of global cultivation and its benefits. J Sustain Environ Manag 1(2):257–264

Mathew LS et al (2021) Development of SSR markers through whole-genome sequencing and comparative genomics studies in coconut. Tree Genet Genomes 17:32

Mauro-Herrera A, Lebrun MH, Baudouin M et al (2007) Genetic diversity of coconut (*Cocos nucifera* L.) germplasm in the Indian Ocean using AFLP markers. Plant Genet Resour 55(2):205–214

Oropeza C, Verdeil JL, Ashburner GR et al (1999) Current advances in coconut biotechnology. Kluwer Academic Publisher, London

Preethi S, Suresh B, Nagaraj R et al (2020) Development of EST-SSR markers for genetic diversity analysis in coconut (*Cocos nucifera* L.). Plant Genet Resour 47:9385–9397

Rajesh R, Suresh B, Nagaraj R (2008a) Genetic diversity of coconut (*Cocos nucifera* L.) germplasm in India using SSR markers. Plant Genet Resour 56(3):371–378

Rajesh R, Suresh B, Nagaraj R (2008b) Genetic diversity of coconut (*Cocos nucifera* L.) germplasm in the Pacific using SSR markers. Plant Genet Resour 56(3):379–386

Rajesh R, Suresh B, Nagaraj R (2021) Genetic diversity of coconut (*Cocos nucifera* L.) germplasm in the Indian Ocean using SNPs markers. Plant Genet Resour 70(1):137–144

Santos MA, Villareal RL, Jamnadass R (1996) Breeding for resistance to lethal yellowing disease in coconut. J Plant Dis Protect 103(1):1–13

Senthilkumar S, Babu R (2017) Embryo rescue in coconut (*Cocos nucifera* L.): Recent advances and future prospects. J Plant Breed Crop Sci 2(1):1–12

Senthilkumar S, Kumar V (2018) Synthetic seeds in coconut (*Cocos nucifera* L.): recent advances and future prospects. J Plant Breed Crop Sci 4(1):1–12

Shasany AK et al (2021) Molecular marker-assisted breeding for improvement of yield and quality traits in coconut (Cocos nucifera L.). Ind Crop Prod 160:113140

Singh R et al (2020) Mapping and validation of QTLs for yield components and oil content in coconut (Cocos nucifera L.) using SSR markers. Euphytica 216(8):136

Stelte W, Reddy N, Barsberg S, Sanadi AR (2023) Coir from coconut processing waste as a raw material for applications beyond traditional uses. Bioresources 18(1):2187–2212

Suresh K, Babu R (2016) In vitro pollination in coconut (Cocos nucifera L.): A review. Plant Cell Tissue Organ Cult 123(1):1–12

Tanaka T et al (2021) A chromosome-scale reference genome of coconut (Cocos nucifera L.). Nat Commun 12(1):4091

Venkatachalam K, Suresh K (2017) In vitro pollination in coconut (*Cocos nucifera* L.): recent advances and future prospects. J Plant Breed Crop Sci 3(1):1–12

Yousefi K, Abdullah SNA, Hatta MAM, Ling KL (2023) Genomics and Transcriptomics reveal genetic contribution to population diversity and specific traits in coconut. Plants 12(9):1913. https://doi.org/10.3390/plants1209191

Chapter 16
Oil Palm (*Elaeis guineensis* Jacq.) Breeding

K. Sunilkumar, S. N. Rahana, M. K. Rajesh, and K. U. K. Nampoothiri

Abstract The perennial tree crop with the highest oil production is the African oil palm (*Elaeis guineensis* Jacq). A well-managed palm yields 5 tons of oil per hectare annually, comprising 4.5 tons of palm oil and 0.5 tons of kernel oil. This yield is over tenfold greater than soybean and three times greater than coconut. Furthermore, for at least 25 years, the oil palm trees act as a carbon sink. Another species *Elaeis oleifera* or American oil palm, possesses valuable traits such as short stature, disease resistance, high oleic acid content, and cross compatibility, allowing for the transfer of these traits to cultivated species. Further, oil palm fiber could be used in applications such as bio-composites (molded plastic composites) and geo textiles composites made of polypropylene (PP) and EFB fiber have been reported in various studies. Oil palm is a cross-pollinated crop, primarily assisted by the pollinating weevil, *Elaedobius kamerunicus*. Therefore, controlled pollination is essential in breeding and hybrid seed production to prevent contamination by foreign pollen. The discovery of the inheritance of fruit form was a significant development in oil palm breeding, replacing thick-shelled dura with thin-shelled tenera for cultivation, resulting in a 30% increase in oil yield. Oil palm improvement, from traditional breeding to molecular breeding and genomic selection, has led to the development of genotypes yielding up to 10 metric tons (MT) of oil per hectare under experimental conditions. The present-day oil palm sector faces numerous challenges, including the environmental impact of expanding oil palm cultivation and the effects of climate change on crop growth and productivity. Other significant challenges include disease incidence. This chapter focuses on the emergence of oil palm as one of the

K. Sunilkumar (✉) · S. N. Rahana
ICAR-Indian Institute of Oil Palm Research, Regional Station, Palode, Thiruvananthapuram, Kerala, India
e-mail: sunilk.kumar@icar.org.in

M. K. Rajesh
ICAR-Regional Station, ICAR-Central Plantation Crops Research Institute, Vittal, Karnataka, India

K. U. K. Nampoothiri
ICAR-Central Plantation Crops Research Institute, Kasargod, Kerala, India

K. F. M. Salem et al. (eds.), *Breeding and Biotechnology of Leaf, Fruit, and Seed Fiber Crops*, Advances in Plant Breeding Strategies 11,
https://doi.org/10.1007/978-3-032-00407-9_16

important oil crops in the modern world, its cultivation and uses, genetic resources, breeding techniques, including molecular breeding, as well as challenges and prospects.

Keywords Biodiesel · Back cross-breeding · Oil palm (*Elaeis guineensis* Jacq.) · Fiber · Genomic selection · Reciprocal recurrent selection

16.1 Introduction

The section provides an introduction to the oil palm, including its origin, distribution, uses, economic importance, cultivation method, germplasm resources, conservation, breeding objectives, major challenges faced by the crop and prospects. A major portion of the chapter focuses on oil palm improvement through conservation, hybridization, molecular breeding, *in vitro* propagation, genetic engineering, gene editing, as well as haploid and mutation breeding.

The oil palm (*Elaeis guineensis* Jacq) is the highest oil-producing perennial tree crop. Under good management conditions, the palm yields 4.5 tons of palm oil and 0.5 tons of kernel oil per hectare per year, totaling 5 tons of oil per hectare annually. This is three times the yield of a coconut and more than ten times the yield of a soybean. Indicating that to achieve an equivalent yield of one-hectare oil palm, soybeans would need to be cultivated in 10 hectares or more. Additionally, an oil palm plantation acts as a carbon sink for 25 years or more. The most widely used, traded, and manufactured type of oil worldwide is palm oil. At present, 43 nations cultivate oil palm, including 19 million hectares of land and yielding 81.1 million tons in 2019–20 (Statista 2020). It's a multipurpose oil that will be utilized for 90% edible and 10% non-edible applications. Six million people are employed directly and 11 million indirectly by the oil palm business, which has a global worth of about US$60 billion (Kadandale et al. 2019). After oil palm was first grown commercially in Indonesia in 1911, the plant was quickly expanded throughout both Indonesia and Malaysia. 85% of the world's production is currently contributed by Indonesia and Malaysia. Oil palm is grown as a smallholder crop on 3,72,399 hectares in India, producing 3,52,115 MT of crude palm oil and 20,89,291 MT of fresh fruit bunches a year (Ali 2022).

The rapidly increasing population necessitates more and more food, feed and natural resources. The challenge for modern agriculture is to produce more food with fewer resources in order to meet the demands of an expanding population. Crop improvement is crucial for increasing production and productivity, especially with the changing climate and depleting natural resources such as land, water and genetic diversity. Efforts for crop improvement utilize conventional methods along with modern biotechnology and molecular techniques to enhance crops by incorporating desirable traits. This chapter focuses on oil palm, the highest oil-yielding crop on earth, by compiling advancements in cultivation, uses, genetic diversity, conventional breeding, biotechnology, conservation and molecular methods. It serves as a source of information on diverse and advanced aspects of crop improvement.

16.1.1 Botany

Within the monocotyledons, the palm family Arecaceae (previously known as the Palmae) is included in the order Arecales (Cronquist 1981). *Elaeis guineensis* Jacq., the oil palm, belongs to the tribe Cocoseae and subfamily Arecoideae, along with other genera that include the coconut (Dransfield et al. 2005).

Kingdom: Plantae
  Clade: Angiosperms
    Clade: Monocots
      Order: Arecales
        Family: Arecaceae
          Subfamily: Arecoideae
            Tribe: Cocoseae
              Genus: *Elaeis* Jacq
                Species: *Elaeis guineensis* Jacq.
                Elaeis oleifera (Kunth) Cortés

There are two primary species in the genus Elaeis, which includes the oil palm. The African oil palm (*E. guineensis*), the first species, is the most oil-producing oil palm that is grown commercially. It is native to West Africa. *Elaeis oleifera*, often referred to as the American oil palm, is the second species. It is native to Central and South America. Better properties, such as reduced acidity, increased levels of carotenoids, tocopherols, and tocotrienols, and a greater percentage of unsaturated fatty acids, are present in the oil that is extracted from the mesocarp of the American oil palm fruit (Cadena et al. 2012; Lopes et al. 2021).

In addition, there are additional traits of the American oil palm that are very interesting for oil palm genetic advancement. These include a slower rate of vertical development of the trunk, which extends the plantations' commercial exploitation period and helps in mechanical harvesting, which lowers harvesting costs. Further, it exhibits resistance to defoliating caterpillars, red ring, bud rot, and other pests and diseases (Barcelos et al. 2015; Gomes et al. 2019; Oteyami et al. 2019; Pinto et al. 2019). The American oil palm is used in the majority of oil palm genetic improvement efforts carried out in Latin America, Africa, and Asia because of its agronomic and industrial importance, making it a main source of genetic diversity (Barcelos et al. 2015).

16.1.2 Origin and Distribution

Evidence suggests that the use of oil palm dates back several thousand years. The first reports of oil palm and its use can be traced back to the fifteenth century, with palm groves being reported in the early sixteenth century. Humans have been the primary agents responsible for the expansion of oil palm, as they have used products made from oil palm.

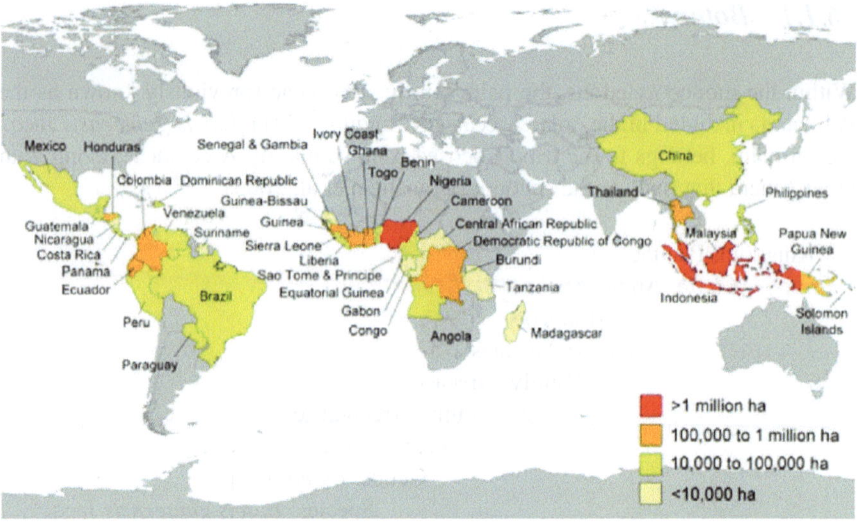

Fig. 16.1 Distribution of oil palm plantations worldwide. (Source: Google Image accessed on 5.12.2023)

Three main areas of the equatorial tropics are home to the oil palm (*Elaeis guineensis*): Africa, Southeast Asia, and South and Central America (Fig. 16.1). Evidence from fossils, history, and language all suggest that Africa is where oil palms (*Elaeis guineensis*) were first domesticated. Pollen dating in Miocene sediments in Nigeria has revealed direct evidence of African ancestry (Zeven 1964). According to Maizura et al. (2006), a study employing 359 oil palm accessions collected from African nations showed that material from Nigeria had the highest percentage of polymorphic loci and the highest number of alleles per locus, suggesting that Nigeria may need to serve as the center of diversity for the wild oil palm.

Schad (1914) and Zeven (1967) provided detailed accounts of the oil palm's distribution, showing that the concentration of palms begins in the highlands of Guinea at 10–11° N, which is the palm's most northern occurrence along the West African coast. The palm belt reaches Cameroon and the Democratic Republic of the Congo after passing through Sierra Leone and all the other nations. Oil palms are native to much of the Congo and Angola in Central Africa, where they are usually found between 3° N and 7° S. Although oil palms rarely grow higher than 300 m, they can grow to heights of 1300 m in the Cameroon mountains and 1000 m in Guinea, East Cameroon, and East Africa with enough rainfall; nevertheless, in these situations, yields are often lower. This suggests that there are palms that can survive in colder areas. Large, semi-wild groves in Africa were the sole economic sources of oil palm before the turn of the century. While there was an exportable surplus of oil and kernel in Africa, the produce from smaller groves was used for local consumption.

16.1.3 Importance and Uses

Mesocarp oil, also referred to as palm oil and palm kernel oil, is the main product of the oil palm. For more than 5000 years, oil palm fruits have been utilized to produce edible oil, and now, palm oil is a major source of biofuels (Corley and Tinker 2016).

16.1.3.1 Major Uses

Deep-frying and fats for cooking are great applications for palm oil. Additionally, it is a component of margarine, ice cream, imitation cream, various dairy products, bakery products, potato chips, and other snacks. The stearin fraction with a high melting point is utilized for bread fats, vanaspati, and shortenings. Conversely, margarine, salad dressings, and cooking oils are made from olein, which has a lower melting point. Owing to the growing demand for organic goods, Ghana, Brazil, and Colombia have begun producing organic palm oil.

These days, the producing nations routinely refine and fractionate crude palm oil. Leading this process, Malaysia now exports a wide range of products, including palm fatty acid distillate (PFAD), biodiesel, neutralized palm oil, refined bleached and deodorized (RBD) palm oil, palm olein, palm stearin, palm kernel oil, palm kernel cake or meal, and various oleochemicals and finished products (Berger 2010; www.mpob.gov.my).

These days, palm oil is used to make a wide range of substances referred to as oleochemicals. The majority of these compounds, which include sulfonated methyl esters, polyols, and polyurethanes, have distinct fatty acid chains linked to a variety of simple functional groups, such as acids, amines, or alcohols. Furthermore, it is possible to extract and use separately a number of the minor components of palm oil, including sterols, vitamin E, and beta-carotene.

The fatty acids in palm oil are converted into methyl esters in order to make palm biodiesel. Due to its characteristics, it may be used in vehicles and buses with success. It is comparable to diesel fuel. The interchangeability of oils means that even while palm oil is used in very modest amounts to make biodiesel, it may be substituted for rapeseed oil in the culinary market. One of the most beneficial aspects of palm oil is its versatility.

16.1.3.2 Other Uses

An estimated 40 t/ha of dry matter is produced annually by the oil palm, which includes stems, leaves (fronds), and Empty Fruit Bunches (EFB). For the manufacturing of palm oil and kernel oil, less than 10 tons of extract is taken out. The residual biomass has multiple uses, such as solid fuel for the production of paper and building boards (Basiron and Salmiah 1994; Jalani et al. 2002). It is not a major problem at the moment, even if the removal of this dry waste could affect the soil's composition and the palm ecosystem. Despite a great deal of research, palm biomass is not yet widely used for several reasons. Contemporary plantations generally recycle biomass on the plantation to make up for the large amount of nutrients that have been taken out of the soil.

In addition, there are many byproducts, including paper and board, as well as structural and building materials. Xylan extracted by acid hydrolysis of EFB (which contains about 24%) is used to obtain xylose for the production of xylitol, a sweetener in diabetic foods. The EFB is disintegrated and used in making paper and boards, although biomass in general has poor inherent strengths (Kamaruddin et al. 1999). Additionally, adding the 20% EFB pulp to the paper improves its mechanical strength.

Additionally, the biomass is utilized as a building and structural material. Oil palm trunk sections are used to make plywood, blackboard, and fiberboard (MDF). By bonding with resins, trunks, leaves, and EFB can be utilized in the production of particle boards. The denser portion of the trunk could be used for furniture after suitable pre-treatments. Seventy percent of EFB is fiber (Shinoj et al. 2011) and this fiber, along with petroleum-derived polymers, could be used in the automotive and building industries. Several of these materials have been tested and usable products have been developed (Koh et al. 1999).

Boiler ash is useful in the partial replacement of masonry blocks for construction work (Rahman et al. 2014). Palm byproducts are also useful as feedstock in producing polyhydroxy alkanoates for use in biodegradable plastics.

The annually pruned leaves are traditional, used for thatching, fence making and other purposes. Additionally, the rachis are used as poles and can be made into ropes by twisting the midribs together.

Palm wine is an important product in African countries. It is obtained by tapping male inflorescence or in the case of felled palms, by tapping the main growing point (Hartley 1988; Lebbie and Guries 2002). It was reported that at some point in Nigeria, the income from palm wine was greater than that from palm oil and kernels.

One of the main byproducts of the processing sector is palm oil mill effluent (POME), which has a lot of potential for use as animal feed. POME contains 12% protein and can be partially substituted as dried feed, although a very high proportion is not tolerated by ruminants. The decanter dryer system (Kanapathy et al. 1981) could yield a product with 10% moisture that can be used in animal feeding trials (Gurmit and Toh 2000). However, the irregularity in the supply of POME has been a major limitation in popularizing this as cattle feed.

16.1.4 Oil Palm Biomass as a Natural Fiber Source

The oil palm plantation and factory produce a significant amount of biomass, including fiber. The fronds are often left on the plantation, creating breeding grounds for insects and reptiles. Additionally, the stalked fronds can obstruct the movement of machinery used for fruit collection and tillage. The trunk waste is typically shredded and incorporated into the soil before replanting. It is important to note that the scattered plantation wastes in farmers' fields add to the economic assessment of any technology aimed at adding value to these wastes, as collection and transportation costs must be considered. On the other hand, the waste generated in palm oil mills, such as EFB and mesocarp waste, is more readily available in bulk and easy to collect. Approximately 250 kg of EFB and 150 kg of mesocarp waste are left per each metric ton of Fresh Fruit Bunches (FFB). It is estimated that 70% of these wastes consist of fibrous materials (Khalil et al. 2008). Empty fruit bunch is the major source of fiber among oil palm residues, leading to a focus on EFB fiber in studies and value addition strategies. Currently, EFB fiber is often disposed of through burning or mulching in young plantations, leading to significant environmental pollution. Finding environmentally friendly ways to utilize these fibers is essential.

16.1.4.1 Value-Added Products from Fresh Fruit Bunches (FFB) Fiber

Value addition of fiber was achieved by making various products from the extracted fibers, such as MDF, mattresses, cooling pads, and yarn (Jayashree et al. 2002). Medium Density Fiber Board (MDFB) of superior quality was prepared by densification of three layers of fiber needled felt mat using the phenol-formaldehyde resin with a concentration of more than 20% at 140 °C and 48 kg/cm^2 pressure. The mattresses prepared from oil palm EFB fiber were graded as extra firm according to Indian Standards based on the indentation hardness index (15.37). Yarn (rope) was prepared in a motorized ratt after blending with coconut fiber, whereas in a traditional Ratt, yarn could be prepared without blending. The fiber length that is selected depends largely on the type of product created. For mattresses, for instance, long fiber is appropriate. Particleboard benefits greatly from short fibers.

Preliminary studies on gasification and densification of the fibers were also attempted. Pulping of the fibers and paper boards from pulped fibers was developed. Yahya et al. (2023) isolated cellulose nanofibres (CNF) from oil palm EFB through modified supercritical CO_2 treatment of the bleached fibers, which has potential in several biomedical applications such as skin regeneration and wound healing.

16.1.4.2 Oil Palm Fiber in Paper and Paper-Based Industries

Darma et al. (2022) reported that biodegradable pots were produced from oil palm EFB and sodium alginate. The pots exhibited sufficient functionality to serve as environmentally sustainable planting containers.

The ever-increasing demand for paper and paper products, coupled with the diminishing stock of wood fibers, makes it necessary to look for alternate sources of natural fibers. Pulping has been attempted with oil palm biomass/residues such as fronds, trunk, and EFB. It was reported that an average of 75% of useful fibrous elements could be recovered and after bleaching, the figure would be 45% (Anon

2000). Removal of the parenchymatous tissue ('depicting') is essential since the parenchyma constitutes 50%, 30% and 5% of the mass of the trunk, frond (leaf stalk) and EFB, respectively. The chemical composition of oil palm biomass indicated that the lignin content (a factor important for chemical pulping) of stem and EFB is similar to exotic woods such as trembling aspen (*Populus tremuloides*) and white birch (*Betula papyrifera*). The solubility in 1% NaOH and water was reported to be comparable to exotic woods like *Acacia mangium*, trembling aspen and white birch. Oil palm fiber contains high ash content, like other nonwood fibers (Table 16.1).

The fiber characteristics were also comparable to those of the woods studied (Kuhoo and Lee 1991). Similar in length to exotic woods, the fiber has a thicker cell wall than wood, which results in a significantly higher rigidity index (as in EFB's case) (Table 16.2). In comparison to their wood counterparts, fibers with a thick cell wall would result in a sheet with a higher bulk and a lesser capability for inter-fiber bonding.

The features of different oil palm biomass concerning pulping and papermaking are described below.

Trunk Studies have been conducted on the chemical pulping of oil palm trunks using a variety of chemicals (Kuhoo and Lee 1991). Oil palm trunk bleached pulp has characteristics akin to those of semi-hardwood kraft. However, because of their comparatively short fibers, these pulps have a low tear index. In printing and writing grades, trunk pulp can be used in place of hardwood kraft components once it has been bleached to the appropriate brightness.

Frond The frond pulp resembles softwood kraft. Its lengthy fibers are most likely the cause of its high tensile and tear indices. Chemi-thermo-mechanical pulping (CTMP) has been studied for fronds. When KOH was utilized as a refiner, whole frond chips without pretreatment produced an 80% pulp yield; when chips were pretreated with alkali prior to refining, the output dropped from 43% to 55%. The pretreatment of chips improved the tensile strength and tear index. CTMP of pretreated frond chips gave a corrugating medium with satisfactory properties.

Empty Fruit Bunch Pulp yield ranges between 45% and 65% and is weaker than trunk pulp in tensile and burst indices. Because of the fiber's high rigidity index and the pulp's long, spiral vessel parts, it has an exceptional tear index.

EFB fiber is clean, biodegradable and compatible with many different fibers from other wood species (Anon 2003). The fiber was reported to be suitable for paper as well as for structural materials like fiberboards (medium density) (Jayashree et al. 2002; Ramli et al. 2002). The performance of MDF was greatly enhanced by pretreatment of the fiber to remove any remaining oils, which also prevented delamination when the panels were being consolidated. The physical characteristics of MDF with EFB fiber are said to be enhanced by the addition of additional resin. EFB pulp can be utilized in newsprint furnishings as a reinforcing element. Although NaOH removed the oil more effectively than water, it produced a lower-quality fiber with a higher bulk density and fewer mechanical qualities. According to Sitinjak et al. (2023), oil palm biomass has the potential to be an abundant source of

Table 16.1 Chemical composition of oil palm biomass compared to some exotic woods

	Lignin	Holocellulose	Alpha-cellulose	Pentosans	Alcohol benzene solubility	1% NaOH solubility	Hot water solubility	Ash
Trunk (whole)	18.8	45.7	29.2	18.8	9.8	40.2	14.2	2.3
Trunk (strands)	22.6	71.8	45.8	25.9	1.2	19.5	2.5	1.63
EFB (strands)	17.2	70	42.7	27.3	0.9	17.2	2.8	0.7
Frond (strands)	15	84.6	50	23.9	1.7	22	–	0.48
Frond (whole)	19.4	67.4	49.6	–	52	36.4	17.5	4.9
Acacia mangium	25.6	–	–	17.4	4.8	16.4	6.16	–
Trembling aspen	18.1	77.8	43.6	20.2	3.7	19.3	2.75	0. 31
White birch	18.4	78.2	41	24.1	3.6	19.1	2.8	0.33

Table 16.2 Fiber characteristics of oil palm and some exotic woods

	Fiber length (mm)	Fiber diameter (D) (μm)	Cell wall thickness (T) (μm)	Fibre coarseness (mg/m)	Rigidity index $(T/D)^3 \times 10$
Whole trunk	1.32	35.3	4.5	–	20.77
Trunk strand	0.96	29.6	4.8	–	42.64
EFB strand	0.99	19.1	3.38	0.107	55.42
Frond strand	1.59	19.7	3.95	–	80.61
Acacia mangium	0.9	22.3	2.9	–	22
Trembling aspen	0.96	20.8	1.93	0.241	7.99
White birch	1.32	21.6	2.44	0.296	14.42

cellulose. Cellulose was isolated from oil palm biomass, including fronds, oil palm fibers and empty fruit bunches (EFBs). The cellulose is hydrolyzed to produce glucose and then fermented to ethanol. The amount of cellulose was higher in oil palm fiber compared to the other two. The nanocellulose synthesized from palm-pressed fiber demonstrated huge potential to be used not only as a fat replacer but also as a rheological modifier for the development of reduced-fat vegan foods (Lee et al. 2022). According to Chungsiriporn et al. (2022), oil palm fiber has the potential to be used as a source for the production of fibrous paper if it is coated with a bee's wax-chitosan solution over a longer period of time. Its use in the preparation of wood pulp in the paper industry could contribute to the fight against deforestation. Ismail et al. (2022) made paper from beaten EFB fiber by incorporating microcrystalline cellulose (MCC). Paper with the addition of 6% MCC had the highest tensile strength. It can make writing or superior printing-grade paper, depending on the physical, mechanical, and optical characteristics of the final product. According to Jarupan et al. (2022), oil palm fronds may be used to make paper and as molded pulp trays for cold-chain fresh product packaging. Souza et al. (2022) compared various pre-treatments for the preparation of cellulose nanocrystals and lignin extraction from mesocarp fibers.

16.1.4.3 Use of Oil Palm Fibre in Biocomposites

Glass fiber composites can be replaced with biopolymers or synthetic polymers bonded with natural or bio-fibers (known as biocomposites). Oil palm fiber could find application in biocomposites such as molded plastic composites (Zakaria and Poh 2002) and geotextile. Composites consisting of polypropylene (PP) and EFB fiber were reported by many workers. According to Karina et al. (2008), glass fiber/

polyester composites and EFB fiber with a 40% volume percentage have comparable flexural strengths. This implies that in many applications where very strong load-bearing qualities are not needed, EFB fiber may be able to substitute for glass. Khalil et al. (2007) investigated the mechanical and physical characteristics of oil palm empty fruit bunch fiber in contrast to glass fiber. The MPOB developed oil palm fiber molded plastic composites by compounding oil palm fiber with plastics (Ridzuan et al. 1999).

The best tensile strength, thermal stability, and water resistance were demonstrated by a thermoplastic starch (TPS) biocomposite that contained 50% powdered mesocarp fiber. The TPS composite, which was gradually broken down by soil microbes, was used to create a biodegradable seedling pot (Saepoo et al. 2023). Oil palm EFB offers enormous potential as an environmentally friendly thermal insulator and sound absorber in commercial buildings, according to Nor-Azlina et al.'s (2021) evaluation of thermal and acoustic qualities.

16.1.4.4 Constraints

The major constraints encountered in the pulping of oil palm biomass and its use in paper-making include:

1. Higher silica and ash content
2. High bulkiness and moisture content
3. Short fiber length
4. High proportion of parenchyma cells (especially in the trunk)
5. Competitiveness
6. Blending of furnish may be necessary to achieve the desired grade of paper.
7. Effluent disposal standards need to be standardized.

16.1.5 History of Oil Palm Research and Development

The first organized professional research on oil palm cultivation was initiated in 1916 at the research station of Algemene Vereniging van Rubber planters Oostkust van Sumatra (AVROS) in Medan, Sumatra (now the Indonesian Oil Palm Research Institute—IOPRI). AVROS pisiferas are still widely used in current seed production (Pamin 1998). In the 1930s, work began in the then Belgian Congo at Yangambi, at Institut National pour l'Étude Agronomique du Congo Belge (INEAC). The connection between tenera, dura, and pisifera palms was an important discovery. Shortly after World War II, the Nigerian Institute for Oil Palm Research (NIFOR), formerly known as the West African Institute for Oil Palm Research, was founded in Benin City, Nigeria. Subsequent research sites were established in Benin and La Mé, Côte d'Ivoire. For many years, individual private enterprises in Malaysia conducted the majority of the research, and their contributions have proven crucial (Chew et al. 1998) and are still being made today.

Although the Department of Agriculture had a modest program, the Malaysian Agricultural Research and Development Institute (MARDI) established a public research program in 1971. The Palm Oil Research Institute of Malaysia (PORIM) took over the program in 1980, and in 2000, it changed its name to the Malaysian Palm Oil Board. A research center called Cenipalma was established in the early 1990s by Fedepalma, the union of Colombian palm farmers. The Empresa Brasileira de Pesquisa Agropecuária (EMBRAPA) has long been conducting research in Brazil. Notable is also the Costa Rican ASD research initiative. Early in the 1960s, an oil palm research station was established in Ecuador at the Instituto Nacional de Investigaciones Agropecuarias (INIAP) in the northwest, close to La Concordia. The oil palm sector in Papua New Guinea established the PNG Oil Palm Research Association. Therefore, national research organizations do exist in some manner in the majority of nations that engage in oil palm agriculture. During the 1970s, oil palm research in India got its modest start at the Central Plantation Crops Research Institute (CPCRI) at the Regional Centre of ICAR in Palode.

16.2 Current Cultivation Practices and Challenges

16.2.1 Current Cultivation Practices

Nursery: Uniform germination of seeds is achieved through preheating and incubation. Germinated seed is the standard planting material and is often raised in two-stage nurseries, primary and secondary. Primary or pre-nurseries seedlings can be planted in reusable plastic trays or polybags; each tray can store 24 seedlings. Less labor, less space, less soil consumption, and cheaper transportation are all benefits of using the trays. Trays are more expensive than polybags, however, their several reuses make up for the initial cost difference. Offering shade is safer in the nursery because direct sunshine can lead to overheating and harm to the leaves. The objective is to gradually remove the shade over 2 weeks, beginning 10 weeks after planting, with the initial reduction in normal light intensity being roughly 60% (Fig. 16.2a, b).

Mulching is a good way to conserve moisture and stop weed growth in seedlings, but it's not very popular in Southeast Asian polybag pre-nurseries. Small nurseries can benefit from hand watering when it comes to irrigation. Because of the tiny soil volume, Mathews et al. (2010) recommended misting plastic trays three times a day with a mist-spray system. Fertilizers may not be required in the prenursery if rich topsoil is employed (Rankine and Fairhurst 1998), but because plastic trays have minimal soil volumes, Mathews et al. (2010) advise weekly foliar treatment of compound fertilizer. Mathews et al. (2010) merely supplemented the water containing the trays with nutrients. Healthy seedlings can be produced in the Far East by filling big black polythene bags with friable topsoil and giving them a lot of water.

Fig. 16.2 Two-stage nursery. (**a**) Primary nursery and (**b**) secondary nursery. (Photo courtesy: Dr. K. Sunilkumar, unpublished)

In the Far East, nurseries generally raise seedlings all year round, starting in April or September and planting them in the field during the wetter months. In West Africa, nurseries are planted the year before since the field-planting season lasts from the start of the rainy season until May or early June. According to Aya (1979) research, the ideal period to set up a nursery in Nigeria is between mid-April and mid-May, when there are early rains. This leads to a decreased incidence of blast disease and better development for up to 10 months. The optimum months to build nurseries and perform transplants in northern Colombia are March through May, which corresponds with the dry season. Before the rainy season, which runs from January to April in Ecuador's Pacific Plain, field nursery planting is best done, bringing an end to the protracted low rainfall but cloudy period from November to December (Hartley 1988).

In conclusion, seedlings should be planted in the nursery 12–16 months ahead of the planned field planting time, or 10–14 months ahead of time if a single-stage nursery is being used. To ensure germination, seeds should be ordered at least 4 months in advance.

16.2.1.1 Climate Requirements

Although oil palm is a typical humid tropical crop, it may thrive in a variety of settings, from semi-arid tropical to wet tropical. The best places to grow oil palm are those that are between 10° N and S latitudes of the equator. In terms of climate, South East Asia has the best conditions for growing oil palm. Temperature affects many aspects of crop yield, including the development of inflorescences, the start of floral primordia, sex differentiation, anthesis, floral abortion, and palm growth.

Oil palm prefers a maximum temperature of 29–33 °C and a minimum temperature of 22–24 °C. Long-term colder weather, below 19 °C, significantly increases male inflorescences while decreasing palm growth and leaf output. Since oil palm grows continually and yields all year long, it needs uniformly distributed yearly rainfall of 1800–2500 mm without a distinct dry season. Being a crop that loves water, oil palm needs moisture all year round for healthy growth and high output. Rainfall must be consistent, with an annual total of at least 1800 mm or 150 mm each month, and a dry spell that lasts no longer than 2 months.

In nations like Malaysia and Indonesia, where the rain falls evenly throughout the year, it is grown as rainfed. Oil palm is grown both in irrigated and rain-fed environments in India. Solar radiation is another important climatic component for the growth of oil palm. Sunlight is a crop that oil palms love. A decent yield needs 1800–2000 h of sunshine every year. The growth and output of oil palm are affected by solar radiation levels below 350 cal cm^{-2} day^{-1}. Oil palm leaves require at least 5 h of direct sunlight each day to absorb this level of radiation.

16.2.1.2 Soil Requirements

Although oil palm can be grown in many types of soils, the best soils are said to be moist, deep loamy or alluvial soils with strong water permeability and a high organic matter content. The pH range of 4–7 is ideal. The majority of oil palm plantations are located in tropical soil orders called ultisols, oxysols, and inceptisols. These soils have little ability to act as a buffer and are extremely acidic. Care must be taken to avoid heavy soils with low drainage, gravelly and coastal sandy soils with poor water holding capacity, as well as saline and alkaline soils. Oil palm is a hardy crop that can tolerate high water and moisture stress.

16.2.1.3 Field Planting

Research shows that, as compared to younger seedlings, polybag seedlings that are 13 months or older yield noticeably more in the first three bearing years. Compared to seedlings that are 8 months old, those that are 14–20 months old bear fruit earlier and generate larger early yields. In the first year of harvest, older seedlings continue to produce more leaves, bear fruit earlier, and produce heavier bunches with a greater fruit/bunch (F/B) ratio and an oil-to-mesocarp ratio. According to Rethinam et al. (2000), seedlings from a single-stage nursery that were 9–15 months old in India grew more quickly than older seedlings that were watered.

Prepare a planting hole that is marginally larger than the diameter of the polybag. After cutting off the bag's base, the entire plant is inserted into the opening. You can cut a slit in the bag's side and take it out, protecting the soil around the roots. Next, topsoil is added to the tiny space that exists between the root ball and the surrounding earth. The plant has to be planted securely, but its young roots also need to be free to spread out into the surrounding soil. It is imperative to plant the seedling with its stem base elevated above the ground. It has been demonstrated that deep planting slows down subsequent growth (WAIFOR 1955), and it could be a specific issue if it is short-term.

In regions without a distinct dry season, planting can occur year-round, although typically the months with higher average rainfall are preferred. Following planting, mulching can be carried out using EFB, biomass from the previous stand, or perforated black polythene sheeting. Following EFB mulching, Loong et al. (1988) reported a 16% increase in yield during thc first 2 years of output. To protect against rodents, wire collars of appropriate sites may be necessary around the palm. Porcupines can also damage your palms, so chicken mesh or wire mesh can be used to cover the seedling up to 1 foot below ground when planting.

Though gaps of up to 3% may arise in locations with an extremely severe dry season, the amount of refilling needed is supposed to be about 0.5%. Polybag seedlings typically grow quickly after being transplanted to the field, which makes "supplying" or gap filling, the process of replacing missing palms, questionable a year after planting and utterly worthless after 2 years. Even in advanced planting material (APM) nurseries, palms grown in a nursery will not be as developed as palms grown in the field at the same age. One or two months after planting is the ideal time to perform an inspection, and a follow-up inspection is advised a year later. In this instance, the seedlings used ought to come from an APM nursery or be of a size appropriate for transplantation. Spacing varies from 7.5 to 10 m, with a triangular arrangement typically used, resulting in 205 to 115 palms per hectare, respectively (Hartley 1988). This planting pattern is referred to as an equilateral triangle pattern.

16.2.1.4 Establishment and Maintenance

Clearing a circular area around each young palm is necessary to ensure that it is never overgrown by competing vegetation. The basin's initial radius should be about 1 m, but it should grow to 2 m in the second year and 3 m in the third year.

Establishing and maintaining a ground cover, especially with leguminous crops, is a recommended practice in young plantations to conserve moisture, reduce runoff and improve soil fertility and physical properties (Fig. 16.3).

Pruning should only involve removing leaves necessary to access a ripe bunch. It is important to leave 33–40 fronds on the crown and only cut a few green fronds. According to Henson (1999), in established plantations, the soil received approximately 9 t/ha of dry matter replenishment annually. It is best to remove epiphytes that are growing on palm trunks and plants with easily caught loose fruits.

A plantation's whole management plan and procedures should be completely integrated with mechanization. A helpful overview of the mechanization and management of United Plantations in Malaysia was given by Bek-Nielsen et al. (2000). Erosion and waterlogging can be mitigated by mechanical weed treatment, which is as simple as slashing the interlines. A typical knapsack sprayer can be used to apply chemical insecticides and herbicides. When paired with GPS/GIS technology, the appropriate dosage can be applied where necessary, provided that digital data regarding weed density or pest infestation is accessible.

Fig. 16.3 Cover crop raised in the juvenile plantation. (Photo courtesy: Dr. K. Sunilkumar, unpublished)

Both soil water shortage and air moisture deficit can affect oil palm. Irrigation could thereby increase the palms' potential production. The mean response varied from 3% to 73% in a series of irrigation tests conducted in Malaysia (Mahamooth et al. 2011). Although there was a notion that drip irrigation would be more effective than other systems, Palat et al. (2000) evaluated drip, sprinklers, micro sprayers, and contour furrows and found no significant variations in yield response among these approaches.

16.2.1.5 Ablation

It is customary to ablate and remove extremely small young bunches at the beginning of the bearing season. It is recommended to remove all female inflorescences from the palm tree 6 months before the start of regular harvesting. After 3 years, young palms can produce rather large bunches, thus, it's critical to halt ablation as soon as a profit is realized.

Cutting the fronds beneath the bunch to be harvested is necessary for tall palms to gain access to the bunch. Harvesting young palms usually requires the use of a chisel with a short handle. The use of a knife and pole has been customary in plantations in Southeast Asia for many years. Since the bunch cannot be closely inspected from the ground, the most practical method of determining maturity is counting the number of loose fruits on the ground. Over the past few decades, the suggested minimum ripeness standard (MRS) has changed. Using the bunch weight from the previous harvest, Ng and Southworth (1973) calculated the ideal MRS to be between two and five loose fruit per kilogram of bunch weight. Harvesting at the correct stage of bunch maturity is equally important for obtaining the best yield from oil palm (Sunilkumar and Sparjanbabu 2013a, b). Overly mature fruits will have higher fatty acids and may be lost due to shattering when the bunch falls during harvest. Under-ripe fruits will have less oil content. Therefore, both yield and oil quality are impacted if harvesting is not done at the proper mature stage (Fig. 16.4a–d).

16.2.2 Digital Farming

Land, soil, and water resources are becoming more scarce due to population growth and rising demand. Precision agriculture (PA) was first proposed in the early 1990s as a solution to the pressing problem of scarce resources. The primary goal of PA is to minimize losses and waste while making the best use of the resources at hand (Luke et al. 2020; Thorp 2014; Yost et al. 2017). PA significantly raises crop profitability, maximizes the use of resources, and reduces loss and waste. Zhang et al. (2002) state that the foundation of PA is a system approach to restructuring the entire agricultural system in the direction of high efficiency, low input, and long-term sustainability.

Fig. 16.4 Fruit types based on color. (**A**) Nigrescence-unripe fruits, (**B**) nigrescence-ripe fruits, (**C**) virescence-unripe fruit, and (**D**) virescence-ripe fruit. (Photo: Dr. K. Sunilkumar)

Expert Systems (ESs) are embraced and employed in a variety of categories to detect, monitor, analyze, and forecast crops in order to achieve these aims, enhancing the effectiveness and dependability of PA (Tan et al. 2022). Expert systems are interactive computer-oriented systems based on knowledge that mimic the decision-making and problem-solving processes performed by human experts (Oyedeji et al. 2019; Saibene et al. 2021). An ES consists of an inference engine, working memory, and a knowledge base (Liao 2005).

According to Oyedeji et al. (2019), an expert system is a knowledge-oriented system that gathers and holds crucial data. (1) Essential information acting as the primary building blocks for a certain problem-solving procedure comprises the core of the ES. (2) Emulation of human decision-makers and problem-solvers' actions (Mirmozaffari 2019). The inference engine uses a set of codified rules derived from the gathered data to accomplish the decision-making process. (3) An application focused on artificial intelligence (AI) that gathers data, models systems, and offers technical and engineering solutions.

Crop, water, and soil management are the three main categories in oil palm PA. Detection, monitoring, analysis, and prediction are all part of crop management. Pests, fruits, trunks, pollination, leaf scale, diseases (such as orange spotted diseases and basal stem rot, or BSR), and ripeness detection can all be detected with ESs

Fig. 16.5 Oil palm crown showing bunches in various stages of ripeness. (Photo: Dr. K. Sunilkumar, unpublished)

(Fig. 16.5). Monitoring is the process of making long-term observations, typically regularly, to compile data and document noteworthy events over time. In this field, monitoring nutrients and landscapes is are common example. In large-scale agricultural settings like oil palm plantations, where UAVs, UGVs, and USVs are mostly used, remote sensing is crucial and common. Many technological solutions have been widely adopted in oil palm PA and incorporated as ESs over the years. The increasing amount of ES-related literature suggests that ESs' relevance and viability for oil palm PA are widely acknowledged. In oil palm PA, prediction is important. Early issues and bottleneck identification help to maximize profits and resources while reducing waste and losses.

With a 25-year production cycle, oil palm is a crop that requires a lot of water. Water management is therefore essential in oil palm PA for the sustainability of crop output and water. Production, usually in fresh fruit bunch production, might be improved by effective water management, which would increase oil yield. Data collection and accurate water footprint identification are possible with ES. Water conservation and yield optimization may result from the use of various ESs in water management.

Crop health, resource efficiency, and oil palm output are all directly impacted by soil management. Accurate soil property data is essential for extending the oil palm area and choosing suitable land. Traditional techniques of evaluating soil, such as soil sampling, are costly, labor-intensive, and require a lot of time from human experts when extracting field samples. The integration of different ESs could facilitate assessment procedures and produce dependable results with the introduction of technological solutions.

Southeast Asian nations, particularly Malaysia and Indonesia, are major contributors to research and investigation in ESs for oil palm PA, given their roles as leaders in the oil palm industry. The majority of oil palm research in Pennsylvania (79.9%) is conducted in the field of crop management.

16.2.3 Major Challenges in Cultivation

The environmental effects of increasing oil palm cultivation are one of the main issues facing the oil palm industry. Since 2000, the conversion of land in Southeast Asia for oil palm production has mostly been driven by the increase in worldwide demand for food (mostly from China and India) as well as for biofuels and other non-food items (largely from Europe). Between 2000 and 2014, Indonesia's oil palm plantations more than tripled in size, from 2.5 million hectares to over 8 million hectares. Concerns have been raised against the clearing of forests and the destruction of habitats of biological diversity, especially of Orangutans in Malaysia and Indonesia. Clearing of peat lands releases more greenhouse gases (CO_2), drainage burning of forest biomass for land clearing, and the application of chemical fertilizers for oil palm plantations releasing Nitrous oxide (NO_3) all contribute to climate change.

The prevalence of disease and pests presents another significant obstacle. Oil palm disease known as Basal Stem Rot (BSR), which was caused by the fungus *Ganoderma boninese*, can result in a 50–80% reduction in output. According to Corley and Tinker (2016), the BSR pathogen can infect oil palm plants at a rate of up to 80% incidence during half of their economic life span. In West and Central Africa, where *Fusarium oxysporum* f. sp. elaeidis is the primary cause of Fusarium wilt, a significant epidemic decimated oil palm fields, where the disease is especially common (Rusli 2017). In South West Colombia, the disease has recently destroyed over 30,000 hectares, and in the Central Zone, over 10,000 hectares. The sharp rise in disease could be linked to climate change.

Pests like *Oryctes rhinoceros*, are causing serious loss to oil palm plantations and are therefore a serious challenge to address. Recently, severe incidence and infestation of invasive rugose spiralling whitefly (RSW) in oil palm was reported from Andhra Pradesh, India (Selvaraj et al. 2020), resulting in serious yield loss.

16.2.4 Objectives of Oil Palm Breeding

The objectives of oil palm breeding are as follows: (1) Breeding for high oil yield: The breeding program primarily focuses on yield enhancement. Breeding for high oil yield involves fine-tuning many factors such as high potential yield, harvestable yield, adaptability to the environment, tolerance to weather and diseases, thinner and longer bunch stalk, nonshedding fruits (thereby reducing the loss of loose fruits while the bunch falls on the ground), low lipase fruits (this can lead to the longer interval between harvest), and virescence fruits (where ripe fruits/bunches can be distinguishable from a distance in case of tall palms. (2) Reduced annual height increment: Under field conditions especially due to labor shortage all bunches are not harvested or bunches are harvested too early or too late. This results in not attaining the potential yield fully. This is also aggravated by the rapid height increase of palms. All these could be addressed by breeding shorter palms. (3) Enhanced oil quality: Through the utilization of *E. oleifera*, oil with high quantities of unsaturated fatty acids can be generated. The desire of consumers for monounsaturated and polyunsaturated dietary oils and fats makes an increase in the oils' unsaturation desirable. Palm oil can penetrate the liquid oil industry because of its higher degree of unsaturation. At room temperature (28 °C), palm oil is currently semisolid and can be separated into 70% liquid olein and 30% solid stearin. By raising the amounts of olein and decreasing the amount of stearin, palm oil quality can be enhanced. (4) Breeding for disease resistance: In West Africa, breeding for resistance to *Fusarium oxysporum* induced vascular wilt is a significant initiative. The Far East is not affected by this disease. Both young and mature palms perish as a result of it. Older palm trees planted in the Far East are particularly vulnerable to basal stem rot (BSR), which is caused by *Ganoderma boninense*. To breed palms resistant to disease, screening methods for identifying tolerant palms are being developed. The American oil palm, *Elaeis oleifera*, was used to produce palms that would not succumb to "sudden wilt" and "little bud rot," which had a detrimental effect on oil palm farms in Central and South America. A small number of hybrids showed resistance to this disease. (5) Breeding for high kernel oil: Due to the significant industrial demand for palm kernel oil, which differs from mesocarp oil in terms of fatty acid composition and its alluringly higher price, efforts are concentrated on producing oil palm hybrids with large kernels. Effective selection may be possible if there is a lot of diversity in the available kernel sizes. (6) Breeding for drought: In their quest to create drought-tolerant ones, Corley and Tinker (2016) have discovered a few Pisifera parents who seem to reliably pass on drought tolerance to their progeny. This information will help produce seeds as well as for future drought-tolerant breeding.

In addition to these, the oil palm improvement project aims to develop planting material with additional qualities such as high bunch index, yield precocity, large kernel, high kernel yield, long stalk, large kernel, selection for natural carotene- and vitamin A-rich palms, efficient usage of fertilizer through physiological and agronomic approaches, low-lipase, palm genotypes s yielding less free fatty acids, and non-shedding fruits.

16.3 Germplasm Biodiversity and Conservation

16.3.1 Germplasm Diversity

Any crop improvement initiative must be successful in assembling a broader genetic variability in the species. Despite being brought to Brazil and other tropical nations by the Portuguese in the fifteenth century, oil palm did not take off in these regions until the nineteenth century. In 1848, four seedlings were planted in Bogor, Indonesia, after the Dutch used Amsterdam to acquire oil palm seeds from West Africa. The offspring of these palms were planted in the Deli as decorative palms, which subsequently gained notoriety as Deli dura. Deli dura was the only plant material used for oil palm growing until the 1960s.

After the inheritance of shell thickness was discovered (Beirnaert and Vanderweyen 1941), tenera began to be used as commercial planting material. Deli dura was continuously used as the female parent and only a limited number of pisifera/tenera palms were involved in oil palm breeding programs on the male side. Therefore, oil palm breeding has a very limited genetic base, which limits the amount and rate of genetic advancement possible through breeding initiatives. As a result, the assembly of genetic diversity has gained attention and various countries have made concerted efforts to collect oil palm genetic resources.

After World War II, INEAC (Institut National Pour I'Etude Agronomique du Congo Belge) started prospecting new oil palm genetic material from commercial plantations. NIFOR (Nigerian Institute for Oil Palm Research) made occasional collections from local markets and through the village chief, establishing collections. Collections were also made from Bamenda, Cameroon, by 1964. Prospection was started in Côte d'Ivoire by IRHO (Institut de Recherche des Oleagineux). NIFOR and MARDI (Malaysian Agriculture Research and Development Institute, then PORIM and currently MPOB) conducted a comprehensive genetic resource prospection in the oil palm region of Nigeria in 1973. Another oil-bearing plant, Pejibaye (*Bactris gasipaes*), was a significant collection that has unusual fatty acids and other desirable traits that can be introduced into new palm populations. In the Republic of Zaire, a collection of elite materials was carried out in 1984. Belgian workers had prospected for the genetic resources of oil palm in the early 1920s and 50s in Congo. Several expeditions were organized by MARDI/PORIM to collect *E. oleifera* resources from Colombia, Panama and Costa Rica in 1967, 1969 and 1980, respectively. In 1981, a joint effort was made by PORIM and IBPGR to collect *E. oleifera* from Central and South America. Collections were made from Surinam, Colombia, Panama, Costa Rica, Nicaragua and Honduras. Oil palm was introduced to India at the National Botanical Garden in Kolkata at the close of the nineteenth century, and it subsequently expanded to other regions of the nation. The first cultivation of oil palm was started by Kerala at Thodupuzha in 1964 using imported materials from Malaysia and Nigeria (Fig. 16.6). Oil palm germplasm's systematic collection and evaluation started in 1979 at the Palode Regional Centre of CPCRI and later all oil palm research and development activities were brought under the ICAR-Indian Institute of Oil Palm Research, Andhra Pradesh.

Fig. 16.6 Oldest plantation in India (estd 1964), Thodupuzha, Kerala, India. (Photo: Dr. K. Sunilkumar, unpublished)

The enrichment of oil palm genetic resources through collection, conservation, and crop development is now essential to the sector's progress. Reports from 29 institutes state that 21,103 oil palm accessions have been gathered worldwide. 1467 oil palm accessions from wild populations have been collected by the Malaysian Palm Oil Board (MPOB) since the 1970s, and these accessions have been preserved (Rajanaidu 1994). According to Okwuagwu et al. (2011), MPOB is home to the largest oil palm germplasm collection in the world. The list of *E. guineensis* and *E. oleifera* collections made by MPOB is shown in Table 16.3. Details of germplasm utilized by various seed producers in Malaysia are shown in Table 16.4.

Genetic material belonging to *Elaeis guineensis* was recently collected from Cameroon and Angola by Indonesia and Colombia. There aren't many collections from Nigeria, Cameroon, or Côte d'Ivoire at CIRAD. Embrapa Brazil, ASD Costa Rica, and CENIPALMA Comprehensive collections can be found in Ecuador and Colombia, as well as in the Brazilian Amazon Basin, Honduras, Nicaragua, Costa Rica, Panama, Surinam, and Tishaa, Ecuador. Field genebanks serve as the primary repository for the genetic resources related to oil palm conservation. 130 accessions have been gathered from primary and secondary sources in India. The diversity of oil palm genetic resources available in India was reported by Sunilkumar et al. (2020), who described genotypes with dwarf, long bunch stalk, large kernel and virescence pisifera available in the collection. Molecular characterization of the same, placed Dura of Nigerian and Tanzanian origins in distinct groups, indicating viability.

Table 16.3 List of *Elaeis guineensis* and *E. oleifera* germplasm collection at Malaysian Palm Oil Board (MPOB)

Species	Originating Country	Collection year	No of accessions
Elaeis guineensis	Nigeria	1973	919
	Cameroon	1984	95
	Zaire	1984	369
	Tanzania	1986	60
	Madagascar	1986	17
	Senegal	1993	104
	Gambia	1993	45
	Sierra Leone	1994	56
	Guinea	1994	61
	Ghana	1996	58
	Angola	1991, 2010	54, 127
Elaeis oleifera	Nicaragua	1982	18
	Honduras	1982	14
	Costa Rica	1982	61
	Colombia	1982, 2004	41
	Suriname	1982	6
	Brazil	1982, 2004	–
	Panama	1982	27
	Peru	2004	–
	Ecuador	2004/2006	5

Murugesan et al. (2015) analyzed the diversity of germplasm resources and indicated maximum diversity in Tanzanian collections among various germplasm accessions available.

16.3.2 Conservation Approaches

Once collected, the germplasm resources need to be conserved without the loss of valuable genes. Insufficient resources in the form of land, labor and money make the field conservation of genetic resources a more daunting task. In this context, *in vitro* conservation methods are very helpful. Among various *in vitro* methods, conservation in liquid nitrogen (around −196 °C) offers the advantage of long-term or indefinite storage of such resources. The chance discovery of glycerol's cryoprotective effect on bull sperm against freezing by Polge et al. (1949) commenced the science of cryopreservation. Since then, many methods have been developed for various types of cells, tissues and organs. Cryopreservation is currently the only technique used to ensure the safe and cost-effective long-term preservation of different types of germplasm, which refers to the storage of biological material at extremely low temperatures, often liquid nitrogen (−196 °C) (Engelmann 2004)

Table 16.4 Utilization of germplasm by various seed producers in Malaysia

Sl. no	Name and location	Germplasm used
1	Felda Agricultural Services Sdn. Bhd. Jerantut, Pahang	Chemara Deli duras IRHO pisiferas & Deli duras IRHO Yangambi ML 161
2	Stesen FASSB Bukit Besar Kulai, Johor	Same as above
3	Ebor Research (Sime Plantations Sdn. Bhd.) Kluang, Johor	Kulai Deli duras MPOB ACROSS
4	Applied Agricultural Resources Sdn. Bhd. Kluang, Johor	Dumpy AVROS (DA) Dumpy Yangambi (DY) DA × DY Chemara Deli Duras Semi-Clonal Seeds & TC
5	Borneo Samudera Sdn. Bhd. Tawau, Sabah	SBP & MPOB duras AVROS pisiferas
6	Ulu Dusun Seeds Sdn. Bhd. Tawau, Sabah	Same as above
7	Kulim Sdn. Bhd. Kota Tinggi, Johor	MPOB Deli duras MPOB AVROS Dami Deli duras Dami AVROS
8	Golden Hope Agrotech Consultancy Sdn. Bhd. Tawau, Sabah	Banting Deli duras Banting AVROS
9	Golden Hope Agrotech Consultancy Sdn. Bhd. Banting, Selangor	Same as above
10	Guthrie Plantation & Agricultural Services Sdn. Bhd. Kluang, Johor	Chemara Deli duras Ex-URT pisiferas
11	Highland Research Unit Sdn. Bhd. Klang, Selangor	Chemara Deli Duras AVROS Dumpy pisiferas
12	IJM Plantations Berhad Labuk, Sabah	MPOB Deli duras MPOB AVROS
13	IOI Corporation Berhad Tampin, Negeri, Sembilan	Chemara Deli duras MPOB AVROS
14	Pamol Plantations Sdn. Bhd. Kluang, Johor	Chemara Deli duras SBP pisiferas
15	Pamol Plantations Sdn. Bhd. Sandakan, Sabah	Same as above
16	RISDA Semaian dan Landskap Sdn. Bhd. Dungun, Terengganu	MPOB Deli duras MPOB AVROS
17	Stesen Penyelidikan Pertanian Ulu Dusun Sandakan, Sabah	SBP duras SBP pisiferas
18	Sasaran Ehsan Utama Sdn. Bhd. Sandakan, Sabah	MPOB Deli duras MPOB AVROS
19	Stesen Penyelidikan MPOB Kluang Kluang, Johor	MPOB Deli duras MPOB AVROS

(continued)

Table 16.4 (continued)

Sl. no	Name and location	Germplasm used
20	United Plantations Berhad Teluk Intan, Perak	Deli Marihat IRHO pisiferas
21	Sarawak Plantation Agriculture Development Sdn. Bhd. Miri, Sarawak	MPOB Deli Duras MPOB AVROS
22	NALFIN Planting Material Sdn. Bhd. Teluk Intan, Perak	MPOB Deli duras MPOB AVROS

within limited space (Li and Pritchard 2009). Rusli (2017) suggested somatic embryos as the most popular source material for cryopreservation.

Even though ultra-freezing at low temperatures is the main process of cryopreservation, pre-growth, freezing, thawing, and post-treatment are the sequential processes of typical cryopreservation procedures, for which ideal conditions must be determined for each material that is chosen. The following section focuses on various efforts and achievements made on the cryopreservation of oil palm genetic resources in the form of somatic embryos, zygotic embryos, kernel and pollen.

16.3.3 Cryopreservation of Somatic Embryos

In 1985, Engelmann et al. successfully cryopreserved oil palm somatic embryos for the first time. Engelmann and Duval (1986) discovered that after being cultured on a medium enhanced with sucrose, finger-like, glossy white clumps of embryos grew in the oil palm and survived freeze storage in liquid nitrogen. An improved procedure for oil palm somatic embryo cryopreservation was reported by Dumet et al. (1993a). For cryopreservation, they utilized clones of conventional somatic embryos from adult hybrid palms with distinct genetic backgrounds (Deli × La Me and Deli × Yangambi), as opposed to embryos that resembled fingers. Standard somatic embryo clones originated from hybrid (Deli × La Me) adult oil palms were used for subsequent cryopreservation studies (Dumet et al. 1993b, c, 1994). Khawnium and Te-chato (2012) evaluated somaclonal variations produced in cryopreservation by the dehydration technique using embryogenic callus initiated from elite "Tenera" oil palm.

In recent years, the successful application of the vitrification technique for cryopreservation of Friable Embryogenic Tissue (FET) developed using young leaves of elite "Tenera" oil palm was developed (Khawnium and Te-chato 2011) and for *in vitro* grown polyembryoids of oil palm (Suranthran et al. 2012). Gantait et al. (2015), using the droplet vitrification technique, achieved a 68% revival in cryopreserved polyembryoids of oil palm for the first time.

16.3.4 Cryopreservation of Zygotic Embryo and Kernel

Within a month of harvest, Grout et al. (1983) successfully cryopreserved zygotic embryos isolated from oil palm seeds. Following this, embryos taken from the dry seeds of the oil palm hybrid were cryopreserved. Cryo preservation of Deli × La Me (Engelmann et al. 1995) and embryos of open-pollinated Dura × Pisifera palm (Hor and Sinnish 2000) were reported. The embryos obtained from the seeds of the outer peripheral bunch were found suitable for cryopreservation (Villa et al. 2007).

Prakash et al. (2019) reported cryopreservation as a tool for the conservation of oil palm genetic resources in India. Suitable moisture content of 10–20% of embryos and/or kernels was achieved using silica gel desiccation. Desiccation of embryos for 4 h in laminar flow was found to be optimum. Furthermore, the use of an endosperm plug as a new explant source for cryopreservation was also suggested, showing 2% germination at 6.1% moisture. Marhalil et al. (2017) suggested silica gel desiccation as an effective step for moisture reduction while handling large-scale germplasm conservation through cryopreservation.

16.3.5 Pollen Cryopreservation

The oldest record of pollen storage dates back to the pollen of date palm used in artificial pollination about 2000 BC (Zirkle 1935). Increased lifetime can be achieved by employing high-quality pollen, desiccating it quickly enough, and then keeping it at extremely low temperatures, according to a defined methodology created for pollen preservation in various plants. Studies regarding the cryopreservation of oil palm pollen are limited and the available information is discussed in this section.

Pollen from oil palms survives in the wild for approximately 1 week following the anthesis of the male inflorescence (Hardon and Turner 1967). Oil palm pollen was effectively cryopreserved for up to 8 years without losing vitality by Tandon et al. (2007). Pollen dried to 5% moisture content using silica gel, cryo-stored for 6 months and used for pollination showed on par fruit set and bunch quality with fresh pollen (Sunilkumar et al. 2013).

For cryopreservation of oil palm pollen, a week before anthesis, immature male inflorescences were surface-sterilized and covered with a bag (Hardon and Davies 1969). The pollen was collected directly into paper sheets when anthers just began dehiscing (Tandon et al. 2007). A 0–5 mm sieve was used to filter the collected pollen to get rid of trash and other foreign objects (Ekaratne and Senathirajah 1983). The same method has been followed in the case of date palms as well (Ateyyeh 2012).

16.4 Traditional Breeding

16.4.1 Improvement Strategies

Breeding is a crucial aspect of producing planting material of high quality and increasing oil palm yield. Traditional hybridizations between dura and pisifera cultivars have significantly increased the productivity of oil palm, especially oil output.

In oil palm, three different types of natural varieties are dura, pisifera, and tenera. Morphologically, they look alike but are distinctly different in their fruit characters. Hence, they are called three different "fruit forms". The existence or lack of a shell and the shell's thickness are the two main distinctions between the three varieties. Pisifera lacks a shell, whereas Dura's kernel is encased in a thick shell. Tenera, the third species, is a hybrid of thick-shell dura and shell-less pisifera, with dura serving as the female parent. Tenera is distinguished by its thin shell as well as a characteristic fiber ring that surrounds and lies in the mesocarp close to the seed (Fig. 16.7a–c). Pisifera Palms are recovered generally from segregating Tenera populations since direct reproduction of this is difficult due to the scarcity of fruits with embryos and the absence of a protective shell. The occurrence of the Tenera hybrid was natural and an explanation of it was given by Beirnaert and Vanderweyen

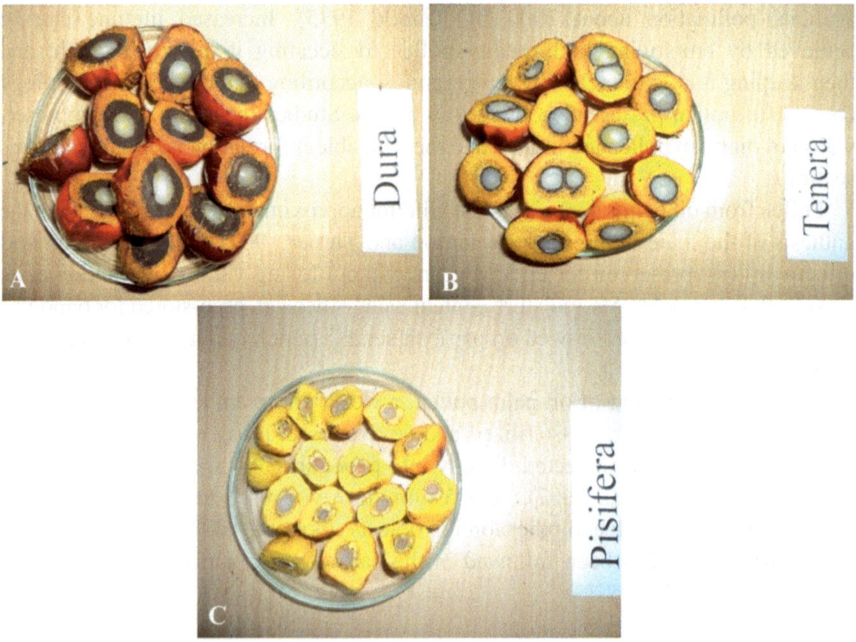

Fig. 16.7 Fruit types based on shell thickness. (**a**) Dura, (**b**) Tenera, and (**c**) Pisifera. (Photo: Dr. K. Sunilkumar and Dr. R. K. Mathur, unpublished)

(1941). Since then, there has been all-around awareness among Oil Palm breeders and planters to deliberately produce these hybrids by artificially crossing Duras and Pisiferas owing to their increased oil yield than that of their planting material all over the world (Hartley 1988).

16.4.2 History of Oil Palm Breeding

The oil palm breeding history can be traced back to the four seedlings of oil palm planted at Bogor in Indonesia and brought via Amsterdam and Reunion (Mauritius). Open-pollinated seeds of these Dura palms were subsequently planted at Deli and hence known as Deli dura and they formed the main population for oil palm breeding throughout the Far East. The thick-shelled Dura × Dura was the commercial planting material in areas with a slower pace of improvement until the 1960s. In 1941, Beirnaert and Vanderweyen reported monogenic inheritance of the shell gene based on their studies in the Congo. They discovered that 100% tenera (D × P) with thicker oil-bearing mesocarp and thin-shelled fruits would result from crossing homozygous dura with homozygous male sterile pisifera. Henceforth, the commercial plantations shifted to D × P (Tenera) material for planting, which resulted in a 30% increase in oil yield. The oil palm area then rapidly expanded throughout the Far East, particularly in Malaysia and Indonesia. Later, advanced materials with 6–9 tons/ha were developed when the yield of wild groves in Africa remained less than 0.1 tons/ha. This was mainly achieved through concerted breeding efforts as well as improved agronomic management of the palms by the government, as well as private stakeholders.

16.4.3 Breeding Methods

Oil palm plant breeding, which is distinguished by wider spacing and longer duration, can take up to 20 years for each cycle of phenotypic selection. Numerous negative effects result from this, including fewer generations that can be reached between two breeding phases and lower breeding intensity.

To circumvent these constraints, oil palm breeding programs have focused on heterosis breeding (through the development of hybrids) and population improvement. The F_1 hybrid produced by the dura × pisifera cross, named tenera, possesses favorable characteristics from both parents.

The breeding method in oil palm was influenced by maize breeding as well as animal breeding programs since the pisifera and dura need to be maintained separately. Like maize, oil palms are monoecious and cross-pollinated. Therefore, the maize breeding system known as Reciprocal Recurrent Selection (RRS) formed the basis for oil palm breeding. Another strategy is based on the principles of animal breeding's Family and Individual Selection (FIS). The fundamental breeding strategies in oil palm are recurrent selection and its variations, such as reciprocal recurrent selection and modified recurrent selection.

16.4.3.1 Modified Recurrent Selection

Since pisiferas are female sterile, they are first chosen based on proven ancestry or their tenera-sib performance in the tenera × tenera/pisifera cross. They are then top-crossed and progeny-tested using a sample of dura mother palms that have been phenotypically chosen. Duras and proven pisiferas that have been phenotypically chosen are used in the commercial manufacture of Tenera hybrid seeds. Parents used for future breeding in dura × dura and tenera × tenera/pisifera crosses are similarly selected phenotypically, although progeny-tested parents are sometimes included. Because family and individual palm selection is often a part of parent selection, this approach is sometimes referred to as the family and individual selection scheme. However, it is also a type of recurrent selection because chosen parents interbreed during each cycle. This scheme highlights general combining ability (GCA) effects and takes advantage of them. The primary benefits of this system are the ability to make and test a greater number of recombinant crossings while conserving time, space and the requirement for thorough progeny testing. The primary drawback is that hybrid progeny testing is typically not done on the original mother palms used for hybrid seed production, nor on the parents selected for additional breeding.

16.4.3.2 Modified Reciprocal Recurrent Selection

The modified recurrent selection approach chooses the dura and pisifera palms for commercial seed selection and further breeding based on family and individual palm performance. In the modified reciprocal recurrent selection method, selected dura and tenera parents are selfed and the crosses between dura and tenera progeny are tested (Fig. 16.8). Commercial D × P hybrid is produced using the Duras from the selves of the dura parent and pisifera from the selves of the tenera parent (T × T segregates into 1D:2T:1P). Both the GCA and SCA (special combining ability) effects are utilized in this technique. The selection of this plan most likely resulted from the assumptions that hybridizing teneras or pisiferas of African origin with Deli duras tended to have hybrid vigor for yields. One of the scheme's benefits is that progeny tests are used to choose the selfed parents for both commercial hybrid creation and subsequent breeding and (ii) there are two separate phases to the plan. Outcrosses are involved in both the "recombinant phase" and the "within-hybrid improvement phase," which occur concurrently. For long-term, sustained improvement, this permits the accumulation of advantageous alleles and preserves genetic variety; also, once the results of the inter-population hybrid test are known, the commercial hybrid can be replicated using duras and pisiferas.

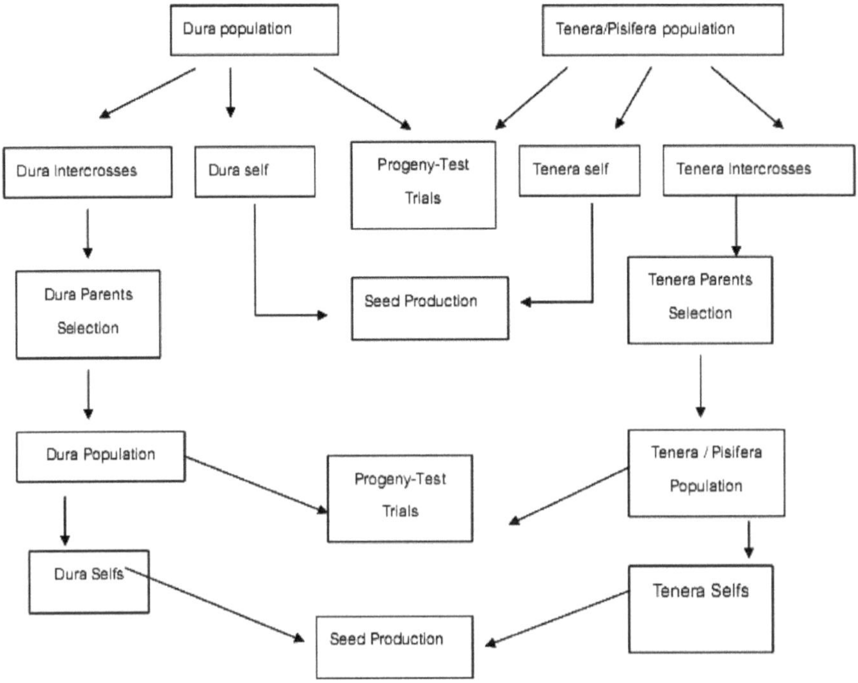

Fig. 16.8 Reciprocal recurrent selection. (Figure courtesy: Dr. K. Sunilkumar)

Mass selection was used to start the genetic enhancement of oil palm output in the 1920s. Subsequently, several revolutionary advancements significantly accelerated the pace of genetic advancement: firstly comprehending fruit form heredity and implementing tenera plantations which have thicker mesocarp; secondly finding hybrid vigor and implementing modified reciprocal recurrent selection; and finally practicing clonal selection, taking advantage of intra-hybrid variability. Furthermore, selection has been more effective with the introduction of linear mixed models to assess genetic values. After a century of breeding, there has been a significant rise in the oil palm yield potential (Nyouma et al. 2019). Promising advancements, including much higher producing cultivars, enhanced oil profiles, increased disease resistance, and more climate resilience, are being brought about by new breeding techniques and management practices (Murphy et al. 2021).

16.4.4 Breeding for Dwarfness

By increasing planting density and extending commercial exploitation, high-yielding, short palms with short leaves and slow vertical growth become a viable substitute for intensive cultivation and enhancing productivity (Escobar and

Alvarado 2004). Plants at Costa Rica's Coto observation plots yielded the original compact palm or OCP. These plots were established using open-pollinated seeds from an *E. oleifera* × *E. guineensis* inter-specific hybrid (O × G) found in Quepos, Costa Rica, in 1966. This hybrid had short leaves and exceptionally slow growth. Despite having small trunks and leaves, the OCP's bunches had not been very attractive. Since 1978, three cycles of backcrossing to *E. guineensis* parents have been carried out to enhance the bunch composition of the OCP. To create marketable seeds, F_1 palms with exceptional traits were chosen and intercrossed to create superior F_2 recombinants following each backcross cycle. A gradual loss of the compact traits of a small stem and short leaves was brought about by the introgression of *E. guineensis* genes through successive backcrossing, especially when backcrossed to AVROS parents. However, compact recombinants with short trunks and small leaves, better bunch parameters and economical bunch yield were created by the second and third backcross cycles, which were on par with conventional commercial seed varieties. Elite palms with distinctive characteristics, such as small leaves, slow stem growth, and a high bunch oil content, were chosen in order to develop clones. Selected progenies from backcross two (BC2) palm selfings were also selected in order to produce compact seeds.

Several other desirable traits of the Pobé Dwarf families have been noted by previous writers: high LAR (Breure 1986), low leaf weight and VDM, and high BI (Corley et al. 1971). Another dwarf genotype found at Elmina Estate in Malaysia is the Deli Dumpy Palm E206, which has a slower rate of height increase. This palm's selfed offspring were all uniformly short and used in numerous breeding initiatives (Jagoe 1952). Furthermore, PS-1 is a dwarf palm developed by crossing MPOB Nigerian dura × AVROS pisifera. Due to the dwarf and compact nature, the planting density of PS-1 could be increased to 170/ha against 143/ha in traditional materials (Kushairi et al. 2001).

16.4.5 Breeding For Drought

With the expanding oil palm industry, there is a need to plant them in less favorable conditions, especially in drought-prone areas. This requires the selection of genotypes tolerant to drought. Corley et al. (2018) exhorted that genotypes should be selected in conditions where they will be planted. Selecting drought-tolerant genotypes in irrigated conditions will not be effective for drought-prone areas and vice versa. It will be helpful for seed production and future breeding to have identified some Pisifera parents who regularly pass on drought tolerance to their progeny. Palat et al. (2008) reported a yield loss of 25–35% due to drought. Compared to mature palms, younger palms are more vulnerable to drought. Several reports have suggested variations in the physiological responses of oil palms to drought (Smith 1993; Suresh et al. 2008; Silva et al. 2017). According to Murugesan et al. (2015), among the genotypes available in India, those introduced from Nigeria, Costa Rica and Colombia have drought-tolerant genotypes.

16.4.6 Breeding for High Yield

Oil palm breeding has significantly increased the oil yield of successive plantations. After four selections, the oil output of Duras increased from 2.8 to 4.5 t/ha. Over the past 40 years, the oil production of teneras has increased from 6.3 to 11.2 t/ha as a result of dura selections that preceded and introgression with selected pisiferas (Jalani et al. 2002).

Sunilkumar et al. (2017) selected high-yielding dura palms and made *interspecific* crosses to produce palms with higher yield potential and the progenies are currently being evaluated in Palode, India. Krishnakumar et al. (2012) reported FFB yields above 50 per ha per year, equivalent to more than 12 MT of oil per ha obtained from farmers' fields in Mysore, India, through the best management of 12-year-old plantations. This indicates the potential for an enhanced yield of 10 MT even with existing planting material under the best management of plantations. According to a study by Hermanto et al. (2023), the best planting material under all rainfall conditions is the highest-yielding and least drought-prone variety, AA86. According to Vera et al. (2023), in two oil palm populations known as C0-Monterrey and C1-Vizcaina, there are larger genetic gains (Δg) for the yield attributes of FFB (19%), for per palm bunch number (18.7%), and for the oil to bunch ratio (6%).

According to Kushairi and Amiruddin (2020), the Deli dura makes up nearly all of the basic population of the maternal line in the majority of breeding operations. The AVROS, La Me, and Yangambi populations were principally responsible for improvements in tenera/pisifera paternal lineages. Modified recurrent selection, or MRS, is a common technique used by oil palm breeders in Malaysia for improvement initiatives. Bunch and oil yields were increased by subsequent parental inbred lines that were created through recurrent selections, crossed, and progeny examined using heterosis. Mathur and Sunilkumar et al. (2015), reported a method to select pisifera palms based on the nursery performance of the D × P, which could help to reduce the evaluation time of pisiferas and, in turn, shorten the breeding cycle.

16.4.7 Interspecific Hybrids

Since it was discovered that *Elaeis oleifera*, also known as *E. melanococca*, can cross with *E. guineensis* to produce interspecific hybrids, the American oil palm has attracted attention. There are three known varieties of *E. oleifera*: Brazilian, Surinamese, and Central American. A few phenotypic traits set the Brazilians apart from the other two varieties, such as the notably larger fruit size. A hybrid plantation has a significantly lower height increment than a classical one, and depending on the combination chosen, it may have an economic lifespan up to three times longer. However, low oil content, high vegetative vigor, short fruit size, low fertility, and natural pollination are the key issues with interspecific hybrids. All varieties of hybrids share certain desirable traits, such as a lower increase in trunk height, an oil richer in carotenes, tocopherols, and unsaturated fatty acids, a brighter fruit color, and a high level of tolerance, or even resistance, to almost all economically significant pests and diseases, such as those found in Latin America.

Murugesan and Sunilkumar (2014) reported five genetic resources, three of *E. oleifera* and two of *E guineensis* with dwarf/compact traits among the collections available in India. The same genotypes have been incorporated into the dwarfness breeding program. The exploitation of oleifera and the production of interspecific hybrids with reduced height increment in India were reported by Sunilkumar et al. (2015). Accordingly, some of the interspecific hybrids are in advanced evaluation trials.

16.4.8 Traditional Breeding: Limitations

The main oil palm breeding objective is to reach a level of selection intensity that allows for advancement while preserving a suitable degree of variability for the next generations. Because oil palms are perennial plants, their breeding cycle lasts around 12 years for each generation, making conventional breeding a labor-intensive technique to increase oil palm productivity. Deli dura and AVROS pisifera populations are extensively bred and selected over many generations to develop hybrid seeds that are used worldwide. These populations are genetically related to one another. Breeding programs incur high expenses due to the substantial manpower investment required for recording and maintaining data. Oil palms are cross-pollinated and propagated only by seed, leading to a high level of heterogeneity in the field.

16.4.9 Haploid Breeding

Because oil palm crops have a long lifespan, breeding progress is very slow. A new oil palm generation needed for commercial seed production usually takes 20 years to grow. Since it would take at least 40 years to produce inbred lines (by eight generations of selfing), there are no accounts of breeders doing so. Therefore, conventional methods are presently the primary means of achieving genetic improvement of oil palm. Breeding perennial crops like oil palms requires enormous acreage for breeding experiments and a lengthy sequence of time-consuming backcrosses, in addition to their extended generation cycle. The development of chromosome doubling strategies in oil palm to produce haploid/double haploid has been the subject of some documented investigations. Although the main focus of these investigations is on doubling diploid material to produce polyploids, the procedure outlined may also help double haploid material (Madon et al. 2005). The process of doubling haploids to create two homozygous elite lines can be crossed to create genetically homogeneous, extremely heterozygous "hybrid" types. However, because of the oil palm's unique breeding system, which makes it difficult to develop inbred lines, there hasn't been much advancement in this field. Oil palm is an outbreeding species because each plant produces either male or female blooms at a given period and differs from maize, which develops both male and female flowers simultaneously on the same plant. Hence, it is difficult for oil palm plants to self-pollinate without controlled pollination with stored pollen. In order to utilize hybrid vigor, oil palm hybrid varieties must be developed. This can only happen when a dependable method for producing homozygous plants is developed. However, no haploid or homozygous diploid oil palm plant has ever been documented. To improve the genetics of the oil palm crop, however, a lot of breeding (Wan et al. 2011), cell culture (Abdullah et al. 2005; Shull 1908; Teixeira et al. 1994), and transformation experiments (US Application 20030159175) have been conducted.

It should be noted that Maluszynski et al. (2003) mistakenly claimed that in 1994, Texeira et al. published a technique for oil palm that doubles haploid production. However, this claim is inaccurate. Somatic embryogenesis from diploid floral tissue is reported here, contrary to the previous study that describes the cultivation of anthers or other reproductive tissue to produce haploid embryos and plants.

Wide hybridization is one method of producing haploids, in addition to the two crucial haploid-origination routes (gynogenesis and androgenesis). Although haploids can sometimes arise organically or by chance, they are uncommon. Many researchers have attempted to search for spontaneous haploids. Dunwell et al. (2010) screened 21,90,000 seedlings using SSR markers to identify natural haploid plants but only found one doubled haploid, indicating a very low frequency of natural doubled haploids. Therefore, a more effective approach may be to develop a sufficient number of haploids through *in vitro* methods (Sunilkumar and Sparjanbabu 2013a, b).

A few significant haploids have been identified in several plants, including rapeseed, barley, wheat, triticale, maize, and rice. Ceballos et al. (2007) state that the advantages of doubled haploids or haploids include producing better hybrids, fixing trait combinations that satisfy industrial needs, improving performance in hybrid combinations, facilitating hybrid or wide breeding, using them in molecular marker research, and increasing the development of mapping populations and trait loci. Crossing or hybridizing one crop cultivar with another one that is incompatible with it is sexually known as "wide hybridization" (Murphy 2006). Several Brassicas, barley, wheat, rice, melon, pepper, and tobacco have all been treated with this method (Ahloowalia et al. 2004). In an attempt to produce haploids, pollen from Cocos nucifera has been crossed with *E. guineensis* (dura), according to Wan et al. (2011). Before being planted in a jiffy, culled oil palm germinated seeds were examined for anomalies, and 12 weeks later, flow cytometry analysis was performed. Genome size and ploidy levels have been estimated using flow cytometry analysis, which is an easy, precise, quick, and practical technique (Dolezel 1991).

Screening of seedlings cultivated from this study's culled germinated seeds revealed a 1.36% possibility of obtaining likely haploids. Unfortunately, the seedlings perished, which might have been brought on by unfavorable planting conditions or by their inherent weakness, a half-sized genome. Ten plantlets were created by the widespread hybridization of *C. nucifera* with *E. guineensis* (dura), from which embryos were extracted from 15-week-old fruits and subcultured in rooting media.

16.4.10 Role of Biotechnology

The application of diverse biotechnological instruments and methodologies has yielded noteworthy outcomes for oil palm improvement initiatives. Important genes and markers linked to desirable features in oil palm have been found through the application of molecular approaches such as marker-assisted selection (MAS), QTL mapping, and genome-wide association studies (GWAS). Clonal propagation and regeneration of oil palm plants have been achieved through the development of tissue culture and somatic embryogenesis procedures for large-scale production and genetic improvement. The identification of genomic regions associated with important oil palm traits has been made easier by the development of genetic linkage maps and high-density SNP genotyping arrays. Oil palm breeding operations may be made more effective and productive through the application of genomic selection, which uses genetic information to forecast an individual's breeding value.

16.4.11 Major Breeding Programmes

Historically, there have been numerous breeding programs, some of which are still ongoing today, while many new programs have been initiated with the development of the oil palm sector. In 2009, above 300 million seeds of oil palm were produced from approximately 40 production centers and this number has continued to increase. A brief description of major breeding programs and seed production is provided below.

16.4.11.1 Africa

RRS was mostly used, which necessitates two base populations. The African program's primary sources have been Deli dura on the one hand and Tenera on the other. One base population consisted of palms producing many small bunches, while the other produced a few large bunches. Country developments are discussed further.

Congo

INEAC developed the initial stage of the scheme after the discovery of the shell thickness inheritance involved a diallel cross of six teneras and was mostly based on RRS. Out of 36 possible crosses, only 20 were ultimately planted due to political unrest in Congo.

Côte d'Ivoire

The origin of the breeding program was the one set up by IRHO (later CIRAD-CP). CIRAD-CP (currently represented by Palm Elit SAS) and the Centre National de Recherche Agronomiques in Côte d'Ivoire, in collaboration with the Institut National de Recherche Agronomiques pour le Development in Cameroon, Socfindo, and the Indonesian Oil Palm Research Institute (IOPRI) in Indonesia, planted trials containing approximately 1300 test crosses from the second cycle RRS (Cochard et al. 1995). By 2009, Côte d'Ivoire and Benin together produced about 16 million seeds from the breeding programs.

Nigeria

At NIFOR, the program was primarily built around RRS. Dura Seeds originated from both the tenera self, as in the INEAC program, and the dura selves and crosses, as in the IRHO program. There are now many excellent tenera palms with shells that are less than 7%; material from Nigeria that was grown in Cote d'Ivoire produced tenera progenies that had palms with 83–90% mesocarp and only 4–9% shell (Hartley 1988).

Ghana

NIFOR materials were brought to Ghana, where genetic values for the parents of the Ghana program was computed using data from the NIFOR program.

Cameroons

There were two primary programs in Cameroon: the Palmol program at Lobe estate and the IRHO program at La Digamba, which contained a lot of the same content as the Ivory Coast Côte d'Ivoire program. In 1953, the Palmol program began producing D × P seeds at the Cowan and Nidian estates. Fertile pisiferas were used in production, and the basis was Ekona material without any Deli duras. Here, too, the approach used was RRS.

16.4.11.2 Asia

Indonesia

World War II had a devastating effect on oil palm breeding in the Far East. However, AVROS continued to breed oil palms under a different name, the Research Institute of the Sumatran Planters Association (RISPA). Additionally, the Research Station at Marihat was established, and both the Marihat and the RISPA programs are now part of the IOPRI program. At least seven different Deli dura subpopulations were present in the sample, along with African teneras such as Yangambi, La Me, NIFOR and AVROS. Working with PT Socfindo, the CIRAD initiative, which originated in Côte d'Ivoire, was duplicated in Sumatra. Nine additional programs in Indonesia were identified by Kushairi et al. (2010), totaling 250 million seeds produced in the nation. Three sites in Sumatra were chosen for a program evaluating 450 D × P crossings. The ASD program in Costa Rica supplied 60 Pisifera clones (Yangambi, AVROS, Calabar La Me, and Ekona) and 225 Deli selfings (all Deli) as the parents of the crosses (Breure 1988). Working with Dami (PNG), another program was developed using Deli × AVROS materials that were originally from Malaysia.

Malaysia

Representative samples of the majority of populations were established by Malaysian breeding's engagement with a number of exchange operations with African centers up until the 1970s. Four of the major Malaysian plantation companies founded the Oil Palm Genetics Laboratory (OPGL), a breeding and genetics center, in 1965 to help partner companies with their selection and seed production programs and to provide a more solid genetic foundation for breeding and selection. Pisifera selection was first based on the performance of dura and tenera sibs in T × T crossings and, if possible, on the outcomes of the D × P progeny tests. The technique used was a system of family and individual mass selection (FIS).

As the Malaysian oil palm sector continues to grow, an increasing number of organizations are beginning to produce seeds and engage in breeding. In 1997, there were 13 commercial seed producers. All nine of them employed Deli Dura as their only female parent and produced Deli × AVROS material. Deli × Yangambi material was also developed by the Felda program, Guthrie, IOI, and United plantations. Some seed manufacturers used Pisifera that were descended from Dumpy × AVROS. Eleven organizations generated 81 million seeds as of 2008 (Kushairi et al. 2010).

Thailand

In Thailand, two breeding initiatives were started in the 1960s. Materials from several origins are included in both programs and Siam Elite Palm is a CIRAD joint venture. About 13 million seeds were produced in 2009 (Kushairi et al. 2010).

Papua New Guinea (PNG)

When the Dami oil palm research station program was first launched in the late 1960s, it used relatively limited samples of Malaysian Deli and AVROS material. Approximately 30 million Deli × AVROS seeds are produced annually by the RRS-based program (Kushairi et al. 2010). Dami took part in the 1980s Combined Breeding program exchanges after seeing the necessity to expand the genetic base.

India

The breeding program was started at a modest level at Central Plantation Crops Research Institute's Regional Station in Kerala, India as early as the 1970s. However, sustained efforts were made only during the mid-1990s by establishing a dedicated research institute for oil palm (presently ICAR-Indian Institute of Oil Palm Research, at Pedavegi in Andhra Pradesh and functioning under the Indian Council of Agricultural Research. The initial breeding program utilized material imported from Malaysia and Nigeria, which predominantly were Deli × AVROS crosses. Subsequently, materials were sourced from Zaire, Nigeria, Cameroon, and Guinea-Bissau Bissau and presently about one million seeds are produced and the remaining requirement is met through imports. It is too brief. We could give our experience on the performance of Indigenous tenera under rainfed and irrigated conditions and the performance of Indigenous materials vs imported ones.

16.4.11.3 Latin America

Most of the initial plantations in Latin America were established using seeds brought from Lancetilla, Honduras.

Costa Rica

In the 1960s, United Fruit Company began collecting *E. oleifera* material, which was traded for cutting-edge breeding lines from multiple research facilities. The current ASD Costa Rica program includes a diverse range of BPROs as a result of these exchanges. Although crosses between Deli × Ekona and Deli × Calabar were also generated, the seed produced is primarily Deli × AVROS. Seeds of about 30 million were produced annually in 2009 (Kushairi et al. 2010).

Columbia

Instituto de Fomento Algodonero began producing deli dura seeds in Colombia in 1958 using materials from Lancetilla. Additionally, Unipalma SA established and is using material from the joint breeding program (as in Congo) in Colombia for the manufacture of seeds.

Ecuador

The INIAP (Instituto Nacional de Investigaciones Agropecuarias) founded an oil palm research facility in the Northwest area in the early 1960s, not far from the first plantation to be grown with Lancetilla open-pollinated seeds.

Brazil

In the 1950s, seeds were shipped to Brazil from Nigeria; in the 1960s, they were sent to Côte d'Ivoire; and in the 1970s, they were sent to Malaysia with AVROS and Deli Dura. Barcelos and Amblard (1999) provided summaries of the breeding programs of Empresa Brasileira de Pesquisa Agropecuaria (EMBRAPA) and Comissão Executiva do Plano da Lavoura Cacaueira (CEPLAC). The majority of the material used in the EMBRAPA program came from IRHO.

16.5 Molecular Breeding

Two approaches mainly followed in oil palm for molecular breeding are Marker Assisted Selection (MAS) and Genomic Selection (GS).

16.5.1 Marker-Assisted Selection

MAS is a breeding process that uses trait-associated DNA markers to pick elites and speed up genetic progress (Xu 2000). Before selected elite varieties can be released, the breeding cycle can be significantly shortened and the number of breeding cycles required can be decreased by using DNA markers for selection in oil palm (Soh et al. 2018). Using this approach, choices could be made right up to the seedling or nursery stage.

16.5.1.1 Types of Molecular Markers

Protein Markers

Molecular markers were first applied to proteins containing isoenzymes or isozymes, which are distinct versions of an enzyme that are separated and whose inheritance is investigated (Ataga and Fatokun 1989; Baudouin 1992; Choong et al. 1996; Ghesquiere 1984, 1985). Because protein markers vary throughout tissues and even within the same tissue, they are generally challenging to deal with. It may also differ depending on the physiological state of the plants, making it challenging to collect equivalent samples from various plants. Additionally, the number of enzymes for which isozymes are readily separable is relatively small (Corley and Tinker 2003). But even with the limitations it can be effectively used for specific purposes and selection processes for the breeding program. The major advantage of the protein markers is found to be its expressivity. Nucleic acid markers showing differences in genotype (even in nucleotide sequence) are of little use in breeding if the difference is not expressed phenotypically. Hence, protein markers can be complementary to DNA markers and still, it is used for the characterization and assessment of the genetic diversity of oil palm (Mandal and Jayanthi 2011).

DNA Markers

DNA markers are used for different purposes where the differentiating palms are required for (a) studying genetic diversity (b) phylogenetic relationships and species interrelationship (c) genetic fidelity assessment of micro propagated plants and identification of somaclonal variants (d) identification of genetically modified plants (e) generation of molecular linkage maps including QTL mapping and tagging useful genes (f) gene pyramiding and (g) marker-assisted selection.

Since DNA is based on the unique structure of genetic material, it is the most dependable type of marker. However, instead of replacing other markers, DNA markers rather complement other marker systems, for example, when a close linkage of an inexpensively scored isozyme for a desirable trait has been identified, this will continue to be a method of choice (Jack and Mayes 1993).

Restriction fragment length polymorphisms (RFLPs) are the oldest known DNA marker technology. The drawback of RLFP was that it required a lot of work and time, but it produced consistent and dependable results. This is a result of the requirement for extracting and purifying the DNA, exposing the photographic film used to detect radioactive probes and digesting it with restriction enzymes. The use of DNA marker systems with special reference to RFLP in oil palm breeding and tissue culture cloning programs was described by Jack and Mayes (1993).

Microsatellites or Simple sequence repeats (SSRs), random amplified polymorphic DNA (RAPD), amplified fragment length polymorphisms (AFLP), and restriction fragment length polymorphisms (RFLPs) are often used in marker systems. A convenient marker system or combination of them is used for different purposes. Some of the variants of DNA markers are gene expression-based, derived basically from RNA, like the expressed sequence tags (ESTs). Cheah and Ives (1998) in their molecular genomic program in Malaysia, examined the oil palm genome using the technique of gene linkage mapping, molecular cytogenetic analysis and cDNA sequencing for the generation of ESTs. Although the primary objective of the program was to enable MAS, the molecular probes developed were also used for fingerprinting the tissue culture clone. Up to date progress towards the development of such specific DNA profiles in oil palm is discussed here.

Marker-assisted selection (MAS) is increasingly employing SSR markers in many crops, including oil palm. Billotte et al. (2001) provided primers, estimates of the allele size range, and predicted heterozygosity in *E. guineensis* and the closely related species *E. oleifera*. An optimal application of the SSR marker was observed. The invention and use of 12 informative SSR markers that were deemed appropriate for oil palm tissue culture clone genome analysis and DNA fingerprinting were documented by Singh et al. (2007).

Similar to other crops, the development of ESTs is progressing in oil palm. The first report of a systematic investigation of genes expressed in oil palm using EST analysis was published by Jouannic et al. (2005). A unidirectional systematic sequence was performed using five distinct cDNA libraries that were created from the shoot apices, zygotic embryos and female as well as male inflorescences. 209 linked sequence groups and 1874 singletons were found using cluster analysis on 2411 valid EST sequences. Out of the non-redundant ESTs collected, 1252 were assigned putative functions. The creation of the EST database marked the beginning of oil palm gene discovery and cDNA array-based expression analysis.

The usage of SNPs (single nucleotide polymorphisms) and small indels (insertions/deletions) as molecular markers has revolutionized since a high-throughput method for detecting them was developed. Candidate SNPs and intron-exon junctions were found by Riju et al. (2007) through the use of 576 ESTs from oil palm databases that were mined. It was composed of 137 indel polymorphisms and 1180 SNP sites, with 1.36 SNPs per 100 bp frequency. SNP frequency in oil palm has been confirmed by the report to use them as markers for genetic research. The oil palm genome is estimated to be 1.8 million bp in size, which is around four times larger than the genomes of rice and maize combined.

16.5.1.2 Markers For Genetic Diversity Analysis

Exploiting genetic diversity is essential if oil palm improvement is to be successful in the long run (Adon et al. 1998). Thus, estimation of the genetic relation of different palm species/varieties/populations is extremely necessary before taking up an oil palm breeding program (Billotte et al. 2001). Traditionally, morphological descriptors were the only information available for the assessment of genetic diversity. Once the molecular markers were developed, they were widely used for detecting the polymorphism and distinguishing the individuals or groups. Reports suggested the use of protein markers in oil palm since 1984 (Ghesquiere), where the enzyme polymorphism was studied in oil palm using nine enzyme systems. Subsequently, Ghesquiere (1985) reported genetic variability of oil palm germplasm accessions from seven different origins using this procedure. An investigation of isozymes from the pollen of seven *E. oleifera* accessions and a hybrid between *E. oleifera* and *E. guineensis* was conducted by Ataga and Fatokun (1989).

Singh et al. (2013a, b) reported that the genomic sequence of the African oil palm *E. guineensis* was 1.8 gigabase (Gb) in size. To anticipate at least 34,802 genes, including those involved in oil production, a homolog of WRINKLED 1 (WRI 1), and other transcriptional regulators that are highly expressed in the kernel, 1.535 Gb of assembled sequencing and transcriptome data from 30 tissue types were employed. The draft sequence of *E. oleifera*, the South American oil palm, was also published. This species makes a fertile interspecific hybrid with *E. guineensis* and has an equal chromosome number (2n = 32), although it seems to have diverged in the New World.

The first publication on the use of RAPD markers for oil palm genetic diversity studies was by Shah et al. (1994). Oil palm germplasm accessions from Africa (Cameroon, Tanzania, Nigeria, and Zaire) have been studied using 20 primers. The results showed significant genetic heterogeneity between the accessions. Then, in order to find RLFP markers that would enable genotype discrimination, a range of DNA probes were employed to screen a different collection of oil palm genotypes (Jack et al. 1995). Mayes et al. (2000) used RFLP markers, which contain 40 probes covering 60% of the oil palm genome, to assess the genetic diversity among 54 palms in a specific oil palm breeding program. To give a larger framework for comparative study, ten samples with a wider range of provenance were added to the 54 palms. After 157 RFLP bands were scored, the data analysis showed that the gene pool of AVROS pisifera is different from the Southeast Asian Deli dura material.

Purba et al. (2000) studied the implications of breeding on a novel facet of the genetic diversity of oil palm collection in Indonesia, as shown by isozymes and AFLP markers. They employed five chosen AFLP primer pairs and four isozyme systems to evaluate 48 parents from the Indonesian oil palm breeding program, which represented four populations. Out of the 158 scorable band levels that were produced, 96 (or 61%) were polymorphic. Crosses within the subpopulation of Africans could be more interesting than crosses between the Deli and African populations, according to the AFLP results.

In their study, Maizura and Rajanaidu (2001) used RFLP to characterize 11 oil palm germplasm collections from Nigeria, Cameroon, Congo DR, Tanzania, Angola, Senegal, Sierra Leone, Guinea, Ghana, Madagascar, and Gambia. They discovered that all of the germplasm collections had greater genetic diversity than the breeding population (Deli dura). Kularatne et al. (2001) used AFLP markers with eight primer combinations to screen 687 accessions from 11 African countries, as well as Deli dura. A total of 377 bands, 97.6% of which were polymorphic, were detected, with an average of 47 bands per primer pair. Oil palm's overall genetic diversity was highest in Nigeria and steadily declined in Senegal, Gambia, Angola, and Tanzania. Three primary groups of oil palm, representing three different locations in Africa, were identified by cluster analysis. The study also demonstrated that Deli dura has a low level of genetic diversity, which may be the consequence of multiple selection cycles. It was discovered that Deli dura is most closely linked to Congo oil palm.

Barcelos et al. (2002) employed RFLP and AFLP to evaluate genetic diversity, its organization, and genetic connections among oil palm germplasm (both *E. oleifera* and *E. guineensis*). The same 37 cDNA probes were used to evaluate 241 *E. oleifera* accessions and 48 *E. guineensis* in the context of the RFLP analysis. Additionally, an AFLP analysis was performed utilizing three pairs of enzymes and primer combinations on a subgroup of 22 *E. guineensis*, 14 *E oleifera* and accessions. The robust structure of diversity in *E. oleifera* that was identified by RFLP was also validated by AFLP.

Using RAPD markers, Moretzsohn et al. (2002) studied the genetic variations in an *E. oleifera* accession that was taken from the Amazon forest. Comparing Brazilian *E. oleifera* accessions to *E. guineensis* accessions, the results indicated that the former possessed moderate levels of genetic diversity. As would be predicted for an allogamous and perennial species, the ANOVA revealed that the majority of the genetic variation was found within the population. Using the standard Deli dura as a control, Maizura et al. (2006) used the RFLP approach to characterize 359 accessions of oil palms (*E. guineensis*) that originated from 11 African nations. Overall, every collection showed more diversity than the average amount. In comparison to natural populations, Deli dura, the standard variety, recorded a loss of 36 alleles, showing a decrease in genetic variability. The majority of loci (67.2%) showed polymorphism and the mean value of alleles per locus was 1.9, found in materials from Nigeria.

A diversity investigation of five dura germplasm accessions in India showed a higher level of DNA polymorphism between the various palms and accessions analyzed. Following that, RAPD was used to examine five exotic accessions of oil palm (ASD Costa Rica, Zambia, Guinea Bissau, Tanzania, Cameroon, and Palode dura) for biochemical and molecular characterization (Mandal and Susmita 2006). In contrast to other groups, the Guinea-Bissau accession's palms were remarkably homogeneous, and they exhibited a greater genetic distance from other groups. Accessions to ASD in Costa Rica and Cameroon are very close to one another.

The effectiveness of SSR markers in displaying the genetic diversity structure of the genus *Elaeis* by known geographic origin and estimated genetic links based on prior molecular investigations was reported by Billotte et al. (2001). Bakoume et al. (2014) used 16 SSR markers to investigate the genetic diversity of 194 oil palms from 49 groups. The genetic diversity was significant, with a total of 209 alleles detected, or an average of 13.1 alleles per locus and a mean predicted heterozygosity of 0.644. The genetic and phenotypic diversity of *E. oleifera* accessions from several South American nations was assessed using agro-morphological features and microsatellite markers (SSRs) (Arias et al. 2015).SSR marker research showed that *E. oleifera* has a high level of genetic diversity ($HT = 0.797$) and that different alleles are present depending on the place of origin. Four distinct *E. oleifera* groups were revealed by the clustering topology, which matched the accession's geographic distribution and demonstrated limited gene flow ($Nm = 0.238$) and strong genetic divergence ($GST = 0.512$) among the participating nations. The findings allowed them to assert that *E. oleifera* has a distinct genetic structure, phenotypic variability, and distinct traits depending on its region of origin. Additionally, the accession of each of these countries has contributed to the overall rise in genetic variety.

16.5.1.3 Marker for Detection of Somaclonal Variation and Clonal Fidelity

Wider adoption of tissue culture technology or clonal propagation of oil palm was hampered by large-scale incidences of bunch failure in some clones. In 1986, it was observed that some of the oil palm clones were not flowering normally after 4 years and produced inflorescences with 'mantled' flowers. There is no fruit development and the mantled flowers are not properly functioning, resulting in serious bunch failure. This kind of abnormality was expected to have a direct effect on oil production and hence considerable efforts were made to understand the causes and possible remedies of this abnormality.

Several factors were attributed to the occurrence of the mantled flower abnormality. Corley et al. (1998) found that, on average, 5–10% of the clones may be affected. The proportion of abnormal flowering increased with increased subculture stages that a clone has undergone (Paranjothy et al. 1995). Eeuwens et al. (2002) described from a 15-year trial that the medium on which embryoids are cultured has a large effect. Rao and Donough (1990) described the preliminary proof of the genetic basis for the abnormalities. The identification of a reliable marker connected with the floral abnormality will be an important asset in screening out abnormal culture and should also be useful in identifying the *in vitro* protocol, if any, that induces abnormalities (Mathius et al. 1998). Various gene isolation and identification methods have been employed in checking the clonal materials, which included RAPD (Paranjothy et al. 1995), RFLP and RAPD markers (Meunier 1995; Mayes et al. 1996), ESTs (Sharifaha et al. 1999). Mathius et al. (1998) reported that the mRNA analysis by RT-PCR has shown potential for the selection of nucleic acid markers in oil palm clones. Singh et al. (2007) demonstrated the use of SSR markers in oil palm tissue culture. They revealed the development of 12 SSR markers that can be used as molecular probes to identify oil palm clones using DNA fingerprinting. The SSR markers are useful for identifying clones, keeping an eye on line uniformity within the clone, and spotting mistakes in culture.

One of the main changes brought about by in vitro cultivation may be a change in the methylation state. It was discovered that the DNA methylation of variant regenerants differed significantly from that of normal regenerants. Significant demethylation was noted by Rival et al. (2002) in heavily mantled plants. According to Shah and Ahmed-Parveez's (1995) research, regenerants with anomalies had significantly higher quantities of 5-methylcytosine than normal regenerants. Jaligot et al. (2000) reported the tendency towards hypomethylation in aberrant plant material, while loss of methylation was detected by Matthes et al. (2001) during tissue culture of oil palm. The incorporation of AFLPs with methylation-sensitive enzymes has made it possible to detect certain tissue culture-induced polymorphisms in oil palm regenerants that were not discovered using other methods. However, only within the same clonal progenies did methylation-sensitive markers distinguish between the two morphologies (Jaligot et al. 2004). It was discovered that the cDNAs of two oil palms, CPH062 and CPH063, showed a distinct methylation pattern in normal and aberrant embryogenic calli (Jaligot et al. 2002). Morcillo et al. (2006) reported that EgM 39A and EgIAA1, two novel oil palm genes were found to have the potential to serve as early markers of clonal conformance. Both of these genes displaced enhanced transcript accumulation in epigenetically aberrant calli. Differential expression studies conducted on oil palm tissue cultures carrying or missing an epigenetic homeotic blooming defect have identified and characterized EGDA1, a gene coding for a potential plant defense (Tregear et al. 2002). During the callus phase of the in vitro regeneration process, EGAD1 transcripts were found to accumulate differently depending on whether the mantled blooming anomaly was present or not. The expression of the EGAD1 gene may therefore serve as a marker for epigenetic somaclonal variant processes.

The EST collection contained homologous genes related to several stages of oil palm flower development, including AGAMOUS-like, CONSTANS-like, AGL2, AGL20, LFY-like, SQUMOSA, and SQUMOSA binding protein (SBP). The MADS-box genes in general are essential for controlling floral structure and reproductive development and hence, Helene et al. (2007) proposed that these genes are likely at the center of many important evolutionary events in plants. The kind of morphological alterations indicates that homeotic MAD box genes, which control floral whorl identity, are involved. Through the use of MAD box-directed mRNA display, several MAD box genes from oil palm inflorescences have been identified.

In India, so far, no work on the genetic fidelity of clonal palms has been reported. However, a variation of esterase isozymes in the embryogenic shoots derived from Tenera seedling explants of oil palm was reported by Anuradha and Karun (2000). RAPD and SSR analysis of clonal oil palm from farmers' fields was carried out (Jayanthi et al. 2008). Out of the ten RAPD primers, three exhibited polymorphism and all ten SSR primers studied produced monomorphic bands with the DNA of the clones. Chan et al. (2014) reported the importance of the reference gene selection for accurate quantification of the extent of gene expression in oil palm. Accordingly, PD00380 and PD00569 were identified as the most stably expressed genes in a total sample of tissue culture lines for RT-qPCR. The somaclonal variation known as the "mantled" fruit anomaly in oil palms originated from tissue culture. It significantly decreased yield, which hindered efforts for cloning elite hybrids for oil output. Many people believed it to be an epigenetic phenomenon. To identify the mantled locus, Abdulla et al. (2015) used epigenome-wide association studies of the African oil palm (*E. guineensis*). In the intron of the homeotic gene deficiencies, a LINE retrotransposon associated with rice Karma had the same DNA hypomethylation in all mantled clones. This was linked to alternative splicing and premature termination.

A normal fruit set was anticipated by dense methylation near the Karma site, also known as the Good Karma epiallele, while homeotic transformation, parthenocarpy, and a significant loss of yield were predicted by hypomethylation, also known as the Bad Karma epiallele. It was discovered that the formation of mantles was related to the loss of Karma methylation and small RNA in tissue culture, whereas non-Mendelian inheritance was explained by restoration in spontaneous revertants. Predicting and eliminating mantling at the plantlet stage will make it easier to introduce more successful crosses and maximize land resources that are sensitive to the environment.

16.5.1.4 Shell Thickness Marker and Identification of Hybrids

Dura (thick-shelled), pisifera (shellless), and tenera (thin-shelled), a hybrid of dura and pisifera, are the three fruit varieties of the oil palm. Typically, female pisifera palms are sterile. The *tenera* palm yields far more oil than *dura* and hence is the only commercial variety. Shell thickness has a monogenic inheritance and hence, the work carried out on shell thickness markers is very important. Singh and Cheah (1999) found that AFLP reveals a high level of polymorphism between Colombian *E. oleifera* (female) and Nigerian *E. guineensis* (male) parents used to generate the inter-specific hybrids. Using only ten primer pair combinations, 674 bands were amplified, and 91 of them showed segregation. Every single marker found was classified as dominant, and the segregation ratio showed that 80% of the markers were inherited in a Mendelian fashion. Since the AFLP method might also be used to identify impurities, it could help determine the fidelity of controlled crosses.

A significant advancement in oil palm marker research was made by Singh et al. (2013a, b) when they mapped and identified the SHELL genes in charge of various fruit morphologies. Thus, they discovered two distinct mutations in the DNA binding domain of a homolog of the MAD box gene SEEDSTICK (STK, also called AGAMOS-like 11), which regulates ovule identity and seed formation in Arabidopsis, utilizing homozygosity mapping by sequencing. Their finding provided a genetic explanation for the heterodimerization-based single-gene hybrid vigor ascribed to SHELL. The tenera phenotype in sub-Saharan African palms, both cultivated and wild, is caused by the SHELL gene. The most significant economic characteristic of oil palm is this mutation.

Three pairs of primers, along with two restriction enzymes, make up the molecular marker system that Ritter et al. (2016) designed to distinguish between the three distinct Sh alleles. Most of the oil palm breeding operations currently in use employ the standard gene pool, which was covered by the verified marker system in dura and pisifera genotypes from various origins.

16.5.2 Linkage Maps and Quantitative Trait Loci (QTLs) Mapping

Finding the markers linked to the desired feature is a fundamental need for MAS. Making a linkage map is required for this. By looking for a wide range of markers in a large population of sibs, it is possible to find pairs or groups of connected markers that are likely to be inherited together. Corley and Tinker (2003) suggested that these groupings are predicted to be on the same chromosome, and the statistically determined proximity of the linkage indicates the markers' relative positions along the chromosome. The basis for constructing physical maps of coconut and oil palm was established as a result of a collaborative project. The consortium consisted of molecular biologists, breeders and producers in Southeast Asia (Philippines, Malaysia, Indonesia) along with European partners. The project established a basis for the construction of physical maps in oil palm as well as coconut. Details of maps, markers and alignments as well as QTLs were presented on the project web page (http://www.NEIKER.net/link2palm).

Jack and Mayes (1993) reported a program using the oil palm RFLP linkage map for the identification of markers associated with shell thickness, Fusarium wilt resistance, oil quality, and yield. The first map that was based on RFLPs was presented by Mayes et al. (1997). By segregating for the shell thickness character (Sh), a mapping population produced through the selfing of a significant breeding material made it possible to map the gene encoding this essential commercial trait. In the mapping population, the closest RFLP marker (pOPgSP1282) was 9.8 cm from Sh, but in a comparable smaller population of 45 palms (A137/30xE80/29), it was 6.6 cm. Jack et al. (1998) updated the map and reported tagging two genes: virescence (Vr) and shell thickness (Sh).

Through the examination of an F_1 tenera × pisifera progeny, two single tree linkage maps were made for a paternal pisifera (sh−sh−) palm and a maternal tenera (sh+sh−) palm using a pseudo-test cross-mapping technique in combination with RAPD markers. RAPD and bulk segregant analysis were used to identify markers nearer the sh+ locus. Two RAPD markers (R11–1282 and T19–1046) were discovered to be connected on both sides of the sh+ locus in linkage group 4. The estimated map distances from sh+ to R11–1282 and T19–1046 were 17.5 and 23.9 cM, respectively, according to Moretzsohn et al. (2000).

Rance et al. (2001) reported an oil palm RFLP marker for quantitative trait loci (QTL) mapping. Utilizing an F_2 population's phenotypic data and information on 153 RFLP markers, the location and effects of QTLs for parameters such as fruit production and its constituents, as well as measurements of vegetative development, were estimated. A linkage map with 22 linkage groups was created by analyzing marker data from the mapping population, which comprised 84 palm trees. Both single-marker and interval mapping analysis revealed significant marker-associated QTL effects for 11 of the 13 characteristics analyzed. The interval mapping analysis of fruit weight, petiole cross-section, and rachis length, along with the ratios of shell: fruit, mesocarp: fruit, and kernel: fruit, identified significant ($P < 0.05$) QTLs at the genome-wide threshold. Using these QTLs in innovative breeding techniques, such as marker-assisted selection for commercial populations, was proposed.

Chua et al. (2001) reported a *dura* map that covered a genetic distance of 555 cM and a *pisifera* map covering a genetic distance of 885 cM. On the three maps generated by the same group, loci for carotene content and clonal abnormalities have been identified. Functional genomics was effective in identifying several differentially expressed genes, yielding 14,537 ESTs from these libraries, from which 2129 singletons and 6464 tentative unique contigs (TUCs) were extracted (Ho et al. 2007; Singh et al. 2001). Gene homologs for enzymes involved in oil production, nitrogen source utilization, and oxygen radical scavenging were discovered among the oil palm ESTs.

A hybrid of two heterozygous parents, a tenera palm from the La Me population and a dura from the Deli population, was used to create a high-density linkage map for oil palm by Billotte et al. in 2005 using microsatellite technology. In oil palm, a set of 390 SSR markers was created. A linkage map comprising 225 microsatellites, 688 AFLPs, and the locus of the Sh gene, which regulates whether an oil palm fruit has a shell or not, was created thanks to a dense and genome-wide microsatellite framework and saturating AFLPs. E-Agg/M-CAA132, an AFLP marker, was mapped 4.7 cm from the Sh locus. Covering 1743 cM, 16 linkage groups (LGs) received distributions of the 944 genomic markers. With 16 distinct linkage groups that correspond to the 16 pairs of homologous chromosomes in the plant, this linkage map was said to be the first in the oil palm species. In the *E. guineensis* species, quantitative trait loci analysis through physical mapping was made possible by this high-density linkage map.

In addition to the direct DNA markers, *in vitro*-synthesized protein from mRNA was reported as a marker for character tagging. Using two-directional gel electrophoresis, Singh and Cheah (2000) discovered a flower-specific protein in oil palm (*E. guineensis*). The inflorescence (flower) and vegetative (leaf and immature etiolated seedlings) tissues showed different banding patterns, which suggested differing gene expression. The pattern of in vitro translation products for mRNA taken from various blooming stages also showed some modifications.

Following the creation of the linkage map, an assembly of markers with a fairly uniform distribution throughout the entire chromosome can be selected. This group of markers can be used to screen other populations and investigate their relationship to relevant traits (Corley and Tinker 2003). It was suggested that heritable traits of economic importance, such as shell thickness, fruit color, height, some fruit and bunch components and fatty acid composition could be tagged with DNA markers (Lim and Rao 2005). The identification of RFLP markers linked to vegetative characteristics, oil/bunch and its constituents, and FFB yield was documented by Rance et al. (2001). Markers linked to the oil's oleic acid and carotene levels were discovered by Singh et al. (2001).

Billotte et al. (2010) tested a quantitative trait loci investigation in oil palm (*Elaeis guineensis* Jacq.) that was intended for a multi-parent population. Using 251 microsatellite loci, an AFLP marker for the Sh major gene that controls the presence of fruit shells, and the Sh major gene locus itself, a consensus genetic map was created. 24 quantitative phenotypic traits were linked to a collection of 76 QTLs. As a result of the linked families, the across-family analysis turned out to be effective. It was proposed that QTL markers for MAS and low progeny numbers could be found.

Using AFLP, RFLP, and SSR markers, Singh et al. (2009) produced a map for an interspecific cross between a Colombian *E. oleifera* (UP 1026) and a Nigerian *E. guineensis* (T 128). Interval mapping and multiple-QTL model (MQM) mapping, also known as composite interval mapping, or CIM, were used to find the QTLs controlling oil quality, which is based on the iodine value (IV) and fatty acid composition. At a 5% genome-wide significance threshold level, QTLs associated with I V, myristic acid (C14:0), palmitic acid (C16:0), palmitoleic acid (18:0), oleic acid (18:1), and linoleic acid (18:2) concentrations were found. Group I appears to have one chromosomal area that affects the content of C14:0, C16:0, 18:0, and 18:1. Around the same locus on Group 15, a significant QTL for C14:0, 16:1, 18:0, and 18:1 content was found, indicating the presence of another significant locus impacting the fatty acid composition in oil palm.

Ting et al. (2013) provided linkage maps enhanced with SSR markers for identification of QTLs linked with callogenesis and embryogenesis, to find QTL associated with oil palm tissue culture amenability. Using AFLP and RFLP markers as a basis, the SSR markers were mapped into previously published parental maps. With 23 linkage groups and 148 markers (33 AFLPs, 71 RFLPs, and 77 SSRs) covering a total of 798 cM map length, the new ENL48 linkage map was created. There were 24 linkage groups and 240 markers on the ML161 map. Two QTLs linked to tissue culture capacity were found for each of the callusing and embryogenesis rates using the updated maps.

For studying the genetics of oil palm, high-density genetic maps of crosses between individuals with various genetic backgrounds are essential resources. They are also helpful for identifying markers that are strongly associated with desirable qualities using comparative mapping analysis. According to Ting et al. (2014), two mapping populations, an intraspecific cross with 87 palms and an interspecific cross with 108 palms, were used to genotype the SNPs and 252 SSRs. Parental maps of 16 linkage groups were created for each of the three *E. guineensis* fruit varieties: dura, pisifera, and tenera. The map resolution was further improved by merging the dura and pisifera maps into an intraspecific integrated map with 1331 markers covering 1867 cM. They also released the first map of a Colombian *E. oleifera*, which covered 471 cM and contained 65 markers and 10 LGs. By directly comparing the parental maps, the 603 transferable markers that were polymorphic in at least two of the parents were identified. Subsequent analysis revealed a significant degree of marker transferability across the intra- and interspecific integrated maps at a distance of 1075 cM. The high-density SNP and SSR-based genetic map has substantially improved marker density and genome coverage in comparison to the original reference map based on AFLP and SSR markers.

Singh et al. (2014) reported on the discovery of oil palm fruit color markers. They claimed to have discovered the Virescens (VIR) gene, which controls the fruit exocarp's color and acts as a maturity indicator. VIR is an R2R3-MYB transcription factor that is homologous to Lilium LhMYB12 and shares characteristics with the anthocyanin pigment1 (pap1) produced by Arabidopsis. In more than 400 accessions from sub-Saharan Africa, they found five distinct mutant alleles of VIR that are responsible for the dominant negative virescent phenotype. Every mutation caused the carboxy-terminal domain of VIR to terminate prematurely, causing it to resemble the maize McClintock's C1-I allele. The discovery of VIR will make it possible to select for the trait 3–6 years before fruit production, at the seed or early nursery stage, significantly accelerating the introgression of the trait into superior breeding material.

Gan (2014) conducted QTL analysis and marker development investigations for use in MAS and genetic linkage mapping. 29 primer pairs were found using the AFLP approach, and these pairs produced 49 putative polymorphic bands associated with shell thickness. A total of 11,675 high-quality DArTSeq polymorphic markers were produced by genotyping of two closely related Tenera self-pollinated F_2 populations using the relatively new Diversity Array Technology Genotyping-by-Sequencing (DArTSeq) platform. The first known DArTSeq-based high-density maps were created using these markers. After the shell thickness (Sh) region was saturated with all available DArTSeq markers and the map integration surrounding the Sh regions for both populations, it was found that 32 SNPs and DArT markers were mapped inside a 5 cM flanking region of the Sh gene. A homology scan of the 64-bp marker sequence tag against the recently released oil palm genome assembly revealed that 23 of the 32 (72%) DArTSeq markers were located on the p5_sc00060 scaffold, which is where the SHELL gene was found.

Pootakham et al. (2015) revealed QTLs linked to agronomic characteristics in oil palm through GBS (Genotyping by Sequencing). They were able to locate 1085 of the 3417 fully informative SNP markers on a linkage map that covered 1429.6 cM and had an average of one marker per 1.26 cM. There were three QTLs influencing trunk height found in linkage groups 10, 14, and 15. Additionally, in linkage group 3, a single QTL linked to fruit bunch weight was found. As a result, the application of the GBS technique in oil palm proved to be quick, affordable, and extremely repeatable.

The height of the trunk is a complex trait that is influenced by a group of genes, the environment, and the interaction of both, according to mapping studies of the QTL for trunk height (Lee et al. 2015; Ong et al. 2018; Pootakham et al. 2015; Theam et al. 2016). These studies also revealed that the QTL for trunk height was found in different linkage groups and had quite different effects in each study. A QTL influencing the trunk height of oil palms was found on chromosome 5 by Babu et al. (2019). Four QTLs for Ganoderma resistance were found by Tisné et al. (2017); two of these QTLs were associated with the onset of the Ganoderma disease and the other two with palm death. For the first time, our study offered useful guidance on choosing oil palm cultivars resistant to Ganoderma illness.

However, most of the QTLs found were family-specific, meaning they couldn't be confirmed in other families with different genetic origins. Therefore, their application was restricted to the families in which they were discovered. It is important to conduct a Genome-Wide Association Study (GWAS) utilizing a sizable population that represents all materials accessible in the world to uncover universal DNA markers that may be used in selecting preferred palms in any community. This is a difficult process. Several GWAS have been published (Babu et al. 2019; Kwong et al. 2016; Ithnin and Kushairi 2020; Ithnin et al. 2017; The et al. 2016). This is a list of genes associated with oil yield. Thanks to the quick development of NGS (next-generation sequencing), millions of SNPs may now be produced economically and quickly, expanding the capacity to provide vast volumes of genotyping data.

Using a revised SNP array, a high-resolution genetic linkage map was created for an oil palm family, enabling more precise mapping of QTLs related to height, height increment, and rachis length. These traits' QTLs were found in certain linkage groups, which account for a sizable amount of the phenotypic variation. Because the detected QTLs were co-located or next to those reported in other populations, they were dependable and consistent. Putative genes and transcription factors involved in conjugation, signaling, cell wall morphogenesis, and phytohormone production were found in the QTL regions. The T128 tenera palm, originating from Nigerian germplasm, showed clear segregation for height, bunch harvestability, and frond length, with shorter frond length, lower height, and slower bunch harvestability compared to other commercial crosses and breeding families (Bai et al. 2018).

16.5.3 Molecular Marker Applications in Other Important Traits of Oil Palm

Molecular markers are helpful not only for breeding certain traits through MAS, but also for illness diagnosis and characterization, genotype/progeny legitimacy determination, and intellectual property rights protection. In most agricultural crops, diagnosis of a disease is based on the presence of causal organisms, which are precisely detected by DNA markers, mainly PCR-based DNA markers. *Ganoderma* spp. caused BSR (basal stem rot) and the *Phytoplasma* caused spear rot disease in oil palm, which can be effectively diagnosed by PCR-based markers. Oil palm is seriously threatened by BSR disease, which Hama-Ali et al. (2015) found to be associated with SSR markers in oil palm. Using 58 SSR markers and three progeny types (KA4G1, KA4G8, and KA14G8), a comparative molecular mapping was performed to determine their relationship to BSR. An average of 5.51 alleles per locus was found among the 319 alleles that were discovered. Through single progeny analysis, five markers, mEgCIR0793: 180, mEgCIR0894: 200, mEgCIR03295: 210, mEgCIR3737: 146, and mEgCIR3785: 299 were shown to be linked to Ganoderma disease. Only two alleles, mEgCIR0804: 213 and mEgCIR3292: 183, were discovered to be connected to Ganoderma illness in pooled data, though. The study found that progeny type KAG1 was resistant to BSR, while the other two were susceptible progenies. They recommended using these markers in subsequent research on oil palm breeding that is resistant to BSR.

Using a cDNA microarray technique, the transcriptional responses of roots treated with *Ganoderma boninense*, a causative agent of BSR in oil palm was investigated. Oil palm roots infected with *G. boninense* at 3 and 6 weeks post-inoculation exhibited 61 of the 3748 transcripts that were examined to be significantly up- or down-regulated in comparison to uninfected roots. Tisné et al. (2017) found four QTLs for Ganoderma resistance; two of these QTLs were associated with the onset of the Ganoderma disease and the other two with palm death. For the first time, our study offered useful guidance on choosing oil palm cultivars resistant to Ganoderma illness.

According to Hama-Ali and Tan (2014), ten numbers of monomorphic SSR markers and two numbers of half-sib families were employed to identify illegitimacy in oil palm. In families 1 and 2, four monomorphic loci, mEgCIR0425, mEgCIR3769, and mEgCIR3902, identified illegitimate offspring IDs 97 and 180. Furthermore, one illegitimate offspring, ID180, was found by five loci. A relatively small number of extremely variable RFLP probes were used in the DNA fingerprinting of 111 elite breeding palms, as described by Mayes et al. (1996).

16.5.4 Functional Genomics

Of late, expressed sequence tags (ESTs) are getting more attention as markers for genetic diversity analysis. Because of their low creation costs, ability to represent transcribed genes, and potential function that can frequently be deduced by homology search, EST-SSR markers are valuable. They could be helpful for marker-assisted selection in oil palm since they have been developed and mapped in several crop species. Ting et al. (2010) developed a set of EST-based SSR markers through *in silico* approaches from a large collection of 19,243 *Elaeis guineensis* ESTs. They combined to produce 10,258 distinct sequences, of which it was discovered that 629 ESTs contained 722 SSRs with a range of patterns. The majority group (45.6%) of dinucleotide repeats was made up of 66.9% AG/CT, 21.9% AT/AT, 10.9% AC/GT, and 0.3% CG/CG motifs.

6103 non-redundant ESTs were annotated and examined cDNA libraries of developing reproductive and vegetative organs for simple sequence repeats (SSRs). Primer pairs from sequences surrounding 289 EST-SSRs were examined to look for polymorphisms in elite breeding parents and their crossovers. SSR polymorphisms were discovered in sequences encoding many RNA-interacting proteins, other transcriptional regulatory proteins, zinc finger, NAC-like, bZIP, and AP2-like transcription factors. Ten EST-SSR primers were created and employed to assess the genetic diversity of 76 oil palm accessions originating from seven African nations as well as the Deli dura population (Singh et al. 2008). There were 2.56 and 1.84 observed and effective alleles on average, respectively. The average polymorphic information content of the EST-SSR markers was found to be 0.53, indicating polymorphism. With a polymorphism (FST) of 0.2492, the group being studied exhibits significant genetic divergence.

Additionally, UPGMA analysis showed a substantial correlation between the population under study's geographic location and genetic distance. It was also discovered that the EST-SSRs were helpful as a novel source of polymorphic markers for phylogenetic research, as a significant proportion of markers exhibited transferability between species and palm taxa.

Taeprayoon et al. (2015) assessed 121 breeding palms from three different populations in Thailand. They employed 96 SSR markers, 20 of which were polymorphic, yielding 109 alleles for the mEgCIR580 and mEgCIR787 markers, respectively, with polymorphism information contents ranging from 0.45 to 0.87.

ANOVA revealed variance across the population and individuals within the population at 33% and 67%, respectively, with a range of three to eight alleles per locus. As a result, the study's use of SSR markers allowed for the classification of the major Thai oil palm genetic stocks according to their genetic backgrounds. Thongthawaee et al. (2010), Zulkifli et al. (2010), Ting et al. (2010), Arias et al. (2013, 2014), and Zaki et al. (2010, 2012) have all investigated the usefulness of microsatellite markers in diversity studies.

16.6 Tissue Culture Applications

Clonal propagation through in vitro multiplication is the answer for producing a huge number of commercial planting materials and obtaining uniform, high-yielding elite oil palms on a wide scale. The production of about 100 million plantlets annually can be sourced commercially through in vitro oil palm growth methods. Following the early successes of plantlet creation by tissue culture in the 1970s, numerous organizations focused on oil palm began experimenting with in vitro production techniques (Jones 1974; Rabechault and Martin 1976). Nevertheless, the somaclonal variety issue led to an average yield advantage in clonal oil palms, as well as mantled fruits and anomalies in the flowers and vegetative stages. Over the past 20 years, oil palm tissue culture techniques have advanced significantly, making it possible to produce clonal palm seedlings with few abnormalities. Furthermore, clonal oil palms outperformed seed-propagated oil palms in terms of yield performance by 25% (Rohani et al. 2000). Clonal propagation in oil palm must therefore continue to be used as a supplement to hybrid seed production. The bulk multiplication of desired hybrids and desirable dura or pisifera of known parentage is made possible by the clonal propagation procedures (Fig. 16.9).

16.6.1 Culture Medium

The most commonly used culture medium for oil palm tissue culture is Murashige and Skoog (1962) and the Y3 medium of Eeuwens (1976). In some cases, the N6 medium was used for oil palm cloning (Thuzar et al. 2012) and the Blaydes medium (Rajesh et al. 2003). Marbun et al. (2015) reported that modified MS medium with extra supplements like Asparagine, biotin, D-Ca-Pantothenate, ascorbic acid, L-cysteine, and Alanine showed significant results in oil palm clonal propagation. Although it is widely known that different labs utilize different techniques to create oil palm clones, the precise media formulas are still unknown due to the strong economic interests in oil palm tissue culture. Every culture medium is kept at a pH of 5.5. In oil palm tissue culture, the growth hormones, often auxins and cytokinins, are typically employed (Eeuwens et al 2002; Ooi et al. 2013).

Fig. 16.9 Tissue culture of oil palm. (**a**) Somatic embryogenesis, (**b**) germination of somatic embryoids, (**c**) shoot in rooting media and (**d**) hardened plants. (Photo: Dr. K. Sunilkumar, unpublished)

16.6.2 Explants and Preparation

The availability of a large number of high-quality, diverse oil palm explants chosen from breeding programs is a requirement for clonal production. To start cultures, a variety of explant sources are utilized, including the immature inflorescences, roots, young leaves, rachis, and apical meristem. The most often utilized explants are the tips of the tertiary roots or immature leaves or inflorescences (Jayanthi et al. 2011; Teixeira et al. 1994; Wooi 1995; Yusnita and Hapsoro 2011). To eradicate the fungus and bacteria from the soil, the root explants must be thoroughly disinfected (Corley et al. 1981). The immature leaf explants require less severe disinfection because it is fully covered by the bases of older leaves. However, swabbing the external closing leaf bases or enclosing spathes with 70% ethanol is to be carried out and the inner tissues have to be removed aseptically to establish sterile explants (Noiret et al. 1985). The excision of immature, unopened leaves severely checks the palm's growth, often with a risk of its death. Hence, it is better to avoid frequent sampling and the palms used in the breeding program should be avoided. According to Wooi et al. (1981), tissues from the young inflorescence at the axil of the tenth leaf produced good results and can be removed without severely harming the palm. Only consistently high-yielding palms should be used to harvest explants. According to Wooi et al. (1981), treating for 15 min with a 0.1% (w/v) mercuric chloride solution ensures disinfection, then washing for 10 min with sodium hypochlorite 10% (w/v) solution and finally rinsing with sterile distilled water. Young leaf explants were successfully surface sterilized using mercuric chloride solution (1%) for 20 min, which effectively prevented contamination (Rabechault and Martin 1976).

16.6.3 Somatic Embryogenesis

Somatic embryogenesis is the method of in vitro regeneration in oil palm that has been documented thus far. Additionally, the embryoids grow shoots after induction and are rooted, toughened, and moved to the nursery. Callus development, induction of embryogenic callus, somatic embryo maturation, and plantlet regeneration are the three primary developmental stages that make up the regeneration process, which includes indirect somatic embryogenesis that is being covered here.

16.6.3.1 Callus Induction

Callus initiation and embryogenesis occur more quickly with leaf explants than with inflorescence or root explants (Eeuwens et al. 2002; Paranjothy and Rohani 1982). While 2,4,5-T (2,4,5-trichlorophenoxy acetic acid) and 2,4,5-TP (2,4,5-trichlorophenoxy propionic acid) were less frequently used, auxins like 2,4-D (2,4-dichlorophenoxy acetic acid), Picloram, and NAA (Naphthalene Acetic Acid) are primarily used for callus initiation in oil palms. However, cytokinins are added at low concentrations and occasionally they are found to be inhibitory (Paranjothy et al. 1990). After 3 weeks of culture with 2,4 D, or NAA, callus formed from the roots of aseptically produced seedlings (Ong 1975). Another study reported that callus initiation was noticed within 4–5 weeks of inoculation of the embryos into the callus induction medium (Rajesh et al. 2003). Yusnita and Hapsoro (2011) found that 2,4 D at 15 μM produced the highest callus induction percentage when oil palm immature leaf explants were cultured in MS media enriched with 2,4 D and/or Picloram and with or without activated charcoal. The appearance of the callus was influenced by the type of auxins used, 2,4 D promoted nodular appearance in the callus, whereas NAA frequently results in rooty callus. Smith and Thomas (1973) and Jones (1974). Significant differences were seen in the growth rates of callus produced from root explants across the several genotypes under investigation (Wooi et al. 1981). Karun and Sajini (1996) showed that nodular calli were produced in 100–120 days from leaf explants of 18-month-old dura seedlings, whereas it took 150–180 days for leaf explants from 6-month-old tenera seedlings. It was reported that maximum callus induction (82%) from tissues of immature male inflorescence was initiated in a modified Y3 medium supplemented with picloram and activated charcoal (3%) (Jayanthi et al. 2015). In another work, Sanputawong and Te-chato (2008) discovered that immature zygotic embryos in MS medium supplemented with 2.5 mg/L Dicamba generated embryogenic callus in 4–5 weeks. Calli from zygotic embryos in Euwens media (Y3) supplemented with 2 mg/L 2,4-D were obtained by Kanchanapoom and Domyoas (1999), who subsequently transferred the embryoids to a medium containing 0.5 mg/L of reduced 2,4-D concentration. After inoculating explants into a culture medium, the callus typically initiates between 2 and 4 months later. It then grows slowly, forming a compact, nodular mass.

On rapidly growing callus, embryogenesis is easily achieved; nevertheless, the clonal plants derived from these embryoids are more prone to experience aberrant flowering. Accordingly, the nodular, slow-growing callus is preferable (Duval et al. 1988). According to some reports, browning caused by phenolic compounds can be stopped by adding activated charcoal to the medium, even if this is not a bad thing for oil palms (Wooi 1990). Approximately 2–60% of leaf explants may produce a callus, and the majority of studied palms had some callus. To maintain callus cultures, the amount of auxin is reduced (Paranjothy and Othman 1982; Smith and Thomas 1973; Thomas and Rao 1985).

Long-term callus culture incubation produces neomorphs, which are embryonal structures that stop growing (Nwankwo and Krikorian 1983). The explant genotype, the growth regulators employed, and the distance of the tissues from the meristematic region (in the case of immature leaf tissues) can all affect the production of callus (Duval et al. 1988; Lioret 1982). In addition, callus development is influenced by the physiological stage of the palm from which explants are taken; callus-generating explants from younger, aggressively growing palms are preferred (Paranjothy 1986). To produce callus in 100–120 days, the leaf explants from oil palm seedlings were cultured in half-strength MS media supplemented with 25 mg/L 2,4 D, 3 mg/L 2-IP, and 30 mg/L adenine sulfate (Karun and Sajini 1996). Correa et al. (2016) proposed that callus might be grown in basal medium supplemented with 9 μM 2,4 D. According to a study by Paranjothy et al. (1990), callus from 3-year-old juvenile palms' leaf and root explants had higher callus frequencies than those from 10-year-old palms. Unfortunately, sampling 3-year-old palms is not practical because it takes several years to discover the best individuals. As a result, samples cannot typically be taken from 3-year-old palms. The veins of immature leaf tissues generate callus, while the pericycle and central stele of the root provide the basis for the root callus (Pannetier et al. 1981; Paranjothy and Othman 1982). In oil palms, Soh et al. (2009) observed a 20% callusing rate. Studies by Ginting et al. also showed that genotypes had an impact on the rate of callus induction (1995). According to Correa et al. (2016), Jayanthi et al. (2015), and Reflini (2017), the cultures are mostly kept in the dark at a temperature of 27 ± 1 °C and relative humidity of 55–60% to induce callus. The callus cultures were then subcultured on the same medium component every month for 16 h with a photoperiod of 2500 lux (Karun and Sajini 1996; Rajesh et al. 2003; Reflini 2017; Yusnita and Hapsoro 2011).

16.6.3.2 Embryoid Formation

Somatic embryogenesis occurs indirectly in oil palm, with an intermediary callus phase. However, the direct embryogenesis (without callus stage) from explants has not been reported except when zygotic embryos were used as explants. Nevertheless, some research revealed that embryoids may be obtained straight from leaf explants or even before the callus stage becomes visible. Jayanthi et al. (2011) observed direct development from cotyledonary nodes of germinated immature zygotic embryos without an intermediate callus phase. Rarely, embryoids may also form on leaf explants right away without going through the callus stage (Wooi 1990).

For embryoid development to occur, calli must be physically separated from explants at the fragmentation lines (Karun and Sajini 1996). Transferring to a medium with no auxin or reduced auxin concentration supplemented with 30 g/L of sucrose and 0.5 g/L of casein hydrolysate stimulates the development of embryoids on the callus. When transferred onto MS medium supplemented with 0.01–0.5 mg/L 2-ip, a nodular callus produced in the presence of 2,4-D becomes embryogenic (Thomas and Rao 1985). Embryoids were generated by Paranjothy and Othman (1982) by lowering the auxin levels and adding cytokinins to the growth medium. According to a different study, embryoid formation occurs when NAA levels are lowered (0–40 mg/L) in the presence of activated charcoal (Nwankwo and Krikorian 1983). The growth of somatic embryos and secondary somatic embryos was significantly more efficient when polyamines such as putrescine and spermine were introduced to the medium rather than zeatin riboside (Rajesh et al. 2003).

When embryoids are developed, they are transferred to the light of intensity 1500 lux with lighting duration varying from 16 to 24 h and temperature 28 ± 1 °C. The embryoids formed can be further proliferated by inoculating them into a fresh proliferation medium. The multiplication medium may contain just auxin, be hormone-free, or have a high ratio of auxin to cytokinin. Previous observations suggest that 50% of embryoid lines proliferate (Rival 2000). The number of subcultures is deliberately avoided in many laboratories to avoid the formation of abnormality, which was partly attributed to the time in culture. In order to halt proliferation and begin somatic embryo development, the cells must be subcultured into an auxin-free medium (Litz 2005).

16.6.3.3 Shoot Development and Rooting

The shoot growth begins on a medium free of hormones or one having less concentration of cytokinin at the same time as other embryoids in the same culture begin to proliferate further. Shoots are removed from the embryoid and placed in the rooting media during each subculture. The shoot development from meristemoids took 90–100 days and after transfer to the rhizogenesis medium, sufficient roots were produced within 3–4 weeks (Karun and Sajini 1996). According to Wooi's (1990) suggestion, a hormone-free baseline medium should be followed by a brief exposure to high NAA or a medium with extremely low NAA to encourage root development. According to reports, the shoot generation from embryogenic lines of re-clones and ortets was 95% and 96% comparable, respectively. Thomas and Rao (1985) demonstrated that in the presence of 1 mg/L GA3 and 20 mg/L NAA, embryoids formed shoots and roots.

The basal medium containing 0.54 μM of NAA, 1000 μM putrescine and active charcoal was used for somatic embryo germination by Correa et al. (2015). According to Jayanthi et al. (2015), the development of shoots and roots was encouraged by regeneration media made up of Y3 basal media with 18 μM BAP (6-benzyladenine), 3.78 μM ABA, and 5.78 μM gibberellic acid. Some studies demonstrated the regeneration of oil palms in cytokinins such as BA and Zeatin riboside (Raju et al. 1989; Karun and Sajini 1996). However, the application of Thidiazuron (TDZ) in the medium resulted in a non-embryogenic callus (Rajesh et al. 2003). According to a study, somatic embryos with distinct bipolar structures formed in Y3 basal media supplemented with BA and ABA under light conditions, and these embryos went on to develop into full plantlets (Jayanthi et al. 2011).

16.6.3.4 Hardening

The well-rooted plantlets were treated with Bavistin (1%) and thereafter with IBA solution (1000 ppm) for an hour after the transfer of the plantlets to pots. The potting mixture containing equal proportions of soil and coir pith was sterilized before use (Karun and Sajini 1996; Rajesh et al. 2003) and stored in the laboratory for 1 or 2 months for initial hardening, covered with a polythene sheet. It is crucial to remember that plants grown in culture lack a typical leaf cuticle, making them extremely vulnerable to dry air exposure and quickly drying out. Thus, to achieve success rates of over 90%, it is important to maintain a high humidity of 60–70% during the first nursery stage, which lasts for 15 days at a temperature of 28 ± 1 °C. In some cases, the acclimatization of transplanted plants done by spraying anti-transpirants to reduce transpiration loss has resulted in unintended effects. Initial survival may be better in sterile soil than in non-sterile soil. The addition of mycorrhizal culture in sterile soil improved the survival of clones by increasing the absorption of nutrients. Due to a shortage of kernel food reserve, plantlets develop more slowly in the nursery at first, but after 12 weeks, clonal plants grow at a rate that is similar to that of seedlings (Wooi et al. 1981). Before being planted in the field, oil palm plantlets are typically raised in a nursery for around a year.

16.6.4 Other In Vitro *Methods*

The liquid culture or suspension culture for oil palm propagation is utilized to improve the uniform growth of callus and somatic embryos. The liquid medium is suitable for micropropagation to reduce production costs, reduce time, and enable automation. This offers uniform culture conditions, and synchronized development, and helps with easier medium replacement. The liquid medium promotes tissues to be in close contact with the medium, which stimulates the uptake of nutrients and phytohormones (Sandal et al. 2001). The suspension culture, innovative shaking flask, and bioreactor technologies are being used to address the poor rate of embryogenesis, which hinders large-scale propagation. Only the embryoid proliferation stage is substituted in suspension culture. By inoculating embryogenic callus cultures derived from oil palm into a liquid medium, suspension cultures are created.

De Touchet et al. (1991) used leaf calli to create the first regenerable oil palm embryogenic cell suspension cultures. According to Monteiro et al. (2018), using larger calli aggregates—that is, those that formed in a liquid culture medium under agitation and did not disaggregate—might be prioritized. Long ago, oil palms were shown to generate somatic embryos from suspension cultures (Teixeira et al. 1994; De Touchet et al. 1991), although the rate of regeneration into plantlets was low. By creating a methodology to raise the standard of shoot formation, this issue was resolved. Friable, embryogenic calli that are separated from nodular compact calluses are used to create embryogenic suspension cultures. In oil palm liquid culture systems, Tarmizi (2002) observed that 1 mg/L 2,4 D and 0.1 mg/L NAA, when supplemented with the liquid medium, were effective for culture proliferation. Oil palm suspension culture grown successfully in MS medium supplemented with 1 mg/L 2,4 D, 1 mg/L NAA, and 100 mg/L glutamine is reported to have multiplied at a rate of 3.5 times in a month in a bioreactor (Gorret et al. 2004). It was claimed that the liquid suspension technology brought the incidence of mantling down to manageable levels in an industrial oil palm tissue culture laboratory in Malaysia (Soh et al. 2001).

Tarmizi et al. (2008) studied the multiplication of liquid palm cultures through bioreactors and found promising results with a 5–12-fold increment in fresh weight of embryogenic callus and further, all the cultures from the bioreactor system were able to regenerate. However, the growth and development vary between clones in bioreactor systems and the optimum parameters for clones have to be standardized. After 6 weeks of transplantation to solid media, the meristems of the shoots and roots differentiate. When benzyl adenine (BA) is added to the maturation medium, root formation halts and shoot development significantly improves.

The bioreactor-mediated somatic embryo production has been extensively discussed earlier (Ibaraki 2001). The effect of dissolved oxygen (DO), CO_2, shear stress and bioreactor design on the viability of plant cells and embryoid regeneration in oil palm has been investigated (Tarmizi et al. 2008). Some parameters can affect the quality of suspension cultures without regard to the species, including nitrogen, the usage of the conditioned medium, and the size of the inoculum (Gorret et al. 2004). Marbun et al. (2015) reported that periodic immersion in the medium improved the embryogenic callus formation and increased the absorption of nutrients efficiently. Further, the somatic embryo of oil palm cultured in a temporary immersion system showed 2.2 times higher fresh weight compared to the somatic embryos cultured in a solid medium (Marbun et al. 2015).

16.6.5 Somaclonal Variation

Somaclonal variation is the genetic variation induced by tissue culturing in plants and it is a considerable problem in many crops. The oil palm is also no exception to somaclonal variation. Many studies observed that some clones showed abnormal flowering and a higher incidence of flowers with the mantled character, i.e., an apparent feminization of male parts in flowers of both sexes. These abnormalities were accompanied by parthenocarpic fruit set (without seeds) and severe bunch failure. According to Corley et al. (1986), plantings of the same three clones in subsequent years showed an increase in the incidence of aberrant flowering. Overall, all varieties of material in a reasonably lengthy transfer interval (8 weeks) and with a high auxin/cytokinin ratio showed the lowest occurrence of mantled blooming. Moreover, utilizing a medium with relatively high kinetin concentrations and low NAA (0 or 0.1 mg/L) concentrations, as well as reducing the transfer interval to 2 or 4 weeks, resulted in a significant incidence of mantled blooming (Eeuwens et al. 2002). However, according to Kushairi et al. (2010), these changes in the medium's composition decreased the somatic embryogenesis's success rate.

In terms of the molecular or biochemical basis, the cause of the abnormality in flowering is not clearly understood. It is widely assumed to be epigenetic modifications induced by the tissue culture process, which results in changes in gene expression level and can be reversible in the field over time by appropriate stimuli. After 9 years in the field, there is also a reversal that results in 50% recovery in highly mantled individuals and 100% recovery in those with mild mantles (Jaligot et al. 2000; Rival et al. 1998). But unlike a straightforward mutation, the anomaly in palms varies in degree and either has a significant impact or none at all. Evidence suggests that the likelihood of abnormalities is increased by short transfer intervals, high levels of cytokinin in the medium, and low levels of auxin. Oil palms generated from somatic embryos exhibit aberrant floral development, with a maximum percentage of 5% deemed commercially viable.

Reliable tissue culture procedures, strict culling at different phases of cloning, and the use of multiple types of ortets (Soh et al. 2003) all contribute to a lower frequency of abnormality (Maheran et al. 1995; Simon et al. 1998). Moreover, many studies were carried out to develop early biomarkers as a diagnostic tool to identify somaclonal variation, thereby screening clonal amenity and conformity. According to certain publications (Jaligot 2002; Jaligot et al. 2000), mantled regenerates' aberrant calluses and leaves have much lower levels of methylation than those of their normal clones. Ong et al. (2015) suggest that DNA hypomethylation of a LINE retrotransposon linked to rice Karma in the intron of the homeotic gene Deficiens is the cause of the somaclonal variation seen in all mantled clones of oil palm.

Though initial success in oil palm cloning was obtained in the 1970s by leading oil palm cultivating countries, they were never disclosed because of commercial interest. In the Indian scenario, few studies were attempted with different explants for oil palm cloning, but only limited success could be achieved with explants like immature zygotic embryos and seedling leaves. Jayanthi et al. (2015) reported that immature inflorescence explants yielded considerable success in oil palm cloning at the laboratory scale but failed to achieve repeatability.

More importantly, the success in commercial production of oil palm clonal planting materials depends on several factors, including the availability of explants from high-quality ortets for year-round production, low contamination in the lab, higher embryogenesis rate and better survival rates in the nursery. Further, the adoption of suspension culture for oil palm tissue culture can bring down the time for plantlet production and automate the methodologies employed. Further, the ability of cloning technologies to produce a large number of palms with specific traits apart from yield traits could potentially improve the demand for clonal planting material. The somaclonal variation is the major problem in clonal planting material faced by leading oil palm growing countries such as Malaysia and Indonesia. The search for floral abnormality (somaclonal variation) markers in oil palms is intensely pursued in many labs.

16.7 Genetic Engineering and Gene Editing

16.7.1 Genetic Transformation

Potential transformation goals for oil palm encompass both quality attributes, including oil composition, and agronomic qualities, like pest or disease resistance (Table 16.5). Modification of the now-well-understood fruit abscission enzymes could be helpful for harvesting, among other agronomic traits. Ethylene stimulates fruit abscission and it was; found that the ethylene synthesis pathway includes a gene called ACC synthase, which is expressed in the mesocarp during the time at which fruit abscission begins, i.e., 160 days after anthesis. It was also found that oil palm fruit has a large number of polygalacturonase genes, one of which was strongly expressed in the abscission zone in response to ethylene.

The MPOB concluded that a higher oleic acid percentage was a means of boosting the market share of palm oil. According to Soh et al. (1994), oleic content may be raised by either utilizing antisense to decrease the activity of C16-ACP thioesterase or by boosting the activity of 3-keto-acyl-ACP synthase II (KAS-II), or by both. Sambanthamurthi et al. discovered associations between KAS-II activity and iodine value, palmitic acid, and oleic acid concentrations in several types of *E. guineensis* and *E. oleifera* genotypes (2000).

Parveez et al. (1994, 2011) identified additional potential transformation targets in addition to an increase in oleic content. These goals included lowering the content of saturated fatty acids by increasing δ9-desaturase activity, raising the content of stearic acids by decreasing δ9-desaturase activity, raising the content of palmitic acid by decreasing KAS-II activity, inserting a palmitoyl-ACP desaturase from a different species to increase palmitoleic acid (C16:1, used in pharmaceuticals), lowering the

Table 16.5 *Agrobacterium*-mediated transformation and direct DNA transfer methods for oil palm genetic transformation

Explants	Transformation/ strain	Vector	Selection marker	Promoter used	Reporter gene	Transgene/ expression	Studied parameters	References
Immature embryo	LBA4404	pCAMBIA 1310	nptII	CaMV35S	Gus	Gus assay	In vitro culture of IE for direct plant regeneration	Abdullah et al. (2005)
Immature embryo	Biolistic	pBI121	nptII	CaMV35S	Gus	Gus assay	In vitro culture of IE for indirect plant regeneration	Abdullah et al. (2005)
Young leaves	LBA4404	*pUBA*	bar	Ubi1	Gus	Southern blot	Study glufosinate-ammonium-resistant transgenic oil palm	Izawati et al. (2012)
Embryonic calli	Biolistic	pMI11, pMI11G, pMI3, pMI3G	pmi gene	Ubi1, CaMV35S	GusA	PCR/ southern blot	To produce transgenic oil palm by using mannose as a selectable agent	Bahariah et al. (2013)
Protoplast, basidiospore, mycelium	LBA4404	pCAMBIA 1300	hyg	GPD fungal promoter	Gus/gfp	PCR	Understanding pathogenicity factor associated with *G. boninense*	Govender et al. (2016)
Embryonic calli	Biolistic	pBINPLUS	hpt	35S (2XCaMV35S)	GFP	PCR/ southern blot	Increase the level of transgene expression and transformation efficiencies in oil palm	Bahariah et al. (2017)
Embryonic calli	LBA4404	pBIDOG	*DOGR1*	CaMV35S	Gus	PCR/ southern blot	Introduction of new selection agent	Yarra et al. (2019)
Calli clump	Electroporation	pCAMBIA 1303	hptII	CaMV35S	GusA/ mgfp5	PCR	An efficient electroporation-mediated transformation method for oil palm calli	Darmawan et al. (2020)

Adopted from Martin et al. (2022)

content of ricinoleic acid (an industrial acid fatty acid usually derived from castor oil) by inserting oleate 12-hydroxylase and lowering the content of lycopene oil by preventing conversion to carotene.

Nurfahisza et al. (2014) reported genetic engineering in oil palm to modify fatty acid composition to increase the production of oleic acid, which is in high demand for use in oleochemical industries. An antisense palmitoyl ACP-thioesterase sequence was inserted into oil palm tissue using a biolistic technique in order to decrease palmitoyl ACP-thioesterase activity and promote the buildup of oleic acid at the expense of palmitic acid. Genomic DNA was extracted from transgenic oil palm, and the presence of the transgene was verified by DNA sequencing. PCR amplification of the chosen marker gene (bar) and gene of interest was effectively completed, as was antisense palmitoyl ACP-thioesterase. Alizadeh et al. investigated the expression profiles of Δ9 stearoyl–acyl carrier protein desaturase (SAD1 and SAD2) and type 3 metallothionein (MT3-A and MT3-B) in oil palm seedlings that had been artificially inoculated with the symbiotic fungus *Trichoderma harzianum* and the pathogenic fungus *Ganoderma boninense*.

Resolving the Intellectual Property Rights issue is another application of DNA markers. The principle is the same as determining the legitimacy of a genotype through DNA fingerprints. It can be used for protecting the IPR of advanced planting materials and breeding stocks.

According to Parveez et al. (2011), plants were regenerated following transformation with genes for high ricinoleic acid, high stearic acid, and high oleic acid; each of these genes included a mesocarp-specific promoter along with the necessary enzymes.

A biosafety greenhouse had hundreds of ostensibly changed palms, but molecular research revealed that not all of them had undergone a stable transformation. Additionally, there has been no data on the oil content of these altered palms as of yet. Additionally, genes for elevated palmitoleic acid and a changed carotene content have been inserted into the cultures. It might also be thought of as enhancing the mesocarp's overall oil content by boosting the activity of the pertinent genes. Although comparable increases in the expression of genes involved in fatty acid synthesis was observed, while oil palms do not collect lipids in their mesocarp, date palms do express majority of genes involved in triglyceride assembly.

Increasing expression of genes on the pathway to C14-ACP appears worth considering, but the idea of a single rate-limiting step is an oversimplification and that attempts to increase fatty acid synthesis by genetic manipulation might simply result in triglyceride assembly becoming limiting. Corley (1998) thought it was possible to get a percentage of 70% oil/fresh mesocarp, although most palms only contain roughly 50%. Another intriguing option is the production of polyhydroxybutyrate (PHB), a biodegradable material. Although bacterial fermentation may manufacture this, the cost is significantly higher than that of polymers like polyethylene (Houmiel et al. 1999).

Therefore, three enzymes are required to create PHB from acetyl coenzyme A, the precursor for fatty acid synthesis. These genes have altered oil palm and the resulting plants are being grown in a biosafety greenhouse (Parveez et al. 2008).

16.7.2 Gene promoters

To manage the expression of the target genes, Soh et al. (1994) noted that mesocarp-specific regulatory genes must be identified. Since lipids make up a significant portion of cell membranes, it will be crucial to make sure that any modifications to the oil composition occur solely in the mesocarp. Alterations to the composition of membranes elsewhere in the plant may have detrimental effects. Parveez et al. (2011) enumerated promoters specific to the mesocarp, leaves, and roots that were identified from oil palm.

16.7.3 Methods of Transformation

The "gene gun," "biolistics," or particle bombardment has been used in the majority of research projects. This is accomplished by bombarding appropriate cells or tissues with tiny (1 μm diameter) tungsten or other metal particles coated with the DNA to be introduced. When the desired DNA is cultivated in a medium containing the antibiotic or herbicide, only the cells that have acquired the resistance gene survive. The marker gene is often present for resistance to antibiotics or herbicides. A marker gene was successfully inserted into oil palm tissue by Parveez et al. (1999), who then successfully grew the altered plants.

Using *Agrobacterium tumefaciens*, also known as the crown gall bacterium, provides an alternative to particle bombardment. This bacteria naturally infects plants by introducing genes that regulate hormone production and promote the growth of galls into the plant cells. The Agrobacterium itself must first transform for the required genes to be inserted instead of the hormone genes. For dicotyledonous species, agrobacterium transformation is the recommended approach; however, monocots are not normally infected by the bacterium. Nonetheless, Chaidamsari et al. (1998) converted cells of oil palm leaf explants, and some success has been had with monocots.

Using Agrobacterium, Izawati et al. (2009) were able to convert the callus produced from oil palm leaves with a marker gene and that gene was expressed in regenerated plants. Abdullah et al. (2005) demonstrated that Agrobacterium or the gene gun may both convert immature embryos. The embryos easily develop callus, and the plants regrow after 4 months. Family selection could be used in conjunction with this technique to produce a sizable population of altered clones for experimentation. Examples of possible transformation problems include unstable transformation, partial or multiple copies of the gene, and gene silence, in which the gene is present but not functioning.

In cases where further breeding is necessary, meiosis stability will also be crucial. As an alternative to transformation (TILLING), targeting induced local lesions in genomes could be tried. This entails applying a mutagen, then using PCR to find mutations in the target genes (for a review, see Wang et al. 2011). It doesn't seem like oil palms have yet been treated with this technique. Although there has been good development, it will likely take many more years before there is a noticeable commercial availability of genetically modified palm oil. To get to that point, the transformed plants need to be examined to see which ones most closely resemble the desired character and to remove any that have had unfavorable side effects from the transformation. After being chosen, the individuals need to be cloned and multiplied until they reach a large enough size for commercial planting. After the changed plants have reached the stage of regeneration, all of this could take up to 15 years (Corley and Stratford 1998).

Further breeding may also be required to increase output unless the clones that were first changed were extremely high-yielding. Clones of changed palms will probably play a bigger role in the oil palm sector of the twenty-first century, despite the lengthy time scale. But it should be remembered that rival annual oil crops have a considerably shorter development period. Conventional breeding has produced high oleic acid cultivars of sunflower, rapeseed, and soybeans that are currently available (Murphy 2006). Annual oilseeds can more easily provide specialized markets than oil palm because annual crop growers can alter crops and varieties from year to year. The fundamental benefit of palm oil will continue to be its comparatively low cost of production; high-lauric rapeseed types are unable to compete with palm kernel or coconut oils on price (Murphy 2006). According to Corley (2001) and Tinker (2000), the conversion of oil palm could have a bigger impact on future yield gains and cost savings than the creation of specialized oils. According to Ong et al. (2015), the transcription of EgDEF1, a MADS-box gene, is impacted by hypomethylation (an epigenetic alteration) in the transposable element Karma, which results in the development of mantled fruit in clones. If verified, this should provide the ability to identify mantled palms before planting, which is crucial for clone planting on a commercial scale.

By using specially designed nucleases having sequence-specific DNA-binding domains for targeting certain DNA sequences, the CRISPR/Cas9 system may be used in oil palm effectively to change desired genes. Young leaves of oil palm can carry the CRISPR/Cas9 transgenes, and tissue culture can be used to restore plants that have undergone a stable transformation. Bahariah et al. (2021) developed a genome editing system for oil palm by using rice as a model system. It was found that there were significant similarities between the rice and oil palm genome sequences. By using specific nucleases with sequence-specific DNA-binding domains to target particular DNA sequences, this might be effectively applied in oil palm to change desired genes. Now that oil palm genome sequencing has been developed, this is now possible.

16.7.4 Nanotechnology

Although presently application of nanotechnology is limited in the oil palm industry, it could offer a sustainable alternative in the areas of food safety and security. Researchers should make full use of nanotechnology as a substitute approach to address the several issues that the oil palm sector faces. Accordingly, there are four major areas where the application of nanotechnology is worth taking up. (1) Use of nano sensing for real-time monitoring of plantations, including crop status, soil, nutrient and water management, and early warning of pest and disease outbreaks. (2) Active ingredients of plant protection chemicals could be incorporated into nanocarrier systems through entrapment or encapsulation for efficient delivery and thus help in integrated pest a disease management. (3) Oil palm biomass could be a great source for the isolation or generation of valuable nanomaterials. Of the total biomass of the palm, only 10% is oil and the rest 90% is generated biowaste and (4) Palm oil may offer a more ecologically acceptable alternative as a capping and stabilizing agent during the biosynthesis of metallic and nonmetallic nanoparticles.

16.8 Mutation Breeding

The scope of use of mutation breeding in crops like oil palm, having a smaller genetic base, is worthwhile to think which could lead to the generation of new genetic variability. The major limitation of the application of mutation breeding in oil palm is its requirement of a lot of resources like land, irrigation and manpower. Mutagenesis methods like gamma rays (Mohammad 2016; Nur et al. 2018; Rohani et al. 2012), and chemical mutagens like EMS (Omar and Novok 1990) have been used to develop desirable phenotypes. However, the mutated palms' final performance has not yet been reported (Yue et al. 2021) because of the lengthy generation gap. The development of various molecular tools now makes it easier to detect and characterize mutants.

16.9 Hybridisation

16.9.1 Technique of Hybridisation

To start with, in any breeding program in oil palm, individual palms are selected by measuring their vegetative as well as bunch yield characters and the bunch is analyzed for oil and kernel contents. To utilize the selected palms for breeding, controlled pollination is to be carried out and the progenies are to be planted in statistically proper experimental designs.

For controlled pollination, the female inflorescence is isolated, and pollen from the male inflorescence is collected, processed, stored and used for pollinating the isolated female inflorescences. The resulting bunches are kept separately throughout the processing of seeds, nursery and in the main field for further evaluation. Controlled pollination is essential to avoid contamination by foreign pollen, either through biotic agents (pollinating weevils) or abiotic forces like wind and rain. With efficient control, it is possible to produce 99% pure crosses. Chin (1999), Donough et al. (1992), and Rao and Kushairi (1999) enumerated the prerequisites for controlled pollination and detailed the quality control methods required to guarantee the production of pure seed. Asmady et al. (2002) described a quality control system with processes prioritized to eliminate the pollinating weevil *Elaeidobius kamerunicus*, backed by an ISO 90001 certification.

Differential longevity of pollen of pisifera and dura was reported by Sunilkumar et al. (2020), which indicated the pollen of pisifera could be stored for a longer time than dura at −5 °C as well as at ambient temperatures. This has a significant advantage in pollination for hybrid seed production and also in breeding programs.

Storing of oil palm pollen in organic solvents like Diethyl ether was found as an effective method to retain viability for more than 200 days (Sunilkumar et al. 2015) under room temperature.

16.9.1.1 Collection and Storage of Pollen

Before the flowers bloom, the male inflorescence needs to be bagged. The entire inflorescence is chopped off while it is still inside the bag and sent to the lab for additional processing when it reaches anthesis, as observed through a window in the bag.

Collecting the inflorescence in the afternoon when the RH is low was found to give better pollen viability. As the anthesis of male and female inflorescence occurs at different times in the same palm, often the pollen needs to be stored for a considerable period. Several studies indicated the storage methods that can extend the viability of pollen. At room temperature, it can be stored for 6–8 weeks. This period could be extended for 1 year or more through storage at 5 °C in the refrigerator (Sunilkumar et al. 2017). Pollen could be cryostored for an extended amount of time without experiencing a significant decrease in viability, as demonstrated by Tandon et al. (2007). The differential longevity of pollen was reported by Sunilkumar et al. (2020), in which Pisifera was found to have more storage life compared to Dura. For long-term storage, vacuum drying and storing in ampules under −18 °C is also recommended. The viability of pollen is to be tested before pollination.

16.9.1.2 Preparation of Female Inflorescence and Pollination

At least 1 week prior to the first flower's anthesis, a female inflorescence should be bagged. Before pollination, the outer surface of the bag covering the female

Fig. 16.10 Oil palm crown showing pollinated bunches for hybrid seed production. (Photo: Dr. K. Sunilkumar, unpublished)

inflorescence is sprayed with a formaldehyde solution. There is a hole cut in the celluloid inspection pane through which pollen is blown in from a glass tube or plastic wash bottle. Adhesive tape is then used to seal the hole. The female flowers' typical receptive period is 36–48 h. The pollen application is repeated after 24 h to ensure maximum pollination and fruit set. The bag should be inspected after 10 days and discarded if damaged. To check the reliability of the pollinator and the efficiency of control measures, blank pollination with chalk powder is done without informing the workers (Fig. 16.10).

16.9.1.3 Breeding Method

Either MRS (Modified Recurrent Selection) or RRS (Reciprocal Recurrent Selection) are the two main breeding programs for oil palm. Programs in the Far East use MRS, whilst West African and Indonesian nations have accepted it through CIRAD cooperation. The modified recurrent selection, as followed by many of the Malaysian breeding programs, consists of crossing individual pisifera to several Duras (Deli). The general and specific combining ability of the pisiferas could be estimated according to the NCM-1 design.

In the case of reciprocal recurrent selection, the parental test crosses and the selves of the parents are planted in the field at the same time and evaluated simultaneously. Here, the land requirement is higher within 12 years, it is possible to produce a limited amount of good-quality D × P seeds. However, MRS takes more time as it involves two stages and requires a minimum of 20 years.

16.9.2 Quality Control of D × P Hybrid Planting Materials

To ensure the authenticity of production and genetic superiority of D × P planting material in oil palm, Malaysia has developed certain standards, known as SIRIM standards (MS 157). It states that the minimum requirement for seed production is as follows:

Requirement	Applicable for
1. Dura yield of a minimum of 160 kg 2. Oil to bunch of 16% 3. Oil yield per palm of 25.6 kg	Deli dura mother palm
4. Minimum progeny yield of 160 kg 5. 24%-Oil to bunch ratio 6. 38.4 kg per palm of progeny oil yield 7. Minimum 3% of kernel yield 8. 5.22 tons of oil yield per ha (136 palms)	D × P progeny

The standard also includes requirements for ortet selection, guidance on production practices, packaging of seed material and related legal matters.

16.10 Conclusion and Prospects

The crop is expected to have a bright future, as evidenced by the following facts. Compared to annual oilseed crops, oil palm has a remarkably small land footprint and is the most important oil crop, accounting for almost 40% of the world's traded vegetable oils. RSPO (Round Table on Sustainable Palm Oil), related to the sustainable production of palm oil, has been instrumental in ensuring the traceability of environmentally sustainable and socially responsible production of oil palm. Breeding new varieties with double yield has been reported, as in the case of Sime Darby, where an average oil yield of 9.9 t/ha was reported over 5 years against the current 4–5 t/ha world average. These palms are part of the plantation company's "Genome Select" genome-based program. In India, a yield of >50 MT FFB was reported, which is equivalent to >10 MT oil from farmers' fields under the best management conditions. Identification of QTLs associated with important traits like plant height can help breed shorter or compact palms for ease of harvest and mechanization, along with high yield. The genomic data available since 2013 and the development of NGS will accelerate the discovery of desirable genes much faster and cheaper. Two *G. boninense* genome assemblies are currently accessible in the NCBI archive, providing access to the transcriptome and genome data for molecular genetic methods to BSR control. Prospects of palm oil as a biofuel has much potential especially with COP 28 Dubai declaration during fag end of 2023 for completely stopping fossil fuels and moving to renewable fuel sources.

In order to meet the continuous increase in global demand for oils and fats in tandem with the increasing population, innovative breeding methods are to be adopted along with environmentally sustainable cultivation practices for maximizing the yield of oil palm. Although breeders have been successful in developing elite genotypes with high yield (as reported by Sime Darby and from India), to sustain the achievements and gain further progress, it is essential to have a multidisciplinary approach involving various fields of plant and life sciences. It also depends on identifying genetic variation for important traits, especially for stress response under unfavorable growing conditions. This, in turn, depends on the exploitation of variability existing in present collections, together with further augmentation of the variability through collections as well as genetic manipulations employing advanced techniques. Oil palm is greatly influenced by climate change concerning growth and productivity and at the same time, the crop also leads to climate change by way of deforestation. The publication of oil palm genomic data was a breakthrough in its improvement. This coupled with NGS technologies, has made it feasible to achieve rapid gains in oil palm breeding. Further, the production technologies assisted with Decision Support systems and Artificial Intelligence will help in a long way in enhancing production through the efficient use of resources.

Appendix 16.1 Research Institutes and Online Resources Available for Oil Palm (*Elaeis guineensis* L.) Breeding

Institute name	Specialization research activities	Contact information and website
Malaysian Palm Oil Board (MPOB), formerly PORIM/MARDI	All aspects	Malaysian Palm Oil Board, PO Box 10620, 50720 Kuala Lumpur, Malaysia. http://mpob.gov.my
Indonesian Palm Oil Research Institute (IOPRI)		Pusat Penelitian Kelapa Sawit Jl. Brigjen Katamso No. 51, Kp. Baru, Kec. Medan Maimun, Kota Medan, Sumatera Utara 20158, Email: Admin@Iopri.Org

(continued)

Institute name	Specialization research activities	Contact information and website
The French Agricultural Research Centre for International Development (CIRAD)		Montpellier, France. www.cirad.fr/en/home-page
Nigerian Institute for Oil palm Research (NIFOR)		PM Bag 1030, Benin City, Nigeria. Email: nifor@infoweb.abs.net
EMBRAPA Brazil		Embrapa Amazônia Ocidental, Rodovia AM-010, km 29, Caixa Postal 319, 69.010-970, Manaus, Amazonas, Brazil
OPRI-Oil Palm Research Institute under the Council for Scientific and Industrial Research (CSIR)	Oil palm improvement	PO. Box 74, Kusi, Kade, Ghana Email: ckusi@opri.csir.org.gh Website: www.csir.org.gh
Applied Agricultural Resources (AAR) Sdn Bhd	Oil palm improvement, interspecific hybrids	Locked Bag 212, Sg Buloh Post office, 47000, Sg Buloh, Selangor, Malaysia
Dami Oil Palm Research Station, Papua New Guinea		www.pngopra.org
Indian Institute of Oil Palm Research under the Indian Council of Agriculture Research (ICAR-IIOPR)	All aspects	Pedavegi, West Godavari Dist, Andhra Pradesh, India https://iiopr.icar.gov.in/
Golden Hope Plantations Bhd. (GHP)	All aspects	13/F Menara PNB, No 201-A Jalan Tun Razak 50400 Kuala Lumpur, Malaysia. http://www.goldenhope.com
United Plantations, Malaysia	All aspects	http://www.unitedplantations.com/Public/Media_room.asp
Oil Palm Research and Development Center – Thailand	All aspects	Website http://www.natres.psu.ac.th/Researchcenter/palm/palm_index.htm
Fedepalma and Cenipalma, Columbia	All aspects	Calle 98 No. 70-91 Piso 14 Centro Empresarial Pontevedra, Bogota, Colombia Phone: 57 (1)313 8600 Email: Infor@fedepalma.org http://web.fedepalma.org/, https://www.cenipalma.org/

Appendix 16.2 Genetic Resources for Oil Palm (*Elaeis guineensis* L.)

Cultivar/ genetic resources	Important traits	Cultivation location
Deli	Deli D is the most common female parent (Dura) in commercial oil palm D × P hybrid seed production programs. It has been widely distributed worldwide and local selection in Malaysia has led to the development of subpopulations of this BPRO. Dumpy Deli of Malaysia and Gunung Melayu Deli are dwarf mutants of the Deli.	Worldwide, the more cultivated areas in Malaysia
AVROS	The majority of the breeding programs in Malaysia, as well as other countries, use AVROS or its derivatives as the male (pisifera) parent. This BPRO descended from the 'Djongo' or 'best' palm found in Zaire features vigorous trunk growth, high yields, big fruits and thick mesocarp	-Do-
Yangambi	The Yangambi BPRO resulted from the breeding work done out at the Yangambi research station in the Republic of the Congo. It features vigorous palm growth, big fruits and thick mesocarp and is popularly known as Ps in D × P hybrid seed production.	Worldwide and more frequent in Indonesia
La Me	The characteristics of this BPRO, which was bred from seeds gathered from Côte d'Ivoire's wild groves, include smaller palm stature, more small fruit bunches, and increased resistance to unfavorable growth conditions.	Indonesia
Ekona	The Ekona BPRO was developed from wild palms found in the Ekona area of Cameroon, and progenies have fruits with high oil content but are smaller in size. Those from CAM2/2311 Ekona palm exhibit resistance to Fusarium wilt and a high yield of oil.	Indonesia
Calabar	The Calabar BPRO was developed from wild palms found in Calabar, and several offspring have been taken to Ghana, Malaysia, Indonesia, and Costa Rica.	West Africa
Interspecific hybrids	Interspecific hybrids with E. guineensis (Iodine Value 20) and *E. oleifera* (IV 80) produced oil with an Iodine Value of 65, are short and compact and tolerant to spear rot, a very serious disease in Latin America. Despite the normally poor oil output, hybrids derived from Ecuadorian *E. oleifera* seem to give yields that are comparable to the current D × P.	Latin America

References

Abdullah R, Zainal A, Heng W et al (2005) Immature embryo: a useful tool for oil palm (*Elaeis guineensis* Jacq.) genetic transformation studies. Electron J Biotechnol 8(1):24–34

Abdulla UF, Naim R, Taib WRW, et al. (2015) Loop-mediated isothermal amplification (LAMP), an innovation in gene amplification: bridging the gap in molecular diagnostics; a review. Indian J Sci Technol 8:1–12. https://doi.org/10.17485/ijst/2015/v8i17/55767

Adon B, Baudouin L, Durand GT et al (1998) Use of non-selected material for oil palm breeding: Angolan germplasm. Plant Rech Dev 5(3):201–207

Ahloowalia BS, Maluszynski M, Nichterlein K (2004) Global impact of mutation-derived varieties, Euphytica 135:187–204

Ali N (2022) My journey with oil palm. Planter 98(1151):133–141

Anonymous (2000) Annual report. Pulp Pap Int 42(7):9

Anonymous (2003) Preparation of cellulose pulp from empty fruit bunches of oil palm bunches. JIRCAS Newsletter 34

Anuradha U, Karun A (2000) Variation of esterase isozyme in regenerated plantlets of oil palm (*Elaeis guineensis* Jacq.). J Plant Crop 28:82–85

Arias D, Carmenza M, Romero H (2013) Molecular characterization of oil palm *Elaeis guineensis* Jacq. materials from Cameroon. Plant Genet Resour 11(2):140–148

Arias D, Ochoa I, Castro F et al (2014) Molecular characterization of oil palm, *Elaeis guineensis* Jacq. of different origins for their utilization in breeding programmes. Plant Genet Resour 12(3): 341–348

Arias D, Gonzalez M, Prada F et al (2015) Genetic and phenotypic diversity of natural American oil palm (*Elaeis oleifera* (HBK) Cortes) accessions. Tree Genet Genomes 11:122

Asmady H, Jacquemard JC, Hayun Z et al (2002) Variety output and oil palm (Elaeis guineensis Jacq.) improvement: the P.T. Socfin Indonesia example. Paper presented at 2002 Int. Oil Palm Conf., Indonesian Oil Palm Res. Inst., Bali, 8–12 July

Ataga CD, Fatokun CA (1989) Disc polyacrylamide gel electrophoresis of pollen proteins in the oil palm (*Elaesis*). Euphytica 40:83–88

Ateyyeh AF (2012) Effect of storage method on date palm and pistachio pollen viability. Jordan J Agric Sci 8(4):573–581

Aya FO (1979) The use of polyethylene bags for raising oil palm seedlings in Nigeria. 2. The influence of bag size, colour, drainage, and planting date on the performance of the seedlings in the nursery. J Nigerian Inst Oil Palm Res 5 (20)

Babu BK, Mathur R, Ravichandran G et al (2019) Genomewide association study (GWAS) for stem height increment in oil palm (*Elaeis guineensis*) germplasm using SNP markers. Tree Genet Genomes 15(3):40

Bahariah B, Ghulam Kadir AP, Mat Yunus AM et al (2013) Biolistic transformation of oil palm using the phosphomannose isomerase (pmi) gene as a positive selectable marker. Biocatal Agric Biotechnol 2:295–304

Bahariah B, Mat Yunus AM, Rasid O et al (2017) Construction of a vector containing hygromycin (HPT) gene driven by double 35S (2XCaMV35S) promoter for oil palm transformation. J Oil Palm Res 29:180–188

Bahariah B, Masani MYA, Rasid OA et al (2021) Multiplex CRISPR/Cas9-mediated genome editing of the FAD2 gene in rice: a model genome editing system for oil palm. J Genet Eng Biotechnol 19:86

Bai B, Wang L, Zhang YJ et al (2018) Developing genome-wide SNPs and constructing an ultrahigh-density linkage map in oil palm. Sci Rep 8(1):691

Bakoume C, Wickneswari R, Siju S et al (2014) Genetic diversity of the world largest oil palm (*E. guineensis* Jacq.) field gene bank accession using microsatellite markers. J Genet Resour Crop Evol 62:349–360

Barcelos E, Amblard P (1999) Oil palm breeding programme at EMBRAPA/Brazil. In: Rajanaidu N, Jalani BS (eds) Seminar science of oil palm breeding, Montpellier, 1992. Palm Oil Research Institute of Malaysia, Kuala Lumpur, pp 190–203

Barcelos E, Amblard P, Berthaud J (2002) Genetic diversity and relationship in American and African oil palm as revealed by RFLP and AFLP molecular markers. Pesq Agrop Brasileira 37: 1105–1114

Barcelos E, Rios SDA, Cunha RNV (2015) Oil palm natural diversity and the potential for yield improvement. Front Plant Sci 6:190. https://doi.org/10.3389/fpls.2015.00190

Basiron Y, Salmiah A (1994) Potential new value-added products from palm oil and palm kernel oil. In: Management for enhanced profitability in plantations (Ed. by K.H. Chee), Incorp. Soc. Planters, Kuala Lumpupp. 409–418

Baudouin L (1992) Utilisation of molecular markers for oil palm breeding. I. Protein marker. Olegineux 47(12):681–691

Beirnaert ADF, Vanderweyen R (1941) Contribution a l'étudegénétique et biométrique des variétés d' *Elaeis guineensis* Jacquin. East Afr Stand 27:101

Bek-Nielsen C, Bek-Nielsen M, Gurmit S (2000) Managing resources resourcefully – a corporate strategy. In: Goh KJ (ed) Managing oil palm for high yields: agronomic principles. Malaysian Society of Soil Science and Param Agricultural Surveys, Kuala Lumpur, pp 1–27

Berger KG (2010) Recommended practices for storage and transport of edible oils and fats. Malaysian Palm Oil Board, pp 1–15

Billotte N, Risterucci AM, Barcelos E et al (2001) Development, characterization and across–taxa utility of oil palm (*Elaeis guineensis* Jacq) microsatellite markers. Genome 44:413–425

Billotte N, Jourjon MF, Marseillac N et al (2010) QTL detection by multiparent linkage mapping in oil palm (*Elaeis guineensis* Jacq.). Theor Appl Genet 120:1673–1687

Breure CJ (1986) Parent selection for yield and bunch index in the oil palm in West New Britain. Euphytica, 35, 65–72

Breure CJ (1988) The effect of palm age and planting density on the partitioning of assimilates in oil palm (Elaeis guineensis Jacq.). Expl Agric., 24, 53–66

Cadena T, Prada F, Perea A et al (2012) Lipase activity, mesocarp oil content and iodine value in oil palm fruits of *Elaeis guineensis*, *Elaeis oleifera* and the interspecific hybrid OxG (*E. oleifera* x *E. guineensis*). J Sci Food Agric 93(3):674–680. https://doi.org/10.1002/jsfa.5940

Ceballos H, Perez VJC, Iglesias CA et al (2007) The use of doubled-haploids in cassava breeding. In: Cassava research and development in Asia: exploring new opportunities for an ancient crop. Centro Internacional de Agricultura Tropical (CIAT), Cassava Office for Asia, Bangkok, pp 150–160

Chaidamsari T, Tahardi JS, Santoso D (1998) Agrobacterium-mediated transformation in leaf explants oil palm. In: Proc. 1998 Int. Oil Palm Conf. 'Commodity of the past, today and the future' (Ed. by A Jatmika et al.), Indonesian Oil Palm Res Inst 602–612

Chan PI, Rose RJ, Munir A et al (2014) Evaluation of reference genes for quantitative real-time PCR in oil palm elite planting materials propagated by tissue culture. PLoS One 9(6):e99774

Chew PS, Soh AC, Goh KJ et al (1998) Role of private sector research in oil palm crop production. In: Proc. Int. Oil Palm Conf. 'Commodity of the past, today and the future' (Ed. by A. Jatmika et al.) Indonesian Oil Palm Res Inst 133–144

Cheah SC, Ives C (1998) The biotechnology of oil palm. In: Bedford BM (ed) Agricultural biotechnology in international development. Palm Oil Research Institute of Malaysia, Kuala Lumpur, pp 161–170

Choong CY, Shah FH, Rajanaidu N et al (1996) Isoenzyme variation of Zairean oil palm (*Elaeis guineensis* Jacq.) germplasm collection. Elaeis 8:45–53

Chua KL, Singh R, Cheah SC (2001) Construction of oil palm (*Elaeis guineensis* Jacq.) linkage maps using AFLP markers. In: Proceedings of the POPOC International Palm Oil Congress, Agriculture conference on 'cutting edge technologies for sustained competitiveness', Kuala Lumpur, Malaysia, 20–22 August 2001, pp 461–467

Chungsiriporn J, Khunthongkaew P, Wongnoipla Y et al (2022) Fibrous packaging paper made of oil palm fiber with beeswax-chitosan solution to improve water resistance. Ind Crop Prod 177(1):114541

Chin CW (1999) Oil palm breeding techniques. In: Proc. Seminar 'Science of oil palm breeding' (Montpellier, 1992) (Ed. by N Rajanaidu, BS Jalani) Palm Oil Res. Inst. Malaysia, Kuala Lumpur pp. 49–64

Cochard B, Noiret JM, Baudouin L et al (1995) Second cycle of reciprocal recurrent selection (RRS) in oil palm, *Elaeis guineensis* Jacq. Results of Deli x La Me hybrid tests. In: Jalani BS et al (eds) Proceedings of the 1993 PORIM International Palm Oil Congress. Palm Oil Research Institute of Malaysia, Kuala Lumpur, pp 24–35

Corley RHV, Tinker PB (2003) The oil palm, World agriculture series, 4th edn. Blackwell Publishing, Oxford, xxviii, 562 pp

Corley RHV, Tinker PB (2016) The oil palm, 5th edn. Wiley Blackwell, Chichester

Corley RHV, Hardon JJ, Tan GY (1971) Analysis of growth of the oil palm (*Elaeis guineensis* Jacq.). 1. Estimation of growth parameters and application in breeding. Euphytica 20:307–315

Corley RHV, Lee CH, Law IM et al (1986) Abnormal flower development in oil palm clones. Planter 62(723):233–240

Corley RHV, Rao V, Palat T et al (2018) Breeding for drought tolerance in oil palm. J Oil Palm Res 30(1):26–35

Correa TR, Motoike SY, de Souza Andrade AP et al (2016) Accelerated in vitro propagation of elite oil palm genotypes (*Elaeis guineensis* Jacq.) by substituting cytokinin with putrescine. Afr J Biotechnol 15(50):2767–2775

Corley, R.H.V. and Breure, C.J. (1981) Measurements in Oil Palm Experiments. Internal Report, Unilever Plantations, London

Corley RHV (1998) What is the upper limit to oil extraction ratio? In: Rajanaidu N, Henson IE, Jalani BS (eds) Proceedings of the 1996 international conference on "Oil and kernel production in oil palm – a global perspective". Palm Oil Research Institute of Malaysia, Kuala Lumpur, pp 256–269

Corley RHV, Stratford R (1998) Biotechnology and oil palm: opportunities and future impact. In: Proc. 1998 Int. Oil Palm Conf. 'Commodity of the past, today and the future' (Ed. by A. Jatmika et al.) Indonesian Oil Palm Res. Inst.80–91

Corley RHV (2001) Oil palm. In: Ecosystems of the world 19 -Tree crop ecosystems (Ed. by F.T. Last), pp. 299–320. Elsevier

Correa TR, Motoike SY, Coser SM, et al (2015) Estimation of genetic parameters for in vitro oil palm characteristics (Elaeis guineensis Jacq.) and selection of genotypes for cloning capacity and oil yield. Industrial Crops and Products, 77, 1033-1038

Cronquist A (1981) An integrated system of classification of flowering plants. Columbia university press

Darma JJ, Muthia E, Sunardi S et al (2022) Physical and mechanical properties of biodegradable pot derived from oil palm empty fruit bunch and sodium alginate. Braz Arch Biol Technol 65:6223

Darmawan C, Wiendi N, Utomo C et al (2020) Electroporation-mediated genetic transformation of oil palm (*Elaeis guineensis*). Biodivers J Biol Divers 21:3720–3726

Donough CR, Ng M, Lai C (1992) Pamol's approach to quality control in controlled pollination for D 3 P seed production. Paper presented at Incorp. Soc. Planters Seminar 'Advances in cocoa and oil palm in meeting future challenges', 16–17 June

Dolezel J (1991) Flow cytometric analysis of nuclear DNA content in higher plants. Phytochem Anal 2(4):143–154

De Touchet B, Duval Y, Pannetier C (1991) Plant regeneration from embryogenic suspension cultures of oil palm (Elaeis guineensis Jacq.). Plant Cell Reports, 10(10), 529-532

Dransfield J, Uhl NW, Asmussen CB et al (2005) A new phylogenetic classification of the palm family, Arecaceae. Kew Bull 60(4):559–569

Dumet D, Engelmann F, Chabrillange N et al (1993a) Cryopreservation of standard oil palm (*Elaeis guineensis* Jacq.) somatic embryos involving a desiccation step. Plant Cell Rep 12:352–355

Dumet D, Engelmann F, Chabrillange N et al (1993b) Development of cryopreservation for oil palm somatic embryos using an improved process. Oleagineux 48:273–278

Dumet D, Engelmann F, Chabrillange N et al (1993c) Importance of sucrose for the acquisition of tolerance to desiccation and cryopreservation of oil palm somatic embryos. CryoLetters 14:243–250

Dumet D, Engelmann F, Chabrillange N et al (1994) Effect of various sugars and polyols on the tolerance to desiccation and freezing of oil palm polyembryonic cultures. Seed Sci Res 4(3): 307–313

Dunwell JM, Wilkinson MJ, Nelson S et al (2010) Production of haploids and doubled haploids in oil palm. BMC Plant Biol 10:218

Duval Y, Durrand-Gasselin T, Konan K et al (1988) Vegetative propagation of oil palm (*Elaeis guineensis* Jacq.) by in vitro culture: strategy and results. Oleagineux 43(2):39–47

Eeuwens CJ (1976) Mineral requirements for growth and callus initiation of tissue explants excised from mature coconut palms (*Cocos nucifera*) and cultured in vitro. Physiol Plant 36(1):23–28

Eeuwens CJ, Lord S, Donough CR (2002) Effects of tissue culture conditions during embryoid multiplication on the incidence of `mantled' flowering in clonally propagated oil palm. Plant Cell Tissue Organ Cult 70(3):311–323

Ekaratne SR, Senathirajah S (1983) Viability and storage of pollen of the oil palm *Elaeis guineensis* Jacq. Ann Bot 51(5):661–668

Engelmann F (2004) Plant cryopreservation: progress and prospects. In Vitro Cell Dev Biol Plant 40(5):427–433

Engelmann F, Duval Y (1986) Cryopreservation of oil palm somatic embryos (*Elaeis guineensis* Jacq.): results and application prospects. Oleagineux 41(4):169–174

Engelmann F, Chabrillange N, Dussert S et al (1995) Cryopreservation of zygotic embryos and kernels of oil palm (*Elaeis guineensis* Jacq.). Seed Sci Res 5(2):81–86

Escobar R, Alvarado A (2004) Strategies in production of oil palm compact seeds and clones. Exp Agric 27:13–26

Gan ST (2014) The development and application of molecular markers for linkage mapping and quantitative trait loci analysis of important agronomic traits in oil palm (*Elaeis guineensis* Jacq.). PhD thesis submitted to the University of Nottingham, UK

Gantait S, Sinniah UR, Suranthran P (2015) Improved cryopreservation of oil palm (*Elaeis guineensis* Jacq.) polyembryoids using droplet vitrification approach and assessment of genetic fidelity. Protoplasma 252(1):89–101

Ghesquiere M (1984) Enzyme polymorphism in oil palm (*Elaeis guineensis* Jacq.) Genetic control of nine enzyme systems. Oleagineux 39:561–574

Ghesquiere M (1985) Enzyme polymorphism in oil palm (*Elaeis guineensis* Jacq.) II. Variability and genetic structure of seven origins of oil palm. Oleagineux 40:529–540

Gomes JR, Rui A, Lopes R et al (2019) Selection gains for bunch production in interspecific hybrids between caiaué and oil palm. Pesq Agrop Brasileira 54:e00819. https://doi.org/10.1590/S1678-3921.pab2019.v54.00819

Gorret N, bin Rosli SK, Oppenheim SF et al (2004) Bioreactor culture of oil palm (*Elaeis guineensis*) and effects of nitrogen source, inoculum size and conditioned medium on biomass production. J Biotechnol 108(3):253–263

Govender N, Wong MY (2016) Detection of Oil Palm Root Penetration by Agrobacterium-Mediated Transformed Ganoderma boninense, Expressing Green Fluorescent Protein. Phytopathology 107, 483–490

Grout BWW, Shelton K, Pritchard HW (1983) Orthodox behaviour of oil palm seed and cryopreservation of the excised embryo for genetic conservation. Ann Bot 52(3):381–384

Gurmit S, Toh TS (2000) Integrating the production of granulated organic fertilisers, compound animal feed and fibre in the palm oil mill. In: Proc. Int. Planters Conf. 'Plantation tree crops in the new millennium: the way ahead' (Ed. by E. Pushparajah), Incorp. Soc. Planters, Kuala Lumpurpp. 843–850

Hama-Ali EO, Tan SG (2014) Using monomorphic microsatellite markers in oil palm (*Elaeis guineensis* Jacq.). Res Rev J Bot Sci 3:1–6

Hama-Ali EO, Panandam JM et al (2015) Association between basal stem rot disease and simple sequence repeat markers in oil palm, *Elaeis guineensis* Jacq. Euphytica 202:199–206

Hardon JJ, Davies MD (1969) Effects of vacuum-drying on the viability of oil palm pollen. Exp Agric 5:59–65

Hardon JJ, Turner PD (1967) Observations on natural pollination in commercial plantings of oil palm (*Elaeis guineensis*) in Malaya. Exp Agric 3(2):105–116

Hartley CWS (1988) The oil palm, 3rd edn. Longman, London

Helene A, Stefan J, Yves O et al (2007) Functional characterization of MADS-box genes involved in the determination of oil palm flower structure. J Exp Bot 58(6):1245–1259

Henson IE (1999) Comparative ecophysiology of oil palm and tropical rainforest. In: Gurmit S et al (eds) Oil palm and the environment – a Malaysian perspective. Malaysian Oil Palm Growers' Council, Kuala Lumpur, pp 9–39

Hermanto A, Gan SH, Mustopa IR et al (2023) Use of multiseasonal oil palm yield data to assess drought tolerance. Sci Hortic 308:111603

Ho CL, Kwan YY, Choi MC et al (2007) Analysis and functional annotation of expressed sequence tags (ESTs) from multiple tissues of oil palm (*Elaeis guineensis* Jacq.). BMC Genomics 8:1–12

Hor YL, Sinnish UR (2000) Conservation of genetic resources of oil palm (*Elaeis guineensis* Jacq.) using cryopreservation. UPM Research Report 1997–2000, Vol II, Section 2-Extended Abstracts, pp 107–108

Houmiel KL, Slater S, Broyles D et al (1999) Poly(beta-hydroxybutyrate) production in oilseed leukoplasts of Brassica napus. Planta 209: 547–550

Ibaraki Y (2001) Automation in somatic embryo production. Prog Biotechnol 18:365–374

Ismail FS, Asaari AZM, Yussof NAM et al (2022) Physical and mechanical properties of paper made from beaten empty fruit bunch fiber incorporated with microcrystalline cellulose. J Nat Fibers 19(3):999–1011

Ithnin M, Kushairi A (eds) (2020) The oil palm genome. Springer, London

Ithnin M, Xu Y, Marjuni M et al (2017) Multiple locus genome-wide association studies for important economic traits of oil palm. Tree Genet Genomes 13(5):103

Izawati AMD, Masani MYA, Ismanizan I et al (2009) Evaluation on the effectiveness of 2-deoxyglucose-6-phosphate phosphatase (DOGR1) gene as a selectable marker for oil palm (*Elaeis guineensis* Jacq.) embryogenic calli transformation mediated by *Agrobacterium tumefaciens*. Front Plant Sci 6:727

Izawati AM, Parveez GK, Masani MY (2012) Transformation of oil palm using *Agrobacterium tumefaciens*. In: Dunwell J, Wetten A (eds) Transgenic plants: methods in molecular biology, vol 847. Humana Press, Totowa, pp 177–188

Jack PL, Mayes S (1993) Use of molecular markers for oil palm breeding. II. Use of DNA markers (RLFPs). Oleagineux 48:1–8

Jack PL, Dimitrijevic TAF, Mayes S (1995) Assessment of nuclear, mitochondrial and chloroplast RFLP markers in oil palm (*Elaeis guineensis* Jacq.). Theor Appl Genet 90:643–649

Jack PL, James C, Price Z (1998) Application of DNA markers in oil palm breeding. In: Jatmika A et al (eds) Proceedings of the International Oil Palm conference on 'commodities of the past, today and future'. Indonesian Palm Oil Research Institute, Medan, pp 315–324

Jagoe RB (1952) The dumpy oil palm. Malay Agric J 35:12

Jalani BS, Basiron Y, Darus A et al (2002) Prospects of elevating national oil palm productivity: a Malaysian perspective. Oil Palm Ind Econ J 2(2):1–9

Jaligot E, Rival A, Beule T (2000) Somaclonal variation in oil palm (*Elaeis guineensis* Jacq.): the DNA methylation hypothesis. Plant Cell Rep 19(7):684–690

Jaligot E, Beule T, Rival A (2002) Methylation-sensitive RFLPs: characterisation of two oil palm markers showing somaclonal variation-associated polymorphism. Theor Appl Genet 104(8): 1263–1269

Jaligot E, Beule T, Baurens FC (2004) Search for methylation-sensitive amplification polymorphisms associated with the "mantled" variant phenotype in oil palm (*Elaeis guineensis* Jacq.). Genome 47:224–228

Jarupan L, Hunsa-Udom R, Bumbudsanpharoke N (2022) Potential use of oil palm fronds for paper making and applications as molded pulp trays for fresh product under simulated cold chain logistics. J Nat Fibers 19(7):2772–2784

Jayanthi M, Mandal PK, Sujatha G et al (2008) Simple sequence repeats (SSR) and RAPD primers for assessment of genetic uniformity among the field-planted clones of oil palm. Paper submitted for presentation in the PLACROSYM-2008, to be held at NRCC, Puttur, Karnataka, India from 4–8 December 2008

Jayanthi M, Mohan NM, Mandal PK (2011) Direct somatic embryogenesis and plantlet regeneration in oil palm. J Plant Biochem Biotechnol 20(2):249–251

Jayanthi M, Susanthi B, Mohan NM (2015) *In vitro* somatic embryogenesis and plantlet regeneration from immature male inflorescence of adult dura and tenera palms of *Elaeis guineensis* (Jacq.). Springerplus 4(1):256

Jayashree E, Mandal PK, Madhava M et al (2002) Development of decorticator for extraction of quality fibre from empty fruit bunches. In: Proceedings of PLACROSYM XV organised by Coffee Board, Mysore, India, pp 678–681

Jones LH (1974) Propagation of clonal oil palms by tissue culture. Oil Palm News 17:1–8

Jouannic S, Argout X, Lechauve F et al (2005) Analysis of expressed sequence tags from oil palm (*Elaeis guineensis*). FEBS Lett 579:2709–2714

Kadandale S, Marten R, Smith R (2019) The palm oil industry and noncommunicable diseases. Bull World Health Organ 97:118

Kamaruddin N, Basri Wahid M, Mohd Isa Z et al (1999) Population studies of Oryctes rhinoceros in an oil palm replant using pheromone traps. In: Preprints, 1999 PORIM Int. Palm Oil Conf, Palm Oil Res Inst Malaysia, Kuala Lumpurpp. 477–496

Kanapathy K, Jorgensen HK, Gurmit S (1981) Preparation and utilisation of dried palm oil mill effluent (POME). In: Proc. Natn. Workshop 'Oil palm by-product utilisation', Palm Oil Res Inst Malaysia, Kuala Lumpurpp. 45–48

Kanchanapoom K, Domyoas P (1999) The origin and development of embryoids in oil palm (*Elaeis guineensis* Jacq) embryo culture. Sci Asia 25:195–202

Karina M, Onggo H, Abdulla AHD et al (2008) Effect of oil palm empty fruit bunch fibre on the physical and mechanical properties of fibre glass reinforced polyester resin. J Biol Sci 8(1): 101–106

Karun A, Sajini KK (1996) Plantlet regeneration from leaf explants of oil palm seedlings. Curr Sci 71:922–926

Khalil HPSA, Hanida S, Kang CW et al (2007) Agro-hybrid composite: the effects on mechanical and physical properties of oil palm fiber (EFB)/glass hybrid reinforced polyester composites. J Reinf Plast Compos 26(2):203–218

Khalil HPSA, Siti AM, Ridzuan R et al (2008) Chemical composition, morphological characteristics and cell wall structure of Malaysian oil palm fibers. Polym Plast Technol Eng 47:273–280

Khawnium T, Te-Chato S (2011) Simple vitrification protocol for cryopreservation of oil palm using embryogenic culture. J Agric Technol 7(2):519–529

Khawnium T, Te-Chato S (2012) Cryopreservation of embryogenic callus of hybrid tenera oil palm by dehydration technique and evaluation of somaclonal variation by SSR marker. J Agric Technol 8(6):2115–2125

Koh MP, Rahim S, Mohd Nor MY et al (1999) Manufacture of building materials from oil palm biomass. In: Oil palm and the environment - a Malaysian perspective (Ed. by Gurmit Singh et al.), Malaysian. Oil Palm Growers' Council, Kuala Lumpur pp. 199–211

Krishnakumar CR, Maheshwar DL, Sunilkumar K (2012) Record yield in oil palm achieved by woman farmer in Karnataka, India. Planter 88(1040):831–833

Kuhoo KC, Lee TW (1991) Pulp and paper from the oil palm. Appita J 44(6):385–388

Kularatne RS, Shah FH, Rajanaidu N (2001) The evaluation of genetic diversity of Deli dura and African oil palm germplasm collection by AFLP technique. Trop Agric Res 13:1–12

Kushairi A, Amiruddin MD (2020) Development of new oil palm cultivars in Malaysia. J Oil Palm Res 32(3):420–426

Kushairi A, Rajanaidu N, Rafi MY et al (2001) Population 12: pollen for the development of PS-1 type oil palm planting materials. MPOB Information Series 15 TT No 100

Kushairi A, Tarmizi AH, Zamzuri I et al (2010) Production, performance and advances in oil palm tissue culture. Paper presented at the International Seminar on Advances in Oil Palm Tissue Culture, in Yogyakarta, Indonesia, vol 29

Kwong QB, Teh CK, Ong AL et al (2016) Development and validation of a high-density SNP genotyping array for African oil palm. Mol Plant 9(8):1132–1141

Lee M, Xia JH, Zou Z et al (2015) A consensus linkage map of oil palm and a major QTL for stem height. Sci Rep 5:8232

Lee ZJ, Tong SC, Tang TK et al (2022) Palm-based cellulose nanofiber isolated from mechano-chemical processing as sustainable rheological modifier in reduced fat. J Food Sci 87(8): 3542–3561

Lebbie AR, Guries RP (2002) The palm wine trade in Freetown, Sierra Leone: production, income, and social construction. Econ Bot 56(3):246–254

Liao SH (2005) Expert system methodologies and applications-a decade review from 1995-2004. Exprt Syst. Appl. 28 (1), 93-103. https://doi.org/10.1016/j.eswa.2004.08.003

Li DZ, Pritchard HW (2009) The science and economics of ex-situ plant conservation. Trends Plant Sci 14(11):614–621

Lim CC, Rao V (2005) DNA marker technology and private sector oil palm breeding. Planter 80:1–12

Litz RE (ed) (2005) Biotechnology of fruit and nut crops, vol 29. CABI, Wallingford

Lioret C (1982) Vegetative propagation of oil palm by somatic embryogenesis. In International Conference on Oil Palm in Agriculture in the Eighties, Kuala Lumpur (Malaysia), 17-20 Jun 1981. PPP (ISP)

Loong SG, Nazeeb M, Letchumanan A (1988) Optimising the use of EFB mulch on oil palms on two different soils. In: Proc. 1987 Int. Oil Palm Conf. 'Progress and prospects' (Ed. by A. Halim Hassan et al.), Palm Oil Res. Inst. Malaysia, Kuala Lumpurpp. 605–639

Lopes R, Antoniassi R, Cunha RNV et al (2021) Physicochemical characterization of bunches and oil composition in a backcross progeny (caiaué x oil palm) x oil palm. Colloq Agrar 17:449

Luke SH, Advento AD, Arywan AAK., etal. (2020) Managing oil palm plantations more sustainably: Large scale Experiments within the Biodiversity and Ecosystem function in tropical Agriculture (BEFTA) programme. Frontiers in Forest and Global Change 2 (January),1-20. https://doi.org/10.3389/ffgc .2019.00075

Madon M, Clyde MM, Hashim H et al (2005) Polyploidy induction of oil palm through colchicines and oryzalin treatments. J Oil Palm Res 17:110–123

Mahamooth TN, Gan HH, Kee KK et al (2011) Water requirements and cycling of oil palm. In: Goh KJ et al (eds) Agronomic principles and practices of oil palm cultivation. Agricultural Crop Trust, Petaling Jaya, pp 89–144

Maheran AB, Abu Zarin O, Aw KT et al (1995) Felda's early experiences with vegetative propagation of the oil palm (*Elaeis guineensis* Jacq.). In: Proceedings of the 1993 PORIM International Palm Oil Congress–update & vision agriculture, pp 99–113

Maizura I, Rajanaidu N (2001) Genetic diversity of oil palm germplasm collections using RFLPs. In: Proceedings of the 2001 PIPOC International Palm Oil Congress, Agriculture conference on 'cutting edge technologies for sustained competitiveness', Kuala Lumpur, Malaysia, 20–22 August 2001, pp 526–535

Maizura I, Rajanaidu N, Zakri AH et al (2006) Assessment of genetic diversity in oil palm (*Elaeis guineensis* Jacq.) using Restriction Fragment Length Polymorphism (RAPD). Genet Resour Crop Evol 53:187–195

Maluszynski M, Kasha KJ, Forster BP et al (2003) Published double haploid protocols in plant species. In: Maluszvanski M et al (eds) Doubled haploid production in crop plants: a manual. Kluwer Academic Publishers, pp 309–335

Mandal PK, Jayanthi M (2011) Oil palm. In: Singh HP, Parthasarathy VA, Nirmal Babu K (eds) Advances in horticulture biotechnology: molecular markers and marker-assisted selection, Vol III Fruit crops, plantation crops and spices. Westville Publishing House, New Delhi

Mandal PK, Susmita D (2006) Characterization of oil palm *Elaeis guineensis* germplasm using biochemical and molecular parameters. J Plant Crop 34:529–533

Marbun CLM, Toruan-Mathius N, Utomo C (2015) Micropropagation of embryogenic callus of oil palm (*Elaeis guineensis* Jacq.) using temporary immersion system. Procedia Chem 14:122–129

Marhalil M, Maizura I, Fadila AM et al (2017) Long-term storage of oil palm germplasm zygotic embryo using cryopreservation. J Oil Palm Res 29:541–547

Martin JJJ, Yarra R, Wei L et al (2022) Oil palm breeding in the modern era: challenges and opportunities. Plants 11(11):1395. https://doi.org/10.3390/plants11111395

Mathews J, Tan TH, Yong KK (2010) Managing oil palm nursery: IOI's experience. Planter 86: 771–785

Mathius NT, Haris N, Ginting G (1998) Use of biomolecular techniques in studies of abnormalities in oil palm clones. In: Proceedings of the International Oil Palm Conference, Nusa Dua, Bali, pp 23–26

Matthes M, Singh R, Cheah SC et al (2001) Variation in oil palm (Elaeis guineensis Jacq.) tissue culture-derived regenerants revealed by AFLPs with methylation-sensitive enzymes. Theor appl Genet, 102

Mayes S, James XM, Horner SF et al (1996) The application of restriction fragment length polymorphism for the genetic fingerprinting of oil palm (*Elaeis guineensis* Jacq.). Mol Breed 2:175–180

Mayes S, Jack PL, Corley RH et al (1997) Construction of a RFLP genetic linkage map for oil palm (*Elaeis guineensis* Jacq.). Genome 40:116–122

Mayes S, Jack PL, Corley RHV (2000) The application of molecular markers in a specific breeding programme for oil palm. Heredity 85:288–293

Meunier J (1995) Biotechnologies: what prospects for tropical tree crops. Plant Res Dev 2:7–18

Mirmozaffari M (2019) Presenting a medical expert system for diagnosis and treatment of nephrolithiasis. Europian Journal of medical and health sciences1 (1),9-11.https://doi.org/10.24018/ejmed.2019.1.1.20

Mohammad AB (2016) Gamma irradiation effects on oil palm (*Elaeis guineensis* Jacq) pollen viability, fruits and bunch formations. JSNM 26(2):1–8

Monteiro TR, Freitas EO, Nogueira GF (2018) Assessing the influence of subcultures and liquid medium during somatic embryogenesis and plant regeneration in oil palm (*Elaeis guineensis* Jacq.). J Hortic Sci Biotechnol 93(2):196–203

Morcillo F, Gagneur C, Adam H et al (2006) Somaclonal variation in micropropagated oil palm. Characterization of two novel genes with enhanced expression in epigenetically abnormal cell lines and in response to auxin. Tree Physiol 26:585–594

Moretzsohn MC, Nunes CDM, Ferreira ME (2000) RAPD linkage mapping of the shell thickness locus in oil palm (*Elaeis guineensis* Jacq.). Theor Appl Genet 100:63–71

Moretzsohn MC, Ferreira MA, Amaral ZJA (2002) Genetic diversity of Brazilian oil palm (*Elaeis oleifera* HBK) germplasm collected in the Amazon Forest. Euphytica 124:35–45

Murashige T, Skoog F (1962) A revised medium for rapid growth and bioassays with tobacco tissue cultures. Physiol Plant 15(3):473–497

Murphy D (2006) Plant breeding and biotechnology. Cambridge University Press, Cambridge, p 33

Murphy DJ, Goggin K, Paterson RRM (2021) Oil palm in the 2020s and beyond: challenges and solutions. CABI Agric Biosci 2:39

Murugesan P, Sunilkumar K (2014) Dwarf palms for future generations. Indian Hortic 59(6):10–12

Murugesan P, Rani KLM, Ramajayam D et al (2015) Genetic diversity of vegetative and bunch traits of African oil palm (*Elaeis guineensis*, Jacq.) germplasm in India. Ind J Agric Sci 85(7): 892–895

Ng KT, Southworth A (1973) Optimum time of harvesting oil palm fruit. In: Wastie RL, Earp DA (eds) Advances in oil palm cultivation. Incorporated Society of Planters, Kuala Lumpur, pp 439–461

Noiret JM, Gascon JP, Pannetier C (1985) La production dc palmier à huile par culture in vitro. Oléagineux 40(7):365–372

Nor-Azlina R, Mohammad J, Safwan IA et al (2021) Evaluation of thermal acoustic properties of oil palm empty fruit bunch/sugarcane bagasse fibers based hybrid composites for wall buildings thermal insulation. Fibers Polym 22(9):2563–2571

Nur F, Forster BP, Osei SA et al (2018) Mutation breeding in oil palm: a manual. CABI

Nurfahisza RA, Aman MDA, Masani MA (2014) Molecular analysis of transgenic oil palm to detect the presence of transgenes. J Oil Palm Res 26(1):96–103

Nyouma A, Bell JM, Jacob F et al (2019) From mass selection to genomic selection: one century of breeding for quantitative yield components of oil palm (*Elaeis guineensis* Jacq.). Tree Genet Genomes 15:69

Nwankwo BA, Krikorian AD (1983) Morphogenetic potential of embryo- and seedling-derived callus of Elaeis guineensisJacq. var. pisiferaBecc. Ann. Bot., 51: 65–76

Okwuagwu CO, Ataga CD, Okoye M (2011) Germplasm collection of highland palms of Afikpo in eastern Nigeria. Bayero J Pure Appl Sci 4:112–114

Omar M, Novok F (1990) In vitro plant regeneration and ethylmethanesulphonate (EMS) uptake in somatic embryos of date palm (*Phoenix dactylifera* L.). Plant Cell Tissue Organ Cult 20(3): 185–190

Ong HT (1975) Callus formation from roots of the oil palm (*Elais guineensis* Jacq.). In: National plant tissue culture symposium, RRIM, Kuala Lumpur, Malaysia, 21 February 1975

Ong AM, Ordway JM, Jiang N (2015) Loss of Karma transposon methylation underlies the mantled somaclonal variant of oil palm. Nature 525(7570):533–537

Ong P, Maizura I, Marhalil M et al (2018) Association of SNP markers with height increment in MPOB-Angolan natural oil palm populations. J Oil Palm Res 30(1):61–70

Ooi SE, Novak O, Dolezal K (2013) Cytokinin differences in vitro cultures and inflorescences from normal and mantled oil palm (*Elaeis guineensis* Jacq.). J Plant Growth Regul 32(4):865–874

Oteyami M, Zannou ET, Coffi A et al (2019) Evaluation of palm trees hybrids (interspecific backcross BC1) *E. oleifera* x *E. guineensis* for their sensitivity to the attacks of the leaf miner. Adv Agric Environ Sci 2(1):1–6. https://doi.org/10.30881/aaeoa.00018

Oyedeji AO, Osifeko MO, Folorunsho OR, Ade-Ikuesan OO (2019) Design and implementation of a medical diagnostic expert system. J. Engg. Sci. 10(2),103-109. https://www2.kuet.ac.bd/JES/

Palat T, Smith BG, Corley RHV (2000) Irrigation of oil palm in Southern Thailand. In: Pushparajah E (ed) Proceedings of the international planters conference. Plantation tree crops in the new millennium: the way ahead. Incorporated Society of Planters, Kuala Lumpur, pp 303–315

Palat T, Chayawat N, Clendon JH et al (2008) A review of 15 years of oil palm irrigation research in Southern Thailand. Planter 84:537–546

Pamin K (1998) A hundred and fifty years of oil palm development in Indonesia: from the Bogor botanic garden to the industry. In: Jatmika A et al (eds) Proceedings of the 1998 international oil palm conference on "Commodity of the past, today and the future". Indonesian Oil Palm Research Institute, Indonesia, pp 3–23

Pannetier C, Arthuis P, Lievoux D (1981) Neo-formation of young *E. guineensis* plantlets from primary calluses obtained on leaf fragments cultured *in vitro*. Oleagineux 36(3):119–122

Paranjothy K (1986) Recent developments in cell and tissue culture of oil-bearing palms. PORIM Occasional Paper, 19

Paranjothy K, Othman R (1982) In vitro propagation of oil palm. In: Fujiwara F (ed) Proceedings of the 5th international congress of plant tissue and cell culture. Japanese Association for Plant Tissue Culture, Tokyo, pp 747–748

Paranjothy K, Saxena S, Banerjee M et al (1990) Clonal multiplication of woody perennials. In: Developments in crop science, vol 19. Elsevier, pp 190–219

Paranjothy K, Ong L, Sharifah SSA (1995) DNA and protein changes in relation to clonal abnormalities in relation to clonal abnormalities. In: Rao V, Henson IE, Rajanaidu N (eds) Recent developments in oil palm tissue culture and biotechnology. PORIM, Kuala Lumpur, pp 86–89

Parveez GKA, Bohari B, Ayub NH et al (2008) Transformation of PHB and PHBV genes driven by maize Ubiquitin promoter into oil palm for the production of biodegradable plastics. J Oil Palm Res (Special Issue) 2:77–86

Pinto SS, Lopes R, Cunha RNV (2019) Produção e composição de cachos e incidência do anelvermelhoemhíbridosinterespecíficos de caiaué com dendezeiro no sul da Bahia. Agrotrópica 1:5–16. https://doi.org/10.21757/0103-3816.2019v31n1p5-16

Polge C, Smith AU, Parkes AS (1949) Revival of spermatozoa after vitrification and dehydration at low temperatures. Nature 164:666

Pootakham W, Jomchai N, Areerate P (2015) Genome-wide SNP discovery and identification of QTL associated with agronomic traits in oil palm using genotyping by sequencing (GBS). Genomics 105:288–295

Prakash K, Sunilkumar K, Rekha Choudhury (2019) Cryopreservation of kernel and zygotic embryos of oil palm (Elaeis guineensis Jacq.) Journal of Plantation Crops. 47 (1):16-23

Purba AR, Noyer JL, Baudouin L (2000) A new aspect of genetic diversity of Indonesian oil palm (*Elaeis guineensis* Jacq.) revealed by isoenzyme and AFLP markers and its consequences to breeding. Theor Appl Genet 101:956–961

Rabechault H, Martin JP (1976) Multiplication vegetative du palmier a huile (*Elaeis guineensis* Jacq.) a l'aide de culture de tissus foliares. C R Acad Sci 283:1735–1737

Rahman ME, Boon AL, Muntohar AS et al (2014) Performance of masonry blocks incorporating Palm Oil Fuel Ash. Journal of Cleaner Production, V 78:195-201. https://doi.org/10.1016/j.jclepro.2014.04.067

Rajanaidu N (1994) Institiut Penyelidikan Minyak Kelapa Sawit Malaysia. PORIM Oil Palm Genebank: collection, evaluation utilization and conservation of oil palm genetic resources. Palm Oil Research Institute of Malaysia, Kuala Lumpur

Rajesh MK, Radha E, Karun A et al (2003) Plant regeneration from embryo-derived callus of oil palm: the effect of exogenous polyamines. Plant Cell Tissue Organ Cult 75(1):41–47

Raju CR, Sajini KK, Balachandran SM et al (1989) Clonal multiplication of oil palm (*Elaeisguineensis* Jacq.). J Plant Crop 16:17–20

Ramli R, Shaler S, Jamaludin MA (2002) Properties of medium density fibre board from oil palm empty fruit bunch fibre. J Oil Palm Res 14(2):34–40

Rance KA, Mayes S, Price Z et al (2001) Quantitative trait loci for yield components in oil palm (*Elaeis guineensis* Jacq.). Theor Appl Genet 103(8):1302–1310

Rankine IR, Fairhurst TH (1998) Field handbook: oil palm series. Volume 1: Nursery. Potash and Phosphate Institute, Singapore

Rao V, Donough CR (1990) Preliminary evidence of a genetic cause for floral abnormalities in some oil palm ramets. Elaeis 2:199–207

Rao V, Kushairi A (1999) Quality of oil palm planting material. In: Proc. 1996 Seminar 'Sourcing of oil palm planting materials for local and overseas joint ventures' (Ed. by N. Rajanaidu & B.S. Jalani), Palm Oil Res. Inst. Malaysia, Kuala Lumpurpp. 188–197

Reflini R (2017) Evaluation of 2,4-D and NAA concentrations for callus and somatic embryos formation in oil palm. J Adv Agric Technol 4(3):215–218

Rethinam P, Suresh K, Reddy VM et al (2000) Effect of age of oil palm seedlings at planting on growth. Int J Oil Palm 1:61–63

Ridzuan R, Jalani BS, Arriffin D (1999) OP-fibre moulded plastic composites. PORIM TT No-68

Riju A, Chandrasekar A, Arunachalam V (2007) Mining for single-nucleotide polymorphisms and insertions/deletions in expressed sequence tag libraries of oil palm. Bioinformatics 2(4): 128–131

Ritter E, de-Armentia EL, Erika P et al (2016) Development of a molecular marker system to distinguish shell thickness in oil palm genotypes. Euphytica 207:367–376

Rival A (2018) The palm oil market: growth and trends Stefano Savi, Roundtable on Sustainable Palm Oil, Malaysia. In: Achieving sustainable cultivation of oil palm, vol 1. Burleigh Dodds Science Publishing, pp 19–32

Rival A, Jaligo E, Laroche A et al (2002) Is DNA methylation underlying somaclonal variation in oil palm? Acta Hortic 575:47–59

Rival A (2000) Somatic embryogenesis in oil palm. In: Somatic embryogenesis in woody plants (Ed. by S.M. Jain, P.K. Gupta R.J. Newton) Kluwer Academic, Dordrecht 249–289

Rohani O, Sharifah SA, Rafii MY et al (2000) Tissue culture of oil palm. Adv Oil Palm Res 1:238–283

Rohani O, Kamal RS, Rajinder S et al (2012) Mutation induction using gamma irradiation in oil palm (*Elaeis guineensis*) cultures. J Oil Palm Res 24:1448–1458

Rusli MH (2017) Disease Epidemiology and Genetic Diversity of *Fusarium oxysporum* f. sp. *elaeidis*, cause of Fusarium wilt of oil palm (Elaeis guineensis Jacq.). J Oil Palm Res.;22 (29):548–61

Saepoo T, Sarak S, Mayakun J et al (2023) Thermoplastic starch composite with oil palm mesocarp fiber waste and its application as biodegradable. Carbohydr Polym 299(1):120221

Sandal I, Bhattacharya A, Ahuja PS (2001) An efficient liquid culture system for tea shoot proliferation. Plant Cell Tissue Organ Cult 65(1):75–80

Sanputawong S, Te-chato S (2008) Effect of genotypes of oil palm as indicator for speed of callus and embryogenic callus formation. J Agric Technol 4:147–156

Saibene A, Assale M and Giltri M (2021) Expert systems: Definitions, advantages, andissues in medical field applications. Expert Syst. Appl. 177 (July 2020).114900. https://doi.org/10.1016/j.eswa.2021.114900

Schad H (1914) Die geographischeVerbreitung der Ölpalme (Elaeis Guineensis). Mittler

Selvaraj K, Vekatesan T, Sumalatha BV (2020) Invasive rugose spiralling whitefly (*Aleurodicusrugioperculatus* Martin). J Oil Palm Res 31(4):651–656

Shah FH, Ahmed-Parveez GK (1995) DNA variation in abnormal tissue culture regenerants of oil palm (*Elaeis guineensis*). Asia Pac J Mol Biol Biotechnol 3:49–53

Shah FH, Rasid O, Simon AJ (1994) The utility of RAPD markers for the determination of genetic variation in oil palm (*Elaeis guineensis*). Theor Appl Genet 89:713–718

Sharifaha SSA, Rajinder S, Cheah SC (1999) Molecular dissection of the floral clonal abnormality in oil palm. In: Proceedings of the PORIM International Palm Oil Congress (unedited) on emerging technologies and opportunities in the next millennium (agriculture), Kuala Lumpur, Malaysia, pp 321–329

Shinoj S, Rangaraju, Visvanathan et al (2011) Oil palm fiber (OPF) and its composites: a review. Industrial Crops and Products33, 7-22

Shull GF (1908) The composition of a field of maize. J Hered 4:296–301

Silva PA, Cosme VS, Rodrigues KCB et al (2017) Drought tolerance in two oil palm hybrids as related to adjustments in carbon metabolism and vegetative growth. Acta Physiol Plant 39:58

Simon S, Hendry T, Chang SW et al (1998) Early yield performance of clonal oil palm (*Elaeis guineensis* Jacq.) plantings in PPB Oil Palms Bhd, Sabah a case study. Planter 74(866):257–269

Singh R, Cheah SC (1999) Analysis of the inheritance of AFLP markers in an interspecific cross of oil palm using the pseudo-test cross strategy. J Oil Palm Res 11:64–73

Singh R, Cheah SC (2000) Identification of flower-specific proteins in the oil palm (*Elaeis guineensis*) by two-dimensional gel electrophoresis. J Oil Palm Res 12:46–54

Singh, R, Cheah SC, Madon M (2001) Genomic strategies for enhancing the value of the oil palm. In: Proceedings of the 2001 PIPOC International Palm Oil Congress, Agriculture conference on 'cutting edge technologies for sustained competitiveness', Kuala Lumpur, Malaysia, pp 3–17

Singh R, Nagappan J, Tan SG et al (2007) Development of simple sequence repeat (SSR) markers for oil palm and their application in genetic mapping and fingerprinting of tissue culture clones. Asia Pac J Mol Biol Biotechnol 15:121–131

Singh R, Zaki NM, Rosli R (2008) Exploiting an oil palm EST database for the development of gene-derived SSR markers and their exploitation for assessment of genetic diversity. Biologia 63:227–235

Singh R, Tan SG, Panandam JM et al (2009) Mapping quantitative trait loci (QTLs) for fatty acid composition in an interspecific cross of oil palm. BMC Plant Biotechnol 9:114

Singh R, Abdullah MO, Low ET et al (2013a) Oil palm genome sequence reveals divergence of interfertile species in old and new world. Nature 500(7462):335–339

Singh R, Low ET, Ooi LC et al (2013b) The oil palm SHELL gene controls oil yield and encodes a homologue of seedstick. Nature 500(7462):340–344. https://doi.org/10.1038/nature12356

Singh R, Low ET, Ooi LC et al (2014) The oil palm VIRESCENS gene controls fruit colour and encodes a R2R3-MYB. Nat Commun 5:4106. https://doi.org/10.1038/ncomms5106

Sitinjak EM, Masmur I, Sitanggan Y et al (2023) A study on cellulose from oil palm biomass as a source of bioethanol production. Cellulose Chem Technol 57(3–4):369–375

Smith BG (1993) Correlations between vegetative and yield characteristics and photosynthetic rate and stomatal conductance in the oil palm (*Elaeis guineensis* Jacq.). Elaeis 5(1):12–26

Smith WK, Thomas JA (1973) The isolation and in vitro cultivation of cells of *Elaeis guineensis*. Oleagineux 28(3):123–127

Soh AC, Wong G, Tan CC et al (2001) Recent advances towards commercial production of elite oil palm clones. In: Cutting-edge technologies for sustained competitiveness: proceedings of the 2001 PIPOC International Palm Oil Congress, Malaysian Palm Oil Board (MPOB), Kuala Lumpur, Malaysia, 20–22 August 2001, pp 33–44

Soh AC, Gan HH, Wong G et al (2003) Estimates of within-family genetic variability for clonal selection in oil palm. Euphytica 133(2):147–163

Soh AC, Wong CK, Ho YW et al (2009) Oil palm. In: Oil crops. Springer, New York, pp 333–367

Soh AC, Mayes S, Roberts JA (2018) Oil palm breeding: genetics and genomics. CRC Press, Boca Raton

Soh AC, Chow SC, Iyama S,Yamada Y (1994) Candidate traits for index selection in choice of oil palm ortets for clonal propagation. Euphytica 76, 23–32

Souza NGF, Almeida JSS, Pinheiro JA et al (2022) Progress in organosolv and steam explosion pretreatments of oil palm fibers for biomacromolecules extraction. J Nat Fibers 19(15): 10708–10722

Statista (2020) Leading producers of palm oil worldwide from 2012/13 to 2019/20.https://www.statista.com/statistics/856231/palm-oil-top-global-producers

Sunilkumar K, Sparjanbabu DS (2013a) Surface color-based prediction of oil content in oil palm (*Elaeis guineensis* Jacq.) fresh fruit bunch. Afr J Agric Res 8(6):564–569

Sunilkumar K, Sparjanbabu DS (2013b) Haploid breeding in palms – a brief review. Adv Crop Sci Technol 1:4. https://doi.org/10.4172/2329-8863.1000113

Sunilkumar K, Mathur RK, Naveenkumar P et al (2013) Proceedings of National Oil Palm Seed Meet – 2013. DOPR, Pedavegi, pp 6–8

Sunilkumar K, Mathur RK, Sparjanbabu DS et al (2015) Evaluation of interspecific oil palm hybrids for dwarfness. J Plant Crop 43(1):29–34

Sunilkumar K, Mathur RK, Sparjanbabu DS (2017) Differential pollen longevity in Dura and Pisifera oil palm (*Elaeis guineensis*) fruit types at storage temperatures. Ind J Agric Sci 87(7): 893–898

Sunilkumar K, Murugesan P, Mathur RK et al (2020) Genetic diversity in oil palm (*Elaeis guineensis* and *Elaeis oleifera*) germplasm as revealed by microsatellite (SSR) markers. Ind J Agric Sci 90(4):741–745

Suranthran P, Gantait S, Sinniah UR et al (2012) Effect of loading and vitrification solutions on survival of cryopreserved oil palm polyembryoids. J Plant Growth Regul 66(2):101–109

Suresh K, Mathur RK, Kochubabu M (2008) Screening of oil palm duras for drought tolerance. J Plant Crop 36:270–275

Taeprayoon P, Tanya P, Lee SH et al (2015) Genetic background of three commercial oil palm breeding populations in Thailand revealed by SSR markers. Aust J Crop Sci 9(4):281–288

Tan XJ, Cheor WL, Yeo KS et al (2022) Expert systems in oil palm precision agriculture: a decade systematic review. J King Saud Univ Sci 34:1569–1594. https://doi.org/10.1016/j.jksuci

Tandon R, Chaudhury R, Shivanna KR (2007) Cryopreservation of oil palm pollen. Curr Sci 92(2): 182–183

Tarmizi AH (2002) Oil palm liquid culture – MPOB Protocol. MPOB Information Series No. 138

Tarmizi AH, Samsul KR, Zaiton R (2008) Multiplication of oil palm liquid cultures in bioreactors. J Oil Palm Res 1:44

Teixeira JB, Sondahl MR, Kirby EG (1994) Somatic embryogenesis from immature inflorescences of oil palm. Plant Cell Rep 13(5):247–250

Theam KL, Islam A, Taufiq-Yap YH, et al (2016) Binary metal-doped methoxide catalyst for biodiesel production from palm stearin. Res Chem Intermed 42: 1943–1963

Thonghawee, S, Tittinutchanon, P, and Volkaert, H (2010) Microsatellites for parentage analysis in oil palm breeding population. Thai journal of genetics, 3(2): 172-181

Thomas V, Rao PS (1985) In vitro propagation of oil palm (*Elaeis guineensis* Jac. var. tenera) through somatic embryogenesis in leaf-derived callus. Curr Sci 54(4):184–185

Thorp K (2014) Precision agriculture.Encyclopedia Earth Sc. Series2050.515-517. https://doi.org/10.1007/978-0-387-36699-9-132

Thuzar M, Vanavichit A, Tragoonrung S (2012) Recloning of regenerated plantlets from elite oil palm (*Elaeis guineensis* Jacq.) cv. Tenera. Afr J Biotechnol 11(82):14761–14770

Tinker PB (2000) Shades of green - a review of UK farming systems. Royal Agricultural Society of England, Stoneleigh Park, U.K

Ting NC, Zaki NM, Rosle R (2010) SSR mining in oil palm EST database: application in oil palm germplasm diversity studies. J Genet 89(2):135–145. https://doi.org/10.1007/s12041-010-0053-7

Ting NC, Jansen J, Nagappan J (2013) Identification of QTLs associated with callogenesis and embryogenesis in oil palm using genetic linkage maps improved with SSR markers. PLoS One 8(1):e53076

Ting NC, Jansen J, Mayes S et al (2014) High-density SNP and SSR-based genetic maps of two independent oil palm hybrids. BMC Genomics 15:309

Tisné S, Pomiès V, Riou V et al (2017) Identification of Ganoderma disease resistance loci using natural field infection of an oil palm multi-parental population. G3 7(6):1683–1692

Tregear JW, Morcillo F, Richaud F (2002) Characterization of a defensin gene expressed in oil palm inflorescences: induction during tissue culture and possible association with epigenetic somaclonal variation events. J Exp Bot 53:1387–1396

Vera AT, Diaz IA, Barrera CF et al (2023) Genetic gains for obtaining improved progenies of oil palm in Colombia. Euphytica 219:38

Villa AL, Jimenez PE, Valbuena RI (2007) Preliminary study of the establishment of cryoconservation protocol for oil palm (*Elaeis guineensis* Jacq.). Agron Colomb 25(2):215–223

WAIFOR (1955) Third Annual Report, 1954–55, West African Inst. Oil Palm Res., Benin City

Wan I, Wan H, Zulkifli Y et al (2011) Obtaining oil palm haploids via screening for naturally occurring haploid (NOH) in oil palm germinated seeds and wide hybridization. In: Proceedings of the PIPOC 2011, International Palm Oil Congress, AP 73, pp 327–330

Wang F, Liu Y, Jin Q et al (2011) Characterization of cocoa butter substitutes, milk fat and cocoa butter mixtures. Eur J Lipid Sci Technol 113: 1145–1151

Wooi KC (1990) Oil palm (*Elaeis guineensis* Jacq.): tissue culture and micropropagation. In: Legumes and oilseed crops. Springer, Berlin, Heidelberg, pp 569–592

Wooi KC (1995) Oil palm tissue culture-current practice and constraints. In: 1993 ISOPB international symposium on recent developments in oil salm tissue culture and biotechnology (No. L-0375), September 24–25, PORIM, Kuala Lumpur, Malaysia

Wooi KC, Wong CY, Corley RHV et al (1981) Tissue culture of palms - a review. In: Proc. COSTED Symp. 'Tissue culture of economically important plants' (Ed. by A.N. Rao),Singapore pp. 138–144
Xu XB (2000) Production of specific-structured triacylglycerols by lipase-catalyzed reactions: a review. Eur J Lipid Sci Technol 102
Yahya EB, Abdul Khalil HPS, Ahmad MI et al (2023) Cleaner approach for preparing antibacterial bioaerogel scaffolds using oil palm waste nanocellulose. Ind Crops Prod 191:115897
Yarra R, Jin L, Zhao Z. Cao, H (2019) Progress in Tissue Culture and Genetic Transformation of Oil Palm: An Overview. Int. J. Mol. Sci. 20, 5353
Yost MA, Kitchen NR, Sudduth K A et al. (2017) Long term impact of precision agriculture system on garin crop production. Precis. Agric.18950823-842. https://doi.org/10.1007/s11119-016-9490-5
Yue GH, Ye BQ, Lee M (2021) Molecular approaches for improving oil palm for oil. Mol Breed 41: 22
Yusnita S, Hapsoro D (2011) In vitro callus induction and embryogenesis of oil palm (*Elaeis guineensis* Jacq.) from leaf explants. Hayati J Biosci 18(2):61–65
Zakaria S, Poh LK (2002) Polyester benzoylated EFB reinforced composites. Polym Plast Technol Eng 41(5):951–962
Zaki NM, Ismail I, Rosli R et al (2010) Development and characterization of *Elaeis oleifera* microsatellite markers. Sains Malays 39(6):909–912
Zaki NM, Singh R, Rosli R et al (2012) *Elaeis oleifera* genomic-SSR markers: exploitation in oil palm germplasm diversity and cross-amplification in arecaceae. Int J Mol Sci 13:4069–4088
Zeven AC (1964) On the origin of the oil palm (Elaeis guineensis Jacq.). Grana Palynol 5: 121–123
Zeven AC (1967) The semi-wild oil palm and its industry in Africa. Agricultural Research Report No 689,178p
Zhang N, Wang M, Wang N (2002) Precision agriculture a world wide over view. Comput. Electron.Agric. 3692-30,113-132.https://doi.org/10.1016/S0168-1699(02)00096-0
Zirkle C (1935) The beginning of plant hybridization. University of Pennsylvania Press, Philadelphia, p 223
Zulkifli, H., Halimah, M., Chan, K. W., et al (2010) Life cycle assessment for oil palm fresh fruit bunch production from continued land use for oil palm planted on mineral soil (Part 2). Journal of Oil Palm Research, 22, 887–894. http://www.palmoilworld.org/PDFs/Environment/9-joprv22dec10-Zulhashim_LCA2.pdf. Accessed 5 Oct 2019

Part III
Seed Fiber Crops

Chapter 17
Tree Cotton (*Gossypium arboreum* L.) Breeding: Current Status and Future Prospects

Surender Kumar Verma, Debashis Paul, Suman Dutta, Sandeep Bagri, Hargopal Dutta, and Nidhi

Abstract The introduction of *Bt* (*Bacillus thuringiensis*) cotton in India resulted in a significant decrease in the cultivation of tree/*desi* cotton. Despite this decline, there is a considerable demand for cotton with short staple lengths, particularly for surgical and denim purposes. Diploid cotton could serve as an alternative to *G. hirsutum* if given equal attention and slightly higher compensation. *Gossypium arboreum*, known as *desi* cotton or Tree cotton, is a native species in the Indian subcontinent and exhibits low input requirements, thriving in high salinity and drought conditions. *Desi* cotton's resistance to sucking pests stems from features such as a greater separation between the midrib's lower epidermis and the phloem, a lower palisade layer, and a dense arrangement of cortex cells in the midrib. By virtue of its tolerance to insect pests and diseases, *G. arboreum* is often chosen as a donor parent in introgressive breeding to enhance other cotton species. This chapter delves into various aspects, such as the genetic diversity of tree cotton (*Gossypium arborerum*), current cultivation practices, challenges faced by tree cotton, unique alleles within *Gossypium arboreum* for crop improvement, genetic diversity for adaptation against biotic and abiotic stress, conservation strategies to preserve genetic diversity in changing environmental conditions, fiber properties, market

S. K. Verma (✉)
ICAR-Indian Institute of Wheat and Barley Research, Seed Research Farm, Hisar, Haryana, India

ICAR-Central Institute for Cotton Research, Regional Station, Sirsa, Haryana, India
e-mail: surendra.verma@icar.gov.in

D. Paul · S. Bagri · Nidhi
ICAR-Central Institute for Cotton Research, Regional Station, Sirsa, Haryana, India
e-mail: debashis.paul@icar.gov.in

S. Dutta
Ramakrishna Mission Vivekananda Educational & Research Institute, Howrah, West Bengal, India

H. Dutta
ICAR-Indian Agricultural Research Institute, New Delhi, India

K. F. M. Salem et al. (eds.), *Breeding and Biotechnology of Leaf, Fruit, and Seed Fiber Crops*, Advances in Plant Breeding Strategies 11,
https://doi.org/10.1007/978-3-032-00407-9_17

opportunities, recent advances in sequencing the *Gossypium arboreum* genome, plant regeneration using somatic embryogenesis, molecular marker-assisted breeding, and the role of genetic engineering and gene editing in conjunction with traditional and mutation breeding. Implementing goal-oriented breeding in *desi* cotton has the potential to increase yield while promoting sustainable livelihoods.

Keywords Breeding strategies · *Desi*/Asiatic cotton · Genetic diversity · Somatic embryogenesis · Tree cotton (*Gossypium arboreum* L.)

17.1 Introduction

Cotton (*Gossypium* sp.), is a natural fiber crop with a remarkable contribution to the textile industry. It is commonly called 'White Gold' for its economic position in the world. Cotton belonging to the family *Malvaceae* (Dongre and Kharbikar 2004; Esmail et al. 2008; Khushboo et al. 2014; Wendel et al. 1992; Brubaker and Wendel 1994), have diploid (26 chromosomes) and allotetraploid (52 chromosomes) species in cultivation. The diploid species *Gossypium herbaceum* and *Gossypium arboreum* are referred to as "Old world or true cotton (*Desi* cotton)", while the allotetraploid species *Gossypium hirsutum* and *Gossypium barbadense* are referred to be "New world cotton" (Wendel et al. 1992).

India's most valuable commercial crop, cotton is grown primarily for its fiber and other valuable byproducts, the most significant of which is edible oil. It has an impact on the major economies of the world, including the United States, Australia, South Asia, and Africa. Both diploid and tetraploid species of cotton are exclusively grown in India. The diploids cover the least area of cultivable land compared to *G. hirsutum* covering more than 90%, however, the diploids show greater resistance to biotic and abiotic stresses, including lower input requirements (Verma et al. 2020). In certain regions of India, farmers continue to cultivate *desi* cotton for its advantage of sustainable yield in marginal soil with minimal management (Mehetre 2015). The cultivation area of *desi* cotton significantly decreased after *Bt* cotton was introduced to India.

However, cotton with short-staple lengths is in great demand for surgical use and denim (Verma et al. 2022). Diploid cotton would be an alternative to that of *G. hirsutum* if provided with equal concern and slightly higher remuneration.

This chapter delves into the fascinating world of tree cotton (*Gossypium arboreum*) embarking on a comprehensive exploration of its genetic diversity, examining the unique alleles that hold immense potential for improving cotton crops. We'll investigate how this genetic variation equips tree cotton to adapt and thrive against various challenges, including harsh environmental conditions and pesky pests and diseases. This ensures a rich pool of genetic resources for future generations. The chapter wouldn't be complete without exploring the cutting-edge advancements in tree cotton research. We'll focus on the exciting world of genome sequencing, unraveling the secrets encoded within the DNA of this remarkable plant. Additionally, this present chapter will also examine the promising technique of somatic embryogenesis for plant regeneration, offering a powerful tool for propagation. Furthermore, this chapter will explore the role of molecular marker-assisted breeding, a technique that harnesses genetic markers to identify and select desirable traits with greater precision. This approach significantly accelerates the breeding process and allows for the development of improved cotton varieties.

Finally, investigation also includes the potential of genetic engineering and gene editing tools, used in conjunction with traditional breeding and mutation breeding methods. This powerful combination could unlock a new era of cotton improvement, leading to higher yields, enhanced fiber quality, and increased resilience to environmental stresses. By implementing these goal-oriented breeding strategies, we can unlock the full potential of tree cotton paving the way for increased yields and promoting sustainable livelihoods for cotton farmers.

17.1.1 Morphology of Tree Cotton Plant

The shrub "*Gossypium arboreum*" typically grows to a height of 1–2 m. Its purple-colored branches are coated with pubescence, and stipules can be found at the base of leaves. These stipules range in shape from linear to lanceolate, sometimes even falcate (sickle-shaped) variation. A 1.5–10 cm petiole holds the leaves to the stem. Resembling the maple leaf, ovate to orbicular-shaped blades with five to seven linear to lanceolate-shaped lobes are present. The sinus is often present with a tooth. Young leaves may be simple or stellate (star-shaped) and have glands along the midrib, while older leaves lose their pubescence. The pedicles are short and carry the flower stalks. Sepals (calyx) are small (5 mm (0.20 in) long), and cupular in shape. The large dentate ovules are cordate (heart-shaped) at the base and acute at the apex. The corolla is small typically 3–4 cm (1.2–1.6 in) long and generally pale yellow with a purple center. The anthers are located on staminal tubes (1.5–2 cm long). The fruit is glabrous, ovoid or oblong in form, and measures 1.5–2.5 cm. It has three to four-celled capsules, a pitted surface and a beak at the very end. Cotton seeds have a globular form and are coated in long white cotton (Abedin and Sultanul 1979).

*17.1.2 Importance and Uses of Desi Cotton (*Gossypium arboreum*)*

Tree/*Desi* cotton (*G. arboreum*) is a native species of the Indian sub-continent. It requires low input due to its ability to endure high saline and dry conditions (Maqbool et al. 2010; Tahir et al. 2011). The greater separation between the midrib's lower epidermis and the phloem, lower palisade layer and dense arrangement of the midrib's cortex cells in *desi* cotton makes it resistant to sucking pests (Nibouche et al. 2008), bollworms (Dhawan et al. 1991), and diseases (Akhtar et al. 2010;

Khushboo et al. 2014; Mehetre et al. 2004; Wheeler et al. 1999). Due to its insect-pest and disease tolerance, *G. arboreum* is chosen as the donor parent in introgressive breeding to enhance the available cotton species (Kulkarni 2002; Ansingkar et al. 2004).

*17.1.3 Fibre and Textile Properties of Tree Cotton (**Gossypium arboreum** L.)*

Cotton fiber is natural, soft, lustrous and fluffy, collected from the plant's seedpods. The main composition of cotton fibers is cellulose a complex carbohydrate consisting of a lengthy chain of glucose molecules joined by chemical bonds. The bonding provides the fiber with strength, flexibility and durability (Nawab and Shaker 2023). Cotton fibers typically ranged from 1 to 1.5 inches in length, sometimes longer. While the fibers are generally white, they can also exhibit variations in shades such as brown, yellow, green, and gray (Ashford 2016). Cotton is recognized as a crucial textile fiber crop by virtue of its breathability, absorbency, and softness. A wide range of products is formed using cotton fibers for textiles, clothing, and households (Jindal and Jindal 2023). Comprised of 91% cellulose and 7.8% water, with the rest consisting of mineral salts, waxes, protoplasm, fatty substances and pectin, cotton fiber is a condensation polymer with a high degree of polymerization (2000–4000), compared to re-generated cellulose like Viscose, which has a polymerization degree of only 300–400 (Khare 2021).

Quality features of cotton fiber, comprising staple length, strength, uniformity and fineness have an intrinsic relationship with yield in textile industry systems (Hovav et al. 2008). The effective processing of *G. arboreum* cotton using high-speed machinery has been significantly hindered by short fiber length, which is considered a key factor in yarn strength and further processing (Chandra and Sreenivasan 2011). The standard staple length limit is specified as nearly 25–30 mm. For instance, the fiber staple length ranged between 16.4 to 27.9 mm, 16 to 23.8 mm and 19 to 29.3 mm with a mean value of 24.1 mm as reported by Chandra and Sreenivasan (2011), Kantartzi and Stewart (2008), Manivannan and Waghmare (2020), respectively.

Another important factor is the uniformity of the fiber which is crucial in determining the utility of lint. The greater the uniformity, the more consistent the size and strength of the yarn spun, reducing manufacturing waste, neps and ends during the spinning process. The uniformity ratio varied from 46 to 56, 76.2 to 77.9 and 45 to 60 with a mean value of 51.7 (Iqbal et al. 2015; Manivannan and Waghmare 2020; Chandra and Sreenivasan 2011), respectively.

Fiber strength varied from 14.4 to 23.3 g/tex, 12.5 to 26.5 g/tex and 13.6 to 27.6 g/tex with a mean value of 19.8 g/tex (Chandra and Sreenivasan 2011; Kantartzi and Stewart 2008; Manivannan and Waghmare 2020), respectively. Elongation ranged from 3.9 to 5.9, 4.5 to 10.5 and 5.0 to 7.7 with a mean value of 5.8

Fig. 17.1 Lint pre-cleaning instrument. (Source: ICAR-Cotton Institute for Cotton Research, Sirsa, Haryana)

(Chandra and Sreenivasan 2011; Kantartzi and Stewart 2008; Manivannan and Waghmare 2020), respectively.

Fiber fineness is an equally important cotton quality parameter next to staple length, determining the texture of cotton fiber as either soft and silky or coarse and rough (Chandra and Sreenivasan 2011). Micronaire ranged from 4.0 to 7.1, 4.1 to 8.0 and 2.7 to 6.8 with a mean value of 5.07 (Chandra and Sreenivasan 2011; Kantartzi and Stewart 2008; Manivannan and Waghmare 2020), respectively (Figs. 17.1 and 17.2).

*17.1.4 Fiber Industry of Tree Cotton (*Gossypium arboreum* L.)*

Cotton is a naturally obtained agricultural commodity that holds great significance in the textile world in various forms such as raw cotton fiber, cotton yarn and goods with added value. Genotypes of cotton with longer staple lengths are in demand for producing higher-quality clothing in the textile industry (Chinchane and Baig 2018). Cotton fibers fall under the category of fibers obtained from seeds known as seed hair which are used to make smooth and slightly porous textile products (Khare 2021). Cotton is hypoallergenic and resistant to wrinkles, making it a popular choice for

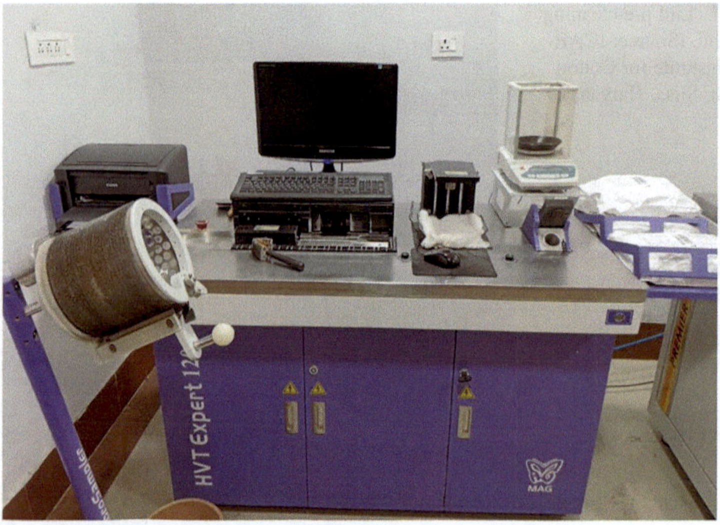

Fig. 17.2 Fiber quality assessment Instrument (HVI technology). (Source: ICAR-Cotton Institute for Cotton Research, Sirsa, Haryana)

clothing, bedding, and industrial products (Jindal and Jindal 2023). Additionally, surgical cotton is in high demand by hospitals, nursing homes, dispensaries, and beauty parlors (Meena et al. 2016).

Following a steady growth rate of surgical cotton, absorbent cotton has gained popularity for sanitary uses in both towns and villages due to its high water-holding capacity (Blaise et al. 2022; Raja et al. 2016). The global market for surgical and absorbent cotton is expanding at an annual rate of 10%, with an estimated annual demand in India of two million bales (each weighing 170 kg). The natural absorbent qualities of Desi cotton (*G. arboreum*) are expected to benefit the medical sector by reducing processing time and costs (Meena et al. 2016). While efforts have been made to increase seed cotton yield, these have resulted in fibers that are too short for spinning at high speeds in the current textile sector (Chinchane and Baig 2018).

17.1.5 Description, Taxonomy Classification and Distribution

The hymn of Rig-Veda written around the fifteenth century BC, mentions the early sighting of cotton. The first recorded use of cotton was reported in Manus Dharma Shastra around 800 B.C. In ancient literature, the Sanskrit word karpas, which resembled cotton, is now and linked to karpas. However, the English word cotton is the appropriate technological and agricultural term used, derived from the Arabic word qutum or kutum. Linnaeus conducted a systematic taxonomic study of cotton (*Gossypium*) in 1953 (https://www.itis.gov/). The source of the Taxonomic Hierarchy of *Gossypium arboreum* can be traced to Fryxell (1969).

Kingdom	Plantae—Planta, plants
Subkingdom	Viridiplantae—Green plants
Infrakingdom	Streptophyta—Land plants
Superdivision	Embryophyta
Division	Tracheophyta—Vascular plants, tracheophytes
Subdivision	Spermatophytina—Spermatophytes, seed plants, phanérogames
Class	Magnoliopsida
Superorder	Rosannae
Order	Malvales
Family	Malvaceae—Mallows, mauves
Genus	*Gossypium* L.—Cotton
Species	*Gossypium arboreum* L.—Tree cotton

The recognized cotton species are approximately 50 in the dicotyledonous genus *Gossypium*, belonging to the family Malvaceae. Out of these, only four are preferable for cultivation (Khadi et al. 2010). *Gossypium arboreum*, also known as Asiatic cotton, is predominantly in Bangladesh, India, Myanmar, and Pakistan. It is also cultivated in limited areas of China, Iran, and Thailand. *Gossypium arboreum* accounts for less than 1% of the total cotton area cultivated worldwide (Edde 2022) (Fig. 17.3).

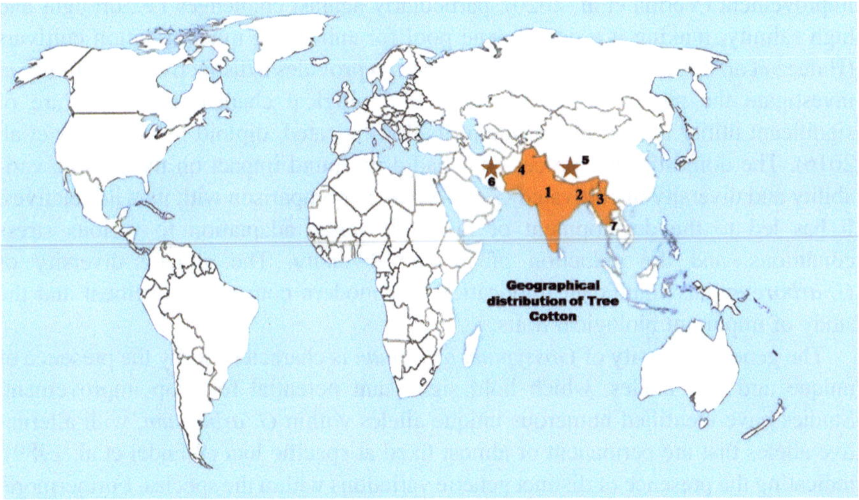

Fig. 17.3 eographical distribution (1: India; 2: Bangladesh, 3: Myanmar, 4: Pakistan, 5: China, 6: Iran and 7: Thailand) of tree cotton (*G. arboreum* L.). (Source: The world map is sourced from the link whereas, the distribution map was prepared by the authors by extensive reviewing of the literature. https://www.google.com/search?sca_esv=cc7d94c02e4b9cbf&rlz=1C1YTUH_enIN1012IN1012&sxsrf=ACQVn0-UtivYfK7N8UZ9HXTqUGVhv304WQ:1710312316118&q=world+map&tbm=isch&source=lnms&prmd=imnsvbtz&sa=X&sqi=2&ved=2ahUKEwiTk7rw0fCEAxWM0KACHWkOCbwQ0pQJegQIDxAB&biw=1440&bih=773&dpr=1#imgrc=ByzJvm8hgfIhQM)

17.1.6 Domestication, Selection and Early Improvements

The domestication process has significantly impacted the genetic diversity of cultivated *Gossypium arboreum* compared to its wild relatives. Recent investigations have clarified the domestication history of cultivated cotton species, including *G. arboreum*, and have revealed the impact of domestication on genetic diversity (Wendel and Cronn 2003). The domestication of *G. arboreum* has led to the development of favorable traits, like drought tolerance, disease and insect pest resilience, making it well adapted to biotic and abiotic challenges (Han et al. 2022; Yuan et al. 2021). Furthermore, studies have shown that *G. arboreum* and *G. herbaceum* domestication were independent of each other, leading to distinct genetic differences between the two species (Renny-Byfield et al. 2016). The repetitive domestication of various wild progenitors is a remarkable example of parallel evolution controlled by humans, resulting in the cultivation of genetically diverse cotton species (Wendel and Grover 2015). The genetic diversity of *G. arboreum* has been influenced by over 2000 years of cultivation in China, contributing significantly to its diversity and adaptation to various stress conditions (Erpelding 2022). Additionally, the domestication process has led to the reduction of genetic diversity in crop species, including *G. arboreum*, compared to their wild relatives (Mehboob-Ur-Rahman et al. 2021). The genetic diversity of *G. arboreum* has been recognized as an important source of genetic variation in modern cotton improvement (Verma et al. 2020), particularly against challenges i.e., drought and high salinity, making it a useful gene pool for enhancing modern cotton cultivars (Hafeez et al. 2021). Furthermore, *G. arboreum* provides a distinctive opportunity to investigate the molecular genetic base of biological characteristics that are of significant utility in agriculture due to its domesticated, diploid nature (Kaur et al. 2016). The domestication process has had a profound impact on the genetic variability and diversity of cultivated *G. arboreum* in comparison with its wild relatives. It has led to the development of favorable traits, adaptation to various stress conditions, and the reduction of genetic diversity. The genetic diversity of *G. arboreum* has significant implications for modern cotton improvement and the study of important biological traits.

The genetic diversity of *Gossypium arboreum* is characterized by the presence of unique and rare alleles, which hold significant potential for crop improvement. Studies have identified numerous unique alleles within *G. arboreum*, with alternative alleles that are permanent or almost fixed at specific loci (Wendel et al. 1989), indicating the presence of distinct genetic variations within the species. Furthermore, genetic diversity analysis using genomic and EST-derived microsatellites has revealed a substantial number of allele variations, with two to four alleles per locus, demonstrating the presence of rare alleles within *G. arborerum* (Kantartzi et al. 2009).

Additionally, *G. arboreum* has been recognized as an important source of genetic diversity in modern cotton improvement, with notable resilience to both biotic and abiotic challenges, particularly drought and salinity, making it a valuable gene pool for enhancing modern cotton cultivars (Hafeez et al. 2021). The identification of unique and rare alleles within *G. arboreum* holds great potential for crop improvement, contributing to enhancing desirable traits like stress resilience, fiber quality, and yield. The existence of a draft genome sequence and the greater allelic diversity in *G. arboreum* provide it as a prime species for studying the genetic foundations for various biological features, especially those governed by orthologous loci in diploid and polyploid species (Kaur et al. 2016). Furthermore, the presence of rare alleles provides opportunities for introgression and breeding programs aimed at developing improved cotton varieties with enhanced genetic diversity and resilience to environmental challenges. The genetic diversity of *G. arboreum* encompasses unique and rare alleles that have significant potential for crop improvement. Utilizing these alleles, cotton cultivars can be developed with enhanced resilience, productivity, and fiber quality, thereby contributing to the sustainability and advancement of cotton agriculture.

17.2 Current Cultivation Practices and Challenges

Cotton is cultivated in *Kharif* season all over the world. It requires 15 °C temperature for better germination in field conditions. The ideal temperature for vegetative growth is between 21 and 27 °C. Temperatures below 21 °C during the vegetative and reproductive phases can damage growth and reduce seed yield. Cotton can withstand temperatures as high as 42 °C. Day temperatures of 27–32 °C and cool nights are necessary during the fruiting stage for better boll development.

In the North Cotton growing zone of India, seedling burning can occur due to high heat during the sowing window, especially Asiatic cotton. The land should and meet the minimum requirements for fertility and texture for the seed crop. Proper rouging and weeding are essential to eliminate weed plants, volunteer plants and other crop plants. The soil in the cultivation area should be free from soil-borne pathogens and insect pests. It is best to avoid planting on land where the same crop was grown in the previous season. If unavoidable, advance irrigation should be provided to break down any remaining seeds from the previous season. Land with good fertility and drainage capabilities should be chosen for cultivation.

17.3 Biodiversity and Conservation of Germplasm

*17.3.1 Genetic Diversity of Tree Cotton (*Gossypium arboreum *L.)*

To understand the genetic diversity of *Gossypium arboreum* across different geographical regions, it is essential to consider various studies that have investigated this aspect. *G. arboreum*, an "Old World" cultivated cotton species, has been grown for more than 2000 years in China, and the diversity of *G. arboreum* is likely influenced by the country's diverse environments (Kantartzi et al. 2009). Studies have shown that *G. arboreum* accessions in China from three geographically distinct genetic groups (Yang et al. 2023). Furthermore, genetic diversity analysis of *G. arboreum* germplasm collections has been assessed using genotyping-by-sequencing, revealing the population structure and genetic variability in the germplasm collection of *G. arboreum* (Li and Erpelding 2016). Additionally, microsatellite analysis was used to assess the genetic diversity of certain germplasm accessions that were taken from various places in China, demonstrating a greater degree of genetic variation at the DNA level among the germplasm accessions of *G. arboreum* (Ma et al. 2008). These findings indicate that *G. arboreum* exhibits significant genetic diversity across different geographical regions, particularly in China.

17.3.2 Genetic Diversity for Adaptation to Biotic and Abiotic Stress

The genetic diversity of *Gossypium arboreum* has been influenced by over 2000 years of cultivation in China, leading to its adaptation to various stress conditions (Jia et al. 2018). *G. arboreum* exhibits advantages in stress resistance, including drought tolerance, and disease and insect pest resilience, making it well adaptable to biotic and abiotic challenges (Chen et al. 2015). Furthermore, these traits possessed by *G. arboreum* are absent in the tetraploid cotton gene pool (Uthirapathy et al. 2023).

The genetic basis of stress tolerance in *G. arboreum* is multifaceted. Unique and rare alleles within *G. arboreum* contribute to its stress resilience, and it has been recognized as a useful source of genetic diversity in advanced cotton improvement, particularly against drought and salinity (Uthirapathy et al. 2023). *G. arboreum* genotypes have been found to express a low level of antixenosis but possess antibiosis and tolerance against cotton-melon aphids, demonstrating its multifaceted resistance to pests (Li et al. 2018). Moreover, the genetic diversity of *G. arboreum* has been associated with resistance to specific pests and diseases.

For example, according to a genome-wide association study, several genes in *G. arboreum* contribute a moderate level of resistance individually to reniform nematode (Stetina and Erpelding 2016). The genetic diversity of *G. arboreum* significantly contributes to its adaptability to both abiotic and biotic stimuli, such as pests and diseases. The presence of unique and rare alleles, multifaceted tolerance to abiotic stresses, and resistance to pests and diseases make *G. arboreum* a valuable genetic resource for improving stress resilience in cotton cultivars.

17.3.3 Conservation Strategies to Preserve Genetic Diversity Under Changing Environmental Conditions

In order to improve and preserve the genetic diversity of *Gossypium arboreum* in the face of changing environmental conditions, several conservation strategies can be implemented. Firstly, *Gossypium arboreum* can be used to develop improved upland cotton cultivars because it has beneficial agronomic traits like disease resistance and drought tolerance (Li and Erpelding 2016). Therefore, conservation efforts should focus on maintaining and utilizing these valuable traits to improve the cotton crop resistance in the face of changing environmental conditions.

Furthermore, *Gossypium arboreum* is known to be a useful gene pool for improving modern cotton cultivars as it possesses a greater ability to withstand abiotic and biotic strains. This highlights the importance of conserving the different germplasm of *Gossypium arboreum* to harness its potential in developing stress-resistant cotton varieties. Genetic diversity analysis using genomic and EST-derived microsatellites has been conducted to assess the genetic diversity in *Gossypium arboreum* cultivars (Hafeez et al. 2021). This research emphasizes the significance of genetic diversity evaluation in *Gossypium arboreum*, which can guide conservation efforts to ensure the preservation of diverse genetic traits within the species.

Moreover, *Gossypium arboreum* provides a significant opportunity to investigate the molecular genetic foundation of important biological characteristics of agronomical importance. Therefore, conservation strategies should include the preservation of diverse biological traits present in *Gossypium arboreum* to sustain and improve the genetic diversity of this valuable species. In addition, microsatellite analysis was used to assess the genetic variability of selected germplasm accessions of *G. arboreum* that were taken from various regions, highlighting the importance of considering geographic variations in conservation efforts. This suggests that conservation strategies should consider the geographic distribution of *Gossypium arboreum* to ensure the preservation of region-specific genetic diversity.

Overall, conservation strategies for *Gossypium arboreum* should focus on maintaining and utilizing its valuable agronomic traits, preserving its genetic diversity through genetic analysis, and considering geographic variations to ensure comprehensive preservation and enhancement of its genetic diversity.

17.3.4 Utilization of Genetic Diversity in Various Breeding Programs

To effectively utilize the identified genetic diversity within *Gossypium arboreum* in breeding programs to improve characteristics like yield, fiber quality, and resistance to pests and diseases, several strategies can be employed. Firstly, *G. arboreum* has been identified as a useful genetic resource for improving modern cotton cultivars (Kantartzi et al. 2009). Genome-wide association studies (GWAS) have revealed significant peak correlations for agronomically significant characters in *G. arboreum*, providing a genetic basis for these traits (Du et al. 2018). Additionally, the genetic diversity of *G. arboreum* has been assessed using microsatellite markers, revealing its potential for increasing the production of nearly all cotton species (Sethi et al. 2015). Furthermore, *G. arboreum* has beneficial agronomic characteristics that can be utilized to create enhanced upland cotton cultivars withstanding biotic and abiotic constraints (Li and Erpelding 2016). In addition, research on the genetic diversity and population structure of *G. arboreum* has revealed important insights for genomic and genetic analysis as well as the systematic utilization of commercially significant traits in this species (Verma et al. 2005; Jia et al. 2018). The capability of *G. arboreum* in improving upland cotton has been demonstrated through genome-wide introgression and quantitative trait locus mapping, which revealed the genomic base of fiber quality and yield-associated characteristics in *G. arboreum* that could benefit upland cotton (Feng et al. 2021; Verma et al. 2017, 2022). Additionally, *G. arboreum* has been identified as an aspect of genetic diversity for upland cotton, offering an opportunity to investigate the molecular genetic base of biological significant characteristics (Kaur et al. 2016; Kebede et al. 2007). Furthermore, the molecular basis of genetic enhancement of *G. arboreum* using molecular markers has been explored, highlighting the need for the development of new molecular markers to enhance breeding programs (Buyyarapu et al. 2011; Sethi et al. 2015). The ability for the development of *G. arboreum* introgressed lines possessing a specific genetic basis of *G. hirsutum* has been suggested, which would simplify the use of certain desirable genes of *G. arboreum* in future cotton breeding schemes (Chen et al. 2015). Additionally, the genetic diversity for cotton seed oil content and seed index in *G. arboreum* and wild species of *Gossypium* has been identified, offering promising opportunities for simultaneous breeding for these traits (Gotmare et al. 2004; Verma et al. 2020). The genetic variation within *G. arboreum* offers significant potential for the enhancement of crucial traits in cotton, like the quality of cotton fiber, yield and resistance to pests and diseases. By leveraging the genetic basis of key agronomic traits, utilizing molecular markers, and exploring the potential for introgression and breeding programs, *G. arboreum* is very crucial in the genetic enhancement of cotton cultivable species.

17.4 Recent Advances in Sequencing *Gossypium arboreum* Genome

There are approximately 45 diploid species and seven tetraploid species in the *Gossypium* genus (Wang et al. 2012). With one of the shortest nuclear gene sequences, *G. raimondii*, also called the D5-genome that originated in Mexico, is thought to be a possible contributor to all polyploid cotton. Collectively referred to as the A-genome, the native *G. herbaceum* of Africa and the native *G. arboreum* of

Asia were designated as the A1 and A2-genomes, respectively (Huang et al. 2021). Similar genetic traits distinguish *G. herbaceum* and *G. arboreum*, which diverged around 0.7 million years ago, a small evolutionary period. These diploid cotton, with their about 1.5 cm fiber cells mainly not suitable for human apparel, have been cultivated for a fair amount of time.

Within the cotton group, it is generally acknowledged that transoceanic dispersion brought together certain diploid cotton species, ultimately resulting in the appearance of at least seven species with allotetraploid AD genomes ($2n = 4x = 52$) (Huang et al. 2020). Approximately 8000 years ago, two of them, *G. hirsutum* [(AD)1-genome] and *G. barbadense* [(AD)2-genome], were introduced for cultivation and commercial purposes. These days, these are the most common cotton cultivars because they yield approximately 3.0 cm long fiber cells that are appropriate for the contemporary textile industry (Wendel et al. 2009).

During diversification, diploid species underwent significant chromosomal evolution, including rearrangements and structural variations. There are 13 chromosomes in every diploid species, yet some species have genomes that are up to three times bigger than those of other species (Chen et al. 2007). The D genome is the smallest and possesses the most basic genetic characteristics. Importantly, *G. hirsutum* and *G. barbadense*, two significant species that produce cotton fiber, are most likely descended from *G. raimondii*, the D genome, and *G. arboreum* (tree cotton) or *G. herbaceum* (levant cotton), the A genome (Wendel 1989).

For this reason, two distinct study groups chose *G. raimondii* as the first species for WGS (Paterson et al. 2012; Wang et al. 2012). The length of the *G. raimondii* genome assembly was estimated in 2012 to be around 740 Mb, with contig N50 values between 44.9 and 135.5 kb. Approximately 60% of the assembly was annotated as repetitive sequences, and thousands of gaps were found in these assemblies (Paterson et al. 2012; Wang et al. 2012).

In 2014, the diploid *G. arboreum* draught genome was made public. Contigs were generated using next-generation sequencing methods, resulting in an assembly fragmented into 40,381 contigs. Despite *G. arboreum*'s assembly size being more than double that of *G. raimondii*, the annotated gene counts from both species are comparable (Li et al. 2014). The cultivar Shixiya1 of tree cotton (A2-genome) after being sequenced and assembled was first reported in 2014 and further clarified in 2018. It is among the close relatives of the tetraploid At-subgenome in terms of phylogeny(Du et al. 2018). The A2-genome's third and final assembly was released recently (Huang et al. 2020). This current version has a more precise genome size of 1637 Mb with 80.1% content of transposable element (TE). For *G. arboreum*, the current assembly serves as a very accurate and comprehensive reference at the chromosomal level (Huang et al. 2020).

A genome mapping of *G. arboreum* was undergone and made available to the social databases (Wang et al. 2019, 2020). In cotton, it reported numerous tissue-specific expressed genes, polyadenylation sites, other splicing hotspots, and additional applications for transcriptional starting regions. In the same paper, an A1-genome of *G. herbaceum* var. *africanum* (A1-0076), was published as a component of comprehensive comparative genomic research (Huang et al. 2020).

A genomic sequence of size 1556 Mb with a 95.7% genome localization rate was obtained by the final assembly (Huang et al. 2020).

Given significant differences in genome size have been seen among *Gossypium* species, even across diploid genomes, cotton provides a great paradigm to examine the evolution of genome size (Wang et al. 2016). The 13 chromosomes found in all diploid cotton species range in size from around 738 Mb in the D5 genome to approximately 2.86 Mb in the K genome of Australia. As a result, approximately 13–15 million years after the beginning of its divergence from a common ancestor was assessed. Furthermore, genome sizes of *Gossypium* have increased by a factor of three; mostly because of the aggregation of long terminal repeat (LTR) retrotransposon activity (Li et al. 2014).

Despite the short period of divergence between the A1 and A2 genomes, these acquired significant genetic differences. For instance, following its speciation, the A1-genome had two large-scale inversions (>40 Mb) in chromosomes 10 and 12, whereas the A2-genome had undergone a reciprocal translocation of chromosomes 1 and 2 (Gerstel 1953; Huang et al. 2020). The two A-genomes have almost 1.0 genetic divergence, indicating that the A1- and A2-genomes are two different cotton species. Although they were originally bred for use in human apparel, these old genomic species have very little economic use today due to their fiber shortness, coarseness, and weakness.

The hypothesis that the exotic ancestor of levant cotton originated from the subspecies *africanum* (found in South Africa) is supported by morphological evidence and genetic data (Huang et al. 2020). The exact origin of tree cotton remains unknown, but there are indications that it was initially found in Madagascar or the Indus Valley and later introduced to South Asia and then to other regions (Huang et al. 2020; Watt 1907). Utilizing 230 and 13 accessions of tree and levant cotton respectively, resequencing was done to create the initial genomic map of diploid cotton (Du et al. 2018). However, one assumption indicated the Chinese origin of *G. arboreum*, which was supported by phylogenetic analysis based on SNPs (Du et al. 2018).

Intense human interference and geographical determines were subjected to *G. arboreum* accessions in China, as evidenced by pairwise comparisons among accessions from the Yangtze River, South China, and Yellow River regions. These comparisons revealed numerous genetically distinct locations coinciding with quantitative trait loci encoding variables for various yield-related parameters (Du et al. 2018). Through interspecific hybridization, significant qualities from diploid cotton may be transmitted to tetraploids, and these discovered domesticated loci constitute a significant genetic resource with high resolution that will aid in the enhancement of complex cotton properties (Huang et al. 2021).

17.5 Regeneration of Plants Using Somatic Embryogenesis

Unfortunately for functional gene analysis and cotton genetic engineering, cultivars or types with a high somatic embryogenesis rate are hard to identify. A significant obstacle to the genetic engineering of desired traits into farmed cotton is the rebellion of *Gossypium* species, particularly the diploid tree cotton (Sakhanokho et al. 2004; Shoemaker et al. 1986). According to Smith et al. (1977), somatic embryos could start and germinate in *G. arboreum*, but there were irregularities in plantlet acclimation, somatic embryo production, and somatic embryo maturation (Sakhanokho et al. 2004). Nevertheless, obtaining regenerated plants by somatic embryogenesis remains an extremely challenging task. Therefore, for cotton functional genomics and genetic improvement, it is highly important to enhance ways to successfully induce genetic transformation in species of diploid cotton (Huang et al. 2020; Satyavathi et al. 2002; Sunilkumar and Rathore 2001).

In diploid cotton species, a variety of factors affecting the formation of callus and somatic embryo development were investigated (Smith et al. 1977); these included in vitro callus initiation conditions and treatment with carbohydrates and desiccation to stimulate development of somatic embryo and germination in *G. arboreum* (Trolinder and Xhixian 1989). Through a variety of culture techniques, the development of somatic embryos and rejuvenation of plants in wild cotton species (*G. klotzschianum*) were also investigated (Finer and Smith 1984; Sun et al. 2003; Yan et al. 2010). Somatic embryogenesis was used to create regenerated plants from certain diploid species basically not in cultivation (Sun et al. 2003, 2006; Wang et al. 2008; Yan et al. 2010). An attempt was made to reliably utilize somatic embryos to rejuvenate plants in *G. arboreum* using the aforementioned techniques, and looked at the appearance of callus at various stages of selecting the callus for culturing (Ke et al. 2022). The accurate process for somatic embryogenesis-based plant regeneration shown here can completely change the genetic transformation of diploid cotton caused by *Agrobacterium tumefaciens* (Ke et al. 2022).

In cotton regeneration, choosing a viable embryogenic callus is an essential stage. Unfortunately, differentiating between embryogenic and non-embryogenic calli in *G. arboreum* is a difficult and inadequate task. To stimulate embryogenic callus induction from non-embryogenic callus, a variety of techniques were employed, such as altering the combination of PGR, suspension culture, inorganic component regulation, inorganic salt stress, and adding amino acids to encourage the development of somatic embryo, maturation, and germination (Tsai and Saunders 1999; Wu et al. 2004). Also, altering the carbon source to carbohydrates to avoid the appearance of cells in brown color (Sun et al. 2006; Kumar et al. 2015).

Therefore, *G. arboreum* and wild cotton might both develop callus when exposed to glucose. While some or all of the strategies were promising to promote somatic embryogenesis and plant regeneration in *G. arboreum*, they were not all useful in this regard for wild cotton species. Crucial stages in somatic culturing and germination were also controlled by PGRs and environmental stressors such as varying major and minor salt concentrations, dehydrating on Whatman paper, and regulating moisture (Firoozabady and DeBoer 1993; Kumria et al. 2003; Sun et al. 2006; Trolinder and Goodin 1987). In the wild species of cotton and *G. arboreum*, a mixture of stressors may be more beneficial than a particular one for the stimulation of somatic embryogenesis. However, the tendency of somatic embryogenesis in wild cotton and *G. arboreum* was reduced due to more abnormalities and lower conversion and maturation tendency of embryos (Sun et al. 2006).

Somatic embryogenesis produced regenerated plants, which were successfully planted in the soil. The technology of culturing somatic cells in *G. arboreum* was continuously refined to decrease the time of regeneration and increase regeneration efficiency. In diploid species, somatic embryogenesis is a viable method of plant regeneration; however, in diploid cotton species, more refinement and optimization are required for gene functional analysis and gene editing (Ke et al. 2022).

17.6 Hybridization

It is the process of breeding two different individuals, or species as parents in a population to combine their qualities and obtain desirable traits in the next generations.

17.6.1 Traditional Breeding

The traditional method of hybridization involves hand emasculation and pollination. Prerequisite requirements for this method are a selection of parent, purity of seed material, emasculation technique for a female parent, bagging, and pollination of a female with pollens of selected males. Maintaining the agronomic requirements, following the general technique of planting ensuring the ratio of male and female (1: 4 or 1:3) and spacing, the crop is to be sown in such a manner that the raised male and female plant should be flowering at the same time. Cotton is often a cross-pollinated crop that allows 6% of out-crossing. The conventional method is uneconomical for Asiatic cotton as the pollen of cotton is heavy and sticky, so only insect pollination is possible in it. Also, the diploid species (Asiatic cotton) have small-sized flowers and brittle pedicel resulting in low boll settings.

There are three crucial steps in the conventional method of hybrid production; (1) Emasculation of the female parent using the doak, surat or copper straw method; (2) Crossing the parents 1 week after flower initiation in the morning; (3) Harvesting only fully opened bolls.

Table 17.1 Hybrids released in cotton genetic male sterility (GMS) based

Hybrids	Species	Releasing year	Released from
G. Cot MDH 11	Intra-*arboreum*	2001	GAU, Surat
AKDH 7	Intra-*arboreum*	2001	PDKV, Akola
CISAA 2	Intra-*arboreum*	2004	CICR RS, Sirsa
DH 7	Intra-*arboreum*	2001	RAU, Sri Ganganagar

17.6.2 Non-Conventional Method: Male Sterility Method

The development of hybrids using male-sterile female parents reduces the cost of emasculation. Male sterility found in cotton is genetic male sterility (GMS) and cytoplasmic genetic male sterility (CGMS). The CGMS system requires a restorer line which is a limitation for its use. However, the GMS system is simple, and more widely accepted, leading to the release of various hybrids (Table 17.1).

17.7 Molecular Breeding

17.7.1 Molecular Marker-Assisted Breeding

Molecular marker-assisted breeding refers to the use of molecular markers in conventional plant breeding, also known as marker-assisted selection (MAS) (Collard and Mackill 2008; Vogel and Aken 2009). Techniques such as Quantitative trait loci (QTL) mapping, marker-assisted selection, and genetic transformations with the use of molecular markers deploying the desired genes in the species are the basis of molecular breeding (Bagri et al. 2020). Conventional breeding is completely dependent on the phenotypic traits and their results are highly manipulated by the favorable and adverse climatic conditions (variation in the environment) (Brumlop and Finckh 2010; Francia et al. 2005; Mohan et al. 2012). Mapping target genes, marker-assisted selection, cloning, gene introgression via backcrossing, germplasm differentiation and phylogenetic study involving the molecular markers make this approach more efficient for breeding (Collard and Mackill 2008; Gupta et al. 2010; Hayashi et al. 2004; Jain et al. 2002; Kumpatla et al. 2012).

Till now, several molecular markers have been identified in many crops. However, limited goals are achieved in the case of cotton due to limited polymorphism in cultivars (Abdurakhmonov et al. 2008; Iqbal et al. 2001; Rahman et al. 2005; Zehr 2012). In order to enhance the "Omics approach", the foundation of molecular markers with genetic purity and involvement in diverse ways for the analysis is further required (Aslam et al. 2013; Joshi et al. 1999; Saha et al. 2003; Sun et al. 2012).

17.7.2 Molecular Markers

Molecular markers are the precise portion of DNA in a sequence on a chromosome targeting a specific trait or variation (Kumar 1999; King and Stansfield 1990; Schulmann 2007). DNA markers are individual flanking fragments of chromosomes inherited (Ortiz 2004). Three kinds of markers: (1) morphological markers, encompassing the morpho-physiological features of plants; (2) biochemical markers, mainly specifying allelic variants of proteins (isozymes); and (3) molecular markers, involving DNA and RNA segments in the genomic sequence (Bagri et al. 2020; Bolek et al. 2016; Joshi and Nguyen 1993; Jones et al. 1997; Winter and Kahl 1995).

17.7.3 Quantitative Trait Loci (QTLs)

Majorly, plant traits follow polygenic inheritance (Bolek et al. 2016; Edwards et al. 1987). Target loci in association with the desired trait (gene of interest) are defined as Quantitative Trait Loci (QTL). Analysis of QTL includes the assessment of genetic diversity (Van Esbroeck et al. 1998). Discrete variability and high recombination frequency of polygenic characters are the main hindrances to QTL mapping (Lee 1995). QTLs allow for the pyramiding and tagging of desired genes in MAB (Bolek et al. 2005; Elshire et al. 2011; Ribaut and Hoisington 1998) and further, transgressive segregants are identified. The superiority of F_2 generation plants over the parental population in one or several traits is transgressive segregation (Abdurakhmonov et al. 2011; Paterson 1997; Tanksley et al. 1989).

QTLs contributing to cotton yield, cotton fiber quality, maturity date (Voss-Fels and Snowdon 2016), biotic (Bolek et al. 2005) and abiotic stress (Elshire et al. 2011; Wang et al. 2015) resistance have been identified.

17.7.4 Breeding for Polygenic Traits

Mapping polygenic traits reveals the hereditary pattern for efficiency in molecular breeding. Distant genetic maps are required to study the impact of several genomic regions on a particular trait in a high population index, with those related to economic value known as QTLs (Collard et al. 2005). The basis for QTLs is homologous sharing at crossing over (Bolek et al. 2016; Paterson 1996) and for QTL mapping the association of the genotype value of markers to phenotype is essential. In the linkage map, the relative distance between markers is estimated with recombination frequency (the frequency of single chromosomal crossing during meiosis). Markers with a recombination ratio of 50% and lower are unlinked and less often transferred to offspring than tightly linked markers (Collard et al. 2005). Reinisch et al. (1994) explored the initial genome mapping of cotton.

17.7.5 DNA Markers in Cotton

Desi cotton is a diploid (2n) species with 26 chromosomes making a complex genome, that requires the allocation of a greater number of evenly and uniformly spaced DNA markers (Khushboo et al. 2014; Park et al. 2005; Qureshi et al. 2004). Highly polymorphic, codominant, zero pleiotropic effect, reproducible, and uniformly distributed are the features of molecular markers (Weising et al. 2005). Three categories of polymorphic markers exist hybridization-based markers, polymerase chain reaction (PCR)-based markers, and DNA sequence-based markers (Kumar et al. 2009). There are numerous kinds of molecular markers available for characterizing crop plant germplasm (Table 17.2).

17.7.5.1 Restriction Fragment Length Polymorphism (RFLP)

RFLP is a category of hybridization-based markers used to detect polymorphic patterns in DNA sequences. Restriction sites are recognized on isolated DNA with restriction endonucleases (Helentjaris et al. 1986; Liu and Cordes 2004; Malik et al. 2014) resulting in fragments on digestion. The resulting fragments are separated via electrophoresis on an agarose gel, transferred to a polyester membrane or nylon, hybridized using probes, and auto-radio graphically scored. RFLP markers are dependable, repeatable and co-dominant. However, it takes longer time to complete and cannot be automated.

Table 17.2 Salient features of various molecular markers

No.	Feature	RFLP	RAPD	AFLP	SSRs	SNPs
1	DNA molecule (μg)	10	0.02	0.5–1.0	0.05	0.05
2	Quality of genomic segment (DNA)	High	High	Medium	Medium	High
3	Polymorphic loci assessed	One to three	One to fifty	Twenty to hundred	One to three	One
4	PCR based	No	Yes	Yes	Yes	Yes
5	Data sequenced	Required	Not required	Not required	Required	Required
6	Inheritance	Codominant	Dominant	Dominant	Codominant	Codominant
7	Allelic interpretation	Yes	No	No	Yes	Yes
8	Utility	Medium	Low to medium	Low to medium	High	Low to medium
9	Assessment price	High	Low	Medium	Low	Low

Source: Lacape et al. (2009); Bolek et al. (2016)

17.7.5.2 PCR-Based Markers

Benefits of using PCR-based markers over hybridization-based markers include (A) a small DNA segment required for genotyping; (B) amplification of fragments even if cells are frozen; (C) high polymorphism; and (D) simultaneously screening of many genes (Bolek et al. 2016; Caeteno-Annolas 1994).

RAPD, AFLP, SSRs and ISSRs are PCR-based markers with high utility (Badigannavar et al. 2012; Bardak 2012; Bolek et al. 2016; Bornet et al. 2002; Essenberg et al. 2002; Khan et al. 2000; Li et al. 2013; Malik et al. 2014; Preetha and Raveendren 2008; Qureshi et al. 2004; Rahman et al. 2002; Reddy et al. 2002; Xiao et al. 2009).

Amplified Fragment Length Polymorphism (AFLP)

AFLP is reliable on a combination of constricted sequences amplified with PCR. The process consists of several steps: Digestion of genomic DNA by restriction enzyme followed by ligation of fragments with oligonucleotide adapters, and then selective fragments are amplified at 3′ ends to complementary primers (Malik et al. 2014). The final step is running gel electrophoresis to detect polymorphic segments (Semagn et al. 2006). AFLPs are highly reproducible, and cost-efficient and analyze more polymorphic loci per primer in a single gel than other markers (Essenberg et al. 2002; Khushboo et al. 2014; Liu and Cordes 2004; Maughan et al. 1996; Russell et al. 1997).

Random Amplified Polymorphic DNA (RAPD)

RAPD is a technique for amplifying short random genome segments of about 10 bp with the use of random primers and PCR (Bardakci 2001). RAPD markers are found to be closely bound to hereditary studies in populations following Hardy-Weinberg assumptions (Lee et al. 2002). The requirement of small fragments of DNA, no prior sequencing, easy to carry out and automation are the main features of the success of RAPDs (Bardakci 2001). These markers are extensively employed for mapping traits and profiling of genotypes. RAPDs show polymorphism and act as dominant markers due to variations in genome sequences at binding sites. The primers used do not contain palindromic sequences and have at least 40% G-C content in the segments (Malik et al. 2014; Preetha and Raveendren 2008).

Several factors like DNA quality and quantity, PCR mixture (primer-template ratio), nature of polymerase and buffer composition contribute to the poor reproducibility of RAPDs (Rafalski 1997; Wolff et al. 1993). Another disadvantage of RAPDs is their inability to distinguish between homozygotes and heterozygotes (Bolek et al. 2016; Liu and Cordes 2004).

Inter Simple Sequence Repeat (ISSRs)

These are modified SSR technologies that involve the utilization of complementary primers to SSR motifs rather than flanking fragments, as stated by Singh et al. (2014). ISSRs are simple, reproducible and highly polymorphic (Tables 17.2 and 17.3).

Table 17.3 Experimental studies with specific marker application

Experimental research/ studies	Markers used	Research outcomes	References
Domestication of *G. hirsutum*	RFLP	Revealed wild ancestor of upland cotton (Yucatan)	Brubaker and Wendel (1994)
Intra-specific populations	RFLP	63 QTLs (A subgenome) and 29 QTLs (D subgenome) at 5 and 3 loci, respectively	Ulloa et al. (2005)
$F_{2:3}$ populations (intra-*hirsutum* accessions)	RFLP	Joint map with 284 polymorphic markers & 49 linked pairs	Ulloa et al. (2002)
MAS for bacterial blight	RFLP	Identified resistant alleles for bacterial blight	Wright et al. (1998)
Genetic diversity of cotton	4 *EcoRI-MseI* primers for AFLP	A total of 17 accessions were assessed viz, 11 of *G. hirsutum* and two of *G. arboreum*, two of *G. herbaceum* and two of *G. barbadense*	Pillay and Myers (1999)
Backcross screening for agronomic traits and fiber quality enhancement	50 AFLPs	E1-M1, M2, M3 & E6-M3 for lint yield %	Wu et al. (2007)
Screening of F_2 population	307 SSR; 72 AFLP oligonucleotides	Genetic map with 27 linkage groups	Bardak (2012)
RILs population by inter-specific introgression (*G. hirsutum* & *G. barbadense*)	AFLP, RFLP, and SSR	A genetic map constructed with 800 markers	Lacape et al. (2009)
Genetic diversity of *G. arboreum* (Pakistan)	RAPD	30 cotton cultivars examined with 50 RAPDs	Yasmin et al. (2008)
Phylogenetic studies	RAPD		Khan et al. (2000); Lan et al. (1999)
Genetic diversity	RAPD		Rahman et al. (2002)
CLCuV disease screening	RAPD		Khanam et al. (2012)
Screening genes for male sterility and fertility restorer traits	RAPD	Mapping of genes R-6592 and UBC607500	Preetha and Raveendren (2008)

(continued)

Table 17.3 (continued)

Experimental research/ studies	Markers used	Research outcomes	References
Resistance to sucking pests	RAPD		Geng et al. (1995)
DNA finger printing	RAPD		Zahra et al. (2011)
Genetic diversity	RAPD	Mapping of primers	Farzaneh et al. (2010); Noormohammadi et al. (2011)
Genetic Characterization of *Desi* Cotton Hybrids	25 RAPD and 9 ISSR	Primer ISSR9–200 bp in Sujay and 550 bp in G.Cot. DH-7; Primer RPI7–450 bp (4011) and RPI8–550 bp (824) in G.Cot.DH-9	Patel et al. (2020)
Screening of F_2 population	26 RAPDs and 17 primers	261 reproducible bands (4.18/primer) and polymorphism (22% in upland and 805 in Opal) in agronomic traits.	Noormohammadi et al. (2013)
Morphological, physico-biochemical and marker-based diversity of *desi* cotton	42 SSR and 10 ISSR	13 SSR and 5 ISSR showed polymorphism; 13 SSRs band size ranged between 94 and 432 bp; 5 ISSRs band size ranged between 85 and 1357 bp.	Sagar et al. (2022)
Genetic diversity of *G. arboreum* (China)	358 SSR	39 accessions examined with 358 SSRs	Liu et al. (2006)
Genetic diversity	60 SSR	108 accessions of Asiatic cotton and 1 *G. herbaceum* assessed	Guo et al. (2006)
Genetic diversity of *G. arboreum* (USA)	115 Genomic and EST-SSRs	A total of 96 accessions were assessed with EST-SSRs	Kantartzi et al. (2009)
Genetic diversity in cotton cultivars	9 ISSR oligonucleotides	62.79% polymorphism	Parkhyia et al. (2014)
Phylogenetic relation in 21 cotton genotypes	12 ISSR	49.6% reproducible rate	Gupta et al. (1994)
Genetic diversity in cotton germplasm	10 ISSR	88.5% polymorphism	Tsumura et al. (1996)
Cotton genome sequencing	SSRs	Reported one SSR per 170 kb of genomic DNA	Abdurakhmonov et al. (2005))
RIL population screening of tri-species hybridization	SSRs (JESPR-13, JESPR-153, and JESPR178)	High association with leaf defoliation trait value	Abdurakhmonov et al. (2005)
Genetic diversity in cotton	300 SSR	SSR-enriched genomic libraries	Reddy et al. (2001)

Simple Sequence Repeats (SSRs)

SSRs also known as microsatellites are short tandem repeats of one to eight nucleotide motifs with a core sequence of a few to a hundred nucleotides. Sequence repetition is assigned as $(AAAC)_n$, where n symbolizes the count of tandem repeats. The SSRs are flanked and the amplification is carried complementary to these regions (Xiao et al. 2009). The extent of variability occurs due to a lapse in DNA polymerization and mispairing at the time of replication and variation in PCR products induces polymorphism. SSRs are more reliable and ideal for studying genomic mapping.

Single Nucleotide Polymorphism (SNPs)

SNPs are unique DNA markers that have emerged by point mutations in single nucleotide base positions. These are mainly base changes from insertions and deletions and transversions (C/G, A/T, C/A, or T/G) to transitions (C/T or G/A) showing single base changes that induce SNPs. SNPs are helpful for genotype diversity where polymorphism is difficult. Reproducibility limits the use of SNP markers in *G. arboreum* (Krawczak 1999).

17.8 Applications of Markers in Breeding Schemes

17.8.1 Marker-Assisted Backcrossing (MAB)

MAB is a common and effective type of molecular marker-assisted breeding scheme. In this scheme, two parents are required, one as donor of the target gene of interest (marker loci) and the other as recipient of the gene. The F_1 generation is developed by hybridization of parents. MAB considers the presence of a molecular marker instead phenotypic value of expression. The F_1 generation is planted to confirm marker loci and selected F_1 is backcrossed with the recurrent parent. BCF_1 generation evaluated and hybridized with the subsidiary recurrent parent that has alternative alleles. The segregation frequency of the F_1 population is observed at the BC_1 generation. By comparing this population with the F_2 generation, efficient mapping is done (Tian et al. 2015). Three to four generations (filials) are developed for marker stability and related traits.

17.8.2 Marker-Assisted Recurrent Selection (MARS)

Marker-Assisted Selection (MAS) plays an important role in developing promising genotypes with desirable traits. The marker-assisted recurrent selection opens up a wider scope for plant improvement programs dealing with the introgression of multiple genes (Jiang et al. 2007). MARS is efficiently used for quantitative traits by selfing and genotyping in a single selection cycle within crop season, resulting in a genetic gain. In some cases, the genetic gain from MARS is doubled.

17.8.3 Genotyping by Sequencing

The need to make reliable SNPs for molecular studies is important and these tools can explore more complex genomics data of polyploid species through genotyping by sequencing (GBS) methodology at economic cost and high-throughput efficiency (Zhu et al. 2003). There are many advantages of GWS including markers availability, quick screening, discovery of more SNPs and validation of mapping sequence (Baird et al. 2008).

17.8.4 Databases

The primary sources of cotton markers and sequenced data are Cotton Gen (Database 1 2016), Comparative Evolutionary Genomics of Cotton (Database 2 2016), National Center for Biotechnology Information (Database 3 2016) for Express sequence tags resource, TropGENE Database (Database 4 2016), the Cotton Diversity Database (Database 5 2016) and the BACMan resources at Plant Genome Mapping Laboratory (Database 6 2016) (Bolek et al. 2016).

17.9 Tissue Culture Applications

Biotechnological techniques applied to cotton include callus culture, somatic embryogenesis, organogenesis, protoplast culture, ovule or embryo culture-mediated interspecific hybridization, somaclonal variation, plant transformation, and other related techniques.

17.9.1 Micropropagation in Plants

Plant tissue culture is a collection of techniques used to grow plant cells or tissues on an artificial culture medium (liquid, semi-solid, or solid growth medium) under sterile conditions. Simply put, it is the cloning of a plant also referred to as micropropagation. This technique of regenerating a complete plant is very helpful in developing transgenic crops. However, cotton is fractious to micropropagation.

Although somatic embryogenesis and shoot tip organogenesis have been successful, they rely on specific genotypes (John 1997; Kalbande and Patil 2016; Solanki et al. 2011; Tian et al. 2010).

17.9.2 Callus Initiation and Culture

Plant part (explant), culture medium, sterile and light/dark conditions are the main requirements for tissue culture. The main steps of the tissue culture procedure in the laboratory: (a) Preparation of culture medium, (b) Inoculation of explant, (c) Incubation at 25–30 °C, (d) Sub-culturing and (e) Acclimatization. Since each part of the plant in cotton such as leaf, stem, anther, ovule, cotyledon, embryo and protoplast, etc. can induce callus however, the achievement is not satisfactory due to lack of further differentiation process of callus. Numerous reports for callus induction and maintenance in cotton; the first callus initiation was reported by Beasley in 1971 (Table 17.4). Since after many researchers appeared but the criteria have not been achieved.

17.9.3 Embryo Rescue Technique

The development of many interspecific hybrids has not been successful due to incompatibility between distant cotton species. In-vitro culture of cotton ovules (Embryo rescue technique) has been developed to overcome these challenges. Success has been achieved in both wild and cultivated cotton species (Bajaj and Gill 1998). Flowers and young bolls should be treated with growth regulators to rescue the embryo which is then cultured on a specific medium. This protocol is preferred for crossing diploid and tetraploid species (Altman et al. 1987; Gill and Bajaj 1987).

17.9.4 Ovule Culture

It is applied to overcome interspecific hybridization problems by understanding the in vitro development of cotton. The necessity of plant hormones is a prerequisite for the success rate in developing the ovule into a whole plant.

17.9.5 In-Vitro Pollination

Seed development followed by stigma pollination of cultured pistils is called in-vitro pollination (Amiteye 2023; Anu et al. 2022). Steps to be taken care of include: (1) determining the time of anthesis, pollination, pollen entry into ovules, and fertilization; (2) emasculation, and bagging and (3) collection of pollens.

Table 17.4 Explants used in callus induction of *G. arboretum*

Year	Part used	Media or regulators	References
1971	Micropylar end of ovules	MS medium	Beasley et al. (1971)
1972	Mesocotyl	0.5 mg/2.4-D, 2 mg p-CPA, 0.1 mg kinetin with sucrose and SH salts	Schenk and Hildebrandt (1972)
1974	Cotyledon tissue	0.1 mg IAA	Davis et al. (1974)
1976	Micropylar end of ovules	Ethylene	Hsu and Stewart (1976)
1976	Hypocotyl and stem	–	Rani and Bhojwani (1976)
1977	Hypocotyl	–	Smith et al. (1977)
1977	The entire outer integument of the cultured ovules	Ethylene and GA	Stewart and Hsu (1977)
1985	Hypocotyl	–	Bajaj and Gill (1985)
1989	Anther		Bajaj and Gill (1989)
2010	Hypocotyl and cotyledon	MS medium, TDZ, 2,4-D + kinetin and 2,4-D	Thakur et al. (2010)
2011	Cotyledon and leaf	MS medium and B5 vitamins, 2,4-D	Anaz and Khanna (2011)
2012	Shoot tip	MS medium, 2 mg/L BA	Sanghera et al. (2012)
2014	Shoot tip	MS medium, 1 mg/L BA, 1 mg/L kinetin	Khatoon et al. (2014)
2018	Shoot tip	MS medium, 2 mg/L BAP, 0.5 mg/L kinetin	Yasin and Yasmin (2018)
2019	Petiole	MS medium with BAP, 2,4-D, NAA	Raut et al. (2019)
2021	Hypocotyl, root and cotyledon	MS medium with B5 vitamins, IBA	Ke et al. (2022)

17.10 Genetic Engineering and Gene Editing

Genetic transformation methodologies are the focus in concern with genetic improvement of desirable characteristics of cotton. The development of Bt cotton has been a great achievement as the first genetically modified crop. In-vitro regeneration has been very difficult in all four cultivated cotton (Cousins et al. 1991; Rajasekaran et al. 1996; Trolinder and Xhixian 1989). The diploids (*G. arboreum*) are less preferable than the tetraploids (*G. hirsutum*) for transgenic transformation and gene introgression. One factor is the morphogenic response of genotypes. Another factor is the genotypes of *G. arboreum* show more tolerance to insects than *G. hirsutum*. Some genotypes are preferable for agronomic and quality traits but are susceptible to insects and pests like DLSa-17. Using a transgenic approach, the cry 1F gene has been transferred into DLSa-17 to make it insect-resilient (Sangannavar 2008). Few transformants with low frequency have been obtained.

By the use of *Agrobacterium* strain GV3101 bearing plasmid pPZP200 vector havingcry1AC gene, limiting factors for transformation efficiency of *Gossypium arboreum* cultivar LD 694 have been analyzed (Sanghera et al. 2011). Subsequent consideration is required to introduce genes for agronomic traits related to yield along with insect, pathogen and virus resilience in *desi* cotton. Cotton is recalcitrant to in vitro regeneration which needs a solution. So, it is reliable to consider the breeding approach rather than genetic transformation.

G. arboreum is reported to be resistant to nematodes, root rot, dehydration and salt, and the Cotton Leaf Curl Virus (CLCuV) (Blaise et al. 2021; Mushtaq et al. 2020). Nevertheless, bollworm, particularly *Helicoverpa armigera*, has a significant negative impact on *G. arboreum* (Nandeshwar et al. 2009). Efforts must be made to genetically modify the genome of *G. arboreumin* in order to increase its resistance to various constraints and enhance quality parameters, as a result of the growing attack of CLCuV on imported types of the plant (Akhtar et al. 2010; Ali et al. 2008). The three most significant pests are *Tetranychus urticae* (spider mite), *Helicoverpa armigera* (spotted bollworm), and *H. punctigera* (Farrell and Johnson 2005). Traditionally, pests are managed by using pesticides, a costly procedure. Approximately, 100 *Bacillus thuringiensis* (Bt) crystal genes, that are encoded for the Cry and Cyt proteins, have been sequenced thus far in order to address these difficulties through biotechnology (Schnepf et al. 1998). The larvae of Diptera, Coleoptera, and Lepidoptera are poisoned by such crystal proteins. Numerous studies have demonstrated that *G. hirsutum* is resistant to lepidopteron and dipteran insects when various *cry* genes (*cry1Ab*, *cry1Ac*, *cry1A*, *cry2A*, *cry1F*, etc.) are altered. Ten cotton genotypes revealing both Cry2A and Cry1Ac endotoxins had been poisonous to *Spodoptra frugiperda*, *H. zea*, *S. exigua*, and *H. armigera* in comparison with genotypes producing a particular toxin (Hernández-Rodríguez et al. 2008; Luo et al. 2007). Delaying the development of resistance in lepidopterans can be achieved by stacking distinct Cry proteins, such as Cry1Ac and Cry2A. Thirteen Stacking *cry1Ac* and *cry2A* genes may be advantageous for managing resistance since the Cry1Ac and Cry2A proteins lack the high-affinity brush border binding sites on midgut membrane vesicles of insects (Hernández-Rodríguez et al. 2008). *G. arboreum* is extremely susceptible to weed competition, which results in 37% of production losses (Latif et al. 2015). In addition to lowering the soil's nutritional value and having a negative impact on cotton harvesting and lint quality, weeds can act as hosts for pests and pathogens. Chemical pesticide sprays provide an alternative to manual weed management for the elimination of weeds, but they cause harm to plants, particularly in cotton fields (Suarez et al. 2016). Glyphosate spray suppresses the enzyme involved in the shikimate pathway (Santos-Sánchez et al. 2019). By inhibiting 5-enolpyruvylshikimate-3-phosphate synthase (EPSPS), an enzyme involved in the shikimate biosynthesis pathway, glyphosate (N-phosphonomethyl glycine) completely damages the crop. Inorganic phosphate and 5-enolpyruvylshikimate-3-phosphate (S3P) are produced by EPSPS from phosphoenolpyruvate and shikimate-3-phosphate (S3P). Numerous secondary compounds, comprising lignins, phytoalexins, folic acids, amino acids, and plastoquinones are produced by this route (Santos-Sánchez et al. 2019). When the EPSPS enzyme is blocked, hydroxybenzoic acid and shikimic acid build up in the leaves or nodes, which kills the plant (Santos-Sánchez et al. 2019). In transgenic crops, EPSPS-encoding bacterial genes are inserted into agricultural plants by stable genetic transformation to provide resistance to glyphosate herbicides. It has been shown that glyphosate tolerance may be successfully induced in upland cotton by transforming a *Calotropis procera EPSPSII* gene (Puspito et al. 2015).

With the current experimental design, our goal is to modify the *cry* genes in *G. arboreum* in conjunction with the Glyphosate Tolerant Gene (*cp4-EPSPS*) in order to generate a transgenic variety that is resistant to both glyphosate and pest assault (Latif et al. 2015).*G. arboreum* naturally resists whiteflies and cotton leaf curl virus (CLCuV), hence the increased fiber length shown in *CpPIP2* transgenic plants makes crop production more financially feasible (Amat-ur-Rasool et al. 2022).

In the current study, an attempt was made to genetically modify the *G. arboreum* cultivar FBD-1 in order to counteract insect and weedicide resistance. This was done by introducing the *cry1Ac*, *cry2A*, and *cp4-EPSPS* genes regulated via *35S* promoter in two distinct cassettes, utilizing *GUS* and kanamycin as markers, by the use of the *Agrobacterium*-mediated stem terminal cut method of cotton (Sufyan Tahir et al. 2021). Prior research on drought-stressed *G. arboreum* leaves identified the universal stress protein gene (*GUSP1*), which was then shown to be expressed in transgenic cotton (Maqbool et al. 2008). In order to examine the effects of site-directed mutagenesis of the *GUSP1* gene in transgenic *Gossypium hirsutum* plants bearing both mutant and non-mutated constructs of *GUSP1*, simulations of *in-silico* involving homology modeling and molecular docking were employed to examine the physiological and biochemical alterations in transgenic plants for drought stress (Hassan et al. 2021).

*17.10.1 Genome Editing in Tree Cotton (***Gossypium arboreum L.***)*

Due to chromosomal doubling, chromosome number revival from repeated backcrossing, and obscurity in the F_1 hybrid generation, *G. hirsutum* has not been successful in obtaining the necessary genes from diploid species (Sattar et al. 2019). Cotton breeders may be able to increase yield, improve seed quality, and improve cotton fiber by combining a variety of breeding techniques with traditional genetics tools. Combining these methods can also result in the early creation of resistant or tolerant cultivars and improved environmental adaptation of cotton crops (Singh et al. 2007). Speeding up the breeding programs, the problems with traditional breeding need to be improved or eliminated. CRISPR/Cas technology may be a useful tool in this regard. Numerous potential genes, including those involved in the manufacture of gossypol, *Verticillium* wilt resistance and fiber growth in cotton, require more study (Sattar et al. 2019). In addition for improves investigation of functional genomics and improves prospects of molecular breeding in cotton (Chen et al. 2017). The CRISPR/Cas system has the potential to investigate these putative genes. Conventional breeding methods might not be able to accomplish these goals quickly. While, the CRISPR/Cas technique has proven effective in creating genetic modifications for the investigation of cotton gene function, concerning the protoplast level and stable generation of transgenic cotton (Chen et al. 2017). In contrast to traditional mutation breeding, the CRISPR/Cas system allows for the creation of targeted mutations at the desired loci as well as the possibility of many off-target loci. Furthermore, there is less unintentional mutation in the crops that have been altered using CRISPR/Cas technology than there is in a traditional mutation breeding program. In the event of an off-target mutation, backcrossing is always an option for further improvement. Unknown and unwanted mutations do, however, develop equally in both conventional breeding and mutation breeding, and they can be eradicated with several crossings in subsequent cycles. CRISPR/Cas, which is assisted by homology recombination, can introduce characteristics for improving cotton. When it comes to creating transgenic goods, the technology is superior to transgene insertion techniques. Therefore, by using natural features found in the cotton crop, CRISPR/Cas is a viable technique for plant breeding with enormous capability that might lead to effective cotton breeding. A proper modification of the cotton genome requires the proper selection of the CRISPR/Cas system, sgRNA promoter, binary vector for sgRNA delivery, and transformation technique, among other things. An appropriate plant promoter and codon-optimized Cas are needed for bacterial Cas expression in cotton. To stimulate the Cas protein in plants, a variety of constitutive and/or inducible promoters have been successfully used, including *CaMV 35*, *CMV*, *EF1A*, *LTR*, and *UBO*. In addition to these, several plant RNAIII-based promoters have also been effectively employed, including *U3p* and *U6p* (Belhaj et al. 2013). After the assessment of the sgRNAs that were created and the choice of CRISPR system, the cassette containing the sgRNA, CRISPR, and Cas/Cpf1 has to be assembled into a plant-compatible binary vector (Sattar et al. 2019). Since the selection of an effective transformation strategy determines the overall performance of the system, delivering the cotton genome with the resulting recombinant vector is an extremely important step. In addition to these methods, successful transformation can also be accomplished by utilizing a variety of plant virus-based vectors. Biolistic inoculations, transit peptides, protoplast transformation, and polyethylene glycol (PEG)-mediated transformation are some of the techniques being used to carry out stable plant transformation (Butler et al. 2016).

17.10.2 Gene Editing for Fiber Quality Improvement

By the use of the *Agrobacterium*-mediated shoot apex cutting method, the *PIP2* gene construct (35SCpPIP2) was used to induce transformation in *G. arboreum*. The increased expression of the *CpPIP2* gene by 35 fold in transgenic plants resulted in an increase of staple length by 150%. FISH method was used to examine the integration of transgene and karyotyping which reported a single transgene present on chromosome 6. Since *G. arboreum* is naturally resilient to whitefly and CLCuV, the evidence of improved fiber length renders the production more economically viable (Amat-ur-Rasool et al. 2022).

17.11 Mutation Breeding

Mutation breeding also known as mutagenesis is the process of inducing phenotypic variations in crop genotypes using mutagens, radiation and enzymes (Engqvist and Rabe 2019). The identification of regulatory genes and selection of desirable traits are key aspects of the mutation breeding scheme for the development of elite genotypes with improved attributes (Zhang and Batley 2020). Site-specific mutation plays an efficient role in crop genetic improvement, involving techniques such as Ethyl methane sulfonate (EMS)-induced mutation, physical and chemical mutation, zinc finger nucleases (ZFNs), transcription activator-like effector nucleases (TALENS), and cluster regulatory interspaced short palindromic repeats (CRISPR-Cas9) nucleases (Khan et al. 2023; Mushtaq et al. 2019).

17.11.1 Conventional Mutagenesis

Conventional methods of mutation comprise physical and chemical mutagenesis. Physical mutagenesis consists of ionizing radiations such as X-rays, gamma rays, neutrons and beam ions (Bado et al. 2015). Chemical mutagenesis consists of hydroxylamine, alkylating agents, nitrous acid, base analogs and acridine dyes (Oladosu et al. 2016). Chemical mutagenesis deals with point mutation and a wider approach than physical mutagenesis.

17.11.2 Insertional Mutagenesis

Insertional mutagenesis comprises deploying an exogenous genomic material (T-DNA, transposable elements, etc.) to create a mutation in a population (Khan et al. 2023). It is a more effective approach than conventional to modify genes and sequencing with PCR techniques. Some other approaches are mutagenesis using genomic and genetic engineering, TILLING, etc. Transposable elements and *Agrobacterium* T-DNA are two forms of insertional mutagenesis that have been extensively and successfully used in plants (Auld et al. 2009).

The Ti plasmid of *agrobacterium* is used in T-DNA technology, which modifies normal gene functioning in plants by inserting it into a gene. T-DNA serves two purposes: first, it disrupts genes to produce mutants; second, it produces a unique tag sequence that can be used to identify physical genome insertions (Krysan et al. 1999). Consequently, in contrast to point mutations produced by EMS/fast neutrons, T-DNA mutant genetic analysis is easy. At Texas Tech University, Xing Qi employed T-DNA to insert into gene-rich areas of the genome (Barakat et al. 2000). It is extensively used in Arabidopsis and other plants where transformation is easier to do.

Transposon mutagenesis, also known as tagging, is a technique for introducing mutations into crops that overcomes the drawbacks of T-DNA technology. Transposons sometimes referred to as jumping genes, were first identified in maize and were found to be involved in the suppression of gene regulation (McClintock 1956). These transposon elements can jump throughout genomes and insert themselves while doing so. Gene function can be altered when these transposons insert themselves into a gene or promoter region. For instance, changes in gene expression were observed when different transposon elements were added or removed from the promoter regions of two transcription factor genes in the cotton At and Dt subgenomes (Wang et al. 2016).

The mutation breeding approach has contributed very little tithe improvement of G. *arboreum* in comparison to upland cotton. However, utilization of various mutagens shows preferable achievements (Table 17.5).

Table 17.5 Mutagens used in cotton improvement

Mutagens	Research outcomes	References
3% Ethyl methanesulfonate (EMS), 0.05% Na azide	*G. arboreum* var. (GA-TAMU, A2-60T, and A2-120W) to enhance the lint percent, lint yields as well fiber quality	Lowery et al. (2007)
Ethyl methanesulfonate (EMS) and gamma rays	EMS has more significant results than gamma rays	Aslam et al. (2013)
Ethyl methanesulfonate (EMS) and gamma rays	Chlorophyll mutants frequency maximum at 0.4% EMS and 30, 40 kR gamma rays	Kalpande et al. (2020)
Gamma rays	Heat-resistant, early maturity	Maluszynski et al. (1995)
60Co-γ-ray	Resistant to glyphosate	Zhao et al. (2022)
Ethyl methanesulfonate (EMS)	Fiber quality	Brown et al. (2015)
Ethyl methanesulfonate (EMS)	Resistance against biotic stresses	Zhang et al. (1998)

Source: Khan et al. (2023)

17.12 Conclusion and Prospectus

Diploid cotton has the potential to act as a substitute for *G. hirsutum*, provided it receives comparable attention and slightly increased compensation. *Gossypium arboreum*, is indigenous to the Indian subcontinent and demonstrates minimal input needs, flourishing particularly in environments characterized by high salinity and drought conditions. *G. arboreum* cotton possesses immense potential for the future of textile production. By implementing innovative breeding strategies like molecular breeding, marker-assisted backcross breeding, and mutation breeding coupled with tissue culture techniques this unique crop can be transformed into a sustainable, cost-effective, and high-quality fiber source, catering to the demands of the modern textile industry. This chapter delves into various facets including encompassing the genetic diversity of *Gossypium arboreum*, prevailing cultivation practices, and existing challenges. The adoption of goal-oriented breeding in *desi* cotton holds the promise of amplifying yields while fostering sustainable livelihoods.

Appendix 17.1 Research Institutes and Online Resources Available for *G. arboreum* Breeding

Country	Institution	Specialization research activities	Website
Bangladesh	Bangladesh Agricultural Research Institute (BARI)	Crop Research and Education	https://bari.gov.bd/
China	Institute of Cotton Research (ICR) of CAAS	Cotton Research	https://cri.caas.cn/en/index.htm
India	ICAR-Central Institute for Cotton Research (ICAR-CICR)	Crop Improvement, Crop Production, Crop Protection, Post Harvest Technology	https://cicr.org.in/
India	ICAR-National Institute of Natural Fibre Engineering and Technology (ICAR-NINFET)	Natural fiber production, processing	http://www.nirjaft.res.in/
India	ICAR-Central Institute for Research on Cotton Technology (ICAR-CIRCOT)	Improvement in the Ginning of cotton; Improvement and Quality Evaluation of Fibre; Yarn and Fabric; Finishing and Dyeing of Cotton with Natural and Environment-Friendly Agents; The utilization of Cotton Plant Residues for Production of Value Added Products; Development of Entrepreneurship for Utilisation of CIRCOT Technologies	https://circot.icar.gov.in/en/

(continued)

Country	Institution	Specialization research activities	Website
India	Punjab Agricultural University	All India Coordinated Research Project on Cotton (AICRP-on Cotton) Center	https://aiccip.cicr.org.in/; https://www.pau.edu/
India	Chaudhary Charan Singh HAU	AICRP-on Cotton Center	https://aiccip.cicr.org.in/; https://hau.ac.in/
India	Swami Keshwanand RAU	AICRP-on Cotton Center	https://aiccip.cicr.org.in/; https://raubikaner.org/
India	Maharana Pratap University of Agri. & Tech.	AICRP-on Cotton Center	https://aiccip.cicr.org.in/; https://www.mpuat.ac.in/
India	Navsari Agricultural University	AICRP-on Cotton Center	https://aiccip.cicr.org.in/; https://nau.in/index
India	Junagadh Agricultural University	AICRP-on Cotton Center	https://aiccip.cicr.org.in/; http://www.jau.in/
India	Mahatma Phule Krishi Vidyapeeth	AICRP-on Cotton Center	https://aiccip.cicr.org.in/; https://mpkv.ac.in/
India	Vasantrao Naik Marathwada Krishi Vidyapeeth	AICRP-on Cotton Center	https://aiccip.cicr.org.in/; https://www.vnmkv.ac.in/
India	Dr. Panjabrao Deshmukh Krishi Vishwa Vidyapeeth	AICRP-on Cotton Center	https://aiccip.cicr.org.in/; https://pdkvu.in/
India	Jawahar Lal Nehru Krishi Vishwa Vidyalaya	AICRP-on Cotton Center	https://aiccip.cicr.org.in/; http://jnkvv.org/
India	University of Agricultural Sciences	AICRP-on Cotton Center	https://aiccip.cicr.org.in/; https://uasd.edu/
India	Acharya N.G.Ranga Agricultural University	AICRP-on Cotton Center	https://aiccip.cicr.org.in/; https://angrau.ac.in/
India	Tamil Nadu Agricultural University	AICRP-on Cotton Center	https://aiccip.cicr.org.in/
Myanmar	Yezin Agricultural University (YAU)	Crops, natural resources, socioeconomics	https://www.yau.edu.mm/
Myanmar	Department of Agricultural Research (DAR)	Crops	http://www.doa.gov.mm/doa/index.php?route=common/home

(continued)

Country	Institution	Specialization research activities	Website
Myanmar	Myanmar Cotton and Sericulture Enterprise (MCSE) Renamed as "Cotton and Sericulture Department (CSD)" under the Ministry of Agriculture, Livestock and Irrigation	Cotton, sericulture	https://www.moali.gov.mm
Pakistan	Cotton Research Institute (CRI), Multan	Cotton Research	https://aari.punjab.gov.pk/cri_multan
Pakistan	Central Cotton Research Institute Pakistan Central Cotton Committee	Cotton Research	https://ccri.gov.pk/Introduction.html
Pakistan	Nuclear Institute for Agriculture and Biology Pakistan Atomic Energy Commission	Cotton and Wheat Research	http://www.nibge.org/Default.aspx

Appendix 17.2 Recommended Varieties/Hybrids for *G. arboreum*

Country	Region	Varieties/Hybrids
India	North India (Irrigated areas)	CICR 1 (CISA 310), CICR 3 (CISA 614), CICR 2 (CISAA 2), HD 123, RG-8, LD 327, LD 649, KR 64, KR 149, KR 151, HD 324, HD 432, LD 1019, LD 949, FDK 124, RG 18, RG 542, AAH 1, RAJ.D.H. 9
	Central India (Rainfed areas)	CNA 1028, CNA 1032, CICR-A Cotton 46 (CNA 1054), CICR-A Cotton 57 (CNA 1091), CICR-A Cotton 56 (CNA 1031), CICR-A Cotton 59 (CNA 17522), JLA 1207, PA 837
	South India (Rainfed areas)	Roja (CNA 1003), CICR-A Cotton 57 (CNA 1091), DWDa 1802
Bangladesh		Hill cotton-1, Hill cotton-2, Comilla
Myanmar		Mahlaing, Wagyi
Pakistan	South-west	Multani Kapas, S.N.R., 15-Mollisoni, 39-Mollisoni, 119-Sanguineum (119-S), 231-R, TD-1, D-9, SKD-10/19, Ravi, Rohi, FDH-170, FDH-228, Bahawalpur Desi, Haroonabad local, Jubilee strains-D.C.17, D.C.26, D.C.37, D.C.40 and D.C.41
China	South China	Million Dollar, Chuan77-1, Acala1517-70, Acala1517-77, LineF, FJA, Zaoshuchangrong7, ITMA96497, Hopical. Now Nanjing, Anyang, Kuerle, Kuche and Shijiazhuang

References

Abdurakhmonov IY, Abdullaev AA, Saha S et al (2005) Simple sequence repeat marker associated with a natural leaf defoliation trait in tetraploid cotton. J Hered 96:644–653

Abdurakhmonov IY, Kohel RJ, Yu JZ et al (2008) Molecular diversity and association mapping of fibre quality traits in exotic *G. hirsutum* L. germplasm. Genomics 92:478–487

Abdurakhmonov IY, Buriev ZT, Shermatov SE et al (2011) Marker-assisted selection for complex fiber traits in cotton. In: 5th World cotton research conference, special session of ICGI, Mumbai, India, 7–12 November 2011

Akhtar KP, Haidar S, Khan MK et al (2010) Evaluation of *Gossypium* species for resistance to cotton leaf curl Burewela virus. Ann Appl Biol 157(1):135–147. https://doi.org/10.1111/j.1744-7348.2010.00416.x

Ali MA, Khan IA, Awan SI et al (2008) Genetics of fibre quality traits in cotton (*Gossypium hirsutum* L.). Aust J Crop Sci 2(1):10–17

Altman DW, Stelly DM, Kohel RJ (1987) Introgression of the glanded plant and glandless, seed trait from *Gossypium sturtianum* Willis into cultivated upland cotton using ovule culture 1. Crop Sci 27:880–884

Amat-ur-Rasool H, Latif A, Yasmeen A et al (2022) Enhanced expression of plasma membrane intrinsic protein 2 improves cotton fiber length in *Gossypium arboreum*. Mol Biol Rep 49(6): 5419–5426

Amiteye S (2023) In vitro embryo rescue techniques and applications in hybrid plant development. In: Advanced crop improvement, volume 2: case studies of economically important crops. Springer International Publishing, Cham, pp 419–456

Anaz MA, Khanna VK (2011) A comparison of *Gossypium hirsutum* and *G. arboreum* cultivars for somatic embryogenesis. Pantnagar J Res 9(1):27–34

Ansingkar AS, Kadke PP, Borikar ST, Bhosle SS (2004) Altering *G. hirsutum*cotton at cellular level to impart multiple sucking pests resistance through interspecific hybridisation. In: Khadi BM, Kategeri IS, Patil SS et al (eds) Proceedings of international symposium on strategies for sustainable cotton production-a global vision: 1 Crop improvement, November 23–25, 2004. University of Agricultural Sciences, Dharwad, pp 101–103

Anu, Yadav S, Singh VK, Bhoyar PI et al (2022) Emerging technologies in plant breeding for fibre crops, cotton, and sunn hemp. In: Technologies in plant biotechnology and breeding of field crops. Springer Nature Singapore, pp 151–180

Ashford B (2016) Fibres to fabrics. AuthorHouse

Aslam U, Khan AA, Cheema HMN et al (2013) Kill curve analysis and response of ethyl methanesulfonate and γ-rays in diploid and tetraploid cotton. Int J Agric Biol Eng 15:11–18

Auld D, Light GG, Fokar M et al (2009) Mutagenesis systems for genetic analysis of *Gossypium*. In: Paterson AH (ed) Genetics and genomics of cotton, plant genetics and genomics: crops and models, vol 3. Springer, New York. https://doi.org/10.1007/978-0-387-70810-2_9

Badigannavar A, Myers GO, Jones DC (2012) Molecular diversity revealed by AFLP markers in upland cotton genotypes. J Crop Improv 26:627–640

Bado S, Forster BP, Nielen S et al (2015) Plant mutation breeding: current progress and future assessment. Plant Breed Rev 39:23–88

Bagri S, Rajat S, Janeja HS (2020) Molecular breeding in maize improvement. Int J Curr Microbiol Appl Sci 9(12):722–729. http://www.ijcmas.com

Baird NA, Etter PD, Atwood TS et al (2008) Rapid SNP discovery and genetic mapping using sequenced RAD markers. PLoS One 3:e3376

Bajaj YP, Gill MS (1985) In vitro induction of genetic variability in cotton (*Gossypium* spp.). Theor Appl Genet 70:363–368

Bajaj YP, Gill MS (1989) Pollen-embryogenesis and chromosomal variation in anther culture of a diploid cotton (*Gossypium arboreum* L.). SABRAO J 21:57–63

Bajaj YP, Gill MS (1998) Interspecific hybridization in cotton through embryo rescue and culture. In: Cotton. Springer, Berlin, Heidelberg, pp 75–97

Barakat A, Gallois P, Raynal M et al (2000) The distribution of T-DNA in the genomes of transgenic Arabidopsis and rice. FEBS Lett 471(2–3):161–164

Bardak A (2012) Mapping of genes related to fiber and fuzz formation in cotton (*G. hirsutum* L.) genome and QTL analysis. PhD thesis, Department of Field Crops, p 141

Bardakci F (2001) Random amplified polymorphic DNA (RAPD) markers. Turk J Biol 25:185–196

Beasley C, Ting I, Feigen L (1971) Improvements in fiber yield and quality may come from test tube cotton. Calif Agric 25(10):6–8

Belhaj K, Chaparro-Garcia A et al (2013) Plant genome editing made easy: targeted mutagenesis in model and crop plants using the CRISPR/Cas system. Plant Methods 9(1):1–10

Blaise D, Kranthi KR et al (2021) High plant density can improve the productivity of rainfed Asiatic cotton (*Gossypium arboreum* L.). Arch Agron Soil Sci 67(5):607–619

Blaise D, Kranthi K, Saxena S et al (2022) Productivity and fibre attributes of absorbent Asiatic cotton (*Gossypium arboreum*) cultivars in rainfed central India. Indian J Agric Sci 92(3): 300–304

Bolek Y, El-Zik KM, Pepper AE et al (2005) Mapping of Verticillium wilt resistance genes in cotton. Plant Sci 168:1581–1590

Bolek Y, Hayat K, Bardak A et al (2016) Molecular breeding of cotton. In: Cotton research. InTech. https://doi.org/10.5772/64593

Bornet B, Muller C, Paulus F, Branchard ML (2002) Highly informative nature of inter simple sequence repeat (ISSR) sequences amplified using tri and tetra-nucleotide primers from DNA of cauliflower (*Brassica oleracea* var. *botrytis* L.). Genome 45:890–896

Brown N, Smith CW, Hague S et al (2015) Within boll yield characteristics and their correlation with fiber quality parameters following mutagenesis of upland cotton, TAM 94L-25. Crop Sci 55(4):1513–1523

Brubaker CL, Wendel JF (1994) Reevaluating the origin of domesticated cotton (*Gossypium hirsutum*; Malvaceae) using nuclear restriction fragment length polymorphisms (RFLPs). Am J Bot 81:1309–1326

Brumlop S, Finckh MR (2010) Applications and potentials of marker-assisted selection (MAS) in plant breeding: final report of the F+E project "Applications and Potentials of Smart Breeding" (FKZ 350 889 0020)-on behalf of the Federal Agency for Nature Conservation. Bundesamt für Naturschutz, Bonn, Germany. http://www.bfn.de

Butler NM, Baltes NJ, Voytas DF, Douches DS (2016) Geminivirus-mediated genome editing in potato (*Solanum tuberosum* L.) using sequence-specific nucleases. Front Plant Sci 7:1045

Buyyarapu R, Kantety RV, Yu JZ et al (2011) Development of new candidate gene and EST-based molecular markers for *Gossypium* species. Int J Plant Genomics 2011:1–9. https://doi.org/10.1155/2011/894598

Caeteno-Annolas G (1994) MAAP: a versatile and universal tool for genome analysis. Plant Mol Biol 25:1011–1026

Chandra M, Sreenivasan S (2011) Studies on improved *Gossypium arboreum* cotton: Part I-Fibre quality parameters. Indian J Fibre Text Res 36:24–34

Chen ZJ, Scheffler BE, Dennis E et al (2007) Toward sequencing cotton (*Gossypium*) genomes. Plant Physiol 145(4):1303–1310

Chen Y, Wang Y, Zhao T et al (2015) A new synthetic amphiploid (AADDAA) between *Gossypium hirsutum* and *G. arboreum* lays the foundation for transferring resistances to Verticillium and drought. PLoS One 10(6):e0128981. https://doi.org/10.1371/JOURNAL.PONE.0128981

Chen X, Lu X, Shu N et al (2017) Targeted mutagenesis in cotton (*Gossypium hirsutum* L.) using the CRISPR/Cas9 system. Sci Rep 7(1):44304

Chinchane VN, Baig KS (2018) Performance of long linted desi cotton (*Gossypium arboreum*) genotypes for yield and fibre quality parameters under rainfed condition. J Pharmacogn Phytochem 7(5):3409–3411

Collard BC, Mackill DJ (2008) Marker-assisted selection: an approach for precision plant breeding in the twenty-first century. Philos Trans R Soc Lond B Biol Sci 363:557–572

Collard BCY, Jahufer MZZ, Brouwer JB, Pang ECK (2005) An introduction to markers, quantitative trait loci (QTL) mapping and marker-assisted selection for crop improvement: the basic concepts. Euphytica 142:169–196

Cousins YL, Lyon BR, Llewellyn DJ (1991) Transformation of an Australian cotton cultivar: prospects for cotton improvement through genetic engineering. Funct Plant Biol 18(5):481–494

Database 1 (2016) CottonGeN [Internet]. Available from: https://www.cottongen.org. Accessed 8 Mar 2016

Database 2 (2016) Comparative Evolutionary Genomics of Cotton [Internet]. Available from: http://cottonevolution.info. Accessed 25 Mar 2016

Database 3 (2016) National Center for Biotechnology Information [Internet]. Available from: http://www.ncbi.nlm.nih.gov. Accessed 25 Mar 2016

Database 4 (2016) TropGENE Database [Internet]. Available from: http://tropgenedb.cirad.fr. Accessed 25 Mar 2016

Database 5 (2016) The Plant Genome Mapping Laboratory [Internet]. Available from: http://www.plantgenome.uga.edu/. Accessed 25 Mar 2016

Database 6 (2016) BACMan resources at Plant Genome Mapping Laboratory [Internet]. Available from: http://www.plantgenome.uga.edu. Accessed 25 Mar 2016

Davis DG, Dusbabek KE, Hoerauf RA (1974) In vitro culture of callus tissues and cell suspensions from okra (*Hibiscus esculentus* L.) and cotton (*Gossypium hirsutum* L.). In Vitro 9:395–398

Dhawan AK, Simwat GS, Sidhu AS (1991) Field reaction of some varieties of Asiatic cotton (*Gossypium arboreum* L.) to sucking and bollworm pests. J Res Punjab Agric Univ 28:57–62

Dongre AB, Kharbikar LL (2004) RAPD-based assessment of genetic diversity in cotton (*Gossypium hirsutum* L.) race stock accessions. Indian J Genet Plant Breed 64:94–96

Du X, Huang G, He S et al (2018) Resequencing of 243 diploid cotton accessions based on an updated A genome identifies the genetic basis of key agronomic traits. Nat Genet 50(6): 796–802. https://doi.org/10.1038/S41588-018-0116-X

Edde PA (2022) Arthropod pests of cotton (*Gossypium hirsutum* L.). In: Field crop arthropod pests of economic importance. Academic Press, London, pp 208–274

Edwards MD, Stuber CW, Wendel JF (1987) Molecular-marker-facilitated investigations of quantitative trait loci in maize. I. Numbers, genomic distribution and types of gene action. Genetics 116:113–125

Elshire RJ, Glaubitz JC, Sun Q et al (2011) A robust, simple genotyping-by-sequencing (GBS) approach for high diversity species. PLoS One 6:e19379

Engqvist MK, Rabe KS (2019) Applications of protein engineering and directed evolution in plant research. Plant Physiol 179(3):907–917

Erpelding JE (2022) Genetic analysis of the Asiatic cotton (*Gossypium arboreum*) petal spot phenotypes. Plant Breed 141(1):71–76. https://doi.org/10.1111/PBR.12984

Esmail RM, Zhang JF, Hamid AM (2008) Genetic diversity in elite cotton germplasm lines using field performance and RAPD markers. World J Agric Sci 4:369–375

Essenberg M, Bayles MB, Samad RA et al (2002) Four near-isogenic lines of cotton with different genes for bacterial blight resistance. Phytopathology 92:1323–1328

Farrell T, Johnson A (2005) Cotton pest management guide 2005/06. NSW Department of Primary Industries, p 99

Farzaneh T et al (2010) Cytogenetic and RAPD analysis of cotton cultivars and their F_1 progenies. Caryologia 63:73–81

Feng L, Chen Y, Xu M et al (2021) Genome-wide introgression and quantitative trait locus mapping reveals the potential of Asian cotton (*Gossypium arboreum*) in improving upland cotton (*Gossypium hirsutum*). Front Plant Sci 12:719371. https://doi.org/10.3389/FPLS.2021.719371

Finer JJ, Smith RH (1984) Initiation of callus and somatic embryos from explants of mature cotton (*Gossypium klotzschianum* Anderss). Plant Cell Rep 3:41–43

Firoozabady E, DeBoer DL (1993) Plant regeneration via somatic embryogenesis in many cultivars of cotton (*Gossypium hirsutum* L.). In Vitro Cell Dev Biol Plant 29:166–173

Francia E, Tacconi G, Crosatti C et al (2005) Marker-assisted selection in crop plants. Plant Cell Tiss Organ Cult 82:317–342

Fryxell PA (1969) A classification of Gossypium L. (Malvaceae). Taxon 1:585–591

Geng CD, Gong ZZ, Huang JQ, Zhang ZC (1995) Identification of difference between cotton cultivars (*G. hirsutum*) using the RAPD method. Jiangsu J Agric Sci 11:21–24

Gerstel DU (1953) Chromosomal translocations in interspecific hybrids of the genus *Gossypium*. Evolution 1:234–244

Gill MS, Bajaj YPS (1987) Hybridization between diploid (*Gossypium arboreum*) and tetraploid (*Gossypium hirsutum*) cotton through ovule culture. Euphytica 36:625–630

Gotmare V, Singh P, Mayee CD et al (2004) Genetic variability for seed oil content and seed index in some wild species and perennial races of cotton. Plant Breed 123(2):207–208. https://doi.org/10.1046/J.1439-0523.2003.00914.X

Guo WZ, Zhou BL, Yang LM et al (2006) Genetic diversity of landraces in *Gossypium arboreum* L. race sinense assessed with simple sequence repeat markers. J Integr Plant Biol 48(9): 1008–1017

Gupta M, Chyi YS, Romero-Severson J, Owen JL (1994) Amplification of DNA markers from evolutionarily diverse genomes using single primers of simple sequence repeats. Theor Appl Genet 89:998–1006

Gupta PK, Kumar J, Mir RR, Kumar A (2010) Marker-assisted selection as a component of conventional plant breeding. Plant Breed Rev 33:145–217

Hafeez MN, Khan MA, Sarwar B et al (2021) Mutant *Gossypium* universal stress protein-2 (GUSP-2) gene confers resistance to various abiotic stresses in E. coli BL-21 and CIM-496-*Gossypium hirsutum*. Sci Rep 11(1):20466. https://doi.org/10.1038/S41598-021-99900-X

Han J, Lopez-Arredondo D, Yu G et al (2022) Genome-wide chromatin accessibility analysis unveils open chromatin convergent evolution during polyploidization in cotton. Proc Natl Acad Sci 119(44):e2209743119. https://doi.org/10.1073/PNAS.2209743119

Hassan S, Ahmad A, Batool F et al (2021) Genetic modification of Gossypium arboreum universal stress protein (GUSP1) improves drought tolerance in transgenic cotton (*Gossypium hirsutum*). Physiol Mol Biol Plants 27(8):1779–1794

Hayashi K, Yoshida H, Ashikawa I (2004) Development of PCR-based SNP markers for rice blast resistance genes at the Piz locus. Theor Appl Genet 108:1212–1280

Helentjaris T, Slocum M, Wright S et al (1986) Construction of genetic linkage maps in maize and tomato using restriction fragment length polymorphisms. Theor Appl Genet 72(6):761–769

Hernández-Rodríguez CS, Van Vliet A, Bautsoens N et al (2008) Specific binding of *Bacillus thuringiensis* Cry2A insecticidal proteins to a common site in the midgut of Helicoverpa species. Appl Environ Microbiol 74(24):7654–7659

Hsu CL, Stewart JM (1976) Callus induction by (2-chloroethyl) phosphonic acid on cultured cotton ovules. Physiol Plant 36(2):150–153

Huang G, Wu Z, Percy RG et al (2020) Genome sequence of *Gossypium herbaceum* and genome updates of *Gossypium arboreum* and *Gossypium hirsutum* provide insights into cotton A-genome evolution. Nat Genet 52(5):516–524

Huang G, Huang JQ, Chen XY, Zhu YX (2021) Recent advances and future perspectives in cotton research. Annu Rev Plant Biol 17:437–462

Iqbal MJ, Reddy OUK, El-Zik KM, Pepper AE (2001) A genetic bottleneck in the evolution under domestication of upland cotton *Gossypium hirsutum* L. examined using DNA fingerprinting. Theor Appl Genet 103:547–554

Iqbal MA, Ammad A, Zafar Y (2015) Characterization of indigenous *Gossypium arboreum* L. genotypes for various fiber quality traits. Pak J Bot 47(6):2347–2354

Jain SM, Brar DS, Ahloowalia BS (2002) Molecular techniques in crop improvement, 2nd edn. Kluwer Academic Publishers, Dordrecht, p 772

Jia Y, Pan Z, He S et al (2018) Genetic diversity and population structure of *Gossypium arboreum* L. collected in China. J Cotton Res 1(1):1–8. https://doi.org/10.1186/S42397-018-0011-0

Jiang GL, Dong Y, Shi J, Ward RW (2007) QTL analysis of resistance to *Fusarium* head blight in the novel wheat germplasm CJ9306. II. Resistance to deoxylevinaeol accumulation and grain yield loss. Theor Appl Genet 115:1043–1052

Jindal A, Jindal R (2023) Textile raw materials. Abhishek Publications

John ME (1997) Cotton crop improvement through genetic engineering. Crit Rev Biotechnol 17(3): 185–208

Jones CJ, Edwaeds KJ, Castaglione S et al (1997) Reproducibility testing of RAPD, AFLP and SSR markers in plants by a network of European laboratories. Mol Breed 3:381–390

Joshi C, Nguyen H (1993) RAPD (random amplified polymorphic DNA) analysis based inter varietal genetic relationships among hexaploid wheats. Plant Sci 93:95–103

Joshi SP, Ranjekar PK, Gupta VS (1999) Molecular markers in plant genome analysis. Curr Sci 25: 230–240

Kalbande BB, Patil AS (2016) Plant tissue culture independent Agrobacterium tumefaciens mediated In-planta transformation strategy for upland cotton (*Gossypium hirsutum*). J Genet Eng Biotechnol 14(1):9–18

Kalpande HV, Borgaonkar SB, Surashe SM, Patil V (2020) Induced mutations in M_2 generation in cotton (*Gossypium arboreum* L.). J Soil Crop 30:326–331

Kantartzi SK, Stewart JM (2008) Association analysis of fibre traits in Gossypium arboreum accessions. Plant Breed 127(2):173–179

Kantartzi SK, Ulloa M, Sacks E, Stewart JM (2009) Assessing genetic diversity in *Gossypium arboreum* L. cultivars using genomic and EST-derived microsatellites. Genetica 136:141–147. https://doi.org/10.1007/s10709-008-9327-x

Kaur B, Andres R, Kuraparthy V (2016) Major leaf shape genes, laciniate in diploid cotton and okra in polyploid upland cotton, map to an orthologous genomic region. Crop Sci 56(3):1095–1105. https://doi.org/10.2135/CROPSCI2015.10.0627

Ke L, Jiang Q, Wang R et al (2022) Plant regeneration via somatic embryogenesis in diploid cultivated cotton (*Gossypium arboreum* L.). Plant Cell Tissue Organ Cult 148:177–188

Kebede H, Burow G, Dani RG, Allen RD (2007) A-genome cotton as a source of genetic variability for upland cotton (*Gossypium hirsutum*). Genet Resour Crop Evol 54:885–895. https://doi.org/10.1007/S10722-006-9157-6

Khadi BM, Santhy V, Yadav MS (2010) Cotton: an introduction. In: Cotton. Biotechnology in agriculture and forestry, vol 65. Springer, Berlin, Heidelberg. https://doi.org/10.1007/978-3-642-04796-1_1

Khan SA, Hussain D, Askari E et al (2000) Molecular phylogeny of *Gossypium* species by DNA fingerprinting. Theor Appl Genet 101:931–938

Khan MK, Hussain M, Ijaz ZA et al (2023) Future of mutation breeding in cotton: a step towards smart cotton. Cotton Innov 3:1–22

Khanam S, Sham A, Bennetzen JL, Aly MAM (2012) Analysis of molecular marker-based characterization and genetic variation in date palm (*Phoenix dactylifera* L.). Aust J Crop Sci 6:1236–1244

Khare AR (2021) Principles of spinning: fibres and blow room cotton processing in spinning. CRC Press

Khatoon S, Akmal M, Sarin NB, Khan JA (2014) Cytokinins improve shoot regeneration efficiency in two Indian cotton (*Gossypium hirsutum* cv. Narashima and *G. arboreum* cv. AKA-7) cultivars. Adv Appl Agric Sci 2(8):25–31

Khushboo S, Priyanka S, Surender KV (2014) Genetic improvement of *Gossypium arboreum* L. using molecular markers: status and development needs. Afr JAgric Res 9(29):2238–2249. https://doi.org/10.5897/AJAR2014.8682

King RC, Stansfield WD (1990) A dictionary of genetics, 4th edn. Oxford University Press, New York, p 188

Krawczak M (1999) Informativity assessment for biallelic single nucleotide polymorphisms. Electrophoresis 20:1676–1681

Krysan PJ, Young JC, Sussman MR (1999) T-DNA as an insertional mutagen in Arabidopsis. Plant Cell 11(12):2283–2290

Kulkarni VN (2002) Hirsutisation of *G. arboreum*cotton and genetic emendation of *G. hirsutum* for sucking pest resistance. PhD thesis submitted to University of Agricultural Sciences, Dharwad, India

Kumar LS (1999) DNA markers in plant improvement: an overview. Biotechnol Adv 17:143–182

Kumar P, Gupta VK, Misra AK et al (2009) Potential of molecular markers in plant biotechnology. Plant Omics 2:141–162

Kumar GP, Subiramani S, Govindarajan S et al (2015) Evaluation of different carbon sources for high-frequency callus culture with reduced phenolic secretion in cotton (*Gossypium hirsutum* L.) cv. SVPR-2. Biotechnol Rep 1:72–80

Kumpatla SP, Buyyarapu R, Abdurakhmonov IY, Mammadov JA (2012) Genomics-assisted plant breeding in the 21st century: technological advances and progress. In: Abdurakhmonov IY (ed) Plant breeding. InTech, Croatia

Kumria R, Sunnichan VG, Das DK et al (2003) High-frequency somatic embryo production and maturation into normal plants in cotton (*Gossypium hirsutum*) through metabolic stress. Plant Cell Rep 21:635–639

Lacape JM, Jacobs J, Arioli T et al (2009) A new interspecific, *Gossypium hirsutum* × *G. barbadense*, RIL population: towards a unified consensus linkage map of tetraploid cotton. Theor Appl Genet 119:281–292

Lan TH, Cook CG, Paterson AH (1999) Identification of a RAPD marker linked to male fertility restoration gene in cotton (*Gossypium hirsutum* L.). J Agric Genomics 1:1–5

Latif A, Rao AQ, Khan MA et al (2015) Herbicide-resistant cotton (*Gossypium hirsutum*) plants: an alternative way of manual weed removal. BMC Res Notes 8(1):1–8

Lee M (1995) DNA markers and plant breeding programs. Adv Agron 55:265–344

Lee SW, Thomas F, Ledig, Johnson DR (2002) Genetic variation at allozyme and RAPD markers in *Pinus longaeva* (Pinaceae) of the White Mountains, California. Am J Bot 89:566–577

Li R, Erpelding JE (2016) Genetic diversity analysis of *Gossypium arboreum* germplasm accessions using genotyping-by-sequencing. Genetica 144:535–545. https://doi.org/10.1007/S10709-016-9921-2

Li C, Wang X, Dong N et al (2013) QTL analysis for early-maturing traits in cotton using two upland cotton (*Gossypium hirsutum* L.) crosses. Breed Sci 63:154–163

Li F, Fan G, Wang K et al (2014) Genome sequence of the cultivated cotton *Gossypium arboreum*. Nat Genet 46(6):567–572

Li R, Erpelding JE, Stetina SR (2018) Genome-wide association study of *Gossypium arboreum* resistance to reniform nematode. BMC Genet 19:1–11. https://doi.org/10.1186/S12863-018-0662-3

Liu ZJ, Cordes JF (2004) DNA marker technologies and their applications in aquaculture genetics. Aquaculture 238:1–37. https://doi.org/10.1016/j.aquaculture.2004.05.027

Liu D, Guo X, Lin Z et al (2006) Genetic diversity of Asian cotton (*Gossypium arboreum* L.) in China evaluated by microsatellite analysis. Genet Resour Crop Evol 53(6):1145–1152

Lowery CC, Auld DL, Bechere E et al (2007) Use of chemical mutagenesis in improving upland cotton. In: World cotton research conference 4, Lubbock, Texas, USA, pp 10–14

Luo S, Wu K, Tian Y et al (2007) Cross-resistance studies of Cry1Ac-resistant strains of *Helicoverpa armigera* (Lepidoptera: Noctuidae) to Cry2Ab. J Econ Entomol 100(3):909–915

Ma XX, Zhou BL, Lü YH et al (2008) Simple sequence repeat genetic linkage maps of a-genome diploid cotton (*Gossypium arboreum*). J Integr Plant Biol 50(4):491–502. https://doi.org/10.1111/J.1744-7909.2008.00636.X

Malik W, Ashraf J, Iqbal MZ et al (2014) Molecular markers and cotton genetic improvement: current status and future prospects. Sci World J 2014:1–15. https://doi.org/10.1155/2014/607091

Maluszynski M, Ahloowalia BS, Sigurbjörnsson B (1995) Application of in vivo and in vitro mutation techniques for crop improvement. Euphytica 85(1):303–315

Manivannan A, Waghmare VN (2020) Assessment of genetic divergence in diploid cotton (*Gossypium arboreum* L.) germplasm using fibre quality traits. Plant Genet Resour 18(5): 351–358

Maqbool A, Zahur M, Irfan M et al (2008) Identification and expression of six drought-responsive transcripts through differential display in *desi* cotton (*Gossypium arboreum*). Mol Biol 42:492–498

Maqbool A, Abbas W, Rao AQ et al (2010) *Gossypium arboreum* GHSP26 enhances drought tolerance in *Gossypium hirsutum*. Biotechnol Prog 26:21–25

Maughan PJ, Seghai MA, Maroof GR et al (1996) Amplified fragment length polymorphism in soybean: species diversity, inheritance and near-isogenic line analysis. Theor Appl Genet 93: 392–401. https://doi.org/10.1007/BF00223181

McClintock B (1956) Controlling elements and the gene. Cold Spring Harb Symp Quant Biol 21: 197–216. Cold Spring Harbor Laboratory Press

Meena RA, Monga D, Venugopalan MV et al (2016) Screening of desi cotton (*G. arboreum*) suitable for surgical properties. J Sci Ind Res 75:570–573

Mehetre SS, Shinde GC, Pardeshi SU (2004) Status and strategies of host plant resistance for biotic stress in cotton. In: Khadi BM, Kategeri IS, Patil SS et al (eds) Proceedings of international symposium on strategies for sustainable cotton production-a global vision: 1 Crop improvement. Universty of Agricultural Sciences, Dharwad, pp 31–47

Mohan M, Nair S, Bhagwat A et al (2012) Genome mapping, molecular markers and marker-assisted selection in crop plants. Mol Breed 3:87–103

Mushtaq M, Sakina A, Wani SH et al (2019) Harnessing genome editing techniques to engineer disease resistance in plants. Front Plant Sci 10:550

Mushtaq R, Shahzad K, Shah ZH et al (2020) Isolation of biotic stress resistance genes from cotton (*Gossypium arboreum*) and their analysis in model plant tobacco (*Nicotiana tabacum*) for resistance against cotton leaf curl disease complex. J Virol Methods 276:113760

Nandeshwar SB, Moghe S, Chakrabarty PK et al (2009) Agrobacterium-mediated transformation of cry 1Ac gene into shoot-tip meristem of diploid cotton *Gossypium arboreum* cv. RG8 and regeneration of transgenic plants. Plant Mol Biol Report 27:549–557

Nawab Y, Shaker K (2023) Textile engineering: an introduction. Walter de Gruyter GmbH & Co KG

Nibouche S, Brevault T, Klassou C et al (2008) Assessment of the resistance of cotton germplasm (*Gossypium* spp.) to aphids (Homoptera, Aphididae) and leafhoppers (Homoptera: Cicadellidae, Typhlocybinae): methodology and genetic variability. Plant Breed 127:376–382. https://doi.org/10.1111/j.1439-0523.2008.01499.x

Noormohammadi Z, Shojaei-Jeshvaghani F, Sheidai M, Farahani F et al (2011) ISSR and RAPD analysis of genetic diversity in Mehr cotton cultivar and its crossing progenies. Afr J Biotechnol 10:11839–11847

Noormohammadi Z, Farahani Y, Sheidai HAM et al (2013) Genetic diversity analysis in Opal cotton hybrids based on SSR, ISSR, and RAPD markers. Genet Mol Res 12:256–269

Oladosu Y, Rafii MY, Abdullah N et al (2016) Principle and application of plant mutagenesis in crop improvement: a review. Biotechnol Biotechnol Equip 30(1):1–16

Ortiz R (2004) Biotechnology with horticultural and agronomic crops in Africa. Acta Hortic 642: 4356

Park YH, Alabady MS, Ulloa M et al (2005) Genetic mapping of new cotton fiber loci using EST-derived microsatellites in an interspecific recombinant inbred line cotton population. Mol Gen Genomics 274:428–441

Parkhyia S, Gohel K, Mehta DDR (2014) Genetic diversity analysis of cotton (*G. hirsutum* L.) genotypes using ISSR markers. Int J Appl Pure Sci Agric 1:7–19

Paterson AH (1996) Making genetic maps. In: Genome mapping in plants, biotechnology intelligence unit. Academic Press, San Diego, pp 23–39

Paterson AH (1997) Molecular dissection of complex traits. CRC Press, Taylor & Francis Group
Paterson AH, Wendel JF, Gundlach H et al (2012) Repeated polyploidization of *Gossypium* genomes and the evolution of spinnable cotton fibres. Nature 492:423–427
Pillay M, Myers GO (1999) Genetic diversity in cotton assessed by variation in ribosomal RNA genes and AFLP markers. Crop Sci 39(6):1881–1886
Preetha S, Raveendren TS (2008) Molecular marker technology in cotton. Biotechnol Mol Biol Rev 3:32–45
Puspito AN, Rao AQ, Hafeez MN et al (2015) Transformation and evaluation of Cry1Ac + Cry2A and GTGene in *Gossypium hirsutum* L. Front Plant Sci 12:943
Qureshi SN, Saha S, Kantety RV et al (2004) Molecular biology and physiology: EST-SSR: a new class of genetic markers in cotton. J Cotton Sci 8:112–123
Rafalski JA (1997) Randomly amplified polymorphic DNA (RAPD) analysis. In: Anolles GC, Gesshoff PM (eds) DNA markers: protocols, applications and overviews. Wiley-Liss, Inc, p 364
Rahman M, Hussain D, Zafar Y (2002) Estimation of genetic divergence among elite cotton (*Gossypium hirsutum* L.) cultivars/genotypes by DNA fingerprinting technology. Crop Sci 42: 2137–2144
Rahman M, Asif M, Ullah I et al (2005) Overview of cotton genomic studies in Pakistan. In: Plant & animal genome conference XIII, San Diego, CA, USA
Raja AS, Arputharaj A, Saxena S, Patil PG (2016) A one bath chemo-enzymatic process for preparation of absorbent cotton. Perspect Sci 8:254–256
Rajasekaran K, Grula JW, Hudspeth RL et al (1996) Herbicide-resistant Acala and Coker cottons transformed with a native gene encoding mutant forms of acetohydroxyacid synthase. Mol Breed 2:307–319
Rani A, Bhojwani SS (1976) Establishment of tissue cultures of cotton. Plant Sci Lett 7:163–169
Raut RV, Patil VM, Rajput JC (2019) A rapid and simple method for in-vitro plant regeneration from petiolar region of diploid *G. arboreum* cotton cultivar (cv. Ambika). J Biotechnol Biochem 5(5):49–55
Reddy OUK, Pepper AE, Abdurakhmonov I et al (2001) New dinucleotide and trinucleotide microsatellite marker resources for cotton genome research. J Cotton Sci 5:103–113
Reddy MP, Sarla N, Siddiq EA (2002) Inter simple sequence repeat (ISSR) polymorphism and its application in plant breeding. Euphytica 128:9–17
Reinisch AJ, Dong JM, Brubaker CL et al (1994) A detailed RFLP map of cotton, *Gossypium hirsutum* × *Gossypium barbadense*: chromosome organization and evolution in a disomic polyploid genome. Genetics 138:829–847
Renny-Byfield S, Page JT, Udall JA et al (2016) Independent domestication of two old world cotton species. Genome Biol Evol 8(6):1940–1947. https://doi.org/10.1093/GBE/EVW129
Russell JR, Fuller JD, Macaulay M et al (1997) Direct comparison of levels of genetic variation among barley accessions detected by RFLPs, AFLPs, SSRs and RAPDs. Theor Appl Genet 95: 714–722
Sagar MR, Kumar S, Patidar D, Sakure AA (2022) Morphological, physico-biochemical and marker-based diversity of *desi* cotton (*Gossypium herbaceum* L.) germplasm. J King Saud Univ Sci 34(8):102336
Saha S, Karaca M, Jenkins JN et al (2003) Simple sequence repeats as useful resources to study transcribed genes of cotton. Euphytica 130:355–364
Sakhanokho HF, Zipf A, Rajasekaran K et al (2004) Somatic embryo initiation and germination in diploid cotton (*Gossypium arboreum* L.). In Vitro Cell Dev Biol Plant 40:177–181
Sangannavar P (2008) Genetic transformation studies in cotton (*Gossypium arboreum* and *G. barbadense*). Accession no: Th9768. MSc thesis submitted to University of Agricultural Sciences, Dharwad AC, Dharwad, Karnataka, India
Sanghera GS, Gill MS, Wani SH, Gosal SS (2011) Factors affecting transformation efficiency of shoot apices of *Gossypium arboreum* and *Gossypium hirsutum* cultivars with *Agrobacterium tumifaciens*. Elixir Bio Tech 37:3934–3940. https://doi.org/10.3906/bot-1209-47

Santos-Sánchez NF, Salas-Coronado R, Hernández-Carlos B, Villanueva-Cañongo C (2019) Shikimic acid pathway in biosynthesis of phenolic compounds. In: Plant physiological aspects of phenolic compounds, vol 1. IntechOpen, pp 1–15

Sattar MN, Iqbal Z, Dangol SD, Bakhsh A (2019) CRISPR/Cas9: a new genome editing tool to accelerate cotton (*Gossypium* spp.) breeding. In: Advances in plant breeding strategies: industrial and food crops, vol 6. Springer International Publishing, pp 61–84. https://doi.org/10.1007/978-3-030-23265-8_3

Satyavathi VV, Prasad V, Lakshmi BG, Sita GL (2002) High-efficiency transformation protocol for three Indian cotton varieties via *Agrobacterium tumefaciens*. Plant Sci 162(2):215–223

Schenk RU, Hildebrandt AC (1972) Medium and techniques for induction and growth of monocotyledonous and dicotyledonous plant cell cultures. Can J Bot 50:199–204

Schnepf E, Crickmore N, Van Rie J et al (1998) *Bacillus thuringiensis* and its pesticidal crystal proteins. Microbiol Mol Biol Rev 62(3):775–806

Schulmann AH (2007) Molecular markers to assess genetic diversity. Euphytica 158:313–321

Semagn K, Bjornstad A, Ndjiondjop MN (2006) An overview of molecular marker methods for plants. Afr J Biotechnol 5:2540–2256

Sethi K, Siwach P, Verma SK (2015) Assessing genetic diversity among six populations of *Gossypium arboreum* L. using microsatellites markers. Physiol Mol Biol Plants 21:531–539. https://doi.org/10.1007/S12298-015-0326-Y

Shoemaker RC, Couche LJ, Galbraith DW (1986) Characterization of somatic embryogenesis and plant regeneration in cotton (*Gossypium hirsutum* L.). Plant Cell Rep 5:178–181

Singh RP, Prasad PV, Sunita K et al (2007) Influence of high temperature and breeding for heat tolerance in cotton: a review. Adv Agron 93:313–385

Singh PK, Sharma H, Srivastava N, Bhagyawant SS (2014) Analysis of genetic diversity among wild and cultivated chickpea genotypes employing ISSR and RAPD markers. Am J Plant Sci 5: 676–682. https://doi.org/10.4236/ajps.2014.55082

Smith RH, Price HJ, Thaxton JB (1977) Defined conditions for the initiation on growth of cotton callus in vitro I. *Gossypium arboreum*. In Vitro 13:329–334

Solanki VH, Khandelwal V, Patel DH, Mahatma M (2011) Short communication *Agrobacterium*-mediated In-planta transformation of *G. hirsutum* CV G COT 10. Indian J Plant Physiol 16:303–308

Stetina S, Erpelding J (2016) Gossypium arboreum accessions resistant to Rotylenchulus reniformis identified. Phytopathology 106(12):38. https://doi.org/10.21307/JOFNEM-2017-030

Stewart JM, Hsu CL (1977) In-ovulo embryo culture and seedling development of cotton (*Gossypium hirsutum* L.). Planta 137:113–117

Suarez LA, Apan A, Werth J (2016) Hyperspectral sensing to detect the impact of herbicide drift on cotton growth and yield. ISPRS J Photogramm Remote Sens 120:65–76

Sufyan Tahir M, Latif A, Bashir S et al (2021) Transformation and evaluation of Broad-Spectrum insect and weedicide resistant genes in *Gossypium arboreum* (Desi Cotton). GM Crops Food 12(1):292–302

Sultanul A (1979) Gossypium arboreum. In: Ali SI, Qaiser M (eds) Flora of Pakistan, vol 130. University of Karachi & Missouri Botanical Garden, St. Louis, p 30

Sun Y, Zhang X, Jin S et al (2003) Somatic embryogenesis and plant regeneration in wild cotton (*Gossypium klotzschianum*). Plant Cell Tissue Organ Cult 75(3):247–253

Sun Y, Zhang X, Huang C et al (2006) Somatic embryogenesis and plant regeneration from different wild diploid cotton (*Gossypium*) species. Plant Cell Rep 25:289–296

Sun FD, Zhang JH, Wang SF et al (2012) QTL mapping for fiber quality traits across multiple generations and environments in upland cotton. Mol Breed 30:569–582

Sunilkumar G, Rathore KS (2001) Transgenic cotton: factors influencing *Agrobacterium*-mediated transformation and regeneration. Mol Breed 8:37–52

Tahir MS, Khan NUI, Sajid Ur R (2011) Development of an interspecific hybrid (Triploid) by crossing *Gossypium hirsutum*and *G. arboreum*. Cytologia 76:193–199. https://doi.org/10.1508/cytologia.76.193

Tanksley SD, Young ND, Paterson AH, Bonierbale MW (1989) RFLP mapping in plant breeding: new tools for an old science. Biotechnology 7:257–264

Thakur T, Dadsena R, Nandeshwar SB, Kumar A (2010) Callus induction from hypocotyls and cotyledonary leaves explants on different media combinations in *G. Arboreum* CV PA-183 & PA-402. Adv Pharmacol Toxicol 11(1):145

Tian ZC, Shen JW, Jun Z et al (2010) Pistil drip following pollination: a simple in planta Agrobacterium-mediated transformation in cotton. Biotechnol Lett 32:547–555

Tian J, Deng Z, Zhang K et al (2015) Genetic analyses of wheat and molecular marker-assisted breeding, vol 1. Springer Netherlands, Dordrecht. https://doi.org/10.1007/978-94-017-7390-4_2

Trolinder NL, Goodin JR (1987) Somatic embryogenesis and plant regeneration in cotton (*Gossypium hirsutum* L.). Plant Cell Rep 6:231–234

Trolinder NL, Xhixian C (1989) Genotype specificity of the somatic embryogenesis response in cotton. Plant Cell Rep 8:133–136

Tsai CJ, Saunders JW (1999) Evaluation of sole nitrogen sources for shoot and leaf disc cultures of sugarbeet. Plant Cell Tissue Organ Cult 59:47–56

Tsumura Y, Ohba K, Strauss SH (1996) Diversity and inheritance of inter-simple sequence repeat polymorphisms in Douglas fir (*Pseudotsuga menziesii*) and sugi (*Cryptomeria japonica*). Theor Appl Genet 92:40–45

Ulloa M, Meredith JWR, Shappley ZW, Kahler AL (2002) RFLP genetic linkage maps from four $F_{2.3}$ populations and a join map of *Gossypium hirsutum* L. Theor Appl Genet 104:200–208

Ulloa M, Saha S, Jenkins JN et al (2005) Chromosomal assignment of RFLP linkage groups harboring important QTLs on an intraspecific cotton (*Gossypium hirsutum* L.) joinmap. J Hered 96:132–144

Uthirapathy P, Marimuthu M, Venkatasamy B et al (2023) Categories of resistance in cotton genotypes, *Gossypium* spp. against cotton-melon aphid, *Aphis gossypii* (Hemiptera: Aphididae). J Econ Entomol 116(4):1411–1422. https://doi.org/10.1093/JEE/TOAD136

Van Esbroeck GA, Bowman DT, Calhoun DS, May OL (1998) Changes in the genetic diversity of cotton in the U.S. from 1970 to 1995. Crop Sci 38:33–37

Verma SK, Tuteja OP, Ahuja SL et al (2005) Identification of potential combiners and combinations from eco-geographical diverse genotypes of Asiatic cotton (*G. arboreum* L.). Cotton Res J (J Indian Soc Cotton Improv) 30(1):39–46

Verma SK, Tuteja OP, Ahuja S (2017) Combining ability estimates for seed cotton yield and quality characters of parents and crosses based on genetic male sterility in Asiatic cotton (*Gossypium arboreum* L.). Electron J Plant Breed 8(4):1046–1052

Verma SK, Tuteja OP, Monga D, Waghmare VN (2020) GMS-CISG 20 – a new genetic male sterile line of diploid cotton (*Gossypium arboreum* L.) with marker-trait. J Cotton Res Dev 34(1):46–49

Verma SK, Paul D, Goyal S et al (2022) Prediction of gene action, heterosis and combining ability to identify superior GMS-based hybrids in Asiatic cotton (*G. arboreum* L.). Int J Plant Soil Sci 34(19):161–174

Vogel B, Aken JV (2009) Smart breeding: marker-assisted selection: a non-invasive biotechnology alternative to genetic engineering of plant varieties. Greenpeace

Voss-Fels K, Snowdon RJ (2016) Understanding and utilizing crop genome diversity via high-resolution genotyping. Plant Biotechnol J 14(4):1086–1094

Wang K, Wang Z, Li F et al (2012) The draft genome of a diploid cotton *Gossypium raimondii*. Nat Genet 44(10):1098–1103

Wang S, Chen J, Zhang W et al (2015) Sequence-based ultra-dense genetic and physical maps reveal structural variations of allopolyploid cotton genomes. Genome Biol 16:108

Wang K, Huang G, Zhu Y (2016) Transposable elements play an important role during cotton genome evolution and fiber cell development. Sci China Life Sci 59:112–121

Wang K, Wang D, Zheng X et al (2019) Multi-strategic RNA-seq analysis reveals a high-resolution transcriptional landscape in cotton. Nat Commun 10(1):4714

Wang D, Fan W, Guo X et al (2020) MaGenDB: a functional genomics hub for Malvaceae plants. Nucleic Acids Res 48(D1):1076–1084

Watt G (1907) The wild and cultivated cotton plants of the world: a revision of the genus *Gossypium*, framed primarily with the object of aiding planters and investigators who may contemplate the systematic improvement of the cotton staple. Longmans, Green, and Company

Weising K, Nybom H, Wolff K, Kahl G (2005) DNA fingerprinting in plants: principles, methods and applications, 2nd edn. CRC Press, Taylor & Francis Group, Boca Raton

Wendel JF (1989) New World tetraploid cottons contain Old World cytoplasm. Proc Natl Acad Sci 86(11):4132–4136

Wendel JF, Cronn RC (2003) Polyploidy and the evolutionary history of cotton. Adv Agron 78: 139–186. https://doi.org/10.1016/S0065-2113(02)78004-8

Wendel JF, Grover CE (2015) Taxonomy and evolution of the cotton genus, *Gossypium*. Cotton 57: 25–44

Wendel JF, Olson PD, Stewart JM (1989) Genetic diversity, introgression, and independent domestication of old world cultivated cottons. Am J Bot 76(12):1795–1806. https://doi.org/10.1002/J.1537-2197.1989.TB15169.X

Wendel JF, Brubaker CL, Percival AE (1992) Genetic diversity in *Gossypium hirsutum* and the origin of upland cotton. Am J Bot 79:1291–1310. https://doi.org/10.2307/2445058

Wendel JF, Brubaker C, Alvarez I et al (2009) Evolution and natural history of the cotton genus. In: Paterson AH (ed) Genetics and genomics of cotton. Plant genetics and genomics: crops and models, vol 3. Springer, New York. https://doi.org/10.1007/978-0-387-70810-2_1

Wheeler TA, Gannaway JR, Keating K (1999) Identification of resistance to *Thielaviopsis basicola* in diploid cotton. Plant Dis 83:831–833. https://doi.org/10.1094/PDIS.1999.83.9.831

Winter P, Kahl G (1995) Molecular marker technologies for plant improvement. World J Microbiol Biotechnol 11:438–448

Wolff K, Schoen ED, Peters-Van RJ (1993) Optimizing the generation of random amplified polymorphic DNA in chrysanthemum. Theor Appl Genet 86:1033–1037

Wright RJ, Thaxton PM, El-Zik KM, Paterson AH (1998) D-subgenome bias of Xcm resistance genes in tetraploid *Gossypium* (cotton) suggests that polyploid formation has created novel avenues for evolution. Genetics 149:1987–1996

Wu J, Zhang X, Nie Y et al (2004) Factors affecting somatic embryogenesis and plant regeneration from a range of recalcitrant genotypes of Chinese cottons (*Gossypium hirsutum* L.). In Vitro Cell Dev Biol Plant 40:371–375

Wu J, Jenkins JN, McCarty JC, Zhong M (2007) AFLP marker associations with agronomic and fiber traits in cotton. Euphytica 153:153–163

Xiao J, Wu K, Fang DD et al (2009) New SSR markers for use in cotton (*Gossypium* spp.) improvement. J Cotton Sci 13:75–157

Yan SF, Zhang Q, Wang JE et al (2010) Somatic embryogenesis and plant regeneration in two wild cotton species belong to G genome. In Vitro Cell Dev Biol Plant 46:298–305

Yang Z, Gao C, Zhang Y et al (2023) Recent progression and future perspectives in cotton genomic breeding. J Integr Plant Biol 65(2):548–569. https://doi.org/10.1111/JIPB.13388

Yasin S, Yasmin A (2018) Optimisation of in vitro propagation of *Gossypium arboreum* L. Pure Appl Biol 7(2):419–426

Yasmin T, Tabbasam N, Ullah I et al (2008) Studying the extent of genetic diversity among *Gossypium arboreum* L. genotypes/cultivars using DNA fingerprinting. Genet Resour Crop Evol 55(3):331–339. https://doi.org/10.1007/S10722-007-9238-1

Yuan D, Grover CE, Hu G et al (2021) Parallel and intertwining threads of domestication in allopolyploid cotton. Adv Sci 8(10):2003634. https://doi.org/10.1002/ADVS.202003634

Zahra N, Shojaei-Jesvaghani F, Sheidai M et al (2011) Inter simple sequence repeats (ISSR) and random amplified polymorphic DNA (RAPD) analyses of genetic diversity in Mehr cotton cultivar and its crossing progenies. Afr J Biotechnol 10:11839–11847

Zehr UB (ed) (2012) Cotton, biotechnology in agriculture and forestry. Springer, Heidelberg

Zhang F, Batley J (2020) Exploring the application of wild species for crop improvement in a changing climate. Curr Opin Plant Biol 56:218–222

Zhang X, Yao M, Shi P, Dong X (1998) Identification of resistance to Fusarium wilt in upland cotton (*Gossypium hirsutum* L.). China Cotton 7:8–9

Zhao Z, Liu Z, Zhou Y et al (2022) Creation of cotton mutant library based on linear electron accelerator radiation mutation. Biochem Biophys Rep 30:101228

Zhu YL, Song QJ, Hyten DL et al (2003) Single-nucleotide polymorphism in soybean. Genetics 163:1123–1134

Chapter 18
Breeding Approaches for Genetic Improvement in Diploid A Genome Levant Cotton (*Gossypium herbaceum* L.)

Aysen Yumurtaci

Abstract Levant cotton (*Gossypium herbaceum* L.), commonly known as Arabian cotton, carries the ancestral A-genome of today's polyploid cotton and is considered a valuable cash crop. Cotton fibers make up 70% of raw materials in the textile industry, while the seeds provide oil and the stalks can be used for fuel. Hybridization between diploid and tetraploid Gossypium species has resulted in the development of allopolyploid commercial cotton varieties. *Gossypium herbaceum* has a haploid chromosome number of n = 13 and a diploid genome size of 1.7 Mbp, which varies among different accessions based on geographic origin. Like other crops, cotton production is impacted by weather conditions, pests, and soil water potential. To address these challenges, recent efforts have focused on breeding cotton for resistant genotypes with high fiber or oil content using advanced technology. *Gossypium herbaceum* represents only 5% of commercially produced diploid cotton varieties globally, limiting the amount of research conducted on this species. However, studies have the potential to yield significant results. This chapter aims to discuss agricultural production, innovative approaches for improving cotton species, and research conducted using biotechnological methods specifically for *Gossypium herbaceum* under various headings.

Keywords Cotton · Biotechnology · Crop · Fiber · Levant cotton (*Gossypium herbaceum* L.)

18.1 Introduction

Cotton (*Gossypium* spp.), a member of *Malvaceae* family is considered an important subtropical crop for its use in the textile industry and provides oil extraction from its seeds and fibers. The genus *Gossypium* has 46 diploids and 5 allotetraploid species that correspond to A-G and K genomes. Among them, independently domesticated

---

A. Yumurtaci (✉)
Marmara University, Institute for Graduate Studies in Pure and Applied Sciences, Biology Department, Goztepe Campus, Istanbul, Turkey

K. F. M. Salem et al. (eds.), *Breeding and Biotechnology of Leaf, Fruit, and Seed Fiber Crops*, Advances in Plant Breeding Strategies 11,
https://doi.org/10.1007/978-3-032-00407-9_18

four species: *G. hirsutum, G. barbadense*, including the two old-world, diploid species *Gossypium herbaceum and G. arboreum* are used in textile production (Hu et al. 2021: Wu et al. 2017). From a botanical perspective, *Gossypium herbaceum* is subdivided into 2 subspecies: the wild form of subsp. *africanum*, and subsp. *herbaceum*. The wild form of *Gossypium herbaceum* subsp. *africanum* is endemic to Southern Africa while *Gossypium herbaceum* is commonly used as a commercial member of the *Gossypium* genus and originated from both Asia and Africa. The genus name comes from an Arabic word which is translated as "soft substance" and *herbaceum* is the Latin description carrying the meaning of herbaceous (https://www.itis.gov/servlet/SingleRpt/SingleRpt?search_topic=TSN&search_value=506096#null). The cotton seed oil extracted from *Gossypium herbaceum* is used for cooking and needs to be refined to remove the specific toxin called "Gossypol" which is harmful to human health. In addition to the food, feed and oil extraction, in Ethiopia, *Gossypium herbaceum* is used for treating fungal infections and as a vaccine for snake bites (Schmidt et al. 2002).

Kingdom Plantae – plantes, Planta, Vegetal, plants
Subkingdom Viridiplantae – green plants
Infrakingdom Streptophyta – land plants
Superdivision Embryophyta
Division Tracheophyta – vascular plants, tracheophytes
Subdivision Spermatophytina – spermatophytes, seed plants, phanérogames Class Magnoliopsida
Superorder Rosanae
Order Malvales
Family Malvaceae – mallows, mauves
Genus Gossypium L. – cotton
Species *Gossypium herbaceum* L. – Levant cotton

All over the world, cotton is produced in 32 million hectares without the need for additional arable land. Between the years 1960–1980, alongside natural cotton, cellulosic cotton also began to gain popularity. Since the 2000s, cellulosic fiber production has tripled. However, fluctuations in economic indicators have negatively impacted cotton fiber production. In the early 1990s, the introduction of commercial "biotech cotton" varieties to the agricultural sector led to an increase in cotton production (www.fao.org/3/cb3269en/cb3269en.pdf).

On a global scale, nearly 25 million tons of cotton are produced annually, with the cotton industry estimated to be worth almost $500 billion in annual utilization. The top ten cotton-producing countries are India, China, the United States, Pakistan, Brazil, Australia, Uzbekistan, Turkey, Turkmenistan, and Burkina Faso (Khan et al. 2020). The first five countries alone account for 75% of global production. Figure 18.1 provides an updated country ranking for cotton production in the 2022–2023 season.

In the past decade, some cotton-producing countries have turned to genetically modified cotton, such as Bt cotton, to address issues faced in production. Bt cotton involves the transfer of a gene from the soil bacterium *Bacillus thuringiensis* into the

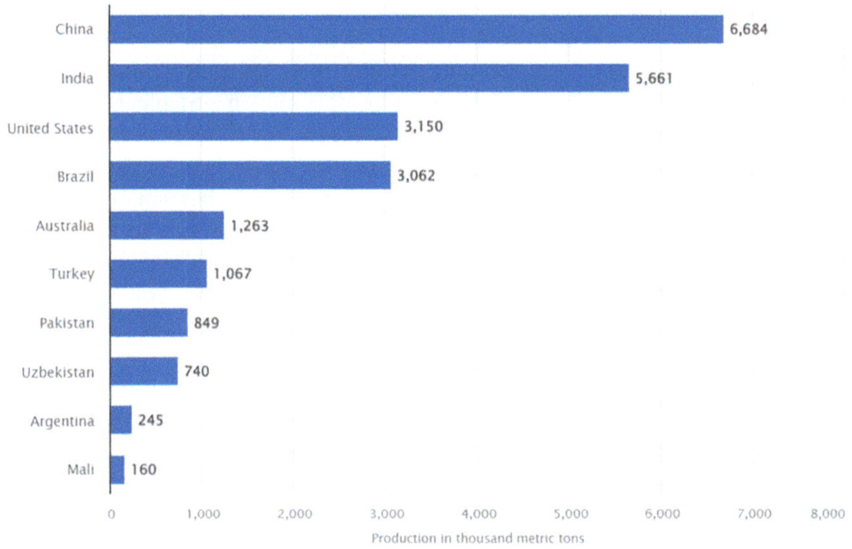

Fig. 18.1 Leading cotton-producing countries in 2022/2023. (Source: statis.com/statistics/263055/cotton–production-worldwide-by-top-countries http://statis.com/statistics/263055/cotton%E2%80%93production-worldwide-by-top-countries)

cotton genome. The Bt gene produces a toxin that is harmful to cotton bollworms, a significant insect pest. Commercial use of Bt cotton began in the mid-1990s in the United States and China.

In 2008, cotton producers in Burkina Faso encountered issues with poor quality end products, specifically with a short staple length, prompting adjustments to their future strategies in line with global market trends (fao.org).

The diploid form of *G. herbaceum* L. (A1A1), (2n = 2x = 26) known as "short-staple cotton", is accepted as the ancestral maternal parent of polyploid cotton (Huang et al. 2020; Khadi et al. 1970). It is grown perennially in the semi-arid regions of Arabia and sub-Saharan Africa and is called Levant cotton in some countries. In 1995, archaeologists from the Institute of Archaeology of Xinjiang-China found 120 well-preserved ancient tombs. From these tombs, almost one hundred grams of seed cotton with boll shells were obtained and fibers were analyzed using radiocarbon dating. The results identified the fibers as belonging to the *Gossypium herbaceum* cotton species. This example may provide evidence for the geographical distribution of *G. herbaceum* from China to Africa and Arabia (Cao et al. 2009).

According to morphological characterization, the main stem is a taproot in the soil. The taproot has a tetrachic structure with a distinct or faint bulge. Lateral roots are arranged in irregular rows on the taproot. Although the exact number of rows is four, they are often folded and irregular, with more than three rows not being well known. The main root either descends vertically into the ground or continues in a zigzag pattern. Lateral roots spread horizontally into the ground, with many of them diving deep into the taproot to participate in its functions. In settled soils, cotton roots cannot extend very deep, resulting in most roots being around 20 cm deep. Under such conditions, the average depth is only 50 cm. Staple length is an important classification factor during quality selection. For *G. herbaceum*, the fibers are short, and the plant features hairy, broadly ovate, 3 to 7-lobed, maple-like leaves with heart-shaped bases and long red-tinged petioles. Cotton fiber makes up approximately 90% of the economic structure of the cotton crop and is a structure called trichomes, formed by the differentiation and elongation of a single cell in the epidermal layer of the seed (Walford et al. 2011). Fiber formation begins with flowering (anthesis) and can grow up to approximately 30 mm within 3 weeks (Tokumoto et al. 2002).

In general, cotton seed composition is defined as the integument layers of the eggs located on the placentas in the compartments inside the cocoon that form the seed coat. As the seed grows, these integuments also expand. Cotton seed has an irregular pear shape, with the ends (micropile part) being very hard and pointed. The weight of 1000 grains is 85–120 g. There are two types of hairs on cotton seeds: short ones called linter and long ones called fiber (lint). In some species, these seedlings are even naked, with tassels of feathers on their pointed parts. Feathering may sometimes be sparse or partial on the seedlings. The wide (chalazal) part of cotton is made of soft and spongy tissue, allowing water to enter the seed for germination. In the cotton seed, the endosperm is pushed aside and remains as a trace, while the embryo fills the seed. The chemical composition of cottonseed (as dry matter) consists of an average of 39% carbohydrates, 23% fat, 22% protein, 12% crude fiber, and 3.5% ash.

Environmental changes, especially temperature fluctuations during flowering, may limit fiber formation and delay the onset of elongation (Ahmed et al. 2018). Although diploid cotton, including *G. herbaceum* (AA) and its close sister *G. arboreum* (DD), have poor fiber characteristics, the "A" genome carries traits of disease and insect resistance, high fiber strength, and plasticity more than other tetraploid cultivated cotton. From this perspective, the tendency towards diploid cotton in the textile sector is increasing due to its production of higher-quality products. Most genomic information, including molecular markers and transcriptomic data, is concentrated on tetraploid cotton species (Malik et al. 2014).

Under environmental stresses, there is a need to identify new cotton genotypes using next-generation biotechnology approaches. Comprehensive sequencing may provide multiple genome data and help select traits at the nucleotide level. With long-read sequencing platforms opening new doors for genomes, Huang et al. (2020) released the first assembly of the Gossypium herbaceum genome (variety Africanum 'Mutema', A1-0076) using the PacBio system and generated almost 225 Gb PacBio single-molecule real-time (SMRT) long reads. In addition to genomic studies, tissue culture is also used to investigate the effects of plant growth regulators on cotton. While there are many studies related to upland cotton species in the literature, studies related to *Gossypium herbaceum* are rare. Chowdhury et al. (2011) tested *Gossypium herbaceum* L. cotyledons for their responses to cytokinins and reported callus formation at high cytokinin concentrations.

In this chapter, biotechnological studies on cotton have been presented under new headings to avoid repetitive information, and the importance of diploid cotton species like *Gossypium herbaceum* has been summarized.

18.2 Current Cultivation Practices and Challenges: Artificial Intelligence and Global Cotton Production

Agriculture has been constantly under the pressure of both subsoil formation and environmental changes, causing stress for plant growth (Ishfaq et al. 2022; Sytar and Kumar 2023). With the development of technology, weather estimations have eased the burden on farmers and provided them with alternative solutions and future reflections during agricultural production. For a farmer, precise yield estimation is an important criterion, and it has been observed that harvesting knowledge as well as harvesting products have positive effects on agricultural production. Artificial intelligence can serve as critical node solvers for the optimization of big data (Stephen et al. 2023).

In addition to production, agricultural storage conditions of cotton are important factors affecting the economic value of this crop. External and internal variables such as temperature, humidity, air quality, and spontaneous combustion during cotton storage require more accurate monitoring and managing systems. To provide precise predictions, the Internet of Things (IoT) and artificial neural network (ANN) models are used in cotton storage with a 99.8% accuracy rate (Shafi et al. 2023). Cotton farming, from seed germination to field observations, involves critical steps. Leading cotton-producing countries have started testing their own systems as well as internationally marketed ones.

In Turkey, AI intelligence-based methods, including artificial neural networks and support vector machines, have been tested to estimate the yield in some cotton farms in Diyarbakır, considering variables such as cotton cultivation area, irrigation periodicity, machinery usage, and fertilizer consumption. According to the results, there was an 84% accuracy rate for the ultimate artificial neural network model (Mikail and Baran 2021). In Pakistan, artificial intelligence is planned to support cotton production with an internationally funded project titled "The Texonomy," conducted by Northumbria University. The textile sector employs almost ten million people in Pakistan and covers 40% of the industrial labor force.

On the other side, climatic pressure is increasing from season to season for all agricultural products, including cotton. In this project, researchers will develop a smart farm application platform and provide this model to farmers, mostly cotton producers, for weather and irrigation-specific alerts (texonomy.org). The results of this project are expected to be obtained by the end of 2023. In India, another major cotton producer, tracking organic cotton farms is planned to be done with the combination of AI and satellite images. Two major companies, including the European Space Agency (ESA) and the Global Organic Textile Standard (GOTS), will work together to quickly classify organic cotton farms in line with cultivation standards, preventing fraudulent applications in cotton production. Hence, organic cotton production certification can be done more easily.

Uzbekistan was selected for the first pilot arable land, and preliminary production was completed in 2021 with a high degree of accuracy by providing discrimination between organic cotton fields and conventional ones (www.thehindu.com/sci-tech/energy-and-environment/satellite-images-and-ai-to-help-detect-organic-cotton-fields/article66927464.ece).

18.3 Germplasm Biodiversity and Conservation

18.3.1 Cotton Genetic Resources

Plant germplasm collections are spread all over the world to preserve wild, endemic, and domesticated genotypes for sustainable plant sources. Endangered plant species may be conserved in vitro under controlled conditions, while others are stocked in situ. Upahyaya (2015) reported that over 1750 gene banks contain a total of 7.4 million plant accessions, with 104,000 accessions belonging to fiber crops. Conservation efforts not only provide a seed source but also improve secondary and tertiary seed bank construction, which can be used as core collections or categorized divisions of specific plant species.

In China, the "Cotton Institute" in Beijing houses more than 8300 accessions, including 600 wild cotton varieties widely used in China. Another cotton collection in Uzbekistan consists of 32,580 *Gossypium* species and 1292 accessions tagged as *G. herbaceum* L. samples. This cotton library contains cultivated and wild relatives from various parts of the world. In Australia, CSIRO (the Commonwealth Scientific and Industrial Research Organisation) holds the largest number of accessions, with cultivated tetraploid cotton dominating and a smaller number of *Gossypium herbaceum* in different germplasm pools divided into secondary and tertiary collections (Egan et al. 2022).

In India, a large cotton collection covers cultivated and wild species, interspecific derivatives, and landraces. Out of 12,335 accessions of Gossypium, the ICAR-CICR Cotton Gene Bank contains 565 *G. herbaceum* accessions, divided into medium-term and short-term storage (cicr.org.in/research-cropimp-genetic-resources/). Another cotton collection is housed in the United States National Cotton Germplasm Collection (NCGC), with stations in different provinces for various public or laboratory-based research purposes. The gene bank's total collection is approximately 10,000 cotton accessions, with high exchange traffic with other cotton-producing countries. Seeds are multiplied every 10 years in this gene bank to ensure the sustainable viability of the collection. *G. herbaceum* is stored with an "A" genome prefix for long-term storage at -20 °C with ~5% seed moisture content.

In the NCGC, two *Gossypium herbaceum* accessions were tested against thrips as "A" genome donors, highlighting the lack of studies on diploid cotton genomes compared to tetraploids commonly found in gene banks (Abdurakhmonov 2014).

*18.3.2 Cytogenetic Studies on Diploid A Genome Levant Cotton (*Gossypium herbaceum *L.)*

Several karyotyping studies have been conducted on the *Gossypium* genus up to the present. In 1933, Davie published cytogenetic measurements of *G. herbaceum*, indicating the length of its chromosomes. This was followed by reports from Arutjunova (1936) and Mikhailova (1938). However, due to the small size of Gossypium chromosomes, a lack of feasible methods and infrastructure hindered progress. For nearly 40 years, there were no acceptable reports on *Gossypium herbaceum*. In 1977, Edwards finally published the karyotype of *G. herbaceum* L.

In another study, Sheidai et al. (1996) screened 42 *Gossypium herbaceum* cultivars available in the Iran National Gene Bank using light microscopy. Unlike Asiatic cotton (*G. herbaceum*), detailed karyotyping results showed a distinct diversity in both chromosome morphology and satellite region settlements in *G. herbaceum* cultivated in Iran. Interestingly, there were no acrocentric chromosome types, and intraspecies-specific chromosome lengths exhibited a wide range of variations. Karyotypic data was analyzed using factorial analysis and a UPGMA dendrogram divided the cultivars into six clusters with different numbers of cultivars. In summary, *G. herbaceum* cultivars in Iran are suggested for future hybridization programs due to their genomic diversity potential.

With technological developments in microscopy and methods, Gossypium karyotypes have been improved with the help of fluorescent dye. Wang Kun-Bo et al. (2008) used FISH-based karyotype analyses with gDNA probes originating from *G. raimondii* to screen *G. herbaceum* chromosomes. The results showed that GISH-NOR probes were functional during mitotic chromosome karyotyping for the *G. herbaceum* genome and could be feasible candidates for locus-specific studies.

In another example using a FISH-based approach, Cui et al. (2015) published the cytogenetic map of *Gossypium herbaceum* chromosome 1 to obtain the genome structure and physical chromosome segmentation, including the satellite regions.

18.4 Molecular Breeding Studies on *Gossypium herbaceum*

Since the beginning of agricultural practices, crops have been bred by humans for food. Later, in line with various needs, the energy and clothing sectors have focused on alternative crops. Cotton, as a versatile crop, has been produced for fiber and oil sources, and has been developed to meet commercial demands. In addition to traditional breeding methods, gene discovery in plant species has been improved with the use of molecular markers, such as RAPD and RFLPs, which have been applied to cotton (Brubaker and Wendel 1994; Multani and Lyon 1995).

Two accessions of *G. herbaceum* were screened using AFLP markers with 4 EcoRI-MseI primer combinations (Pillary and Myers 1999). Generally, microsatellite-enriched genomic libraries are mined to capture feasible molecular marker sources. Jena et al. (2012) extracted 1970 expressed SSRs and 263 gSSRs using this approach for the *G. herbaceum* genome and tested 150 gSSRs and 50 expressed SSRs in four species of cotton. Additionally, 15 genotypes of Levant cotton were assessed with these 200 SSRs. However, there is limited sequence information available for *G. herbaceum*, so major marker sources from tetraploid cotton were used in phylogenetic studies. Parekh et al. (2016) used a set of 60 SSR primers associated with fiber development on a panel of 23 *G. herbaceum* genotypes. A set of 13 SSR and five ISSR markers were used for molecular diversity evaluation of 48 cotton (*Gossypium herbaceum*), and morphological traits including yield parameters, seed physical properties, fiber quality, and seed chemical parameters were also analyzed (Sagar et al. 2022). Surveying polymorphism with SNP markers (Single nucleotide polymorphism) among different *G. herbaceum* accessions was used for the characterization of heterozygosity rates (Grover et al. 2021). Molecular marker studies supported by genome sequencing efforts were also used to identify SNPs in *G. herbaceum* (Du et al. 2018). Cotton full genome sequencing efforts began in 2012 in Gossypium species, with the first sequenced cotton genome being *G. raimondii* (Wang et al. 2012). Additionally, the *G. hirsutum* "TM-1" genome was the first sequenced allotetraploid cotton (Li et al. 2015). Ramaraj et al. (2023) used a high-quality genome sequencing approach for the *G. herbaceum* cultivar (cv.) Wagad. From this perspective, the development of functional molecular markers is at the core of cotton molecular breeding and has accelerated in the last decade. Especially, cotton breeding has mostly focused on yield characteristics of commercial varieties, but improving resistance to environmental stresses has become an important parameter (Yang et al. 2023). Cotton fiber quality is a very important criterion for the textile industry and is controlled by genetic factors (Paterson et al. 2003). Genes involved in fiber development and elongation were identified as the corresponding functions. Especially "A" genome-related ones are cellulose synthase CELA1-(Genbank Accession number: GHU58283) and CELA1-Cellulose Synthase catalytic subunit, cellulose biosynthesis in developing cotton fibers (Genbank Accession number: AF150630) (Zhu et al. 2012). Also, transcriptome sequencing of *Gossypium herbaceum* fibers from different stages showed that some transcription factors play an important role in the development of cotton fiber elongation (Parekh et al. 2018). In general, fiber formation-related genes were mostly studied in *Gossypium hirsutum* and some of these genes cloned in upland cotton species; glutamine synthetase (GhGS) Yajun et al. (2008) brassinosteroid synthetase GhDWF1 (Ming et al. 2007), Myb109 and MYB2, transcription factors expressed during fiber formation (Suo et al. 2003), GhKCH2 (Xu et al. 2009), Sucrose Synthase (SuSy) 5 (Notle et al. 1995), transporter protein GhWBC1 (Zhu et al. 2003) and GhRac1 gene is highly expressed during the fiber elongation phase (Kim and Triplett 2004).

Drought is a significant environmental stressor that greatly impacts the development of cotton from seed to fiber. In this study, leaf and root tissues from 1-month-old cotton plants (*Gossypium herbaceum*) were analyzed using comparative transcriptomic profiling under drought conditions. The study also identified specific key regulatory genes expressed in both leaves and roots using the 454 pyrosequencing platform. Results indicated that cotton roots exhibited higher expression levels under drought stress compared to leaves (Ranjan and Sawant 2015). Marker-assisted selection merged with the mutation breeding approaches provides rare allele selection from advanced cultivars. However, mutation breeding is not a new method, it was used by Kumar et al. (2022) to develop *G. herbaceum* mutant lines. In this research, a chemical mutagen ethyl methane sulfonate known as EMS was used to develop *G. herbaceum* mutant population and randomly selected 150 M3 *G. herbaceum* mutants were further screened with RAPD markers.

18.5 Hybridization Studies on *Gossypium herbaceum*

Generally, hybridization studies on *G. herbaceum* began at the intra and interspecies levels. Gennur et al. (1986) reported a hybridization study on two closely related *Gossypium* species, *Gossypium herbaceum* L. and *Gossypium arboreum*. According to hybridization studies on *Gossypium herbaceum* × *G. raimondii* F_1 hybrids, Wu et al. (2017) selected 500 F_1 hybrid pollen grains and found that almost 58% of pollen grains showed normal meiosis with dyad and tetrad formation. However, abnormal pollen formations such as oval-shaped or cracked ones were observed in a new synthetic cotton line, which could be the reason for sterile plant formation. Yin et al. (2020) used both *Gossypium herbaceum* L. and *Gossypium nelsonii* Fryx. to obtain a new hybrid F_1 generation for providing disease-resistant cotton genotypes. A distant hybridization method was used to create a new allotetraploid cotton germplasm (A1A1G3G3). Chromosome doubling was achieved through colchicine treatment, and these F1 lines (as parental lines) were crossed with the TM1 line to obtain fertile tetraploid cotton lines. The S2 generation resulting from these crosses was screened for Verticillium resistance. According to the results, the new hybrid line can be used as a germplasm source for cotton breeding studies. Badigannavar and Myers (2015) studied fiber quality in two diploid cotton species hybrid lines, one belonging to *G. herbaceum*, and identified 19 QTLs covering traits such as cotton fiber length, strength, elongation, and seed index. Thus, mapping of important traits in tetraploids could potentially start with the hybridization of diploid cotton species. To develop drought-tolerant varieties in *G. herbaceum*, a wild species belonging to the 'E' genome, *G. stocksii*, was hybridized with *G. herbaceum* using the embryo-rescue method (Gill and Bajaj 1996).

18.6 Conclusions and Prospects

In many countries, cotton, also known as "white gold," holds strategic importance for the industrial sector as it is used as a raw material for food, feed, oil production, and clothing. Almost no waste is produced in all processes involving cotton, from agriculture to production and processing. An extraordinary example of this is the use of cotton filaments in the filaments of 3D printers due to their resistance to heat and wetness. Consequently, cotton has become a core material for multiple applications and economies in most countries, whether developed or underdeveloped. Breeding studies on cotton plants have entered a new phase with advancements in plant biotechnology. Various cotton species with different genome sizes are being sequenced using high-technology sequencing platforms with high accuracy. Since the early 2000s, efforts have been made to improve cotton plants by targeting new traits of interest through molecular breeding methods using wild germplasm sources. Genotyping for trait selection requires more accessions to conduct comparative genomic analysis. Therefore, various molecular marker types, from RFLP to SNPs, are used to facilitate selection at both intra and interspecies levels.

However, the molecular bottleneck effect and the limited sequence information obtained from different cotton species, including wild and country-specific varieties, remain challenges for the Gossypium genus. Cotton breeding is supported by germplasm sources that encompass wild and domesticated forms of cotton. In addition to enhancing commercial varieties and selecting more qualified genotypes, there is a focus on developing biotic and abiotic stress tolerance in cotton. Factors such as absorbable water in the soil and soil-borne pathogens can have negative physiological effects on cotton plant development, reducing yields of fiber and oil content.

On the other hand, rapid changes in weather conditions, such as severe droughts, flooding, or irregular rainfall, pose barriers for cotton producers. Additionally, shrinking farming areas are limiting crop production lands and affecting the mineral content of the soil, which in turn impacts the amount of fertilizer needed. Therefore, biotechnological approaches should be applied using multiple factors to improve commercial cotton species.

Research on cotton at different genome levels plays a crucial role in the development of today's commercial varieties. Fiber quality is universally recognized as a key trait of interest. As environmental conditions continue to change, cotton development is being shaped by next-generation tools and approaches. A-genome-originated cotton provides valuable genomic resources for cotton genetic improvement. While studies on *Gossypium herbaceum* are not sufficient, gene bank resources will aid in screening new cotton varieties for future agricultural production. Omics techniques will help shorten breeding times and enhance the accuracy of selecting viable cotton genotypes.

References

Abdurakhmonov IY (2014) World cotton germplasm resources. BoD–books on demand; 2014 Jul 2. Intech ISBN: 978-953-51-1622-6

Ahmed M, Shahid AA, Din SU et al (2018) An overview of genetic and hormonal control of cotton fiber development. Pak J Bot 50(1):433–443

Arutjunova LG (1936) An investigation of chromosome morphology in the genus *Gossypium*. C R (Doklady) Aca Sci URSS 3:37–40

Badigannavar A, Myers GO (2015) Construction of genetic linkage map and QTL analysis for fiber traits in diploid cotton (*Gossypium arboreum* x *Gossypium herbaceum*). J Cotton Sci 19(1): 15–26

Brubaker CL, Wendel JF (1994) Reevaluating the origin of domesticated cotton. *Gossypium hirsutum*. Am J Bot 81:1309–1326

Cao Q, Zhu S, Pan N, Zhu Y, Tu H (2009) Characterization of archaeological cotton (*G. herbaceum*) fibers from Yingpan. Techn Brief Hist Archaeol 4:18–28

Chowdhury AN, Rahman MZ, Samad A, Alam AKMS, Khaleda S (2011) In vitro plant regeneration from cultured cotyledons of cotton (*Gossypium herbaceum* L.). Bangladesh. J Sci Ind Res 46(3):359–364

Cui X, Liu F, Liu Y et al (2015) Construction of cytogenetic map of *Gossypium herbaceum* chromosome 1 and its integration with genetic maps. Mol Cytogenet 8(1):1

Davie JH (1933) Cytological studies in the *Malvaceae* and certain related families. Genet 28:33–67

Du X, Huang G, He S et al (2018) Resequencing of 243 diploid cotton accessions based on an updated A genome identifies the genetic basis of key agronomic traits. Nat Genet 50(6):796–802

Edwards GA (1977) The Karyotype of *Gossypium herbaceum* L. Caryologia 30(3):369–374. https://doi.org/10.1080/00087114.1977.10796710

Egan LM, Conaty WC, Stiller WN (2022) Core collections: is there any value for cotton breeding? Front Plant Sci 13:895155. https://doi.org/10.3389/fpls.2022.895155

Gennur MN, Habib AF, Kadapa SN, Goud JV (1986) Cytogenetic studies in interspecific and intraspecific hybrids of Gossypium herbaceum L. and *Gossypium arboreum* L. Caryologia 39: 1:65–68. https://doi.org/10.1080/00087114.1986.10797768

Gill M, Bajaj Y (1996) Hybridization between *Gossypium herbaceum* and *Gossypium stocksii* through embryo rescue. Cell Res 6:85–90. https://doi.org/10.1038/cr.1996.10

Grover CE, Arick MA, Thrash A, Sharbrough J, Hu G et al (2021) Dual domestication, diversity, and differential introgression in old world cotton diploids. bioRxiv. https://doi.org/10.1101/2021.10.20.465142

Hu G, Grover CE, Yuan D, Dong Y, Miller E, Conover JL, Wendel JF (2021) Evolution and diversity of the cotton genome. In: Rahman M-U, Zafar Y, Zhang T (eds) Cotton precision breeding. Springer, New York, pp 25–78

Huang G, Wu Z, Percy RG et al (2020) Genome sequence of *Gossypium herbaceum* and genome updates of *Gossypium arboreum* and *Gossypium hirsutum* provide insights into cotton A-genome evolution. Nat Genet 52(5):516–524. https://doi.org/10.1038/s41588-020-0607-4

Ishfaq M, Kiran A, ur Rehman H et al (2022) Foliar nutrition: potential and challenges under multifaceted agriculture. Environ Exp Bot 200:104909

Jena SN, Srivastava A, Rai KM, Ranjan A, Singh SK, Nisar T, Srivastava M, Bag SK, Mantri S, Asif MH, Yadav HK (2012) Development and characterization of genomic and expressed SSRs for levant cotton (*Gossypium herbaceum* L.). Theor Appl Genet 124:565–576

Khan M, Wahid A, Ahmad M et al (2020) World cotton production and consumption: an overview. https://doi.org/10.1007/978-981-15-1472-2_1

Kim HJ, Triplett BA (2004) Characterization of GhRac1 GTPase expressed in developing cotton (*Gossypium Hirsutum* L.) fibers. Biochim Biophys Acta 1679:214–221

Khadi B, Venoor S, Yadav M (1970) Cotton: An Introduction. https://doi.org/10.1007/978-3-642-04796-1_1

Kumar U, Chavan NS, Sawant SV, Kumar YH (2022) Evaluation and characterization of EMS-induced mutant population of *Gossypium herbaceum*. Vegetos 35(4)1036–1046

Kun-Bo W, Guo-Li S, Chun-Ying W, San-Hong L, Fang L, Mao-Xue L, Shao-Hui L, Xiang-Di Z, Yu-Hong W (2008) FISH-based Karyotype of *Gossypium herbaceum* generated with 45S rDNA and gDNA of *Gossypium raimondii* as probes[J]. Cotton Sci 20(4):264–273. https://doi.org/10.11963/cs080404

Li F, Fan G, Lu C et al (2015) Genome sequence of cultivated Upland cotton (*Gossypium hirsutum* TM-1) provides insights into genome evolution. Nat Biotechnol 33(5):524–530

Malik W, Ashraf J, Iqbal MZ et al (2014) Molecular markers and cotton genetic improvement: current status and future prospects. Sci World J 2014:15

Mikail N, Baran MF (2021) Application of artificial intelligence methods to predict cotton production in Turkey. Turkish J Agric Natl Sci 8(4):1018–1027. https://doi.org/10.30910/turkjans.947978

Mikailova KA (1938) Chromosome morphology of cotton. C R (Doklady) Acad Sci URSS 19:181–184

Ming L, Zhong-Yi X, Yue-Hua X et al (2007) Cloning and expression analysis of a brassinosteroid biosynthetic enzyme gene, GhDWF1, from cotton (*Gossypium hirsutum* L.). Agric Sci China 6(11):1297–1305

Multani DS, Lyon BR (1995) Genetic fingerprinting of Australian cotton cultivars with RAPD markers. Genome 38(5):1005–8

Notle KD, Hendrix DL, Radin JW, Koch KE (1995) Sucrose synthase localization during initiation of seed development and trichome differentiation in cotton ovules. Plant Physiol 109:1285–1293

Parekh MJ, Kumar S, Zala HN et al (2016) Development and validation of novel fiber relevant dbEST–SSR markers and their utility in revealing genetic diversity in diploid cotton (*Gossypium herbaceum* and *G. arboreum*). Ind Crop Prod 83:620–9

Parekh MJ, Kumar S, Fougat RS et al (2018) Transcriptomic profiling of developing fiber in levant cotton (*Gossypium Herbaceum* L.). Funct Integr Genomics 18:211–223

Pillay M, Myers GO (1999) Genetic diversity in cotton assessed by variation in ribosomal RNA genes and AFLP markers. Crop Sci 39(6):1881–1886. https://doi.org/10.2135/cropsci1999.3961881x

Paterson AH, Bowers JE, Peterson DG, Estill JC, Chapman BA (2003) Structure and evolution of cereal genomes. Curr Opin Genet Dev 13:644–650

Ramaraj T, Grover CE, Mendoza AC et al (2023) The *Gossypium herbaceum* L. Wagad genome as a resource for understanding cotton domestication. G3 13(2):jkac308

Ranjan A, Sawant S (2015) Genome-wide transcriptomic comparison of cotton (*Gossypium herbaceum*) leaf and root under drought stress. 3 Biotech 5:585–596

Sagar MR, Kumar S, Patidar D, Sakure AA (2022) Morphological, physico-biochemical and marker-based diversity of desi cotton (*Gossypium herbaceum* L.) germplasm. J King Saud Univ Sci 34(8):102336

Schmidt E, Lötter M, McCleland W (2002) Trees & shrubs of Mpumalanga and Kruger National Park. Jacana, Johannesburg

Shafi UF, Bajwa IS, Anwar W, Sattar H, Ramzan S, Mahmood A (2023) Sensing spontaneous combustion in agricultural storage using IoT and ML. Inventions 8(5):122. https://doi.org/10.3390/inventions8050122

Sheidai M, Vojdani P, Alishah O (1996) Karyological studies in *Gossypium herbaceum* cultivars of Iran. Cytologia 61:365–374

Stephen A, Arumugam P, Arumugam C (2023) An efficient deep learning with a big data-based cotton plant monitoring system. Int J Inf Technol 13:1–7

Suo J, Liang X, Pu L, Zhang Y, Xue Y (2003) Identification of GhMYB109 encoding a R2R3 MYB transcription factor that expressed specifically in fiber initials and elongating fibers of cotton (*Gossypium hirsutum* L.). Biochim Biophys Acta 1630:25–34

Sytar O, Kumar A (2023) Editorial: the adaptation strategies of plants to alleviate important environmental stresses. Front Plant Sci 14:1307399. https://doi.org/10.3389/fpls.2023.1307399

Tokumoto H, Wakabayashi K, Kamisaka S, Hoson T (2002) Changes in the sugar composition and molecular mass distribution of matrix polysaccharides during cotton fiber development. Plant Cell Physiol 43(4):411–418

Upahyaya H (2015) Establishing core collections for enhanced use of germplasm in crop improvement. Ekin J Crop Breed Gene 1:1–12

Walford SA, Wu Y, Llewellyn DJ, Dennis ES (2011) GhMYB25-like: a key factor in early cotton fibre development. Plant J 65(5):785–797

Wang K, Wang Z, Li F et al (2012) The draft genome of a diploid cotton *Gossypium raimondii*. Nat Genet 44(10):1098–1103

Wu Y, Chen D, Zhu S, Zhang L, Li L (2017) A new sythetic hybrid (A1D5) between *Gossypium herbaceum* and *G. raimondii* and its morphological, cytogenetic, molecular characterization. PLoS One 12(2):e0169833. https://doi.org/10.1371/journal.pone.0169833

Yajun H, Wangzhen G, Xinlian S, Tianzhen Z (2008) Molecular cloning and characterization of a cytosolic glutamine Synthetase gene, a fiber strength associated gene in cotton. Planta 228:473–483

Yang Z, Gao C, Zhang Y, Yan Q, Hu W, Yang L, Wang Z, Li F (2023) Recent progression and future perspectives in cotton genomic breeding. J Integr Plant Biol 65(2):548–569

Yin X, Zhan R, He Y, Song S, Wang L, Ge Y, Chen D (2020) Morphological description of a novel synthetic allotetraploid (A1A1G3G3) of *Gossypium herbaceum* L. and *G. nelsonii* Fryx. suitable for disease-resistant breeding applications. PLoS One 15(12):e0242620. https://doi.org/10.1371/journal.pone.0242620. eCollection 2020

Zhu YQ, Xu KX, Luo B, Wang JW, Chen XY (2003) An ATP-binding cassette transporter GhWBC1 from elongating cotton fibers. Plant Physiol 133:580–588

Zhu H, Lv JL, Zhao X et al (2012) Molecular evolution and phylogenetic analysis of genes related to cotton fibers development from wild and domesticated cotton species in *Gossypium*. Mol Phylogenet Evol 63(3):589–597 [illegible] Kumar A (2023) [illegible] Exploration strategies of plant [illegible] [illegible] Plant [illegible] https://doi.org/[illegible]

[illegible] (2002) [illegible] changes in the sugar composition and [illegible] during fiber development [illegible]

[illegible]

Chapter 19
Recent Advances in Upland Cotton (*Gossypium hirsutum* L.) Breeding

Waleed M. B. Yehia

Abstract Cotton is one of the most significant fiber commodities in the world, and it has significant economic and monetary returns that significantly impact the national income of cotton-producing countries in large quantities and across large areas. There are about 50 types of cotton in the world, but only four types of these fifty types represent the spread of cotton and its cultivation in the world. We find that American cotton or Upland cotton (*Gossypium hirsutum* L.) is spread in an area of up to about 90% of the total areas planted with cotton in the world, which amounts to 2.5% of the cultivated area on the globe. Then comes Indian cotton and African cotton (*Gossypium herbaceum* L.) and *Gossypium arboreum* L.), which together represent about 7% of the total cotton cultivated area in the world. Then comes Egyptian cotton (*Gossypium barbadense* L.), which represents the best and softest types of cotton in the world and represents approximately 3% of the total cotton cultivated area in the world. We find that one of the most important features of American cotton is that it is tolerant to different weather conditions and tolerant to temperatures, whether high or low temperatures and contains genetic genes specific to drought tolerance and salinity tolerance, whether soil salinity or soil salinity. In addition, the presence of a group of descriptive characteristics and the possibility of applying genetic engineering techniques to produce insect-resistant plants, represent the economic dilemma of cotton production in the world. Therefore, its spread over a very wide range of latitudes and over a wide range of temperatures, whether low temperatures or high temperatures and its ability to spread over a wide range of lands is what creates a place for it to spread as the largest cotton cultivated area in the world. This chapter elucidates the properties and potential applications of fibers obtained from upland cotton species leaves, the techniques for extracting fibers, germplasm biodiversity and conservation, the goals and stages of beetroot breeding initiatives, and traditional breeding methodologies. It also covers contemporary methods of plant breeding, including genetic alteration and marker-assisted selection.

W. M. B. Yehia (✉)
Cotton Research Institute, Agriculture Research Center (ARC), Giza, Egypt
e-mail: waleed.yehia@arc.sci.eg

K. F. M. Salem et al. (eds.), *Breeding and Biotechnology of Leaf, Fruit, and Seed Fiber Crops*, Advances in Plant Breeding Strategies 11,
https://doi.org/10.1007/978-3-032-00407-9_19

Keywords Upland cotton (*Gossypium hirsutum* L.) · Breeding · Salinity · Drought tolerance · Germplasm · Selection

19.1 Introduction

Gossypium hirsutum L. is referred to as "cotton" in this chapter. In areas and nations that cultivate cotton, this species is the most prevalent and extensively distributed. However, the term "cotton" generally refers to four species of the genus *Gossypium* (Malvacea) that have been independently adapted as major sources of textile fibers: *Gossypium hirsutum* L. (upland cotton), *Gossypium barbadense* L. (Egyptian cotton), *Gossypium arboreum* L., and *Gossypium herbaceum* L. (Brubaker et al. 1999). With *Gossypium hirsutum* L. making up 92% of global cotton production, *Gossypium hirsutum* L. and *Gossypium barbadense* L. are currently the most extensively grown cotton species worldwide (Jenkins 2003). Egyptian cotton, or *Gossypium barbadense* L., is mostly farmed in Egypt, Peru, Sudan, the United States, Uzbekistan, China, India, and other small nations. It accounts for roughly 3–5% of global cotton production (Wu et al. 2005). The *Gossypium arboreum* L. is primarily grown in India, but the *Gossypium herbaceum* L. allied genus is found in the *Gossypieae*, either slightly more recently in the Pliocene (Cronn et al. 2002) or at least 12.5 million years ago in the Miocene (Seelanan et al. 1997 and Wendel and Albert 1992). The origins of New World cotton are still hotly contested (see Wendel and Cronn 2003). According to some scholars, hybridization of the A and D genomes took place between 60 and 100 million years ago, which is a long time ago, before South America and Africa split apart. According to the knowledge that is currently available, hybridization with D-genome species most likely happened lately, within the last 6000 years, even though there is evidence that human transfer of A-genome African or Asian plants may have occurred followed by unintentional or intentional hybridization. Molecular evidence, such as DNA sequence data that have suggested and documented an origin originating from the middle Pleistocene (1–2 million years ago) (Wendel 1989; Fryxel 1979b), does not, however, definitively support any of these views.

19.1.1 Origin, Taxonomy and Distribution

Upland cotton (*Gossypium hirsutum* L.) is a flowering plant in the class Magnoliopsida, order Malvales and family Malvaceae. Many edible plants including various *Gossypium* species, cacao (*Theobroma cacao*), linden (Tilia species), durian (Durio species), Hibiscus, and okra (*Abelmoschus esculentus*) are also in the Malvaceae family (https://www.itis.gov). Upland cotton (*Gossypium hirsutum* L.) belongs to the genus *Gossypium* and species.

Kingdom Plantae – plantes, Planta, Vegetal, plants
Subkingdom Viridiplantae – green plants
Infrakingdom Streptophyta – land plants
Superdivision Embryophyta
Division Tracheophyta – vascular plants, tracheophytes
Subdivision Spermatophyta – spermatophytes, seed plants, phanérogames
Class Magnoliopsida
Superorder Rosanae
Order Malvales
Family Malvaceae – mallows, mauves
Genus *Gossypium* L. – cotton
Species *Gossypium hirsutum* L. – Upland cotton, algodón

Historical evidence and information indicate that cotton fibers have been used in the manufacture of threads and textiles since 6000 BC. The *Gossypium* thread used in the copper bead thread in Pakistan has been historically confirmed to date back to the sixth millennium BC (Moulherat et al. 2002). It is not certain or known whether this thread was domesticated or wild cotton. However, the available information indicates that cotton fibers and textiles were known, used and widespread at this time. The cotton may have been used for preparation and arrangement for use in spinning into threads and then into textiles and fabrics (Smith 1995). However, the most important thing is that these fibers and threads have been identified as New World cotton *Gossypium pp.* From before 2500 BC to after 1000 BC, seed size and fiber diameter increased, according to historical finds and documents (Stephens and Moseley 1973). Historical data indicates that North America, Central America, and the Mexico region are *G. hirsutum's* main geographic centers of origin (Jenkins 2003). Pre-Columbian people or those who conquered the Yucatan Peninsula likely were the first to produce *Gossypium hirsutum* L. (Brubaker and Wendel 1993). These plants and semi-domesticated buildings proliferated throughout the Caribbean, northern South America, and the remainder of Central America. Selection subsequently reduced seed dormancy, growth traits, and flowering processes, resulting in the production of genotypes and other genotypes that are more similar to the new cultivars currently distributed in different regions of the world. Modern North American genotypes and cultivars of *Gossypium hirsutum* L. have very limited gene content, suggesting that they are genetically predisposed to adaptation pressure (Iqbal et al. 2001). It is believed that Indians in the southwestern United States domesticated *Gossypium hirsutum* L as early as the first century AD (Fryxell 1979a). The majority of wild cotton species have short photoperiods, brief flowering seasons, and short days. Cotton cultivars that are less sensitive to photoperiods have therefore been chosen for evaluation (Lee 1984).

Annuals are also unknown among wild cotton species (Fryxell 1979a). Growth characteristics and the natural response associated with flowering are key evaluation steps that depend on human selection and the extent to which these plants can be grown outside the region. Wild cotton species have a high degree of seed dormancy, which can persist in the seed bank before germination (Jenkins 2003). The data and results collected indicate that doubling the seed size resulted in a 3-fold increase in the amount of fiber on the seed and an 80% increase in the average fiber length in the upper half of the staple length during the measurement and evaluation system (Stephens 1965). This increased fiber length was achieved by the growth of fibers in the new cultivars compared to wild cotton *Gossypium hirsutum* L.. *Gossypium hirsutum* L. is now found in many parts of the Caribbean, Central and South America, and a few Pacific islands (Wendel and Cronn 2003). This chapter clarifies the characteristics and possible uses of fibers derived from upland cotton species leaves, the methods for fiber extraction, germplasm biodiversity and preservation, the objectives and phases of beetroot breeding programs, and conventional breeding procedures. Additionally, it examines contemporary plant breeding methodologies, including marker-assisted selection and genetic engineering.

19.1.2 Cotton Cultivation and Uses

One of the most significant fiber crops in the world, cotton is farmed commercially in more than 50 nations in both tropical and subtropical climates (Smith 1999). At least 2.4% of the world's arable land is thought to be used for cotton production (Blaise 2006). The United States, India, China, Pakistan, Uzbekistan, Tajikistan, the Middle East, and Australia are among the countries that have been designated as production land for cotton cultivation. These countries have climatic conditions that are conducive to the natural growth of cotton production, as well as periods of hot and dry weather and the availability of moisture, which is typically obtained through irrigation. According to the data at our disposal, the average global production of cotton in 2005 and 2006 was 24.88 million tons (Wood et al. 2007). The average yearly production of cotton lint was 764 kg per hectare, with the highest country averages being Australia (2011), China 1892, Turkey 1827, Brazil 1719, Mexico 1592, and Greece 1121 kg per hectare. In 2021 and 2022, the world produced approximately 24.96 metric tons of cotton, with China 5730, India 5270, United States 3815, Brazil 2360, Australia 1277, and Pakistan 1266 metric tons. 2022 ICAC Data Book.

The short hair, or "lint," that covers the ginned seeds of the cotton plant *G. hirsutum*. Before the seeds can be planted or used to crush cottonseed oil, this lint needs to be eliminated. Either mechanical de-linting or acid de-linting produces first-cut or second-cut lint. The medium-length upper half of first-cut lint is used to make mops, furniture upholstery, and mattresses. The medium-length upper half of second-cut lint, which is a significant source of cellulose, is created by sulphuric acid or mechanical brushing. It serves as a thickening in toothpaste and ice cream and as a cellulose foundation in goods like high-fiber foods. In the chemical industry, cellulose products including cellulose acetate, nitrocellulose, and a variety of other compounds are made from this second-cut or acid-treated lint in combination with other materials (Gregory et al. 1999). Upland or American cotton yields cottonseed that is composed of 28% husks, 10% fluff, 45% meal, and 17% crude oil (Smith 1995). Oil, meal, and husks are produced from delinted cottonseed (Cherry and Leffler 1984). Heating, adding sodium hydroxide, bleaching, filtering, and vacuum-steaming are some of the processes used to prepare cottonseed oil (OECD 2004).

Since the middle of the 19th century, cottonseed oil has been a very popular and widely used oil. Because of its extensive use before 1958, it was accorded GRAS classification under the US Federal Food, Drug, and Cosmetic Act (ANZFA 2002). The product left over after the oil has been crushed away is called post-oil cottonseed residue, and it can have as much as 41% protein (Smith 1995). Because cottonseed contains natural toxins (cyclopropane fatty acids and gossypol), its application in animal feed or for fattening and feeding animals is limited. In the United States and other nations, cottonseed meal has been allowed for use as human food when made from gossypol-free cultivars or after gossypol is removed during processing, even though it is not used for human consumption in Australia or New Zealand. While the FDA has a minimum limit of 450 ppm, the FAO and WHO permit up to 0.6 μg/mg (600 ppm) of free gossypol in edible cottonseed products (Lusas and Jividen 1987; Frank 1987). To increase the efficiency of animal manure composting, cottonseed waste can be utilized as a filler (Brampton 2001). The fact that cotton stalks contain a specific energy that is comparable to cottonwood (17.1–18.1 μJ/kg) (Coates 2000) is also quite intriguing. This has led to the idea that the residual cottonwood or stalks could be utilized as an industrial energy source for a power plant (Gomes et al. 1997). Additionally, there has been some interest in fermenting cotton plant waste and leftovers to generate large amounts of ethanol (Jeoh and Agblevor 2001). It is also noteworthy that gossypol, which is mostly found in cotton plant extracts from the stalks, leaves, seeds, and roots, has been utilized as a medicine (Sawyer 1955).

19.2 Current Cultivation Practices

The majority of the world's countries (more than 70) grow cotton, a staple crop that yields around 26 million tons annually, with only five generating more than 70% of the total: the United States accounts for 18%, India for 24%, China for 22%, Pakistan for 7%, and Brazil for 5%. Most countries that produce short-staple cotton do not manufacture a large part of it, which causes the supply to always exceed what is manufactured, and thus its prices globally (selling prices) to decrease. Short staple cottons are used in the textile industry such as (jeans, casual clothes, children's clothes, etc.). They are low-quality cottons that produce thick numbers not exceeding 30 English. Some countries that produce medium cotton (lowest in quality) in large quantities (such as India) resort to purchasing large quantities of Egyptian cotton (highest in quality) and mixing them to produce threads with specific and good specifications that cannot be produced from short-staple cotton alone and exporting them in large quantities in global markets and achieving very high profits.

The prices of medium-quality cotton during the past 20 years ranged from 70 to 90 cents.

19.2.1 Digital Farming

The productivity of American cotton in many areas where it is grown is still low and small, despite the rapid progress and use of smart agriculture and high technology currently. These previous reasons and others still require a set of agricultural transactions, methods and agricultural practices that focus on different areas of crop management at the global level. Therefore, when following the growth of cotton and the use of smart agriculture and modern technology, we must be aware of an overview of the current developments in the knowledge, skills and technologies needed to produce sustainable cotton on a small-scale basis. Specifically, the areas of the cotton chain that have been studied and their methods and methods have been developed are three: (1) exploring sound, simple and effective agricultural practices that can be given to the production of sustainable cotton on a small-scale basis; (2) studying and identifying the existing social and economic conditions that show the disparities in the contribution to the production of sustainable cotton on the small-scale basis; and (3) exploring the existing and feasible institutional policies that should be tested and farmers trained on them in order to produce sustainable cotton on the small-scale basis, which represents most of the American cotton areas in the world. The study of production conditions, quantity and reasons for its decline shows that cotton production worldwide involves traditional and organic systems which represent the traditional production system.

The intensive use of nitrogen and mineral fertilizers and pesticides are common practice in traditional cotton production systems worldwide. We find that the adoption of precision agriculture practices, smart agriculture, modern methods and genetically modified varieties represents a major shift in cotton production and increased productivity. Awareness should also be raised about the importance of intercropping with other crops, crop rotation systems and the use of other early maturing crops as viable options to improve farmers' attitudes towards adopting modern and smart technologies for cotton production. These methods and applications have the greatest impact in the social and economic field as they lead to improving farmers' livelihoods by increasing income from cotton sales and increasing productivity per unit area and labor in cotton production and manufacturing. In addition, modern systems and new initiatives in cotton production achieve the principle of gender equality in the cotton industry. Studies show that females as a group enjoy a higher and better level of technology adoption when compared to males.

19.2.2 Artificial Intelligence

Climate change has a significant impact on cotton cultivation through a wide range of factors, causes and mechanisms. All of this creates a set of important and major challenges for agricultural sustainability, environmental sustainability and cotton production sustainability. For example, we find that high or low global temperature, along with patterns of globally variable rainfall rates, in addition to an increase in extreme and unstable weather events (such as drought, floods, heat waves, humidity, sometimes low temperatures, floods and frost) lead to a transformation of the agricultural reality. These previously mentioned causes and changes ultimately lead to a decrease in cotton yields and threaten the sustainability of production (Habib et al. 2022; Kumar et al. 2022). Adaptation systems and strategies, transforming agricultural operations and changing prevailing systems, for example, changing planting dates, crop rotation, developing crop management systems, and breeding and producing varieties resistant to drought, salinity and high temperatures. The use of new methods and approaches to improve water productivity through modern irrigation systems is important and necessary in order to mitigate the severity and impact of these effects, in addition to maintaining and ensuring sustainable agricultural productivity of cotton (Habib et al. 2022; Garflo et al. 2023).

Climate change can significantly affect the yield and quality of cotton (*Gossypium hirsutum* L.) fibers (Dabbert and Gore 2014). Given the global importance of upland cotton, these effects and changes resulting from climate change are of particular concern. Although cotton is generally considered a drought-resistant crop, prolonged water stress can eventually lead to a decline in the quality and deterioration of cotton fibers. In addition, mono-crop farming methods require high inputs, especially water, chemicals, fertilizers, and other production requirements, to improve and increase productivity (Celik 2023; Zhang et al. 2023). Studies show that the negative impact of drought stress on cotton yields, which has increased due to the impact of climate change, depends on several important factors, such as the time of water stress and its severity, in addition to the type and variety of cotton grown (Abdelrheem et al. 2020; Abro et al. 2022).

Research and studies show that thermal and water changes can lead to decreased productivity and fiber characteristics, such as decreased fiber length and strength (Pettigrew 2024; Wu et al. 2022). Therefore, artificial intelligence and the availability of methods that allow rapid monitoring of water levels in cotton, irrigation, and its quantity in plant cells can help avoid many of the negative effects and damages resulting from drought and climate change, improve irrigation water provision, and increase water unit productivity. We find that the traditional methods for measuring and determining the water status of cotton are using measurements of soil moisture, the extent of water reaching the plant leaves, the state of the stomata, and the state of water in the cotton stems. Therefore, we find that artificial intelligence provides sensors that can measure soil moisture. Such as devices for measuring the state of stomata and estimating field capacity, as well as data on the water content in soil particles (Byrd et al. 2015). Some devices provide a complete picture of the water status inside the leaves and the extent of its impact on environmental conditions, daily variations, and temperatures (Maldera et al. 2024).

Modern devices have shown the possibility of using the water status of cotton stems as a more stable and documented indicator of the water status in the plant, in addition to images of the total water balance of the cotton plant (Myburgh and Howell 2022). This method provides great importance and a direct assessment of the water status in the plant, as recent research has shown that stem water is very sensitive to water stress, making it an important and documented tool and guide for managing irrigation in fields and reducing the impact of drought (Pedrero et al. 2015). Therefore, it may be very difficult to make these measurements to a sufficient degree for the water status of the stem to be representative of all places within the field and throughout the entire growing season to regulate and schedule irrigation water. However, we find that the integration of remote sensing and artificial intelligence technologies can allow for rapid monitoring of many useful indicators for crop irrigation management (Sellami et al. 2022; Garofalo et al. 2024). We also find that through remote sensing technology and artificial intelligence applications, it is possible to monitor and analyze soil surface properties by measuring the amounts of radiation reflected or emitted by soil particles (Weiss et al. 2020).

Satellites, which offer high-precision findings at varying costs based on the latest technologies used, are then used to collect the data primarily and permanently. According to studies and research conducted recently, the agricultural industry has become much more dependent on artificial intelligence and remote sensing applications and technology, particularly in the cotton sector (Weiss et al. 2020). Sentinel-2 is a mission of the European Space Research Agency's Copernicus program that uses two satellites, 2A and 2B. Sentinel-2 provides very high-resolution multispectral photos that can be used to enhance crop monitoring and precision agriculture.

Three methods parametric, non-parametric, and physics-based are utilized to employ remote sensing data in crop monitoring and management programs that leverage artificial intelligence technology. The direct correlation between crop attributes and data from remote sensing is investigated using parametric techniques. Non-parametric methods, like machine learning algorithms, can handle non-normal distributions and capture complicated and non-linear interactions. Although physical characteristics form the basis of physics-based techniques, their practical use in the agricultural system is frequently limited by their complexity (Jiang et al. 2022).

Machine learning in agriculture, artificial intelligence, and remote sensing applications can help monitor crop indicators and potentially improve productivity and increase the efficiency of using available resources (Biswas and Banik 2024). For instance, a drone was utilized to assess water stress indicators for cotton crops under various irrigation schemes in order to apply artificial intelligence applications to monitor the water status in cotton fields (Ballester et al. 2019). By measuring vegetation cover indicators across the field area, drone photos were used to evaluate the impacts of water stress on cotton plants.

19.2.3 Current Agricultural Challenges

Continuous cotton cultivation under the no-tillage system is always challenging to accomplish because of numerous agricultural operations, including tearing and cutting the plant roots, according to studies and research. Moreover, plowing is necessary for the use of flood irrigation, which eliminates and destroys insect pupae (CRDC 2017). It can also be added that there are some disadvantages represented by the presence of finished or retained crop residues, which can interfere with the operation of agricultural machinery and equipment, which can also lead to a significant increase in the spread of pests and diseases, in addition to the increased reliance on chemical herbicides to control harmful weeds (Scott et al. 2010). Given the high costs, small farmers face numerous obstacles and challenges in implementing this system, particularly in nations like China and India where farms are dispersed and farms are small (Kassam et al. 2009).

Furthermore, it is crucial to use subsurface plows and deep plowing to alleviate impermeable layers and soil compaction (Geller et al. 2009). Additionally, investigations, research, and farmer follow-up reveal a significant disconnect between the implementation of sustainable agriculture and the attainment of the intended and anticipated outcomes (Bhan and Behera 2014). The lack of willingness of farmers to take risks, in addition to the high costs of purchasing new equipment and the low availability of capital required for investment, are among the reasons behind the lack of adoption of sustainable agriculture (Knoller and Bradshaw 2007). Therefore, farmers, especially small farmers, must confront the decline in yield by adopting sustainable agriculture after traditional agriculture (FAO 2018). In addition, to the development and spread of cotton weeds in abundance, in addition to being resistant to glyphosate, here plowing is the main and best choice for controlling weeds resistant to glyphosate (Price et al. 2011).

19.2.4 Breeding Objectives

Cotton is grown for its fiber, mainly for use in the textile industry. Moreover, its seed may contribute an adequate amount of oil. Losses due to pests and diseases in cotton cause significant reductions in yield and quality of lint and seed and increase production costs. Therefore, one of the major breeding objectives for the cotton breeder is to develop or produce cultivars with some level of insect resistance, while having desirable yield and fiber quality.

Essential Characters: lint yield and fiber quality improvements: studies have shown a strong, negative genetic correlation between lint yield and fiber length and strength. In the last decades, cotton breeders have emphasized transferring genes for extra fiber strength to improve cotton cultivars. Several breeding successes in combining both high lint yield and fiber quality suggested that several genetic linkages between lint yield, and fiber, quality had been broken through intercrosses, outcrosses backcrosses and selection. Hybridization is required for the recombination of genes for high lint yield and fiber strength or length, and selection is necessary to locate the desirable recombinants in early generations.

19.2.4.1 Breeding for Insect and Disease Resistance

The cotton plant is a preferred host to many pests. Consequently, the greatest expense in most cotton production operations is the insecticide applications required to prevent serious economic loss. Therefore, the breeding of improved cotton cultivars with high levels of heritable resistance to pests is essential Desirable characters Morphological traits (resistance to pests): The association of several morphological traits with insect and disease resistance has been identified in cotton. These traits are; okra-shaped leaf, frego bracts, nectarless, glabrous (smooth leaf) and early maturity. No single trait provides adequate resistance to all pests as can be seen in Okra leaf and super okra both confer some resistance to white flies, pink bollworms and boll weevils. Plains with okra leaves are early mature and also produce less trash.

The frego bract trait is associated with a high level of resistance to boll weevil and reduces ball rot damage in humid areas. The nectariless trait provides resistance to plant bugs and some pathogens such as black rot and boll rot. Many insects prefer to feed on nectar and lay eggs on Cottons that have an extra 4. The glabrous (smooth leaf) cotton has less damage from the bollworm and pink bollworm. Out it is traditionally attacked by jassids.5. Varieties that mature their crop insignificantly in fewer days than normal can avoid prolonged exposure to pest attack and thereby need fewer pesticide applications.

Physiological resistance: Two classes of secondary metabolites, terpenoid aldehydes (e.g. gossypol) and flavonols (e.g. condensed tannins) are important sources of resistance to insects in cotton. Increasing gossypol and cotton from insects and diseases.

One disadvantage of the character is the increase in the gossypol content of the seeds which aggravates processing difficulties at the oil mills, where gossypol is regarded as a contaminant. It could be concluded that remarkable advances have been made in breeding concepts and methodologies for developing plants resistant to pests Breeding for salinity stress: In irrigated fields, dissolved salts in the water remain after evapotranspiration and salt concentration increases to Levels which harmful to both plants and the soil structure. There are, however, conflicting statements about the effects of salinity on cotton It could be concluded that both yield and quality of cotton can be affected by increases in salinity levels. There is some progress in developing cultivars with salinity tolerance.

19.2.4.2 Germplasm Biodiversity and Conservation

Among other crops, cotton (*Gossypium spp.*) is the primary source of fiber, making it one of the most significant crops in the world. Especially high-quality and high-tensile fibers used in textiles. In addition to many other important additional products such as oil, meal, seed husks, and lint. Continuous research and studies and continuous improvement processes using Gossypium germplasm are essential requirements in order for cotton to remain a sustainable, competitive and constantly improving crop. Plant introductions and collection of germplasm from all over the world have been a necessary and important and vital part of the history of cotton improvement and modernization processes and the introduction of new genes to it and the domestication of wild types.

All recent studies show that the introduction of desirable traits from wild or domesticated germplasm sources into adapted varieties is an ongoing process in most cotton breeding programs and the improvement and modernization of varieties. Useful genetic variation for several traits has been found in the collection of primitive lines of Gossypium hirsutum. The response to long photoperiods and failure to complete flowering under long days in temperate growing seasons represent a major obstacle to the use of more primitive cotton germplasms or wild varieties, which can have high flowering responses but whose genes are not readily available for use, growth, reproduction, or introduction into cotton breeding programs in different parts of the world.

There are two possible ways to overcome this problem: (1) hybridizing non-domesticated wild germplasm with commercially grown varieties in a tropical garden or in a greenhouse, in which case both varieties will flower simultaneously, or (2) introducing day-neutrality genes scientifically and systematically in accordance with sound scientific rules into the germplasm through a reverse hybridization scheme, in addition to making the germplasm available for use in temperate regions. Here we compile information on wild specimens in the American cotton collection and begin a program to estimate the variation between wild and adapted genetic populations and determine their relationships. Cotton Seedlings The genetic collection trips of 1946, 1947, and 1948 established the presumed origin (Mexico and Central America) of the upland cotton species, *G. hirsutum*, by bringing more than 600 wild specimens to the United States. When these specimens were grown and germinated in College Station, Texas, more than half of the specimens remained wild and undomesticated (Lewis and Richmond 1957). Hutchinson (1951) cultivated a large number of wild species in Africa and proposed further collection and germination of geographic species of *G. hirsutum*. Although the term "species" is not a valid taxonomic category, names are helpful for discussing and collecting species and cultivars (Fryxell 1976). Approximately 2500 primitive strains from collecting expeditions over the previous 50 years are currently included in the *Gossypium* collection (Percival 1987; Anonymous 1997).

19.3 Traditional Breeding of Upland Cotton

19.3.1 Improvement Strategies

The most common and well-known pollination system in cotton is self-pollination, although there is a percentage of cross-pollination that can reach 5% in the presence of bees. Also, we find that the majority of cotton varieties of the type *G. hirsutum* grown in all countries of the world focus on reaching maximum productivity and the best fiber qualities in addition to wide adaptation to environmental conditions. In 2000, research on cotton breeders in the United States revealed that less than 3% of the breeding material came from sources other than *G. hirsutum* cotton. This was due to the practice of depending on parental hybridization followed by backcrossing or re-selection from existing hybrids (Bowman 2000). Although American, Russian, and other cotton varieties may still be employed in breeding operations, bred parental genes are currently being used in many breeding, hybridization, and selection procedures (Constable et al. 2001).

19.3.2 Traditional Breeding Methodologies

Many breeding programs for upland cotton have produced and shown the production of new high-yielding hybrids, genotypes or genes with fiber quality and physical traits, in addition to breeding programs for biotic and abiotic stresses, and selection of drought-tolerant genotypes, salinity-tolerant genotypes with the ability or more tolerance to current climate changes. Through studies, reports and results, it was found that the cotton production system in drylands, which has high production potential, can improve the quality of water use efficiency and contain the best irrigation water use quality system (Stiller et al. 2005). The results indicate that cotton breeding for upland cotton has contributed more than 45% to the improvement of yield and crop components since 1983 (Constable et al. 2001). Furthermore, through the results and studies, quantitative traits between *G. hirsutum* and *G. barbadense* have been improved, such as the number of days until the first flower opens or the first five flowers open, and breeding to produce disease-resistant strains and new strains that tolerate insect infestations and produce genetic combinations that can tolerate both high and low temperatures through plant selection programs (Brubaker et al. 1999; Wang et al. 1995).

Upland cotton *G. hirsutum* shares tetraploid AD genomes (Gerstel and Sarvella 1956), and can be crossed to produce fertile or highly fertile first generation (F_1). However, the second generation (F2) plants show greater improvement in gene pools and genetic isolations to obtain in the following generations better and higher yielding and quality structures with high genetic stability (Gerstel 1954; Stephens and Phillips 1972).

19.3.3 Molecular Breeding

Genetic studies have also shown that American cotton has different sequences of ribosomal DNA and chloroplast genomes, which allows for progress in selection (Lee et al. 2006; Wendel et al. 1995). Genetic, physical and morphological isolation mechanisms have evolved to maintain the distinctiveness of existing genotypes (Stephens 1946; Stephens 1950a, b; Stephens and Phillips 1972), in addition to maintaining the characteristics of differences in the timing of pollen dispersal in the flower, and the date and dates of flower opening (Stephens and Phillips 1972), in addition to Pollination, fertilization, pollen viability and fertility (Brubaker et al. 1999; Kearney and Harrison 1932).

However, despite all of the above, these differences can be overcome through directed breeding programs to produce a genetic composition that carries certain desirable traits. Modern programs and research have developed in the processes of pollination, hybridization and selection for the cotton variety *G. hirsutum*, followed by backcrossing or repeated hybridization programs, in addition to repeated selection programs for the possibility of performing QTL (quantitative trait loci) analysis to increase fiber length and staple quality (Chee et al. 2005a), fiber softness and strength (Draye et al. 2005), fiber length (Chee et al. 2005b), in addition to improving cotton crop traits and components (Saha et al. 2006).

Wild relatives of *Gossypium* of cultivated tetraploid cotton are being investigated as an important source of novel genes and for the production of novel genotypes, for example for resistance to Fusarium wilt (McFadden et al. 2004). Crosses between *G. sturtianum* and the upland cotton cultivar *G. hirsutum* have shown improved wilt resistance and significantly better offspring yields compared to the parent, suggesting that *G. sturtianum* may have a useful source of resistance that could be improved through successive crosses and selection programs (Becerra Lopez-Lavalle et al. 2007).

However, many generations of backcrossing and selection are needed to produce genotypes with high quality, disease resistance and environmental tolerance. It appears from the results of research and studies that *G. Raimondi* is resistant to jassid and plants of this species have been used as a genetic source to transfer the resistance trait to *G. hirsutum* varieties. In addition, hybridization between *G. Raimondi* and *G. hirsutum* with successive selection programs has produced new genetic combinations with resistance to jassid. After using colchicine to restore fertility to the resulting combinations, backcrossing is done with parental plants of *G. hirsutum* to produce new varieties carrying the parental traits in addition to the jassid resistance trait (Saravanan et al. 2007).

19.3.4 Genetic Transformation Methodologies, Gene Editing Methodologies and Transgenic Cultivars

The tissue culture technique was initially used to grow cotton crops in 1983 (Davidonis and Hamilton 1983). Since then, cotton plants have transformed, primarily through the use of tissue culture and regeneration-ready Coker cultivars (*Gossypium hirsutum*). After that, this species' commercial cultivars were all backcrossed. The findings of the research, however, indicated that while many cultivars of this cotton would develop into somatic embryos and create a callus, not all of them are equally successful or successfully regenerate into whole plants (Sakhanukou et al. 2004). However, as science has advanced, regeneration techniques for all commercial cotton *G. hirsutum* types have been established (Gould et al. 1991; Sakhanokho et al. 2001, 2004).

The creation of genetically engineered cotton plants with strong resistance to insects and herbicides, particularly to insect pests of the order Lepidoptera like cotton leafworms, has been the primary emphasis. These were primarily made with *Bacillus thuringensis* gene transfer systems (cry1Ac in INGARD® and cry1Ac and cry2Ab in Bollgard II®). When introducing herbicide resistance genes, Roundup Ready® *G. hirsutum* is produced using a single gene from *Agrobacterium spp.* CP4 (cp4) Esps., and Liberty Link® *G. hirsutum* is produced using an exogenous gene from *Streptomyces hygroscopius* to produce glyphosate ammonium herbicide resistance.

Research approvals in later stages are still focusing on different agronomic characteristics of cotton plants. After that, field trials were gradually approved in many countries for the genetically modified cotton type *G. hirsutum*, its use, cultivation of its varieties, and increasing the cultivated areas of it, in addition to introducing new traits to it, for example, increasing its ability to withstand water stress, increasing its ability to use water efficiently, tolerating water stress, and the ability to resist insects, in addition to improving its ability to resist fungi and fungal diseases. The same studies and programs, for instance, are being carried out in the United States of America and include characteristics like enhancing resistance to cold and high temperatures, resistance to water stressors (drought), and boosting plants' capacity to withstand salinity, whether it be soil or water salinity. There are also many studies and research being conducted on genetically modified cotton to improve fiber quality and various quality traits (USDA 2007).

There are numerous reports and studies on genetically modified cotton plants with increased yield, fiber length, and good fiber quality from upland cotton, even though genetic engineering advancements to improve cotton fiber yield, seed quality, cotton lint yield, and fiber quality traits are steadily advancing daily. There is constant interest in lowering the amount of gossypol in the seeds so that they can be used as human food or to feed and fatten farm animals. Cotton is one of the most significant high-quality oil crops in the world, and cottonseed meal is a very high nutritional value food source for animals that feed on it (Wilkins et al. 2000). It is currently possible to produce genetically modified cotton plants of the type *G. hirsutum* with very low levels of gossypol in their seeds while still maintaining their high productivity and fiber quality characteristics through the efforts of international research centers (Sunilkumar et al. 2006).

19.4 Morphology of Upland Cotton

Regarding the Upland cotton plant's morphology, it can be defined as a perennial that has been tamed and acclimated to grow as an annual, reaching a height of roughly 1.5–2 meters. All varieties of this type are grown as annual plants in all countries of the world that produce it. With the removal of plants and plant remains after harvesting the crop to obtain seeds and fibers. Cotton plants are characterized by their growth usually being unlimited. This means that the plant can grow and produce leaves, stems, flowers, fruits (almonds) and seeds at the same time, and its growth is true-axis and not false-axis, meaning that the growth of the stem does not end with a flower or an almond that stops its growth, but rather ends with a green bud that continues to grow.

Vegetative and fruiting branches are the two categories into which the cotton plant's current branches fall. While fruiting branches have several stem tissues, each beginning after the previous fruiting bud and displaying a zigzag development tendency, vegetative branches have only one stem tissue and grow long and straight as a primary stem. Fruiting branches start to form after the first five to six nodes of the main stem sustain primarily vegetative growth. As seen in Fig. 19.1, fruiting branches have a 3⁄8-reciprocal branching arrangement (Oosterhuis and Jernstedt 1999; and Ritchie et al. 2007). In order to optimize light absorption throughout the day, *G. hirsutum's* leaves are heliotropic, meaning they are typically flat and follow

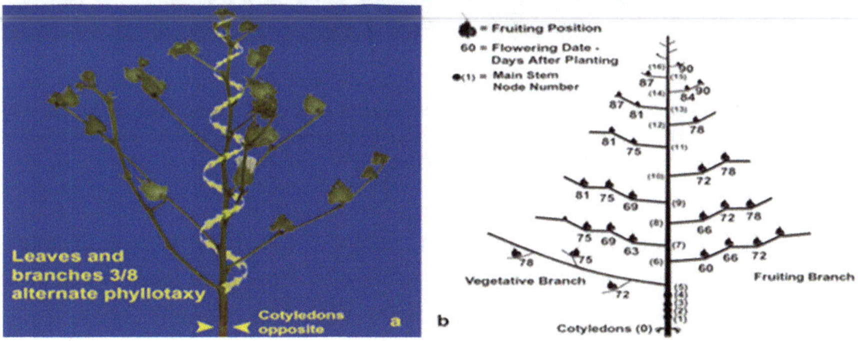

Fig. 19.1 The morphology of cotton plants (**a**) The 3\8 alternating phyllotaxy of branches is visible on a cotton plant that has lost its leaves. From the branch below it, each branch is 3/8 of a circle around the stem. The main stem leaves' axils give rise to the branches. (**b**) A schematic showing when flowers generally emerge from buds on fruiting branches according to fruiting position. (Reproduced from Ritchie et al. 2007 with permission)

Fig. 19.2 *Gossypium hirsutum* L. is heliotropic, its leaves are generally flat and track the sun to maximize light adsorption throughout the day, (Sassenrath-Cole 1995; Wise et al. 2000; Lu et al. 1997; Wise et al. 2000). Generally, leaves on the main stem axis have five for *G. hirsutum*, whereas those on the fruiting branches have three lobes (Gore 1935)

the sun (Sassenrath-Cole 1995; Wise et al. 2000; Lu et al. 1997; Wise et al. 2000). For *G. hirsutum*, leaves on the fruiting branches often have three lobes, while those on the main stem axis typically have five (Gore 1935) (Fig. 19.2 and Table 19.1).

19.4.1 Flowering and Reproductive Form

The botanical description of cotton indicates that its blooms are pentamerous (i.e., grouped in five segments), large (5–9 cm), and complete (i.e., including both male

Table 19.1 Description of cotton (*Gossypium hirsutum* L.*)* Fryxell plant morphology

Gossypium hirsutum L.	
Habit	Shrubs that are at least one meter tall, typically have many branches, are about stellate-pubescent, and have glands all over them.
Leaves	Broadly triangular to ovate, cordate, long-petiolate, weakly three to five-lobed, and sharp to acuminate lobes
Stipules	subulate, 5–15 (rarely to 20) mm long
Distribution	Originally from Middle America, the Antilles, and several Pacific Islands (Socorro, the Marquesas, Samoa, etc.), it is now grown almost anywhere in the world.

and female organs in the same flower). Additionally, we discover that they simultaneously contain floral nectar on the inside and the outside (Moffett 1983). The stigma measures 0.5:1 cm in length, while the style measures about 2.5 cm. We also discover that each almond has 3:5 divisions or gaps in the ovary, each of which contains 5:10 ovules. The majority of the style is encased in a sheath that contains several 0.5:1 cm long stamen, each of which terminates in an anther that often yields copious amounts of self-pollinating pollen grains that are viable for roughly 24 hours (McGregor 1976). Additionally, tests and studies have revealed that each blossom contains roughly 20,000 pollen grains (Ter Ava Nesian 1978). *Gossypium hirsutum* L. flowers are distinguished by their appearance and the way they communicate to pollinators their desire for food (see Fig. 19.3). The blooms of *Gossypium hirsutum* L. release a lot of nectar and have cream-colored pollen grains (McGregor 1976; Moffett 1983), which may influence the likelihood of cross-pollination (Table 19.2).

19.4.2 Flowering and Reproductive Development in Upland Cotton (**Gossypium hirsutum** *L.)*

The growth of the cotton plant can be specifically divided into three main and fundamental developmental stages if we look at it from an agricultural perspective. The first stage is known as the reproductive and dispersal stage, the second is known as the germination and seedling establishment stage, and the third is known as the leaf area and canopy development stage. From the germination stage to the maturity of the first fruit, the entire plant life or developmental and growth stage of the cotton varieties of the species *Gossypium hirsutum* L. typically lasts 15–17 weeks, although this time frame may vary depending on the current temperatures and the overall environmental conditions surrounding the plant (Oosterhuis and Jernstedt 1999; Ritchie et al. 2007).

Fig. 19.3 Cotton flowers. (**a**, **b**) *Gossypium hirsutum* L. flower. (Source: Ritchie et al. (2007))

19.4.3 Reproduction in Cotton

We find that upland cotton plants reproduce sexually in general and mainly, although there are reports and studies on the possibility of rooting cuttings or asexual reproduction as explained below.

Table 19.2 description of upland cotton (*Gossypium hirsutum* L.) flowering

Gossypium hirsutum L.	
Flowers	flowers usually in sympodial inflorescences, the pedicels 20–40 mm long, surmounted by three involucellar nectaries
Bracts	bracts of the involucel inserted above each nectary, foliaceous (enclosing the bud), ovate, three to 19-laciniate
Calyx	truncate or five-toothed, 5–6 mm long (excluding teeth)
Petals	up to 50 mm long, cream-colored or pale yellow, with or without a dark spot at the base; androecium included
Style	single with decurrent stigmatic lobes, more or less enclosed by androecium or somewhat exceeding androecium
Capsule	three to five-celled, glabrous, smooth, broadly ovoid or subglobose
Seeds	several per locule, lanate, the seed fibers white, tan, or red-brown

19.4.4 Asexual Reproduction or Vegetative Reproduction

Cotton plants do not reproduce vegetatively and do not respond to this sort of reproduction in their natural environment and growth. However, *Gossypium hirsutum* L. plants and certain hybrids between *Gossypium hirsutum* L. x *Gossypium barbadense* have a rooting system under experimental conditions. These hybrids have been shown to root semi-hardwood cuttings, but only when plant hormones like NAA and indole butyric acid are added (Sheelavantar et al. 1975). Additionally, grafting *Gossypium hirsutum* L. plants onto several rootstocks was accomplished with success, completing the process of asexual reproduction in upland cotton plants (Rea 1931, 1933). It should be remembered, nonetheless, that the grafts need to be finished less than an hour after the pieces are cut in order for vegetative propagation in cotton to be completely successful.

19.4.5 Sexual Reproduction

According to research on the process of sexual reproduction in cotton plants, the plant reaches complete reproductive maturity 4–5 weeks after seeding. This is when flower buds (also known as "squares") start to form. As seen in (Fig. 19.4), these flower buds first take the form of a little pyramid, which is surrounded by three enormous green cones that entirely enclose the budding flower. The actual flowering phase, or the opening of the flower, occurs approximately 25 days after the square first appears (Oosterhuis and Jernstedt 1999; Ritchie et al. 2007).

Cotton plants of *Gossypium hirsutum* L. start flowering, flower production, growth and flowering curve progression after about 60 days from planting (Bang et al. 2002). Generally speaking, if *Gossypium hirsutum* L. is planted in March and early April, flowering and flowering normally occurs after about 60 days. Additionally, research has demonstrated that while all contemporary cotton cultivars are

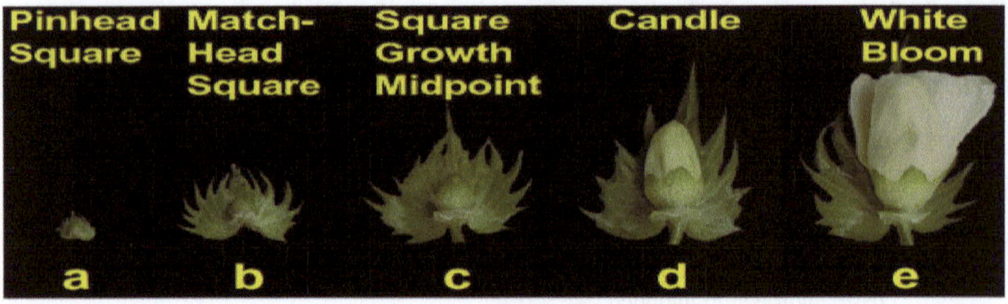

Fig. 19.4 Developmental stages of floral buds. (**a**) initiation of the floral bud, (**b**) early bud enlargement, (**c**) mid-stage of bud development, (**d**) petals start to elongate and become visible and (**e**) fully opened flower with visible white petals. (Source: Ritchie et al. (2007))

unaffected by the length of the day or the number of light hours, they may exhibit a preference for fruiting on cool nights and moderate water stress, as seen by an increase in fruit set during these circumstances (Hearn 1981). Studies and research have shown that under normal crop conditions, natural shedding of up to 60% of the immature fruits occurs before they open and produce their own crop or are completed. Studies have also shown that mature, uninfected flowers do not usually fall before pollination and are fully developed and open properly (Oosterhuis and Jernstedt 1999).

As for the flower opening system in cotton plants, we find that the flowers open in a sequence that can be predicted, studied, and graphically drawn, from which we can judge the varieties, their yield potential, and the extent to which they are affected by environmental conditions. The first bloom, or number 1, opens at the base of the plant, which is closest to the stem, as shown in Fig. 19.3. The next flower will then open in the same relative position on the next highest branch after roughly 3 days, followed by the next flower on the lowest branch after another 3 days, and so on, until all of the cotton plant's mature bolls are fully developed.

Also, through the information and studies, it is clear that the flowering system progresses and takes place in an ascending spiral pattern (Oosterhuis and Jernstedt 1999). According to research findings, upland cotton flowers open at or close to dawn, releasing pollen first when the bloom is closed. At sunrise, the flower opens and stays open all day and for just 1 day. According to Beasley (1975), over 90% of flowers that open in a single day do so within an hour. At flowering, the petals of *Gossypium hirsutum* L. are creamy white, turning reddish pink within 1 day of pollination as shown in Fig. 19.5 from (a) to (d). The petals then fall after turning reddish brown. Cotton is characterized by an indeterminate flowering pattern, so flowers appear over several weeks, but in some cotton, it can extend to three months (Cherry and Leffler 1984). At the peak of flowering, there are usually several flowers open along the plant each day (McGregor 1976).

Fig. 19.5 American cotton (*Gossypium hirsutum* L.). (**a–d**) shows the Patil color from the opening flower to pollination and hybridization and drop the petals. Source: Yehia, W.M.B. 2023

Table 19.3 Pollen size and spine length of American cotton (*Gossypium hirsutum* L.)

Gossypium hirsutum	Species Size (μm)	Spines (μm)	Spine density	References
	85–88	7.5	–	El Nagger
	100.9	12.1	8.3×10^{-3} spines/μm^2	Kakani et al.
	103 ± 6.2	–	–	Saad

19.4.6 Pollination and Pollen Spread

Immediately after the flowering of highland cotton plants and the flower opens, the stamens in cotton flowers open and burst, causing pollen to fall. After that, and after the pollen grains fall, the cotton plants grow after that and after the opening of the corolla becomes large enough or enough for pollinators to reach it and enter the flower and flower petals, which are bees, and they represent the most important pollinators. The stigmas become receptive shortly after that for pollen to fall and pollination and fertilization to occur. Therefore, flowers in cotton plants and species are mostly self-pollination, with a percentage that almost reaches 100% most of the time, except for a small percentage of external cross-pollination in the presence of bees, which reaches 3–5%, especially since there are no mechanisms of self-incompatibility in cotton plants. Cotton pollen grains are relatively large and have long thorns. The specifications of pollen grains for American cotton are shown in Table 19.3.

We find from previous studies and research that the viability and survival of pollen grains decrease rapidly after 8 hours for the *Gossypium hirsutum* L. species according to (Govila and Rao 1969; Richards et al. 2005). Studies have confirmed that high temperatures to which *Gossypium hirsutum* L. flowers are exposed to full sunlight for long periods of time lead to a significant decrease in the vitality and germination of pollen grains (Burke et al. 2004; McGregor 1976). The results obtained from a research study on the minimum, maximum and optimum temperatures for survival of 12 varieties of Upland cotton (*Gossypium hirsutum* L.) as the optimum and 43 °C as the maximum temperature (Kakani et al. 2005). From botanical descriptions and studies, we find that pollen grains germinate 30 minutes after falling on the stigma, and then ovules are fertilized within 24–48 minutes after pollination (Pundir 1972). For complete fertilization, which results in the presence of all the seeds in the boll, about 35 ovules must be fertilized, and thus at least 35 live, fully vital and viable pollen grains must touch the stigma (McGregor 1976). According to studies, the pollen tube grows more quickly when there are a lot of pollen grains on the stigma (Ter Avanesian 1978). The nucleus of the pollen tube shifts a few microns ahead of the sperm as it develops along the style. After roughly 15 hours, the sperm and all of its contents are deposited into the ovule's germinal sac, and fertilization is finished 24–30 hours after the flower opens (Gore 1932).

19.4.7 Pollination

We find from studies and research that the cotton plant is characterized by the presence of self-pollination, which is the predominant system in the first place with large, sticky and heavy pollen grains that cannot be easily spread by the wind (McGregor 1976; Moffett 1983). The flowers in cotton are large and clear and have attractive colors for bees, which are the most important pollinators (Green and Jones 1953). Therefore, the general description of cotton is that it is a self-pollinating and fertilizing plant and does not hybridize externally except when honey bees are present (Oosterhuis and Jernstedt 1999). Honey bees are thought to be the most probable insects and the main cause of any percentage of cross-pollination in cotton, according to all prior research and in many nations (Mungomery and Glassop 1969; Thomson 1966).

In many studies, honey bees were the best, most important and sometimes the only pollinator as a major pollinator (Mungomery and Glassop 1969). However, with this information and acceptance, in a similar study in the Ord River Valley in the absence of honey bees in Western Australia, a small amount of cross-pollination was observed (Thomson 1966). Pollen beetles (Carpophilus sp.) were proposed as the primary cross-pollinators in cotton outcrossing experiments near Narrabri, New South Wales, even though no bees or honey bees were found in the area, despite the presence of a few wasps and flies (Llewellyn and Fitt 1996; McGregor 1976). Since cotton is a prominent honey plant and is present in large quantities, honey bees are seen to visit cotton blossoms largely for the sole purpose of collecting nectar (Vansell 1944). Although honey bees hardly ever gather pollen from cotton, pollen adheres to their body hairs and influences pollination, resulting in cross-pollination or a certain amount of it in American cotton (Moffett et al. 1975; Vaissière and Vinson 1994). Additionally, since pollen obtained in pollen baskets is unavailable for the cotton pollination process, we discover that bees' incapacity to gather cotton pollen and transfer it to their colonies has nothing to do with their capacity to cross-pollinate cotton blossoms.

19.4.8 Percentage of Outcrossing in Upland Cotton

According to scientific studies and research, the prevalence of insects and pollinators in cotton production fields (Llewellyn et al. 2007; Phelung 2001), as well as other natural conditions like location, timing, temperature, and wind, have a significant impact on the amount and percentage of cross-pollination in cotton (Moffett et al. 1975, 1976). Furthermore, since many pollinator or insect species prefer to collect nectar over pollen grains, it can be argued that the extent or volume of insect visits to fields or plants influences the estimation of cross-pollination rates (Moffett et al. 1975; Rao et al. 1996). Cross-pollination has been estimated to be 10% or less in a number of field studies conducted in Australia, the USA, other nations, and abroad (Gridley 1974; Llewellyn and Fitt 1996; Smith 1976; Meredith and Bridge 1973; Umbeck et al. 1991). According to independent research, plants sown in nearby rows of cotton exhibit 1–2% cross-pollination (Llewellyn and Fitt 1996; Mungomery and Glassop 1969; Thomson 1966). Considering what has been noted in certain other nations, this estimate of cross-pollination is comparatively low. Low or high rates of cross-pollination, especially the presence or absence of bees, which are regarded to be the finest pollinators, may be caused by differences in insect and pollinator species (Llewellyn and Fitt 1996).

Several studies of cotton pollen distribution have shown that when cross-pollination occurs, the pollen source, honeybees, and the distance between pollinators and the cotton fields have been studied. These studies have recorded that the rate of cross-pollination decreases significantly as the distance between insect pollinators and the cotton crop increases (Chauhan et al. 1983; Llewellyn and Fitt 1996; Thomson 1966; Umbeck et al. 1991). To study the rate and proportion of cross-pollination in cotton fields in Australia, researchers observed that using plots of genetically modified *Gossypium hirsutum* L. cotton surrounded by some rows of non-*Gossypium hirsutum* L. cotton plants, a significant flow of pollen to the non-*Gossypium hirsutum* L. cotton occurred (Llewellyn and Fitt 1996). The results obtained from research and studies on the rates of cross-pollination showed that there are large differences in the rates of cross-pollination according to the seasons and wind direction and speed. It was estimated that the highest level of cross-pollination was (0.9%) in the first row at a distance of 10 meters, while the rates of cross-pollination were generally rare as the distance increased, reaching 0.01% up to 16 meters. The results also showed that no cross-pollination was detected between 16 and 20 meters as a separation distance. The results of other experiments confirmed that cross-pollination rarely occurs after a distance of 20 meters (Llewellyn et al. 2007).

The expansion of cotton cultivation in the northern tropical regions has led to the tendency to make estimates of the rate of cross-pollination in these regions, which have high numbers of insects and different environmental conditions (Llewellyn et al. 2007). It is believed that the presence of higher rates and percentages of cross-pollination in cotton fields is due to the increase in the number of pollinators resulting from the spread of bee hives in the fields adjacent to cotton (Llewellyn et al. 2007). Cross-pollination between neighboring plants as a percentage of red-leaf cotton was in the range of 0–5%, with mean values of 1.6% and 1.0% in both seasons, according to a previous cross-pollination experiment of red-leaf cotton *Gossypium hirsutum* L. planted in a field of green-leaf *Gossypium hirsutum* L. A (Thompson 1966) in the Ord River Valley, Washington. No cross-pollination was found at distances other than 3–8 m, and very little cross-pollination happened at distances greater than 3 m (mean <0.01%).

Results from other studies conducted in Mississippi, USA, to measure and estimate the extent of pollen dispersal from genetically modified cotton *G. hirsutum* to a field of non-GM cotton was found to be as high as 5.7% in the first isolation row, with less than 1% at 7 m, but lower levels of cross-pollination were observed at distances of 25 m and no cross-pollination was observed at distances greater than 25 m (Van Deynze et Umbeck et al. 1991, 2005). In a field where bees were present, cross-pollination was estimated at 0.32% at a distance of 30 m. A study conducted in Greece using phenotypic features to analyze cross-pollination in cotton revealed a rate of 2.2% at 1 m and zero at 10 m. Another experiment revealed a slightly higher rate of 3.8% at 1 m and zero at 20 m (Van Deynze et al. 2005; Xanthopoulos and Kechagia 2000).

19.4.9 Fruiting and Seed Development in Upland Cotton: Fruiting Development

After the flower appears and opens and the pollination and fertilization processes are completed in approximately 5–7 days, the flower petals dry up and fall from the plants to reveal the cotton fruit that grows and increases in size or boll (Ritchie et al. 2007). In Greece, a study that analyzed cross-pollination in cotton using phenotypic traits found that the rate was 2.2% at 1 m and zero at 10 m. A slightly higher rate of 3.8% at 1 m and zero at 20 m was found in another experiment (Van Deynze et al. 2005; Xanthopoulos and Kechagia 2000). The growth and development processes of the boll can be discussed in three stages. The first stage is at the beginning where growth occurs in the length of the cotton fibers until the maximum size of the boll and the seeds contained in it is reached. The second stage occurs after approximately 3 weeks, and here we find that the filling stage begins, where cellulose layers are deposited inside the hollow cotton fibers at a rate of 2 layers per day for 3 weeks.

After roughly 6 weeks, the boll reaches its third and last stage, when it starts to mature and dry up in preparation for opening (Ritchie et al. 2007). According to Yasuor et al. (2007), each mature berry produces 29–34 seeds per mature, opened boll and splits into four or five tufts, each of which contains several seeds encased in fibers (Berardi and Goldblat 1980).

The ripe fruits dry quickly and are thick and leathery. Frequently, these bolls fracture, revealing the fibers and seeds within. Additionally, many of the opening fruits stay in the field for a long period before being collected because cotton is often harvested only once or twice (Cherry and Leffler 1984). Seed quality can deteriorate once the boll opens and the white cotton fiber-covered seeds are exposed to weather and climate, which can occasionally result in a loss of strength and a decreased ability to germinate (Hopper and McDaniel 1999). Since cotton fibers are longitudinal growths from epidermal cells, they are among the finest, strongest, and softest plant fibers (Smith 1995). Lint growth starts as soon as the flower opens and is not dependent on the completion of fertilization or pollination (Gore 1932).

After flowering, about 20% of each seed's epidermal cells start to lengthen and get long enough to spin into fibers. About 6 days after flowering, epidermal cells start to lengthen and create lint, which are thick, short fibers. The fibers are made up of a primary wall, a secondary wall, a layer of protoplasm, and a cavity known as the central vacuole during the elongation or lengthening phase. Additionally, we discover that during the cellulose filling stage, cellulose microfibrils are deposited on the cavity's inner side and visible under a microscope as a daily growth ring during the last stage of fiber maturation. When the fibers dry out, the cavity collapses, giving the appearance of a twisted or semi-twisted ribbon (Smith 1995). The ripe cotton seed is oval, pointed at the tip, and between 8 and 12 mm long. It is primarily made up of a kernel and pericarp, with a thin membrane between the two. Furthermore, gossypol glands are found throughout the kernel's tissues as oval-shaped patches that range in length from 100 to 400 μm (Berardi and Goldblatt 1980) (Fig. 19.6).

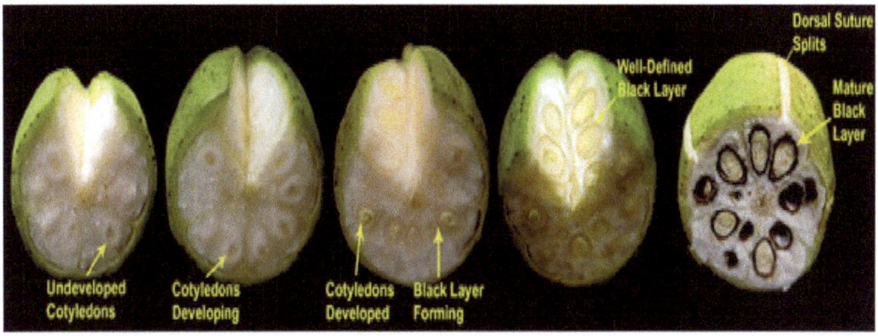

Fig. 19.6 Development in cotton (*Gossypium hirsutum* L.) bolls

19.5 Classification of Upland Cotton

High-volume precision instruments assess the following characteristics of American cotton (Upland cotton): fiber length, uniformity, strength, Micronair, color or depth of fiber color, waste, and degree of lint. This method is known as "high-volume instrument grading" (Fig. 19.7).

19.5.1 Fiber Length

The average length of the longer half of the fiber (the average upper half length) is the definition of fiber length for cotton. Both 100 and 32 inches are noted (Table 19.4). An optical sensor is used to measure the length of a "beard" of parallel cotton fibers. When fibers from a cotton sample are automatically held in place with a clip and then combed and rubbed in a parallel direction, a beard is created. The fibers' lengths and limitations in centimeters and their corresponding lengths in inches are displayed in the following Table 19.4.

Studies have shown that fiber length is greatly affected by the presence of diversity, in addition to the exposure of cotton plants to high temperatures for long periods or exposure of cotton to water stress or a deficiency in the necessary nutrients for the plant and for plant nutrition, which may lead to shorter fibers and greatly affect the optimum length of the cultivated variety. Also, excessive cleaning of cotton flowers or extensive drying at the ginning site or in the ginning mills may lead

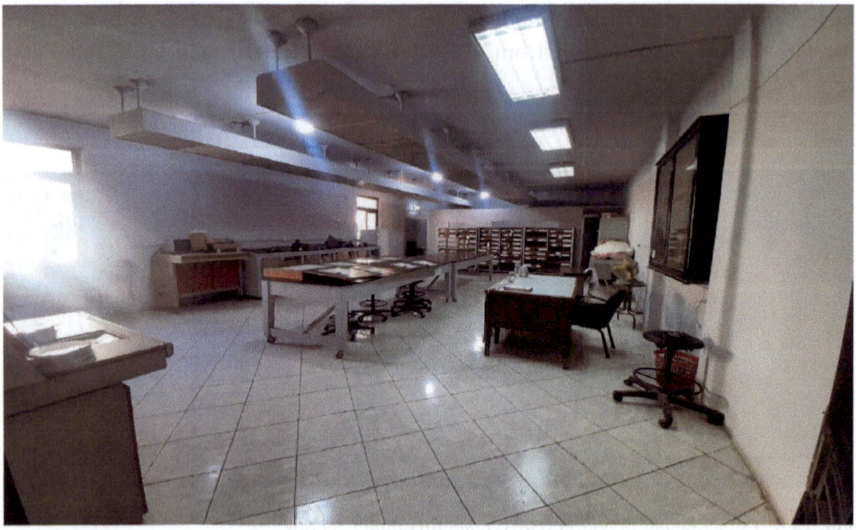

Fig. 19.7 Classing methodology is based on both grades. (Source, W.M.B. Yehia Cotton Research Institute)

Table 19.4 Upland cotton fiber length conversion for Upland cotton (*Gossypium hirsutum* L.)

Upland cotton fiber length conversion			
Inches	32nds	Inches	32nds
0.79 & shorter	24	1.11–1.13	36
0.80–0.85	26	1.14–1.17	37
0.86–0.89	28	1.18–1.20	38
0.90–0.92	29	1.21–1.23	39
0.93–0.95	30	1.24–1.26	40
0.96–0.98	31	1.27–1.29	41
0.99–1.01	32	1.30–1.32	42
1.02–1.04	33	1.33–1.35	43
1.05–1.07	34	1.36–longer	44 & linger
1.08–1.10	35		

Table 19.5 Interpreting length uniformity for Upland cotton (*Gossypium hirsutum* L.)

Interpreting length uniformity	
Description of the degree of uniformity	Length uniformity index (percent)
Very high	Very high
high	83–85
Intermediate	80–82
low	77–79
Very low	Below 77

to shorter fibers. It can be said that fiber length affects the strength and durability of the resulting yarn and the efficiency of the spinning process itself. Fiber length is also affected by the fineness of the yarn that can be successfully produced from the ideal fibers.

19.5.2 *Length Uniformity*

The relationship or ratio of the average length to the average length of the fibers, or the ratio of the average length of the fiber to the average length of the upper half, given as a percentage, is known as length uniformity. The average length and the average length of the upper half of a bale would be the same if every fiber in it had the same length, resulting in 100% length uniformity. However, length uniformity is always less than 100% since cotton fibers naturally vary in length. A reference for evaluating length uniformity measures can be found in Table 19.5. According to research and findings, the degree of length uniformity influences the strength and durability of the finished yarn, its uniformity, and the scope and effectiveness of the spinning process itself. The degree of short fiber content (fiber content less than 1/2 inch) has also been found to be closely associated with this. There is no doubt that cotton with a low uniformity index has a significant proportion of short fibers.

Furthermore, processing such cotton is challenging, and it is a given that the resulting yarn will be of poor quality.

19.5.3 Fiber Strength

Grams per Tex are used to quantify cotton strength. The weight of 1000 meters of fiber in grams is called tex. Thus, the concept or meaning is that the force in grams needed to break a bundle of fibers of one text size equals the resulting or measured force. The same cotton skeins used to measure fiber length were also used to test force. The amount of power needed to break the fibers was measured while the skeins were held in two sets of jaws, separated by 1/8 inch. Guidance for evaluating fiber strength measurements is given in the accompanying table. Since every cotton variety has a unique fiber strength that sets it apart from other variations, the variety has a significant influence on fiber strength. However, deficiencies or excesses of plant nutrients, nutrients required for the plant, and the surrounding weather can all have an impact on fiber strength. One could argue that there is a strong or tight relationship between the strength of yarns and fibers. Cotton with a high fiber strength will also be able to resist breaking during production (Table 19.6 and Fig. 19.8).

Table 19.6 Interpreting fiber strength for Upland cotton (*Gossypium hirsutum* L.)

Interpreting fiber strength	
Description of degree of strength	Strength (grams per tex)
Very strong	31 & above
strong	29–30
average	26–28
Intermediate	24–25
weak	23 & below

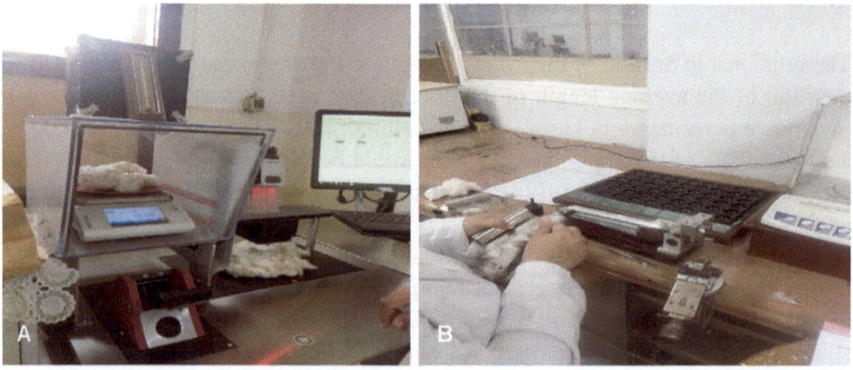

Fig. 19.8 Fiber length and strength measurements are made on the same. (Source W.M.B. Yehia Cotton Research Institute)

19.5.4 Micronair: Fiber Fineness

The fineness and maturity of cotton fibers are gauged by the micronair reading value. The plan is to test the air permeability of a fixed mass of cotton fibers compacted to a fixed size using an airflow instrument through cotton. The graph displays the micronair measurements and readings. According to studies and research, the weather and environmental factors surrounding the plant during the growth period, such as temperature and humidity during the growing season, cellulose deposition, sunlight, the presence or absence of plant nutrients, and the type of cotton grown, can all have an impact on the micronair reading value. It's also critical to understand that the softness and fineness of the fibers have multiple effects on the quality of the finished product after spinning as well as the effectiveness of chemical processing.

For instance, it is crucial to use cotton with a low micronair value throughout the opening, cleaning, and combing operations. Additionally, slower processing speeds are required for fine fibers to avoid fiber breakage. We discover that threads composed of softer and finer fibers have more fibers per cross-section, making them stronger and more resilient. Additionally, research indicates that the maturity of the fibers influences the yarn's capacity to absorb dyes and the degree of their retention. The better the capacity to absorb and hold onto dyes, the higher the maturity values (Fig. 19.9).

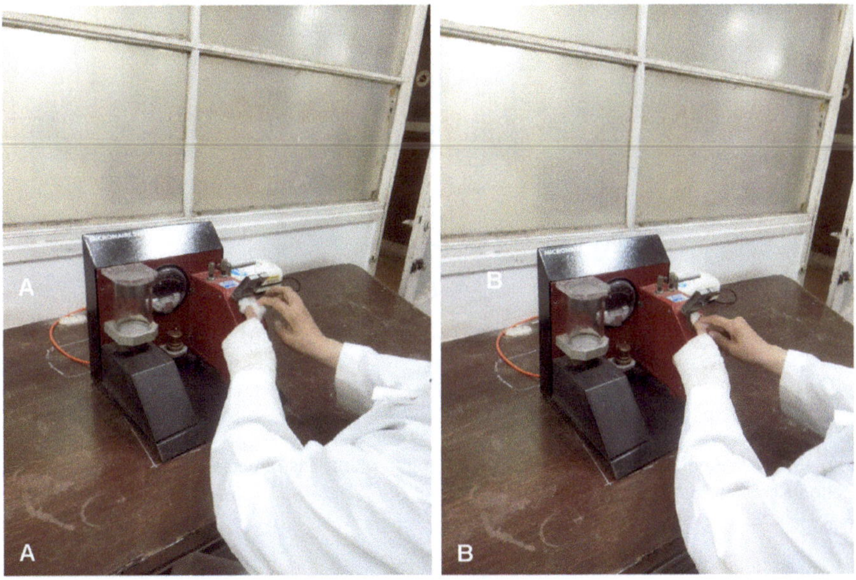

Fig. 19.9 Upland cotton with micronair of (**a**, **b**). (Source: W. M. B. Yehia Cotton Research Institute)

19.5.5 Color Estimation

In American cotton (Upland cotton) and the color characteristic and its estimation, we find that the color characteristic of the fibers is determined by estimating the degree of reflection (Rd) and estimating the degree of yellowness (+b), according to what is stipulated in the official and standard standards for estimating the color characteristic. The measurement is done using a large-sized instrument. The measured reflectance values indicate the brightness or dullness of the existing cotton sample, while the measured yellowness values indicate the degree of dyeing or dyeability. After that, a color code was determined. This code consists of three numbers and is output or estimated by determining the point where the Rd and +b values intersect on the color chart for the type of American high-quality cotton.

The amount of rainfall, low temperatures or snowfall, the spread of insects and fungi, and the spread of spots in cotton hair due to contact with soil, grass, cotton plant leaves, other plants, or weeds can all have an impact on the color of cotton fibers, according to observations and findings from earlier research on estimating color values. Numerous studies also show that excessive moisture and storage temperature, whether before or after ginning, can have an impact on color. It is crucial to remember that environmental factors that cause color deterioration impact fibers' capacity to absorb and hold dyes (Fig. 19.10).

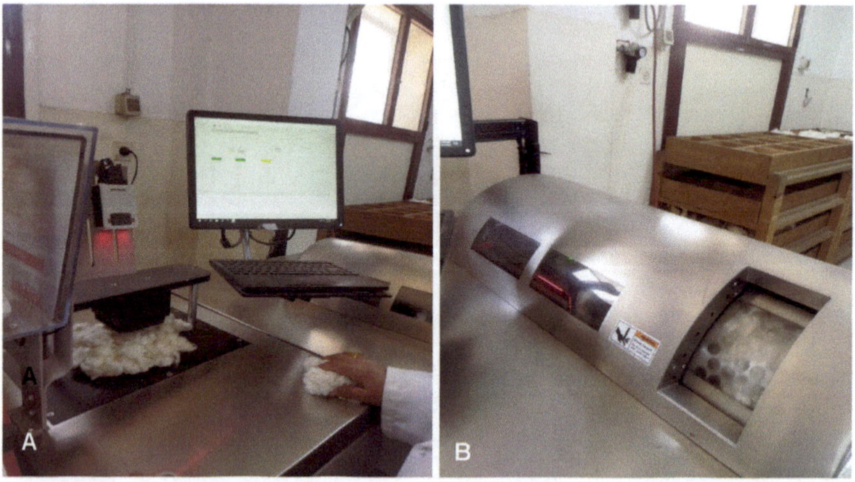

Fig. 19.10 The color of cotton graded by the high-volume instruments. (Source, W.M.B. Yehi Cotton Research Institute)

19.5.6 Trash

One may argue that the trash % is a crucial indicator of the quantity of wastes, materials, and non-lint leftovers found in cotton, such as leaves, bark, trees, and plant residues. A digital camera is used to scan the cotton hair sample's surface, and the digital images that are produced are then examined. Next, we determine the number of visible litter particles (particle count) and the percentage of the surface area occupied by trash particles (percentage area). The data that results is then summarized and explained. Through the measurements, we discover that the average size of foreign or non-lint particles in the cotton sample can be accurately predicted by the computed ratio of the relationship between the percentage of trash area and the number of litter particles. An estimated lower average particle size is indicated, for instance, by a low percentage of area combined with a high number of particles as opposed to a high percentage with a low number of particles.

The research's findings, which provide reasons, show that a high percentage of trash area raises the amount of waste generated during the textile spinning mill's processing and lowers the quality and value of the yarn that is produced. Additionally, we discover that tiny waste particles, also known as "seed waste," are more unwanted since they are more challenging for the mill to extract from the cotton lint than larger waste particles.

19.5.7 Leaf Grade

When talking about the degree and spread of leaves in a cotton sample, we find that it is a measure of the extent of the spread of plant leaves in the resulting cotton. Recent research, studies and intensive developments have shown the acceptance of mechanical leaf grades. In addition to the number of particles, we discover that a good estimate of the proportion of big mechanical waste area may be used to predict the degree, distribution, and quantity of leaves. Based on the American Pima Grade and Universal Upland Grade standards, it can be said that the extent and dispersion of leaves have been evaluated and computed from these factors.

It is worth noting that the content of leaves, their quantity and spread are affected by the type of plant present or variety, in addition to the harvesting methods and harvesting conditions present in the field. We discover that the number of leaves that remain in the lint after ginning is determined by the number of leaves that were present and dispersed in the cotton crop prior to ginning, the amount of cleaning, and the kind of drying and cleaning equipment that was used in the gins to remove these leaves. A small percentage of leaves are still present in the cotton lint, even the most precise harvesting and ginning techniques. From the perspective of manufacturing, the paper's composition is waste, and extracting it from the cotton requires charges or expenses. Furthermore, tiny particles might eventually lower the quality of the finished product even though they are not always completely or successfully eliminated.

19.5.8 Extraneous Matter

Anything in cotton other than fibers or leaves is referred to as extraneous matter. Bark, grass, weeds, broken seed residues, spindle fragments, seed coat fragments, dust, oil, plastic, grease, textiles, textile residues, metals, and leather are a few examples of foreign matter. According to research on the dispersion of extraneous matter in cotton and its varieties, plastic foreign matter is not evenly dispersed throughout the plastic-contaminated bale, in contrast to plant foreign matter (such as particles of bark, grass, or seed coat). As a result, plastic foreign matter may or may not be present in a classification sample taken from a bale contaminated with plastic.

By analyzing and evaluating cotton samples, we discover that unusual, subpar, or non-conforming preparation is another element mentioned in the classification and grading record and classified under "foreign matter." For instance, "preparation" refers to the level of roughness or softness of the ginned cotton pile. We discover that variations in the mechanical and manual techniques of harvesting cotton, as well as the machinery employed for the process, lead to variations in the preparation's smoothness or roughness, which are frequently quite noticeable. Improved harvesting and ginning techniques have led to a dramatic reduction in the unnatural preparation of high-altitude cotton in recent years.

19.5.9 Classification Standard Averages of Upland Cotton

It is important to talk about what is called the Standard Average, which is a voluntary program that has been offered since 1991 to customers of the Cotton Program in the United States of America at no additional cost to the cotton product. In practice, though, we discover that this approach enhances the repeatability of high-volume instrument measurements of cotton's strength, length, homogeneity, micronair, and other physical characteristics. All stakeholders in the system can trade American cotton with more assurance and higher quality standards thanks to this program's reproducibility and increased accuracy, which raise the value of American cotton grading. Because the Standard Average employs the measurements made through the current approach to acquire a sample from each side of each existing bale, it does not require additional sampling procedures, particular methodologies, the addition of duties, or special measurement techniques.

This new standard average will allow us to assign an average or overall mean value to each bale in the unit by averaging all individual bale fiber quality measurements inside a unit or trailer. The average strength for the unit, for instance, is calculated by adding the individual strength readings for each bale in the unit and dividing the result by the number of bales in the unit. This strength value is then used and allocated to each bale in the unit. This average functions as both a general average and a gauge of the bale's or unit's ultimate quality.

19.6 Conclusions and Prospects

The most commonly planted cotton species worldwide is *Gossypium hirsutum* L, sometimes referred to as Mexican cotton or Upland cotton. Cultivars of this species account for almost 90% of all cotton produced worldwide. Upland cotton is a subshrub belonging to the mallow family that can grow in a variety of ways. It typically grows to a height of five feet and has two different types of branches, the fruiting branches of which bear the fruits and flowers. It typically grows to a height of five feet and has two different types of branches, the fruiting branches of which bear the fruits and flowers. It's also the most common type of cotton in the US, making up 95% of the cotton planted on American soil. Fiber length ranges from 15 to 35 mm. Grown mostly in China, India, Brazil, Argentina, USA etc. The adaptability of Upland Cotton is quite astounding. It is the chameleon of crops, changing from the sterility of medical gauze to the coziness of your favorite t-shirt. Upland cotton is used for more than just clothes; it's a mainstay in home furnishings and even has applications in medicine.

Postharvest processing, crop management, and genetics can all be used to increase the quality of cotton fiber. To create improvement methods, one must understand how fiber qualities affect processing, as well as how they are inherited, relate to one another, and are influenced by the environment. Enhancing the longest fibers or fiber strength for ring yarn manufacturing systems has historically been the main goal of breeding to increase fiber quality (Liody 2008). The majority of upland cotton cultivars are closely related, and the germplasm base is somewhat small. Over time, advancements in cotton breeding have been delayed by links like the well-known inverse relationship between fiber strength and yield. However, fiber strength has increased as a result of transgressive segregation, which is brought about by genetic recombination. Due to practical considerations, cotton breeders choose fiber traits that affect processing instead of directly selecting for yarn properties. This technique is known as indirect selection. The inherent variety in fiber qualities caused by the environment makes it difficult to improve them through breeding or biotechnological methods. Future breeding and biotechnological efforts should concentrate on improving fiber qualities and decreasing variability at the same time since variability in fiber properties poses a challenge to fiber processing (Liody 2008).

Sustainability in agriculture and the environment is seriously threatened by climate change. Extreme weather conditions brought on by climate change will eventually lower crop yield. According to earlier projections, the temperature of the atmosphere has risen by 0.74 °C and is expected to continue rising until the end of this century, reaching 1.8–4.0 °C. Cotton production is affected by climate change in both positive and bad ways, depending on the location. Climate change has a favorable impact on the majority of regions. Climate change has a generally positive effect on cotton production. For the majority of the cotton-growing regions, rainfall and evapotranspiration estimates are questionable. Temperature increases and unpredictable rains shorten crop duration, increase evapotranspiration and mineralization, and attract pest populations. Cotton production will consequently drop by as much as 40–50%. In addition, higher carbon dioxide levels will promote biomass production, photosynthesis, and ultimately cotton yield. Cotton production benefits from rising carbon dioxide, while it suffers from rising temperatures (Iqbal et al. 2022).

Cotton has vertical tap roots giving it some resistance to drought and extreme temperatures. However, the crop is vulnerable to water availability, especially when flowering and boll formation are at their peak. Unless daytime temperatures are higher than 32 °C, rising temperatures encourage the growth of plants. Overall, less water available for irrigation is one of the negative effects of climate change on cotton production, especially in Xinjiang (China), Pakistan, Australia, and the western United States. In Pakistan specifically, heat stress could result in lower yields, while in other nations, modest temperature increases could encourage cotton plant development and extend the growing season. The Yellow River region of China, India, the southeast United States, and south-eastern Anatolia (Turkey) will probably benefit from climate change's effects on rainfall. It's unknown how this may affect rainfall in Brazil and West and Central Africa.

Therefore, some adaptation methods will be used to reduce these impacts and maintain cotton output. Furthermore, modeling techniques must be used to quantify adaption strategies. Consequently, in order to lessen the effects of climate change, adaptive techniques are required. The creation of cultivars that can withstand biotic and abiotic stresses and adapt to rising temperatures and carbon dioxide levels would lessen the adverse effects of climate change. Additionally, changes in production technologies that give growers, researchers, and policymakers meaningful information will support cotton output in the face of a changing environment in the future.

Cotton's basic needs for extended growing times and particular seasonal temperatures restrict the amount of arable land that can be used for cotton cultivation worldwide. Early-maturing varieties help to optimize the cropping system by reducing competition between cotton and other crops for arable land and mitigating yield losses caused by unfavorable weather in early spring and late autumn. This is achieved by delaying the sowing dates without affecting the boll-opening time. This review outlines research and breeding initiatives for early-maturing cotton that effectively uses pleiotropic genes to combine early maturity, high-quality, multi-resistance characteristics, and appropriate plant architecture. An attempt is also made to provide a summary of what we currently know about the molecular mechanisms underpinning early maturation, which include a variety of routes such as hormone signaling, circadian clock, and epigenetics. Additionally, fresh approaches and practical solutions are suggested for the fine-scale breeding of early-maturing crops in order to guarantee the agricultural sector's sound growth (Zhao et al. 2023).

Appendix 19.1 Research Institutes and Online Resources Available for Upland Cotton (*Gossypium hirsutum* L.)

Institution name	Specialization and research activities	Address	Contact information and website
ACSA International Cotton Institute, USA	Basic education on all aspects of cotton		https://bf.memphis.edu/cotton/index.php/main/instructions
Cotton Development Board (CDB)	Cotton Development Broad or CDB is a government organization responsible for cotton production and research in Bangladesh located in Dhaka.	Khamarbari Rd, Dhaka, Bangladesh.	Email: ict.cdb.gov.bd@gmail.com www.Cdb.gov.bd
Cotton Research Institute (CRI)	Research focused on breeding improved, high-yielding varieties, based on the selection of desirable mutations or genetic segregation of earlier natural hybridization. As of 1921, the Cotton Research Council approved the utilization of artificial hybridization to develop new cotton varieties. The aim was to incorporate into a new strain such desirable traits from parent plants. The technique was further developed to become the major approach for developing new varieties. Cotton technology research was also developed to complement agronomy research The Experimental Spinning mill	9, El-Gamma St. 12619, Giza, Egypt.	Email: cri@arc.eg. http://www.arc.sci.eg/InstsLabs/Default.a
Central Cotton Research Institute, Multan (CCRI)	The institute has contributed significantly by advancing and generating knowledge in cotton research and development since its establishment in 1970.	4CXQ+CRX, OLD Shuja Abad Rd, ShershahTwon, Multan, Punjab 60000, Pakistan.	Email: mnaagr@gmail.com www.ccri.gov.pk

(continued)

Institution name	Specialization and research activities	Address	Contact information and website
	Over the years, with many achievements of high-yielding varieties with standard fiber quality characteristics like staple length, fineness and strength, the institute increased cotton productivity from 370 kg/ha in 1970 to 772 kg/ha during the current years.		
Central Institute for Cotton Research (CICR)	Research on Cotton at ICAR-CICR continues to focus on innovations in crop improvement, crop production and crop protection aspects to enhance cotton productivity, develop technologies for resource conservation, increase resource use efficiency and eco-friendly crop protection.	Post Bag No.2, Shankar Nagar Post Office, Nagpur - 440 010, Maharashtra, India	Email:cicrnagpur@gmail.com https:www.cicr.org.in
Cotton Egypt Association (CEA)	CEA is a non-profit association, established in 2005. CEA exclusively manages and promotes the registered Egyptian Cotton™ logo trademark, owned by the Ministry of Trade & Industry and Alexandria Cotton Exporter's Association. CEA is also responsible for the licensing program. Our mission is to protect and promote the value-add of the Egyptian Cotton™ identity in supply chains end-to-end. CEA is supporting sustainability by collaborating with the Better Cotton Program launched in Egypt by UNIDO in 2020.	22, Syria St. 3rd Floor, Mohandessine, Giza, 12411, Egypt.	Email: info@cottonegyptassociation.com https://www.cottonegyptassociation.com

(continued)

Institution name	Specialization and research activities	Address	Contact information and website
Nazili Cotton Research Institute, Turkey	Cotton research and development		https://www. administrator@nazilli.tagem. gov.tr
Uzbekt Research Institute, Uzbekistan	Cotton germplasm resources.		https://en.yellowpages.uz/ company/uzbek-scientific-researchinstitute-of-cotton

Appendix 19.2 Genetic Resources of Long and Extra-Long Staple for Upland Cotton (*Gossypium hirsutum* L.) Varieties

	Fiber quality traits			
Variety names	Fiber length mm	Fiber fineness MIC	Fiber strength g/tex	Cultivation location
CB-9	28.0	4.6	30.1	Bangladesh
CB-12	29.0	4.0	30.5	Bangladesh
CB-13	29.5	3.8	29.0	Bangladesh
CB-14	30.5	3.9	30.4	Bangladesh
CB-15	30.7	4.2	29.5	Bangladesh
Xinjiang,	27.8–29.7	3.6–4.9	26.0–30.9	China
ICS-202 J-34 SG	25.0–28.0	4.0–4.5	23.0–27.0	India
ICS-105 Shankar-6	27.0–30.0	3.5–4.4	26.0–29.0	India
ICS-102 V-797	22.0–24.0	4.6–5.5	20.0–23.0	India
ICS-105- H-4 / MECH-1	27.0–29.0	3.5–4.9	26.0–28.0	India
ICS-105- LRA-5166	25.0–27.0	3.5–4.9	24.0–27.0	India
ICS-104- Y-1	23.0–25.0	4.0–5.5	22.0–24.0	India
Varamin	30.0	4.1	27.5	Iran
Bakhtegan	29.8	4.1	28.0	Iran
Sahel	30.0	4.1	28.0	Iran
Golestan	29.5	4.3	27.0	Iran
Sepid	30.8	4.2	27.2	Iran
Armaghan	29.4	4.3	26.5	Iran
Lydia	29–32	4.0–4.5	34–38	Turkey
Flash	29–31	4.6–4.8	33–35	Turkey
Carisma	28–30	4.4–4.9	30–32	Turkey
PG 2018	28–30	4.6–4.8	31–33	Turkey
Gloria	30–32	4.1–4.4	33–36	Turkey
NC18-11	32.2	4.3	36.7	USA
NC18-12	31.0	4.7	35.6	USA

(continued)

Variety names	Fiber quality traits			Cultivation location
	Fiber length mm	Fiber fineness MIC	Fiber strength g/tex	
NC18-13	32.3	4.4	34.5	USA
NC18-14	29.1	5.7	29.7	USA
DP393	30.7	5.2	32.8	USA
Sure-Grow 747	29.5	5.3	29.3	USA
UA48	33.2	5.3	37.3	USA

References

Abdelraheem A, Adams N, Zhang J (2020) Effects of drought on agronomic and fiber quality in an introgressed backcross inbred line population of Upland cotton under field conditions. Field Crop Res 254:107850

Abro S, Rizwan M, Deho ZA, Abro SA, Sial MA (2022) Identification of heat tolerant cotton lines showing genetic variation in cell membrane thermostability, stomata, and trichome size and its effect on yield and fiber quality traits. Front Plant Sci 12:804315

Anonymous (1997) Preservation and utilization of germplasm in cotton. Southern Co-operative Series Bulletin No. 386

ANZFA (2002) Final assessment report (Full assessment - S.15) Application A436: Oil and linters derived from insect-protected cotton containing event 15985. Report No. Full assessment - S.15 Application A436, Australia New Zealand Food Authority Canberra, Australia

Ballester C, Brinkhoff J, Quayle WC, Hornbuckle J (2019) Monitoring the effects of water stress in cotton using the green red vegetation index and red edge ratio. Remote Sens 11:873

Bange M, Marshall J, Stiller WN (2002) Agronomy. Chapter 8. In: Australian Cotton Cooperative Research Centre (ed) Australian dryland cotton production guide, 3rd edn. Cotton Research & Development Corporation Narrabri NSW 2390, pp 41–52

Beasley CA (1975) Developmental morphology of cotton flowers and seed as seen with the scanning electron microscope. Am J Bot 62:584–592

Becerra Lopez-Lavalle L, McFadden H, Brubaker C (2007) The effect of Gossypium C-genome chromosomes on resistance to fusarium wilt in allotetraploid cotton. TAG Theor Appl Genet 115:477–488

Berardi, L.C., and L.A. Goldblatt. 1980. Gossypol. Pp. 184-238 in I.E. Liener, ed., Toxic Constituentsof Plant Foodstuffs, 2nd ed. Academic Press, New York

Bhan S, Behera UK (2014) Conservation agriculture in India–problems, prospects and policy issues. Int Soil Water Conserv Res 2(4):1–12

Biswas A, Banik R (2024) Machine learning integration in agriculture domain: concepts and applications. In: Fog computing for intelligent cloud IoT systems. Wiley, Hoboken, pp 71–97

Blaise D (2006) Yield, Boll distribution and fibre quality of hybrid cotton (*Gossypium hirsutum* L.) as influenced by organic and modern methods of cultivation. J Agron Crop Sci 192:248–256

Bowman DT (2000) Contemporary issue: attributes of public and private cotton breeding programs. J Cotton Sci 4:130–136

Brampton M (2001) Research turns trash to treasure. USQ News 2001[21 February] University of Southern Queensland

Brubaker CL, Wendel JF (1993) On the specific status of *Gossypium lanceolatum* Todaro. Genet Resour Crop Evol 40:165–170

Brubaker CL, Bourland FM, Wendel JE (1999) The origin and domestication of cotton. Chapter 1.1. In: Smith CW, Cothren JT (eds) Cotton: origin, history, technology, and production. Wiley, New York, pp 3–31

Burke JJ, Velten J, Oliver MJ (2004) In vitro analysis of cotton pollen germination. Agron J 96: 359–368

Byrd SA, Rowland DL, Bennett J, Zotarelli L, Wright D, Alva A, Nordgaard J (2015) The relationship between Sap flow and commercial soil water sensor readings in irrigated potato (Solanum tuberosum L.) production. Am J Potato Res 92:582–592

Çelik S (2023) Assessing drought tolerance in a large number of upland cotton plants (Gossypium hirsutum L.) under different irrigation regimes at the seedling stage. Life 13:2067

Chauhan DS, Singh P, Singh P (1983) Isolation distance in upland cotton under Nagpur conditions. Cotton Dev 13:23

Chee P, Draye X, Jiang CX, Decanini L, Delmonte TA, Bredhauer R, Smith CW, Paterson AH (2005a) Molecular dissection of interspecific variation between Gossypium hirsutum and Gossypium barbadense (cotton) by a backcross-self approach: I. Fiber elongation. TAG Theor Appl Genet 111:757–763

Chee PW, Draye X, Jiang CX, Decanini L, Delmonte TA, Bredhauer R, Smith CW, Paterson AH (2005b) Molecular dissection of phenotypic variation between Gossypium hirsutum and Gossypium barbadense (cotton) by a backcross-self approach: III. Fiber length. TAG Theor Appl Genet 111:772–781

Cherry JP, Leffler HR (1984) Seed. Chapter 13. In: Kohel RJ, Lewis CF (eds) Cotton, agronomy monograph No. 24, 24th edn. ASA-CSSA-SSSA, Madison, pp 511–558

Coates W (2000) Using cotton plant residue to produce briquettes. Biomass Bioenergy 18:201–208

Constable GA, Reid PE, Thomson NJ (2001) Approaches utilised in breeding and development of cotton cultivars in Australia. Chapter 1. In: Jenkins JN, Saha S (eds) Genetic improvement of cotton: emerging technologies. Science Publishers Inc., pp 1–15

CRDC (2017) Australian cotton production manual, CRDC, Australia. Department of Agriculture and Fisheries 2018, Cotton growing, https://www.daf.qld.gov.au/business-priorities/plants/field-crops-and-pastures/broadacre-field-crops/cotton/growing/varieties-and-planting

Cronn RC, Small RL, Haselkorn T, Wendel JF (2002) Rapid diversification of the cotton genus (*Gossypium*: Malvaceae) revealed by analysis of sixteen nuclear and chloroplast genes. Am J Bot 89:707–725

Dabbert TA, Gore MA (2014) Challenges and perspectives on improving heat and drought stress resilience in cotton. J Cotton Sci 18:393–409

Davidonis GH, Hamilton RH (1983) Plant regeneration from callus tissue of Gossypium hirsutum L. Plant Sci Lett 32:89–93

Draye X, Chee P, Jiang CX, Decanini L, Delmonte TA, Bredhauer R, Smith CW, Paterson AH (2005) Molecular dissection of interspecific variation between *Gossypium hirsutum* and *G. barbadense* (cotton) by a backcross-self approach: II. Fiber fineness. TAG Theor Appl Genet 111:764–777

FAO (2018) Conservation Agriculture http://www.fao.org/conservation-agriculture/in-practice/soil-organic-cover/en

Frank AW (1987) Food uses of cottonseed protein. Chapter 2. In: Hudson BJF (ed) Development in food proteins, vol 5. Elsevier Applied Science, New York, pp 31–80

Fryxell PA (1976) A nomenclator of Gossypium – The botanical names of cotton. USDA Tech. Bull, p 1491

Fryxell PA (1979a) The influence of man. Chapter 6. In: The natural history of the cotton tribe (Malvaceae, Tribe Gossypieae), 1st edn. Texas A&M University Press, College Station, pp 159–177

Fryxell PA (1979b) The origin and spread of the tribe. In: The natural history of the cotton tribe (Malvaceae, tribe Gossypieae). Texas A & M University Press, College Station, pp 131–218

Garofalo SP, Intrigliolo DS, Camposeo S et al (2023) Agronomic responses of grapevines to an irrigation scheduling approach based on continuous monitoring of soil water content. Agronomy 13:2821
Garofalo SP, Giannico V, Costanza L et al (2024) Prediction of stem water potential in Olive Orchards using high-resolution planet satellite images and machine learning techniques. Agronomy 14:1
Gerstel DU (1954) A new lethal combination in interspecific cotton hybrids. Genetics 39:628–639
Gerstel DU, Sarvella PA (1956) Additional observations on chromosomal translocations in cotton hybrids. Evolution 10:408–414
Giller KE, Witter E, Corbeels M, Tittonell P (2009) Conservation agriculture and smallholder farming in Africa: the heretics' view. Field Crop Res 114(1):23–33
Gomes RS, Wilson N, Coates WE, Fox RW (1997) Cotton (*Gossypium*) plant residue for industrial fuel.: An economic assessment. Ind Crop Prod 7:1–8
Gore UR (1932) Development of the female gametophyte and embryo in cotton. Am J Bot 19:795–807
Gore UR (1935) Morphogenetic studies on the inflorescence of cotton. Bot Gaz 97:118–138
Gould J, Banister S, Hasegawa O, Fahima M, Smith RH (1991) Regeneration of *Gossypium hirsutum* and *G. barbadense* from shoot apex tissues for transformation. Plant Cell Rep 10: 12–16
Govila OP, Rao CH (1969) Studies on the in vitro germination and storage of cotton pollen. J Palynol 5:37–41
Green JM, Jones MD (1953) Isolation of cotton for seed increase. Agron J 45:366–368
Gregory SR, Hernandez E, Savoy BR (1999) Cottonseed processing. Chapter 4.5. In: Smith CW, Cothren JT (eds) Cotton: origin, history, technology and production. Wiley, New York, pp 793–819
Gridley HE (1974) Natural and artificial crossing in upland cotton at Namulongu, Uganda. Cotton Growing Re
Habib-ur-Rahman M, Ahmad A, Raza A, Hasnain MU, Alharby HF, Alzahrani YM, Bamagoos AA, Hakeem KR, Ahmad S, Nasim W et al (2022) Impact of climate change on agricultural production; Issues, challenges, and opportunities in Asia. Front Plant Sci 13:925548
Hearn AB (1981) Cotton nutrition. Field Crop Abstr 34:11–34
Hopper WM, McDaniel RG (1999) The cotton seeds. Chapter 2.4. In: Smith CW, Cothren JT (eds) Cotton: origin, history, technology and production. Wiley, New York, pp 289–317
Hutchinson JB (1951) Intra-specific differentiation in Gossypium hirsutum. Heredity 5:161–193
ICAC (2022) ICAC data book 2022
Iqbal MJ, Reddy OUK, El Zik KM, Pepper AE (2001) A genetic bottleneck in the 'evolution under domestication' of upland cotton *Gossypium hirsutum* L. examined using DNA fingerprinting. Theor Appl Genet 103:547–554
Iqbal A, Mazhar I, Madeeha A, Dong Q, Wang X, Gui H, Zhang H, Pang N (2022) Climate change and cotton production improvement of plant production in the era of climate change. CRC Press, Taylor & Francis Group
Jenkins JN (2003) Cotton. In: Traditional crop breeding practices: an historical review to serve as a baseline for assessing the role of modern biotechnology. OECD, pp 61–70
Jeoh T, Agblevor FA (2001) Characterization and fermentation of steam exploded cotton gin waste. Biomass Bioenergy 21:109–120
Jiang J, Johansen K, Stanschewski CS, Wellman G, Mousa MA, Fiene GM, Asiry KA, Tester M, McCabe MF (2022) Phenotyping a diversity panel of quinoa using UAV-retrieved leaf area index, SPAD-based chlorophyll and a random forest approach. Precis Agric 23:961–983
Kakani VG, Reddy KR, Koti S, Wallace TP, Prasad PVV, Reddy VR, Zhao D (2005) Differences in in vitro pollen germination and pollen tube growth of cotton cultivars in response to high temperature. Ann Bot 96:59–67
Kassam A, Friedrich T, Shaxson F, Pretty J (2009) The spread of conservation agriculture: justification, sustainability and uptake. Int J Agric Sustain 7(4):292–320

Kearney TH, Harrison GJ (1932) Pollen antagonism in cotton. J Agric Res 44:191–226
Knowler D, Bradshaw B (2007) Farmers' adoption of conservation agriculture: a review and synthesis of recent research. Food Policy 32(1):25–48
Kumar L, Chhogyel N, Gopalakrishnan T, Hasan MK, Jayasinghe SL, Kariyawasam CS, Kogo BK, Ratnayak ES (2022) Climate change and future of agri-food production. In: Future foods. Academic Press, Cambridge, MA, pp 49–79
Lee JA (1984) Cotton as a world crop. Chapter 1. In: Kohel RJ, Lewis CF (eds) Cotton, pp 1–25
Lee S-B, Kaittanis C, Jansen RK, Hostetler JB, Tallon LJ, Town CD, Daniell H (2006) The complete chloroplast genome sequence of Gossypium hirsutum: organization and phylogenetic relationship to other angiosperms. BMC Genomics 7:61–72
Lewis CF, Richmond TR (1957) The genetics of flowering response in cotton. I. Fruiting behaviour of *Gossypium hirsutum* var.marie-galante in a cross with a variety of cultivated American upland cotton. Genetics 42:499–509
Lioyd OM (2008) Quality improvement of upland cotton (*Gossypium hirsutum* L.). J Crop Production 5(1-2):371–394
Llewellyn D, Fitt G (1996) Pollen dispersal from two field trials of transgenic cotton in the Namoi Valley, Australia. Mol Breed 2:157–166
Llewellyn DJ, Tyson C, Constable GA et al (2007) Containment of regulated genetically modified cotton in the field. Agr Ecosyst Environ 121:419–429
Lu Z, Chen J, Percy RG, Zeiger E (1997) Photosynthetic rate, stomatal conductance and leaf area in two cotton species (*Gossypium barbadense* and *Gossypium hirsutum*) and their relation with heat resistance and yield. Funct Plant Biol 24:693–700
Lusas EW, Jividen GM (1987) Glandless cottonseed: a review of the first 25 years of processing and utilization research. J Am Oil Chem Soc 64:839–854
Maldera F, Garofalo SP, Camposeo S (2024) Ecophysiological recovery of micropropagated Olive cultivars: field research in an irrigated super-high-density Orchard. Agronomy 14:1560
McFadden H, Beasley D, Brubaker C (2004) Assessment of *Gossypium sturtianum* and *G. australe* as potential sources of Fusarium wilt resistance to cotton. Euphytica 138:61–72
McGregor SE (1976) Crop plants and exotic plants - cotton. Chapter 9.10. In: Insect pollination of cultivated crop plants. USDA, Agricultural Research Service, Washington, D. C, pp 171–190. (now available at http://gears.tucson.ars.ag.gov/book/chap9/cotton.html)
Meredith WRJ, Bridge RR (1973) Natural crossing in cotton (*Gossypium hirsutum* L.) in the Delta of the Mississippi. Crop Sci 13:551–552
Moffett JO (1983) Pollination of entomophilous hybrid seed parents - hybrid cotton. Chapter 8. In: Jones CE, Little RJ (eds) Handbook of experimental pollination biology. Van Nostrand Reinhold, New York, pp 508–514
Moffett JO, Stith LS, Burkhart CC, Shipman CW (1975) Honey bee visits to cotton flowers. Environ Entomol 4:203–206
Moffett JO, Stith LS, Burkhardt CC, Shipman CW (1976) Fluctuation of wild bee and wasp visits to cotton flowers. Arizona Acad Sci 11:64–68
Moulherat C, Tengberg M, Haquet JF, Mille B (2002) First evidence of cotton at neolithic Mehrgarh, Pakistan: analysis of mineralized fibres from a copper bead. J Archaeol Sci 29: 1393–1140
Mungomery VE, Glassop AJ (1969) Natural cross-pollination of cotton in central Queensland. Queensland J Agric Anim Sci 26:69–74
Myburgh PA, Howell CL (2022) Determining a midday stem water potential threshold for irrigation of table grapes. S Afr J Enol Vitic 43:96–102
OECD (2004) Consensus document on compositional considerations for new varieties of cotton (*Gossypium hirsutum* and *Gossypium barbadense*): Key food and feed nutrients and anti-nutrients. Report No. 11, OECD Environment, Health and Safety Publications Paris.
Oosterhuis DM, Jernstedt J (1999) Morphology and anatomy of the cotton plant. In: Smith CW, Cothren JT (eds) Cotton: origin, history, technology and production. Wiley, New York, pp 175–206

Pedrero F, Maestre-Valero JF, Mounzer O et al (2015) Response of young 'Star Ruby' grapefruit trees to regulated deficit irrigation with saline reclaimed water. Agric Water Manag 158:51–60

Percival AE (1987) The national collection of Gossypium germplasm. Southern Cooperative Series Bulletin No. 321

Pettigrew WT (2024) Physiological consequences of moisture deficit stress in cotton. Crop Sci 44: 1265–1272

Pheloung PC (2001) Weed risk assessment for plant introductions to Australia. In: Groves RH, Panetta FD, Virtue JG (eds) Weed risk assessment. CSIRO Publishing, Melbourne, pp 83–92

Price AJ, Balkcom KS, Culpepper SA et al (2011) Glyphosate-resistant Palmer amaranth: a threat to conservation tillage. J Soil Water Conserv 66(4):265–275

Pundir NS (1972) Experimental embryology of *Gossypium arboreum* L. and *G. hirsutum* L. and their reciprocal crosses. Bot Gaz 133:7–26

Rao GM, Nadre KR, Suryanarayana MC (1996) Studies on the utility of honey bees on production of foundation seed of cotton cv. NCMHH-20. Indian Bee J 58:13–15

Rea HE (1931) Grafting experiments with cotton. Plant Physiol 6:193–196

Rea HE (1933) Field propagation of cotton by means of grafts. Plant Physiol 8:171

Richards JS, Stanley JN, Gregg PC (2005) Viability of cotton and canola pollen on the proboscis of *Helicoverpa armigera*: implications for spread of transgenes and pollination ecology. Ecol Entomol 30:327–333

Ritchie GL, Bednarz CW, Jost PH, Brown SM (2007) Cotton growth and development. 1–14. http://pubs.caes.uga.edu/caespubs/pubcd/B1252.htm. Accessed on 14 August 2007

Saha S, Jenkins JN, Wu J, McCarty JC, Gutierrez OA, Percy RG, Cantrell RG, Stelly DM (2006) Effects of chromosome-specific introgression in Upland Cotton on fiber and agronomic traits. Genetics 172:1927–1938

Sakhanokho HF, Zipf A, Rajasekaran K, Saha S, Sharma GC (2001) Induction of highly embryogenic calli and plant regeneration in Upland (*Gossypium hirsutum* L.) and Pima (*Gossypium barbadense* L.) cottons. Crop Sci 41:1235–1240

Sakhanokho HF, Ozias-Akins P, May OL, Chee PW (2004) Induction of somatic embryogenesis and plant regeneration in select Georgia and Pee Dee cotton lines. Crop Sci 44:2199–2205

Saravanan NA, Ram SG, Thiruvengadam V, Ravikesavan R, Raveendram TS (2007) Production and fertility restoration of an interspecific hybrid between *Gossypium hirsutum* L. and G. raimondii U. Cytologia 72:195–203

Sassenrath-Cole GF (1995) Dependence of canopy light distribution on leaf and canopy structure for two cotton (*Gossypium*) species. Agric For Meteorol 77:55–72

Sawyer WH (1955) Medicinal uses of plants by native Inaguans. Sci Mon 80:371–376

Scott BJ, Eberbach PL, Evans J, Wade LJ (2010) Stubble retention in cropping systems in Southern Australia: benefits and challenges. EH Graham Centre Monograph No. 1. www.csu.edu.au/research/grahamcentre

Seelanan T, Schnabel A, Wendel JF (1997) Congruence and consensus in the cotton tribe (Malvaceae). Syst Botan 22:259–290

Sellami MH, Albrizio R, Čolović M et al (2022) Selection of hyperspectral vegetation indices for monitoring yield and physiological response in sweet maize under different water and nitrogen availability. Agronomy 12:489

Sheelavantar MN, Prabhakar AS, Patil SV (1975) Propogation of hybrid cotton through cuttings. Indian J Agric Sci 45:91–92

Smith CW (1976) Natural cross-pollination of cotton. Arkansas Farm Res 25:6

Smith CW (1995) Cotton (Gossypium hirsutum L.). Chapter 6. In: Crop production: evolution, history, and technology. Wiley, New York, pp 287–349

Smith WC (1999) Production statistics. Chapter 3.1. In: Smith WC, Cothren JT (eds) Cotton: origin, history, technology and production. Wiley, pp 435–449

Stephens SG (1946) The genetics of 'corky' I. The New World alleles and their possible role as an interspecific isolating mechanism. J Genet 47:150–161

Stephens SG (1950a) The genetics of 'corky' II. Further studies on its genetic basis in relation to the general problem of interspecific isolating mechanisms. J Genet 50:9–20
Stephens SG (1950b) The internal mechanisms of speciation in Gossypium. Botan Rev 16:115–149
Stephens SG (1965) The effects of domestication on certain seed and fiber properties of perennial forms of cotton, Gossypium hirsutum L. Am Nat 99:355–372
Stephens SG, Moseley ME (1973) Cotton remains from archeological sites in central Coastal Peru. Science 180:186–188
Stephens SG, Phillips LL (1972) The history and geographical distribution of a polymorphic system in New World cottons. Biotropica 4:49–60
Stiller WN, Read JJ, Constable GA, Reid PE (2005) Selection for water use efficiency traits in a cotton breeding program: Cultivar differences. Crop Sci 45:1107–1113
Sunilkumar G, Campbell LM, Puckhaber L, Stipanovic RD, Rathore KS (2006) Engineering cottonseed for use in human nutrition by tissue-specific reduction of toxic gossypol. Proc Natl Acad Sci 103:18054
Ter Avanesian DV (1978) The effect of varying the number of pollen grains used in fertilization. Theor Appl Genet 52:77–79
Thomson NJ (1966) Cotton variety trials in the Ord Valley, North Western Australia: 4. Natural crossing of cotton. Empire Cotton Grow Rev 43:18–21
Umbeck PF, Barton KA, Nordheim EV, McCarty JC, Parrot WL, Jenkins JN (1991) Degree of pollen dispersal by insects from a field test of genetically engineered cotton. J Econ Entomol 84: 1943–1950
USDA (2007) Information Systems for Biotechnology. Field Test Releases in the US. http://www.isb.vt.edu/2002menu/regulatory_information.cfm
Vaissière BE, Vinson SB (1994) Pollen morphology and its effect on pollen collection by honey bees, Apis mellifera L. (Hymenoptera: Aphidae), with special reference to upland cotton, Gossypium hirsutum L. (Malvaceae). Grana 33:128–138
Van Deynze AE, Sundstrom FJ, Bradford KJ (2005) Pollen-mediated gene flow in California cotton depends on pollinator activity. Crop Sci 45:1565–1570
Vansell GH (1944) Cotton nectar in relation to bee activity and honey production. J Econ Entomol 37:528–530
Wang L, Dong M, Paterson AH (1995) The distribution of Gossypium hirsutum chromatin in *G. barbadense* germplasm: molecular analysis of introgressive plant breeding. TAG Theor Appl Genet 91:1153–1161
Weiss M, Jacob F, Duveiller G (2020) Remote sensing for agricultural applications: A meta-review. Remote Sens Environ 236:111402
Wendel JF (1989) New World tetraploid cottons contain Old World cytoplasm. Proc Natl Acad Sci U S A 86:4132–4136
Wendel JF, Albert VA (1992) Phylogenetics of the cotton genus (Gossypium): Character-State Weighted Parsimony Analysis of chloroplast DNA restriction site data and its systematic and biogeographic implications. Syst Botan 17:115–143
Wendel JF, Cronn RC (2003) Polyploidy and the evolutionary history of cotton. Adv Agron 78: 139–186
Wendel JF, Schnabel A, Seelanan T (1995) An Unusual Ribosomal DNA Sequence from *Gossypium gossypioides* Reveals Ancient, Cryptic, Intergenomic Introgression. Mol Phylogenet Evol 4:298–313
Wilkins TA, Rajasekaran K, Anderson DM (2000) Cotton biotechnology. Crit Rev Plant Sci 19: 511–550
Wise RR, Sassenrath-Cole GF, Percy RG (2000) A comparison of leaf anatomy in field-grown *Gossypium hirsutum* and *G. barbadense*. Ann Bot 86:731–738
Wood AW, Drum F, Shaw I, Ashton D (2007) Natural Fibres - outlook for wool and cotton to 2011-12. Aust Commod 14:52–61
Wu Z, Soliman KM, Zipf A, Saha S, Sharma GC, Jenkins JN (2005) Isolation and characterization of genes differentially expressed in fiber of Gossypium barbadense L. J Cotton Sci 9:166–174

Wu N, Yang J, Wang G, Ke H, Zhang Y, Liu Z, Ma Z, Wang X (2022) Novel insights into water-deficit-responsive mRNAs and lncRNAs during fiber development in *Gossypium hirsutum*. BMC Plant Biol 22:6

Xanthopoulos FP, Kechagia UE (2000) Natural crossing in cotton (*Gossypium hirsutum* L.). Aust J Agr Res 51:979–983

Zhang Z, Huang J, Yao Y, Peters G, Macdonald B, La Rosa AD, Wang Z, Scherer L (2023) Environmental impacts of cotton and opportunities for improvement. Nat Rev Earth Environ 4: 703–715

Zhao H, Chen Y, Liu J, Wang Z, Li Z, Ge Z (2023) Recent advances and future perspectives in early-maturing cotton research. New Phytol 237:1100–1114. https://doi.org/10.1111/nph.18611

Chapter 20
Recent Molecular Genetics and Breeding Strategies in Egyptian Cotton (*Gossypium barbadense* L.)

Ahmed M. Abdelmoghny, Reham H. A. O. Gibely, Abeer S. Arafa, and Khaled F. M. Salem

Abstract Egyptian cotton, Pima cotton or Sea Island cotton (*Gossypium barbadense* L.) is a member of the Malvaceae family and is a self-pollinating crop with 15–20% cross-pollination. It is one of the most important and economically significant cultivated fiber crops globally. It is used in a wide range of clothing from jackets to everyday shirts, as well as in home textiles like bedsheets and curtains. The seed oil is utilized in food and cosmetics, and the fibers are even used in coffee filters. The seeds are fed to cattle and processed to make oil, rubber and plastics. Numerous genetic resources of cotton, including wild species, have been maintained in gene banks. There are multiple species with varying chromosome numbers and cotton diversity is influenced by geographic regions and morphological features. Introducing new alleles through crossing different genetic resources such as crossing common varieties with locally developed ones or wild genotypes, can enhance genetic diversity and select for beneficial traits. Ensuring adequate natural variety in phenotypes is crucial. The main goals of Breeders and geneticists' are to improve new cotton varieties with desired traits, including high yield, good quality, and tolerance to both biotic and abiotic stress. Enhancing traditional breeding strategies with contemporary new genetic technologies can aid in achieving these objectives. This chapter gives a summary of the biodiversity and conservation of Egyptian cotton germplasm, as well as the goals and phases of Egyptian cotton breeding projects and traditional breeding techniques. Modern plant breeding methods, such as genetic engineering and marker-assisted breeding, are also covered.

A. M. Abdelmoghny (✉) · R. H. A. O. Gibely · A. S. Arafa
Cotton Research Institute (CRI), Agricultural Research Center (ARC), Giza, Egypt
e-mail: ahmed.abdelbaky@arc.sci.eg; reham.omar@arc.sci.org; abeer.hassan@arc.sci.org

K. F. M. Salem
Department of Biological Sciences, Shaqra University, Ad-Dawadimi, Saudi Arabia

Department of Plant Biotechnology, Genetic Engineering and Biotechnology Research Institute (GEBRI), University of Sadat City, Sadat City, Egypt
e-mail: khaled.salem@gebri.usc.edu.eg; khaledfathi@su.edu.sa

K. F. M. Salem et al. (eds.), *Breeding and Biotechnology of Leaf, Fruit, and Seed Fiber Crops*, Advances in Plant Breeding Strategies 11,
https://doi.org/10.1007/978-3-032-00407-9_20

Keywords Biodiversity · Molecular breeding · Egyptian cotton · *Gossypium barbadense* L. · Fiber

20.1 Introduction

About 2.5% of the world's arable land is used for the cultivation of this crop, which is grown in almost 100 nations with a wide range of agroclimatic conditions and agricultural techniques. It is of great importance for the economies of these countries (Fig. 20.1) (Scarpin et al. 2022). China, India and the USA are the leading countries in production. Cotton fiber is a major source for many industries such as textile production and healthcare. Cotton is used in the production of various products including edible oil, cottonseed cake for animal feed, gloves, mufflers, bags, socks, garments, jackets, underwear, towels, carpets, t-shirts, fishnets, coffee filters, book-binding, archival paper, beds, curtains, bed sheets and others. Cotton oil is used as vegetable oil, margarine and soaps among other things. About 7% of the overall labor force in developing countries is engaged directly or indirectly in the cotton sector, encompassing about 350 million individuals from basic farming to textile production. The global cotton market is estimated to be valued at over USD 44 billion. In the USA alone, the cotton market provides $21 billion and employs more than 125,000 people, while in China, it employs 300 million people across 35 provinces (Adeleke 2024). The cotton crop is key to improving the lives of people in many underprivileged countries especially in West Africa.

Cotton is cultivated in tropical and subtropical areas between 46° N and 36° S latitude as a perennial species, although it is primarily harvested as an annual crop. Cultivated cotton is grown as an annual plant, while the species originally had a perpetual growth habit. This region provides favorable climatic conditions, with an

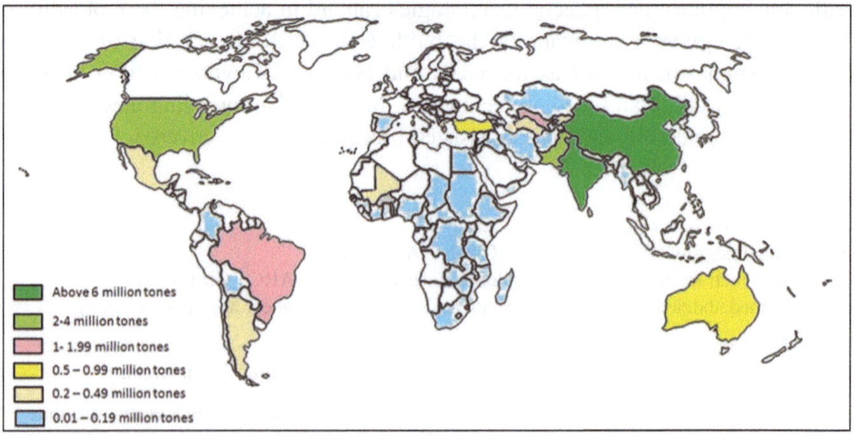

Fig. 20.1 The world map of cotton production. The map was prepared as per food and agriculture statistics. (Source: FAO (2016) FAO and Rauf et al. (2020))

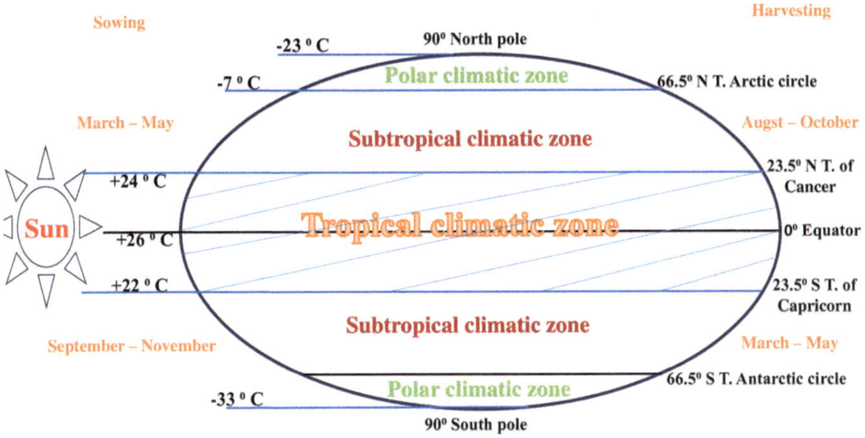

Fig. 20.2 The relationship between latitude, irradiation, and average regional temperature. (Source: Prepared by Dr. A. Abdelmoghny)

annual quantity of received solar radiation (Fig. 20.2). Globally conventional cotton production is a major consumer of freshwater for irrigation. Approximately 73% of worldwide cotton cultivation relies on freshwater irrigation, whilst 27% depends on rainfed conditions. Furthermore, cotton agronomy is a significant source of ground and surface water pollution and land degradation due to the use of enormous quantities of pesticides, herbicides, and heavy-duty machinery for tillage, management, and harvesting (Adeleke 2024). The cotton crop has a wide range of suitable soil for growing from deep black clay to sandy soils. It is tolerant to high temperatures and drought but sensitive to water logging and excessive moisture. The global cultivated cotton is distributed across the continents of the world with 70% in Asia, 20% in Americans, 6% in Africa and 2% in Europe.

Cotton (*Gossypium barbadense* L.) is among the most significant and ancient non-food crops to be domesticated. Four *Gossypium* species have been domesticated: two tetraploid, or new-world, cotton (*G. hirsutum* and *G. barbadense*) and two diploid, or old-world, cotton (*G. arboreum* and *G. herbaceum*). A tiny shrub, cotton is grown for its fiber. A cotton "boll" is the pod that remains after a bloom withers. A fruit with tiny seeds encased in white fibres is called a boll. The boll turns brown as it ripens. The boll cracks, the fibres swell, and the cotton shows through. Cotton is produced via ginning, which separates the fiber from the seed. *Gossypium barbadense* is long (LS) and extra-long staple fibers (ELS) over 35 mm in length, such as Egyptian cotton, Pima cotton and Sea Island cotton (ICAC 2023). Archaeological evidence indicates that Gossypium barbadense domestication began as early as 2500 BCE on Peru's central coast and was tied to the fishing economy and marine-based culture. Interestingly, *G. barbadense* also has medical uses to treat high blood pressure, abdominal cramps or skin rashes. Its growth requires full sun and high humidity making *G. barbadense* fibers rare and expensive. This chapter presents an overview of cotton's origin, distribution, taxonomic status, production methods, cultivar characterization, and germplasm biodiversity. It also discusses recent developments in biotechnology, molecular markers and their application for crop improvement in conjunction with tissue culture, mutation, and traditional cotton breeding.

20.1.1 Origin and Classification

Egyptian cotton (*Gossypium barbadense* L.) is a flowering plant in the class Dicotyledonae, order Malvales and family Malvaceae. Many edible plants including various *Gossypium* species, cacao (*Theobroma cacao*), linden (Tilia species), durian (Durio species), Hibiscus, and okra (*Abelmoschus esculentus*) are also in the Malvaceae family (https://www.itis.gov). Egyptian cotton (*Gossypium barbadense* L.) belongs to the genus *Gossypium* and species.

Kingdom	Plantae—Plants
Subkingdom	Tracheobionta—Vascular plants
Superdivision	Spermatophyta—Seed plants
Division	Magnoliophyta—Flowering plants
Class	Magnoliopsida—Dicotyledons
Subclass	Dilleniidae
Order	Malvales
Family	Malvaceae—Mallow family
Genus	*Gossypium* L.—Cotton
Species	*G. barbadense* $(AD)_2$
Common names:	Egyptian cotton, Pima cotton, Sea Island cotton, Tanguish cotton

The spice is a tropical, frost-sensitive perennial with yellow blooms and black seeds. It grows as a bush or tiny tree, producing cotton with extraordinarily long, silky fibers. This is a description of Egyptian cotton (*Gossypium barbadense* L.) plants (Table 20.1 and Figs. 20.3, 20.4 and 20.5). The Egyptian cotton (*Gossypium barbadense* L.) originated in southwest Ecuador and northwest Peru. It is now cultivated around the world, including China, Egypt, India, Australia, Peru, Israel, the southwestern United States, Tajikistan, Turkmenistan, and Uzbekistan. It accounts for about 1.2% of the world's cotton production in 2022/2023 (ICAC 2023).

Table 20.1 Description of Egyptian cotton (*Gossypium barbadense* L.) plants

Plant part	Description of *barbadense*
Plant height	Perennial, shrub/annual, sub-shrub, reaching approximately 1–3 m tall annually
Root	Taproot can reach depths of 1–3 m depending on age, soil characteristics and management with many lateral roots (Fig. 20.3c).
Leaves	Leaf lobes are broader at the base and pointed at the top (long triangular), mostly 8–20 cm long, with three to seven measuring lobes, 7.5–15 × 7.5–15 cm (Fig. 20.3a, b). Phyllotaxis is 3/8 back over the last leaf. Mature leaves are larger and thinner than *G. hirsutum*. With higher stomatal density and smaller stomata surface area per leaf compared to *G. hirsutum*.
Lobes	Lobes are more deeply cut, about ⅔ the length of the leaf.
Flowers	Yellow with maroon nectar guide, with five petals that partially open and reach up to 8 cm long. Petals are usually yellow or creamy yellow with a red spot at the base during anthesis, turning rose pink (Fig. 20.4). Pollen is orange and large with long spines, producing more nectar of a lower sugar concentration and larger grains than *G. hirsutum* (Fig. 20.4c).
Bolls	Egg-shaped or conform, with three to four locules, each producing five to eight seeds (Figs. 20.4d and 20.5a). They are darker green, conspicuously pitted with numerous glands. Tapering longer than broad and weight 3–5 g.
Seeds	8–10 mm long, smaller than *G. hirsutum* seeds, mostly naked or less fuzz (Fig. 20.5c, d). Rich in protein (10–20%) and oil (up to 25%), with higher oil content than *G. hirsutum*.
Ginning out turn	Less than *G. hirsutum*.
Fiber length	36–50 mm
Linters	Docs not produce linters.
Resistance	Shows some resistance to Earias spp., jassids and spider mites, due to higher gossypol content than *G. hirsutum*.
Gossypol glands	Higher amount of compared to *G. hirsutum*.
Cyclopropenoid fatty acids (CPFAs)	Lower than *G. hirsutum*.

Prepared by Dr. A. Abdelmoghny

20.1.2 *World Cotton Production for Long and Extra-Long Staple Cottons*

Cotton cultivation began over 7000 years ago for its fiber. Cotton crops cover nearly 2.5% of the world's cultivated area across more than 75 countries, with over 150 countries involved in the cotton trade and textile industry (Abdullah and Khan 2022). However, *G. barbadense* is cultivated in restricted production zones in some countries, including China, India, Uzbekistan, Tajikistan, Turkmenistan, Spain,

Fig. 20.3 Cotton parts. Cotton leaves (**a**) front side and (**b**) upper side and (**c**) cotton tap root. (Source: Prepared by Dr. A. Abdelmoghny)

Egypt, Israel, Sudan, Peru, and the United States, because of its low yield potential and need for warm and dry weather (Table 20.2). The total world cotton production in 2021/2022 was 25 million tonnes, with ELS and LS cotton production around 308,000 tonnes, comprising 1.2% of total world production. The USA, Egypt and India individually accounted for 0.3% of total world cotton production. World production was 20% lower than in the 2020/2021 season (Fig. 20.6). This decrease was mainly due to the reduction in cultivated area of ELS cotton in the United States, although other major producers like India, China and Israel also experienced a decrease in production. The top producing countries for this category of cotton are the USA, Egypt, India, China, Israel, Turkmenistan, Turkey and Peru. India, Pakistan and China, together consume more than 80% of the world's LS and ELS production. In the 2021/2022 season, total exports were 181,000 tonnes, 32% lower than the previous season. The major exporters were the USA and Egypt, with both countries exporting almost their entire production every year, comprising more than 90% of total world ELS and LS exports in 2021/2022 (Figs. 20.6 and 20.7). On the

Fig. 20.4 (**a**) Cotton flower, (**b**) opening flower, (**c**) orange pollen grains, (**d**) flower after anthesis, (**e**) cotton bud, and (**f**) mature cotton boll. (Source: Prepared by Dr. A. Abdelmoghny)

other hand, the leading importers in 2021/2022 were Pakistan at 135,000 tonnes, followed by India and China (ICAC 2023).

Long and extra-long staple cotton is used to make durable, softer household items like blankets, sheets, and towels. The cultivated cotton varieties of *G. barbadense* are categorized into the following categories based on fiber quality parameters (Table 20.3).

20.1.2.1 Sea Island Cotton

Extra-long staple cotton with a smooth appearance and outstanding luster (length $>$ 1–3/8). Superior fiber strength, consistent growth, and vibrant color grown on South Carolina's and Georgia's Sea Islands. The finest cotton yarn is produced by weaving it into high counts. Its hairlessness and uniformity make it suitable for softer wear, preserving color and quality even after frequent washings and use. The cost of this cotton kind is higher than that of other varieties.

Fig. 20.5 (**a**) Opening cotton boll with three locules, (**b**) seed cotton yield, (**c**) lint cotton yield and seeds after ginning and (**d**) cottonseeds. (Source: Prepared by Dr. A. Abdelmoghny)

Table 20.2 Long and extra-long staple world cotton production and its percentage of total world production

Cotton production of 000 tonnes				
	Season of 2020/2021	Percent of total world production	Season of 2021/2022	Present of total world production
Total world cotton production	23,989		25,176	
Total output of both long and extra-long staple cotton	387	1.6	308	1.2
Production countries for LS and ELS				
USA	119	0.50	72	0.30
Egypt	73	0.30	76	0.30
India	85	0.32	70	0.30
China	65	0.25	40	0.20
Israel	6	0.03	5	0.02
Peru	6	0.06	9	0.04
Turkmenistan	15	0.07	15	0.06
Turkey	17	0.00	15	0.06
Uzbekistan	1	0.03	4	0.02
Tajikistan	0		1	0.00
Thailand	0		1	0.00
Total production	387		308	

Source: ICAC (2023)

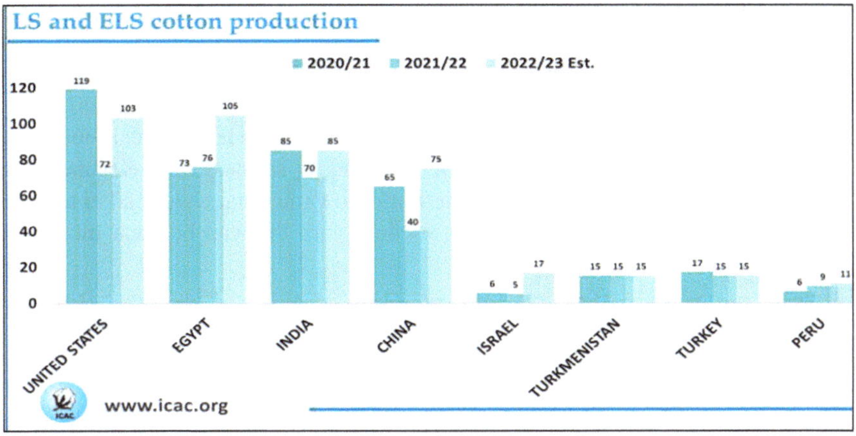

Fig. 20.6 World cotton production for long and extra-long staple cotton during the last three seasons. (Source: ICAC (2023))

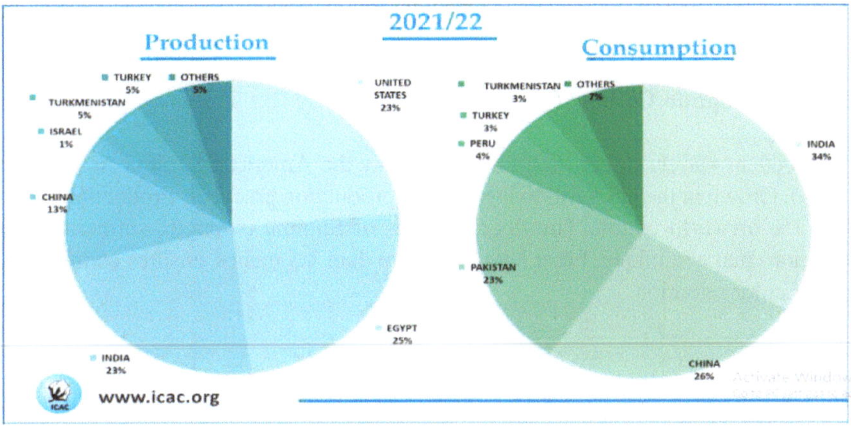

Fig. 20.7 World cotton production and consumption for long and extra-long staple cotton during the 2021/2022 season. (Source: ICAC (2023))

Table 20.3 Classification of cotton fiber based on fiber quality characteristics

Classification	Fiber length		Spinning count	Fiber fineness (μm)
	mm	inch		
Short	<24	15/16–1	Coarse <20s	Over 5
Medium	24–28	1–1/32 to 1–3/32	Medium count 20s–34s	3.6–4.9
Long	28–34	1–3/32 to 1–3/8	Fine count 34s–60s	3.6–2.8
Extra-long	34–40	1–3/8 to 1–9/16	Superfine count 80s–140s	Less than 2.8

Prepared by Dr. Abdelmoghny

20.1.2.2 Egyptian Cotton

Staple cotton fibers that are long or extra long. Hand-picked used to reduce stress on fibers, keeping them straight, unbroken and spun into fine yarn of 200 counts or more for good strength, luster, and low hairiness. Grown in a few African countries, including Egypt and Sudan. Known for its strength, softness, and elegance, this cotton is regarded as one of the best in the world (low micronaire value). For a softer, more flexible fabric, Egyptian cotton uses finer yarn that can be spun into every square inch.

20.1.2.3 Pima Cotton: American Extra-Long Staple Variety

In the early 1790s, South Carolina began growing Pima cotton. Cotton fibers with a long or extra-long staple. Longer, silkier fibers are appropriate for finer fabrics. Pima cotton is widely used and considered the best cotton for garments, towels, and bedding due to its strength, softness, durability, and absorbency. Fine yarn spun between 200 and 300 counts. The quality of Pima cotton is comparable to Egyptian cotton.

20.1.2.4 Supima Cotton

An excellent variety of certified Pima cotton by the American Supima Association (ASA). Grown in the USA following organic production practices. Production is less than 1% of world cotton. The fiber qualities of Supima (strength, softness, color retention) make it unique. Fiber length of more than 1.5 inches, ensures comfort and longer color retention.

20.1.3 Importance and Uses

Around the world, 100 million people work in cotton production, particularly in areas where poverty and hunger are prevalent, with over 250 million working in related jobs such as transportation, ginning, and processing, and several million in textile manufacturing, agriculture inputs, and cottonseed crushing. Cotton oil derived from seeds is utilized in the cosmetics and paint industries. If the gossypol concentration is very low, cotton oil can be utilized for consumption (cooking), and the leftover extraction is termed cottonseed meal, which is used in dairy feed (Katageri et al. 2020). To satisfy global edible oil demand, cotton breeders must improve oil percentage by selecting cultivars with superior oil content and quality. To satisfy global edible oil demand, cotton breeders must improve oil percentage by selecting cultivars with superior oil content and quality (Majeed et al. 2020). Refined oil cotton is free of phenolic chemicals and can be used for cooking. Cotton oil after removing phenolic and gossypol contents is considered one of the best edible oils due to its low cholesterol content (Rojo-Gutiérrez et al. 2020). The composition of cottonseed oil is genetically controlled and varies from variety to variety, but generally, it contains 26% palmitic acid, 15% oleic acid, and 58% linoleic acid (Gao et al. 2020). With low-density lipoprotein cholesterol-raising qualities, the oil's high palmitic acid content stabilizes it and makes it perfect for high-temperature cooking.

After extracting oil from cotton seeds, what remains is cottonseed meal, which is used in the livestock industry as a high source of protein for animal feed. The issue lies in the gossypol content (a natural phenol) produced in the cotton plant, making it unsuitable for human consumption and non-ruminant feed. Cotton breeders have developed cultivars with low gossypol content by producing glandless cotton varieties with no gossypols on leaves and seeds. However, the gossypol content in leaves acts as a natural resistance against insects or pests. Researchers at the University of Texas A&M recently developed cotton cultivars with extraordinarily low gossypol content in the seed, without affecting the gossypol levels in other plant components (Prasad and Blaise 2020).

Cotton fiber quality is defined by a variety of sophisticated qualitative features in the trade sector. Reflectance regulates cotton brightness, which is expressed in Rd (a unit of reflectance). The variety, hand picking, ginning processes, and environmental factors all affect the fiber's properties. The spinning industry wants longer, finer, white, bright, and completely mature lint, although the price of cotton is principally determined by fiber staple length, micronaire, grade, and color (Abdullah and Khan 2022). Cotton breeders strive to integrate the greatest features into a single variety while solving additional production issues such as biotic and abiotic environments. Cotton is often divided into divisions depending on fiber properties (Table 20.3), with this category accounting for 3% of total global cotton production. Increasing commercial output requires certain crop management strategies and a suitable environment. The largest cotton-producing countries are China, the United States, India, and Egypt, which use the majority of global production. Small-scale production takes place in nations such as Australia, Israel, Tajikistan, Uzbekistan, and Turkmenistan. International cotton commerce is generally related to fiber quality parameters such as strength, uniformity, maturity, fineness, elongation, neps, short fiber content, spinning performance, dyeing ability, and cleanliness (Abdullah and Khan 2022).

Cotton prices are influenced by several fiber quality attributes (Fig. 20.8), as well as physical aspects such as litter and non-plant contamination, which are important contributors in raising cotton weight (Abdullah and Khan 2022). Reflectance (Rd) is used to determine bright cotton, as brighter cotton results in better dyeing and a more lustrous finished product. Fiber length, measured in millimeters (mm), affects yarn fineness and quality. Fiber fineness measured as micronaire value (MIC), indicates the starch content in the fiber. Fiber strength and neps in the fiber are also important parameters that impact machine efficiency in spinning and weaving.

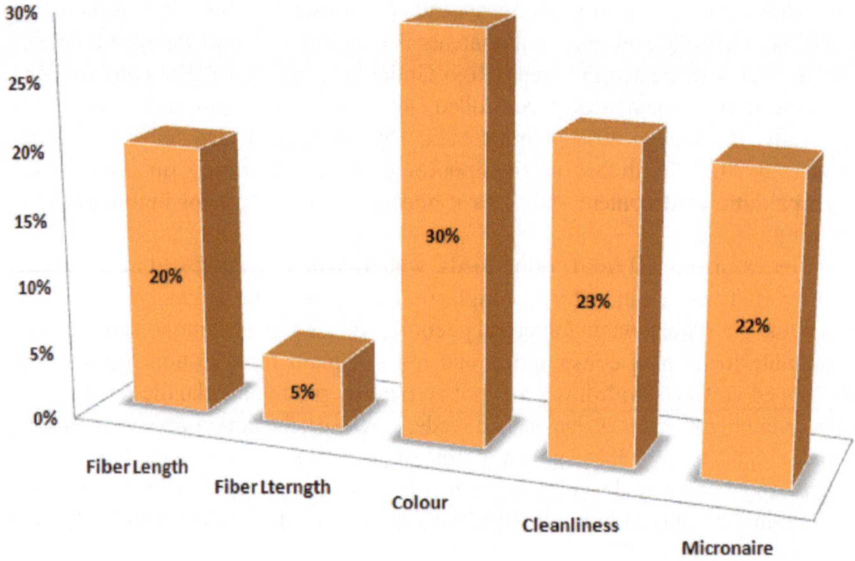

Fig. 20.8 Average price contribution of fiber quality attributes. (Source: Prepared by Dr. A. Abdelmoghny)

Table 20.4 Composition of cotton fiber at different maturity levels

Constituent	Percentage of dry weight		
	Typical	Low	High
Cellulose	94.0	88.0	96.0
Protein (N × 6.25)	1.3	1.1	1.9
Pectin substances	1.2	0.7	1.2
Ash	1.2	0.7	1.6
Wax (principally higher alcohol and long chain organic acids)	0.6	0.4	1.0
Total sugars	0.3	–	–
Pigment	(2)	–	–
Others	1.4	–	–

Prepared by Dr. Abeer Arafa

20.1.4 Fiber and Textile Properties

Cotton belonging to the genus *Gossypium* is considered a primary fiber crop, grown in tropical, subtropical, and semiarid regions worldwide. Fiber is the main product of cotton and is extensively used in the textile industry (Shahzad et al. 2022). Cotton is a soft, fluffy staple fiber that grows as a protective casing around cotton seeds and accounts for 30% of global textile fiber production. The fiber is formed from the epidermal layer of the ovules consisting mainly of pure cellulose which holds significant economic value due to its use in the textile industry. Nearly 90% of cotton fibers are cellulose, a polymer made up of glucose (Table 20.4). Cotton seeds produce two types of fibers: lint, which grows from the seed coat's epidermal cells and detaches readily, and fuzz, which grows for a shorter length and is thicker than lint fiber. Some cotton types feature bare seeds, without lint or fuzz. Fuzz fiber has a thinner cell wall and larger diameters than lint fiber and it comes in various colors like green, gray, creamy and brown, depending on the variety. Fuzz fiber is used in cotton batting, rayon and various types of plastics.

20.1.4.1 Cotton Fiber Development and Structure

A deep study of cotton fiber growth, initiation and inner structure is of primary importance to all cotton breeders, researchers and manufacturers. This is because the quality properties of cotton fiber are largely determined by its growing conditions and chemical composition. Cotton fiber development follows biologically determined stages (Fig. 20.9). After fertilization, cotton fibers undergo tip refining and shift to the elongation phase (Jareczek et al. 2023).

Fig. 20.9 Depiction of overlapping stages of fiber growth that displays *G. hirsutum* fiber from both macro (**a**) and microscopic (**b**) perspectives at various significant junctures, as detailed in (**c**) Fibre initials are visible on the ovule's surface at 0 DPA. Fiber tips are polished and elongation starts at 1–2 DPA. All of the fibers are elongated by 7 DPA. When secondary wall synthesis starts at the transition period (~14 DPA, depending on species and/or accession), elongation continues and then starts to taper. The fiber is actively involved in secondary cell wall deposition by 30 DPA, at which point elongation has stopped. Around 50 DPA is when the fruit reaches its final maturity, at which point the boll dies and the mature fibers have died. Scale bars for the scanning electron images are 5 mm; all other scale bars are 1 cm. (Source: Lee et al. (2007))

Elongation Phase

Cotton fiber is defined as the elongation of the outer layer of the seed coat (epidermal layer). Beginning on the day of anthesis, cells in the outer layer of the ovule start to elongate outward until as late as 25 days post-anthesis (DPA) in domesticated accessions. During this stage, the fiber is expandable and grows in length up to 6 cm. The cotton fiber reaches its final length between 16 and 25 DPA depending on variety and environmental conditions. Elongation starts slow, speeds up for a few days, then slows down towards the end of the period Fig. 20.10. Fiber elongation is a genetically controlled trait that is influenced by environmental conditions like water scarcity and high temperature Cross-section shows that the fiber is circular during this stage and the boll size may not match.

Transition Phase

The secondary cell wall deposition stage begins approximately 15–20 days following anthesis. The transition phase lasts roughly 15–20 DPA. The thickness of the secondary wall continues to increase until approximately 40 DPA or maturity.

20.1.5 Cotton Fiber Classification

Scientists have determined that cotton quality is defined by a set of characteristics that describe the physical, mechanical and structural properties of the cotton. These properties are measured using special instruments according to international principles, methodologies and calibration under consistent conditions. Known as main quality properties (Commercial parameters). Cotton fiber is classified based on various traits including length, strength, length uniformity, fineness, color, trash, leaf grade, extraneous materials and module averaging (Kumar et al. 2022). The classification of cotton is based on these fiber characteristics (Table 20.3).

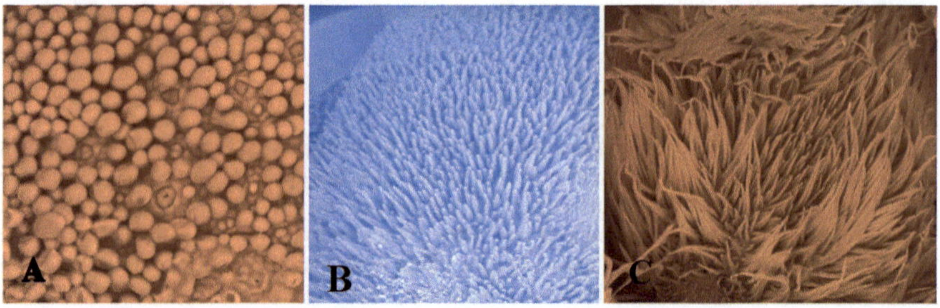

Fig. 20.10 Development sequence of lint and fuzz fibers on (**a**) the day of anthesis, (**b**) 3 days after anthesis (**c**) 6 days after anthesis. (Source: Dr. Abeer Arafa)

20.1.5.1 Fiber Length

Fiber length is the average length of the longest half of the fibers (upper half mean length). It is a sensitive index that can be used to detect changes in fiber length during processing particularly combing. Fiber length is largely influenced by variety, extreme temperatures, water stress and nutrients. Fiber length is measured using different methods: (1) Manual methods such as halo length and staple length, (2) Sorters like Shirley Comb Sorter, Balls Sorter and Sutter Webb Sorter, and (3) Electronic methods including Fibrograph, High Volume Instrument (HVI), Advanced Fiber Information System (AFIS) and fiber classification system FCC, Beyer Tex. Lab (Fig. 20.11a, d).

Fiber length affects spinning machinery efficiency, spinning performance and yarn quality (Kumar et al. 2022). Cotton is classified into five classes based on fiber length: Short Staple: Staple length shorter than $^{13}/_{16}$ inch (14.4 mm). Medium Staple: Staple length ranging from $^{13}/_{16}$ to 1 inch (14.4–25.4 mm). Medium Long Staple: Staple length ranging from $1^1/_{32}$ to $^{13}/_{32}$ inch (26.2–27.7 mm). Long Staple: Staple length ranging from $1^1/_4$ to $1^5/_{16}$ inch (28.6–33.3 mm) and Extra Long Staple: Staple length longer than $1^3/_8$ inch (34.9 mm).

20.1.5.2 Fiber Fineness

Fiber fineness and maturity are two of the most important fiber properties because they have an impact on cotton processing and the quality of the final product. Fiber fineness affects both the strength and appearance of yarn. Fineness determines how

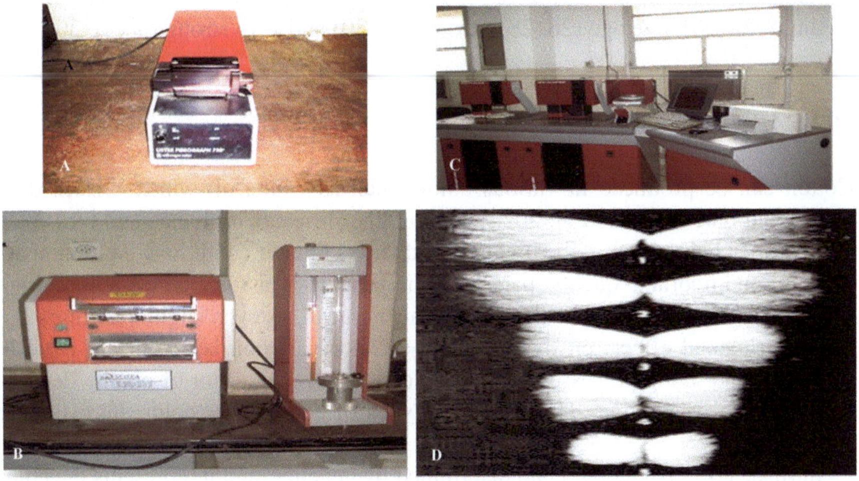

Fig. 20.11 Cotton fiber quality instruments. (**a**) Fibrograph to test fiber length. (**b**) Micronaire instrument to test fiber fineness. (**c**) High volume instrument (HVI). (**d**) Fiber length. (Source: Prepared by Dr. Abeer Arafa)

many fibers may be packed into a cross-section of yarn at a given count. The finer the fibers, the bigger the quantity, and hence the greater the fiber surface area per cross-section of yarn. As the yarn size becomes finer, fiber fineness becomes increasingly important. Finer fibers are usually of higher quality than coarser ones. The finer fibers that are incorporated in a fabric, the greater the number of individual reflecting surfaces per unit area of that fabric. Thus, fiber fineness influences yarn and fabric luster and appearance. Micronaire instrument measures fiber fineness and maturity using an airflow instrument to gauge the air permeability of compressed cotton fibers (Table 20.4 and Fig. 20.11b). Moisture, temperature, sunlight, plant nutrients, and harsh growing circumstances all have an impact on micronaire, which can range from fine to coarse (Kumar et al. 2022).

20.1.5.3 Fiber Strength

Strength and elongation can be measured for single fibers or bundles using various methods, devices and instruments such as Favimat and Instron for measuring single fiber and bundle strength, Pressley flat bundle Tester, Stelometer flat bundle tester and High-Volume Instrument (HVI) (Fig. 20.11c). The factors that primarily determine the strength and other mechanical properties of cotton fibers are fiber structure including spiral angles percentage of crystalline and amorphous cellulose, degree of polymerization, convolutions, structural reversals and weak points. As well as test conditions including the rate of loading, length of the test specimen and atmospheric conditions, particularly humidity. Fiber strength is typically measured in grams per tex, indicating the force in grams required to break a bundle of fibrils per tex. Tex is defined as the weight of fiber in grams of 1000 m of fibril. The international classification of fiber strength using grams per tex is as follows: weak (below 23), medium (23–24), average (26–28), strong (29–30) and very strong (above 30).

20.1.5.4 Fiber Color

Cotton can be produced in various colors such as white, off-white, brown, dark brown and blue (Kumar et al. 2022). Cotton fiber loses its brilliance and turns pale yellow after being exposed to light and adverse climatic conditions continuously. This is an important quality parameter because it is associated with bleaching efficiency. The uniformity of lint color is also crucial, as even slight variations can have detrimental effects on dyeing quality. Color variation is a constant issue in bleaching and dying processes and lint color is a good indicator of quality deterioration due to weathering. There are two types of cotton fiber color: intrinsic and extrinsic.

Intrinsic color is inherited and determined by genetics, and it represents the color of newly opened bolls before exposure to environmental degradation. Bleaching is an option for removing these hues. Extrinsic or acquired color, on the other hand, is caused by external weathering forces following boll opening, such as frost, rain, microbial action, and insect infestation. The extrinsic color shows fiber breakdown and is closely associated with cotton grading. It is tough to remove during bleaching and degrades color quality. The primary characteristics that describe the color of cotton fiber are hue, brightness, and chroma. Hue is the property of color that permits it to be classified as green or blue. Brightness describes the lightness or darkness of a color, ranging from bright to dull. Chroma is a color's intensity, measured in degrees of yellowness. Lint color is measured by two parameters: Rd % (reflectance degree) which measures the brightness and lightness of the cotton and +b, which describes the yellowness or brownish tint of the color.

20.1.5.5 Trash

Trash refers to any additional noncotton materials mixed with cotton fiber such as fragments, grass, sand, and dust. Removing small fragments is more challenging than removing larger ones (Kumar et al. 2022).

20.1.5.6 Leaf Grade

Leaf grade represents the amount of leaf content in cotton fiber as evaluated by a garbage meter. Leaf grade criteria vary from "1" to "7" and are influenced by foreign fragments, harvesting conditions, and techniques (Kumar et al. 2022).

20.1.6 Domestication, Selection and Early Improvements

Domesticated cotton has more plentiful and longer fibers that are more easily removed from the seed than wild varieties. Crop evolution combines the taming of wild plants with the gradual adaptation of crops to new environments outside their ancestor's origins as they expand over the planet. Cotton fibers are the world's most widely used plant textile and serve as a model for plant cell wall biosynthesis. The genus Gossypium has four domesticated species that generate cotton fiber from solitary epidermal seed trichomes (Jareczek et al. 2023). The genus *Gossypium* consists of 50 species distributed across tropical and subtropical regions in the world (Fig. 20.12). These 50 species are grown to generate varied degrees of fiber (long, strong seed hairs used in seed dispersal and textile manufacture) and/or fuzz (shorter hairs unsuitable for textiles) and are the global greatest source of renewable natural textile fiber. The genus *Gossypium* is divided into 8 monophyletic diploid genome groups ($2n = 26$) and a single monophyletic allotetraploid ($2n = 4x = 52$). The genus originated nearly 5–10 million years ago (MYA), while the two allotetraploid species (A and D) diverged from 1 to 2 million years ago the two species (A and D) were reunited to form the allotetraploid (AD), which diverged into seven polyploid species (AD1–AD7) (Hu et al. 2021). Approximately 5000–8000 years

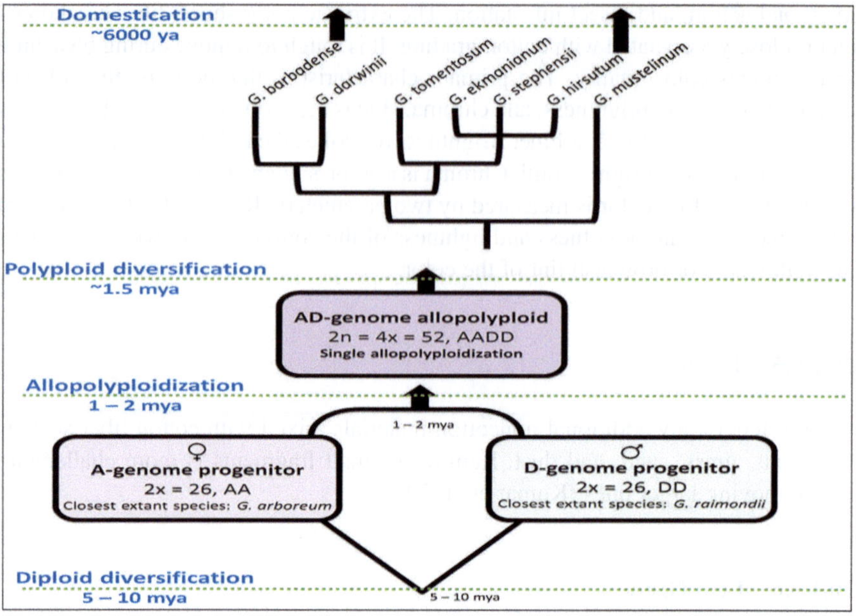

Fig. 20.12 *Gossypium* allopolyploid evolutionary model. The now-extinct A-(maternal) and D-(paternal) progenitors of the allopolyploid clade followed distinct evolutionary trajectories as a result of cotton's diversification around 5–10 mya. These lineages then hybridized and experienced genome doubling, giving rise to the allotetraploid AD clade (~1 mya). Two of the seven polyploid species that emerged from this clade's diversification underwent improvement and domestication, i.e., *G. hirsutum* and *G. barbadense*. (Source: Jareczek et al. (2023))

ago, humans domesticated four wild cotton species four times resulting in two allotetraploid *G. hirsutum*, (AD1) and *G. barbadense* (AD2) and the two diploid species *G. herbaceum* (A1) and *G. arboreum* (A2). All the wild cotton species produce fiber that arises from the seed epidermal layer. The domestication process led to the production of longer, stronger, finer, whiter hairs and increased susceptibility to different biotic and abiotic stresses. After domestication and human selection, the wild cotton germplasm evolved into an annual plant habit, photoperiod insensitivity, early blooming, increased boll size, yield, and fiber attributes such as greater fiber strength, fineness, and length. However, some of these selection mechanisms generated genetic bottlenecks in cultivated cotton, leading to increased loss of genetic diversity (Campbell et al. 2010).

The two major forms of domesticated cotton are the two species that originated in the New World. Pima or Egyptian cotton belonging to *Gossypium barbadense* was domesticated in the Peruvian Andes between 4000 and 5000 years ago. It was used to produce a well-known luxury textile with a lower yield and not as well-adapted to diverse environments. On the other hand, Upland cotton (*Gossypium hirsutum*) was domesticated between 4000 and 5000 years ago in the Yucatan peninsula. It produces coarser and weaker fiber with high yield productivity, is widely adapted to different environments and is more resistant to pests and diseases. During the domestication process, both species underwent many phenotypic and gene expression changes, such as a shift to compact plants, reduction of seed dormancy, and loss of photoperiod sensitivity. The most significant change is in seed trichomes basis cotton fiber, which became longer, finer and stronger than wild species (Gross and Strasburg 2010). Cytoskeletal function genes were up-regulated during some developmental stages and are associated with increasing fiber length and quality. In 1730 the first spinning machine was developed in England. Eli Whitney from Massachusetts, invented the first ginning machine in 1793, which increased the quantity of cotton fiber production (Aslam et al. 2020).

20.1.7 *Current Cultivation Practices and Challenges*

Cotton crop management is a vital aspect of the cotton industry that extends beyond just varieties. Cotton researchers are actively working to develop and enhance crop management strategies and technologies to maximize the combined economic, environmental and social benefits to address today's challenges and those in the future. This approach is known as better or Best Management Practices (BMPs), which are defined as agricultural practices aimed at optimizing the three pillars of sustainability: social responsibility, environmental integrity and economic viability. By integrating financial requirements for agriculture, such as high-yields with environmental and social concerns such as water, nutrients and pesticide use. BMPs seek to establish more ecologically friendly practices, encourage the use of local resources, and increase input efficiency. In addition to enhanced agronomic methods, these practices can be divided into five major categories: soil fertility management, water management, insect and disease management, proper harvesting and storage, and crop residue management. Furthermore, boosting profits for all players in the cotton supply chain while conserving wealth eventually leads to capital formation (Mehdi et al. 2022). The key to decreasing production costs and increasing cotton profitability is more efficient production practices to sustain production and protect the environment.

20.1.7.1 Current Cultivation Practices

Cotton thrives in temperatures between 21 and 27 °C and can tolerate temperatures up to 43 °C but anything below 21 °C can harm the crop. Ideal conditions for cotton growth include warm days, cool nights and significant diurnal temperature variations during the fruiting period which are conducive to good boll and fiber development. Cotton is cultivated in a range of soils, from well-drained deep alluvial soils to black clayey and mixed black and red soils. Cotton needs well-drained soils with high water retention and can be cultivated in both irrigated and rain-fed situations.

Time of Sowing

The timing of cotton sowing in subtropical regions depends on soil and air temperatures, while in tropical regions, it is dictated by the rainfall season. Planting typically occurs from the beginning of February to the beginning of June in the Northern Hemisphere, with harvests taking place between August and October. In the Southern Hemisphere cotton is sown from September to November, with harvests occurring from March to May. Cotton is usually ready for harvest approximately 6 months after sowing. In subtropical regions, the sowing date is determined when the soil temperature stabilizes at 15 °C for 10 consecutive days at a depth of 20 cm by 8 a.m. Adequate soil preparation, including plowing, cultivating and rotavating, is essential to create a suitable environment for good root development and soil conservation. In tropical regions, cotton is sown at the beginning of the rainy season to ensure sufficient moisture during the first half of vegetative growth and dry conditions during ripening and harvesting. The depth for planting cotton seeds typically ranges from 2.5 to 5 cm depending on the seedbed and weather conditions (Afzal et al. 2020).

Variety Selection

Cotton varieties must be bred for high yields under high-input conditions such as fertilizers, pesticides, and irrigation. They must also have good fiber quality, disease and pest resistance and adaptation to local conditions. The most interesting varieties for farmers are those that combine traits like drought resistance and high yield with long fibers, especially in areas with limited or no irrigation. When selecting a variety, farmers should consider factors such as site conditions (soil properties, rainfall amount and time, availability of irrigation water), farm conditions (availability of manure and fertilizers, ability to manage pests) and buyers' requirements for fiber quality (length, strength, fineness) at a high price.

Field Preparation

Field tillage is often carried out four to five times to loosen the soil, eradicate weeds, raise soil temperature and prevent moisture loss. Ridge creation is particularly significant since it improves irrigation and drainage while also helping to reduce lodging. Ridges and furrows should be approximately 10 m long with suitable spacing based on the variety being grown. Ridge plows or bund formers can be used to help save on cultivation costs. In fields with ragi stubbles, cotton seeds can be dibbled at specified spacing. It is worth noting that different varieties/hybrids may require specific spacing between ridges (Farooq et al. 2020; Kumar et al. 2023).

Sowing

Planting technique is a significant factor that determines the healthy vegetative and reproductive growth of cotton crops ultimately affecting lint yield. It also helps achieve proper crop stand and root development by improving nutrient uptake in the root zone by the cotton plant (Adeleke 2024). Sowing cotton requires careful soil preparation to ensure adequate moisture for high germination and root growth rate. Soil pre-preparation and ridge establishment are crucial for optimal drainage and temperature maintenance. Soil temperature plays a vital role in determining the sowing date (Robertson et al. 2007). When sowing, dibble the seeds at a depth of 3–5 cm on the side of the ridge 2/3 of the way from the top, above the band where fertilizers and insecticides are applied. Maintain correct spacing and cover the seeds with soil. In the case of intercropping, sow the intercrop seeds between the paired rows of cotton, with rows 5 cm apart and cover the seeds. Sowing involves adding the required number of seeds in each hole (Datta et al. 2020).

The most conventional cotton sowing technique is the raised seedbed which is widely used in the USA and other countries and remains one of the most sophisticated and profitable methods. This technique involves using raised-seedbed shapers/planters (Fig. 20.13) and planting, depending on the soil conditions, at a depth of 12.7–38 mm (0.5–1.5 in.) The advantages of this technique include reduced crust formation, increased germination rate, optimal crop stands, improved lint output, easier drainage of rainfall runoff, improved water and nutrient use efficiency, comparatively less lodging of insect or disease stress, and improved root propagation (Adeleke 2024).

Seed Rate and Spacing Between Holes

The spacing of cotton plants varies depending on soil, type, region, variety and irrigated conditions. Row spacing can range from 70 to 120 cm, with wide spacing

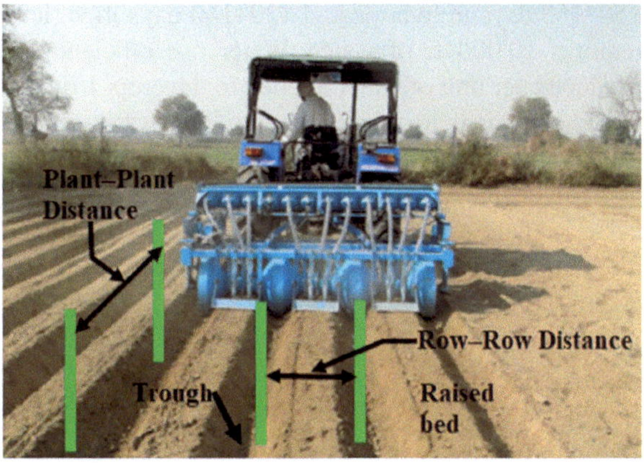

Fig. 20.13 An image showing seedbed design parameters for a multi-crop raised seedbed planter in action on a field. The machine plants the seeds in its path after first forming the raised trapezoidal seedbeds. (Source: Kumar et al. (2023))

recommended for deep black soils. Plant spacing can vary from 30 to 60 cm, with hybrids typically requiring greater spacing than varieties. The seed rate depends on factors such as seed type, growth habits and vigor (Farooq et al. 2020).

Water

The cotton crop is not water intensive because cotton is tolerant to drought and high temperatures, making it suitable for cultivation in regions with limited water supply. Additionally, cotton is often grown and cultivated under rainfed conditions. Approximately 53% of global cotton production occurs in dry, semi-arid and arid regions (such as Pakistan, Australia, Uzbekistan, Egypt, the US, and some provinces of China) using irrigation systems. The irrigation water source comes from freshwater sources in Uzbekistan, Egypt, and Xinjiang province of China, while 31% of cotton production in Pakistan relies on groundwater sources. Most of these irrigated areas use traditional flood irrigation techniques, which involve delivering fresh water from dams, lakes, rivers, or reservoirs to the surface of the cotton fields through open channels. However, these traditional irrigation systems are inefficient due to high seepage and evaporation rates, making them unsustainable. Modern irrigation systems such as plastic film mulching, drip and center pivot irrigation methods have proven to be effective in water conservation (Fig. 20.14). These new methods have led to the development of fertigation technology, which integrates solid or liquid fertilizer application with irrigation. Despite their effectiveness, modern irrigation systems can be costly, especially in developing countries that rely on groundwater, and measures need to be taken to prevent the depletion of groundwater sources to avoid future water crises (Adeleke 2024).

As a general rule, the cotton plant requires a minimum of 500 mm of water between germination and boll formation. The cotton plant cycle is divided into five stages, with moisture playing a crucial role in plant growth and overall health. Water stress during these stages can significantly reduce yield. In black soils, the cotton plant requires 5–6 irrigations, while in red loamy or red sandy soils, they need 10–12 irrigations. During the boll bursting period, no irrigation is required (after 150–160 days in hybrids and 130–140 days in varieties). Producing 1 kg of cotton requires 10,000 L of water. Water use efficiency (WUE) is the total lint yield produced per unit of water applied to the crop. Irrigated cotton has shown improved WUE compared to rainfed cotton. The water requirement varies throughout the growing stages, with critical stages for water availability being squaring, flowering, and boll development. Improved water management throughout the cotton plant life cycle leads to an effective root system, optimal vegetative growth, less boll shedding, and uniform bursting in a shorter period, resulting in enhanced yield (Datta et al. 2020).

Fig. 20.14 Center pivot (also known as linear move machine). Compared to drip irrigation, which is better suited for smaller farms, it is more appropriate for variable rate irrigation water application on a larger scale. (Source: Valley Irrigation)

Fertilizer

NPK fertilizers should be applied to soil based on soil analysis, at the plants base. Nitrogen (N) should be applied during the plant vegetative stage, divided into three equal doses at 30, 60 and 90 days after sowing. Phosphorus (P) should be applied during soil preparation at the time of the final plow. Potassium (K) should be applied in a single dose at the beginning of the flowering stage. *G. barbadense* is more sensitive to nitrogen fertilizer than *G. hirsutum* with excessive nitrogen causing delayed maturity and reduced yield. Barbadense requires more N, P and K per unit to produce lint. Micronutrient elements are required depending on soil content and environmental conditions. Recommended doses may vary among different farming systems and soil types and should be confirmed with local agricultural research stations. Nutrient use efficiencies (NUE) including macronutrients and micronutrients can be enhanced through selective breeding to increase high lint yield (Xiao and Yin 2020). The balance between macro and micronutrients, as well as organic and inorganic nutrients, is crucial for cotton production and fiber quality.

Harvesting

Cotton harvesting is a boring and extremely laborious activity. In tropical cotton countries, harvesting is typically done by hand, in three rounds to prevent deterioration of lint quality from open bolls left in the field. The first picking usually begins approximately 10 days after the fruits have opened; with subsequent rounds following at intervals of slightly over 2–3 weeks. Sorting the cotton during harvesting into white cotton, stained cotton and waste is essential. The cotton should be as dry as possible during harvesting, and in forest regions or after late rains, additional drying on racks may be necessary. Hand-picking seed cotton requires a significant amount of labor but results in cleaner cotton and higher recovery compared to mechanical harvesting. The adult person will collect 15–20 kg of cotton per day (Mishra et al. 2023).

Mechanical harvesting is common in the United States and Australia, due to the high cost of hand-picking. In these areas, low-growing, weakly-branching cultivars are planted at narrow spacing and sprayed with defoliants before harvesting. Mechanically harvested fiber requires heating and shaking to remove impurities, making the fiber more brittle (Majumdar et al. 2020). When John Rust created the first cotton harvester prototype, known as the harvesting locomotive, in the 1930s, the development of the cotton harvesting machine began. However, it needed regular maintenance and was costly and brittle (Fig. 20.15). The first spindle-type

Fig. 20.15 The development of the cotton harvesting machine. (**a**) A second-generation cotton picker M manufactured in the 1940s that was commercially successful A 7760 JD cotton harvester with an onboard module building feature, which eliminates the need for additional human labor, costly equipment and tractors to construct cotton modules, improved (**b**) a boll buggy receiving seed cotton from a cotton picker, an approach that required multiple equipment types and is labor intensive (**c**) is John Deere machine. (Sources: (**a**) Mississippi State University Extension, (**b**) West Tennessee Historical Society, (**c**) John Deere, Inc)

mechanical cotton piker was introduced in the mid-1940s by International Harvester (IH), which revolutionized cotton farming from a labor-intensive to a capital-intensive system. Also, in the 1940s John Rust succeeded in redesigning his spindle-type harvester to be more durable and commercially viable. The major revolution happened between the 1950s and 1980s when John Deere (JD) transformed the cotton harvester into a fully functional cotton picker with their 4-row model that worked in cooperation with tractors. Subsequently, Jhon Deere invented the cotton piker with 6-rows in the 1990s, which increased the speed and decreased the time required for cotton harvesting (Adeleke 2024).

20.1.8 Germplasm Biodiversity and Conservation

Germplasm conservation is an efficient technique for preserving the genetic traits of vulnerable and economically valuable species. For all of the genes found in a plant, germplasm acts as a live information source. It can also be preserved for extended periods, enabling future regeneration as required. Germplasm conservation helps retain knowledge about extinct, wild, and living species of crop plants, as human interference has led to the erosion of genetic diversity by favoring certain genes and eliminating fewer desirable ones, ultimately leading to the extinction of historic genetic material.

The genus *Gossypium* contains 45 diploid and 5 allotetraploid species mainly grown in tropical and subtropical regions worldwide (Percival et al. 1999). Out of the 50 species, only two diploid and two tetraploid species are cultivated (Paterson et al. 2012). The most genetically diverse genus, *G. barbadense*, is found in the western Andes, South America, southern Mesoamerica, and the Caribbean basin (Fryxell 1979; Percy and Wendel 1990). *G. barbadense* was first cultivated in the United States coastal islands and lowlands as Sea Island cotton before being introduced to Egypt's Nile Valley as Egyptian cotton. This type of cotton is distinguished by its long and extra-long staple, fine and stronger fibers (Abdalla et al. 2001). This genus accounts for around 9% of total world cotton production. Nine nations hold major cotton germplasm collections: India, Brazil, France (CIRAD), the United States, Australia, Uzbekistan, Russia, China, and Pakistan (Table 20.5) (Campbell et al. 2010; Abdullaev et al. 2013). These collections comprise breeding resources, germplasm from wild species, as well as genetic and cytogenetic stocks for cotton gene pools.

20.1.8.1 Genetic Diversity

Genetic diversity is the biological variation that occurs within species, allowing them to adapt to climate changes (Alqahtani et al. 2024; El Rabey et al. 2013; Elabd and Salem 2009; El-Absawy et al. 2012, 2015; El-Zanaty et al. 2011, 2013; Esmail et al. 2009; Ghazy et al. 2020; Ibrahim et al. 2024; Mattar et al. 2016; Salem 2004, 2009;

Table 20.5 Total number of major cotton countries collected genus *Gossypium*

Country	No. of collected accessions	
	G. barbadense	*G. hirsutum*
Uzbekistan	3019	13,241
USA	1584	6302
Brazil	1509	1660
Russia	1057	4503
China	633	7712
India	534	7633
France	483	2103
Australia	104	1518
Pakistan	109	1025

Salem and El-Zanaty 2015; Salem and Ibrahim 2024; Salem and Mattar 2014a, b; Salem and Sallam 2016; Salem et al. 2008a, b, 2010, 2015a, 2021b, c, 2022, 2024; Varshney et al. 2008). The region where a plant species has the greatest amount of genetic variation is known as its center of diversity, and it encompasses both wild relatives and gene variants as well as cultivated species (or subspecies). Genetic diversity has three sources: mutation, recombination and gene immigration. *Gossypium* for example, contains three major biological diversity hotspots: Mexico, Australia, and Africa and the Arabian Peninsula. South America, Mesoamerica (including Mexico and Central America), and Africa and Asia are the three centers of domestication.

The major hindrance of cotton cotton-sustaining productivity worldwide is a narrow genetic base. During the last three decades, the cultivar's improvement has slowed down due to Limited genetic diversity and the low efficiency of traditional selection methods. The small genetic foundation results in high levels of homogeneity across broad areas of cropland. The major factors for low genetic variability in cotton; The cotton breeding programs depend on adapted cotton genotypes because the cotton breeders avoid wild germplasm because of positive linkage between desired and unwanted traits. Lack of innovative ways to introduce beneficial genetic variety into breeding cultivars from several exotic cotton species in the Gossypium genus. Together, these elements caused evolution's genetic bottleneck (Bardak et al. 2018). The cotton breeder must investigate the genetic variability and genetic relationship between breeding materials to select the best parents which could improve breeding programs. Many statistical techniques are used to determine the extent of genetic diversity between breeding materials, including the Percent Disagreement Values (PDVs) distance matrix, tree clustering diagram, and Polymorphism Information Content (PIC), which is used to evaluate polymorphisms via primers or markers.

Using 26,952 polymorphic SNPs, assessments of genetic diversity and population structure demonstrated gradually decreasing diversity in US Upland cotton compared to landraces (Gowda et al. 2023). Thus, cotton breeding efforts require a broad genetic base to develop new varied populations with appropriate yield and fiber quality attributes, which are then used to increase genetic gains and produce hybrids with better heterotrophy.

20.1.8.2 Cultivars Characterization and Phylogeny

The characteristic features of the phylogenetic tree include a rooted tree with a lineage of a common ancestor and an unrooted tree that illustrates the organism with relativeness that does not have a common ancestor. The branches denote the occurrence of divergence character-based approaches and compare every sequence at once, taking into account a single character or site at a time. They are maximum likelihood and maximum parsimony. These techniques take into account variation in a set of sequences and employ probability. Phylogenetic tree construction techniques include Bayesian inference, maximum likelihood, parsimony, and distance-based techniques. Phylogenies help structure classifications, organize knowledge of biological diversity, and shed light on evolutionary events. Recently, a large number of released varieties characterized by narrow genetic bases, while the major target for cotton breeders is increasing productivity (Amiteye 2021; Kalivas et al. 2011). A relatively new field of study called molecular phylogeny compares the amino acid and structural characteristics of proteins and the nucleotide sequences of genes to infer evolutionary histories, relationships, and in certain situations, functions. DNA sequencing reveals genetic similarities and differences to infer evolutionary relationships among species.

The cotton breeder used morphological and biochemical markers for new variety identifications, but these techniques are influenced by the environment. The molecular breeder used molecular markers based on polymerase chain reaction (PCR-based DNA) or hybridization, which are widely used in the cotton genome. The most widely used molecular markers are random amplified polymorphic DNA (RAPD), Allozymes, restriction fragment length polymorphism (RFLP) and amplified fragment length polymorphism (AFLP). However, since the new cotton varieties are allotetraploid with large genomes, the efficiency of genotyping cotton genome using these markers is very low (IAEA, Vienna 2002; Kalivas et al. 2011; Nadeem et al. 2018; Amiteye 2021). Therefore, the molecular breeder has turned to using simple sequence repeat (SSR) or microsatellites for genotyping the cotton genome (Kalivas et al. 2011). SSR markers are commonly used in studies to investigate genetic diversity due to their ability to amplify genomic loci specifically, a high degree of polymorphism and distribution across the whole genome (IAEA, Vienna 2002; Kalivas et al. 2011; Nadeem et al. 2018; Amiteye 2021).

The goal of contemporary breeding methods is to create new cotton cultivars with traits like upright leaf angles, erect, compact stem architecture, decreased branching, high harvest indices, and higher seed output (Li et al. 2013). The ideal type for optimal light interception can be achieved by having vertical branches near the meristem that are more horizontal at the lower canopy, which is associated with high yield (Sun et al. 2022). Currently, cotton breeders are using genome editing tools like CRISPR to modify plant architecture, including branch and leaf orientation to optimize light exposure and water uptake (Kangben et al. 2024).

20.1.8.3 Genetic Resources Conservation Approaches

Germplasm refers to the complete genetic material, or alleles of distinct genes, found in crop species and their wild counterparts. This is also known as genetic stock, gene pool, or genetic resources (Börner et al. 2009, 2010; Elabd et al. 2020; Saleh et al. 2020; Salem et al. 2004a). The following are the main characteristics of plant genetic resources: The genetic pool refers to the entire amount of genetic variability or diversity found in crop species. Germplasm is gathered in a variety of locations, including centers of diversity, gene banks, gene sanctuaries, farmer's fields, markers, and seed corporations (Abulaila et al. 2025; Abdullaev et al. 2025; Akparov et al. 2025; Aladele et al. 2025; Al Lawati 2025; Almarri et al. 2025; Arab et al. 2025; Békefi et al. 2025; Breidy 2025; Cañar et al. 2025; de Galarreta et al. 2025; Desheva et al. 2025; Díez et al. 2025; Galea-Gragera and Llera Cid 2025; Ghidan and Abdulmajid 2025; Gudaye et al. 2025; Harzli et al. 2025; Hidayatun and Kurniawan 2025; Ihsan and Prihatini 2025; Jatoi and Siddiqui 2025; Laghetti et al. 2025; Mallor and Luna 2025; Monteros-Altamirano et al. 2025; Osterc et al. 2025; Popović et al. 2025; Rocha et al. 2025; Wang et al. 2025). Germplasm includes both cultivated and wild species and crop plant relatives, as well as landraces, modern cultivars, obsolete cultivars, breeding stocks, wild forms, and wild species of cultivated crops. It provides the foundation for starting a crop development program, and it may be indigenous (collected within the country) or exotic (collected from foreign countries).

Genetic resources are the raw materials utilized to develop new cultivars with desirable features that permit resistance to biotic and abiotic challenges. Many crop species' genetic diversity has gradually reduced as a result of increased human activities. Germplasm conservation, on the other hand, refers to the protection of crop plant genetic diversity against genetic degradation. Germplasm conservation, on the other hand, refers to the protection of crop plant genetic diversity against genetic degradation. In-situ and ex-situ conservation are two crucial techniques for conserving or preserving germplasm. Eight nations-Australia, Brazil, China, France, India, Russia, the United States, and Uzbekistan have the majority of the world's cotton germplasm resources (Abdullaev et al. 2013). A program for the conservation and storage of germplasm is in effect in these eight countries.

Genetic Resources at ICAR-CICR, India

The ICAR-CICR Cotton Gene Bank houses over 12,335 Gossypium accessions, including *G. hirsutum* (8851), *G. barbadense* (536), *G. arboreum* (2053), *G. herbaceum* (565), Wild Species (24), interspecific derivatives (40), Perennials and landraces (254), and races and derivatives of cultivated species (12).

Additionally, 69 desi cotton germplasms from various states in the North Eastern Hill Region (NEH Region) were characterized and evaluated. Breeders, scientists at CICR, State Agricultural Universities, and private seed companies received 48 germplasms for their cotton enhancement program, which included *G. hirsutum*, *G. arboreum*, and wild species. Medium-term storage (MTS) was used to store 52 *G. herbaceum* lines multiplied in Surat and 565 accessions multiplied in Nagpur (https://cicr.org.in/reserch-cropimp-genetic-resources).

Mexico

Mexico is the centre of origin of the Gossypium genus, with 11 of its 13 diploid species (*G. armourianum*, *G. lobatum*, *G. gossypioides*, *G. aridum*, *G. laxum*, *G. shwendimanii*, *G. thurberi*, *G. trilobum*, *G. davisonii*, *G. turneri* and *G. harknesii*) and one tetraploid species (*Gossypium hirsutum*). Together, these species constitute a genetic pool that is useful for the utilization and improvement of this genus. The greatest amount of data has been collected for *G. hirsutum*, while other species such as *G. armourianum*, *G. shwendimanii*, *G. laxum* and *G. arboreum* have less data available and are stored in germplasm banks in the United States, China and Russia (Mendoza et al. 2016). The World Information Network on Biodiversity (CIRAD) is another research center that preserves one of the largest collections of genetic cotton resources, with more than 3000 accessions of cotton from over a hundred countries.

Uzbekistan

The Institute of Genetics and Plant Experimental Biology at the Academy of Sciences of Uzbekistan (IGPEB) houses approximately 7500 cotton accessions, which include 4500 *G. hirsutum*, 971 *G. barbadense*, 857 *G. herbaceum* and 547 *G. arboreum*, as well as 200 photoperiod-converted mutants of photoperiodic allotetraploid species and numerous intra- and inter-specific hybrids. They are all maintained vegetatively in the germplasm unit greenhouse (Abdullaev et al. 2013).

20.1.8.4 Cytogenetics

Cultivated cotton exists in both diploid ($2n = 26$) and allotetraploid ($2n = 4x = 52$) varieties. Cotton has eight distinct genome types, called A through G and K. The most significant commercial variety in the US is tetraploid cotton, *G. hirsutum* and *G. barbadense*. The cotton cytogenetics studies began in the early 1930s. The chromosomes were very small ranging from 1 to 4 μm, quite similar in morphology and very difficult to identify (Muravenko et al. 1998). The F genome had the largest chromosome length 3.5 μm, while the D genome contains the smallest one 1.5 μm.

The species *Gossypium barbadense* L. is a tetraploid cotton with an (AD2) genome and consists of Ab and Db homoeologous sub-genomes. The (AD2) genome is divided into sub-genomes Ab and Db. The *G. barbadense* karyotype and the idiograms of the two sub-genomes Ab and Db are given in Figs. 20.16 and

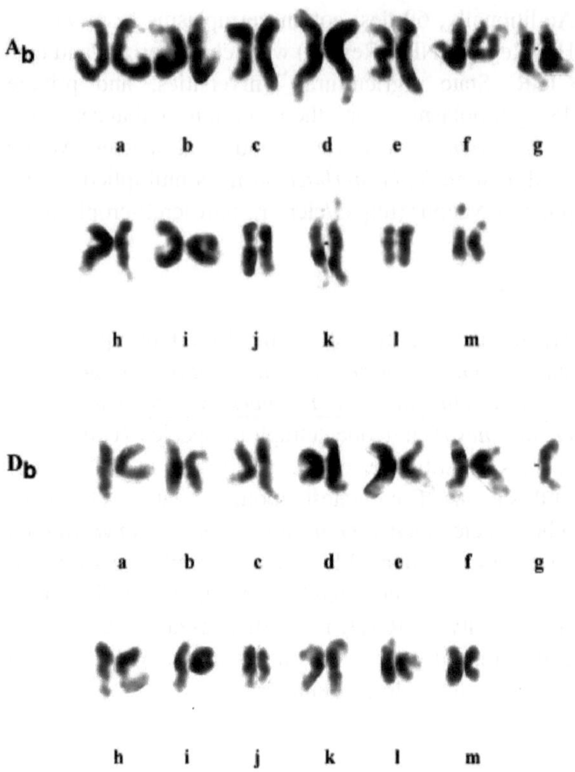

Fig. 20.16 Egyptian cotton (*Gossypium barbadense* L.) karyotype. (Source: Muravenko et al. (1998))

20.17, respectively. The results showed that the chromosomes of the A genome are larger than the D-genome and there are no significant differences in homoeologous chromosomes between the A1 genome and the Ab sub-genome. However, there are more pronounced differences between the Ab and Db sub-genomes. The two genomes A and D had different sizes and molecular structures due to the differences in both repetitive DNA and repeat tandem families.

20.1.9 Cotton Breeding Methods

The term conventional cotton breeding refers to techniques that do not involve genetic modifications or are more appropriate to be a reference for all methods not involving recombinant DNA. This process involves crossing parents with desired traits and selecting offspring with the desired combination of genes inherited from the parents (Abdel-Hamid and Salem 2021). The breeding cycle is considered complete when the top lines are issued as improved pure-line cultivars or germplasm. The farmer grows pure-line cultivars. As a result, the fundamental goal of

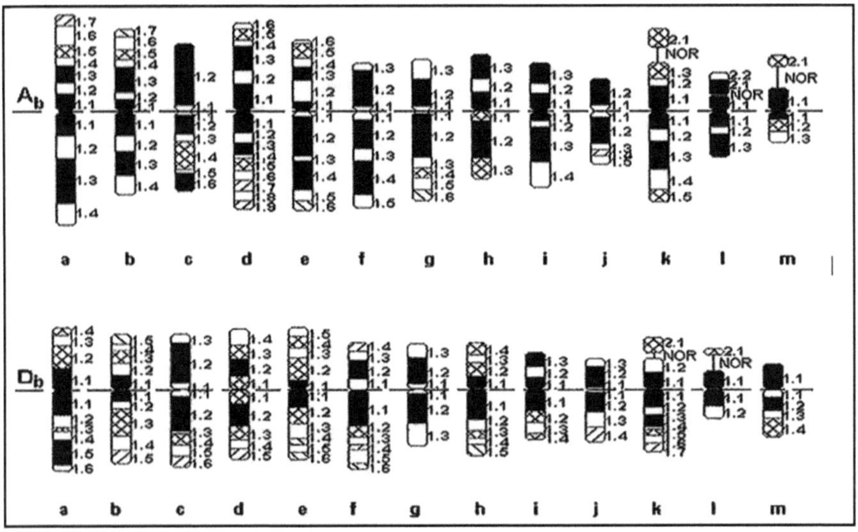

Fig. 20.17 Egyptian cotton (*Gossypium barbadense* L.) ideogram of the two sub-genomes Ab and Db. (Source: Muravenko et al. (1998))

conventional breeding methods is to increase the genetic variety and breed for better production, fiber quality, and disease resistance (Waheed et al. 2022).

20.1.9.1 Cotton Breeding Objectives

Cotton breeders' primary goal is to develop and improve cultivars with high yield potential, good fiber quality (longer, stronger, and finer), a high capacity to tolerate changes in climatic conditions, early maturity, high nutrient use efficiency (NUE), water use efficiency (WUE), and resistance to diseases and pests (Abdelbary et al. 2021, Abdelmoghny et al, 2022; Razzaq et al. 2021; Percy et al. 2009; Ulloa et al. 2009; Ur-Rahman et al. 2014) as shown in Fig. 20.18.

20.1.9.2 Traditional Cotton Breeding Methods

Plant breeding entails modifying plant species to develop plants with desirable characteristics such as increased yield and disease resistance (Aldahak et al. 2021; Hassan et al. 2021; Ibrahim et al. 2023; Mourad et al. 2019; Salem et al. 2015b, c, 2023). Crop improvement is achieved primarily by introduction, selection, hybridization, and mutant breeding (Fig. 20.19). Cotton breeders have effectively improved productivity, fiber quality, disease resistance, and stress tolerance in several cotton-producing countries using conventional breeding methods (Bourland and Myers 2015). Most cotton breeding programs in various countries utilize

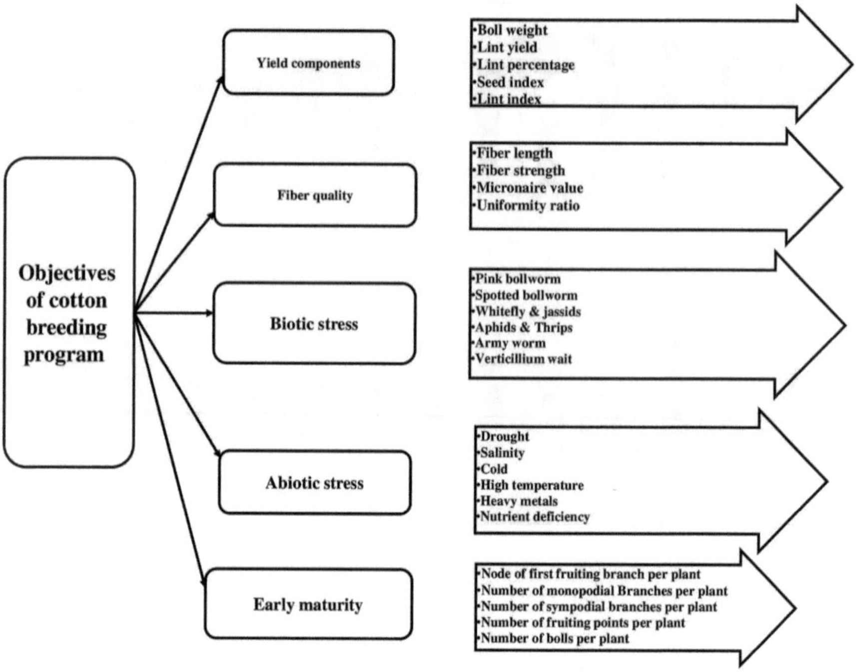

Fig. 20.18 Main objective of the cotton breeding program. (Source: Prepared by Dr. A. Abdelmoghny)

hybridization and selection breeding techniques to develop new cotton varieties in Pakistan (Farzana et al. 2022; Razzaq et al. 2021; Ur-Rahman et al. 2014), Egypt (Abdelbary et al. 2021; Abdelmoghny et al. 2022; El-Adly et al. 2018) and the USA for Pima cotton (Percy et al. 2009; Ulloa et al. 2009). All breeding methods follow similar steps to reach the commercial variety, starting with hybridization, selection, preliminary yield trials, multi-location replicated yield trials, registration and DUS tests, and finally seed multiplication and release of the commercial variety (Fig. 20.20). The difference between breeding methods lies in the selection methodology, which determines the breeding years.

Hybridization

Hybridization is the process of crossing genetically distinct parents from the same or other species to increase variability and produce desirable changes in organisms (Aboshama et al. 2015; Emara et al. 2011, 2013; Nawara et al. 2016, 2017; Salem et al. 2021a). This approach can be used on both plants and animals. There are two kinds of sexual hybridization techniques: intra-specific and inter-specific. Cotton, being a self-pollinated crop with 15–20% cross-pollination, relies on insects like honey bees and bumble bees for pollination. Most cotton breeders use the pedigree

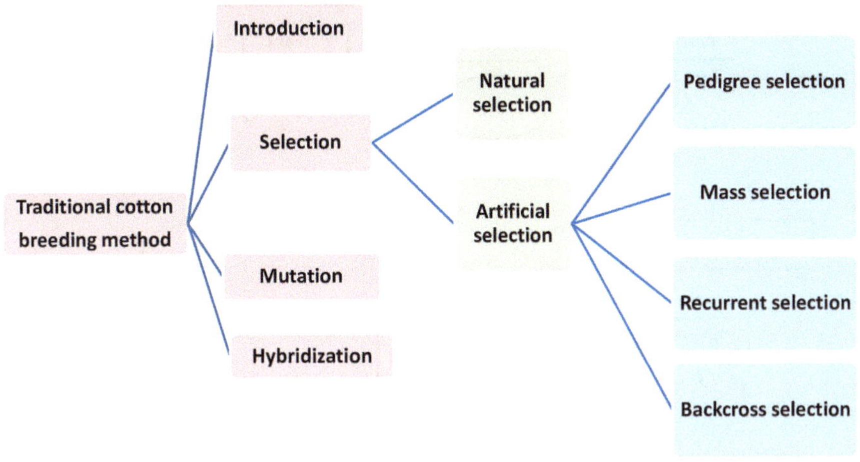

Fig. 20.19 Traditional cotton breeding methods. (Source: Prepared by Dr. A. Abdelmoghny)

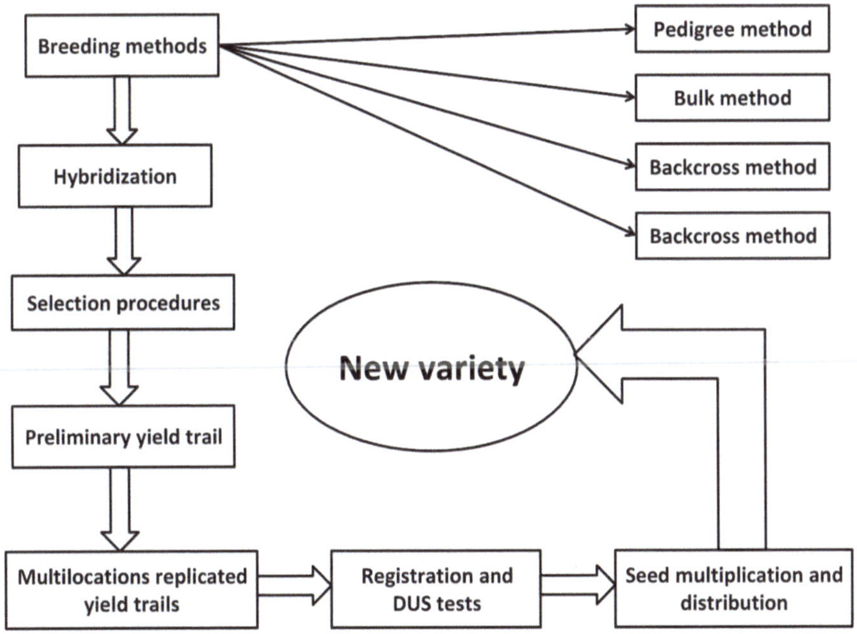

Fig. 20.20 Main steps of different cotton breeding methods. (Source: Prepared by Dr. A. Abdelmoghny)

breeding approach to produce pure-line cultivars (Anwar et al. 2022). Breeders select parents with various desirable traits for cross-hybridization, evaluate the segregating generation population in the field to select individual plants with desirable traits, and evaluate seeds from selected plants in progeny rows and inbred lines that outperform check cultivars in multi-location yield trials (Anwar et al. 2022).

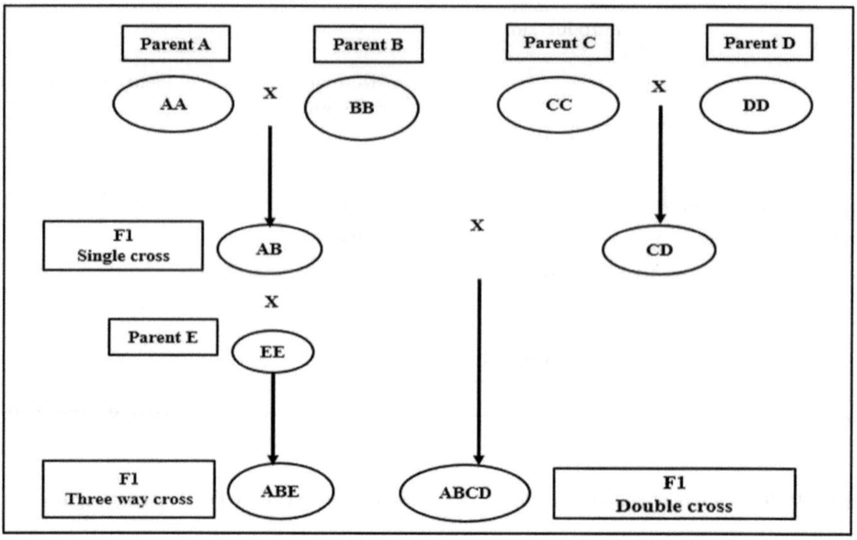

Fig. 20.21 Different types of cotton crosses depending on the number of parents. (Source: Prepared by Dr. A. Abdelmoghny)

Cotton hybrids (F_1 Filial) can be simple or complex according to the number of parents involved (Fig. 20.21). A simple cross or single cross involves crossing between only two parents to create F_1 seeds. Complex cross, also known as convergent cross, is the process of crossing more than two parents to produce F_1 seeds, which combine genes from many parents into a single variety. Three-way cross: When F_1 is crossed with a third parent. Double-cross happens when two F_1s are crossed with different parents.

In cotton, common hybridization types include intra-specific (inter-varietal hybridization) and inter-specific hybridization. (A) Intra-specific hybridization; intra-specific crosses are used to create commercial cotton varieties between pure lines or varieties within the same species. This includes crossing *G. barbadense* species in Egyptian and Pima breeding programs and crossing *G. hirsutum* species to release commercial Upland cotton varieties at the tetraploid level. Mating within *G. arboreum* species is particularly beneficial for tolerance to pests and drought conditions at the diploid level. (B) Plant breeders utilize inter-specific hybridization when genetic variation within a species is restricted. They cross various species of the same genus to convey beneficial genes into the genome of farmed cotton (Nijagum and Khadi 2001). These approaches have been used to transfer the CLCuV resistance introgressed resistant B6 genes against bacterial blight caused by *Xanthomonas malvacearum* (Brinkerhoff 1970; Knight 1956; Nazeer et al. 2014). Inter-specific hybridization results in three types of crosses; fully fertile, partially fertile and fully sterile in different crop species (Table 20.6).

Table 20.6 The results of some inter-specific cotton crosses between different species

No.	Cross between	Results in F_1
1	*G. hirsutum* × *G. barbadense*	Fertile
2	*G. arboreum* × *G. herbaceum*	Fertile
3	*G. hirsutum* × *G. arboreum*	Partially fertile
4	*G. hirsutum* × *G. herbaceum*	Partially fertile
5	*G. hirsutum* × *G. raimondii*	Partially fertile
6	*G. barbadense* × *G. arboreum*	Partially fertile
7	*G. barbadense* × *G. herbaceum*	Partially fertile
8	*G. barbadense* × *G. raimondii*	Partially fertile
9	*G. barbadense* × *G. thurberi*	Partially fertile
10	*G. barbadense* × *G. gossypioides*	Sterile
11	*G. africanum* × *G. raimondii*	Sterile
12	*G. herbaceum* × *G. thurberi*	Sterile
13	*G. arboreum* × *G. stocksii*	Sterile
14	*G. arboreum* × *G. raimondii*	Sterile

Source: (Singh 2013)

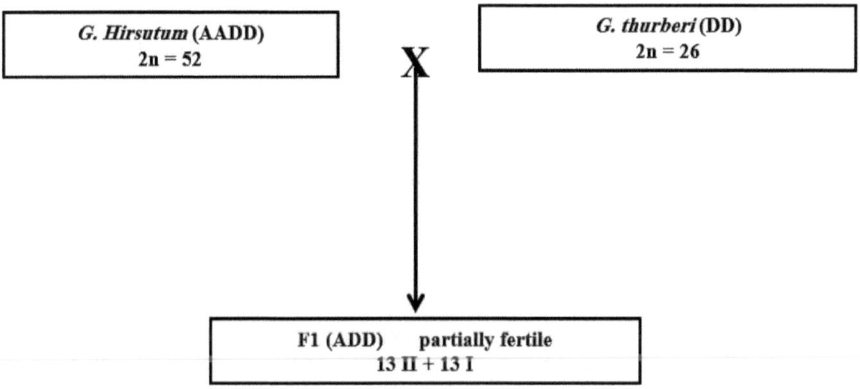

Fig. 20.22 Hybridization between *G. hirsutum* and *G. thurberi*. (Source: Prepared by Dr. A. Abdelmoghny)

Fully fertile crosses occur between species with complete chromosomal homology resulting in fully fertile F_1 plants. Inter-specific hybrids are fully fertile especially between new-world cotton *G. hirsutum* and *G. barbadense* at the tetraploid level (2n = 52) and between the old-world cotton *G. herbaceum* and *G. arboreum* at the diploid level (2n = 26). Partially fertile crosses: Occur between species with differing chromosome numbers but some chromosomes in common, leading to partially fertile and partially sterile F_1 plants. The cross between American cultivated cotton (*G. hirsutum*, 2n = 52) and American wild diploid (*G. thurberi*, 2n = 26) is partially fertile due to a shared chromosomal D genome. Meiosis in F_1 leads to the formation of 13 bivalents and 13 univalent (Fig. 20.22).

Fully sterile crosses occur between species that have no chromosomal homology, meaning their chromosome numbers may or may not be similar. This prevents pairing between the chromosomes of the two species during meiosis, resulting in F_1 plants that are fully self-sterile. However, these crosses can be made self-fertile by doubling the chromosome number through colchicine treatment.

Pedigree Method

The pedigree selection method refers to individual plants that begin with F2 and progress through succeeding generations, where their progenies are formed and a record of the parent-progeny relationship (pedigree) is kept. As a result, each progeny in every generation may be traced back to the F_2 plant from which it originated, forming a pedigree record. Selection is practiced between and within progeny families. The pedigree method was first outlined by Lowe (1927). Plant breeders are always looking for a more efficient and successful way to choose plants. Cotton breeders have used several selection methods to improve various cotton traits. Pedigree selection eventually became the most prominent breeding approach in cotton breeding programs in many cotton-producing countries. This technique is used to select from the segregating generation of crosses in self-pollinated crops. Figure 20.23 represents an example of the pedigree selection method from the Egyptian cotton breeding program.

Until the population is close to homozygosity (from F_2 to $F_{4:5}$), this strategy depends on selecting individual plants. After that, selection is carried out among families. According to Abdelmoghny et al. (2021) and Abdelmoghny et al. (2016), each F_2 plant is unique. Segregation would still be observed within families, even if many loci would become homozygous in the F_3 and F_4 generations. This is why the selection is based on the superior plants in the best families from F_3 through F_4 and F_5 generations (Falconer and Mackay 1996; Singh et al. 2021). Most families should be homozygous at most loci by $F_{5:6}$ generations, which means there will be little genetic variation within the family. As a result, selection becomes less effective within the family and more effective among families (Abdelmoghny 2016; Abdelmoghny et al. 2021; Gibely 2021a, b).

Pedigree Selection Procedure

The procedures of the pedigree method are as follows (1) Hybridization: Select two parents for crossing to produce F_1 seeds (15–30 plants are selected for the next generation). One parent should be a commercial variety and the other should complement the first one. (2) F_1 generation: Plant F_1 seeds to produce the maximum number of F_2 seeds. (3) F_2 generation: This generation exhibits the greatest range of genetic variability and offers the first opportunity to select individual plants. A significant number of plants are planted to be evaluated. Field selection of individual plants is carried out based on their behavior, the breeding program's purpose, and the breeder's experience. Individual plants are harvested, and a comprehensive record of

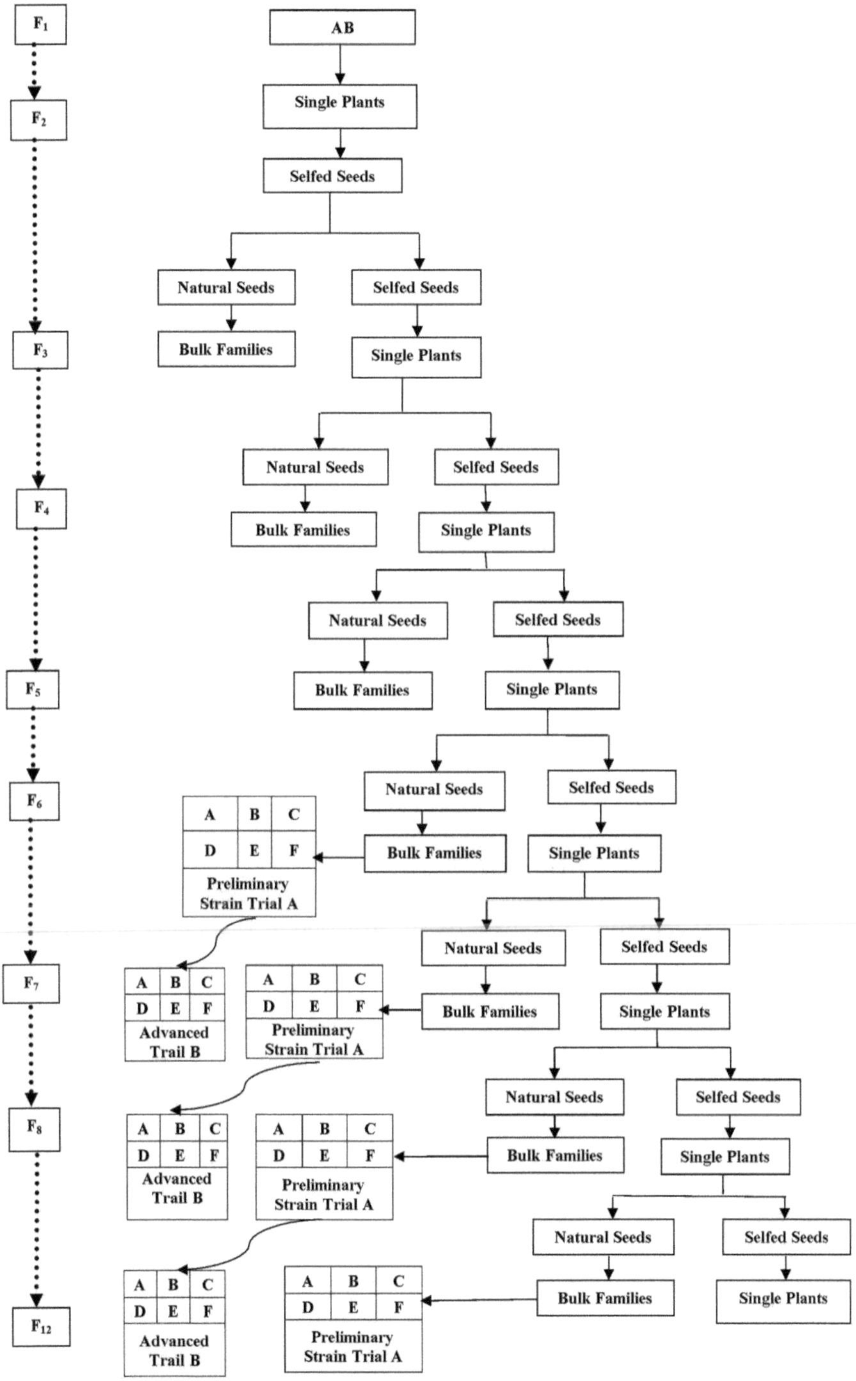

Fig. 20.23 Pedigree selection breeding method. (Source: Prepared by Dr. A. Abdelmoghny)

all genetic, agronomic, and fiber quality data is kept. (4) F_3 generation: the selected F_2 plants are raised alongside their parents as check variety. The plants are meticulously inspected, visually compared with their parents and tested against diseases and pests. The selected plants are harvested individually, and all genetic, agronomic, and fibre quality data are documented. (5) F_4 generation: Selected F_3 plants are propagated to form F_4 families. The F_4 family is made up of offspring from a single F_3 plant. These families are checked for diseases and pests, bulk harvested, and all genetic, agronomic, and fiber quality attributes are documented. Selection of attractive families and outstanding plants within each family is carried out. Selection occurs within and between families. (6) F_5 generation: Selected families from the F_4 generation are grown to produce F_5 families and individual plants of the same families are grown. All processes from the F_4 generation are repeated. Many families become homozygous and are bulk-harvested. The breeder assesses the preliminary yield trial in several replicates to evaluate and test the best families against commercial varieties of the region. (7) F_6 generation: More than 99% homozygosity is obtained. Selected lines from the preliminary yield trial that exceed the commercial varieties are evaluated in multilocation yield trials with several replications under different locations. Commercial varieties of these locations are used as check varieties and screening for resistance to diseases and pests is conducted. (8) F_7–F_{10} generations: The best lines are examined in multiple-location yield trails for genetic, agronomic, and fiber quality attributes, as well as disease and insect resistance. Superior lines should have a high yield potential along with higher stability or a low genotype × environment interaction (GEI). (9) F_{11} generation: The superior line superior to commercial varieties is identified for seed multiplication and the variety registration process to release as a new variety.

This breeding method is widely used to release commercial *G. barbadense* cotton varieties in different countries. Using pedigree selection, the USDA–ARS and New Mexico State Agricultural Experiment Station collaborated to create and release two Pima cotton Germplasm Lines in 2008: PSI 113 (Reg.No.GP-916, PI655939) and PSI 425 (Reg.No.GP-917, PI655940) (Percy et al. 2009). These lines outperformed commercial check cultivars in terms of fiber lengths and strength, but the yield was comparable. Furthermore, in 2008, the USDA–ARS and New Mexico State University Agricultural Experiment Station jointly released four Pima cotton germplasm lines: SJ-07P-FR01 (Reg.No.GP-910, PI 654065), SJ-07P-FR02 (Reg.No.GP-911, PI 654066), SJ-07P-FR03 (Reg.No.GP-912, PI 654067), and SJ-07P-FR04 (Reg. No.GP-913, PI654068) (Ulloa et al. 2009). The new Pima lines produced through artificial hybridization had high resistance to Fusarium wilt race 4 (*Fusarium oxysporum f. sp. Vasinfectum* Atk. Sny & Hans) coupled with moderate yields and good fiber quality.

In Egypt all Egyptian cotton varieties released using the pedigree selection method from Super Giza 1 to Super Giza 98 except, Super Giza 36 (Dandara) created by direct selection from the field of Ashmouni variety, without employing artificial hybridization (Al Didi 1972, 1982). The new variety must have sustainable yield, early maturity, and resilience to pests and diseases, along with good fiber quality as compared to the commercial one. El-Adly et al. (2018) noted that Giza

95 had a higher increasing percentage of 6.90% for seed cotton yield and 16.32% for lint yield compared to the commercial variety Giza 90. Super Giza 94 was released in 2016 (Abdelmoghny et al. 2022) and is distinguished from the commercial variety Giza 86 by its early maturity, high yield potential, fluffy opening and easy picking, high lint percentage (40.2%), fiber length (34.1 mm), fiber strength (43.4 g/tex), micronaire reading (4.2), uniformity ratio (86.9%), yellowness or +b (8.3), brightness Rd (79.8%), and white lint color. In comparison with the two commercial varieties (Super Giza 94 and Super Giza 86), the new variety Super Giza 97 has dramatically enhanced yield and lint percentage in all sites. The fiber quality traits of this long staple cotton category include yarn strength (2400), yellowness or +b (8.9), micronaire value (4.4), fiber length (33.8 mm), uniformity index (85.6%), and fiber strength (44.6 g/tex) (Abdelbary et al. 2021).

Bulked Breeding Method

Nilsson Ehle, a Swedish plant geneticist, developed the bulk method in 1908 while studying winter wheat. Natural selection serves as the foundation for this approach. The main idea of this method is to rely on natural selection from F_2 to F_4 without conducting artificial selection and bulking of the seeds. The breeder begins to apply artificial selection from F_5 or F_6 when maximum homozygosity is achieved (Fig. 20.24). The steps of the bulk breeding method are as follows; (1) Choose desirable parents for making crosses. (2) Harvest seeds in bulk from F_2 to F_4 generations without applying artificial selection. (3) Start from F_5 or F_6 by applying

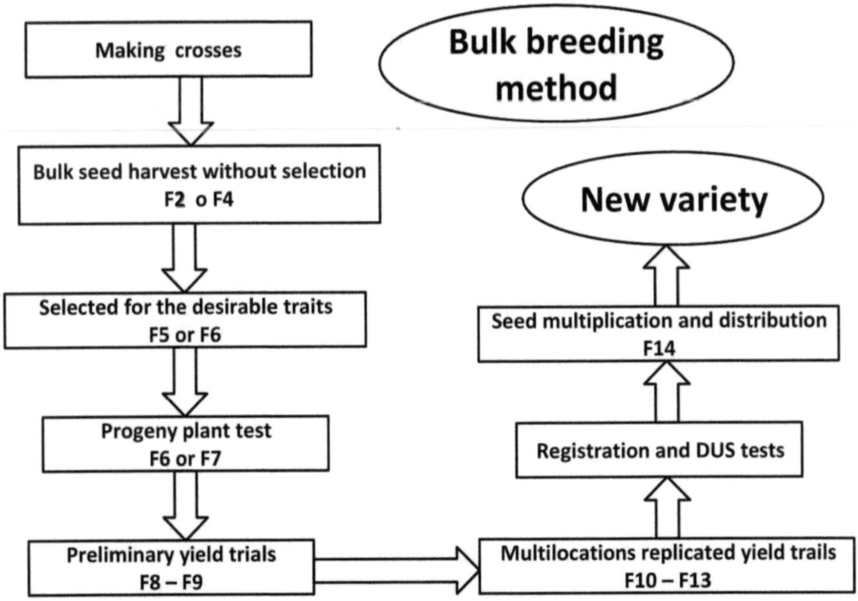

Fig. 20.24 Bulked breeding method. (Source: Prepared by Dr. A. Abdelmoghny)

artificial selection to select a large number of desirable plants (single plant selections, PSS). (4) Progeny test from F_6 or F_7 to select the most superior progenies. (5) Establish preliminary yield trial in F_8. (6) Conduct multilocation replicated yield trials from F_8 to compare the new lines with check varieties. Evaluation could be stopped from F_{10} to F_{13}. (7) Registration and DUS tests. (8) Seed multiplication and distribution of the new variety could be accomplished in F_{14}.

Backcross Breeding Method

Backcross breeding involves transferring a desired attribute from one variety (donor parent, DP) into the favored genetic background of the adapted variety (recurrent parent, RP), which is regulated by a small number of genes (Harlan and Pope 1922). The result of this crossing method is the progeny increasingly resembles the parent to which the backcrosses are made (Fig. 20.25). By BC_6F_1 the plants will have 98.438% genes from the donor parent. This breeding approach produces predictable results and is commonly used to transfer qualitative features like disease resistance. Unfortunately, in cotton, this strategy has not been successful in transferring disease-resistant genes.

In order to improve the fiber strength of cotton variety (T1) (*Gossypium hirsutum* L.) in Stoneville, Mississippi, William Meredith Jr (1977) employed three cycles of backcross breeding, resulting in favorable yield combinations. FTA 263-20 (FTA) was the source of higher T1, and "Deltapine 16" (DPL 16) was the recurrent parent.

Requirements of the Backcross Breeding Program

The backcross breeding method requires (1) a recurrent parent: a popular variety lacking one or two traits, (2) a donor parent, which possesses a parent having the lacking trait in the recurrent parent in high intensity, (3) high heritability of the lacking trait, (4) a large enough number of backcrosses to retrieve the recurrent parent's genetic makeup (RPG).

The Procedure of the Backcross Method

The backcross breeding method involves several steps; (1) Hybridization: Select two parents for crossing to produce the F_1 seeds. One parent should be a commercial variety and the other should complement the first one. (2) F_1 generation: Backcross F_1 plants to the recurrent parent (aa). (3) BC_1F_1: Resistant and Susceptible genotypes in BC1 would be 1:1. Select resistant plants and backcross to the recurrent parent (aa). (4) BC_2F_1–BC_5F_1: In each backcross generation, selected resistant plants are backcrossed to the recurrent parent (aa). (5) BC_6 F_1 generation: The BC_6 F_1 plants will have 98.438% of their genes from the recurrent parent (aa). Select plants and self, their seeds and harvest them separately. (6) BC_6F_2: Grow individual plant progenies from the selfed seeds. Select resistant plants that are similar to the plant type of the recurrent parent and harvest their selfed seeds separately. (7) BC_6F_3:

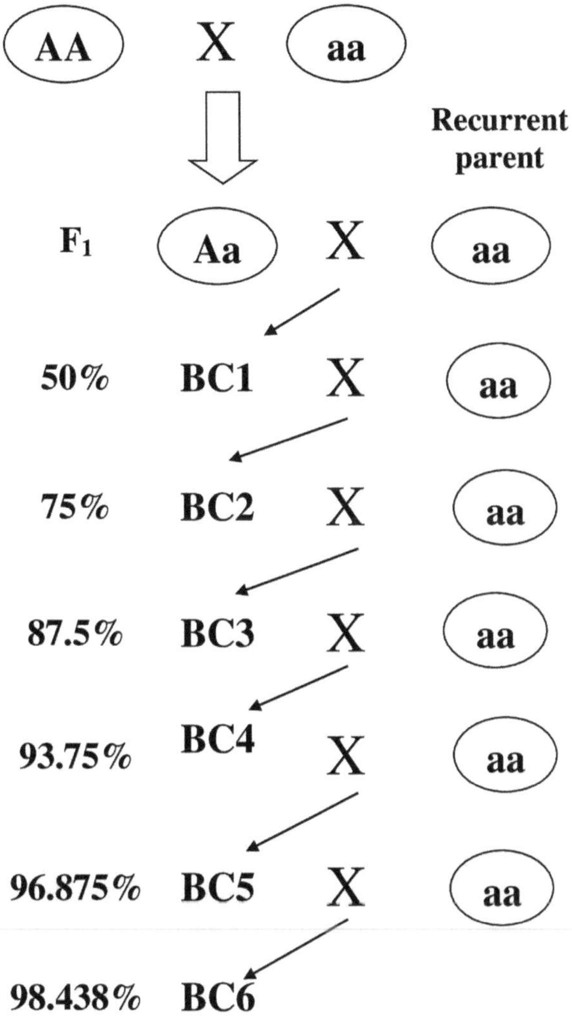

Fig. 20.25 Backcross breeding method. (Source: Prepared by Dr. A. Abdelmoghny)

Grow individual plant progenies. Harvest progenies that are homozygous for resistance and similar to the plant type of the recurrent parent in bulk. (8) Yield test: Test the new variety in a primary yield trial (PYT) along with a check variety. Typically, the new variety should be identical to the recurrent parent.

20.1.10 Molecular Breeding

Molecular breeding is the practice of increasing desirable plant traits by genetic manipulation at the DNA level. Additionally, it could involve genomic selection, gene and genome mapping, marker-assisted selection (MAS), gene editing, and

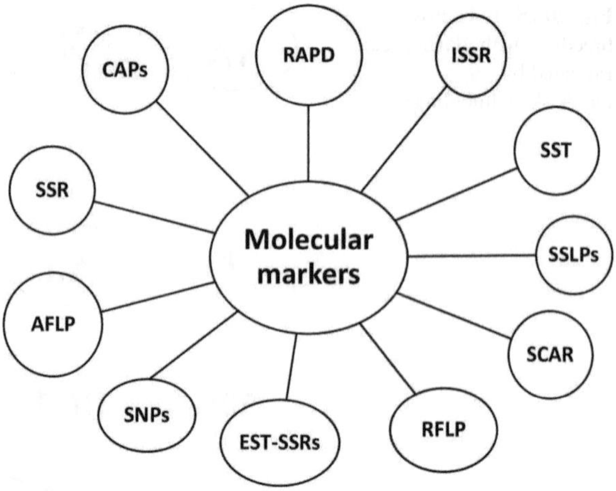

Fig. 20.26 Different types of molecular markers in cotton breeding programs. (Source: Prepared by Dr. A. Abdelmoghny)

genetic engineering (Eltaher et al. 2017, 2018, 2021, 2022; Ismail et al. 2013; Salem 2015; Salem et al. 2003, 2004b, 2007a, b). Breeding using molecular markers followed by selection is known as molecular breeding. The major distinction between classical and molecular breeding strategies is that classical breeding modifies the plant's genetics indirectly by selecting plants with specified characteristics. In contrast, molecular breeding modifies traits by making direct modifications to the DNA of the desired traits only. Molecular markers are DNA segments that can be identified throughout the genome. They are located in certain regions of the genome and are phenotypically neutral. Molecular markers are used to 'flag' the location of a particular gene or the inheritance of specific features. Molecular markers come in various forms, including Microsatellites or simple sequence length polymorphisms (SSLPs), Cleaved Amplified Polymorphic Sequence (CAPS), Random Amplified Polymorphic DNA (RAPD), Restriction Fragment Length Polymorphism (RFLP), Amplified Fragment Length Polymorphism (AFLP), and Inter Simple Sequence Repeat (ISSR) markers. All of these distinct molecular markers are employed to improve cotton (Fig. 20.26). The benefits of utilizing molecular markers include boosting selection efficiency, and precision, and speeding up the breeding cycle to release a new cultivar with the desired feature (Collard and Mackill 2008). Cotton breeders have discovered that these markers are also linked to the most desirable qualities, such as yield components, fiber quality, boll size, and genes that respond to biotic and abiotic challenges in cotton (Majeed et al. 2019). The first linkage map of cotton was established by Reinisch et al. (1994) using RFLP molecular marker. To break the correlation between trait pairs, multiple strategies have been applied, including maker-assisted selection based on QTLs and GWAS, multi-trait selection, and development of new germplasm (Darmanov et al. 2022; Niu et al. 2022). A comparison of the key properties of important molecular markers utilized in cotton breeding programs (Table 20.7).

Table 20.7 Comparing some salient characteristics of major molecular marker techniques

Marker	PCR based	Mode of inheritance	Locus specificity	Level of polymorphism	Reproducibility
RFLP	No	Co-dominant	Yes	Low-medium	Low
RAPD	Yes	Dominant	No	Medium-high	Low
SCAR	Yes	Co-dominant	Yes	High	High
AFLP	Yes	Dominant	No	High	Medium
SSR	Yes	Co-dominant	Yes	High	High
ISSR	Yes	Dominant	Yes	High	Medium
CPSSR	Yes	Co-dominant	Yes	Low	High
SAMPL	Yes	Co-dominant	Yes	High	High
SRAP	Yes	Dominant	No	Medium-high	Low
SSCP	Yes	Co-dominant	Yes	High	High
CAPS	Yes	Co-dominant	Yes	High	High
SNP	Yes	Co-dominant	Yes	Extremely high	High
DArt	No	Dominant	Yes	High	High
EST	Yes	Dominant	Yes	High	High
STS	Yes	Co-dominant	Yes	High	High
RAMP	Yes	Co-dominant	Yes	High	High

Modified form IAEA, Vienna (2002) and Nadeem et al. (2018)

20.1.10.1 Marker-Assisted Selection Breeding Method (MAS)

The entire cotton breeding procedure was more scientific and efficient than ever thanks to the cotton breeder's use of contemporary breeding equipment. Marker-assisted selection breeding method (MAS) or marker aided selection (MAS) helped the cotton breeder discover the genes and quantitative trait loci (QTLs) in a short duration. According to Ribaut and Hoisington (1998), marker-assisted selection (MAS) is the process of indirectly selecting a certain characteristic by manipulating genomic areas implicated in the desired phenotype using DNA markers. Because MAS links the trait to the molecular marker, it increases the selection effectiveness of the targeted trait (Jiang 2013) by lowering the environmental effect (Tester and Langridge 2010), allowing the cotton breeder to rely less on visual phenotypic selection. MAS is an indirect selection mechanism that selects the target characteristic via the connected marker (Fig. 20.27). MAS can be effective for traits that are difficult to assess, have poor heredity, and/or emerge late in development. The MAS approach is used to characterize genetic material and select individuals in the early segregating generation, which speeds up the breeding cycle and improves accuracy (Nadeem et al. 2018). The most favorable molecular indicators for MAS are those with significant recombination in gene pool entry. MAS provides various advantages, including quick speed, consistent findings, biosafety, and high efficiency.

The MAS method in molecular breeding consists of the following steps: the quality and quantity of DNA extraction, the identification and creation of DNA markers closely associated with the desired traits, the mapping of target genes or quantitative trait loci (QTLs) using DNA markers, and the selection of traits based on

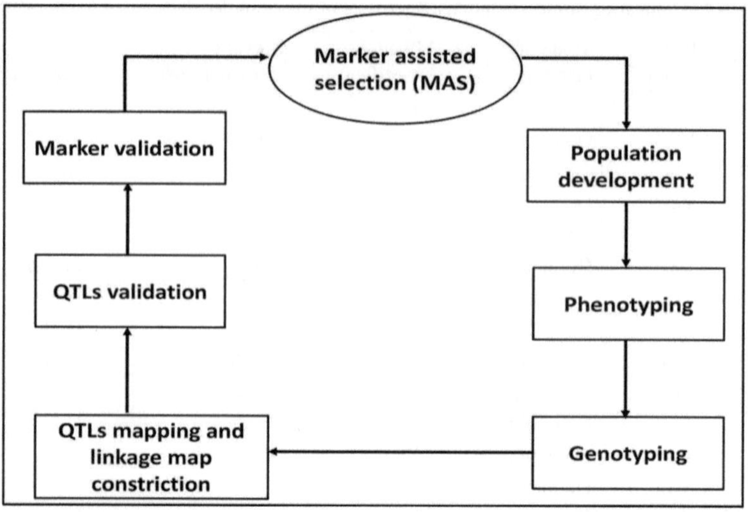

Fig. 20.27 Marker-assisted selection (MAS) used in cotton breeding programs. (Source: Prepared by Dr. A. Abdelmoghny)

DNA markers. Notwithstanding consistent procedures, the particular approach may differ based on the characteristic's complexity and the history of the target gene or genes, often known as quantitative trait loci (QTLs).

Methods of Marker-Assisted Selection (MAS)

Marker-assisted selection (MAS) utilizes various approaches in cotton breeding to identify genetic materials and select superior plants in early segregating generations, ultimately reducing breeding cycles (Kumawat et al. 2020). These methods include genomic selection (GS), gene pyramiding, marker-assisted recurrent selection (MARS), and marker-assisted backcrossing (MABC).

Achievements Made Through Marker-Assisted Selection (MAS)

Despite many QTL map studies that have been conducted on inter-specific (*G. hirsutum* × *G. barbadense*) and intra-specific populations in *G. hirsutum*, few studies have been done on *G. barbadense* (Wang et al. 2013 and 2015). To establish the genetic basis of economic features in the *G. barbadense* genus (Table 20.8). Some QTL maps were created using marker-assisted selection (MAS), which demonstrated the technique's effectiveness in releasing new cotton varieties.

Table 20.8 Reported characterization of QTLs in Egyptian cotton *G. barbadense* genus for various breeding traits

Targeted trait	Cotton species	No. QTL	Markers	Source of genotyping	Cotton cross	No. chromosomal location of major QTLs	References
Number of seeds	*G. hirsutum* X *G. barbadense*	22	AFLPs RFLPs SSRs	120 F_2 Individuals	Acala.44 X Pima S-7	Chr01–Chr04 Chr06–Chr09 Chr10–Chr12 Chr14–Chr18 Chr20–Chr22 Chr25–Chr26	Mei et al. (2004)
Fiber length		14					
Seed weight		22					
Fiber tenacity		5					
Fiber fineness		35					
Fiber elongation		6					
Hundred seed weight	*G. barbadense*	5	SSRs EST-SSRs SRAPs SSCP-SNPs	124 F_2 Individuals	Hai7124 X 3–79	Chr03–Chr05 Chr06–Chr07 Chr08–Chr10 Chr11–Chr12 Chr13–Chr14 Chr15–Chr17 Chr19–Chr20 Chr21–Chr22 Chr23–Chr24 Chr25–Chr26	Wang et al. (2013)
Lint percentage		5					
Lint weight/boll		3					
Seed cotton yield/boll		3					
Lint index		3					
Seed index		2					
Micronaire value		4					
Fiber length		1					
Fiber strength		3					
Fiber uniformity		1					
Fiber elongation		3					
Nulliplex-branch gene (gb_nb1)	*G. hirsutum* X *G. barbadense*		SNPs	168 F_2 population	TM-1 X Pima (Xinhai-18)	Chr16	Chen et al. (2015)
Root-knot nematode (RKN)	*G. hirsutum* X *G. barbadense*	2	SSRs	174 F_2 plants	CIR1348 X FiberMax 966	Chr11	Gomez et al. (2016)

(continued)

Table 20.8 (continued)

Targeted trait	Cotton species	No. QTL	Markers	Source of genotyping	Cotton cross	No. chromosomal location of major QTLs	References
Fiber length	*G. hirsutum* X *G. barbadense*	4	SNPs	146 Backcross inbred lines (BILs)	NMGA-062 X NMGA-105	Chr05–Chr11 Chr12–Chr21	Li et al. (2017)
Days to the first flower	*G. barbadense*	2	63K SNP chip	185 F_2 plants	Suvin X BCS23-18-7 (BCS)	Chr02–Chr03 Chr05–Chr06 Chr08–Chr09 Chr10–Chr14 Chr17–Chr18 Chr23–Chr24 Chr26	Kumar et al. (2019)
Plant height		2					
Monopodial/plant		4					
Sympodial/plant		1					
Bolls per plant		1					
Seed index		2					
Ginning outturn		1					
Boll weight		2					
Fiber length		3					
Fiber strength		3					
Fibre fineness		2					
Maturity ratio		3					
Elongation ratio		3					
Fusarium wilt	*G. barbadense*	9	SNPs	Recombinant inbred line (RIL) population of 110 lines	Susceptible Xinhai 14 X Resistant 06-146	ChrA01–ChrA05 ChrA07–ChrA09 ChrD03–ChrD05 ChrD09–ChrD03	Han et al. (2022)

Source prepared by Dr. A. Abdelmoghny

20.1.10.2 Cotton Genomics

Both diploid (2n = 2x = 26) and allotetraploid (2n = 4x = 52) forms of cotton are known to exist. Eight distinct genome types-known as A and K genomes-appear in diploid cotton. Around 0.7 million years ago, the two diploid A genomes (A1 and A2) separated from a shared ancestor, A0. Between 1.0 and 1.6 million years ago, the A0 and D5 genomes hybridized to produce allotetraploid cotton (AADD) (Iqbal et al. 2022). In order to produce natural, renewable textile fibers, two allotetraploid cotton cultivars have long been grown around the world: Upland cotton *Gossypium hirsutum* (AD)1 and extra-long staple (ELS) cotton *Gossypium barbadense* (AD)2. Both species' draft genome sequences are available, although they are quite fragmented and lacking for *G. barbadense* and *G. hirsutum* (Liu et al. 2015; Wang et al. 2019; Zhang et al. 2015). Extra-long fibers for exceptional textiles are produced by *G. barbadense*. In *barbadense*, the average of the D-genome is 3.29 μm in chromosome length (and ranges from 1.76 to 4.25 μm), whereas the average chromosome length for the A-genome is 4.20 μm, with the largest seven measuring 4.34 μm or more (the entire is ranged from 2.23 to 5.81 μm). Because retrotransposons proliferate according to lineage, genome sizes differ between groups. A synopsis of the *G. barbadense* genus' genomic sequence (Table 20.9). Contigs N50 of 72 kilobases (kb) and scaffold N50 of 503 kb, which cover 1.395 gigabases (Gb) of the A sub-genome (At) and 0.776 Gb of the D sub-genome (Dt), are features of the genome of *G. barbadense*. K-mer estimate accounted for about 88% of the 2.470 Gb genome overall. The genetics of complex cotton plants have been transformed by next-generation sequencing (NGS) technologies, which have also made it easier to analyze agronomic aspects. The extensive cotton data set that NGS offers has aided researchers in standardizing cotton's

Table 20.9 Statistics of *G. barbadense* genome features

Genomic feature	*G. barbadense* genome
Total length of contigs	2,222,525,789
Total length of assemblies	2,266,656,771
Estimated gap size, bp	44,130,982
% of anchoring, bp	97.68%
% of anchoring and ordering, bp	96.35%
Number of contigs	4930
Counting L50, bp	2,151,565
Number of scaffolds	3032
Scaffold L50, bp	92,880,876
GC content	34.2%
% of repeat sequences	69.83%
Number of genes	71,297
Number of transcripts	109,778

Source: Wang et al. (2019)

Table 20.10 Germplasm resources of *G. barbadense* collected by different countries

Country	Type				
	Accessions	Cultivars	Landraces	Hybrids	Unclassified
China	831	–	–	–	–
Australia	51	45	3	–	–
USA	1584	–	–	–	–
Pakistan	109	–	–	–	–
India	530	3	1	–	–
Russia	360	603	132 + 35	–	27
Uzbekistan	–	1799	41	881	1183
Brazil	1470	–	39	–	–
France	214	269	–	–	–

Table 20.11 Genome assembly statistics of *G. barbadense*

Cultivar	3–79	Hai7124	3–79	3–79
Number of contigs	4766	6902	4930	–
N50 of contigs (kb)	1800	7766	2151.56 (150)	–
No. of scaffolds	2048	29	3032	29.751
N50 of scaffolds (Mb)	93.8	23.44	92.88 (L50)	0.260
Total assembled genome size (Mb)/scaffold length (Mb)	2195.8	2224.98	2266.65	2573.19
Number of annotated protein coding genes	74.561	75.071	71.297	80.876

functional genomics resources (Table 20.10). The majority of cotton genomics databases are created and maintained by the National Centre for Biotechnology Information (NCBI) and the International Cotton Genome Imitative (ICGI) (https://www.cottongen.org), the National Center for Biotechnology Information (NCBI) (https://www.ncbi.nlm.nih.gov/genome?term=Gossypium+barbadense%5Borgn%5D&cmd=DetailsSearch), and the Cotton Functional Genomics Database (https://cottonfgd.org). To aid cotton growers in identifying allele variants and determining how to apply the favorable alleles in cotton breeding programs using marker-assisted selection (MAS), the genomic data found in all of these databases makes fine QTL mapping easier. Additionally, several genomic, transcriptomic, genetic variation, and epigenetic databases, including CottonGen, CottonFGD, ccNET, MaGenDB, CottonGVD, and GRAND, have been created and made available for the cotton genus (Yang et al. 2023). In CottonGen, four *G. barbadense* assemblies have been reported (Table 20.11). Numerous researchers have shared their findings on cottonGen, which has led to the creation of guide RNAs for gene editing, linkage maps, genome-wide association studies (GWAS), Validation of linkage maps, expression investigations, and genome-wide characterization studies for gene families (Katageri et al. 2020).

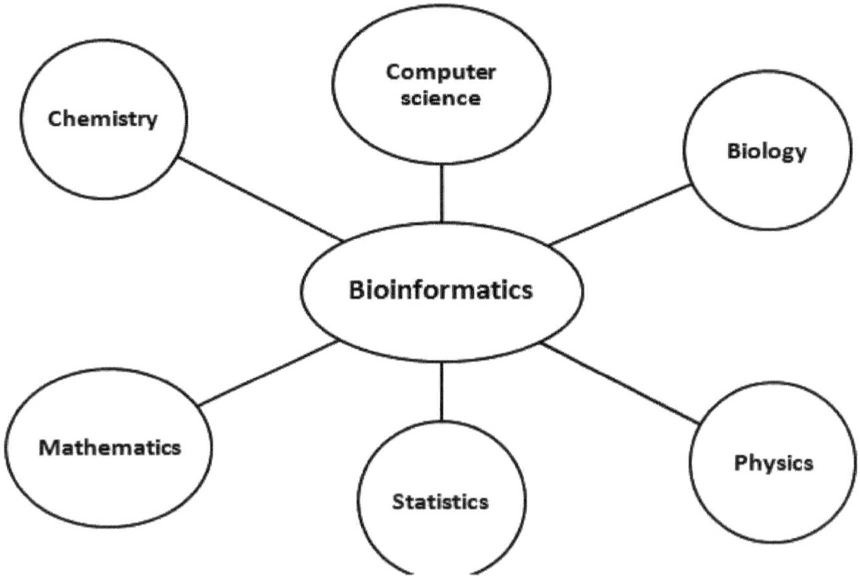

Fig. 20.28 Science related to bioinformatics. (Source: Prepared by Dr. A. Abdelmoghny)

20.1.10.3 Bioinformatics

Genetics and genomics are related topics, and bioinformatics is a scientific subfield that employs computer technology to gather, store, analyze, and communicate biological data and information, such as amino acid and DNA sequences or annotations on those sequences (Fig. 20.28). Biotechnology (BT) and information technology (IT) are combined to develop bioinformatics.

In 1979, Paulien Hogeweg was the first to refer to this discipline as bioinformatics. Because it enables scientists to arrange vast amounts of data, it is especially helpful in the study of genomes and DNA sequencing. It also aids in the comprehension and identification of trends in biological data. The Human Microbiome Project and the Human Genome Project are two instances of bioinformatics. The order of base pairs in the human genome and related microbial genomes was ascertained by both projects using genome sequencing technology. These days, bioinformatics is utilized to develop algorithms and database improvements that use statistical and computational methods to solve issues arising from biological data. Scientists can now easily analyze data computationally thanks to computational algorithms. Bioinformatics software and tools range from basic command-line tools to intricate databases and applications. There are numerous uses and applications for SOAP and REST-based interfaces in bioinformatics.

20.2 Tissue Culture Applications

Under controlled conditions, plant tissue culture is the process of developing plant parts in vitro from a single cell, tissue, or organ. Cotton is one of the many commercially significant crops that commonly use tissue culture techniques. In cotton tissue culture, organogenesis and embryogenesis are the primary techniques for regeneration (Sakhanokho and Rajasekaran 2016). Organogenesis involves inducting callus using various plant parts as explants, such as hypocotyl, stems, leaves, shoot apices, roots, young inflorescences, ovular tissue, and embryos. Because somatic embryos are produced from single cells, somatic embryogenesis is the favored way of regeneration over organogenesis. However, compared to other main crops, cotton has a restricted capacity for large-scale in vitro regeneration of plantlets, which makes somatic embryogenesis more difficult to accomplish (Zhang et al. 2001). To success of cotton regeneration procedures, particularly in obtaining genetically modified plants, depends on factors such as Cotton genotype, media composition, plant growth hormones, and environmental variables. Commonly used parts for regeneration in *G. barbadense* include leaf sections, shoot apices, nodes, and immature zygotic embryos (Guo et al. 2019; Li et al. 2019;). The type and age of explants as well as plant growth hormones play a crucial role in the successful regeneration potential of cotton, with explants exhibiting high meristematic activity being more responsive. Except for Coker genotypes, which are susceptible to gene transfer, the majority of cotton cultivars regenerate via callus. Therefore, using traditional techniques like the backcross method, the genes are first transferred to the Coker variety before being transferred to other genotypes. Transferring target features requires the creation of tissue culture methods that effectively promote proliferation without regard to genotype (Bakhsh et al. 2015).

20.2.1 Somatic Embryogenesis

Somatic embryogenesis is used in plant biotechnology and genetic engineering to produce changed plants using either Agrobacterium-mediated transformation or particle bombardment (Waheed et al. 2022). Price and Smith (1979) observed cotton somatic embryogenesis in *Gossypium koltzchianum*, but did not report any plantlet regeneration. Davidonis and Hamilton (1983) demonstrated the first successful plant regeneration from a 2-year-old callus of *Gossypium hirsutum* L. cv Coker 310 via somatic embryogenesis. Since then, major advances have been made in cotton tissue cultivation (Fig. 20.29). Highly efficient somatic embryogenesis techniques have been developed to regenerate plantlets from various cotton cultivars using different plant parts as explants such as shoot apices, hypocotyls and nodes of 10 days seedlings.

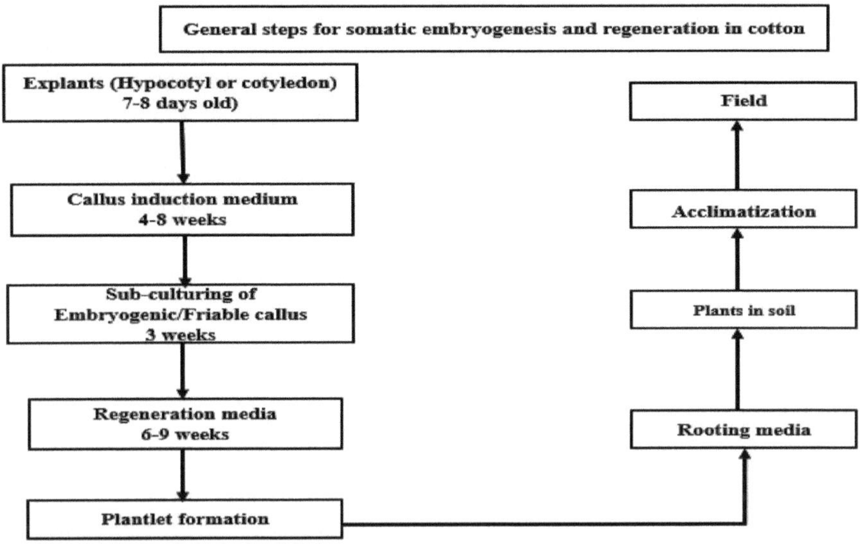

Fig. 20.29 General steps for somatic embryogenesis and regeneration in cotton. (Source: Prepared by Dr. A. Abdelmoghny)

20.2.2 *Organogenesis*

Organogenesis is used to produce somatic embryos from somatic tissues Which resemble true zygotic embryos. This can be accomplished via direct or indirect somatic embryogenesis approaches. The primary distinction between the two procedures is that the direct method converts single cells straight into somatic embryos without passing through the dedifferentiation (callus) phase. Three organs are used to regenerate the whole plant from callus culture: explants and the axillary bud.

20.3 Genetic Engineering and Gene Editing

Genome editing or gene editing procedures are a sort of genetic engineering that produces genetically modified organisms (GMOs). Genome editing refers to a set of procedures that allow scientists to change the genetic material (DNA) of plants, animals, and microbes by adding, removing, or altering genetic material at a specific position in the genome. Genetic engineering aims to modify genes to enhance the capabilities of organisms beyond what is normal. Genetically engineered organisms (GEOs) are organisms changed by human via direct tampering with parental cells and their genomes. This technique requires the delivery of DNA into parental cells (Platani et al. 2022). Recombinant DNA is one of the major tools of genetic engineering that involves both snipping and splicing DNA fragments from different organisms and recombining them to produce a new genetic sequence. Another technique of genetic engineering is gene editing, which involves making targeted alterations to particular genes using methods like CRISPR-Cas9 (Patil et al. 2022). The development of DNA editing and engineering tools advanced our fundamental understanding of biology, allowing us to develop societally relevant applications. CRISPR-Cas editing tools have revolutionized the field of genome editing due to their simplicity, accuracy, and efficiency across all forms of life.

20.3.1 Genetic Transformation Methodologies

Genetic transformation is an important strategy for enhancing cotton genotypes because it transfers genes from one genotype to another. Various approaches for genetic transformation such as *Agrobacterium*-mediated transformation, particle bombardment transformation and Shoot Apical Meristem (SAM) cells-mediated Transformation System (SAMT), are successful techniques for introducing desirable traits to improve cotton (Ge et al. 2023; Waheed et al. 2022). For example, resistance to verticillium wilt (anti-fungal) has been transferred to *G. barbadense*, cv. H7124 using the Gbve1 gene through the Agrobacterium-mediated method (Li et al. 2017; Zhang et al. 2016). Cotton transformation is essential for addressing losses due to biotic (insect pests, pathogens), and abiotic stresses in cotton production. *Agrobacterium*-mediated and bombardment transformation strategies are widely utilized to introduce genes into cotton (Zhang 2023). The cotton transformation has become a major bottleneck for introducing genes into cotton.

20.3.1.1 *Agrobacterium*-Mediated Transformation

The earliest applications of *Agrobacterium*-mediated cotton transformation happened in the late 1980s and early 1990s. *Agrobacterium tumefaciens* transfers target genes from the bacterium to the plant nuclear genome utilizing a large tumor-inducing (Ti) or rhizogenic (Ri) plasmid residing in *Agrobacterium*. (Waheed et al. 2022). The first successful cotton transformation was reported by Firoozabady et al. (1987) in *G. hirsutum* var. model Coker-312. This optimal donor model is most favorable for cotton somatic embryogenesis, resulting in a reliable cotton transformation process. This approach is distinguished by a low copy number and consistent, intact gene expression. This approach is used to introduce novel genes into cotton cultivars to improve tolerance to biotic pressures (pests and herbicides), increase resistance to environmental challenges, and improve fiber quality (Waheed et al. 2022; Zhang 2023).

20.3.1.2 Particle Bombardment Transformation

The particle bombardment method involves coating DNA with gold particles and delivering them into host tissue at a high velocity. The first successful stable transformation in the Upland cotton variety (Coker) was reported by Finer and McMullen (1990), with a transformation efficiency rate of 0.7%. Challenges of this method include unstable integration of transgenes, gene silencing, and insertion of multiple copy numbers (Waheed et al. 2022).

20.3.1.3 Pollen Tube Pathway-Mediated Transformation

The pollen tube pathway-mediated transformation is a straightforward, genotype-independent, and cost-effective technique for agricultural output. In 1978 was the first used in cotton. Three techniques are used: microinjection, foreign gene cultivation and pollination with pollen, and direct DNA drop to stigma. The method of microinjection yields stable transgenic plants and has the best transformation efficiency ratio. In order to insert exogenous DNA directly into the plant genome during self-pollination, Zhou et al. (1983) used pollen tube pathway-mediated transformation. This approach is rarely employed for many plant species, has a modest transformation rate, requires competent personnel, and does not need tissue culture conditions (Waheed et al. 2022).

20.3.1.4 The Shoot Apical Meristem (SAM) cells-Mediated Transformation System (SAMT)

Ge et al. (2023) at the Chinese Academy of Agricultural Sciences' Institute of Cotton Research created a novel cotton transformation technique. In addition to dicot crops, this innovative method has demonstrated potential in changing other cotton genotypes and species, including Asiatic cotton (*Gossypium arboreum* L.), Sea Island cotton (*Gossypium barbadense* L.), and Upland cotton (*Gossypium hirsutum*). This method's benefits include speed, ease of use, and transformation efficiency that are on par with or better than those of other approaches. Transgenic cotton may be produced from seeds to seeds in an average of 9 months, as opposed to 12–24 months for traditional approaches. Furthermore, compared to conventional techniques, the chimaera ratio is significantly reduced (Zhang 2023). Two foreign genes, GFP and RUBY, were introduced into the SAM cells of certain resistant cotton genotypes using SAMT. Transgenic adventitious shoots formed from the axillary meristem zone could be harvested and grown into full cotton plants in 2–3 months (Zhang 2023).

20.3.2 Gene Editing

In agriculture, genome editing usually tries to boost a positive characteristic or deactivate an undesirable trait inside a plant to raise output, improve resilience to increasingly harsh weather conditions, or reduce the need for inputs such as insecticides or fungicides. Genome editing methods engineered from proteins include zinc-finger nucleases (ZFNs), transcription activator-like effector nucleases (TALENs), and meganucleases. In comparison to CRISPR/Cas9, these approaches did not gain popularity because of their low efficiency, specificity, engineering practicality, and design simplicity (Katageri et al. 2020). An additional method is called clustered regularly interspaced short palindromic repeats, also known as CRISPR/Cas9. Cotton breeders can use gene editing to make modifications to plants more rapidly and precisely than with traditional plant breeding approaches. CRISPR/Cas-9 genome editing involves three steps: recognition, cleavage, and repair. CRISPR is easier, faster, cheaper, and more accurate than other genome editing approaches. CRISPR/Cas-9 is used to precisely alter plant DNA sequences related to particular traits, in order to breed new plants with desired traits and avoid many of the negative side effects of conventional plant breeding techniques. Cotton genetic improvement is tough because of issues such as ploidy level, crop duration, crop genetics, and regeneration issues (Khan et al. 2022b) (Table 20.12). The CRISPR genome editing workflow is as follows: (1) Design: Choose your Cas enzyme. (2) Deliver: To deliver the Cas enzyme and guide RNA to cells, it is recommended to complex the enzyme and gRNA to form a ribonucleoprotein (RNP). (3) Repair. (4) Analyze.

Table 20.12 Improving cotton traits through gene editing

No.	Targeted trait	Method	Targeted gene	References
1	Root length	CRISPR/Cas-9	Arginase	Wang et al. (2017)
2	Color	CRISPR/Cas-9	*GhCLA*	Chen et al. (2017)
3		CRISPR/Cas-9	*GhPDS*, *GhCLA1* and *GhEF1*	Gao et al. (2017)
4		CRISPR/Cas-9	*GhCLA*	Wang et al. (2018)
5	Fiber development	CRISPR/Cas-9	*ALARP*	Zhu et al. (2018)
6	CLCuV resistance	TALEN	*GhSWEET10*	Mubarik et al. (2021)
7	Bacterial blight resistance	TALEN	*GhSWEET10*	Cox et al. (2017)
8	Metastasis	ZFN	*HAb18G/CD147*	Li et al. (2015)

Source: Khan et al. (2022a, b)

20.3.3 Enhanced Traits

All cotton breeding initiatives strive to generate new varieties with high yield, fiber quality, and resilience to diseases and insects, exhibiting minimal variance under varied circumstances. Cotton breeder aims to concentrate on the selection of staple varieties that have less interaction with environments. Variety stability is known as a genetic trait. The selection of high-stability varieties would help in sustainable production over a large area (Yasar 2023). The term "cotton" describes the boll, the portion of the cotton plant that produces the fluffy cotton fibers. The cotton plant's seedpod, known as a boll, reaches a size comparable to a golf ball. The boll begins to grow after the cotton plant has finished flowering. The cotton fibers that resemble seed hairs are linked to each of the seven or eight seeds that are inside each boll. There could be up to 20,000 fibers in each seed, which translates to up to 150,000 fibers per boll. After being picked, cleaned, and spun into yarn, these bolls are woven into the familiarly soft and sturdy fabric. Genetic enhancement is necessary in cotton for several traits, which are listed in Table 20.13.

20.3.4 Transgenic Cultivars

All traditional plant breeding methods result in genetic changes, making all offspring technically genetically modified compared to their parents. Gene transfer is the process of artificially transferring desirable features or genes from one genotype to another. The transgenic procedure entails introducing desired genes into the plant genome and then regenerating the entire plant from the changed tissue/cell (Fig. 20.30). The most prevalent ways for transferring genes into plants are *Agrobacterium*-mediated Transformation and particle bombardment. Other techniques, such as polyethylene glycol (PEG)-mediated transformation and electroporation are also used (Katageri et al. 2020). These methods are also known as systems of DNA delivery for genetic transformation. Transgenic cotton was among the first transgenic plants to be commercially adopted around the world. Since its introduction in the 1990s, transgenic cotton has been rapidly embraced by cotton farmers in both developed and developing countries as an important and effective tool for cotton improvement. Most Particle bombardment transformations and *Agrobacterium*-mediated transformations have proven to be successful methods of introducing desirable traits (Waheed et al. 2022). *G. barbadense* cotton cultivars have been improved using a range of genetic transformation approaches to impart desirable traits (Table 20.14). Cotton is a difficult crop to regenerate from in vitro tissue cultures because it requires more effort to produce somatic embryogenesis, shoot multiplication, and plant regeneration than other crops (Katageri et al. 2020).

Table 20.13 Definition of enhanced breeding traits in cotton breeding programs across the world cotton production countries

Traits	Abb.	Measured unit	Definition
Morphological traits			
Taproot length	TRL	cm	Length from rhizome separation to the farthest point of the main root.
Total root length	TRH	cm	The total length of all roots per plant
Root volume	RVE	cm^3	The total volume of all roots per plant
Root average diameter	RAD	mm	The average diameter of all roots (including lateral roots) per plant
Root surface area	RSA	cm^2	Total root surface area of roots per plant
Root tips number	RTN	Num	Average tips number of all roots per plant
Root tissue density	RTD	gcm^{-3}	Total root dry mass divided by root volume
Specific root tip density	SRTD	g^{-1}	Average root tips number divided by root dry mass
Specific root length	SRL	cmg^{-1}	Total root length divided by root dry mass
Specific root area	SRA	cm^2g^{-1}	Root surface area divided by root dry biomass
SPAD Value	SPAD	Num	The relative value of chlorophyll content in leaves
Shoot fresh weight	SFW	$gplant^{-1}$	The weight measured immediately after the fresh shoots are collected
Root fresh weight	RFW	$gplant^{-1}$	The weight measured immediately after the fresh roots are collected
Shoot dry biomass	SDB	$gplant^{-1}$	Total shoot dry mass per plant
Root dry biomass	RDB	$gplant^{-1}$	Total dry mass of all roots per plant
Total dry biomass	TDB	$gplant^{-1}$	constant whole plant weight
Root to shoot ratio	R/S	%	The ratio of the dry matter weight of the aboveground part and the underground part
Plant height	Ph	cm	The measurement from the ground level to the tip of the main shoot at maturity.
Physiological traits			
Total N concentration	TNC	mg/g	N concentration in plant aboveground and underground tissues
Total N accumulation	TNA	g/g	Average N content per unit weight of seedlings
N use efficiency	NUE	g^2/mg^{-1}	Weight of dry matter produced by unit N absorption
N uptake efficiency	NUtE	mgg^{-1}	Total N uptake of plants

(continued)

Table 20.13 (continued)

Traits	Abb.	Measured unit	Definition
Total P concentration	TPC	mg/g	P concentration in total plant
Total P accumulation	TPA	g/g	Average P content per unit weight of seedlings
P use efficiency	PUE	g^2/mg^{-1}	Weight of dry matter produced by a unit of phosphorus absorption
P uptake efficiency	PUtE	mgg^{-1}	Total phosphorus uptake of plants
Maturity traits			
Node of first fruiting branch per plant	NFB/P	Num	Total number of nodes above zero nodes on each plant.
Number of monopodial branches per plant	NMB/P	Num	Count of vegetative branches and indirect fruiting branches that arise from the main for each plant.
Number of sympodial branches per plant	NSB/P	Num	Count of fruiting branches that bear direct fruiting in each plant.
Number of fruiting points per plant	NFB/P	Num	All the open and shaded bolls were counted and added in order to calculate the total number of fruiting points per plant.
Number of bolls per plant	NB/P	Num	The count numbers of opening bolls for each plant.
Cotton yield traits			
Boll retention percentage	BR%	%	Calculated by dividing the total number of bolls from each plant of every cross on the fruiting points per plant and multiplied by a hundred. Boll retention percentage $= \frac{\text{Total number of bolls per plant}}{\text{Fruiting points per plant}} \times 100$
Boll weight	BW	g	Average boll weight was recorded by dividing the total seed cotton weight from each plant of every cross by the total number of mature/opened bolls that were picked. Average boll weight per plant $= \frac{\text{Total seed cotton yield per plant}}{\text{Number of bolls per plant}}$
Seed cotton yield per plant	SCY/P	g	Weigh of total seed cotton of each recorded plant.
Seed cotton yield per plot	SCY/P	g	Weigh of total seed cotton of all plants in the plot.
Lint cotton yield per plant	LY/P	g	Weigh of total lint yield of each recorded plant after gaining.

(continued)

Table 20.13 (continued)

Traits	Abb.	Measured unit	Definition
Lint cotton yield per plot	LY/P	g	Weigh of total lint yield of all plants in the plot after gaining.
Lint percentage	L%	%	Line percentage $= \frac{\text{Weight of lint sample}}{\text{Weight of seed cotton sample}} \times 1$
Lint index	LI	g	Lint index = (Seed index × ginning %)/(100 − ginning %).
Seed index	SI	g	Weight of 100 seeds in grams.
Number of seeds per boll	NS/B	Num	The total number of seeds per boll.
Number of locules per boll	NL/B	Num	The number of locules for each boll.
Harvest index	HI	%	The ratio of the harvested product (lint and seed) to the above-ground plant dry weight or biomass (stem, leaves and fruits).
Fiber quality traits			
Fiber length	FL	mm	Fiber length is the average length of the longer half of the fibers (upper half mean length) measured in millimeters by the tuft method.
Uniformity ratio (%)	UR%	%	Length uniformity is the ratio between the mean length and the upper-half mean length of the fibers, expressed as a percentage.
Fiber strength	FS	g/tex	The enforcement in grams is necessary to crack the bunch of fibrils per Tex. While Tex is defined as the weight of fiber in grams of 1000 m of fibril.
Elongation (%)	E%	%	
Micronaire reading	MIC	Num	Micronaire is a measure of fiber fineness and maturity. An airflow instrument is used to measure the air permeability of a constant mass of cotton fibers compressed to a fixed volume. Varies from fine to coarse.

20.4 Mutation Breeding Method

Conventional cotton breeding methods improve cotton yield while also making the crop more susceptible to a variety of insects and pests, resulting in a major loss of natural genetic variability. Induced mutation can induce innovative modifications in functional genes to develop greater germplasm resources (Abid et al. 2020). Mutation breeding methods utilized both chemical and ionizing radiation mutagens in cotton breeding programs during the 1960s and 1970s in some cotton-producing countries but they are now not commonly used. The primary goal of utilizing mutagenesis is to generate novel enhanced features that do not exist in common

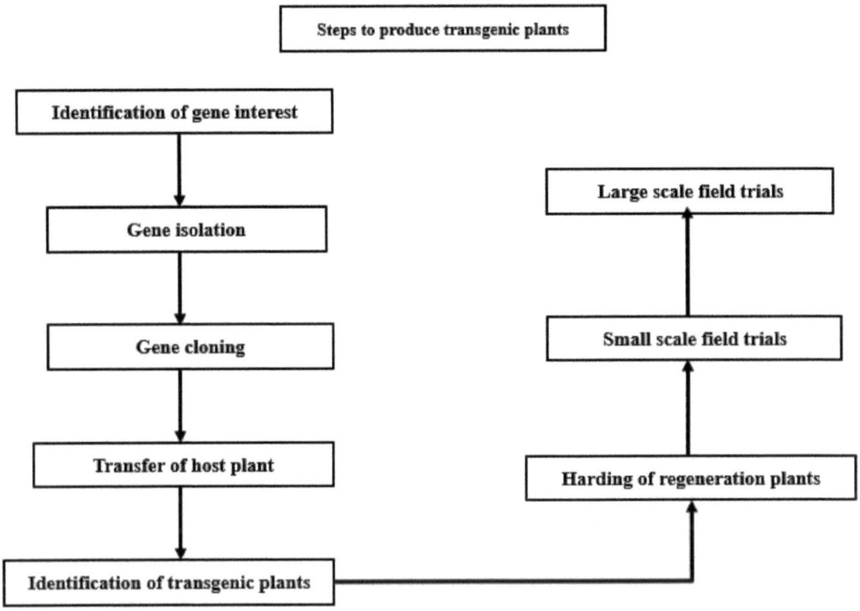

Fig. 20.30 Steps of transformation to produce a transformed cotton plant. (Source: Prepared by A. Abdelmoghny)

Table 20.14 List of gene transformation methods in *G. barbadense*

Transformation method	Variety	Gene	Function	Reference
Agrobacterium	Hai7124	Gbvdr3	Resistant to *Verticillium*	Zhang et al. (2016) Li et al. (2017)
Agrobacterium		LOS5/ ABA3	Increase drought stress tolerance	Bakhsh et al. (2015)
Agrobacterium		GHSP26	Genes for heat shock proteins	Al-Shafeay et al. (2011)

variations (Farzana et al. 2022). More than 3400 mutant varieties of industrial crops, vegetables, oilseeds, and cereals have been introduced globally since the 1920s (https://nucleus.iaea.org/sites/mvd/SitePages/Home.aspx). These breeding strategies focused on early maturity, disease resistance, high fruiting, and reduced monopodial growth habits (Rauf et al. 2020) and (Tach et al, 2022). However, the mutation breeding method has been used to release 48 cotton varieties worldwide during the last few decades according to IAEA and Khan et al. (2023) (Table 20.15). This technique is used to create or increase genetic variation in common varieties or lines. Different parts of cotton (seeds, pollen, shoot tips, and ovule) are treated with different doses of radiation or chemical mutagens, and then the desirable mutants are selected during segregating generations based on high yield, good fiber quality,

Table 20.15 Forty-eight cotton varieties released via mutational breeding methods worldwide during the last decades

No.	Variety name	Country	Released year
1	113	China	1985
2	Agdash 3	Russian Federation	1983
3	AN-20	Uzbekistan	1986
4	AN-401	Uzbekistan	1971
5	Badnawar-1	India	1961
6	C-3585	Uzbekistan	1979
7	C-7503	Uzbekistan	1980
8	C-7510	Uzbekistan	1991
9	Chandi 95	Pakistan	1995
10	Chuanpei 1	China	1982
11	DS-1	India	1985
12	Emian 15	China	1991
13	Helius	Bulgaria	2007
14	Indore-2	India	1950
15	Jimian 8	China	1984
16	Karisma 1	Indonesia	2008
17	Khandwa-2	India	1971
18	Lumian 1	China	1976
19	M.A.9	India	1948
20	NCU 10	India	1982
21	MCU 7	India	1972
22	ML-11	Uzbekistan	1979
23	Mutant 1	Uzbekistan	1966
24	Mutant 5	Uzbekistan	1969
25	NIAB 111	Pakistan	2004
26	NIAB 777	Pakistan	2009
27	NIAB 999	Pakistan	2003
28	NIAB Karishma	Pakistan	1996
29	NIAB-2008	Pakistan	2016
30	NIAB-26N	Pakistan	1992
31	NIAB-78	Pakistan	1983
32	NIAB-846	Pakistan	2008
33	NIAB-852	Pakistan	2012
34	NIAB-86	Pakistan	1990
35	Oktyabr	Russian Federation	1984
36	Pusa Ageti	India	1978
37	Rasmi	India	1976
38	Sohni	Pakistan	2004
39	Trakia	Bulgaria	2007
40	VNIISSKH-101	Uzbekistan	1986
41	Xinhai 2	China	1979

(continued)

Table 20.15 (continued)

No.	Variety name	Country	Released year
42	Yanmian 48	China	1985
43	Yunfu 885	China	1977
44	Zhong 279-5	China	2005
45	Zhong 297-5	China	2005
46	Zhongmiansuo 42	China	2002
47	Zhongmiansuo 50	China	2007
48	Zhongmiansuo 69	China	2008

Source: IAEA and Khan et al. (2023)

disease resistance, and early maturity in comparison to the parent line (Rauf et al. 2019). The challenge of using the mutation breeding method is to select desirable mutations that do not have any negative linkage. Mutagens can be split into two major groups. Physical and chemical mutagens. These conventional mutagenesis techniques have some limitations such as, creating random mutations with mutagens differing for every crop, requiring proper checks and balances. However, new approaches are evolving, such as antisense technology, zinc finger nucleases (ZFNs), transcription activator-like effector nucleases (TALENS), and cluster regulatory interspaced short palindromic repeats (CRISPR-Cas9) nucleases (Khan et al. 2023). The most commonly employed chemical mutagens in plant breeding are ethyl methanesulfonate (EMS), sodium azide (SA), and N-methyl-N-nitrosourea (MNU), which mostly cause point mutations. The most often employed physical mutagens, on the other hand, are gamma or X-rays, which frequently result in massive deletions that cause DNA breaks, chromosome abnormalities, and the loss of several genes (Winkler et al. 2023).

The majority of mutation research is to find the appropriate mutagen and dose combination that will produce the best response. Finding the ideal mutagen to produce a high mutation rate while causing plant harm and avoiding an impact on germination and plant growth is thus the first step in mutation breeding (Winkler et al. 2023). Mutants are generated by exposing plant tissues such as seed (30 kR), pollen (10 Gy gamma rays), shoot tips (5–50 Gy gamma rays), ovule (50 Gy gamma rays) or the chemical mutagen 1–5 mM of ethyl methane sulfonate (EMS) (Iqbal et al. 1994; Bechere et al. 2013; Aslam et al. 2018). Lower doses and concentrations increase genetic variation in variables like early blooming, plant height, number of bolls, seed cotton yield, ginning percentage, seed index, harvest index, and fiber traits (Muthusamy and Jayabalan 2011, 2014). Ethyl methanesulfonate (EMS) is an excellent mutagenesis agent that triggers the mutagenic potential of many agricultural plants, among other physical and chemical mutagenic causes. In order to enhance genotypic and phenotypic variety, EMS mutagenesis produced point mutations dispersed throughout the entire genome. Through 3-ethyl adenine pairing defects, the process replaces G: C pairing at the nitrogenous base level, causing 2–10 mutations per Mb of diploid DNA (Abid et al. 2020).

The procedure of the mutation breeding program to improve disease resistance in cotton includes the following steps: (1) Seed treatment with selected mutagen (X-ray, gamma rays, or chemical mutagens) using the recommended dose and duration of treatment. (2) Rising the M_1 generation using wider spacing (individual plants). (3) Recording observations on morphological variants and fertility. (4) Selfing all M_1 plants to avoid contamination. (5) Harvesting all the M_1 plants separately. (6) Rising the M_2 generation from selfed seeds obtained from M_1 using wider spacing. (7) Identifying and selecting disease-resistance mutants. (8) Harvesting seeds of the mutant plants separately. (9) Rising the M_3 generation separately for each M_2-selected plant. (10) Evaluating for homozygosity for disease resistance. (11) Bulking of homozygous disease-resistant M_3 progeny. (12) Planting the M_4 and M_5 generation in a replicated trial using local check varieties for comparison. (13) Multilocation yield experiments are evaluated in disease resistance trials that are coordinated. The disease-resistant pure line is propagated and distributed as a new variety.

Cotton breeders can assess the harm done to the M1 generation by counting the number of seeds obtained that can be utilized to create M2 or M3 populations. Different parameters are used by the breeder to identify the somatic effects of mutagenic effect (Winkler et al. 2023); (1) Delay in seed germination. (2) The severity of abnormalities in the cell cycle. (3) The incidence of chromosomal abnormalities in meristematic tissues. (4) Reduced seedling emergence. (5) Minimised seedling and plant growth. (6) Evidence of chlorophyll deficiency. (7) Decreased fertility, plant survival at maturity, and sterility.

20.4.1 In Vitro-Mutagenesis and Selection

In vitro mutagenesis is the process of causing a mutation in a segment of cloned DNA, putting it into a cell or organism, and then studying the outcomes. This method is a crucial component of genetic analysis. In a procedure known as "reverse genetics," controlled mutagenesis has revealed the function of novel genes and made it possible to analyze the mechanisms underlying well-known proteins. The insertional mutagenesis technique is used to modify gene function and investigate gene sequence to recover the sequence of the flanking region of the genome directly via employing various PCR-based techniques. There are two types of insertional mutagenesis; transposable elements and T-DNA of Agrobacterium. The genes impacted via this method might be discovered rapidly, characterize (Khan et al. 2023). Furthermore, site-directed mutagenesis has emerged as a crucial biotechnology tool. Numerous in vitro mutagenesis procedures, including haploidization, protoplast fusion, gene transfer, and the use of somaclonal variation, are used for crop improvement (Muthusamy and Jayabalan 2011). In order to get beyond the natural obstacles to fertilization, ovule and embryo cultures have recently been used in in vitro fertilization. To generate cottonseeds with low gossypol content, which improves the shelf life of cotton seed oil, and high gossypol content in cotton plants, which boost disease resistance through transspecies bridge crosses, both ovule and embryo culture are now required (Bi et al. 1999). The majority of cotton plant in vitro cultural research was successful when they used somatic embryogenesis or call cultures, particularly in Cocker types.

20.4.2 Molecular Analysis

Molecular markers are a very useful tool for molecular analysis or characterization in cotton breeding to identify genetic variability. Polymerase chain reaction (PCR)-based markers are the helpful types utilized in cotton because of their high effectiveness and utilization, and they are not linked to phenotypic expression. The molecular markers used for population mapping to investigate genetic similarity between exotic germplasm and common varieties. The genetic similarity ranges from 81.45% to 90.59% (Razzaq et al. 2021). There are many molecular analyses used such as:

20.4.3 MutMap Approach

The MutMap approach is a new genetic technique that depends on next-generation high-throughput sequencing techniques. This technique is cost-effective and efficient. The steps of this technique as outlined by Wang et al. (2022) are as follows (1) Crossing an interesting (desirable) mutant plant with the parental line which was used for mutagenesis. (2) Selfing of M_1 individuals to develop M_2 progenies. (3) If the phenotypic changes are due to the single recessive mutation, then it is expected that the F_2 population will segregate 3:1 for wild/type and mutant progeny. (4) From the M_2 population at least 20 individuals confirming mutations through phenotypic screening are used for DNA extraction. (5) The high-quality extracted DNA is pooled in equal ratios and subjected to a newly developed sequencing technology such as PacBio for whole genome sequencing with more than 10× coverage (Sugihara et al. 2022). (6) Following sequencing, the short/small reads are matched with the reference genome to find SNPs. (7) The mutant lines from the M_2 population that carry mutant SNPs, must have nucleotide that differs from the reference genome to confirm the SNP associated with the phenotypic change. (8) The desirable mutants are selected from M_3 to M_5 generations. Finally, following sequencing, the short/small reads are matched with the reference genome to find SNPs (Fig. 20.31). Crossing the selected mutants with parental lines and then evaluating through phenotypic changes in the F_2 segregating generation.

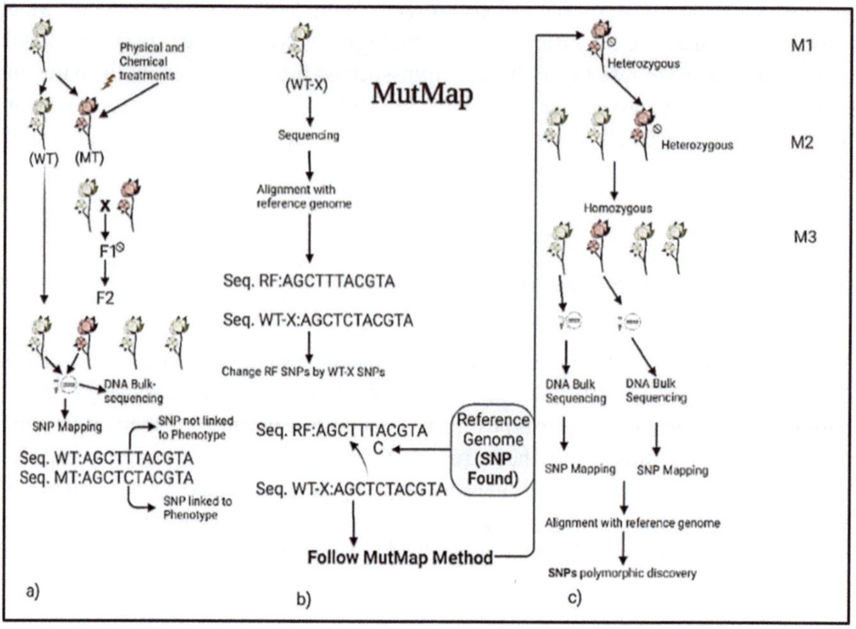

Fig. 20.31 Pictorial illustration of MutMap approach, (**a**) mutation induction via chemical/physical mutagens, (**b**) exploitation of NGS approach for identification of mutant, (**c**) identification of SNPs associated with the mutant phenotype. (Source: Khan et al. (2023))

20.4.3.1 MutChromSeq Approach

The MutChromSeq approach is a widely used method in different crop species to identify novel genes. This method enables the quick detection of mutations and the identification of DNA sequences and genes that regulate gene expression in polyploid plants and complicated genomes (Khan et al. 2023). In this approach: (1) Chromosome flow sorting and traditional mutagenesis are coupled to filter the region of interest, where the gene of interest can be identified. (2) The chromosomes containing the desired gene are filtered based on fluorescent marker tags. (3) Sequencing of both non-mutant and mutant chromosomes is then performed to identify the targeted/desirable gene (Chaudhary et al. 2019).

20.4.3.2 Whole Genome Sequencing Based Mapping

Recently, the cotton geneticist has a high-quality reference genome for cotton species, which makes resequencing of mutant cotton genotypes easier and more cost-efficient to discover the variations (SNPs) at the nucleotide level over the entire genome (Khan et al. 2023). Whole Genome Sequencing Mapping (GWAS) is a new molecular technique used for all crops to identify the desirable SNPs that correlate with beneficial phenotypic changes. Then, mapped all the changes across genes or chromosomes. The results of GWAS using whole cotton genome sequencing (WGS) to identify millions of SNPs to generate high-quality WAS maps (Li et al. 2022). Constructing a WGS map using the GWAS technique has limitations such as; not identifying the association between phenotypes and complete genetic variations (SNPs), even after alignment with the reference genome (Su et al. 2020). This is caused by the hidden or missing variation in the reference genome.

20.4.3.3 Tilling Approach

TILLING is a new breeding technique used to discover various mutations in crops. TILLING is used to identify lesions in a desired sequence (Khan et al. 2022a, b). TILLING is a reverse genetic technique that combines conventional mutagenesis and molecular screening techniques to detect point mutation/single nucleotide polymorphism (SNP). This method yields good results for crop species that have a narrow genetic basis (Khan et al. 2023). Among conventional mutation methods, EMS is the best mutagenesis used with the TILLING technique, because it creates point mutation or single nucleotide polymorphisms (SNPs). Changing the coding region of the target gene will cause changes in both transcriptional and translational levels will lead to a change in gene expression. The limitations of this method are expensive and require skillful labor. On the other hand, is a strong tool for producing point mutations that can be used to examine gene function.

20.4.3.4 Enhanced Traits and Improved Cultivars

Mutation breeding techniques have been used by cotton breeders in recent decades to add additional variability to cotton plants. With this technique, plant material such as seeds or somatic embryos is combined with chemicals like EMS, SA, gamma, or X-rays. These investigations were utilized to select pure lines and perform morpho-physiological assessments that improved agronomical variables like yield, fiber quality, stress tolerance, and disease resistance (Bechere et al. 2013; Aslam et al. 2018; Abid et al. 2020; Scarpin et al. 2022; Zhao et al. 2022; Winkler et al. 2023). The cotton breeders succeeded in releasing new varieties using the mutation breeding method (Table 20.16).

20.5 Hybridization

The process of hybridization creates new genetic combinations by combining the DNA of selected parents (pure lines, strains, varieties, or races) (Farzana et al. 2022). It is a key activity in any breeding program, aiming to select the genotype with the most desired gene combinations from both parents. When two pure lines are crossed,

Table 20.16 Enhanced traits and released cultivars using mutation breeding method

Mutant	Mutation dose	Enhanced traits	Reference
NIAB-78	Irradiating the F1 cross of (Deltapine × AC-134).	Early maturing and with high yield potential. Suitable for integration into the double cropping system (wheat-cotton). Increased cotton production in Pakistan from 3 to about 12.8 million bales. Abandoned due to its susceptibility to CLCuVD during the mid-1990s.	www.niab.org.pk
NIAB-92	Irradiating seed of Stoneville 231 with gamma rays at 30 kR from 60CO.	A semi-hairy. Compact canopy with 0–2 monopodial branches as compared to its parental cultivar.	Iqbal et al. (1994)
NIAB-846	A cross between NIAB 78 × REBA 288 (pollen irradiated with 10 Gy of gamma rays).	Resistant to cotton leaf curl virus disease (old strain). Better tolerant to CLCuV-B (Burewal strain). Good plant type. Desirable leaf foliage. Heat tolerance Avoidance of pink bollworm.	http://www.niab.org.pk/mutation.htm
NIAB-999, NIAB-111		High yielding. Heat tolerant. CLCuVD resistant.	Khan et al. (2023)
NIAB-777	Developed by crossing NIAB-78 with REBA-288. The pollen of REBA-288 was irradiated with 10 Gy of gamma rays before pollinating NIAB-78.	High yield. Better fiber quality. Disease resistance during segregation resulted in the evolution of cv. NIAB-777.	Aslam et al. (2018)
MD 15 M4	3.2% v/v of ethyl methanesulfonate.	A mutant with higher fiber quality (fiber strength = 40.5) than that of the parent (38.7).	Bechere et al. (2013)

the result is the F_1 generation, which is identical and heterozygous. F_2 seeds, the first segregating generation with 50% lower heterozygosity, are produced when F_1 plants self. The heterozygosity drops by 50% with each selfed generation, until homozygosity is reached at F_5 or F_6. These plants are characterized by superior combinations and are known as pure lines. Their performance is evaluated in multilocation yield trials compared to commercial varieties. Cotton breeders have two methods to produce cotton hybrids: the conventional or manual method and the male-sterile system. Successful hybridization in cotton involves emasculation and dusting techniques. Emasculation refers to the physical removal of the male organ (stamen column) from the female parent, followed by hand pollination with the selected male parent (Fig. 20.32).

Fig. 20.32 Steps of cotton hybridization. (**a**) Select female parent, (**b**) emasculation, (**c**) female parent only, (**d**) the flower is covered with a color butter paper bag for easy identification the next day morning, (**e**) select male parent and make self-pollination, (**f**) mature pollen grains, (**g**) artificial pollination and (**h**) cover with color butter paper bag for easy identification. (Source: Prepared by Dr. A. Abdelmoghny)

20.5.1 Hybrid Cultivars

First-generation (F1) hybrids are generated by crossing two-line, inbred, pollination, clone, or other genetically distinct populations to provide more features or yield than either the parent or the normal seed variety. Most commercial hybrid varieties are F_1 from two or more pure lines or inbreds. India is one of the first nations in the world to

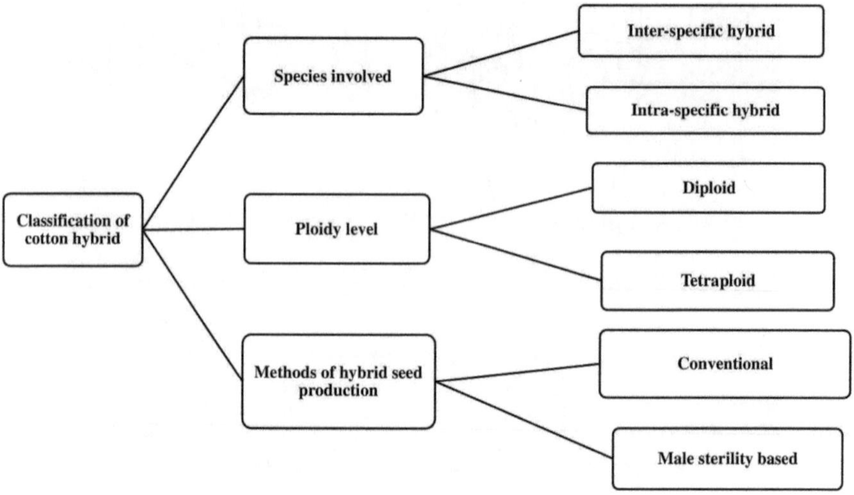

Fig. 20.33 Classification of hybrid cotton. (Source: Prepared by Dr. A Abdelmoghny)

commercially grow hybrid cotton. Dr. Patel is rightly called the father of hybrid cotton and Gujarat is the home of hybrids. The first intra-specific *G. hirsutum* cotton hybrid (H4) was produced in India in 1970 by C. T. Patel from Cotton Research Station, Surat for commercial cultivation at Navasari Agricultural University in Gujarat State. B. H. Katarki developed the first inter-specific hybrid (Varalaxmi) between (*G. hirsutum* × *G. barbadense*) in 1972 utilizing hybrid vigor for higher productivity in the Indian subcontinent in Karnataka state (Katageri et al. 2020). Thus, there are six different types of cotton hybrids from the four farmed species: four intra-specific hybrids (one in each species) and two inter-specific hybrids, one diploid and the other tetraploid (Fig. 20.33). India is the only country in the world to grow all four of the cultivated *Gossypium* species, India holds a special distinction.

Only four varieties of hybrids have been created for commercial production, nevertheless. These comprise both intra-specific hybrids and interspecific hybrids between diploid and tetraploid species in *G. hirsutum* and *G. arboreum*. There are currently no intra-specific hybrids of *G. herbaceum* and *G. barbadense* available (Singh 2007). In cotton, a 20% heterosis over the popular hybrid and a 50% heterosis over the popular variety are deemed significant for hybrid development. In India, 90% of hybrid cotton is *Gossypium hirsutum*, and all of the Bt cotton hybrids that are now in use are either *Gossypium hirsutum* or interspecific hybrids with *Gossypium barbadense*. *Gossypium hirsutum* is most commonly grown due to its excellent yield potential and broad range of adaption.

Cotton is frequently cross-pollinated. To create hybrid cotton, three processes must be followed: identifying and developing the male and female parents, emasculating the female parent, and pollinating a female parent with the identified male parent (Zhang et al. 2023).

The traditional method is very inefficient in diploid cotton because of the small flower size and fragile pedicel, which result in a low boll setting. Using male sterility to create hybrid seeds removes the necessity for emasculation. Pollination is done by hand, and the female parent's anthers are sterile and devoid of pollen. This procedure reduces the price of hybrid seeds. Cotton seed production uses two basic types of male sterility: genetic male sterility and cytoplasmic genetic male sterility.

20.6 Conclusions and Prospects

Cotton fiber has been used by humans for thousands of years. Several tactics were used to boost the growth and fiber yield of cotton plants. To get around several obstacles in the manufacturing of cotton, conventional techniques have been used. Traditional breeding methods have demonstrated potential in strengthening cotton's resistance to different stressors and boosting yield and fiber quality. Three methods are usually included in conventional plant breeding: introduction, selection, and hybridization. Nowadays, a lot of crops are improved and new types are created through hybridization. Conventional approaches make trait introgression difficult, however, hybridization provides a faster and easier way to do it. To improve cotton's resistance to abiotic stress, new genes from wild species can be successfully incorporated into local cultivars through interspecific hybridization and backcrossing. The development of certified cotton cultivars has been greatly aided by conventional breeding.

Cotton breeders are now able to increase breeding accuracy and precision through the development of biotechnology and plant breeding. Cotton breeding has been transformed by molecular tools, while there is still an opportunity for improvement in controlling the spatiotemporal expression of transgenes. Although molecular-assisted selection (MAS) methods are still useful, their effectiveness can be increased by using genome-editing tools. To enhance MAS and increase its accuracy, the most recent genetic and bioinformatics technologies ought to be applied. Cotton breeding may benefit from the combination of QTL mapping and transcriptome analysis, especially for qualitative qualities.

Modern molecular markers have aided in the mapping of genes on cotton chromosomes. AFLP, RFLP, ESTs, RAPD, and SSR markers were used to identify high-yield, stress-tolerant, and fiber-quality genes in the cotton A and D genomes. Next-generation sequencing technology may allow for the identification of major and minor genes across the cotton genome, perhaps resulting in the development of the best plants for current and future farming systems around the world.

Climate change has made heat stress a major problem for cotton cultivation. Microarray technology is an effective tool for studying gene expression levels in response to heat exposure. For aggressive selection, DNA markers are necessary for tracking genetic variation and detecting QTLs. Limited genetic variation in germplasm that is impacted by different conditions can be addressed using molecular markers. To fulfill the demands of an expanding population, cotton production can be increased through the combination of biotechnological and molecular breeding techniques. To produce cotton sustainably, heat-tolerant germplasm must be developed.

Sustainable yield is a key goal in agriculture, with water stress being a major limiting factor. Water deficit stress affects plant growth, morpho-physiological characteristics, fiber quality and seed cotton yield. Cotton, both cultivated and wild, can be sequenced and re-sequenced to reveal fiber-specific genes, high-density markers, and epigenetic research. The identification of mutation through reverse genetics techniques can lead to improved phenotypes and crop productivity under moisture deficit stress. Cotton faces challenges from lepidopteran pests with Bt technology offering a safe and effective alternative to chemical insecticides. However, evolving pests pose risks, necessitating the development of new technologies like RNAi and CRISPR/Cas9.

Cotton was the first commercially utilized herbicide-resistant transgenic plant made when recombinant DNA technology emerged. Biotechnology primarily used in cotton, has led to the creation of Bt cotton, known for its insect resistance and reduced pesticide usage, making it eco-friendly. However, public acceptance of genetically modified plants remains a major concern. There is a need to bridge the gap between scientists and the general people by raising awareness. Concerns about health and the environment must be treated seriously. Breeders and biotechnologists face challenges as the world's population grows, demand for food and fiber increases, production land shrinks, and climate changes. To solve these issues with cotton output and fiber quality, contemporary breeding initiatives are required, such as participatory breeding employing novel biotechnological techniques. Cotton genetics research has been primarily focused on acquiring genomic resources and tools for basic and applied genetics and genomics research. Despite the availability of several techniques and resources, research on cotton genetics lags behind that of other significant crops. These materials have enhanced our understanding of cotton biology by assisting in the discovery of relevant DNA sequences linked with yield, fiber quality, and stress tolerance. However, further work is needed to enhance these resources and tools for more effective use in cotton genetic improvement.

Cotton crops are also at risk from weed infestations, which reduce crop yield and fiber quality by competing with plants for nutrients, light, and moisture. There are several ways to control weeds, such as chemical applications like herbicides, mechanical control, allelopathy, and cultural practices. Cotton fields may benefit from more efficient weed control if these methods are applied in tandem. When compared to other ways, the use of herbicides—chemicals used to control weeds in almost all crops, including cotton—is the most effective, rapid, and simple. However, because there are various classes and types of herbicides, each with unique functions and modes of action, it is necessary to apply them carefully. One of the most effective ways to make plants resistant to herbicide application is to alter the genetic composition of cotton. Farmers prefer to cultivate resistant genotypes to herbicides, and the majority of genetic modification effort in cotton is done for herbicides. Today, scientists all around the world are successfully transmitting this characteristic to local cotton types by including GM herbicide-resistant cotton in their traditional breeding operations. In addition to saving time, herbicide-resistant cotton provides a low-cost and easy technique to control weeds in cotton.

Over the last two decades, the commercialization of biotech crops has shown considerable agronomic, ecological, economic, social, and health benefits. Unlike chemical pesticides, which affect pollinators, wild animals, and other non-target invertebrate pests, genetically modified cotton is the first transgenic non-food crop that provides a focused, safe, and effective weapon against lepidopterans. Nevertheless, despite its benefits, opinions on the biosafety of genetically modified crops have split scientists, policymakers, and other stakeholders. Regulators have guaranteed that a sizable portion of the yearly crop yield is contributed by genetically modified crops. After years of selection and the necessary regulatory mechanisms, these crops are judged safe and suited for usage. We think that sensitivity and misinterpretation are the reasons behind some of the questions that are brought up in print and electronic media. Since the first genetically modified crop was brought to market about 25 years ago, there hasn't been a single unfavorable report about its detrimental effects. However, some research groups expressed concerns and found that genetically modified plants had detrimental effects at the feeding level. On the other hand, the scientific community has adequately addressed the following issues: experimental designs, errors, reproducibility, and misleading findings.

The goal of the cottonseed system is to provide farmers with seeds so they can plant the crop for the following growing season. To guarantee varietal purity and identity, a formal system will incorporate breeder seed increases and maybe a seed certification procedure. Cottonseed farmers may not choose to participate in the seed certification system, but they do follow their quality management system to ensure seed quality and varietal purity. An informal seed system varies according to the setting. In a typical informal seed system, the seed producer keeps his seed and sells it to other local farmers so that they can grow the crop the next season. Other informal systems include insufficient information on seed production, handling, storage, and shipment, as well as unknown seed sources. The seed being sold cannot be properly traced in an informal seed system. Since such seed is typically not checked, its quality is uncertain, which results in a low crop stand and a significant risk of weed seed production. Before purchasing seed through an unofficial system, farmers are advised to get the seed checked at a respectable laboratory. Based on the data, regional policymakers and cotton stakeholders can have a productive discussion to improve the cotton production system by determining the problems that require attention and coming up with solutions. Even though cotton production and trade are powerless because of environmental variables and price volatility, the chain of cotton stakeholders will profit over time from the variety of income-raising alternatives. Cotton production zones are constrained not only by rainfall patterns, immature picking, environmental degradation, and a lack of labor during the decade but also by increased pressure from livestock and population growth, as well as potential adverse effects from cotton cultivation itself. The only way to revive the system with a viable diversification chance in the current situation is to devote time and money to cotton research and development.

Cotton breeding is a complementing aspect of related businesses like textile, oil, and production agriculture, and it is not limited to just generating high-quality seed. These sectors are crucial to the nation's overall economic growth since they account for a sizable portion of the agricultural GDP. To achieve the goal of sustainable economic development, it is critical that the private sector's cotton breeding enterprise shift from government-focused cottonseed production to self-motivated entrepreneurial aspiration. This can be accomplished by legislators creating an environment that encourages the private sector to participate and perform its constructive purpose.

Appendix 20.1 Research Institutes and Online Resources Available for Egyptian Cotton (*Gossypium barbadense* L.) Breeding

Institution name	Specialization and research activities	Address	Contact information and website
Australian Cotton Research Institute (ACRI)	Located between Narrabri and Wee Waa in the center of New South Wales' main cotton production areas. Scientists at ACRI are assisting the cotton industry to develop a sustainable future based on productivity and best practice natural resource, disease and pest management. It includes glasshouses, laboratories, cotton gin, controlled environment growth rooms, insectaries, workshops and conference rooms.	Narrabri – Australian Cotton Research Institute, Locked Bag 1000, Narrabri, NSW 2390	https://www.dpi.nsw.gov.au/
ACSA International Cotton Institute, USA	Basic education on all aspects of cotton	ACSA International Cotton Institute, USA	https://bf.memphis.edu/cotton/index.php/main/instructions
Cotton Development Board (CDB)	Cotton Development Broad or CDB is a government organization responsible for cotton production and research in Bangladesh located in Dhaka.	Khamarbari Rd, Dhaka, Bangladesh.	Email: http://www.ict.cdb.gov.bd@gmail.com www.cdb.gov.bd

(continued)

Institution name	Specialization and research activities	Address	Contact information and website
Cotton Research Institute (CRI)	Research focused on breeding improved, high-yielding varieties, based on the selection of desirable mutations or genetic segregation of earlier natural hybridization. As of 1921, the Cotton Research Council approved the utilization of artificial hybridization to develop new cotton varieties. The aim was to incorporate into a new strain such desirable traits from parent plants. The technique was further developed to become the major approach for developing new varieties. Cotton technology research was also developed to complement agronomy research The Experimental Spinning mill	9, El-Gamma St. 12619, Giza, Egypt.	Email: cri@arc.eg. http://www.arc.sci.eg/InstsLabs/Default.a
Central Cotton Research Institute, Multan (CCRI)	The institute has contributed significantly by advancing and generating knowledge in cotton research and development since its establishment in 1970. Over the years, with many achievements of high-yielding varieties with standard fiber quality characteristics like staple length, fineness and strength, the institute increased cotton productivity from 370 kg/ha in 1970 to 772 kg/ha during the current years.	4CXQ+CRX, OLD Shuja Abad Rd, Shershah Town, Multan, Punjab 60000, Pakistan.	Email: mnaagr@gmail.com www.ccri.gov.pk

(continued)

Institution name	Specialization and research activities	Address	Contact information and website
Central Institute for Cotton Research (CICR)	Research on Cotton at ICAR-CICR continues to focus on innovations in crop improvement, crop production and crop protection aspects to enhance cotton productivity, develop technologies for resource conservation, increase resource use efficiency and eco-friendly crop protection.	Post Bag No. 2, Shankar Nagar Post Office, Nagpur – 440 010, Maharashtra, India	Email: cicrnagpur@gmail.com https://www.cicr.org.in
Central Institute for Research on Cotton Technology (CRICOT)	Is engaged in carrying out basic and strategic research in the processing of cotton & its agro-residues, development of value-added products and quality assessment of cotton. The Institute extends effective technological support to the country's cotton breeding program for varietal development with improved productivity and quality that suits industry needs.	Central Institute for Research on Cotton Technology, Adenwala Road, Matunga (East, Mumbai – 400 019).	Email: director.circot@icar.gov.in https://www.circot.icar.gov.in/en/
Cirad Agriculture Research Institute, France	Cotton germplasm, databases.	Cirad Agriculture Research Institute, France	https://www.cirad.fr/en
Cotton Development Trust (CDT)	To contribute to the increase of farm yields and seed cotton production in Zambia. CDT's main objectives are: to promote, develop, and improve cotton production through research, extension and training of farmers and extension staff and seed multiplication. To contribute to the	PO. Box 670057, Mazabuka City, Lusaka, Zambia	Email: cdt@zamtel.zm https://www.asti.cgiar.org/zambia/directory/cotton-development-trust-cdt

(continued)

Institution name	Specialization and research activities	Address	Contact information and website
	agricultural development of Zambia through innovative commercial production and enhanced partnerships in technology development and information dissemination with local and international institutions		
Cotton Egypt Association (CEA)	CEA is a non-profit association, established in 2005. CEA exclusively manages and promotes the registered Egyptian Cotton™ logo trademark, owned by the Ministry of Trade & Industry and Alexandria Cotton Exporter's Association. CEA is also responsible for the licensing program. Our mission is to protect and promote the value-add of the Egyptian Cotton™ identity in supply chains end-to-end. CEA is supporting sustainability by collaborating with the Better Cotton Program launched in Egypt by UNIDO in 2020.	22, Syria St. 3rd Floor, Mohandessine, Giza, 12411, Egypt.	Email: info@cottonegyptassociation.com https://www.cottonegyptassociation.com
Cotton Research Institute, Nanjing University China	Cotton genomics and genetics	Cotton Research Institute, Nanjing University China	http://mascotton.njau.edu.cn
Cotton Research Institute, Zimbabwe	Cotton varietal development, Integrated Pest Management	Cotton Research Institute, Zimbabwe	https://www.gfar.net/organizations/cotton-research-institute-1
International Cotton Advisory Committee (ICAC)	The International Cotton Advisory Committee is an association of governments of cotton-producing,	1629-K Street, NW, Suite 702, Washington DC-20006-1635-USA	Email: secretariat@icac.org https://www.icac.org

(continued)

Institution name	Specialization and research activities	Address	Contact information and website
	consuming and trading countries that acts as the international commodity body for cotton and cotton textiles		
International Cotton Researchers Association (ICRA)	The concept of an International Cotton Researchers Association (ICRA) was the initiative of the International Cotton Advisory Committee (ICAC), Washington DC, stimulating research collaboration across the world, including assistance with organizing World Cotton Research Conferences (WCRC).	1629-K Street, NW, Suite 702, Washington DC-20006-1635-USA	Email: secretariat@icac.org https://www.icracotton.org/
International Cotton Association Ltd (ICA)	ICA is the world's leading cotton trade association and arbitral body. From cotton growing and spinning to insurance and shipping, any firm or individual involved in the cotton trade can apply to become a member of ICA. With over 500 members including the world's major cotton firms, we aim to promote safe trading whilst protecting the interests of the global cotton trade.	6th floor, Walker House, Exchange Flags, Liverpool L2 3YL, United Kingdom.	Email: info@ica-itd.org https://www.ica-itd.org
Institute of Cotton Research, China	Genetic breeding, germplasm resources, farming cultivation, plant protection, molecular biology.	Institute of Cotton Research, China	http://www.caas.cn/en
Pakistan Central Cotton Committee (PCCC)	Leading cotton research and development body in the public sector, conducting research on variety; nutritional	PTV Station, Shershah Town, Multan, Punjab 60000, Pakistan	Email: dr@pccc.gov.pk http://www.pccc.gov.pk/

(continued)

Institution name	Specialization and research activities	Address	Contact information and website
	requirements; crop agronomy; water management; insect-pests and disease management; and resistance management in its seven research facilitates across the country. Also, maintains a germplasm bank with over 6143 accessions of all four cultivated cotton species.		
National Cotton Council (NCI)	The National Cotton Council of America's mission is to ensure the ability of all U.-S. cotton industry segments to compete effectively and profitably in the raw cotton; oilseed and U.S manufactured product markets at home and abroad.	National Cotton Council of America, 1521 New Hampshire Ave NW, Washington, DC 20036-1205	https://www.cotton.org/about/index.cfm
Nazili Cotton Research Institute, Turkey	Cotton research and development	Nazili Cotton Research Institute, Turkey	https://www.administrator@nazilli.tagem.gov.tr
Uzbek Research Institute, Uzbekistan	Cotton germplasm resources.	Uzbek Research Institute, Uzbekistan	https://en.yellowpages.uz/company/uzbek-scientific-researchinstitute-of-cotton

Appendix 20.2 Genetic Resources of Long and Extra-Long Staple for Egyptian Cotton (*Gossypium barbadense* L.) Varieties

Cultivars	Important fiber quality traits				Cultivation location
	Fiber length (mm)	Fiber fineness (μm)	Fiber strength (g/tex)	Yellowness (+b)	
Australian Pima	35.6	–	39.0	–	Australia
Xinjiang China	35.3	–	39.3	–	China
Giza 45	35.6	3.0	45.8	8.9	Egypt
Giza 86	33.5	4.3	46.0	8.2	Egypt

(continued)

Cultivars	Important fiber quality traits				Cultivation location
	Fiber length (mm)	Fiber fineness (μm)	Fiber strength (g/tex)	Yellowness (+b)	
Giza 92	3.5	3.0	48.6	7.7	Egypt
Giza 93	35.8	3.0	47.2	11.1	Egypt
Giza 94	34.0	4.2	43.7	8.0	Egypt
Giza 95	30.5	4.4	37.7	11.8	Egypt
Giza 96	35.8	4.0	47.4	8.2	Egypt
Giza 97	33.7	4.3	46.1	9.0	Egypt
Giza 98	30.7	4.3	38.1	11.6	Egypt
West Indian	35.3	–	36.8	–	India
Pima	36–38	3.7–4.5	40–42	–	Israel
Acalpi	34–35	3.4–4.1	34.37	–	Israel
Peru Pima	36.8	–	36.0	–	Peru
Yoloten-14	40.0–41.0	3.3–4.0	–	–	Turkmenistan
Yoloten-7	33–34	3.8–4.8	–	–	Turkmenistan
Pima S3	35.3	3.4	–	–	USA
Pima S4	34.3	2.9	–	–	USA
Pima S6	36.5	3.5	–	–	USA
Supima	36.1	–	41.5	–	USA
Sea Island	36.8	3.3	39.0	–	USA
Mehnat	34–35	4.5	29.7	8.2	Uzbekistan
Gulbahor-2	34–35	4.6	26.4	–	Uzbekistan
AN-514	34–35	4.5–4.6	27.6	6.3	Uzbekistan
AN-517	34–36	4.4–4.6	27.7	7.0	Uzbekistan
AN-516-DV	36–37	4.1–4.3	27.7	7.3	Uzbekistan
Ishonch	33.5–34.5	4.3–4.4	28.0	6.5	Uzbekistan
UzFA-703	34.5–35.5	4.0–4.4	29–31	7.0	Uzbekistan
748-B	39.2–41.2	–	32.6–38.6	–	Tajikistan

References

Abdalla AM, Reddy OUK, El-Zik KM, Pepper AE (2001) Genetic diversity and relationships of diploid and tetraploid cottons revealed using AFLP. Theor Appl Genet 102:222–229

Abdelbary AM, Hamoud HM, Yahiya WM et al (2021) Giza 97 a new Egyptian long staple cotton variety. Egypt J Agric Res 99(3):284–295. https://ejar.journals.ekb.eg/article_199748.html

Abdel-Hamid AME, Salem KFM (2021) Breeding strategies of garden pea (*Pisum sativum* L.). In: Al-Khayri JM, Jain SM, Johnson DV (eds) Advances in plant breeding strategies: vegetable crops. Springer, Cham. https://doi.org/10.1007/978-3-030-66969-0_9

Abdelmoghny AM (2016) Genetic analysis and prediction of new recombination in some cotton (*G. barbadense* L.) crosses. J Agric Res Kafr El-Sheikh Univ 42(3):319–335

Abdelmoghny AM, Max SM, Orabi MHM (2016) Genetic variability in segregating generations of some cotton crosses. Egypt J Plant Breed 20(3):507–527

Abdelmoghny AM, Max SM, Hassan SS (2021) Heritability and genetic advance as genetic indicators for improvement in two cotton crosses. J Gene Eng Bio Res 3(2):26–39. https://doi.org/10.33140/IGEBR.03.02.03

Abdelmoghny AM, Abdelbary AM, Hamoud HM et al (2022) A new high yielding and long staple Egyptian cotton (*Gossypium barbadense* L.) variety super Giza 94. New Valley J Agric Sci 2(6):552–564. https://doi.org/10.21608/NVJAS.2022.174385.1112

Abdullaev A, Abdumavlon A, Abdullaev A et al (2013) Cotton germplasm collection of Uzbekistan. Asian Australas J Plant Sci Biotechnol 7(2):1–15

Abdullaev FK, Rustamov NSU, Yakubov MD (2025) Uzbekistan genebank. In: Al-Khayri JM, Salem KFM, Jain SM (eds) Plant gene banks. Springer, Singapore. https://doi.org/10.1007/978-981-99-4236-7_57-1

Abdullah K, Khan Z (2022) Breeding cotton for international trade. In: Khan AA (ed) Cotton breeding and biotechnology challenges and opportunities. Taylor & Francis Group, p 368. https://doi.org/10.1201/9781003096856

Abid MA, Wang P, Zhu T et al (2020) Construction of *Gossypium barbadense* mutant library provides genetic resources for cotton germplasm improvement. Int J Mol Sci 21(1):13. https://doi.org/10.3390/ijms21186505

Aboshama HM, Salem KFM, Hassan HMF, Tahoun MH (2015) Estimating heterosis and combining ability for *in vivo* and *in vitro* traits using diallel cross on *tomato* (*Solanum lycopersicum* L.). Minufiya J Agric Res 70:757–770

Abulaila K, Al Sane KO, Tahabsom Z (2025) Jordan seed bank. In: Al-Khayri JM, Salem KFM, Jain SM (eds) Plant gene banks. Springer, Singapore. https://doi.org/10.1007/978-981-99-4236-7_45-1

Adeleke A (2024) Technological advancements in cotton agronomy: a review and prospects. Preprints.org. www.preprints.org, https://doi.org/10.20944/preprints202402.1342.v1

Afzal MN, Tariq M, Ahmed M et al (2020) Managing planting time for cotton production. In: Ahmad S, Hasanuzzaman M (eds) Cotton production and uses agronomy, crop protection, and postharvest technologies. Springer Nature Singapore Pte Ltd. https://doi.org/10.1007/978-981-15-1472-2

Akparov ZI, Mammadova SA, Kalantarova NS, Mirzaliyeva IA, Mammadov AT (2025) National genebank of Azerbaijan. In: Al-Khayri JM, Salem KFM, Jain SM (eds) Plant gene banks. Springer, Singapore. https://doi.org/10.1007/978-981-99-4236-7_31-1

Al Didi MA (1972) History of Egyptian cotton varieties. Egypt Cotton Gaz 58:36–56

Al Didi MA (1982) The Egyptian cotton crop 1950–1980. XXXXI Plenary meeting of the International Cotton Advisory Committee (ICAC), Cairo, Egypt

Al Lawati AH (2025) Plant genetic resource conservation of Oman. In: Al-Khayri JM, Salem KFM, Jain SM (eds) Plant gene banks. Springer, Singapore. https://doi.org/10.1007/978-981-99-4236-7_50-1

Aladele SE et al (2025) Plant genetic resources conservation and utilization in Nigeria. In: Al-Khayri JM, Salem KFM, Jain SM (eds) Plant gene banks. Springer, Singapore. https://doi.org/10.1007/978-981-99-4236-7_17-1

Aldahak L, Salem KFM, Al-Salim SHF, Al-Khayri JM (2021) Advances in chicory (*Cichorium intybus* L.) breeding strategies. In: Al-Khayri JM, Jain SM, Johnson DV (eds) Advances plant breeding strategies: vegetable crops. Springer, Cham. https://doi.org/10.1007/978-3-030-66969-0_1

Almarri NB, Ahmad S, Alsubaie GTG, Elshal MH (2025) Saudi Arabia gene bank. In: Al-Khayri JM, Salem KFM, Jain SM (eds) Plant gene banks. Springer, Singapore. https://doi.org/10.1007/978-981-99-4236-7_54-1

Alqahtani MM, Saleh MM, Alwutayd KM et al (2024) Performance and genotypic variability in diverse date palm (*Phoenix dactylifera* L.) cultivars for fruit characteristics. Genet Resour Crop Evol 71:1759–1772. https://doi.org/10.1007/s10722-023-01719-0

Al-Shafeay AF, Ibrahim AS, Nesiem MR, Tawfik MS (2011) Establishment of regeneration and transformation system in Egyptian sesame (*Sesamum indicum* L.) cv Sohag 1. GM Crops 2(3): 182–192

Amiteye S (2021) Basic concepts and methodologies of DNA marker systems in plant molecular breeding. Heliyon 7:e08093. https://doi.org/10.1016/j.heliyon.2021.e08093

Arab S et al (2025) National gene bank of Egypt. In: Al-Khayri JM, Salem KFM, Jain SM (eds) Plant gene banks. Springer, Singapore. https://doi.org/10.1007/978-981-99-4236-7_5-1

Aslam M, Haq MA, Bandesha AA, Haidar S (2018) NIAB-777: an early maturing, high yielding and better-quality cotton mutant developed through pollen irradiation technique – suitable for high density planting. J Anim Plant Sci 28(2):636–646

Aslam S, Khan SH, Ahmed A, Dandekar AM (2020) The tale of cotton plant: from wild type to domestication, leading to its improvement by genetic transformation. Am J Mol Biol 10:91–127

Anwar M, Iqbal MZ, Abro AA, Memon S, Bhutto L.A, Memon, SA, Peng Y (2022) Inter-Specific Hybridization in Cotton (Gossypium hirsutum) for Crop Improvement. Agronomy 2022, 12, 3158. https://doi.org/10.3390/agronomy12123158

Bakhsh A, Anayol E, Özcan SF et al (2015) An insight into cotton genetic engineering (*Gossypium hirsutum* L.): current endeavors and prospects. Acta Physiol Plant 37(8):171

Bardak A, Khezir H, Oktay E, Zahid M et al (2018) Genetic mapping in cotton. In: Past, present and future trends in cotton breeding. IntechOpen, pp 93–112

Bechere E, Meredith WR, Boykin JC (2013) Registration of mutant population MD 15 M4 Gossypium hirsutum L. with enhanced fiber quality. J Plant Regist 7(2):216–219

Békefi Z, Szilágyi S, Desiderio F, Demku T, Varga J (2025) Prunus gene bank in Hungarian University of Agriculture and Life Sciences. In: Al-Khayri JM, Salem KFM, Jain SM (eds) Plant gene banks. Springer, Singapore. https://doi.org/10.1007/978-981-99-4236-7_71-1

Bi IV, Baudoin JP, Hau B, Mergeai G (1999) Development of high gossypol cotton plants with low-gossypol seeds using trispecies bridge crosses and in vitro culture of seed embryos. Euphytica 106:243–251

Börner A, Landjeva S, Salem KFM, Lohwasser U (2009) Pflanzengenetische Ressourcen als Grundlage für die Züchtung klimatoleranter Sorten. Proceedings of a German Conference "Weizenforum = Wheat Forum" Norddeutsches Weizenforum Schriftenreihe des Inst. Für Pflanzenbau und –züchtung der CAU Kiel, Germany 63:23–27

Börner A, Landjeva S, Salem KFM, Lohwasser U (2010) Plant genetic resources: a prerequisite for drought tolerance breeding in cereals. In: Proceedings of the Jahrestagung der Vereinigung der Pflanzenzüchter und Saatgutkaufleute Österreichs, 24–26 November 2009, Raumberg-Gumpenstein, pp 11–13

Bourland F, Myers GO (2015) Conventional cotton breeding. In: Fang DD, Richard GP (eds) Agronomy monographs. ASA, CSSA, SSSA, Madison, pp 1–25. https://doi.org/10.2134/agronmonogr57.2013.0025

Breidy J (2025) Lebanon gene bank. In: Al-Khayri JM, Salem KFM, Jain SM (eds) Plant gene banks. Springer, Singapore. https://doi.org/10.1007/978-981-99-4236-7_48-1

Brinkerhoff L (1970) Variation in *Xanthomonas malvacearum* and its relation to control. Annu Rev Phytopathol 8(1):85–110

Campbell BT, Saha S, Percy R et al (2010) Status of the global cotton germplasm resources. Crop Sci 50(4):1161–1179

Cañar DY et al (2025) Colombia gene bank. In: Al-Khayri JM, Salem KFM, Jain SM (eds) Plant gene banks. Springer, Singapore. https://doi.org/10.1007/978-981-99-4236-7_95-1

Chaudhary J, Alisha A, Bhatt V, Chandanshive S et al (2019) Mutation breeding in tomato: advances, applicability and challenges. Plants 8(5):128

Chen W, Yao J, Chu L et al (2015) Genetic mapping of the nulliplex-branch gene (gb_nb1) in cotton using next-generation sequencing. Theor Appl Genet 128(3):539–547. https://doi.org/10.1007/s00122-014-2452-2

Chen X, Lu X, Shu N et al (2017) Targeted mutagenesis in cotton (*Gossypium hirsutum* L.) using the CRISPR/Cas9 system. Sci Rep 7:44304

Collard BC, Mackill DJ (2008) Marker assisted selection: an approach for precision plant breeding in the twenty first century. Philos Trans R Soc B Biol Sci 363(1491):557–572

Cox KL, Meng F, Wilkins KE et al (2017) TAL effector driven induction of a SWEET gene confers susceptibility to bacterial blight of cotton. Nat Commun 8(1):1–14

Darmanov MM, Makamov AK, Ayubov MS et al (2022) Development of superior fiber quality upland cotton cultivar series "Ravnaq" using marker-assisted selection. Front Plant Sci 13: 906472

Datta A, Ullah H, Ferdous Z et al (2020) Water management in cotton. In: Jabran K, Chauhan BS (eds) Cotton production. Wiley

Davidonis GH, Hamilton RH (1983) Plant regeneration from callus tissue of Gossypium hirsutum L. - Plant Sci. Lett. 32:89–93

de Galarreta JIR, Barandalla L, Ritter E, Velázquez ME, Rios D (2025) Spanish potato germplasm bank. In: Al-Khayri JM, Salem KFM, Jain SM (eds) Plant gene banks. Springer, Singapore. https://doi.org/10.1007/978-981-99-4236-7_114-1

Desheva G, Valchinova E, Deshev M (2025) Bulgarian gene bank. In: Al-Khayri JM, Salem KFM, Jain SM (eds) Plant gene banks. Springer, Singapore. https://doi.org/10.1007/978-981-99-4236-7_66-1

Díez MJ, Valcárcel JV, Solbes E, Muñoz E (2025) Forty years conserving Spanish biodiversity in the gene bank of the Universitat Politècnica de València, Spain. In: Al-Khayri JM, Salem KFM, Jain SM (eds) Plant gene banks. Springer, Singapore. https://doi.org/10.1007/978-981-99-4236-7_86-1

El Rabey H, Salem KFM, Mattar MZ (2013) The genetic diversity and relatedness of eight rice (*Oryza sativa* L.) cultivars as revealed by AFLP and SSRs markers. Life Sci J 10(1):1471–1479

Elabd AB, Salem KFM (2009) Detection of genetic diversity in rice (*Oryza sativa* L.) varieties using microsatellite DNA-markers and quantitative traits. In: 6th Egyptian Plant Society of breeding conference, 4–7 May 2009, Ismailia, Egypt

Elabd AB, Saleh MM, Aldahak L, Salem KFM (2020) Identification of superior rice (Oryza sativa L.) genotypes under irrigated conditions based on multivariate analysis. In: 7th International scientific conference of genetic and environment from 8th–9th December 2020, Baghdad, Iraq

El-Absawy EA, Salem KFM, Abdellatif KF, ElShamy H (2012) Estimation of genetic diversity among some Egyptian barley (*Hordeum vulgare* L.) cultivars using RAPD markers and agronomic traits. In: 1st International conference on biotechnology applications in agriculture, Moshtohor-Hurghada, 18–22 February 2012, Faculty of Agriculture, Benha University, Egypt

El-Absawy EA, Nowar AA, Salem KFM et al (2015) Estimation of genetic diversity among some bread wheat (*Triticum aestivum* L.) genotypes using molecular markers and agronomic traits. Minufiya J Agric Res 40(3):223–233

El-Adly HH, Eissa AEM, Khalif HS et al (2018) Giza 95 a long staple Egyptian cotton variety for middle and upper Egypt. Egypt J Agric Res 96(2):717–726

Eltaher S, Sallam A, Belamkar V, Emara HA, Nower A, Salem KFM, Poland J, Baenziger PS (2017) Genome-wide association mapping for drought tolerance traits at seedling stage using Nebraska winter wheat. In: International conference on productive crop design (genome to phenome), April 2017, USA

Eltaher S, Sallam A, Belamkar V et al (2018) Genetic diversity and population structure of F3:6 Nebraska winter wheat genotypes using genotyping-by-sequencing. Front Genet 9:76. https://doi.org/10.3389/fgene.2018.00076

Eltaher S, Baenziger S, Belamkar V et al (2021) GWAS revealed effect of genotype x environment interactions for grain yield of Nebraska winter wheat. BMC Genomics 22:1–14

Eltaher S, Sallam A, Emara HA et al (2022) Genome-wide association mapping revealed SNP alleles associated with spike traits in wheat. Agronomy 12:1469. https://doi.org/10.3390/agronomy12061469

El-Zanaty AM, Salem KFM, Esmail RM (2011) Detection of genetic diversity in Egyptian cotton (*Gossypium barbadense* L.) varieties using RAPD markers and morphological traits. J Am Sci 7(12):1107–1115

El-Zanaty AM, Salem KFM, Ismail M, Esmail RM (2013) Impact of plant breeding on genetic diversity of the Egyptian cotton (Gossypium barbadense L.) varieties as revealed by RAPD markers. J Agric Chem Biotechnol 4:217–227

Emara HA, Nower AA, Salem KFM, Eltaher S (2011) Estimating heterosis and combining ability for *in vivo* and *in vitro* traits using diallel cross in Maize (*Zea mays* L). Int J Prod Dev 16:101–120

Emara HA, Salem KFM, Abdel-Halim AA, Abd El-Kowy RNF (2013) An *in vivo* and *in vitro* traits analysis of a diallel cross in rice (*Oryza sativa L.*). Egypt J Plant Breed 17(2):591–610

Esmail RM, Salem KFM, El-Zanaty AM (2009) Genetic diversity studies of Egyptian cotton (Gossypium barbadense L.) varieties using RAPD markers and agronomic traits. In: 6th Egyptian Plant Society of breeding conference, 4–7 May 2009, Ismalia, Egypt

Falconer DS, Mackay TFC (1996) Introduction to quantitative genetics, 4th edn. Addison Wesley Longman, Harlow

FAO (2016) Food and Agriculture data. Retrieved from http://www.fao.org/faostat/en/#home. Accessed 20 Feb 2016

Farooq O, Mubeen K, Khan AA, Ahmad S (2020) Sowing methods for cotton production. In: Ahmad S, Hasanuzzaman M (eds) Cotton production and uses agronomy, crop protection, and postharvest technologies. Springer Nature Singapore Pte Ltd. https://doi.org/10.1007/978-981-15-1472-2

Farzana AN, Iqbal W, Nazeer ZU et al (2022) Conventional breeding of cotton. In: Khan Z, Ali Z, Khan AA (eds) Cotton breeding and biotechnology challenges and opportunities, 1st edn. Taylor & Francis Group, p 368. https://doi.org/10.1201/9781003096856

Finer JJ, McMullen MD (1990) Transformation of cotton (Gossypium hirsutum L.) via particle bombardment. Plant Cell Rep 8:886–889

Firoozabady E, DeBoer DL, Merlo DJ et al (1987) Transformation of cotton (Gossypium hirsutum L.) by Agrobacterium tumefaciens and regeneration of transgenic plants. Plant Mol Biol 10:105–116. https://doi.org/10.1007/BF00016148

Fryxell PA (1979) Natural history of the cotton tribe. Texas A&M University Press, College Station, p 267

Galea-Gragera FA, Llera Cid F (2025) Spain germplasm bank of annual forage legumes seeds at the Scientific and Technological Research Center of Extremadura (CICYTEX). In: Al-Khayri JM, Salem KFM, Jain SM (eds) Plant gene banks. Springer, Singapore. https://doi.org/10.1007/978-981-99-4236-7_102-1

Gao W, Long L, Tian X et al (2017) Genome editing in cotton with the CRISPR/Cas9 system. Front Plant Sci 8:1364

Ge X, Xu J, Yang Z et al (2023) Efficient genotype independent cotton genetic transformation and genome editing. J Integr Plant Biol 65(4):907–917. https://doi.org/10.1111/jipb.13427

Ghazy MI, Salem KFM, Sallam A (2020) Utilization of genetic diversity and marker-trait to improve drought tolerance in rice (*Oryza sativa* L.). Mol Biol Rep 48:157–170. https://doi.org/10.1007/s11033-020-06029-7

Ghidan WF, Abdulmajid DA (2025) Egyptian rice gene bank. In: Al-Khayri JM, Salem KFM, Jain SM (eds) Plant gene banks. Springer, Singapore. https://doi.org/10.1007/978-981-99-4236-7_106-1

Gibely RH (2021a) Estimation of genetic variance components and identification of transgressive segregants for two intraspecific extra-long staple cotton crosses. Menoufia J Plant Prod 6(1): 53–70

Gibely RH (2021b) Variability, genetic components and selection response in segregating generations among some cotton crosses. J Sustain Agric Sci 47(1):1–12

Gomez GM, Moretzon MC, da Silva EH et al (2016) Mapping of QTLs identified in a novel source of resistance to the root-knot nematode (*Meloidogyne incognita* race 3) in cotton (*Gossypium barbadense* L.). ICGI. Goiânia, ICGI, Résumé, World Cotton Research Conference 6, Goiânia, Brésil, 2–6 May 2016, p 2

Gowda SA, Bourland FM, Kaur B et al (2023) Genetic diversity and population structure analyses and genome-wide association studies of photoperiod sensitivity in cotton (*Gossypium hirsutum* L.). Theor Appl Genet 136(11):230. https://doi.org/10.1007/s00122-023-04477-w
Gross BL, Strasburg JL (2010) Cotton domestication: dramatic changes in a single cell. BMC Biol 8:137
Gudaye TY, Lakew BF, Selamo MM, Tessema ES (2025) Ethiopian crop gene bank. In: Al-Khayri JM, Salem KFM, Jain SM (eds) Plant gene banks. Springer, Singapore. https://doi.org/10.1007/978-981-99-4236-7_7-1
Guo H, Guo H, Zhang L et al (2019) Metabolome and transcriptome association analysis reveals dynamic regulation of purine metabolism and flavonoid synthesis in transdifferentiation during somatic embryogenesis in cotton. Int J Mol Sci 20(9):2070
Gao LH, Chen W, Xu XY, Zhang J, Singh TK, Liu SM, et al (2020) Engineering trienoic fatty acids into cottonseed oil improves low-temperature seed germination, plant photosynthesis and cotton fiber quality. Plant Cell Physiol 61:1335–1347. https://doi.org/10.1093/pcp/pcaa062
Han W, Zhao J, Deng X et al (2022) Quantitative trait locus mapping and identification of candidate genes for resistance to Fusarium wilt race 7 using a resequencing-based high density genetic bin map in a recombinant inbred line population of *Gossypium barbadense*. Front Plant Sci 13:1–13. https://doi.org/10.3389/fpls.2022.815643
Harlan HV, Pope MN (1922) The use and value of back-crosses in small-grain breeding. J Hered 13(7):319–322
Harzli I, Mostafa K, Abdulla MF, Mehmood M (2025) Sustaining Turkish genetic resources. In: Al-Khayri JM, Salem KFM, Jain SM (eds) Plant gene banks. Springer, Singapore. https://doi.org/10.1007/978-981-99-4236-7_88-1
Hassan MN, Mekkawy SA, Mahdy M, Salem KFM et al (2021) Recent molecular and breeding strategies in lettuce (Lactuca spp.). Genet Resour Crop Evol 68:3055–3079. https://doi.org/10.1007/s10722-021-01246-w
Hidayatun N, Kurniawan H (2025) Conservation of agricultural genetic resources in Indonesia. In: Al-Khayri JM, Salem KFM, Jain SM (eds) Plant gene banks. Springer, Singapore. https://doi.org/10.1007/978-981-99-4236-7_40-1
Hu G, Grover CE, Yuan D et al (2021) Evolution and diversity of the cotton genome. In: Rahman M-U, Zafar Y, Zhang T (eds) Cotton precision breeding. Springer International Publishing, Cham, pp 25–78
IAEA (International Atomic Energy Agency), Vienna (2002) Mutant germplasm characterization using molecular markers. A manual. IAEA-TCS-19
Ibrahim AA, Salem KFM, Abdein MA, Ramadan SM (2023) Advances in summer squash (*Cucurbita pepo* L.) molecular breeding strategies. In: Singh S, Sharma D, Sharma SK, Singh R (eds) Smart plant breeding for vegetable crops in post-genomics era. Springer, Singapore. https://doi.org/10.1007/978-981-19-5367-5_8
Ibrahim AA, Abd-Ellatif S, Abdel Razik ESS, Salem KFM (2024) Biodiversity of cereal crops and utilization in food and nutritional security. In: Al-Khayri JM, Jain SM, Penna S (eds) Sustainable utilization and conservation of plant genetic diversity, Sustainable development and biodiversity, vol 35. Springer, Singapore. https://doi.org/10.1007/978-981-99-5245-8_2
ICAC (2023) Specialty cotton report. International Cotton Advisory Committee, New Your
Ihsan F, Prihatini R (2025) Indonesian tropical fruit gene bank. In: Al-Khayri JM, Salem KFM, Jain SM (eds) Plant gene banks. Springer, Singapore. https://doi.org/10.1007/978-981-99-4236-7_41-1
Iqbal SRM, Chaudhry MB, Aslam M, Bandesha AA (1994) Development of a high yielding cotton mutant, NIAB-92 through the use of induced mutations. Pak J Bot 26:99–104
Iqbal MU, Khan SH, Mubarik MS et al (2022) Cotton genetics and genomics. In: Khan AA, Khan AA (eds) Cotton breeding and biotechnology challenges and opportunities. Taylor & Francis Group, LLC, p 368

Ismail M, Salem KFM, El-Zanaty AM, Esmail RM (2013) Association mapping of fiber and agronomic characters in Egyptian cotton (*Gossypium barbadense* L.) genotypes. J Agric Chem Biotechnol 4:205–216

Jareczek JJ, Grover CE, Wendel JF (2023) Cotton fiber as a model for understanding shifts in cell development under domestication. Front Plant Sci 14:1146802. https://doi.org/10.3389/fpls.2023.1146802

Jatoi SA, Siddiqui SU (2025) National gene bank of Pakistan. In: Al-Khayri JM, Salem KFM, Jain SM (eds) Plant gene banks. Springer, Singapore. https://doi.org/10.1007/978-981-99-4236-7_51-1

Jiang GL (2013) Molecular markers and marker-assisted breeding in plants. In: Andersen SB (ed) Plant breeding from laboratories to fields, vol 22. IntechOpen, Rijeka, pp 45–83. https://doi.org/10.5772/52583

Kalivas A, Xanthopoulos F, Kehagia O, Tsaftaris AS (2011) Agronomic characterization, genetic diversity and association analysis of cotton cultivars using simple sequence repeat molecular markers. Genet Mol Res 10(1):208–217

Kangben F, Kumar S, Li Z et al (2024) Phylogenetic and functional analysis of tiller angle control homeologs in allotetraploid cotton. Front Plant Sci 14:1320638. https://doi.org/10.3389/fpls.2023.1320638

Katageri IS, Gowda SA, Prashanth BN et al (2020) Prospects for molecular breeding in cotton *Gossypium* spp. In: Abdurakhmonov IY (ed) Plant breeding current and future views. IntechOpen, London, p 350. https://doi.org/10.5772/intechopen.95004

Khan MKR, Liu F, Wang B et al (2022a) Breeding cotton for heat tolerance. In: Cotton breeding and biotechnology challenges and opportunities. CRC Press, pp 113–138

Khan Z, Ali Z, Khan AA (2022b) Molecular cotton breeding. In: Khan Z, Ali Z, Khan AA (eds) Cotton breeding and biotechnology challenges and opportunities. Taylor & Francis Group, LLC, p 368

Khan MKR, Hussien M, Anwar Z et al (2023) Future of mutation breeding in cotton: a step towards smart cotton. Cotton Innov 3(1):1–22

Knight R (1956) Blackarm disease of cotton and its control. In: The plant protection conference. Butterworth, London

Kumar NVM, Katageri IS, Gowda SA et al (2019) 63K SNP chip-based linkage mapping and QTL analysis for fiber quality and yield component traits in *Gossypium barbadense* L. cotton. Euphytica 215(6):1–16. https://doi.org/10.1007/s10681-018-2326-9

Kumar P, Cherala SR, Jay PS et al (2022) Synthesis of cotton fiber and its structure. In: Natural and synthetic fiber reinforced composites. Wiley, pp 17–36. https://doi.org/10.1002/9783527832996.ch2

Kumar N, Upadhyay G, Choudhary S et al (2023) Resource conserving mechanization technologies for dryland agriculture. In: Naorem A, Machiwal D (eds) Enhancing resilience of dryland agriculture under changing climate. Springer, Singapore. https://doi.org/10.1007/978-981-19-9159-2_33

Kumawat G, Kumawat CK, Chandra K et al (2020) Insights into marker assisted selection and its applications in plant breeding. In: Abdurakhmonov IY (ed) Plant breeding-current and future views. IntechOpen, London, p 350. https://doi.org/10.5772/intechopen.95004

Laghetti G, Piergiovanni AR, De Paola D, Direnzo P, Bucci G, Sonnante G (2025) Italy IBBR gene bank. In: Al-Khayri JM, Salem KFM, Jain SM (eds) Plant gene banks. Springer, Singapore. https://doi.org/10.1007/978-981-99-4236-7_73-1

Lee JJ, Woodward AW, Chen ZJ (2007) Gene expression changes and early events in cotton fibre development. Ann Bot 100:1391–1401. https://doi.org/10.1093/aob/mcm232

Li Y, Zhao SC, Ma JX, Li D, Yan L, Li J et al (2013) Molecular footprints of domestication and improvement in soybean revealed by whole genome resequencing. BMC Genomics 14(1):579. https://doi.org/10.1186/1471-2164-14-579

Li HW, Yang XM, Tang J et al (2015) Effects of HAb18G/CD147 knockout on hepatocellular carcinoma cells in vitro using a novel zinc-finger nuclease-targeted gene knockout approach. Cell Biochem Biophys 71(2):881–890

Li T, Ma X, Li N et al (2017) Genome-wide association study discovered candidate genes of Verticillium wilt resistance in upland cotton (*Gossypium hirsutum* L.). Plant Biotechnol J 15(12):1520–1532

Li J, Wang M, Li Y et al (2019) Multi-omics analyses reveal epigenomics basis for cotton somatic embryogenesis through successive regeneration acclimation process. Plant Biotechnol J 17(2): 435–450

Li W, Liu J, Zhang H et al (2022) Plant pan-genomics: recent advances, new challenges, and roads ahead. J Genet Genomics 49(9):833–846. https://doi.org/10.1016/j.jgg.2022.06.004

Liu X, Zhao B, Zheng HJ et al (2015) *Gossypium barbadense* genome sequence provides insight into the evolution of extra-long staple fiber and specialized metabolites. Sci Rep 5:14139. https://doi.org/10.1038/srep14139

Lowe HH (1927) Dosage: Pedigree and Performance. The Russell Meerdink Company Ltd

Majeed S, Ahmad RI, Muhammad AR et al (2019) Role of SNPs in determining QTLs for major traits in cotton. J Cotton Res 2–5:1–13. https://doi.org/10.1186/s42397-019-0022-5

Majeed S, Mubarik MS, Iqbal MU, Azhar MT (2020) Breeding cotton for value-added traits. In: Khan Z, Ali Z, Khan AA (eds) Cotton breeding and biotechnology challenges and opportunities. Taylor & Francis Group, LLC, p 368. https://doi.org/10.1201/9781003096856-18

Majumdar G, Singh SB, Shukla SK (2020) Seed production, harvesting, and ginning of cotton. In: Jabran K, Chauhan BS (eds) Cotton production, Wiley

Mallor C, Luna E (2025) The genetic diversity of the Spanish vegetable genebank BGHZ-CITA: conservation and utilization. In: Al-Khayri JM, Salem KFM, Jain SM (eds) Plant gene banks. Springer, Singapore. https://doi.org/10.1007/978-981-99-4236-7_109-1

Mattar MZ, Salem KFM, El-Abed AB (2016) Identification of candidate microsatellite markers associated with agronomic traits in rice (*Oryza sativa* L.). J Bot 15(5):213–232

Mehdi M, Raza MH, Azeem MT (2022) Cotton breeding and entrepreneurship challenges and opportunities. In: Khan Z, Ali Z, Khan AA (eds) Cotton breeding and biotechnology challenges and opportunities. Taylor & Francis Group, LLC, p 368. https://doi.org/10.1201/9781003096856-18

Mei M, Syed NH, Gao W et al (2004) Genetic mapping and QTL analysis of fiber-related traits in cotton (*Gossypium*). Theor Appl Genet 108(2):280–291. https://doi.org/10.1007/s00122-003-1433-7

Mendoza CP, Gómez MRT, Gonzalez QO et al (2016) Genetic resources of cotton in Mexico: ex situ and in situ conservation and use. Rev Mex Cienc Agríc 7(1):5–16

Meredith W Jr (1977) Backcross breeding to increase fiber strength of cotton. Crop Sci 17(1): 172–175. https://doi.org/10.2135/cropsci1977.0011183X001700010045x

Mishra PK, Sharma A, Prakash A (2023) Current research and development in cotton harvesters: A review with application to Indian cotton production systems. Heliyon. 9(5):e16124. https://doi.org/10.1016/j.heliyon.2023.e16124. PMID: 37234621; PMCID: PMC10208840

Monteros-Altamirano Á et al (2025) Plant genetic resources in Ecuador. In: Al-Khayri JM, Salem KFM, Jain SM (eds) Plant gene banks. Springer, Singapore. https://doi.org/10.1007/978-981-99-4236-7_99-1

Mourad AMI, Alomari DZ, Alqudah AM et al (2019) Recent advances in wheat (*Triticum* spp.) breeding. In: Al-Khayri J, Jain S, Johnson D (eds) Advances in plant breeding strategies: Cereals. Springer, Cham. https://doi.org/10.1007/978-3-030-23108-8_15

Mubarik MS, Wang X, Khan SH et al (2021) Engineering broad-spectrum resistance to cotton leaf curl disease by CRISPR-Cas9 based multiplex editing in plants. GM Crops Food 12(2):647–658

Muravenko OV, Fedotov AR, Punina EO et al (1998) Comparison of chromosome BrdU–Hoechst–Giemsa banding patterns of the A1 and (AD)2 genomes of cotton. Genome 41:616–625. https://doi.org/10.1139/g98-049

Muthusamy A, Jayabalan N (2011) In vitro induction of mutation in cotton (*Gossypium hirsutum* L.) and isolation of mutants with improved yield and fiber characters. Acta Physiol Plant 33(5): 1793–1801

Muthusamy A, Jayabalan N (2014) Radiation and chemical mutagen induced somaclonal variations through in vitro organogenesis of cotton (Gossypium hirsutum L.). Int J Radiat Biol 90(12): 1229–1239

Nadeem MA, Nawaz MA, Shahid MQ et al (2018) DNA molecular markers in plant breeding: current status and recent advancements in genomic selection and genome editing. Biotechnol Biotechnol Equip 32(2):261–285

Nawara HM, Mattar MZ, Salem KFM, Eissa OA (2016) Estimating heterosis and combining ability for agronomic and tissue culture traits using diallel cross on bread wheat (*Triticum aestivum* L.). Egypt J Bot 2016:177–200

Nawara HM, Mattar MZ, Salem KFM, Eissa OA (2017) Diallel study on some in vitro traits of bread wheat (*Triticum aestivum* L.) under salt stress. Int J Agric Environ Res 3(1):1988–2006

Nazeer W, Tipu AL, Ahmad S et al (2014) Evaluation of cotton leaf curl virus resistance in BC_1, BC_2, and BC_3 progenies from an interspecific cross between *Gossypium arboreum* and *Gossypium hirsutum*. PLoS One 9(11):e111861

Nijagum H, Khadi B (2001) Progeny analysis of fiber characteristics of DCH 32-an interspecific cotton hybrid (*Gossypium hirsutum* L.-*Gossypium barbadense* L.-India). J Genet Breed 55(3): 209–216

Niu H, Ge Q, Shang H, Yuan Y (2022) Inheritance, QTLs, and candidate genes of lint percentage in upland cotton. Front Genet 13:855574

Osterc G, Gostinčar K, Hudina M, Usenik V, Luthar Z (2025) Slovene fruit gene bank. In: Al-Khayri JM, Salem KFM, Jain SM (eds) Plant gene banks. Springer, Singapore. https://doi.org/10.1007/978-981-99-4236-7_84-1

Paterson AH, Wendel JF, Gundlach H (2012) Repeated polyploidization of *Gossypium* genomes and the evolution of spinnable cotton fibres. Nature 492(7429):423–427

Patil R, Sivaramand A, Patil N (2022) Gene isolation methods: beginner's guide. In: A complete guide to gene cloning: from basic to advanced. Springer, pp 43–55

Percival AE, Stewart JM, Wendel JF (1999) Taxonomy and germplasm resources. In: Smith CW, Cothren JT (eds) Cotton: origin, history, technology and production. Wiley, New York, pp 33–63

Percy RG, Wendel JF (1990) Allozyme evidence for origin and diversification of *Gossypium barbadense* L. Theor Appl Genet 79:529–542

Percy RG, Ulloa M, Zhang J (2009) Registration of PSI 113 and PSI 425 germplasm lines of Pima cotton possessing superior fiber quality traits. J Plant Regist 3(3):297–299. https://doi.org/10.3198/jpr2009.02.0060crg

Platani M, Sokefun O, Bassil E, Apidianakis Y (2022) Genetic engineering and genome editing in plants, animals and humans: facts and myths. Gene 856:147141. https://doi.org/10.1016/j.gene.2022.147141

Popović V et al (2025) Economic seed production and storage of fiber crops in gene banks in Serbia. In: Al-Khayri JM, Salem KFM, Jain SM (eds) Plant gene banks. Springer, Singapore. https://doi.org/10.1007/978-981-99-4236-7_108-1

Prasad R, Blaise D (2020) Low gossypol containing cottonseed: not only a fibre but also a food crop. Natl Acad Sci Lett 43:599–602. https://doi.org/10.1007/s40009-020-00931-1

Price HJ, Smith RH (1979) Somatic embryogenesis in suspension cultures of Gossypium klotzschianum Anderss. - Planta 145:305–307

Rauf S, Shehzad M, Al-Khayri JM et al (2019) Cotton (*Gossypium hirsutum* L.) breeding strategies. In: Al-Khayri JM, Jain SM, Johnson DV (eds) Advances in plant breeding strategies: industrial and food crops. Springer, pp 29–59. https://doi.org/10.1007/978-3-030-23265-8_2

Rauf S, Shehzad M, Al-Khayri JM (2020) Cotton (*Gossypium hirsutum* L.) breeding strategies. In: Al-Khayri JM, Jain SM, Johnson DV (eds) Advances in plant breeding strategies: industrial and food crops, vol 6. Springer. https://doi.org/10.1007/978-3-030-23265-8

Razzaq A, Zafar MM, Ali A et al (2021) Cotton germplasm improvement and progress in Pakistan. J Cotton Res 4:1. https://doi.org/10.1186/s42397-020-00077-x

Reinisch AJ, Dong JM, Brubaker CL et al (1994) A detailed RFLP map of cotton *Gossypium hirsutum* × *Gossypium barbadense*: chromosome organization and evolution in a disomic polyploidy genome. Genetics 138:829–847

Ribaut JM, Hoisington D (1998) Marker assisted selection: new tools and strategies. Trends Plant Sci 3(6):236–239

Robertson B, Bednarz C, Burmester C (2007) Growth and development first 60 days. Cotton Physiol Today 13(2):1–5

Rocha F et al (2025) Portugal gene bank. In: Al-Khayri JM, Salem KFM, Jain SM (eds) Plant gene banks. Springer, Singapore. https://doi.org/10.1007/978-981-99-4236-7_79-1

Rojo-Gutiérrez E, Buenrostro-Figueroa JJ, López-Martínez LX et al (2020) Biotechnological potential of cottonseed, a by-product of cotton production. In: Zakaria Z, Aguilar C, Kusumaningtyas R, Binod P (eds) Valorisation of agro-industrial residues, Volume II: Non-biological approaches, applied environmental science and engineering for a sustainable future. Springer, Cham, pp 63–82. https://doi.org/10.1007/978-3-030-39208-6_3

Saleh MM, Salem KFM, Elabd AB (2020) Definition of selection criterion using correlation and path coefficient analysis in rice (*Oryza sativa* L.) genotypes. Bull Natl Res Cent 44(1):1–6

Salem KFM (2004) The inheritance and molecular mapping of genes for post-anthesis drought tolerance (PADT) in wheat. Martin-Luther-University, Halle-Wittenberg, 124 pp

Salem KFM (2009) Relationship between genetic diversity based on SSRs markers with heterosis and combining ability in diallel cross of bread wheat (*Triticum aestivum* L.). Minufiya J Agric Res 6(34):2159–2178

Salem KFM (2015) Allelic state at the microsatellite locus *Xgwm261* marking the dwarfing gene *Rht8* in Egyptian bread wheat (*Triticum aestivum* L.) genotypes released from 1947 to 2004. Genetika 47:741–750

Salem KFM, El-Zanaty AM (2015) Assessing genetic diversity of old and modern Egyptian rice (Oryza sativa L.) using microsatellite markers. Minufiya J Agric Res 40(6):1507–1518

Salem KFM, Ibrahim AA (2024) Plant biodiversity in the context of food security under climate change. In: Al-Khayri JM, Jain SM, Penna S (eds) Sustainable utilization and conservation of plant genetic diversity, Sustainable development and biodiversity, vol 35. Springer, Singapore. https://doi.org/10.1007/978-981-99-5245-8_1

Salem KFM, Mattar MZ (2014a) Genetic diversity in old and modern Egyptian bread wheat (*Triticum aestivum* L.) varieties revealed by simple sequence repeats. Egypt J Genet Cytol 43: 143–156

Salem KFM, Mattar MZ (2014b) Identification of microsatellite alleles for salt tolerance at seedling stage in wheat (*Triticum aestivum* L.). Life Sci J 11(12s):1064–1073

Salem KFM, Sallam A (2016) Analysis of population structure and genetic diversity of Egyptian and exotic rice (*Oryza sativa* L.) genotypes using SSR markers. C R Biol 339:1–9. https://doi.org/10.1016/j.crvi.2015.11.003

Salem KFM, Röder MS, Börner A (2003) Mapping of quantitative trait loci (QTLs) for post-anthesis drought tolerance in bread wheat (Triticum aestivum L.). In: Gesellschaft für Pflanzenzüchtung, GPZ-Tagung, 11. Vortragsveranstaltung "Harnessing genetic diversity: genomics and allele mining", 16–17 September 2003, Institut für Pflanzengenetik und Kulturpflanzenforschung (IPK), Gatersleben, Germany

Salem KFM, Röder MS, Börner A (2004a) Drought tolerance in wheat germplasm – evaluation and molecular mapping. In: Institut für Pflanzengenetik und Kulturpflanzenforschung, IPK Tagung, 22–24/08/2004, Ag Ressourcengenetik und Reproduktion, Abteilung Genbank, Gatersleben, Germany

Salem KFM, Röder MS, Börner A (2004b) Molecular mapping of quantitative trait loci (QTLs) determining post-anthesis drought tolerance in hexaploid wheat (Triticum aestivum L.). In: 7 Gesellschaft für Pflanzenzüchtung, GPZ-Tagung, Vortragsveranstaltung zum Thema: "Klimatische und edaphische Sortenanpassung und Züchtung für Nachwachsende Rohstoffe", 3–5 March 2004, Halle/Saale, Germany. Vort Pflanzenzüchtung 64:21–24

Salem KFM, Röder MS, Börner A (2007a) Evaluation of some barley varieties for the presence of thermostable alleles of *ßamylase*. In: 8th African Crop Science Society conference, El-Minia, Egypt, 27–31 October 2007, pp 643–648
Salem KFM, Röder MS, Börner A (2007b) Identification and mapping quantitative trait loci for stem reserve mobilisation in wheat (*Triticum aestivum* L.). Cereal Res Commun 35:1367–1374
Salem KFM, Varshney RK, Röder MS, Börner A (2008a) Detection of genetic diversity in Egyptian barley genotypes using barley expressed sequence Tag-SSR markers. In: 10th International Barley Genetics Symposium (IBGS), Alexandria, Egypt, 5–10 April 2008. The International Center for Agricultural Research in the Dry Areas (ICARDA) and the Bibliotheca Alexandrina, Alexandria
Salem KFM, El-Zanaty AM, Esmail RM (2008b) Assessing wheat (*Triticum aestivum* L.) genetic diversity using morphological characters and microsatellite markers. World J Agric Sci 4:538–544
Salem KFM, Varshney RK, Röder MS, Börner A (2010) EST-SSR based estimates on functional genetic variation in a barley (*Hordeum vulgare* L.) collection from Egypt. Genet Resour Crop Evol 57(4):515–521
Salem KFM, Röder MS, Börner A (2015a) Assessing genetic diversity of Egyptian hexaploid wheat (*Triticum aestivum* L.) using microsatellite markers. Genet Resour Crop Evol 62:377–385
Salem KFM, Yaseen YM, El-Abd AB, Ayoub A (2015b) Inheritance of some yield components and grain quality traits in rice (*Oryza sativa* L.). J Prod Dev 20:161–177
Salem KFM, Yaseen YM, El-Abd AB, Ayoub A (2015c) Breeding studies on some agronomic and quality traits in rice (Oryza sativa L.). J Prod Dev 20(2):211–226
Salem KFM, Elabsawy A, Aldahak L et al (2021a) Evaluation of stem reserve mobilization in Egyptian bread wheat (Triticum aestivum L.) genotypes and F1 hybrids under post-anthesis chemical desiccation stress. J Genet Environ Resour Conserv 9(1):176–182
Salem KFM, El-Absawy EA, Khidr YA, El-Shamy H (2021b) Assessment genetic diversity among some Egyptian bread wheat (*Triticum aestivum* L.) cultivars using RAPD markers. J Genet Environ Resour Conserv 9(1):168–175
Salem KFM, Saleh MM, Aldahak L, Elabd AB (2021c) Assessment phenotypic diversity of rice (*Oryza sativa* L.) genotypes by multivariate analysis. J Arid Agric 7:52–59. https://doi.org/10.25081/jaa.2021.v7.6516
Salem KFM, Alghuthaymi MA, Elabd AB et al (2022) Prediction of heterosis for agronomic traits in half-diallel cross of rice (*Oryza sativa* L.) under drought stress using microsatellite markers. Plants 11:1532. https://doi.org/10.3390/plants11121532
Salem KFM, Tawfik E, Mahdy M et al (2023) Advances in lettuce (*Lactuca* spp.) molecular breeding strategies. In: Singh S, Sharma D, Sharma SK, Singh R (eds) Smart plant breeding for vegetable crops in post-genomics era. Springer, Singapore. https://doi.org/10.1007/978-981-19-5367-5_11
Salem KFM, Safhi FA, Alwutayd KM et al (2024) Analysis of genetic diversity, population structure and phylogenetic relationships of rice (*Oryza sativa* L.) cultivars using simple sequence repeat (SSR) markers. Genet Resour Crop Evol 71:2213–2227. https://doi.org/10.1007/s10722-023-01789-0
Sakhanokho HF, Rajasekaran K (2016) Cotton Regeneration In Vitro. K.G. Ramawat and M.R. Ahuja (eds.), Fiber Plants, Sustainable Development and Biodiversity 13, https://doi.org/10.1007/978-3-319-44570-0_6. © Springer International Publishing Switzerland
Scarpin GJ, Dileo PN, Winkler HM, Cereijo AE et al (2022) Genetic progress in cotton lint and yield components in Argentina. Field Crops Res 275:108322. https://doi.org/10.1016/j.fcr.2021.108322
Shahzad K, Mubeen I, Zhang M et al (2022) Progress and perspective on cotton breeding in Pakistan. J Cotton Res 5(29):1–17. https://doi.org/10.1186/s42397-022-00137-4
Singh P (2007) Breeding hybrids cotton, 2nd edn. Kalyani Publishers, New Delhi
Singh DPA, Singh K, Singh A (2021) Plant breeding and cultivar development. Elsevier Science Publishing Co Inc. https://doi.org/10.1016/C2018-0-01730-2

Singh P (2013) Cotton Genetics. Kalyani Publishers. New Delhi, India
Su J, Wang C, Ma Q et al (2020) An RTMGWAS procedure reveals the QTL alleles and candidate genes for three yield-related traits in upland cotton. BMC Plant Biol 20(1):1–15
Sugihara Y, Young L, Yaegashi H et al (2022) High-performance pipeline for MutMap and QTL-seq. PeerJ 10:e13170
Sun Q, Xie Y, Li H et al (2022) Cotton GhBRC1 regulates branching, flowering, and growth by integrating multiple hormone pathways. Crop J 10(1):75–87. https://doi.org/10.1016/j.cj.2021.01.007
Tcach MA, Landau AM, Montenegro A et al (2022) Isolation and characterization of a new imidazolinone-tolerant mutant in cotton. Crop Sci 62:2222–2232. https://doi.org/10.1002/CSC2.20814
Tester M, Langridge P (2010) Breeding technologies to increase crop production in a changing world. Science 327(5967):818–822
Ulloa MR, Percy R, Zhang J et al (2009) Registration of four Pima cotton germplasm lines having good levels of Fusarium wilt race 4 resistance with moderate yields and good fibers. J Plant Regist 3:198–202. https://doi.org/10.3198/jpr2008.09.0530crg
Ur-Rahman M, Rahmat Z, Mahmood A et al (2014) Cotton germplasm of Pakistan. In: Abdurakhmonov IY (ed) World cotton germplasm resources. IntechOpen, London, pp 137–166. https://doi.org/10.5772/58620
Varshney RK, Salem KFM, Röder MS et al (2008) Assessment and comparison of genetic diversity in barley germplasm collection using gSSR, eSSR and SNP markers. Plant Genet Res 6:167–174
Waheed U, Shabir S, Mazhar F et al (2022) Cotton transformation and regeneration. In: Khan Z, Ali Z, Khan AA (eds) Cotton breeding and biotechnology challenges and opportunities. Taylor & Francis Group, LLC, p 368. https://doi.org/10.1201/9781003096856-7
Wang X, Yu Y, Sang J et al (2013) Intraspecific linkage map construction and QTL mapping of yield and fiber quality of *Gossypium barbadense*. Aust J Crop Sci 7(9):1252–1261
Wang S, Chen J, Zhang W et al (2015) Sequence-based ultra-dense genetic and physical maps reveal structural variations of allopolyploid cotton genomes. BMC Genome Biol 16(1):108. https://doi.org/10.1186/s13059-015-0678-1
Wang Y, Meng Z, Liang C et al (2017) Increased lateral root formation by CRISPR/Cas9-mediated editing of arginase genes in cotton. Sci China Life Sci 60(5):524–527
Wang P, Zhang J, Sun L et al (2018) Highly efficient multisets genome editing in allotetraploid cotton (*Gossypium hirsutum*) using CRISPR/Cas9 system. Plant Biotechnol J 16(1):137–150
Wang M, Tu L, Yuan D et al (2019) Reference genome sequences of two cultivated allotetraploid cottons, *Gossypium hirsutum* and *Gossypium barbadense*. Nat Genet 51:224–229. https://doi.org/10.1038/s41588-018-0282-x
Wang X, Zhang X, Fan D et al (2022) AAQSP increases mapping resolution of stable QTLs through applying NGS-BSA in multiple genetic backgrounds. Theor Appl Genet 135(9):3223–3235
Wang J et al (2025) China national gene bank (CNGB). In: Al-Khayri JM, Salem KFM, Jain SM (eds) Plant gene banks. Springer, Singapore. https://doi.org/10.1007/978-981-99-4236-7_103-1
Winkler HM, Cereijo AE, Scarpin GJ et al (2023) Phenotypic effects of different doses of physical and chemical mutagens in cotton plants. Rev Investig Agropecu 49:71–84
Xiao J, Yin X (2020) Nutrient management in cotton. In: Jabran K, Chauhan BS (eds) Cotton production. Wiley
Yang Z, Wang J, Huang Y et al (2023) Cotton MD: a multi-omics database for cotton biological study. Nucleic Acids Res 51:1446–1456. https://doi.org/10.1093/nar/gkac863
Yasar M (2023) Yield and fiber quality traits of cotton (*Gossypium hirsutum* L.) cultivars analyzed by biplot method. J King Saud Univ Sci 35:1–10
Zhang HA (2023) A breakthrough in cotton transformation technology. J Cotton Res 6:8. https://doi.org/10.1186/s42397-023-00147-w

Zhang BH, Feng R, Liu F, Wang Q (2001) High frequency somatic embryogenesis and plant regeneration of an elite Chinese cotton variety. Bot Bull Acad Sin 42:9–16

Zhang T, Hu Y, Jiang W et al (2015) Sequencing of allotetraploid cotton (*Gossypium hirsutum* L. acc. TM-1) provides a resource for fiber improvement. Nat Biotechnol 33:531–537. https://doi.org/10.1038/nbt.3207

Zhang BJ, Zhang HP, Chen QZ et al (2016) Molecular cloning and analysis of a receptor-like promoter of *Gbvdr3* gene in sea island cotton. Genet Mol Res 15(2):15028636

Zhang T, Xuan L, Mao Y, Hu Y (2023) Cotton heterosis and hybrid cultivar development. Theor Appl Genet 136(4):89. https://doi.org/10.1007/s00122-023-04334

Zhao Z, Liu Z, Zhou Y et al (2022) Creation of cotton mutant library based on linear electron accelerator radiation mutation. Biochem Biophys Rep 30:101228. https://doi.org/10.1016/j.bbrep.2022.101228

Zhou GY, Weng J, Zeng Y et al (1983) Introduction of exogenous DNA into cotton embryos. In: Methods in enzymology, vol 101. Academic Press, pp 433–481

Zhu S, Yu X, Li Y et al (2018) Highly efficient targeted gene editing in upland cotton using the CRISPR/Cas9 system. Int J Mol Sci 19(10):3000

Chapter 21
Kapok (*Ceiba pentandra* L.) Breeding

Essam Fathy El-Hashash and Karima Mohamed El-Absy

Abstract *Ceiba pentandra* (L.) Gaertn. (*C. pentandra*), also known as kapok tree, silk cotton tree or Java cotton, belongs to is a member of the Malvaceae (Bombacoideae) family and the Ceiba genus. It is found in Mexico, the Caribbean, India, Thailand, Indonesia, and South and Central America. The two nations that produce the most kapok worldwide are Indonesia and Thailand. The fiber generated by tree fruits and its distinct fiber structure are the main reasons for the *C. pentandra* fame. Kapok fiber is a naturally occurring plant fiber (cellulose fiber) that is abundant, biocompatible, and biodegradable. It's interesting to note that this plant is utilized for food and fodder, as well as that its parts have therapeutic properties in addition to having a particular fiber, wood, and oils with various applications. Although *C. pentandra* populations may face threats from climate change throughout their existing areas of distribution, there is still potential for the in situ persistence of the most genetically diverse and unique populations. A summary of current knowledge and understanding about *C. pentandra* is given in this chapter. Many topics have been covered including taxonomy, economic importance, climate change, domestication, selection and early improvements, current cultivation practices and challenges, germplasm biodiversity and conservation, breeding strategies, and conclusions and prospects. Because of its important economic benefits, the kapok tree needs more studies on traditional and biotechnology breeding methods as well as the applications of artificial intelligence to the management of agricultural and industrial practices of the kapok tree.

E. F. El-Hashash (✉)
Department of Agronomy, Faculty of Agriculture, Al-Azhar University, Cairo, Egypt
e-mail: dressamelhashash@azhar.edu.eg

K. M. El-Absy (✉)
Biology Department, University College of Tayma, University of Tabuk, Tabuk, Saudi Arabia

Eco-physiology Unit, Plant Ecology and Ranges Department, Desert Research Center, Cairo, Egypt
e-mail: k.alabssi@ut.edu.sa

K. F. M. Salem et al. (eds.), *Breeding and Biotechnology of Leaf, Fruit, and Seed Fiber Crops*, Advances in Plant Breeding Strategies 11,
https://doi.org/10.1007/978-3-032-00407-9_21

Keywords Biodiversity · Biotechnology · Breeding techniques · Climate change · Economic importance · Germplasm · Kapok (*Ceiba pentandra* L.)

21.1 Introduction

Notably, the Malvaceae family includes three of the primary fiber-producing plants: cotton, jute, and kapok (Gómez-Maqueo and Gamboa-deBuen 2022). Within the family Malvaceae, the Bombacoideae comprises trees that are exceptionally significant ecologically in tropical regions, which include roughly 160 species and 17 genera, with 90% of the species being found in the Neotropics (Carvalho-Sobrinho et al. 2016). Among the 17 genera included in the Malvaceae (Bombacoideae) family, the Ceiba genus is considered the most important economically. The genus Ceiba has eighteen species in the Neotropics within the taxonomic sections Ceiba and Campylanthera (Pezzini et al. 2021). *Ceiba pentandra* (L.) Gaertn. (*C. pentandra*) is the tallest and most widely distributed of these species, where it holds the most prominent position in Neotropical cosmovisions (Tareau et al. 2022). According to Gómez-Maqueo and Gamboa-deBuen (2022), the term "kapok" is commonly used to refer to a type of "seed fiber" (such as cotton) that is primarily produced by two plant species known as "kapok trees": *C. pentandra* (Malvaceae, also known as "silk-cotton" or "Java cotton") and *Bombax ceiba* L. (Malvaceae Juss., also known as "red cotton tree" or "red silk-cotton").

Large and hermaphrodite, the kapok tree (*C. pentandra*) is indigenous to North West Africa, the Caribbean, Central and South America, and the Caribbean (Hending et al. 2021). The tree is widely distributed across the tropics and its seeds likely made their way to Africa by floating over the Atlantic Ocean, thanks to its multiple commercial uses that have brought it to several nations in Africa and Asia (Dick et al. 2007). While the other species are more common in seasonally dry woods, *C. pentandra* is primarily found in rainforests (Pezzini et al. 2021). Similar to numerous other neophytes, *C. pentandra* grows quickly and thrives in arid environments with subpar soil (Immanuel and Ganapathy 2020).

The kapok tree is known for growing quickly and having a very unique appearance (Fig. 21.1). *C. pentandra* is a large and tall tree, that often rises to up to 50 meters or even more (reach to 70 meters) above the ground and develops massive buttresses covered in conical prickles. With five to eight leaflets and either a serrated or plain border, the leaves are digitate composite or palmate. Aculeate trunks and branches with spines that are characteristic. The flowers are bisexual and arranged in sets of three to eight. The actinomorphic, pentamerous flowers have short (2–3 cm) leaves that are glabrous and a corolla composed of five white or pink petals. The size and color of corolla vary greatly; they can be brightly colored like pink or red, or they might be pale like white, ivory, yellow, or light pink. Occasionally, the base of the corolla will have yellow tones. There are instances where the petals exhibit striations of a black hue. The black seeds emerge in fruits (elongated woody capsules), 12-to 24-cm-long fruits, which have modified endocarps that form long, tubular trichomes that make up the fiber of kapok. The long, woolly down (fluff),

Fig. 21.1 Morphological description of the kapok tree. (Source: Nkouam et al. (2017))

non-wettable hairs known as kapok envelop seeds. It originates from the pericarp's inner layer. This raw material is fluffy and light, and it doesn't rot or absorb water. 500–4000 seedpods are produced by a kapok tree. This fiber aids in the wind-borne dissemination of seeds. Many species of bats, butterflies, bees, and hummingbirds pollinate Ceiba species because of the variety of their flower morphology (Baraniak and Kania-Dobrowolska 2023; Dick et al. 2007; Gibbs and Semir 2022; Gómez-Maqueo and Gamboa-deBuen 2022; Mori et al. 2002; Rijavec 2008; Tareau et al. 2022). About 1 to 2 centimeters in length, the fiber of kapok is a light fiber with a hollow tube structure. Cotton fibers have a mean external diameter of 16.8 μm and a wall thickness of 3.9 μm, whereas kapok fibers are composed of microtubes with a mean external diameter of about 10 μm and a wall thickness of 0.1 μm (Graham and Haigler 2021).

For its floss, or fiber, kapok is grown. Kapok fiber's technical and economic benefits are apparent (Tareau et al. 2022). It's interesting to note that the plant is consumed as food, a source of wood and that some of its parts have therapeutic properties in addition to being a particular fiber with several applications (Baraniak and Kania-Dobrowolska 2023). In addition to being used in textiles, and biotechnological industries, as an exchange for cotton-seed oil, and as a filler for pillows and

Table 21.1 Statistics data of Kapok fiber, raw, Kapok fruit, and Kapokseed in the shell in the world, as average from 2010 to 2020

Years	Kapok fiber, raw (production t)	Kapok fruit		Kapokseed in shell (production t)
		Area harvested (ha)	Production (t)	
2010	76,956	197,587	268927.08	174212.2
2011	93,919	194,259	324623.82	209468.6
2012	92,455	186,934	319090.39	205789.6
2013	89,663	179,902	309337.03	199474.3
2014	96,000	177,069	293018.1	197,606
2015	95,790	190,836	269862.18	181,011
2016	93,240	182,603	290739.1	171,493
2017	91,670	181,707	294846.91	165,766
2018	90,520	185,049	283873.47	160,186
2019	78,657	183,120	282232.55	187502.59
2020	78,674	183,292	281416.96	187536.71

Source: FAOSTAT (2023)
t tons, *ha* hectares

cushions, it also plays an importance role in the ecosystems in which it is found, particularly in conservation and restoration efforts aimed at promoting sustainable management of agro-forest systems (Gómez-Maqueo and Gamboa-deBuen 2022).

Production of kapok is exclusively restricted to the Asia-Pacific region, namely Indonesia and Thailand. According to FAOSTAT (2023), the yearly global kapok fruit harvested area, yield, and production in 2021 were approximately 183,820 (ha), 15,285 (100 g/ha), and 280970.94 (t), respectively. World production of kapok fiber, raw, kapok fruit, and kapok seed in shell fluctuated significantly in recent years through the 2010–2020 period (Table 21.1). During this period, production is observed to rise, then gradually decline, and then climb again until declining in 2021, all leading up to the current period. Indonesia and Thailand are the top two countries that account for 100.00% of kapok production in the world. Indonesia is the top country in kapok production in the world, followed by Thailand. Based on average 1994–2021, kapok production in Indonesia accounts for 66% of the world's, followed by Indonesia with 34% (Fig. 21.2), according to FAOSTAT (2023).

This chapter provides a summary of the existing information and understanding of *C. pentandra*. Numerous subjects have been addressed, including taxonomy, economic significance, climatic change, domestication, selection, and first enhancements, contemporary cultivation techniques and problems, germplasm biodiversity and conservation, breeding methodologies, and conclusions and future possibilities. The kapok tree requires further research on conventional and biotechnological breeding techniques, as well as the application of artificial intelligence in the administration of its agricultural and industrial activities, due to its significant economic advantages.

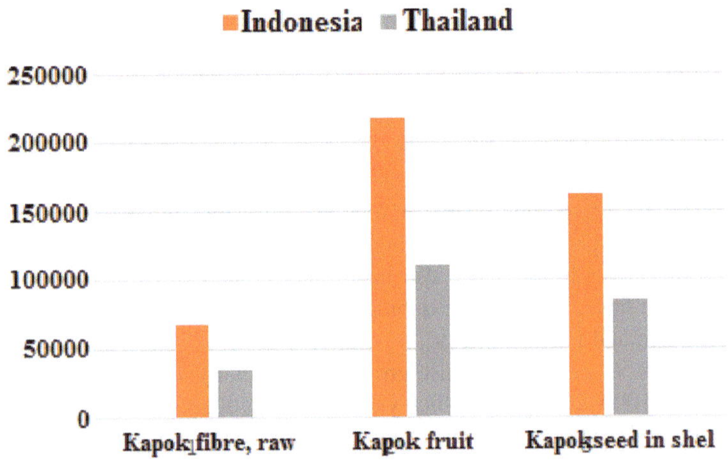

Fig. 21.2 Top two producers for production of kapok fiber, raw, kapok fruit, and kapok seed in shel in the world (Average 1994–2021). (Source: FAOSTAT (2023))

Fig. 21.3 Types of natural fibers found in nature. (Prepared by: E.F. El-Hashash and K.M. El-Absy)

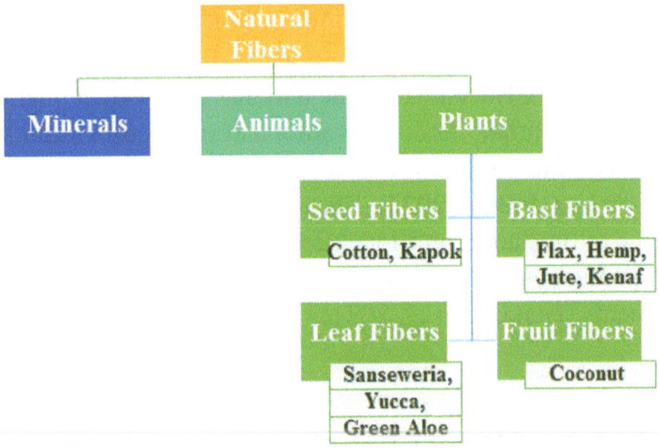

21.1.1 Economic Importance

There are three types of natural fibers found in nature: those derived from plants, animals, and minerals (Fig. 21.3). Kapok fiber is a typical cellulosic fiber (including 69% cellulose and nanocellulose). Kapok fiber is affordable, readily biodegradable, and has a pale yellow to light brown tint. The characteristics of kapok fiber include a thin cell wall, a big lumen, softness, lightness (more than eight times lighter than cotton), low density, extreme moisture resistance, and hydrophobic-oleophilic qualities. Fiber of kapok is a naturally occurring plant fiber that is abundant, biocompatible, and biodegradable. It has garnered raising attention in both academic and industrial regions for its full exploration and potential applications (Baraniak and Kania-Dobrowolska 2023; Sartika et al. 2021; Zheng et al. 2015). It has been discovered that this fiber contains antibacterial and anti-mite qualities in addition to being biodegradable and environmentally beneficial. The kapok has been utilized in the paper, textile, aviation, and upholstery industries because of the advantageous qualities of its fibers (Baraniak and Kania-Dobrowolska 2023). According to reports from many parts of the world, the most popular application for the fiber generated by any of the Ceiba species is filling for pillows and cushions (Gómez-Maqueo and Gamboa-deBuen 2022). Nonetheless, because of their hollow structure, clumps of kapok fibers have important characteristics like porosity and superhydrophobicity, which make them perfect for life-saving supplies because of their increased buoyancy and maneuverability. They also have other qualities that artificial buoyancy materials don't have, like biodegradability, acid/alkali resistance, and natural abundance (Graham and Haigler 2021).

For its seeds, *C. pentandra* is also grown commercially. Every year, every kapok tree produces 1000–2000 pods, which yield 15–25 kg of seeds. The oil derived from the kapok seeds is extracted and used to make soap, replacing cottonseed oil (Gómez-Maqueo and Gamboa-deBuen 2022). In addition, it is utilized in food preparations (young shoots, oil, leaves, and fruit). Its wood is used by people to build houses and canoes (Hending et al. 2021). The seeds, bark, and roots are also conventionally utilized as a diuretic, painkiller, or to cure wounds, and to treat a variety of ailments, like skin infections, kidney problems, gastritis, and blood sugar reduction (Avendaño et al. 2006). Additionally, according to Gómez-Maqueo and Gamboa-deBuen (2022), the bark has antioxidant qualities. Numerous Ceiba species are appropriate for conservation and restoration initiatives since it has been demonstrated that they are significant members of their ecosystems (Gómez-Maqueo and Gamboa-deBuen 2022). Because *C. pentandra* grows quickly, it could be a good fit for reforestation and forest restoration projects, as it promotes seed dissemination, spontaneous forest regeneration, and animal migration across nearby forest fragments (Hending et al. 2021).

21.1.2 Fiber of Kapok

The fiber of kapok, a naturally occurring grain derived from kapok fruits, is renowned for its high hollowness and low density, which enable it to effectively absorb and retain oil (Abdullah et al. 2010). Single cells grouped in a narrow hollow tube shape with a thin wall, a large lumen (wide inner channel filled with air), and a thick covering of wax that gives the fibers a smooth surface make up kapok fibers, as shown in Fig. 21.4 (Baraniak and Kania-Dobrowolska 2023; Xiang et al. 2013; Zheng et al. 2015). Natural vegetable waxes that are hydrophobic and oleophilic coat the surface of kapok fiber. Van der Waals forces and hydrophobic interactions between the oil and the waxy surface allow oil to be absorbed by the kapok fiber (Chen et al. 2022). Each pod contains 12–15 g of fiber, with each strand measuring 10–35 mm in length (Smole et al. 2013). The fiber diameter is between 15 and 35 μm, and the wall thickness is approximately 1–2 μm. Approximately 80–90% of the fiber surface is taken up by the channel (Liua et al. 2022; Xiang et al. 2013). Ito

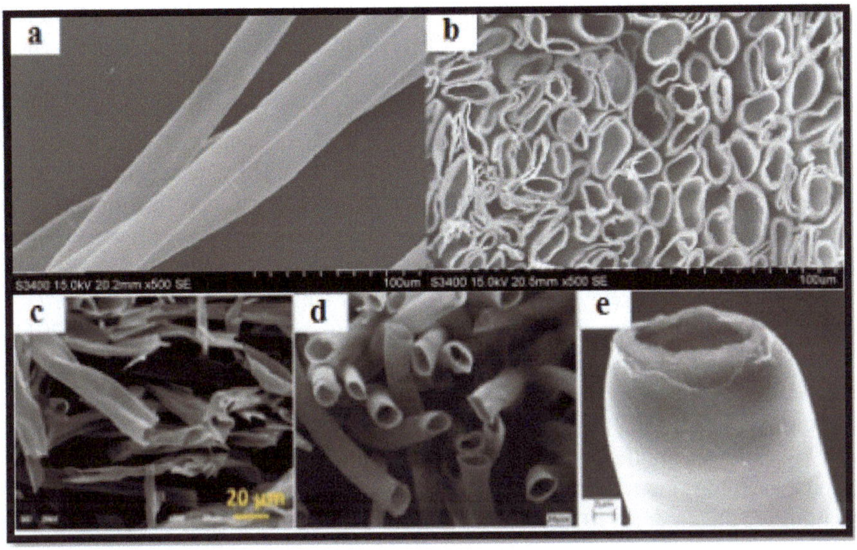

Fig. 21.4 A microphotograph was taken with a scanning electron microscope: (**a**) longitudinal and (**b**) cross-sectional view of kapok fibers with magnification (Baraniak and Kania-Dobrowolska 2023), (**c**) fiber body (Tang et al. 2021), (**d**) hollow structure, and (**e**) tail of kapok fiber (Nyankson et al. 2015)

et al. (2020) reported that the kapok fiber has a surface roughness ranging from 140 to 155 nm.

Regarding the chemical composition, Kapok fiber's chemical makeup includes a high cellulose percentage and significant acetylation. As part of the natural fiber family, the fiber of kapok is mostly made up of cellulose, lignin, pectin, and trace amounts of inorganic substances (Futalan et al. 2022). The locale, the age of the fruit used, and the preparation technique all affect how much cellulose is present (Chaiarrekij et al. 2012; Zheng et al. 2015). The chemical composition of kapok fiber is different according to multiple sources. Approximately 22% lignin, 22% hemicellulose, and 35% cellulose make up kapok fiber (Yıldızhan et al. 2018). According to another study, the biological composition of kapok fiber was found to be 64% cellulose, 23% pentosan, and 13% lignin by mass (Zerga and Tahir 2022). Zheng et al. (2015) found that, in contrast to conventional fiber, kapok fiber had a higher ratio of guaiacyl/syringyl units (6–4) and a significant number of acetyl groups (14.0%). It was also constituted of 21.5% lignin, 22% xylan, and 35% cellulose. Variations in the origins and preparation methods of kapok may be the cause of the differences. Nonetheless, because of its enormous lumen, kapok fiber's specific refraction is 0.017, bulk density is 0.30 g/cm3, and crystallization level has been calculated to be 35.90% (Zerga and Tahir 2022).

Two main layers of kapok fiber are the internal and outer layers, each of which has a unique microfibrillar structure. Cellulose microfibrils on the outside are oriented longitudinally with respect to the fiber axis. On the other hand, fibrils nearly perpendicular to the fiber axis are present in the inner layer (Nilsson and Björdal 2008). According to Mahadeva et al. (2011) and Naharudin et al. (2019), These fibers are cylinder-shaped, have a color ranging from yellow to brown, and are biodegradable, lightweight, abundant, and easy to work with. Because of its hardness, low cohesion, and fragility, the raw fiber of kapok cannot be woven like the fiber of cotton; nevertheless, it may be successfully blended with the fiber of cotton to create yarns (Zerga and Tahir 2022). Moreover, a higher kapok fraction increases yarn flexibility and decreases the uniformity and toughness of the fibers, lowering the fibers' total manufacturing costs (Dauda and Kolawole 2003). Furthermore, kapok fiber has less dyeing efficacy because of its large aperture and waxy surface, which prevent aqueous watercolor additives or pigments from entering (Lou 2011). The combination of these properties improves the hydrophobic-oleophilic properties for the fiber of kapok, enabling its use as both an oil-absorbing and buoyant material (Zerga and Tahir 2022). Fibers of kapok are being utilized in the industry as a nonwoven material filler and as bedding and upholstery fillers. But kapok fiber may also be useful for the adsorptive removal of dyes and heavy metal ions from aqueous solutions (Futalan et al. 2022).

These days, there is a major risk of oil contamination to both the environment and human health. Fiber-based oil sorption materials are reusable and eco-friendly solutions that can help mitigate this issue. For oil sorption materials, it is still difficult to retain both enhanced mechanical qualities and a high sorption rate (Zhou et al. 2023). The fiber of kapok has been used for oil and water separation and spill cleanup because of its remarkable hydrophobic and lipophilic properties, which are attributed to the wax coating on its upper layer (Zhang et al. 2023), as well as in packaging paper that needs to be waterproof and durable (Zerga and Tahir 2022). Wastewater treatment, solar-powered saltwater desalination, and oily wastewater treatment are just a few of the many uses for kapok fibers-MXene, an MXene composite aerogel (Chen et al. 2022). The fiber of kapok's inherent hollow structure makes it a good model or support material for applications like catalyst transporters (Fan et al. 2012). Moreover, according to Chung et al. (2013) and Tye et al. (2012), when creating flexible activated carbon fibers or a second-generation bioethanol generator, this fiber is a suitable place to start. Using the fiber of kapok as a bio-template, the simultaneous hydrothermal growth of C-doped g-C_3N_4 and TiO_2 in a single pot is a simple, environmentally benign, and economically viable synthesis method (Zerga and Tahir 2022). From kapok fiber, carbon microtubes have been created and are used as adsorbents for things like organic dyes, materials for sound absorption, and catalysis. Furthermore, the carbon microtube generated from kapok fiber exhibits significant promise for use in broadband microwave attenuation (Long et al. 2022). An effective way to clean up oil spills is to use natural fiber assemblies or blend them with synthetic fibers (Zhou et al. 2023).

21.1.3 Climate Change

Temperature variations and heat waves, rainfall, stronger and more frequent storms, longer droughts, and a higher risk of wildfires are all examples of climate change. Within the next 20 years, it is predicted that all of these consequences will hurt crop yields and food security globally (Jägermeyr et al. 2021; Li et al. 2019; Zhao et al. 2017). Global warming and enhanced planetary heat-trapping are the results of greenhouse gas (GHG) accumulation in the atmosphere, like carbon dioxide and methane (Eckardt et al. 2023). According to the IPCC's sixth assessment report (IPCC 2022), severe climate change effects can only be avoided by keeping global warming to 1.5 °C over pre-industrial levels. This calls for removing an extra 2–10 billion metric tons (Gt) of carbon dioxide (CO_2) annually, cutting global CO_2 emissions by half by 2030, and reaching net zero by 2050 (Eckardt et al. 2023). Climate changes pose significant challenges to humid tropical forests, affecting essential ecosystem attributes like nutrient cycling, carbon storage, and variations in temperature and precipitation patterns. These challenges will ultimately have adverse effects on ecosystem services and diversity (Cusack et al. 2016).

Major CO_2 fluxes like photosynthesis and respiration primarily govern the global carbon budget of terrestrial ecosystems, but several smaller flows also have a significant impact on carbon stocks and their turnover (Ito 2019). According to González-Meler et al. (2004), respiration in plants and ecosystems contributes significantly to the global carbon cycle because CO_2 is a quantitatively significant byproduct of respiration. Also, Sudarshan et al. (2014) reported that tree biomass can also be gathered and transformed into long-lasting plant products. Because they store about 60% of the terrestrial organic carbon above ground and 40% below ground, trees are significant to the global carbon cycle (IPCC 2001). Many nations have chosen to use fast-growing plantation trees as a sustainable supply of tree products and to lessen the strain on wild forests (Widyorini et al. 2009). The atmospheric CO_2 is absorbed by vegetation and retained as tree biomass during the productive season (Losi et al. 2003; Phat et al. 2004). Globally, plantation forests are growing quickly to provide a variety of products and services while also reducing deforestation and degradation of wild forests. Because of resource scarcity and climate change, mixed-species plantations are growing in popularity even though monoculture plantations have historically dominated operations and have been the subject of substantial research. For forestry plantations to effectively address the concerns of biodiversity and climate change, a variety of genotypes are required (Liu et al. 2018).

Although *C. pentandra* populations may face threats from climate change throughout their existing areas of distribution, there is still potential for the in situ persistence of the most genetically diverse and unique populations (Bocanegra-González et al. 2018). The most significant agroforestry tree species, kapok, is crucial for carbon storage. A 12-year-old kapok plantation was found to retain 38.7 tons of carbon per hectare (Sudarshan et al. 2014). Drought stress had a deleterious effect on both the photosynthetic and total biomass of *C. pentandra*, with the latter being less responsive to increased CO_2. Reforestation trials with a limited water supply could assess the *C. pentandra* tree's performance in response to high CO_2 and water stress (Silveira et al. 2023).

21.1.4 Domestication, Selection and Early Improvements

One well-known agroforestry species that smallholders frequently give priority to developing in their agroforestry systems is *C. pentandra* (Gunasena and Roshetko 2000; Roshetko and Evans 1999). As a non-invasive tree, *C. pentandra* has spread around the world (Chan et al. 2023). This tropical tree is native to the Caribbean, Mexico, tropical West Africa, Central America, and Northern South America (as *C. pentandra var. guineensis*), although it is also widely grown across the tropics, particularly in southeast Asian rainforests (Ravi Kiran et al. 2015). *C. pentandra* is found in dry forests, gallery forests, and various moist evergreen and deciduous woods. It is primarily found in secondary woods as a pioneer species. This tree is considered exotic in Cambodia, Ethiopia, Eritrea, Ghana, Gambia, Kenya, South Africa, Thailand, Tanzania, Zanzibar, and Uganda, but it is local in Indonesia, India, and the USA (Orwa et al. 2009). Many countries in southern Asia, such as Indonesia, Malaysia, India, South America, Central Africa, and Mexico, cultivate *C. pentandra* (Aji et al. 2022).

Drought tolerance and ecosystem roles as windbreaks are provided by *C. pentandra* (Brown 2012). In contrast to other more commercially valued commodities like rubber, coffee, and cocoa, their domestication is limited. *C. pentandra* gives farmers extra money, especially when other more economically desirable species are less productive due to droughts. Data on growth and productivity are useful in assessing a species' potential for increased domestication, which in turn improves farmers' lives (Harja et al. 2017). From the Neotropics, the species *C. pentandra* expanded to West Africa, where populations now exist in the wild from Senegal's coast to Angola (Herzog 1994). Cash crops like cocoa and coffee were planted alongside *C. pentandra* in Côte d'Ivoire. The species was brought to tropical Asia from Africa. *C. pentandra*, also referred to as kapok, was initially reported from Java, Indonesia, and subsequently from nearby nations including Malaysia and Thailand (Sahid and Zeven 2003). Along with other tree species like sengon, teak, and mahogany, kapok was planted in Java as part of a community-based forestry project (Irawanti et al. 2017). A species of tree native to West Africa is the red kapok tree. The majority of its organs and plant parts have therapeutic, culinary, and commercial uses (Assogba et al. 2018).

Aspects such as animal use, growth characteristics, and human use may be taken into account when selecting a species of *C. pentandra* (Hending et al. 2021). *C. pentandra* has a poor rate of regeneration in the province of Banwa, which poses a problem due to the species' low rate of regeneration. This indicates that its settlements have not been revitalized. Numerous factors affect the adult settlement's regeneration and selection (Zoma et al. 2022). For tree breeders who would like to begin selecting as soon as possible, there should be a significant variance in tree heights among *C. pentandra* accessions. One could draw the conclusion that choosing fast-growing accessions of *C. pentandra* would result in higher gains in plantation yield. Thus, the early identification of genetic variations would be highly beneficial for agroforestry tree improvement initiatives such as *C. pentandra* (Abengmeneng et al. 2015).

21.2 Current Cultivation Practices and Challenges

In order to maintain soil conservation, increase organic matter, and facilitate effective nutrient cycling, trees must be integrated into marginal and fallow croplands. Concerns regarding the sustainability of woody perennial-based land use practices are developing in light of ecological and economic issues. In these circumstances, diverse farming methods like agroforestry become crucial (Gawali et al. 2015). Only if the developed varieties are released in tandem with the best practices for sowing and field management will the genetic refinement of important plants offer reliable solutions to the issues brought about by climate change. This will allow for the full exploitation of all the traits that plant breeders have discovered during the phase of experimentation.

21.2.1 Current Cultivation Practices

The deciduous *C. pentandra* tree grows enormously quickly, reaching heights of 25–70 m and a diameter of 1–3 m. It has been cited as one of the global tallest trees (Adeniji et al. 2019). According to Olajide et al. (2022), the study proved that *C. pentandra* may be cultivated regularly without losing its nutritional value. At first, plant seeds that are still fresh can be planted. Before planting *C. pentandra* seeds, they need to be pre-treated by soaking them in lukewarm water for a full day, or until they start to swell. Direct sowing of the seeds takes place in 20 × 10 cm polythene bags, or 10 mm deep, in trays or pots filled with moist, high-quality seed compost. Daily watering is required to keep the plants at the ideal temperature of 25 ° C. After seeding, seeds begin to germinate, and this process takes 15 days to complete. Because germination can be highly unpredictable, plant sparingly and prick out the seedlings as soon as they are big enough to handle, with as little disturbance to the remaining compost as possible. Generally, seedlings that are 6 months old are used for planting. *C. pentandra* seedlings should be planted during the rainy season or whenever there is access to irrigation. For the first 5 years of life, the species is usually grown with agricultural crops at a planting spacing of 7 m × 7 m in a tangy system; after that, additional crop cultivation becomes unfeasible due to canopy closure. During the first 2 years, when *C. pentandra* develops above the level of weeds, all rank growth in the intervening strip that is likely to suppress the seedlings should be trimmed once or twice. In the third and fifth years, general caring can be carried out, which includes pruning the climbers and getting rid of unhealthy and dead trees (Orwa et al. 2009).

Irrigation is essential in places with less than 1000 mm of annual rainfall, especially in the summer when it should be done every 10 days. For the first 2 years, it ought to be completed weeding. *C. pentandra* uses 6% nitrogen, 12% phosphorus, and 12% potassium as fertilizer. Applying 20 t ha^{-1} or 5 t ha^{-1} of poultry manure supplemented with 200 kg ha^{-1} NPK might be suitable for cultivating *C. pentandra*, as the leaves of these plants showed higher concentrations of zinc and calcium. The use and domestication of *C. pentandra* for commercial cultivation may be guided by this information (Olajide et al. 2022). *C. pentandra's* growth features were enhanced by the exclusive application of poultry manure at a rate of 20 t ha^{-1} (Olajide and Baiyeri 2021). Using liquid fertilizer makes the most sense most of the time, especially if you feed your kapok tree while watering it. Furthermore, this tree needs to be protected from pests and diseases. *Eulepidotis modestula, Ephyriades arcas, Pericalia ricini* and *Oiketicus kirbiyi* are examples of insects that defoliate foliage. Additionally, parasitic plants like *Dendropthoe falcata* and *Loranthus spp.* host on the tree. Also, pathogenic fungi that affect trees include Armillaria, Calonectria, Camillea, and Cercospora (Orwa et al. 2009). Regarding maturity and harvesting, kapok trees start producing pods 3–4 years after they are planted. Kapok is typically harvested in the dry months. Achieving a full lactation phase takes 3–7 years. For the first 2 years, pulse crops can be raised as an intercrop. A 6 m spacing between each bundle can be used, and agricultural crops can be grown on the main field. Any crop's yield is based on its nutritional needs, maintaining the quality and yield of a new plant necessitates the use of suitable crop management techniques, particularly soil fertility management (Ani and Baiyeri 2008; Ndukwe and Baiyeri 2020). According to El-Khateeb et al. (2010), *C. pentandra* may tolerate salt levels of up to 6000 ppm and even benefit from them in certain situations.

21.2.2 Digital Farming

In the coming years, global agriculture will face several challenges in satisfying the need for food and fiber while maintaining total productivity without further harming soil, water, and other agroecological systems (Cole et al. 2018; Finger et al. 2019). Digital farming, sometimes referred to as smart farming or agriculture 4.0, is expected to address these problems by leveraging information and communication technology to collect and process data in order to support profitable farming methods (Bacco et al. 2019). There are large datasets generated that can be processed using cutting-edge computing to yield even more precise insights on decisions about agricultural management, such as when to plant, treat with chemicals, irrigate, and so on. The phrase "digital agriculture" refers to the use of big data in food production in connection with blockchain, artificial intelligence (AI), machine learning, cloud computing, robotics, unmanned aerial vehicles (UAVs), and remote sensing technologies. Understanding the processes of acceptance and dissemination of digital farming technologies is especially crucial given the emergence of these technologies and the potentially disruptive effects of smart digital agriculture (Gugganig and Bronson 2022; Shang et al. 2021).

21.2.3 Current Agricultural Challenges

Although growing *C. pentandra* can be a lucrative endeavor, growers may encounter a number of difficulties. These difficulties may affect the kapok plantation's overall performance and yield (AGC 2023), as shown in Fig. 21.5. The seasonally dry tropical forest is currently one of the most endangered ecosystems due to deforestation made essential to distribute arable land, which is a result of diverse land uses and the fertility of the soil (Bullock et al. 1995). The intense exploitation of *C. pentandra* wild populations in the Peruvian and Brazilian Amazon is a concern posed by the plywood industry (Brondani et al. 2003). Furthermore, the fundamental ecosystem attributes of humid tropical forests, nutrient cycling, carbon storage, temperature changes, and rain patterns are affected by climate change, posing significant challenges that will ultimately have a detrimental effect on ecosystem services and diversity (Cusack et al. 2016). A further difficulty in land-use planning is striking a balance between the requirements and demands of local stakeholders and scientific and technical factors (such as those pertaining to soil capabilities, site appropriateness, terrain, watersheds, ecological importance, proximity to cities, and minerals), according to FAO (2016). Even with high-resolution satellite images, there are ongoing gains and losses in forests, and obtaining accurate data on these dynamics can be difficult (FAO 2016). In Australia, kapok is recognized as a weed of the natural environment, and in other places where it has been introduced, it is considered a low-level weed. But it doesn't seem to be mentioned anywhere as a dangerous weed or invasive species.

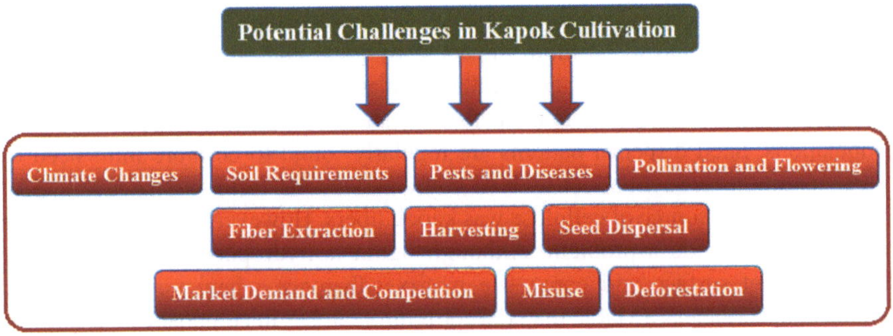

Fig. 21.5 Potential Challenges in Kapok Cultivation. (Prepared by: E.F. El-Hashash and K.M. El-Absy)

Lower fiber production might result from kapok tree growth and development being hampered by unsuitable climate conditions or poor soil quality (AGC 2023). The main theory underlying the auger hole planting method is the control of the plant root zone by the soil environment modification at a restricted quantity of amendments to a deeper soil profile, which is essential for tree establishment. Deeper rooting was promoted and the kankar layers were penetrated by the performance of *C. pentandra* cultivated using the auger hole method. As a result, the trees can resist harsh drought conditions during times of moisture deficit and reach deeper soil layers in search of water and nutrients (Immanuel and Ganapathy 2019).

Like any other crop, kapok trees are vulnerable to a variety of pests and diseases that can seriously harm the trees. Bees and bats are two examples of pollinators that kapok trees depend on for effective fertilization and fruit production. Reduced fruit sets can lead to insufficient pollination, which will ultimately result in lower fiber production. Pollinator populations can be significantly impacted by factors including habitat loss and pesticide use, which makes kapok production challenging (AGC 2023). The right methods and precise time are needed to harvest kapok fibers. The seed pods that encase the fibers must be harvested when they are fully developed but before they split open. It can take a lot of time and labor to separate the fibers from the seeds during the extraction process. The competitiveness and demand of the market affect kapok cultivation as well. The profitability of kapok cultivation may be impacted by changes in market demand or the introduction of substitute materials (AGC 2023).

Seed dispersal patterns and seedling mobilization vary throughout most plant species; these variations can be attributed to seed predation and post-dispersal seed movement (Gribel et al. 1999). These factors may interact to limit seed germination and have an adverse effect on the long-term performance of plant populations. In addition to seed predation and dispersal mechanisms, human activity is a major factor in how plant population dynamics are influenced. First, across its range of distribution, *C. pentandra* has frequently faced severe pressure due to deforestation caused by the harvest of lumber (Gentry and Vasquez 1988). According to Rija et al. (2013), the species is under increased pressure due to its use as animal fodder and for medicinal purposes in the surrounding area. The performance of tree populations has generally been found to be more impacted by local and spatial seed distribution patterns. For instance, Nathan and Muller-Landau (2000) analyze how seed density and the distance a seed travels after dissemination affect tree recruitment in plant populations. Seed dispersal patterns are regularly found to be more influenced by local conditions brought about by both biotic and abiotic factors (Rija et al. 2013).

One potential source of cellulose was kapok fibers. However, kapok's high lignin content can provide a problem. Although lignin is hydrophobic, it may prevent reinforcement surface and composite matrix adherence when present in kapok, which is nevertheless a viable material for reinforcement in composites. As a result, a barrier is created on the composite, which lowers its quality (Rezekinta et al. 2023). Usually, the only uses for kapok are as filler in dolls, mattresses, and pillows. What's left are pillows and mattresses filled with kapok. Kapok consequently became plentiful and was underutilized. To boost the added economic value, more research on the sustainable use of kapok is required (Sartika et al. 2020). Farmers can maximize the productivity and profitability of their kapok plantations by addressing these issues through effective planning, disease and pest management strategies, scientific land management with agro techniques, and ongoing learning. After all, these are the farmers' primary resources and they help to meet the needs of society (AGC 2023; Immanuel and Ganapathy 2019). Also, overcoming these obstacles can be facilitated by keeping up with industry developments and developing effective marketing plans (AGC 2023).

21.2.4 Breeding Objectives

Cuttings from branches stems, and seeds can all be used to grow *C. pentandra*. Research conducted over 6 years on the phenology, floral biology, pollination ecology, and breeding system of *C. pentandra* has shown that 17 of the 21 trees studied bloomed once or twice. In the Brazilian Amazon, a wide variety of nocturnal (like bats and hawk moths) and diurnal (like bees, wasps, and hummingbirds) species visit the blossoms of *C. pentandra* (Gribel et al. 1999). P*hyllostomus hastatus* and *P. discolor*, the only nocturnal phyllostomid bats that visit flowers, are efficient pollinators that promote cross-pollination. Pollen tubes did not pass through the style to reach the ovary, making diurnal floral visitors ineffectual at pollinating (Chan et al. 2023; Gribel et al. 1999). This species may exhibit varying degrees of self-incompatibility (Chan et al. 2023). Additionally, self-incompatibility was noted in *C. pentandra* in the southeast of Costa Rica, where cross-pollination produced a 25% fruit set while self-pollination produced very little or no fruit set (Rojas-Sandoval et al. 2008). Producing long-fiber plants that can provide significant volumes of high-quality fiber in a variety of soil and climate conditions is the primary objective of kapok breeding. Additionally, the plant needs to be disease- and pest-resistant. Even though grown kapok is quite important, there are still certain qualities that might be enhanced, such as low stature, regular yielding, resilience to drought, and fiber qualities (Brightness, Vibrantness, and Sensuality). To guarantee that the appropriate steps are taken for the sustainable management of *C. pentandra*, the study of variability within the species using both morphological and genetic markers is being conducted. Additionally, it would make it possible to compare phenotypic traits with genetic markers in order to better understand the variances in the genotypes and provide enough data to identify superior genotypes (Abengmeneng 2013). According to Fu et al. (2020), compared to selfing species, outcrossing species should be predicted to have lower population differentiation values and more intrapopulation genetic variation.

21.3 Germplasm Biodiversity and Conservation

There has been a discernible decline in plant biodiversity during the last 10 years, which can be linked to shifting climatic circumstances, the destruction of ecosystems to make way for farms, and ongoing selective breeding for certain features. Plant biologists around the world are facing a major obstacle in their efforts to find sustainable solutions to the existing obstacles to agricultural productivity, and that obstacle is the loss of biodiversity. An important part of the solution to this issue can come from plant genetic resource centers or genebanks that preserve plant germplasm (Subramanian et al. 2023). The term "germplasm" refers to a plant's genetic resources, such as tissues, cells, and seeds, as well as plant parts and trees kept in nurseries for research, conservation, and the acquisition of desirable breeding traits (Quazi et al. 2021). Genetic resources must be carefully collected, stored, analyzed, documented, and exchanged in order to be used successfully. Among the methods being used to conserve germplasm are slow-growing cultures, pollen, cryopreservation, and DNA or gene banks, genetic reserves, botanical gardens, and farmers' fields (Priyanka et al. 2021). The collection of germplasm typically includes genes intended to capture plant traits resulting from natural selection as well as wild species (Bonner 1990). Since the majority of plant species are becoming endangered over time, protecting plant resources is crucial. One of the key areas where knowledge needs to be continuously improved and changed is plant genetic conservation. The preservation of the genetic foundation required for reproduction is aided by the phenomena of conservation (Quazi et al. 2021). A priceless natural resource, germplasm is essential for preserving plant diversity and can reveal information about a species' genetic makeup. Strategies for protecting germplasm not only involve saving endangered plant species but also aid in maintaining all necessary plants, as the life of all organisms depends on them (Priyanka et al. 2021).

21.3.1 Germplasm Diversity

To ensure the long-term viability of tree planting activities, especially in light of climate change, it is imperative to understand the genetic diversity and structure of tree species. This understanding will drive conservation measures as well as the source of planting materials (Bocanegra-González et al. 2018). One of the most defining features of many neotropical tropical woods that are periodically dry is ceiba trees. It is a useful case study to look into biome-specific variations in the characteristics of species and their diversification paths, though, as it also includes species that are exclusive to the Amazon rainforest (Pezzini et al. 2021). Although *C. pentandra* populations may face threats from climate change throughout their existing areas of distribution, there is still potential for the in situ persistence of the most genetically diverse and unique populations (Bocanegra-González et al. 2018). There are still few *C. pentandra* plants, and the majority are not good seed sources

(Hefzi et al. 2023). Because this promotes the adaptability and sustainability of the species, Plantations of kapok require seeds from populations with a high level of genetic diversity (Ingvarson and Dahlberg 2019). Because pollinators may spread pollen over quite large distances and promote outcrossing, their movement also influences genetic variation within populations (Hefzi et al. 2023).

Establishing seed orchards from different populations is necessary to maintain genetic variety within and among populations for breeding methods. To generate high-quality, genetically superior plants that can adapt to environmental change and preserve biodiversity, the source is carefully chosen trees or seeds (better seeds) (Kuswantoro et al. 2020). The genetic variability of the kapok tree is an important genetic resource that will help with future efforts to conserve and reproduce this species (Hefzi et al. 2023). Significant genetic heterogeneity in *C. pentandra* was found by research using the RAPD marker on 25 accessions in West Sumatra (Hefzi et al. 2023) and 36 genotypes in Ghana (Abengmeneng et al. 2016). To choose better seeds, information regarding the genetic variety of *C. Pentandra* found on crucial land is utilized (Hefzi et al. 2023). Given its recent origin, this degree of variation may appear remarkable, but it should be remembered that many plant species have recently undergone swift adaptive radiations. For instance, the almost 300 species of Neotropical rainforest species that make up the species-rich tree genus Inga most likely had a common ancestor fewer than two million years ago (Lavin 2006).

21.3.2 Cultivars Characterization and Phylogeny

There are three forms of *C. pentandra*. In Africa, *C. pentandra* occurs in two distinct native forms and one hybrid. The enormous tree with the prickly trunk is the rainforest type. This study focuses on this particular form, which is the only one discovered in the Neotropics. The savanna form is found only in the savannas of West Africa. It is a smoother-trunked, smaller tree that doesn't reproduce at the same time as the rainforest form. The cultivated type is a hybrid/cross of the savanna and rainforest varieties. The low branches and indehiscent fruits of the cultivated variety make it easier to harvest the kapok fibers, which are used to stuff pillows, beds, and life preservers. The planted version, such as "java kapok," is widely cultivated outside of the range of the parental types (species), although it does not appear to have developed natural populations. Many physical and cytological differences could lead to the classification of the savanna and rainforest forms as distinct species under a more sophisticated taxonomy (Baker 1965; Dick et al. 2007). According to Abengmeneng (2013), Baker (1965) and Dick et al. (2007), the morphological and cytological character differences between them are shown in Fig. 21.6. Morphological characteristics, such as stem height, diameter, color, leaf stalk color, and number of spines, can be used to describe the variability among *C. pentandra* accessions (Abengmeneng 2013). Furthermore, the degree to which environmental factors influence tree features varies, and genetically inferior trees might appear phenotypically acceptable if cultivated in an exceptionally favorable microenvironment.

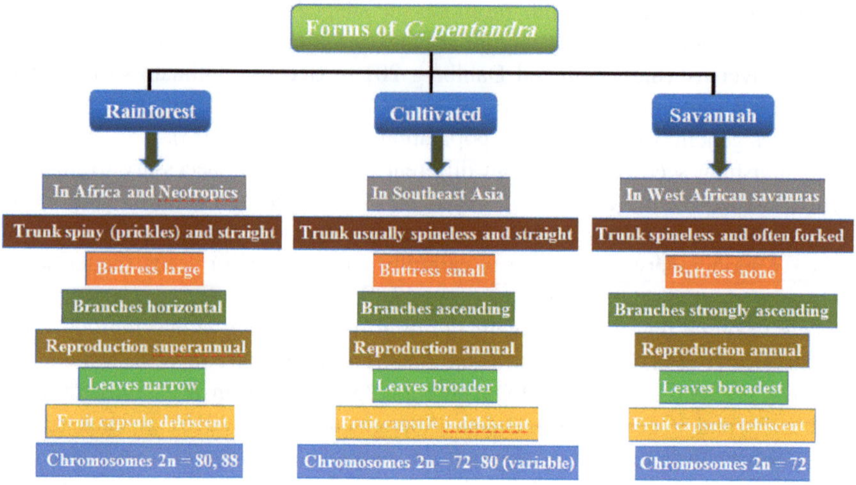

Fig. 21.6 Distinctions in morphological and cytological characterization among the three forms of *C. pentandra*. (Prepared by: E.F. El-Hashash and K.M. El-Absy)

Similarly, unfavorable environmental circumstances might make genetically better plants seem phenotypically unattractive (Frampton 1996). According to Abengmeneng (2013), accessions with lower performance levels would do better in an alternative testing setting, while accessions with higher performance levels might fare worse. Because *C. pentandra* is highly adapted to growing on barren terrain and water lack, it is imperative to develop it by paying attention to prospects, especially in critical regions (Hefzi et al. 2023).

21.3.3 Genetic Resources Conservation Approaches

Although it is not currently listed as endangered, habitat loss and deforestation pose risks to the kapok tree. To guarantee the survival of this species, conservation activities are crucial. Opportunities for the in situ persistence of the most genetically diverse and unique kapok populations persist beyond their current distribution areas (Bocanegra-González et al. 2018). A good genetic conservation level for *C. pentandra* is indicated by the high genetic diversity seen in all populations studied. Despite being an introduced species, conservation and restoration initiatives in Africa and Asia suggest the species as a viable tree. Developing in situ and ex-situ conservation techniques may benefit from a deeper comprehension of the genetic relationships between various *C. pentandra* accessions. Additionally, by identifying the proper seed zones, the study of this species' genetic variation distribution patterns will aid in the restoration of its deteriorated populations (Abengmeneng 2013). To make sure that this species' natural harvesting is sustainable, monitoring is required. This species is reported to exist in more than 100 ex situ collections at botanic gardens throughout the globe (BGCI 2017). For kapok, seed storage is required to preserve germplasm. Kapok seeds exhibited typical storage characteristics (Zheng et al. 2016).

The management of *C. pentandra* genetic resources is impacted by molecular information, and managing the species carries a significant risk to both its sustainable use and conservation (Abengmeneng et al. 2016). Many modifications and techniques, including micropropagation, slow-growth storage, somatic embryogenesis, and cryopreservation, are part of the vast array of in vitro plant propagation technologies that can be employed in integrated schemes (Kulak et al. 2022). One practical use of cryobiology, or the study of life at low temperatures, typically below −100 °C, is cryopreservation. In particular, plant cryopreservation is the technique of freezing and preserving plant structures such as tissues, cells, or organs in liquid nitrogen (LN; −196 °C) or LN vapor (−150 °C to −180 °C). This technique permits indefinite storage and guarantees the preservation of samples' viability during thawing (Benelli 2021; Chaudhury and Malik 2017). The vitrification (glass formation) phenomenon is the basis for new cryotechniques. Samples are exposed to concentrated cryoprotective solutions or dried by air to accomplish cell desiccation (Chaudhury and Malik 2017).

They have a long shelf life at subzero temperatures and low moisture content. Light-brown seeds were shown to retain their viability and vigor best when stored at 4 degrees Celsius (Zheng et al. 2016). Preserving the grandeur of the Kapok Tree requires intensive cultivation and conservation efforts. Preserving the Kapok Tree is an investment in maintaining not just a particular species but also the abundant biodiversity and ecological balance of tropical rainforests.

21.3.4 Cytogenetics

Numerous tiny chromosomes characterize the paleopolyploid ancestry of the genus Ceiba, which has been shown to have varied intra- and interspecific chromosomal numbers (Figueredo et al. 2016). It is discovered that *C. pentandra* has 88 somatic chromosomes by Tigo (1948). According to Baker (1965), African rainforest trees ($2n = 80, 88$) and Neotropical trees ($2n = 88$) have comparable and maybe overlapping chromosomal numbers. The chromosomal count of the African savanna variety is $2n = 72$, while the farmed Ceiba have varying chromosome counts ($2n = 72, 74, 75, 76, 80$). The offspring of selfing exhibited a variation in the chromosomal count, with the mother *C. pentandra* belonging to the indica subtype. It appears that $2n = 72$ is the minimum chromosomal number required for a plant to thrive, suggesting that this species is probably polyploid (Zeven 1969). Nucleotide divergence is very minor, and *C. pentandra* has significant cytological variation (Dick et al. 2007).

For the pan-tropical forest tree *C. pentandra*, microsatellite markers were created in 2003. A total of 112 alleles were found, with an average of 14 alleles per locus. The genetic information richness of every microsatellite locus was quite high, with anticipated heterozygosity ranging from 0.814 to 0.895 (Brondani et al. 2003). The genetic relationships between the 36 genotypes of *C. pentandra* in Ghana were discovered in 2016 using RAPD and ISSR markers. This information is advised to all forestry stakeholders, particularly those that use *C. pentandra* as a main species in the growth of their plantations (Abengmeneng et al. 2016). Every species had a karyotype that was primarily made up of metacentric chromosomes, with 2n = 86. Technical issues related to large chromosomal counts and/or smaller Bombacoideae chromosomes may be the cause of chromosomal count inaccuracies (Figueredo et al. 2016). Regarding 2018, a study on the genetic diversity of *C. pentandra* in the seasonally dry tropical forest of Colombia revealed that heterozygosity scores at the majority of the 12 places tested were in the range of what is expected by Hardy-Weinberg. Boncanegra-González et al. (2018) determined the priority regions for the in-situ conservation of *C. pentandra* in the Colombian SDTF based on this conclusion. There is still a dearth of genetic data and molecular research on *C. pentandra* (Hefzi and Mansyurdin 2023).

21.4 Breeding Strategies

21.4.1 Traditional Methods

During repeated cycles of actions like breeding, selection, and genetic testing, tree breeding programs create genetically superior types for forestation (White 2004). To produce better seeds and seedlings for reforestation, tree improvement techniques include provenance trials for testing and selection of seed sources, progeny tests for testing and parent selection, and the establishment and upkeep of seed orchards (White 1987). The part of tree improvement that involves breeding or crossing suggests a multigenerational program that, with each breeding cycle, produces a new generation of improved trees (chosen parents for improved orchards), according to Nelson (2023).

The chiropterophilic flowers of the pantropical tree *C. pentandra* and the breeding system vary depending on where in its range the tree is found (Jorge et al. 2005). While stamens and stigmas of neighboring flowers on the same tree can come into contact to cause self-pollination, bisexual blooms open at night and are rich in pollinators (cross-pollination). Breeding systems vary in their degree of selfing; they can be completely self-incompatible or mixed (Jorge et al. 2005). It has been said to be completely self-compatible in Africa and Southeast Asia (Baker 1965), but as self-incompatible in Samoa and Brazil (Gribel et al. 1999). It is considered to have a mixed breeding system in Panama (Murawski and Hamrick 1992). Studying the adaptability of the breeding system in relation to environmental factors like pollinator diversity and abundance is made possible by the diversity of the *C. pentandra* breeding system (Jorge et al. 2005). One of the primary determinants of plant mating systems and reproductive success is the identity and behavior of pollinators (Jorge et al. 2005).

It will be easier to select for height growth in tree breeding if the genetics of its constituents are well understood (Rweyongeza et al. 2003). The genetic gain and heritability in height growth of kapok are poorly understood, making it difficult to make informed decisions about selection and future breeding. Additionally, dieback of kapok seedlings and saplings has been noted in plantations as well as nurseries (Apetorgbor et al. 2003). This phenomenon is quite dangerous for the large-scale commercial establishment of this species and jeopardizes significant investments made in this direction. As a result, it is necessary to identify and employ resistant stock. The selection of genetically resistant genotypes also requires the calculation of genetic factors, such as breeding value prediction (Abengmeneng et al. 2015). While, the genetic gain and heritability in height growth in kapok are not well understood, making it difficult to make informed decisions about selection and subsequent breeding (Abengmeneng et al. 2015).

To facilitate even better selections in the ensuing selection cycle, breeding reconstructs and repackages genetic variation. Enhancing the frequency of advantageous alleles of a few essential qualities is the goal of selection (White 2004). Line selection must be adequate in order to produce useful material from the original indigenous Southeast Asian material (Zeven 1969). *C. pentandra* material can be multiplied in seed gardens with a multiplication factor of 100, which allows for a rapid rise in seeds without the need for precautions against out-pollination (Toxopeus 1948). During every breeding cycle, genetic testing is used to rank the parents in the breeding population and maximize the benefits of selection. This breeding method is the basis for nearly all programs that use recurrent selection for general combining ability (White 2004).

Natural selection may operate as a result of increased genetic variety produced by hybridization events (Gaskin 2016). Hybridization is a useful technique to help restoration and conservation efforts by making tree plants more resilient to stress, particularly from pests and diseases (Brennan et al. 2020). To achieve additional *C. pentandra* enhancements, crosses ought to be made. The following breeding protocol should be followed after the parents have been chosen: several crossings between different parents should be made, and all F_1s should be examined. It is recommended to propagate the best offspring asexually and to grow and examine the F_2 offspring of the chosen F_1 offspring (Huitema 1937). It is best to backcross when using a parent with wild (Caribaea) features. The seedlings should be budded on the rootstocks of established trees in order to shorten the generation period. But keep in mind that it takes 14 years to choose a BC_2 progeny under such a procedure (Zeven 1969).

New breeding technologies like tissue culture, which enables the operational planting of tested clones, marker-assisted breeding, which increases gain from selection, and genetic modification, which inserts novel genes, have the potential to improve the benefits of tree breeding, though they still need to be tested for efficacy, safety, and public acceptance (White 2004).

21.4.2 Biotechnology Techniques

The fields of forest genetics and tree improvement employ techniques such as biotechnology and tree breeding. The conventional method of genetically enhancing trees is commonly perceived as tree breeding, whereas biotechnology is considered the more contemporary and cutting-edge technique (Nelson 2023). Based on their utilize in preserving, evaluating, or generating genetic variability, the foundation of genetic improvement, the biotechnologies can be divided into three groups about tree improvement, as depicted in Fig. 21.7. Some of the modern biotechnologies utilized in tree breeding include tissue culture-based propagation, molecular genetic markers, genome mapping and sequencing, gene and genome editing, gene cloning, and genetic transformation/modification (Nelson 2023).

In tropical climates, ceiba species are commonly farmed due to their high ornamental value. Still unknown, though, is the genetic heritage of domesticated Ceiba plants (Han et al. 2023). Morphological and molecular factors should be taken into account when choosing *C. pentandra* accessions for conservation (Abengmeneng 2013). The management of *C. pentandra* genetic resources is impacted by molecular information, and managing the species carries a significant risk to both its sustainable use and conservation (Abengmeneng et al. 2016). The lack of a molecular genetic linkage map and the limited understanding of *C. pentandra* stem dieback quantitative trait loci (QTL) may have significantly impeded the efficacy of the used markers (Abengmeneng 2013).

According to Said et al. (2013), analysis of the amplified fragments produced by RAPD reactions indicates that the biotype of kapok yields distinct molecular patterns from members of the family that have been studied previously. According to

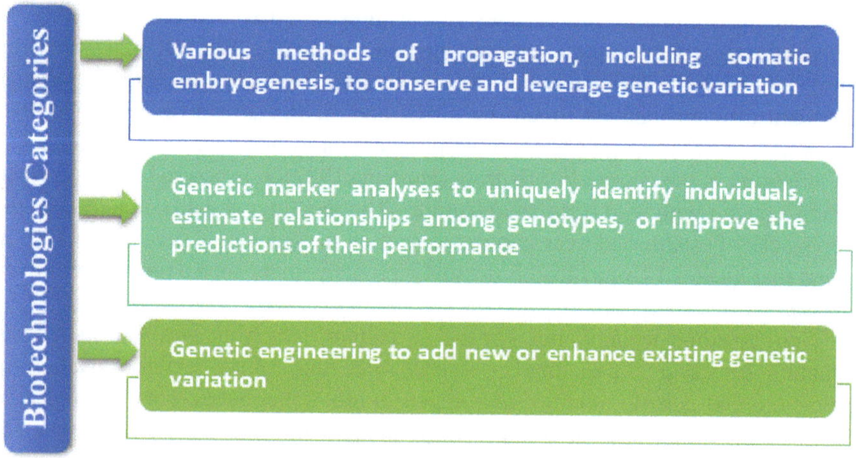

Fig. 21.7 The three categories of combined biotechnologies for tree improvement. (Prepared by E. F El-Hashash and K.M. El-Absy)

Darwish et al. (2015), there are primers available that can be used to identify and genetically differentiate *C. pentandra* from other morphologically associated types in the future, where the generated fragments' molecular sizes showed that a variety of sequences between 114 and 1904 bp were present. *C. pentandra* from five ecological zones in Ghana was characterized using five morphological features and three molecular marker technologies: ISSR and RAPD, as well as the combination of ISSR and RAPD (Abengmeneng 2013).

To comprehend the genetic basis of significant horticultural features and the genetic relatedness of cultivated Ceiba plants, Han et al. (2023) looked into the genetic relatedness of 153 accessions of Ceiba plants grown in Southern China. The researchers found SNPs linked to five horticultural qualities based on 11,704 SNPs obtained using double-digest restriction-site associated DNA sequencing (ddRAD-seq). The genome-wide association analysis (GWAS) revealed that two, four, two, three, and four SNPs were linked to petal color, petal striation number, flowering duration, trunk morphology, and prickles on the trunk and branches, respectively. Near the SNPs that were highly connected to these characteristics, one or two candidate genes were discovered. This paper showed the genetic relatedness of the Ceiba plants cultivated in Southern China and presented the first GWAS analysis for five horticultural features, laying the groundwork for phenotype-related marker selection and molecular breeding. Hefzi et al. (2023) carried out another study employing the descriptive approach and molecular data utilizing the molecular marker RAPD (random amplified polymorphic DNA) for twenty-five individuals of *C. pentandra*. The findings demonstrated that polymorphisms may be found using the primers OPA-01, OPA-02, and OPB-10. In plant breeding projects, the parent can be chosen based on *C. pentandra* genetic variety.

21.5 Conclusions and Prospects

A recent study published in Nature Plants suggests that massive trees in tropical forests, which have witnessed centuries of civilization, might be caught in a hazardous feedback cycle. In fully developed tropical forests, the largest trees hold half of the carbon stored in the atmosphere. However, due to climate change, these trees may soon die and release vast amounts of carbon back into the sky. Changes in climate are causing the most frequent and severe droughts as well as more powerful storms in many parts of the tropics. Big trees may be more vulnerable to wind or lightning damage because they tower over other trees. They are more susceptible to drought than other trees because they must draw groundwater uphill.

Although *C. pentandra* populations may face threats from climate change throughout their existing areas of distribution, there is still potential for the in situ persistence of the more genetically various and unique populations. Kapok is a multipurpose tree that is cultivated in afforestation and agroforestry operations in most nations. The species is widely distributed around the world and is non-invasive. In addition to providing wood, textiles, oils, and fodder that support local livelihoods, the tree also supplies pollen to beekeepers in the area. It may be possible to create beneficial compounds with commercial potential from *C. pentandra*. They consist of absorbents made from kapok fibers and biodiesel. It is possible to produce both medicinal and non-medicinal compounds from *C. pentandra* seed oil. An excellent, biodegradable organic fiber is kapok. Kapok fiber is frequently utilized. Paper, textiles, buoyancy, acoustical and support materials, thermal insulation, oil-absorbing materials, biofuel, and activated carbon fibers derived from kapok are only a few applications for it. Both medication and food are made from the plant and its preparations. *C. pentandra* is an extremely valued plant with numerous additional advantageous biological qualities. Due to its combination of hydrophobic and oleophilic qualities, as well as its increased fiber hollowness, this plant is becoming more and more interesting. The newly released scientific studies demonstrate the great interest in the plant and its fiber qualities. Research on this topic should be prioritized, nevertheless, as more in-depth study is still required to fully understand this plant's industrial relevance, its characteristics, and new technologies.

Predicted climate change scenarios based on several models were covered in Sect. 21.1.2 along with potential effects on the kapok tree. Major obstacles related to climate change affect the basic ecosystem functions of mid-tropical forests, such as nutrient cycling, carbon storage, temperature changes, and precipitation patterns. These challenges will eventually harm ecosystem services and diversity.

Fighting climate change and lowering greenhouse gas emissions are two of the UN's main objectives for its "green legacy" effort. Planting and growing trees is one way to combat climate change, which is currently a top global priority. The UN urges nations to embrace sustainable transportation systems, energy-efficient practices, and renewable energy sources. The United Nations states that in addition to its ability to lower atmospheric carbon dioxide levels, trees have other environmental advantages. They enhance the quality of the air and water, aid in stopping soil erosion, and supply habitat for wildlife.

Globally, there is a growing movement to reduce total reliance on fossil fuels by using renewable energy sources like biofuel, as a result of the rising deterioration of fossil fuels. Recent years have seen an increase in the popularity of green renewable resources due to the development of environmental technological solutions. A type of cellulosic fiber called fiber of kapok is made from the seeds of the *C. pentandra* tree. The greatest option for supplying clean and renewable energy sources is provided by kapok fiber, which acts as a bio-template. Compared to other local species, seedlings exhibit greater resilience to high temperatures. Given the challenges posed by climate change, these attributes may be crucial in guiding the implementation of conservation strategies.

The majority of research on kapok fibers has concentrated on unstructured fibrous formations; further investigation into structured fibers is required to enable more practical applications. Most research should be done to enhance the spectrum of applications for the fiber of kapok by utilizing its many good vacuousness and hydrophobic–oleophilic qualities, given the rising focus on this sustainable cellulosic fiber. Early genetic difference discovery would be highly beneficial for agroforestry tree enhancement initiatives. Molecular trials are still desperately needed to investigate and confirm the genetic basis underlying these important attractive features of kapok, despite potential connections between several traits and candidate genes having been explored.

Given the significant economic value of the kapok tree, further research is needed on germplasm diversity, taxonomy, cytogenetics, genetic resource conservation, conventional breeding techniques, conventional and somatic cell hybridization, genetic engineering, molecular breeding, and mutation breeding. Also, investigate the application of artificial intelligence to the management of agricultural practices, the observation of environmental variables, the enhancement of resources and their sustainability, the enhancement of crop monitoring, and the advancement of kapok tree-based industries. Biotechnology techniques, like molecular breeding (molecular marker-assisted breeding, genomics, and bioinformatics), gene editing, and genetic engineering, can be used in kapok breeding to facilitate certain tasks and avoid introducing foreign genes into the developed crop.

Appendix 21.1 Research Institutes and Online Resources Available for Kapok Breeding

Institute name	Specialization and research activities	Address	Contact information and website
Institute of Natural Fibres and Medicinal Plants	The Institute of Natural Fibres and Medicinal Plants conducts cutting-edge research for the advancement of the bioeconomy (textiles and paper industry, etc.).	Wojska Polskiego 71B, 60-630 Poznań, Poland	https://iwnirz.pl/
The International Union for Conservation of Nature and Natural Resources	It is the most complete knowledge repository on the planet regarding the risk of extinction for plant, animal, and fungal species.	The David Attenborough BuildingPembroke Street, Cambridge CB2, 3QZ, United Kingdom	https://www.iucnredlist.org/species/61782438/61782442
Feedipedia	It is a joint project of INRAE (formerly INRA-Institut National de la Recherche Agronomique), CIRAD (Centre de Coopération Internationale en Recherche Agronomique pour le Développement), AFZ (Association Française de Zootechnie) and FAO		https://www.feedipedia.org/node/48

References

Abdullah MA, Rahmah AU, Man Z (2010) Physicochemical and sorption characteristics of Malaysian *Ceiba pentandra* (L.) Gaertn. as a natural oil sorbent. J Hazard Mate 177(1–3): 683–691. https://doi.org/10.1016/j.jhazmat.2009.12.085

Abengmeneng CS (2013) Evaluation of *Ceiba pentandra* for stem dieback disease resistance and characterization by molecular markers. Thesis Ph.D. Faculty of Renewable Natural Resources, College of Agriculture and Natural Resources, Ghana

Abengmeneng CS, Ofori DA, Kumapley P et al (2015) Estimation of heritability and genetic gain in height growth in *Ceiba pentandra*. Afr J Biot 14(22):1880–1885. https://doi.org/10.5897/AJB2014.14287

Abengmeneng CS, Ofori D, Kumapley P et al (2016) Genetic relationship among 36 genotypes of *Ceiba pentandra* (L.) as revealed by RAPD and ISSR markers. Am J Agric Forest 4(4):86–96. https://doi.org/10.11648/j.ajaf.20160404.13

Adeniji IT, Olomola DB, Jegede OC (2019) Seed germination and seedling growth of *Ceiba pentandra* (L) as influenced by different soil types in Ibadan, Southwest Nigeria. J Res Forest Wildlife Environ 11(3):316–320

AGC (2023). Agriculture.Gov.Capital. What are the potential challenges in kapok cultivation?. https://agriculture.gov.capital/what-are-the-potential-challenges-in-kapok-cultivation/

Aji A, Oktafiani D, Yuniarto A et al (2022) Biosynthesis of gold nanoparticles using Kapok (*Ceiba pentandra*) leaf aqueous extract and investigating their antioxidant activity. J Mol Str 1270: 133906. https://doi.org/10.1016/j.molstruc.2022.133906

Ani JU, Baiyeri KP (2008) Impact of poultry manure and harvest season on juice quality of yellow passion fruit (*Passiflora edulis* var. flavicarpa Deg.) in the sub-humid zone of Nigeria. Fruits 63: 239–247

Apetorgbor MM, Siaw DEKA, Gyimah A (2003) Decline of *Ceiba pentandra* seedlings, a tropical timber species in nurseries and plantations. Ghana J For 11(2):51–53

Assogba GA, Fandohan AB, Gandji K et al (2018) *Bombax costatum* (Malvaceae): state of knowns, unknowns and prospects in West Africa. Biotechnol Agron Soc Env 22(4):267–275

Avendaño A, Casas A, Dávila P et al (2006) Use forms, management and commercialization of "pochote" *Ceiba aesculifolia* (HB & K.) Britten & Baker f. subsp. parvifolia (Rose) PE Gibbs & Semir (Bombacaceae) in the Tehuacán Valley, Central Mexico. J Arid Environ 67:15–35. https://doi.org/10.1016/j.jaridenv.2006.02.004

Bacco M, Barsocchi P, Ferro E et al (2019) The digitisation of agriculture: a survey of research activities on smart farming. Array 3–4:100009. https://doi.org/10.1016/j.array.2019.100009

Baker HG (1965) The evolution of the cultivated kapok tree: a probable West African product. In: Brokensha D, Baker HG (eds) Ecology and economic development in Africa. Institute of International Studies, University of California, Berkeley, pp 185–217

Baraniak J, Kania-Dobrowolska M (2023) Multi-purpose utilization of Kapok fiber and properties of *Ceiba pentandra* tree in various branches of industry. J Nat Fibers 20:1. https://doi.org/10.1080/15440478.2023.2192542

Benelli C (2021) Plant cryopreservation: a look at the present and the future. Plants 10(12):2744. https://doi.org/10.3390/plants10122744

BGCI (2017) Plant search. Botanic Gardens Conservation International, London. Available at: www.bgci.org/plant_search.php

Bocanegra-González KT, Thomas E, Guillemin ML (2018) Genetic diversity of *Ceiba pentandra* in Colombian seasonally dry tropical forest: implications for conservation and management. Bio Conserv 227:29–37. https://doi.org/10.1016/j.biocon.2018.08.021

Bonner FT (1990) Storage of seeds: potential and limitations for germplasm conservation. Forest Ecoand Manag 35(1–2):35–43

Brennan AN, McKenna JR, Hoban SM et al (2020) Hybrid breeding for restoration of threatened forest trees: evidence for incorporating disease tolerance in *Juglans cinerea*. Front Plant Sci 11: 580693. https://doi.org/10.3389/fpls.2020.580693

Brondani RPV, Gaiotto FA, Missiaggia AA et al (2003) Microsatellite markers for *Ceiba pentandra* (Bombacaceae), an endangered tree species of the Amazon Forest. Mol Ecol Notes 3:177–179. https://doi.org/10.1046/j.1471-8286.2003.00389.x

Brown SH (2012) *Ceiba pentandra*. IFAS extension Lee County, Southwest Florida. Gainsville, USA: University of Florida, USA

Bullock SH, Mooney HA, Medina E (1995) Seasonally dry tropical forests (No. 574.52642 S4). In: Mooney HA, Bullock SH, Medina E (eds) Seasonally dry tropical forests. Cambridge University Press, Cambridge

Carvalho-Sobrinho JG, Alverson WS, Alcantara S et al (2016) Revisiting the phylogeny of Bombacoideae (Malvaceae): novel relationships, morphologically cohesive clades, and a new tribal classification based on multilocus phylogenetic analyses. Mol Phylogenet Evo 101:56–74. https://doi.org/10.1016/j.ympev.2016.05.006

Chaiarrekij S, Apirakchaiskul A, Suvarnakich K et al (2012) Characteristcs of kapok fiber as a potential pulp source for papermaking. Bio Res 7:475–488

Chan EWC, Yeong SW, Wong CW et al (2023) *Ceiba pentandra* (L.) Gaertn.: an overview of its botany, uses, reproductive biology, pharmacological properties, and industrial potentials. J App Bio Biot 11(1):1–7. https://doi.org/10.7324/JABB.2023.110101

Chaudhury R, Malik SK (2017) Cryopreservation of plant species: practical approaches from handling to Cryobanking. Indian Council of Agricultural Research (ICAR) National Bureau of Plant Genetic Resources, Pusa Campus, New Delhi, p 52

Chen L, Mu X, Guo Y et al (2022) MXene-doped kapok fiber aerogels with oleophobicity for efficient interfacial solar steam generation. J Colloid and Interface Sci 626:35–46. https://doi.org/10.1016/j.jcis.2022.06.143

Chung J-T, Hwang K-J, Shim W-G et al (2013) Synthesis and characterization of activated hollow carbon fibers from *Ceiba pentandra* (L.) Gaertn. (kapok). Mater Lett 93:401–403. https://doi.org/10.1016/j.matlet.2012.09.016

Cole MB, Augustin MA, Robertson MJ et al (2018) The science of food security. NPJ Sci Food 2(1):14. https://doi.org/10.1038/s41538-018-0021-9

Cusack DF, Karpman J, Ashdown D et al (2016) Global change effects on humid tropical forests: evidence for biogeochemical and biodiversity shifts at an ecosystem scale. Rev Geophys 54(3): 523–610. https://doi.org/10.1002/2015RG000510

Darwish FMM, Orabi MAA, Abdelkader MSA et al (2015) Botanical study and DNA fingerprint of *Ceiba pentandra* (L.) Gaertn. var. pentandra cultivated in Egypt. Bull Pharm Sci Assiut Univ 38:61–90

Dauda BMD, Kolawole EG (2003) Processibility of Nigerian kapok fibre. Indian J Fibre Text Res 28:147–149

Dick CW, Bermingham E, Lemes MR et al (2007) Extreme long-distance dispersal of the lowland tropical rainforest tree *Ceiba pentandra* L. (Malvaceae) in Africa and the Neotropics. Mol Ecol 16(14):3039–3049. https://doi.org/10.1111/j.1365-294X.2007.03341.x

Eckardt NA, Ainsworth EA, Bahuguna RN et al (2023) Climate change challenges, plant science solutions. Plant Cell 35(1):24–66. https://doi.org/10.1093/plcell/koac303

El-Khateeb M, Nabih A, Nasr A et al (2010) Growth and chemical constituents of *Ceiba pentandra* L. plant in response to different levels of saline irrigation water. Egyptian J Agric Sci 61(2): 214–221. https://doi.org/10.21608/ejarc.2010.215406

Fan H, Yu X, Long Y et al (2012) Preparation of kapok–polyacrylonitrile core–shell composite microtube and its application as gold nanoparticles carrier. Appl Surf Sci 258:2876–2882. https://doi.org/10.1016/j.apsusc.2011.10.151

FAO (2016) State of the World's Forests 2016. Forests and agriculture: land-use challenges and opportunities, Rome

FAOSTAT (2023) Food and agriculture data, crops. https://www.fao.org/faostat/en/#data/QCL/visualize. Accessed 7 Nov 2023

Figueredo A, de Oliveira ÁWL, Carvalho-Sobrinho JG et al (2016) Karyotypic stability in the paleopolyploid genus Ceiba Mill. (Bombacoideae, Malvaceae). Braz J Bot 39:1087–1093. https://doi.org/10.1007/s40415-016-0296-5

Finger R, Swinton SM, El Benni N et al (2019) Precision farming at the nexus of agricultural production and the environment. Ann Rev Res Econ 11(1):313–335. https://doi.org/10.1146/annurev-resource-100518-093929

Frampton J (1996) The tree improvement process. Limbs Needles 23(4):10–14

Fu Q, Lu G, Fu Y et al (2020) Genetic differentiation between two varieties of *Oreocharis benthamii* (Gesneriaceae) in sympatric and allopatric regions. Ecol Evo 10:7792–7805

Futalan CM, Choi AES, Soriano HGO et al (2022) Modification strategies of Kapok fiber composites and its application in the adsorption of heavy metal ions and dyes from aqueous solutions: a systematic review. Int J Env Res Pub Health 19(5):2703. https://doi.org/10.3390/ijerph19052703

Gaskin JF (2016) The role of hybridization in facilitating tree invasion. AoB PLANTS 9(1):plw079. Advanced online publication. https://doi.org/10.1093/aobpla/plw079

Gawali A, Puri S, Swamy SL (2015) Evaluation growth and yield of wheat varieties under *Ceiba pentandra* (L) based agrisilviculture system. Univ J Agric Res 3(6):173–181. https://doi.org/10.13189/ujar.2015.030601

Gentry AH, Vasquez R (1988) Where have all the Ceibas gone? A case history of mismanagement of a tropical forest resource. Forest Ecol Manag 23:73–76. https://doi.org/10.1016/0378-1127(88)90015-1

Gibbs P, Semir J (2022) A taxonomic revision of the genus Ceiba Mill. (Bombacaceae). Anales Jard Bot Madrid 60:259–300. https://doi.org/10.3989/ajbm.2002.v60.i2.92

Gómez-Maqueo X, Gamboa-deBuen A (2022) The biology of the genus Ceiba, a potential source for sustainable production of natural fiber. Plants 11(4):521. https://doi.org/10.3390/plants11040521

Gonzalez-Meler MA, Taneva L, Trueman RJ (2004) Plant respiration and elevated atmospheric CO_2 concentration: cellular responses and global significance. Ann Bot 94(5):647–656. https://doi.org/10.1093/aob/mch189

Graham BP, Haigler CH (2021) Microtubules exert early, partial, and variable control of cotton fiber diameter. Planta 253(2):47. https://doi.org/10.1007/s00425-020-03557-1

Gribel R, Gibbs PE, Queiroâz AL (1999) Flowering phenology and pollination biology of *Ceiba pentandra* (Bombacaceae) in Central Amazonia. J Trop Ecol 15:247–263. https://doi.org/10.1017/S0266467499000796

Gugganig M, Bronson K (2022) Perspective: digital agriculture. In: Ottawa ON (ed) Food studies: matter, meaning, movement. Food Studies Press, Canada. https://doi.org/10.22215/fsmmm/bk05

Gunasena HPM, Roshetko JM (2000) Tree domestication in Southeast Asia: results of a regional study on institutional capacity. International Centre for Research in Agroforestry, Bogor, Indonesia

Han B, Huang S, Huang G et al (2023) Genetic relatedness and association mapping of horticulturally valuable traits for the Ceiba plants using ddRAD sequencing. Hortic Plant J 9(4):826–836. https://doi.org/10.1016/j.hpj.2022.01.006

Harja D, Martini E, Munawir A et al (2017) Agroforestry and forestry in Sulawesi series: modelling tree production based on farmers' knowledge: case for kapok (*Ceiba pentandra*) and candlenut (*Aleurites mollucana*) under various agroforestry scenarios. Working Paper 269. World Agroforestry Centre (ICRAF) Southeast Asia Regional Program, Bogor, Indonesia. https://doi.org/10.5716/WP17361

Hefzi F, Mansyurdin M (2023) The biology and molecular studies of *Ceiba Pentandra* (L.) gaertn: a review. Int J Prog Sci Tech 36(2):74–79

Hefzi F, Mansyurdin M, Maideliza T (2023) Analysis of genetic variations of *Ceiba pentandra* (L.) gaertn on several critical lands in west Sumatra using RAPD molecular markers. Biosaintifika: J Biol Bio Edu 15(1):26–35

Hending D, Randrianarison H, Holderied M et al (2021) The kapok tree (*Ceiba pentandra* (L.) Gaertn, Malvaceae) as a food source for native vertebrate species during times of resource scarcity and its potential for reforestation in Madagascar. Aust J 46(8):1440–1444. https://doi.org/10.1111/aec.13076

Herzog F (1994) Multipurpose shade trees in coffee and cocoa plantations in Côte d'Ivoire. Agrofor Syst 27:259–267

Huitema WK (1937) Het selectiewerk bij kapok (Breeding work in kapok). Mimeograph, pp 20

Immanuel RR, Ganapathy M (2019) Agro-techniques for afforestation of degraded coastal agricultural lands with silk cotton (*Ceiba pentendra* (L.) Gaertn.). J Pharmacog Phytochem 8:1587–1590

Immanuel RR, Ganapathy M (2020) Restoration of soil microbial biomass using multipurpose trees. Plant Arch 20:2641–2645

Ingvarson PK, Dahlberg H (2019) The effects of clonal forestry on genetic diversity in wild and domesticated stands of forest trees. Scand J Res 34:370–379

IPCC (2001) Climate change: Working Group I: the scientific basis. Cambridge University Press, New York

IPCC (2022) Climate change 2022: impacts, adaptation, and vulnerability. In: Pörtner HO, Roberts DC, Tignor M et al (eds) Contribution of working group II to the sixth assessment report of the intergovernmental panel on climate change. Cambridge University Press, Cambridge, UK/-New York, p 3056

Irawanti S, Race D, Stewart H (2017) Understanding the timber value chain in community-based forestry in Indonesia: analysis of Sengon in Central Java. J Sustain For 36:847–862

Ito A (2019) Disequilibrium of terrestrial ecosystem CO_2 budget caused by disturbance-induced emissions and non-CO_2 carbon export flows: a global model assessment. Earth Syst Dynam 10: 685–709. https://doi.org/10.5194/esd-10-685-2019

Ito EN, Macedo MJP, Silva GS et al (2020) Surface modification of Kapok fibers by cold plasma surface treatment. J Mater Res Technol 9:2467–2476. https://doi.org/10.1016/j.jmrt.2019.12.077

Jägermeyr J, Müller C, Ruane AC et al (2021) Climate impacts on global agriculture emerge earlier in new generation of climate and crop models. Nat Food 2:873–885

Jorge A, Lobo JA, Quesada M et al (2005) Effects of pollination by bats on the mating system of *Ceiba pentandra* (Bombacaceae) populations in two tropical life zones in Costa Rica. Am J Bot 92(2):370–376

Kulak V, Longboat S, Brunet ND (2022) In vitro technology in plant conservation: relevance to biocultural diversity. Plants 11(4):503. https://doi.org/10.3390/plants11040503

Kuswantoro H, Artari R, Iswanto R et al (2020) Family structure of F_5 soybeans lines derived from soybean varieties with the main differences on seed size and maturity traits. Biodiversitas 21(6): 2576–2585

Lavin M (2006) Floristic and geographical stability of discontinuous seasonally dry tropical forests explains patterns of plant phylogeny and endemism. In: Pennington RT, Lewis GP, Ratter JA (eds) Neotropical savannas and seasonally dry forests: plant diversity, biogeography and conservation. CRC Press, Boca Raton, pp 433–447

Li Y, Guan K, Schnitkey GD et al (2019) Excessive rainfall leads to maize yield loss of a comparable magnitude to extreme drought in the United States. Glob Chang Biol 25:2325–2337

Liu CLC, Kuchma O, Krutovsky KV (2018) Mixed-species versus monocultures in plantation forestry: development, benefits, ecosystem services and perspectives for the future. Global Ecoand Cons e00419:2351–9894. https://doi.org/10.1016/j.gecco.2018.e00419

Liua X, Zhua Y, Dengb Z et al (2022) The normal incidence sound absorption of the thin band yarn woven fabric with kapok fiber. App Acoustics 194:108805. https://doi.org/10.1016/j.apacoust.2022.108805

Long A, Zhao P, Liao L et al (2022) Sustainable Kapok fiber-derived carbon microtube as broadband microwave absorbing material. Materials 15(14):4845. https://doi.org/10.3390/ma15144845

Losi CJ, Siccama TG, Condit R et al (2003) Analysis of alternative methods for estimating carbon stock in young tropical plantations. Forest Ecol Manag 184(1–3):355–368

Lou L (2011) Structure, properties of kapok fiber and test on heat retention. Shanghai Text Sci Technol 39:15–17

Mahadeva SK, Yun S, Kim J (2011) Flexible humidity and temperature sensor based on cellulose–Polypyrrole nanocomposite. Sens Actuat A Phys 165:194–199. https://doi.org/10.1016/j.sna.2010.10.018

Mori SA, Cremers G, Gracie C et al (2002) Guide to the vascular plants of central French Guiana: part 2. Dicotyledons, memoirs of the New York Botanical Garden. New York Botanical Garden Press, New York

Murawski DA, Hamrick JL (1992) Mating system and phenology of *Ceiba pentandra* (Bombacaceae) in Central Panama. J Hered 83:401–404

Naharudin AU, Shaarani SHN, Rou LM et al (2019) Kapok as an adsorbent for industrial wastewater. J Chem Eng Ind Biot 5:48–54. https://doi.org/10.15282/jceib.v5i2.3727

Nathan R, Muller-Landau HC (2000) Spatial patterns of seed dispersal, their determinants and consequences for recruitment. Trends Ecol Evol 15:278–285. https://doi.org/10.1016/S0169-5347(00)01874-7

Ndukwe OO, Baiyeri KP (2020) Agronomic evaluation of two passion fruit genotypes in containers (megapot) predicted corresponding performances under field conditions. Trop Subtrop Agroeco 23(2):1–15

Nelson CD (2023) Tree breeding, a necessary complement to genetic engineering. New For 54: 721–738. https://doi.org/10.1007/s11056-022-09931-z

Nilsson T, Björdal C (2008) The use of kapok fibres for enrichment cultures of lignocellulose-degrading bacteria. Int Biodeterior Biodegradation 61:11–16. https://doi.org/10.1016/j.ibiod.2007.06.009

Nkouam GB, Adjoh G, Leudeu CBT et al (2017) Local uses of kapok (*Ceiba pentandra* Gaertn.) tree from the northern part of Cameroon. Int J Envion Agric Biotech 2(4):2214–2219. https://doi.org/10.22161/ijeab/2.4.82

Nyankson E, Rodene D, Gupta RB (2015) Advancements in crude oil spill remediation research after the deepwater horizon oil spill. Water Air Soil Pollut 227:1–22. https://doi.org/10.1007/s11270-015-2727-5

Olajide K, Baiyeri KP (2021) Early growth response of *Ceiba pentandra* to organic and inorganic fertilizers under field conditions in southeastern Nigeria. J Agric Res 6(3):273. https://doi.org/10.23880/oajar-16000273

Olajide K, Baiyeri KP, Ndubuaku UM (2022) Nutritional qualities of kapok (*Ceiba pentandra* L. Gaertn.) in response to fertilizer management. J Bio Res 20(3):1676–1684. https://doi.org/10.4314/br.v20i3.5

Orwa C, Mutua A, Kindt R et al (2009) Agroforestree Database: a tree reference and selection guide version 4.0. 1–5. http://www.worldagroforestry.org/sites/treedbs/treedatabases.asp

Pezzini FF, Dexter KG, de Carvalho-Sobrinho JG et al (2021) Phylogeny and biogeography of Ceiba Mill. (Malvaceae, Bombacoideae). Front Biogeogr 13(2):e49226. https://doi.org/10.21425/F5FBG49226

Phat NK, Knorr W, Kim S (2004) Appropriate measures for conservation of terrestrial carbon stocks analysis of trends of forest management in Southeast Asia. Forest Ecol Manag 191(1–3): 283–299

Priyanka V, Kumar R, Dhaliwal I et al (2021) Germplasm conservation: instrumental in agricultural biodiversity—a review. Sustainability 13(12):6743. https://doi.org/10.3390/su13126743

Quazi S, Golani T, Martino Capuzzo A (2021) Germplasm conservation. IntechOpen. https://doi.org/10.5772/intechopen.96184

Ravi Kiran C, Rao DB, Sirisha N et al (2015) Assessment of phytochemicals and antioxidant activities of raw and germinating *Ceiba pentandra* (kapok) seeds. J Biomed Res 29(5):414–419. https://doi.org/10.7555/JBR.27.20120145

Rezekinta FA, Kasim A, Syafri E et al (2023) Lignin degradation of kapok fiber (*Ceiba pentandra*, L.) with different times of pulping. J App Agric Sci Tech 7(3):329–336. https://doi.org/10.55043/jaast.v7i3.186

Rija A, Said A, Mwita J et al (2013) Floss release, seed fall and germination of kapok seeds (*Ceiba pentandra*) in an urban environment. Open J Ecol 3:423–430. https://doi.org/10.4236/oje.2013.36048

Rijavec T (2008) Kapok in technical textiles. Tekstilec 51(10–12):319–331

Rojas-Sandoval J, Budde K, Fernández M et al (2008) Phenology and pollination biology of *Ceiba pentandra* (Bombacaceae) in the wet forest of south-eastern Costa Rica. Stapfia 88:539–545

Roshetko JM, Evans DO (1999) Domestication of agroforestry trees in Southeast Asia. Working group reports from a regional workshop on the domestication of agroforestry trees in Southeast Asia, 4–7 November, Gajah Mada University, Yogyakarta, Indonesia. Bogor, Indonesia: International Centre for Research in Agroforestry. http://www.worldagroforestry.org/sea/Publications/files/report/RP0059-04.pdf

Rweyongeza DM, Francis C, Yeh FC et al (2003) Genetic variation in stem growth components in white spruce seedlings and its implications to retrospective early selection. For Genet 10(4): 299–308

Sahid M, Zeven AC (2003) *Ceiba pentandra* (L.) gaertn. In: Brink M, Escobin RP (eds) Plant resources of South-East Asia No. 17: fiber plants. PROSEA Foundation, Bogor, Indonesia

Said WM, Ehsan NO, Khalifa NS (2013) Comparative study of three species of Malvatheca (Bombacoideae and Malvoideae "*Malvaceae sensu lato*") using morphological, anatomical and RAPDPCR analyses. Adv Environ Biol 7(2):415–426

Sartika D, Syamsu K, Warsiki E et al (2020) Isolation of microfiber cellulose from kapok fiber (*Ceiba pentandra*) by using chemical-hydrothermal treatment. Ecol Envion Conserv 26(2): 654–662

Sartika D, Syamsu K, Warsiki E et al (2021) Nanocrystalline cellulose from kapok fiber (*Ceiba pentandra*) and its reinforcement effect on alginate hydrogel bead. Starch-Stärke 79:2100033. https://doi.org/10.1002/star.202100033

Shang L, Heckelei T, Gerullis MK et al (2021) Adoption and diffusion of digital farming technologies- integrating farm-level evidence and system interaction. Agric Syst 190:103074. https://doi.org/10.1016/j.agsy.2021.103074

Silveira AMF, Coelho Netto RA, Marenco RA (2023) Biomass allocation in *Ceiba pentandra* (Malvaceae) under water stress and high CO_2 concentration. Sci Fores 51:e3955. https://doi.org/10.18671/scifor.v51.10

Smole MS, Hribernik S, Kleinschek KS et al (2013) Advances in agrophysical research. Plant fibres for textile and technical applications. Intech, p 369

Subramanian P, Kim SH, Hahn BS (2023) Brassica biodiversity conservation: prevailing constraints and future avenues for sustainable distribution of plant genetic resources. Front Plant Sci 14:1220134. https://doi.org/10.3389/fpls.2023.1220134

Sudarshan A, Goroji PT, Lokesh SL et al (2014) Carbon sequestration potential of kapok (*Ceiba pentandra*) plantations in Theni district of Tamil Nadu, India. Ecol Envion Conserv 20(3): 1287–1292

Tang B, Chen X, He Y et al (2021) Fabrication of kapok fibers and natural rubber composites for pressure sensor applications. Cellulose 28:2287–2301. https://doi.org/10.1007/s10570-020-03647-z

Tareau MA, Alexander Greene A, Odonne G et al (2022) *Ceiba pentandra* (Malvaceae) and associated species: spiritual keystone species of the Neotropics. Botany 100(2):127–140. https://doi.org/10.1139/cjb-2021-0099

Tjio JH (1948) Notes on nucleolar conditions *in Ceiba pentandra*. Hereditas 34(1–2):204–208. https://doi.org/10.1111/j.1601-5223.1948.tb02836.x

Toxopeus HJ (1948) Over de bloembiologie, de spontane hybridisatie en de veredeling van de Java kapok (*Ceiba pentandra var. indica*) (On floral biology, spontaneous hybrization and breeding of Javanese kapok (*Ceiba pentandra* var. indica)). Med Alg Proefstat voor de Landbouw 20: 365–390

Tye YY, Lee KT, Abdullah WNW et al (2012) Potential of *Ceiba pentandra* (L.) Gaertn. (kapok fiber) as a resource for second generation bioethanol: effect of various simple pretreatment methods on sugar production. Bioresour Technol 116:536–539. https://doi.org/10.1016/j.biortech.2012.04.025

White TL (1987) A conceptual framework for tree improvement programs. New For 1:325–342 (1987). https://doi.org/10.1007/BF00031742

White TL (2004) Tree breeding, principles, breeding theory and genetic testing. In: Burley J (ed) Encyclopedia of forest sciences. Elsevier, pp 1551–1561. https://doi.org/10.1016/B0-12-145160-7/00095-8

Widyorini R, Kawai S, Subiyanto B et al (2009) Evaluation of biomass production of plantation forest in tropical areas: A case study of *Acacia mangium* plantation Forest, PT Musi Hutan Persada, Indonesia. The JSPS Global COE Program (E-04), In search of sustainable Humanosphore in Asia and Africa

Xiang H, Wang D, Liua H et al (2013) Investigation on sound absorption properties of kapok fibers. Chin J Polym Sci 31:521–529. https://doi.org/10.1007/s10118-013-1241-8

Yıldızhan Ş, Çalık A, Özcanlı M et al (2018) Bio-composite materials: a short review of recent trends, mechanical and chemical properties, and applications. Eur Mech Sci 2(3):83–91. https://doi.org/10.26701/ems.369005

Zerga AY, Tahir M (2022) Biobased Kapok fiber Nano-structure for energy and environment application: a critical review. Molecules 27(22):8107. https://doi.org/10.3390/molecules27228107

Zeven AC (1969) Kapok tree (*Ceiba pentandra* Gaertn.). In: Ferwerda FP, Wit, Veenman F (eds) Outlines of perennial crop breeding in the tropics, miscellaneous papers 4. The Netherlands, pp 269–288

Zhang M, Bu X, Xu X et al (2023) Miniaturized kapok fiber-supported liquid extraction for convenient extraction of pesticide residues in vegetable oils: determination of organochlorine pesticides as a proof-of-concept study. Talanta 253:123982. https://doi.org/10.1016/j.talanta.2022.123982

Zhao C, Liu B, Piao S (2017) Temperature increase reduces global yields of major crops in four independent estimates. Proc Natl Acad Sci USA 114(35):9326–9331. https://doi.org/10.1073/pnas.1701762114

Zheng Y, Wang J, Zhu Y et al (2015) Research and application of kapok fiber as an absorbing material: a mini-review. J Environ Sci 27:21–32. https://doi.org/10.1016/j.jes.2014.09.026

Zheng YL, Zhao G, Ma HC (2016) Effects of desiccation and storage temperature on seed germination in kapok. Hortic Tech 26(1):83–88. https://doi.org/10.21273/HORTTECH.26.1.83

Zhou Y, Zhu B, Yang T et al (2023) Hollow polyester/kapok/hollow polyester fiber-based needle punched nonwoven composite materials for rapid and efficient oil sorption. RSC Adv 13(39): 27077–27087. https://doi.org/10.1039/d3ra03695b

Zoma V, Tarama WJI, Traore A et al (2022) Potentialities and local conservation strategy of *Ceiba pentandra* (L.) Gaertn in the Banwa province in Burkina Faso. Int J Envion Ecol Res 4(2):1–9

Index

K. F. M. Salem et al. (eds.), *Breeding and Biotechnology of Leaf, Fruit, and Seed Fiber Crops*, Advances in Plant Breeding Strategies 11,
https://doi.org/10.1007/978-3-032-00407-9

H

N

O

P

Q